# NETTER'S
# INFECTIOUS
# DISEASES

## 2nd EDITION

# NETTER'S INFECTIOUS DISEASES

**Elaine C. Jong, MD, FIDSA, FASTMH**
Clinical Professor of Medicine Emeritus
Division of Allergy and Infectious Diseases
University of Washington School of Medicine
Seattle, Washington

**Dennis L. Stevens, MD, PhD**
Director
NIH Center of Excellence in Emerging/Reemerging Infectious Diseases
Associate Chief of Staff
Research & Development Service
Veterans Affairs Medical Center
Boise, Idaho
Professor of Medicine
Department of Medicine
University of Washington
Seattle, Washington

*Illustrations by* **Frank H. Netter, MD**

**CONTRIBUTING ILLUSTRATORS:**
Carlos A.G. Machado, MD
John A. Craig, MD
Tiffany S. DaVanzo, MA, CMI
Anita Impagliazzo, MA, CMI
Kristen Wienandt Marzejon, MS, MFA
James A. Perkins, MS, MFA

ELSEVIER

Elsevier
1600 John F. Kennedy Blvd.
Ste 1800
Philadelphia, PA 19103-2899

NETTER'S INFECTIOUS DISEASES, SECOND EDITION

ISBN: 978-0-323-71159-3

---

**Notice**

Practitioners and researchers must always rely on their own experience and knowledge in evaluating and using any information, methods, compounds or experiments described herein. Because of rapid advances in the medical sciences, in particular, independent verification of diagnoses and drug dosages should be made. To the fullest extent of the law, no responsibility is assumed by Elsevier, authors, editors or contributors for any injury and/or damage to persons or property as a matter of products liability, negligence or otherwise, or from any use or operation of any methods, products, instructions, or ideas contained in the material herein.

---

Previous edition copyrighted 2012.

**Library of Congress Control Number: 2021932970**

*Content Strategist:* Marybeth Thiel
*Publishing Services Manager:* Catherine Jackson
*Senior Project Manager:* Daniel Fitzgerald
*Designer:* Patrick Ferguson

Printed in the United States of America.

Last digit is the print number:   9   8   7   6   5   4   3

## Frank H. Netter, MD

Frank H. Netter was born in 1906, in New York City. He studied art at the Art Students League and the National Academy of Design before entering medical school at New York University, where he received his MD degree in 1931. During his student years, Dr. Netter's notebook sketches attracted the attention of the medical faculty and other physicians, allowing him to augment his income by illustrating articles and textbooks. He continued illustrating as a sideline after establishing a surgical practice in 1933, but he ultimately opted to give up his practice in favor of a full-time commitment to art. After service in the United States Army during World War II, Dr. Netter began his long collaboration with the CIBA Pharmaceutical Company (now Novartis Pharmaceuticals). This 45-year partnership resulted in the production of the extraordinary collection of medical art so familiar to physicians and other medical professionals worldwide.

In 2005, Elsevier, Inc. purchased the Netter Collection and all publications from Icon Learning Systems. There are now more than 50 publications featuring the art of Dr. Netter available through Elsevier, Inc.

Dr. Netter's works are among the finest examples of the use of illustration in the teaching of medical concepts. The 13-book *Netter Collection of Medical Illustrations*, which includes the greater part of the more than 20,000 paintings created by Dr. Netter, became and remains one of the most famous medical works ever published. *Netter's Atlas of Human Anatomy*, first published in 1989, presents the anatomical paintings from the Netter Collection. Now translated into 16 languages, it is the anatomy atlas of choice among medical and health professions students the world over.

The Netter illustrations are appreciated not only for their aesthetic qualities, but, more important, for their intellectual content. As Dr. Netter wrote in 1949, "…clarification of a subject is the aim and goal of illustration. No matter how beautifully painted, how delicately and subtly rendered a subject may be, it is of little value as a medical illustration if it does not serve to make clear some medical point." Dr. Netter's planning, conception, point of view, and approach are what inform his paintings and what makes them so intellectually valuable.

Frank H. Netter, MD, physician and artist, died in 1991.

Learn more about the physician-artist whose work has inspired the Netter Reference collection: https://netterimages.com/artist-frank-h-netter.html.

## Carlos A.G. Machado, MD

Carlos Machado was chosen by Novartis to be Dr. Netter's successor. He continues to be the main artist who contributes to the Netter collection of medical illustrations.

Self-taught in medical illustration, cardiologist Carlos Machado has contributed meticulous updates to some of Dr. Netter's original plates and has created many paintings of his own in the style of Netter as an extension of the Netter collection. Dr. Machado's photorealistic expertise and his keen insight into the physician–patient relationship inform his vivid and unforgettable visual style. His dedication to researching each topic and subject he paints places him among the premier medical illustrators at work today.

Learn more about his background and see more of his art at: https://netterimages.com/artist-carlos-a-g-machado.html.

# ABOUT THE EDITORS

**Elaine C. Jong, MD, FIDSA, FASTMH,** is Clinical Professor of Medicine Emeritus at the University of Washington School of Medicine (UWSOM) in Seattle, Washington. She was born in New York City, New York, graduated from Wellesley College with a bachelor of arts degree in biological sciences, and received her medical degree at the University of California-San Diego School of Medicine in La Jolla. After completing her internal medicine residency and a fellowship in Infectious Diseases at UWSOM, she joined the faculty in the Department of Medicine-Division of Allergy and Infectious Diseases and also served as an attending physician in the Division of Emergency Medicine. Engaging in basic research on the role of eosinophils in the host defense against schistosomiasis early in her career led to a lifelong interest in parasitic diseases, exotic infections, refugee and immigrant health, infections associated with international travel and outdoor activities, and vaccine-preventable diseases. Publications include more than 100 journal articles and book chapters and serving as editor or co-editor of more than a dozen books. At UW, Dr. Jong founded and directed courses and training clinics for the practice of refugee and immigrant medicine, and travel and tropical medicine. She served as the Director of the UW Student Health Center at the UW Hall Health Center and was the first Medical Director of the UW Campus Health Service, implementing programs for campus public health, occupational health, and employee health. Dr. Jong is a Fellow of the Infectious Diseases Society of America (IDSA) and a Fellow of the American Society of Tropical Medicine and Hygiene (ASTMH); she is a past President of the Clinical Group of the ASTMH and recently was awarded the Society's first Martin S. Wolfe Mentoring Award. Dr. Jong has served for many years as Chair of the International Medical Advisory Board for the International Association for Medical Assistance to Travellers (IAMAT).

**Dennis L. Stevens, MD, PhD,** has been Chief of the Infectious Diseases Section at the Veterans Affairs Medical Center in Boise, Idaho for 40 years and is currently Professor of Medicine at the University of Washington School of Medicine in Seattle. He is currently Program Director of a National Institutes of Health (NIH) Center of Excellence for Emerging and Re-emerging Pathogens in Boise, Idaho. Dr. Stevens obtained a Bachelor of Arts degree in microbiology from the University of Montana, a doctoral degree in microbiology from Montana State University, and a medical degree from the University of Utah. He completed an internal medicine residency at the University of Utah and performed his fellowship in infectious diseases at Brooke Army Medical Center. Dr. Stevens' major research interests have been the pathogenesis of serious infections caused by toxin-producing gram-positive pathogens including *Clostridium perfringens, Clostridium sordellii,* group A streptococcus, and methicillin-resistant *Staphylococcus aureus* (MRSA). Dr. Stevens recently received the IDSA Citation for his work on group A streptococcal infections and the William Altemeier Award from the Surgical Infections Society and was elected to membership in the Association of American Physicians. In 2018 he received the Veterans Affairs (VA) Infectious Disease Practitioners Lifetime Achievement Award. He has published more than 185 articles and 120 book chapters on serious invasive infections caused by gram-positive organisms and has been visiting professor at more than 70 national and 30 international institutions. He has been a member of the Centers for Disease Control and Prevention Working Group on Invasive Streptococcal Infections and a consultant to the World Health Organization, and he has been an invited participant to the National Institutes of Health Workforce on severe group A streptococcal infection. He has testified twice before the US Congress on the importance of basic science research in infectious diseases and on invasive group A streptococcal infections. Dr. Stevens is the current chairman of the IDSA's Guideline Committee for the Treatment of Skin and Skin Structure Infections.

# ACKNOWLEDGMENTS

The invitation to edit a new book on infectious diseases was an honor and a challenge almost a decade ago when the two of us, colleagues in the Division of Allergy and Infectious Diseases at the University of Washington School of Medicine in Seattle, decided to combine our complementary academic interests and clinical experiences to create the first edition of *Netter's Infectious Diseases*. We both shared a deep admiration for the amazing medical artwork of the late Frank H. Netter, MD; as medical educators, our goal was to place the original Netter illustrations into the context of modern medical science, judiciously augmented by additional artwork and photographic images to create memorable moments of understanding and insight for medical students, trainees, and clinicians in practice, generalists and specialists alike. This remains our goal for the second edition: the content has been thoroughly reviewed, updated, and augmented; new contributing authors add fresh perspective to classic and emerging infectious diseases; and the addition of Clinical Vignettes to most chapters in this edition will further enhance the book as an educational resource and reference.

We acknowledge with gratitude the dedication and precious time contributed to this book by our Section Editors and Contributing Authors. As we reached out to our colleagues and friends, recognized subject experts in the broad field of infectious diseases (ID), to contribute chapters to the second edition, the Coronavirus Disease pandemic of 2019 (COVID-19) emerged midway into the book project timeline. The tidal wave of seriously ill patients stressed healthcare systems across the United States and around the world, and ID specialists in particular took on burdens of added responsibilities in response to the crisis. Despite all this, the contributing authors came through with their best; their chapters are readable, up-to-date, and unforgettable.

Production of this book in the Netter Medical Series of books is a concerted team effort, and we want to acknowledge and thank Marybeth Thiel, Content Strategist–Education Content, and Daniel Fitzgerald, Project Manager–Clinical Solutions, at Elsevier for their expert guidance, assistance, and collaboration—it was our pleasure to work with them. The encouragement of our families, especially our spouses, Dr. Britt Litchford and Dr. Amy E. Bryant, was essential to the completion of the book, and we thank them for understanding the toll on our family time.

A special thanks to the late Dr. Frank H. Netter for his artistic skills that incorporated anatomy, clinical signs, and pathogenesis into remarkable images that have improved patient care for over 5 decades. We would like to dedicate this book to those colleagues who are no longer with us but contributed greatly to modern infectious diseases: Merle Sande, Richard Root, William Kirby, Seymour Klebanoff, Walter Stamm, Alan Bisno, Robert Moellering, Sydney Finegold, and John Bartlett.

# PREFACE

As longtime colleagues in the Division of Allergy and Infectious Diseases at the School of Medicine, University of Washington, Seattle, we were honored and challenged by the unique opportunity to create a new textbook of infectious diseases, with the goal of utilizing the beautiful medical artwork created by the late Dr. Frank H. Netter to teach and clarify important concepts in infectious diseases. Mindful of existing textbooks, such as Mandell's *Principles and Practice of Infectious Diseases*, that are considered authoritative as well as standards of excellence in the field, we set out to create a new resource with a strongly clinical orientation; the first edition of *Netter's Infectious Diseases* made its debut in 2012. This second edition continues our purpose: to provide healthcare providers, both generalists and specialists, with up-to-date clinical approaches to the broad spectrum of infectious diseases from the perspective of how these various infections may impact patients as individuals, members of communities, and as citizens of global society.

Reflecting the rapid accumulation of advances in the medical sciences, the infectious diseases specialty has many branches or subspecialties, and we were fortunate in recruiting outstanding Section Editors: Patrick W. Hickey (Vaccine-Preventable Diseases in Children and Adolescents), Thomas M. File, Jr. (Respiratory Tract Infections), E. Patchen Dellinger (Surgical Infections), Jeanne M. Marrazzo (Sexually Transmitted Infections), Vernon Ansdell (Parasitic Diseases), and M. Patricia Joyce (Emerging Infectious Diseases and Pandemics). We also participated as Section Editors: Skin and Soft-Tissue Infections (Dennis L. Stevens), Systemic Infections (Dennis L. Stevens), and Infections Associated With International Travel and Outdoor Activities (Elaine C. Jong).

A superb roster of talented contributing authors, recruited from our extended network of colleagues and friends, contributed their valuable time and expertise to writing concise, highly readable, and clinically relevant chapters. In each chapter, Dr. Netter's medical illustrations, in some cases revised to reflect new advances, are used to illustrate key points from the text, augmented by radiographic and photographic images, tables, and graphs. For some topics, there was creation of new and updated artwork by Carlos A. G. Machado, Tiffany S. DaVanzo, and Anita Impagliazzo, talented medical artists carrying on Dr. Netter's mission of using art as an educational icon.

Although our academic interests, clinical activities, and teaching commitments have kept us on different paths within our large Division in the past, we were drawn together to work on this book by our mutual admiration for the work of Dr. Frank H. Netter. His understanding of anatomy, physiology, pathogenesis, and clinical signs of disease are translated by his incredible artistic talent into visual images that are so powerful that they reduce complexities into simple concepts that remain embedded in our memories for decades. It is difficult to express the appreciation we have for how much his art contributed to our own enjoyment in medical education. We both had a strong desire to extend this experience to our peers, trainees, and students through the creation of a new up-to-date resource for learning about infectious diseases. We hope that we have succeeded in our goal and welcome your feedback, seeking further improvements for future editions.

**Elaine C. Jong, MD, FIDSA, FASTMH**
*Clinical Professor of Medicine Emeritus*
*Division of Allergy and Infectious Diseases*
*University of Washington School of Medicine*
*Seattle, Washington*

**Dennis L. Stevens, MD, PhD**
*Director*
*NIH Center of Excellence in Emerging/Reemerging Infectious Diseases*
*Associate Chief of Staff*
*Research & Development Service*
*Veterans Affairs Medical Center*
*Boise, Idaho*
*Professor of Medicine*
*Department of Medicine*
*University of Washington*
*Seattle, Washington*

## EDITORS

**Elaine C. Jong, MD, FIDSA, FASTMH**
Clinical Professor of Medicine Emeritus
Division of Allergy and Infectious Diseases
University of Washington School of
    Medicine
Seattle, Washington

**Dennis L. Stevens, MD, PhD**
Director
NIH Center of Excellence in Emerging/
    Reemerging Infectious Diseases
Associate Chief of Staff
Research & Development Service
Veterans Affairs Medical Center
Boise, Idaho
Professor of Medicine
Department of Medicine
University of Washington
Seattle, Washington

## SECTION EDITORS

**Vernon Ansdell, MD, FRCP, DTM&H**
Associate Clinical Professor
Department of Tropical Medicine, Medical
    Microbiology and Pharmacology
University of Hawaii
Honolulu, Hawaii

**E. Patchen Dellinger, MD**
Professor, Emeritus
Department of Surgery
University of Washington
Seattle, Washington

**Thomas M. File, Jr., MD, MSc**
Professor, Chair
Infectious Disease Section
Northeast Ohio Medical University
Rootstown, Ohio
Chair
Infectious Disease Division
Summa Health
Akron, Ohio

**Patrick W. Hickey, MD**
Associate Professor and Chair
Pediatrics
Uniformed Services University
Bethesda, Maryland

**M. Patricia Joyce, MD**
Medical Epidemiologist
Retired
Centers for Disease Control and Prevention
Tucker, Georgia

**Jeanne M. Marrazzo, MD, MPH, FACP, FIDSA**
Professor of Medicine
Director-Division of Infectious Diseases
University of Alabama at Birmingham
Birmingham, Alabama

## CONTRIBUTORS

**Daniel J. Adams, MD**
Pediatric Infectious Disease Physician
Department of Pediatrics
Naval Medical Center Portsmouth
Portsmouth, Virginia
Associate Professor
Department of Pediatrics
Uniformed Services University
Bethesda, Maryland

**Paul C. Adamson, MD, MPH**
Fellow
Division of Infectious Diseases
David Geffen School of Medicine at UCLA
Los Angeles, California

**Jan M. Agosti, MD, CTropMed**
Clinical Faculty
Division of Allergy and Infectious Diseases
University of Washington School of
    Medicine
Seattle, Washington

**Ibne Karim M. Ali, PhD**
Biologist
Free-Living and Intestinal Amebas
    Laboratory
Waterborne Disease Prevention Branch
Division of Foodborne, Waterborne, and
    Environmental Diseases
U.S. Centers for Disease Control and
    Prevention
Atlanta, Georgia

**Daniel A. Anaya, MD**
Professor and Chief
GI Surgery & Head
Hepatobiliary Section
Gastrointestinal Oncology
Moffitt Cancer Center
Tampa, Florida

**Margaret C. Bash, MD, MPH**
Medical Officer, Principal Investigator
Laboratory of Bacterial Polysaccharides
Center for Biologics Evaluation and
    Research
U.S. Food and Drug Administration
Silver Spring, Maryland

**Sky R. Blue, MD**
Infectious Diseases Physician
Sawtooth Epidemiology and Infectious
    Diseases
Boise, Idaho

**Dean A. Blumberg, MD**
Professor
Division of Pediatric Infectious Diseases
Department of Pediatrics
University of California, Davis School of
    Medicine
Sacramento, California

**John R. Bower, MD**
Associate Professor of Microbiology/
    Immunology
Department of Integrative Medical Sciences
Northeast Ohio Medical University
Rootstown, Ohio
Attending Physician
Division of Pediatric Infectious Diseases
Akron Children's Hospital
Akron, Ohio

**David M. Brett-Major, MD, MPH**
Professor
Department of Epidemiology
College of Public Health
University of Nebraska Medical Center
Omaha, Nebraska

**Amy E. Bryant, PhD**
Research Professor
Biological and Pharmaceutical Sciences
Idaho State University
Meridian, Idaho

**Anthony P. Cardile, DO, FACP**
Research Physician
Medicine
USAMRIID
Fort Detrick, Maryland

**Anna Cervantes-Arslanian, MD**
Boston University School of Medicine
Departments of Neurology, Neurosurgery,
    and Medicine
Boston Medical Center
Boston, Massachusetts

**Ritu Cheema, MD, FAAP**
Clinical Assistant Professor
Division of Pediatric Infectious Diseases
Department of Pediatrics
University of California, Davis School of
    Medicine
Sacramento, California

**Anthony W. Chow, MD, FRCPC, FACP**
Professor Emeritus
Internal Medicine/Infectious Diseases
University of British Columbia
Honorary Consultant
Internal Medicine/Infectious Diseases
Vancouver Coastal Health Acute
Vancouver, British Columbia, Canada

**Blaise L. Congeni, AB, MD**
Director
Division of Infectious Diseases
Department of Pediatrics
Akron Children's Hospital
Akron, Ohio

**Bradley A. Connor, MD**
Clinical Professor
Medicine
Weill Cornell Medical College
New York, New York

**James L. Cook, MD**
Clinical Professor of Medicine
Division of Infectious Diseases
Loyola University Medical Center
Maywood, Illinois
Staff Physician and Research Scientist
Section of Infectious Diseases
Edward Hines, Jr. Veterans Administration
   Hospital
Hines, Illinois

**Jennifer R. Cope, MD, MPH**
Medical Epidemiologist
Waterborne Disease Prevention Branch
Division of Foodborne, Waterborne, and
   Environmental Diseases
U.S. Centers for Disease Control and
   Prevention
Atlanta, Georgia

**Christina M. Coyle, MD, MS**
Professor of Medicine-Infectious Diseases
Assistant Dean for Faculty Development
Albert Einstein College of Medicine
Bronx, New York

**Birgitt L. Dau, MD**
Infectious Disease Physician
Sawtooth Epidemiology and Infectious
   Disease
Boise, Idaho

**Katherine L. DeNiro, MD**
Acting Assistant Professor
Department of Medicine
Division of Dermatology
University of Washington
Seattle, Washington

**Shireesha Dhanireddy, MD**
Professor
Medicine
University of Washington
Seattle, Washington

**Megan L. Donahue, MD**
Pediatric Infectious Diseases Fellow
Division of Pediatric Infectious Diseases
Walter Reed National Military Medical
   Center
Bethesda, Maryland

**Dimitri Drekonja, MD, MS**
Chief
Infectious Disease Section
Minneapolis VA Health Care System
Associate Professor of Medicine
University of Minnesota Medical School
Minneapolis, Minnesota

**Claire Panosian Dunavan, MD, DTM&H
(London)**
Clinical Professor of Medicine Emeritus-
   Recalled
Division of Infectious Diseases
David Geffen School of Medicine at UCLA
Los Angeles, California

**Eileen F. Dunne, MD, MPH**
Medical Officer
Division of HIV/AIDS Prevention CDC
Captain
U.S. Public Health Service
Bangkok, Thailand

**Matthew D. Eberly, MD**
Associate Professor of Pediatrics
Department of Pediatrics
Uniformed Services University
Bethesda, Maryland

**Jasmina Ehab, BS, MS**
H. Lee Moffitt Cancer Center
Gastrointestinal Oncology
University of South Florida Morsani College
   of Medicine
Tampa, Florida

**Seth I. Felder, MD**
Assistant Member
Gastrointestinal Oncology
Moffitt Cancer Center
Tampa, Florida

**Marc Fischer, MD, MPH**
Medical Epidemiologist
Arboviral Diseases Branch
Centers for Disease Control and Prevention
Fort Collins, Colorado

**Paul Froom, MD**
Head
Department of Clinical Utility
Sanz Medical Center, Laniado Hospital
Netanya, Israel, School of Public Health
University of Tel Aviv
Tel-Aviv, Israel

**Hector H. Garcia, MD, PhD**
Head
Cysticercosis Unit
Instituto Nacional de Ciencias Neurologicas
Director
Center for Global Health
Universidad Peruana Cayetano Heredia
Lima, Peru

**William M. Geisler, MD, MPH**
Professor
Medicine
The University of Alabama at Birmingham
Birmingham, Alabama

**Mark D. Gershman, MD**
Medical Epidemiologist
Division of Global Migration and Quarantine
Centers for Disease Control and Prevention
Atlanta, Georgia

**Radhika Gharpure, DVM, MPH**
Epidemic Intelligence Service Officer
Waterborne Disease Prevention Branch
Division of Foodborne, Waterborne, and
   Environmental Diseases
U.S. Centers for Disease Control and Prevention
Atlanta, Georgia

**Sandra G. Gompf, MD, FACP**
Associate Professor of Medicine
Division of Infectious Disease and
   International Medicine
USF Health Morsani College of Medicine
Chief
Infectious Disease Section
James A. Haley Veterans Hospital
Tampa, Florida

**Fernando B. Guerena, MD, MPH, FACP,
FIDSA, FACPM**
Director of Core Sciences USAMRIID
US Army Medical Research Institute of
   Infectious Diseases
Frederick, Maryland

**Martin Haditsch, MD, PhD**
Medical Head
Microbiology
Labor Hannover MVZ GmbH
Hannover, Germany
Medical Head
TravelMedCenter
Leonding, Austria

**Alison M. Helfrich, DO**
Assistant Professor
Pediatrics
Uniformed Services University of the Health
    Sciences
Bethesda, Maryland

**Susan L. Hills, MBBS, MTH**
Medical Epidemiologist
Arboviral Diseases Branch
Centers for Disease Control and Prevention
Fort Collins, Colorado

**Natasha S. Hochberg, MD, MPH**
Associate Professor of Medicine
Section of Infectious Diseases
Boston University School of Medicine
Attending Physician and Director of
    Tropical Medicine
Boston Medical Center
Associate Professor of Epidemiology
Boston University School of Public Health
Boston, Massachusetts

**Christopher M. Hull, MD**
Associate Professor
Dermatology
University of Utah School of Medicine
Salt Lake City, Utah

**Abir Hussein, MD**
Senior Fellow
Division of Allergy and Infectious Disease
University of Washington
Seattle, Washington

**Nathan K. Jansen, DO**
Infectious Disease Physician
United States Army Medical Corps
Fort Detrick, Maryland

**James R. Johnson, MD**
Staff Physician
Infectious Diseases
Minneapolis VA Health Care System
Professor
Medicine
University of Minnesota
Minneapolis, Minnesota

**Milissa U. Jones, MD, MPH**
Pediatric Infectious Diseases Physician
Pediatrics
Uniformed Services University
Bethesda, Maryland

**Stephen J. Jordan, MD, PhD**
Assistant Professor
Medicine
Indiana University
Indianapolis, Indiana

**Carol A. Kauffman, MD**
Chief
Infectious Diseases
Veterans Affairs Ann Arbor Healthcare System
Professor
Internal Medicine
University of Michigan
Ann Arbor, Michigan

**Maryam Keshtkar-Jahromi, MD, MPH**
Assistant Professor
Medicine-Infectious Diseases
Johns Hopkins University, School of Medicine
Baltimore, Maryland

**Amber K. Kirby, MD**
Department of Pediatrics
Naval Medical Center Portsmouth
Portsmouth, Virginia

**Natalie Kirilcuk, MD, MS**
Clinical Assistant Professor
Surgery
Stanford University
Palo Alto, California

**Patricia Kissinger, PhD**
Professor of Epidemiology
Department of Epidemiology
Tulane University School of Public Health
    and Tropical Medicine
New Orleans, Louisiana

**Jeffrey D. Klausner, MD, MPH**
Professor of Medicine and Public Health
UCLA David Geffen School of Medicine and
    Fielding School of Public Health
Los Angeles, California

**Marin H. Kollef, MD**
Professor of Medicine
Division of Pulmonary and Critical Care
    Medicine
Washington University School of Medicine
St. Louis, Missouri

**John D. Kriesel, MD**
Associate Professor
Internal Medicine/Infectious Diseases
University of Utah
Salt Lake City, Utah

**Kevin P. Labadie, MD**
General Surgery Resident
University of Washington
Seattle, Washington

**Alexa R. Lindley, MD, MPH**
Acting Assistant Professor
Department of Family Medicine
University of Washington
Seattle, Washington

**Nicole Lindsey, MS**
Epidemiologist
Division of Vectorborne Diseases
Centers for Disease Control and Prevention
Fort Collins, Colorado

**Fabiana Simão Machado, PhD**
Associate Professor
Department of Biochemistry and
    Immunology
Universidade Federal de Minas Gerais
Belo Horizonte, Minas Gerais
Brazil

**Douglas W. MacPherson, MD,
MSc(CTM), FRCPC**
Adjunct Professor
Department of Medicine
Chief of Infectious Diseases
Schulich School of Medicine and Dentistry
Western University
London, Ontario, Canada

**Lionel A. Mandell, MD, FRCPC**
Professor Emeritus
Medicine
McMaster University
Hamilton, Ontario, Canada

**Lauri E. Markowitz, MD**
Medical Epidemiologist
Division of Viral Diseases
Centers for Disease Control and Prevention
Atlanta, Georgia

**Blaine A. Mathison, BS, M(ASCP)**
Scientist
Institute for Clinical and Experimental
    Pathology
ARUP Laboratories
Salt Lake City, Utah

**Nancy McClung, PhD, RN**
Epidemiologist
Division of Viral Diseases
Centers for Disease Control and Prevention
Atlanta, Georgia

**Emily McDonald, MD, MPH**
EIS Officer
Arboviral Diseases Branch
Centers for Disease Control and Prevention
Fort Collins, Colorado

**Peter P. McKellar, MD**
Associate Professor of Medicine
Department of Medicine
Banner University Medical Center
Phoenix, Arizona

**Graham W. McLaren, MD**
Resident Physician
Department of Surgery
Western Michigan University
Kalamazoo, Michigan

**Eric A. Meyerowitz, MD**
Assistant Professor
Montefiore Medical Center
Albert Einstein College of Medicine
Bronx, New York

**Andrew P. Michelson, MD**
Assistant Professor
Pulmonary & Critical Care
Division of Pulmonary and Critical Care
    Medicine
Washington University School of Medicine
St. Louis, Missouri

**David W. Miranda, MD**
General Surgery Resident
University of Washington
Seattle, Washington

**Arden M. Morris, MD, MPH**
Professor
Surgery
Stanford University
Palo Alto, California

**Sapna Bamrah Morris, MD, MBA**
Medical Officer, Team Lead
Division of Tuberculosis Elimination
Centers for Disease Control and Prevention
Atlanta, Georgia

**Nicholas J. Moss, MD, MPH**
Director
HIV STD Section
Division of Communicable Disease Control
    and Prevention
Alameda County Public Health Department
Oakland, California

**Christina A. Muzny, MD, MSPH**
Associate Professor of Medicine &
    Epidemiology
Division of Infectious Diseases
University of Alabama at Birmingham
Birmingham, Alabama

**Natasha A. Nakra, MD**
Associate Professor
Division of Pediatric Infectious Diseases
University of California, Davis School of
    Medicine
Sacramento, California

**Lori M. Newman, MD**
Medical Officer
Division of Microbiology and Infectious
    Diseases
National Institute of Allergy and Infectious
    Diseases
Rockville, Maryland

**Michael S. Niederman, MD**
Associate Division Chief
Clinical Director, Pulmonary and Critical
    Care Medicine
NYP-Weill Cornell Medical Center
Professor of Clinical Medicine
Weill Cornell Medical College
New York, New York

**Arthur C. Okwesili, DO, MPH**
Chief
Occupational & Environmental Medicine
Department of Preventive Medicine
Brooke Army Medical Center
JBSA–Fort Sam Houston, Texas

**Winnie W. Ooi, MD, DMD, MPH**
Director
Travel and Tropical Medicine
Division of Infectious Diseases
Lahey Hospital and Medical Center
Co-Director
Hansen's Disease Outpatient Clinic
Burlington, Massachusetts
Assistant Professor of Medicine
Tufts School of Medicine
Boston, Massachusetts

**Martin G. Ottolini, MD**
Director
Capstone Student Research Program
Office of the Dean
F. Edward Hebert School of Medicine
USUHS
Consultant
Pediatric Infectious Diseases
Pediatrics
Walter Reed National Military Medical Center
Professor of Pediatrics
Pediatrics
F. Edward Hebert School of Medicine
USUHS
Bethesda, Maryland

**James O. Park, MD**
Professor
Department of Surgery
University of Washington
Seattle, Washington

**Elizabeth H. Partridge, MD, MPH**
Pediatric Infectious Disease
UC Davis Children's Hospital
Sacramento, California

**Benjamin C. Pierson, DO, MPH**
Research Physician
Division of Medicine
United States Army Medical Research
    Institute of Infectious Diseases
    (USAMRIID)
Fort Detrick, Maryland

**Phillip R. Pittman, MD, MPH**
Chief
Department of Clinical Research
United States Army Medical Research
    Institute of Infectious Diseases
Fort Detrick, Maryland
Director
Military Vaccine Clinical Research Center
USAMRIID
Fort Detrick, Maryland

**Paul S. Pottinger, MD, DTM&H**
University of Washington
Seattle, Washington

**R. Douglas Pratt, MD, MPH**
Division of Vaccines and Related Products
    Applications
Office of Vaccines Research and Review
U.S. Food and Drug Administration
Silver Spring, Maryland

**Bobbi S. Pritt, MD, MSc, (D)TMH**
Professor; Chair
Division of Clinical Microbiology
Department of Laboratory Medicine and
    Pathology
Mayo Clinic
Rochester, Minnesota

**Latha Rajan, MD, MPHTM, MBA, CTropMed, FASTMH**
Clinical Associate Professor
Department of Tropical Medicine
Tulane University School of Public Health &
    Tropical Medicine
New Orleans, Louisiana

**Michael Rajnik, MD**
Chief
Infectious Diseases Division
Pediatrics
Uniformed Services University of the Health
    Sciences
Bethesda, Maryland

**Mark S. Riddle, MD, DrPH**
Associate Dean of Clinical Research
Department of Medicine
University of Nevada Reno School of
    Medicine
Associate Chief of Staff–Research
VA Sierra Nevada Health Care System
Reno, Nevada

**Caitlin K. Rochester, MD**
Department of Pediatrics
Naval Medical Center Portsmouth
Portsmouth, Virginia

**Edward T. Ryan, MD**
Director
Global Infectious Diseases
Massachusetts General Hospital
Professor of Medicine
Harvard Medical School
Professor of Immunology
Harvard T.H. Chan School of Public Health
Boston, Massachusetts

**Rabeeya Sabzwari, MD**
Assistant Professor of Medicine
Division of Infectious Diseases
Loyola University Medical Center
Maywood, Illinois
Staff Physician and Research Scientist
Section of Infectious Diseases
Edward Hines VA Medical Center
Hines, Illinois

**Rebecca Sainato, MD, MTM&H**
Pediatric Infectious Diseases
Madigan Army Medical Center
Tacoma, Washington

**Christopher A. Sanford, MD, MPH, DTM&H**
Associate Professor
Family Medicine, Global Health
University of Washington
Seattle, Washington

**David L. Saunders, MD, MPH**
Chief
Division of Medicine
US Army Research Institute of Infectious Disease
Fort Detrick, Maryland

**Robert G. Sawyer, MD**
Professor and Chair of Surgery
Department of Surgery
Western Michigan University Homer Stryker MD School of Medicine
Adjunct Professor of Engineering and Applied Sciences
Engineering and Applied Sciences
Western Michigan University
Kalamazoo, Michigan
Adjunct Professor of Surgery
Surgery
University of Virginia
Charlottesville, Virginia

**April Schachtel, MD**
Acting Instructor
Department of Medicine
Division of Dermatology
University of Washington
Seattle, Washington

**Elizabeth R. Schnaubelt, MD**
Clinical Assistant Professor
Division of Infectious Diseases
University of Nebraska Medical Center
Omaha, Nebraska
U.S. Air Force School of Aerospace Medicine
Wright Patterson Air Force Base, Ohio

**Sanjay Sethi, MD**
Professor
Medicine
University at Buffalo, SUNY
Buffalo, New York

**Zvi Shimoni, MD**
Clinical Assistant Professor
Faculty of Medicine
Ruth and Bruce Rappaport School of Medicine
Haifa, Israel
Department of Internal Medicine B
Sanz Medical Center, Laniado Hospital
Netanya, Israel

**J. Erin Staples, MD, PhD**
Medical Epidemiologist
Arboviral Diseases Branch
Centers for Disease Control and Prevention
Fort Collins, Colorado

**Angela Starks, PhD**
Branch Chief
Division of Tuberculosis Elimination/Laboratory Branch
Centers for Disease Control and Prevention
Atlanta, Georgia

**Russell W. Steele, MD**
Division Head
Pediatrics
Ochsner Children's Health Center
New Orleans, Louisiana

**Michael J. Tan, MD, FACP, FIDSA**
Professor of Internal Medicine
Northeast Ohio Medical University
Rootstown, Ohio
Lead Physician
Infectious Disease
Summa Health
Akron, Ohio

**Stephanie N. Taylor, MD**
Professor of Medicine
Section of Infectious Diseases
Louisiana State University Health Sciences Center
New Orleans, Louisiana

**Derek T. Tessman, DO**
Resident Physician
Department of Surgery
Western Michigan University
Kalamazoo, Michigan

**John F. Toney, MD**
Professor of Medicine
Division of Infectious Disease and International Medicine
USF Health Morsani College of Medicine
Assistant Chief
Infectious Disease Section
James A. Haley Veterans Hospital
Tampa, Florida

**Christina Topham, MD**
School of Medicine
University of Utah
Salt Lake City, Utah

**Alan F. Utria, MD**
Pediatric Surgery Research Fellow
University of Washington
Seattle, Washington

**Anna Wald, MD, MPH**
Head of Allergy and Infectious Diseases Division
Professor of Medicine, Laboratory Medicine & Pathology, & Epidemiology
University of Washington
Seattle, Washington

**Ana A. Weil, MD, MPH**
Acting Assistant Professor
Department of Medicine
Division of Allergy and Infectious Diseases
University of Washington
Seattle, Washington

**Louis M. Weiss, MD, MPH**
Professor of Pathology
Division of Parasitology and Tropical Medicine
Professor of Medicine
Division of Infectious Diseases
Albert Einstein College of Medicine
Bronx, New York

**Patrick S. Wolf, MD, FACS**
Surgical Oncology
The Surgical Clinic
Nashville, Tennessee

**Kate R. Woodworth, MD, MPH**
Medical Officer
Division of Birth Defects and Infant
  Disorders
Centers for Disease Control and Prevention
Atlanta, Georgia

**Kimberly A. Workowski, MD, FACP,
FIDSA**
Professor of Medicine
Infectious Diseases
Emory University
Atlanta, Georgia

**Jonathan M. Wortham, MD**
Medical Officer, Team Lead
Division of Tuberculosis Elimination
Centers for Disease Control and Prevention
Atlanta, Georgia

**Casi M. Wyatt, DO, FIDSA**
Infectious Diseases Physician
Sawtooth Epidemiology and Infectious Diseases
Boise, Idaho

**Johnnie A. Yates, MD**
Medical Director
Travel & Tropical Medicine
Hawaii Permanente Medical Group
Honolulu, Hawaii

**Sylvia H. Yeh, MD**
Professor of Pediatrics
Pediatric Infectious Diseases
Harbor-UCLA Medical Center
David Geffen School of Medicine at UCLA
Torrance, California

**Sarah S. Zhu, MCMSc, PA-C**
Inpatient Surgical Physician Assistant
Gastrointestinal Surgical Oncology
H. Lee Moffitt Cancer Center
Tampa, Florida

# CONTENTS

# ONLINE CONTENTS

Visit your ebook (see inside front cover for details) for the following printable patient education brochures from *Ferri's Netter Patient Advisor*, 3rd edition.

# NETTER'S INFECTIOUS DISEASES

# Vaccine-Preventable Diseases in Children and Adolescents

*Patrick W. Hickey*

# Introduction to Vaccine-Preventable Diseases in Children and Adolescents

*Patrick W. Hickey*

 **ABSTRACT**

This section provides information about the vaccines currently recommended in the United States for routine immunization of children and adolescents, as well as the epidemiology and clinical manifestations of these diseases. This chapter provides an overview of how vaccine schedules are developed, the types of products used, and considerations related to safety and usage. Although overall vaccine coverage remains high in the United States and other industrialized nations and is expanding globally, sustaining access to and uptake of vaccines requires continued attention from policy makers and public health officials.

## GEOGRAPHIC DISTRIBUTION AND EPIDEMIOLOGICAL TRENDS

Worldwide in 2018, the percentage of children immunized with three doses of diphtheria, tetanus, and pertussis (DTP) and oral polio vaccines, and a measles-containing vaccine exceeds 85%. Although vaccines have proven safe and cost-effective, saving $16.00 in healthcare costs, lost wages, and lost productivity due to illness and death for every dollar spent, many developing countries do not have adequate and consistent access to available or affordable vaccines, particularly for newer vaccine products, some of which have narrow geographic use, such as protein-conjugate pneumococcal, typhoid, and meningococcal vaccines, and Japanese encephalitis.

During the time period of 2016 to 2018, the immunization coverage rate in the United States at age 24 months with the full seven-vaccine series (4:3:1:3:3:1:4 series): four diphtheria toxoid and tetanus toxoid with acellular pertussis vaccine (DTaP); three polio; one measles, mumps, rubella vaccine (MMR); three (or four) *Haemophilus influenzae* type b vaccine [Hib]; three hepatitis B, one varicella, and four pneumococcal conjugate vaccines was only 68.5%. However, coverage rates exceeded 90% for polio, MMR, hepatitis B, and, varicella on an individual vaccine basis. Although only 1.3% of children had received no vaccines, vaccine coverage in the United States varies widely based on both geography and family demographics. Low vaccine coverage is associated with lack of private health insurance, poverty, and being of the Black or American Indian/Alaska Native race. Twenty states have MMR coverage less than 90%, with pockets in some communities significantly lower. In many of these localities, parental reluctance to vaccinate and acceptance by local governance of nonmedical exemptions to vaccine requirements are highly prevalent, putting them at risk for outbreaks of vaccine-preventable diseases, as represented by the more than 1200 measles cases that occurred nationwide in 2019. Vaccine hesitancy is an emerging problem for the United States and other industrialized nations that historically have had high coverage rates and had eliminated many of these diseases, only to see reemergence of the diseases with decreasing coverage. Engaging parents reluctant to vaccinate requires careful consideration of the underlying concerns and a thoughtful communication strategy, preferably from a trusted source.

## GENERAL PRINCIPLES

### Schedules

Synchronized immunization schedules for the United States are developed by the Advisory Committee on Immunization Practices (ACIP) of the Centers for Disease Control and Prevention (CDC), the American Academy of Pediatrics (AAP) Committee on Infectious Diseases (Red Book), and the American Academy of Family Practitioners (AAFP) and are posted annually in January. Two immunization schedules are posted for the pediatric age groups: one for children younger than 7 years of age and one for individuals 7 through 18 years of age. A separate schedule is available for immunizations for adults over the age of 18 years. The World Health Organization (WHO) Expanded Program on Immunization (EPI) publishes immunization schedules for all of the countries in the world.

### Immunizations Received in Other Countries

Healthy individuals immunized in countries outside of the United States, now living in the United States should receive vaccines according to the recommended schedule for healthy infants, children, and adolescents. In general, only written documentation should be accepted as proof of previous vaccination. Written, dated, and appropriate records (e.g., correct age, dates, intervals, and number of doses) may be considered as valid, and immunizations may resume according to the US schedule. If vaccination status is uncertain, the options include vaccinating or performing serologic testing for antibodies against the selected vaccine antigen, if testing is available.

## TYPES OF IMMUNIZATIONS

The two major types of immunizations are active and passive.

### Passive Immunization

Passive immunization refers to receipt of preparations of preformed antibodies, usually as immune globulin (IG). IG may be a general formulation or hyperimmune IG developed with high concentrations of antibodies against a specific disease, such as hepatitis B immune globulin (HBIG).

Administration of IG may be useful for (1) prophylactic immunization for a host who is not able to make antibodies (e.g., an infant with congenital immunodeficiency), and (2) immediate preexposure or postexposure protection of individuals, especially when there is not sufficient time for the host to mount a protective antibody response (e.g., in acute exposure to hepatitis A in an immune compromised individual, or an infant too young to receive active immunization).

## Active Immunization

With active immunization, a vaccine antigen is given to the host to elicit a protective immune response (e.g., antibodies and cellular immunity). The vaccine antigen may be composed of whole microorganisms, partial microorganisms, or a modified product (e.g., toxoid or purified component) of microorganisms. Whole organisms may be inactivated or live-attenuated. The elicited immune response usually mimics the response seen with natural infection, and ideally this occurs with no or minimal risks to the recipient.

## VACCINE RECIPIENTS

### Healthy Pediatric Populations

In the United States, all licensed vaccines have undergone review by the US Food and Drug Administration (FDA) and have been proven safe and effective for the targeted population. Most of the routinely recommended pediatric vaccines are targeted for healthy children and adolescents (Fig. 1.1). Alternative schedules that are delayed or staggered have not been systematically studied for safety and efficacy and pose increased risk of disease acquisition.

### Adolescents

Since 2005, several vaccines have become available for routine use in adolescents. These include tetanus toxoid with reduced-dose diphtheria toxoid and reduced-dose acellular pertussis vaccine (Tdap), human papillomavirus (HPV) vaccine, and meningococcal conjugate vaccines; all of these are discussed in detail in the chapters in this section. In addition, some existing vaccines for use in children, such as influenza and varicella vaccines, were given new recommendations for routine or "catch-up" indications in adolescents. The AAP has recommended a routine health visit at 11 to 12 years, and this visit can be used to ensure that the adolescent has received all recommended immunizations, as well as to afford the opportunity to provide anticipatory guidance for safe and healthy living for the teen years.

### Immunocompromised Children

A growing number of children and adolescents have congenital or acquired immune dysfunction and should not receive immunizations as routinely recommended. Special accommodations may be needed for immunizing these individuals, such as adjusting the schedule or possibly not administering some agents. However, there are no indications for giving decreased or partial doses of vaccines. The plan for vaccination of an immunocompromised child should be determined by the nature and degree of the immunosuppression, weighing the risks and benefits of vaccination with those of exposure to natural infection. Efforts that support high rates of vaccine uptake among the general population are important to providing protective herd immunity for those with contraindications.

### Preterm (<37 Weeks of Gestation) and Low–Birth-Weight (<2000 g) Infants

In general, medically stable premature and low–birth-weight infants may be immunized at the same dose, schedule, and postnatal age as full-term infants. One notable exception is the use of hepatitis B vaccine in infants who weigh less than 2000 g; details are provided in Chapter 15.

### International Adoptees, Travelers, Immigrants, and Refugees

All routinely recommended vaccines should be up to date for age, as many families travel abroad without recognizing the possible exposures to vaccine-preventable diseases. In addition, traveling children and teens should receive vaccinations, as well as other preventative measures (e.g., malaria prophylaxis), targeted for their destination. There may be a need for an accelerated schedule—for example, early immunization with MMR and hepatitis A for infants 6 to 12 months of age traveling to endemic regions. Use of IG prophylaxis can be considered for some individuals susceptible to hepatitis A (e.g., infants under 6 months of age, short-notice travelers with chronic liver disease). Current recommendations for travelers are posted on the CDC website.

## ADVERSE EVENTS AND VACCINE INFORMATION

### Adverse Events

Safety information about vaccines for healthcare providers and laypersons is available from several reliable resources including the AAP, CDC, FDA, and WHO. A select list of internet resources for vaccine information is provided in Table 1.1. The vaccine manufacturer's package insert provides safety and tolerability data from the clinical trials for each specific vaccine. As with any medication, no vaccine is completely free of adverse effects (AEs), and the known AEs should be discussed with vaccinees (nonminors) and/or parents or legal guardians. Most AEs observed after routine immunizations are local injection-site reactions such as erythema, edema, and pain and systemic reactions such as fever or irritability. Although the majority of AEs are mild and self-limiting, some may be associated with transient impairment for the vaccinee, such as limited limb mobility because of pain. Serious AEs, which may lead to permanent disability or life-threatening illness, are rarely observed after routine pediatric vaccinations. The occurrence of an AE after immunization proves not that the vaccine is the cause of the event but that there is a temporal relationship. If a serious AE occurs after administration of a vaccine (especially within 30 days of receipt), a complete evaluation for all plausible causes, including the role of the vaccine antigen, should be performed. All serious AEs and clinically significant AEs should be reported to the Vaccine Adverse Event Reporting System (VAERS), which is maintained by the CDC and FDA. Reporting AEs is valuable because it helps identify events that are infrequent or unexpected and not observed in the prelicensure clinical trials.

### Informing Vaccine Recipients and Parents

Vaccine recipients and parents or legal guardians should be informed about the risks and benefits of vaccination and about the natural disease that the vaccine is designed to prevent. The National Childhood

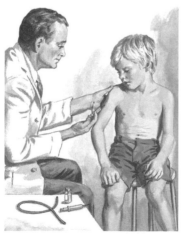

**Fig. 1.1** Vaccination.

| TABLE 1.1 | Select Internet Resources for Vaccine Information | |
|---|---|---|
| **Resource** | **For Healthcare Providers** | **For Lay Persons** |
| American Academy of Pediatrics (AAP) | https://www.aap.org/en-us/advocacy-and-policy/aap-health-initiatives/immunizations/Pages/Immunizations-home.aspx | https://www.healthychildren.org/english/safety-prevention/immunizations/ |
| Centers for Disease Control and Prevention (CDC) and the Advisory Committee on Immunization Practices (ACIP) | https://www.cdc.gov/vaccines/hcp/ https://www.cdc.gov/vaccines/hcp/acip-recs/ | https://www.cdc.gov/vaccines/parents/ https://www.cdc.gov/vaccines/vac-gen/ |
| CDC travelers' health recommendations | https://wwwnc.cdc.gov/travel/yellowbook/2020/table-of-contents | https://wwwnc.cdc.gov/travel |
| CDC and AAP communication toolkits, messaging strategies | https://www.cdc.gov/vaccines/hcp/vis/ https://www.cdc.gov/vaccines/partners/childhood/index.html | https://www.healthychildren.org/english/safety-prevention/immunizations/ |
| US Food and Drug Administration (FDA) | https://www.fda.gov/vaccines-blood-biologics/vaccines | https://www.fda.gov/vaccines-blood-biologics/resources-you-biologics/consumers-biologics |
| Vaccine Adverse Event Reporting System (VAERS) | https://vaers.hhs.gov/professionals | http://vaers.hhs.gov/ |
| World Health Organization (WHO) Expanded Programme on Immunization | https://www.who.int/immunization/programmes_systems/supply_chain/benefits_of_immunization/en/ | www.who.int/vaccine_safety/en |

Vaccine Injury Act of 1986 requires that parents receive a Vaccine Information Statement (VIS) each time a child receives a vaccine covered under this legislation, regardless of the funding source used to purchase the vaccine. The VISs are available from the CDC at the National Immunization Program site. The vaccine manufacturer, lot number, date of administration, and that the VISs were provided should be documented.

Particularly in an era when many adults have not seen or experienced the diseases for which childhood vaccination is offered, nor the morbidity and mortality they cause, there is a risk of hesitancy or refusal to vaccinate children. In situations where this occurs, it is important to both identify the source of this reluctance and then to have an informed discussion with parents that demonstrates both respect and prioritizing the child's welfare. Some caregivers will express religious reasons for vaccine refusal, although these situations are in fact quite rare among major religious faiths. Common misconceptions related to vaccines include such beliefs as, "natural immunity is better," "too many vaccines can overload the immune system," "vaccines are not effective anyway," "vaccines cause autoimmune diseases and/or autism," and "spreading out the vaccine series is safer." There is strong evidence to counter these false beliefs, and the AAP and CDC offer a number of resources (see Table 1.1) to help address these concerns with parents and caregivers. Parents should be informed of legal requirements for vaccination to access school and childcare services. Although some state laws allow for philosophical/nonmedical waivers to be provided, both the AAP and CDC discourage their provision. When a vaccine is refused, a "vaccine refusal" document should be signed by the caregiver.

## ACKNOWLEDGMENT

The author acknowledges ChrisAnna M. Mink for her contribution on this chapter in the previous edition. The assertions expressed herein are those of the author and do not reflect the official policy or position of the Uniformed Services University or the Department of Defense.

## ADDITIONAL RESOURCES

American Academy of Pediatrics (AAP): Vaccine information. In Pickering LK, Baker CJ, Kimberlin DW, Long SS, eds: *Red Book: 2018 Report of the Committee on Infectious Diseases*. 31st ed. Elk Grove Village, IL: AAP; 2009: pp 1-111.

Cattaneo R, Engert SF, Gray D, and Vineyard C. *Immunization Training Guide & Practice Procedure Manual*. 3rd ed. Elk Grove, IL: American Academy of Pediatrics; 2016. Available at: https://www.aap.org/en-us/Documents/immunizations_training_guide.pdf.

Centers for Disease Control and Prevention. *Epidemiology and Prevention of Vaccine-Preventable Diseases*. Hamborsky J, Kroger A, Wolfe S, eds. 13th ed. Washington, D.C. Available at: https://www.cdc.gov/vaccines/pubs/pinkbook/.

Centers for Disease Control and Prevention (CDC): Vaccines and immunizations: immunization schedules. Available at: https://www.cdc.gov/vaccines/schedules/.

Edwards KM, Hackell JM; Committee on Infectious Diseases, Committee on Practice and Ambulatory Medicine. Countering vaccine hesitancy. *Pediatrics*. 2016 Sep;138(3). pii: e20162146. Available at: https://pediatrics.aappublications.org/content/138/3/e20162146#sec-7.

Robinson CL, Bernstein H, Romero JR, Szilagyi P. Advisory Committee on Immunization Practices Recommended Immunization Schedule for Children and Adolescents Aged 18 Years or Younger—United States, 2019. *MMWR Morb Mortal Wkly Rep*. 2019 Feb 8;68(5):112-114. Available at: https://www.cdc.gov/mmwr/volumes/68/wr/mm6805a4.htm.

# Diphtheria and Tetanus

*Megan L. Donahue, Matthew D. Eberly*

##  ABSTRACT

Diphtheria and tetanus are bacterial diseases mediated by extremely potent toxins. Diphtheria is a communicable infection of the upper respiratory tract, skin, and rarely other mucous membranes caused by *Corynebacterium diphtheriae,* whereas tetanus is a neurotoxin-mediated disease resulting from anaerobic wound infections caused by *Clostridium tetani.* These diseases can be life-threatening, and early recognition and intervention are essential for effective management. Immunization against both diseases is usually performed with combination vaccines containing diphtheria and tetanus toxoids, which induce toxin-neutralizing antibodies that are protective.

Diphtheria and tetanus are very different diseases in their clinical presentation. Nevertheless, the two diseases are commonly considered together because they share a common history, as well as key elements of pathogenesis and prevention. Potent toxins are central to the pathogenesis of diphtheria and tetanus. Diphtheria toxin (DipT) and tetanus neurotoxin (TeNT) were among the earliest recognized bacterial toxins, and early immunology was stimulated by the discovery of protective, toxin-neutralizing serum antibodies. The science of vaccination was promoted by the revelation that DipT and TeNT can be chemically treated to produce toxoids, namely, molecules that have lost toxicity but retain their ability to induce protective antibodies.

For both diseases the key to prevention is to maintain adequate concentrations of toxin-neutralizing antibodies. As a result of successful immunization programs, diphtheria and tetanus are now rare in the United States and in other developed nations. If diphtheria or tetanus is suspected, state and local health departments should be contacted for guidance because both diseases require treatment with specific antitoxin.

# DIPHTHERIA

## ✳ CLINICAL VIGNETTE

A 5-year-old child with a history of incomplete immunizations developed sore throat and low-grade fever a few days after returning from visiting family in rural India. She presented to an urgent care center, where rapid streptococcal testing was negative. She was diagnosed with viral pharyngitis and discharged home. Two days later, she complained of worsening sore throat and difficulty swallowing. She presented to an emergency department where examination revealed a thick, adherent, grayish membrane across the posterior oropharynx, involving the tonsillar pillars and uvula. Attempts to remove the membrane resulted in bleeding. She also had notable swelling of her anterior cervical lymph nodes. Based on her clinical presentation, immunization status, and recent travel history, a clinical diagnosis of diphtheria was made. A swab of the tonsillar exudate was sent for culture, and the Centers for Disease Control and Prevention (CDC) were contacted for release of diphtheria antitoxin (DAT). She was started on IV erythromycin while awaiting arrival of DAT from the CDC.

## Geographic Distribution and Magnitude of Disease Burden

*Corynebacterium diphtheriae* is a species of aerobic, nonencapsulated, non–spore-forming, mostly nonmotile pleomorphic, gram-positive rods, and humans are the only natural host. Strains are found in four biotypes known as *gravis, intermedius, mitis,* and *belfanti;* all four types can cause human disease. Spread occurs through contact with respiratory secretions or infected skin lesions. The genes responsible for DipT production are carried on a chromosomally integrated bacteriophage. Strains of *C. diphtheriae* that do not carry the phage commonly colonize the human respiratory tract but cannot cause clinical diphtheria. Asymptomatic carriers of both toxigenic and nontoxigenic strains have been reported. Nontoxigenic strains are increasingly reported in several countries and have been associated with systemic disease in immunocompromised individuals. The bacteriophage can also be carried by *Corynebacterium ulcerans* or *Corynebacterium pseudotuberculosis,* and diphtheria-like illness has been observed in patients infected with *C. ulcerans.*

Before widespread vaccination, diphtheria was a leading cause of morbidity and mortality in the United States. In the prevaccine era, approximately 70% of diphtheria cases occurred in children younger than 15 years old. Disease was less common in infants younger than 6 months of age, presumably because of the protection provided by maternal antibodies acquired transplacentally. Asymptomatic infections were common. Clinical disease was less common in adults because most had immunity as a result of natural exposure. Immunity from exposure does not appear to be lifelong; however, immunity was maintained by frequent boosting through natural exposure.

In the United States, there were two cases of diphtheria reported between 2004 and 2017. Cases of cutaneous diphtheria continue to occur but are not reportable in the United States. However, respiratory diphtheria continues to cause disease globally in countries with poor routine vaccination coverage, with the annual number of reported diphtheria cases remaining largely unchanged for more than a decade. In the states of the former Soviet Union, a large outbreak of diphtheria involving 157,000 cases and 5000 deaths, primarily in adults, occurred from 1990 to 1998. This outbreak was found to be associated with declines in the public health infrastructure and vaccination coverage rates and demonstrated the importance of maintaining high immunization coverage in all populations. More recently, India has had the largest amount of reported cases annually, with 18,350 cases reported between 2011 and 2015. During this same time period, Southeast Asia accounted for 55% to 99% of all reported diphtheria cases, according to reports made to the World Health Organization (WHO). The majority of cases worldwide occur in adolescents and adults, likely due to increasing vaccination coverage in children and incomplete vaccination or waning immunity in adolescents and adults.

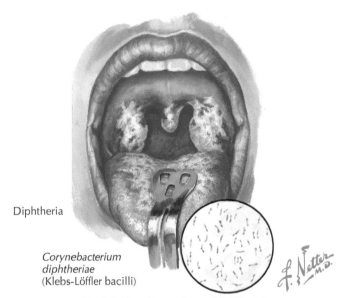

Diphtheria

*Corynebacterium diphtheriae* (Klebs-Löffler bacilli)

**Fig. 2.1** Pseudomembrane (diphtheria).

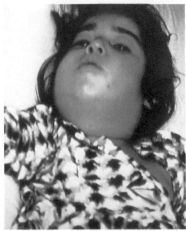

**Fig. 2.2** Classic bull-neck appearance. (From Centers for Disease Control and Prevention, Public Health Image Library, 1995.)

Travelers to regions where diphtheria remains endemic due to inadequate vaccine coverage should ensure that they are up to date with diphtheria immunizations. Similarly, diphtheria should be considered in the differential diagnosis for symptomatic individuals coming from those regions.

## Risk Factors

Increased risk is directly associated with inadequate serum concentrations of DipT-neutralizing antibodies. Notably, immunity from vaccine or natural disease does not appear to be lifelong. In the United States, coverage with diphtheria booster immunization decreases with age, and studies have suggested that 40% to 70% of adults older than 40 years of age are susceptible. In the United States and Canada, circulation of toxigenic strains is uncommon, except in some areas within the northern plains region. Although the opportunity exists for introduction of toxigenic strains into the general population, outbreaks are rare when high immunization rates in infants and children are maintained.

## Clinical Features

The initial diagnosis is based on the observation of classic clinical features. Two main types of infection can occur: a benign, self-limited, and nonspecific skin infection, and a respiratory form that can manifest as a localized nasal infection or a more serious pharyngeal or laryngeal disease. The incubation period is usually 2 to 5 days but can range from 1 to 10 days. The respiratory disease begins gradually with nonspecific symptoms such as fatigue, sore throat, anorexia, and low-grade fever. Approximately 2 to 3 days after onset of symptoms, patients develop the classic pseudomembrane, which is an adherent grayish-white membrane which can cover the tonsils, posterior pharynx, uvula, and/or posterior tongue and will bleed with attempts to remove it (Fig. 2.1). As the disease progresses, patients may develop difficulty swallowing and a hoarse voice. In the most severe forms of respiratory diphtheria, extensive membrane formation and edema can result in airway obstruction; therefore patients should be monitored closely for respiratory compromise. Associated extensive cervical lymphadenopathy and soft-tissue swelling may cause the classic "bull-neck" appearance (Fig. 2.2). Absorption of DipT into the bloodstream can cause serious systemic complications, notably myocarditis with heart block and cranial and peripheral neuropathies.

## Diagnosis

Diphtheria is initially a clinical diagnosis, and laboratory confirmation requires isolation of toxigenic strains of *C. diphtheriae* from the site of infection. Proper culture requires special collection techniques and growth media; few laboratories have maintained this capability. Therefore, in addition to discussion with the local laboratory, an experienced public health microbiology laboratory should be contacted for further guidance. For respiratory diphtheria, cultures should be obtained from the involved nasal or pharyngeal mucosa and should include both the pseudomembrane and the material beneath the membrane. Because there are asymptomatic carriers of nontoxigenic *C. diphtheriae,* confirmation also requires detection of DipT. The Elek immunoprecipitation assay is the traditional way to detect toxin production but is time consuming. The diphtheria toxin gene *(tox)* can be detected by polymerase chain reaction (PCR) testing, but actual production of toxin should be performed by the Elek test.

The differential diagnosis includes pharyngitis from more common causes, including bacterial pharyngitis caused by group A *Streptococcus* or *Arcanobacterium* and viral pharyngitis (e.g., caused by adenoviruses and enteroviruses); infectious mononucleosis from Epstein-Barr virus; and more unusual diseases such as acute necrotizing ulcerative gingivitis (Vincent angina) and severe oropharyngeal candidiasis.

## Clinical Management

Respiratory diphtheria mandates prompt treatment with both antitoxin and antibiotics, supplemented with intensive supportive care. When diphtheria is suspected, treatment with equine DAT should begin before laboratory confirmation is obtained. No licensed product is available in the United States; however, DAT can be obtained from the CDC under an investigational new drug (IND) protocol. The local or state public health departments should be contacted for public health investigations. Antitoxin only neutralizes circulating DipT and has no effect on intracellularly bound toxin; therefore early use is required to minimize the severity of the disease. Because the antitoxin is made from horse serum, it carries the risk of hypersensitivity reactions or serum sickness, and patients should be tested for sensitivity before administration. Individuals with hypersensitivity should receive the antitoxin according to the desensitization procedure provided by the CDC protocol and only in settings equipped for treatment of anaphylaxis.

Antibiotics are also an important aspect of therapy, but they do not replace the use of antitoxin. Although antibiotics have no effect on existing DipT, they will help to prevent further bacterial growth, slow toxin production, and decrease the risk of transmission. Antibiotic treatment consists of a 14-day course of either erythromycin (40 mg/kg/day; maximum 2 g/day given orally or by injection) or penicillin G (300,000 units IM every 12 hours for those weighing 10 kg or less; 600,000 units every 12 hours for those weighing more than 10 kg). Intravenous medications should be used initially but can be transitioned to oral medications as soon as the patient is able to tolerate oral therapy.

Supportive care involves careful respiratory and cardiac monitoring because patients are at risk for airway obstruction as well as arrhythmia and cardiac compromise from myocarditis. Droplet precautions should be maintained for patients with suspected respiratory diphtheria until completion of antibiotic regimens and until two cultures, separated by at least 24 hours, are negative. Contact precautions are recommended for individuals with cutaneous diphtheria, and lesions should be covered. Treatment should include antibiotics and routine management for skin ulcers; DAT is rarely needed. Active diphtheria infection may not induce protective immunity, and all patients should receive an appropriate vaccination series after resolution of the acute illness.

Immunization status should be assessed for all contacts of diphtheria cases, with full catch-up immunization recommended for incompletely immunized contacts. Swabs for diphtheria culture should be taken from all contacts, and a course of erythromycin or penicillin should be administered for 7 days.

## Prognosis

Prognosis depends on the severity of the pharyngeal disease, the extent of respiratory compromise, the duration of disease before initiation of treatment, and the presence of myocarditis. Duration of illness depends on the severity of the disease and resulting complications and can range from a few days to several months. The overall case fatality rate of respiratory diphtheria is 5% to 10%, although it can be higher in children younger than 5 years and adults older than 40 years. Cutaneous diphtheria is rarely fatal.

# TETANUS

## Geographic Distribution and Magnitude of Disease Burden

*Clostridium tetani* is a gram-positive, spore-forming, strictly anaerobic bacterium that typically exhibits a terminal spore and can infect wounds. The spores are found in soil in nearly all areas of the world. Tetanus is typically associated with deep or penetrating wounds that create the anaerobic conditions that facilitate germination, growth of spores, and release of TeNT (often called *tetanospasmin*), the toxin responsible for the clinical manifestations of disease.

TeNT, one of the most toxic molecules known, achieves its toxicity through a series of complex steps that includes migration from the periphery to the central nervous system (Fig. 2.3). TeNT binds to neuronal cells at the site of infection and then is transported centrally, where it interferes with release of inhibitory neurotransmitters. Once the inhibitory control is lost, motor neurons undergo sustained excitation leading to the muscular stiffness and spasms characteristic of tetanus.

In the United States there is an average of 30 tetanus cases reported annually. Tetanus first became a reportable disease in 1947, and since then there has been more than a 95% reduction in cases, attributed in large part to immunization, as well as hygienic improvements in wound management and childbirth practices. Cases have been reported in all age groups; however, the incidence rates tend to increase with increasing age—for example, from 2001 to 2008, 30% of the cases reported to the CDC were in individuals older than 65 years of age, with less than 10% of cases reported in individuals less than 20 years of age. The case fatality rate over the same period was 13%; however, at least 75% of deaths occurred in individuals older than 65 years old.

Worldwide, a steady increase in tetanus toxoid vaccinations has been associated with a decline in cases and deaths. However, tetanus remains a problem in parts of the world with low immunization coverage, with maternal and neonatal tetanus responsible for the great majority of deaths. Maternal tetanus is associated with inadequate vaccination and unhygienic obstetric practices. Neonatal tetanus results from infection of the umbilical stump, particularly in infants born to mothers with inadequate immunization. Since 1989, the WHO has encouraged efforts to reduce maternal and neonatal tetanus (i.e., a reduction of neonatal tetanus incidence less than 1 case per 1000 live births per year in every district); however, as of July 2019, there are still 12 countries who have not achieved Maternal and Neonatal Tetanus Elimination (MNTE) status. In 2017, the WHO estimated that 30,848 newborns died from neonatal tetanus, an 85% reduction from the year 2000. Both maternal and neonatal tetanus are preventable via immunization of mothers, with neonatal protection resulting from transplacentally acquired TeNT-neutralizing antibodies.

## Risk Factors

Mirroring the situation described earlier for diphtheria, higher risk is associated with inadequate concentrations of neutralizing antibodies, usually through a failure to stay up to date with recommended immunizations. Most cases of tetanus occur in individuals who lack TeNT-neutralizing antibodies and incur a wound that promotes the anaerobic growth of *C. tetani*. Conditions that encourage an anaerobic environment include deep puncture wounds, coinfection with other bacteria, devitalized tissue, or presence of a foreign body. Other risk factors include intravenous drug use and diabetes. Most nonneonatal cases occur after a penetrating injury; however, in approximately 30% of cases, the site of infection cannot be identified.

## Clinical Features

Tetanus disease occurs in one of three clinical patterns: generalized (including neonatal), localized, or cephalic. The severity is influenced by the amount of toxin produced as well as the presence of preexisting, albeit not fully protective, concentrations of antibody. The incubation period from the time of inoculation is typically approximately 1 week but can be as rapid as 1 to 3 days or as long as many months. In general, rapid disease progression is associated with a more severe course. The most common and severe form of tetanus is generalized tetanus, accounting for 88% of tetanus cases in the United States. Generalized tetanus is characterized by diffuse tonic contraction of skeletal muscles as well as intermittent, painful muscular spasms (see Fig. 2.3). The disease typically begins with localized muscle spasms of the jaw known as trismus (lockjaw) or other cranial nerve involvement, which progress to include a stiff neck, opisthotonus (spasms and arching of the back), risus sardonicus (sustained muscle spasm of the face), a rigid abdomen, dysphagia, or apnea (caused by the contraction of the thoracic muscles and/or the glottal or pharyngeal muscles). Generalized tetanic spasms lead to a characteristic posture: clenching of the fists, arching of the back, flexion and abduction of the arms, and extension of the legs, often accompanied by apnea. Spasms can be exacerbated by noxious stimuli, including loud noises. Toxin damage to autonomic nerves also leads to autonomic instability, characterized early by irritability and restlessness, sweating, and tachycardia and later by labile blood pressures, cardiac arrhythmias, and fever.

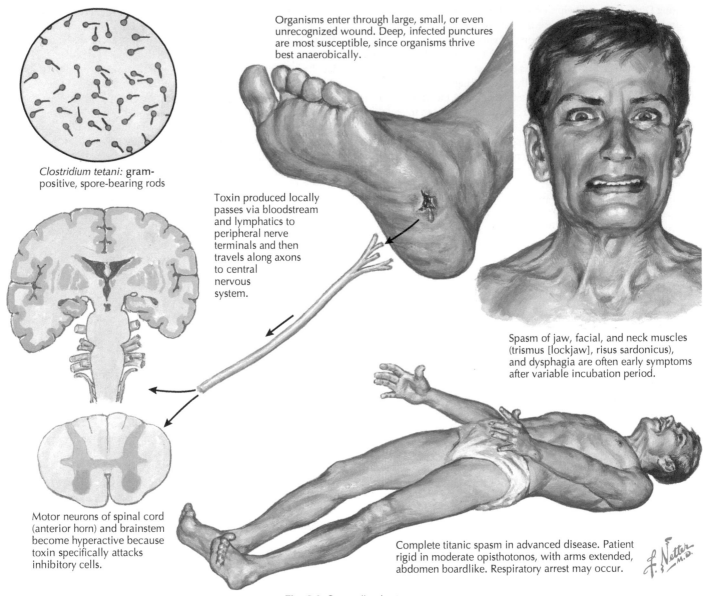

Clostridium tetani: gram-positive, spore-bearing rods

Organisms enter through large, small, or even unrecognized wound. Deep, infected punctures are most susceptible, since organisms thrive best anaerobically.

Toxin produced locally passes via bloodstream and lymphatics to peripheral nerve terminals and then travels along axons to central nervous system.

Spasm of jaw, facial, and neck muscles (trismus [lockjaw], risus sardonicus), and dysphagia are often early symptoms after variable incubation period.

Motor neurons of spinal cord (anterior horn) and brainstem become hyperactive because toxin specifically attacks inhibitory cells.

Complete titanic spasm in advanced disease. Patient rigid in moderate opisthotonos, with arms extended, abdomen boardlike. Respiratory arrest may occur.

**Fig. 2.3** Generalized tetanus.

Neonatal tetanus is a subset of generalized tetanus. Disease onset is very rapid and generally occurs within the first 2 weeks of life. Symptoms are similar to those of generalized tetanus, including diffuse rigidity, muscle spasms, and trismus, leading to complications of apnea and the inability to suck. In addition, seizures have been observed in infants with tetanus.

Localized tetanus is a rare clinical pattern characterized by muscle contraction in a single extremity or body region, accounting for 12% of tetanus cases in the United States. Cephalic tetanus is similar but involves only the cranial nerves, most commonly the facial nerve. It may be associated with otitis media or head lesions and may result in cranial nerve palsies. Many cases of both localized and cephalic tetanus progress to generalized tetanus and may represent an early stage of the disease.

## Diagnosis

Tetanus is diagnosed solely on clinical grounds. Tetanus should be considered in the differential diagnosis of muscle spasms, particularly in patients with a history of an antecedent injury and inadequate

immune status. The differential diagnosis includes drug-induced dystonias, dental infections, neuroleptic malignant syndrome, strychnine poisoning, and stiff person syndrome.

## Clinical Management and Drug Treatment

Effective treatment of tetanus requires a multipronged approach, including neutralization of circulating toxin, reduction in toxin production, medical control of muscle spasms, management of autonomic instability, aggressive supportive care, and immunization. Neutralization of unbound, circulating antibody is achieved by the use of human tetanus immune globulin (TIG), a commercially available product licensed by the US Food and Drug Administration (FDA). TIG is administered intramuscularly as a one-time dose. The optimal therapeutic dose has not been established but experts recommend 500 international units. Prompt administration helps to minimize disease severity. If TIG is not available, human intravenous immune globulin (IVIG) can be used at a dose of 200 to 400 mg/kg; however, the FDA has not approved IVIG for this use. Aggressive wound debridement, as well as antibiotic treatment with metronidazole (30 mg/kg/day divided every 6 hours), can reduce further TeNT production. Penicillin G (100,000 unit/kg/day given in 4- to 6-hour intervals) is an alternative treatment. Antibiotic treatment should continue for 7 to 10 days. Muscle spasms, in addition to being intensely painful, can be life-threatening when they lead to apnea. Limiting stimulation can minimize the frequency of spasms. Additional management includes the use of sedatives such as benzodiazepines or, in severe, refractory cases, neuromuscular blockade. Autonomic instability is often managed with magnesium sulfate, morphine, or beta-blockade. Supportive care is a vital part of the management of tetanus, as full recovery may take several weeks. Many patients require respiratory support, and early tracheostomy should be considered to avoid complications of prolonged intubation. Early enteral nutrition and prompt initiation of physical therapy may speed recovery. Because of the extreme potency of small amounts of tetanus toxin, tetanus disease does not confer lasting protective immunity, and all patients should immediately receive appropriate immunization.

## Prognosis

The prognosis of people with tetanus is dependent on the availability of supportive care, as well as the age and underlying health of the patient. In general, the shorter the incubation period and the time to onset of spasms, the worse the prognosis. Recovery from tetanus requires regrowth of axonal nerve terminals, and therefore the duration of disease can be prolonged (typically 4 to 6 weeks). Because of the availability of supportive care, most patients in developed countries recover. In contrast, in developing countries, without access to intensive care, the case fatality rates vary from 10% to 70%, nearing 100% in the youngest and oldest patients. Most adults and children who survive recover fully, but neonates may have varying degrees of neurologic damage, including intellectual deficits and cerebral palsy.

## PREVENTION AND CONTROL OF DIPHTHERIA AND TETANUS

Vaccination with diphtheria and tetanus toxoids is the basis of prevention and control of these two diseases. Measurement of antitoxin antibodies is possible; however, interpretation of the antibody concentrations is complex and may not readily predict the protection status of the individual. More important than a specific antibody concentration, the key to protection is assurance of the appropriate series, including booster doses, of vaccinations.

In addition to routine vaccinations, tetanus immunizations have a role in management after potential exposure from an injury. Anyone with a clean, minor wound should receive a dose of a tetanus toxoid–containing vaccine if he or she has not had a booster dose in the past 10 years. For all other injuries, including but not limited to dirty, penetrating, or burn wounds, a dose of tetanus toxoid–containing vaccine should be given if the individual has not received a booster in the past 5 years, has received fewer than three vaccinations, or has unknown vaccination status. In addition, prophylactic administration of TIG is recommended for underimmunized patients with serious wounds.

Tetanus toxoid is available as a single antigen (TT) or in combination with pediatric or adult formulations of diphtheria toxoid (DT or Td, respectively) and acellular pertussis antigens (DTaP and Tdap). The diphtheria toxoid component of childhood vaccines has a higher toxoid content (represented by capital D) and is designed to induce a good antibody response in immunologically naïve individuals younger than 7 years of age. For older and previously immunized individuals, a lower toxoid content vaccine (designated by lowercase d) is used because it induces a good antibody response while yielding fewer reactions. Products available vary by country and age, and national agencies should be consulted for the currently recommended products. In the United States the current recommendation is for an initial five-dose series in childhood, with doses at 2, 4, and 6 months, a fourth dose at 15 to 18 months, and a fifth dose at 4 to 6 years of age. At least three doses of each toxoid are required for development of immunity, and additional doses are recommended to maintain this immunity. In the United States, booster doses of diphtheria and tetanus toxoid vaccines are recommended at age 11 to 12 years (combined with reduced dose acellular pertussis vaccines, Tdap) and then every 10 years to maintain protection. For individuals at least 11 years of age who are eligible to receive acellular pertussis vaccine, one dose of Tdap is recommended instead of Td; this can be given earlier than the 10-year mark. Since 2012, the CDC's Advisory Council on Immunization Practices (ACIP) has recommended Tdap immunization in the third trimester of every pregnancy and in October of 2019, the ACIP voted to update vaccine recommendations for booster doses to include Tdap immunization at any point when Td might be administered, anticipated to be incorporated into the CDC's 2020 Immunization Schedules. Anyone with uncertain vaccination history should be considered unvaccinated and receive the recommended age-appropriate series.

## ACKNOWLEDGMENT

The authors acknowledge the work of Bruce D. Meade and Kristin E. Meade on this chapter in the previous edition.

## EVIDENCE

Centers for Disease Control. Tetanus Surveillance—United States, 2001-2008. *MMWR*. 2011;60:365-396. *This resource describes characteristics of recently reported tetanus cases in the United States.*

Dittmann S et al. Successful control of epidemic diphtheria in the states of the former Union of Soviet Socialist Republics: Lessons learned. *J Infect Dis.* 2000;181(1):S10-S22. *This resource describes the largest recent outbreak of diphtheria.*

Liang JL et al. Prevention of pertussis, tetanus, and diphtheria with vaccines in the United States: Recommendations of the Advisory Committee on Immunization Practices (ACIP). *MMWR Recomm Rep.* 2018;67(RR-2):1-44. *This resource provides the latest ACIP recommendations for tetanus, diphtheria, and pertussis vaccinations in the United States.*

World Health Organization. Diphtheria vaccine: WHO position paper—August 2017. *Wkly Epidemiol Rec* 2017;92:417-436. *This resource provides a summary of the recent global epidemiology, discussions of vaccination strategies, and recommendations for prevention of diphtheria.*

World Health Organization. Tetanus vaccines: WHO position paper—February 2017. *Wkly Epidemiol Rec* 2017;92:53-76. *This resource provides a summary of the recent global epidemiology, discussions of vaccination strategies, and recommendations for prevention of tetanus.*

## ADDITIONAL RESOURCES

American Academy of Pediatrics (AAP): Diphtheria. In Kimberlin DW, Brady MT, Jackson MA, Long SS, eds. *Red Book: 2018 Report of the Committee on Infectious Diseases*, ed 31, Itasca, IL: AAP; 2018: 319-323. *This resource provides up-to-date information about diagnosis, treatment, and control of diphtheria.*

American Academy of Pediatrics (AAP): Tetanus. In Kimberlin DW, Brady MT, Jackson MA, Long SS, eds. *Red Book: 2018 Report of the Committee on Infectious Diseases*, ed 31, Itasca, IL: AAP; 2018: 793-798. *This resource provides up-to-date information about diagnosis, treatment, and control of tetanus.*

Brook I. *Clostridium tetani* (Tetanus). In Long SS, Pickering LK, Prober CG, eds. *Principles and Practice of Pediatric Infectious Diseases,* ed 4: Elsevier; 2012: 966-970. *This resource provides information about the diagnosis, treatment and control of tetanus.*

Centers for Disease Control and Prevention (CDC): *Diphtheria.* Available at: www.cdc.gov/diphtheria. *This resource provides up-to-date information about epidemiology and prevention of diphtheria.*

Centers for Disease Control and Prevention (CDC): *Tetanus.* Available at: http://www.cdc.gov/vaccines/vpd-vac/tetanus. *This resource provides up-to-date information about epidemiology and prevention of tetanus.*

Daskalaki I. *Corynebacterium diphtheriae.* In Long SS, Pickering LK, Prober CG, eds. *Principles and Practice of Pediatric Infectious Diseases,* ed 4: Elsevier; 2012: 754-759. *This resource provides information about the diagnosis, treatment and control of diphtheria.*

World Health Organization (WHO): *Health topics. Diphtheria.* Available at: http://www.who.int/topics/diphtheria/en. *This resource provides information about the global epidemiology and prevention of diphtheria.*

World Health Organization (WHO): *Health topics. Tetanus.* Available at: http://www.who.int/topics/tetanus/en. *This resource provides information about the global epidemiology and prevention of tetanus.*

# 3

# Bordetella pertussis and Pertussis (Whooping Cough)

*Sylvia H. Yeh*

## ABSTRACT

Despite the availability of vaccines against pertussis in most developed and developing countries, pertussis remains a significant cause of morbidity and mortality worldwide due to immunity following vaccination and clinical disease waning with time. Routine use of pertussis vaccines has shifted the burden of disease from middle childhood to young infants and older children, adolescents, and adults. The changing epidemiology dictates the need for new vaccines and new vaccine programs targeting these age groups.

## CLINICAL VIGNETTE

A 9-year-old child presents with approximately 2 weeks of coughing. She reports that coughing comes on suddenly and has been increasingly forceful and prolonged, to the point where she feels she cannot breathe. She is anxious about going to sleep for fear of having a coughing fit. She is presently without rhinorrhea and has had no fever or other associated symptoms. There have been no ill contacts at home. Her immunizations are current, and there have been no new exposures to second-hand smoke, pets, or other allergens. There has been no travel outside the area she lives in. On exam, she is afebrile with normal vital signs. No abnormalities are noted on examination.

COMMENT: Due to changing pertussis epidemiology, children 7 to 10 years of age increasingly present with pertussis symptoms, presumably due to waning immunity from their last diphtheria toxoid and tetanus toxoid (DTaP) vaccination and before the recommended adolescent vaccination of tetanus toxoid and reduced-dose diphtheria toxoid (Tdap). Frequently, the illness progresses insidiously over time before it is recognizable as pertussis.

## GEOGRAPHIC DISTRIBUTION AND DISEASE BURDEN

The World Health Organization (WHO) estimates, based on modeling data, that in 2014 worldwide there were 24.1 million pertussis cases and 167,000 deaths in children younger than 5 years of age. This is significantly greater than the 150,000 passively reported cases and approximately 80,000 to 90,000 annual deaths. Global vaccination coverage rate with three doses of pertussis vaccine stands at 86%. Pertussis is endemic globally, with epidemic peaks every 2 to 5 years. This cycling is unchanged even in the postvaccine era, likely because of both an accumulation of susceptible individuals, and in part a result of waning immunity.

## RISK FACTORS

*Bordetella pertussis* is strictly a human pathogen and is readily transmitted by aerosolized droplets. It is the primary cause of clinical whooping cough, with *B. parapertussis*, *B. bronchiseptica*, and *B. holmseii* creating a similar, although typically milder, spectrum of illness. The attack rate for susceptible individuals is estimated to be more than 80%. Globally, as vaccination coverage has increased, the overall disease burden has declined (Fig. 3.1). In the United States, after the introduction of whole-cell pertussis vaccines in the 1940s, disease burden was the lowest in 1976 but has had a steady increase since 1980 (Fig. 3.2). The incidence of disease has increased in infants, especially those too young to have completed the primary immunization series, and in adolescents and adults who experience waning immunity from either vaccination or natural exposure. In the mid-2000s, following the transition to acellular pertussis vaccine, there was an age-specific increase in pertussis clustering among those 7 to 10 years of age, presumably due to waning immunity from acellular pertussis vaccine (Fig. 3.3). This was then followed by an increased disease burden among adolescents 13 to 14 years of age in 2012.

Infants younger than 12 months of age have the greatest risk of morbidity and mortality due to pertussis compared with all other age groups, with those younger than 2 months of age with the highest pertussis-related hospitalization and deaths. Studies from industrialized countries have reported rates of hospitalization because of pertussis ranging from 17 (6- to 11-month-old infants) to 280 (0- to 5-month-old infants) per 100,000 population. From 2004 to 2016, the Centers for Disease Control and Prevention (CDC) reports that among infants hospitalized for pertussis, 54.4% were younger than 2 months of age, and among infants who died, 85.5% were younger than 2 months of age. A study of pertussis deaths in the 1990s revealed a higher than expected rate of death in Hispanic infants and infants born at less than 37 weeks of gestation. Some preliminary data suggest this is due to limited maternal immunity in Hispanic mothers, and for preterm infants it is speculated that the cause is due to decreased placental transfer of maternal immunoglobulin G (IgG). Data suggest that administration of two to three doses of pertussis vaccine within the first 6 months of life is protective against severe disease. With the changes in pertussis epidemiology over time, mothers were originally the most common source of pertussis transmission to infants, whereas siblings are currently the most common source of pertussis infection for infants.

Since 2013, there have been reports of disease-causing strains of *B. pertussis* being deficient in a specific antigen, pertactin, which is one of the common antigenic targets of acellular pertussis vaccine. Subsequent vaccine efficacy studies suggest that the currently available acellular vaccines remain effective regardless of current circulating strains, even those deficient in pertactin. In addition, data do not suggest the pertactin-deficient strains differ in severity of disease compared with non–pertactin-deficient strains.

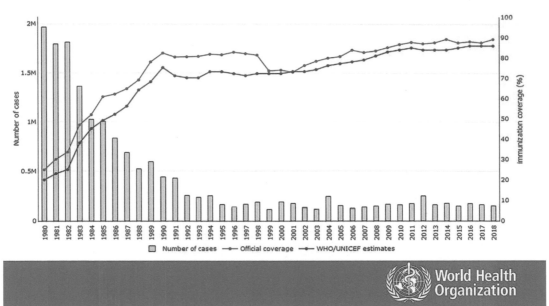

**Fig. 3.1** World Health Organization (WHO) pertussis global annual reported cases and DTP3 coverage 1980–2018. (Reused with permission from WHO/UNICEF, Joint Reporting From July 8, 2019, Immunization, Vaccines, and Biological [IVB], World Health Organization. https://www.who.int/immunization/monitoring_surveillance/burden/vpd/surveillance_type/passive/pertussis_coverage_2018.jpg?ua=1.)

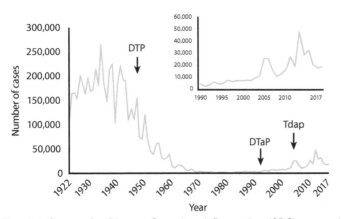

**Fig. 3.2** Centers for Disease Control and Prevention (CDC) reported National Notifiable Diseases Surveillance System (NNDS) pertussis cases: 1922–2017. *DTaP,* Diphtheria toxoid, tetanus toxoid, acelluar pertussis; *DTP,* diphtheria toxoid, tetanus toxoid, whole-cell pertussis *Tdap,* tetanus toxoid, reduced dose diphtheria toxoid, reduced dose acellular pertussis. (Reused with permission from CDC, National Notifiable Diseases Surveillance System and 1922–1949, passive reports to the Public Health Service. https://www.cdc.gov/pertussis/images/incidence-graph-2017.jpg.)

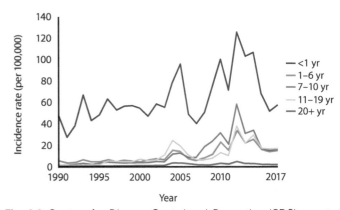

**Fig. 3.3** Centers for Disease Control and Prevention (CDC) reported pertussis incidence by age group: 1990–2017. (Reused with permission from CDC. National Notifiable Diseases Surveillance System, 2017. https://www.cdc.gov/pertussis/images/incidence-graph-age-2017.jpg.)

## CLINICAL FEATURES

After exposure to the bacteria, the average incubation period is 7 to 10 days, which is then followed by onset of symptoms. The clinical illness is divided into three stages: catarrhal, paroxysmal, and convalescent.

### Catarrhal Stage

The catarrhal stage appears similar to the "common cold," with mild cough and coryza, and generally lasts 1 to 2 weeks. Fever is uncommon, and if present it is usually low grade. Unlike a common viral upper respiratory infection, which resolves quickly, in pertussis the cough gradually increases, and infected individuals are most contagious during this phase.

### Paroxysmal Stage

In the paroxysmal stage the coughing persists and gradually increases in severity, resulting in the classic paroxysmal attacks. Whooping may be observed during the paroxysmal phase, which is characterized by the noise of the forced inspiratory effort after the coughing attacks (Fig. 3.4). Posttussive emesis may be observed. This stage may last 6 to 12 weeks and is the period in which complications are most likely to occur. The occurrence of complications is inversely related to age, with the youngest infants having the highest rate of complications. The most common complications include hospitalizations, apnea, pneumonia (primary and secondary), barotrauma events (e.g., subconjunctival hemorrhages, umbilical hernias, pneumothorax), and failure to thrive in infants who are unable to feed due to persistent coughing. Seizures, encephalopathy (likely related to generalized hypoxia), and death occur in less than 2% of patients and are generally seen in the youngest infants, although these events

**Fig. 3.4** Child coughing.

have been reported in adults. Pneumonia is the most common complication of pertussis (from either primary pertussis or secondary bacterial infection) and also the most common cause of pertussis-associated deaths. Infants younger than 4 months of age who died from pneumonia have been found to have histopathologic findings consistent with pulmonary hypertension.

### Convalescent Stage

During the convalescent stage the cough continues to decrease gradually over several weeks to months. Sporadic coughing paroxysms may reappear with subsequent upper respiratory infections during convalescence.

### Atypical Pertussis

Infants younger than 6 months of age and previously immunized older children, adolescents, and adults are less likely to have the features of classic whooping cough. Young infants may have apnea as their only presenting feature, and a history of a coughing household member is often the diagnostic clue. Older children, teens, and adults may have chronic cough of varying severity as their primary sign. Persistent coughing has led to erroneous diagnoses such as asthma or reflux aspiration.

## DIAGNOSTIC APPROACH

### Clinical Diagnosis

Pertussis is a clinical diagnosis, and a high index of suspicion is essential in making the correct diagnosis. When the classic features of whooping cough are present, the diagnosis may be readily entertained. However, because the spectrum of illness is quite varied, pertussis may not be considered. The following clinical case definitions for pertussis have been established by the WHO and the CDC in conjunction with the Council of State and Territorial Epidemiologists:

- In the absence of a more likely diagnosis, a cough illness lasting 2 weeks or longer with one of the following symptoms: paroxysms of coughing, inspiratory "whoop," posttussive vomiting, or apnea with or without cyanosis for infants younger than 1 year of age.
- A probable case is one that meets the clinical case definition and has neither laboratory confirmation nor epidemiologically linkage to a laboratory confirmed cases. For infants younger than 1 year of age, probable cases are defined by meeting the clinical definition and having either a polymerase chain reaction (PCR) positive for pertussis or being a contact with a laboratory-confirmed case of pertussis.

- A confirmed case is one that meets the clinical case definition and is PCR positive for pertussis, or had a contact older than 1 year of age with a laboratory-confirmed case of pertussis, or is any acute cough illness of any duration with isolation of *B. pertussis* from a culture of a clinical specimen.

### Laboratory Studies

Although bacterial culture remains the "gold standard" for laboratory confirmation of pertussis infection, the sensitivity is variable and takes a long time to result. In the past decade, many more cases are being diagnosed via PCR. Serology, preferably paired, is an option for epidemiologic evaluation or confirmation of pertussis cases 2 to 8 weeks following cough onset. Although direct fluorescent antibody (DFA) testing was previously used, it has largely been replaced with PCR due to rapid turnaround and improved sensitivity and specificity of PCR compared with DFA.

### Bacterial Culture

Growth of *B. pertussis* on specialized media (Bordet-Gengou or Regan-Lowe charcoal agar) is considered the gold standard, and although it is 100% specific, the sensitivity is variable and may be quite low. Factors that contribute to the poor sensitivity of culture include (1) inadequate nasopharyngeal (NP) sample, (2) samples obtained late in the illness, (3) prior antibiotic therapy, (4) previous pertussis immunizations, and (5) difficulty growing and identifying this fastidious organism in the clinical microbiology laboratory. Growth of *B. pertussis* may take 3 to 5 days, but generally microbiology laboratories will maintain the culture plates for 7 to 14 days.

### Polymerase Chain Reaction

PCR is an amplified molecular testing tool that often detects sequences in the pertussis toxin gene. In multiple trials evaluating NP samples, PCR has demonstrated higher sensitivity than culture, as well as a more rapid turnaround time. There are commercial assays that include *B. pertussis* in a panel of pathogens detected by multiplex PCR. The limitation of PCR is the potential for contamination yielding false-positive results. Testing via PCR can be done on NP samples up to 4 weeks after cough onset.

### Serology

The most widely available serologic assay is an enzyme-linked immunosorbent assay (ELISA) for antibodies to purified antigens of *B. pertussis*. However, there is no established serologic correlate of immunity or serologic marker of acute infection, nor standardization between commercially available tests. Currently, serologic testing is most informative when simultaneous testing of acute and convalescent serum samples is performed; however, fold-rise in titers may not be observed if the first sample is obtained late in the illness. Serologic testing is most useful for laboratory diagnosis late in the disease, usually 2 to 12 weeks after cough onset.

### Other Laboratory Findings

During the paroxysmal stage, leukocytosis and lymphocytosis are the characteristic hematologic findings, although these may not be seen, especially in older individuals. The degree of leukocytosis can exceed 50,000 cells/mm³, and the absolute lymphocyte count may exceed 10,000 cells/mm³. The degree of lymphocytosis has been shown to correlate with the severity of illness and may result in a hyperviscosity syndrome. Hypoglycemia has been reported in young infants, which may be related to their inability to eat because of coughing or possibly secondary to toxins produced by the organism. Chest radiographs may be normal or may have subtle abnormalities such as peribronchial cuffing, perihilar infiltrates, or atelectasis. Pulmonary consolidation occurs in 20% of hospitalized patents.

## CLINICAL MANAGEMENT

### Supportive Care

Supportive care is the mainstay for management of B. pertussis infection in patients of all ages. Young infants are more likely to need hospitalization because of their risks for apnea, hypoxia, failure to thrive, and other complications. However, more than 3% of pertussis-associated hospitalizations in the United States have been in adults older than 20 years of age. Infants with severe or frequent coughing paroxysms or apnea may require assisted ventilation. Some infants benefit from oxygen supplementation during coughing spells, feedings, and other exertions. Although there are no established criteria for hospitalization for infants, inability to maintain $O_2$ saturation during feeding and paroxysms warrants hospitalization.

For infants younger than 3 months of age, some experts recommend hospitalization for close monitoring. Severe pertussis and death have been associated with extreme leukocytosis (>30,000 white blood cells [WBCs]/mm$^3$). Monitoring WBC counts closely over time is recommended by some experts, and they would recommend double volume exchange transfusion or leukapheresis if there is continuing increase or if counts are more than 30,000 cells/mm$^3$. Although there have been no controlled studies, data suggest that for exchange transfusion to be successful, it needs to be done before evidence of multiorgan failure has occurred. Extracorporeal membrane oxygenation (ECMO) may be required in some cases.

Maintaining adequate hydration and caloric intake may be difficult for infants because of the paroxysms, which often prohibit them from eating and may cause increased caloric expenditure. Thus close attention to fluid and nutritional status is imperative. In addition, whenever possible, known triggers (e.g., exercise, cold temperatures) that cause coughing paroxysms in the child should be avoided.

### Antibiotic Therapy

Antibiotic therapy should be initiated based on a high degree of clinical suspicion, even without laboratory confirmation. Antibiotic treatment is recommended for all infected individuals, regardless of their age or immunization status. The duration of symptoms before treatment seems to be an important factor for the impact of antibiotics. Early treatment (e.g., during the first 7 days of symptoms) may decrease the severity of symptoms, as well as decrease the risk of spread to other susceptible individuals. After 21 days of symptoms, antibiotics may still decrease spread to contacts but likely do little to alter the clinical course. Unfortunately, most patients do not come to medical attention until the paroxysmal phase, when there is likely to be less of an impact on the disease progression.

Azithromycin taken for 5 days is the first line choice for treatment and postexposure prophylaxis. Azithromycin should be used with caution in persons with prolonged QT interval and any proarrhythmic conditions. Oral erythromycin and possibly azithromycin have been associated with infantile hypertrophic pyloric stenosis (IHPS) in infants younger than 1 month of age. In infants younger than 1 month old, the risk of pertussis and its complications outweighs the risks of azithromycin therapy. Therefore the American Academy of Pediatrics (AAP) currently recommends azithromycin for infants younger than 1 month of age even though this is not a US Food and Drug Administration (FDA)-approved indication. Depending on the age of the individual, alternatives may include doxycycline, fluoroquinolones, and trimethoprim-sulfamethoxazole (TMP-SMX). Neither doxycycline nor the fluoroquinolones are routinely recommended for use in young children.

Although it has excellent in vitro activity, TMP-SMX has not been systematically evaluated, and its clinical efficacy is largely unproven. TMP-SMX should not be used in the first 6 weeks of life because of the risks of bilirubin displacement and kernicterus. The β-lactam antibiotics have variable activity against B. pertussis and are not recommended.

### Adjunctive Therapies

Treatment with agents such as antitussives, steroids, and aerosolized bronchodilators have not demonstrated efficacy or benefit for pertussis and are not routinely recommended. Initial studies of B. pertussis immunoglobulin (BPIG) showed an improvement in cough paroxysms; however, development of this product has not been pursued and BPIG is not available.

## PROGNOSIS

As described earlier, infants have the greatest risk of morbidity and mortality, although complications from pertussis have also been reported in adolescents and adults. The prognosis of infants and children appears to be related to the occurrence and severity of complications. For adults, although the illness may be prolonged, recovery is usually complete.

## PREVENTION AND CONTROL: VACCINES

The original vaccines for pertussis were killed whole-cell pertussis organisms, which were later available in combination with diphtheria toxoid and tetanus toxoid (DTwP). These vaccines were effective in reducing the overall burden of pertussis. However, reported temporal associations of adverse events led to a decrease in acceptance of these vaccines in many industrialized nations. Subsequently, pertussis vaccines that were more purified or composed of acellular pertussis antigens (aP) were developed. Currently, several aP vaccines in combination with diptheria toxoid and tetanus toxoid (DTaP) are widely available for use in infants and young children. The aP vaccines have a reduced rate of adverse events compared with whole-cell pertussis vaccines.

Throughout the world, pertussis vaccination generally begins around 2 months of age and consists of three to five doses by age 5 years. In the United States, DTaP is recommended for routine administration to infants at 2, 4, 6, and 12 to 15 months, with a booster dose at 4 to 6 years of age.

Because of the increasing recognition of pertussis in persons 10 years of age and older, who may expose young infants, vaccines targeted to the older age groups have been developed. Reduced-dose acellular pertussis (ap) vaccines combined with tetanus toxoid and reduced dose diphtheria toxoid (Tdap) have been available for individuals 10 years of age and older in the United States since 2005. In the United States, Tdap is recommended for routine administration in early adolescents (11 to 12 years of age), and for one-time use to replace Td booster vaccination in older teens and adults. Tdap vaccines are also recommended for persons (e.g., pediatric healthcare workers; household contacts, and caregivers of young infants) who have contact with individuals at risk for severe pertussis.

For pregnant women, Tdap is recommended to be administered with each pregnancy regardless of prior Tdap receipt history. Ideally, Tdap should be given between 27 and 36 weeks gestation, with earlier in the time frame preferred to maximize the transfer of placental antibody. However, Tdap can be given at any time during pregnancy. For women who did not get Tdap during pregnancy, it is recommended that they receive Tdap in the postpartum period if there is no prior history of Tdap receipt.

After pertussis exposure, DTaP vaccine should be administered as soon as possible for unimmunized children 7 years of age and younger with incomplete pertussis vaccine series or for those with four or fewer doses and for whom more than 3 years have elapsed since the last pertussis vaccine. Tdap should be administered to exposed individuals older than 10 years of age who are unimmunized. Tdap can be administered at intervals as short as 2 years after previous vaccines containing tetanus, diphtheria, or pertussis, during a community outbreak, or in persons with a high risk of complications with pertussis.

Universal immunization of infants and children with the whole-cell and acellular pertussis vaccines has been effective in reducing the disease burden of pertussis. However, with the shifting epidemiology, it has become apparent that new vaccines and delivery programs are necessary to control this human pathogen. Recently in many industrialized countries, routine immunization programs that included only vaccination of infants and children with DTwP or DTaP have been revised to include Tdap for adolescents and adults. The goals of expanding pertussis vaccinations to include individuals 10 years of age and older are not only to reduce infections in the older age groups but ultimately, by decreasing pertussis circulation in the community, to provide protection for infants too young to be immunized.

## EVIDENCE

Breakwell L, Kelso P, Finley C, et al. Pertussis vaccine effectiveness in the setting of pertactin-deficient pertussis. *Pediatrics*. 2016;137(5):e20153973. *This paper details vaccine efficacy in light of pertactin-deficient pertussis.*

Faulkner AE, Skoff TH, Tondella L, Cohn A, Clark TA, Martin SW. Trends in pertussis diagnostic testing in the United States, 1990 to 2012. *Pediatr Infec Dis J.* 2016;35:39-44. *This paper provides the trends in diagnostic testing for pertussis over time.*

Nieves D, Bradley JS, Gargas J, et al. Exchange blood transfusion in the management of severe pertussis in young infants. *Pediatr Infect Dis J.* 2013;32:698-699. *This paper provides clinical experience of 10 infants where exchange transfusion was done.*

## ADDITIONAL RESOURCES

Centers of Disease Control and Prevention (CDC): Pertussis (Whooping Cough)—Clinicians. Available at: https://www.cdc.gov/pertussis/clinical/index.html. Accessed September 2019. *Provides updated information and recommendations for clinicians related to pertussis.*

Centers for Disease Control and Prevention (CDC): Prevention of pertussis, tetanus, and diphtheria with vaccines in the United States: Recommendations of the Advisory Committee on Immunization Practices (ACIP). *MMWR Recomm Rep* 2018;67:RR-2 April 27, 2018. *This resource provides updated information for the epidemiology recommendations for the prevention of pertussis in the United States.*

Cherry JD, Heininger U: Pertussis and other Bordetella infections. In Feigin RD, Cherry JD, eds: *Textbook of pediatric infectious diseases*, ed 4, Philadelphia, 1998, WB Saunders, p 1423. *This resource provides a review of pertussis epidemiology, pathogenesis, clinical manifestations, and management.*

# 4

# *Haemophilus influenzae* Type b

*Sylvia H. Yeh*

## ABSTRACT

*Haemophilus influenzae* type b (Hib) was a leading cause of bacterial sepsis and meningitis in children before the widespread availability of conjugate vaccines. Globally, the burden and incidence of Hib disease has dramatically declined with increases in vaccine uptake.

## CLINICAL VIGNETTE

A 3-year-old boy presents to the emergency department with high fever and difficulty breathing for 1 day. There is mild rhinorrhea but no rash. The breathing problems began suddenly, and there are no ill contacts. His parents have previously refused routine vaccinations. On exam, he is febrile, tachycardic, and in obvious respiratory distress with his upper body leaning forward and head tilted up. There is audible stridor at rest.

COMMENT: With the widespread use of Hib vaccination, epiglottitis is a very rare clinical presentation in the United States. When present, it is a medical emergency, necessitating rapid availability of sedation and intubation to minimize airway collapse.

## GEOGRAPHIC DISTRIBUTION AND DISEASE BURDEN

*Haemophilus influenzae* is a pleomorphic, gram-negative coccobacillary human pathogen that can be transmitted person to person via respiratory droplets. It is the causative agent for a wide range of infections whose severity is related to the presence or lack of a bacterial capsule. Encapsulated (typeable) strains of *H. influenzae* account for the vast majority of invasive disease, with Hib being the predominant serotype, especially in children younger than 5 years of age. Before widespread availability of conjugate vaccine in 1991, Hib was a leading cause of bacterial meningitis, occult bacteremia, epiglottitis, pneumonia, empyema, cellulitis, septic arthritis, and pericarditis. Invasive Hib infections occurred in 1 in 200 children in the United States during the first 5 years of life. This incidence has dramatically declined to approximately 1 per 100,000 over the past 15 years, reflecting the success of Hib conjugate immunization. Among all age groups in the United States, nontypeable *H. influenzae* currently causes the majority of invasive *H. influenzae* disease. In 2012, the World Health Organization (WHO) estimated that Hib was the cause of 2% of child deaths younger than 5 years of age due to all causes. As of 2018, the WHO estimated the global vaccination coverage with three doses of Hib vaccine to be 72%.

## RISK FACTORS

Antibodies against the Hib polysaccharide capsule, polyribosylribitol phosphate (PRP), are protective against invasive disease. Young age is a primary risk factor, with infants 6 to 18 months of age having the greatest risk for invasive Hib disease because of their paucity of PRP antibodies. Infants younger than 6 months of age likely have some protection from passively acquired maternal antibodies. Underlying conditions such as human immunodeficiency virus (HIV), sickle cell anemia, functional asplenia, antibody or complement deficiency syndromes, and malignancy are also risk factors for Hib disease. Environmental exposures such as crowding, household size, daycare attendance, low family income, and low parental education level are also risk factors. Conversely, breastfeeding has been demonstrated to be protective against invasive Hib disease. In the prevaccine era, Hib invasive disease was rare in children older than the age of 5 years, primarily related to antibody acquisition from natural exposure to the organism.

Significantly increased risk of invasive Hib disease has been reported in indigenous populations, including Australian Aboriginal children, Native Alaskan Eskimos, Native Americans, and Canadian First Nations children. These population differences potentially relate to a variety of factors, including early exposure, crowding, microbiologic differences in the circulating strains, socioeconomic factors, and possibly genetics. In adults, underlying conditions such as chronic obstructive pulmonary disease, smoking, HIV infection, alcoholism, pregnancy, splenectomy or functional asplenia, and malignancy increase the risk of invasive Hib disease.

## CLINICAL FEATURES

Although Hib can cause a variety of respiratory tract infections, in the prevaccine era it was notorious for causing significant invasive disease in infants and young children and may occur simultaneously at multiple sites. Although these infections are currently rare in immunized populations, practitioners should be familiar with their clinical manifestations because of possible resurgence of disease related to vaccine shortages and the potential for caring for unimmunized children, either due to vaccine refusal or coming from areas of the world without access to Hib vaccines. The following features are described as classically seen in the prevaccine era.

### Cellulitis

Cellulitis is a somewhat unique manifestation of Hib disease seen in infants younger than 1 year of age. Hib cellulitis rarely occurs on the extremities, with most cases (74%) occurring at buccal, periorbital, or cervical areas. Facial cellulitis in infants often manifests with acute fever, with a unilateral area of induration, warmth, and tenderness, which may progress to have a violaceous hue. An aspirate of the point of maximal swelling will typically yield organisms. Facial cellulitis from Hib is often associated with bacteremia, and 10% to 20% of children will have a secondary focus, including meningitis.

## Epiglottitis

Classically, epiglottitis from Hib occurs in older children, 2 to 7 years of age. Signs may begin with abrupt onset of high fever and drooling, with rapid progression to significant respiratory distress with children assuming a tripod position (sitting leaning forward, with mouth open and tongue and jaw protruding) (Fig. 4.1) to allow air entry. Approximately 70% to 90% of patients with epiglottitis have positive blood cultures. The mortality rate is 5% to 10%, and death is usually related to abrupt airway obstruction. Children in respiratory distress exhibiting the tripod position should not have their oropharynx examined without the presence of suitable personnel and equipment for rapid intubation. In children younger than 2 years of age, Hib epiglottitis typically has a rapid onset and progression leading to severe blockage of the airway.

## Pneumonia

Hib pneumonia was common, often severe, and frequently associated with empyema, pericarditis, and multilobe involvement. Hib pneumonia is usually preceded by an upper respiratory infection such as a "common cold" with fever and cough and is clinically indistinguishable from pneumonia caused by other bacterial agents. Frequently a peripheral leukocytosis with a polymorphonuclear predominance is observed. Culture of blood, pleural fluid, tracheal aspirate, or lung aspirate may be positive in 75% to 90% of cases. The disease process is usually acute and not associated with long-term pulmonary dysfunction.

## Septic Arthritis and Osteomyelitis

Hib was the leading cause of septic arthritis for children younger than 2 years of age and can be associated with contiguous osteomyelitis. Presenting signs include fever and limited limb use. Approximately 90% of septic arthritis cases involve a single large joint such as the knee, ankle, elbow, or hip. Long-term cartilage damage may result from Hib arthritic degeneration. Hib antigen detection is often positive from infected articular fluid.

**Fig. 4.1** Tripod position assumed by child with epiglottitis.

## Bacteremia and Sepsis

Hib was the second leading cause of occult bacteremia in febrile children 6 to 36 months of age. Unlike bacteremia with *Streptococcus pneumoniae* (the leading cause), which may spontaneously resolve, bacteremia with Hib is commonly associated with disseminated infections. Of children with occult Hib bacteremia, 30% to 50% had a focus (e.g., meningitis, pneumonia, or cellulitis).

## Meningitis

Meningitis derives from high-grade bacteremia that seeds the cerebrospinal fluid (CSF) through the choroid plexus and eventually reaches the arachnoid villi to cause meningitis. There are no clinical features that differentiate Hib meningitis from meningitis from other bacterial causes (see Fig. 4.2A). Typically in bacterial meningitis the child demonstrates fever, irritability, and nuchal rigidity. Classical signs of involuntary flexion of the knees with passive flexion of the neck (Brudzinki sign) and inability to extend a flexed leg (Kernig sign) may be seen in older children and adults (see Fig. 4.2B). Altered mental status, vomiting, or seizures may also be presenting symptoms and signs. Hib meningitis can have a fulminant course with clinical sepsis, disseminated intravascular coagulation (DIC), and rapid neurologic deterioration leading to increased intracranial pressure, seizure, coma, and respiratory arrest. Approximately 10% to 20% of children with Hib meningitis will have additional foci of infection (e.g., cellulitis, arthritis, or pneumonia) with their concomitant bacteremia.

In fulminant meningitis, 80% of CSF Gram stain analyses will demonstrate the organism. However, the ability to identify the organisms on Gram stain is diminished with prior antibiotic administration. Although previous antibiotic administration may negatively affect the ability to isolate the organism from culture, it will not affect the overall CSF cell count differential, glucose, or protein, which have the usual abnormalities associated with bacterial meningitis. Peripheral blood may reveal anemia, leukocytosis, thrombocytosis, or thrombocytopenia.

## DIAGNOSTIC APPROACH

Evaluation of a young child with fever should be directed by history, including vaccinations and risk factors for serious invasive disease (e.g., age or underlying condition), and physical findings. In young infants with fever without a specific source, complete blood count, urinalysis, and potentially a chest radiograph and lumbar puncture to assess for CSF pleocytosis should be considered, along with appropriate cultures of blood, urine, CSF, and any nidus found on physical examination. The diagnosis of infection due to Hib is usually made by isolation of the organism by culture. Most invasive Hib disease is associated with bacteremia; therefore blood cultures should be performed if Hib is a consideration, even if the probable primary focus of infection is thought to have been identified (e.g., pneumonia or buccal cellulitis). Culture specimens should be processed quickly because the organism is fastidious. Once *H. influenzae* has been isolated, it can be serotyped, usually in a public health laboratory facility.

Nucleic acid amplification tests (NAAT) have been developed for detection of *H. influenzae* organisms but are not serotype specific. At least one multiplex assay is licensed in the United States for the detection of *H. influenzae* in CSF samples. The American Academy of Pediatrics Committee on Infectious Diseases recommends NAATs be done in conjunction with bacterial culture to allow for serotype testing and surveillance and for monitoring antimicrobial susceptibilities. However, NAAT may be useful in the setting of antibiotic pretreatment when cultures may become negative before sampling.

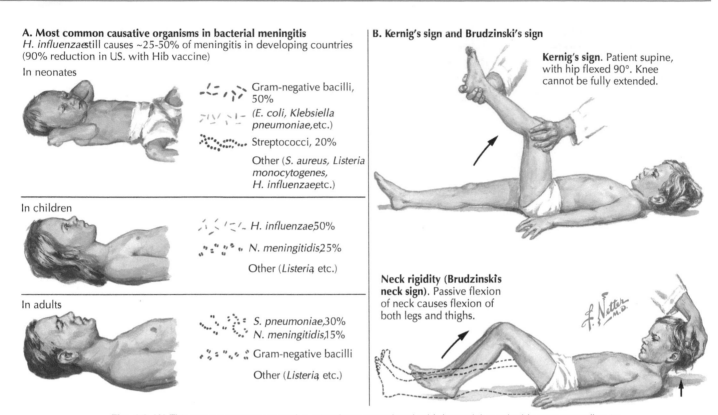

**A. Most common causative organisms in bacterial meningitis**
*H. influenzae* still causes ~25-50% of meningitis in developing countries (90% reduction in U.S. with Hib vaccine)

In neonates

Gram-negative bacilli, 50%
(*E. coli, Klebsiella pneumoniae,* etc.)

Streptococci, 20%

Other (*S. aureus, Listeria monocytogenes, H. influenzae,* etc.)

In children

*H. influenzae* 50%

*N. meningitidis* 25%

Other (*Listeria,* etc.)

In adults

*S. pneumoniae,* 30%
*N. meningitidis,* 15%

Gram-negative bacilli

Other (*Listeria,* etc.)

**B. Kernig's sign and Brudzinski's sign**

**Kernig's sign.** Patient supine, with hip flexed 90°. Knee cannot be fully extended.

**Neck rigidity (Brudzinski's neck sign).** Passive flexion of neck causes flexion of both legs and thighs.

**Fig. 4.2** (A) The most common causative organisms associated with bacterial meningitis vary according to age group. (B) Kernig's sign and Brudzinski's sign, if present during examination of older children and adults, have a high predictive value for bacterial meningitis.

## CLINICAL MANAGEMENT

### General Management

Increasing rates of Hib resistant to multiple antimicrobial agents have been reported worldwide. Approximately 50% of Hib isolates in the United States have evidence of plasmid-mediated β-lactamase production (greater for nontypeable strains than type b). This β-lactamase production renders the organism resistant to ampicillin and amoxicillin. Use of β-lactamase inhibitors, such as found in amoxicillin-clavulanic acid, restores the antibiotic killing activity against such strains. Second-, third-, and fourth-generation cephalosporins are generally active against Hib, including strains with β-lactamase production. Extended-spectrum macrolide antibiotics such as clarithromycin and azithromycin are also generally active against most strains of Hib, including those with resistance due to β-lactamase. Trimethoprim-sulfamethoxazole is active against Hib, but rates of resistance to this agent are increasing. Levofloxacin and other quinolones also have potent activity against Hib but are inappropriate for use in pediatric patients when there are safer alternative medications. The selection of antibiotic for treatment of Hib disease should be based on (1) suspected site of infection, (2) need for penetration of the blood-brain barrier and achievable bactericidal activity, (3) local antibiotic susceptibility of invasive isolates, and (4) duration needed for sterilization of both primary and secondary foci.

### Invasive Disease

Pending determination of cause and susceptibility, the empiric treatment of suspected Hib meningitis should include a third-generation cephalosporin (cefotaxime or ceftriaxone) because of the ability of these agents to cross the blood-brain barrier and their resistance to β-lactamase activity. For other types of invasive Hib disease (pneumonia, septic arthritis, periorbital cellulitis), cefuroxime or a third-generation

cephalosporin can be used for empiric therapy. Aminoglycosides are not recommended despite in vitro susceptibility.

Studies of pediatric patients with Hib meningitis suggest that adjunctive use of dexamethasone decreases the inflammatory response and may lessen the likelihood of hearing loss. The dose of dexamethasone is 0.6 mg/kg/day divided every 6 hours for a total of 4 days. Ideally, the first dose of dexamethasone should be administered just before or concurrently with the first antibiotic dose.

For abscesses, subdural empyema, pleural empyema, and pericardial effusions, percutaneous or surgical drainage is often necessary. In cases of septic arthritis, aspiration of infected joint fluid often is necessary for both diagnosis and reduction of intraarticular pressure. Repeated aspiration or placement of a surgical drain may also be necessary.

In addition, adjunctive and supportive therapies are important in the management of children with invasive Hib disease. For meningitis, careful evaluation and anticipation of potential complications, such as shock, syndrome of inappropriate secretion of antidiuretic hormone (SIADH), seizures, subdural empyema, and evaluation for development of secondary foci are important. Fever may persist for prolonged periods in Hib meningitis, with approximately 10% of children remaining febrile for at least 10 days. For epiglottitis, airway management is critical, often dictating the need for intubation before the occurrence of airway obstruction.

Duration of therapy is determined by the site of infection, clinical response, and underlying host factors. For bacteremia, sepsis, and uncomplicated Hib meningitis, 10 days of antibiotics is generally sufficient. For cellulitis treatment, transition to oral antibiotics after a period of parenteral antibiotics can be made if there is a good clinical response. Similarly, for septic arthritis, pericarditis, empyema, and osteomyelitis, although they may require a longer duration of antibiotics (3 to 6 weeks), antibiotics may be switched from parenteral to

## TABLE 4.1   *Haemophilus influenzae* Type b Conjugate Vaccines

| Hib CONJUGATE | PRP-OMP | PRP-T |
|---|---|---|
| Carrier protein | *Neisseria meningitides* group B outer membrane protein | Tetanus toxoid |
| Trade name (manufacturer) | PedvaxHIB (Merck) | ActHIB (Sanofi Pasteur) Hiberix (GlaxoSmithKline) |
| Dosing schedule in United States (age in months) | 2, 4, 12–15 | 2, 4, 6, 12–15 |
| Available combinations (trade name) | | DTaP/PRP-T/IPV (Pentacel, Sanofi Pasteur for 2, 4, 6, 12–15 months of age) |

*DTaP*, Diphtheria, tetanus, acellular pertussis; *Hib, Haemophilus influenzae* type b; *IPV*, inactivated poliovirus vaccine; *PRP-OMP*, polyribosylribitol phosphate conjugated to outer membrane protein; *PRP-T*, polyribosylribitol phosphate conjugated to tetanus toxoid.

oral administration after documentation of susceptibility, good therapeutic response, adequate antimicrobial blood levels, and ensured compliance.

## PROGNOSIS

With the exception of meningitis and fulminant sepsis, patients with invasive Hib disease, if treated early with appropriate antibiotics, may recover with minimal to no long-term sequelae. However, even with prompt intensive care, mortality from Hib meningitis is approximately 5%. Complications occur early in the disease course and include seizure, cerebral edema, subdural effusions or empyema, SIADH, cortical infarction, cerebritis, intracerebral abscess, hydrocephalus, and, rarely, herniation. Routine imaging such as head computed tomography or magnetic resonance imaging is not necessary but can help to clarify focal neurologic findings or complications that occur during the clinical course, especially subdural empyema. Small sterile subdural effusions are common findings on imaging but are usually of no clinical significance.

Long-term sequelae occur in approximately 15% to 30% of meningitis survivors. These sequelae include sensorineural hearing loss, delay in language acquisition, developmental delay, gross motor abnormalities, vision impairment, and behavior abnormalities. A substantial proportion of these abnormalities may resolve over time, and therefore long-term monitoring is necessary. Evaluation of hearing during the initial hospitalization and follow-up if a hearing loss is detected are necessary to provide the earliest interventional services necessary should a hearing deficit become permanent.

## PREVENTION AND CONTROL

### *Haemophilus influenzae* Type b Immunoprophylaxis

The first vaccine developed for the prevention of invasive Hib was a purified Hib capsular PRP polysaccharide vaccine. However, polysaccharide vaccines are poorly immunogenic in young children, especially those younger than 18 months of age, owing to the polysaccharide's inability to induce T cell–dependent response. Linking a polysaccharide antigen to an immunogenic carrier protein allows for T-cell recognition, leading to inducible antibody responses and long-term protection in young infants.

Four Hib conjugate vaccines have been developed, but only two conjugate formulations are used in the United States, with three manufacturers and one available in a combination vaccine (Table 4.1). In general, it is ideal to complete the primary series with the same Hib conjugate vaccine; however, if this is not possible, the vaccines can be interchanged. In this instance three doses of the primary series are required. The booster dose at 12 to 15 months of age can be with any Hib conjugate vaccine, regardless of the type used in the primary series. Unimmunized children older than 59 months of age with underlying conditions that increase the individual risk of Hib disease should receive a single dose of Hib conjugate vaccine. For individuals who are younger than 59 months of age

with HIV or immunoglobulin G2 (IgG2) subclass deficiency, and who are unimmunized, two doses of Hib conjugate vaccine given 4 to 8 weeks apart are suggested.

### Chemoprophylaxis

Postexposure chemoprophylaxis is recommended for all household members of cases, if there is a contact younger than 4 years of age who has not received all Hib vaccinations appropriate for his or her age or if there is an immunocompromised child contact regardless of his or her immunization status. Prophylaxis should be initiated as soon as possible and ideally within 2 weeks of the onset of disease in the index case. Rifampin is the drug of choice for chemoprophylaxis in the dose of 20 mg/kg once daily (maximum daily dose of 600 mg) for 4 days. Other agents such as ampicillin, trimethoprim-sulfamethoxazole, erythromycin-sulfisoxazole, and cefaclor have been shown to be ineffective for chemoprophylaxis. Recommendations for treatment of contacts in daycare centers are controversial, but most experts recommend prophylaxis if two or more cases of Hib disease have occurred among attendees within 60 days.

## EVIDENCE

Cochi SL, Broome CV: Vaccine prevention of H. influenzae type b disease: past, present and future, *Pediatr Infect Dis J* 1986;5:12-19. *This reference provides information about the development and activity of Hib vaccines.*

Koomen I, Grobbee DE, Roord JJ, et al: Hearing loss at school age in survivors of bacterial meningitis: assessment, incidence, and prediction, *Pediatrics* 2003;112:1049. *This reference provides data about the sequela of hearing loss after bacterial meningitis.*

## ADDITIONAL RESOURCES

Allen CH: Fever without a source in children 3 to 36 months of age. In Kaplan S, Fleischer G, eds: *2009 UptoDate Version 17.2* (Revised), Wellesley, MA, 2009. Available at: www.uptodate.com. *This resource provides a review of evaluation of infants and young children with fever, as it relates to identification and management of invasive bacterial diseases, including Hib infection.*

American Academy of Pediatrics (AAP) Committee on Infectious Diseases: Haemophilus influenzae infections. In Kimberlin DW, Brady MT, Jackson MA, Long SS, eds: *Red Book: 2018-2021 Report of the Committee on Infectious Diseases*, ed 31, Elk Grove Village, IL, 2018, AAP, pp 367-378. *This resource provides an overview of the diagnosis, treatment, and prevention of Haemophilus influenzae infections.*

Centers for Disease Control and Prevention (CDC): Hib vaccination recommendations. Available at: www.cdc.gov/vaccines/vpd-vac/hib/default.htm. *This site provides the current recommendations for prevention of Hib infection.*

Ward JI, Zangwill KM: *Haemophilus influenzae* vaccines. In Plotkin SA, Orienstein WA, eds: *Vaccines*, ed 3, Philadelphia, 1999, WB Saunders. *This resource provides a review of Hib epidemiology, vaccine history, and immunology.*

World Health Organization. Immunizations, Vaccines and Biologicals: Haemophilus influenza type b (Hib). Available at: https://www.who.int/immunization/diseases/hib/en/. Accessed September 2019.

# 5

# Pneumococcal Disease: Infections Caused by *Streptococcus pneumoniae*

*R. Douglas Pratt*

## ABSTRACT

*Streptococcus pneumoniae* (pneumococcus) is a gram-positive encapsulated bacterium that causes significant morbidity and mortality across all age groups. *S. pneumoniae* can be carried asymptomatically in the nasopharynx, and it can cause a wide range of diseases from upper respiratory infections including sinusitis and otitis media, lower respiratory infections (most commonly pneumonia), and invasive disease, including bacteremia and meningitis. Children younger than age 2 years, the elderly, and individuals with immunocompromise or anatomic or functional asplenia are most susceptible to invasive disease. Treatment is guided by severity of disease, site of infection, and susceptibility to antimicrobials. A 23-valent polysaccharide vaccine has been available in the United States for use in adults and high-risk children ages 2 years and older since 1983. A polysaccharide-protein conjugate vaccine covering 7 serotypes became available in the United States in 2000, followed by a 13-valent vaccine in 2010. Conjugate pneumococcal vaccines have been highly effective in preventing invasive disease in infants and young children, as well as affording indirect protection of the elderly population via herd immunity. The 13-valent vaccine was also shown to prevent pneumococcal pneumonia and invasive disease by active immunization in an elderly population. Pneumococcal conjugate vaccines are recommended for routine immunization of infants and young children globally.

## CLINICAL VIGNETTE

A 22-month-old child was admitted to the hospital with complaints of high (39.4°C) fever, headache, vomiting, and impaired consciousness. On the basis of findings from physical examination and initial laboratory results, a working diagnosis of bacterial meningitis was made and empirical ceftriaxone and vancomycin therapy were initiated. The cerebrospinal fluid culture yielded penicillin-susceptible pneumococci, and the isolate was identified as serotype 35F by quellung reaction. The patient fully recovered with 14 days of tailored therapy without any complications during follow-up.

COMMENT: Following the introduction of the 13-valent pneumococcal conjugate vaccine (PCV13) into mass infant vaccination programs, invasive pneumococcal disease (IPD) due to the vaccine serotypes has tended to decrease in both vaccinated young children and nonvaccinated age groups due to herd immunity. However, IPD remains a risk, particularly in vaccinated children younger than 2 years of age, children with primary/secondary immunodeficiencies, and adults older than the age of 65. PCV13 serotypes currently comprise a minority of all IPC cases, with non-PCV13 serotypes now predominant.

## GEOGRAPHIC DISTRIBUTION AND BURDEN OF DISEASE

The World Health Organization (WHO) estimates that pneumococcal infections accounted for approximately 5% of all-cause child mortality in children younger than 5 years of age in 2008. Countries implementing routine use of pneumococcal conjugate vaccines in infancy have seen a marked reduction in the incidence of serious diseases due to pneumococcal serotypes in the vaccines. Still, pneumonia is the single largest infectious cause of death in children worldwide. The WHO estimated that pneumonia accounted for 15% of all deaths of children younger than 5 years old, *S. pneumoniae* being the most common cause of bacterial pneumonia in children. Among adults living in Europe and the United States, *S. pneumoniae* is the most common cause of community-acquired bacterial pneumonia.

Distribution of serotypes varies temporally and geographically. Outbreaks of pneumococcal disease caused by the same serotypes have been reported in institutional settings; however, epidemics in the general population are rare in developed countries.

In temperate regions, invasive pneumococcal disease (IPD), defined as an infection of a normally sterile body site, is more common during winter. Close proximity indoors and spread of viral respiratory pathogens facilitate transmissibility and invasiveness.

In low-income countries, data about the burden of pneumococcal disease are limited. Using information from hospital-based studies and vaccine efficacy trials and inferring from disease patterns among native populations, the estimated burden of disease is high. Rates of IPD among young children have been reduced substantially in countries in which the conjugate vaccines are in widespread use. After introduction of conjugate vaccines in the United States, annual rates of IPD in children young decreased from 100 cases per 100,000 people in 1998 to 9 cases per 100,000 in 2015, and IPD caused by the 13 conjugate vaccine serotypes decreased from 91 cases per 100,000 people in 1998 to 2 cases per 100,000 people in 2015.

*S. pneumoniae* is a common pathogen in middle ear aspirates from children with acute otitis media (AOM). AOM is also the leading reason for prescribing antibiotics during childhood, and this use contributes substantially to increased antimicrobial resistance.

## MICROBIOLOGY AND PATHOGENESIS

*S. pneumoniae* is a strictly human pathogen consisting of gram-positive, encapsulated, lancet-shaped bacteria occurring in pairs (called *diplococci*) and chains (Fig. 5.1). At least 90 serotypes have been identified based on differences in the polysaccharide capsule, as observed with the quellung reaction, in which serotype-specific antibodies bind to the bacterial capsule causing the bacteria to appear opaque and enlarged on microscopy (see Fig. 5.1).

The polysaccharide capsule is considered the primary virulence factor of *S. pneumoniae*. Nonencapsulated strains are less likely to cause severe disease. The capsule contributes to pathogenesis by protecting the organism from phagocytosis. Antibody binding to the capsule can facilitate phagocytosis and bacterial killing by phagocytic cells. Some pneumococcal proteins, including hyaluronate lyase, pneumolysin,

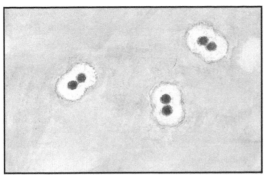

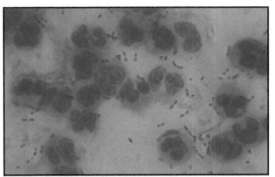

**Quellung reaction.** Swelling of bacterial capsule when exposed to antibody

Purulent sputum with pneumococci (Gram stain)

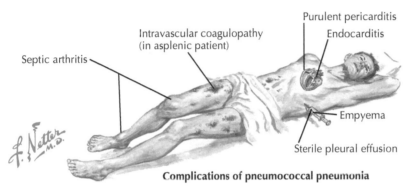

Complications of pneumococcal pneumonia

Fig. 5.1 Pneumococcal disease.

neuraminidase, major autolysin, choline-binding protein A, and pneumococcal surface protein A (PspA) have functions that also contribute to pathogenicity. These virulence proteins have also been considered for development as vaccine antigen candidates.

## RISK FACTORS

Children younger than age 2 years carry the highest burden of *S. pneumoniae* disease worldwide. Other groups at increased risk of invasive disease include the elderly, people with impaired immunity (e.g., humoral immunodeficiency, complement deficiency, human immunodeficiency virus [HIV] infection, and anatomic or functional asplenia [e.g., sickle cell disease]), cigarette smokers, those with cochlear implants and cerebrospinal fluid (CSF) leaks, and individuals with chronic diseases such as diabetes and heart, kidney, and lung disorders. IPD occurs at higher rates among certain ethnic groups; for example, prior to widespread use of conjugate pneumococcal vaccines, Native Americans and Eskimos had rates of IPD that were several-fold higher than in the general population in the United States.

## CLINICAL PRESENTATION

*S. pneumoniae* organisms are transmitted person to person via contact with infected respiratory droplets. Asymptomatic nasopharyngeal (NP) carriage is common in children 6 months to 5 years of age, and they are a source of transmission among close contacts by projecting droplets (>5 μm) across a short distance (1 m).

### Invasive Pneumococcal Disease

*S. pneumoniae* infection can manifest as fever and bacteremia without a focus or as an infection of any organ system, including AOM, sinusitis,

conjunctivitis, periorbital cellulitis, soft-tissue infection, pyogenic arthritis, osteomyelitis, community-acquired pneumonia (CAP), empyema, endocarditis, peritonitis, sepsis, and meningitis (see Fig. 5.1).

Onset of fever is typical for infections in all tissues and may be the only sign in young children with bacteremia. Otitis media typically causes pain in the ears. Pneumonia can cause fever, cough, pleuritic chest pain, and purulent or blood-tinged sputum.

Pneumococcal meningitis is characterized by high fever, headache, neck stiffness, and altered mental status. Other symptoms include nausea, vomiting, photophobia, and lethargy. Infants and young children may have nonspecific symptoms, including irritability or poor feeding. In later stages, patients of any age may have seizures or focal neurologic findings or may be comatose.

The differential diagnosis for IPD includes other bacterial pathogens, such as *Haemophilus influenzae* type b (Hib), *Neisseria meningitidis*, and *Staphylococcus aureus*. In the United States and other areas using Hib vaccines, most cases of bacterial meningitis are caused by *S. pneumoniae* and *N. meningitidis*. Viral meningitis caused by enteroviruses can manifest like bacterial meningitis but is generally less severe.

Before the widespread use of the conjugate vaccine in the United States, *S. pneumoniae* was the most common bacterial cause of pneumonia in young children. Other bacterial causes of CAP in children include *H. influenzae* (nontypeable), *Moraxella catarrhalis*, *Streptococcus pyogenes*, and *S. aureus* (including methicillin-resistant *S. aureus* [MRSA]). However, respiratory viruses (e.g., respiratory syncytial virus [RSV], influenza, human metapneumovirus) cause most cases of pneumonia in children.

## DIAGNOSTIC APPROACH

When *S. pneumoniae* is suspected as a cause of invasive disease, specimens should be obtained from the site(s) of infection for culture and

Gram stain. Cultures provide highly specific diagnostic information, the ability to test for antibiotic susceptibility, and specimens for serotype-specific epidemiology. However, because pneumococci can colonize the upper respiratory tract, recovery of *S. pneumoniae* from the nasopharynx does not confirm it as the causative agent.

Most patients with IPD have leukocytosis (>12,000 cells/μL) and elevated markers of inflammation (e.g., C-reactive protein, erythrocyte sedimentation rate). Leukopenia may be seen with meningitis and other severe pneumococcal infections.

Diagnosing pneumococcal pneumonia can be challenging and treatment is typically empiric. Typical chest radiographs show consolidation of a segment, an entire lobe, or multiple lobes (see Fig. 26.1). In hospitalized patients with pneumonia, blood cultures are positive in 10% of children and up to 25% of adults. Respiratory specimens, such as sputum (in adolescents and adults) and endotracheal or bronchoalveolar lavage samples, showing gram-positive diplococci and many polymorphonuclear neutrophils suggest a pneumococcal cause pending culture results. Thoracentesis may be needed to drain pleural effusions or empyemas and to obtain specimens for culture.

In cases of fever without a focus and suspected IPD, blood should be obtained for culture. Depending on the signs and symptoms, specimens of CSF should be obtained for laboratory evaluation including cultures. Gram stain of the CSF sediment may reveal the characteristic gram-positive diplococci. The CSF will usually show a pleocytosis with a predominance of polymorphonuclear leukocytes (PMNs), although early in the disease lymphocytes may predominate. Typically, CSF protein is elevated and glucose is low relative to blood glucose levels.

A rapid in vitro diagnostic test based on the presence of pneumococcal capsular C polysaccharide in urine has been approved by the US Food and Drug Administration (FDA) for diagnosing pneumococcal pneumonia in adults. The test is 70% to 80% sensitive and greater than 90% specific when compared with conventional methods and is not affected by antibiotics. However, this test is not for use in children as it lacks specificity, probably because of higher rates of NP carriage. This rapid test is also used for detecting pneumococci in CSF of patients with meningitis. Multiplex polymerase chain reaction (PCR) tests have been approved by the FDA for detection of respiratory pathogens, including *S. pneumoniae*, in sputum and bronchioalveolar specimens, and a similar test is available for CSF specimens, making this the preferred non–culture-based approach to diagnosis.

## TREATMENT

Adjunctive and supportive treatments of IPD are similar to therapies used for Hib infections, as described in Chapter 4.

Pneumococcal infections are treated with antimicrobials to which the organism is susceptible. All cultures of *S. pneumoniae* from sterile body sites should be evaluated for antimicrobial susceptibilities, but in most cases empirical therapy should begin before susceptibilities are known.

The prevalence of *S. pneumoniae* strains that are not fully susceptible to penicillin varies by region, but the proportion of such strains has decreased in the United States since introduction of the conjugate vaccine. Penicillin-resistant strains are defined as intermediately resistant (minimum inhibitory concentration [MIC] > 0.1 to 1 mcg/mL) or highly resistant (MIC ≥ 2 mcg/mL). *S. pneumoniae* strains that are resistant to penicillin are often resistant to other antimicrobials, including cephalosporins and macrolides, and these multidrug-resistant strains have been identified throughout the world. *S. pneumoniae* strains resistant to vancomycin have not been identified in the United States.

Bacterial meningitis is treated with higher doses of antibiotics than used for other infections to ensure adequate drug levels in the CSF.

The initial regimen for suspected pneumococcal meningitis in all ages should include vancomycin and ceftriaxone or cefotaxime until the antimicrobial susceptibilities are known. Meropenem may be an alternative. Corticosteroids have been used in the treatment of bacterial meningitis to reduce intracerebral inflammation; however, evidence of improved outcomes in children is equivocal. If used, corticosteroids should be given before or concurrently with the first dose of antimicrobials.

In CAP, the specific pathogen is usually not known and patients are treated presumptively with antimicrobials that are effective against *S. pneumoniae,* as well as other common bacterial pathogens. For inpatient adults, combination therapy with a cephalosporin and a macrolide or fluroquinolone antibiotic is recommended while awaiting specific bacterial diagnosis. In children with mild-to-moderate CAP suggestive of bacterial infection, amoxicillin can be used for empirical treatment in an outpatient setting. If the clinical presentation is consistent with both bacterial and atypical pneumonia, a macrolide may be considered. For hospitalized children, empirical treatment with intravenous ampicillin or a third-generation parenteral cephalosporin (e.g., cefotaxime, ceftriaxone) is effective in most settings. Cefotaxime or ceftriaxone is recommended for children hospitalized with pneumonia caused by pneumococci suspected or proven to be penicillin-resistant strains, for serious infections including empyema, or in those not fully immunized with PCV13. Vancomycin should be included in those with life-threatening infection.

First line treatment for uncomplicated AOM is amoxicillin (80 to 90 mg/kg/day). If initial therapy fails, antibiotics active against penicillin-nonsusceptible pneumococci and β-lactamase–producing *H. influenzae* and *M. catarrhalis* should be used (e.g., amoxicillin-clavulanic acid, second- and third-generation oral cephalosporins). In severe cases and when second line therapy fails, parenteral ceftriaxone may be given and/or tympanocentesis may be used to drain infected middle ear fluid, obtain cultures, and guide therapy, as well as to provide pain relief.

## PROGNOSIS

Early diagnosis and treatment with appropriate antimicrobials are keys to better clinical outcomes.

Nearly all children with ear infections recover, although recurrent infections can lead to hearing loss and delayed language development. In the United States and other high-income countries, pneumococcal pneumonia may result in hospitalization, although the mortality rate is low. In children who have bacteremia without a focus, 10% will develop focal complications, 3% to 6% will develop meningitis, and approximately 1% will die. Of children younger than 5 years of age with pneumococcal meningitis, approximately 5% will die and 25% of survivors may have long-term problems such as hearing loss or learning disability. Sequelae in patients with meningitis are associated with the presence of coma and low CSF glucose level (<0.6 mmol/L).

The incidence of IPD and mortality resulting from bacterial infections in sickle cell disease have been declining due to use of penicillin prophylaxis, implementation of new vaccination strategies using conjugate vaccines, and improved medical care.

## PREVENTION

Good respiratory hygiene and active immunization are effective prevention strategies. Risk of IPD may be reduced by improvement in predisposing conditions such as diabetes and HIV, smoking cessation, and avoidance of crowded living conditions.

## Immunoprophylaxis

The first vaccines developed against *S. pneumonia* were composed of polysaccharides extracted from common invasive serotypes. In the United States a tetravalent polysaccharide vaccine was licensed in 1946 but was discontinued in 1951 owing to the increasing use of penicillin. In 1977 a 14-valent polysaccharide vaccine became available, and a 23-valent polysaccharide vaccine that covers serotypes responsible for more than 90% of pneumococcal disease has been available since 1983.

The pneumococcal polysaccharide vaccine provides only limited protection in children younger than 2 years of age. Polysaccharide antigens do not elicit T-cell help for antibody production efficiently, particularly in young children. Protein-polysaccharide conjugate vaccines, manufactured by chemical linkage of bacterial polysaccharides to protein antigens such as diphtheria and tetanus toxoids, are able to elicit T-cell help and induce protective immune responses, even in infants and young children. In clinical trials, a conjugate vaccine composed of seven serotypes (4, 6B, 9V, 14, 18C, 19F, and 23F), which accounted for approximately 80% of invasive disease in young children in the United States at the time of the trial, was proven highly effective in preventing invasive disease. Since licensure in the United States in 2000, high rates of vaccine coverage have resulted in a marked reduction in IPD in infants and small children, as well as adult and elderly populations, likely because of decreased transmission and herd immunity. In addition, rates of antimicrobial resistance have fallen because the conjugate vaccine includes serotypes associated with high rates of resistance.

As disease caused by conjugate vaccine serotypes diminished with wide use of the vaccine, some nonvaccine serotypes emerged as important causes of disease in children. Pneumococcal conjugate vaccines formulated with additional serotypes, including serotypes important globally, became available in 2009 and 2010. A 13-valent vaccine containing the seven original serotype plus serotypes 1, 3, 5, 6A, 7F, and 19A was licensed in the United States for use in children 6 weeks through 5 years of age. Subsequently, the 13-valent vaccine was evaluated in adults and shown to be effective in preventing invasive disease, and nonbacteremic and bacteremic pneumonia due to vaccine serotypes. A 10-valent conjugate vaccine containing additional serotypes 1, 5, and 7F is available in Europe, Canada, and elsewhere. Investigational vaccines targeting proteins common to most pneumococci and conjugate vaccines with additional serotypes hold promise in providing broad protection against pneumococcal disease.

The WHO recommends that all infants and young children receive at least three doses of a pneumococcal conjugate vaccine. In the United States, the Centers for Disease Control and Prevention (CDC) recommends that all infants and toddlers receive the conjugate pneumococcal vaccine at 2, 4, 6, and 12 to 15 months of age. Both the conjugate and 23-valent polysaccharide vaccines are recommended for certain groups of children (ages 2 years and older) and adults who are at increased risk of pneumococcal disease.

CDC vaccine recommendations may be viewed at: https://www.cdc.gov/vaccines/schedules/index.html.

## Chemoprophylaxis

Chemoprophylaxis is not routinely recommended for contacts of individuals with IPD or for travelers.

Penicillin chemoprophylaxis is recommended by the CDC for persons with functional or anatomic asplenia because of the high risk of severe infections. For children with sickle cell hemoglobinopathy, oral penicillin V (125 mg, twice daily) is recommended beginning before 4 months of age. The optimal duration of penicillin prophylaxis in these children has not been determined; however, stopping at age 5 years in children who are fully vaccinated and who have been free of severe pneumococcal infections has not resulted in increased infections.

## EVIDENCE

Black S, Shinefield H, Fireman B, et al: Efficacy, safety and immunogenicity of heptavalent pneumococcal conjugate vaccine in children. Northern California Kaiser Permanente Vaccine Study Center Group, *Pediatr Infect Dis J* 19:187-195, 2000. *This reference reports on the efficacy trial in US children of the pneumococcal conjugate vaccine for prevention of invasive pneumococcal disease in children.*

Bonten MJ, Huijts SM, Bolkenbaas M, Webber C, et al. Polysaccharide conjugate vaccine against pneumococcal pneumonia in adults. *N Engl J Med* 12:1114-1125, 2015. *This reference reports on the effectiveness of the 13-valent pneumococcal conjugate vaccine in preventing pneumonia in elderly adults.*

Hava DL, LeMieux J, Camilli A: From nose to lung: the regulation behind *Streptococcus pneumoniae* virulence factors, *Mol Microbiol* 50:1103-1110, 2003. *This reference provides information about the virulence factors of S. pneumoniae.*

Swartz MN: Bacterial meningitis: a view of the past 90 years, *N Engl J Med* 351:1826-1828, 2004. *This reference provides a comprehensive review of evolving epidemiology, diagnosis, management, and treatment.*

## ADDITIONAL RESOURCES

American Academy of Pediatrics (AAP): Pneumococcal infections. In Kimberlin DW, Brady MT, Jackson MA, et al., eds: *Red Book: 2018 Report of the Committee on Infectious Diseases*, ed 31, Elk Grove Village, IL, 2009, AAP, pp 639-651. *This resource provides a concise summary of the diagnosis, treatment, and prevention of pneumococcal infections in children.*

Centers for Disease Control and Prevention (CDC): Licensure of a 13-valent pneumococcal conjugate vaccine (PCV13) and recommendations for use among children—Advisory Committee on Immunization Practices (ACIP), *MMWR Morb Mortal Wkly Rep* 2010;59:258-261. *This reference provides information about the use of chemoprophylaxis and vaccines against pneumococcal infections.*

Matanock A, Lee G, Gierke R, Kobayashi M, Leidner A, Pilishvili T. Use of 13-valent pneumococcal conjugate vaccine and 23-valent pneumococcal polysaccharide vaccine among adults aged ≥65 years: Updated recommendations of the Advisory Committee on Immunization Practices. *MMWR* 2019;68:1069-1075. *This reference provides recommendations from CDC on vaccination of elderly adults.*

Varmann M, Chatterjee A, Wick NA, John CC: Diagnosis and treatment of adults with community-acquired pneumonia; an official clinical practice guideline of the American Thoracic Society and Infectious Diseases Society of America, *Am J Respir Crit Care Med* 2019;200(7):e45-e67. *This reference provides current diagnosis and treatment of community acquired pneumonia in adults.*

World Health Organization (WHO): Immunization, Vaccines and Biologicals, Monitoring and Surveillance. Available at: www.who.int/immunization_monitoring/diseases/en. *This WHO website contains information on vaccine-preventable diseases globally and links to related topics.*

# 6

# Infections Caused by *Neisseria meningitidis*

*Margaret C. Bash*

## ABSTRACT

*Neisseria meningitidis* is both a commensal organism of the upper respiratory tract and a significant pathogen for humans. It causes devastating disease with significant morbidity and mortality worldwide. Infections caused by *N. meningitidis* can have various clinical presentations ranging from asymptomatic carriage or a mild upper respiratory illness to purulent meningitis and/or disseminated disease with fulminant sepsis. Strains with capsule types belonging to groups A, B, C, Y, W, and recently X are associated with invasive disease. The majority of patients infected with *N. meningitidis* are children, usually younger than 5 years old, adolescents, young adults, and adults older than 65 years of age. Effective polysaccharide and polysaccharide conjugate vaccines are available for prevention of serogroup A, C, Y, and W disease. The serogroup B polysaccharide is poorly immunogenic, and outer membrane protein–based vaccines have been developed and recently licensed.

## CLINICAL VIGNETTE

A 17-year-old previously healthy female university student was brought to the emergency room with symptoms of fever, headache, and malaise. She had been well until the prior evening, when her symptoms began with generalized fatigue causing her to leave her friends early to go to sleep in her dormitory. She felt too ill to join them for breakfast, and when they returned to her room they were concerned to find her confused and febrile.

Her exam was significant for a temperature of 102°F, photophobia, and positive Kernig and Brudzinski signs (pain with hip flexion and knee extension, and flexion of the hips and knees when the neck is flexed by the examiner, respectively). Rare, fine petechiae were present on her trunk and upper extremities, and the rash progressed during the evaluation. Laboratory investigations revealed a total leukocyte count of 19000/mm³ and a platelet count of 10000/mm³. Blood cultures and lumbar puncture were performed. Cerebrospinal fluid (CSF) white blood cell (WBC) count was 15 with 80% neutrophils. Gram stain revealed gram-negative diplococci.

The patient was initially treated with vancomycin (1 g intravenously every 12 hours) and ceftriaxone (1 g intravenously every 12 hours). The patient showed early signs of hemodynamic instability (systolic blood pressure of 80 mm Hg and heart rate of 140/min) in spite of an initial intravenous fluid bolus, and she was transferred to the intensive care unit for continued fluid and inotropic support. Norepinephrine was required at a rate of 1 mcg/kg/min during the first 8 hours of admission but was successfully discontinued during the first hospital day. Dexamethasone was administered for the first 4 days because the CSF gram stain and subsequent culture confirmed *Neisseria meningitidis* meningitis in addition to sepsis. Vancomycin was stopped following culture confirmation of *N. meningitidis*.

COMMENT: The patient recovered rapidly and continued antibiotics for 7 days. Close monitoring for sequelae of meningitis including hearing loss and cognitive dysfunction was planned. Antibiotic prophylaxis was administered to her roommates and boyfriend.

## GEOGRAPHIC DISTRIBUTION AND BURDEN OF DISEASE

*N. meningitidis* has a worldwide distribution, but there are distinct regional characteristics of the microbiology and epidemiology of disease. The Centers for Disease Control and Prevention (CDC) has reported the worldwide incidence of invasive meningococcal disease to be 0.5 to 5 per 100,000 population per year. In the United States the overall incidence ranged from 0.5 to 1.5 per 100,000 population per year in the 1990s but has declined over the past two decades to 0.11 cases per 100,000 persons in 2017. Disease is primarily caused by serogroup B, although cases of C, Y, and W occur. European countries have a wide range of disease incidence (0.2 to 14 cases per 100,000). The majority of these cases are caused by serogroup B, especially in countries where meningococcal C conjugate (MCC) vaccines are used routinely. An epidemic of serogroup B meningococcal disease persisted for almost a decade in New Zealand (17.4 cases per 100,000 in 2001), but attack rates declined to 2.6 per 10,000 in 2007, at least in part as a result of a national vaccination campaign using an outer membrane vesicle (OMV) vaccine made from the outbreak strain.

Before World War II, serogroup A was the most common serogroup identified from disease isolates in industrialized countries. In the second half of the 20th century, this serogroup was primarily associated with epidemic meningitis, especially in the sub-Saharan region of Africa known as the *meningitis belt*. Countries in this region experienced recurrent epidemics every 5 to 10 years with incidence rates reaching 1000 per 100,000. Even in interepidemic years, serogroup A disease rates were as high as 25 per 100,000. Following introduction and widespread use of a serogroup A meningococcal conjugate vaccine (MenAfriVac) in national vaccination campaigns, serogroup A epidemics have disappeared from this region. Serogroup A epidemics also occurred in India, China, and Russia. Other areas of Africa have yearly rates of invasive serogroup A disease higher than in industrialized nations. Travel to the Hajj pilgrimage has been a significant risk factor for exposure to serogroup A meningococci.

Asymptomatic nasopharyngeal carriage of *N. meningitidis* is common and varies by age. Infants have a carriage rate of 1% to 2%, whereas adolescents and young adults have a carriage rate of 15% to 25%. Carriage of highly virulent strains occurs in less than 5% of the general population, even during outbreaks. Carriage isolates are often unencapsulated, or less virulent, than strains that cause invasive disease. Transmission of *N. meningitidis* is primarily through respiratory droplets. Invasive meningococcal disease is associated with recent acquisition of a new pathogenic strain rather than after extended colonization. Studies of military recruits in the 1970s helped to determine the infectivity and route of spread of the organism and established that preexisting antibodies that trigger complement-mediated killing of the bacteria provide protection against invasive disease.

The prevalence of bactericidal antibodies in the population increases with age and is inversely related to the rates of invasive disease. Infection rates are highest in children younger than age 5 years, especially 6 months to 1 year of age. A second period of increased risk is observed during adolescence and young adulthood. College freshman living in dormitories and military recruits are at moderate risk, whereas and individuals with close or intimate contact with an index case are at 500 times increased risk of invasive disease. Individuals with terminal complement component deficiencies or those who are receiving complement inhibitor therapies are at high risk of recurrent invasive meningococcal infections.

## PATHOGENESIS

*N. meningitidis,* a gram-negative diplococcus, is an obligate human pathogen. Endemic sporadic disease is caused by highly diverse strains, but most disease isolates can be grouped by genetic analysis of housekeeping genes (multilocus sequence typing [MLST]) into one of several hypervirulent lineages. In contrast, local outbreaks and sustained epidemics can be caused by a single strain type and are considered clonal.

Surface structures of the bacterium are important for strain characterization, disease pathogenesis, and immunity. Capsular polysaccharide type defines the serogroup of a strain. Strains expressing serogroup A, B, C, Y, W, and more recently serogroup X capsules are associated with invasive disease, whereas unencapsulated strains and strains of the remaining serogroups are usually associated with asymptomatic nasopharyngeal colonization. The polysaccharide capsule enhances efficient transmission and inhibits phagocytosis. Lipooligosaccharide (LOS) is an endotoxin that induces an inflammatory cascade that leads to the clinical features seen in septicemia, septic shock, and meningitis.

*N. meningitidis* has extensive mechanisms for the uptake and incorporation of deoxyribonucleic acid (DNA), a process known as *horizontal exchange,* which allows the organism to adapt to the host and evade natural immunity. This process results in antigenic diversity of many surface proteins and occasionally capsule type switching. In addition, phase variation of surface structures such as Opa proteins and LOS also contribute to phenotypic diversity of the organism.

Genetic polymorphisms of the human host affect susceptibility to meningococcal infections and may affect disease outcomes. Genetic deficiencies of complement factors 5 to 9 (late complement component deficiency) are well recognized risk factors for recurrent or familial meningococcal disease. In addition, some case control studies have shown an association between meningococcal disease and polymorphisms of interleukin-1 receptor agonist (IL1RA), carcinoembryonic antigen cell adhesion molecules 3 and 6, surfactant proteins A and D, and factor H.

## CLINICAL PRESENTATION

The spectrum of *N. meningitidis* infections can range from asymptomatic to fulminant septicemia and/or meningitis. The progression of invasive disease can be rapid, with circulatory collapse and death occurring within hours of presentation. In contrast, resolution of unsuspected culture-positive meningococcemia without treatment has been documented in adults and infants with fever and upper respiratory symptoms.

The hallmark features of meningococcal disease are fever and rash, classically a nonblanching petechial rash, which can progress to purpura associated with disseminated intravascular coagulation (DIC). The rash is not diagnostic; a macular or maculopapular rash can also be observed, and of note, rash is absent in almost one-third of culture-proven disease in children. Approximately 11% to 15% of children presenting with petechiae have meningococcemia.

The most common presentation of invasive meningococcal disease is meningitis. Typical clinical findings of meningitis are the result of inflammation in the subarachnoid space and include fever, headache, meningismus, photophobia, and lethargy. Older children may also have positive Kernig and Brudzinski signs (see Fig. 4.2B). Infants may not have these classic symptoms of meningitis but usually demonstrate irritability, inconsolable crying, poor feeding, and lethargy. Meningococcal meningitis can occur with or without associated septicemia.

Septicemia accounts for 15% to 20% of invasive meningococcal disease and results from significant levels of bacteria and endotoxin in the bloodstream. The onset of disease is rapid, and clinical decline with significant morbidity or mortality can occur within 24 to 48 hours. Bacteremia and increased endotoxin production initiate an intravascular inflammatory cascade that causes endothelial damage resulting in DIC and multiorgan failure (Fig. 6.1). The vascular damage initially results in petechiae and purpura and can ultimately lead to autoamputation of digits or entire limbs. The average duration from onset of symptoms to admission for patients with sepsis is 12 hours, which is less than half the time for patients with meningitis.

Other less common presentations of invasive disease include pneumonia, pyogenic arthritis, purulent pericarditis, osteomyelitis, and endophthalmitis. These clinical presentations are more often associated with serogroups Y or W-135 and are usually seen in older patients. Recently, outbreaks of meningococcal urethritis caused by unencapsulated strains that have acquired some genetic features of *N. gonorrhoeae* have been reported.

Chronic meningococcemia is a rare condition that manifests with recurrent intermittent fever, rash, and arthralgia or arthritis and may last for months. The presentation can be like that of many other viral, rheumatologic, or vasculitic conditions. *N. meningitidis* can be cultured from serum during acute episodes, and in the absence of appropriate treatment, some patients eventually progress to disseminated disease. The pathogenesis of chronic meningococcemia is not known; however, it is more commonly seen in patients with underlying terminal complement deficiency.

## DIAGNOSTIC APPROACH

Invasive meningococcal disease should be considered in any patient with fever and signs or symptoms that are consistent with bacterial sepsis or meningitis. The initial evaluation includes cultures of blood and cerebrospinal fluid (CSF) and, in some instances, biopsies of skin lesions. The CSF cell count, protein level, and glucose concentrations are usually abnormal, although normal CSF profiles have been reported. CSF should also be examined for gram-negative diplococci. Peripheral leukocytosis with a predominance of polymorphonuclear cells is typical, but leukopenia, thrombocytopenia, and anemia may be present. Coagulopathy and metabolic abnormalities, such as hyponatremia, hypoglycemia, and metabolic acidosis, can complicate the clinical management of patients with invasive disease.

Viable *N. meningitidis* can be obtained from 40% to 75% of blood and 90% of CSF cultures when obtained before the administration of antibiotics. The microbiologic evaluation should not delay therapy when invasive meningococcal disease is suspected; in some circumstances, antibiotics must be administered empirically before cultures can be obtained. Blood and CSF can be rapidly sterilized after antibiotic administration; however, organisms have been identified in skin lesions up to 12 hours after antibiotics have been given. *N. meningitidis* antigen testing has been replaced by nucleic acid amplification–based

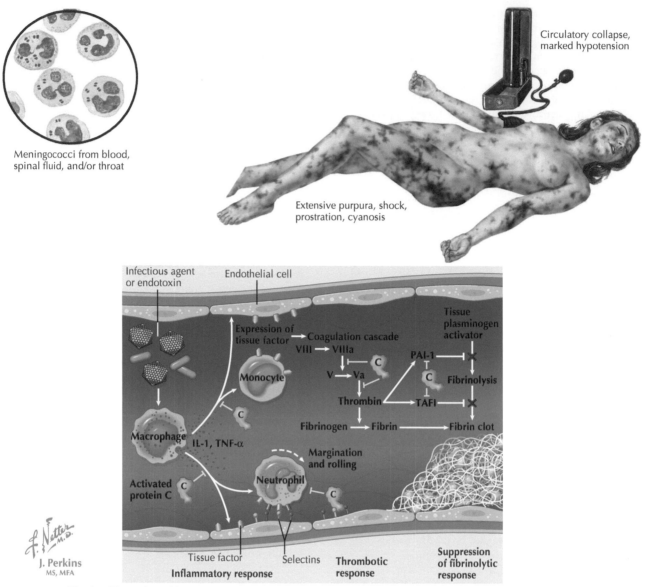

Meningococci from blood, spinal fluid, and/or throat

Circulatory collapse, marked hypotension

Extensive purpura, shock, prostration, cyanosis

**Fig. 6.1** Disseminated intravascular coagulation (DIC). *IL-1*, Interleukin-1; *PAI-1*, plasminogen activator inhibitor-1; *TAFI*, thrombin activatable fibrinolysis inhibitor; *TNF-α*, tumor necrosis factor alpha.

diagnostic methods, and these diagnostic tests can be informative, especially when treatment has been initiated prior to obtaining blood or CSF for culture; in locations where antigen testing is still performed, it is important to note that cross-reaction occurs between the serogroup B capsular antigen and *Escherichia coli* K1.

The differential diagnosis is influenced by the age and epidemiologic history of the patient and includes other causes of bacterial sepsis and meningitis including *Streptococcus pneumoniae, Haemophilus influenzae* type b, *Staphylococcus aureus,* and gram-negative enteric pathogens, as well as rickettsial disease, Henoch-Schönlein purpura, and other noninfectious causes of vasculitis.

## TREATMENT

Prompt initiation of parenteral antibiotics is the mainstay of treatment for invasive meningococcal disease. Third-generation cephalosporins, such as ceftriaxone and cefotaxime, are bactericidal, have excellent central nervous system penetration, and are most often used as initial therapy of meningitis until culture results and susceptibility profiles are determined. Most meningococcal isolates are susceptible to

penicillin G (250,000 units/kg/day IV or IM divided every 4 to 6 hours [maximum: 24 million units/day]).

Penicillin-resistant *N. meningitidis* strains have been described worldwide; however, the rate of penicillin resistance in the United States has remained low at approximately 3%. Ciprofloxacin-resistant isolates and clinical treatment and prophylaxis failures have been reported. Routine susceptibility testing is recommended for all *N. meningitidis* isolates from sterile body sites. The recommended duration of therapy for meningococcemia or meningitis is 7 to 10 days; however, there are reports of clinical and microbiologic resolution with shorter courses of therapy.

Emergent management of *N. meningitidis* infections involves fluid resuscitation, maintenance of blood pressure with pressor support, and early intubation if indicated to prevent circulatory collapse or increases in intracranial pressure. Adjuvant therapy with corticosteroids for patients with septic shock and meningitis has shown some benefit in adults and in pediatric patients with *H. influenzae* meningitis; however, there has been no proven benefit in meningococcal sepsis or meningitis. Activated protein C (APC), a component in the coagulation pathway, may be considered in adult patients with invasive meningococcal disease and severe sepsis who present early in their disease course. In pediatric patients, the risk and benefits of this therapy should be weighed, although it is generally not recommended.

## PROGNOSIS

The overall mortality rate from meningococcemia is approximately 10% in the United States, with most cases of death occurring in young infants and adolescents. Factors associated with poor outcome and high mortality rates include young age, absence of meningitis, presence of shock, clinical signs of ischemic damage, leukopenia, or thrombocytopenia. In less severe disease with prompt management, most patients have significant resolution of symptoms by the second day of illness and complete recovery within 1 week. After acute illness, less than 10% of patients develop postinfectious complications as a result of immune complex–induced inflammation such as arthritis, iritis, myocarditis, pericarditis, and vasculitis. These conditions usually manifest 4 or more days after onset of infection and can be managed with nonsteroidal antiinflammatory agents. Sequelae from invasive meningococcal disease occur in up to 19% of patients and can include hearing loss, neurologic dysfunction or motor deficits, digit or limb amputation, and skin scarring.

## PREVENTION AND CONTROL

Prevention of secondary cases of invasive meningococcal disease involves patient isolation, chemoprophylaxis of exposed individuals, and in some situations vaccination. Respiratory droplet precautions should be used until the patient has received 24 hours of appropriate antimicrobial therapy to eradicate carriage. Patients who are treated with antibiotics other than ceftriaxone or cefotaxime should receive chemoprophylaxis to eradicate nasopharyngeal colonization. Chemoprophylaxis is essential for close contacts because their attack rate is 500 to 800 times higher than in the general population. This includes all individuals who have been directly exposed to oral secretions from the index case, all childcare or preschool contacts, and all individuals who slept in the same dwelling as the index case in the 7 days before onset of disease. Individuals who were seated next to the index case on a prolonged airplane flight (>8 hours) are also eligible for chemoprophylaxis. Routine prophylaxis is not given to medical personnel unless they had significant exposure to respiratory secretions. Ideally, chemoprophylaxis should be given within 24 hours after diagnosis in the index patient. Rifampin, ceftriaxone, ciprofloxacin, and azithromycin are effective chemoprophylactic medications. Rifampin or ceftriaxone can be used in younger children, but ciprofloxacin is recommended only for nonpregnant adults older than 18 years of age. Ciprofloxacin is not recommended in regions where resistant strains have been identified. In addition, vaccination of close contacts is recommended if the index case was caused by a serogroup included in current vaccines.

Vaccines are available for prevention of invasive disease caused by *N. meningitidis* groups A, C, Y, and W. Historically, the polysaccharide tetravalent meningococcal vaccine (MPSV4) was used for control of outbreaks and recommended for routine use in high-risk populations such as military recruits, individuals with terminal complement component deficiencies or asplenia, travelers to geographic regions with high rates of disease, and more recently college freshman living in dormitories. Polysaccharide vaccines are not generally immunogenic in children younger than 2 years of age except serogroup A meningococcal polysaccharide vaccine, which was used in infants 6 months of age and older in a two-dose series during serogroup A outbreaks.

Because polysaccharides do not induce memory and immunity wanes after 3 to 5 years, the MPSV4 has been replaced by conjugate vaccines. Chemical conjugation of polysaccharides to a protein carrier creates T cell–dependent antigens that are usually highly immunogenic in infant populations. After the licensure of a tetravalent glycoconjugate meningococcal vaccines (MCV4) in the United States, routine immunization of all children 11 years of age and older, and all individuals who are at increased risk of invasive meningococcal disease was recommended. A booster dose of MCV4 is administered 3 to 5 years after the initial immunization for high-risk children and routinely at 16 years of age for all adolescents who were initially immunized at 11 years of age. Monovalent serogroup C conjugate vaccines have been incorporated into routine infant immunization schedules in some countries where serogroup C disease was common. These vaccines have been shown to decrease colonization and provide herd immunity.

Management of epidemic serogroup A disease in Africa previously depended on a reactive vaccination strategy initiated when disease rates exceed certain thresholds. This approach provided some benefit in controlling outbreaks; however, it was not useful in their prevention. The Meningitis Vaccine Project, a public-private partnership, successfully developed and incorporated into widespread use an affordable effective serogroup A conjugate vaccine, eliminating the recurrent epidemics of this disease in the meningitis belt region of Africa.

Vaccines for the prevention of serogroup B meningococcal disease target subcapsular protein antigens because the serogroup B polysaccharide is poorly immunogenic. Vaccines using OMVs depleted of LOS have been used in the Netherlands, Cuba, Brazil, Chile, and most recently in New Zealand to address serogroup B epidemics. These OMV vaccines were made from the epidemic strain and, especially in young children or infants, are thought to have provided strain specific immunity. New vaccines that are designed to be broadly protective against the diverse strains associated with endemic disease have been developed. Two serogroup B vaccines have been approved for use in individuals 10 through 25 years of age in the United States. One vaccine, 4CMenB (Bexerso), contains four antigenic components: recombinant factor H binding protein (Fhbp), recombinant neisserial adhesin A (NadA), recombinant neisserial heparin-binding protein (NHBP), and OMVs from strain NZ98/254 that include a porin protein PorA (serosubtype P1.4). The other licensed vaccine, MenBFHbp (Trumenba), contains two lipidated recombinant FHbp antigens, one from each of two genetic subfamilies. In clinical studies, both vaccines elicited bactericidal antibodies against selected MenB strains measured with the human complement serum bactericidal activity. Serogroup B vaccines are recommended for people 10 years of age and older who are at increased risk for meningococcal disease. Although not routinely recommended for adolescents, these vaccines may be considered for immunization of individuals 16 through 23 years of age.

## ACKNOWLEDGMENT

The author acknowledges Anjali N. Kunz for contributions to the previous edition.

## EVIDENCE

Goldschneider I, Gotschlich E, Artenstein M: Human immunity to the meningococcus. II: Development of natural immunity, *J Exp Med* 1969;129:1327. *This classic article provides the first data about the immune responses to the meningococcus.*

Stephens DS, Greenwood B, Brandtzaeg P: Epidemic meningitis, meningococcemia, and Neisseria meningitidis, *Lancet* 2007;369:2196-2210. *This review article highlights the distinguishing clinical and epidemiologic aspects of meningococcal disease and preventive measures.*

Wong VK, Hitchcock W, Mason WH: Meningococcal infections in children: a review of 100 cases, *Pediatr Infect Dis* 1989;8:224-227. *This article explores the variety of clinical presentations of meningococcal disease in pediatric patients and assesses the factors involved in their outcomes.*

## ADDITIONAL RESOURCES

Acevedo R, Bai X, Borrow R, et al. The Global Meningococcal Initiative meeting on prevention of meningococcal disease worldwide: Epidemiology, surveillance, hypervirulent strains, antibiotic resistance and high-risk populations. *Expert Rev Vaccines* 2019 Jan;18(1):15-30. doi: 10.1080/14760584.2019.1557520. Epub 2018 Dec 27. Review. PubMed PMID:30526162. *This journal supplement provides a collection of review articles regarding epidemiology, pathogenesis, clinical disease, and vaccine development and prevention for meningococcal disease.*

American Academy of Pediatrics (AAP): Meningococcal infections. In Kimberlin DW, Brady MT, Jackson MA, eds: *Red Book: 2018 Report of the Committee on Infectious Diseases*, ed 31, Elk Grove Village, IL, 2009, AAP, pp 455-463. *This resource provides an abbreviated summary of the diagnosis, management, and prevention of meningococcal infections.*

Centers for Disease Control and Prevention (CDC): Meningitis. Available at: https:// www.cdc.gov/meningitis/clinical-resources.html and Centers for Disease Control and Prevention (CDC): Meningococcal Disease. Available at: www.cdc.gov/menincoccal/clinical-info.html. Accessed January 15, 2020. *These websites are available to use as a reference, to review current statistics and epidemiologic information, and to obtain clinical updates regarding treatment and management.*

Cohn AC, MacNeil JR, Clark TA, et al; Centers for Disease Control and Prevention (CDC). Prevention and control of meningococcal disease: recommendations of the Advisory Committee on Immunization Practices (ACIP). *MMWR Recomm Rep* 2013 Mar 22;62(RR-2):1-28. PubMed PMID: 23515099. *This is a collaborative report reviewing the recommendations of the ACIP regarding meningococcal vaccination and preventive strategies.*

# 7

# Poliomyelitis (Polio) and Polioviruses

*Alison M. Helfrich, Michael Rajnik*

## ABSTRACT

In the early 20th century, poliomyelitis was one of the most feared illnesses of humans, in part because it affected previously healthy individuals with little or no warning and could result in devastating paralysis. Although the disease was once endemic worldwide, with vaccines eradication has been achieved in all but two countries. In current times, acute flaccid paralysis (AFP) is associated with vaccine-derived poliovirus strains or other viral agents more commonly than wild-type polio virus. However, with disruption of proper sanitation or maintenance of vaccination within the population, outbreaks can quickly reemerge.

## CLINICAL VIGNETTE

A 3-year-old male from Syria presents to the local health clinic for progressive leg weakness for 2 days. Yesterday, he collapsed after getting out of bed and has not been able to stand. His parents report a recent febrile illness with diarrhea that self-resolved. Prior to this illness, he appeared healthy. Due to recent unrest and violence in the region, the family relocated to a rural area where healthcare, including vaccines, is limited due to the civil war. His parents recall no vaccines administered in the past 2 years.

Exam revealed a thin child who appeared in pain but was without acute distress. Cranial nerves were intact, and he was controlling oral secretions with normal breathing. He was unable to move his left leg, and he only had 3/5 strength in his right leg. The lower extremity patellar and Achilles reflexes were 0/4 on the left and 2/4 on the right. He was transferred to a referral hospital for management. Stool polymerase chain reaction (PCR) ultimately confirmed poliomyelitis due to a circulating vaccine-derived poliovirus strain. After 6 months of physical therapy, he started to regain strength and function.

COMMENT: Oral polio vaccine (OPV) is used in mass vaccination campaigns worldwide because of ease in administration and fecal shedding of the vaccine viruses by vaccine recipients can promote herd immunity in the community. Rarely, the circulating vaccine-derived polioviruses (cVDPVs) cause paralytic polio.

## EPIDEMIOLOGY, GEOGRAPHIC DISTRIBUTION, AND DISEASE BURDEN

Polioviruses are positive-sense, single-stranded ribonucleic acid (RNA) viruses that belong to the *Enterovirus C* species of the Picornaviridae family. They include three antigenically distinct serotypes (1, 2, and 3), with serotype 1 as the most common serotype leading to infection. The polioviruses were previously ubiquitous throughout the world and are highly infectious, spreading through contaminated fecal or respiratory secretions via the fecal-oral route. Humans are the only known natural hosts and reservoir. Although replication in other primates can occur, only humans are infected. Outbreaks were most common in summer and fall in temperate climates, whereas tropical climates have episodes year-round.

Poliomyelitis, often referred to as polio, was recognized by ancient Egyptians, as evidenced in hieroglyphics depicting people with deformed limbs. Throughout history, a majority of the infections were subclinical. Most of these infections occurred in infants and were attenuated by maternal antibodies, leading to subclinical disease but widespread immunity. With the improvement of sanitation in the United States, fewer infants were exposed to the virus and therefore not immune. This created a pool of susceptible individuals, specifically older children and adults, resulting in sporadic epidemics that occurred every few years, quickly increasing in frequency and size. The largest outbreak in the United States occurred in 1952, when nearly 58,000 cases of polio were diagnosed. In developing countries, polio remained endemic into the latter half of the 20th century. With improved sanitation in these regions, epidemics began to occur similar to the pattern observed in the turn-of-the-century United States.

The introduction of polio vaccines dramatically reduced the incidence of polio worldwide. The last case of naturally occurring polio in the United States was reported in 1979. The Western Hemisphere was certified free of wild-type poliovirus (WTPV) in 1994. In October 2019, the Polio Global Eradication Initiative announced that polio is on the verge of eradication, with only two countries (Pakistan and Afghanistan) reporting indigenous WTPV. Of the three serotypes of poliovirus, only serotype 1 remains endemic.

The OPV is a live-attenuated vaccine that has been used worldwide, but there is a risk in 1 in 2.5 million doses to develop vaccine-associated paralytic poliomyelitis (VAPP). The use of OPV also led to the development of cVDPVs, which have been the cause of numerous outbreaks of polio illnesses in parts of Africa, the Middle East, and Southeast Asia.

## RISK FACTORS

In the prevaccine era, virtually all infants were exposed to the polioviruses by the age of 6 months, although very few developed the paralytic disease. Risk factors for the infection progressing to paralytic disease include young age, pregnancy, antibody deficiency states, male gender (prepuberty), strenuous exercise, and preceding limb injury within 4 weeks of infection. In adults, women have an increased risk of infection but do not necessarily have an increased risk of paralysis.

## CLINICAL FEATURES

Even in the postvaccine era, approximately 70% of infections are asymptomatic. Clinically inapparent infections are defined as isolation of a poliovirus from stool or throat with concomitant fourfold rise in antibody titers. Less than 1% of infections progress to the weakness

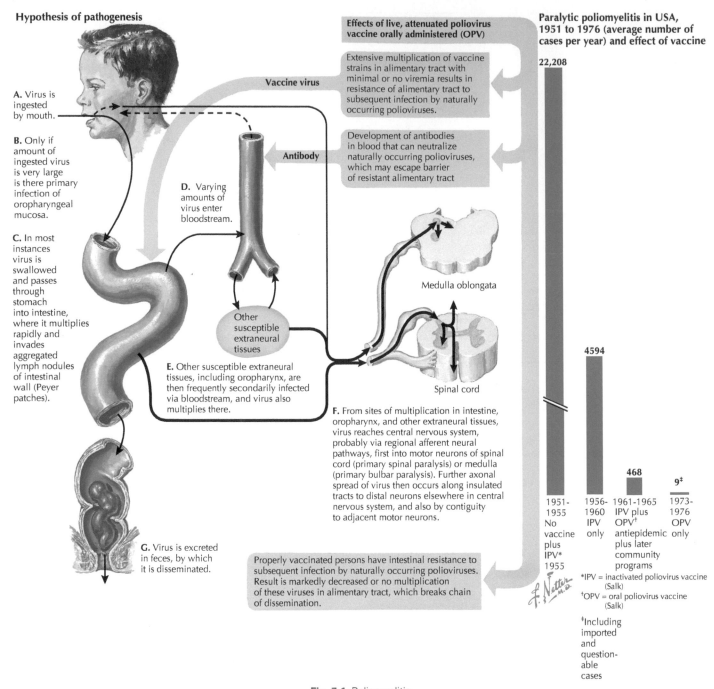

**Hypothesis of pathogenesis**

**A.** Virus is ingested by mouth.

**B.** Only if amount of ingested virus is very large is there primary infection of oropharyngeal mucosa.

**C.** In most instances virus is swallowed and passes through stomach into intestine, where it multiplies rapidly and invades aggregated lymph nodules of intestinal wall (Peyer patches).

**D.** Varying amounts of virus enter bloodstream.

Vaccine virus

Antibody

Other susceptible extraneural tissues

**E.** Other susceptible extraneural tissues, including oropharynx, are then frequently secondarily infected via bloodstream, and virus also multiplies there.

**G.** Virus is excreted in feces, by which it is disseminated.

**Effects of live, attenuated poliovirus vaccine orally administered (OPV)**

Extensive multiplication of vaccine strains in alimentary tract with minimal or no viremia results in resistance of alimentary tract to subsequent infection by naturally occurring polioviruses.

Development of antibodies in blood that can neutralize naturally occurring polioviruses, which may escape barrier of resistant alimentary tract

Medulla oblongata

Spinal cord

**F.** From sites of multiplication in intestine, oropharynx, and other extraneural tissues, virus reaches central nervous system, probably via regional afferent neural pathways, first into motor neurons of spinal cord (primary spinal paralysis) or medulla (primary bulbar paralysis). Further axonal spread of virus then occurs along insulated tracts to distal neurons elsewhere in central nervous system, and also by contiguity to adjacent motor neurons.

Properly vaccinated persons have intestinal resistance to subsequent infection by naturally occurring polioviruses. Result is markedly decreased or no multiplication of these viruses in alimentary tract, which breaks chain of dissemination.

**Paralytic poliomyelitis in USA, 1951 to 1976 (average number of cases per year) and effect of vaccine**

22,208

4594

468

9‡

| 1951- 1955 No vaccine plus IPV* 1955 | 1956- 1960 IPV only | 1961-1965 IPV plus OPV† antiepidemic plus later community programs | 1973- 1976 OPV only |

*IPV = inactivated poliovirus vaccine (Salk)

†OPV = oral poliovirus vaccine (Salk)

‡Including imported and question-able cases

**Fig. 7.1** Poliomyelitis.

and paralysis seen in paralytic poliomyelitis. The incubation period is 3 to 6 days; however, symptom onset varies depending on clinical presentation.

## Nonparalytic Poliomyelitis

During the incubation period, the virus replicates in the lymphatic tissue, first in the tonsils and lymph nodes of the neck, then in the Peyer patches of the small intestine. After several days, a primary minor viremia occurs, with systemic spread of the poliovirus in the blood to muscle, fat, bone marrow, liver, and spleen (Fig. 7.1). This period is typically asymptomatic, but up to 24% of patients have a mild illness with gastrointestinal symptoms or a respiratory influenza-like illness. These individuals will typically recover in 5 to 10 days, without the development of antibodies to neutralize the virus. In 1% of patients,

a secondary major viremia occurs, and the poliovirus spreads to the central nervous system (CNS) and causes an aseptic meningitis. This presents with a severe headache and meningeal irritation but also has a full recovery in 5 to 10 days.

## Paralytic Poliomyelitis

Polioviruses replicate within neurons, with a strong predilection for the anterior horn cells of the spinal cord. Rarely, the posterior horn cells and dorsal root ganglia are infected also (Fig. 7.2). Once inside the neurons, the poliovirus will replicate and cause selective destruction of motor neurons; this results in paralysis of the muscle fibers of the infected neuron. It can spread within the CNS laterally to nearby neurons, or retrograde along the axonal transport and travel superiorly along the spine, even as far as the brainstem.

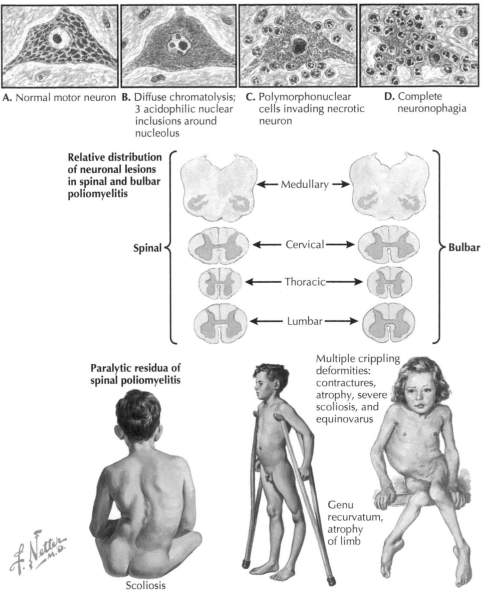

**Stages in destruction of a motor neuron by poliovirus**

**A.** Normal motor neuron

**B.** Diffuse chromatolysis; 3 acidophilic nuclear inclusions around nucleolus

**C.** Polymorphonuclear cells invading necrotic neuron

**D.** Complete neuronophagia

**Relative distribution of neuronal lesions in spinal and bulbar poliomyelitis**

Spinal ⟨ — Medullary → / — Cervical → / — Thoracic → / — Lumbar → ⟩ Bulbar

**Paralytic residua of spinal poliomyelitis**

Multiple crippling deformities: contractures, atrophy, severe scoliosis, and equinovarus

Genu recurvatum, atrophy of limb

Scoliosis

**Fig. 7.2** Poliomyelitis.

Approximately 0.1% to 0.2% of the infected, symptomatic individuals develop paralytic poliomyelitis. The hallmark of paralytic poliomyelitis is asymmetric flaccid paresis or paralysis with areflexia. In children, classic paralytic poliomyelitis has a biphasic course, with an initial minor illness (as previously discussed), then a 7- to 10-day symptom-free period of presumed recovery. It is followed by a rapid, abrupt onset of a major illness characterized by severe back, neck, and muscle pain with the development of muscle weakness, loss of reflexes, and asymmetric flaccid paralysis. The maximal loss of function is typically seen within 3 to 5 days of onset but may progress for up to 1 week.

Lower extremities are more often involved than upper extremities, with the proximal muscles more often involved than distal muscles (see Fig. 7.2). Paralysis may diminish over time, with maximal improvement by 6 months after infection; some patients make a full recovery. Unfortunately, if motor dysfunction persists beyond 12 months, the patient will have lifelong disability. It is estimated up to 60% patients with poliomyelitis will have some residual deficit. Sensory loss with polio is extremely rare. Spastic paralysis suggesting upper motor

neuron disease may be observed with polioencephalitis, which is a rare manifestation of infants.

## Bulbar Poliomyelitis

During epidemics, bulbar poliomyelitis accounts for 5% to 35% of paralytic cases and is the most severe form of polio. Due to its brainstem involvement, including cranial nerves 9, 10, and 12, it results in respiratory compromise. The patient may have nasal speech, inability to swallow secretions, and dyspnea, often appearing anxious or distraught, attributable to difficulty breathing and the associated hypoxia. Prior to modern ventilators, it was associated with a 60% mortality; in contrast, paralytic poliomyelitis affecting only the spinal cord had a less than 7% mortality rate.

## DIFFERENTIAL DIAGNOSIS

The World Health Organization defines AFP as any case of "sudden onset of paralysis/weakness in any part of the body of a child younger than 15 years of age." This definition is utilized in the global surveillance of poliomyelitis, because it is the most common category of diagnoses that includes paralytic polio in the differential. Infectious agents are the leading identified cause of AFP, with up to 90% of cases occurring after a viral infection. Other nonpolio enteroviruses such as enterovirus A71, enterovirus D68, coxsackievirus A7, and coxsackievirus A16 cause AFP. Noninfectious causes of AFP include multiple sclerosis and spinal cord disorders (e.g., transverse myelitis and cord compression). The most common cause of AFP is Guillain-Barré syndrome (GBS), and it can easily be confused with poliomyelitis. However, GBS manifests as a bilateral process, with ascending paralysis, loss of sensory functions, and motor neuron dysfunction. The cerebrospinal fluid (CSF) in GBS generally has elevated protein, with minimal or no pleocytosis. Identifying GBS is important because management differs from poliomyelitis.

## Laboratory Findings

Laboratory findings, such as the complete blood count and serum chemistry values, are usually normal or have nonspecific mild abnormalities. CSF findings with poliovirus infection are similar to those of other forms of viral/aseptic meningitis. The CSF pleocytosis is mild (20 to 300 cells/mm$^3$) with a lymphocyte predominance, although initially there may be a polymorphonuclear predominance. The total protein may be normal or mildly elevated, and the CSF glucose is normal.

Poliovirus is rarely detected in the CSF via culture, due to low viral loads and the presence of neutralizing antibodies. However, it can be rapidly and easily isolated from a throat swab or stool specimens. The poliovirus persists in the throat for 1 to 2 weeks post illness and is shed in the stool for 3 to 6 weeks. Viral culture from a throat swab or stool specimen is the preferred diagnostic tool to confirm diagnosis, preferably performed early in disease course. Molecular methods with real-time reverse-transcriptase polymerase chain reaction (RT-PCR) have a slightly higher sensitivity than viral cell culture and provide results quicker; RT-PCR can also differentiate between WTPV and vaccine-associated poliovirus, whereas viral culture cannot. Confirmatory testing requires two samples obtained greater than 24 hours apart. In addition, confirmation of polio infection may be obtained with serologic testing demonstrating a fourfold rise in antibody titers in paired acute and convalescent sera. Serum neutralizing antibodies develop about 1 week after infection and are lifelong. Immunity is type specific; however, serology cannot differentiate between infection with wild-type or vaccine-associated poliovirus.

## CLINICAL MANAGEMENT AND DRUG TREATMENT

Supportive care is the mainstay of treatment, which may include pain management and physical therapy, as well as ventilation for patients who progress to respiratory failure. Surgical management may be required for long-term sequelae such as contractures. The role of antiviral therapy for polioviruses remains unclear. Pocapavir is a novel antiviral targeting picornaviruses, which acts as a capsid inhibitor to prevent virion uncoating upon entry into the cell. A double-blind placebo study in adults found that Pocapavir decreased the length of time of poliovirus shedding after OPV administration. This finding suggests that Pocapavir could be used to decrease transmission of the poliovirus and may be clinically useful in select groups such as patients with primary B-lymphocyte immune deficiencies, who were found to excrete the vaccine-derived poliovirus for up to 25 years.

## PROGNOSIS

During the time of polio epidemics, approximately 50% mortality was observed in individuals with respiratory failure, but the overall mortality rate was approximately 5% to 10% of people who developed paralytic poliomyelitis. Approximately two-thirds of patients with AFP do not regain full strength and/or function. Approximately 25% to 40% of children infected with polio were at risk of developing a secondary complication called postpolio syndrome. This syndrome was a new onset of functional deterioration after a prolonged period of stability, occurring 15 to 40 years after the initial poliomyelitis infection. Clinical presentation is characterized by muscle weakness, atrophy, and fatigue in the same muscles that were involved in the original illness. Even without meeting the criteria for postpolio syndrome, many polio survivors have long-term sequelae including muscle weakness, chronic pain, contractures, fatigue, depression, and other disorders associated with a lower quality of life.

## PREVENTION AND CONTROL

The polio vaccine was heralded as one of the greatest successes of medical research in the 20th century. In 1955, the inactivated polio vaccine (IPV) was introduced by Dr. Jonas Salk, followed by rapid decline in polio cases. In 1961 to 1962, the OPV, a live attenuated virus vaccine, was introduced by Dr. Albert Sabin and quickly replaced IPV, because of the ease of administration and potential benefits of herd immunity through fecal shedding of the vaccine viruses. After the introduction of both polio vaccines, the rates of wild-type polio dropped dramatically.

In the latter half of the 20th century, VAPP was observed more frequently than wild-type polio in the United States. This led to adoption of a sequential vaccine regimen, with two doses of IPV followed by two doses of OPV from 1997 through 1999. Ultimately, this was changed to an all-IPV dosage regimen in 2000. IPV is safe and generally well tolerated by recipients. It is contraindicated in individuals with previous adverse reaction to the vaccine or its components. Several combination vaccines for infants and children containing IPV are available and offer the benefit of fewer injections per round of vaccinations.

Due to VAPP, adjustments have been made to the OPV to decrease the risk of cVDPV. Serotype 2 WTPV was declared eradicated in 1999, but many of the cVDPV were related to serotype 2. In 2016 the trivalent-OPV was switched to the bivalent-OPV, to eliminate serotype 2 from the vaccine in efforts to decrease the cVDPV. With the recent eradication of serotype 3, further changes may occur and lead to the development of a monovalent OPV.

Currently, both versions of OPV are not routinely produced nor used in the United States; however, an emergency stockpile of OPV is maintained in the event of a polio outbreak. OPV is recommended as the vaccine of choice during a mass vaccination campaign. OPV should not be used in immunocompromised individuals, unvaccinated adults, or in those requiring vaccination but who have an immunocompromised household contact.

## EVIDENCE

Centers for Disease Control and Prevention (CDC): Progress toward polio eradication—worldwide January 2017–March 2019. *MMWR* May 24, 2019;68(20);458-462. *This reference provides data about polio strains involved in current outbreaks and strategies for control efforts.*

Collett MS, Hincks JR, Benschop K, et al. Antiviral activity of Pocapavir in a randomized, blinded, placebo-controlled human oral poliovirus vaccine challenge model. *J Infect Dis* 2017;215(3):335-343. *This reference discusses the clinical trial for the novel antiretroviral Pocapavir and poliovirus.*

World Health Organization (WHO) Global Polio Eradication Initiative News Stories: Two out of three wild poliovirus strains eradicated. 24 October 2019. http://polioeradication.org/news-post/two-out-of-three-wild-poliovirus-strains-eradicated. *This article announced the eradication of wildtype poliovirus 3.*

Zangwill KM, Yeh SH, Wong EJ, Marcy SM, Eriksen E, Huff KR, et al. Paralytic syndromes in children: epidemiology and relationship to vaccination. *Pediatr Neurol* 2010;42(3):206-12. doi: 10.1016/j.pediatrneurol.2009.10.012. *This article reviews efficacy of polio vaccine and rates of acute flaccid paralysis.*

## ADDITIONAL RESOURCES

American Academy of Pediatrics. Poliovirus infections. In: Kimberlin DW, Brady MT, Jackson MA, eds: *Red Book: 2018 Report of the Committee on Infectious Diseases*, ed 31, Elk Grove Village, IL, 2009, AAP, pp 541-545. *This resource provides abbreviated "standard of care" information about diagnosis, management, and prevention.*

Mehndiratta MM, Mehndiratta P, Pande R. Poliomyelitis: historical facts, epidemiology, and current challenges in eradication. *Neurohospitalist* 2014;4(4):223-229. doi:10.1177/1941874414533352. *This article reviews the history and epidemiology of poliomyelitis.*

Troy SB, Moldanado YA. Polioviruses. In: Long SS, Prober CG, Fischer M, eds. *Principles and Practices of Pediatric Infectious Diseases*, ed 5, Philadelphia, 2018, Elsevier, pp 1201-1205. *This resource provides a review of polio pathogenesis, clinical manifestations, and management.*

# 8

# Influenza

*Sylvia H. Yeh*

## ABSTRACT

Influenza is a viral infection characterized by abrupt onset of fever, chills, myalgias, and respiratory symptoms such as cough, sore throat, and rhinitis. Influenza viruses, types A and B, cause annual epidemics worldwide leading to a substantial morbidity and mortality. Vaccination is the most cost-effective means of prevention. Antiviral therapy is available, but development of strain-specific resistance is becoming more common and requires continued global monitoring and vigilance.

## CLINICAL VIGNETTE

An 8-year-old child presents with fever, cough, rhinorrhea, and muscle aches for 1 day. His 5-year-old sibling had similar symptoms 5 days prior. The child has no significant past medical history. His routine immunization status is up to date; however, the family has not seen any health provider since July of this year. Exam reveals a febrile child with otherwise normal vital signs. The child appears tired and the exam is significant for rhinorrhea, mild pharyngeal erythema, and breath sounds without focal changes.

COMMENT: School-aged children are important sources of influenza transmission within a community. Although influenza infection in this age group is associated with a low risk for complications, persons with underlying diseases such as asthma, renal insufficiency, or significant immunosuppression, as well as pregnant women, very young infants, and persons older than 65 years of age, are at increased risk for significant morbidity and mortality. All persons 6 months of age and older are recommended to receive influenza vaccination annually.

## GEOGRAPHIC DISTRIBUTION AND MAGNITUDE OF DISEASE BURDEN

Influenza infections occur in worldwide epidemics annually and are estimated to cause 5.3/1000 episodes of lower respiratory illness across all ages and 290,000 to 650,000 deaths annually. In the United States, for the period 2010 to 2018, seasonal influenza epidemics accounted for a range of 4.4 to 23 million medical visits, 140,000 to 900,000 hospitalizations, and 12,000 to 79,000 respiratory and circulatory deaths each year. Pandemic influenza is defined as the emergence and global spread of a new influenza A subtype to which the population has little or no immunity and that spreads rapidly from human to human. Pandemics may cause greater numbers of hospitalizations and deaths. During the 1918 to 1919 pandemic, an estimated 21 million deaths occurred globally. In general, lower rates for morbidity and mortality were associated with the 2009 to 2010 H1N1/09 pandemic, with approximately 14,500 deaths reported globally from April 2009 through January 2010.

In temperate regions, the disease is prominent during the winter months, with the peak usually occurring in January and February. In tropical areas, influenza viruses may circulate year-round with one or two peaks of activity, usually associated with the regional rainy season. During pandemic influenza, the typical seasonality may not be observed because of sustained transmission among immune-naïve populations. The 2009 pandemic started in spring 2009, spread globally, and had a second observed peak of cases in late October 2009. A similar pattern was observed in the 1918 to 1919 pandemic.

Three types of influenza viruses—A, B, and C—have been described; however, types A and B are the main causes of human disease. Influenza type C is associated with sporadic outbreaks of mild illnesses, primarily in children. Influenza viruses are single-stranded ribonucleic acid (RNA) viruses in the Orthomyxoviridae family and are structurally similar but vary antigenically. Influenza A viruses are further classified based on antigenic variations on their surface proteins, hemagglutinin (HA) and neuraminidase (NA) (Fig. 8.1). Recently the dominant circulating A strains have been H1N1 and H3N2. Both influenza A and B viruses undergo rapid antigenic changes. Minor changes account for the yearly seasonal epidemics (antigenic drifts) wherein previous infections with influenza may provide some protection against disease. Novel antigenic changes (antigenic shifts) typically involve reassortment of antigens between human and nonhuman influenza viruses. These antigenic shifts can lead to pandemics, in which individuals are immune naïve and thus susceptible to more severe disease and widespread transmission. Humans may be infected with nonhuman influenza A strains, including swine and avian strains.

## RISK FACTORS FOR ENDEMIC DISEASE

With seasonal influenza epidemics, children have the highest rates of infections (accounting for 6 to 15 per 1000 pediatric outpatient visits). In contrast, the highest risk for complications and hospitalizations is seen in persons 65 years of age and older, infants and children younger than 2 years of age, and persons with chronic medical conditions. These conditions include congenital or acquired immunodeficiencies, chronic pulmonary, cardiac, renal, hepatic, neurologic or neuromuscular, hematologic, and metabolic diseases (including diabetes), and receipt of long-term aspirin therapy. Pregnant women and residents of long-term care facilities are also at increased risk of complications.

With pandemic influenza, the rate of infections and complications may disproportionately affect healthy adults and others with no known risk factors. In the 2009 pandemic, excess mortality was highest in the 45 to 54 age group among males and females, with fewer than expected deaths in those older than 85 years of age.

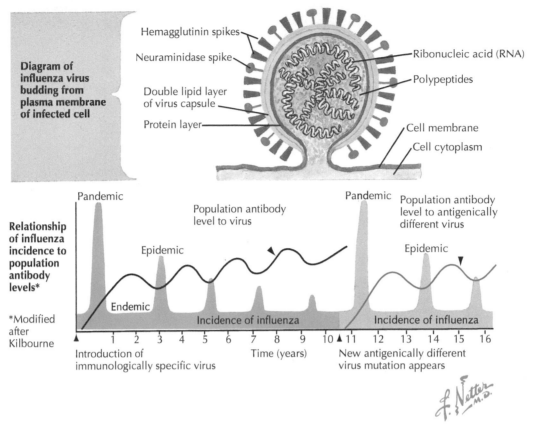

**Diagram of influenza virus budding from plasma membrane of infected cell**

Hemagglutinin spikes
Neuraminidase spike
Double lipid layer of virus capsule
Protein layer
Ribonucleic acid (RNA)
Polypeptides
Cell membrane
Cell cytoplasm

**Relationship of influenza incidence to population antibody levels***

*Modified after Kilbourne

Pandemic
Population antibody level to virus
Epidemic
Endemic
Incidence of influenza

Pandemic
Population antibody level to antigenically different virus
Epidemic
Incidence of influenza

1  2  3  4  5  6  7  8  9  10  11  12  13  14  15  16
Time (years)

Introduction of immunologically specific virus
New antigenically different virus mutation appears

**Fig. 8.1** Influenza virus and epidemiology.

## CLINICAL FEATURES

Influenza viruses are readily spread person to person by infected respiratory droplets, and spread may also occur via fomites contaminated with infected droplets. Viral shedding may occur for 24 hours before onset of symptoms, continues for 5 to 10 days, and correlates directly with the degree of fever. Shedding may be longer in young children and immunocompromised individuals.

After exposure to the virus, the incubation period is 1 to 4 days, with a mean of 2 days. Influenza in older children and adults typically manifests with abrupt onset of fever, chills, myalgia, malaise, headache, and upper respiratory infection (URI) symptoms such as rhinorrhea, cough, and pharyngitis. Younger children may have only fever or may demonstrate URI symptoms indistinguishable from those of other viral respiratory infections. Conjunctival injection, abdominal pain, vomiting, and diarrhea are less commonly observed. Very young children and infants may have symptoms ranging in severity from nonspecific URI, to moderate respiratory illness, to a sepsis-like picture. Individuals who have preexisting immunity from previous natural infections or vaccination may have milder symptoms.

## COMPLICATIONS

Although most influenza illnesses are self-limited, complications are not infrequent. Influenza virus may directly cause manifestations in any part of the respiratory tract, including otitis media, croup, bronchiolitis, and pneumonia (see Fig. 30.2). Approximately 10% to 50% of children younger than 5 years of age develop viral bronchopneumonia with symptoms that are typically mild and resolve without treatment.

The most common complications are bacterial superinfection, which may occur anywhere along the respiratory tract, including otitis media, sinusitis, and secondary bacterial pneumonia. Bacterial sepsis may also be seen in association with influenza. Secondary bacterial infections typically cause a recrudescence of fever with associated localizing symptoms (e.g., ear pain for otitis media, cough or shortness of breath for pneumonia). The most common bacterial pathogens include *Staphylococcus aureus* (including methicillin-resistant *S. aureus*), *Streptococcus pneumoniae*, Group A streptococci, and *Haemophilus influenzae*. *S. aureus* is a prominent player for severe, progressive pneumonia as well as bacterial tracheitis and toxic shock syndrome in a setting of preceding influenza infection.

Myositis is not an uncommon feature, more often seen after influenza B infection, at the time that respiratory symptoms are resolving. Patients frequently exhibit calf tenderness and refusal to walk. The myositis often self-resolves, although rarely rhabdomyolysis and renal failure can occur. Rare extrapulmonary complications caused by direct invasion of an organ by the influenza virus include myocarditis and encephalitis. Complications involving the central nervous system (CNS) range from febrile seizures to severe encephalopathy. Postinfluenza CNS events, presumably caused by the host response to the virus, include encephalopathy, Guillain-Barré syndrome, postinfluenza encephalitis, and transverse myelitis. Although rarely reported since the cessation of aspirin's use in children, Reye syndrome manifests as acute hepatic encephalopathy and carries an approximately 30% mortality rate.

In children, it is estimated that influenza accounts for 50 to 85 per 1000 medical office visits and 6 to 27 per 1000 emergency room visits each year in the United States. In addition, a 10% to 30% increase in the number of prescribed antimicrobial agents has been reported. Hospitalization rates for young children rival those for the elderly, with 108 compared with 190 hospitalizations per 100,000 person-years in children younger than 5 years of age and persons older than 65 years of age, respectively.

| Method | Influenza Virus Types Detected | Types of Specimens to Be Tested | Test Time | CLIA Waived | Performance Characteristics | Notes |
|---|---|---|---|---|---|---|
| Rapid molecular assay (influenza viral RNA or nucleic acid detection) | A & B | NP swab, nasal swab | 15–30 min | Varies by product | High sensitivity, high specificity | Preferred over rapid influenza diagnostic tests for treatment/management decisions |
| Rapid influenza diagnostic tests (antigen detection) | A & B | NP swab, aspirate or wash, nasal swab, aspirate or wash, throat swab | <15 min | Varies by product | Low to moderate sensitivity, high specificity | |
| Molecular assays (RT-PCR or other RNA-based, single or multiplex) | A & B | NP swab or wash, bronchial wash, nasal or endotracheal aspirate | Varies (1–8 h depending on assay) | No | High sensitivity, high specificity | RT-PCR or other molecular assays preferred for hospitalized patients; multiplex RT-PCR targeting multiple pathogens preferred in hospitalized immunocompromised patients |
| Immunofluorescence, direct (DFA) or indirect (IFA) fluorescent antibody staining (antigen detection) | A & B | NP swab or wash, bronchial wash, nasal or endotracheal aspirate | 1–4 h | No | Moderate sensitivity, high specificity | |
| Shell vial tissue culture | A & B | NP swab, throat swap, NP or bronchial wash, nasal or endotracheal aspirate, sputum (specimens placed in viral transport media) | 1–3 days | No | High sensitivity, high specificity | Yields live virus |
| Conventional viral tissue culture | A & B | NP swab, throat swab, NP or bronchial wash, nasal or endotracheal aspirate, sputum (specimens placed in special viral transport media) | 3–10 days | No | High sensitivity, high specificity | Yields live virus. Useful for testing for antiviral resistance. Not useful for diagnostic decision making |

**TABLE 8.1   Influenza Diagnostic Tests**

*NP*, Nasopharyngeal; *RNA*, ribonucleic acid; *RT-PCR*, real-time polymerase chain reaction.
Adapted from CDC Table of Influenza Testing Methods (https://www.cdc.gov/flu/professionals/diagnosis/table-testing-methods.htm), and IDSA Guidelines (Uyeki TM, Bernstein HH, Bradley JS, et al. Clinical practice guidelines by the Infectious Diseases Society of America: 2018 Update on diagnosis, treatment, chemoprophylaxis, and institutional outbreak management of seasonal influenza. *CID* 2018;68:e1-47).

## DIAGNOSTIC APPROACH

During an influenza epidemic, the diagnosis of influenza may be made clinically. However, the clinical presentation may be difficult to distinguish from that of infections from other respiratory pathogens, and laboratory identification of the causative agent can be useful. In addition, specific viral testing and pathogen identification may be important for epidemiologic reasons (e.g., public health surveillance), for consideration of empiric antiviral therapy, and for management of individuals at increased risk of complications.

Viral culture requires 1 to 3 days of incubation if using the shell vial technique but has generally been supplanted in clinical practice by use of molecular and antigen detection assays. There has been a surge of technologic advances in rapid diagnosis of influenza in recent years, with several methods currently available (Table 8.1). The current recommendations from the Infectious Diseases Society of America and Centers for Disease Control and Prevention (CDC) favor use of rapid molecular assays over rapid influenza antigen detection tests for both outpatient and inpatient settings (https://www.cdc.gov/flu/professionals/diagnosis/consider-influenza-testing.htm). Point of care access to tests may provide reassurance and limit the administration of unnecessary antibiotics and further diagnostic testing; however, rapid antigen detection tests typically have a sensitivity of 50% to 70%. The positive predictive value of rapid antigen detection tests is also highly dependent on the prevalence of influenza in the community; they are of limited value when the prevalence is less than 10%. In addition, the rapid antigen detection tests may not be able to detect influenza viruses with significant antigenic changes, such as in a pandemic setting.

## CLINICAL MANAGEMENT

For most individuals with influenza, the care is supportive, emphasizing patient comfort and fever control with antipyretics. Acetaminophen is the preferred antipyretic for most infants and young children. For most individuals, ibuprofen is also acceptable with assurance of good hydration. Aspirin is contraindicated for children with influenza because of the association with Reye syndrome.

Specific antiinfluenza medications for treatment of influenza include adamantanes and NA inhibitors. These antivirals, when used within 48 hours of illness onset, may shorten the duration of symptoms. The NA inhibitors may reduce the risk of morbidity from influenza in certain high-risk hosts. Due to increasing rates of influenza strains with resistance to adamantanes, NA inhibitors are currently the only class of antiinfluenza medications recommended when treatment for influenza is indicated. Treatment recommendations may change based on ongoing monitoring of susceptibility data of circulating strains.

## Adamantanes

Available drugs in the adamantane class are amantadine and rimantadine, which are active only against influenza A. These drugs target the M2 protein, which is present on influenza A, but not type B. When the adamantanes are used within 48 hours of symptom onset, the duration of fever, systemic symptoms, and viral shedding is shortened by 1 to 2 days. The most common adverse drug effects are insomnia, anxiety, nausea, and loss of appetite, and these are more commonly associated with amantadine than rimantadine. Infrequent CNS side effects include agitation and an increase in seizure activity in persons with epilepsy; these are also more common with amantadine. This class of medications is not licensed for use in children younger than 1 year of age.

## Neuraminidase Inhibitors

NA inhibitors inhibit the NA protein present on both influenza A and B, blocking the influenza virus from fusing with the host cell membrane. Three drugs in this class are currently available: oseltamivir, which is administered orally; zanamivir, which is inhaled; and peramivir, which is given as a single intravenous dose. Use of these medications within 24 to 48 hours of symptoms onset is associated with shortened duration of fever and constitutional symptoms but does not necessarily shorten viral shedding. Use of oseltamivir within 24 hours is associated with reduced risk of asthma exacerbations due to influenza. The most common adverse effects from oseltamivir are nausea and vomiting. Reports of self-injurious behavior and delirium, associated with oseltamivir, have been determined to be more likely caused by influenza disease itself, but advising parents to monitor for abnormal behavior may be prudent. Bronchospasm has been reported with zanamivir administration, and therefore this drug should be used with caution in persons with asthma or other preexisting pulmonary dysfunction. Current studies suggest that the rate of adverse events following peramivir are similar to placebo.

Routine use for mild illnesses in immunocompetent hosts is not recommended. However, antiviral treatment should be given as soon as possible for adults and children with documented or suspected influenza with any of the following: (1) any person hospitalized with influenza regardless of duration of illness, (2) outpatients of any age with severe or progressive illness regardless of duration, (3) outpatients who are at high risk of complications from influenza including those with chronic medical conditions or immunocompromised, (4) children less than 2 years of age and adults 65 years of age or older, and (5) pregnant women and those within 2 weeks postpartum. For individuals who are not at high risk for complications from influenza, but have documented or suspected influenza, treatment may be considered in the following circumstances: (1) outpatients with illness 2 days or less at presentation, (2) persons with symptomatic influenza who are household contacts of persons who are at high risk of developing complications from influenza, or (3) symptomatic health care workers who care for patients at high risk of developing complications from influenza.

## PROGNOSIS

In most immunocompetent hosts, influenza is self-limited illness. However, in very young children and the elderly, rates of complications and death are higher than seen in other age groups. Hospitalization rates are much higher in very young children, with rates of 240 to 720 per 100,000 children per year in infants younger than 6 months of age compared with 20 per 100,000 children per year in children 2 to 5 years of age. Although death is rare, the rate is estimated at 0.4 per 100,000 children younger than 5 years of age per year. Data from 2003 to 2005 reveal that, although children with underlying medical conditions are at greater risk of death, most (51%) pediatric deaths resulting from laboratory-confirmed influenza occurred in children with no known risk factors.

Individuals older than 65 years of age with underlying medical conditions have the highest rates of hospitalization, estimated at 560 per 100,000 persons per year (compared with 190 per 100,000 for persons older than 65 years of age who are healthy). Persons with immunodeficiency conditions, such as human immunodeficiency virus (HIV), and pregnant women also have demonstrable increased risk of hospitalization and death compared with persons in similar age groups without these conditions.

## PREVENTION AND CONTROL
### Immunoprophylaxis

It is estimated that influenza vaccination prevented 1.6 to 6.7 million illnesses, 39,000 to 87,000 hospitalizations, and 3000 to 10,000 deaths each influenza season between 2010 and 2016. Several influenza vaccines are currently available for the prevention of influenza A and B (Table 8.2). All persons 6 months of age or older without contraindications to influenza vaccinations should be vaccinated against influenza annually. When vaccine supply is limited, vaccination strategies should target groups that are at increased risk for developing complications due to influenza (Box 8.1). Influenza vaccines typically contain either four strains (quadrivalent with two influenza A strains and two influenza B strains) or three strains (trivalent with two influenza A strains and one influenza B strain). The strains included are based on the predicted likelihood of what strains will circulate in the hemisphere in the upcoming season. These predictions are based on patterns of global circulation and are determined by experts from the World Health Organization, and from the US Food and Drug Administration and CDC in the United States approximately 9 months ahead of the influenza season. The efficacy of inactivated influenza vaccine is dependent on the age of the host, underlying host factors, and the antigenic match between the vaccine strain and the circulating strain. The efficacy of influenza vaccines varies from 60% to 95% against culture-confirmed influenza when the vaccine strain matches the circulating strain.

Children 6 months of age to 8 years of age require two doses of vaccine, separated by at least 4 weeks, during the first season they are immunized, and a single annual dose thereafter. In this age group, if prior to the current influenza season the child has received less than two doses of influenza vaccine, then two doses of the current season's vaccine is recommended separated by a minimum of 4 weeks. Persons 9 years of age or older need only one dose of influenza vaccination regardless of prior influenza vaccination history.

Two vaccine formulations have been developed to improve efficacy in those older than 65 years of age and include a high-dose trivalent inactivated influenza vaccine and an adjuvanted trivalent inactivated influenza vaccine. Most vaccines are egg based in development, and these formulations may have trace amounts of egg protein. However, there are formulations that are developed in cell culture or using

**TABLE 8.2  Influenza Vaccine Types**

| Vaccine Type | Age Indication | Route |
|---|---|---|
| **Quadrivalent (Contains 2 Influenza A Strains, 2 Influenza B Strains)** | | |
| Inactivated Influenza Vaccine 4 (IIV4)—Egg Based—Standard Dose | Varies based on manufacturer & formulation | IM |
| Inactivated Influenza Vaccine 4 (IIV4)—Cell Culture Based—Standard Dose | ≥4 years old | IM |
| Live Attenuated Influenza Vaccine (LAIV4)—Egg Based | 2 through 49 years | NAS |
| Recombinant Influenza Vaccine (RIV4) | ≥18 years old | IM |
| **Trivalent (Contains 2 Influenza A Strains, 1 Influenza B Strain)** | | |
| Inactivated Influenza Vaccine (IIV3)—Egg Based With MF59 Adjuvant—Standard Dose | ≥65 years old | IM |
| Inactivated Influenza Vaccine (IIV3)—Egg Based—High Dose | ≥65 years old | IM |

*IM,* Intramuscular; *NAS,* nasal spray.
Adapted from CDC. Prevention and control of seasonal influenza with vaccines: Recommendations of the Advisory Committee on Immunization Practices—United States, 2019-20 influenza season. *MMWR Recomm Rep* 2019;68(3):1-21.

**BOX 8.1  Centers for Disease Control and Prevention Advisory Committee on Immunization Practices Specific Groups for Influenza Vaccinations 2019**

- All persons 6 months of age or older should be vaccinated annually if there are no contraindications for vaccination.
- When the vaccine supply is limited, vaccination efforts should focus on vaccinating the following groups:
  - Persons at higher risk for medical complications from severe influenza:
    - All children aged 6 to 59 months
    - All persons 50 years of age or older who have chronic pulmonary (including asthma), cardiovascular (except hypertension), renal, hepatic, cognitive, neurologic or neuromuscular, hematologic, or metabolic disorders (including diabetes mellitus)
    - are immunosuppressed (including immunosuppression caused by medications or by human immunodeficiency virus)
    - are receiving long-term aspirin therapy and therefore might be at risk for experiencing Reye syndrome after influenza virus infection
    - are residents of long-term care facilities
    - will be pregnant during the influenza season
    - are American Indians/Alaska Natives
    - are extremely obese (body mass index ≥ 40 for adults)
  - Persons who live with or care for persons at higher risk for influenza-related complications
    - healthcare personnel
    - household contacts and caregivers of children age younger than 5 years and adults age 50 years or older, with particular emphasis on vaccinating contacts of children age younger than 6 months
    - household contacts and caregivers of persons with medical conditions that put them at higher risk for severe complications from influenza

Adapted from Centers for Disease Control and Prevention (CDC). Prevention and control of seasonal influenza with vaccines: Recommendations of the Advisory Committee on Immunization Practices—United States, 2019-20 influenza season. *MMWR Recomm Rep* 2019;68(3):1-21 (https://www.cdc.gov/mmwr/volumes/68/rr/rr6803a1.htm, accessed January 2020).

or required epinephrine or other emergency medical intervention may still receive any licensed recommended influenza vaccine, but it is recommend that the vaccine be administered in a medical setting supervised by a health provider able to recognize and manage severe allergic reactions. Prior history of severe allergic reaction to influenza vaccination is a contraindication for future influenza vaccination.

Live attenuated influenza vaccine (LAIV) is indicated for healthy persons 2 to 49 years of age. LAIV should not be administered to persons with reactive airway disease or asthma, known or suspected immunodeficiency, pregnant women, persons receiving salicylates, or persons with conditions considered to be high risk for severe influenza, for whom inactivated influenza vaccine is recommended. LAIV should be used with caution in persons with a history of Guillain-Barré syndrome within 6 weeks after a previous dose of influenza vaccination. The efficacy of LAIV is approximately 90% against culture-confirmed influenza cases when the vaccine strain matches the epidemic strain. In addition, available data suggest that LAIV may provide cross-protection against mismatched strains owing to antigenic drift.

## Chemoprophylaxis

In the event that annual influenza vaccination is not possible or is contraindicated, chemoprophylaxis with antiviral medication is second line protection. Studies of antiviral chemoprophylaxis have demonstrated 20% to 40% reduction of secondary cases within households. However, chemoprophylaxis has potential drug toxicity, as well as potential for promoting development of antiviral resistance, and should not be considered equivalent to vaccination. Chemoprophylaxis may be considered (1) for unimmunized individuals at high risk of complications from influenza or those who were vaccinated less than 2 weeks before high-risk exposure to influenza virus, (2) for unimmunized close contacts of high-risk persons, (3) for immunized high-risk children when the vaccine strain poorly matches the circulating strain, and (4) for control of an influenza outbreak in a close setting (such as an institution). Because of the emergence of resistance to the antivirals, recommendations regarding chemoprophylaxis need to be adjusted based on susceptibility of circulating strains. For the United States, these recommendations are available through the CDC (https://www.cdc.gov/flu/professionals/antivirals/index.htm, accessed January 2020).

## EVIDENCE

Ambrose CS, Yi T, Walker RE, Conner EM: Duration of protection provided by live attenuated influenza vaccine in children, *Pediatr Infect Dis J* 2008;27:7444-7448. *This reference provides a clinical trial of live attenuated*

recombinant technology. Persons who have a history of egg allergy with only urticaria may receive any influenza vaccine that is appropriate for their age and health history. Persons who have a history of egg allergy with symptoms other than urticaria (e.g., angioedema or swelling, respiratory distress, lightheadedness, or recurrent vomiting)

influenza vaccine examining length of immunity and possible cross-protection from non-vaccine influenza strains.

Finelli L, Fiore A, Dhara R, et al: Influenza-associated pediatric mortality in the United States: increase of *Staphylococcus aureus* coinfection, *Pediatrics* 2008;122:805-811. *This reference provides an epidemiologic study of influenza in children and associated disease with* S. aureus.

Nguyen AM, Noymer A. Influenza mortality in the United States, 2009 pandemic: Burden, timing and age distribution. *PLOS One* 2013;8:e64198. *This article lists the excess US mortality during the 2009 pandemic.*

Walter ND, Taylor TH, Shay DK, et al: Influenza circulation and the burden of invasive pneumococcal pneumonia during a non-pandemic period in the United States, *Clin Infect Dis* 2010;50:175-183. *This reference provides an epidemiologic study of influenza in adults and risk of concomitant disease with pneumococcal pneumonia.*

## ADDITIONAL RESOURCES

Centers for Disease Control and Prevention (CDC): Influenza. Available at: www.cdc.gov/flu. *This CDC site provides the most up-to-date data about epidemiology of circulating influenza strains.*

Centers for Disease Control and Prevention (CDC). Prevention and control of seasonal influenza with vaccines: Recommendations of the Advisory Committee on Immunization Practices—United States, 2019-20 influenza season. *MMWR Recomm Rep* 2019;68(3):1-21. *Provides the recent recommendations related to influenza vaccination recommendations in the United States.*

Derlet RW, Sandrock SE: Influenza. Available at: http://emedicine. medscape.com/article/219557-overview. Accessed April 18, 2010. *This resource provides a summary for diagnosis and management of influenza.*

Fiore AE, Shay DK, Broder K, et al: Prevention and control of seasonal influenza with vaccines. Recommendations of the Advisory Committee on Immunization Practices (ACIP), *MMWR Recomm Rep* 2009;58:1-52. *This provides the recent Centers for Disease Control and Prevention (CDC) recommendations for prevention and control of seasonal influenza.*

Long SS, Pickering LK, Prober C, eds: *Principles and practices of pediatric infectious diseases*, ed 3, St Louis, 2009, Elsevier, pp 1176-1179. *This reference provides a comprehensive review of the current clinical approach to influenza in pediatrics.*

Uyeki TM, Bernstein HH, Bradley JS et al. Clinical practice guidelines by the Infectious Diseases Society of America: 2018 update on diagnosis, treatment, chemoprophylaxis, and institutional outbreak management of seasonal influenza. *CID* 2018;68:e1-47. *Provides the recent recommendations regarding diagnosis and treatment of influenza in the United States.*

# 9

# Rotavirus Infection

*Rebecca Sainato*

## ABSTRACT

Rotavirus infections are the most common cause of severe dehydrating gastro-enteritis worldwide. In developing countries, virtually all children have been infected by 2 to 3 years of age. Rotaviruses are segmented, nonenveloped, double-stranded ribonucleic acid (RNA) viruses that belong to the family Reoviridae. There are eight distinct groups (A through H). Group A viruses cause the majority of rotavirus diarrhea worldwide; groups B and C are also associated with gastro-enteritis. Among group A viruses, the antigens residing on the outer capsid of VP7 (G type) and VP4 (P type) proteins determine the genotype; five strains account for more than 90% of infections and these serve as the primary vaccine targets. Clinical presentations are characterized by varying severity of vomiting, diarrhea, and fever, either alone or in combination. Illness may be mild or asymptomatic, particularly in neonates, older children, and adults. Infection is self-limited and treatment is supportive, directed at maintaining fluid and electrolyte balance and nutrition. Natural infection is generally protective against subsequent severe illness. Immunization has markedly reduced the rates of illness and death previously associated with rotavirus gastroenteritis. Two highly effective rotavirus vaccines are recommended for routine use.

## CLINICAL VIGNETTE

A 6-month-old male infant is brought to the emergency department (ED) for a second time because of increasing lethargy. He was seen in the ED less than 24 hours ago with a febrile seizure; his temperature was 103.4°F. The child's mother had also reported vomiting and diarrhea for the past 48 h. At this initial visit, he was observed over several hours and reportedly was a playful, active, and well-appearing infant without evidence of dehydration. No labs were obtained. Since then he has continued to vomit occasionally and has had multiple watery stools, without blood or mucus. He continues to have low-grade fevers, maximum body temperature 101.2°F. He has been too tired to take formula. Mom cannot recall the last time he urinated.

He was previously well until onset of symptoms. He is an otherwise healthy baby without history of prematurity or other complicated infections. He attends daycare, and the family was just notified that multiple other infants have similar symptoms.

On exam you note that the infant is limp in his mother's arms. He has a weak cry and does not withdraw from venous sampling. His eyes are sunken and mucous membranes are dry. Serum sodium level is 166 mEq/L, bicarbonate 10.6 mmol/L, anion gap 32.4 mEq/L, and creatinine 2.4 mg/dL. Multiplex gastrointestinal polymerase chain reaction panel is positive for rotavirus and negative for all other pathogens.

COMMENT: Due to severe dehydration the infant was hospitalized and started on parenteral fluids. After 48 h, he was switched to oral rehydration and then advanced to his regular formula. Upon discharge home his mother was given instructions on hygienic measures to prevent transmission to others in the household.

## GEOGRAPHIC DISTRIBUTION AND DISEASE BURDEN

Rotavirus gastroenteritis is the most common cause of severe dehydrating diarrhea in infants and young children worldwide. Before the availability of rotavirus immunization almost all children acquired infection by age 3 years, irrespective of their geographic location or living conditions. In the United States, infection was responsible for a significant number of pediatric office and ED visits as well as hospitalizations. Based on the Centers for Disease Control and Prevention (CDC) surveillance from 2007 to 2018, the median annual percentage of rotavirus positive tests declined from 25.6% to 6.1% during the postvaccine period. Rotavirus still causes approximately 200,000 child deaths each year, although this is a remarkable improvement over previous estimates. Still, 75% of the world's infants remain unvaccinated. Shifts in strain prevalence have been observed with introduction of current vaccine options. However, vaccine efficacy remains high, likely due to robust cross-protection.

## RISK FACTORS

Transmission of infection is primarily fecal-oral via person-to-person spread and contact with contaminated environmental surfaces where this virus can remain infectious for weeks to months. Contamination of water and food has caused outbreaks; respiratory droplet transmission may play a minor role. Rotavirus is shed in the stool of infected children not only during the acute illness but also several days before and after. The high rate of transmission and prevalence of early childhood infection is likely potentiated by a combination of intense viral shedding during clinical disease (up to $10^{12}$ virions per gram of stool) together with a small oral infective dose (as few as 100 particles).

Risk factors for illness in infancy include use of formula rather than breastfeeding, low birth weight, residence in a household with one or more children younger than 2 years of age, attending a childcare facility, male gender, and young maternal age (Fig. 9.1).

## CLINICAL FEATURES

Historically, outbreaks of rotavirus have been seasonal, typically during the cooler and drier months. Widespread use of rotavirus vaccine in infancy has resulted in a markedly diminished incidence of illness, with cases still emerging in late winter/early spring. The peak age of illness in developed countries is from 6 to 24 months. In developing countries most cases occur in the first year of life.

Illness begins from 1 to 2 days after infection, most often with the abrupt onset of vomiting and fever, followed shortly by watery diarrhea (Fig. 9.2). Any of these findings can occur alone or in combination; however, 95% of children have vomiting, diarrhea, or both.

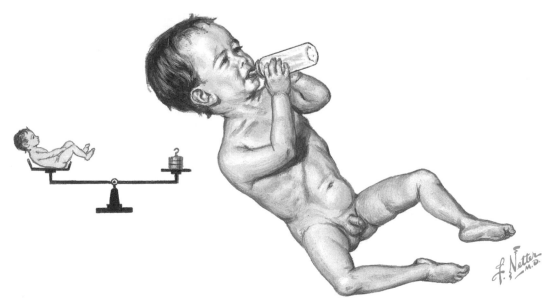

**Fig. 9.1** Low birth rate and use of formula rather than breastfeeding are risk factors for rotavirus infection in infants. The bottle-fed boy shows signs of dehydration from rotavirus infection.

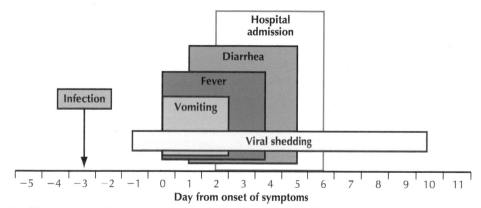

**Fig. 9.2** Clinical course of rotavirus infections. (Data from Marshall G, Dennehy P, Matson DO, Staat MA: Rotavirus: prevention and vaccination strategies to address burden of disease, Thorofare, NJ, 2008, Vindico Medical Education.)

Diarrhea can be both profuse and frequent, occurring 8 to 20 times per day. Vomiting is usually brief in duration, and diarrhea resolves in most cases after 3 to 7 days. Although a third of children manifest a fever of 39°C (102.2°F), up to 25% are afebrile at the onset of clinical disease.

Age is an important determinant of the clinical course. In the neonatal period, illness is usually asymptomatic or very mild. The first infection after 3 months of age is typically the most severe. Infected infants and toddlers manifest a broad spectrum of signs and symptoms, ranging from a single episode of watery diarrhea or emesis to fulminant gastroenteritis leading rapidly to electrolyte imbalance, hypovolemia, and shock. Fatalities are almost exclusively related to dehydration, acidosis, and electrolyte imbalance. With rare exceptions, adolescents and adults have asymptomatic infection or mild disease, probably as a result of protection derived from previous infection(s).

Children and adults who are immunocompromised as a result of congenital or acquired immunodeficiency may experience more severe or prolonged gastroenteritis. Rotavirus infection has been associated with necrotizing enterocolitis, particularly in preterm infants. Infection has also been associated with encephalitis, meningitis, pancreatitis, myocarditis, myositis, and lower respiratory tract infection.

In cases of rotavirus-associated seizures, it has been identified within the cerebrospinal fluid. Rotavirus infection is unlikely to be a principal cause of intussusception.

## DIAGNOSTIC APPROACH

No clinical features reliably distinguish rotavirus disease from that caused by many other gastrointestinal pathogens. In hospitalized children, it is common to observe electrolyte abnormalities and an elevation in serum hepatic enzymes. Fecal leukocytes are generally not observed. A number of assays are used to detect rotavirus. Enzyme-linked immunosorbent assay (ELISA) or latex agglutination tests are available in most clinical laboratories to detect group A virus antigens. Molecular diagnostic methods improve sensitivity, although increased use of these modalities has led to frequent detection along with identification of multiple pathogens, sometimes in asymptomatic patients.

## TREATMENT

Treatment of rotavirus gastroenteritis is supportive, directed at restoration and maintenance of vascular volume, electrolyte, and caloric balance. Illness is self-limited; there is no antiviral therapy. In most cases a few days of oral rehydration therapy (ORT) with appropriate electrolyte-carbohydrate solutions are sufficient. Severe dehydration, persistent vomiting, intractable watery diarrhea, or refusal of oral fluids may necessitate parenteral therapy. Breastfeeding should be continued or reinstituted as quickly as possible. In older infants and children, early refeeding with solids should be encouraged. Use of probiotics may provide some benefit in shortening the course of illness.

## PREVENTION AND CONTROL

Breastfeeding, hand washing, maintaining surface cleanliness (particularly in daycare centers), and diligently using contact precautions with hospitalized children are helpful in preventing rotavirus infection. In the United States, two rotavirus vaccines (RotaTeq, Merck; Rotarix, Glaxo SmithKline) are recommended for routine infant immunization in the first 8 months of life, including preterm infants. Both are live, oral vaccines. They are comparably effective and have resulted in an 80% reduction in rotavirus gastroenteritis of any severity and more than 95% reduction against severe dehydrating gastroenteritis. In addition to providing direct protection in immunized infants, there is epidemiologic evidence that reduction in the community incidence of rotavirus has occurred, likely related to development of herd immunity. For unclear reasons, vaccine efficacy appears to be reduced in the developing world. Infants living in households with immunocompromised people can be immunized against rotavirus but highly immunocompromised people should avoid handling diapers of infants for at least 4 weeks following vaccination. Severe combined immunodeficiency (SCID) and history of intussusception are both contraindications for use of rotavirus vaccines. With current vaccine formulations and schedules, there likely remains a small attributable risk of vaccine-associated intussusception with 1 excess case per 20 to 100,000 vaccinated infants, most often during the first week following the first or second vaccine dose.

## FUTURE DIRECTIONS

Many countries have not yet introduced a rotavirus vaccine into their national immunization programs. Cold-chain capacity, cost, and product availability remain as obstacles to the expanded use of rotavirus vaccines within the developing world. Current vaccine options also have not demonstrated equivalent efficacy within the developing world as observed in industrialized countries, although this may be confounded by a lower average age of infection.

## ACKNOWLEDGMENT

The author acknowledges S. Michael Marcy and Susan Partridge for their contribution to the previous edition chapter.

## EVIDENCE

Bernstein DI, ed: The changing epidemiology of rotavirus gastroenteritis, *Pediatr Infect Dis J* 2009;28:S49-S62. *This reference provides data about the changing epidemiology of rotavirus in recent times.*

Chiu M, Bao C, Sadarangani M. Dilemmas with rotavirus vaccine: The neonate and immunocompromised. *Pediatr Infect Dis J* 2019 Jun;38(6S Suppl 1):S43-S46. *This articles reviews current research and consensus statements on the safety of the rotavirus vaccine.*

Kuehn B. Reductions in rotavirus infections, *JAMA* 2019;322(6):496. *This article reviews current U.S. surveillance data on incidence and seasonality of positive test results.*

Lee B. Rotavirus epidemiology and vaccine effectiveness: Continuing successes and ongoing challenges, *Pediatrics* 2019 [Epub ahead of print]. *This reference provides information about how use of current vaccines impacts circulating strains and rates of infections.*

World Health Organization (WHO): Oral rehydration salts (ORS): a new reduced osmolarity formulation. Available at: www.who.int/child-adolescent-health/New_Publications/NEWS/Statement.htm. *This reference provides information about the preparation and use of ORS for diarrhea and dehydration in infants and children.*

Zhou B, Niu W. Efficacy of live oral rotavirus vaccines, *Lancet Infect Dis* 2019;19(9):929. *This meta-regression study estimates the efficacy of live oral rotavirus vaccines.*

## ADDITIONAL RESOURCES

American Academy of Pediatrics (AAP): Rotavirus. In Kimberlin DW, Brady MT, Jackson MA, Long SS, eds: *Red Book: 2018-2021 Report of the Committee on Infectious Diseases*, ed 31, Elk Grove Village, IL, 2018, AAP, pp 700-704. *This resource provides an abbreviated summary of the diagnosis, management, and prevention of rotavirus infections.*

Cortese MM, Parashar UD, Centers for Disease Control and Prevention (CDC): Prevention of rotavirus gastroenteritis among infants and children: recommendations of the Advisory Committee on Immunization Practices (ACIP), *MMWR Recomm Rep* 2009;58:1-37. *This provides a review of the preventative measures available for rotavirus.*

Long SS, Pickering LK, Prober CG: Rotavirus. In *Principles and Practice of Pediatric Infectious Diseases*, ed 4, Philadelphia, PA, 2012, pp 1094-1097. *This resource provides a more in depth review of the diagnosis, management, and prevention of rotavirus infections.*

Magill AJ, Ryan ET, Hill DR, Solomon T. *Hunter's Tropical Medicine and Emerging Infectious Diseases*, ed 9, Elsevier, 2013, pp 276-279. *This resource provides a global perspective to epidemiology, management, and prevention of rotavirus infections.*

# 10

# Measles

*Milissa U. Jones, Martin G. Ottolini*

## ABSTRACT

Measles (rubeola) is an extremely contagious, prolonged respiratory and systemic viral illness characterized by high fever, an erythematous maculopapular rash, cough, coryza, and conjunctivitis. Measles infects the immune system and leads to continued susceptibility to other pathogens long after recovery. A variety of complications lead to serious morbidity and mortality, particularly in under- and malnourished populations, where death rates can be up to 30%. Vitamin A supplementation may reduce the morbidity and mortality of complications such as pneumonia, encephalitis, and prolonged debilitating diarrhea. A live attenuated vaccine has been available since the late 1960s, and a two-dose immunization strategy has proven safe and highly effective. The global incidence of disease and associated childhood mortality has been greatly reduced through extensive immunization programs, but further progress is hampered by economic and logistic challenges, which are frequently exacerbated by natural and human-made disasters. Additionally, substantial progress in many regions is threatened by an increasing distrust of recommended immunizations.

## CLINICAL VIGNETTE

A 20-month-old child presents to her pediatrician's office because of fever, malaise, cough, and a rash. The family has just returned from a month-long trip to Europe to visit relatives. The child's symptoms began 4 days earlier with a brassy cough, runny nose, and conjunctivitis. Additionally, she has had a fever ranging from 104°F to 105°F as well as fatigue. Her cough is worsening and she is sensitive to light. On the previous day, she developed an erythematous maculopapular rash that began with a few lesions but today covers her face and seems to be moving to her chest and back.

The child's past history is unremarkable. She has received only her 2-month immunizations as her parents have chosen to delay the subsequent vaccines to avoid "overwhelming" her immune system.

Physical exam reveals a febrile child who is exquisitely irritable. Her temperature is 103.5°F. Examination of the oropharynx reveals fading white lesions with a bluish hue on the buccal mucosa. Conjunctivitis and photophobia are present on examination of the eyes. Diffuse rhonchi are present upon auscultation of the lungs. The dermatologic exam reveals a confluent erythematous maculopapular rash, covering the face with a few scattered lesions on the neck that were not noted previously by the parents.

The pediatrician makes a clinical diagnosis of measles. The local health department is contacted and supportive treatment is prescribed, including 200,000 international units (IU) of vitamin A for 2 days. Arrangements are then made for an expedited serology and a respiratory polymerase chain reaction (PCR) panel.

COMMENT: The immediate diagnosis of measles is clinical, based on the constellation of classic symptoms as well as potential exposure to measles virus. Confirmatory diagnostic tests are critical for the management of contacts and to control this public health emergency. This patient was likely exposed during the latter half of her trip to Europe. She demonstrates the classic prodrome of illness including high fever and malaise as well as the "classic 3 Cs": "cough, coryza, and conjunctivitis." The oral lesions described, also known as Koplik spots, are pathognomonic for measles; however, these are often missed on exam or may be fading or absent by the time the rash has become more prominent. Although there is a highly effective live virus vaccine against the measles virus, problems in eliminating disease remain in some low-resource countries and in regions suffering from humanitarian crises; the growing trend of "vaccine refusal" in both Europe and the Unites States also hampers these efforts.

## GLOBAL EPIDEMIOLOGY AND CONTAGIOUSNESS

The number of naive contacts potentially infected by exposure to one contagious individual, defined as the measles basic reproduction number ($R_0$), has been estimated to be 12 to 18; this is among the highest numbers for any known pathogen. The World Health Organization (WHO), in concert with several partners, has supported the Global Measles and Rubella Strategic Plan in an attempt to greatly reduce measles by the year 2020. Global numbers of measles cases decreased from over 28 million in 2000 (with 535,000 annual deaths) to fewer than 8 million (with less than 90,000 annual deaths) by the end of 2016. Since that nadir, however, measles cases have risen; more than 140,000 deaths were reported in 2018, with significant measles outbreaks currently occurring in every part of the globe. The strategic goal of measles elimination in five of six WHO regions was not met by the year 2020; in fact, "The Americas," which was the only region to eliminate endemic measles, recently lost that status due to outbreaks aggravated by the collapse of the health care infrastructure in Venezuela. Multiple factors continue to challenge further reductions and the goal of measles elimination. These factors include the virus's intrinsic extreme contagiousness in an era of increasing travel, the financial and logistic burdens of the two-dose schedule, the increasing number and severity of human-made and natural disasters that regionally weaken health care infrastructure, and the emergence of elective vaccine avoidance.

## CLINICAL FEATURES

Rubeola virus is a single-stranded ribonucleic acid (RNA) virus belonging to the family Paramyxoviridae. Infection is transmitted by direct contact with infected respiratory droplets and less commonly by airborne spread. It is one of the most communicable infectious diseases, with an attack rate well over 90% for susceptible persons. Infected individuals are most contagious during the late prodromal phase, when cough and coryza are maximal but the specific diagnosis has not yet been made. The incubation period ranges from 7 to 21 days from the time of exposure to the onset of symptoms. Unlike most paramyxoviruses, measles virus directly infects the dendritic cells of the respiratory or conjunctival mucosa, with rapid migration to regional lymph nodes and ultimately infection of CD150 (SLAM)-expressing T and B cells. Over time, the respiratory epithelium becomes infected, resulting in damage and sloughing; this induces coughing, which expels infectious secretions and restarts the infectious cycle.

The prodromal phase begins approximately 10 days following exposure and can last for 3 to 6 days. The prodrome begins as a severe upper respiratory tract infection, with malaise, fever, and the classic "3 Cs" (cough, coryza, and conjunctivitis) of measles. Profuse nasal discharge and a brassy cough are typically maximal on day 5. This represents the most infectious phase of illness. In addition to conjunctivitis, patients may also develop photophobia. About 2 to 3 days before the appearance of the rash, Koplik spots (bluish-gray specks, at times described as appearing like coarse salt crystals, on an erythematous base) appear on the buccal mucosa, usually opposite the molars (Fig. 10.1). Koplik spots are pathognomonic for measles; however, in the absence of close examination and clinical experience, they are often overlooked or misdiagnosed as thrush. As the rash emerges, Koplik spots begin to fade, contributing to the difficulty of less experienced providers in diagnosing measles correctly.

The characteristic morbilliform rash usually begins as blanching macules behind the ears and at the hairline (see Fig. 10.1); it then spreads to involve the face and proceeds down the body to the extremities and last to the palms and soles. The rash usually lasts 5 days, often becoming confluent, and then fades. A brownish discoloration of the skin may remain for 10 days after the disappearance of the rash. The skin may desquamate, but this usually spares the hands and feet. In uncomplicated illness, the duration from late prodrome to resolution of fever and rash is 7 to 10 days. Infected persons remain contagious to others until approximately 4 days following the appearance of the rash.

## ATYPICAL PRESENTATIONS

Measles can present atypically as a result of a variety of patient factors, which often results in a delayed diagnosis. For example, the rash may be more difficult to discern in patients with darker skin. Malnourished patients may have a prolonged desquamating dermatitis rather than the classic maculopapular rash. The rash may not be present at all in immunocompromised individuals despite severe systemic illness.

### Modified Measles

A mild form of measles has been observed in persons with some degree of passive immunity, such as infants less than 6 months of age who have passively acquired maternal antibodies and some individuals who have received measles immune globulin. The illness is usually mild and abbreviated, although the incubation period may be prolonged up to 21 days. Patients with modified measles do not typically experience complications, and transmission to others is extremely rare.

### Black Measles

Very rarely, disseminated intravascular coagulation occurs as a systemic complication, which—with measles virus infection of endothelium—results in bleeding from epithelial and mucosal surfaces (*black measles*). Hemorrhagic measles is often severe and has a high mortality rate.

### Measles in Immunocompromised Hosts

Severe measles infection may occur in persons with deficient cellular immunity, such as those being treated for malignancy, acquired immunodeficiency syndrome (AIDS), or congenital immunodeficiency. Often, these individuals do not have a rash; however, there is a significant incidence of pneumonitis and encephalitis.

### Measles in Malnourished Hosts

Malnutrition, which impairs cellular immunity, appears to contribute to higher morbidity and mortality in underresourced settings. Vitamin A plays a role in the stability of certain epithelial cells and has a positive role in immune modulation. Patients with vitamin A deficiency who become infected with measles experience a more severe clinical course, which has led the WHO to strongly promote both therapeutic and preventive supplementation.

### Measles in Pregnancy and Congenital Infection

Measles during pregnancy may be severe, mainly related to primary measles pneumonitis. During pregnancy, measles is associated with a higher risk of miscarriage and premature delivery; however, it is not known to cause congenital anomalies of the fetus. Congenital measles, in which the rash is present at birth or appears in the first 10 days of life, varies from mild to severe, with mortality approaching 30%. Postnatally acquired (neonatal) measles is rare, because generally infants are protected by the higher levels of transplacentally acquired antibodies from their mothers.

## COMPLICATIONS

The most common complications of measles virus infection are pneumonia (primary viral or secondary bacterial), acute otitis media, and diarrhea due to the combination of immunosuppression and mucosal

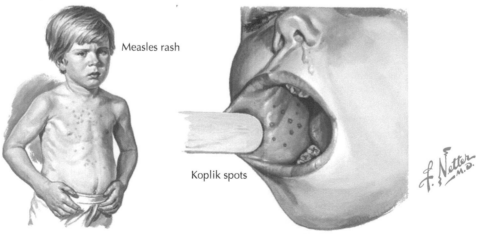

Measles rash

Koplik spots

**Fig. 10.1** Clinical features of measles.

injury. An infection with measles virus can result in death, with standard fatality rates ranging from 2 to 3 per 1000 persons; they are highest among infants and in adults over 30. Mortality increases in populations with malnutrition and vitamin A deficiency.

## Immune Amnesia

Measles virus infection results in a temporary yet significant suppression of the immune system. Recent immunologic studies have demonstrated the action of cytotoxic T cells upon infected B and T cells that express CD150 (SLAM). It is estimated that measles virus can lead to the depletion of 11% to 73% of the immune repertoire. Decimation of lymphocytes in both the respiratory and gastrointestinal epithelium results in increased frequency and severity of infection, ultimately leading to a greater risk for all-cause mortality over many years following infection with measles virus.

## Respiratory Infections

Infection with measles virus can lead to respiratory pathology as a direct result of the virus's effects on the respiratory epithelium or from secondary viral or bacterial pathogens during the period of transient immune amnesia and mucosal injury. Viral infection or bacterial superinfection can occur in any area of the respiratory tract, clinically appearing as pneumonia, laryngotracheobronchitis (croup), and bronchiolitis. Among patients with measles, radiographic evidence of pneumonia is common, even with uncomplicated disease. Bacterial pneumonia is commonly caused by *Streptococcus pneumoniae*, *Haemophilus influenzae*, or *Staphylococcus aureus* and usually occurs 5 to 10 days after onset of the rash. Pneumonia is the most common cause of measles-associated deaths in infants. In adults, a primary measles virus pneumonia is more common than secondary bacterial pneumonia and is seen clinically as increasing respiratory distress and hypoxia along with the rash.

## Central Nervous System Manifestations

Cerebrospinal fluid (CSF) pleocytosis and electroencephalographic (EEG) abnormalities may occur in up to 50% of individuals with measles, even without overt central nervous system (CNS) symptoms. Acute measles encephalitis occurs during the convalescent phase and is characterized by a resurgence of fever, headaches, and altered consciousness. The incidence of measles encephalitis is 1 to 3 per 1000 cases, with an estimated fatality rate of 10%. The condition ranges from mild to severe, with permanent neurologic sequelae present in 25% of survivors.

Acute postinfectious measles encephalomyelitis or acute disseminated encephalomyelitis (ADEM) may also be seen after measles. It is rare in children younger than 2 years of age but has an incidence of approximately 1 in 1000 in older children. It generally develops after the first week of rash and manifests with an abrupt onset of fever, headache, and seizures proceeding to obtundation and coma.

Subacute sclerosing panencephalitis (SSPE) is a chronic, degenerative CNS disorder attributed to continuing infection with an altered measles virus that persists despite a robust host immune response. SSPE is a rare complication occurring in approximately 1 in 25,000 cases of measles. It occurs several years after the initial measles infection and young children seem to be most at risk for sequelae. The progression of SSPE is variable; some patients can experience remission, but it is usually fatal within 3 years.

## DIAGNOSIS

### Clinical Diagnosis

The diagnosis of measles can be difficult, as many childhood and adult diseases display similar "morbilliform" rashes and systemic findings. The serious implications of the diagnosis of measles require an immediate clinical judgment from an experienced clinician who also has weighed both the patient's immunization history and his or her risk for recent exposures to infected individuals. This is critical not only for isolating and supporting the infected patient, but also for identifying and evaluating all recent contacts for consideration of postexposure prevention measures for outbreak control.

### Laboratory Diagnosis

Confirmatory diagnostics should always be performed when feasible. Simple blood counts often reveal leukopenia and lymphopenia. Molecular diagnostic methods (e.g., real-time polymerase chain reaction [RT-PCR]) with samples from sites of shedding—including the nasopharynx, oropharynx, conjunctiva, and even urine—can make diagnosis possible within hours but require access to a modern laboratory with this capability. Naive infected individuals usually develop an elevated IgM antibody as soon as the onset of rash; RT-PCR—with its high sensitivity and specificity—therefore remains one of the most valuable clinical and public health tools. Acute and convalescent serology can detect a fourfold or greater rise in measles-specific antibody but is less relevant for acute management.

## TREATMENT

### Supportive Therapy

No specific antiviral agents have proven to be effective in treating individuals with measles. Supportive therapy includes reducing fever, maintaining hydration and nutrition, and treatment of any of the individual complications previously discussed. Hospitalization becomes problematic due to the need for negative-air-pressure isolation rooms to protect other patients as well as staff. All of this requires expensive medical resources that are often absent in the regions where measles outbreaks are occurring.

### Administration of Vitamin A

Immediate vitamin A supplementation is used to reduce the risk of xerophthalmia, corneal scarring, and blindness as well as the debilitating prolonged diarrhea that may exacerbate malnutrition and increase mortality postinfection. Dosing is simple: 50,000 units for infants older than 6 months of age, 100,000 units for those 6 to 11 months of age, and 200,000 units for those older than 12 months of age given on the first and second days of presentation. An additional dose can be given 4 to 6 weeks later to patients with ophthalmic findings of vitamin A deficiency.

### Prevention of Disease in Immediate Contacts

If a risk of exposure has been identified early, contacts can be identified and screened for natural immunity or adequate prior immunization. If inadequate protection is identified, a dose of measles vaccine can be delivered within 72 hours of exposure. Alternatively, if it is more than 3 days and fewer than 6 or if an exposed subject is pregnant, has late-stage HIV, or other contraindications to vaccination, passive immunoprophylaxis with human immunoglobulin preparations can be given, which will achieve protective levels of antimeasles antibody at recommended doses.

## REDUCTION OF GLOBAL DISEASE BURDEN

### Immunization With Live Virus Vaccines

The introduction of an effective live-attenuated vaccine during the late 1960s has led to a dramatic global reduction in the disease burden and resultant mortality. The vaccine is often combined with rubella and mumps virus (measles, mumps, and rubella, or MMR) and varicella

(MMRV) and is equally effective regardless of combination or source of manufacture. Although vaccine schedules vary subtly by region, the universal goal is to deliver two immunizations at least 1 month apart when the protective yet also vaccine-inhibitory effects of passively transferred maternal antibodies have diminished. The first dose is routinely administered at 9 to 12 months of age and the second can vary from 1 month to years later. The first can be given as early as 6 months if the infant is born during an outbreak or is in a high-risk environment or traveling to such a region, which may necessitate a third dose to ensure that two are given after the loss of maternal antibody. Immunization strategies designed to diminish measles propagation by protecting both the individual and the surrounding population illustrate an epidemiologic principle known as *herd immunity*. This has been estimated to be between 89% and 94% but is possibly higher in regions of great population density. As the world's population shifts to relying on vaccine-induced immunity, it is possible that additional doses may be indicated in later life.

## Vaccine Safety

The modern measles vaccine is highly safe, with minimal side effects. Common reactions include fever in less than 15% of recipients 7 to 12 days postimmunization and transient rashes or lymphadenopathy in less than 5% of children and 20% of adults. Up to 25% of naive postpubertal females have transient arthralgia 1 to 3 weeks postimmunization, which rarely persists or recurs. Febrile seizures occur in 1 per 3 to 4000 infants 6 to 14 days postimmunization. Very rare occurrences of anaphylaxis or thrombocytopenia have also occurred. Extensive independent literature reviews by the Institute of Medicine, the Cochrane Database, and WHO experts found no evidence supporting a causal relationship between the administration of measles vaccine and autism spectrum disorders or autoimmune disorders including Guillain-Barré syndrome, inflammatory bowel disease, and type 1 diabetes.

## Immunization of Special Populations

MMR vaccines are contraindicated in pregnant women or those who may become pregnant in the subsequent 28 days primarily to avoid exposure to the rubella component, although vaccine-associated problems for any component have been extraordinarily rare. As a principle, all live vaccines, including the MMR or MMRV, should be avoided in primary or acquired immunodeficiency, including late-stage HIV or ongoing systemic immunosuppressive therapy, or in those who have recently received antibody-containing products (IG, or IG-containing blood components) that might inhibit the vaccines. Immunization is encouraged for early-stage HIV-infected individuals or those successfully managed with antiviral therapy, and reimmunization should be considered for children who are immune reconstituted later in life. Measles vaccine should be planned around the timing of induction and recovery from immunosuppression for the growing group of individuals receiving targeted immunotherapies or after recovery from cancer chemotherapy.

## Vaccine Avoidance

Since the introduction of vaccines in modern medicine, there has always been a degree of suspicion regarding their use to prevent disease. Voluntary refusal or intentional delays from the recommended schedule of measles immunization has gained momentum in recent years, primarily in developed countries. Public distrust was further inflamed after a now-discredited publication in 1998 attempted to associate measles immunization with autism spectrum and gastrointestinal disorders. Providers need to be advocates who are willing to discuss the scientific evidence supporting the individual and community benefits of measles vaccine with their patients and families and to recognize that their concerns do not reflect a single issue. Recent studies have found a diversity of concerns about perceived toxins, distrust of government and pharmaceutical companies, and a desire for a "natural" lifestyle. The WHO's 2012 Strategic Plan recognizes the need to "Communicate and engage to build public confidence and demand for immunization," as this challenge to reduce the global burden of measles may be among the most difficult to overcome.

## EVIDENCE

Bello S, Meremikwu MM, Ejemot-Nwadiaro RI, Oduwole O. Routine vitamin A supplementation for the prevention of blindness due to measles in children. *Cochrane Database Syst Rev* 2016;8:1-28. doi: 10.1002/14651858. CD007719.pub4. *This is a comprehensive analysis of the evidence supporting the current recommendations for vitamin A supplementation during measles infection.*

Demicheli V, Rivetti A, Debalini MG, Di Pietrantonj C. Vaccines for measles, mumps, and rubella in children. *Cochrane Database Syst Rev* 2012;2:1-168, doi: 10.1002/14651858.CD004407.pub3. *This is a comprehensive review of the evidence to date which refutes the concerns of the association of the MMR vaccine with autism spectrum and immune disorders, but also honestly discusses the insufficiencies of good systematic data collection.*

Furuse Y, Oshitani H. Global transmission dynamics of measles in the measles elimination era. *Viruses* 2017;9(4):1-10. doi: 10.3390/v9040082. *Through use of modern viral sequencing the authors present a "phylogenetic analysis" of the increasing complexities of measles outbreaks and transmission across the globe over the past 60 years.*

Guerra F, Bolotin S, Lim G, et al. The basic reproduction number ($R_0$) of measles: a systematic review. *Lancet Infect Dis* 2017;17(12):e420-e428. doi: 10.1016/S1473-3099(17)30307-9. *Fundamental epidemiologic tools describe the sobering evidence of measles' tremendous infectivity.*

Hoffman BL, Felterc EM, Chua KH, et al. It's not all about autism: The emerging landscape of anti-vaccination sentiment on Facebook. *Vaccine* 2019;37:2216-2223. *This is a fascinating insight into the prevailing beliefs posted on social medial by individuals promoting vaccine avoidance. The authors used qualitative analytic tools to evaluate the direct content of nearly 200 sites.*

Lewis PE, Burnett DG, Costello AA, Olsen CH, Tchandja JN, Webber BJ. Measles, mumps, and rubella titers in Air Force recruits: Below herd immunity thresholds? *Am J Prev Med* 2015;49(5):757-760. *This is a recent cross-sectional study of the percentages of protective titers in sub-groups of younger adults against the three pathogens addressed by the MMR vaccine, which indicates that we may need to further vaccinate populations which rely solely on vaccine-induced protection.*

Mina MJ, Kula T, Leng Y, et al. Measles virus infection diminishes preexisting antibodies that offer protection from other pathogens. *Science* 2019; 366(6465):599-606. doi:10.1126/science.aay6485. *This is a prospective study of 77 children evaluating their immune repertoire before and after measles virus infection and showed that preexisting antibodies were significantly diminished.*

## ADDITIONAL RESOURCES

Alter SJ, Bennett JS, Koranyi K, Kreppel A, Simon R. Common childhood viral infections. *Cur Prob PediatrAdolesc Health Care* 2015;45:21-53. *A recent extensive review of the clinical findings and complications from measles infection.*

Griffith DE, The immune response in measles: Virus control, clearance and protective immunity. *Viruses* 2016;8(10):1-8. doi:10.3390/v8100282. *An extensive description of the intricate host response to measles infection and the resulting impact on the human immune system.*

Kutty P, Rota J, Bellini W, et al, ed. *Manual for the Surveillance of Vaccine-Preventable Diseases*, Atlanta, GA, 2012, Centers for Disease Control and Prevention, pp 7.1-7.21. *This is the standard reference for current laboratory diagnostic tests available for measles and other important infectious disease pathogens.*

World Health Organization. *Global measles and rubella—strategic plan: 2012-2020*. Geneva, Switzerland: 2012, WHO. *This is the foundational document describing the current global epidemiology of measles and the ongoing plan for its reduction and elimination.*

Zachariah P, Stockwell MS. Measles vaccine: Past, present, and future. *J Clin Pharmacol* 2016;56(2):133-140. doi: 10.1002/jcph.606. *This is an extensive review of the biology of measles and the components, immunogenicity, and safety of the modern measles vaccine—a useful update for providers who need to explain the safety and risks of vaccine use to recipients.*

# 11

# Mumps

*Amber K. Kirby, Daniel J. Adams*

##  ABSTRACT

Mumps is an acute, highly contagious systemic viral infection and the leading cause of parotitis worldwide. The most common clinical manifestation is nonsuppurative inflammation of the salivary glands, most notably the parotid glands. In most prepubertal children and adults, the illness is benign and self-limited, with one third of infections being subclinical. Postpubertal children are at higher risk for severe disease and more frequently experience extrasalivary gland manifestations of mumps, including orchitis (15%–30%), oophoritis (5%), meningitis (1%–10%), and rarely encephalitis or pancreatitis. Although sterility is a feared complication of mumps orchitis, it rarely occurs. Treatment for mumps is supportive; the primary means of prevention is use of live-attenuated measles, mumps, and rubella (MMR) vaccine. Approximately 88% of vaccine recipients who complete a two-dose MMR series will develop immunity, and use of this vaccine has led to a marked decline in mumps cases worldwide.

##  CLINICAL VIGNETTE

A 19-year-old male college student presents to his university's health clinic with 1 day of jaw pain and "swollen cheeks." Three days prior to presentation, he began experiencing fatigue, anorexia, and a low-grade fever. He reports tenderness over the swollen areas of his jaw and pain with chewing. He is otherwise healthy and takes no medications. He is up to date on routine age-appropriate immunizations. His exam reveals marked bilateral painful swelling of both parotid glands, obscuring the angle of the mandible. Oral examination showed erythema surrounding the Stensen duct bilaterally and healthy dentition. There are no other abnormal findings. Laboratory evaluation reveals a mildly elevated serum amylase and positive serum mumps IgM. A reverse transcription polymerase chain reaction (RT-PCR) panel from a buccal swab is positive for mumps virus.

## GEOGRAPHIC DISTRIBUTION

Mumps virus is an RNA virus of the *Paramyxoviridae* family, genus *Rubulavirus*. Mumps infection occurs throughout the world, being clustered in areas with poor vaccine coverage. In the prevaccine era, epidemics occurred every 2 to 5 years, peaking in late winter and early spring in temperate climates and appearing year round in tropical climates. In the United States, use of the MMR vaccine has led to a 99% decrease in mumps incidence; however, sporadic mumps outbreaks still occur, mostly in close-contact environments such as college dormitories and military barracks.

## RISK FACTORS

Absence of mumps-specific IgG is the primary risk factor for infection. Mumps is uncommon in infants younger than 1 year of age, likely due to passive immunity from maternal antibodies. In the prevaccine era,

mumps occurred most commonly in children 5 to 9 years of age, and more than 90% of children younger than 15 years of age had a history of mumps infection. More recently, in countries with universal childhood immunization programs, outbreaks of mumps have been reported among adolescents in universities and within military units. In 2015, a total of 301 students at the University of Iowa were diagnosed with acute mumps infection and 86% of these had completed a two-dose MMR series, suggesting a waning of vaccine-induced immunity over time.

## CLINICAL FEATURES

The mumps virus is transmitted through contact with infectious respiratory droplets. The incubation period ranges from 12 to 25 days with a mean of 16 to 18 days. Mumps virus can be isolated from saliva 5 days before the onset of parotitis until 5 days after. Mumps is highly contagious, and peak infectivity occurs just before the onset of parotitis.

In symptomatic individuals, a prodrome of low-grade fever, anorexia, malaise, and headaches occurs. In primary infections, approximately 25% of patients initially have unilateral parotid involvement; however, the second parotid gland usually enlarges within a few days. Patients commonly complain of earache, and parotid gland tenderness is noted on palpation. The parotid gland or glands may be painful and visibly enlarged. The ear on the affected side may be displaced superiorly and outward. The enlarged parotid gland may obscure the angle of the mandible and should not be confused with cervical adenopathy, which does not usually obscure the mandible. After the parotid swelling peaks (at 3 to 5 days), the fever and tenderness rapidly resolve and the parotid gland or glands return to normal size, usually within a week. Involvement of other salivary glands may occur, and submandibular gland swelling may be confused with anterior cervical lymphadenitis. Involvement of sublingual glands is less common but can occur. The orifices of the Stensen and Wharton ducts are often erythematous and edematous (Fig. 11.1). Patients may complain of trismus and difficulty speaking and drinking. Acidic foods may induce parotid pain.

Natural infection does not confer lifelong immunity. Mumps reinfections, however, are milder, often lacking the typical parotitis seen in primary mumps infection.

### Central Nervous System Involvement

Central nervous system (CNS) involvement is the most common extrasalivary gland manifestation of mumps and is due to neurotropism of the virus. For unclear reasons, this is more prevalent in males. CNS manifestations range from asymptomatic cerebrospinal fluid (CSF) pleocytosis (50% of patients) to a very rare fulminant and potentially fatal encephalitis. Clinical meningitis occurs in 10% to 30% of cases and may occur before, during, after parotitis or even in its absence. Typical clinical

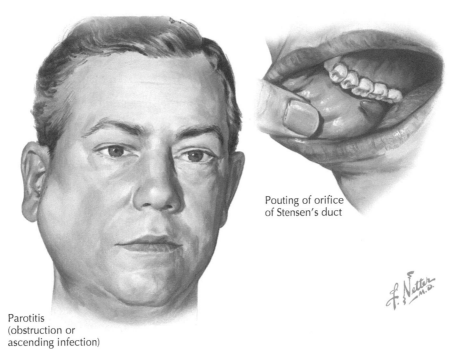

Pouting of orifice
of Stensen's duct

Parotitis
(obstruction or
ascending infection)

**Fig. 11.1** Mumps parotitis.

features include headache, vomiting, fever, and nuchal rigidity. The CSF may contain up to 2000 cells/mm³, predominantly lymphocytes, although a neutrophil predominance may occur early in the disease. Protein levels are normal to mildly elevated, and the glucose concentration may be low in up to 30% of patients, which can lead clinicians to consider bacterial meningitis in the differential diagnosis. Mumps meningitis is generally benign without any neurologic sequelae.

Encephalitis, a rare but serious CNS manifestation of mumps infection, manifests as altered mental status, convulsions, paresis, aphasia, and involuntary movements. It has a bimodal distribution of onset: early encephalitis, in which onset coincides with the presence of parotitis, and late encephalitis, in which the condition develops 7 to 10 days after the onset of parotitis. Early-onset encephalitis represents direct damage to neurons as a result of viral invasion, whereas late-onset disease is considered to be a postinfectious process related to the host's immune response. The CSF findings are similar to those of mumps meningitis. The neurologic manifestations usually resolve over a period of 1 to 2 weeks; however, a small minority of patients have significant neurologic sequelae, and death can occur.

Cerebellar ataxia, facial palsy, transverse myelitis, Guillain-Barré syndrome, and a poliomyelitis-like syndrome have rarely been associated with mumps infection. Aqueductal stenosis resulting in hydrocephalus has developed after CNS infection caused by mumps.

## Epididymo-orchitis and Oophoritis

Epididymo-orchitis is a common mumps complication in the postpubertal period, developing in 20% to 30% of infected postpubertal males. Most cases occur during the first week of parotitis, but it may precede parotitis or be the only manifestation of mumps. Physical examination typically reveals a single warm, swollen, tender testicle and erythema of the scrotum accompanied by fever, headache, nausea, and vomiting. The testicular swelling usually resolves in 7 to 10 days, but residual tenderness may persist for several weeks. Although mumps orchitis has been associated with sexual impotence and sterility, these complications rarely occur. Some degree of testicular atrophy occurs in about half of patients with epididymo-orchitis months to years after

the infection. In men with unilateral epididymo-orchitis, a slight cosmetic imbalance in testicular size may occur.

Oophoritis develops in about 5% of postpubertal females with mumps. Symptoms of fever, nausea, and lower abdominal pain are frequently reported. Rarely, reduced fertility and premature menopause have been reported. Some postpubertal women with mumps complain of swelling and pain in their breasts consistent with mastitis.

## Pancreatitis

Pancreatitis is an uncommon (<1% of cases) and severe complication of mumps that presents with fever, nausea, vomiting, and severe epigastric pain. There is some controversy surrounding the relationship between mumps infection and juvenile diabetes mellitus. Although mumps virus has been shown to infect beta cells from pancreatic islets in vitro, no evidence of infection has been demonstrated in vivo and no clear causal link has been established.

## Arthritis and Arthralgia

Mumps arthritis or arthralgia are seen infrequently in adults and very rarely in children. Migratory polyarthritis of large or small joints is the most common manifestation, but monoarticular arthritis has been reported. Symptoms usually appear 10 to 14 days after the onset of parotitis and may last several weeks. The etiology of the mumps arthritis is unclear, but it is typically self-limited and without any permanent joint damage.

## Myocarditis

Transient electrocardiographic changes occur in up to 15% of patients with mumps, the most common being ST-segment depression. Other findings include flattened or inverted T waves and prolonged PR intervals. Clinical myocarditis is rare; however, deaths due to rapidly progressing myocarditis have been reported.

## Deafness

Sensorineural hearing loss caused by cochlear damage from the mumps virus does occur, albeit infrequently, with an incidence of 0.5 to 5 cases per 100,000 cases of mumps. Those with mumps-associated hearing loss generally present with acute symptoms. Deafness may have an abrupt or gradual onset, is often unilateral, and may be transient or permanent. Vestibular symptoms, such as vertigo, commonly accompany mumps-associated hearing loss.

### Other Associated Conditions

Other rare associations with mumps infection include nephritis, hepatitis, thyroiditis, thrombocytopenia, and ocular complications (e.g., iritis, keratitis, and central retinal vein thrombosis).

### Congenital Infection

An increase in fetal death and spontaneous abortion has been documented in pregnant women who acquire mumps in the first trimester of pregnancy. No significant increase in fetal loss has been shown with infections during the second or third trimesters. Mumps infection during pregnancy does not result in congenital abnormalities of the fetus. During active infection, mumps virus is excreted into breast milk; however, breastfeeding during maternal mumps infection is considered safe and helpful in protecting the infant from infection.

## DIAGNOSIS

### Clinical Findings

Diagnosis of mumps infection is usually based on the history of mumps exposure combined with the physical examination finding of parotitis. However, in the post-MMR era, a variety of other infections and noninfectious conditions are much more common causes of acute parotid gland swelling. Parainfluenza 3, coxsackievirus, Epstein-Barr virus, cytomegalovirus, and influenza have all caused parotitis. Bilateral parotid swelling can also occur in human immunodeficiency virus infection. Noninfectious causes of bilateral parotid swelling include salivary gland stones, Sjögren syndrome, medications, diabetes mellitus, malnutrition, and metabolic disorders. Acute bacterial infection of the parotid gland as well as tumors, cysts, or ductal obstruction typically cause unilateral parotid swelling.

### Laboratory Diagnosis
#### Blood Abnormalities

Serum amylase is often elevated in mumps, indicating inflammation of the salivary glands or pancreas. The origin of the amylase can be determined by isoenzyme analysis or by determining pancreatic lipase. Amylase in the presence of parotitis may remain elevated for 2 to 3 weeks after resolution of symptoms.

The peripheral white blood cell and differential counts are usually normal; however, mild leukopenia with a relative lymphocytosis may be seen. In patients with evidence of meningitis, epididymo-orchitis, or pancreatitis, a leukocytosis with polymorphonuclear predominance can occur.

### Serology

Laboratory confirmation of a mumps case is important given its outbreak potential and public health reporting requirements. The diagnosis of mumps is confirmed with serology by detection of mumps immunoglobulin M (IgM) antibodies. Alternatively, seroconversion from negative to positive or a fourfold rise in mumps IgG antibody titers between acute and convalescent sera can confirm the diagnosis. Serologic diagnosis is challenging in previously immunized individuals given that they have lower rates of detectable mumps IgM and may already have a high mumps IgG titer in the acute phase.

### Virology

Buccal swabs obtained less than 3 days after onset of parotitis are the preferred specimens for RT-PCR to confirm mumps infection. Early testing is critical in MMR-immunized individuals, who shed virus for a shorter duration. Mumps RT-PCR can also be performed on samples from throat washings or CSF. Given its sensitivity for detecting mumps and fast results, RT-PCR testing has replaced the use of viral cultures.

## CLINICAL MANAGEMENT

No antiviral therapy for the treatment of mumps infection is available. Treatment is supportive, including the use of analgesics, warm compresses, and sialagogues to relieve the pain associated with parotitis.

## PREVENTION AND CONTROL

### Infection Control Precautions

Infected individuals should be considered contagious for 5 days after the onset of parotid swelling. To prevent nosocomial spread, hospitalized patients with mumps should be maintained with droplet precautions for 5 days after the onset of parotid swelling.

### Vaccination

Active immunization with MMR vaccine elicits protective levels of mumps-neutralizing antibodies in more than 78% of recipients after one dose and 88% after two doses. Significant adverse reactions to the mumps component of the MMR vaccination are rare, but parotitis, orchitis, and fever have been reported. Anaphylactic reactions to the MMR vaccine are exceedingly rare and thought to be caused by small amounts of neomycin or gelatin. Individuals with egg allergies are at very low risk for anaphylactic reactions as MMR vaccine is produced in chicken embryo cells, which contain little ovalbumin.

Routine MMR vaccination is recommended for children at 12 to 15 months of age, with a second dose at 4 to 6 years of age. A majority of mumps-infected individuals in recent outbreaks were considered completely immunized, having received the recommended two doses of MMR prior to their infection. In 2018 this prompted the Advisory Committee on Immunization Practices to recommend a third MMR dose for individuals at increased risk due to an active mumps outbreak.

MMR vaccine should not be administered to pregnant women or immunocompromised individuals and should be delayed for 3 to 11 months in those with recent receipt of immunoglobulin or blood products (depending on the amount of IgG in the product).

Mumps vaccine given to mumps-exposed individuals does not prevent mumps infection or alter the clinical course but will provide protection from subsequent exposures. Immunoglobulin is not recommended for mumps-exposed patients.

## NOTES

**Disclaimer.** The views expressed in this chapter are those of the authors and do not necessarily reflect the official policy or position of the Department of the Navy, US Air Force, Department of Defense, or the United States Government.

**Copyright statement.** This work was prepared as part of our official duties as military service members or employees of the United States Government. Title 17 U.S.C. 105 provides that "Copyright protection under this title is not available for any work of the United States Government." Title 17 U.S.C. 101 defines a United States Government work as a work prepared by a military service member or employee of the United States Government as part of that person's official duties. Daniel J. Adams, MD, is a major in the US Air Force Medical Corps and Amber K. Kirby, MD, is a captain in the US Air Force Medical Corps.

**Financial support.** None

**Potential conflicts of interest.** All authors: No reported conflicts.

## ACKNOWLEDGMENT

The authors acknowledge the work of Alison Margaret Kesson on the previous edition chapter.

## EVIDENCE

Lewnard J, Grad YH: Vaccine waning and mumps re-emergence in the United States. *Sci Transl Med* 2018;10(433):eaao5945. *This reference provides data on MMR vaccine effectiveness during recent mumps outbreaks in the United States.*

Ramanathan R, Voigt EA, Kennedy RB, Poland GA: Knowledge gaps persist and hinder progress in eliminating mumps. *Vaccine* 2018;36(26):3721-3726. *This resource provides data on recent mumps outbreaks and potential contributing factors.*

Shah M, Quinlisk P, Weigel A, Riley J, James L, Patterson J: Mumps outbreak response team: Mumps outbreak in a highly vaccinated university-affiliated setting before and after a measles-mumps-rubella vaccination campaign-Iowa July 2015-May 2016. *Clin Infect Dis* 2018;66(1):81-88. *This reference provides information and clinical data on the epidemiology, clinical presentations and complications that occurred during a mumps outbreak at the University of Iowa in 2015.*

## ADDITIONAL RESOURCES

American Academy of Pediatrics: Mumps. In Kimberlin DW, Brady MT, Jackson MA, Long SS, eds: *Red Book: 2018 Report of the Committee on Infectious Diseases*, ed 31, Itasca, IL, 2009, AAP, pp 567-573. *An essential resource from leading experts in the pediatric infectious disease field, summarizing the clinical features, diagnosis, management, and prevention of mumps.*

Bellini W, Icenogle J, Hickman C: Measles, mumps and rubella viruses. In Loeffelholz M, Hodinka R, Young S, Pinsky B, eds: *Clinical Virology Manual*, ed 5, Washington, DC, 2016, ASM Press, pp 293-310. *This resource provides a comprehensive review of the virology and pathogenesis of mumps virus.*

Cherry J, Quinn K: Mumps virus. In Cherry J, Harrison G, Kaplan S, Steinbach W, Hotez P, eds: *Textbook of Pediatric Infectious Diseases*, ed 8, Philadelphia, 2019, Elsevier, pp 1771-1779. *This valuable textbook chapter provides a broad scope of mumps virus, including a detailed description of pathophysiology and immune responses.*

Maldonado Y, Shetty A: Mumps virus. In Long SS, Prober CG, Fischer M, eds: *Principles and Practice of Pediatric Infectious Diseases*, Philadelphia, 2018, Elsevier, pp 1157-1162. *This expertly written textbook chapter includes all important features of mumps virus infection, including a thorough review of the epidemiology of recent mumps outbreaks.*

# 12

# Rubella

Caitlin K. Rochester, Daniel J. Adams

## ABSTRACT

Rubella (German measles), a leading cause of vaccine-preventable birth defects, is an acute viral infection that can affect people of all ages. When acquired postnatally, rubella is most often characterized by a generalized maculopapular rash, fever, and lymphadenopathy resembling a mild case of measles (rubeola). Many postnatal rubella infections (20%–50%) are very mild or even subclinical, occurring without the prototypical rash. In adolescent and adult females, up to 70% of infections will be associated with joint symptoms, including arthritis or arthralgias. When infection occurs just before conception or in early pregnancy, it has the potential to cause severe infection of the developing fetus with resultant birth defects, known as congenital rubella syndrome (CRS). No specific antiviral treatment is available for rubella; therefore, prevention through use of a live-attenuated vaccine administered during childhood has been the focus of a worldwide campaign to eliminate rubella.

## CLINICAL VIGNETTE

A 2-day-old male infant is being evaluated in the newborn nursery after failing his newborn hearing screen twice. He is a term infant born via uncomplicated spontaneous vaginal delivery, with an Apgar score of 8/9 (points deducted for color only). His mother is an immigrant and came late to care, not receiving routine obstetric care until the third trimester because she spent the first part of her pregnancy visiting family in Africa. Maternal blood type is A+. She was group B *Streptococcus* negative with unremarkable serologies except for not being immune to rubella. Examination of the infant is notable for a holosystolic murmur best appreciated at the left upper sternal border and bilateral leukocoria; echocardiography confirms the presence of a patent ductus arteriosus and ophthalmology confirms the presence of cataracts. CRS is suspected and a serum reverse transcription polymerase chain reaction (RT-PCR) confirms the diagnosis.

## GEOGRAPHIC DISTRIBUTION

In the prevaccine era, rubella occurred globally, with minor epidemics every 5 to 9 years. The last rubella pandemic occurred from 1962 to 1965, just prior to the release of the live-attenuated rubella vaccine in the United States in 1969. Due to large-scale vaccination efforts, the United States was able to eliminate rubella in 2004, a status that was reconfirmed in 2011 and 2014. In 2015, the World Health Organization (WHO) verified elimination of rubella from the entire region of the Americas. Based on data collected from the 194 WHO member states, 69% of the world's infants had been vaccinated against rubella in 2018, an increase of 21% from 2000. Furthermore, the number of rubella cases reported in 2000 decreased by 96% in 2018, despite the number of countries reporting their cases having increased from 53% to 91%. The largest incidence of CRS in 2018 was reported in the WHO's South-East Asia Region, while the only region free of CRS cases in 2018 was the Region of the Americas.

## PATHOGENESIS

Humans are the only known host for rubella virus. Rubella virus, an enveloped, positive-sense, single-stranded ribonucleic acid (RNA) virus belonging to the family *Togaviridae*, is the only member of the genus *Rubivirus*. Rubella virus replicates in the nasopharyngeal mucosa and local lymph nodes and is then transmitted primarily through direct or droplet contact from respiratory secretions. The incubation period ranges from 12 to 23 days, with an average of 18 days. Initial infection of the nasopharyngeal epithelium is followed by spread to regional lymph nodes and transient viremia 5 to 7 days after exposure. Persons with rubella are most infectious when the rash is erupting, but they can shed the virus from 7 days before to 14 days after rash onset. Maximal viral shedding occurs 1 to 5 days after rash onset.

When rubella infection occurs during pregnancy, especially in the first trimester, serious consequences, including miscarriages, fetal demise, and a constellation of well-documented birth defects, can result. In general, the younger the fetus when infected, the more severe the injuries observed. When the mother is infected during the first 12 weeks of gestation, there is up to an 85% chance that her fetus will have congenital defects or that a spontaneous abortion will occur. By the end of week 16 of gestation, the risk of developing a single defect, such as deafness or congenital heart disease, is approximately 25%.

Rubella infection confers lifelong protection in most people. However, despite the persistence of specific immunoglobulin G (IgG) immunity to rubella virus, reinfection can occur. The majority of reinfections are asymptomatic. Rubella reinfection during pregnancy has been documented to cause CRS in the fetus; however, this is an extremely rare event.

## CLINICAL FEATURES

### Postnatal Rubella

Generally, postnatal rubella is a mild self-limited illness that often occurs in childhood but can affect all ages. Characteristic clinical features include lymphadenopathy, which may last for several weeks, as well as fever and rash. The lymphadenopathy—of the postauricular, posterior cervical, and occipital chains—typically precedes the rash and lasts 5 to 10 days. The rash of rubella begins on the face and then moves down the body. It is maculopapular, like many viral exanthems. Although unlike the measles rash specifically, the rash of rubella infection is not confluent (Fig. 12.1). It occurs in 50% to 80% of cases; when it is absent, the presence and distribution of lymphadenopathy may suggest the diagnosis. Other common findings of this illness include malaise, mild conjunctivitis, and headache, though these prodromal

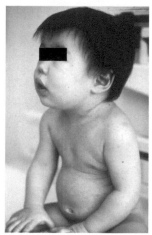

Fig. 12.1 Child with rubella rash. (Courtesy Centers for Disease Control and Prevention Public Health Image Library.)

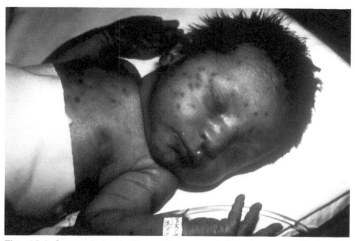

Fig. 12.2 Congenital rubella syndrome and "blueberry muffin" rash. (Courtesy Centers for Disease Control and Prevention Public Health Image Library, Dr. Andre J. Lebrun.)

symptoms are more common in adults than in children. An enanthem consisting of petechial lesions of the soft palate (Forschheimer spots) has been described with rubella, but this is not diagnostic. Arthritis and arthralgia have been reported in as many as 70% of adult women with rubella, but joint symptoms are much less common in children and in men. The arthritis tends to involve fingers, wrists, and knees and may be slow to resolve, lasting up to 1 month, but it rarely becomes chronic.

Sometimes called *three-day measles*, rubella infection clinically may resemble a mild case of measles. In addition to measles, other diagnostic considerations include scarlet fever, roseola, parvovirus B19 infection, infectious mononucleosis, toxoplasmosis, some enteroviral infections, Kawasaki disease, rheumatologic disorders, and allergic reactions.

The majority of postnatal rubella cases resolve without complications. Rare but potentially serious complications include postinfectious encephalitis and thrombocytopenia. Although both are uncommon, the former is more frequently encountered when rubella infects adults whereas the latter occurs more often in children.

## Congenital Rubella Syndrome

The connection between maternal rubella infection and certain birth defects was first described in 1941. Specific signs and symptoms of CRS can be classified as either temporary (e.g., low birth weight), permanent (e.g., deafness), or progressive (e.g., myopia). The most common manifestations are ophthalmic (cataracts, glaucoma, microphthalmia, pigmentary retinopathy), auditory (sensorineural deafness), cardiac (peripheral pulmonary stenosis, patent ductus arteriosus), and neurologic (behavioral disorders, developmental delay, microcephaly). Although many infants are afflicted with several symptoms, they can present with a single defect, hearing impairment being the most common of these. Additionally, CRS can present with neonatal manifestations including growth restriction, interstitial pneumonitis, radiolucent bone disease, hepatosplenomegaly, thrombocytopenia, and dermal erythropoiesis. Although it is classically associated with rubella, the characteristic "blueberry muffin rash" of dermal erythropoiesis occurs in only about 5% of CRS cases (Fig. 12.2). CRS should not be considered a static disease. Delayed manifestations include a progressive encephalopathy, which resembles the subacute sclerosing panencephalitis of measles; developmental delay, including autism; and endocrinopathies. The development of insulin-dependent diabetes mellitus in late childhood has been observed in approximately 20% of individuals with CRS. Autoimmune thyroid dysfunction has also been described.

## DIAGNOSIS

As rubella becomes rarer and in consideration of its nonspecific presenting features, both congenital and postnatal rubella infection can be difficult to diagnose clinically. All suspected cases of rubella should therefore be laboratory confirmed. Historically, the majority of rubella serologic testing was done using hemagglutination inhibition (HAI) assay, which is considered to be the assay standard. However, more recently, latex agglutination or enzyme-linked immunosorbent assay (ELISA) techniques, which detect either rubella IgG or IgM antibodies, have become widely available and are technically easier to perform, making this method the screening assay of choice.

Acute postnatal infection may be diagnosed by the presence of rubella IgM or a minimal fourfold increase in rubella IgG antibody titer between acute and convalescent serum specimens performed 4 to 6 weeks apart. Some groups advise repeat IgM testing in 5 to 10 days for all positive rubella IgM antibody tests in conjunction with avidity testing to help differentiate false positives from acute infections. Rheumatoid factor, parvovirus IgM, and heterophile antibodies may all result in false-positive IgM tests for rubella. Reportedly, rubella IgM can be detected 4 to 30 days after symptom onset; however, samples for serologic testing should be collected as soon as the diagnosis is suspected. Detection of IgG antibodies indicates immunity to rubella.

Although viral culture was common historically, testing with RT-PCR on swab samples from the throat or nasal passages is becoming increasingly common. Rubella can also be isolated from the urine and cerebrospinal fluid. Molecular typing now plays a critical role in identifying specific rubella isolates during outbreaks. All suspected, probable, and confirmed cases of rubella or CRS should be reported within 24 hours to local health departments, who then notify the Centers for Disease Control and Prevention (CDC), in order to establish proper control measures and further surveillance.

For CRS, the detection of rubella IgM antibodies in newborn infant serum is diagnostic of congenital infection, as this antibody does not cross the placenta. Congenital rubella infection can also be diagnosed by placental biopsy at 12 weeks' gestational age with demonstration of rubella RNA by PCR. It may also be diagnosed by the presence of specific IgM in fetal blood, but this is usually not detectable until at least 22 weeks of gestation. Infants with CRS can shed the virus for up to 1 year of age.

## TREATMENT

No specific antiviral therapy is available to treat rubella. Treatment of acute rubella is therefore supportive. Follow-up treatment of CRS is targeted to specific CRS manifestations—for example cataract surgery as appropriate and audiology screening for hearing loss.

## PREVENTION AND CONTROL

### Infection Control

Hospitalized patients with rubella should be isolated with contact and droplet precautions to prevent nosocomial spread. Individuals with postnatal rubella should be isolated for 7 days after the onset of rash. If an outbreak occurs, all unvaccinated persons should receive the measles, mumps, rubella (MMR) vaccine as part of the outbreak control measures. Exposed individuals without evidence of immunity should be excluded from school or childcare until 21 days after the onset of rash of the last case in the outbreak. Infants with CRS should be considered infectious until they are at least 1 year of age or have demonstrated two negative RT-PCRs after 3 months of age, which indicates that any hospital admission during the first year of life warrants contact and droplet isolation precautions. Notably, breastfeeding for rubella-infected mothers is safe.

### Vaccination

Immunization against rubella in the United States is achieved with the provision of a live-attenuated vaccine in either the MMR form or the tetravalent measles, mumps, rubella, and varicella (MMRV) form. The goal of vaccination primarily is to prevent congenital rubella infection. The first dose of the rubella vaccine can be administered between 12 and 15 months of age; a single dose of the combination vaccine is sufficient to achieve greater than 95% effective immunity against rubella infection. To achieve and maintain immunization coverage and thereby decrease the risk of CRS in a community, the WHO recommends a minimum of 80% vaccination coverage of a population. Immunization with rubella vaccine may cause a mild transient viremia. Although uncommon, the main complications of vaccination are fever, lymphadenopathy, arthritis, and arthralgia.

No cases of CRS have been attributed to rubella vaccination; however, vaccination during pregnancy is contraindicated. If a pregnant woman is inadvertently vaccinated with rubella vaccine, termination of pregnancy is not recommended because of the very low theoretical risk of CRS. Live-attenuated vaccines are not recommended to persons with severe immunodeficiency. However, vaccination of close contacts of these persons is safe, since the risk of transmission of rubella due to the vaccine is negligible.

## NOTES

**Disclaimer.** The views expressed in this chapter are those of the authors and do not necessarily reflect the official policy or position of the Department of the Navy, US Air Force, Department of Defense, or United States Government.
**Copyright statement.** This work was prepared as part of our official duties as military service members or employees of the United States Government. Title 17 U.S.C. 105 provides that "Copyright protection under this title is not available for any work of the United States Government." Title 17 U.S.C. 101 defines a United States Government work as a work prepared by a military service member or employee of the United States Government as part of that person's official duties. Daniel J. Adams, MD, is a major in the US Air Force Medical Corps and Caitlin K. Rochester, MD, is a captain in the US Air Force Medical Corps.
**Financial support.** None.
**Potential conflicts of interest.** All authors: No reported conflicts.

## ACKNOWLEDGMENT

The authors acknowledge Alison Margaret Kesson for her contributions as author of a prior version of this chapter.

## EVIDENCE

Brown KE, Rota PA, Goodson JL, et al. Genetic characterization of measles and rubella viruses detected through global measles and rubella elimination surveillance, 2016–2018. *MMWR Morb Mortal Wkly Rep* 2019;68:587-591. doi: https://dx.doi.org/10.15585/mmwr.mm6826a3. *This reference reveals the most recent data collected from the surveillance efforts of the World Health Organization on rubella virus.*

Grant GB, Desai S, Dumolard L, Kretsinger K, Reef SE. Progress toward rubella and congenital rubella syndrome control and elimination—worldwide, 2000–2018. *MMWR Morb Mortal Wkly Rep* 2019;68:855-859. doi: https://dx.doi.org/10.15585/mmwr.mm6839a5. *This reference provides the most current data on global progress towards rubella elimination.*

World Health Organization: Rubella Vaccines: WHO Position Paper, Weekly Epidemiological Record. 2011 Jul 15;86(29):301-316. *This reference outlines the importance of rubella vaccination campaigns.*

## ADDITIONAL RESOURCES

American Academy of Pediatrics: Rubella. In Kimberlin DW, Brady MT, Jackson MA, Long SS, eds. *Red Book: 2018 Report of the Committee on Infectious Diseases*, ed 31, Itasca, IL, 2018, American Academy of Pediatrics, pp 705-711. *This resource provides a brief overview of the clinical manifestations, diagnosis and prevention of rubella.*

Centers for Disease Control and Prevention: Chapter 14: Rubella. In Lanzieri T, Redd S, Abernathy E, Icenogle J. *Manual for the Surveillance of Vaccine-Preventable Diseases: Centers for Disease Control and Prevention;* May 18, 2018. *The resource provides an overview of rubella and surveillance measures, including explanations for laboratory screening basics.*

Centers for Disease Control and Prevention: Chapter 15: Congenital Rubella Syndrome. In Lanzieri T, Redd S, Abernathy E, Icenogle J. *Manual for the Surveillance of Vaccine-Preventable Diseases: Centers for Disease Control and Prevention;* June 12, 2018. *This resource provides an overview of congenital rubella syndrome and surveillance measures.*

# 13

# Varicella-Zoster Virus Infections

*Sylvia H. Yeh*

## ABSTRACT

Varicella-zoster virus (VZV) causes primary varicella, a common childhood illness called *chickenpox*. This virus establishes latency and may reactivate later in life, causing herpes zoster, which is commonly called *shingles* or simply *zoster*. Although both chickenpox and zoster usually resolve without event, significant complications may develop, even in healthy individuals. In the United States, widespread use of a VZV vaccine in children has diminished the burden of disease. Although the World Health Organization (WHO) recommends that countries implement varicella vaccination in their immunization programs, not all countries have done so, and VZV infections remain a significant cause of morbidity and mortality in those areas.

## CLINICAL VIGNETTE

A 12-year-old child presents with a pruritic rash on his torso. The rash started 2 days earlier with papules on the chest and abdomen. There has been no fever. He had visited his aunt, whose cat has new kittens, a few days before. There are no known ill contacts at home, but a notice was sent home from school about a potential chickenpox exposure. His immunizations are current and up to date. On examination, he is afebrile and his vital signs are normal for age. There are approximately 10 erythematous papular lesions scattered over his trunk, about 1 to 5 mm in diameter; several have central vesicles and none appear to be umbilicated. The remainder of the exam is unremarkable.

COMMENT: With widespread varicella vaccination coverage in the United States, including the addition of a preschool-age second varicella vaccine dose to the childhood immunization schedule, the overall incidence of chickenpox has decreased by approximately 95%. The highest proportion of cases are now seen between the ages of 10 and 14 years, manifesting as breakthrough disease among those who have already received either one or two doses of vaccine. Breakthrough chickenpox is typically milder than primary chickenpox in those who are unvaccinated, with less frequent occurrence of fever and respiratory symptoms; the skin lesions may be difficult to recognize, as the papules may not have the characteristic vesicle on an erythematous base, the classic "dewdrops on a rose petal" appearance.

## GEOGRAPHIC DISTRIBUTION AND DISEASE BURDEN

VZV is a member of the Herpesviridae family, and initial infection (varicella) is commonly called *chickenpox*. Varicella is present worldwide and is common in areas without routine vaccination. In temperate regions without high vaccine coverage, seasonal epidemics with peaks in late winter and spring occur. In the United States, before the introduction of routine vaccination in 1996, approximately 4 million cases of chickenpox occurred each year. Most individuals were infected by adolescence, and the peak age of onset was 4 to 5 years of age. Attack rates are lower and less seasonal in tropical areas, where a greater proportion of adolescents and adults are susceptible.

In the United States, the overall incidence of chickenpox is much lower than in the prevaccine period. Breakthrough varicella (occurrence of wild-type VZV infection in a vaccinated host 42 days or more after receipt of varicella vaccination) now makes up a greater proportion of chickenpox disease. One study estimates an incidence of 2.2 cases of breakthrough varicella per 1000 person years in children vaccinated with two doses of varicella vaccine.

Following primary chickenpox infection, the virus remains latent in the dorsal root ganglia and can reactivate, causing herpes zoster, commonly referred to as *shingles*. Herpes zoster is infrequent in immunocompetent children, but persons older than 60 years of age have an incidence of 7.2 to 11.8 cases per 1000 population per year, which increases with advancing age.

## RISK FACTORS

VZV is highly transmissible, with attack rates of nearly 90% among susceptible household contacts but only 10% to 30% among susceptible persons with casual contact (e.g., classroom or hospitals). Individuals with impaired immunity are at increased risk for severe disseminated disease caused by VZV, including complications of pneumonitis and hepatitis, which rarely may lead to hepatic failure. Adolescents and adults, pregnant women, and neonates born within 2 days before or 4 days after onset of maternal varicella are also at risk for progressive disease.

## CLINICAL FEATURES

### Varicella (Chickenpox)

The incubation period for primary VZV infection ranges from 10 to 21 days. Once infected, persons are contagious (via respiratory droplets and direct contact) for 1 to 2 days before the onset of rash and until all lesions have crusted (usually 5–7 days). For 24 to 48 hours before the rash, constitutional symptoms such as fever, malaise, anorexia, headache, and mild abdominal pain may be present. Cutaneous lesions develop first on the scalp, face, or trunk and then spread to the lower portions of the body. Lesions start as erythematous macules, which then develop into clear fluid–filled vesicles, often described as resembling "dewdrops on a rose petal" (Fig. 13.1). These lesions are usually pruritic. After 24 to 48 hours, the vesicle fluid becomes cloudy, and eventually the lesions crust over. Lesions develop at different intervals, such that at any one time lesions at multiple stages are present. In unimmunized children, primary infection produces fewer than 300 lesions. Primary varicella infection tends to be more severe in children infected from a household contact (likely because of more intense exposure) as well as in adolescents and adults. Individuals with

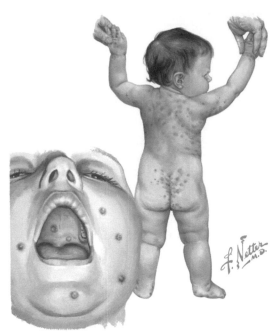

**Fig. 13.1** Child with chickenpox.

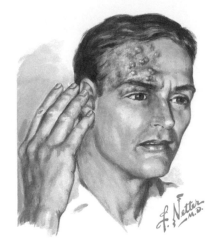

**Fig. 13.2** Zoster in the ophthalmic branch of the trigeminal nerve.

underlying skin conditions such as eczema, trauma, and sunburn may have higher concentrations of lesions at the affected sites.

Primary varicella usually resolves without event and healthy children rarely develop complications. Bacterial superinfection (commonly with *Staphylococcus aureus* or *Streptococcus pyogenes*) is the most common cause of morbidity in healthy children. These children may have impetigo-like lesions, cellulitis, lymphadenitis, subcutaneous abscesses, bacteremia, osteomyelitis, or necrotizing fasciitis. Neurologic complications such as meningoencephalitis or cerebellar ataxia are typically immune mediated and generally resolve without treatment. Meningoencephalitis is associated with rapid recovery, whereas cerebellar ataxia may take weeks to resolve. Other rare complications include hepatitis, Reye syndrome, thrombocytopenia, nephritis, and arthritis.

Breakthrough varicella is usually mild, with much fewer lesions (<50), and a shorter illness duration compared with varicella in unvaccinated hosts. However, it is estimated that approximately 25% to 35% of persons who have received only one dose of varicella vaccine will develop chickenpox, with features similar to those in unvaccinated hosts.

## Herpes Zoster (Shingles)

Emergence of latent VZV as herpes zoster manifests with an eruption of vesicular lesions in a sensory dermatomal distribution (Fig. 13.2). Lesions may erupt over a course of 3 to 7 days, with complete resolution of the rash usually within 2 weeks. The rash is pruritic and may be associated with local pain and hyperesthesia. Zoster is unusual in childhood, but when present, the extent of the lesions is limited and with minimal neuropathic symptoms. Risk factors for developing zoster during childhood include having primary varicella infection in the first year of life and immunocompromising conditions, especially those affecting cellular immunity (e.g., acquired immunodeficiency syndrome, chemotherapy, or prolonged high-dose steroids). In compromised hosts, zoster may disseminate, with lesions occurring outside of the dermatome and involvement of internal organs. Other complications of zoster depend on the dermatome affected.

Some unique manifestations of herpes zoster occur when the reactivated virus involves nerves of the head and neck. Ramsay-Hunt

syndrome occurs when VZV reactivates in the geniculate ganglion and consists of unilateral facial palsy, pain and vesicles in the auditory canal, and loss of taste in the anterior two-thirds of the tongue. When zoster involves the ophthalmic branch of the trigeminal nerve (herpes zoster ophthalmicus), sight-threatening keratitis may develop, necessitating emergent ophthalmologic consultation. Zoster involving the maxillary or mandibular branch of the trigeminal nerve may cause intraoral lesions affecting the palate, tonsillar fossa, tongue, and floor of the mouth.

Pain preceding the rash, acute neuritis, and postherpetic neuralgia is more common with zoster in adults than in children. Approximately 80% to 85% of persons over 50 years of age with zoster will develop postherpetic neuralgia. The pain can be quite debilitating and may last for weeks to months. Early antiviral treatment can ameliorate the severity and duration.

## Disease in High-Risk Hosts

In the absence of antiviral therapy in children with underlying immunodeficiency (congenital or acquired), primary varicella is often progressive, manifesting with an extended period of new lesion eruptions, pneumonitis, hepatitis, encephalitis, and/or disseminated intravascular coagulation (DIC) (Fig. 13.3). The mortality rate of untreated primary varicella in these hosts ranges from 7% to 17%. Presentations of potentially life-threatening varicella include respiratory symptoms, hemorrhage into a vesicular lesion (indicative of potential DIC), or severe abdominal or back pain.

Adolescents and adults are at increased risk for having complicated primary varicella despite being otherwise healthy. In this population the most common complication is varicella pneumonia, which may manifest with cough, pleuritic chest pain, and hemoptysis with or without cyanosis. This occurs on average 3 days (range of 1–6 days) after the onset of the rash. Resolution usually occurs within 24 to 72 hours without antiviral treatment, but varicella pneumonia can progress to respiratory failure.

Nonimmune pregnant women are at high risk for adverse outcomes with primary varicella, with pneumonia being the major cause of morbidity and mortality. In addition, severe sequelae to the fetus, including spontaneous abortion, fetal demise, or premature delivery, can occur. Rarely (less than 2% of cases), maternal varicella can lead to congenital varicella syndrome, the severity of which is dependent on gestational age at the time of infection. Infection during the first 20 weeks of gestation carries the highest risk of severe embryopathy. Congenital varicella can affect the skin (cutaneous defects, cicatricial scars, hypopigmentation, bullous lesions), the extremities

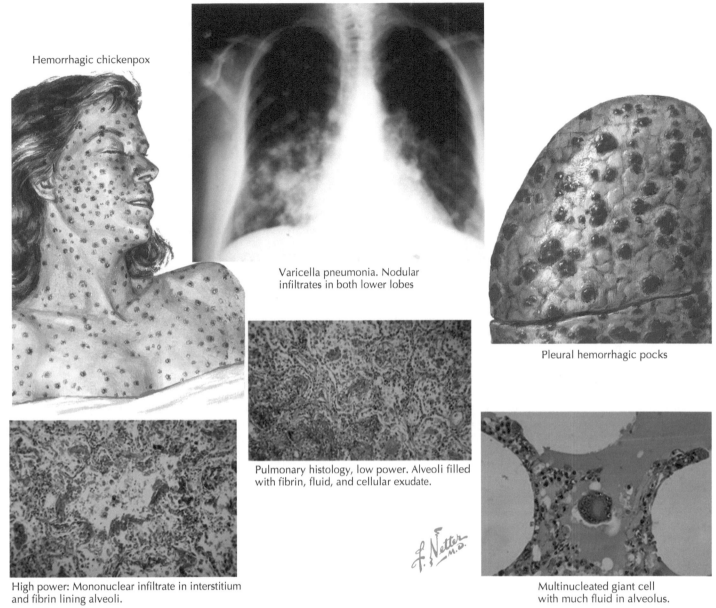

Hemorrhagic chickenpox

Varicella pneumonia. Nodular infiltrates in both lower lobes

Pleural hemorrhagic pocks

Pulmonary histology, low power. Alveoli filled with fibrin, fluid, and cellular exudate.

High power: Mononuclear infiltrate in interstitium and fibrin lining alveoli.

Multinucleated giant cell with much fluid in alveolus.

**Fig. 13.3** Varicella pneumonia.

(hypoplastic limb, muscular atrophy and denervation, joint abnormality, absent or malformed digits), eye (chorioretinitis, microphthalmia, anisocoria), the central nervous system (CNS) (intrauterine encephalitis with cortical atrophy, seizures, mental retardation), or other organs (e.g., hydronephrosis or hydroureter and esophageal dilatation or reflux).

Varicella may also be severe for the neonate born within 2 days before or 4 days after the onset of maternal varicella, with a mortality rate of 30% if untreated. This is a result of the lack of transfer of maternal antibodies with the primary viremia. Neonates born 5 days after the onset of maternal varicella are not at increased risk for severe disease because maternal antibodies have had the opportunity to cross the placenta.

Herpes zoster in immunocompromised hosts can lead to severe dermatomal infection, extension beyond the dermatome, and dissemination. Pneumonia, hepatitis, encephalitis, and DIC may result from hematogenous spread of reactivated virus. Severely immunocompromised hosts such as those with human immunodeficiency virus (HIV)

infection may have chronic or relapsing cutaneous disease, retinitis, and CNS infection with or without skin findings.

## DIAGNOSTIC APPROACH

In healthy individuals, primary varicella and zoster are clinical diagnoses. In atypical presentations, suspected breakthrough disease, or in patients with other possible causes of vesicular lesions, laboratory confirmation may be necessary to facilitate initiating appropriate therapy. Vesicular fluid or lesions can be tested for VZV by polymerase chain reaction (PCR), which is the most sensitive and specific of the available testing methods. Serologic tests for VZV-specific antibodies are often not useful in the acute phase because immune globulin M (IgM) anti-VZV antibodies are inconsistently present. If present, IgM anti-VZV antibodies indicate current or recent infection. Acute and convalescent serum specimens may demonstrate a rise in IgG VZV antibodies and may improve sensitivity. Direct fluorescent antibody (DFA) and Tzanck smear are not recommended due to their limited sensitivity.

Complete blood count and liver function tests may be indicated when disseminated disease is suspected; however, lymphocytosis and mild hepatic enzymatic elevations are common in uncomplicated disease. When neurologic complications exist, mild cerebrospinal pleocytosis (<100 cells/mm³) with moderate protein elevation (<200 mg/mL) may be present. Patients with uncomplicated zoster may also have cerebrospinal fluid pleocytosis.

## CLINICAL MANAGEMENT

For most healthy children, symptomatic therapy for chickenpox and herpes zoster is all that is required. Antiviral therapy is recommended for adolescents, adults, immunocompromised hosts, and individuals with disseminated disease but is not routinely recommended for healthy children. The optimal timing of therapy is within the first 72 hours of the onset of the rash. Acyclovir or valacyclovir may be used when oral administration is an option, but immune compromised patients at risk of severe disease should receive intravenous acyclovir therapy (Table 13.1).

Aspirin is contraindicated during varicella infections owing to the risk of Reye syndrome. Symptomatic relief of fever with acetaminophen may be undertaken with careful attention to dosage. Ibuprofen use has been associated with increased rates of necrotizing fasciitis in isolated reports; therefore this drug should be used cautiously. After weighing risks and benefits, antihistamines and topical calamine lotion may be used for pruritus treatment. Should bacterial superinfection develop, antibiotics that include coverage for group A *Streptococcus* and *S. aureus* (including methicillin-resistant *S. aureus*), based on local epidemiology and sensitivity patterns, should be instituted.

For zoster in all age groups, acyclovir reduces the development of new lesions from over 1 week to less than 3 days. In adults, acyclovir or valacyclovir treatment may also reduce the likelihood of developing acute neuritis.

## PROGNOSIS

For healthy children with primary varicella, the prognosis is good, with complete recovery. Although most lesions heal without scarring, many children have one or two scars, typically on the face, as these earlier lesions may involve deeper layers of the skin. For adolescents, adults, pregnant women, and immunocompromised hosts, there is greater risk of morbidity and mortality, but this may be reduced with prompt antiviral therapy. After herpes zoster, most children do not incur long-term sequelae. Adults, however, may experience postherpetic neuralgia, which may be debilitating and may require treatment targeting pain control.

## PREVENTION AND CONTROL

Individuals with varicella should be considered contagious until all lesions are crusted. Hospitalized patients should be maintained with airborne precautions. Children should be excluded from school or day care. Persons with zoster do not need to be isolated if the lesions can be covered. Health care workers without evidence of immunity who have been exposed to varicella should not be in the hospital setting from days 8 to 21 after exposure or for 28 days if they received hyperimmune VZV immune globulin (see later).

### Active Immunoprophylaxis: Vaccination

Currently in the United States, a live-attenuated varicella vaccine (Varivax, Merck & Co., Inc., Whitehouse Station, NJ) is recommended for use in all children; a two-dose regimen is used, with vaccine given at 12 to 15 months and 4 to 6 years of age. After licensure in 1995, widespread vaccination at 12 to 15 months of age led to a dramatic decline in primary varicella. However, from 2001 to 2005, breakthrough cases of chickenpox began to occur, peaking in children aged 7 to 10 years. Although most of these cases were mild, typically with fewer than 50 lesions, they posed a risk to the children and their susceptible contacts. Therefore in 2007 the addition of the second dose of vaccine at 4 to 6 years of age was recommended after postlicensure data suggested a reduction of breakthrough disease in this preadolescent age group.

Because it is a live virus, VZV vaccine is contraindicated in persons with altered immunity and in pregnant women. Varicella vaccines should not be given to persons on high-dose immunosuppressive therapy including systemic steroids (≥2 mg/kg of prednisone or total ≥20 mg/d of prednisone or equivalent for ≥2 weeks). Quadrivalent measles, mumps, rubella, and varicella (MMRV) is available in the United States and licensed for use in healthy children 12 months to 12 years of age.

## TABLE 13.1    Recommendation and Doses of Acyclovir or Valacyclovir for Varicella

| Indication | Dosage | Timing |
| --- | --- | --- |
| Heathy persons with increased risk of moderate to severe varicella:<br>Unvaccinated persons >12 years of age<br>Persons with chronic cutaneous or pulmonary disorders<br>Persons on long-term salicylate therapy<br>Persons receiving short or intermittent courses of corticosteroids<br>Secondary cases of household contacts | **Valacyclovir:** 20 mg/kg per dose, max 1000 mg given PO three times a day orally for 5 days (2 to younger than 18 years old)<br>**Acylovir PO:** ≤40 kg: 80 mg/kg per day in four divided doses for 5 days; max daily dose 3200 mg/d; >40 kg: 3200 mg, in four divided doses for 5 days<br>**Acylovir IV** (if hospitalized):<br><2 years old: 30 mg/kg per day in three divided doses for 7–10 days; ≥2 years old: 1500 mg/m² per day in three divided doses for 7–10 days | Ideally initiate within 24 hours after initial lesions appear for maximum benefit |
| Immunocompromised hosts (including persons on high-dose corticosteroids >14 days) | **IV Acyclovir:** <2 years old: 30 mg/kg per day in three divided doses for 7–10 days; ≥2 years old: 1500 mg/m² per day in three divided doses for 7–10 days | Initiate as soon as possible after initial lesions appear |
| Immunocompromised hosts possibly at low to moderate risk of developing severe varicella:<br>Persons with HIV with normal CD4+ T lymphocytes and selected children with leukemia, with careful follow-up | **Valacyclovir:** 20 mg/kg per dose, max 1000 mg, given PO three times a day orally for 5 days | Initiate as soon as possible after initial lesions appear |

For otherwise healthy young children, antiviral treatment of varicella is not routinely recommended. However, for specific indications or concerns, the recommended dosage and timing of most benefit as listed above.
Data from American Academy of Pediatrics. In: Kimberlin DW, Brady MT, Jackson MA, Long SS, eds. *Red Book: 2018 Report of the Committee on Infectious Diseases.* Elk Grove Village, IL: Elk Grove Village, IL: American Academy of Pediatrics; 2018.

In 2006, zoster vaccine live (ZVL) (Zostavax, Merck & Co., Inc., Whitehouse Station, NJ) was licensed for adults 60 years of age and older to reduce the risk of zoster and expressly to prevent postherpetic neuralgia (67% efficacy in preventing postherpetic neuralgia and 73% efficacy in preventing postherpetic pain lasting more than 6 months). Zostavax differs from Varivax in that it contains up to a 14 times greater amount of live-attenuated VZV. Zostavax is contraindicated in persons with primary or acquired immunodeficiency.

In 2017, zoster vaccine-recombinant, adjuvanted (RZV) (Shingrix, GlaxoSmithKline Biologicals, Rixensart, Belgium, distributed by GlaxoSmithKline, Research Triangle Park, NC) two-dose subunit vaccine was licensed in the United States for use in healthy adults 50 years of age or older. This vaccine contains a recombinant glycoprotein E (important for the replication and intercellular viral diffusion of VZV) with a novel adjuvant (AS01B). In vaccine trial studies, Shingrix was 96.6% efficacious in preventing herpes zoster and 88% to 91% efficacious in preventing postherpetic neuralgia in persons 50 years of age or older. There have been no head-to-head comparative trials of Zostavax versus Shingrix. However, due to the higher reported efficacy in the Shingrix studies and evidence that immunity following Shingrix is more durable over time, in 2018 the Advisory Committee on Immunization Practices stated a preference of Shingrix over Zostavax and recommends Shingrix for use in persons 50 years of age or older for the prevention of herpes zoster.

## Postexposure Prophylaxis
### Vaccination

If not otherwise contraindicated, VZV vaccine can be given to susceptible individuals after exposure to varicella. Postexposure vaccination should be given within 3 days but can be given up to 5 days after exposure and may reduce the likelihood of disease or its severity. A second dose should be administered after an age-appropriate interval.

### Passive Immunoprophylaxis

Varicella-Zoster Immune Globulin (VariZIG, Saol Therapeutics, Inc., Roswell, GA) is approved by the US Food and Drug Administration and available in the United States for postexposure prophylaxis to prevent or mitigate varicella infection in high-risk individuals (Box 13.1). VariZIG should be administered within 10 days of exposure but ideally within 96 hours. If VariZIG is not available, intravenous immune globulin (IVIG) may be substituted, although the amount of varicella-specific antibody in different preparations of IVIG varies.

### Chemoprophylaxis

Preemptive use of acyclovir or valacyclovir may help to prevent or attenuate varicella disease for susceptible immunocompromised individuals when neither active nor passive immunization is possible. Preemptive therapy is begun 7 to 10 days after exposure and continued for 7 days. Age-appropriate vaccination is still recommended but should not be given while antiviral therapy is being given.

## EVIDENCE

Coffin SE, Hodinka RL. Utility of direct immunofluorescence and virus culture for detection of varicella-zoster virus in skin lesions. *J Clin Microbiol* 1995;33:2792-2795. *This reference provides comparative data for the DFA and culture of VZV.*

Gabutti G, Bolognesi N, Sandri F, Florescu C, Stefanati A. Varicella zoster virus vaccines: an update. *Immuno Targets and Therapy* 2019;8:15-28. *Provides a review of vaccines against varicella-zoster virus.*

Kuter B, Matthews H, Shinefield H, et al. Ten-year follow-up of healthy children who received one or two dose injections of varicella vaccine. *Pediatr Infect Dis J* 2004;23:132-137. *This reference provides data about the postlicensure follow-up of clinical trial participants for varicella vaccine and incidence of breakthrough varicella.*

Zhu S, Zeng F, Xia L, He H, Zhang J. Incidence rate of breakthrough varicella observed in healthy children after 1 or 2 doses of varicella vaccine: results from a meta-analysis. *Amer J Infect Control* 2018;28:e1-7. *This article provides data for incidence of breakthrough varicella.*

## ADDITIONAL RESOURCES

American Academy of Pediatrics (AAP) Committee on Infectious Diseases: Varicella-zoster infections. In Kimberlin DW, Brady MT, Jackson MA, Long SS, eds: *Red Book: 2018-2021 Report of the Committee on Infectious Diseases,* ed 31, Elk Grove Village, IL, 2018, AAP, pp 869-883. *This resource provides an overview of the diagnosis, treatment, and prevention of VZV infections.*

Arvin A: Varicella-zoster virus. In Long SS, Pickering LK, Prober CG, eds: *Principles and Practice of Pediatric Infectious Diseases,* ed 3, Philadelphia, 2008, Churchill Livingstone, pp 1021-1029. *This resource provides a review of the virology, diagnosis, and treatment of VZV infections.*

Centers for Disease Control and Prevention. Chickenpox (Varicella): for healthcare professionals. Available at: https://www.cdc.gov/chickenpox/hcp/index.html accessed Sept 2019. *This website provides US guidelines and recommendations for chicken pox.*

Dooling KL, Guo A, Patel M, et al. Recommendations of the Advisory Committee on Immunization Practices for use of herpes zoster vaccines. *MMWR Weekly Rep* 2018;67(3):103-108. *This resource provides recommendations for the prevention of herpes zoster.*

Marin M, Güris D, Chaves SS, et al; Advisory Committee on Immunization Practices, Centers for Disease Control and Prevention (CDC). Prevention of varicella: recommendations of the Advisory Committee on Immunization Practices (ACIP). *MMWR Weekly Rep* 2007;56:1-39. *This resource provides recommendations for the prevention of varicella infections.*

---

### BOX 13.1  Situations for Consideration of Postexposure Antibody Prophylaxis

**Candidates for Varicella-Zoster Immune Globulin (VariZIG) With Significant Exposure**

- Susceptible immunocompromised children
- Susceptible pregnant women
- Newborn infant whose mother had onset of chickenpox within 5 days before delivery or within 48 hours after delivery
- Hospitalized preterm infant (≥28 weeks' gestation) of mother without evidence of varicella immunity
- Hospitalized preterm infant (<28 weeks' gestation or ≤1000 g birth weight) regardless of maternal immunity status

**Situations Qualifying for Significant Exposure**

- Residing in the same household
- Face-to-face indoor play lasting at least 5 min (some experts use 1 h)
- Hospital varicella exposure: same two- to four-person bedroom or adjacent beds in a large ward
- Hospital zoster exposure: intimate contact (e.g., touching or hugging) a person considered to be contagious
- Newborn infant

# 14

# Hepatitis A Infection and Prevention

*Elizabeth H. Partridge, Dean A. Blumberg*

## ABSTRACT

Hepatitis A infection is the most common cause of viral hepatitis worldwide and a commonly reported vaccine-preventable disease in the United States. Clinical illness—characterized by fever, malaise, jaundice, and nausea—caused by hepatitis A virus (HAV) is similar to hepatitis caused by other viral pathogens. Although most cases of hepatitis A are self-limited, fulminant hepatitis resulting in death may occur. Widespread routine childhood vaccination against hepatitis A has resulted in a significant decline in hepatitis A infections, with a notable diminishment of previous racial or ethnic and geographic disparities.

## CLINICAL VIGNETTE

A 14-year-old female presents to her local emergency room with a history of fatigue and intermittent abdominal pain with poor appetite for 1 week. This morning her mother noticed that the whites of her eyes looked yellow. In the emergency room she has a temperature of 101.5°F, heart rate 130/min, blood pressure 110/68 mm Hg, and respiratory rate 16/min. On exam, she is tired appearing with dry mucosal membranes and faint scleral icterus. Her abdomen is distended and tender to palpation; her liver is 3 cm below the costal margin. The girl's mother reports that she doesn't believe in vaccines and wants her children to "build their own immunity." The patient denies sick contacts. She and the rest of the family have been busy welcoming the arrival of her new 3-year-old adopted sister, who arrived from China 1 month earlier.

COMMENT: The patient presented with signs and symptoms compatible with a diagnosis of acute hepatitis. Given the recent addition to the family of a 3-year-old child adopted from China, the most likely diagnosis is acute hepatitis A. However, given the history of the mother's vaccine hesitancy, the patient could lack other standard childhood immunizations and be susceptible to hepatitis B, as there is also a risk of transmission of hepatitis B to close contacts of an adopted orphan coming from a highly endemic country. Both hepatitis A and B serologic tests were sent to the laboratory; the IgM anti–hepatitis A virus antibody test was reported back positive.

## ETIOLOGY

Hepatitis A infection is caused by HAV, a ribonucleic acid (RNA) virus in the genus *Hepatovirus* within the family Picornaviridae. HAV is an icosahedral nonenveloped virus that replicates in the cytoplasm. The genome consists of single-stranded RNA that encodes information for 11 individual proteins. There are four closely related genotypes of human HAV strains but only one major serotype.

## GEOGRAPHIC DISTRIBUTION AND MAGNITUDE OF DISEASE BURDEN

Approximately 1.5 million cases of hepatitis A are reported each year worldwide, and the prevalence of antibodies (indicating previous infection) in the population can vary from 15% to nearly 100% in some countries. Hepatitis A infection remains highly endemic in Africa, India, Central and South America, the Middle East, and Asia. In developing areas of the world, most adults show serologic evidence of past infection. Nordic countries show the lowest prevalence of infection, approximately 15%. In other parts of Europe, Australia, Japan, and the United States, the seroprevalence of antibodies to HAV is 30% to 70% in adults (Fig. 14.1).

In the United States, the incidence of hepatitis A has significantly declined since the late 1990s, when hepatitis A vaccines came into wider use. In the prevaccine era, the incidence of hepatitis A in the United States was cyclic, with nationwide increases occurring every 10 to 15 years, often linked to foodborne outbreaks. Historically, higher rates of hepatitis A have occurred in the western United States, but this geographic variation is no longer present. The last peak incidence occurred in 1995. Rates decreased more than 95% from 1995 to 2011, when the rate of infection was at a historic low of 0.4/100,000 population. Rates increased by 140% from 2011 to 2017 due to large person-to-person outbreaks occurring primarily among homeless people and users of illicit drugs. In 2017, there were an estimated 6700 hepatitis A cases in the United States (Fig. 14.2). From 2016 to 2018, reports of hepatitis A cases increased by 294%, with approximately 15,000 hepatitis A infections reported from US states and territories. Since 2008, the highest rates of hepatitis A infections have been seen in the Asian and Pacific Islander populations (Fig. 14.3).

## RISK FACTORS FOR INFECTION

HAV is most commonly contracted via ingestion. After a brief viremia, the primary site of replication is the liver; the virus is then shed into the bile with subsequent passage to the intestines and feces. The highest concentration of virus is in the stool, and the period of greatest infectivity occurs during the 2 weeks before onset of symptoms. Infected stools result in transmission to others through either ingestion of contaminated water or food or person-to-person contact. An outbreak of hepatitis A infection can be caused by a combination of these factors—for example, contaminated food may infect restaurant patrons and employees, who then become the sources for further transmission within their larger communities.

Children are at increased risk for acquiring infection from fecal/oral pathogens, including HAV, because they have limited proficiency with personal hygiene and toileting habits and because of their propensity to explore the environment with their mouths. Because most infected children have asymptomatic or unrecognized infection, they can play a role in community outbreaks.

Epidemics may occur locally or nationwide, sometimes related to foodborne outbreaks. With more frequent international travel and global food production, hepatitis A can readily spread beyond

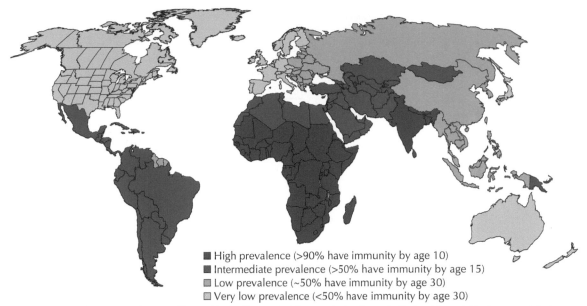

High prevalence (>90% have immunity by age 10)
Intermediate prevalence (>50% have immunity by age 15)
Low prevalence (~50% have immunity by age 30)
Very low prevalence (<50% have immunity by age 30)

Fig. 14.1 Global distribution of hepatitis A. (Reused with permission from Jefferies M et al. Update on global epidemiology of viral hepatitis and prevention strategies. *World J Clin Cases* 2018 Nov 6;6[13]:589-599. https://www.ncbi.nlm.nih.gov/pmc/articles/PMC6232563/figure/F1/. Accessed January 2020.)

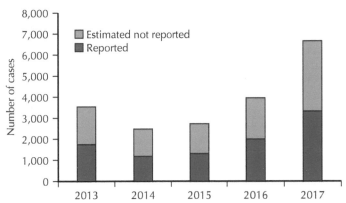

Fig. 14.2 Actual number of hepatitis A cases submitted to the Centers for Disease Control by states and estimated number of hepatitis A cases (note: estimated cases were determined by multiplying the number of reported cases by a factor that adjusted for underascertainment and underreporting)—United States, 2013–2017. (Data from Centers for Disease Control and Prevention. Available at: https://www.cdc.gov/hepatitis/statistics/2017surveillance/TablesFigures-HepA.htm, Fig. 2.1. Accessed January 2020.)

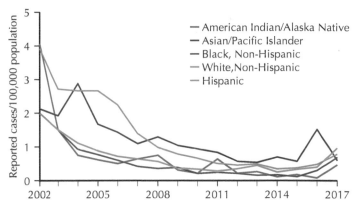

Fig. 14.3 Rates of reported hepatitis A by race/ethnicity—United States, 2002–2017 (Data from Centers for Disease Control and Prevention. Available at: https://www.cdc.gov/hepatitis/statistics/2017surveillance/TablesFigures-HepA.htm, Fig. 2.5. Accessed January 2020.)

the regions of known high endemicity to any area of the world. In communities with high rates of hepatitis A, most people are infected before reaching young adulthood. In the United States, communities with high rates of infection include Asian/Pacific Islander, Alaskan Native, American Indian, and selected Hispanic, migrant, and religious communities, although in recent times these elevated rates are less pronounced, likely because of widespread vaccination (see Fig. 14.3). Recent outbreaks of hepatitis A among persons reporting homelessness, due to unsafe sanitary conditions, represent a shift in HAV infection epidemiology in the United States, with over 7000 outbreak-associated cases reported from 12 states in 2018.

Hepatitis A may also be transmitted through sexual contact and in rare instances through transfusion of blood products collected from donors during the viremic period. Adults most at risk for infection with hepatitis A include household and sexual contacts of infected individuals,

men who have sex with men, close contacts of newly arriving international adoptees, users of illicit drugs, homeless people, and individuals traveling to countries where hepatitis A is common. In the United States, 85% of travel-related cases are associated with travel to Central and South America. Of note, a specific risk factor cannot be determined for more than half of all individuals who acquire hepatitis A infection.

## CLINICAL FEATURES

HAV can cause symptomatic or asymptomatic infection. The necessary infectious dose is not known but presumably is 10 to 100 viral particles. Clinical features usually appear after an average incubation period of 28 days, with a range of 15 to 50 days. Symptomatic infection is characterized by a relatively acute onset of symptoms that may include jaundice, weakness, fatigue, myalgia, anorexia, nausea, abdominal pain, fever, clay-colored stools, and/or dark-colored urine. Pruritus is less common. Diarrhea is common in young children (60%) but infrequent in adults. Physical findings can include jaundice, abdominal tenderness, hepatomegaly, or splenomegaly (Fig. 14.4).

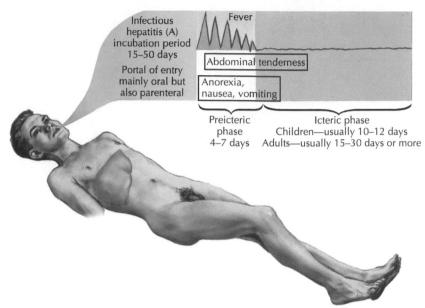

Infectious hepatitis (A) incubation period 15–50 days

Portal of entry mainly oral but also parenteral

Fever

Abdominal tenderness

Anorexia, nausea, vomiting

Preicteric phase 4–7 days

Icteric phase
Children—usually 10–12 days
Adults—usually 15–30 days or more

**Fig. 14.4** Clinical course of hepatitis A.

The symptoms and severity of illness correlate with age. In children younger than 6 years of age, 70% of infections are asymptomatic, and less than 10% of patients will have jaundice. Older children and adults usually develop symptoms. Generally, more than 75% of infected adults are symptomatic, with 50% to 85% exhibiting jaundice.

## COMPLICATIONS

No chronic or long-term infection is associated with hepatitis A; however, prolonged or relapsing symptoms for up to 9 months occur in approximately 10% of infected individuals. Infrequent complications of hepatitis A include pancreatitis, autoimmune hepatitis, and cholestatis hepatitis, which is characterized by fever, pruritis, and prolonged elevated bilirubin that can persist for months.

Hepatitis A is a rare but important cause of fulminant hepatitis, manifested by increasingly severe jaundice and deteriorating liver function. This may result in hepatic encephalopathy and coagulopathy, particularly in older adults and individuals with chronic liver disease.

## DIAGNOSTIC APPROACH

Clinically, infection with HAV may be indistinguishable from other causes of hepatitis. Hepatitis A can be differentiated from other types of viral hepatitis only through laboratory testing. Serologic testing detects immunoglobulin M antibody to HAV (IgM anti-HAV) starting 5 to 10 days before symptom onset, and these antibodies may be detectable for up to 6 months after infection. Immunoglobulin G antibody to HAV (IgG anti-HAV) also appears early in the course of infection and is detectable for life, denoting immunity to hepatitis A infection.

Other laboratory abnormalities include elevated liver enzymes. Serum alanine aminotransferase (ALT), aspartate aminotransferase (AST), and γ-glutamyltranspeptidase (GGT) concentrations are elevated before symptom onset, and bilirubin elevation follows. ALT and AST elevations usually range from 200 to 5000 IU/L and peak at 3 to 10 days after onset of symptoms, though these findings are not specific for HAV infection. Elevated liver function tests generally return to normal by 2 to 3 months after the onset of illness. Patients with acute HAV infection typically have a mild lymphocytosis, and occasional atypical mononuclear cells may be seen.

Although polymerase chain reaction testing of serum and stool for HAV can be helpful in investigating common-source outbreaks of hepatitis A, such tests are not widely available. Liver biopsy is generally not indicated for HAV infection. However, when performed, the typical pathology of viral hepatitis is observed, including hepatocellular necrosis, inflammatory cell infiltration, and regeneration of hepatocytes.

Other diagnostic considerations for HAV infection include other viral agents, most commonly hepatitis B and C but also Epstein-Barr virus (EBV), cytomegalovirus (CMV) and other enteroviruses, and bacterial infections (e.g., leptospirosis). Toxic hepatitis caused by prescribed medications that are metabolized by the liver (e.g., isoniazid, acetaminophen), alcohol, illicit drugs or poisons (e.g., carbon tetrachloride), and nonspecific injury (e.g., ischemia or shock) may also be considered.

## CLINICAL MANAGEMENT

There is no specific antiviral treatment for patients with hepatitis A; supportive care is the mainstay. Alcohol avoidance is recommended during the acute stage. No other dietary restrictions are recommended. Patients are encouraged to monitor their energy level and increase activities judiciously. Medications that might cause liver damage or are metabolized hepatically should be used with caution (Fig. 14.5). Patients with prolonged jaundice and cholestatic hepatitis may benefit from a short course of rapidly tapered corticosteroids, resulting in improved symptoms and resolution of disease.

In general, most individuals infected with HAV can be managed as outpatients. Because nausea and vomiting may lead to inadequate fluid intake, patients should be monitored for dehydration. If dehydration occurs, intravenous fluid administration and hospitalization may be necessary. Infected individuals with signs or symptoms of acute liver failure require management in an intensive care setting for appropriate monitoring of liver function as well as supportive care. Liver transplantation may be considered for patients with fulminant hepatic failure.

## PROGNOSIS

Most infected individuals recover within 2 months. Hepatitis A does not cause chronic infection. Relapsing hepatitis, in which exacerbations may

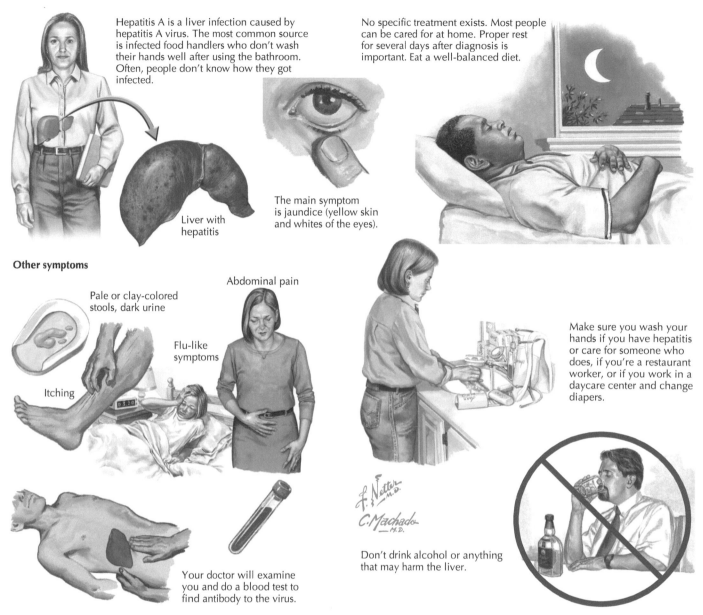

Hepatitis A is a liver infection caused by hepatitis A virus. The most common source is infected food handlers who don't wash their hands well after using the bathroom. Often, people don't know how they got infected.

Liver with hepatitis

The main symptom is jaundice (yellow skin and whites of the eyes).

No specific treatment exists. Most people can be cared for at home. Proper rest for several days after diagnosis is important. Eat a well-balanced diet.

**Other symptoms**

Pale or clay-colored stools, dark urine

Abdominal pain

Flu-like symptoms

Itching

Make sure you wash your hands if you have hepatitis or care for someone who does, if you're a restaurant worker, or if you work in a daycare center and change diapers.

Your doctor will examine you and do a blood test to find antibody to the virus.

Don't drink alcohol or anything that may harm the liver.

**Fig. 14.5** Management of hepatitis A.

occur weeks or months after apparent recovery, occurs in 10% to 15% of cases. Fulminant hepatitis is rare but can result in death. The overall case fatality rate of hepatitis A is 0.8% in the United States; it rises to 2.6% in adults 60 years of age and older. In the United States, more than 70% of deaths from hepatitis A occur in individuals 50 years of age and older.

## PREVENTION AND CONTROL

Hepatitis A vaccine provides the best protection against hepatitis A infection, but other important preventive measures include improved sanitation and meticulous personal hygiene practices, such as good hand washing and proper food-handling techniques.

### Hepatitis A Vaccines

The first hepatitis A vaccine was licensed in the United States in 1995. Vaccine guidelines have evolved over the years from target-group recommendations (i.e., vaccine for specific ages, communities, and groups with risk factors) to the current recommendations for routine childhood vaccination.

The hepatitis A vaccines currently available in the United States are prepared from purified cell culture–grown HAV that is formalin-inactivated; these vaccines are licensed for intramuscular use in persons 1 year of age and older. HAV vaccines are highly immunogenic, resulting in excellent efficacy (≥95% efficacy starting 4 weeks after one dose and >99% 1 month after the second dose). Two inactivated whole-virus vaccines are available: HAVRIX (GlaxoSmithKline, Brentford, Middlesex, UK) and VAQTA (Merck & Co., Kenilworth, NJ). Both have pediatric formulations approved for persons 12 months to 18 years of age and adult formulations have been approved for persons 19 years of age and older. These vaccines are given in a two-dose series consisting of an initial dose with a second dose 6 to 18 months later. The duration of protection after vaccination is not known but, based on surveillance data and kinetic models of antibody decline, is predicted to be at least 20 years. Adverse events after immunization are generally mild local reactions and are self-limited. Hepatitis A vaccines are contraindicated in those with a severe allergic reaction to a previous dose of hepatitis A vaccine or known allergy to any of the vaccine components.

Hepatitis A vaccine is also available in combination with hepatitis B vaccine. The combined hepatitis A–hepatitis B vaccine is licensed in the United States for those 18 years of age and older and is administered as either a three- or four-dose series.

## Preexposure Prophylaxis With Hepatitis A Vaccine

Hepatitis A vaccine is preferred for preexposure protection in all populations unless contraindicated and for postexposure prophylaxis for most individuals 1 year of age or older.

Since 2005 in the United States, routine childhood vaccination against hepatitis A has been recommended beginning at 1 year (12 to 23 months) of age. Vaccination is also recommended for unvaccinated individuals 1 year of age and older who are traveling to a hepatitis A endemic area, all individuals 1 year of age and older during an outbreak, men who have sex with men, users of illicit drugs, homeless persons, those with occupational exposure, persons who anticipate close contact with an international adoptee from a country of high or intermediate HAV endemicity, individuals receiving clotting factor concentrates, those with chronic liver disease, and anyone wishing to be immunized.

The Advisory Committee on Immunization Practices now recommends that HepA vaccine be administered to infants aged 6 to 11 months traveling outside the United States to areas with increased rates of hepatitis A. The travel-related dose for infants aged 6 to 11 months should not be counted toward the routine two-dose series. For infants younger than 6 months of age traveling to HAV endemic areas, prophylactic immune globulin (IG) is recommended (see further on).

## Postexposure Prophylaxis With Hepatitis A Vaccine

Postexposure prophylaxis should be given to persons who have recently been exposed to hepatitis A if they have not been previously vaccinated. They should receive a hepatitis A vaccine and/or intramuscular IG (0.1 mL/kg) as soon as possible or within 2 weeks of exposure. Hepatitis A vaccine is preferred if they are older than 12 months of age. Administration of IG in addition to HepA vaccine should be considered for persons with special risk factors for HAV infection or increased risk of complication from HAV exposure, including those who are over 40 years of age, immunocompromised, have chronic liver disease, are close contacts (household contacts, caretakers, sexual contacts) of persons with HAV infection or those with an occupational risk (persons working with nonhuman primates or with HAV in a research laboratory). For children younger than 12 months of age or persons who have a vaccine contraindication, IG should be administered. Measles, mumps, rubella, and varicella (MMRV) vaccines should not be administered sooner than 3 months after IG administration.

## Preexposure Prophylaxis With Immune Globulin

Intramuscular IG confers greater than 85% protection against HAV infection if administered before or within 2 weeks of HAV exposure. Preexposure prophylaxis with IG should be provided to susceptible individuals traveling to an area with an increased rate of hepatitis A

and in whom use of HAV vaccine is not permissible. Administration of hepatitis A vaccine is generally recommended; however, there are situations in which prophylaxis with IG instead of or in addition to the hepatitis A vaccine is appropriate. Older adults, immunocompromised persons, and persons with chronic liver disease or another chronic medical condition may not have a brisk immune response to hepatitis A vaccine. If these individuals are planning to depart in less than 2 weeks, they should receive IG in addition to hepatitis A vaccine (simultaneously but at separate anatomic sites). In addition, travelers with a contraindication to hepatitis A vaccine should receive a single dose of IG. The dose depends on the anticipated length of exposure: 0.1 mL/kg provides protection for up to 1 month; for travel up to 2 months, the dose is 0.2 mL/kg; if the travel period is 2 months or more, a dose of 0.2 mL/kg should be repeated every 2 months for the duration of travel.

## EVIDENCE

Advisory Committee on Immunization Practices (ACIP), Fiore AE, Wasley A, Bell BP. Prevention of hepatitis A through active or passive immunization: recommendations of the Advisory Committee on Immunization Practices (ACIP). *MMWR Recomm Rep* 2006;55:1-23. *This reference provides information about the efficacy of hepatitis A vaccine and immunoglobulin for preventing hepatitis A infections.*

Foster MA, Hofmeister MG, et al; Centers for Disease Control and Prevention (CDC). Increase in hepatitis A virus infections—United States, 2013-2018. *MMWR* 2019;68(18):413-415. *This reference provides the most recent data on hepatitis A in the United States.*

## ADDITIONAL RESOURCES

Centers for Disease Control and Prevention (CDC): Hepatitis A information for health professionals. Available at: www.cdc.gov/hepatitis/HAV/index.htm. Accessed September 24, 2019. *This resource provides an overview of hepatitis A for medical professionals.*

Centers for Disease Control and Prevention (CDC): Update: Recommendations of the Advisory Committee on Immunization Practices for Use of Hepatitis A Vaccine for Postexposure Prophylaxis and for Preexposure Prophylaxis for International Travel. *MMWR Morb Mortal Wkly Rep* 2018;67:1216-1220. *This resource provides updated recommendations for prevention of HAV infections.*

Kimberlin DW, Brady MT, Jackson MA, eds: *Red Book: 2018 Report of the Committee on Infectious Diseases*, ed 31, Elk Grove Village, IL, 2009, AAP, pp 392-400. *This resource provides an overview of the diagnosis, treatment, and prevention of HAV infections.*

Recommendations of the Advisory Committee on Immunization Practices for Use of Hepatitis A Vaccine for Persons Experiencing Homelessness. *MMWR Morb Mortal Wkly Rep* 2019;68:153-156. *This resource provides updated recommendations for prevention of HAV infections.*

U.S. Food and Drug Administration (FDA): Hepatitis A virus. In: *Bad Bug Book, Foodborne Pathogenic Microorganisms and Natural Toxins*, ed 2, 2012. Available at: www.fda.gov/Food/FoodSafety/FoodborneIllness/FoodborneIllnessFoodbornePathogensNaturalToxins/BadBugBook/ucm071294.htm. *This resource provides an overview of food-related hepatitis A infections.*

# 15

# Hepatitis B Infection

*Ritu Cheema, Dean A. Blumberg*

## ABSTRACT

Hepatitis B infection is one of the most common global health problems. Persons with acute hepatitis B infection can be asymptomatic or present with hepatitis with or without jaundice. Most acute infections resolve without sequelae; however, progression to chronic infection may occur and the risk is inversely related to age at the time of acquisition of infection. This progression to chronic hepatitis B infection may result in cirrhosis or hepatocellular carcinoma. Treatment of acute hepatitis B infection is primarily supportive. Patients with chronic hepatitis B infection should undergo routine monitoring for disease progression and to determine appropriate treatment initiation. Treatment includes nucleos(t)ide analogues and interferon and is usually managed by liver specialists. Hepatitis B is preventable with vaccination, which is universally recommended for infants in most industrialized nations as well as in many low-resource areas.

## CLINICAL VIGNETTE

A 12-year-old boy who recently immigrated from a low-resource country presented to the clinic with fatigue, vague abdominal pain, and nausea without fever. Physical examination was notable for right-upper-quadrant tenderness without jaundice. Laboratory testing revealed elevated liver enzymes, positive HBsAg, negative anti-HBs antibody, positive total anti-HBc antibody, negative IgM anti-HBc antibody, positive anti-HBe antibody, and a high load of hepatitis B virus (HBV). Does he have acute or chronic hepatitis? Does he need to be started on treatment?

COMMENT: He has chronic hepatitis B supported by the presence of HBsAg and negative IgM anti-HBc. Both total anti-HBc and anti-HBe can be present in resolved and chronic infection, but lack of anti-HBs antibody supports chronic infection. His need for treatment is based on the degree of elevation of liver enzymes.

## ETIOLOGY

HBV is a partially double-stranded deoxyribonucleic acid (DNA) virus of the *Hepadnaviridae* family with hepatitis B core antigen (HBcAg), surrounded by hepatitis B surface antigen (HBsAg). Humans are the only known host, and people with chronic hepatitis B infection are the primary reservoir, with the liver being the primary site of HBV replication.

## GEOGRAPHIC DISTRIBUTION AND MAGNITUDE OF DISEASE BURDEN

In the United States, the Centers for Disease Control and Prevention (CDC) estimates the actual number of acute hepatitis B cases annually as close to 22,100. In the United States, chronic hepatitis B has a reported estimated prevalence (2011 to 2016) around 862,000. Globally, approximately 257 million people have chronic hepatitis B.

The majority of persons with chronic HBV infections in the United States are immigrants, have immigrant parents, or became exposed through other close household contacts. Immunosuppression increases the risk of progression to chronic infection as well. Since 1990 the incidence of HBV infection has declined in all age groups in the United States, with the largest decline (approximately 98%) occurring in children younger than 15 years of age, largely attributable to use of hepatitis B vaccination during infancy.

## RISK FACTORS FOR INFECTION

In an infected person, HBV is present in all body fluids, although semen, vaginal secretions, and serum are primarily infectious. HBV is highly infectious, can be transmitted in the absence of visible blood, and remains infectious on environmental surfaces for at least 7 days. Hepatitis B is transmitted parenterally (e.g., by needlestick or sharing of needles), perinatally, sexually, and less commonly horizontally. Infants are most often infected from exposure to the blood of an infected mother during the birth process. Horizontal transmission is less well defined but appears to occur rarely in households where one or more members have chronic infection. It is important to remember that HBV is not spread through food or water, the sharing of eating utensils, breastfeeding, hugging, kissing, hand holding, coughing, or sneezing.

Other risk factors for infection include high-risk sexual activity (e.g., multiple partners, traumatic contact, men who have sex with men) and injection drug use (IDU). Transmission of HBV from the transfusion of blood or blood products is rare because of donor screening and viral inactivation procedures. Even after intensive investigation, up to 40% of patients with acute hepatitis B infection have no identifiable risk factors. Many international adoptees come from geographic areas with a high prevalence of HBV infection. Because infected children may have few or no symptoms, all internationally adopted children should be screened.

## CLINICAL FEATURES

### Acute Infection

The incubation period is generally 3 months but ranges from 6 weeks to 6 months. The clinical features of acute HBV infection depend partly on age. Most children under age 5 years and newly infected immunosuppressed adults are generally asymptomatic, whereas 30% to 50% of persons 5 years of age or older have signs and symptoms. The manifestations of infection range from being asymptomatic to nausea, vomiting, abdominal pain, fever, dark urine, clay-colored stools, hepatomegaly, splenomegaly, and jaundice (Fig. 15.1). Malaise and anorexia may precede jaundice by 1 to 2 weeks. Fulminant HBV infection is uncommon (<1%) but often results in death or liver failure, necessitating liver transplantation. Recovery is gradual, with fatigue and malaise usually resolving within a few weeks or months.

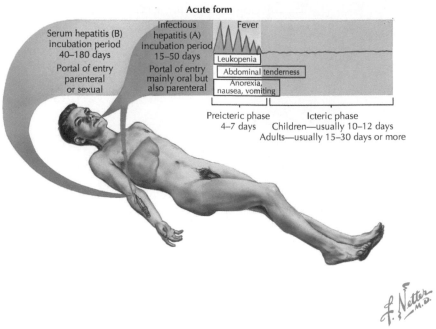

**Fig. 15.1** Hepatitis B clinical presentation.

## Chronic Infection

Chronic HBV infection is defined as persistence of HBsAg in serum for at least 6 months. It may cause no symptoms or may result in chronic active hepatitis and cirrhosis (Fig. 15.2). Individuals with chronic active hepatitis often have fever, abdominal pain, and malaise. Advanced disease may lead to spider nevi, ascites, coagulopathy, and esophageal varices if progression to cirrhosis occurs. Chronic HBV carriers are at risk of developing hepatocellular carcinoma, the most common infectious cause of liver cancer.

Extrahepatic manifestations can occur in 10% to 20% of chronic HBV infections and may include glomerulonephritis/membranous nephropathy, serum sickness, arthralgias and arthritis, and rashes such as urticaria, papular acrodermatitis (Gianotti-Crosti), and polyarteritis nodosa.

## DIAGNOSTIC APPROACH

Patients with acute HBV infection have elevated transaminases and bilirubin. The diagnosis is most commonly confirmed by the use of HBV-specific antigens and antibodies, as described in Table 15.1. The presence of hepatitis E antigen (HBeAg) is a marker of increased infectivity in those with acute or chronic infection.

Polymerase chain reaction (PCR) assays are available to detect and quantify hepatitis B DNA, which may be useful for monitoring disease progression.

## DIFFERENTIAL DIAGNOSIS

The diagnostic considerations vary with age, vaccination history, and exposures such as travel, sexual activity, and known infectious contacts. The clinical presentation can be similar due to hepatitis A, C, D, and E (if there is a history of travel to endemic regions), cytomegalovirus, herpes simplex virus, Epstein-Barr virus, and yellow fever. Bacterial causes of hepatitis are less common but may include sepsis, leptospirosis, and syphilis. Noninfectious considerations include Wilson disease, malignancies, toxins, illicit drugs, and ingestions (accidental or intentional).

## CLINICAL MANAGEMENT

### Acute Infection

Treatment for acute infection is primarily supportive, with particular attention to hydration, nutrition, and patient comfort. Antiviral therapy (entecavir, tenofovir) is indicted only for acute liver failure or protracted severe courses (Fig. 15.3).

### Chronic Infection

Patients diagnosed with chronic HBV infection need baseline testing of alanine aminotransferase (ALT), alpha fetoprotein, HBeAg, anti-HBe, and viral load; they should also have a liver ultrasound scan and an assessment of the family history for liver disease. If the ALT is normal, alpha fetoprotein is below 10 ng/mL, and family history is negative, periodic follow-up blood tests are recommended (ALT and DNA every 3 to 6 months; HBeAg every 6 to 12 months). The development of anti-HBe antibody is a good prognostic sign and may appear before resolution of HBsAg. Liver biopsy may be considered to assess the disease stage. Patients with elevated ALT or alpha fetoprotein or a positive family history of liver disease should have consultation with a hepatologist.

In patients of all age groups, coinfection should be assessed by checking antibodies to human immunodeficiency virus (HIV) and hepatitis A (HAV), hepatitis C (HCV), and hepatitis D virus (HDV) (if the patient is from an endemic area or has a history of IDU). To prevent further hepatic injury, patients should be vaccinated against hepatitis A if they are nonimmune. Heavy alcohol use as well as use of medications that are metabolized by the hepatic system should be avoided.

Indications for treatment are based on disease progression, viral load, liver biopsy results; they are generally initiated and monitored by liver specialists. For chronic HBV infection, nucleos(t)ide therapy is preferred (i.e., entecavir, tenofovir, lamivudine, adefovir, telbivudine) over interferon α-2b/pegylated interferon α-2a. These nucleos(t)ide analogues work by inhibiting the reverse transcription of pregenomic RNA to DNA. Entecavir and tenofovir (tenofovir disoproxil fumarate, tenofovir alafenamide) are

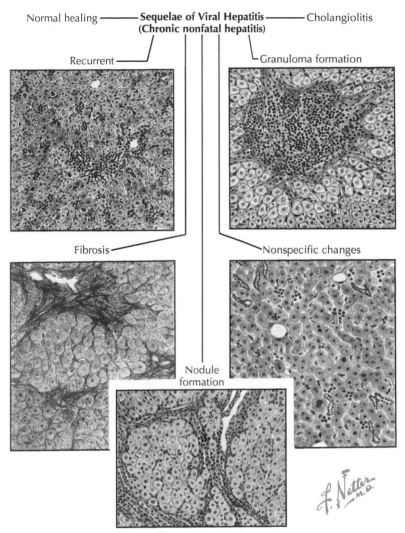

Fig. 15.2 Chronic hepatitis B sequelae.

| TABLE 15.1 | **Antibody/Antigen Interpretation for Hepatitis B Infection** | | | | | |
|---|---|---|---|---|---|---|
| **Clinical State** | **HBsAg** | **Anti-HBs** | **IgM (anti-HBc)** | **Total Anti-HBc** | **HBe** | **Anti-HBe** |
| Susceptible | – | – | – | – | – | – |
| Immunized | – | + | – | – | – | – |
| Acute infection | + | – | + | + | + | – |
| Resolved infection | – | + | – | + | – | + |
| Chronic infection | + | – | – | + | +/– | + |

Adapted from Mast EE, et al. A comprehensive immunization strategy to eliminate transmission of hepatitis B virus infection in the United States: recommendations of the Advisory Committee on Immunization Practices (ACIP) part 1: immunization of infants, children, and adolescents. *MMWR Recomm Rep* 2005;54(RR-16):1-31.

the most commonly used oral agents (preferred) and less likely to cause the virus to develop resistance compared with lamivudine, adefovir, and telbivudine. Interferon has been hypothesized to interfere with the entry of virus, uncoating of virion, and transcription/translation and assembly of nucleocapsids along with the additional role of promoting the clearance of HBV-infected hepatocytes by augmenting cell-mediating immunity. It is usually given for a finite duration, unlike oral antivirals, which require treatment for years until the desired response is achieved.

## PROGNOSIS

### Acute Infection

Recovery without sequelae occurs in 90% to 95% of infants with horizontal transmission and 80% to 85% of adults, with case fatality rates below 2%. Fulminant hepatitis may occur in 1% to 2% of infants and 1% of adults. Coinfection with HDV may result in a more severe clinical course of acute HBV infection.

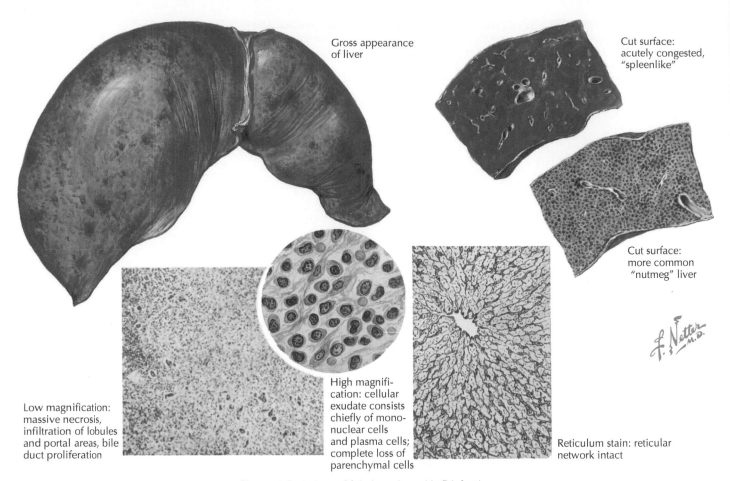

Gross appearance of liver

Cut surface: acutely congested, "spleenlike"

Cut surface: more common "nutmeg" liver

Low magnification: massive necrosis, infiltration of lobules and portal areas, bile duct proliferation

High magnification: cellular exudate consists chiefly of mononuclear cells and plasma cells; complete loss of parenchymal cells

Reticulum stain: reticular network intact

**Fig. 15.3** Pathology of fulminant hepatitis B infection.

## Chronic Infection

Age of infection is inversely correlated with the risk of chronic infection. The risk of developing chronic infection is 90% among newborns infected perinatally, 50% for children infected at 1 year of age, 20% for children infected at 1 to 4 years of age, and 5% to 7% for older children and adults. Chronic infection may lead to cirrhosis and hepatocellular carcinoma, both of which occur more commonly in those infected as infants as compared with adults. Patients with chronic HBV infection are susceptible to superinfection with HDV, which increases the risks of development of cirrhosis and of mortality.

## PREVENTION AND CONTROL

### Nonvaccine Measures

Blood and body fluids from patients with acute or HBV infection are infectious, and appropriate measures should be implemented to prevent transmission to other people. Standard precautions should be followed for hospitalized patients. HBV-infected individuals should not share items that may be contaminated with blood, such as toothbrushes or razors; they should also not donate blood, sperm, or organs for transplantation.

### Screening for Hepatitis B Infection

Foreign-born individuals—including international adoptees, immigrants, and refugees from areas with high prevalence of hepatitis B infection (>2%)—should be screened for infection. Serologic testing may also be considered for other high-risk groups, but testing should not delay the initiation of immunizations.

To prevent perinatal HBV transmission, all pregnant women should be tested for HBsAg at an early prenatal visit. If clinical hepatitis develops, an exposure occurs, or other risk factors are present (e.g., IDU), testing should be repeated in late pregnancy. If no prenatal testing is performed or the results are not available, the mother should be tested at the time of delivery. Maternal antiviral therapy during pregnancy further reduces perinatal HBV transmission; hence antiviral therapy has been suggested when maternal HBV DNA is greater than 200,000 IU/mL. The management of infants based on the mother's HBV status is shown in Table 15.2.

### Passive Immune Prophylaxis

Hepatitis B immune globulin (HBIG) is prepared from hyperimmunized donors whose plasma has high concentrations of anti-HBs and may be used to provide short-term protection in specific postexposure settings. Indications include perinatal exposure of infants to mother with unknown status or known hepatitis B infection and cases of specific discrete exposure to HBsAg-positive or unknown-status individuals.

### Active Vaccination to Prevent Infection

Hepatitis B vaccination via the intramuscular route is available for pre- and postexposure prophylaxis and provides long-term protection. Hepatitis B vaccines are produced recombinantly in yeast cell systems. The vaccines contain noninfectious HBsAg, a small amount of yeast protein, and aluminum hydroxide as an adjuvant. Pediatric formulations contain trace or no thimerosal. Adverse effects are generally mild and mainly consist of local tenderness and low-grade fever. After a vaccine

## TABLE 15.2   Infant Hepatitis B Immunoprophylaxis and Vaccine Administration Based on Infant's Birth Weight and Mother's Hepatitis B Status

| Maternal HBsag Status | Infant's Birth Weight ≥2000 g | Infant's Birth Weight <2000 g |
|---|---|---|
| HBsAg positive | Hepatitis B vaccine + HBIG (within 12 hours of birth). Continue vaccine series beginning at 1–2 months of age. | Hepatitis B vaccine + HBIG (within 12 hours of birth). Continue vaccine series beginning at 1–2 months of age. Immunize with four doses of vaccine; do not count the birth dose as part of the series. |
| | Check anti-HBs and HBsAg after completion of vaccine series at typically 9–12 months of age. HBsAg-negative infants with anti-HBs levels ≥10 mIU/mL are protected and need no further interventions. HBsAg-negative infants with anti-HBs levels ≤10 mIU/mL at 9 months of age should be reimmunized with fourth dose and retested. If the infant continues to have anti-HBs levels <10 mIU/mL, two additional doses should be administered followed by retesting. If, after the sixth dose, the anti-HBs levels are <10 mIU/mL, no additional doses are indicated. HBsAg-positive infants need follow-up and monitoring for chronic liver disease. | Check anti-HBs and HBsAg after completion of vaccine series. HBsAg-negative infants with anti-HBs levels ≥10 mIU/mL are protected and need no further interventions. HBsAg-negative infants with anti-HBs levels ≤10 mIU/mL at 9 months of age should be reimmunized with fourth dose and retested. If the infant continues to have anti-HBs levels <10 mIU/mL, two additional doses should be administered followed by retesting. If, after the sixth dose, the anti-HBs levels are <10 mIU/mL, no additional doses are indicated. HBsAg-positive infants need follow-up and monitoring for chronic liver disease. |
| HBsAg unknown | Test the mother for HBsAg immediately on admission for delivery. Hepatitis B vaccine within 12 hours of birth. HBIG within 7 days if mother tests positive for HBsAg; if mother's status remains unknown, some experts recommend HBIG within 7 days. Continue vaccine series beginning at 1–2 months of age based on mother's HBsAg status. | Test the mother for HBsAg immediately on admission for delivery. Hepatitis B vaccine within 12 hours of birth. HBIG within 12 hours if mother tests positive for HBsAg; if mother's status remains unknown, recommend HBIG within 12 hours. Continue vaccine series beginning at 1–2 months of age based on mother's HBsAg status. Immunize with four doses of vaccine; do not count the birth dose as part of the series. |
| HBsAg negative | Hepatitis B vaccine at birth (within 24 hours of birth) before hospital discharge. Continue vaccine series beginning at 1–2 months of age. Follow-up HBsAg and anti-HBs not needed. | Hepatitis B vaccine dose 1 at 30 days of chronologic age if medically stable or at hospital discharge if <30 days. Continue vaccine series beginning at 1–2 months of age. Follow-up HBsAg and anti-HBs not needed. |

anti-HBs, Antibody to HBsAg; HBIG, hepatitis B immune globulin; HBsAg, hepatitis B surface antigen.
Adapted from Pickering LK, Baker CJ, Kimberlin DW, Long SS, eds: *Red Book: 2018-2021 Report of the Committee on Infectious Diseases*, ed 31, Elk Grove Village, IL, 2018, AAP.

series, more than 95% seroconversion is achieved, which results in greater than 90% efficacy, and immunity may last for up to 20 years.

Three HBV single-antigen vaccines are available in the United States: Recombivax HB, Engerix-B, and Heplisav-B (adults only). Pediatric vaccines contain 5 to 10 μg per 0.5 mL dose and adult formulations contain 10 to 20 μg per 1.0 mL dose. High-dose vaccines are available for adult hemodialysis and immunocompromised patients. Pediatric vaccines are given in a three-dose series (at 0, 1 to 2, and 6 months) and are generally interchangeable. A fourth dose may be given if a birth dose was administered. The birth dose must be a single-antigen hepatitis B vaccine formulation. Hepatitis B vaccine is also available as a component of combination vaccines: a diphtheria-tetanus–acellular pertussis–hepatitis B–inactivated polio vaccine (DTaP-HBV-IPV, [Pediarix]) and hepatitis B–hepatitis A vaccine (HBV-HAV, [Twinrix for adults]).

In the United States and most industrialized nations as well as many low-resource countries, routine immunizations are started at birth and completed during infancy. Hepatitis B vaccines are also recommended for all unvaccinated individuals, especially those at increased risk of exposure to HBV, including travelers to countries where HBV infection is endemic. Health care workers should have documented HBV immunization because of the increased risk of percutaneous exposure in healthcare settings. All close contacts of individuals with chronic HBV infection should receive hepatitis B vaccinations.

"Catch-up" immunization for HBV may be performed at any age for unimmunized individuals. Booster doses of HBV vaccine beyond the initial series are generally not recommended. The long incubation period of HBV theoretically allows for the development of a protective anamnestic immune response after exposure.

### Postexposure Prophylaxis

Postexposure prophylaxis may be administered to unvaccinated individuals after an identified percutaneous or sexual exposure to HBV. The regimen includes prompt initiation and completion of the appropriate doses of the hepatitis B vaccine series for those previously unimmunized as well as the prompt (ideally within 24 hours) administration of HBIG.

### EVIDENCE

AAP Committee on Infectious Diseases and AAP Committee on Fetus and Newborn Policy Statement. Elimination of perinatal hepatitis B: providing the first vaccine dose within 24 hours of birth. *Pediatrics* 2017;

140(3):e20171870. *This resource provides information on efforts towards elimination of perinatal hepatitis B.*

Kowdley KV, Wang CC, Welch S et al. Prevalence of chronic hepatitis B among foreign-born persons living in the United States by country of origin. *Hepatology* 2012;56(2):422. *This resource provides information on prevalence of chronic Hepatitis B in the United States.*

Schille S, Murphy T, Sawyer M et al. CDC guidance for evaluating health-care personnel for hepatitis B virus protection and for administering postexposure management. *MMWR Recomm Rep* 2013;62(RR-10):1. *This resource provides additional information about recommendations for evaluating immunity in health care professionals and managing accidental exposures.*

## ADDITIONAL RESOURCES

Blach S, Razavi H, Razavi-Shearer D et al, World Health Organization. Global hepatitis report, 2017. Geneva, Switzerland: World Health Organization; 2017. *This resource provides worldwide epidemiology of hepatitis B.*

Pickering LK, Baker CJ, Kimberlin DW, Long SS. *Red Book: 2018-2021 Report of the Committee on Infectious Diseases*, ed 31, Elk Grove Village, IL, 2018, American Academy of Pediatrics. *This resource provides abbreviated information about the diagnosis, management, and prevention of hepatitis B infections in infants, children, and adolescents.*

Schillie S, Vellozzi C, Reingold A et al. Prevention of hepatitis B virus infection in the United States: Recommendations of the Advisory Committee on Immunization Practices. *MMWR* 2018;67(1):1-31. *This resource provides recommended strategies towards hepatitis B elimination.*

Terrault N, Lok A, McMahon B et al, American Association for the Study of Liver Diseases (AASLD) Practice guidelines are available for the treatment of chronic hepatitis B and can be found on this site: http://www.aasld.org/publications/practice-guidelines-0external icon. *This resource provides recommendations on indications of treatment and follow up needed in chronic hepatitis B patients.*

# 16

# Human Papillomavirus Infections and Prevention

*Natasha A. Nakra, Dean A. Blumberg*

## ABSTRACT

Human papillomavirus (HPV) is the most common sexually transmitted infection (STI) worldwide and is the etiologic agent of cervical cancer, the majority of anogenital and oropharyngeal cancers, and genital warts. In the United States, vaccination to prevent HPV infection was first introduced in 2006 and is now routinely recommended for adolescent males and females aged 11 to 12 years to prevent morbidity and mortality from cervical and anogenital cancers and genital warts. Initial studies since vaccine implementation have demonstrated a reduction in prevalence of HPV vaccine types, genital warts, and cervical cancer precursors, despite moderate vaccine coverage rates.

## CLINICAL VIGNETTE

A 12-year-old healthy girl presents to her pediatrician's office for a well child visit prior to 7th grade entry. Her physical exam is normal, and she has no complaints. The pediatrician discusses with her mother that she should receive her routine adolescent vaccinations at today's visit. Her mother expresses concern about the human papillomavirus (HPV) vaccine, stating that she is worried that her daughter may exhibit more risk-taking sexual behaviors if she is vaccinated. She is also worried about the safety of the vaccine, due to reports she has heard on social media. The pediatrician responds that she understands the mother's concerns but that she strongly recommends the vaccine. She informs the girl's mother that studies have shown effectiveness at reducing infection with HPV types that cause cervical cancer and precursors to cervical cancer, but the vaccine must be given prior to sexual activity for it to be effective. The pediatrician reassures her that studies have not demonstrated increased risk-taking behaviors in vaccinated adolescents and that the vaccine is highly safe, with the most common reactions being fever and local reactions. The girl's mother ultimately consents to the vaccine, and the girl tolerates it without any significant adverse events.

## ETIOLOGY

HPV is a nonenveloped, double-stranded deoxyribonucleic acid (DNA) virus with an icosahedral nucleocapsid composed of structural proteins L1 and L2. The L1 protein comprises 85% of the nucleocapsid and is the basis for the current HPV vaccine. Viral proteins E6 and E7 are implicated in malignant cellular transformation through inactivation of proteins encoded by host tumor-suppressor genes *p53* and retinoblastoma *(RB),* respectively.

There are more than 200 types of HPV, and different types demonstrate tropism for specific tissues. They can be classified as high risk, indicating association with cancer, or low risk, which are not associated with malignancy. HPV types 1 to 4 are low-risk types that cause skin warts. HPV types 6 and 11 are low-risk types that cause 90% of genital warts, as well as respiratory tract papillomas. Types classified as high risk cause premalignant lesions and may progress to cause cervical and

other anogenital cancers, as well as head and neck cancers. High-risk types 16 and 18 cause the majority of cervical and other cancers. Other high-risk types include types 31, 33, 35, 39, 45, 51, 52, 56, 58, and 59.

## GEOGRAPHIC DISTRIBUTION AND MAGNITUDE OF DISEASE BURDEN

HPV infection occurs worldwide and is the most common STI. Globally, approximately 500,000 cases of cervical cancer are caused by HPV every year, representing approximately 7% of all cancers of women. Cervical cancer is the fourth most common cause of cancer mortality in women worldwide and is the second most common cause of cancer mortality in women in developing countries. In addition, HPV causes most oropharyngeal, vulvar, vaginal, penile, and anal cancers.

In the United States, approximately 79 million people are actively infected with HPV, with more than 14 million new infections occurring every year. HPV causes approximately 35,000 cancers annually in the United States, of which 21,000 occur in females and 14,000 occur in males. HPV causes approximately 11,000 cases of cervical cancer every year, with a mortality rate of 4000/year. Cervical cancer incidence has been decreasing steadily since the 1970s with the advent of screening. Invasive cervical cancer incidence and mortality rates are higher in Black and Hispanic women as compared with white women. Oropharyngeal cancers are currently the most common HPV-associated malignancy in the United States, with 13,000 cases attributable to HPV annually. HPV is additionally responsible for 6000 anal cancers among men and women per year.

Approximately 350,000 cases of genital warts and 1000 cases of recurrent respiratory papillomatosis (RRP) occur every year in the United States. Direct medical costs related to HPV disease, detection, and management are estimated to be $8 billion annually, second in cost only to human immunodeficiency virus (HIV) among STIs.

Since the introduction of HPV vaccine in 2006, there has been a significant reduction in the proportion of high-grade cervical lesions caused by HPV types 16 and 18, which are included in all three US Food and Drug Administration (FDA)-approved vaccines, as well as the overall incidence of high-grade cervical lesions in women aged 15 to 24 years. Due to the prolonged time between HPV infection and development of cancer, decreases in cancer incidence have not been documented at this time but are likely to be seen in the future.

## RISK FACTORS FOR INFECTION

HPV causes cutaneous and mucosal infections. Primary HPV infection occurs in the basal layer of the epidermis. Minor trauma of the epidermis allows HPV to access the basal layer, where replication initially occurs. Infectious virions are then assembled and released from

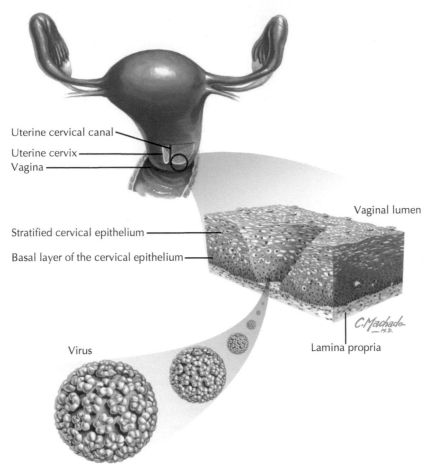

**Fig. 16.1** Human papillomavirus (HPV) cervical infection. HPV initially replicates in the basal cell layer of the cervical epithelium, and HPV progeny virions are then released from cells at the mucosal surface.

the squamous cells of the epidermis, which can then infect others. Naturally thin mucosal layers, such as occur in the cervix or anus, are particularly susceptible to infection (Fig. 16.1). Infection may result from an infectious contact or autoinoculation. Minor trauma may not be clinically apparent and can result from normal sexual intercourse or other skin-to-skin contact activities.

The lifetime risk of cervical HPV infection in women is estimated to be as high as 80%. Sexual behavior influences risk of infection, with increased risk noted in individuals with young age at sexual initiation, increasing number of sex partners, and increasing number of partner's sex partners. Sexual intercourse, genital-to-genital contact, and contact with HPV-contaminated fomites can all lead to anogenital HPV transmission. Oral sex is associated with oropharyngeal acquisition. In the United States, more than 50% of men and women acquire HPV infection within 5 years of initiation of sexual activity. Men who have sex with men are at increased risk for infection and for developing anogenital cancer. Other factors that have been associated with HPV infection include alcohol use, smoking, hormonal contraceptives, inconsistent condom use, uncircumcised male partner, bacterial vaginosis, and coinfection with other STIs, including HIV.

## CLINICAL PRESENTATION

Most HPV infections are asymptomatic, and many associated lesions are self-limited. However, persistent HPV infections, most notably anogenital and head and neck lesions, can lead to cancer. Risk factors for persistent infection include infection with high-risk HPV types,

infection with multiple types, high viral load, host genetic susceptibility, immune compromise, higher parity, smoking, and infection with other STIs.

### Warts

Cutaneous HPV infection causes common skin warts, plantar warts, and flat warts. Mucosal infection with low-risk types, most commonly types 6 and 11, causes genital warts (see Fig. 57.1). If mothers have genital warts or asymptomatic genital HPV infection, newborns may be exposed to the virus during the birth process. Infection of the neonatal respiratory tract can result in a condition known as juvenile-onset RRP, which is benign but associated with significant morbidity. Respiratory papillomas most commonly occur in the laryngeal area, and affected children generally present before age 5 with hoarseness, noisy breathing, and symptoms of airway obstruction.

### Cancers

Virtually all cervical cancers are caused by HPV. HPV also causes approximately two-thirds of vaginal, vulvar, and penile cancers; 90% of anal cancers; and 70% of oropharyngeal cancers, including squamous cell carcinoma of the oral cavity and oropharynx. Persistent cervical HPV infection can lead to mild cervical dysplasia, designated as cervical intraepithelial neoplasia (CIN) I. This lesion may then progress to CIN II (moderate dysplasia) and further to CIN III (severe dysplasia or carcinoma in situ). Most early CIN lesions regress spontaneously, but CIN III is the immediate precursor to invasive cervical cancer. Progression from initial infection to invasive cervical cancer usually occurs over 20 or more years. This prolonged interval provides the opportunity for effective screening and treatment to prevent invasive cervical cancer. Of note, it appears that HPV infection alone does not cause malignant transformation of the infected tissue, but cofactors (e.g., tobacco use, folate deficiency, increased parity, presence of other STIs, and immune suppression including HIV) are essential for progression to malignancy.

## DIAGNOSTIC APPROACH

Please see discussion in Chapter 57.

## TREATMENT

There is no specific antiviral therapy available for HPV infections. In general, treatment is directed against HPV-associated conditions such as warts, cervical dysplasia, or cancer.

### Management of Warts

For treatment of warts, reduction of symptoms is usually the primary goal. Discussion of the specific treatment of warts is included in Chapter 57.

Juvenile-onset RRP is a rare condition due to infection of the neonatal respiratory tract with HPV, which may require multiple surgical procedures throughout childhood to relieve airway obstruction. Recent studies have examined the use of HPV vaccine as a therapeutic agent for treatment of RRP, and a meta-analysis concluded that the need for surgical interventions is significantly reduced following HPV vaccination.

### Prevention of Progression to Cervical Cancer

Periodic Papanicolaou testing (Pap tests or smears) or high-risk HPV polymerase chain reaction (PCR) testing is recommended in women to detect precancerous lesions that may progress to cervical cancer. With the initiation of routine screening with Pap smears in the 1950s, detection and removal of these lesions has resulted in a 70% reduction in cervical cancer mortality in the United States. Further discussion of screening and treatment is included in Chapter 57.

## PROGNOSIS

More than 90% of HPV infections are self-limited. High-risk HPV types that cause persistent infection (detectable for more than 6 to 12 months) are the most important predictor of high-grade cervical cancer precursors. Persistent active infection causes continued cellular proliferation leading to inhibition of cellular differentiation and the development of premalignant and malignant lesions.

## PREVENTION AND CONTROL

### Nonvaccine Measures to Prevent Infection

Abstinence from sexual activity or lifetime mutual monogamy may prevent genital infection. Consistent male condom use results in an approximately 70% reduction in female HPV infection and up to a 50% reduction in male HPV acquisition. Because HPV may infect areas outside of those covered by a condom, condom use does not fully protect against infection of external anogenital areas.

### Vaccination to Prevent Infection

The goal of vaccination against HPV is to prevent HPV infection and thus reduce HPV-associated morbidity and mortality. Currently licensed HPV vaccines consist of virus-like particles (VLPs): empty viral shells composed of recombinant L1 (viral capsid) protein that contain no infectious genetic material. L1 is produced in a heterologous system (yeast), and the recombinant proteins self-assemble into VLPs that mimic the structure of natural virions. The vaccines induce neutralizing antibodies against L1 that prevent HPV cell entry. Immunity to HPV is predominantly type specific, although cross-protection has been documented against other nonvaccine types.

Currently, there is only a single HPV vaccine that is in use in the United States (Gardasil 9, Merck & Co., Kenilworth, NJ). This is a 9-valent HPV vaccine (9vHPV) that contains VLPs against HPV types 16, 18, 31, 33, 45, 52, and 58, which are implicated in precancerous lesions and cancer, as well as types 6 and 11, which are implicated in genital warts. 9vHPV was first approved by the FDA in 2015 and has effectively replaced the previously used quadrivalent vaccine (included types 6, 11, 16, and 18) and bivalent vaccine (included types 16 and 18).

The Advisory Committee on Immunization Practices (ACIP) of the Centers for Disease Control and Prevention (CDC) recommends routine HPV vaccination of all girls and boys at 11 to 12 years of age, with catch-up immunization recommended for individuals aged 13 to 26 years of age who were not adequately vaccinated previously. A two-dose series (0 and 6 to 12 months) is approved for girls and boys who initiate the series at age 9 to 14 years, due to studies demonstrating more robust antibody responses in this younger age population. Adolescent boys and girls who initiate the vaccine series at age 15 or older should receive three doses, on a schedule of 0, 1 to 2 months, and 6 months. In addition, individuals with some primary and secondary immune deficiencies, including HIV infection, should receive a three-dose series. Immunosuppressed individuals may not have as robust of an immune response as immunocompetent subjects but may still benefit from vaccination.

For maximal benefit, vaccination should be administered prior to onset of sexual activity because HPV acquisition typically occurs quickly after sexual debut. Nevertheless, HPV vaccination is recommended for individuals with an equivocal or abnormal Pap test result, positive HPV DNA test result, or genital warts, because past infection is not likely with all HPV types included in the vaccines; therefore vaccination should provide protection against some HPV types.

In 2018 the FDA expanded the approved age range to 9 to 45 years. However, catch-up vaccination is not routinely recommended for adults aged 27 to 45 years, because HPV acquisition generally occurs earlier in life, leading to lower vaccine effectiveness and population benefit in this older age group. The ACIP currently recommends that providers engage in shared decision-making with adult patients aged 27 to 45 as to the potential benefit of HPV vaccination.

Precautions and contraindications for HPV vaccine administration include moderate or severe illness (defer until clinical improvement) and a history of immediate hypersensitivity or severe allergic reaction to yeast or any vaccine component. HPV vaccines should not be administered during pregnancy, although no adverse events have been reported following inadvertent vaccination of a pregnant woman.

9vHPV is generally well tolerated, with the most common adverse events being local reactions (injection site–related pain, swelling, and erythema). These adverse events are generally mild to moderate and self-limited. Systemic events, such as fever, headache, and nausea, occur in approximately 30% of vaccinees. Syncope is a rare but potentially serious outcome following vaccination and is not specific to HPV vaccine; observation for at least 15 minutes after vaccination is recommended to prevent injury from falling.

Studies of quadrivalent and bivalent HPV vaccine have shown efficacy in preventing the acquisition and persistence of high-risk HPV types. In addition, vaccine studies in countries with high vaccine uptake have demonstrated a reduction of greater than 90% of precancerous cervical lesions caused by vaccine types, even up to 10 years after initial immunization. Quadrivalent vaccine has been shown to be effective in reducing incidence of genital warts in males and females; in one US study, there was a decrease in the incidence of genital warts of 77% with three vaccine doses as compared with unvaccinated individuals. Of note, studies in countries with high uptake have shown that unimmunized individuals also have decreased rates of HPV acquisition, genital warts, and HPV-associated precancerous cervical lesions, likely due to herd immunity. In Australia, which has had a national vaccination program since 2007 with high vaccine uptake in both girls and boys, current modeling suggests that cervical cancer elimination (defined as ≤4 new cases/100,000 women each year) will be achieved by the year 2028.

9vHPV has been shown to be noninferior to quadrivalent vaccine in eliciting antibody responses against types 6, 11, 16, and 18 and additionally is 97% effective in generating antibody responses against the remaining five types. Based on estimates, 92% of cancers attributable to HPV in the United States are due to types included in 9vHPV.

Despite the known benefits of HPV vaccine, 2018 estimates from the CDC indicate that only 68% of adolescents in the United States have received one or more HPV vaccines, and only 51% have completed the HPV series. Uninsured status, white race, and residence in a rural area are associated with lower rates of vaccination. Studies suggest that parents are more willing to accept vaccination for their children if health care providers deliver a "clear and unambiguous" message about the benefits of HPV vaccine for both boys and girls. In addition, the American Academy of Pediatrics (AAP) recommends that providers should present the HPV vaccine as part of the routine adolescent vaccine schedule, rather than an optional or special vaccine.

Despite initial public concern that administration of HPV vaccine to young adolescents would result in increased risk-taking sexual behaviors, this has not been borne out over time.

## EVIDENCE

Arbyn M, Xu L. Efficacy and safety of prophylactic HPV vaccines. A Cochrane review of randomized trials. *Expert Rev Vaccines* 2018;17:1085-1091. *This article critically reviews the evidence demonstrating that HPV vaccines protect against cervical precancerous lesions, including CIN grade 2 and 3 and adenocarcinoma in situ.*

Markowitz LE, Gee J, Chesson H, Stokley S. Ten years of human papillomavirus vaccination in the United States. *Acad Pediatr* 2018;18:S3-S10. *This article reviews HPV vaccination policy, implementation, safety, and effectiveness in the U.S. since introduction of HPV vaccine in 2006.*

## ADDITIONAL RESOURCES

Petrosky E, Bocchini JA, Hariri S et al. Use of 9-valent human papillomavirus (HPV) vaccine: updated HPV vaccination recommendations of the Advisory Committee on Immunization Practices. *MMWR Morb Mortal Wkly Rep* 2015;64:300-304. *This resource reviews the evidence demonstrating non-inferior efficacy and immunogenicity of 9-valent HPV vaccine as compared to quadrivalent and bivalent vaccines, and provides specific recommendations regarding 9vHPV vaccine use, including in special populations.*

World Health Organization. Human papillomavirus vaccines: WHO position paper, May 2017. *Weekly Epidemiologic Record* 2017;92:241-268. *This is an excellent overview of the worldwide epidemiology of HPV-related diseases and HPV vaccine efficacy, safety, and cost-effectiveness.*

# Skin and Soft-Tissue Infections

*Dennis L. Stevens*

# 17

# Introduction to Skin and Soft-Tissue Infections

*Dennis L. Stevens*

## ABSTRACT

Skin and soft-tissue infections are among the most common human afflictions, and descriptions from the past few centuries suggest that many of these have the same clinical presentations today as historically described. Regardless of the cause, location, or chronicity, the cardinal manifestations of inflammation (heat, swelling, erythema, and tenderness or pain) are invariably present. Infections of the skin and/or subcutaneous tissues are highly diverse in respect to cause, location, incidence, systemic manifestations, severity, and complications. They may occur as single or recurrent episodes. Many cases are mild or self-limited, but some progress to cause scarring, loss of digits or limbs, or even death. Although skin and soft-tissue infections are common, the causes are diverse, and therefore the clinician faces an immense challenge in establishing a specific diagnosis and prescribing definitive treatment. Important considerations in evaluating patients are the following:

- The patient's symptoms
- The general appearance of the infected site
- Historical clues such as contact with insects or animals, especially involving bites, travel to specific geographic areas, occupation, use of a hot tub, or having a home aquarium
- The immune status of the host
- Chronicity
- Anatomic distribution

Anatomic considerations are paramount to the correct diagnosis. In general, vesicular lesions (herpes simplex) and crusted lesions are the most superficial and may be either impetigo (*Staphylococcus aureus* or group A *Streptococcus*) or dermatomycosis caused by a variety of fungi. Erysipelas is also superficial, caused by group A *Streptococcus*, and characterized by brilliant red color, well-demarcated edges, and pain. None of these superficial infections is associated with scarring. Localized soft-tissue infections (folliculitis, furuncles, carbuncles, and abscesses) involve the epidermis and dermis and are usually purulent in nature and caused by *S. aureus,* including methicillin-resistant *S. aureus* (MRSA). Cellulitis is caused by a variety of streptococci and affects the subcutaneous tissues and is characteristically a pinkish erythema with less distinct borders than erysipelas. Necrotizing infections invariably involve the fascia (necrotizing fasciitis) or muscle (gas gangrene and clostridial myonecrosis). In general, these necrotizing infections may ultimately involve all the layers of the soft tissue, particularly if there is a delay in recognition and treatment.

If the diagnosis cannot be established based on the history, symptoms, and signs, then needle aspiration, biopsy, or surgical exploration may be necessary to obtain specimens for appropriate staining and culture. This is extremely important in those patients with systemic signs of infection or those with historical or observational evidence of rapid progression. With the emergence of antimicrobial resistance among many of these microbes, treatment (particularly for severe infections) should be based on the results of Gram stain and cultures and antimicrobial susceptibility.

## EPIDEMIOLOGY

Although the exact incidence of these infections in the general population is unknown, they are among the most common infections occurring in all age groups. Some skin and soft-tissue infections are age related—for example, impetigo is more common in children, and erysipelas is more common in older adults. Infections of the skin and soft tissues can be caused by bacteria (including rickettsiae), fungi, viruses, parasites, and spirochetes. There are hundreds of possible causative agents, with two common species of gram-positive cocci being the predominant causes of skin and soft-tissue infections—*S. aureus* and *Streptococcus pyogenes*—although other streptococci, such as groups B, C, and G streptococci, are being recognized more commonly. Staphylococcal skin infections are most commonly associated with infections that begin in hair follicles and cause a variety of infections ranging from folliculitis, furuncles, carbuncles, and subcutaneous abscesses. Staphylococcal infections usually have a focal central collection of pus with surrounding erythema. Some practitioners have advocated referring to these types of infections as *purulent cellulitis*. It is unclear if the surrounding erythema is indeed cellulitis (infection in the skin) or merely sympathetic inflammation. In contrast, streptococci more commonly cause erysipelas or cellulitis. Both cause a diffuse erythematous infection not associated with purulent drainage. Thus streptococcal infections should be called *nonpurulent cellulitis*. It should be noted that staphylococcus and group A *Streptococcus* can both cause impetigo, which is superficial infection of the keratin layer of the skin associated with crusting lesions that occasionally weep golden brown fluid. In the past, many terms have been used to describe rapidly destructive lesions of the skin and soft tissues, and such nomenclature has not been helpful to the clinician. Gas gangrene and necrotizing fasciitis are examples of such devastating infection. Although gas gangrene is generally also referred to as *clostridial myonecrosis*, in fact most cases involve all the layers of the soft tissue, including muscle, fascia, subcutaneous tissue, and skin. Similarly, necrotizing fasciitis can be caused by mixed aerobic and anaerobic bacteria (polymicrobial) or monomicrobial. The latter type is most commonly caused by group A *Streptococcus*, but cases involving *Aeromonas hydrophila* and *Vibrio vulnificus* have also been described. As with gas gangrene, necrotizing fasciitis commonly destroys not only fascia but also muscle, subcutaneous tissue, and skin. Thus rightfully these should both be reclassified as simply necrotizing soft-tissue infections. Because these necrotizing soft-tissue infections are rapidly progressive and cause extensive mortality and morbidity, an aggressive approach to diagnosis and treatment is mandatory.

Skin and soft-tissue infections caused by newly recognized or previously rarely encountered microbes are continually being described in immunocompromised patients, especially those with acquired immunodeficiency syndrome (AIDS).

Several noninfectious diseases can mimic infection of the soft tissues—for example, contact dermatitis, pyoderma gangrenosum, gout, psoriatic arthritis with distal dactylitis, Reiter syndrome, relapsing polychondritis, or mixed cryoglobulinemia secondary to immune complex disease from chronic hepatitis C or B virus infection. All may manifest with erythematous rashes, with or without fever.

## PATHOGENESIS

The integument is an organ that reacts to noxious, infectious, external, and internal stimuli in a limited number of ways. It is therefore not surprising that infection can be mimicked by the noninfectious inflammatory conditions listed previously. The rich plexus of capillaries beneath the dermal papillae provides nutrition to the stratum germinativum and the dermatocytes, which are bound together by tight junctions and form the barrier to microbial invasion. Once microbes have penetrated this barrier through a hair follicle, cut, or bite, the dermal plexus of capillaries delivers the components of the host's defense—oxygen, complement, immunoglobulins, macrophages, lymphocytes, and granulocytes—to the site of infection. At some locations, factors such as pressure, thrombosis, or drugs may reduce or stop blood flow, resulting in inadequate oxygenation. If tissue perfusion is moderately attenuated, tissues may remain viable, but the threshold for progression of infection may be lowered. Predisposing conditions in this category include the following:

- Peripheral vascular disease affecting large arteries
- Diabetes mellitus causing microvascular disease
- Chronic venous stasis causing postcapillary obstruction

Necrosis of the skin and deeper tissue may occur if there is severe hypoxia. The following are two examples:

- Pressure necrosis resulting in decubitus ulcers
- Compartment syndromes resulting in hypoxia and then necrosis in muscles confined within tight fascial bundles

When the host is physiologically, structurally, and immunologically normal, only certain pathogens such as *S. aureus* and group A streptococci are able to cause disease by virtue of their potent virulence factors, such as toxins, capsules, or dermonecrotic enzymes. This statement is supported by the observation that normal skin, although constantly exposed to many indigenous and exogenous microbes, rarely becomes infected. In contrast, patients who have compromised skin integrity (e.g., patients with burns), vascular defects (e.g., those who have diabetes mellitus or pressure ulceration), or immunologic deficits may become infected with either virulent organisms (e.g., staphylococci or streptococci) or microbes that are usually saprophytic, such as *Pseudomonas aeruginosa, Escherichia coli,* enterococci, or *Fusarium* species.

## PREVENTION

Avoidance of cuts, scratches, and other forms of trauma that disrupt the natural barrier function of the skin helps to prevent skin and soft-tissue infections. For example, stopping shaving may prevent recurrent folliculitis in the beard area (sycosis barbae). Prompt cleansing, debridement, and disinfection of such lesions are important for preventing infection, particularly in the case of bite wounds. Treatment of eczema reduces the risk of secondary bacterial infection.

Prevention of recurrent folliculitis or furunculosis is difficult to achieve, but there has been some success using intranasal applications of bacitracin or mupirocin ointment. Hexachlorophene or chlorhexidine soaps may be tried to eliminate or reduce staphylococcal carriage in adults. Prophylaxis with systemic antibiotics is of doubtful efficacy and can result in the emergence of resistant strains; it should be tried only for severe cases. Finally, recurrent bacterial cellulitis of the lower extremities can often be prevented by topical antifungal treatment for dermatophyte infections such as tinea pedis because even minor or inapparent superficial fungal infection can serve as a portal of entry for gram-positive cocci.

# 18

# Impetigo

*Dennis L. Stevens*

## ABSTRACT

Impetigo is a common crusted and superficial infection of the skin that occurs in individuals throughout the world. From the historical perspective, impetigo has occurred most frequently among economically disadvantaged children in tropical or subtropical regions, but it is also prevalent in northern climates during the summer months. Its peak incidence is in children aged 2 to 5 years, although older children and adults, including the homeless and migrant farm workers, may also be affected. There is no sex predilection, all races are susceptible, and impetigo is nearly always caused by beta-hemolytic streptococci and/or *Staphylococcus aureus*. In recent times impetigo caused by *S. aureus*, including methicillin-resistant *S. aureus* (MRSA), has become problematic.

## CLINICAL VIGNETTE

A 6-year-old boy from the Red Lake Indian reservation in the northern part of the United States developed dry scaly lesions over his face and upper left extremity. They have grown in size over the course of 2 weeks. By the time he sought medical care he had also noted that his urine had turned reddish brown in color. A urine test showed that he had red cell casts and high protein levels; blood studies demonstrated elevated serum creatinine, low serum complement, and low albumin content. A kidney biopsy demonstrated enlarged glomeruli with massive influx of monocytes and polymorphonuclear leukocytes. Immunofluorescence studies revealed a coarse deposition of IgM, IgG, and complement on the capillary walls of the glomeruli. He was treated with oral penicillin. Over the course of 10 days, the impetiginous lesions resolved, as did the hematuria. Six months later the creatinine returned to normal, proteinuria resolved, and serum albumin and complement levels returned to normal. Subsequent typing of the group A streptococcus (GAS) isolate revealed that it was M-type 49. Five years later, there was no evidence of kidney dysfunction.

## GEOGRAPHIC DISTRIBUTION

Impetigo occurs in all regions of the world but is most common among children in the tropics. In temperate areas, impetigo has been described in Native American children living on Indian reservations and in children in poor economic conditions in cities. Recent studies in the South Pacific Islands have documented that 65% of children with impetigo had scabies. In adults, impetigo has recently been described in the homeless, migrant farm workers, and travelers returning from tropical areas.

## RISK FACTORS

Studies of streptococcal impetigo performed at the Red Lake Indian Reservation in the 1960s demonstrated that personal hygiene, humidity, and geographic location influence the incidence of disease. Predisposing factors also include minor trauma, abrasions, and insect bites. Colonization of unbroken skin with particular streptococcal strains such as M-type 49 precedes the development of impetiginous lesions by an average of 10 days. Inoculation or colonization of surface organisms into the skin by abrasions, minor trauma, or insect bites then ensues. Over the course of 2 to 3 weeks, proliferation of streptococcal strains results in the classic crusted lesions with yellow serosanguineous drainage. At this stage streptococci may be transferred from the skin and/or impetigo lesions to the upper respiratory tract. In contrast, in patients with staphylococcal impetigo, nasal colonization usually precedes cutaneous disease.

## CLINICAL FEATURES

The cutaneous distribution of impetigo is largely on exposed areas of the body, most often the face and extremities (Fig. 18.1). Interiginous areas are spared. Individual lesions are well localized, but multiple cutaneous lesions may develop and may be either bullous or nonbullous in appearance. The former appear initially as superficial vesicles that rapidly enlarge to form flaccid bullae filled with clear yellow fluid, which later becomes darker, more turbid, and often purulent. Bullae may rupture, leaving a thin brown crust resembling lacquer. The lesions of nonbullous impetigo begin as papules that rapidly evolve into vesicles surrounded by an area of erythema and then into pustules that gradually enlarge and then break down over a period of 4 to 6 days to form characteristic thick-crusted cutaneous lesions. Impetiginous lesions heal slowly and leave depigmented areas but do not form scars. A deeply ulcerated form of impetigo is known as *ecthyma*. Regional lymphadenitis may occur, but systemic symptoms are usually absent.

The appearance of impetiginous lesions provides some clues regarding the etiology. For example, bullous impetigo is usually caused by strains of *Staphylococcus aureus* that produce a toxin causing cleavage in the superficial skin layer. In the past, nonbullous lesions were usually caused by streptococci. Today 50% of cases are caused by staphylococci alone or in combination with streptococci. Streptococci isolated from lesions are primarily group A, but occasionally other serogroups such as C and G are responsible.

## DIAGNOSTIC APPROACH

Impetigo is most commonly diagnosed based on the characteristic dry, crusted skin lesions that weep golden-colored fluid. Gram stain of vesicular fluid or exudates beneath the crusts reveals gram-positive cocci resembling *S. aureus* or GAS. Cultures are invariably positive for one or the other or both. Cultures are more important than ever based on the evolving resistance of *S. aureus* (MRSA) and high-level resistance in specific geographic areas of GAS to erythromycin. The anti–streptolysin O response is weak in patients with

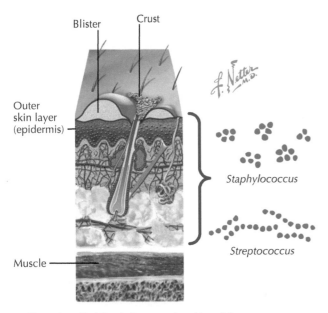

Blister Crust

Outer
skin layer
(epidermis)

*Staphylococcus*

*Streptococcus*

Muscle

Bacteria called *Staphylococcus* (staph) and *Streptococcus* (strep) cause impetigo. They normally live on the skin surface but can go deeper and cause infection.

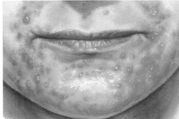

Impetigo starts with painless blisters, usually on the face, especially near the nose and mouth. Blisters fill with clear or yellow fluid and crust over.

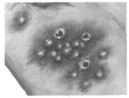

Your doctor makes a diagnosis from the look of the blisters and sores.

**Fig. 18.1** Managing impetigo.

streptococcal impetigo, presumably because skin lipids suppress the streptolysin O response, but anti–DNase B levels are consistently elevated.

Although assays of streptococcal antibodies are of limited value in the diagnosis and management of impetigo, they provide helpful supporting evidence of recent streptococcal infection in patients suspected of having poststreptococcal glomerulonephritis. It must be emphasized that the greatest consequence of impetigo caused by GAS is not the impetiginous lesion but the postinfection sequela, poststreptococcal glomerulonephritis. In times past, there was a strong association between specific strains of GAS that caused impetigo and the development of poststreptococcal glomerulonephritis. In the last 20 years, the incidence of this complication has markedly declined to less than 1 case per 100,000 population. This is likely related to the rarity of the classic M-49 strain (Red Lake strain) in modern times compared with the 1950s era. There is some evidence that impetigo may also be associated with acute rheumatic fever, although it is unclear whether those with impetigo caused by GAS may also be colonized with GAS in the throat.

## CLINICAL MANAGEMENT

In the past, therapy directed primarily at group A streptococci (e.g., penicillin) was successful, both in healing the lesions and decreasing recurrences of nonbullous impetigo for at least several weeks. Because *S. aureus* currently accounts for most cases of bullous impetigo as well as a substantial portion of nonbullous infection, penicillinase-resistant penicillins or first-generation cephalosporins are preferred, although impetigo caused by MRSA is increasing in frequency. Erythromycin has been a mainstay of pyoderma therapy, but its utility may be lessened in areas where erythromycin-resistant strains of *S. aureus,* or more recently *Streptococcus pyogenes,* are prevalent. Topical therapy with mupirocin is equivalent in effect to oral systemic antimicrobials and may be used when lesions are limited in number. It is expensive, however, and resistance not only has developed but is in part related to the general usage of mupirocin in the community. Retapamulin ointment was recently approved by the US Food and Drug Administration

(FDA) for the treatment of impetigo caused by both GAS and *S. aureus.* Similarly, ozenoxacin 1% cream for 5 days has been approved by the FDA and is an effective treatment of impetigo. Bullous impetigo caused by MRSA poses a difficult problem for clinicians. Although empiric treatment with dicloxacillin, amoxicillin-clavulanate, or an oral cephalosporin may be the easiest choice, patients should be observed carefully for evidence of clinical failure as a result of MRSA. If this occurs, trimethoprim-sulfamethoxazole or clindamycin may be effective, although resistance to both agents has been described. In patients with extensive impetigo caused by MRSA, parenteral antibiotics such as vancomycin, daptomycin, or linezolid would occasionally be a reasonable choice based on susceptibility testing. All these agents have been approved by the FDA for complicated skin and soft tissue infections; however, no clinical trials have been performed with these agents for impetigo per se.

## PROGNOSIS

Impetiginous lesions resolve spontaneously with or without antibiotics and do not leave a scar because the infection occurs in the outer keratin layer of skin. Antibiotic treatment topically or orally speeds the rate of resolution. Suppurative complications of streptococcal impetigo are uncommon and, for as yet unexplained reasons, rheumatic fever has never occurred after streptococcal impetigo. On the other hand, cutaneous infections with nephritogenic strains of GAS are the major antecedent of poststreptococcal glomerulonephritis in many areas of the world. No conclusive data indicate that treatment of streptococcal pyoderma prevents nephritis.

## PREVENTION AND CONTROL

General body hygiene is important in preventing impetigo. In addition, minor skin trauma and abrasions in children should be quickly cleansed with soap and water and appropriately covered with sterile dressings. Antibiotic ointment probably aids in prevention because it is useful in shortening the course of established impetigo, yet such studies have not been performed. Topical or oral antibiotics are also important as an epidemiologic intervention to eradicating nephritogenic strains from the community and for reducing transmission within families and day care settings.

## EVIDENCE

Barton LL, Friedman AD, Sharkey AM, et al. Impetigo contagiosa III. Comparative efficacy of oral erythromycin and topical mupirocin. *Pediatr Dermatol* 1989;6:134-138. *First study to demonstrate the efficacy of mupirocin in the treatment of impetigo.*

Bisno AL, Nelson KE, Waytz P, Brunt J. Factors influencing serum antibody response in streptococcal pyoderma. *J Lab Clin Med* 1973;81:410-420. *This clinical study describes low titers of anti–streptolysin O in patients with impetigo.*

Dagan R, Bar-David Y. Comparison of amoxicillin and clavulanic acid (augmentin) for the treatment of nonbullous impetigo. *Am J Dis Child* 1989;143:916-918. *Forty-nine patients with impetigo were treated with either amoxicillin or amoxicillin clavulanate. All patients had S. aureus isolated, and 29% had group A streptococcus. Responses were 71% for amoxicillin and 95% for amoxicillin clavulanate, and the recurrence rate was 26%.*

Demidovich CW, Wittler RR, Ruff ME, et al. Impetigo: current etiology and comparison of penicillin, erythromycin, and cephalexin therapies. *Am J Dis Child* 1990;144:1313-1315. *Excellent study demonstrating the efficacy of different classes of antibiotics in the treatment of impetigo.*

Ferrieri P, Dajani AS, Wannamaker LW. A controlled study of penicillin prophylaxis against streptococcal impetigo. *J Infect Dis* 1974;129:429-438. *An important paper that demonstrated that penicillin prophylaxis could interrupt transmission of impetigo to susceptible patients.*

Ferrieri P, Dajani AS, Wannamaker LW, Chapman SS. Natural history of impetigo. Site sequence of acquisition and familial patterns of spread of cutaneous streptococci. *J Clin Invest* 1972;51:2851-2862. *A classic study among Native Americans of the factors leading to impetigo and the dynamics of spread among family members.*

Kaplan EL, Wannamaker LW. Suppression of the anti–streptolysin O response by cholesterol and by lipid extracts of rabbit skin. *J Exp Med* 1976;144:754-767. *This study demonstrated that the reason that anti–streptolysin O titers are not useful in the diagnosis of impetigo is because skin lipids, especially cholesterol, neutralize the streptolysin O hemolysin.*

Schachner L, Andriessen A, Bhatia N, et al. Topical ozenoxacin cream 1% for impetigo: a review. *J Drugs Dermatol* 2019;18(7):655-661. *A meta-analysis of Ozenoxacin 1% cream in the treatment of impetigo.*

Weinstein L, Le Frock J. Does antimicrobial therapy of streptococcal pharyngitis or pyoderma alter the risk of glomerulonephritis? *J Infect Dis* 1971;124:229-231. *This study reviewed the evidence that antibiotic treatment of impetigo does not affect the development of poststreptococcal glomerulonephritis, whereas treatment of streptococcal pharyngitis does.*

Yang LP, Keam SJ. Spotlight on retapamulin in impetigo and other uncomplicated superficial skin infection. *Am J Clin Dermatol* 2008;9:411-413. *First study to demonstrate that retapamulin is an effective topical agent in the treatment of impetigo caused by staphylococci and GAS.*

Yun HJ, Lee SW, Yoon GM, et al. Prevalence and mechanisms of low- and high-level mupirocin resistance in staphylococcal isolates from a Korean hospital. *J Antimicrob Chemother* 2003;51:619. *An in-depth study demonstrating the emergency of mupirocin resistance among staphylococci and correlating this resistance to the pressure of mupirocin usage.*

## ADDITIONAL RESOURCES

Derrick CW Jr, Dillon HC Jr. Impetigo contagiosa. *Am Fam Physician* 1971;4:75-81. *A classic review of the cause, clinical course, and epidemiology of impetigo.*

Hirschmann JV. Impetigo: etiology and therapy. *Curr Clin Top Infect Dis* 2002;22:42-51. *This is an excellent, in-depth review article on impetigo.*

Parks T, Smeesters PR, Steer AC. Streptococcal skin infection and rheumatic heart disease. *Curr Opin Infect Dis* 2012;25(2):145-153. https://doi.org/10.1097/QCO.0b013e3283511d27. *A potential association with GAS impetigo and rheumatic fever in the South Pacific Islands.*

Osti MH, Sokana O, Phelan S, et al. Prevalence of scabies and impetigo in the Solomon Islands: a school survey. *BMC Infect Dis* 2019;19(1):803. https://doi.org/10.1186/s12879-019-4382-8. *A strong association with scabies and impetigo in the South Pacific Islands.*

Stevens DL, Bisno AL, Chambers HF, et al. Practice guidelines for the diagnosis and management of skin and soft tissue infections: 2014 update by the Infectious Diseases Society of America. *Clin Infect Dis* 2014;59(2):147-159. https://doi.org/10.1093/cid/ciu296. *This is a comprehensive, evidence-based guideline for the treatment of all skin and soft-tissue infections including impetigo.*

# 19

# Erysipelas and Cellulitis

*Dennis L. Stevens*

## ABSTRACT

Pathologically, *cellulitis* is defined as a diffuse area of soft tissue infection characterized by leukocytic infiltration of the dermis, capillary dilatation, and proliferation of bacteria. Clinically, cellulitis is recognized as an acute infection of the skin characterized by localized pain, pinkish erythema, swelling, heat, and a diffuse, indistinct border. Erysipelas is similar to cellulitis but characteristically has fiery red erythema and a distinct border. Cellulitis can be caused by microbes colonizing the skin or by bacteria introduced through animal contact, including bites from dogs, cats, and humans. In the last case, causative microbes are normal inhabitants of the oral flora of the specific animal. Endogenous flora of the human skin that most commonly cause cellulitis are streptococcal species such as groups A, B, C, and G streptococci. Cultures of cellulitic skin are generally positive in bite wound cases but positive in only approximately 25% of cases in which the infection is caused by endogenous flora. This is likely because the streptococci produce a variety of readily diffusible protein toxins that mediate the inflammatory reaction. Thus, relatively few microbes may actually be present in the skin. Clearly, better diagnostic reagents are needed to improve the management of patients with cellulitis.

## CLINICAL VIGNETTE

A 45-year-old woman presented with redness and swelling of her left upper arm that had been present for 2 days. She had chills and a temperature of 39°F. On examination the arm was swollen from the elbow to the axilla. There was diffuse redness and tenderness; her heart rate was 110 beats/min, and blood pressure 120/80 mm Hg. Laboratory tests demonstrated a white blood cell (WBC) count of 11,400/mm$^3$ with increased band forms, normal creatinine, a C-reactive protein (CRP) of 5.6, serum bicarbonate of 24, and a lactic acid level within normal limits. She had had a left mastectomy with axillary node resection. Once she had been admitted to the hospital, supportive measures and intravenous antibiotic treatment with cefazolin were given. Over the course of 2 days, she clinically improved and was discharged on cephalexin orally for 7 days. Indications for admission were fever and a WBC count with left shift. Two months later she returned with the same signs and symptoms, indicating recurrent cellulitis. Blood cultures remained negative. The most common cause is group A streptococcus. The patient was again treated with cefazolin and responded to this treatment over the course of 2 days. She was treated once a week with benzathine penicillin for 4 weeks and had no further episodes. Recurrent cellulitis is a significant problem in female patients with axillary node dissection for breast cancer, male patients with prostate cancer with inguinal node dissection, and patients with elephantiasis. Prolonged treatment with appropriate antibiotics can reduce the chance of recurrence. Group A streptococcus or occasionally group C or G streptococci are usually the cause.

## DISEASE BURDEN, EPIDEMIOLOGY, AND MICROBIOLOGY

The epidemiology of cellulitis is poorly defined. It is not a reportable disease, and establishment of a specific causative diagnosis may be difficult, as described previously. Currently, there are nearly 1 million hospital admissions for skin and soft tissue infections (SSTIs) in the United States annually, and this figure has increased by 29% since 2000. Figures for outpatient cases are not available. Cellulitis is clearly among the more common diagnoses within the category of SSTIs that necessitate hospitalization. Recent clinical trials suggest that cellulitis represents about 30% of complicated SSTIs; however, this figure is likely an underrepresentation, because these trials generally exclude patients who have not had a specific microbe isolated. In addition, such trials enhance the representation of localized abscesses and carbuncles owing to the ease of obtaining culture-positive material. Thus, these trials uniformly suggest that *Staphylococcus aureus* is the most common cause of "culturable" SSTI. Cellulitis and erysipelas are most commonly caused by *Streptococcus pyogenes* and occasionally by streptococci of groups B, C, and G. Establishment of the correct disease burden of cellulitis will require prospective population-based studies and more sensitive microbial detection methods (Fig. 19.1).

## PATHOGENESIS AND RISK FACTORS

Intact healthy skin in nonimmunocompromised hosts is a nearly perfect barrier to infection. Thus cellulitis occurs only when one or more risk factors are present. Minute breaks in the skin barrier are most commonly caused by abrasions, insect bites, burns, splinters, dermatophyte infections particularly of the toe webs, animal or human bites, or surgery. In addition, certain conditions appear to predispose to infection without providing a portal of entry, including chronic venous insufficiency and lymphedema (Table 19.1). Thus cellulitis and erysipelas caused by streptococci are much more common in patients with stasis dermatitis from venous insufficiency and chronic lymphedema from elephantiasis, radical mastectomy, or prostatectomy with regional node dissection.

Virtually nothing is known regarding the host response and microbe virulence factors responsible for pathogenesis. Clearly all the cardinal manifestations of acute inflammation are present in diffuse spreading cellulitis. Because cytokines such as tumor necrosis factor (TNF) and interleukin-1 (IL-1) are important acute response cytokines in general, they likely play an important role in mediating the pain, redness, and heat that are so characteristic of cellulitis and erysipelas. In addition, diffusion of a number of extracellular toxins from a nidus of infection to surrounding tissues undoubtedly also contributes to these signs and symptoms of disease.

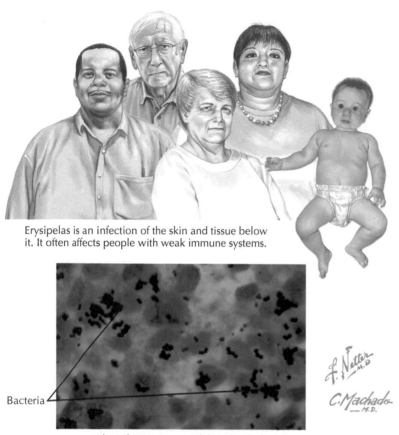

Erysipelas is an infection of the skin and tissue below it. It often affects people with weak immune systems.

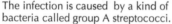

Bacteria

The infection is caused by a kind of bacteria called group A streptococci.

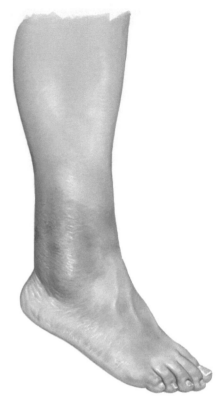

The leg is usually affected. The skin becomes red and swollen, feels tender and hot, and can have blisters. Other symptoms are fever, chills, headache, tiredness, loss of appetite, abdominal pain, and swollen glands.

**Fig. 19.1** Erysipelas.

| TABLE 19.1 | **Cellulitis Associated With Specific Exposures or Underlying Medical Conditions** |
|---|---|
| **Risk Factors** | **Likely Causative Agent** |
| Cat bite | *Pasteurella multocida* |
| Dog bite | *P. multocida, Capnocytophaga canimorsus* (DF-2), *Staphylococcus intermedius* |
| Human bite | *Streptococcus pyogenes, Eikenella corrodens, Staphylococcus aureus, Fusobacterium, Peptostreptococcus, Prevotella, Porphyromonas* |
| Tick bite + erythema chronicum migrans | *Borrelia burgdorferi* |
| Hot tub exposure | *Pseudomonas aeruginosa* |
| Diabetes mellitus | Group B streptococci |
| Periorbital cellulitis (children) | *Haemophilus influenzae* |
| Saphenous vein donor site | Group C and G streptococci |
| Freshwater laceration | *Aeromonas hydrophila* |
| Seawater exposure, cirrhosis, raw oysters | *Vibrio vulnificus* |
| Chronic stasis dermatitis | Group A, C, and G streptococci |
| Lymphedema | Group A, C, and G streptococci |
| Cat scratch | *Bartonella henselae, Bartonella quintana* |
| Tilapia fish | *Streptococcus iniae* |
| Fishmongering, bone rendering | *Erysipelothrix rhusiopathiae* |
| Fishtank exposure | *Mycobacterium marinum* |
| Compromised host + ecthyma gangrenosum | *P. aeruginosa* |
| Compromised host | *Stenotrophomonas hydrophila* |

## CLINICAL FEATURES

### Erysipelas

Erysipelas may be a specific variant of cellulitis but can be distinguished by its precise margin and bright color (Fig. 19.2). Erysipelas involves the outer layer of the epidermis, whereas cellulitis extends into the subcutaneous tissues, which probably explains the more diffuse margins and pinkish color of cellulitis. Erysipelas is characterized by the abrupt onset of fiery red swelling of the face or extremities. It is most common in elderly adults, and the severity of systemic toxicity can vary from region to region. Distinctive features are well-defined margins, particularly along the nasolabial fold; rapid progression; and intense pain. Flaccid superficial bullae may develop during the second to third day of the illness, but extension to deeper soft tissues is rare. Desquamation of the involved skin occurs after 5 to 10 days. Whereas most studies suggest that erysipelas has become less common and less severe now than in the past based on hospitalization records, a recent study from Belgium demonstrated an increase from 1.88 cases per 1000 population to 2.49 cases per 1000 population between 1994 and 2004. Erysipelas is most commonly caused by *S. pyogenes* and less commonly by group C or G *Streptococcus*. The erythema, pain, tenderness, and swelling of erysipelas are likely caused by potent streptococcal toxins and/or the host responses to these toxins. Thus cultures may be negative owing to the relatively low number of microbes that is sufficient to cause infection. Occasionally bullous lesions and throat cultures are positive for group A streptococci even when skin cultures are negative. A rare form of erysipelas is caused by *Campylobacter jejuni* and *Campylobacter fetus* in patients with agammaglobulinemia and occasionally in those with acquired immunodeficiency syndrome (AIDS). Surgical debridement is rarely necessary, and treatment with penicillin is effective. Swelling may progress for a time despite appropriate treatment, even while fever, pain, and the intense

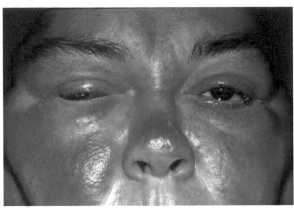

Fig. 19.2 Erysipelas. Note the bright red salmon color of the skin and the well demarcated edge along the nasolabial fold. Erysipelas is a superficial infection and scarring is rare. Infection can occur on the face or lower extremity.

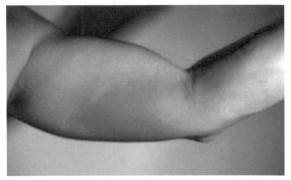

Fig. 19.3 Cellulitis. Note the pinkish color and less distinct edge of the infection. This type of subcutaneous infection is usually caused by streptococci, including groups A, B, C, and G (nonpurulent cellulitis). There is usually no portal of entry. If a purulent center is present, the surrounding erythema is referred to as a purulent cellulitis and the cause is most commonly *Staphylococcus aureus*.

red color are diminishing. A hemorrhagic form of erysipelas caused by group A streptococci has been described in six patients who developed purpura, bullae, and petechiae despite antibiotics but who responded to a combination of antibiotics and corticosteroids.

## Cellulitis

The term *cellulitis* is commonly used by physicians but is not well defined in the literature. Pathologically it is a diffuse area of soft tissue inflammation associated with leukocytic infiltration of the dermis, capillary dilatation, and proliferation of bacteria. Clinically, cellulitis is recognized as an acute inflammatory condition of the skin characterized by localized pain, erythema, swelling, and heat. The area of erythema is a paler pink than the flaming red of erysipelas and has indistinct margins (Fig. 19.3).

Cellulitis is most commonly caused by indigenous flora such as *S. pyogenes* followed by groups B, C, and G streptococci, which colonize the skin and appendages. Bacteria may gain access to the epidermis through cracks in the skin, abrasions, cuts, burns, insect bites, surgical incisions, and intravenous catheters. The erythema surrounding a central localized infection such as a furuncle, carbuncle, or abscess caused by *S. aureus* is in reality likely just an inflammatory reaction to the focal infection but is sometimes said to be cellulitis. Drainage of these abscesses involves opening the central part of the lesion, and usually the incision is not extended into the surrounding erythema. Cultures of the purulent center are invariably positive for staphylococcal species (Fig. 19.4), but cultures of the areas of centripetal erythema are usually negative. The genesis of these localized staphylococcal abscesses is usually secondary to folliculitis or a foreign body (e.g., a splinter, prosthetic device, or intravascular catheter) or surgical procedures. In contrast, streptococcal cellulitis may begin spontaneously with no defined portal of entry and is a more rapidly spreading diffuse process commonly associated with lymphangitis and fever.

### Recurrent Cellulitis and Erysipelas

Recurrent streptococcal cellulitis of the lower extremities may be caused by group A, C, or G streptococci in association with skin lesions such as chronic venous stasis, saphenous venectomy for coronary artery bypass surgery, and lymphedema. A retrospective study recently demonstrated that a recurrence of cellulitis or erysipelas was found in 47% of 171 patients and was positively associated with chronic edema of the affected leg. Streptococci also cause recurrent cellulitis or erysipelas in patients with chronic lymphedema resulting from irradiation, lymph node dissection (breast and prostatic cancer), Milroy disease, or elephantiasis. Finally, among 47 patients with an average of 4.1

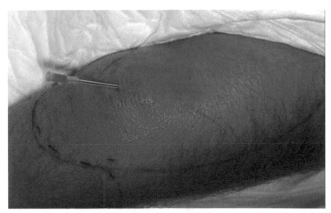

Fig. 19.4 Purulent cellulitis. The focus of infection in this elbow is the olecranon bursa, and the organism is methicillin-resistant *Staphylococcus aureus*.

recurrent episodes of erysipelas, disruption of the cutaneous barrier was found in 81% and had usually led to intertrigo (60%).

In a recent study, patients with at least three recurrences of erysipelas of the upper extremity after radical mastectomy for breast cancer were treated with 2.4 million units of benzathine penicillin given intramuscularly every 2 weeks. The mean erysipelas-free period was 2.7 years and the estimated recurrence rate at 1 year was 26%. Thus there was a therapeutic benefit, but recurrences did arise despite parenteral prophylaxis. A second study investigating recurrences after an initial episode of erysipelas identified a recurrence rate of around 8% of patients despite a variety of antibiotic prophylactic treatments. Noncompliance and inappropriate antibiotics were the major causes of failure.

## DIAGNOSIS

The cause of cellulitis and erysipelas can be suspected on the basis of the epidemiologic data supplied earlier. If there is drainage, an open wound, or an obvious portal of entry, Gram stain and culture can often provide a definitive diagnosis. In the absence of these findings, the bacterial cause of cellulitis may be difficult to establish. Even with needle aspiration from the leading edge or punch biopsy of the cellulitis itself, cultures are positive in only 20% of cases. This suggests that relatively low numbers of bacteria may cause cellulitis and that the expanding area of erythema within the skin may be the direct result of extracellular toxins or the soluble mediators of inflammation elicited by the host.

## DIFFERENTIAL DIAGNOSIS

Staphylococcal infections of deeper tissues may also cause superficial redness, warmth, and swelling of the skin, even though the skin itself is not infected. Examples include olecranon bursitis, septic arthritis, osteomyelitis, and staphylococcal parotitis. Acute gout can also be associated with red, hot, and tender skin overlying the affected joint and can therefore mimic cellulitis. Other deep infections of the head and neck—such as anaerobic infections, actinomycosis, and tooth abscesses—can also be confused with cellulitis. Deep infections such as staphylococcal or streptococcal myositis, necrotizing fasciitis, or gas gangrene may resemble cellulitis initially, but in these severe infections marked evidence of systemic toxicity is usually present. The bite of the brown recluse spider resembles acute infection soon after the bite but rapidly progresses to localized tissue destruction and central necrosis from the action of its dermonecrotic toxins. These infections may resemble pyoderma gangrenosum or may become secondarily infected with skin organisms. Finally, cutaneous allergic reactions to soaps, topical agents, insect bites, and so on can have the appearance of cellulitis and, of course, may serve as a portal of entry for indigenous flora and become secondarily infected.

## MANAGEMENT AND ANTIBIOTIC TREATMENT

In contrast to staphylococcal cutaneous abscesses, surgical debridement of streptococcal cellulitis is rarely necessary and treatment with penicillin is effective. Swelling may progress for 24 to 36 hours despite appropriate treatment. However, fever, pain, and the intense red color diminish. Commonly, the red to pink color of cellulitis becomes more of a reddish blue color after 24 hours of appropriate antibiotic treatment. Thus these findings are indicative of a good response to treatment. In contrast, patients who manifest continued swelling, increased pain, and persistent fever and have evidence of systemic toxicity—such as rising serum creatinine, leukocytosis, tachycardia, or hypotension—should be emergently evaluated for deeper infection such as septic joint, necrotizing fasciitis, or gas gangrene.

### Empiric Choices of Antibiotics for the Treatment of Cellulitis

Because many different microbes can cause cellulitis, the choice of initial empiric antibiotic therapy depends on the risk factors, exposures, and clinical features described earlier. Once cultures and sensitivities have become available, the choice is easier and more specific. The physician must first decide whether the patient's illness is severe enough to require parenteral treatment, either in the hospital or on an outpatient basis.

### Presumed Streptococcal or Staphylococcal Cellulitis Requiring Hospitalization

For presumed streptococcal or staphylococcal cellulitis, nafcillin, cephalothin, cefuroxime, vancomycin, or erythromycin is a good choice. Cefazolin and ceftriaxone have less activity against *S. aureus* than cephalothin, although clinical trials have shown a high degree of efficacy. Ceftriaxone is a useful choice for outpatient treatment because it can be given once daily. Similarly, teicoplanin and vancomycin have excellent activity against *S. pyogenes* and *S. aureus*, but teicoplanin, unlike vancomycin, may be given once daily by intravenous or intramuscular injection. Because of the virtual epidemic of methicillin-resistant *S. aureus* (MRSA) infections worldwide, severe soft tissue infections in patients who are toxic, in those who have been recently hospitalized, or in those who previously received antibiotics should be treated with agents that have high-level activity against these strains. Reasonable choices would include vancomycin, tigecycline, linezolid, tedizolid,

daptomycin, teicoplanin, and telavancin. Appropriate cultures and sensitivities as well as local antibiograms are thus crucial for treatment rationale.

### Empiric Choices of Antibiotics for the Outpatient Treatment of Cellulitis

For patients with less severe infections, treatment with an oral antibiotic such as dicloxacillin, cefuroxime axetil, cefpodoxime, erythromycin, clarithromycin, or azithromycin is effective.

For known group A, B, C, or G streptococcal infections, penicillin or erythromycin should be used orally or parenterally. For serious group A streptococcal infections such as necrotizing fasciitis or streptococcal toxic shock syndrome, clindamycin is more efficacious than penicillin. This is probably because in this type of infection, in which there are large numbers of bacteria, streptococci are in a stationary growth phase and do not express a full complement of penicillin-binding proteins. In contrast, the activity of clindamycin is not affected by inoculum size or growth phase. In addition, clindamycin suppresses the synthesis of many streptococcal exotoxins and surface proteins. In situations in which MRSA coverage is necessary, clindamycin has been used effectively in children, although clindamycin resistance is increasingly becoming problematic. Although trimethoprim and tetracycline have been used for the oral treatment of minor MRSA infections after incision and drainage (lesions less than 5 cm in diameter), these antibiotics do not have predictable in vitro activity against streptococci. Thus, if empiric treatment demands coverage for both MRSA and streptococci, trimethoprim (Septra) plus penicillin or an oral cephalosporin may be necessary. Linezolid and tedizolid had excellent activity against both and they are oral agents, but the expense is a major consideration.

### Treatment of Cellulitis Caused by Unusual Microbes

For cellulitis associated with *Eikenella corrodens*, useful antibiotics are penicillin, ceftriaxone, trimethoprim-sulfamethoxazole, tetracyclines, and fluoroquinolones. Interestingly, this organism is resistant to oxacillin, cefazolin, clindamycin, and erythromycin. Cellulitis associated with cat bites may fail to respond to treatment with oral cephalosporins, erythromycins, and dicloxacillin. Reasons for failure include resistance of *Pasteurella multocida* to oxacillin and dicloxacillin and the inadequate serum and tissue levels attained with older oral cephalosporins and erythromycins. *P. multocida* is resistant to dicloxacillin and nafcillin but sensitive to all other β-lactam antimicrobials as well as quinolones, tetracycline, and erythromycin. Ampicillin-clavulanate, ampicillin-sulbactam, and cefoxitin are good choices for treating animal or human bite infections. *Aeromonas hydrophila* is sensitive to aminoglycosides, fluoroquinolones, chloramphenicol, trimethoprim-sulfamethoxazole (co-trimoxazole), and third-generation cephalosporins but is resistant to ampicillin. Rifampin (rifampicin) plus ethambutol has been an effective treatment for *Mycobacterium marinum*, although no comprehensive studies have been carried out. In addition, some strains of *M. marinum* are also susceptible to tetracycline or trimethoprim-sulfamethoxazole. The gram-positive aerobic rod *Erysipelothrix rhusiopathiae*, which causes cellulitis in bone renderers and fishmongers, remains susceptible to erythromycin, clindamycin, tetracycline, and cephalosporins but is resistant to sulfonamides and chloramphenicol.

Soft tissue infections caused by *Pseudomonas aeruginosa* are seen most often in compromised hosts, burn patients, or those with hot tub exposure. In addition, *P. aeruginosa* may be introduced into the deep tissues by stepping on a nail, a scenario referred to as the "sweaty tennis shoe syndrome." Treatment includes surgical inspection and drainage, particularly if the injury also involves bone or the joint capsule.

Choices for the empiric treatment of *P. aeruginosa* SSTIs pending antimicrobial susceptibility data include aminoglycosides, third-generation cephalosporins such as ceftazidime, cefoperazone, or cefotaxime, semisynthetic penicillins such as piperacillin-tazobactam, penem compounds, and fluoroquinolones. (The quinolones are not approved for use in children younger than 13 years of age.)

*Stenotrophomonas maltophilia* has recently emerged as an important cause of nosocomial cellulitis in patients who have cancer. The bacterium has been isolated from incubators, nebulizers, humidifiers, and tap water in hospitals. The cellulitis may be related to intravenous catheters and in some circumstances may be metastatic via the bloodstream. Trimethoprim-sulfamethoxazole or ticarcillin–clavulanic acid with or without ciprofloxacin is a reasonable treatment choice, although cultures and sensitivities are important because of the high prevalence of antibiotic-resistant organisms in the health care environment.

Further information regarding the diagnosis and treatment of common and uncommon SSTIs, including those in compromised hosts, can be found in "Practice Guidelines for Diagnosis and Management of Skin and Soft-Tissue Infections." See Additional Resources.

# EVIDENCE

Bartholomeeusen S, Vandenbrouchke J, Truyers C, Buntinx F. Epidemiology and cormorbidity or erysipelas in primary care. *Dermatology* 2007;215:118-122. *Among outpatients in Belgium there was an increase in the incidence of erysipelas from 1.88 per 1000 patients to 2.49 per 1000 patients between 1994 and 2004.*

Cox NH. Oedema as a risk factor for multiple episodes of cellulitis/erysipelas of the lower leg: a series with community follow-up. *Br J Dermatol* 2006;155:947-950. *Among 171 patients with cellulitis or erysipelas, 47% had recurrent episodes and 46% had chronic leg edema.*

Ezzine Sebai N, Hicheri J, Trojjet S, et al. Systemic corticosteroids and their place in the management of hemorrhagic erysipelas. *Tunis Med* 2008;86:49-52. *In six patients, erysipelas associated with bullae, ecchymoses, and petechiae progressed despite antibiotics alone but responded when corticosteroids were added.*

Koster JB, Kullberg BJ, van der Meer JW. Recurrent erysipelas despite antibiotic prophylaxis: an analysis from case studies. *Neth J Med* 2007;65:89-94. *There were approximately 8% recurrences in 117 patients with erysipelas despite antibiotic prophylaxis. Noncompliance or inappropriate antibiotic with improper dosage was the main reason for failure.*

Leclerc S, Teixeira A, Hahe E, et al. Recurrent erysipelas: 47 cases. *Dermatology* 2007;214:52-57. *These authors demonstrated that the major risk factor for recurrence of erysipelas or cellulitis was interruption of skin integrity, usually but not exclusively as a result of intertrigo.*

Tirupathi R, Areti S, Salim SA, et al. Acute bacterial skin and soft tissue infections: new drugs in ID armamentarium. *Journal of Community Hospital Internal Medicine Perspectives* 2019;9(4):310-313. https://doi.org/10.1080/20009666.2019.1651482. *A review of new antimicrobial agents approved for treatment of bacterial infections.*

Vignes S, Dupuy A. Recurrence of lymphoedema-associated cellulitis (erysipelas) under prophylactic antibiotherapy: a retrospective cohort study. *J Eur Acad Dermatol Venereol* 2006;20:818-822. *This study showed that recurrence of erysipelas of the upper extremity could be reduced in frequency but not absolutely prevented by use of benzathine penicillin given intramuscularly every 2 weeks in women who had undergone radical mastectomy.*

Yamamoto Y, Yamamoto S. Achenbach's syndrome. *The New England Journal of Medicine* 2017;376(26):e53. https://doi.org/10.1056/nejmicm1610146. PMID:28657879. *Clinical trial for treatment of minor purulent infections comparing incision and drainage alone versus incision and drainage plus either clindamycin or trimethoprim/sulfamethoxazole for abscesses less than 5 cm in diameter. Fewer recurrences with clindamycin compared to incision and drainage alone.*

# ADDITIONAL RESOURCES

Bisno AL, Stevens DL. Streptococcal infections in skin and soft tissues. *N Engl J Med* 1996;334:240-245. *A comprehensive review of all skin and soft tissue infections caused by group A Streptococcus.*

Stevens DL, Bisno AL, Chambers HF, et al. Practice guidelines for the diagnosis and management of skin and soft tissue infections: 2014 update by the Infectious Diseases Society of America. *Clin Infect Dis* 2014;59(2):147-159. https://doi.org/10.1093/cid/ciu296. PMID:24947530. *Consensus statement of the Infectious Diseases Society of America on the diagnosis, clinical manifestations, management, and antibiotic treatment of skin and soft tissue infections.*

Stevens DL, Bryant AE. Necrotizing soft-tissue infections. *N Engl J Med* 2017;377(23):2253-2265. https://doi.org/10.1056/NEJMra1600673. PMID: 29211672. *Review article describing necrotizing soft tissue infections, diagnosis, differential diagnosis, laboratory features, pathogenesis, microbial etiology and treatments.*

# 20

# Folliculitis, Furuncles, and Carbuncles

*Dennis L. Stevens*

## ABSTRACT

Localized purulent infections of the skin are extremely common in all parts to the world, in all age groups, and in both sexes. *Staphylococcus aureus* is the single most common cause of these infections; most are minor, requiring only local treatment such as drainage, unless systemic effects are evident, in which case surgical incision and drainage as well as appropriate antibiotics are necessary. In the past a wide variety of antibiotics were effective against these bacteria, but over the last 5 to 6 years there has been a virtual explosion of skin and soft tissue infections caused by methicillin-resistant *S. aureus* (MRSA). MRSA infections pose two major problems for the clinician: first, resistance to all β-lactam antibiotics including cephalosporins, and second, an apparent increase in severity.

## CLINICAL VIGNETTE

A 25-year-old patient noted a small papule on her left arm 2 days earlier. The papule has gotten red and painful and has reached a size of 5 cm. She has never had such lesions and reports to her physician. The area looks like a volcano with reddened skin and fluctuance. The patient denies having had a fever or chills and has no known underlying illnesses. The physician decides to do an incision and drainage, obtain cultures, and have the patient return in 1 month.

The patient returns with a recurrent abscess at the same site and a second lesion on her right arm.

Cultures from the previous incision and drainage demonstrate MRSA. The organism is sensitive to linezolid, vancomycin, trimethoprim-sulfamethoxazole, and clindamycin. The physician again performs incision and drainage and prescribes a 5-day course of clindamycin taken orally.

A nasal swab was obtained, which grew the same microbe. This is instructive to the patient that the source of these recurrent infections is the anterior nares and that placing a finger in the nares and then scratching her skin inoculates the organism wherever she scratches. With a positive nasal culture, topical agents to the anterior nares could include mupirocin, bacitracin, or other agents. In addition, oral treatment for staphylococcal abscesses beyond incision and drainage have been associated with reduced recurrence.

## DISEASE BURDEN

*S. aureus* is once again reemerging as a major threat to human health and well-being the world over. Folliculitis, furuncles, and carbuncles have plagued *Homo sapiens* throughout their evolution. *S. aureus* too has evolved and adapted to a wide variety of human conditions and medical innovations. After the introduction of antibiotics, *S. aureus* developed resistance to penicillin in the 1940s; in the 1950s it emerged as an important cause of serious nosocomial infections. With the development and widespread use of chloramphenicol and tetracycline in the 1960s, a wide variety of superinfections with antibiotic-resistant *S.*

*aureus* strains occurred. The timely discovery of β-lactamase–resistant cephalosporins and later the semisynthetic β-lactam antibiotics (methicillin, oxacillin, and nafcillin) provided effective treatments for the next 10 to 15 years. Still, as early as the 1970s, sporadic reports of MRSA began to appear. Epidemics of MRSA were reported in some unique facilities with extremely ill patients and with intense antibiotic usage. Over the subsequent 20 to 30 years, the widespread emergence of MRSA infections has been observed in certain regions of Europe, throughout the United States, as well as in Japan and the Western Pacific. Until very recently, these MRSA strains have largely been associated with hospital-acquired MRSA (HA-MRSA). Only recently have reports of true community-acquired MRSA (CA-MRSA) infections begun to emerge. Empiric treatment of some of these cases with conventional antibiotic agents was inadequate; it resulted in disastrous outcomes before it was determined that the causative agent was MRSA. CA-MRSA has been named the USA300 strain.

## PATHOGENESIS OF CUTANEOUS INFECTION AND THE GENETICS OF MRSA STRAINS

Staphylococci colonize the skin and nasal mucosa of approximately 20% of the population. Proliferation of these bacteria within the epidermis or dermis can occur after minor trauma that carries the microbes to deeper structures. Thus a sliver can carry staphylococci into the dermis and result in a felon or subcutaneous abscess (Fig. 20.1). In addition, alteration of the barrier function of intact skin after water immersion, thumb sucking, or the removal of a hangnail can result in paronychia (Fig. 20.2). Similarly, obstruction of hair follicles can result in the development of folliculitis, furuncles, and carbuncles in the skin (see Fig. 20.1) or in the anterior nares (Fig. 20.3).

The earliest tissue response in staphylococcal infection is acute inflammation with a vigorous exudation of polymorphonuclear leukocytes. Vascular thrombosis and tissue necrosis quickly lead to abscess formation. As a result of the development of a fibrin meshwork and later fibroblast proliferation, these abscesses become walled-off zones of loculated infection and tissue destruction, with dying leukocytes and viable bacteria at the center. Fibrosis and scarring are often prominent in healing. CA-MRSA strains appear to have the ability to cause necrosis of not only soft tissue structures but also of lung tissue. Although the precise toxin responsible is debated, the histopathology suggests disruption of vascular integrity, reduced tissue perfusion, and subsequent tissue necrosis.

There is now clear genetic-based evidence that CA-MRSA strains are distinct from HA-MRSA. In fact, Daum and colleagues (2004) demonstrated that there are at least four different types of *mecA* (methicillin resistance) gene cassettes. Interestingly, types I, II, and III are associated with strains causing HA-MRSA infections, whereas type

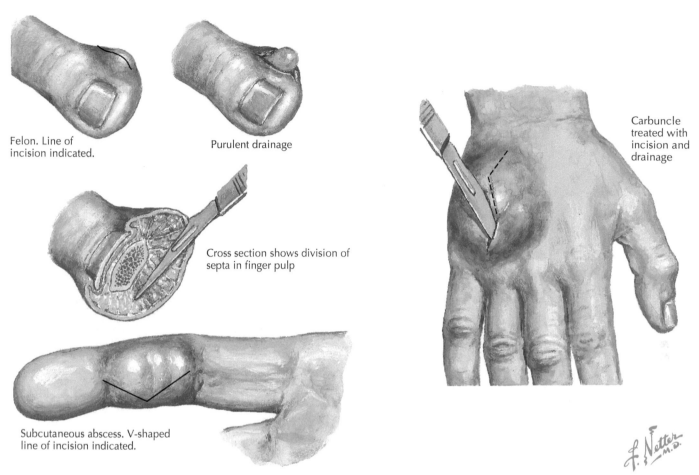

Felon. Line of
incision indicated.

Purulent drainage

Cross section shows division of
septa in finger pulp

Subcutaneous abscess. V-shaped
line of incision indicated.

Carbuncle
treated with
incision and
drainage

**Fig. 20.1** Cellulitis and epidermal access.

IV distinguishes CA-MRSA strains. The type IV *mecA* gene cassette is much smaller (23 kD) compared with types I, II, and III, which are 95, 80, and 55 kD, respectively. It has been postulated that if a smaller gene cassette size is in fact associated with a greater likelihood of transfer of antibiotic resistance to sensitive strains, then many strains of *S. aureus* in the community could acquire *mecA* gene cassette type IV and the prevalence of CA-MRSA infections would increase dramatically. Indeed, this phenomenon has come to fruition, and recent estimates document that 59% of staphylococcal isolates causing skin and soft tissue infections in the outpatient setting are in fact CA-MRSA. Whereas most of the infections associated with CA-MRSA are folliculitis, furuncles, carbuncles, and cutaneous abscesses, more severe infections have been described, including necrotizing fasciitis and hemorrhagic necrotizing pneumonia. In each case, these novel syndromes have occurred after *S. aureus* strains acquired mobile genetic elements, usually bacteriophages carrying genes coding for certain virulence factors. Subsequently, specific toxin genes, including the Panton-Valentine leukocidin (PVL) gene, were incorporated into the CA-MRSA genome. Although there is clearly an association between CA-MRSA strains, severity of infection, and presence of the PVL gene, Hamilton and colleagues (2007) showed no correlation between the severity of human diseases and the level of PVL production in various CA-MRSA clinical strains. Additional studies to identify the role of virulent extracellular toxins of *S. aureus* in human diseases are warranted. Whereas PVL toxin has gotten much attention, clearly alpha-hemolysin– and fibronectin-binding proteins, among others, are well-recognized virulence factors in a variety of *S. aureus* infections.

## RISK FACTORS

The importance of the granulocyte in host defense is supported by the enhanced susceptibility to staphylococcal infections seen in patients with neutropenia or various disorders of neutrophil function, such as chronic granulomatous disease, Chediak-Higashi syndrome, and various disorders of neutrophil chemotaxis. Although it has been suggested that diabetic patients are especially prone to develop boils and carbuncles, few data support this concept. In contrast, it is well established that patients with Job syndrome, who have high levels of serum immunoglobulin E (IgE) antibody and often a congenital defect of the STAT3 signaling pathway, are strongly predisposed to these focal *S. aureus* infections. However, the most important factors predisposing to staphylococcal infections are not immunologic defects but mechanical defects. Minute skin abrasions, other minor trauma, and puncture wounds from slivers, to mention only a few, provide the portal of entry for most staphylococcal skin infections.

## CLINICAL FEATURES OF FOLLICULITIS, FURUNCLES, AND CARBUNCLES

### Evolution of Folliculitis to Furuncles, Carbuncles, and Abscesses

Pustules or abscesses can develop when microorganisms that permanently or transiently reside on skin surfaces are introduced into deeper tissues after even minor skin trauma, as described earlier. Pathogens can also seed the skin hematogenously after bacteremia secondary to a

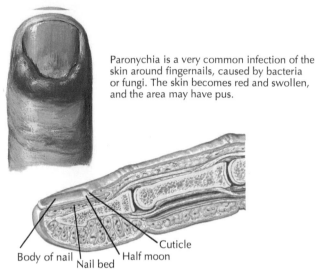

Paronychia is a very common infection of the skin around fingernails, caused by bacteria or fungi. The skin becomes red and swollen, and the area may have pus.

Body of nail   Nail bed   Half moon   Cuticle

The nail body lies on top of the nail bed. An infection can happen when the seal between the body and bed is broken.

Treatment can be simply soaking in water with liquid antibacterial soap. Your doctor may prescribe a topical antibiotic cream or lotion (if bacteria are the cause) or an antifungal medicine given by mouth (if a fungus is the cause).

Your doctor will diagnose paronychia by examining your fingernails.

Your doctor may need to drain the pus.

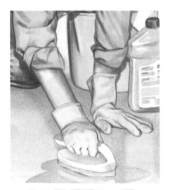

Wear vinyl gloves to prevent contact with irritating substances, such as water, soap, detergent, scouring pads, and chemicals.

You may be more prone to developing the infection if your hands are often wet (such as dishwashers and bartenders). Biting nails or hangnails, thumb sucking, penetrating injuries (e.g., from splinters), and exposure to harsh chemicals, acrylic nails, or nail glue can lead to infection.

**Fig. 20.2** Managing paronychia.

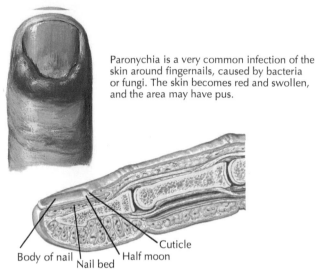

variety of infections, including endocarditis, or by contiguous spread from infectious foci in the lung or gastrointestinal tract. Most commonly, small focal abscesses develop in the superficial layers of the skin, where hair follicles serve as the portal of entry. Such lesions are called *folliculitis*. *S. aureus* accounts for most of these infections, but many different bacterial species can cause localized folliculitis on occasion.

Folliculitis can progress to form subcutaneous abscesses called *furuncles* or *boils*, which usually drain and resolve spontaneously but may progress to form a large, exquisitely painful group of contiguous furuncles, called a *carbuncle*. Felons are subcutaneous abscesses that develop on the pulp of digits (see Fig. 20.1), and paronychia develops between the nail bed and the overlying skin of a digit (see Fig. 20.2).

### Recurrent Furunculosis

The greatest predisposing factor for developing recurrent furunculosis has been the colonization of the anterior nares with *S. aureus*. Small abscesses or furuncles can develop around hair follicles of the anterior nares as well (see Fig. 20.3). Thus touching the nose or nasal secretions

**Fig. 20.3** Furuncle of nasal mucosa.

and then rubbing or scratching the skin results in autoinoculation and abscess formation. Interpersonal contact with contaminated fingers may result in the spread of *S. aureus* carriage or infection to others, especially in situations where there is close physical contact. CA-MRSA strains that harbor PVL have been associated with epidemics of these types of infections among prisoners, athletes, and children in day care centers. Breaking the cycle can be useful to prevent recurrences; however, it is first important to document nasal colonization by appropriate culture techniques. Topical application of intranasal mupirocin or bacitracin ointment for the first 5 days of each month has been shown to decrease colonization and reduce the frequency of recurrent infection by about 50%. Treatment of recurrent furunculosis may require surgical incision and drainage as well as antistaphylococcal antibiotics such as oral dicloxacillin or parenteral nafcillin for methicillin-sensitive *S. aureus* (MSSA) or alternative treatments for MRSA, as described later.

## Diffuse Folliculitis

Diffuse folliculitis occurs in two distinct settings. The first, "hot tub folliculitis," is caused by *Pseudomonas aeruginosa* bacteria, which can flourish in water maintained at a temperature of 98.6°F to 104°F (37°C–40°C) that is insufficiently chlorinated. The infection is usually self-limited, although serious complications of bacteremia and shock have occasionally been reported.

The second form of diffuse folliculitis, "swimmer's itch," occurs when the skin is exposed to bodies of freshwater contaminated with avian schistosomes. Warm water temperatures and alkaline pH provide a suitable habitat for snails, which are the intermediate hosts between birds and humans. Free-swimming parasitic larvae, called *cercariae*, which are shed by the snails into freshwater lakes and rivers, readily penetrate human hair follicles or pores but quickly die. This triggers a brisk allergic reaction, causing intense itching, erythema, and sometimes a papular rash. The infestation is self-limited, secondary infection is uncommon, and oral antihistamines and topical corticosteroid cream usually relieve the symptoms promptly.

## DIAGNOSTIC APPROACH

*S. aureus* is clearly the most common microbe causing focal cutaneous infections. However, the empiric treatment of such infections has become increasingly complex owing to the rapid increase in the prevalence of both CA-MRSA and HA-MRSA strains. Clinical decisions regarding diagnostic tests, selection of empiric antibiotic therapy, route of antibiotic administration, and surgical intervention must be based on the seriousness of the infection, the trends in staphylococcal resistance patterns (i.e., percentage of strains that are MRSA) in the specific geographic area, and the established risk factors for MRSA infection in a given patient. More than ever, making a correct diagnosis and obtaining good culture material for bacterial isolation and antibiotic susceptibility testing are crucial for rational antibiotic selection, as switching from the initial empiric therapy may be warranted by the results of susceptibility testing on the culture isolate. Should a CA-MRSA strain be isolated, careful follow-up, monitoring of the patient's clinical course, and making changes in therapy promptly according to the antibiotic susceptibility pattern may be necessary to ensure good outcomes.

## TREATMENT

### Treatment of Minor Staphylococcal Infections

Minor staphylococcal infections of the skin—such as folliculitis, furunculosis, paronychia, and styes—generally respond well to the topical application of warm soaks. Larger focal infections, such as carbuncles or deep abscesses (particularly those over 5 cm in diameter), often require incision and drainage (see Fig. 20.1). Concomitant antibiotic treatment is indicated in the following situations:

- When the infection is associated with systemic signs of infection (e.g., fever or tachycardia)
- When lesions are large, numerous, or recurrent
- When surgical drainage alone has failed
- If patients have underlying medical problems (e.g., valvular heart disease or implanted prosthetic devices)
- When the nose or face is involved

Cloxacillin (Cloxapen) or dicloxacillin (Dynapen) in an oral dose of 250 to 500 mg every 6 hours is generally sufficient for MSSA. Alternatives would be an oral cephalosporin, erythromycin, clindamycin, or one of the newer macrolides. A recent randomized trial in patients with skin abscesses less than 5 cm compared incision and drainage alone with incision and drainage plus either trimethoprim-sulfamethoxazole or clindamycin given orally. The cure rate was higher in groups receiving either antibiotic and recurrences at 1 month were less in both antibiotic groups.

### Treatment of Major Staphylococcal Infections With No Risk Factors for MRSA

For serious staphylococcal skin infections—such as large abscesses with fever, deep-wound sepsis, or necrotizing fasciitis—parenteral antibiotics are mandatory. Oxacillin (Bactocill) or nafcillin in a dose of 1 to 2 g every 4 hours and parenteral cephalosporins—such as cephalothin, cefuroxime axetil (Ceftin), or cefuroxime sodium (Zinacef)—are excellent alternatives.

### Treatment of Serious Staphylococcal Infections With a High Likelihood of MRSA

Empiric treatment of severe staphylococcal soft tissue infections should involve an agent that has proven efficacy against MRSA, at least until antibiotic susceptibilities are available. Constitutive resistance to erythromycin is common among HA-MRSA strains and in certain geographic regions. Inducible clindamycin resistance is increasing in prevalence in both CA-MRSA and HA-MRSA strains. In addition, the use of ciprofloxacin (Cipro, Proquin XR) and even the newer quinolones is limited by the emergence of resistance. Interestingly, TMP-SMX (Bactrim, Septra) remains active against both MSSA and MRSA strains. Anecdotally, TMP-SMX is also commonly used to treat MRSA-complicated skin and skin structure infections. In situations where streptococcal infections cannot be excluded, cephalexin, penicillin, or clindamycin should be administered.

Vancomycin has been traditionally used as the workhorse antibiotic to treat all MRSA infections; however, numerous problems have emerged. With the increased use of vancomycin, strains of *S. aureus* have appeared that are resistant to vancomycin, with minimum inhibitory concentrations (MICs) of greater than 16 μg/mL (vancomycin-resistant *S. aureus* [VRSA]). Although these organisms are uncommon in the United States, vigorous control measures are required to prevent them from joining vancomycin-resistant enterococci (VRE) as major nosocomial pathogens. Vancomycin-intermediate *S. aureus* (VISA) with MICs of 2 to 4 μg/mL and heteroresistant strains with MICs of 4 to 16 μg/mL have also appeared. Heteroresistance can be detected only in broth culture in the presence of a very large inoculum ($10^7$ colony-forming units [CFUs] per milliliter). There is also mounting evidence that over the course of 3 to 4 decades of vancomycin use, MICs have gradually increased, even in sensitive strains. This development has relevance to the treatment of skin and soft tissue infections because tissue levels of vancomycin may reach only 2 to 4 μg/mL owing to the limited tissue penetration of this antibiotic. In the past, MRSA strains commonly had MICs of 0.1 to 0.5 μg/mL, values considerably below achievable tissue levels of vancomycin (Vancocin). As suggested, at the present time and in the future, we may encounter strains with MICs of 1 to 2 μg/mL, which may account for the greater failure rate of vancomycin in the treatment of a variety of infections including those involving the skin and soft tissue.

### New Antibiotics to Treat MRSA Infections of Skin and Soft Tissue

Recent clinical trials have documented the efficacy of linezolid, tedizolid, daptomycin, quinupristin-dalfopristin, tigecycline, omadacycline, and telavancin in the treatment of skin and soft tissue infections. One other antibiotic—teicoplanin—has excellent activity against MRSA, has proven activity in skin and soft tissue infections, and is available in other countries but not in the United States.

Linezolid and tedizolid inhibit protein synthesis but, unlike the other antibiotics in this class, they prevent the formation of the complex of transfer ribonucleic acid (tRNA), messenger ribonucleic acid (mRNA), and ribosome. As a result, cross resistance to other types of protein synthesis inhibitors is not possible. Linezolid and tedizolid have been approved in the United States for skin and soft tissue infections as well as nosocomial and community-acquired pneumonias (CAPs) caused by MRSA, and evidence is accumulating that, like clindamycin, these agents are a potent suppressors of toxin production by Staphylococcus strains. Because of this characteristic, they may be more suitable agents to use in patients with staphylococcal toxic shock syndrome (TSS) or necrotizing fasciitis caused by PVL-producing strains. In vitro, linezolid and daptomycin have shown excellent activity against VISA and VRSA.

Daptomycin and tigecycline have been approved for treatment of skin and soft tissue infections, including MRSA, in the United States. Daptomycin's mechanism of action involves the formation of pores in the cell membranes of bacteria with rapid loss of intracellular potassium. Tigecycline is a minocycline derivative that inhibits protein synthesis.

Quinupristin-dalfopristin is a semisynthetic derivative of pristinamycin that is at least as active as vancomycin against MRSA and may be a useful alternative when resistance phenotypes for erythromycin, ciprofloxacin, rifampicin, or gentamicin are detected. However, its administration has been associated with a high incidence of phlebitis and myopathy. As a result, its use requires the placement of a central line.

Telavancin and dalbavancin are semi-synthetic lipoglycopeptide antibiotics that contain a heptapeptide core enabling them to inhibit cell wall synthesis by interfering with cross-linking. In addition, telavancin and dalbavancin affect the integrity of bacterial cell membranes, increasing membrane permeability. Both drugs have activity against VISA strains but have poor activity against VRSA. Telavancin requires daily administration, but dalbavancin has an extremely long half-life (147 to 258 hours) and is given by intravenous (IV) injection of a 1-g dose and, 7 days later, 500 mg intravenously. Both are approved by the US Food and Drug Administration (FDA) for uncomplicated skin and soft tissue infections. Oritavancin is a third lipoglycopeptide, and phase III clinical trials have demonstrated noninferiority to conventional agents. Iclaprim is a folic acid antagonist with four- to tenfold higher activity against staphylococci than TMP-SMX; it has been approved for treatment of skin and soft tissue infections including those caused by MRSA. Omadacycline has been approved by the FDA for uncomplicated skin and soft tissue infections including those caused by MRSA.

## PREVENTION AND CONTROL

Epidemiologic control of staphylococcal infections requires the ongoing surveillance and reporting of infections. The dramatic increase in the prevalence of MRSA in community and hospital environments is proof of the difficulty of controlling the spread of these microbes. Contact precautions should be followed in the management of patients with active infections of skin or wounds. There has been a movement among infection-control practitioners to culture the nares of all patients admitted to hospitals in an effort to define patients at risk for MRSA infection. However, the detection and treatment of these nasal carriers is labor intensive, and decreased infection rates may not be achieved, particularly if hospital personnel include nasal or rectal carriers with recurrent furunculosis. Topical treatment with germicidal soaps, povidone-iodine solution, or antibiotic ointments has been advocated, but long-term results have been disappointing. Topical mupirocin 2% ointment (Bactroban, Centany) can reduce the MRSA carrier rate, but, because recolonization is common, mupirocin is not recommended for extended use in long-term-care facilities. A placebo-controlled trial of nasal mupirocin (Bactroban) in 34 patients who were S. aureus carriers found that a monthly course of nasal mupirocin reduced the incidence of nasal colonization and skin infections for at least 1 year. However, because resistance to mupirocin and recolonization after therapy can occur, indiscriminate use of mupirocin should be avoided. Orally administered antibiotics, including rifampin (Rifadin, Rimactane), TMP-SMX, and ciprofloxacin have failed to live up to initially promising findings.

Bacterial interference, which attempts to replace epidemiologically virulent strains of staphylococci with strains that have been deliberately attenuated to be less virulent, has generally been abandoned, in part because infections have been caused by these supposedly less virulent strains. Attempts to develop staphylococcal vaccines are continuing. In a recent clinical trial in dialysis patients, a staphylococcal surface carbohydrate conjugated to Pseudomonas exotoxin A significantly reduced the incidence of bacteremia, although the protective antibodies lasted only 8 months.

## SUMMARY

The MRSA epidemic has evolved rapidly, and new data are needed regarding the efficacy of older antibiotic agents, such as tetracycline, clindamycin, and TMP-SMX in the treatment of MRSA infections. Because of the high prevalence of MRSA in hospitals and in the community, we are probably at the tipping point where practitioners should select antibiotic agents with proven efficacy in clinical trials for the treatment of MRSA infections. This recommendation can be stated with more confidence in regard to seriously ill patients. As a result, a practitioner should select one of the newer antibiotic agents discussed in this chapter to treat a seriously ill patient. Because of the higher cost of those agents, it is more important than ever to make a correct microbial diagnosis and obtain antibiotic susceptibilities on clinical isolates. Based on antibiotic susceptibility testing, it may then be possible to step down to an older and cheaper antibiotic to complete the course of therapy.

## EVIDENCE

Arbeit RD, Maki D, Tally FP, et al. The safety and efficacy of daptomycin for the treatment of complicated skin and skin-structure infections. *Clin Infect Dis* 2004;38:1673-1681. *Daptomycin was noninferior to vancomycin in the treatment of complicated skin and soft tissue infections.*

Bogdanovich T, Ednie LM, Shapiro S, Appelbaum PC. Antistaphylococcal activity of ceftobiprole, a new broad-spectrum cephalosporin. *Antimicrob Agents Chemother* 2005;49:4210-4219. *Description of the in vitro activity of ceftobiprole.*

Centers for Disease Control (CDC). From the Centers for Disease Control and Prevention: four pediatric deaths from community-acquired methicillin-resistant Staphylococcus aureus—Minnesota and North Dakota, 1997-1999. *JAMA* 1999;282:1123-1125. *An early report describing the deaths of four children who developed shock and organ failure caused by MRSA acquired in the community.*

Daum RS, Ito T, Hiramatsu K, et al. A novel methicillin-resistance cassette in community-acquired methicillin-resistant Staphylococcus aureus isolates of diverse genetic backgrounds. *J Infect Dis* 2002;186:1344-1347. *This important study demonstrated differences in the genetic cassette containing the mecA gene between hospital-associated strains and community-acquired strains.*

Daums RS, Miller LG, Immergluck L, et al. A placebo-controlled trial of antibiotics for smaller skin abscesses. *N Engl J Med* 2017;376(26):2545-2555. https://doi.org/10.1056/NEJMoa1607033. PMID: 28657870. *A comparative trial of incision and drainage alone compared to either oral trimethoprim/sulfamethoxazole or clindamycin plus incision and drainage for cutaneous abscesses less than 5 cm in diameter. Fewer recurrences in 1 month were seen with groups receiving antibiotics.*

Duong M, Markwell S, Peter J, Barenkamp S. Randomized, controlled trial of antibiotics in the management of community-acquired skin abscesses in the pediatric patient. *Ann Emerg Med* 2010;55:401-407. *Septra was no better than placebo in treating MRSA cutaneous abscesses in children. It should be noted that all patients underwent surgical debridement. Subsequent development of new MRSA abscesses was less in the Septra group.*

Ellis-Grosse EJ, Babinchak T, Dartois N, et al. The efficacy and safety of tigecycline in the treatment of skin and skin-structure infections: results of 2 double-blind phase 3 comparison studies with vancomycin-aztreonam. *Clin Infect Dis* 2005;41(suppl 5):S341-S353. *Describes two double-blind studies that evaluated the efficacy and safety profile of tigecycline in the treatment of complicated skin and soft tissue infections.*

Graffunder EM, Venezia RA. Risk factors associated with nosocomial methicillin-resistant Staphylococcus aureus (MRSA) infection including previous use of antimicrobials. *J Antimicrob Chemother* 2002;49:999-1005. *An epidemiologic study that demonstrated that hospitalization, antibiotics, and prior surgery were risk factors for developing hospital-associated MRSA infection.*

Hamilton SM, Bryant AE, Carroll K, et al. In vitro production of Panton-Valentine leukocidin (PVL) among strains of methicillin-resistant Staphylococcus aureus causing diverse infections. *Clin Infect Dis* 2007;45:1550-1558. *The quantity of PVL toxin produced in vitro by strains of MRSA did not correlate with the severity of infection in patients.*

Jauregui LE, Babazadeh S, Seltzer E, et al. Randomized, double-blind comparison of once-weekly dalbavancin versus twice-daily linezolid therapy for the treatment of complicated skin and skin structure infections. *Clin Infect Dis* 2005;41:1407-1415. *A small comparative trial that demonstrated the efficacy of weekly dalbavancin in the treatment of complicated skin and soft tissue infections.*

Kotilainen P, Routamaa M, Peltonen R, et al. Elimination of epidemic methicillin-resistant Staphylococcus aureus from a university hospital and district institutions, Finland. *Emerg Infect Dis* 2003;9:169-175. *An aggressive approach to reduce nosocomial spread of MRSA in hospitals is described.*

Lina F, Piemont Y, Godail-Gamot F, et al. Involvement of Panton-Valentine leukocidin–producing Staphylococcus aureus in primary skin infections and pneumonia. *Clin Infect Dis* 2003;29:1128-1132. *These authors describe an association between poor outcome in staphylococcal infections and the presence of the PVL gene.*

Miller LG, Perdreau-Remington F, Rieg G, et al. Necrotizing fasciitis caused by community-associated methicillin-resistant Staphylococcus aureus in Los Angeles. *N Engl J Med* 2005;352:1445-1453. *First description of a series of patients with necrotizing soft tissue infections caused by PVL containing CA-MRSA.*

Moran GJ, Krishnadasan A, Gorwitz RJ, et al. Methicillin-resistant S. aureus infections among patients in the emergency department. *N Engl J Med* 2006;355:666-674. *This study demonstrated that MRSA was the most common cause of culturable skin and soft tissue infections in patients in a variety of emergency departments throughout the United States.*

Stevens DL, Bisno AL, Chambers HF, et al. Practice guidelines for the diagnosis and management of skin and soft tissue infections: 2014 update by the Infectious Diseases Society of America. *Clin Infect Dis* 2014;59(2):147-159. https://doi.org/10.1093/cid/ciu296. PMID: 24947530. *Practice guidelines from the Infectious Disease Society of America.*

Stevens DL, Herr D, Lampiris H, et al. Linezolid versus vancomycin for the treatment of methicillin-resistant Staphylococcus aureus infections. *Clin Infect Dis* 2002;34:1481-1490. *Results of a double-blind clinical trial comparing the efficacy of vancomycin versus linezolid for the treatment of a variety of MRSA infections.*

Stryjewski ME, Chu VH, O'Riordan WD, et al. Telavancin versus standard therapy for treatment of complicated skin and skin structure infections caused by gram-positive bacteria: FAST 2 study. *Antimicrob Agents Chemother* 2006;50:862-867. *Comparative trial results demonstrated that telavancin was not inferior to standard treatment for skin and soft tissue infection.*

Weigelt J, Itani K, Stevens D, et al. Linezolid versus vancomycin in treatment of complicated skin and soft tissue infections, *Antimicrob Agents Chemother* 2005;49:2260-2266. *Linezolid was noninferior to vancomycin in the treatment of skin and soft tissue infections and was superior to vancomycin for a subset of surgical infections caused by MRSA.*

## ADDITIONAL RESOURCE

Lowy FD. *Staphylococcus aureus* infections. *N Engl J Med* 1998;339:520. *An excellent review article describing virulence factors of S. aureus.*

# 21

# Life-Threatening Skin and Soft-Tissue Infections

*Dennis L. Stevens, Amy E. Bryant*

## ABSTRACT

This chapter discusses those soft-tissue infections that are truly life threatening. As such, early clinical recognition is the most important aspect of their clinical management. Staphylococcal and streptococcal toxic shock syndromes (referred to here as StaphTSS and StrepTSS, respectively) have similar clinical features once established, yet their prognosis and management are quite different.

Necrotizing soft-tissue infections occur in three distinct settings. The first, necrotizing fasciitis (NF) type I, occurs when mucosal barriers are breached such that mixed aerobic and anaerobic microbes are leaked into the deep soft tissues, resulting in rapidly progressive necrotizing infections. These infections are usually associated with gas-producing microbes. Surgical inspection of suspicious lesions is paramount, and if necrosis is found, adequate debridement is absolutely necessary.

The second, type II NF, is usually monomicrobic; group A *Streptococcus* (GAS) is clearly most common, though *Vibrio vulnificus*, *Aeromonas hydrophila*, and methicillin-resistant *Staphylococcus aureus* (MRSA) may also cause extensive soft-tissue necrosis. Surgical intervention is also of major importance and also provides a definitive etiologic diagnosis. These infections are usually not characterized by gas in the tissue.

The third type of necrotizing soft-tissue infection is gas gangrene (also known as *clostridial myonecrosis*). This infection is always associated with gas in the tissue and, like other forms of necrotizing soft-tissue infection, is rapidly progressive. Causative organisms include *Clostridium perfringens*, *Clostridium histolyticum*, *Clostridium septicum*, *Clostridium novyi*, and *Clostridium sordellii*. All these species are agents of gas gangrene after penetrating trauma; however, the more aerotolerant *C. septicum* can also cause spontaneous gas gangrene in patients with adenocarcinoma of the colon or neutropenia. *C. sordellii* and *C. novyi* have been associated with gas gangrene after skin injection of black tar heroin. *C. sordellii* has also recently been associated with a toxic shock–like syndrome in women after abortion or childbirth.

## CLINICAL VIGNETTE

A 25-year-old weight lifter strained his trapezius muscle. He had immediate pain and took ibuprofen. After 24 h, the pain subsided but later became severe and he went to an emergency department (ED). He was found to have nausea, a pulse rate of 124 beats/min, blood pressure of 130/65 mm Hg, and oral temperature of 99.3°F. There was no evidence of swelling or erythema around the trapezius although the area was tender to palpation. He was given intravenous morphine and discharged on oral narcotics and ibuprofen. He then developed vomiting, shaking chills, and dizziness. He returned to the ED 18 h later and was found to have a heart rate of 150 beats/min and blood pressure of 120/50 mm Hg. There was massive swelling over the trapezius along with erythema and purple bullae. Lab tests showed a white blood cell (WBC) count of 11,000/mm³ with 30%

bands, 5% metamyelocytes, 4% myelocytes, and a platelet count of 70,000/mm³. A serum metabolic panel demonstrated a bicarbonate of 15, creatinine of 2.5 mg/dL, albumin of 2.0 g/dL, and creatine phosphokinase (CPK) of 2500 U/dL. Computed tomography (CT) demonstrated edema extending from the trapezius to the deltoid and pectoralis muscles. Within 2 hrs he developed profound hyotension with evidence of diffuse capillary leak. He remained hypotensive despite the intravenous administration of 10 L of saline and died 30 h after admission.

This is an example of a necrotizing soft-tissue infection and StrepTSS presenting with no portal of entry. Infection begins deep in the muscle at the exact site of injury. There is a trophism of GAS to injured muscle, which is mediated by vimentin. A biphasic pain pattern is usually present; ibuprofen or other nonsteroidal antiinflammatory drugs (NSAIDs) may predispose to worse outcomes. Evidence of renal failure can precede hypotension by several hours. Severe pain is the greatest clue and, because the infection begins deep, the classic signs of a necrotizing process—such as skin sloughing, ecchymosis, and violaceous bullae—may not be present until later in the course. Other clues are a normal or slightly elevated WBC count, but with a dramatic increase in immature cells and a markedly elevated CPK, indicating muscle destruction.

## DISEASE BURDEN

Life-threatening skin and soft-tissue infections are relatively uncommon. Comprehensive epidemiology has been performed only for invasive group A streptococcal infections, including NF. The incidence is approximately 3.5 cases per 100,000 population per year in the United States. Of these, approximately 50% are NF or myonecrosis, of which half are associated with TSS. Before the 1900s, the incidence of malignant scarlet fever was as high as 25 cases per 100,000 population, but it has dramatically declined both in frequency and severity since the advent of antibiotics. Gas gangrene reached epidemic proportions during the Civil War and World Wars I and II; however, antibiotics, rapid transport to evacuation hospitals, and immediate vascular reconstruction have contributed to its declining incidence among active duty military personnel. Mixed aerobic and anaerobic infections are most common in diabetic patients but may also occur after a variety of surgical procedures, such as cholecystectomy, colonic surgery, and gynecologic procedures. Although relatively uncommon, these infections can be devastating and often affect otherwise healthy individuals.

## STAPHYLOCOCCAL TOXIC SHOCK SYNDROME

StaphTSS was first reported in 1978; by 1990, more than 3300 cases had been reported in the United States, 90% of which occurred during menstruation in women who were using tampons. A specific toxin, staphylococcal toxic shock syndrome toxin-1 (TSST-1), was implicated in these cases. The incidence of StaphTSS declined precipitously after superabsorbable tampons were withdrawn from the market.

**Etiology and Pathogenesis**

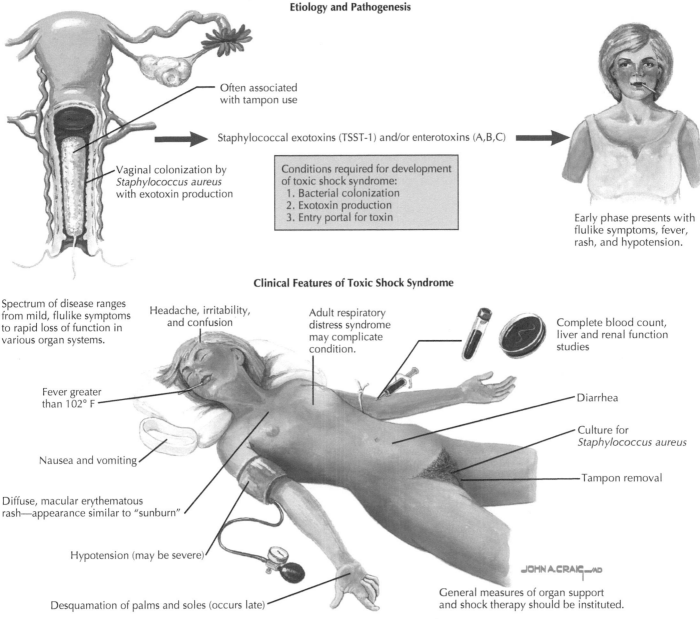

Often associated with tampon use

Staphylococcal exotoxins (TSST-1) and/or enterotoxins (A,B,C)

Vaginal colonization by *Staphylococcus aureus* with exotoxin production

Conditions required for development of toxic shock syndrome:
1. Bacterial colonization
2. Exotoxin production
3. Entry portal for toxin

Early phase presents with flulike symptoms, fever, rash, and hypotension.

**Clinical Features of Toxic Shock Syndrome**

Spectrum of disease ranges from mild, flulike symptoms to rapid loss of function in various organ systems.

Headache, irritability, and confusion

Adult respiratory distress syndrome may complicate condition.

Complete blood count, liver and renal function studies

Fever greater than 102° F

Diarrhea

Culture for *Staphylococcus aureus*

Nausea and vomiting

Tampon removal

Diffuse, macular erythematous rash—appearance similar to "sunburn"

Hypotension (may be severe)

General measures of organ support and shock therapy should be instituted.

Desquamation of palms and soles (occurs late)

JOHN A. CRAIG—AD

Fig. 21.1 Staphylococcal toxic shock syndrome.

Currently, fewer than 100 cases of StaphTSS occur each year, and most of these nonmenstrual cases are nosocomially acquired, often as a result of postoperative staphylococcal wound infections when packing material has been used (e.g., rhinoplasty). It is interesting to note that these cases are associated with strains that produce staphylococcal enterotoxin B (SEB).

## Clinical Presentation and Diagnostic Approach

StaphTSS is a multisystemic disease with diverse clinical manifestations (Fig. 21.1). A characteristic sunburn-type rash is present in 90% of patients; it is most prominent on the face and trunk and in intertriginous areas. It blanches with digital pressure and is often confused with a drug rash (Fig. 21.2, *right*). Other early signs of StaphTSS include fever, confusion, nausea, vomiting, diarrhea, tachycardia, and hypotension. Sudden onset and skin rash are the best clues to the diagnosis.

Laboratory tests are helpful to confirm the diagnosis. Leukocytosis with a left shift and thrombocytopenia (with a platelet count of

$<100,000/mm^3$) are common findings. Urinalysis may show mild pyuria and, occasionally, microscopic hematuria. Blood urea nitrogen and creatinine levels are elevated in more than 50% of patients. Serum bilirubin and hepatic enzyme levels are raised in about half of patients. Serum creatine kinase levels are high in more than one-third of patients, and myoglobinuria has developed in some patients. Elevated serum amylase levels are also found but may be related to the azotemia rather than to clinically evident pancreatitis. Unexplained marked hypocalcemia is often observed. The drop in serum calcium level is out of proportion to the degree of hypoalbuminuria noted in some patients and may be caused by elevated serum calcitonin levels.

Blood cultures are negative in most cases, but a positive blood culture should not exclude the diagnosis. The original Centers for Disease Control and Prevention (CDC) definition of StaphTSS excluded patients with bacteremia, but it has been shown that some strains that cause bacteremia produce either TSST-1 or SEB. Thus some patients with bacteremia may have more complicated septic shock caused by

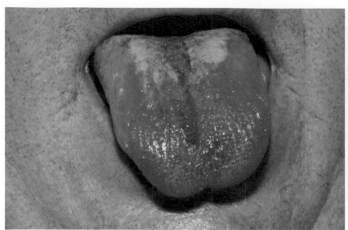

Desquamation of the tongue with a strawberry appearance.

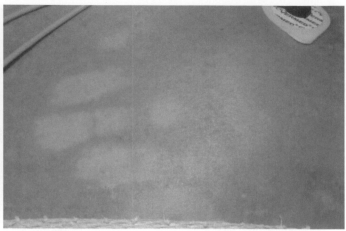

The diffuse sunburn-type rash of toxic shock syndrome blanches under digital pressure.

**Fig. 21.2** Staphylococcal toxic shock syndrome.

the presence of these toxin genes. Group A streptococci can produce a severe form of TSS that resembles StaphTSS except for the paucity of cutaneous manifestations and a marked difference in mortality (see the discussion of StrepTSS later and also Table 21.1).

## Treatment

The management of StaphTSS demands immediate treatment of hypotension and shock with vigorous fluid replacement (and supplemental catecholamines if needed), attention to the site of *S. aureus* colonization or infection (e.g., removal of packing material, drainage of any abscess), and systemic antimicrobial therapy with an antistaphylococcal agent. Because this infection is commonly associated with a diffuse capillary leak syndrome, administration of 10 to 12 L of normal saline may also be necessary. If the albumin drops below 1.5 g/dL, replacement with colloid rather than crystalloid should be considered. One retrospective analysis of 45 patients suggested that glucocorticoids may assist in recovery, but more data are needed before this can be recommended for all patients with StaphTSS. With proper management and support, most patients recover in 1 to 2 weeks; mortality is less than 5%.

## STREPTOCOCCAL TOXIC SHOCK SYNDROME

Like StaphTSS, StrepTSS is a toxin-mediated disease. Although multiple exotoxins have been implicated in experimental studies, streptococcal pyrogenic exotoxin type A (SPEA) has been linked epidemiologically. The increased incidence of invasive streptococcal infections may also be linked to the global spread of toxigenic strains of streptococci, particularly M-types 1 and 3, containing the gene for SPEA. Like the staphylococcal toxins TSST-1 and SEB, SPEA enters the circulation and functions as a superantigen, stimulating both lymphocytes and macrophages to produce cytokines that mediate the shock syndrome.

People of all ages are vulnerable to StrepTSS; about half of the patients have diabetes or alcoholism. When a primary focus is identified, it is most often a necrotizing soft-tissue infection; respiratory infections are the next most common focus.

### Clinical Presentation and Diagnostic Approach

The onset of StrepTSS can be subtle, with fever, chills, nausea, vomiting, and diarrhea for 12 to 36 hours before the sudden onset of hypotension, which occurs in all patients and is often severe. Other clinical

| TABLE 21.1 **Staphylococcal Versus Streptococcal Toxic Shock Syndrome** | | |
|---|---|---|
| **Feature** | **Staphylococcal** | **Streptococcal** |
| Age | 15–35 | 0–80 |
| Severe pain | Rare | Common |
| Hypotension | 100% | 100% |
| Rash | Very common | Less common |
| Renal failure | Common | Common |
| Bacteremia | Low | 60% |
| Tissue necrosis | Rare | Common |
| Risk factors | Tampons, packing | Trauma, varicella |
| Thrombocytopenia | Common | Common |
| Mortality | <3% | 30%–70% |

features include a generalized erythematous rash (10% of cases), which desquamates 7 to 10 days into the illness, acute respiratory distress syndrome (60% of cases), renal failure (80% of cases), and soft-tissue necrosis, such as NF or myonecrosis (see Table 21.1). In patients with NF or myonecrosis, the initial complaint is often severe, unrelenting pain out of proportion to the clinical findings (see the discussion of necrotizing fasciitis, later). In many such cases, infection begins at the site of antecedent trauma that does not break the skin (e.g., muscle strain, bruise). Without a portal of bacterial entry to provide clinical clues, the diagnosis is often missed or delayed, causing mortality in these "cryptic" infections to approach 85%. Further, experimental evidence suggests that administration of nonsteroidal antiinflammatory agents may predispose to worse outcomes in these infections. Laboratory evidence of multiorgan involvement can typically be found and characteristically includes evidence of renal impairment, hepatic abnormalities, and laboratory evidence of disseminated intravascular coagulation, though clinical evidence of coagulopathy is rarely present.

### Treatment

Even after aggressive treatment—including antibiotics, circulatory and respiratory support, and appropriate surgical debridement—mortality rates range from 30% to 70% (see the discussion of streptococcal NF, later, for antibiotic treatment and the use of intravenous immunoglobulin [IGIV]).

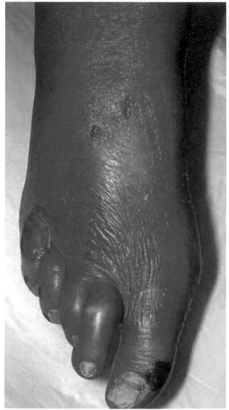

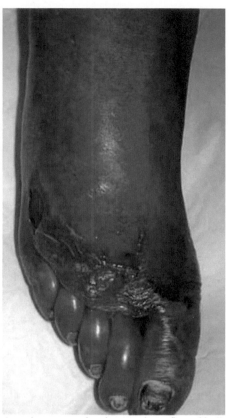

The appearance of the foot on the day the patient was treated with oral dicloxacillin for presumed cellulitis.

Over the course of 48 hours, a large flaccid bulla developed and cultures grew methicillin-resistant *Staphylococcus aureus* (MRSA). This disease is more common in neonatal intensive care units but can also occur in adults. A toxin (exfolitin) causes degradation of desmosomes, which serve as intracellular bridges holding skin intact.

**Fig. 21.3** Staphylococcal scalded skin syndrome.

## SCARLET FEVER

The incidence of scarlet fever has declined sharply in the antibiotic era. When it does occur, it most commonly accompanies acute streptococcal pharyngitis. The initial symptoms are fever and sore throat. Within 1 to 5 days the characteristic fine red sandpaper-like eruptions appear on the skin, often beginning on the chest and rapidly spreading to other parts of the body. Although the tongue and buccal mucosa are classically involved, the perioral area may be spared, thus accounting for the typical circumoral pallor. The rash is caused by hyperemia and capillary damage produced by erythrogenic (scarlatina) toxins. In areas of trauma, such as the antecubital fossae, punctate hemorrhages (Pastia sign) may occur. Nausea and vomiting may be present, and fever and prostration may be severe. Desquamation of skin and mucous membranes is prominent during healing; one characteristic feature is the strawberry tongue (see Fig. 21.2, *left*). Therapy is the same as that for the underlying streptococcal infection.

## STAPHYLOCOCCAL SCALDED SKIN SYNDROME

Epidemics of staphylococcal scalded skin syndrome (SSSS) have been reported among children, usually in neonatal intensive care units, but sporadic cases have also been described among the elderly. A particular strain of *S. aureus* belonging to phage group II is responsible, and these strains produce an extracellular toxin called *exfolatin* that degrades the intercellular desmosomes, which provide tight binding between adjacent epithelial cells. Thus flaccid fluid-filled bullae develop. The shear plane is very superficial and does not result in scarring, although considerable fluid can extravasate if lesions are extensive. The propensity

of skin to slough is demonstrated by a positive Nikolsky sign, which is elicited by placing a thumb on skin and applying lateral pressure. Despite fluid loss, infection usually remains superficial and skin slippage can occur at sites remote from the primary source of infection. Thus the mortality is low.

This disease must be distinguished from toxic epidermal necrolysis (TEN), which also results in skin sloughing and is associated with a positive Nikolsky sign. TEN is more common in adults, is usually associated with drug reactions, and carries a higher mortality rate; the cleavage plane in TEN is much deeper, at the stratum germinativum layer. Thus a skin biopsy or frozen section readily distinguishes SSSS from TEN. Fig. 21.3 illustrates the evolution of SSSS in an elderly adult. Note the flaccid bullae.

## NECROTIZING FASCIITIS

NF is a surgical diagnosis demonstrating friable deep fascia and an accumulation of inflammatory fluid that resembles dishwater. Although this is a very specific surgical diagnosis, it is quite apparent that NF is also associated with necrosis of the skin, subcutaneous tissue, fascia, and in some cases muscle. Thus a better term might be *necrotizing soft-tissue infection*.

Under the current definition, NF encompasses two microbiologic entities. Type I disease is caused by mixed anaerobes (e.g., *Clostridium, Bacteroides, Prevotella,* and *Peptostreptococcus* organisms), streptococci, and enteric gram-negative bacilli (e.g., *Escherichia coli, Klebsiella,* and *Proteus* organisms). Infection most often complicates deep wounds, including those resulting from surgical procedures involving the

This patient developed a group A streptococcal infection at the site of minor non-penetrating trauma. Severe pain, fever, and localized tenderness quickly evolved to hemorrhagic bullae, skin sloughing, and expanding necrosis of the skin, fascia, and muscle.

**Fig. 21.4** Necrotizing fasciitis.

gastrointestinal tract or genitourinary systems. Predisposing factors include a compromised vascular supply, either generalized or diabetes related. Type II NF (hemolytic streptococcal gangrene, or "flesh-eating disease") is most commonly caused by *Streptococcus pyogenes* and may occur at the site of a cut, burn, or insect bite. It may also follow surgical wounds or childbirth or may occur spontaneously at sites of antecedent trauma that does not break the skin.

## Disease Burden

NF per se is not a reportable disease, and few data exist regarding its prevalence. However, the widespread publicity given to streptococcal NF has fueled popular concern about invasive group A streptococcal infections. The CDC estimates that 10,000 to 15,000 cases of invasive group A streptococcal infections occur each year in the United States; of these, NF occurs in 5% to 10% of patients and carries a case fatality rate of about 30%. In patients with StrepTSS, a mortality rate as high as 70% to 85% is the norm. Group A streptococcal NF is usually community acquired and sporadic in nature, and many patients have predisposing conditions such as varicella infection, trauma, diabetes, and immunosuppressive disorders. Most group A streptococci that cause invasive disease produce one or more pyrogenic (formerly erythrogenic) toxins, but the genetic heterogeneity of causative strains does not support a clonal basis for the resurgence of invasive streptococcal infections.

## Clinical Presentation and Approach to the Diagnosis

Patients with NF generally have a history of the rapid onset of severe pain in a limb, along with malaise, chills, and fever. The affected area is red, hot, shiny, swollen, and exquisitely tender. Late findings of NF are a blue-black discoloration indicative of superficial necrosis near the center, blistering or bulla formation, and edema that extends beyond the margins of skin erythema. Crepitus may be palpable or audible in patients with NF resulting from gas-forming organisms such as *Clostridium* species or mixed aerobic and anaerobic bacteria. The margins of erythema may progress visibly over a matter of hours.

Approximately 50% of patients with NF caused by group A streptococci do not manifest the classic cutaneous manifestations described earlier; these patients may have only fever, chills, and severe pain. In roughly half of these patients, NF begins at the site of cutaneous penetrating trauma such as burns, insect bites, splinters, abrasions, chickenpox vesicles, or surgical incisions. In the remaining cases, infection begins at the site of nonpenetrating deep trauma, such as hematoma, ankle sprain, tendon rupture, or muscle tear. Most likely the organism translocates from the pharynx to the site of injury via the bloodstream.

Because this type of infection begins deep within fascia and muscle, cutaneous changes are initially less prominent than pain, swelling, and systemic toxicity. Fig. 21.4 demonstrates the appearance of fascia and hypoxic muscle in a patient with StrepTSS and NF.

In many cases, hypotension and evidence of renal impairment precede the skin manifestations of streptococcal NF. In these patients, elevated serum creatinine, hypotension, marked tachycardia, and severe pain should prompt surgical consultation to inspect the deep tissues and to obtain culture material. Although radiographs, CT, and magnetic resonance imaging may indicate swelling and edema in NF, they do not provide a definitive diagnosis. Thus these imaging studies can help to define the depth and extent of infection; however, in those with preexisting trauma, infection may be difficult to distinguish from trauma alone. An elevation in serum CPK may provide additional clues facilitating surgical intervention.

## Treatment

The Gram stain result of an aspirate or biopsy should be used to guide therapy. Mixed gram-positive and gram-negative bacteria indicate type I disease, and broad-spectrum coverage for gram-negative and anaerobic organisms is needed. Because of the risk of antimicrobial resistance in gram-negative organisms, using at least two antibiotics active against enteric gram-negative pathogens is recommended until culture results are obtained. A carbapenem or a third- or fourth-generation cephalosporin with metronidazole, in combination with ciprofloxacin or an aminoglycoside, is a reasonable option. In the absence of a Gram stain, patients should receive treatment for type II disease until a causative diagnosis has been made.

GAS remains highly susceptible to penicillin and cephalosporins; however, erythromycin resistance has emerged in Japan, Finland, Sweden, and recently the United States. In addition, penicillin and other β-lactam antibiotics have been shown to fail in the treatment of severe group A streptococcal infections in humans and experimental animals, whereas protein synthesis inhibitor antibiotics (e.g., clindamycin) have been protective. For these reasons, the currently recommended antibiotic treatment for severe group A streptococcal NF or StrepTSS is penicillin plus clindamycin. Clindamycin's efficacy is largely related to its ability to directly suppress bacterial toxin production at the ribosomal level; some studies have also demonstrated that it can beneficially modulate the host's immune response.

For patients with severe staphylococcal infections such as NF, recommendations for therapy have changed dramatically in the last few years. With the prevalence of MRSA approaching 50% in many communities, empiric treatment of severe *S. aureus* infections must now be based on the clinician's judgment of severity. Specifically, suspected staphylococcal infections in patients in a toxic condition should be treated with vancomycin, linezolid, or daptomycin. Stepping down to a more conventional agent could be accomplished once susceptibility results are known.

Irrespective of the cause of NF, early surgical intervention clearly reduces morbidity and mortality rates, and surgery for diagnosis and debridement of involved tissue should be performed as soon as possible—within the first hours—after admission. Surgical consultation should also be considered for infections of the hand—particularly infected cat, dog, or human bite wounds—and for patients with a history of intravenous drug abuse. These patients are at particular risk for soft-tissue abscesses, septic arthritis, and osteomyelitis.

In addition to antibiotics and early surgical intervention, meticulous metabolic and circulatory support is critically important for patients with NF. Bacteremia and StrepTSS often complicate NF and portend worse outcomes.

## Ancillary Treatment Modalities

Neutralization of circulating streptococcal virulence factors would be a useful therapeutic modality, and studies suggest that some batches of IVIG have neutralizing activity against some streptococcal exotoxins. One observational cohort study supports the use of a single dose of 2 g/kg of IVIG; however, a recent double-blind clinical study did not demonstrate improved survival or diminution of NF, although the study was stopped prematurely because of low enrollment. In contrast, IVIG was shown to be useful in GAS NF only in patients receiving clindamycin. The latest multicenter trial in 2020 showed the patients who did not receive IVIG had the greatest mortality.

All persons with traumatic wounds or infections associated with the gastrointestinal tract require assessment of tetanus immunization status.

The use of hyperbaric oxygen (HBO) for the treatment of NF is controversial. There is no indication for its use in the treatment of group A streptococcal NF. It still is being advocated for the treatment of Fournier gangrene and is indicated in gas gangrene. It should never take priority over surgery.

## CLOSTRIDIAL MYONECROSIS (GAS GANGRENE)

Clostridial species such as *C. perfringens, C. histolyticum, C. septicum, C. novyi,* and *C. sordellii* cause aggressive necrotizing infections of the skin and soft tissues, largely because of the elaboration of bacterial proteases, phospholipases, and cytotoxins. Necrotizing clostridial soft-tissue infections (gas gangrene) are rapidly progressive and characterized by marked tissue destruction, gas in the tissues, shock, and frequently death. Gas gangrene can develop after trauma, injection of either legal or illicit drugs, or pregnancy; it can also develop spontaneously in patients with carcinoma of the colon or neutropenia.

### Traumatic Gas Gangrene Caused by Clostridia

Clostridial myonecrosis (gas gangrene) is one of the most fulminant gram-positive infections of humans. Predisposing conditions include crush type injury, laceration of large or medium-sized arteries, and open fractures of long bones that are contaminated with soil containing the bacterial spores. Gas gangrene of the abdominal wall and flanks occurs after penetrating injuries such as knife or gunshot wounds sufficient to compromise intestinal integrity, with resultant leakage of bowel contents into the soft tissues. In the last few years, cases of cutaneous gas gangrene caused by *C. perfringens, C. novyi* type A, and *C. sordellii* have been described in the United States and northern Europe

among drug abusers injecting "black tar heroin" subcutaneously; these cases involve the rapid development of a foul-smelling wound with a thin serosanguineous discharge and gas bubbles. Brawny edema and induration develop and give way to cutaneous blisters containing bluish to maroon fluid. Later, such tissue may become liquefied and slough. The margin between healthy and necrotic tissue often advances several inches per hour despite appropriate antibiotic therapy, and radical amputation remains the single best lifesaving treatment. Shock and organ failure frequently accompany gas gangrene; when patients become bacteremic, the mortality exceeds 80%.

Diagnosis is not difficult because the infection always begins at the site of significant trauma, is associated with gas in the tissue, and is rapidly progressive. Gram stain of drainage or tissue biopsy specimen is usually definitive, demonstrating large gram-positive rods and an absence of inflammatory cells. Using experimental models, we have recently demonstrated that the severe pain, rapid progression, marked tissue destruction, and absence of neutrophils in *C. perfringens* gas gangrene are caused by the alpha toxin–induced occlusion of blood vessels by large heterotypic aggregates of platelets and neutrophils. Because these aggregates were caused in large part by alpha toxin–induced activation of the platelet fibrinogen receptor GPIIb-IIIa, it implies that agents targeting this receptor (e.g., eptifibatide, abciximab) may be therapeutic for maintaining tissue blood flow in gas gangrene. This notion is currently being tested.

### Spontaneous, Nontraumatic Gas Gangrene Caused by *Clostridium septicum*

The first symptom of spontaneous *C. septicum* gas gangrene may be confusion, followed by the abrupt onset of excruciating pain and rapid progression of tissue destruction with demonstrable gas in the tissue. Swelling increases and bullae appear filled with clear, cloudy, hemorrhagic, or purplish fluid. The surrounding skin has a purple hue, perhaps reflecting vascular compromise resulting from bacterial toxins diffusing into surrounding tissues. Mortality from spontaneous gangrene ranges from 67% to 100%, with the majority of deaths occurring within 24 hours of onset. Predisposing host factors include colonic carcinoma, diverticulitis, gastrointestinal surgery, leukemia, lymphoproliferative disorders, cancer chemotherapy, radiation therapy, and, more recently, acquired immunodeficiency syndrome (AIDS). Cyclic, congenital or acquired neutropenia is also strongly associated with an increased incidence of spontaneous gas gangrene caused by *C. septicum*; in such cases necrotizing enterocolitis, cecitis, or distal ileitis is commonly found. These gastrointestinal pathologies permit bacterial access to the bloodstream; consequently, aerotolerant *C. septicum* can proliferate in normal tissues. Patients surviving bacteremia or spontaneous gangrene caused by *C. septicum* should have aggressive diagnostic studies to rule out gastrointestinal pathology.

### Gynecologic Infections Caused by *Clostridium sordellii*

Historically, gas gangrene of the uterus, especially that caused by *C. sordellii* and *C. perfringens*, occurred after illegal or self-induced abortions, but in modern times it also follows spontaneous abortion, normal vaginal delivery, and cesarean section. In addition to infections caused by *C. perfringens* and *C. sordellii*, GAS can also result in puerperal sepsis and uterine myonecrosis. Recently, *C. sordellii* has also been implicated in medically induced abortion. Fatal postpartum *C. sordellii* infections in young, previously healthy women involve a unique clinical picture of little or no fever, lack of a purulent discharge, refractory hypotension, extensive peripheral edema and effusions, hemoconcentration, and a markedly elevated WBC count with left shift.

## Other Clostridial Skin and Soft-Tissue Infections

Crepitant cellulitis, also called *anaerobic cellulitis*, is seen principally in diabetic patients, characteristically involves subcutaneous tissues or retroperitoneal tissues, and can progress to fulminant systemic disease; the muscle and fascia are not involved. Cases of *C. histolyticum* infection with cellulitis, abscess formation, or endocarditis have also been documented in injecting drug users. *C. sordellii* was responsible for endophthalmitis after suture removal from a corneal transplant.

## FUTURE DIRECTIONS

Progression of necrosis can be alarmingly rapid in all the life-threatening infections discussed in this chapter, and shock and multiorgan failure are common. Despite the introduction of highly active antibiotics and aggressive surgical intervention, the morbidity and mortality rates of NF and gas gangrene remain high. This is in part because of the failure of treatment strategies to alter the physiologic processes responsible for the local tissue destruction and systemic toxicity. Clearly, to reduce the need for radical surgery and amputation, new translational research is sorely needed.

## EVIDENCE

Bilton BD, Zibari GB, McMillan RW, et al. Aggressive surgical management of necrotizing fasciitis serves to decrease mortality: a retrospective study. *Am Surg* 1998;64:397-400. *Important article that demonstrates that early and aggressive debridement of necrotic tissue improves outcomes.*

Bruun T, Rath E, Bruun Madsen M, et al. Risk factors and predictors of mortality in Streptococcal necrotizing soft-tissue infections: a multicenter prospective study [published online ahead of print, 2020 Jan 10]. *Clin Infect Dis* 2020;ciaa027. https://doi.org/10.1093/cid/ciaa027. *This study provides evidence that patients who did not receive IVIG had increased mortality.*

Bryant AE, Bayer CR, Aldape MJ, et al. Clostridium perfringens phospholipase C–induced platelet/leukocyte interactions impede neutrophil diapedesis. *J Med Microbiol* 2006;55(Pt 5):495-504. *Alpha toxin, a phospholipase C, causes dramatic reduction in neutrophil diapedesis related to induction of platelet-neutrophil complexes. This may explain the absence of neutrophils at the site of clostridial gas gangrene.*

Bryant AE, Bayer CR, Chen RY, et al. Vascular dysfunction and ischemic destruction of tissue in Streptococcus pyogenes infection: the role of streptolysin O–induced platelet/neutrophil complexes. *J Infect Dis* 2005;192:1014-1022. *Streptolysin O causes immediate reduction in blood flow, as measured by laser Doppler, caused by toxin-induced platelet-neutrophil aggregates.*

Davies HD, McGeer A, Schwartz B, et al. Invasive group A streptococcal infections in Ontario, Canada. *N Engl J Med* 1996;335:547-554. *Prospective review of the clinical aspects of invasive group A streptococcal infections, which demonstrates a higher mortality in patients that have NF associated with StrepTSS.*

Hamilton SM, Bayer CR, Stevens DL, et al. Muscle injury, vimentin expression, and nonsteroidal anti-inflammatory drugs predispose to cryptic group A streptococcal necrotizing infection. *J Infect Dis* 2008;198:1692-1698. *Experimental study demonstrating that minor muscle trauma associated with vimentin expression provides a niche for homing of group A Streptococcus to the exact site of muscle injury. In addition, administration of nonsteroidal antiinflammatory agents enhances the magnitude of streptococci homing to the site of injury.*

Kaul R, McGeer A, Low DE, et al. Population-based surveillance for group A streptococcal necrotizing fasciitis: clinical features, prognostic indicators, and microbiologic analysis of seventy-seven cases, Ontario Group A Streptococcal Study. *Am J Med* 1997;103:18-24. *Comprehensive review of the clinical features and epidemiology of group A streptococcal NF in Ontario, Canada.*

Majeski J, Majeski E. Necrotizing fasciitis: improved survival with early recognition by tissue biopsy and aggressive surgical treatment. *South Med J* 1997;90:1065-1068. *This study documents the importance of an aggressive surgical approach to the diagnosis and treatment of necrotizing soft-tissue infections.*

## ADDITIONAL RESOURCES

Aldape MJ, Bryant AE, Stevens DL. Clostridium sordellii infections: epidemiology, clinical findings and current perspectives on diagnosis and treatment. *Clin Infect Dis* 2006;43:1436-1446. *A recent review article describes the clinical presentation, epidemiology, and pathogenic mechanisms involved in devastating C. sordellii infections.*

Bisno AL, Stevens DL. Streptococcal infections of skin and soft tissues. *N Engl J Med* 1996;334:240-245. *A comprehensive review of the spectrum of group A streptococcal infections including impetigo, cellulitis, erysipelas, NF, and myonecrosis.*

Chapnick EK, Abter EI. Necrotizing soft-tissue infections. *Infect Dis Clin North Am* 1996;10:835-855. *Comprehensive review of the clinical spectrum and etiology of necrotizing soft-tissue infections.*

Nguyen HB, Rivers EP, Abrahamian FM, et al. Severe sepsis and septic shock: review of the literature and emergency department management guidelines. *Ann Emerg Med* 2006;48:28-54. *A guideline for the management of patients with severe sepsis in the emergency department. Comprehensive review of parameters for fluid resuscitation and use of pressors, corticosteroids, antibiotics, and activated protein C.*

Stevens DL. The toxic shock syndromes. *Infect Dis Clin North Am* 1996;10:727-746. *Comprehensive review of the epidemiology, clinical characteristics, pathogenesis, and treatment of streptococcal and staphylococcal TSSs.*

Stevens DL, Bisno AL, Chambers HF, et al. Practice guidelines for the diagnosis and management of skin and soft tissue infections: 2014 update by the Infectious Diseases Society of America. *Clin Infect Dis* 2014;59(2):e10-52. *This article provides guidelines for the diagnosis and treatment of all skin and soft-tissue infections.*

Stevens DL, Bryant AE. Necrotizing soft-tissue infections. *N Engl J Med* 2017;377(23):2253-2265. *An in-depth review of the clinical manifestations, pathogenesis and treatment of all types of necrotizing infections.*

Working Group on Severe Streptococcal Infections. Defining the group A streptococcal toxic shock syndrome: rationale and consensus definition. *JAMA* 1993;269:390-391. *This article describes the case definition for StrepTSS.*

# 22

# Superficial Dermatophyte Infections of the Skin

*April Schachtel, Katherine L. DeNiro*

## ABSTRACT

Fungal infections of the skin can be divided into superficial and deep infections. The vast majority are superficial and are caused by dermatophytes, which invade only fully keratinized tissues (stratum corneum, hair, and nails). They are classified into three genera: *Epidermophyton*, *Microsporum*, and *Trichophyton*. Less frequent causes include nondermatophyte fungi (e.g., *Malassezia furfur* in tinea versicolor) and *Candida* species. In naming clinical dermatophyte infections, traditionally *tinea* precedes the Latin name of the involved body region. Superficial fungal infections are divided into tinea barbae, tinea capitis, tinea corporis, tinea faciei, tinea manus, tinea pedis, tinea cruris, tinea unguium, tinea versicolor, and cutaneous candidiasis.

## DISEASE BURDEN

Dermatophyte infections are among the most common dermatologic conditions and are frequently refractory to treatment, leading to recurrence. In 2010, fungal skin diseases were the fourth most common of all diseases worldwide. The prevalence varies across the age spectrum; men are more likely to be affected than women. With the exception of tinea capitis, which primarily affects prepubertal children, the vast majority of dermatophytoses occur in adult hosts. Although dermatophytes rarely undergo deep local invasion and systemic dissemination, they can cause considerable morbidity, particularly in immunocompromised hosts.

## PATHOGENESIS AND RISK FACTORS

Dermatophytes are not endogenous pathogens. Transmission can occur directly from person to person (anthropophilic organisms), animal to person (zoophilic organisms), and soil to person (geophilic organisms) as well as from fomites like hair brushes and hats. Infective spores in dermal scales and hair can remain viable for months to years. Once the infectious elements of the fungus (arthroconidia) enter the skin, they germinate and invade the superficial skin layers. Dermatophytosis may cause defects in the skin barrier, which serve as a portal of entry to bacterial infections such as cellulitis. Disease severity is affected by several factors: sebum (which has an inhibitory effect on dermatophytes), the cutaneous barrier, genetic susceptibility to certain fungal infections, and the host's immune system.

## CLINICAL FEATURES

### Tinea Barbae

Tinea barbae, also known as *tinea sycosis* and *barber's itch*, is an uncommon infection that involves the terminal hairs and skin in the beard and mustache area of adult men. The lesions can be divided into deep and superficial. The deep type, sometimes known as tinea barbae profunda, tends to be more severe, nodular, and suppurative and is often caused by zoophilic

dermatophytes, namely *Trichophyton mentagrophytes* and *Trichophyton verrucosum*. The superficial type may cause crusted, partially bald patches with folliculitis; *Trichophyton rubrum* and *Trichophyton violaceum* are usually the causative organisms. Reversible alopecia may sometimes be seen accompanying these lesions. Farm workers are most often affected because the causative organisms are usually zoophilic dermatophytes. Autoinoculation may also occur after minor trauma or from a razor blade while shaving. The use of topical steroids may play a role as well.

The differential diagnosis includes bacterial folliculitis (due to *Staphylococcus aureus*), furuncle or carbuncle, perioral dermatitis, pseudofolliculitis barbae, contact dermatitis, and herpes simplex. The clinical diagnosis can be confirmed using microscopy to examine potassium hydroxide (KOH) of plucked hair or a biopsy specimen.

Tinea barbae is treated with systemic antifungal therapy. Common regimens in adults include terbinafine 250 mg/day, itraconazole 200 mg/day, griseofulvin 500 mg twice daily, and fluconazole 150 mg once weekly. Treatment is continued for 2 to 3 weeks after lesion resolution. Topical agents are helpful only as adjunctive therapy.

### Tinea Capitis

Tinea capitis is an infection of the scalp and hair shafts. It is most common in children and more common in boys than girls. Poor hygiene and overcrowding can foster transmission, which can occur through inanimate objects such as hats, brushes, and pillowcases. In the United States, 90% to 95% of tinea scalp infection is caused by the anthropophilic dermatophyte *Trichophyton tonsurans*, which does not fluoresce. Before 1950, fluorescent *Microsporum* species were the most common cause. KOH, fungal culture, and Wood lamp examination can be used for diagnostic testing. Tinea capitis begins with a small papule, which spreads to form scaly, irregular, or well-demarcated areas of alopecia. "Black dot" alopecia is produced when swollen hairs fracture a few millimeters from the scalp. Cervical and occipital lymphadenopathy may be prominent. A boggy, sterile inflammatory mass known as a kerion may also be seen, which can result in permanent alopecia if not treated promptly.

There are many scalp and hair conditions that lead to scaling or alopecia (scarring and nonscarring) that are not caused by fungal infections. When scaling and inflammation are prominent, diagnoses to be ruled out include atopic dermatitis, seborrheic dermatitis, and psoriasis. When alopecia is prominent, other diagnoses to consider include alopecia areata, traction alopecia, and trichotillomania.

Systemic antifungal therapy is required to penetrate the hair follicles. Griseofulvin 20 to 25 mg/kg/day for 8 weeks is commonly recommended (Fig. 22.1). Several other agents—including itraconazole, terbinafine, and fluconazole—have been reported as effective and safe. They allow a shorter course of treatment and can be used in patients with griseofulvin allergy or in recalcitrant cases. Terbinafine may be more effective in cases due to *T. tonsurans*, whereas griseofulvin is likely more effective in cases due to *M. canis*. Preventive measures are also important, because

Tinea capitis (ringworm) is an infection of the scalp with mold-like fungi called *dermatophytes*. It's usual name is *ringworm* because of its round lesions. It's the most common fungal infection in children, usually 2 to 10 years old, often African American boys.

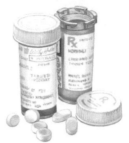

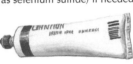

Oral medicines (such as griseofulvin or terbinafine) can cure tinea capitis, but treatment usually takes weeks to months. It's very important to finish the full course of treatment and follow the doctor's instructions. Otherwise, the infection can return.

The doctor may also prescribe special creams or shampoos (such as selenium sulfide) if needed.

The cause is the common fungus named *Trichophyton*. People get it by direct contact with infected people or infected animals, especially cats and dogs.

Avoid contact with infected people. Check family members for the disorder. Don't share combs, brushes, or hats with other people.

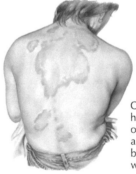

Check pets (such as cats, dogs, hamsters, and guinea pigs) for skin infections or irritations, and take your pet to your veterinarian if infections are present.

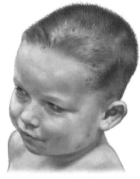

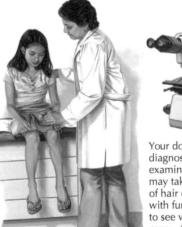

Symptoms may be mild and first include redness and swelling of the scalp, followed by hair loss. Infected hairs become brittle and broken. Pus-filled blisterlike lesions called *kerions* may occur.

Your doctor makes a diagnosis by a skin examination. The doctor may take small samples of hair or infected skin with fungi for culture, to see whether fungi grow. A microscope is used to study them.

Call your doctor if you have fever, pus drainage, or swelling, or if other areas of the scalp or body become infected even with treatment.

**Fig. 22.1** Managing tinea capitis.

the disease is contagious; all individuals residing with the infected patient should be evaluated and appropriately treated. Adjunctive topical therapy with antifungal shampoo, such as 2% ketoconazole or 2.5% selenium sulfide, may be useful to decrease shedding of viable fungi and spores.

### ✳ CLINICAL VIGNETTE

A 52-year-old man presented with an itchy rash on his left wrist, which had been present for 3 months. He had treated it with over-the-counter hydrocortisone with minimal improvement, and the lesions appeared to be spreading. His skin exam was notable for a 5-cm area of affected skin containing erythematous scaly papules with admixed pustules. KOH examination of the affected area reveals fungal hyphae. A diagnosis of Majocchi granuloma, a superficially invasive dermatophyte infection usually caused by *T. rubrum*, was made. Patients with Majocchi granuloma often have a history of corticosteroid use, minor trauma, immunosuppression, and/or of onychomycosis or tinea corporis. A KOH examination is not always positive; if it is negative, a skin biopsy with fungal staining may be required to make the diagnosis. Systemic antifungal treatment is necessary. Example regimens include terbinafine 250 mg PO daily for 4 weeks.

### Tinea Corporis

Tinea corporis typically presents on the trunk or extremities as a single lesion or multiple scaly annular lesions with a slightly raised, scaly erythematous edge and central clearing (Fig. 22.2). The border of the lesion may contain papules, pustules, or vesicles. The intensity of itch is variable. The disease is more common in tropical climates and can occur at any age. Although it can be caused by any of the dermatophytes, *T. rubrum*, *M. canis*, and *T. mentagrophytes* are the common organisms found in the United States. Risk factors for transmission include occupational or recreational exposure, contact with contaminated clothing and furniture, and a personal history of or close contact with tinea capitis, tinea pedis, or an affected pet. The diagnosis is typically based on clinical appearance and KOH microscopy of scrapings from the active edge. The differential diagnosis includes nummular eczema, granuloma annulare, psoriasis, contact dermatitis, pityriasis rosea, and tinea versicolor.

Clinical variants of tinea corporis include tinea incognito, tinea profunda, Majocchi granuloma, and tinea imbricata. *Tinea incognito* is a term applied to atypical clinical lesions produced from previous

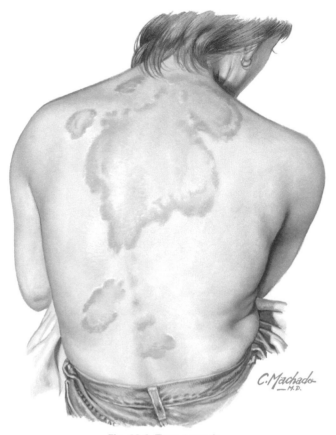

**Fig. 22.2** Tinea corporis.

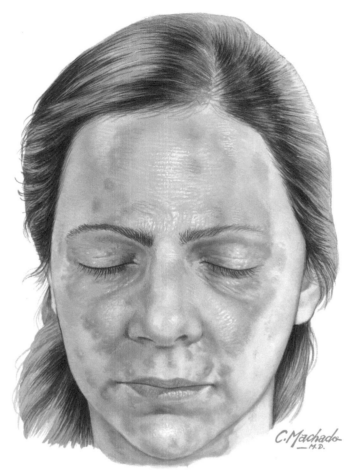

**Fig. 22.3** Tinea faciei.

topical corticosteroid use and, less commonly, calcineurin inhibitors. *Tinea profunda* refers to granulomatous or verrucous appearance from an excessive inflammatory response to a dermatophyte that is analogous to a kerion on the scalp. Majocchi granuloma is characterized by perifollicular pustules, papules, or granulomatous plaques that are typically caused by *T. rubrum*. It is more commonly seen in men. Tinea imbricata is caused by the anthropophilic dermatophyte *Trichophyton concentricum* and is limited to southwest Polynesia, Melanesia, Southeast Asia, India, and Central America. The clinical presentation consists of characteristic concentric rings of scales forming extensive patches with polycyclic borders resembling erythema gyratum repens.

Tinea corporis is usually treated with topical antifungal creams. Twice-daily application of topical terbinafine, clotrimazole, ketoconazole, or ciclopirox is usually very effective. Topical nystatin is ineffective against dermatophytes. The recurrence rate is high for those with extensive infections, who may require systemic antifungal therapy. Appropriate systemic agents for extensive disease include oral terbinafine 250 mg/day for 1 to 4 weeks, itraconazole 200 mg/day for 1 to 2 weeks, and fluconazole 150 mg to 300 mg once weekly for 2 to 4 weeks.

## Tinea Faciei

Tinea faciei is seen on the nonbearded areas of the face. The patient usually experiences itching and burning, which may become worse after exposure to sunlight (Fig. 22.3). Some lesions may have classic features (e.g., scale, annular configuration, pustules in the border), whereas others can be more difficult to diagnose clinically and require a high index of suspicion. Owing to the atypical presentation, tinea faciei is often confused with other skin disorders that affect the face, including contact dermatitis, rosacea, discoid lupus erythematosus, and seborrheic dermatitis. Usually the infection is caused by *T.*

*rubrum, T. mentagrophytes,* or *M. canis.* KOH microscopy of scrapings from the leading edge of the skin change may help to establish the diagnosis. Treatment is similar to that for tinea corporis.

## Tinea Manus and Tinea Pedis

Tinea manus is a superficial dermatophyte infection of the palm and/or interdigital spaces of one or both hands. It is a relatively uncommon infection and accounts for less than 2% of all of the superficial mycoses. *T. rubrum* is the most commonly associated pathogen, followed by *Microsporum canis* and other *Trichophyton* species. Occupations at high risk for tinea manus include massage therapists, hair stylists, and veterinarians. The clinical presentation of tinea manus is similar to that of other superficial dermatophytes and includes scaly erythematous papules and plaques that are often annular and pruritic. Tinea manus is frequently unilateral. Tinea manus isolated to one hand often occurs in concert with tinea pedis on both feet and is referred to as "two feet–one hand syndrome." Autoinoculation from feet to hands commonly occurs.

Tinea pedis refers to a superficial dermatophyte infection of the feet and is the most common form of dermatophytosis, as some estimates suggest that more than half of the population will be infected at some point in their lives (Fig. 22.4). *T. rubrum*, followed by *T. interdigitale* and *E. floccosum*, is the most frequently implicated dermatophyte. On examination, erythema and scale are invariably present. The interdigital presentation of tinea pedis is characterized by maceration, most commonly found within the interdigital spaces, particularly between the fourth and fifth toes; it can spread to the soles and instep of the foot with scaling and erythema. The maceration causes a skin barrier defect that predisposes to lower extremity cellulitis, particularly due to *S. aureus*

The two most common forms of tinea pedis are interdigital and moccasin.

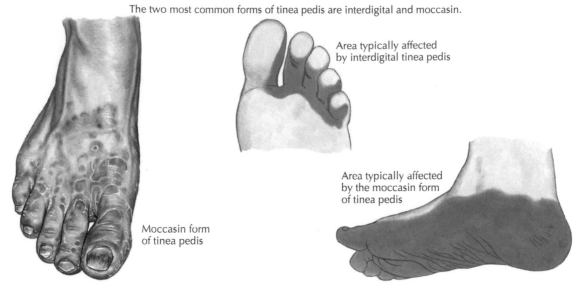

Area typically affected by interdigital tinea pedis

Area typically affected by the moccasin form of tinea pedis

Moccasin form of tinea pedis

**Dermatologic conditions that may mimic tinea pedis**
Other infectious and inflammatory disorders share some of the clinical features of tinea pedis, including scaling, pruritus, and (as seen in inflammatory forms of the condition) pruritic vesicles and bullae. Psoriasis, lichen simplex chronicus, eczema, and interdigital erythrasma should be considered in the differential diagnosis. A simple KOH examination of scale or vesicle roof can provide the correct diagnosis.

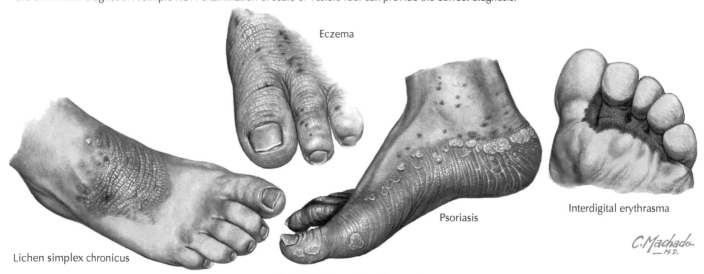

Eczema

Psoriasis

Interdigital erythrasma

Lichen simplex chronicus

**Fig. 22.4** Diagnosis of tinea pedis.

and beta-hemolytic streptococci. *Moccasin-type tinea pedis* refers to diffuse scale in a moccasin distribution along the sole; significant erythema may or may not be present. Vesicular or bullous tinea pedis is relatively uncommon. Acute ulcerative type of tinea pedis is characterized by ulcers and erosions in the web spaces with frequent extension onto the balls of the feet. This form is commonly secondarily infected with gram-negative bacteria such as *Pseudomonas aeruginosa* and is referred to as "gram-negative bacterial toe web infection (GNBTWI)" or "dermatophytosis complex." GNBTWI presents as exudative erythematous maceration of the interdigital spaces with a well-demarcated punched-out edge associated with malodor, itching and pain. The presence of the dermatophyte can be challenging to document during the acute inflammatory phase of GNBTWI, but swabs demonstrating gram-negative bacteria are invariably positive. Treatment of GNBTWI includes debridement of macerated skin, especially around the border; application of ciclopirox 0.77% gel twice daily for 8 weeks; and Burow solution soaks (aqueous solution of aluminum triacetate) to dry the moist skin. Prevention of recurrence with treatment of the predisposing tinea pedis with topical antifungal creams is necessary.

Tinea pedis may mimic other dermatologic diseases of the foot, including atopic dermatitis, contact dermatitis, or psoriasis. Confirmation of the diagnosis can be obtained by KOH prep showing branching hyphae or fungal culture. Tinea pedis is a predisposing factor in cellulitis, which should prompt treatment in high-risk individuals such as those with poorly controlled diabetes, recurrent cellulitis, or venous insufficiency. Treatment of both tinea manus and pedis can be achieved in most cases with topical antifungal creams such as terbinafine used once daily for 4 weeks. Recurrences are common; the use of topical antifungals prophylactically may help to prevent relapse.

## Tinea Cruris

*Tinea cruris* refers to a dermatophyte infection of the groin or crural fold (Fig. 22.5). Common dermatophytes implicated include *T. rubrum*, *E. floccosum*, and *T. interdigitale*. The infection is far more common in men, and risk increases with comorbid tinea pedis (through autoinoculation from the feet), obesity, immunosuppression, diabetes, and hyperhidrosis. The clinical exam demonstrates an erythematous annular plaque with scale most commonly within the inguinal fold, which may spread to the

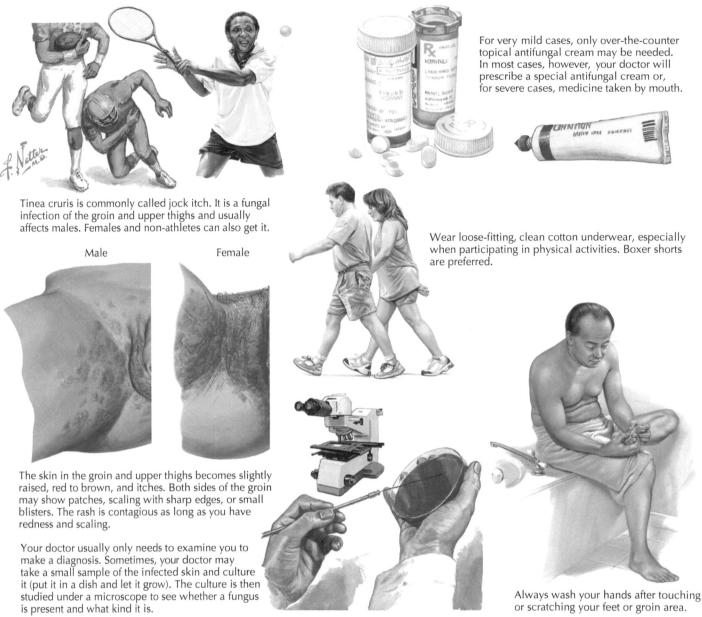

Tinea cruris is commonly called jock itch. It is a fungal infection of the groin and upper thighs and usually affects males. Females and non-athletes can also get it.

Male                    Female

The skin in the groin and upper thighs becomes slightly raised, red to brown, and itches. Both sides of the groin may show patches, scaling with sharp edges, or small blisters. The rash is contagious as long as you have redness and scaling.

Your doctor usually only needs to examine you to make a diagnosis. Sometimes, your doctor may take a small sample of the infected skin and culture it (put it in a dish and let it grow). The culture is then studied under a microscope to see whether a fungus is present and what kind it is.

For very mild cases, only over-the-counter topical antifungal cream may be needed. In most cases, however, your doctor will prescribe a special antifungal cream or, for severe cases, medicine taken by mouth.

Wear loose-fitting, clean cotton underwear, especially when participating in physical activities. Boxer shorts are preferred.

Always wash your hands after touching or scratching your feet or groin area.

Fig. 22.5 Managing tinea cruris.

perineum and buttocks; the scrotum is typically spared, perhaps due to the cooler temperature of this region. The differential diagnosis includes erythrasma (which is due to *Corynebacterium* infection and glows coral red with Wood lamp examination), seborrheic dermatitis, inverse psoriasis, and candidal intertrigo. The diagnosis of tinea cruris can be confirmed with a KOH examination of a sample taken from the scale along the active border. Treatment is similar to that for tinea corporis. Typical regimens include topical therapy with azoles, allylamines (such as terbinafine), butenafine, ciclopirox or tolnaftate. Oral antifungals are reserved for extensive or recalcitrant infections and regimens are similar to that of tinea corporis. Recurrence of tinea cruris is common and may be prevented by treating occurring tinea pedis and/or onychomycosis.

## Tinea Unguium

Tinea unguium is defined as an infection of the nail unit caused by dermatophytes. *Onychomycosis* is a broader term denoting an infection of the nail caused by either dermatophytes or nondermatophyte molds or yeasts. Tinea unguium is common, with a worldwide prevalence of 5%,

and is associated with reduced quality of life, pain, and social embarrassment. Risk factors for its acquisition include increasing age, immunosuppression, diabetes, tinea pedis, and infection of family members and/or household contacts.

The most common dermatophyte causing tinea unguium is *T. rubrum*, followed by *T. mentagrophytes*, *Epidermophyton floccosum*, and other *Trichophyton* species. Tinea unguium most commonly presents as thickening and discoloration of the nail plate with prominent subungual debris and onycholysis. Other less common presentations include longitudinal melanonychia (caused by *T. rubrum var nigricans*), a diffuse white/yellowish longitudinal streak (referred to as a dermatophytoma), and proximal superficial white onychomycosis, which can be seen in HIV or other causes of immunosuppression.

The differential diagnosis of multiple dystrophic nails includes psoriasis and lichen planus (both of which can occasionally affect only the nails without other cutaneous involvement), trauma, yellow nail syndrome, subungual or periungual warts, and paronychia. Given the broad differential diagnosis, it is important to confirm the diagnosis of dermatophyte infection prior to treatment in order to avoid treatment failures, incorrect diagnosis, medication side effects, and patient frustration. The diagnosis can be confirmed by either KOH prep of subungual debris, histology from a nail clipping, or fungal culture from subungual debris. Of these techniques, histology has the highest sensitivity and specificity but is also the most costly. Fungal culture is relatively inexpensive and highly specific but with lower sensitivity. Nondermatophyte molds, such as *Aspergillus* spp., *Scopulariopsis brevicaulis, Acremonium* and *Fusarium* spp, and yeasts such as *Candida* spp. can also cause onychomycosis. Differentiating a nondermatophyte organism recovered from culture as a colonizer versus causative agent of the infection can be difficult. In the case of suspected nondermatophyte onychomycosis, our approach is to perform histology on the nail plate and confirm the presence of hyphae, followed by two separate cultures taken at different times from subungual debris showing the same nondermatophyte organism.

Treatment of tinea unguium is challenging and relapse or reinfection can occur in up to 25% of individuals within the first year after treatment. Topical treatment options include ciclopirox 8% lacquer, efinaconazole

10% solution, and tavaborole 5% solution, but all have rates of complete cure in the range of only 8% to 15%. Furthermore, efinaconazole and tavaborole are expensive ($575 and $1576 per 4-mL bottle, respectively). An additional barrier to topical treatments is that they must be applied for 48 weeks, and patient compliance may wane over time. Topical antifungals may be more effective in children than in adults. Terbinafine and itraconazole are two oral medications approved by the US Food and Drug Administration (FDA) for the treatment tinea unguium, and both have higher complete cure rates than topical medications. Terbinafine is preferred over itraconazole, given fewer drug-drug interactions, fewer side effects, and higher complete cure rates. A typical treatment regimen for tinea unguium with terbinafine is 250 mg/day for 12 weeks. The necessity of laboratory monitoring for terbinafine is controversial.

## Tinea Versicolor

Tinea versicolor, also known as pityriasis versicolor, is a common superficial fungal infection caused by the overgrowth of nondermatophyte yeasts in the *Malassezia* genus (Fig. 22.6). *Malassezia* is ubiquitous; it is a component of normal skin flora that thrives in sebum-rich

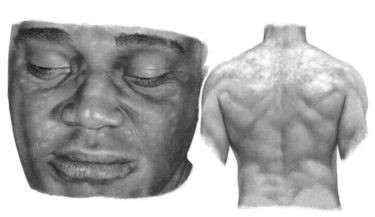

Tinea versicolor is a rash caused by fungus that usually lives on the skin. Teenagers and young adults can get it, more often in warmer weather, because fungus grows easily in heat and humidity. Unlike similar infections, tinea versicolor won't be passed among people.

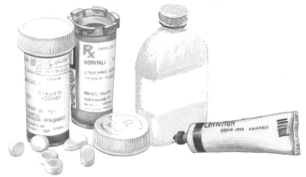

Your doctor may prescribe a cream, lotion, or pills to treat the rash. Treatment may last up to 1 month, but your skin may take several months to return to normal. So don't be discouraged if it doesn't look normal after a few weeks.

Sunlight may help your rash, but ask your doctor whether being in the sun is good for you, and if so, how long you should stay in the sun. Using sunscreen is also important.

Don't scratch at the rash. If the itching really bothers you, talk to your doctor about medicine to help relieve it.

The rash occurs on the upper arms, chest, back, and neck and sometimes the face. The rash can be various colors and has small, white-to-pink or tan-to-dark spots with sharp edges and scales.

Your doctor makes a diagnosis by seeing the usual appearance of the rash on the skin. Your doctor may take a small scraping of a patch for study with a microscope if the diagnosis isn't clear.

**Fig. 22.6** Managing tinea versicolor.

areas of the skin. The trunk and proximal extremities are the most commonly affected locations, and the eruption is most frequently seen in otherwise healthy adolescents or young adults. Living in a warm humid climate, physical activity with excessive sweating, diabetes, or immunosuppression may contribute to the overgrowth of *Malassezia*. The skin exam of tinea versicolor is notable for hyperpigmented or hypopigmented faintly scaly thin patches and plaques. Although most patients are asymptomatic, some may report distress from the uneven skin color, scaling, and sometimes pruritus. Patients often report that the involved skin fails to tan in the summer. This is because the yeast produces azelaic acid, which inhibits pigment transfer to keratinocytes, leading to accentuated demarcation of the uninfected skin. The diagnosis can be confirmed by a KOH prep showing short hyphae and yeast forms, commonly referred to as "spaghetti and meatballs."

Treatment of tinea versicolor with an antidandruff shampoo such as selenium sulfide 2.5% or ketoconazole 2% is often effective; patients should be counseled to apply the shampoo daily as a body wash, lather, and let sit for 5 to 10 minutes before rinsing; this should be repeated daily for 2 weeks. Using these shampoos as a body wash once weekly indefinitely may help to prevent recurrences. Patients should also be counseled that it is relatively straightforward to treat the overgrowth of yeast with topical medications. However, it can take many months for the skin color to return to normal, and this does not represent treatment failure. Should documented treatment failure occur or when, with extensive disease, patient compliance is difficult, oral azole antifungals such as itraconazole and fluconazole are effective. A typical regimen for fluconazole for tinea versicolor is 300 mg once weekly for 2 weeks. Itraconazole is given as 200 mg/day for 5 days.

## Cutaneous Candidiasis

Candidiasis of the skin is an infection caused most commonly by *Candida albicans*. *Candida* is a normal colonizer of the gastrointestinal and genitourinary tracts in humans, but it can thrive opportunistically in warm moist areas of the skin such as intertriginous regions of the groin, axillae, and both inframammary and abdominal folds. The clinical presentation of cutaneous candidiasis is characterized by beefy red pustules, papules, and plaques. Satellite papules and pustules adjacent to the larger primary lesions are hallmarks of cutaneous candidiasis. Light peripheral scale may be present. The diagnosis can be confirmed by KOH examination, which displays short hyphal elements and yeast forms. Treatment of candidiasis involves making the local environment less hospitable to yeast by avoiding tight-fitting clothing, keeping cool, and thoroughly drying intertriginous regions after bathing. Nystatin cream is commonly used for treatment. Topical azole antifungal therapy applied once or twice daily for approximately 2 weeks is also quite effective.

## DIAGNOSTIC METHODS

### Potassium Hydroxide Microscopy

KOH microscopy is the hallmark test for a quick and reliable diagnosis of superficial cutaneous dermatophyte infections. The technique for tinea infections on intact skin involves scraping scale onto a glass microscope slide. Our preferred technique is to wipe the skin with a wet alcohol pad to dampen the scale and then use a number 15 scalpel blade to superficially scrape the wet scale until it forms small damp pile on the side of the blade that can be wiped onto the microscope slide. For tinea unguium, the subungual debris should be collected, and for tinea capitis, any scale present should be obtained in addition to plucking several hairs in order to examine the hair bulb and shaft. A cover slip is then placed over the scale, hair or debris and a drop of

10% KOH solution is added, which will wick under the cover slip by capillary action. The underside of the slide is then heated briefly, after which it is examined under a light microscope. At first it can be difficult for the novice to identify the hyphae, but over time one can learn to distinguish the small fine branching hyphae from other debris and keratinocytes. KOH microscopy is fast, inexpensive, can be done in the office, and has a relatively high sensitivity and specificity when performed by a trained clinician. For the clinician who is unsure of his or her diagnostic abilities with a KOH prep, special stains such as calcofluor white can be used to help highlight the fungal elements. Finally, many laboratories can perform a KOH examination on submitted scale.

### Fungal Culture

The advantage of a fungal culture for the diagnosis of dermatophyte infections is its high specificity. However, the sensitivity is relatively low, as organisms can be challenging and slow to grow. Scale should be submitted for suspected tinea infection of intact skin, and subungual debris and several plucked hairs submitted for tinea unguium and tinea capitis, respectively.

### Wood Lamp Examination

Most dermatophytes do not fluoresce under a Wood lamp. The main benefit of using a Wood lamp lies in ruling out erythrasma (caused by the bacterium *Corynebacterium minutissimum*), which fluoresces a brilliant coral red, whereas tinea cruris and cutaneous candidiasis do not fluoresce. Tinea versicolor produces a pale white-yellow fluorescence.

### Histology

Submission of a skin biopsy specimen or nail clipping for histology and PAS stain to aid in the diagnosis of a dermatophyte infection has a high sensitivity and specificity but is relatively expensive and requires expertise in the biopsy technique.

## ACKNOWLEDGMENTS

The authors wish to acknowledge the work of previous edition authors Gregory Raugi and Thao U. Nguyen.

## EVIDENCE

Gupta AK, Skinner AR, Cooper EA. Evaluation of the efficacy of ciclopirox 0.77% gel in the treatment of tinea pedis interdigitalis (dermatophytosis complex) in a randomized, double-blind, placebo-controlled trial. *Int J Dermatol* 2005;44(7):590-3. *Randomized controlled trial for the treatment of dermatophytosis complex (GNBTWI)with ciclopirox.*

Hay RJ, et al. The global burden of skin disease in 2010: an analysis of the prevalence and impact of skin conditions. *J Invest Dermatol* 2014;134(6):1527-1534. *A study examining the most frequent diseases worldwide with a focus on skin diseases and the disability caused by each.*

Ilkit M, Durdu M. Tinea pedis: the etiology and global epidemiology of a common fungal infection. *Crit Rev Microbiol* 2015;41(3):374-88. *Epidemiological study of tinea pedis.*

Veraldi S, Schianchi R, Benzecry V, Gorani A. Tinea manuum: a report of 18 cases observed in the metropolitan area of Milan and review of the literature. *Mycoses* 2019;62(7):604-608. *Identifies occupations at highest risk of developing tinea manus.*

## ADDITIONAL RESOURCES

Boral H, Durdu M, Ilkit M. Majocchi's granuloma: current perspectives. *Infection and drug resistance* 2018;11:751–760. *A review of updated factors predisposing to Majocchi granuloma.*

Chen X, Jiang X, Yang M, et al. Systemic antifungal therapy for tinea capitis in children: an abridged Cochrane Review. *J Am Acad Dermatol* 2017;76(2):368-374. *A review of randomized trials comparing the effectiveness of different systemic treatments for tinea capitis.*

Ely JW, Rosenfeld S, Seabury Stone M. Diagnosis and management of tinea infections. *Am Fam Physician* 2014;90(10):702-710. *A review of diagnostic methods and treatments for common dermatophyte infections.*

Hainer BL. Dermatophyte infections. *Am Fam Physician* 2003;67(1):101-108. *A review of diagnostic methods and treatments for common dermatophyte infections.*

Hay R. Therapy of skin, hair and nail fungal infections. *J Fungi* 2018;4(3):99. *An overview of updated treatments for common superficial fungal infections.*

Lipner SR, Scher RK. Onychomycosis: clinical overview and diagnosis. *J Am Acad Dermatol* 2019;80(4):835-851. *Review article with up-to-date summary of the presentation and diagnosis of onychomycosis, with useful table summarizing sensitivity, specificity and cost of diagnostic methods for onychomycosis.*

Lipner SR, Scher RK. Onychomycosis: treatment and prevention of recurrence. *J Am Acad Dermatol* 2019;80(4):853-867. *Review article of the treatment of onychomycosis.*

Weidner T, Tittelbach J, Illing T, Elsner P. Gram-negative bacterial toe web infection—a systematic review. *J Eur Acad Dermatol Venereol* 2018;32(1):39-47. *Excellent review article covering terminology, risk factors, and treatment of dermatophytosis complex (GNBTWI).*

# 23

# Herpes Simplex Virus Infection

*John D. Kriesel, Christopher M. Hull, Christina Topham*

## ABSTRACT

Herpes simplex viruses (HSVs) are double-stranded deoxyribonucleic acid (DNA) viruses that cause lifelong infection and frequent reinfections or reactivations. There are two types of HSV: HSV-1, the cause of human cold sores, and HSV-2, the usual cause of genital herpes. Each virus type has different clinical manifestations, modes of transmission, and epidemiologies. The development of type-specific serologic assays has allowed for differentiation between the two strains of viruses. These sensitive and specific serologic assays have expanded knowledge about the geographic distribution, the burden of disease, and risk factors for HSV-1 and HSV-2. Understanding of these diseases is growing, and there have been advances in both the treatment and the prevention of HSV infections with antiviral medications.

## CLINICAL VIGNETTE

A 20-year-old male presented to his dermatology clinic with a rash on his lower lip. He also endorsed fever, malaise, myalgias, and headaches. The day prior, he had complained of a tingling sensation involving his lower lip. He denies any history of similar rashes, herpes infections, or recent exposures. He admits to a sunburn of his lips after a boating trip the weekend prior. Physical exam demonstrated grouped vesicles, erosions, and ulcers on a faintly red base localized to the vermillion border. A vesicular lesion was unroofed and swabbed for viral polymerase chain reaction (PCR).

COMMENT: This patient is a classic example of herpes labialis occurring after ultraviolet (UV) exposure. Lesions of herpes labialis tend to occur on the outer third of the lips, and the lower lip is more frequently involved than the upper lip. Lesions classically progress through distinct stages, as seen in the patient presented—a prodrome of localized tingling, erythema, papule, vesicle, ulcer, crust, and healing. Diagnosis can be confirmed through viral culture of a young vesicle or PCR for herpes simplex virus DNA. Antiviral therapy can be used to treat acute episodes of herpes labialis or used for prophylaxis prior to high intensity UV exposure. Prophylaxis options include acyclovir 400 mg bid, valacyclovir 500 mg daily, or famciclovir 250 mg bid and should be initiated 24 to 48 hours before exposure, with continuation until removal of the UV exposure.

## GEOGRAPHIC DISTRIBUTION AND MAGNITUDE OF DISEASE BURDEN

HSVs are common and ubiquitous worldwide pathogens. There are two identified strains of HSVs: HSV-1 and HSV-2. The most common clinical presentation of HSV-1 infection is herpes labialis, commonly known as *cold sores* or *fever blisters*. Herpes labialis occurs in approximately 20% to 40% of the general population and manifests as recurrent vesicular herpetic lesions on the lips or around the mouth.

Serologic studies show that approximately 60% of the adult population in the United States is infected with HSV-1, although many infected persons are asymptomatic (i.e., do not have recognized herpes labialis or regular outbreaks).

HSV-2 is the primary cause of genital herpes, which manifests as recurrent genital vesicular and ulcerative lesions. Genital herpes is one of the most common sexually transmitted infections worldwide. The seroprevalence of HSV-2 infection varies widely within the US adult population, but overall it is approximately 20%. This translates to some 60 million infected persons in this country, approximately half of whom have recognized genital herpes disease.

## RISK FACTORS FOR INFECTION AND DISEASE

Most HSV-1 reactivations are mild, although uncomfortable and cosmetically disfiguring. In persons with an underlying immunosuppressing disease, active facial and intraoral HSV-1 infection can be persistent and may spread to cause major morbidity. The same is true for HSV-2 infections. Patients with acquired immunodeficiency syndrome (AIDS), for instance, sometimes experience continuous genital or perirectal ulcerations because of persistent replication of HSV-2.

Primary orofacial infection with HSV-1 is predominantly acquired during childhood and is often asymptomatic. Age, socioeconomic status, and geographic location affect the frequency of HSV-1 infection. Women are somewhat more susceptible to genital herpes than men. Other risk factors for HSV-2 infection include a high number of lifetime sexual partners, a history of sexually transmitted diseases, and early age of first intercourse.

## CLINICAL FEATURES

Primary infection with HSV occurs via inoculation of the oral or genital mucosa. The virus must contact mucosal surfaces or abraded or eroded skin, where it replicates and initiates infection. HSV-1 is spread primarily via direct contact with contaminated secretions or saliva. HSV-2 is usually spread via sexual contact from infected genital skin or secretions. The spectrum of primary infection is variable, ranging from asymptomatic infection to fulminant gingivostomatitis or genital infection, often with associated constitutional symptoms including fever, malaise, headache, and myalgia. More important, the virus has the ability to ascend through sensory nerve axons to establish chronic, lifelong, latent infection within the trigeminal, geniculate, vagal, and sacral ganglia (Fig. 23.1). These ganglia contain the sensory neuron cell bodies. HSV-1 and HSV-2 do not integrate into the human genome but rather are maintained in an episomal (or extrachromosomal) state within the sensory neuron nucleus.

On reactivation of HSV from its latent state, the virus descends along sensory neurons to reinvade cutaneous tissues, where it appears

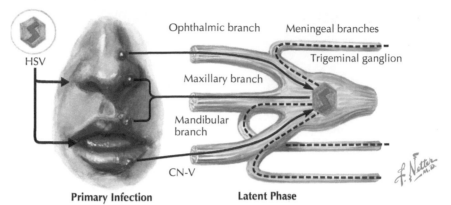

**Primary Infection**

Virus enters via cutaneous or mucosal surfaces to infect sensory or autonomic nerve endings with transport to cell bodies in ganglia.

**Latent Phase**

Virus replicates in ganglia before establishing latent phase.

**Fig. 23.1** Herpes simplex. *HSV,* Herpes simplex virus.

as vesicles and ulcers. Oral HSV outbreaks are often preceded by a pro-drome consisting of localized burning, tenderness, tingling, or pain near the site of reactivation. Other associated symptoms may include headache, fever, malaise, lymphadenopathy, and fatigue. Genital her-pes outbreaks are also commonly preceded by prodromal symptoms including dysesthesias and muscle cramping.

## Herpes Labialis

Recurrent HSV-1 lesions usually develop on the lower lip but can also be found on the upper lip, nose, cheek, chin, and eyelid (Fig. 23.2). The resultant epithelial cell death and inflammatory response leads to the characteristic vesicular (sometimes even pustular) and ulcerative or necrotic lesions. The lesions of herpes labialis tend to occur on the outer third of the lips, and the lower lip is more frequently involved than the upper lip. Lesions on the nose, chin, or cheeks account for less than 10% of cases. Lesions are most often single; "secondary lesions," those appearing one or more days after the first sore, develop in one-fifth of cases. From episode to episode, lesions commonly cross the midline of the face or move from one lip to the other if the patient experiences frequent episodes; in patients with infrequent recurrences, the lesion location generally remains the same. The healing time for herpes labialis lesions is variable, with the majority healed within 7 or 8 days. Most patients will have one to three outbreaks per year, but approximately 10% will have more frequent recurrences (more than six lesions per year).

The severity of recurrent herpes labialis is variable, ranging from prodromal symptoms without the development of any signs to exten-sive disease of both lips and cheeks after severe sunburn. Lesions that do not progress beyond the papule stage have been called "aborted" lesions. Aborted genital outbreaks also occur commonly, especially with early antiviral treatment. Episodes in which there is complete destruction of the epithelium, manifested by the development of the vesicle, ulcer, and/or crust stages, have been called "classical" lesions. Aborted lesions are the outcome of incipient episodes in approxi-mately 25% of cases. Of these lesions, approximately half are limited to prodromal symptoms, with or without erythema, and the others progress to the papule stage before resolving. Although the former are often termed "false" prodromes because of failure of the episode to progress, they are associated with a 60% rate of HSV excretion in the oral cavity and thus appear to be caused by reactivation of the virus. The classic herpes virus lesion progresses through distinct and identi-fiable stages from a prodrome (localized tingling or burning at the site of herpes reactivation), erythema, papule (edema), vesicle, ulcer, crust (soft debris then hard eschar), and healed (loss of crust).

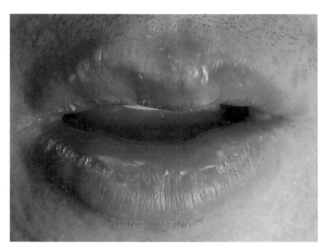

**Fig. 23.2** Herpes labialis. (Courtesy Salt Lake City/County STD Clinic.)

Well-documented stimuli that appear to induce HSV-1 recurrences in humans include UV exposure (sunburn) of the lips, febrile illness, and menstruation. Some medical procedures also induce herpes labi-alis, including surgical manipulation of the trigeminal nerve (to treat trigeminal neuralgia), hyperthermia, laser-assisted in situ keratomil-eusis (LASIK), epidural procedures, chemical or laser resurfacing of the face and others. The mechanisms by which trigger factors induce reactivation of HSV-1 are not completely understood, but they seem to relate to interferon effectors in the skin and/or ganglion.

## Genital Herpes

Primary genital herpes is most often caused by HSV-2 infection, although up to 25% of primary genital outbreaks may be caused by HSV-1, presumably transmitted by oral-genital contact. A patient's first genital herpes outbreak is typically more severe than subsequent recurrences. Patients with primary genital herpes due to HSV-2 should understand that they are very likely to experience one or more genital herpes recurrences in the next few months. Those with HSV-1 genital disease suffer fewer, and sometimes no, recurrences.

Recurrent genital herpes lesions need not follow a definite primary outbreak; for unknown reasons, some previously HSV-2–infected patients simply begin to have recurrent genital herpes lesions. In men, genital herpes lesions are often vesicular (blisters with clear fluid) or pustular. They are usually painful and most often occur on the shaft of the penis but also are common on the pubis, groin, thigh, buttocks, and scrotum. In women, genital herpes lesions are more likely to manifest

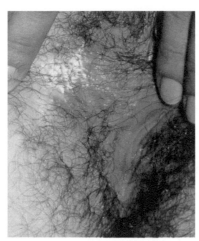

Fig. 23.3 Genital herpes. This is a typical genital herpes recurrence with erythema in the labial folds of a young woman. Genital recurrences are usually painful and may present manifest as bumps, fissures, or vesicles. (Courtesy Salt Lake City/County STD Clinic.)

as a small ulcer or fissure on the vulva or perineum (Fig. 23.3). Women may also experience genital herpes outbreaks on the pubis, groin, buttocks, or lower back. Men who have sex with men are more likely to have perirectal genital herpes outbreaks. Primary genital herpes may manifest as proctitis in men who have sex with men.

Genital herpes outbreaks are often heralded by a prodrome. Prodromal symptoms include itching, burning, dysuria, and other abnormal sensations in the genital area. Some genital herpes patients may have buttock or leg pain, occasionally mimicking sciatica, as a prodromal symptom. Occasionally, typical genital herpes prodromal symptoms are not followed by a visible outbreak. These "aborted" genital herpes episodes are more common when antiviral treatment is started promptly.

The frequency of genital herpes outbreaks is highly variable, ranging from one every few years to almost monthly, and HSV-2 infection that is completely asymptomatic is not uncommon. Triggering factors are difficult to identify because the outbreaks occur randomly. Some patients believe that stress, vigorous sex, lack of sleep, or sun exposure leads to outbreaks. Most lesions heal in 1 to 2 weeks. Genital herpes lesions can be very severe and persistent in immunocompromised individuals, including those with AIDS.

### Asymptomatic Shedding and Unrecognized Infection

Infectious HSV is shed from the genital or oral mucosa on approximately 2% to 3% of days. Shedding of HSV DNA at these sites is much more common, occurring on 25% to 40% of days. Studies show that only 10% to 25% of persons with HSV-2 antibodies identify themselves as having genital herpes disease. It follows that most genital herpes transmissions occur either from persons without recognized disease or during periods when the transmitting partner did not have a recognized outbreak.

### Other Manifestations of Herpes Simplex Virus Infection

HSV infections can occur elsewhere and can present with life-threatening infection in some cases. *Eczema herpeticum* (also referred to as *Kaposi varicelliform eruption*) is a term used for disseminated cutaneous infection by HSV in patients with other chronic dermatologic conditions, including atopic dermatitis. Herpes gladiatorum is a unique cutaneous infection with HSV seen primarily in individuals who wrestle. Transmission of virus occurs from infected individuals to susceptible persons during contact. Lesions usually develop on the

lateral neck, face, and forearms, areas in direct contact with the face of the infected wrestling partner. *Herpetic whitlow* refers to infection of the digits, often the index finger. Whitlow caused by HSV-1 was more common before the widespread use of protective gloves in dental and medical personnel. Most cases of herpetic whitlow now are caused by HSV-2 infection, presumably acquired from sexual activity.

Erythema multiforme (EM) is an acute, self-limited, cutaneous eruption characterized by the development of dusky erythematous macules. These are often referred to as "targetoid" because of their central dusky or purple zone and outer erythematous rim. Occasionally these are associated with central vesiculation or ulceration. The lesions of EM are commonly found on the hands, forearms, and oral mucosa. The development of EM is most commonly precipitated by HSV infection. Preceding herpes labialis will be seen in approximately 50% of subjects with EM. The herpes labialis lesions may develop before, simultaneously, or after the lesions of EM. The pathogenesis of HSV-associated EM is unclear but is likely related to an HSV-specific host response. HSV-encoded proteins and HSV DNA can be identified within lesional skin of EM, and virally encoded antigens have been detected on keratinocytes.

HSV-1 may also infect and reactivate in the cornea. There are approximately 25,000 new cases of ocular herpes per year in the United States, with a total case burden of approximately 400,000. These cases are particularly problematic in that ocular recurrences are difficult to prevent, treatment is not entirely satisfactory, and frequent ocular reactivations can overwhelm the natural antiinflammatory capacity of the cornea, leading to blindness.

Herpes encephalitis can be caused by either HSV-1 or HSV-2, although HSV-1 predominates. This disease is rare and may be insidious or rapid in onset, often heralded by personality changes or mood swings. Virtually all patients proven to have HSV encephalitis have fever. Mollaret (aseptic) meningitis is a related condition caused by recurrent HSV-2 reactivation from the sacral ganglia into the spinal cord and meninges, often without accompanying genital lesions. Neonatal herpes is a rare but particularly feared complication of genital HSV infection. These cases are usually caused by HSV-2 passed to the child either during gestation or, more commonly, at the time of delivery.

## DIAGNOSTIC APPROACH

There are a number of laboratory tests that can be used to confirm the diagnosis of HSV. PCR for HSV-1 and HSV-2 has now supplanted viral cultures at many institutions. Viral cultures are still used on occasion (e.g., with resistance to antiviral drugs is suspected). The virus and/or its DNA should be obtained from the surface of infected tissue or fluid (vesicles, ulcerations, cornea, throat, and other sites). Cultures are most sensitive within a few days of the onset of the outbreak, and PCR can detect viral DNA for up to 2 weeks. If possible, the lesion to be tested (vesicle or pustule) should be unroofed or carefully broken with a sterile needle. The resulting fluid should be collected on a Dacron swab, placed in viral transport medium, and sent to the laboratory for testing. Typing of cultured virus or PCR amplicons distinguishes between HSV-1 and HSV-2. HSV typing is strongly recommended because this often affects the patient's treatment and prognosis.

Direct florescence antigen (DFA) testing is a rapid, sensitive, and reproducible test for both HSV-1 and HSV-2. The test is performed by swabbing the base of a vesicle with a 15 blade or sterile cotton-tipped swab and transferring the cells to the glass slide from a DFA kit. Following fixing of the slide, the specimen is then sent for immunofluorescence analysis.

HSV antibody tests have been developed to take advantage of antigenic differences between HSV-1 and HSV-2. The type-specific enzyme-linked immunosorbent assay (ELISA) test is based on differences in the HSV envelope glycoprotein G between HSV-1 and HSV-2. This relatively simple serologic assay has sensitivity and specificity exceeding 90%, similar to the more difficult and expensive Western blot method. It is crucial to be sure one is ordering a type-specific assay when performing a herpes serology. The older non–type-specific assays frequently lead to confusion because they cannot distinguish between HSV-1 and HSV-2. There are several US Food and Drug Administration (FDA)-approved HSV type-specific serologic assays available in the United States. These blood tests cannot determine whether HSV infection is oral, genital, or both. A recent systematic review found that HSV-2–specific serologic tests have low specificity and high false-positive rates when used for screening at-risk asymptomatic individuals. Therefore serologic screening for genital herpes in asymptomatic individuals is not recommended given the potential for psychosocial harm with a false-positive result.

## DIFFERENTIAL DIAGNOSIS

The differential diagnosis for herpes labialis should include folliculitis, aphthous stomatitis, oral candidiasis, medication-induced stomatitis, pharyngitis, Stevens-Johnson syndrome, and hand, foot, and mouth disease (coxsackievirus infection). Intraoral lesions of HSV, clusters of tiny (1 to 3 mm) vesicles and ulcers, develop exclusively on the gingivae and anterior hard palate in immunocompetent patients, are difficult to see, and are less symptomatic than their cutaneous counterpart. Recurrent HSV lesions seldom appear on the soft palate or posterior pharynx. Aphthous stomatitis, or "canker sores," are common intraoral ulcerations of unknown cause. In contrast to intraoral HSV lesions, aphthous ulcers are located on the tongue or buccal mucous membranes and are generally larger, more painful, single lesions.

Discrete genital or anal ulcers in sexually active young adults in most parts of the United States are genital herpes until proven otherwise. Chancroid, syphilis, and lymphogranuloma venereum can also cause genital ulcerative lesions. Primary syphilis may be distinguished from other ulcers by the presence of an indurated, nonpurulent ulcer (that may be painful). Other ulcer characteristics are not helpful in distinguishing infectious causes. Furunculosis, often caused by *Staphylococcus aureus*, can also appear similar to genital herpes. Shaving the genital area appears to predispose to furunculosis, which can be diagnosed by a Gram stain of the lesion exudate and/or a negative HSV culture. Diagnostic testing of suspicious genital lesions, including recurring bumps and fissures, is critical to prevent a missed diagnosis of genital herpes.

## TREATMENT

Major progress has been made in recent years in the understanding of HSV infections and the development of safe and effective antiviral drugs. Acyclovir is the prototype antiviral medication and functions as a nucleoside analogue that competitively inhibits viral (but not human) DNA polymerase. Acyclovir must first be phosphorylated by the herpes-specific thymidine kinase and then phosphorylated two more times by host cell enzymes to the active form. The active drug then competes for binding of the viral DNA polymerase in virus-infected cells. This limits replication and further spread of the virus but does not prevent death of the infected cells. Acyclovir is available in topical, oral, and intravenous formulations. Valacyclovir is the l-valyl ester prodrug of acyclovir. After oral absorption, it is metabolized to acyclovir and has the same mechanism of action and safety profile as acyclovir. The oral bioavailability of valacyclovir is threefold to fivefold

**TABLE 23.1** **Treatment Recommendations for Herpes Simplex Virus Infections**

| Infection | Recommended Treatment |
|---|---|
| Herpes labialis (recurrence) | VACV 2 g PO bid × 1 day<br>FACV 1.5 g PO × 1 dose |
| Genital herpes (first episode) | ACV 400 mg tid × 7–10 days<br>ACV 200 mg 5× a day × 7–10 days<br>FACV 250 mg tid × 7–10 days<br>VACV 1 g PO bid × 7–10 days |
| Genital herpes (recurrence) | ACV 400 mg tid × 5 days<br>VACV 500 mg bid × 3 days<br>FACV 1 g bid × 1 day |
| Genital herpes (suppression) | ACV 400 mg bid<br>VACV 500–1000 mg once daily<br>FACV 250 mg bid |
| Neonatal herpes | ACV 10 mg/kg IV q8h × 10–21 days |
| Immunocompromised | ACV 400 mg tid × 5–10 days<br>FCV 500 mg bid × 5–10 days<br>VACV 1000 mg bid × 5–10 days<br>IV ACV 5 mg/kg q8h × 5–7 days |

*ACV*, Acyclovir; *bid*, twice per day; *FACV*, famciclovir; *IV*, intravenously; *PO*, by mouth; *tid*, three times per day; *VACV*, valacyclovir.

higher than that of acyclovir. Famciclovir is the prodrug of penciclovir. Penciclovir (Denavir) is available only in a topical formulation because of poor oral bioavailability. Famciclovir, like valacyclovir, has a higher oral bioavailability and is metabolized to famciclovir after oral absorption. Famciclovir and penciclovir have a mechanism of action similar to that of acyclovir and valacyclovir and inhibit viral DNA polymerase. Table 23.1 gives specific treatment recommendations.

### Herpes Labialis

Episodic or prophylactic treatment with antiviral drug therapy is the current standard of care for recurrent herpes labialis infections. A variety of topical over-the-counter (OTC) preparations are available, but in the majority of cases the mechanism of action is not clear and rigorous clinical trials to define efficacy have not been performed. Abreva (10% docosanol cream) has been the most intensively studied product and is approved for OTC sale in the United States. Two controlled human studies showed modest effects from treatment.

Studies of acyclovir ointment in immunocompetent subjects have provided little or no evidence of efficacy. However, acyclovir ointment was effective for herpes labialis in immunocompromised patients and was approved for this indication in the United States. Acyclovir was shown to penetrate human skin more effectively from cream than ointment formulation. Accordingly, data supporting the efficacy of acyclovir cream have been more readily obtained than for the ointment. Acyclovir 5% cream is applied topically to the affected area five times per day for 4 days. Penciclovir 1% cream is an alternative topical treatment that is applied topically to the affected area every 2 hours for 4 days. Acyclovir/hydrocortisone cream is a newer option topically available for recurrent herpes labialis. The addition of hydrocortisone (a topical corticosteroid) improved healing times and lesion severity compared to topical acyclovir alone.

In most instances, oral antiviral therapy for herpes labialis is preferred because of the limited efficacy of topical treatments. Acyclovir, valacyclovir, and famciclovir have all been used for episodic treatment of herpes labialis. With early initiation of therapy, acyclovir, valacyclovir, and famciclovir reduce lesion healing time, viral shedding, and

## TABLE 23.2   Prophylaxis of Herpes Simplex Infections

| Condition or Stimulus | Treatment | Comments |
|---|---|---|
| Ultraviolet radiation | ACV 400 mg bid<br>VACV 500–1000 mg daily | Start at least 1 day before ultraviolet exposure and continue for 7 days. |
| Facial resurfacing: laser resurfacing, chemical peels, dermabrasion | ACV 400 mg bid<br>VACV 500 mg daily, bid<br>FACV 250 mg bid | Start 1–2 days before procedure and continue for 7 days or until reepithelialization. |
| Recurrent erythema multiforme | ACV 400 mg bid | Suppressive therapy needed.<br>Episodic therapy does not appear to be helpful. |
| Frequent herpes labialis recurrences (six or more per year) | ACV 400 mg bid<br>VACV 500 mg daily | May also be considered in patients with less frequent outbreaks whose appearance is very important or those who experience severe anxiety with outbreaks. |
| Herpes gladiatorum | VACV 500–1000 mg daily | Any wrestler with a history of herpes labialis or gladiatorum should be considered for suppressive therapy during periods of training and competition. |
| Ocular herpes | ACV 400 mg bid | Important to consider in individuals with a history of recurrent ocular HSV and in patients undergoing ocular procedures. Prevents only approximately 50% of outbreaks. |
| Genital herpes | ACV 400 mg bid<br>VACV 500–1000 mg daily<br>FACV 250 mg bid | Highly effective for the prevention of HSV-2 outbreaks.<br>If using VACV, start with 1000 mg daily and then may dose reduce after 1 year. |

*ACV*, Acyclovir; *bid*, twice per day; *FACV*, famciclovir; *HSV*, herpes simplex virus; *VACV*, valacyclovir.

pain. High-dose, short-course therapy with valacyclovir and famciclovir have become the most commonly used treatments for episodic herpes labialis. Valacyclovir is given at a dosage of 2 g twice a day for 1 day, and famciclovir is given as a single 1500-mg dose at the earliest signs of herpes labialis recurrence.

Three previous exploratory studies have investigated the effect of topical corticosteroids on the severity of recurrent herpes labialis. In these studies, the addition of a topical corticosteroid to an antiviral compound reduced lesion size and increased the number of aborted lesions. In all three studies, the antiviral-corticosteroid combinations were well tolerated, and there were no increased adverse events attributable to topical corticosteroids. The expanded clinical activity demonstrated in these studies (more rapid healing and increased aborted lesions) constitutes an improvement compared with what is seen with antiviral medications alone (more rapid healing) and supports the concept that corticosteroids are a valuable new therapeutic modality in this disease. Larger controlled studies are needed to expand and confirm these findings.

Prophylactic oral antiviral therapy for certain herpes labialis recurrences may be an effective and appropriate management strategy for selected patients. Patients who may benefit from this approach include individuals with frequent recurrences (more than six per year), patients with a history of HSV-associated EM, individuals anticipating intense sun exposure, patients undergoing certain surgical procedures, immunocompromised patients, and wrestlers with a history of herpes gladiatorum (Table 23.2).

## Genital Herpes

For some patients, genital herpes is considered to be nothing more than a nuisance that occurs uncommonly. For these patients, episodic treatment or no treatment at all may be appropriate. However, for other patients, particularly those with very frequent or very severe outbreaks, genital herpes is a stressful and difficult disease. There is an undeserved stigma attached to genital herpes—an infection borne by nearly 20% of the adult population—and most patients require reassurance and appropriate counseling. Patients with genital herpes caused by HSV-1

usually experience fewer outbreaks (0 to 1 per year) than those with HSV-2 genital herpes (3 or more per year). Therefore viral typing of genital herpes lesions is important and should be performed whenever possible. For patients with at least four recurrences of genital herpes per year or those with actual or perceived severe genital herpes, prophylactic antiviral therapy is appropriate.

Episodic treatment of genital herpes outbreaks with oral antivirals decreases the duration of the outbreaks by approximately one-third. The duration of live viral shedding is similarly decreased. Initiation of therapy immediately on recognition of the outbreak is recommended, to maximize the effect of antiviral therapy. Patients are sometimes disappointed by the limited efficacy of this treatment strategy, but by the time antiviral therapy is taken and absorbed, in most instances, HSV has already reactivated within the sensory neuron, traveled down the axon, and infected the innervated skin, creating a visible lesion. Antiviral medications act by preventing further viral replication; they cannot by themselves affect the natural inflammation already set in motion by the existing viral infection.

The current Centers for Disease Control and Prevention (CDC)-recommended regimens for episodic treatment of genital herpes outbreaks include acyclovir 400 mg three times per day for 5 days, acyclovir 800 mg twice a day for 5 days, acyclovir 800 mg three times a day for 2 days, famciclovir 125 mg twice a day for 5 days, famciclovir 1000 mg twice a day for 1 day, valacyclovir 500 mg twice a day for 3 days, and valacyclovir 1000 mg daily for 5 days. The cost of generically available acyclovir versus the more expensive valacyclovir and famciclovir regimens often figures into the decision about which antiviral to prescribe. On the other hand, the ease and convenience of the less frequent valacyclovir and famciclovir regimens favor compliance in patients who can afford them.

Patients experiencing their first (primary) episode of genital herpes should receive 10 days of oral antiviral therapy (acyclovir 400 mg three times per day, acyclovir 200 mg five times per day, valacyclovir 1000 mg orally twice a day or famciclovir 250 mg three times a day). Treating primary genital herpes promptly reduces the duration of symptoms. Topical antivirals are not effective in this setting.

Prophylactic therapy with daily antivirals is remarkably effective, preventing 95% of HSV-2–induced lesions (see Table 23.2). The cost and inconvenience of daily therapy should be weighed against the expected effects of prophylaxis.

Shorter courses of antiviral medications have also been studied and proven to be effective for the treatment of genital herpes. These studies have been patient-initiated to maximize the effect of antiviral treatment. In addition to the CDC-recommended 3-day valacyclovir course, acyclovir 800 mg three times a day for 2 days is effective. This high-dose regimen resulted in aborted lesions in 27% of the patients without decreasing the time to next recurrence. A shorter-course (2-day) treatment with famciclovir 500 to 250 mg twice a day was recently shown to be as effective as a full 5-day treatment course.

## PROGNOSIS

Herpes simplex infections cannot be cured with existing antiviral medications. Many patients, including those with herpes labialis and genital herpes, find that outbreaks become less common over time. Patients with genital herpes caused by HSV-1 infection usually have far fewer outbreaks than those with disease caused by HSV-2. Fortunately, neither oral nor genital herpes recurrences typically cause scarring or anesthesia at the site of the outbreaks.

## PREVENTION AND CONTROL

Attempts to vaccinate against herpes simplex infections have generally been unsuccessful. During the 1990s, one genital herpes vaccine candidate failed to protect its recipients while another showed promise, but only among women who were uninfected with HSV-1. A promising HSV-2 subunit vaccine was recently studied in more than 7000 young women. Unfortunately, it failed to prevent both acquisition of HSV-2 infection and genital diseased caused by HSV-2. Other trials of HSV-2–protective vaccines have been performed, but none has proven efficacious. Therapeutic vaccines aimed to improve natural immunity and decrease recurrences in patients with HSV-2 have also been tested. These HSV vaccines were either ineffective or not effective enough to lead to an approved and marketed drug.

In recent years, studies have demonstrated decreased risk of HSV-2 acquisition in patients using pericoital and/or daily preexposure prophylactic tenofovir gel. Additional studies are needed prior to recommending tenofovir gel for prevention in routine care.

Unlike many other sexually transmitted diseases (e.g., gonorrhea, chlamydia), genital herpes infections are not reported to the CDC. Therefore the usual epidemiologic method of finding, contacting, testing, and treating partners is not used for genital herpes patients. Nevertheless, recent serologic data show that HSV-2 infections are decreasing gradually over time in the United States.

## EVIDENCE

Abdool Karim SS, Abdool Karim Q, Kharsany AB et al. Tenofovir gel for the prevention of herpes simplex virus type 2 infection. *N Engl J Med* 2015; 373:530-9. *This study demonstrated the efficacy of daily tenofovir gel for preexposure prophylaxis against HSV-2 in high risk individuals.*

Ashley RL, Wald A. Genital herpes. review of the epidemic and potential use of type-specific serology. *Clin Microbiol Rev* 1999;12:1-8. *Evidence for the utility of type-specific HSV serologies.*

Bernstein DI, Wald A, Warren T et al. Therapeutic vaccine for genital herpes simplex virus-2 infection: findings from a randomized trial. *J Infect Dis* 2017;215:856-64. *Therapeutic vaccine trial of GEN-003 demonstrating reduced genital HSV shedding and lesion rates.*

Crespi CM, Cumberland WG, Wald A, et al. Longitudinal study of herpes simplex virus type 2 infection using viral dynamic modelling. *Sex Transm Infect* 2007;83:359-364. *This study shows the decreased rate of outbreaks over time among genital herpes patients.*

Evans TG, Bernstein DI, Raborn GW, et al. Double-blind, randomized, placebo-controlled study of topical 5% acyclovir–1% hydrocortisone cream (ME-609) for treatment of UV radiation-induced herpes labialis. *Antimicrob Agents Chemother* 2002;46:1870-1874. *This study details the striking effects of topical steroid when combined with an antiviral against herpes labialis.*

Feltner C, Grodensky C, Ebel C, et al. Serologic screening for genital herpes: an updated evidence report and systematic review for the US Preventive Services Task Force. *JAMA* 2016;316:2531-43. *Systematic review including 10 studies that demonstrated a high false-positive rate in patients screened for genital herpes with serologic testing.*

Habbema L, De Boulle K, Roders GA, Katz DH. n-Docosanol 10% cream in the treatment of recurrent herpes labialis: a randomised, double-blind, placebo-controlled study. *Acta Derm Venereol* 1996;76:479-481. *Several clinical trials have been performed evaluating topical n-docosanol cream for herpes labialis. The results have been mixed and the benefits in healing time are modest. This medication is available OTC in the U.S. and many other countries.*

Heslop R, Roberts H, Flower D, et al. Interventions for men and women with their first episode of genital herpes. *Cochrane Database Syst Rev* 2016;CD010684. *Review demonstrating the benefit of oral antivirals and lack of benefit of topical antivirals for first outbreak of genital herpes.*

Hull C, McKeough M, Sebastian K, et al. Valacyclovir and topical clobetasol gel for the episodic treatment of herpes labialis: a patient-initiated, double-blind, placebo-controlled pilot trial. *J Eur Acad Dermatol Venereol* 2009;23:263-267. *This study details the striking effects of topical steroid when combined with an antiviral against herpes labialis.*

Le Cleach L, Trinquart L, Do G, et al. Oral antiviral therapy for prevention of genital herpes outbreaks in immunocompetent and nonpregnant patients. *Cochrane Database Syst Rev* 2014;CD009036. *Review demonstrating effectiveness of suppressive antiviral therapy in patients experiencing at least four genital herpes outbreaks per year.*

Marrazzo JM, Rabe L, Kelly C, et al. Tenofovir gel for prevention of herpes simplex virus type 2 acquisition: findings from the VOICE trial. *J Infect Dis* 2019;219:1940-1947. *This study demonstrated the efficacy of pericoital tenofovir gel for preexposure prophylaxis against HSV-2 in high risk individuals.*

Sacks SL, Thisted RA, Jones T, et al. Clinical efficacy of topical docosanol 10% cream for herpes simplex labialis: a multicenter, randomized, placebo-controlled trial. *J Am Acad Dermatol* 2001;45:222-230. *Two studies showing an effect of topical docosanol on herpes labialis episodes.*

Schofield JK, Tatnall FM, Brown J, et al. Recurrent erythema multiforme: tissue typing in a large series of patients. *Br J Dermatol* 1994;131:532-535. *Evidence for HSV as the primary cause of erythema multiforme.*

Spruance S, Kriesel J. Treatment of herpes simplex labialis. *Herpes* 2002;9:6-11. *Review of evidence showing a lack of efficacy of topical acyclovir on herpes labialis.*

Spruance SL, McKeough MB. Combination treatment with famciclovir and a topical corticosteroid gel versus famciclovir alone for experimental ultraviolet radiation–induced herpes simplex labialis: a pilot study. *J Infect Dis* 2000;181:1906-1910. *This study details the striking effects of topical steroid use when combined with an antiviral agent against herpes labialis.*

Spruance SL, Rea TL, Thoming C, et al. Penciclovir cream for the treatment of herpes simplex labialis: a randomized, multicenter, double-blind, placebo-controlled trial. Topical Penciclovir Collaborative Study Group. *JAMA* 1997;277:1374-1379. *This study shows the effect of topical penciclovir on herpes labialis.*

Stanberry LR, Spruance SL, Cunningham AL, et al. Glycoprotein-D-adjuvant vaccine to prevent genital herpes. *N Engl J Med* 2002;347:1652-1661. *This study documents the efficacy of the SmithKline Beecham vaccine against HSV-2 in double-seronegative females.*

Wald A. Genital HSV-1 infections. *Sex Transm Infect* 2006;82:189-190. *This study outlines the prognosis of patients with HSV-1 induced genital herpes.*

## ADDITIONAL RESOURCES

American Social Health Association (ASHA): ASHA website. Available at: www.ashastd.org. *A comprehensive website that is informative and easy to understand.*

Centers for Disease Control and Prevention (CDC), Workowski KA, Berman SM. Sexually transmitted diseases treatment guidelines, 2006. *MMWR Recomm Rep* 2006;55:1-94. *This document provides specific CDC recommendations for genital infections, including HSV.*

Cernik C, Gallina K, Brodell RT. The treatment of herpes simplex infections: an evidence-based review. *Arch Intern Med* 2008;168:1137-1144. *An evidence-based review of therapies for genital and labial herpes.*

Gupta R, Warren T, Wald A. Genital herpes. *Lancet* 2007;370:2127-2137. *A review of the risk factors and epidemiology of genital herpes.*

International Herpes Management Forum (IHMF): IHMF website. Available at: www.ihmf.org. *The IHMF is a medical and scientific research forum that was established to improve the awareness and understanding of herpes viruses and the counseling and management of people with these infections.*

Sacks SL. Genital herpes simplex infection and treatment. In Sacks SL, Straus SE, Whitley RJ, Griffiths PD, eds: *Clinical management of herpes viruses,* Amsterdam, 1995, IOS Press, pp 3-42. *Excellent and thorough review of genital herpes by a leading authority.*

Spruance SL. Herpes simplex labialis. In Sacks SL, Straus SE, Whitley RJ, Griffiths PD, eds: *Clinical management of herpes viruses,* Amsterdam, 1995, IOS Press, pp 55-67. *Excellent and thorough review of herpes labialis by a leading authority.*

Westover Heights Clinic: Updated herpes handbook. Available at: www. westoverheights.com/genital_herpes/handbook.html. *Link to an excellent free publication about genital herpes.*

Whitley RJ. Herpes simplex encephalitis: adolescents and adults. *Antiviral Res* 2006;71:141-148. *A review of the findings among patients with HSV encephalitis.*

Xu F, Sternberg MR, Kottiri BJ, et al. Trends in herpes simplex virus type 1 and type 2 seroprevalence in the United States. *JAMA* 2006;296:964-973. *The most complete reference outlining the prevalence and trends among HSV infections in the United States.*

Yeung-Yue K, Brentjens MH, Lee PC, Tyring SK. Herpes simplex viruses 1 and 2. *Dermatol Clin* 2002;20:249-266. *This is a thorough review of HSV-1 and HSV-2 infections.*

# 24

# Nontuberculous Mycobacterial Skin and Soft-Tissue Infections

*Rabeeya Sabzwari, James L. Cook*

## ABSTRACT

Nontuberculous mycobacteria (NTM) are a group of acid-fast bacteria that are ubiquitous in the environment. They have been isolated from water and soil, and, when it is possible to identify the etiology, related infections are almost always associated with environmental sources. NTM skin and soft-tissue infections (SSTIs) have been reported with slowly growing NTM species (e.g., avium complex, kansasii, marinum, haemophilum), but rapidly growing NTM species (e.g., abscessus, chelonae, fortuitum) are more commonly involved. NTM SSTIs typically result from a breach in the skin barrier and can be multifocal in immunocompromised hosts (underlying autoimmune disease, immunosuppressive therapy or human immunodeficiency virus [HIV]/acquired immunodeficiency syndrome [AIDS]). Localized NTM SSTIs are usually the result of direct inoculation of the organism after trauma (e.g., penetrating wounds, fish tank exposure, compound fractures) or contamination of surgical sites with NTM-colonized water sources. Hematogenous dissemination of infection from NTM SSTIs is almost always associated with immunosuppressive therapy or underlying defects in cellular immunity. A strong clinical suspicion is necessary for diagnosis, appropriate pathologic and laboratory testing, and management of NTM SSTIs. Effective care usually requires a combination of long-term, multidrug, antibiotic therapy and debridement and, where possible, reduction or elimination of immunosuppression.

## EPIDEMIOLOGY

In adults, NTM infections are usually manifested as pulmonary disease. In children, lymphadenitis comprises 75% to 85% of total infections. Extrapulmonary presentations in adults include SSTIs, tenosynovitis, septic arthritis, osteomyelitis, and keratitis. Disseminated NTM disease, occasionally involving the central nervous system, can occur, especially with *Mycobacterium avium* or *Mycobacterium abscessus* infections in immunocompromised hosts.

NTM SSTIs are usually caused by exposure to environmental water and soil sources of contamination. Such infections have been reported with traumatic skin lesions, exposure to contaminated gentian violet, nail salon pedicures, and tattoos. Few surveillance data are available on the public health burden of NTM SSTIs, because they are not reportable in most states.

NTM SSTIs are more commonly caused by rapidly growing mycobacteria (RGM) species, including *Mycobacterium fortuitum*, *M. abscessus*, and *Mycobacterium chelonae* and the slowly growing mycobacteria (SGM), *Mycobacterium marinum*, *Mycobacterium ulcerans*, *Mycobacterium chimaera*, and *Mycobacterium haemophilum*. *M. haemophilum* and *Mycobacterium genavense* SSTIs are usually diagnosed in immunocompromised patients. However, virtually all NTM species can cause SSTIs.

Immunocompromised hosts, including patients with advanced HIV infection, organ or stem cell transplantation and rare individuals with mendelian susceptibility to mycobacterial disease (MSMD) or anti–interferon-γ antibody are at greatest risk of disseminated NTM disease to multiple sites, including skin and soft tissues.

## PATHOGENESIS

*Direct NTM inoculation*—either traumatic or iatrogenic—can breach the skin barrier and cause SSTIs. Skin punctures by foreign bodies such as wood splinters, metal fragments, fish spines, or associated with intravascular catheters or devices can introduce environmental NTM into wounds. Infections can arise through inadvertent contamination of surgical wounds, usually from water sources. Skin and soft-tissue involvement can also result from hematogenous dissemination of NTM (e.g., from primary pulmonary infection), resulting in multifocal lesions, mostly in immunosuppressed hosts.

*Biofilm formation* is another factor in NTM pathogenesis. Mycobacterial biofilms form both in environmental sites of colonization, such as water distribution systems, aquariums, hot tubs, swimming pools, and water-containing medical devices, and after insertion of various biomaterials, such as catheters and implantable prosthetic devices. Biofilms are a factor in antimicrobial resistance of bacterial populations, protecting the NTM against antibiotics that might otherwise be effective against bacteria growing in the absence of biofilms. RGM are the most common NTM associated with foreign body biofilms, but virtually any NTM species can be involved in such infections. Formation of NTM biofilms on foreign bodies and implanted biomaterials underscores the importance of foreign body removal and debridement in managing NTM SSTIs, as an adjunct to antibiotic therapy.

*Impaired cell-mediated immunity* can increase susceptibility to and alter the pathogenic response to NTM SSTIs, for example as associated with AIDS, malignancy, solid organ or hematopoietic stem cell transplantation, or treatment with biologic response modifiers or other immunosuppressive therapy. The NTM species and the quality of the host inflammatory response to NTM SSTIs can affect the clinical presentation and suspicion of mycobacterial infection and therefore the timeliness of diagnosis. For example, RGM NTM species can elicit more of an acute inflammatory response than the typical granulomatous response seen with slowly growing NTM species during chronic infection, and both types of inflammatory responses can be rendered less intense and characteristic by immunosuppressive conditions or therapies. Therefore it is important to use a combination of tissue histopathology, acid-fast bacilli (AFB) stain and mycobacterial culture to define the cause of SSTI lesions, especially in immunocompromised hosts.

## CLINICAL PRESENTATION

NTM SSTIs have varied clinical presentations, including cellulitis, non-healing ulcers, nodular skin lesions, abscesses, sinus tracts, and lymphadenitis. The most common exposures that result in such infections involve

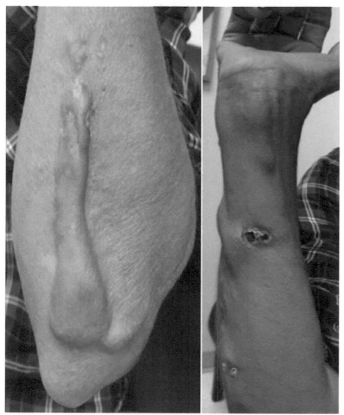

Fig. 24.1 Nodular skin lesions and an ulcer involving the arms in a patient with disseminated *M. haemophilum* infection.

direct inoculation via penetrating trauma or some other compromise in the skin barrier (e.g., work or play related accidents, injections, cosmetic piercings, acupuncture, or surgical procedures) associated with environmental soil or water contamination. Transient bacteremia can occur during NTM SSTIs in immunocompetent patients but would usually be missed because mycobacterial blood cultures are not obtained, except in immunosuppressed patients where disseminated infection is suspected. Conversely, diffuse skin and soft-tissue lesions can be caused by bloodborne NTM dissemination, usually in immunosuppressed individuals. Patients receiving systemic corticosteroid therapy (e.g., for Crohn disease or rheumatoid arthritis [RA]), chemotherapeutic agents, and tumor necrosis factor (TNF)-α blockers, and those with defects in cell-mediated immunity, including transplant recipients, are at greatest risk of disseminated, multifocal infection (Fig. 24.1, Clinical Vignette 1). Therefore a high index of suspicion is required for timely diagnosis and antimycobacterial therapy.

### ✳ CLINICAL VIGNETTE 1

A 58-year-old woman with RA, treated with TNF alpha blockers and oral corticosteroids, had progressive nodular lesions and ulcers involving the hands and forearms and a left lower extremity abscess. She had frequent exposure to contaminated lake water while living on a boat. A skin lesion was culture-positive for *M. haemophilum*. She was treated with rifampin, azithromycin, clofazimine, and an oral fluoroquinolone, and immunosuppressive therapy was placed on hold, with RA management using nonsteroidal anti-inflammatory drugs (NSAIDs).

COMMENT: Repeated exposure of this patient to contaminated water likely caused the initial SSTI that became disseminated, as a result of immunosuppression. Reducing immunosuppressive therapy and treating with a multidrug antimycobacterial regimen resulted in marked improvement of her multiple skin lesions and abscesses.

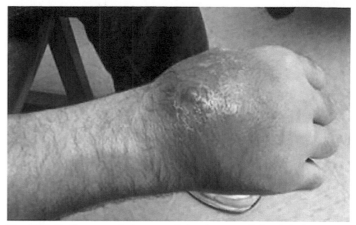

Fig. 24.2 Erythema and abscess of the dorsum of the right hand in a patient with *M. fortuitum* infection.

NTM SSTIs involving the extremities are more common in adults because of soil or water contamination during work- or play-related trauma but have occasionally been observed in children (e.g., associated with contaminated public swimming pools). Young children, and less commonly adults, can present with subacute cervicofacial lymphadenitis that can progress to abscess and sinus tract formation, in the absence of any apparent trauma, possibly because of an oral route of infection. Lack of positive routine bacterial cultures, along with persistent infection after empirical antibiotic treatment, in both extremity SSTIs and lymphadenitis should raise the suspicion for NTM infection.

Patients with RA or gout and NTM infection can be initially misdiagnosed as having a flare of their underlying joint disease and treated with systemic or injected corticosteroid therapy. Patients with joint problems or tenosynovitis following trauma are also commonly treated with corticosteroid injections. Delays in diagnosis during trials of antiinflammatory therapy and using routine bacterial cultures can be avoided by early addition of AFB cultures as part of the evaluation (Fig. 24.2, Clinical Vignette 2).

### ✳ CLINICAL VIGNETTE 2

A 52-year-old male sheet metal worker, with a work history of innumerable nicks of his hands, developed pain and swelling of right-hand proximal interphalangeal joint (PIP) and metacarpophalangeal joint (MCP) joints and was treated with multiple corticosteroid injections for the presumed diagnosis of carpal tunnel syndrome. He developed dorsal hand erythema and swelling with tenosynovitis and operative cultures positive for *M. conceptionense* (an *M. fortuitum* group RGM species). He was initially treated with intravenous (IV) cefoxitin and IV amikacin therapy, followed by long-term, multidrug oral antimycobacterial therapy, along with repeated debridement of the hand and forearm lesions.

COMMENT: This patient received multiple corticosteroid injections in the hand that resulted in progression of the *M. fortuitum* SSTI to extensive tenosynovitis and soft-tissue infection in the arm, requiring combined medical/surgical therapy.

*M. marinum* is a slowly growing species found in fresh, salt, and brackish water. SSTIs have been observed following direct inoculation of this pathogen from a contaminated source, after minor skin trauma in individuals handling fish or cleaning fish tanks. A sporotrichoid form of disease, with lesions extending proximally in the affected extremity, suggestive of lymphocutaneous spread, is often noted (Fig. 24.3, Clinical Vignette 3). Prolonged treatment with combination antimycobacterial agents usually results in cure. Because the bacterial burden is usually low, a single antibiotic can often be effective in patients with localized, superficial *M. marinum* infection.

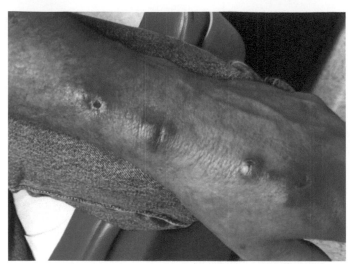

Fig. 24.3 *M. marinum* associated "sporotrichoid-form" of disease involving the hand and forearm of an immunocompetent patient.

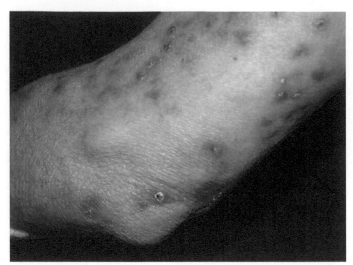

Fig. 24.4 Nodular skin lesions associated caused by *M. abscessus* infection associated with a central venous line placed in that arm.

 **CLINICAL VIGNETTE 3**

A 60-year-old, immunocompetent male developed hand pain, swelling and nodular lesions a few weeks after cleaning his home fish tank. He had an incision and drainage (I&D), with negative, routine bacterial cultures and had progression of the initial lesion, with ascending, sporotrichoid nodules in the forearm, despite empirical trimethoprim-sulfamethoxazole (TMP-SMX) therapy. Repeat biopsy revealed a positive AFB stain and a positive culture for *M. marinum*. Six months therapy with clarithromycin resulted in complete resolution of his lesions.

COMMENT: Minor skin trauma associated with fish-tank cleaning is the classical history of patients with *M. marinum* SSTIs. Failure to include AFB stains and cultures during initial evaluation can result in delayed diagnosis and treatment.

Central line–associated NTM SSTIs have been reported in the hospital setting, sometimes prompting an investigation into possible environmental sources. These line-associated infections may be localized to the affected extremity or insertion site. This presentation is more common in immunosuppressed patients (Fig. 24.4, Clinical Vignette 4). NTM bacteremia can be seen in injection drug users who access central venous lines (sometimes using drugs injected in tap water) at home or in hospital. NTM SSTIs associated with cardiac or other endovascular devices, including exit site infections and pacemaker pocket infections, require prompt removal of the catheter and/or cardiac device, to prevent further complications and possible bacteremic spread to cardiac valves and other sites.

 **CLINICAL VIGNETTE 4**

A 54-year-old female with Crohn's disease, on prednisone therapy, and with a left arm central venous line, who developed numerous nodules with variable inflammatory response involving the left hand and upper arm. Skin biopsy was culture-positive for *M. abscessus*. She was treated with combined IV and oral antimycobacterial therapy and prednisone dose reduction, with slow improvement in the lesions over 6 months but also with periodic appearance of new subcutaneous nodules during therapy.

COMMENT: Central line insertion can predispose to NTM SSTIs in immunosuppressed patients. Skin nodules can begin as indurated, minimally or noninflamed lesions that progress to increasingly inflamed skin and subcutaneous lesions that may drain serosanguinous fluid. Skin biopsy with AFB stain and culture and pathology is usually required for diagnosis. Reduction of immunosuppression, if possible, along with prolonged, multidrug antimycobacterial therapy is required for resolution, and larger lesions might require debridement.

Cosmetic surgery procedures, both domestic and associated with medical tourism, have been associated with NTM SSTIs. Face-lift, blepharoplasty, breast augmentation, abdominoplasty, liposuction, and buttock-lift have all been associated with NTM infections. The most commonly identified NTM species in this context are the RGM, *M. fortuitum*, *M. abscessus*, and *M. chelonae*. Suboptimal sterilization techniques, contaminated water sources, and formation of biofilms by RGM are involved in these surgical infections. Abdominal surgeries, liposuction, or procedures that require mesh placement can cause RGM SSTIs that require removal of infected material and serial debridements, in addition to antibiotic therapy. Periprosthetic NTM infections can occur following augmentation mammoplasty and breast reconstruction. Most patients present with a delayed onset of signs of surgical site inflammation and serosanguinous exudates, negative routine bacterial cultures, and persistent skin and soft-tissue lesions despite multiple courses of empirical antibacterial treatment. Patients with idiopathic granulomatous mastitis can present with superimposed infections with RGM, and it can be difficult to differentiate the inflammatory pathologic findings of the underlying disease from those cause by superimposed RGM infection (Clinical Vignette 5).

**CLINICAL VIGNETTE 5**

A 36-year-old female with metabolic syndrome was referred for management of left breast abscess. A breast biopsy revealed granulomatous mastitis, with negative special stains for AFB. AFB cultures were eventually positive for *M. abscessus*. Given the extent of the granulomatous reaction and the absence of any AFB on histopathology, it was concluded that the patient had a primary problem of idiopathic granulomatous mastitis, with a mycobacterial superinfection—not a simple distinction.

The patient was treated with IV cefoxitin and IV amikacin and oral clofazimine, in the initial intensive phase of treatment. She responded well but required intermittent course of oral corticosteroid therapy for flares of granulomatous mastitis, despite serial negative breast cultures for mycobacteria. Given a multidrug-resistant *M. abscessus* isolate, the patient was maintained on long-term suppressive therapy with clofazimine and continued to improve, during more than 1 year of follow-up with negative cultures.

COMMENT: NTM infection in the context of underlying, granulomatous disease creates difficulty in interpreting the cause and effect relationship between the infection and the underlying (usually autoimmune) process. In this setting, both immunosuppressive and antibiotic therapy may be required, to reach the best balance between control of the underlying disease and the mycobacterial superinfection. Long-term, suppressive, antimycobacterial therapy may be required while trying to reduce the immunosuppressive therapy to the lowest possible level.

Outbreak-associated cases of NTM SSTIs have been reported in tattoo studios, due to poor infection control practices and contamination of tap water and of graywash ink at the point of manufacture. Similar outbreaks, usually caused by RGM, can also occur in patients visiting spas and nail salons and using foot baths, as the result of minor skin trauma and exposure to contaminated water sources.

Outbreaks with *M. chimaera* infections have been linked to factory associated, contaminated water in heater-cooler devices, used during the cardiovascular surgeries. Resulting presentations can include sternal wound infections, mediastinitis, vascular graft infections, osteomyelitis, and dissemination to remote bone and soft-tissue sites. Other postthoracotomy NTM SSTIs are uncommon, usually only seen after wound contamination during resectional lung surgery for NTM disease. Such postoperative infections can result in persistent surgical site infections and sinus tracts and can be associated with underlying bronchopleural fistula and empyema.

Leprosy and Buruli ulcer are two other forms of NTM SSTIs that are beyond the scope of this review. Buruli ulcer is caused by *M. ulcerans* and is essentially only seen in patients in or from the tropics. Infection associated with this species results in development of ulcers with a predilection for the lower extremities. The ulcers are usually painless and progress rapidly to involve bone and result in permanent disability and loss of function, if treatment is not initiated promptly.

## DIAGNOSTIC APPROACH

Because NTM SSTIs are uncommon, they may not be considered in the differential diagnosis of inflammatory skin or soft-tissue lesions. Therefore patients who are eventually diagnosed with NTM SSTIs have often not had initial AFB studies (cultures/pathology) and have usually failed treatment with other empirical antibiotic therapy. In cases with underlying autoimmune disease, treatment with corticosteroids, injected locally, in oral, or IV form, is often reported. Having a high index of suspicion of a possible NTM etiology and a low threshold for doing AFB studies of skin and soft-tissue lesions of unknown etiology is a good general recommendation.

Immunologic testing is of little value in diagnosing NTM SSTIs. Mycobacterial skin testing and interferon-γ response assays are rarely useful in diagnosis.

Tissue biopsies and collection of purulent material and debrided tissues are much more sensitive than swab cultures for NTM infection diagnosis and have the added advantage of providing material for pathology with special stains. It is important to communicate with pathologists and microbiology laboratory personnel about the possibility of NTM infection, so that processing of stain and culture methods can be optimized. Culture, using a combination of solid and liquid media systems, can improve the sensitivity of diagnosis. Incubation at temperatures other than 37°C is required to grow some NTM species; for example, *M. marinum*, *M. chelonae*, and *M. haemophilum* grow optimally at 28°C to 31°C, whereas *M. xenopi* grows optimally at 40°C to 42°C.

Pathologic studies are important complements to culture. Inflammatory responses to NTM/SSTIs, as evidenced in hematoxylin and eosin (H&E)-stained tissues, can vary from acute and chronic inflammatory reactions to granulomatous lesions, depending on the age of the lesion and the pathogen involved. Granulomatous reactions can suggest mycobacterial infection, but, without a positive special stain for AFB, can also be consistent with other chronic, inflammatory process. Even with special stains for AFB, the numbers of organisms detected can be small, or the stain might be negative because of the low bacterial burden in some cases; thus it is important to obtain simultaneous tissue specimen cultures.

Molecular methods are being used increasingly for NTM diagnosis in both culture-based and pathology-based studies. Nucleic acid probes, PCR, matrix-assisted laser depolarization/ionization time of flight (MALDI TOF) mass spectrometry, and high performance liquid chromatography (HPLC) are methods of speciation of growing mycobacterial cultures that were initially limited to reference laboratories but that have become widely available in medical center clinical microbiology and send-out support laboratories. A few reference laboratories also offer methods for direct detection of mycobacterial DNA and speciation from tissue specimens. The sensitivity of such analysis is greater with fresh-frozen tissue but can also be attempted with formalin-fixed specimens.

The key message for all of these diagnostic methods is the same. Because AFB testing is not routine for SSTIs, the clinician must include mycobacterial infection in the differential diagnosis, especially for unusual SSTIs, and should discuss that possibility with colleagues in clinical microbiology and pathology laboratories to optimize utilization of the array of diagnostic methods and susceptibility testing to guide antimycobacterial therapy.

## TREATMENT

The mainstay of therapy for NTM SSTIs includes consideration of surgical management, coupled with prompt initiation of pathogen-directed antimycobacterial therapy. In the case of NTM lymphadenitis, complete surgical excision of the affected lymph node(s) is usually curative. When complete lymph node excision is not possible (e.g., with facial or other nerve involvement), a combined medical and surgical approach will often result in cure. For other NTM SSTIs, aggressive and repeated surgical drainage/debridement, as needed, and removal of infected foreign material are critical adjuncts to antibiotic therapy.

The choice and duration of antimycobacterial therapy depends on the NTM species, extent of disease, antibiotic susceptibility, ability to achieve source control and overall response to treatment. A combination of oral and IV antibiotics should be considered in the initial, intensive phase of therapy for extensive disease. Macrolides (azithromycin, clarithromycin) are the core of treatment for most such NTM infections, including those caused by both slowly growing (SGM) and rapidly growing (RGM) mycobacterial species. The exception is that some RGM species, such as *M. abscessus* ssp. abscessus, have a functional *erm* gene, which causes inducible macrolide resistance by encoding a methylase that alters the bacterial ribosomal binding site for macrolides. Therefore, as a general rule, treatment of all RGM should be based upon the results of antibiotic susceptibility testing. The choice of other antibiotics can generally follow NTM species–related guidelines, as published by ATS/IDSA. The following are selected comments, to complement those guidelines.

IV cefoxitin or imipenem-cilastatin in combination with IV amikacin (10 to 15 mg/kg daily or 20 to 25 mg/kg thrice-weekly) should be considered for the initial control of progressive RGM SSTIs, before transition to a long-term oral regimen. It is important to note that *M. chelonae* is inherently resistant to cefoxitin. The use of IV tigecycline should be reserved for patients who are allergic to or intolerant of other antibiotics, given the common, debilitating gastrointestinal (GI) side effects reported with tigecycline, including rare cases of pancreatitis. Dual β-lactam antibiotic combinations, such as ceftazidime plus imipenem, have revealed synergistic activity against RGM during in vitro studies and therefore might provide an option for initial, parenteral treatment. Periodic monitoring for drug-related side effects and central line complications are important during follow up. Cefoxitin-induced rash, amikacin-associated nephrotoxicity and ototoxicity, macrolide-related hearing loss, and leukopenia associated with β-lactam therapy are among the treatment issues to be considered.

It can be challenging to define an effective oral antimycobacterial regimen for the long-term maintenance phase of RGM SSTIs and macrolide-resistant SGM SSTIs. Fluoroquinolone antibiotics may be considered as components of a multidrug regimen for selected infections, but require EKG QTc monitoring, especially when used in combination with macrolides for patients who have a history of cardiac arrhythmias. Linezolid use is limited by the adverse effects of bone marrow suppression and neurotoxicity when used for more than a few weeks. Tedizolid, with its improved side effect profile and generally lower minimum inhibitory concentration (MIC) values than linezolid, and the newer tetracycline, omadacycline, provide additional options for the oral maintenance phase of RGM SSTI therapy.

Oral clofazimine can be considered as part of the initial intensive phase or prolonged maintenance phase of therapy for RGM SSTIs or therapy of SGM SSTIs, because the combination of clofazimine with macrolides, quinolones, or amikacin demonstrates in vitro synergy against both RGM and SGM species. Skin hyperpigmentation is expected after weeks to months of clofazimine therapy, because of drug deposition. GI associated adverse effects might occur with this drug but can usually be managed by temporarily holding therapy and resuming at a lower dose, after GI symptoms abate. The drug's long half-life (~70 days) should be considered in dosing and management during long-term therapy.

Emerging therapies for NTM infections are being considered for newly discovered anti-tuberculosis (TB) drugs that have collateral anti-NTM activity. One example is bedaquiline, a bacteriostatic agent that targets mycobacterial ATP synthase. Whether this drug and others evolving from the international TB drug discovery initiative will be clinically useful for NTM SSTIs remains to be determined.

In summary, a high index of suspicion, prompt diagnosis, adequate source control with debridement when necessary, and a focus on removal of the infected hardware/prostheses/devices, minimizing or eliminating the use of immunosuppressive therapy when feasible, along with the use of combination antimycobacterial therapy, provide the basis of successful treatment outcomes for NTM SSTIs. Expected cure rates depend upon the mycobacterial species isolated, host factors, extent of disease, antibiotic efficacy and tolerance, and long-term monitoring for signs of infection recurrence.

## EVIDENCE

Cusumano LR, Tran V, Tlamsa A, et al. Rapidly growing Mycobacterium infections after cosmetic surgery in medical tourists: the Bronx experience and a review of the literature. *Int J Infect Dis* 2017;63: 1-6. *A case series that describes the presentation and management of patients with RGM infections after undergoing cosmetic surgeries abroad that raises the awareness of this potential problem associated with medical tourism and the more general problem of NTM SSTIs in patients undergoing cosmetic surgery.*

Ferro BE, Meletiadis J, Wattenberg M, et al. Clofazimine prevents the regrowth of Mycobacterium abscessus and Mycobacterium avium type strains exposed to amikacin and clarithromycin. *Antimicrob Agents Chemother* 2016;60(2): 1097-1105. *One of several reports on the synergistic activity of clofazimine with other antimycobacterial antibiotics, specifically amikacin and clofazimine. This study provides detailed pharmacodynamic analysis of the synergy of clofazimine and amikacin and clofazimine and clarithromycin against M. avium complex and M. abscessus.*

Griffin I, Schmitz A, Oliver C, et al. Outbreak of tattoo-associated nontuberculous mycobacterial skin infections. *Clin Infect Dis* 2019;69(6): 949-955. *One of several reports of outbreaks of NTM SSTIs occurring in patients who received tattoos at commercial facilities, where the source of infection was traced to contaminated ink and tap water used for dilution. Awareness of this problem can sensitize clinicians to consider NTM infection in the differential diagnosis of tattoo-related SSTIs.*

Nash KA, Brown-Elliott BA, Wallace RJ Jr., et al. A novel gene, erm(41), confers inducible macrolide resistance to clinical isolates of Mycobacterium abscessus but is absent from Mycobacterium chelonae. *Antimicrob Agents Chemother* 2009;53(4): 1367-1376. *Characterization of the bacterial genetic mechanism of inducible macrolide resistance in certain RGM subspecies of the M. abscessus group.*

Pandey R, Chen L, Manca C, et al. Dual beta-lactam combinations highly active against Mycobacterium abscessus complex in vitro. *MBio* 2019;10(1). *A detailed in vitro study documenting the synergistic activities of ceftazidime and imipenem or ceftaroline against multiple isolates of Mycobacterium abscessus that offers a possible, new strategy for initial, parenteral treatment of progressive infection, when conventional options cannot be used.*

Sax H, Bloemberg G, Hasse B, et al. Prolonged outbreak of Mycobacterium chimaera infection after open-chest heart surgery. *Clin Infect Dis* 2015;61(1): 67-75. *One of the earliest reports of outbreaks of disseminated M. chimaera infection associated with the use of heater-cooler devices during open-chest cardiac surgery. This provides a beginning reference to this literature that can be used for perspective in cases of M. chimaera sternal wound, mediastinal and related, disseminated infections that might present initially as SSTIs with positive blood cultures.*

Winthrop, KL, Albridge K, South D, et al. The clinical management and outcome of nail salon-acquired Mycobacterium fortuitum skin infection. *Clin Infect Dis* 2004;38(1): 38-44. *The largest published case series of RGM infections associated with nail salon whirlpool footbaths. The epidemiologic data, clinical courses and treatment responses provide information useful for considerations of early diagnosis and effective management.*

## ADDITIONAL RESOURCES

Esteban, J. and M. Garcia-Coca. Mycobacterium biofilms. *Front Microbiol* 2017;8: 2651. *A review of the history, characteristics, and environmental and clinical implications of NTM biofilms that also considers the therapeutic implications of biofilm formation.*

Franco-Paredes C, Marcos LA, Henao-Martinez AF, et al. Cutaneous Mycobacterial infections. *Clin Microbiol Rev* 2018;32(1). *A review that compares the epidemiological and clinical features of four different categories of mycobacterial skin infections: tuberculosis, leprosy, Buruli ulcer and NTM disease. Included are several photographs of the clinical presentation of each type of mycobacterial skin infection.*

Griffith DE, Aksamit T, Brown-Elliott BA, et al. An official ATS/IDSA statement: diagnosis, treatment, and prevention of nontuberculous mycobacterial diseases. *Am J Respir Crit Care Med* 2007;175(4): 367-416. *A comprehensive guide about the wide range of clinically relevant NTM species. This is an excellent, well-referenced, searchable reference with information on clinical presentation, diagnosis, and treatment.*

Henkle E, Winthrop KL. Nontuberculous mycobacteria infections in immunosuppressed hosts. *Clin Chest Med* 2015;36(1): 91-99. *An overview of the diagnosis, prevention and treatment of NTM disease associated with a variety of immunosuppressing conditions, including HIV/AIDS, corticosteroid and biologic therapies, solid tumors and hematologic malignancies, solid organ and hematopoietic stem cell transplants and primary immunodeficiency diseases.*

Misch EA, Saddler C, Davis JM. Skin and soft tissue infections due to nontuberculous Mycobacteria. *Curr Infect Dis Rep* 2018;20(4): 6. *A review of epidemiologic trends, pathogenesis and advances in diagnosis and treatment of NTM SSTIs.*

# Respiratory Tract Infections

*Thomas M. File, Jr.*

# Introduction to Respiratory Tract Infections

*Thomas M. File, Jr.*

Respiratory tract infections (RTIs) are the most common type of infection managed by healthcare providers, and they are of great consequence because of cost, mortality, emergence of antibiotic-resistant pathogens, and potential for epidemic dissemination.

Acute RTIs are the greatest single cause of death in children worldwide. Community-acquired pneumonia is the eighth leading cause of death and the most common cause of death caused by infectious disease in the United States, whereas nosocomial pneumonia is a leading cause of death from infections acquired in the hospital.

Indicators of overall burden of RTIs suggest that the morbidity and mortality rates attributed to these diseases are increasing, as are the proportion of hospitalizations attributed to these infections. Many factors, such as the emergence of acquired immunodeficiency syndrome (AIDS), the increased association of underlying conditions (e.g., diabetes and heart or lung disease), the emergence of new pathogens (e.g., community-acquired methicillin-resistant *Staphylococcus aureus* [CA-MRSA], H5N1 influenza A, and 2009 H1N1 influenza A), and increased antibiotic resistance, probably contribute to this increase.

RTIs are also the cause of most antibiotic use. Approximately three-quarters of all outpatient antimicrobial use is for respiratory infections—especially for acute bacterial sinusitis for adults and acute bacterial otitis for children. Although many RTIs require antimicrobial therapy for optimal management, most outpatient respiratory infections (e.g., acute bronchitis, nasal pharyngitis, cold, nonspecific upper respiratory tract infection [URI]) are caused by respiratory viruses for which antibiotic use is not warranted. Overuse of antibiotics for both community- and hospital-acquired respiratory infections is a source of great antibiotic abuse and increases the likelihood of further hindering the already high level of antibiotic resistance. The emergence of CA-MRSA in the community and multidrug-resistant gram-negative pathogens (e.g., multiple drug–resistant *Acinetobacter* or *Pseudomonas*; carbapenem-resistant Enterobacteriaceae [CRE]) in the hospital setting further challenges our ability to successfully treat these infections. The emergence of the pandemic of H1N1 influenza A has illustrated the importance of viral infections as well as bacterial causes of these infections (see Section IX: Emerging Infectious Diseases and Pandemics). The pathogenic mechanisms for secondary bacterial pneumonia in patients with influenza is of major importance because these coinfections are frequently associated with high mortality.

There have been several new developments concerning RTIs since the prior version of *Netter's Infectious Diseases*; these include: advances in molecular diagnostic methods to identify the causative pathogen; increasing awareness of viral etiology of community-acquired pneumonia; increase of macrolide-resistant *S. pneumoniae*; awareness of the role of the lung microbiome; and elimination of the healthcare-associated pneumonia (HCAP) category for pneumonia (discussed in Chapter 29).

Critical to the appropriate management of RTIs is an understanding of the pathogenesis, microbiology, diagnosis, and treatment of these diseases. This section reviews the current concepts of RTIs, including both upper and lower tract infections, in light of recent information and published guidelines. It provides a useful source of essential information for healthcare providers with the goal of achieving optimal care of patients.

# Community-Acquired Pneumonia, Bacterial

*Thomas M. File, Jr.*

## ABSTRACT

There have been several new developments concerning community-acquired pneumonia (CAP) since the prior version of *Netter's Infectious Diseases*; these include: advances in molecular diagnostic methods to identify the causative pathogen; increasing awareness of viral etiology; decreasing rate of *Streptococcus pneumoniae* (although still the most common bacterial cause); increase of macrolide-resistant *S. pneumoniae*; awareness of the role of the lung microbiome; and elimination of the healthcare-associated pneumonia (HCAP) category for pneumonia.

## CLINICAL VIGNETTE

A 58-year-old male presents in January to an urgent care center with a 2-day history of fever, chills, nonproductive cough, and dyspnea. He has a history of CHF and was immunized previously with both influenza and *S. pneumoniae* vaccines. Exam: slightly confused per wife; T 38.3°C; blood pressure (BP) 140/80; resp rate 24; pulse 100; pulse oximetry 91%; chest x-ray (CXR) reveals right lower lobe (RLL) reticular, patchy infiltrate. Labs: white blood cell (WBC) 9000; influenza polymerase chain reaction (PCR): negative.

Patient administered ceftriaxone and azithromycin and transferred to hospital.

FOLLOW-UP: Blood cultures obtained prior to antimicrobial administration were no growth and urinary antigen tests for *S. pneumoniae* and *Legionella* were not detected. No sputum was produced; a multiplex PCR respiratory panel was positive for human metapneumovirus. Procalcitonin on day 2 was very low at less than 0.1 ng/mL. Based on the positive test for a respiratory virus and negative studies for bacterial coinfection and a low procalcitonin level suggesting bacterial infection was unlikely, the antibacterial agents were discontinued on day 2. The patient was clinically improved at 48 hours and discharged. At 30 days follow-up, he was back to baseline status.

## INTRODUCTION

CAP is defined as an acute infection of the pulmonary parenchyma in a patient who has acquired the infection outside of the hospital setting. CAP is a common and potentially serious illness, particularly in elderly patients and those with significant comorbidities. The term HCAP was previously used to identify patients at risk for infection with multi-drug-resistant pathogens since they had some exposure to a healthcare setting (e.g., prior hospitalization or patient of a long-term care facility, dialysis, etc.). However, this categorization may have been overly sensitive, leading to increased, inappropriately broad antibiotic use and was thus retired. Patients previously classified as having HCAP should be managed in a similar way to those with CAP (assessing risks for drug resistant organisms) because patients with HCAP frequently present from the community and are initially cared for in emergency departments.

An update of the American Thoracic Society/Infectious Diseases Society of America (ATS/IDSA) guidelines on the management of CAP has recently been published. This chapter will focus on the general approach to CAP in adults, with a concentration primarily on typical bacterial causes. Other chapters focus specifically on pneumonia caused by aspiration, viruses, or atypical organisms.

## DISEASE BURDEN

CAP is associated with significant morbidity and mortality and considerable costs of care. In the United States, CAP is the most frequent cause of death resulting from infectious diseases and is the eighth leading cause of death overall. The mortality rate of patients ranges from less than 1% for patients treated on an outpatient basis to 30% for those with severe CAP who require admission to the intensive care unit (ICU). Data from the Centers for Medicare and Medicaid Services database estimate the 30-day mortality rate of CAP patients (mostly those >65 years of age) requiring admission to the hospital in the United States to be approximately 12%.

The overall rate of CAP ranges from 8 to 23 per 1000 persons per year, depending on the population considered; the highest rates are at the extremes of age. More cases occur during the winter months. In the United States, approximately 30% of patients with CAP are hospitalized. The economic cost exceeds $17 billion a year.

## PATHOGENESIS AND RISK FACTORS

Historically, the lung had been considered sterile. However, newer molecular and gene sequencing methods have identified diverse communities of microbes that reside within the alveoli. This resident flora may play a role in the development of pneumonia, either by modulating the host immune response to infecting pathogens or through direct overgrowth of specific pathogens within the alveolar microbiome.

Respiratory pathogens are primarily transmitted from person to person via droplets or, less commonly, via aerosol inhalation. Following colonization of the nasopharynx microorganisms may reach the lung alveoli via microaspiration. When the inoculum size is sufficient and/or host immune defenses are impaired, infection results. Replication of the pathogen, the production of virulence factors, and the host immune response lead to inflammation and damage of the lung parenchyma, resulting in pneumonia.

Once bacteria reach the lungs, they can cause an inflammatory response that results in disease. This is best studied with *S. pneumoniae*, which in the absence of opsonizing antibodies, rapidly multiplies in the alveolar spaces, leading to local hyperemia, edema, and mobilization of neutrophils. The filling of alveoli with bacteria, red cells, and fluid leads to significant increase in weight of the lung in this early phase of consolidation (Fig. 26.1). Subsequently this leads to advanced consolidation with increased neutrophils, pulmonary cells, and fibrin.

In some cases, CAP might also arise from uncontrolled replication of microbes that normally reside in the alveoli. Hypothetically, exogenous insults such as a viral infection or smoke exposure might alter the composition of the alveolar microbiome and trigger overgrowth of certain microbes. Because organisms that comprise the alveolar microbiome typically cannot be cultivated using standard cultures, this hypothesis might explain the low rate of pathogen detection among patients with CAP.

There are several predisposing conditions to development of pneumonia (Box 26.1).

## MICROBIOLOGY

Although numerous pathogens have been associated as a cause of CAP, a limited range of key pathogens cause the majority of cases (Table 26.1). The predominant bacterial pathogen continues to be *S. pneumoniae* (pneumococcus). Although *S. pneumoniae* (pneumococcus) is the most commonly detected bacterial cause of CAP in most studies, the overall incidence of pneumococcal pneumonia is decreasing in the United States. This likely is in part due to widespread use of pneumococcal vaccination, which results in both a decline in the individual rates of pneumococcal pneumonia and herd immunity in the population. Other causative agents include (but are not limited to) *Haemophilus influenzae, Mycoplasma pneumoniae, Chlamydophila pneumoniae, Legionella* species, enteric gram-negative bacteria

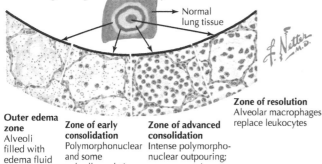

**A. Lobar pneumonia; right upper lobe.** Mixed red and gray hepatization (transition stage); pleural fibrinous exudate

**B.** Right upper lobe and segment of right lower lobe pneumonia

**C. Pathologic changes in zones of the pneumonic lesion**

Normal lung tissue

**Outer edema zone**
Alveoli filled with edema fluid containing pneumococci

**Zone of early consolidation**
Polymorphonuclear and some red cell exudation

**Zone of advanced consolidation**
Intense polymorphonuclear outpouring; pneumococci phagocytized and destroyed

**Zone of resolution**
Alveolar macrophages replace leukocytes

**Fig. 26.1** Pneumococcal pneumonia.

### BOX 26.1  Predisposing Conditions of Community-Acquired Pneumonia

- Alterations in the level of consciousness, which predispose to both macroaspiration of stomach contents (because of stroke, seizures, drug intoxication, anesthesia, and alcohol abuse) and microaspiration of upper airway secretions during sleep
- Smoking
- Alcohol consumption
- Toxic inhalations
- Pulmonary edema
- Uremia
- Malnutrition
- Administration of immunosuppressive agents (solid organ or stem cell transplant recipients or patients receiving chemotherapy)
- Mechanical obstruction of a bronchus
- Being elderly (there is a marked increase in the rate of pneumonia in persons ≥65 years)
- Cystic fibrosis
- Bronchiectasis
- Chronic obstructive pulmonary disease (COPD)
- Previous episode of pneumonia or chronic bronchitis
- Uncontrolled comorbidities (e.g., congestive heart failure, diabetes)

### TABLE 26.1  Most Common Causes of Community-Acquired Pneumonia

| Ambulatory Patients | Hospitalized (Non-ICU) | Severe (ICU) |
| --- | --- | --- |
| Streptococcus pneumonia | S. pneumonia | S. pneumonia |
| Mycoplasma pneumonia | M. pneumonia | Staphylococcus aureus |
| Haemophilus influenza | C. pneumonia | Legionella species |
| Chlamydophila pneumonia | H. influenza | Gram-negative bacilli |
| Respiratory viruses[a] | Legionella species | H. influenzae |
| | Aspiration | |
| | Respiratory viruses[a] | |

[a]Influenza A and B, adenovirus, respiratory syncytial virus, parainfluenza.
*ICU,* Intensive care unit.
Adapted from Mandell LA, Wunderink RG, Anzueto A, et al: Infectious Diseases Society of America/American Thoracic Society consensus guidelines on the management of community-acquired pneumonia in adults, *Clin Infect Dis* 2007; 44(suppl 2):S27-S72 based on collective data from recent studies.

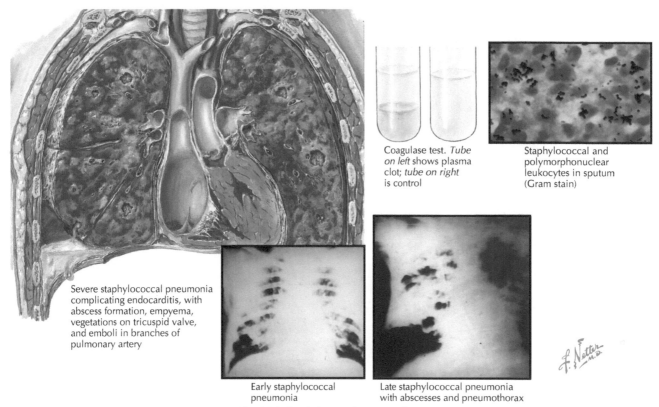

Severe staphylococcal pneumonia complicating endocarditis, with abscess formation, empyema, vegetations on tricuspid valve, and emboli in branches of pulmonary artery

Coagulase test. *Tube on left* shows plasma clot; *tube on right* is control

Staphylococcal and polymorphonuclear leukocytes in sputum (Gram stain)

Early staphylococcal pneumonia

Late staphylococcal pneumonia with abscesses and pneumothorax

**Fig. 26.2** Staphylococcal pneumonia.

(Enterobacteriaceae), *Pseudomonas aeruginosa, Staphylococcus aureus (both methicillin susceptible and resistant)*, anaerobes (aspiration pneumonia), and respiratory viruses (influenza, adenovirus, respiratory syncytial virus, parainfluenza, coronavirus). Respiratory viruses have been detected in approximately one-third of cases of CAP in adults when using molecular methods. The extent to which respiratory viruses serve as single pathogens, cofactors in the development of bacterial CAP, or triggers for dysregulated host immune response has not been established.

S. aureus (Fig. 26.2) and gram-negative bacilli (such as *Klebsiella* species; Fig. 26.3) are less frequently isolated and are the cause in selected patients (e.g., patients with severe CAP requiring intensive care admission or those who have recently received antimicrobial therapy or have pulmonary comorbidities). The frequency of other causes—for example, *Mycobacterium tuberculosis, Chlamydia psittaci* (psittacosis), *Coxiella burnetii* (Q fever), *Francisella tularensis* (tularemia), and endemic fungi (histoplasmosis, coccidioidomycosis, blastomycosis)—varies with epidemiologic setting. The alveolar microbiome is similar to oral flora and primarily comprised of anaerobic bacteria (e.g., *Prevotella* and *Veillonella*) and microaerophilic streptococci.

Antimicrobial resistance, especially the emergence of multidrug-resistant *S. pneumoniae*, has escalated worldwide over the past two decades. Risk factors for drug-resistant *S. pneumoniae* include very young age (≤2 years) and old age (≥65 years); β-lactam, macrolide, or fluoroquinolone therapy within the previous 3 months; alcoholism; medical comorbidities; immunosuppressive illness or therapy; and exposure to a child in a daycare center. Available data suggest that a clinically relevant level of penicillin resistance is a minimum inhibitory concentration (MIC) of 4 mcg/mL or greater. At this breakpoint the rate of penicillin resistance for pneumonia for most locations in North America is 10% or less. Since the prior version of *Netter's Infectious Diseases*, there continues to be an increase of macrolide-resistant

*S. pneumoniae*. In many regions of the world, including the United States, the overall prevalence of macrolide-resistant *S. pneumoniae* (high level and low level) is now greater than 25%; in the United States, it is greater than 40%.

## CLINICAL FEATURES

The host immune response to microbial replication within the alveoli plays an important role in determining disease severity. For some patients, the local inflammatory response in the lung is sufficient to localize the infection; in others, a systemic immune response may be necessary to control infection and prevent spread or complications, such as bacteremia. In a minority, the systemic response can become dysregulated, leading to tissue injury, sepsis, acute respiratory distress syndrome, and/or multiorgan dysfunction.

Symptoms and signs of CAP include cough (either productive or nonproductive), pleuritic chest pain, shortness of breath, temperature greater than 38°C, and crackles on auscultation. Mucopurulent sputum production is more frequently found in association with bacterial pneumonia, whereas scant or watery sputum production is more suggestive of an atypical pathogen. Although there are classic descriptions of certain types of sputum production and particular pathogens (e.g., pneumococcal pneumonia and rust-colored sputum), these clinical descriptions usually do not help with initial clinical decision-making regarding treatment because the clinical presentations of the specific pathogens are variable. Gastrointestinal symptoms (nausea, vomiting, diarrhea) and mental status changes may accompany respiratory manifestations. Signs and symptoms of pneumonia can also be subtle in patients with advanced age and/or impaired immune systems, and a higher degree of suspicion may be needed to make the diagnosis.

On physical examination the majority of patients are febrile, although this finding is frequently absent in older patients. Increased

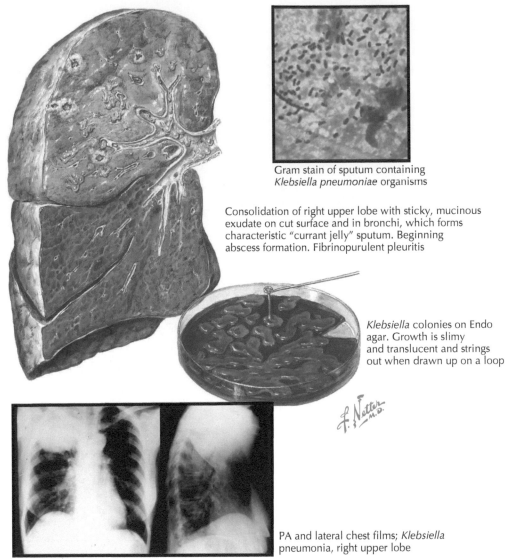

Gram stain of sputum containing
*Klebsiella pneumoniae* organisms

Consolidation of right upper lobe with sticky, mucinous
exudate on cut surface and in bronchi, which forms
characteristic "currant jelly" sputum. Beginning
abscess formation. Fibrinopurulent pleuritis

*Klebsiella* colonies on Endo
agar. Growth is slimy
and translucent and strings
out when drawn up on a loop

PA and lateral chest films; *Klebsiella*
pneumonia, right upper lobe

**Fig. 26.3** *Klebsiella* (Friedländer) pneumonia.

respiratory rate is frequently noted, and this may be the most sensitive sign in elderly patients; tachycardia is also common. Chest examination reveals audible rales in most patients, and approximately one-third have evidence of consolidation. However, no clear constellation of symptoms and signs has been found to accurately predict whether or not the patient has pneumonia.

The major blood test abnormality is leukocytosis (typically 15,000 to 30,000 per mm$^3$) with a leftward shift. Leukopenia can occur and generally is associated with a poor prognosis.

## DIAGNOSTIC APPROACH

Adult patients who are immunocompetent should be evaluated for pneumonia if they demonstrate signs including cough, sputum production, labored breathing (including altered breath sounds and rales), and/or fever. These symptoms (especially cough) are nonspecific and may also be present in patients with other diseases that should be considered in the differential diagnosis (Box 26.2; Fig. 26.4).

The presence of an infiltrate on plain chest radiograph is considered the "gold standard" for diagnosing pneumonia when clinical features are supportive. Radiologists cannot reliably differentiate bacterial from

nonbacterial pneumonia on the basis of the radiographic appearance. Computed tomography (CT) scans are significantly more sensitive in detecting pulmonary infiltrates in patients hospitalized with CAP, but the clinical significance of this finding is unclear.

### Determining Severity of Illness and Site of Care

Site of care in patients with CAP affects the overall cost of treatment, the intensity of diagnostic testing, and the empirical antimicrobial(s). A general consensus is that the majority of patients can be safely treated as outpatients. However, selected patients should be hospitalized based on the requirements of care (e.g., need for close observation, respiratory support, intravenous antibiotics, or other concerns). The advantages of not admitting patients for treatment of CAP include decreased cost, patient preference, and avoidance of iatrogenic complications in the hospital. For elderly patients, particularly, a reduction in time in a hospital bed can facilitate better convalescence. Hospitalization should be considered when (1) patients have preexisting conditions that may compromise the safety of home care, (2) patients have hypoxemia, (3) patients are unable to take oral medications, or (4) psychosocial factors can potentially influence effective treatment (such as an unstable home environment or psychiatric disorders that may hinder adherence

## BOX 26.2   Differential Diagnosis of Cough and Fever

**Infectious**
- Upper respiratory tract
  - Acute or chronic sinusitis
- Lower respiratory tract
  - Acute or chronic bronchitis
  - Acute exacerbation of chronic obstructive pulmonary disease
  - Bronchiectasis
  - Tuberculosis

**Noninfectious**
- Pulmonary embolism or infarction
- Pulmonary neoplasm
- Radiation pneumonitis
- Interstitial lung disease
- Sarcoidosis
- Collagen vascular disease
- Drug-induced pulmonary disease
- Hypersensitivity pneumonitis
- Granulomatous vasculitis
- Eosinophilic pneumonitis

to therapy). Mortality prediction tools can also help to guide clinicians in determining the requirement of hospitalization.

The pneumonia prediction rule, developed more than 20 years ago, offers important insights into the risk of mortality. This technique uses a combination of demographic variables, comorbidities, physical observations, and laboratory and radiographic variables to assign patients to one of five classes. Those belonging to pneumonia severity index (PSI) class 1 or 2 have a low risk of mortality (<1%) and can be treated as outpatients. Those in PSI class 3 have a slightly higher risk of mortality (<5%) and may require a brief observational stay in a hospital. Those in PSI class 4 or 5 have the highest mortality risk (8% to 40%) and will require hospitalization; those in PSI class 5 should be admitted to an ICU. Although the Pneumonia Patient Outcomes Research Team (PORT) prediction rule is effective in determining mortality risk, it is not the most practical approach in the clinical setting as it is partly based on laboratory evaluations that can be time consuming.

The Confusion, Uremia, Respiratory Rate, Blood Pressure (CURB)-65 rule uses only five aspects in making a clinical determination—confusion, urea concentration, respiratory rate, blood pressure, and age. Those meeting two or more of these criteria should be considered for hospitalization. However, this method requires a blood sample and laboratory analysis for urea concentration. In response to this, the Confusion, Respiratory Rate, Blood Pressure (CRB)-65 was designed. It omitted the blood urea measurement and was practical for office-based settings. In CRB-65, a score of 0 equates to home treatment, a score of 1 to hospital-supervised treatment, and a score of 2 or more to hospitalization.

Recommendations regarding admission to the ICU are provided by the IDSA/ATS guidelines for management of CAP. According to the IDSA/ATS guidelines, direct admission to the ICU is essential for patients with septic shock requiring vasopressor or for patients with acute respiratory failure requiring intubation and mechanical ventilation. ICU admission should also be considered for patients who have three or more of the following: confusion or disorientation, uremia (blood urea nitrogen ≥20 mg/dL), respiratory rate of 30 breaths/min or greater, hypotension requiring aggressive fluid resuscitation, arterial oxygen pressure ($PaO_2$)/fractional inspired oxygen ($FIO_2$) ratio of 250 or less, multilobar infiltrates, leukopenia (white blood count <4000 cells/mm$^3$), thrombocytopenia (platelet count <100,000 cells/mm$^3$), or hypothermia (core temperature <36°C).

## Determining Causative Diagnosis

The benefit of obtaining a microbiologic diagnosis should be balanced against the time and cost associated with an extensive evaluation in each patient. Advancements in molecular testing methods are offering more rapid results and can provide an increasing rate of pathogen identification. The cost benefit of these tests requires further evaluation. Nevertheless, there is good rationale for establishing a causative diagnosis (i.e., to permit antibiotic selection that permits optimal selection of agents against a specific pathogen, to limit the consequences of antibiotic misuse, and to identify pathogens of potential epidemiologic significance such as Legionella or tuberculosis [TB]).

Routine microbiologic tests are not recommended by most of the recent guidelines for patients managed in the community. However, if a patient has purulent sputum, it is reasonable to send a sample to the laboratory for Gram stain and culture on the basis that the information may be of value for directing specific therapy if the patient's condition is failing to respond to initial empirical therapy.

Investigations that are variably recommended for patients requiring admission include blood cultures, Gram stain and culture, urinary antigen tests, and thoracentesis if there is significant pleural fluid present. Approximately 5% to 11% of patients with CAP admitted to the hospital will have positive blood cultures, more commonly associated with severe illness. Because false-positive blood cultures may be more common than pathogen-positive blood cultures from patients admitted to a general ward, they may have limited benefit for such patients and are considered "optional" in recent guidelines. In more severely ill patients requiring ICU admission, blood cultures may increase the likelihood of finding a pathogen not covered by customary antimicrobial therapy and are recommended for all such patients. The value of performing a sputum Gram stain and culture is limited by the facts that many patients cannot produce a good specimen and the validity of the Gram stain is related directly to the experience of the interpreter; however, when stringent criteria are applied, the specificity for pneumococcal pneumonia can approach 90%. Sputum culture for other pathogens (e.g., *Legionella* species, fungus, virus, *Mycobacterium* species) should be obtained based on epidemiologic considerations. *M. tuberculosis* should be considered when patients are from endemic areas, when the chest radiograph shows an upper lobe cavitary infiltrate, when patients have had prolonged symptoms of cough (often with weight loss and night sweats), when patients are homeless, or when they have a history of exposure to TB.

Because the early administration of therapy is important for the outcome of CAP, an attempt to obtain expectorated sputum should never delay prompt initiation of antimicrobial therapy.

Other tests considered helpful for patients admitted to the hospital include the urinary antigen assays for *Legionella* and *S. pneumoniae* and a direct stain (i.e., acid-fast) for detection of mycobacterial infections in patients with epidemiologic risks for TB.

Many rapid diagnostic tests such as nucleic acid amplification tests (i.e., PCR) assays are currently available, offer promise for rapid diagnosis, and will likely become increasingly used in the future. The commercial availability of the multiplex PCR methods provides clinicians the ability to assess for a wide range of possible pathogens.

Serologic tests are generally not helpful in the early management of CAP because the determination of acute and convalescent titers is required before the cause can be ascribed to a specific pathogen. Percutaneous transthoracic needle aspiration (PTNA) or other invasive testing (including bronchoscopy and biopsy) is not routinely recommended for the evaluation of patients but may be valuable in immunocompromised hosts, suspected TB in the absence of productive cough, selected cases of chronic pneumonia, pneumonia associated with suspected neoplasm or foreign body, suspected *Pneumocystis*

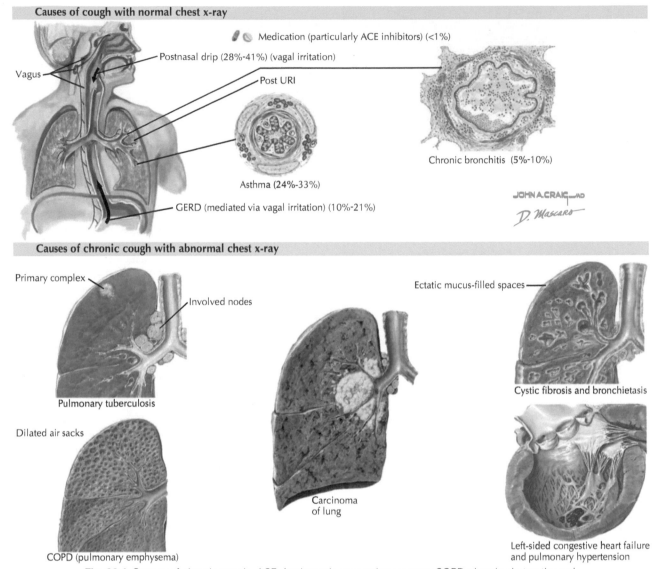

**Causes of cough with normal chest x-ray**

Medication (particularly ACE inhibitors) (<1%)

Postnasal drip (28%-41%) (vagal irritation)

Vagus

Post URI

Chronic bronchitis (5%-10%)

Asthma (24%-33%)

GERD (mediated via vagal irritation) (10%-21%)

JOHN A. CRAIG—MD

D. Mascaro

**Causes of chronic cough with abnormal chest x-ray**

Primary complex

Involved nodes

Ectatic mucus-filled spaces

Pulmonary tuberculosis

Dilated air sacks

Cystic fibrosis and bronchietasis

Carcinoma of lung

COPD (pulmonary emphysema)

Left-sided congestive heart failure and pulmonary hypertension

**Fig. 26.4** Causes of chronic cough. *ACE,* Angiotensin converting enzyme; *COPD,* chronic obstructive pulmonary disease; GERD, gastroesophageal reflux disease; URI, upper respiratory infection.

*carinii* pneumonia, some cases in which intubation is required, and suspected conditions that require lung biopsy. During influenza season (e.g., late fall to early spring in the northern hemisphere) testing for influenza by PCR is recommended.

## CLINICAL MANAGEMENT AND DRUG TREATMENT

### Empirical Antimicrobial Therapy

For most patients with CAP and until point-of-care diagnostic test methods become more available, empiric treatment directed at the most likely pathogens is appropriate. The pathogens most likely to cause CAP vary with severity of illness, local epidemiology, and patient risk factors for infection with drug-resistant organisms. Recently published guidelines from North America give specific recommendations for empirical therapy for CAP (Box 26.3).

For outpatients, empirical therapy with a macrolide, doxycycline, amoxicillin, antipneumococcal fluoroquinolone (e.g., levofloxacin, gemifloxacin, moxifloxacin), or the combination of a β-lactam plus macrolide is a recommended treatment option based on risk factors of recent antimicrobial use and comorbidities. In regions such as the United States where the prevalence of macrolide resistance is high but

doxycycline resistance is usually less than 20%, I typically treat with doxycycline for patients without comorbidities such as chronic heart, lung, liver, or renal disease; diabetes mellitus; alcoholism; malignancy; or asplenia and without recent use of antimicrobials (the strongest risk factor for resistance). Because doxycycline resistance rates are not well established, I generally follow up with patients treated with doxycycline monotherapy approximately 48 to 72 hours after start of therapy to ensure improvement. Another option as listed in the recent ATS/IDSA guidelines is amoxicillin (1 g three times daily), which will cover almost all *S. pneumoniae* but will not be effective of organisms associated with atypical pneumonia. In regions where the prevalence of macrolide- and doxycycline-resistant *S. pneumoniae* is greater than 25% or who have major comorbidities and/or have recently used antibiotics, I treat with either combination therapy with a β-lactam (e.g., high-dose amoxicillin [1 g orally three times daily], extended-release amoxicillin-clavulanate [2 g orally twice daily], cefpodoxime, cefuroxime) **PLUS** either a macrolide **OR** doxycycline or with a fluoroquinolone (levofloxacin or moxifloxacin) monotherapy. Modifications to these regimens may be needed for antibiotic allergy, potential drug interactions, specific exposures, and other patient-specific factors. In particular, during influenza season, patients may warrant antiviral therapy.

For general inpatient treatment, combination therapy with a β-lactam (e.g., cefotaxime, ceftriaxone, ceftaroline, ertapenem, or ampicillin-sulbactam) plus azithromycin or doxycycline or monotherapy with a respiratory fluoroquinolone is recommended. For patients with severe CAP requiring ICU admission, recommendations are given based on risks for methicillin-resistant *S. aureus* (MRSA) and/or *Pseudomonas*.

If MRSA is a consideration, linezolid or vancomycin should be added to the regimen. Although methicillin-resistant strains of *S. aureus* are still the minority, the excess mortality of inappropriate antibiotic therapy would suggest that empirical coverage should be considered when community-acquired (CA)-MRSA is a concern. The best indicator of *S. aureus* is the presence of gram-positive cocci in clusters in a tracheal aspirate or adequate sputum sample (see Fig. 26.2). Clinical risk factors for *S. aureus* CAP include end-stage renal disease, intravenous drug abuse, prior influenza, and prior antibiotics (especially fluoroquinolones). Risk factors for CAP due to *Pseudomonas* as well as resistant gram-negative bacilli include previous antibiotic therapy, recent hospitalization, immunosuppression, pulmonary comorbidity (e.g., cystic fibrosis, bronchiectasis, or repeated exacerbations of chronic obstructive pulmonary disease that require frequent glucocorticoid and/or antibiotic use), and the presence of multiple medical comorbidities (e.g., diabetes mellitus, alcoholism). For patients with risk factors for *Pseudomonas*, I typically use combination therapy with both an antipseudomonal β-lactam (e.g., piperacillin-tazobactam, cefepime, ceftazidime, meropenem, or imipenem) plus an antipseudomonal fluoroquinolone (e.g., ciprofloxacin or levofloxacin). Antiviral treatment (e.g., oseltamivir) should be given as soon as possible for any hospitalized patient with known or suspected influenza. (See "Treatment of seasonal influenza in adults.")

Two newer US Food and Drug Administration (FDA)-approved options of therapy for CAP include omadacycline and Lefamulin, each of which can be used as monotherapy and cover both standard bacterial and atypical pathogens. Both have in vitro activity for MRSA but neither for *Pseudomonas*. Both are available as intravenous and oral formulations.

## Pathogen-Directed Therapy

Treatment options are obviously simplified if the causative agent is established or strongly suspected (Table 26.2). Diagnostic procedures that provide identification of a specific cause within 24 to 72 hours can still be useful for guiding continued therapy. If, for example, an appropriate culture reveals the isolation of penicillin-susceptible *S. pneumoniae*, therapy can be specified by selecting a narrow-spectrum agent (e.g., penicillin or amoxicillin). This will hopefully reduce the selective pressure for resistance, as well as cause less disturbance of the microbiota, the disruption of which may impairment the immune system. This information is often available at the time of switch from parenteral to oral therapy and may be used to direct specific antimicrobial choices.

## Duration of Therapy

The duration of therapy should be based on the patient's clinical response to therapy. For all patients, treatment should be given until the patient has been afebrile and clinically stable for at least 48 hours and for a minimum of 5 days. Patients with mild infection generally require 5 to 7 days of therapy. Patients with severe infection or chronic comorbidities generally require 7 to 10 days of therapy. Extended courses may be needed for immunocompromised patients, patients with infections causes by certain pathogens (e.g., *P. aeruginosa*), or those with complications.

For patients for whom there is uncertainty of duration, I often use procalcitonin thresholds as an adjunct to clinical judgment to help guide antibiotic discontinuation. If the level is less than 0.25 μg/mL or if there has been an 80% decrease from baseline level, there is greater confidence of being able to discontinue antibiotic therapy.

## PROGNOSIS

In most patients, fever is reduced within 3 to 4 days of therapy; however, cough and fatigue will often persist for several weeks even in

**TABLE 26.2   Epidemiologic Conditions and/or Risk Factors Related to Specific Pathogens in Community-Acquired Pneumonia**

| Condition | Commonly Encountered Pathogen(s) |
|---|---|
| Alcoholism | *Streptococcus pneumoniae*, oral anaerobes, *Klebsiella pneumoniae*, *Acinetobacter* species, *Mycobacterium tuberculosis* |
| COPD and/or smoking | *Haemophilus influenzae*, *Pseudomonas aeruginosa*, *Legionella* species, *S. pneumoniae*, *Moraxella catarrhalis*, *Chlamydophila pneumoniae* |
| Aspiration | Gram-negative enterics, oral anaerobes |
| Lung abscess | CA-MRSA, oral anaerobes, endemic fungal pneumonia, *M. tuberculosis*, atypical mycobacteria |
| Exposure to bat or bird droppings | *Histoplasma capsulatum* |
| Exposure to birds | *Chlamydia (Chlamydophila) psittaci* (if poultry—avian influenza) |
| Exposure to rabbits | *Francisella tularensis* |
| Exposure to farm animals or parturient cats | *Coxiella burnetii* (Q fever) |
| HIV (early) | *S. pneumoniae*, *H. influenzae*, *M. tuberculosis* |
| HIV (late) | Above plus *Pneumocystis jiroveci*, *Cryptococcus*, *Histoplasmosis*, *Aspergillus*, atypical mycobacteria (especially *Mycobacterium kansasii*), *P. aeruginosa*, *H. influenzae* |
| Hotel or cruise ship stay in previous 2 weeks | *Legionella* |
| Travel to or residence in southwestern United States | *Coccidioides* species, hantavirus |
| Travel to or residence in Southeast and East Asia | *Burkholderia pseudomallei*, avian influenza, SARS |
| Influenza active in community | Influenza, *S. pneumoniae*, *Staphylococcus aureus*, *H. influenzae* |
| Cough >2 weeks with whoop or post tussive vomiting | *Bordetella pertussis* |
| Structural lung disease (e.g., bronchiectasis) | *P. aeruginosa*, *Burkholderia cepacia*, *S. aureus* |
| Injection drug use | *S. aureus*, anaerobes, *M. tuberculosis*, *S. pneumoniae* |
| Endobronchial obstruction | Anaerobes, *S. pneumoniae*, *H. influenzae*, *S. aureus* |
| In context of bioterrorism | *Bacillus anthracis* (anthrax), *Yersinia pestis* (plague), *F. tularensis* (tularemia) |

*CA-MRSA*, Community-associated methicillin-resistant *Staphylococcus aureus*; *COPD*, chronic obstructive pulmonary disease; *HIV*, human immunodeficiency virus; *SARS,* severe acute respiratory syndrome.
Adapted from Mandell LA, Wunderink RG, Anzueto A, et al: Infectious Diseases Society of America/American Thoracic Society consensus guidelines on the management of community-acquired pneumonia in adults, *Clin Infect Dis* 44(suppl 2):S27-S72, 2007.

patients with mild disease. Symptoms can be expected to last longer in more seriously ill patients. Such information should be imparted to patients for better awareness of their illness and anticipated clinical course.

Up to 15% of patients may not respond appropriately to initial antibiotic therapy. Nonresponse can be defined as absence or delay in achieving clinical stability within several days or actual clinical deterioration. A number of possibilities, both infectious and noninfectious, should be considered (Table 26.3). A systematic assessment of patients should take these considerations into account. The most common causes of treatment failure are lack of or delayed response by the host despite appropriate antibiotics and infection with an organism that is not covered by the initial antibiotic regimen. Antibiotic changes in this time period should be considered only for patients with deterioration or in whom new culture data or epidemiologic clues suggest alternative causes.

## PREVENTION

Vaccines targeting pneumococcal disease and influenza are a mainstay for preventing CAP. Pneumococcal vaccine and inactivated or live, attenuated influenza vaccine should be considered for all patients according to recent recommendations from the Centers for Disease Control and Prevention (CDC; refer to most recent CDC recommendations). Chemoprophylaxis (i.e., oseltamivir or zanamivir) can be used as an adjunct to influenza vaccination. Clinicians should also

intervene to modify some of the associated risk factors for pneumonia in adults. Because smoking is a significant risk factor for CAP, smoking cessation should be attempted; this is particularly important and relevant when patients are hospitalized for pneumonia. In addition, stabilization of underlying conditions (e.g., congestive heart failure, diabetes) and promotion of appropriate nutrition may help to reduce the risk of CAP and thereby promote longer and healthier lives.

## EVIDENCE

Dickson RP, Erb-Downward JR, Martinez FJ, Huffnagle GB. The microbiome and the respiratory tract. *Annu Rev Physiol* 2016;78:481. *Description of the population dynamics of microbiome of the respiratory tract, both in health and as altered by lung disease.*

File TM Jr, Goldberg L, Das Ak, Sweeney C, Saviski J, et al. Efficacy and safety of IV-to-oral lefamulin, a pleuromutilin antibiotic, for treatment of community-acquired bacterial pneumonia: the phase 3 LEAP 1 Trial. *Clin Infect Dis* 2019;69(11):1856–1867. *Randomized clinical trial documented efficacy and safety of lefamulin for community-acquired pneumonia.*

File TM Jr, Marrie TJ. Burden of community-acquired pneumonia in North American adults, *Postgrad Med* 2010;122:130-141. *An update of economic cost associated with community-acquired pneumonia.*

Gadsby, NJ, Russell CD, McHugh MP, Mark H, Morris, AC., et al. Comprehensive molecular testing for respiratory pathogens in community-acquired pneumonia. *Clin Infect Dis* 2016;62:817-23. *Comprehensive molecular testing significantly improves pathogen detection*

## TABLE 26.3 Recommended Antimicrobial Therapy for Specific Bacterial Pathogens

| Organism | Preferred Antimicrobial(s)[a] | Alternative Antimicrobial(s)[a] |
|---|---|---|
| *Streptococcus pneumoniae*, penicillin nonresistant (MIC <2 mcg/mL) | Penicillin G; amoxicillin | Macrolide; cephalosporins (oral—cefpodoxime, cefprozil, cefuroxime, cefdinir, cefditoren; parenteral—cefuroxime, ceftriaxone, cefotaxime); clindamycin; doxycycline; respiratory fluoroquinolone[b] <br> Newer agents: omadacycline; Lefamulin |
| *S. pneumoniae*, penicillin resistant (MIC ≥2 mcg/mL) | Agents based on susceptibility, including cefotaxime, ceftriaxone, fluoroquinolone | Vancomycin, linezolid, high-dose amoxicillin (3 g/day with penicillin MIC ≤4 mcg/mL) <br> Newer agents: omadacycline; Lefamulin |
| *Haemophilus influenzae* | Non–β-lactamase producing: amoxicillin <br> β-lactamase producing: second- or third-generation cephalosporin; amoxicillin-clavulanate | Fluoroquinolone; doxycycline; azithromycin[c]; clarithromycin[d] <br> Newer agents: omadacycline; Lefamulin |
| *Mycoplasma pneumoniae* or *Chlamydophila pneumoniae* | Macrolide; a tetracycline | Fluoroquinolone <br> Newer agents: omadacycline; Lefamulin |
| *Legionella* species | Fluoroquinolone; azithromycin | Doxycycline <br> Newer agents: omadacycline; Lefamulin |
| *Chlamydia psittaci* | A tetracycline | Macrolide |
| *Coxiella burnetii* | A tetracycline | Macrolide |
| *Francisella tularensis* | Doxycycline | Gentamicin, streptomycin |
| *Yersinia pestis* | Streptomycin, gentamicin | Doxycycline, fluoroquinolone |
| Anthrax (inhalation) | Ciprofloxacin, levofloxacin, doxycycline (usually with second agent) | Other fluoroquinolones; β-lactam, if susceptible; rifampin; clindamycin; chloramphenicol |
| Enterobacteriaceae | Third-generation cephalosporin; carbapenem[d] (drug of choice if extended-spectrum β-lactamase producer) | β-lactam or β-lactamase inhibitor[e]; fluoroquinolone |
| *Pseudomonas aeruginosa* | Antipseudomonal β-lactam[f] *plus* ciprofloxacin or levofloxacin[g] or aminoglycoside | Aminoglycoside *plus* (ciprofloxacin or levofloxacin[g]) |
| *Burkholderia pseudomallei* | Carbapenem, ceftazidime | Fluoroquinolone, TMP-SMX |
| *Acinetobacter* species | Carbapenem | Cephalosporin-aminoglycoside, ampicillin-sulbactam, colistin |
| *Staphylococcus aureus:* <br> Methicillin susceptible <br> Methicillin resistant | Antistaphylococcal penicillin[h] <br> Vancomycin or linezolid | Cefazolin; clindamycin <br> TMP-SMX |
| *Bordetella pertussis* | Macrolide | TMP-SMX |
| Anaerobe (aspiration) | β-lactam or β-lactamase inhibitor[e]; clindamycin | Ertapenem |

[a]Choices should be modified based on susceptibility test results and advice from local specialists. Refer to local references for appropriate doses.
[b]Levofloxacin, moxifloxacin (not a first line choice for penicillin-susceptible strains); ciprofloxacin is appropriate for *Legionella* and most gram-negative bacilli (including *H. influenzae*).
[c]Azithromycin more active in vitro than clarithromycin for *H. influenzae*.
[d]Imipenem-cilastatin, meropenem, ertapenem.
[e]Piperacillin-tazobactam for gram-negative bacilli; ampicillin-sulbactam or amoxicillin-clavulanate.
[f]Piperacillin-tazobactam, ceftazidime; cefepime, aztreonam, imipenem, meropenem.
[g]750 mg qd.
[h]Nafcillin, oxacillin, flucloxacillin.
*MIC*, Minimum inhibitory concentration; *TMP-SMX*, trimethoprim-sulfamethoxazole.
Adapted from Mandell LA, Wunderink RG, Anzueto A, et al: Infectious Diseases Society of America/American Thoracic Society consensus guidelines on the management of community-acquired pneumonia in adults, *Clin Infect Dis* 2007;44(suppl 2):S27-S72.

in CAP. It also has the potential to enable early de-escalation from broad spectrum empirical antimicrobials to pathogen-directed therapy.

Schuetz P, Wirz Y, Sager R, Christ-Crain M, Stolz D, et al. Effect of procalcitonin-guided antibiotic treatment on mortality in acute respiratory infections: a patient level meta-analysis. *Lancet Infectious Dis.* 2018;18: 95-107. *Use of procalcitonin to guide antibiotic treatment in patients with acute respiratory infections, including pneumonia, has the potential to improve antibiotic management with positive effects on clinical outcomes and on the current threat of increasing antibiotic multiresistance.*

Stets R, Popescu M, Gonong JR, Mitha I, Nseir W. et al. Omadacycline for community-acquired pneumonia. *NEJM* 2019;380:517-27. *Randomized*

*clinical trial documented efficacy and safety of omadacycline for community-acquired pneumonia.*

## ADDITIONAL RESOURCES

Faner R, Sibila O, Agustí A, et al. The microbiome in respiratory medicine: current challenges and future perspectives. *Eur Respir J* 2017;49. *Although the healthy lung had previously been considered to be sterile, culture-independent techniques report that large numbers of microorganisms coexist in the lung. Dysbiosis of the lung microbiota has been related to disease of the lung.*

Jain S, Self WH, Wunderink RG, et al. Community-acquired pneumonia requiring hospitalization among U.S. adults. *N Engl J Med* 2015;373:415. *A multicenter study coordinated by The Center for Disease Control and Prevention to identify the etiology of adults with community-acquired pneumonia.*

Metlay JP, Waterer G, Long AC, et al. Diagnosis and treatment of adults with community-acquired pneumonia: an official clinical practice guideline of the American Thoracic Society and Infectious Diseases Society of America. *Am J Respir Crit Care Med* 2019;200:e45-e67. *A comprehensive, updated, evidence-based set of recommendations regarding the diagnosis and management of adults with community-acquired pneumonia.*

National Institute for Health and Care Excellence. Pneumonia in adults: Diagnosis and management. https://www.nice.org.uk/guidance/cg191 (Accessed on July 26, 2018). *This guideline provides recommendations for the management of suspected and confirmed community- and hospital-acquired pneumonia in adults. By National Institute for Health and Care Excellence of United Kingdom.*

Ramirez JA, Wiemken TL, Peyrani P, et al. Adults hospitalized with pneumonia in the United States: incidence, epidemiology, and mortality. *Clin Infect Dis* 2017;65:1806. *The estimated US burden of CAP is substantial, with greater than 1.5 million unique adults being hospitalized annually, 100 000 deaths occurring during hospitalization, and approximately 1 of 3 patients hospitalized with CAP dying within 1 year.*

# 27

# Hospital-Acquired Pneumonia

*Andrew P. Michelson, Marin H. Kollef*

## ABSTRACT

Hospital-acquired pneumonia (HAP) is a new infection of the lung parenchyma that develops more than 48 hours after hospital admission. Epidemiologic data suggest that HAP occurs in up to 1.6% of patients, prolongs hospital stay by 2 to 3 days, and represents the most common hospital-acquired infection leading to death in critically ill patients. The morbidity and mortality attributed to HAP are significantly increased if the infection is caused by multidrug-resistant pathogens. The subset of HAP occurring more than 48 hours after initiation of mechanical ventilation is termed *ventilator-associated pneumonia* (VAP). The concept of a healthcare-associated pneumonia (HCAP), referring to a pneumonia diagnosed outside of the acute care setting, was purposefully removed from the 2016 Infectious Disease Society of American (IDSA)/American Thoracic Society (ATS) guidelines as more recent evidence suggested HCAP pathogens more closely resemble those seen in community-acquired pneumonia than those seen in HAP. Risk factors for HAP/VAP are diverse and include both patient-specific and treatment-associated elements. The diagnostic criteria for HAP/VAP require a new infiltrate on chest radiography in conjunction with typical clinical and laboratory findings of pneumonia. Although microbiologic confirmation is ideal, the optimal diagnostic strategy is debated. Recent trends in the microbial resistance rates for pathogens causing HAP/VAP have led to significant changes in the recommendations for empirical therapy, with an increased reliance on patient-specific factors and local antimicrobial resistance patterns. Limited new antibiotic development and high attributable mortality rates highlight the importance of evidence-based HAP/VAP prevention strategies.

## CLINICAL VIGNETTE

A 72-year-old-man with a history of chronic obstructive pulmonary disease (COPD) sought attention for progressive shortness of breath and wheezing. Two days before presentation, the patient developed subjective fevers, chills, and myalgias. The following morning, his chronic cough had significantly worsened and he was producing more phlegm than usual. By the afternoon, he developed progressive shortness of breath with wheezing that would not respond to his rescue inhalers. His wife called the paramedics, and he was brought into the emergency department.

On exam, the patient had a temperature of 38.4°C, respiratory rate of 32, blood pressure of 147/86 and an oxygen saturation of 86% on 6 L/min of supplemental oxygen. He was dyspneic and unable to speak in full sentences with notable diaphoresis. His pulmonary examination revealed use of accessory muscles with minimal air movement and diffuse expiratory wheezing. The patient was urgently intubated and transferred to the medical intensive care unit (ICU). His complete blood count (CBC) showed a white blood cell count (WBC) of 11,000/mm³, and his chest radiograph showed hyperexpanded lungs with no focal consolidations, pleural effusions, or pneumothorax. A respiratory viral panel revealed infection with influenza A and the patient was treated for both his viral infection and a COPD exacerbation with azithromycin,

methylprednisone, bronchodilators, and oseltamivir. The patient subsequently defervesced, and his oxygen requirements became minimal.

Five days later the patient developed a new fever and had an increase in his respiratory rate and oxygen requirements. A CBC revealed an increase in his WBC, and a repeat chest radiograph showed a new right lower lobe consolidation. The patient was diagnosed with a VAP and had a tracheal aspirate sent for Gram stain and culture. He was started on empiric broad-spectrum antibiotics, and 2 days later his cultures returned positive for methicillin-resistant *Staphylococcus aureus* (MRSA). His antibiotics were narrowed to vancomycin monotherapy, and he was successfully extubated 2 days later.

Patients with COPD are prone to developing respiratory failure when exposed to respiratory viruses, such as influenza. This patient's age and underlying lung disease increased his predilection to develop a VAP, which is a known complication of endotracheal intubation with mechanical ventilation. His severity of illness merited initiation of antimicrobial therapy that covered MRSA, which is a common postinfluenza superinfection.

## EPIDEMIOLOGY AND IMPACT

In the United States, pneumonia accounts for 22% of all hospital-acquired infections and is the leading cause of death from nosocomial infection. Although most cases of HAP arise outside the ICU, HAP rates are highest in ICU patients—particularly those who are mechanically ventilated. In this setting, approximately 10% of patients will be diagnosed with a VAP. These patients often experience a prolonged duration of endotracheal intubation, have up to 13% higher rates of mortality, and incur an estimated $40,000 in additional healthcare costs.

## PATHOPHYSIOLOGY

HAP/VAP—like any other pneumonia—results when microbes penetrate the normally sterile lower respiratory tract, overwhelm local host defenses, and establish infection. Although pathogens are most often introduced in aspirated oropharyngeal secretions, rarely they may enter the lung hematogenously in patients with bacteremia. Intubated patients develop a VAP as a direct consequence of the endotracheal tube acting as a foreign body that bypasses key barriers to infection. VAP ultimately results from varying degrees of aspiration of secretions pooled above the endotracheal tube cuff and/or direct inoculation from the biofilm that forms on the endotracheal tube surface.

Some of the risk factors that predispose patients to develop HAP/VAP are patient specific and include male sex, advanced age, chronic underlying disease (especially pulmonary disease),

**Patient-specific risk factors:**

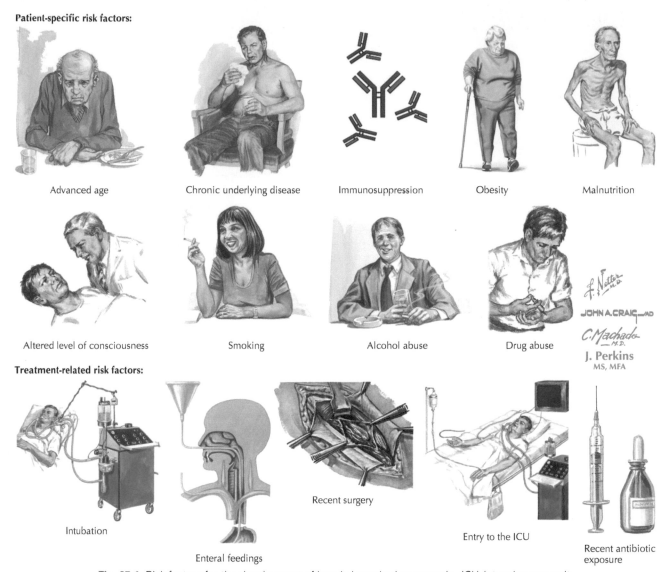

Advanced age

Chronic underlying disease

Immunosuppression

Obesity

Malnutrition

Altered level of consciousness

Smoking

Alcohol abuse

Drug abuse

**Treatment-related risk factors:**

Intubation

Enteral feedings

Recent surgery

Entry to the ICU

Recent antibiotic exposure

**Fig. 27.1** Risk factors for the development of hospital-acquired pneumonia. *ICU,* Intensive care unit.

immunosuppression, obesity or malnutrition, altered level of consciousness, smoking, and alcohol or drug abuse. Other risk factors are treatment related, including intubation, enteral feedings, recent surgery, entry to the ICU, and recent antibiotic exposure (Fig. 27.1).

## MICROBIOLOGY

Although causative agents include a wide variety of bacterial pathogens and may be polymicrobial, anaerobes are uncommon in HAP/VAP (Table 27.1). The bacteria that most frequently cause HAP/VAP are stratified into those causing early- and late-onset disease. Early-onset HAP/VAP occurs before hospital day 5 and is often caused by community-acquired organisms such as *Streptococcus pneumoniae* and *Haemophilus influenzae*. In contrast, late-onset HAP occurs beyond hospital day 5 and is more frequently caused by hospital-acquired gram-negative bacilli and *S. aureus* (including methicillin-resistant strains). HAP/VAP caused by *S. aureus* is more common in patients with head trauma, diabetes, or admission to an ICU. Multidrug resistant pathogens are a growing concern for both community and nosocomial pneumonia, and patients at highest risk are those with higher rates of healthcare contact (Table 27.2). Even though fungi are infrequently implicated as causal pathogens in immunocompetent patients,

| TABLE 27.1　**Common Nosocomial Pneumonia Pathogens** | |
|---|---|
| **Gram-Positive Pathogens** | **Gram-Negative Pathogens** |
| *Staphylococcus aureus* | *Enterobacter* species |
| *Streptococcus pneumoniae* | *Escherichia coli* |
| | *Klebsiella pneumoniae* |
| | *Pseudomonas aeruginosa* |
| | *Acinetobacter* species |

the use of antifungal therapies has also started to induce a concerning shift in echinocandin resistance among candida species. This dynamic pattern of pathogen resistance evolves over time and varies by geographic location, hospital system, and patient population, highlighting the need for regularly updated local surveillance data.

## CLINICAL PRESENTATION

The presenting signs and symptoms of HAP/VAP are nonspecific and vary in intensity based on the severity of the infection and the patient's underlying pulmonary reserve. The most common signs and symptoms include fever, cough, increasing sputum or purulent sputum

## TABLE 27.2 Patient-Specific Risk Factors for Antimicrobial-Resistant Pneumonia

| | |
|---|---|
| Treatment-related risk factors | Gastric acid suppression |
| | Hemodialysis |
| | Immunosuppression |
| | Home wound care |
| | Hospitalization for ≥2 days within the past 90 days |
| | Residence in a long-term care facility |
| Patient-specific risk factors | MRSA colonization |
| | *Pseudomonas aeruginosa* colonization |
| | Prior multidrug resistant infection |
| | Recurrent skin infections |
| | Poor functional status |
| | Structural lung disease (severe COPD, bronchiectasis) |
| | Tracheostomy |
| Antibiotic-induced resistance | Systemic antibiotics within the past 6 months |

*COPD,* Chronic obstructive pulmonary disease; *MRSA,* methicillin-resistant *Staphylococcus aureus.*

production, dyspnea, and pleuritic chest pain. Pneumonia should also be considered in patients who have an increase in their oxygen requirements, respiratory rate, or work of breathing. The presentation of HAP in the elderly population may be more subtle, with a predominance of vague complaints. Geriatric patients are less likely to have a significant fever but are more likely to have altered mental status. Evaluation of the patient with VAP is particularly challenging because patients are unable to provide any history and innumerable ICU-associated conditions are easily confused with pneumonia.

Physical examination findings of HAP/VAP are similarly nonspecific and fall along a spectrum of severity that is associated with the physiologic effects of infection on the patient. Initially insignificant abnormalities may progress to cyanosis and progressive respiratory failure in severe cases. As with any respiratory infection, the most common findings of HAP/VAP include hyperthermia, tachycardia, and tachypnea. Findings of consolidation—crepitations, bronchial breath sounds, tactile fremitus, dullness to percussion, reduced motion of the thorax, and egophony—may all be present. However, many of these findings will be absent or difficult to elicit in critically ill or mechanically ventilated patients.

Given the nondescript nature of many findings associated with HAP/VAP, alternative causes of fever or clinical deterioration must be evaluated in hospitalized patients (Fig. 27.2). Possible considerations

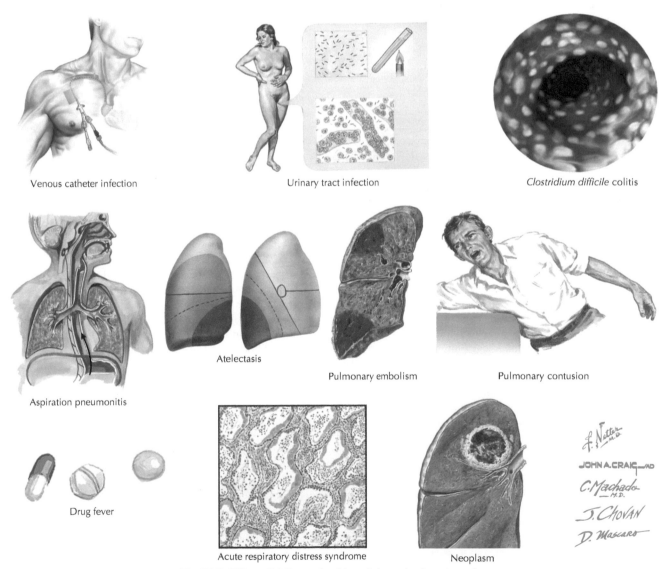

Venous catheter infection · Urinary tract infection · *Clostridium difficile* colitis · Aspiration pneumonitis · Atelectasis · Pulmonary embolism · Pulmonary contusion · Drug fever · Acute respiratory distress syndrome · Neoplasm

**Fig. 27.2** Differential diagnosis of hospital-acquired pneumonia.

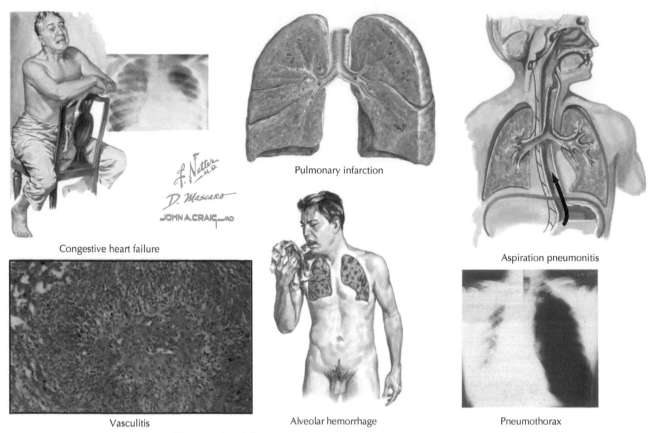

Congestive heart failure

Pulmonary infarction

Aspiration pneumonitis

Vasculitis

Alveolar hemorrhage

Pneumothorax

**Fig. 27.3** Noninfectious causes of an abnormal chest radiograph.

include central venous catheter infections, urinary tract infections, *Clostridium difficile* colitis, aspiration pneumonitis, atelectasis, pulmonary embolism, pneumothorax, pulmonary contusion, drug fever, acute respiratory distress syndrome, and neoplasms. Each of these entities has clinical characteristics that overlap with HAP/VAP and merits a high level of suspicion.

## DIAGNOSTIC APPROACH

The 2016 IDSA/ATS and European Respiratory Society (ERS)/European Society of Intensive Care Medicine (ESICM)/European Society of Clinical Microbiology and Infectious Diseases (ESCMID)/Asociación Latinoamericana del Tórax (ALAT) no longer endorse the use of scoring systems, such as the Clinical Pulmonary Infection Score (CPIS), for the diagnosis of HAP/VAP. Instead, the diagnosis of HAP/VAP is based on clinical criteria and incorporates a combination of the aforementioned symptoms in combination with radiographic and laboratory data. Radiographic demonstration of a new infiltrate is required for the diagnosis of HAP/VAP but is nonspecific and must be accompanied by other evidence supporting an active infection. When possible, microbiologic evidence of infection should be aggressively pursued to exclude noninfectious causes of an abnormal chest radiograph such as congestive heart failure, pulmonary infarction, aspiration pneumonitis, vasculitis, and alveolar hemorrhage (Fig. 27.3). Although posteroanterior and lateral radiographs are usually adequate, computed tomography scans are useful in assessing pneumonia-related complications such as empyema or lung abscess. False-negative chest radiographs are very uncommon but may be seen in patients with neutropenia, acquired immunodeficiency syndrome, tuberculosis, or profound dehydration.

Laboratory studies are used to confirm the diagnosis of pneumonia, to establish a causative pathogen, and to assess severity of illness.

Appropriate testing includes a CBC, two sets of blood cultures drawn from independent sites, and sputum for Gram stain with culture and sensitivity testing. The CBC gauges the leukocyte response to infection and gives information regarding anemia that may further impair oxygen delivery. Positive blood cultures serve two purposes: they identify a causative pathogen in up to 20% of cases, and they identify a subset of patients with increased mortality. A Gram stain that is timely, properly prepared, and correctly interpreted is useful for guiding the empirical therapeutic regimen. Any sputum sample showing leukocytes without organisms should raise suspicion for atypical organisms such as *Legionella*, mycobacteria, or viruses.

Arterial blood gas sampling is indicated in all critically ill patients and in those with pulse oximetry saturations less than 92% while breathing room air. Additional tests to assess concurrent end-organ dysfunction include measurements of electrolytes and serum glucose, hepatic function tests, assessment of renal parameters, and measurement of lactic acid. The role of nonspecific inflammatory markers such as procalcitonin and sedimentation rates is not well defined. Diagnostic thoracentesis must be promptly performed to exclude empyema in all patients with significant parapneumonic effusions. Serum immunoglobulins and human immunodeficiency virus testing should be offered to younger patients who develop HAP/VAP. Because these patients may rapidly decompensate, a focused yet comprehensive diagnostic regimen is mandatory (Fig. 27.4).

Intubated patients with suspected VAP are unique in that the endotracheal tube provides convenient, immediate access to uncontaminated tracheal specimens. All such patients should have endotracheal aspirates sent for Gram stain and quantitative cultures. Quantitative endotracheal aspirates (QEAs) containing $10^5$ or more organisms per milliliter correlate with diagnostic thresholds by more invasive techniques such as bronchoalveolar lavage (BAL) or protected specimen

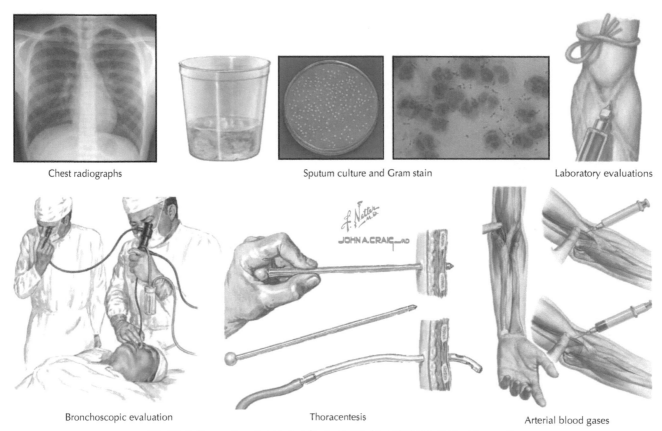

Chest radiographs          Sputum culture and Gram stain          Laboratory evaluations

Bronchoscopic evaluation          Thoracentesis          Arterial blood gases

**Fig. 27.4** Comprehensive testing for suspected hospital-acquired pneumonia.

brush (PSB) sampling. Incorporating QEA with clinical diagnostic criteria improves diagnostic certainty beyond clinical criteria alone and leads to outcomes similar to those seen with more invasive strategies (BAL or PSB). The added expense and inherent risks of BAL or PSB may be justified in patients with severe initial illness, patients not responding to initial therapy, and those suspected to have infections caused by unusual pathogens.

## TREATMENT

A key initial decision in managing patients with HAP is whether their severity of illness merits transfer to the ICU. Indications for ICU monitoring and supportive care include hemodynamic instability, impending respiratory failure, and alteration in mental status that may lead to inability to maintain a patent airway. Other factors that justify more rigorous monitoring include advanced age, limited physiologic reserve because of comorbid chronic illnesses, need for intensive nursing care, evidence of new end-organ injury, and rapid rate of clinical worsening.

Regardless of the site of care, the patient's oxygen saturations should be closely monitored and supplemental oxygen administered as needed to maintain values at or greater than 92%. Adequate pulmonary toilet and regular reassessment of the patient's ability to maintain a patent airway are critical. Clinicians must also ensure adequate nutrition, hydration, and deep venous thrombosis prophylaxis. In complicated cases, early consultation with an infectious disease or pulmonary specialist is prudent.

Because the causative pathogen is typically unknown at the time of HAP/VAP diagnosis, initial antibiotic therapy is empirical. Both delayed initiation of antibiotic therapy and/or inadequate empirical therapy are associated with increased morbidity, mortality, and hospital costs. Accordingly, it is recommended that intravenous (IV)

antibiotics with an appropriate spectrum of activity be initiated as soon as possible after the diagnosis of HAP/VAP has been established. Although sputum and blood cultures should ideally be obtained before antibiotic therapy is initiated, treatment should not be delayed to pursue diagnostic measures.

The current treatment strategy endorsed by the IDSA and ATS is to select empiric antibiotics based on each patient's risk for multidrug resistant organisms, their severity of illness and the local antimicrobial resistance patterns (Table 27.3). Current guidelines recommend that patients without risk factors for resistant pathogens be treated with IV antibiotics to cover *S. aureus*, *Pseudomonas aeruginosa*, and other gram-negative bacilli (Table 27.4). In patients with risk factors for resistant organisms, the guidelines recommend double coverage for gram-negative organisms (an antipseudomonal carbapenem, antipseudomonal cephalosporin, or antipseudomonal β-lactam and β-lactamase inhibitor in combination with an antipseudomonal fluoroquinolone or an aminoglycoside) with consideration of an agent having anti-MRSA activity. Optimal administration of these antibiotics includes adequate doses administered at proper intervals with particular attention to adjustments for renal and/ or hepatic insufficiency. Additional influencing factors include risk of drug interactions, drug allergies, and cost.

Adoption of routine empirical broad-spectrum therapy requires a concurrent commitment to culture-based deescalation of the initial antibiotic prescription. This strategy mandates aggressive sampling of respiratory tract specimens followed by frequent communication with the microbiology laboratory. Once the pathogen has been identified and the in vitro susceptibility demonstrated, the onus falls on the clinician to stop therapies that are ineffective or redundant. Adjuvant therapies such as granulocyte colony-stimulating factors and steroids are of unproven benefit in patients with HAP/VAP and should not be routinely used.

**TABLE 27.3  Infectious Disease Society of America and American Thoracic Society Criteria for Expanded Antimicrobial Coverage**

| | Cover for MRSA If: | Cover *Pseudomonas* With Two Antibiotics From Different Classes If: |
|---|---|---|
| HAP | • Prior intravenous antibiotics within the preceding 90 days<br>• Hospitalization in a unit with a >20% prevalence of MRSA<br>• Hospitalization in a unit with an unknown prevalence MRSA<br>• High mortality risk (mechanical ventilation or septic shock) | • Prior intravenous antibiotics within the preceding 90 days<br>• High mortality risk (mechanical ventilation or septic shock)<br>• Structural lung disease (i.e., bronchiectasis and cystic fibrosis) |
| VAP | • Prior intravenous antibiotics within the preceding 90 days<br>• Hospitalization in a unit with a >10%–20% prevalence of MRSA<br>• Hospitalization in a unit with an unknown prevalence MRSA<br>• Septic shock at the time of VAP diagnosis<br>• ARDS before VAP<br>• Acute renal replacement therapy before VAP<br>• Five or more days of hospitalization prior to VAP onset | • Prior intravenous antibiotics within the preceding 90 days<br>• Septic shock at the time of VAP<br>• ARDS before VAP<br>• Acute renal replacement therapy before VAP<br>• Five or more days of hospitalization prior to VAP onset<br>• Hospitalization in a unit with >10% of gram-negative isolates resistant to monotherapy<br>• Hospitalization in a unit with unknown antimicrobial resistance patterns |

*ARDS,* Acute respiratory distress syndrome; *HAP,* hospital-acquired pneumonia; *MRSA,* methicillin-resistant *S. aureus*; *VAP,* ventilator-associated pneumonia.

**TABLE 27.4  Infectious Disease Society of America and American Thoracic Society Empiric Antibiotic Regimens for Hospital-Acquired and Ventilator-Associated Pneumonia**

| HAP/VAP Coverage[a]: | If MRSA Coverage Is Needed, Add[a]: | If 2 Classes of Anti-Pseudomonal Antibiotics Are Needed, Choose 2 Different Classes From[a,b,c]: |
|---|---|---|
| Piperacillin-tazobactam 4.5 g IV every 6 h<br>OR<br>Cefepime 2 g IV every 8 h<br>OR<br>Levofloxacin 750 mg IV daily<br>OR<br>Imipenem 500 mg IV every 6 h<br>Meropenem 1 g IV every 8 h | Vancomycin 15 mg/kg every 12 h<br>OR<br>Linezolid 600 mg IV every 12 h | Piperacillin-tazobactam 4.5 g IV every 6 h<br>OR<br>Cefepime 2 g IV every 8 h<br>Ceftazidime 2 g IV every 8 h<br>OR<br>Levofloxacin 750 mg IV daily<br>Ciprofloxacin 400 mg IV every 8 h<br>OR<br>Imipenem 500 mg IV every 6 h<br>Meropenem 1 g IV every 8 h<br>OR<br>Amikacin 15-20 mg/kg IV daily[d]<br>Gentamicin 5–7 mg/kg IV daily[d]<br>Tobramycin 5–7 mg/kg IV daily[d]<br>OR<br>Aztreonam 2g IV every 8 h |

[a]Recommended doses assume normal renal and hepatic function.
[b]Select two different classes of antibiotics, avoiding two β-lactams.
[c]If MRSA coverage is not going to be used, cover for MSSA with piperacillin-tazobactam, cefepime, levofloxacin, imipenem, meropenem.
[d]Recommended trough levels: gentamicin <1 mcg/mL; tobramycin <1 mcg/mL; amikacin 4–5 mcg/mL; and vancomycin 15–20 mcg/mL.
*HAP,* Hospital-acquired pneumonia; *IV,* intravenous; *MRSA,* methicillin-resistant *Staphylococcus aureus*; *MSSA,* methicillin-sensitive *Staphylococcus aureus*; *VAP,* ventilator-associated pneumonia.

Assessment of the clinical response to initial therapy requires ongoing monitoring of the patient's temperature, sputum volume and characteristic, radiographic findings, oxygenation parameters, WBC count, and other pertinent laboratory results. The course of HAP/VAP is variable, but clinical response should be apparent within 72 hours of antibiotic initiation. Failure to improve or clinical deterioration may be caused by isolation of the wrong organism, inadequate antibiotic administration, an incorrect diagnosis of pneumonia, or development of a pneumonia-related complication. In such instances a comprehensive work-up should immediately be initiated. Repeat microbiologic sampling is essential, and persistent negative cultures should prompt consideration of atypical infections and noninfectious pulmonary processes. Detailed radiographs—including a combination of lateral decubitus films, ultrasound, and computed tomography scans—are necessary to exclude local complications such as cavitation, abscess formation, and empyema (Fig. 27.5).

The standard duration of therapy for HAP/VAP is 7 days; however, prolonged antibiotic courses may be necessary in patients with delayed antimicrobial coverage, isolation of a resistant pathogen, those with critical illness or immunosuppression at the time of diagnosis, and those with complications such as bacteremia or empyema.

## PROGNOSIS

Despite high absolute mortality rates in HAP patients, the mortality that is directly attributable to HAP is controversial. Crude mortality rates as high as 70% have been reported in patients with nosocomial pneumonia, but many of these critically ill patients die from their underlying

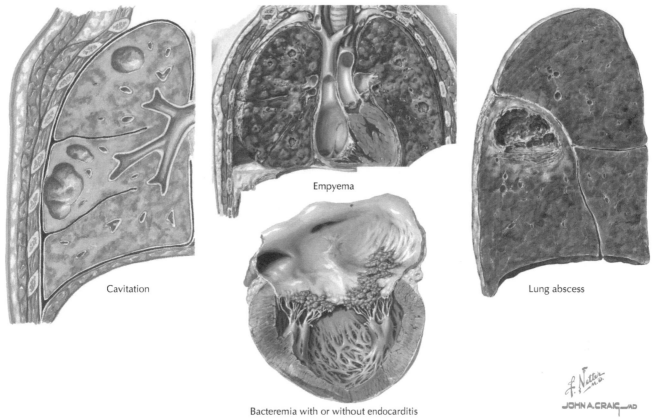

Cavitation

Empyema

Lung abscess

Bacteremia with or without endocarditis

**Fig. 27.5** Complications of hospital-acquired pneumonia.

disease and not because of their pneumonia. Case-control studies estimate that HAP increases mortality by 33% to 50% while VAP increases mortality 2- to 10-fold. Variables associated with increased mortality in nosocomial pneumonia include serious illness at the time of diagnosis (increased acute physiology and chronic health evaluation [APACHE] score, shock, coma, respiratory failure, acute respiratory distress syndrome); bacteremia; severe underlying comorbid disease; infection caused by an organism associated with multidrug resistance (*P. aeruginosa*, *Acinetobacter* species); radiographic infiltrates that are multilobar, cavitating, or rapidly progressive; and ineffective empirical therapy.

## PREVENTION

HAP/VAP-specific prevention strategies focus on reducing risk factors for aerodigestive tract colonization and aspiration, particularly in intubated patients. Simple and inexpensive measures such as minimizing the use of sedatives and maintaining the patient in a semi-upright position (head of bed at greater than 30 degrees from horizontal) reduce aspiration and decrease the risk of HAP/VAP. Indiscriminate antibiotic use must be avoided because it may result in mucosal colonization with nosocomial pathogens, including antibiotic-resistant strains. Cytotoxic drugs and immunosuppressive agents should be used judiciously, given that they impair the host response to infection. Current evidence suggests that a multipronged approach that includes early mobilization, incentive spirometry, early identification and treatment of dysphagia, and the prevention of viral cross-infection together can reduce the rate of HAP development.

Because a nosocomial pneumonia is 20 times more likely to occur in ventilated patients, intubation and mechanical ventilation should be avoided when possible and discontinued as soon as feasible. In intubated patients, meticulous attention should be given to the ventilator circuit, endotracheal tube, and suction apparatus. Orotracheal and orogastric tubes are preferred to nasotracheal and nasogastric tubes because they are associated with reduced rates of sinusitis and clinically diagnosed VAP. Ventilator circuits should be changed whenever visibly soiled; more frequent changes are not associated with reduced VAP rates. Heat-moisture exchangers are used to eliminate condensate accumulation, but the data on attributable VAP reduction are conflicting. Secretion management using closed suctioning systems is recommended to minimize interruptions of the ventilator circuit. Drainage of subglottic secretions and silver-coated endotracheal tubes are novel VAP-prevention strategies that attempt to reduce or sterilize biofilms. These devices have significant acquisition costs, but they appear to be cost effective in patients receiving more than 72 hours of mechanical ventilation. Regardless of the tube employed, adequate endotracheal cuff pressures must be maintained to prevent aspiration and tracheal damage.

Selective decontamination of the digestive tract is a prevention strategy that involves administration of nonabsorbable antibiotics to the oropharynx and stomach with or without systemic antibiotics. A similar strategy involves regular rinsing of the oral cavity with chlorhexidine. Conflicting results regarding the efficacy of aerodigestive decontamination in the prevention of VAP, the lack of a mortality benefit, contrasting results in different subpopulations, and concerns over long-term selection of multidrug-resistant pathogens currently limit the use of these practices.

In 2014, the Society for Healthcare Epidemiology of America (SHEA) in conjunction with the IDSA released updated guidelines on the prevention of VAP. These guidelines propose the establishment of a high-value care bundle that includes the minimization of sedation, daily interruptions in sedation, daily readiness for extubation assessment, early exercise and mobilization, subglottic suctioning if duration of mechanical ventilation is longer than 48 hours, and elevation of the head of bed at least 30 degrees. With implementation of these bundles, hospitals have reduced the rate of VAP occurrence by 25% to 40%. The implementation of these VAP prevention bundles significantly reduced the rate of VAP occurrence and, with higher levels of compliance, could have an even greater impact.

## FUTURE DIRECTIONS

With an aging and more medically complex population, patients are using more healthcare resources than ever before. As a result, more patients are increasingly exposed to antibiotics driving up pathogen resistance rates. To combat this trend, there is a national antibiotic stewardship effort to minimize the utilization of unnecessarily broad-spectrum antibiotics and determine the shortest course of therapy required for pathogen clearance. To aid in this effort expeditious molecular sequencing tests are being developed to rapidly determine the pathogens of significance and their projected antimicrobial resistance patterns. Nevertheless, the best treatment for HAP/VAP remains good prevention and continued surveillance of VAP bundle compliance.

## EVIDENCE

Chastre J, Wolff M, Fagon JY, et al: Comparison of 8 vs 15 days of antibiotic therapy for ventilator-associated pneumonia in adults: a randomized trial. *JAMA*. 2003;290(19):2588-2598. doi:10.1001/jama.290.19.2588. *A critically important clinical trial comparing the clinical outcomes of patients treated with shorter courses of antibiotic therapy with those receiving treatment for a more conventional duration.*

Gross AE, Van Schooneveld TC, Olsen KM, et al. Epidemiology and predictors of multidrug-resistant community-acquired and health care-associated pneumonia. *Antimicrob Agents Chemother* 2014;58(9):5262-5268. doi:10.1128/AAC.02582-14. *This retrospective analysis evaluated the association between the individual components of the HCAP definition and nicely demonstrated that the key patient characteristics associated with a multi-drug resistant organism are antimicrobial use in the past 90 days, admission from a nursing home, and duration of hospitalization within the previous 90 days.*

Kollef MH, Shorr A, Tabak YP, et al. Epidemiology and outcomes of health-care–associated pneumonia: results from a large US database of culture-positive pneumonia, *Chest* 2005;128:3854-3862. *This retrospective study clearly demonstrated that the distribution of pathogens causing hospital-acquired and VAP is distinct from that seen with community-acquired pneumonia. The observed trend of increased multidrug-resistant pathogens causing infections in patients with nosocomial pneumonia was associated with worse clinical outcomes.*

Magill SS, Edwards JR, Bamberg W, et al. Multistate point-prevalence survey of health care-associated infections. *N Engl J Med* 2014;370(13):1198-1208. doi:10.1056/NEJMoa1306801. *A large multi-institutional US-based survey developed in collaboration with the Centers for Disease Control and Prevention designed to assess the most prevalent types of hospital-acquired infections and their most frequently associated causative organisms. This study showed that pneumonias are one of the most common types of hospital-acquired infections and that the most common pathogens are Staphylococcus aureus, Klebsiella species, and Pseudomonas aeruginosa.*

Magill SS, O'Leary E, Edwards JR. Changes in prevalence of health care-associated infections. Reply. *N Engl J Med* 2019;380:1085-1086. *This reply to a high-impact article that documented the changes in antimicrobial resistance patterns over time extends the manuscript by highlighting the growing echinocandin resistance patterns seen in candida species.*

Pileggi C, Mascaro V, Bianco A, Nobile CGA, Pavia M. Ventilator bundle and its effects on mortality among ICU patients. *Crit Car Med* 2018;46(7):1167-1174. doi:10.1097/CCM.0000000000003136. *This meta-analysis definitively demonstrated that deploying a VAP prevention bundle significantly reduces patient mortality.*

Torres A, Chalmers JD, Cruz Dela CS, et al. Challenges in severe community-acquired pneumonia: a point-of-view review. *Intensive Care Med* 2019;45(2):159-171. doi:10.1007/s00134-019-05519-y. *This high impact review concisely summarizes the patient-specific risk factors associated with drug-resistant pneumonia.*

## ADDITIONAL RESOURCES

Chalmers JD, Rother C, Salih W, Ewig S. Healthcare-associated pneumonia does not accurately identify potentially resistant pathogens: a systematic review and meta-analysis. *Clin Infect Dis* 2013;58(3):330-339. doi:10.1093/cid/cit734. *This high-impact meta-analysis reviewed the definition of HCAP and demonstrated that the proposed criteria do not accurately predict pneumonia due to multi-drug resistant organisms.*

Kalil AC, Metersky ML, Klompas M, et al. Management of adults with hospital-acquired and ventilator-associated pneumonia: 2016 clinical practice guidelines by the Infectious Diseases Society of America and the American Thoracic Society. *Clin Infect Dis* 2016;63(5):e61-e111. *Clinically impactful guideline for the diagnosis and treatment of hospital-acquired and VAP.*

Pugh R, Grant C, Cooke RP, Dempsey G. Short-course versus prolonged-course antibiotic therapy for hospital-acquired pneumonia in critically ill adults. *Cochrane Database Syst Rev* 2015;7(8):e41290-64. doi:10.1002/14651858.CD007577.pub3. *This high-quality study demonstrated that limiting treatment for VAP to a fixed duration (7 to 8 days) would not increase the risk for adverse outcomes and may reduce the emergence of resistant pathogens.*

Resar R, Pronovost P, Haraden C, et al. Using a bundle approach to improve ventilator care processes and reduce ventilator-associated pneumonia. *Jt Comm J Qual Saf* 2005;31(5):243-248. doi:10.1016/S1553-7250(05)31031-2. *A set of guidelines released by the Institute for Healthcare Improvement in partnership with the Joint Commission that established a bundled care pathway to prevent VAP, which is now the standard of care in many ICUs.*

Torres A, Niederman MS, Chastre J, et al. International ERS/ESICM/ESCMID/ALAT guidelines for the management of hospital-acquired pneumonia and ventilator-associated pneumonia: guidelines for the management of hospital-acquired pneumonia (HAP)/ventilator-associated pneumonia (VAP) of the European Respiratory Society (ERS), European Society of Intensive Care Medicine (ESICM), European Society of Clinical Microbiology and Infectious Diseases (ESCMID) and Asociacion Latinoamericana del Torax (ALAT). *Eur Respir J* 2017;50(3). *Updated international and multidisciplinary guidelines on the diagnosis and treatment of hospital-acquired and VAP.*

# 28

# Atypical Pneumonia

*Thomas M. File, Jr.*

 **ABSTRACT**

The term *atypical pneumonia* was first used more than 50 years ago to describe cases of pneumonia caused by an unknown agent(s) and that appeared clinically different from pneumococcal pneumonia. Although the original distinction between atypical and typical pneumonia arose from the perception that the clinical presentation of patients was different, recent studies have shown that there is excessive overlap with clinical manifestations from specific causes, which does not permit empirical therapeutic decisions to be made solely on this basis. Thus the scientific and clinical merit of the designation *atypical pneumonia* is controversial, and many authorities have suggested that the term *atypical* be discontinued. However, the term remains popular among clinicians and investigators and prevalent in recent literature regardless of its clinical value. Moreover, options for appropriate antimicrobial therapy for the most common causes are similar, which is considered justification by some for lumping these together.

**✳ CLINICAL VIGNETTE**

A 68-year-old male with history of smoking and congestive heart failure presents to the emergency department with a 2-day history of fever, cough with sputum, and dyspnea. He has stage 3 chronic renal disease. He recently moved into an older home which required extensive plumbing maintenance. Exam: blood pressure (BP)-130/90; temperature 38.6°C; pulse 110; respiratory rate 24; auscultation of lungs revealed evidence of consolidation of right lung field; pulse oximetry-93%; laboratory-white blood cell (WBC)-18,000; sodium level slightly low; slight elevation of transaminases.

Atypical pneumonia was initially characterized by constitutional symptoms, often with upper and lower respiratory tract symptoms and signs, a protracted course with gradual resolution, a lack of typical findings of consolidation on chest radiograph, failure to isolate a pathogen on routine bacteriologic methods, and a lack of response to penicillin therapy. In the 1940s an agent that was believed to be the principal cause was identified as *Mycoplasma pneumoniae*. Subsequently other pathogens have been linked with atypical pneumonia because of similar clinical presentation, including a variety of respiratory viruses, *Chlamydia psittaci*, *Coxiella burnetii*, and *Chlamydophila* (also known as *Chlamydia*) *pneumoniae*. Less common causative agents associated with atypical pneumonia include *Francisella tularensis* and *Yersinia pestis* (plague), although these agents are often associated with a more acute clinical syndrome. Finally, pneumonia caused by *Legionella* species, albeit often more characteristic of pyogenic pneumonia, is also included because it is not isolated using routine microbiologic methods.

## BURDEN OF DISEASE

*M. pneumoniae*, *C. pneumoniae*, and *Legionella pneumophila* are the most common causes of atypical pneumonia (Figs. 28.1 and 28.2). The results of recent studies indicate that they cause from 15% to as much as 50% (in selected outpatient populations) of cases of community-acquired pneumonia (CAP). However, until recently these pathogens (with the exception of *L. pneumophila*) have not often been identified in clinical practice because of lack of specific, rapid, or standardized tests for their detection. The recent availability of US Food and Drug Administration (FDA)-approved molecular testing methods will likely provide increased identification of these pathogens. The other causes of atypical pneumonia occur with much less frequency.

## PATHOGENESIS

*M. pneumoniae* infections are ubiquitous and can affect people in all age groups. *M. pneumoniae* are extracellular pathogens that adhere to the respiratory epithelium by means of specialized protein attachments. They are unique among bacteria because they do not have a cell wall; this property renders the organism resistant to β-lactam antimicrobial agents. *M. pneumoniae* are transmitted from person to person by respiratory droplets with a usual incubation period of several weeks. It is estimated that only 3% to 10% of infected persons develop pneumonia. Many of the pathogenic features of infection are believed to be immune mediated rather than induced directly by bacteria (antibodies produced against the glycolipid antigens of *M. pneumoniae* may cross-react with human red cells and brain cells).

*C. pneumoniae* are very small bacteria (once considered viruses) that are obligate intracellular parasites and are unique among bacteria for their developmental cycle, forming infectious forms (elementary bodies) and noninfectious forms (reticulate bodies). Infections are often acquired early in life, and the bacteria may remain in a latent form afterward. Reinfections or recrudescent processes, both referred to as *recurrent infection*, may occur throughout one's lifetime. Most adults who are hospitalized with *C. pneumoniae* pneumonia have recurrent infection.

*Legionella* species are small bacilli that have special growth requirements in the laboratory. They do not stain with common reagents but can be seen in tissues stained with Dieterle silver stain (see Fig. 28.2). They are intracellular organisms that are engulfed by alveolar macrophages via phagocytosis. More than 49 different *Legionella* species have been identified. The most common to infect humans is *L. pneumophila*, which contains 16 different serogroups (serogroup 1 causes most cases of infection in North America).

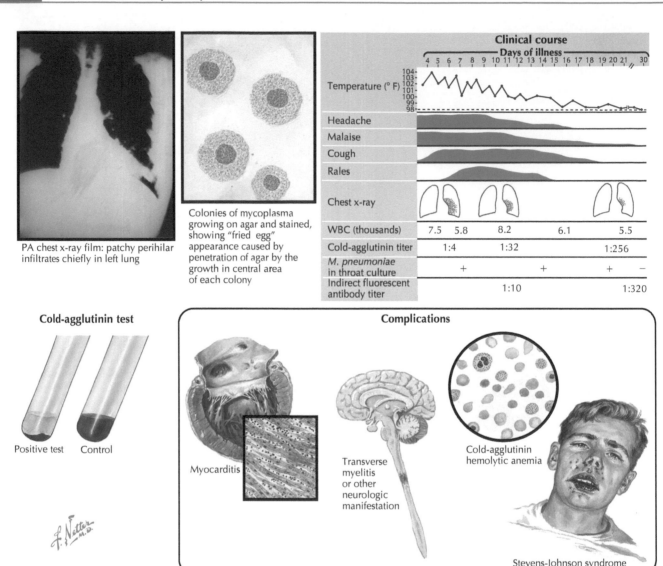

PA chest x-ray film: patchy perihilar infiltrates chiefly in left lung

Colonies of mycoplasma growing on agar and stained, showing "fried egg" appearance caused by penetration of agar by the growth in central area of each colony

| Clinical course — Days of illness | | | | |
|---|---|---|---|---|
| | 4 5 6 7 8 9 10 11 12 13 14 15 16 17 18 19 20 21 30 | | | |
| Temperature (° F) | | | | |
| Headache | | | | |
| Malaise | | | | |
| Cough | | | | |
| Rales | | | | |
| Chest x-ray | | | | |
| WBC (thousands) | 7.5   5.8 | 8.2 | 6.1 | 5.5 |
| Cold-agglutinin titer | 1:4 | 1:32 | | 1:256 |
| *M. pneumoniae* in throat culture | + | + | + | − |
| Indirect fluorescent antibody titer | | 1:10 | | 1:320 |

Cold-agglutinin test

Positive test      Control

Complications

Myocarditis

Transverse myelitis or other neurologic manifestation

Cold-agglutinin hemolytic anemia

Stevens-Johnson syndrome

**Fig. 28.1** Mycoplasma (Eaton agent) pneumonia (primary atypical pneumonia).

*Legionella* organisms produce virulence factors that enhance intracellular survival and growth within the macrophages; chemokines and cytokines released by the infected macrophages trigger an inflammatory response that is often severe and can lead to a rigorous influx of neutrophils within the alveoli (see Fig. 28.2). *Legionella* is not spread person to person but usually by exposure to water. Outbreaks may be associated with infected water sources. The incubation period is 10 days.

## CLINICAL FEATURES

Although the diagnosis of these specific pathogens is difficult to establish on clinical manifestations alone, there are several generalizations that may be helpful to the clinician in considering these infections.

### *Mycoplasma pneumoniae*

Although commonly perceived as a cause of CAP, predominantly in young healthy patients, the incidence of *M. pneumoniae* pneumonia increases with age, highlighting the importance of this pathogen in the elderly as well. The onset is usually insidious over several days to a week. Constitutional symptoms including headache (usually worse with cough), malaise, myalgias, and sore throat are frequently present. Cough is typically initially dry, may be paroxysmal, is frequently worse at night, and may become productive of mucopurulent sputum. Sinus

and ear pain are occasionally reported. *M. pneumoniae* pneumonia is often associated with extrapulmonary manifestations including rash, neurologic involvement (i.e., aseptic meningitis, meningoencephalitis, cerebral ataxia, Guillain-Barré syndrome, and transverse myelitis), hemolytic anemia (associated with cold agglutinins), myopericarditis, polyarthritis, and pancreatitis.

The physical findings are often minimal, seemingly disproportionate to the patient's complaints. Auscultation of the lungs usually reveals variable scattered rales or wheezes. Bullous myringitis, first described in volunteer subjects infected with *M. pneumoniae,* has been infrequent in naturally occurring infection and is not a diagnostic sign. Chest radiograph findings are variable. Most common are patchy opacities, which can be unilateral or bilateral, and parabronchial pneumonia. Dermatologic manifestations may range from a mild maculopapular or vesicular rash to Stevens-Johnson syndrome (see Fig. 28.1).

The course of *M. pneumoniae* pneumonia is usually mild and self-limiting. However, significant pulmonary complications may occur and include pleural effusion, pneumatocele, lung abscess, pneumothorax, bronchiectasis, chronic interstitial fibrosis, respiratory distress syndrome, and bronchiolitis obliterans. Some strains of *M. pneumoniae* exhibit expression of community-acquired respiratory distress syndrome (CARDS) toxin, which has been shown to be a virulence factor in severe infection.

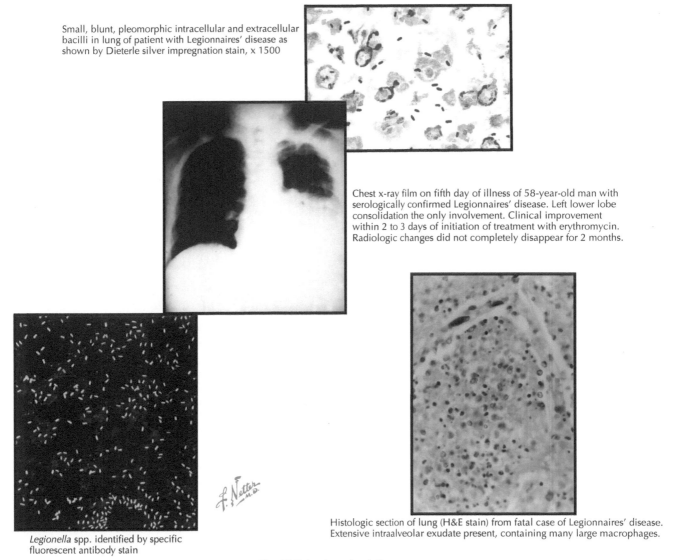

Small, blunt, pleomorphic intracellular and extracellular bacilli in lung of patient with Legionnaires' disease as shown by Dieterle silver impregnation stain, x 1500

Chest x-ray film on fifth day of illness of 58-year-old man with serologically confirmed Legionnaires' disease. Left lower lobe consolidation the only involvement. Clinical improvement within 2 to 3 days of initiation of treatment with erythromycin. Radiologic changes did not completely disappear for 2 months.

*Legionella* spp. identified by specific fluorescent antibody stain

Histologic section of lung (H&E stain) from fatal case of Legionnaires' disease. Extensive intraalveolar exudate present, containing many large macrophages.

Fig. 28.2 Legionnaires' disease.

## *Chlamydophila pneumoniae*

Pneumonia caused by *C. pneumoniae* may be sporadic or epidemic. The clinical manifestations of *C. pneumoniae* pneumonia remain somewhat unclear because of the lack of a gold standard of diagnosis and the contributing effect of co-pathogens. The onset is usually insidious. Infections often manifest initially with sore throat, hoarseness, and headache as important nonclassical pneumonic findings. Cough is prominent but unproductive and may last, if not treated early and effectively, for weeks or even months. Clinical characteristics, however, are generally not predictive of *C. pneumoniae* as a cause. Chest radiographs of patients with *C. pneumoniae* pneumonia tend to show less extensive opacifications in relation to clinical findings than other processes.

The clinical characteristics associated with primary infection may be difficult to distinguish from those of reinfection because of the confounding effect of comorbid conditions on age. However, patients with primary infection are usually younger and tend to have higher fever. For older patients with reinfection, the presence of comorbid illness and the requirement for supplemental oxygen therapy are often the reason for hospital admission.

## *Legionella pneumophila*

Legionellosis is primarily associated with two clinically distinct syndromes: Legionnaires' disease, a potentially fatal form of pneumonia, and Pontiac fever, a self-limited, nonpneumonic illness. Legionellae are environmental organisms often found in water and infection is often associated with exposure to water sources (plumbing of water pipes, showers, pools, fountains, cooling towers, etc.). Many of the clinical features of Legionnaires' disease are more typical of pyogenic (bacterial) pneumonias than the previously described atypical pneumonia. However, as Legionnaires' disease has become increasingly recognized, less severely ill patients are seen earlier in the course of disease, and thus clinical manifestations of unusual severity once considered to be distinctive Legionnaires' disease are now known to be less specific.

The onset is often acute, with high fever, myalgias, anorexia, and headache. Temperature often exceeds 40°C. Gastrointestinal symptoms are prominent, especially diarrhea, which occurs in 20% to 40% of cases. Relative bradycardia, which had been purported to be a common finding in earlier studies, has been overemphasized as a diagnostic finding. Hyponatremia, elevated lactate dehydrogenase (LDH) levels, and elevated transaminase enzymes are common abnormal laboratory findings.

## Other Causes of Atypical Pneumonia *Coxiella burnetii* (Q fever), Psittacosis, Tularemia, Plague

Several of the less common causes of the atypical pneumonia syndrome are infections transmitted from animals to humans. In such cases epidemiologic clues may be very important; and although

**TABLE 28.1    Common Characteristics and Therapy for the Other Atypical Pneumonias**

| Pathogen | Epidemiologic or Underlying Condition | Clinical Features | Recommended Therapy |
|---|---|---|---|
| Chlamydophila (Chlamydia) psittaci | Exposure to birds | Headache, myalgia prominent, liver involvement, Horder spots (macular rash) | Tetracycline, doxycycline, macrolide |
| Coxiella burnetii (Q fever) | Exposure to farm animals (especially parturient) | Headache prominent, liver involvement | Tetracycline, doxycycline; macrolides |
| Francisella tularensis (tularemia) | Exposure to rabbits | Headache, chest pain prominent; hilar adenopathy | Streptomycin or gentamicin considered as drug of choice; doxycycline effective for most cases (especially if not severe) |
| Yersinia pestis (pneumonic plague) | Exposure to infected animals (rodents, cats, squirrels, chipmunks, prairie dogs) | For inhalation, acute onset with rapidly severe pneumonia; blood-tinged sputum | Streptomycin, gentamicin; tetracycline, doxycycline |

**TABLE 28.2    Diagnostic Studies for Pathogens Associated With Atypical Pneumonia**

| Pathogen | Rapid Test | Standard Culture or Microbiologic Test(s) | Serology, Other Tests[d] |
|---|---|---|---|
| Mycoplasma pneumoniae | PCR [95][b]; as part of multiplex PCR respiratory infection panel | Culture is possible but requires specialized media and takes approximately 2–3 weeks to grow. Thus most clinical laboratories do not attempt to culture this organism | ELISA, CF[c] [75–80] (IgM may be present after 1 week but can persist 2–12 months) Diagnostic criteria:<br>• Definite: fourfold titer rise<br>• Possible: IgG ≥ 1:64 (CF); IgM ≥ 1:16 (ELISA)<br>• Cold agglutinin [50] (less than 50% specificity; takes several weeks to develop) |
| Chlamydophila pneumoniae | PCR [80–90]; as part of multiplex PCR respiratory infection panel | Cell culture; however, this method is rarely used as most clinical laboratories are not equipped to culture and can take a week [80–90] | MIF[c] (IgM may take up to 4–6 weeks to appear in primary infection) Diagnostic criteria:<br>• Definite: fourfold titer rise<br>• Possible: IgG ≥ 1:512; IgM ≥ 1:32 |
| Legionella pneumophila | Urine antigen[e] [60–70] PCR, as part of multiplex PCR respiratory infection panel; DFA[f] [25–75] | Sputum, bronchoscopy [75–99] (selective media required, 2–6 days) | IFA[c] [40–75] Diagnostic criteria:<br>• Definite: fourfold titer rise<br>• Possible: IgG or IgM ≥ 1:512 (titer of 1:256 has positive predictive value of only 15%) |
| Chlamydia psittaci | PCR[a] | Usually not done (considered laboratory hazard) | CF (presumptive IgG > 1:32) MIF for IgM |
| Coxiella burnetii | PCR[a] | Usually not done (considered laboratory hazard) | ELISA, IFA, CF |
| Francisella tularensis | | Culture (selective media) | ELISA preferred Passive hemagglutination |
| Yersinia pestis | Gram stain, morphology, gram-negative coccobacillus exhibiting bipolar staining ("safety pin"); PCR | Culture | Serology available |

[a]Available in selected laboratories; reagents are not cleared by the US Food and Drug Administration.
[b][ ] = relative % sensitivity of test.
[c]Rarely done; requires specialized culture techniques.
[d]Paired sera generally required.
[e]Only for *L. pneumophila* serogroup 1 (≈60% to 70% of cases); can be positive for months.
[f]DFA; primarily for *L. pneumophila* serogroup 1; some false-positive results with other species; technically demanding.
*CF*, Complement fixation; *DFA*, direct fluorescence antibody; *ELISA*, enzyme-linked immunosorbent assay; *IFA*, indirect fluorescence antibody; *IgG*, immunoglobulin G; *IgM*, immunoglobulin M; *MIF*, microimmunofluorescence; *NP*, nasopharyngeal; *PCR*, polymerase chain reaction.

specific manifestations cannot be considered diagnostic of a specific cause, there are general findings that are characteristic of these diseases (Table 28.1). Respiratory viruses may also cause an atypical pneumonia syndrome (see Chapter 30: Viral Respiratory Infections).

## DIAGNOSTIC APPROACH

Laboratory tests used for the diagnosis of the causative agents associated with atypical pneumonia are listed in Table 28.2. Serologic tests are the most common tests for diagnosis of *M. pneumoniae* and

**TABLE 28.3  Recommended Antimicrobial Therapy for *Mycoplasma pneumoniae* and *Chlamydophila pneumoniae* (Adult Doses[a])**

| Antimicrobial | Dose | Duration (Days) |
|---|---|---|
| Erythromycin[c] | 500 mg qid | 10–14 |
| Clarithromycin (Biaxin) | 500 mg bid | 7–10 |
| Azithromycin (Zithromax)[c] | 500 mg initially | |
| | then 250 mg qd | 5 |
| | (alternative 500 mg qd) | (3) |
| Dirithromycin (Dynabac) | 500 mg qd | 10 |
| Tetracycline | 500 mg qid | 10–14 |
| Doxycycline[c] | 100 mg bid | 7–10 |
| Omadacycline | 300 mg daily (fast for at least 4 h) | 7 |
| Lefamulin[b] | 600 mg bid | 5 days |
| Levofloxacin (Levaquin)[c] | 500 mg qd | 7–14 |
| | 750 mg qd | 5 (data are limited) |
| Moxifloxacin (Avelox)[c] | 400 mg qd | 7–14 |
| Delafloxacin[b] | | 5–7 days |

[a]Oral except where noted.
[b]Approval pending at time of submission.
[c]Also, can be administered intravenously in equivalent dose.
*bid*, Twice per day; *qd*, once per day; *qid*, four times per day.

**TABLE 28.4  Parenteral Therapy for Serious *Legionella* Infections[a]**

| Preferred Antimicrobial | Alternative Antimicrobial |
|---|---|
| Fluoroquinolone | Erythromycin 1 g IV q6h ± rifampin[b,c] |
| Levofloxacin (Levaquin) 750 mg IV q24h (750 mg qd for 5 days possible for immunocompetent patients) | |
| Moxifloxacin (Avelox) 400 mg IV q24h | Doxycycline (Vibramycin) 100 mg IV q12h ± rifampin |
| Azithromycin (Zithromax) 500 mg IV q24h | |

[a]Requiring hospitalization or in immunocompromised patients; can change to oral when clinically stable and can take orally.
[b]300–600 mg IV every 12 hours.
[c]Not approved by US Food and Drug Administration for this indication.
*IV*, Intravenously; *qd*, once daily.

*C. pneumoniae* but are of limited value in the clinical evaluation of a patient given the requirement for measurement during acute and convalescent specimens. Recently nucleic acid amplification tests (most commonly polymerase chain reaction, PCR) have been developed, and several multiplex PCR panels are now FDA approved and available for diagnosis of respiratory infections. A rapid urinary antigen test is available for *L. pneumophila* serogroup 1.

## ANTIMICROBIAL THERAPY

Antimicrobial agents generally considered effective for the atypical pathogens are included in Tables 28.1, 28.3, and 28.4. Because most cases of atypical pneumonia are treated empirically, clinicians must also consider the possibility of other standard pathogens (e.g., *Streptococcus pneumoniae*, *Haemophilus influenzae*) when deciding on antimicrobial therapy. Antimicrobial therapy for *M. pneumoniae*, *C. pneumonia*, or Legionella generally include a macrolide, tetracycline, or a fluoroquinolone. However, macrolide-resistant *M. pneumoniae* is now very common in parts of the world, especially Asia, where doxycycline or a fluoroquinolone is preferred. Newer approved agents omadacycline and lefamulin are also effective for *M. pneumoniae*, *C. pneumoniae*, and *Legionella* spp.

Therapy for *Mycoplasma* and *Chlamydia* has been the subject of some conjecture. A common view is that it really does not matter whether antibiotics are given for most of these infections because the mortality is low, these infections are often self-limiting, there may be ambiguity of diagnosis (especially for *C. pneumoniae*), co-pathogens may have confounding effects, and antimicrobial efficacy is questioned. However, there are data indicating that treatment (especially for *M. pneumoniae*) reduces the morbidity of pneumonia and shortens the duration of symptoms.

There is little debate concerning the need for therapy for *Legionella* pneumonia. Delay in instituting appropriate antimicrobial therapy for *Legionella* pneumonia significantly increases mortality. Therefore empirical anti-*Legionella* therapy should be included in the treatment of severe CAP. Erythromycin had initially been accepted as the treatment of choice for Legionnaires' disease; however, intracellular models as well as animal models of *Legionella* infection indicate that the systemic fluoroquinolones and the newer macrolides (especially azithromycin) show superior activity compared with erythromycin. These newer agents have better pharmacokinetic properties: better bioavailability, longer half-life (resulting in fewer doses per day), better intracellular penetration into macrophages, and better tolerability. On the basis of greater activity in intracellular models and several observational studies, the quinolones may produce a superior clinical response compared with macrolides. The addition of rifampin to erythromycin has been suggested for patients who are severely ill; however, there is no convincing laboratory data to show that adding rifampin to fluoroquinolones or the more active macrolide therapy improves bacterial killing. Doxycycline has also been shown to be effective in limited, well-documented cases. Recommendations for initial parenteral therapy are listed in Table 28.4. Oral therapy for less serious cases or for step-down from intravenous therapy includes the oral macrolides and fluoroquinolones as well as doxycycline.

The duration of therapy for optimal response of *C. pneumoniae* and *M. pneumoniae* has not been well established. In initial descriptions of *C. pneumoniae* pneumonia, observers found that respiratory symptoms frequently recurred or persisted after short courses (5 to 10 days) of erythromycin or tetracycline. In recent recommendations the usual duration of therapy for *C. pneumoniae* or *M. pneumoniae* using more recently approved agents has been 7 to 10 days (shorter for azithromycin because of the longer half-life); however, recent studies (mostly with the fluoroquinolones) have suggested that a minimum of 5 days may be adequate for immunocompetent patients if the patient has had a good clinical response within 48 to 72 hours. Similarly, the usual duration of therapy for Legionnaires' disease in immunocompetent adults has been 7 to 10 days; one recent study showed good efficacy of 750 mg per day of levofloxacin for 5 days. For therapy of immunocompromised patients or more severe disease, longer duration is recommended.

Therapy for the other atypical pneumonias is included in Table 28.1. The tetracyclines are generally considered the drugs of choice for *C. psittaci*, with the macrolides as appropriate alternatives (similar duration as for *C. pneumoniae*). The newer fluoroquinolones are active in vitro and in animal models, but their efficacy for human infection is unknown. For *C. burnetii* the tetracyclines and macrolides are both considered effective (usually for 10 days). In one small prospective study, doxycycline was slightly more effective than erythromycin, but most cases were benign and self-limiting. Combination therapy (e.g., doxycycline plus ciprofloxacin or rifampin) has been used for Q fever endocarditis. No prospective controlled clinical trials have defined optimal antimicrobial therapy for *F. tularensis*. The traditional choice of therapy for pneumonic tularemia is streptomycin (1 g every 12 hours if patient is severely ill or 500 mg every 12 hours in milder disease) or gentamicin (3 to 5 mg/kg/day) for 7 to 14 days. Doxycycline (100 mg intravenously or by mouth twice daily) has often been used with good success, particularly in less severe pneumonia, and is easier to administer.

## PREVENTION

Because the source of *Legionella* is often the water supply, prevention of nosocomial legionellosis is possible by disinfection of the water source. Methods include use of copper-silver ionization units, superheating of the water to a temperature of 70°C (158°F) and flushing distal outlets, and treatment of the water supply with chloride dioxide.

---

### ✳ CLINICAL VIGNETTE

The patient was admitted to a general ward and empirically administered ceftriaxone and azithromycin. Sputum and blood cultures were no growth. The urinary antigen was positive for Legionella pneumophila serogroup 1. As he was clinically improved at 48 h, ceftriaxone was discontinued and azithromycin continued. He was discharged on hospital day 4 on oral azithromycin to complete a total course of 7 days. Recent plumbing of older water pipes is a risk factor for Legionella.

## EVIDENCE

Centers for Disease Control and Prevention. Avian Influenza. www.cdc.gov/flu/avianflu/. *Excellent source for up-to-date information regarding this new viral infection.*

Centers for Disease Control and Prevention. Middle East Respiratory Syndrome Coronavirus (MERS-CoV). http://www.cdc.gov/coronavirus/mers/index.html. *Excellent source for up-to-date information regarding this new viral infection.*

File TM Jr., Marrie TJ. Does empiric therapy for atypical pathogens improve outcomes for patients with CAP? *Infect Dis Clin N Am* 2013;27: 99-114. *A critical review of the significance of empirical therapy of atypical pneumonia, concluding that available evidence supports treatment for atypical pathogens empirically.*

Gramegna A, Sotgiu G, Di Pasquale M, et al. Atypical pathogens in hospitalized patients with community-acquired pneumonia: a worldwide perspective. *BMC Infect Dis* 2018;18:677. *Among 3702 CAP patients 1250 (33.8%) underwent at least one test for atypical pathogens. Patients with CAP due to atypical pathogens were significantly younger and showed less cardiovascular, renal, and metabolic comorbidities in comparison to adult patients hospitalized due to non-atypical pathogen CAP.*

Jain S, Self WH, Wunderink RG, et al., CDC EPIC Study Team. Community-acquired pneumonia requiring hospitalization among U.S. adults. *N Engl J Med* 2015;358:415-427. *A multicenter study of the etiology of CAP requiring hospitalization. Of note viruses accounted for approximately 25% as sole identified pathogen.*

Shefet D, Robenshtok E, Paul M, Leibovici. Empirical atypical coverage for inpatients with community-acquired pneumonia: a systematic review of randomized controlled trials. *Arch Intern Med* 2005;165:1992-2000. *One of a few "meta-analyses" that suggest there was no significant difference in mortality or clinical response using a standard endpoint (e.g., 7 to 10 days following end of therapy) for assessment. However, regimens with coverage of atypical pathogens showed a trend toward clinical success. Subgroup analysis in patients with Legionella species found a significantly lower failure rate in those who were treated with antibiotics active against atypical pathogens.*

## ADDITIONAL RESOURCES

Arnold FW, Summersgill JT, Lajoie AS, et al. A worldwide perspective of atypical pathogens in community-acquired pneumonia. *Am J Respir Crit Care Med* 2007;175:1086-1093. *Observational study from a large international cohort showing patients with atypical pneumonia treated empirically with antimicrobials with coverage for atypical pathogens had faster time to clinical stability.*

Eljaaly K, Alshehri S, Aljabri A et al. Clinical failure with and without atypical bacterial coverage in hospitalized adults with community-acquired pneumonia: systemic review and meta-analysis. *BMC Infect Dis* 2017;17:385. *Empiric atypical coverage was associated with a significant reduction in clinical failure in hospitalized adults.*

File TM Jr. Atypical pneumonia. In Schlossberg D and Cunha C, eds: *Current therapy of infectious diseases*, 3rd edition. St Louis, pending publication, Elsevier. *A concise review that discusses in greater detail the clinical aspects of the common causes of atypical pneumonia.*

Marrie TJ, Costain N, La Scola B, et al. The role of atypical pathogens in community-acquired pneumonia. *Semin Respir Crit Care Med* 2012;33:244-256. *Review of atypical pneumonia.*

Metlay JP, Waterer G, Long AC et al. Diagnosis and Treatment of Adults with Community-Acquired Pneumonia: An Official Clinical Practice Guideline of the American Thoracic Society and Infectious Diseases Society of America. *Am J Respir Crit Care Med* 2019;200:e45-e67. *A comprehensive, updated, evidence-based set of recommendations regarding the diagnosis and management of adults with CAP.*

# 29

# Aspiration Pneumonia

*Lionel A. Mandell, Michael S. Niederman*

##  ABSTRACT

While microaspiration of oropharyngeal secretions is usually the cause of most cases of community- and hospital-acquired pneumonias, macroaspiration of oropharyngeal or upper gastrointestinal contents is the cause of aspiration pneumonia (AP). Chemical pneumonitis (CP) is triggered by aspiration of acidic gastric material.

A variety of different microbial communities exist in the lung. It is postulated that a macroaspiration event may trigger an inflammatory response resulting in epithelial or endothelial injury and subsequent activation of feedback loops which may lead to emergence of a dominant pathogen such as *Streptococcus pneumoniae* or *Pseudomonas aeruginosa*.

Conditions which allow either enhanced access or reduced clearance of oral or gastrointestinal (GI) contents to or from the lower airways are risk factors. These include conditions such as upper airway and esophageal neoplasms, esophageal strictures, motility disorders, seizures, impaired consciousness, and certain neurodegenerative diseases.

Clinical manifestations range from mild to severe and may appear within hours to days and can include coughing, wheezing, shortness of breath, and fever. With foreign body aspiration obstruction and subsequent postobstructive pneumonia may occur.

Chest radiographic findings may initially be normal but may reveal an infiltrate within 24 to 48 hours.

Diagnosis depends upon a history of a witnessed aspiration event, risk factors, clinical assessment, and chest radiographic findings. At times quantitative lung-lavage cultures may help to distinguish infection from noninfection situations.

Our treatment approach to AP is based on site of acquisition of the infection; that is community, hospital, or long-term care facility (LTCF); and initial chest radiograph appearance. Consideration should also be given to the risk of infection with resistant organisms and to potential adverse events including *Clostridium difficile* infection.

Treatment regimens are outlined in the chapter. The duration of treatment for those with a good clinical response is 5 to 7 days but longer in complicated cases.

For CP, antibiotics are usually unnecessary unless bacteria are present as in patients with elevated gastric pH secondary to acid suppressing medication or in those with small bowel obstruction. Supportive therapy is important in both AP and CP but glucocorticoids do not play a role in the routine management of these patients.

A number of interventions may be used to prevent AP and CP and are outlined in the Prevention section of the chapter.

## ✳ CLINICAL VIGNETTE

A 68-year-old gentleman with mild type 2 diabetes and GERD underwent a left total hip replacement for severe osteoarthritis. He experienced a significant amount of postoperative pain and was given narcotics. The drugs caused some dysphoria and nausea and on the second day he vomited and aspirated. This was a witnessed event and within several hours he developed shortness of breath and cough.

On examination his temperature was 38.4°C, respiratory rate 22/min, heart rate 105/min and regular, and blood pressure 135/85. A chest radiograph showed an infiltrate in the superior segment of the right lower lobe.

COMMENT: This patient experienced a macroaspiration event likely caused by the narcotics which induced both nausea and a reduced level of consciousness. The aspiration of gastric contents usually results in a CP secondary to inflammation triggered by the acidic gastric contents. Typically, in such cases infection is not an issue initially. In this case however, because of his GERD and use of acid-suppressing medications for several months the pH of the gastric contents was higher than usual and this could result in bacterial overgrowth. This in turn could lead to an infection (AP) rather than a purely inflammatory process (CP). It was felt that the patient was ill enough to warrant initial empiric antimicrobial therapy. The early appearance of an infiltrate on the chest radiograph, however, suggests that infection may not yet be established and serial chest radiographs should be done. If the infiltrate clears rapidly and the patient is doing well clinically consideration may be given to stopping the antibiotics. Unfortunately, pro-calcitonin does not help in differentiating AP from CP.

In selecting a treatment regimen, one should consider the site of acquisition, chest radiograph findings, and other variables such as the state of dentition, the risk of resistance, and the severity of clinical presentation.

Antibiotic treatment is usually not required for CP but in this particular case infection is certainly possible and antibiotics should be initiated. Steroids do not play a role routinely in the initial management of either AP or CP.

## BURDEN OF DISEASE

Microaspiration of oropharyngeal secretions is the main pathogenic mechanism of most cases of community- and hospital-acquired pneumonias. Macroaspiration (large volume) of oropharyngeal or upper gastrointestinal contents is the cause of AP.

AP is an important part of the pneumonia continuum that includes both community- and hospital-acquired infections. Unfortunately, unlike community-acquired pneumonia (CAP) and hospital-acquired pneumonia (HAP), diagnostic criteria for AP are unavailable. It is thought that up to 15% of CAP cases are AP and they are associated with a higher mortality rate than other forms of CAP (29.4% vs. 11.6%).

Patients with AP usually have risk factors for macroaspiration and the resultant syndromes may involve the airways or pulmonary parenchyma. Macroaspiration can lead to either AP, an infection, or to CP which represents an inflammatory reaction secondary to aspiration of gastric contents.

## MICROBIOLOGY AND PATHOGENESIS

Until recently it was thought that in healthy individuals the lower respiratory tract was sterile. Genomic methods, however, have demonstrated numerous bacterial phylotypes. A complex taxonomic bacterial landscape that includes diverse microbial communities exists in the lung and studies are helping to define the role of the lung microbiome in health and disease and in the pathogenesis of pneumonia. In healthy individuals it is believed that the lung microbiome helps calibrate the immune tone of the airways and alveoli.

To explain the role of the microbiome in the pathogenesis of pneumonia various models have been proposed. An interplay between movement of bacteria into and out of the lungs and local feedback loops may be central to the development of pneumonia. Bacteria may gain access to the lungs by microaspiration and are eliminated by coughing and ciliary clearance. Positive and negative feedback loops can promote or suppress inflammation. An inflammatory event resulting in epithelial or endothelial injury may create a positive loop which can further increase inflammation and local susceptibility to infection.

The complex adaptive system model may help explain the interplay between the lung microbiome and AP. A change in the lung microbiome (dysbiosis) may occur as the result of illness and may then impair local defenses. A significant macroaspiration event could then possibly overwhelm the host's ability to clear the bacteria further disrupting bacterial homeostasis thereby triggering an increase in a positive feedback loop.

Substances such as cytokines, neurotransmitters, and hormones (e.g., glucocorticoids) can promote growth of bacteria such as *S. pneumoniae*, and certain gram-negative rods and may result in a shift from a diverse microbiome to one with a dominant species (e.g., *pneumococcus* or *P. aeruginosa*).

With illness or aging, cell surface fibronectin may be lost from the surface of airway epithelial cells exposing receptors on these cells to gram-negative rods which may adhere to them, thereby shifting the oral microbial flora to more virulent bacteria. Colonization of the oropharynx by these new potential pathogens may be followed by their micro- or macroaspiration to the lung.

Historically anaerobes with or without aerobes were typically associated with AP. CP on the other hand is an inflammatory response to acidic gastric contents or bile acids and is not the result of infection. A shift has been noted away from anaerobes to bacteria typically seen with CAP and HAP. In one study, pathogens in community-acquired cases were *S. pneumoniae*, *Staphylococcus aureus*, *Haemophilus influenzae*, and *Enterobacteriaceae* while in hospital-acquired cases gram-negative rods including *P. aeruginosa* predominated. Anaerobes are more likely to occur in patients with lung abscess but are far less common in those with AP but no abscess or empyema.

Less importance of anaerobes has also been noted in AP among the elderly. It is postulated that this may be the result of a shift in patient demographic characteristics and to earlier sampling for cultures in the course of infection rather than after development of empyema or abscess.

## RISK FACTORS

Risk factors for AP are those resulting in either increased access of oral or gastric contents to the lower airways or reduced clearance or both (Fig. 29.1). Swallowing dysfunction and impaired cough reflex can occur in a number of settings including neoplasms involving the head, neck, or esophagus, esophageal strictures and motility disorders, chronic obstructive pulmonary disease (COPD), and seizures. Other swallowing difficulties may arise with central nervous system dysfunction, for example multiple sclerosis, parkinsonism, dementia, and impaired consciousness. Reduced consciousness may be secondary to a variety of drugs including narcotics, anesthetic agents, antidepressants, and alcohol. Stroke and cardiac arrest may also be significant. Many of

the above may impair the cough reflex which is the primary clearance mechanism.

In some patients, multiple risk factors may exist significantly increasing the risk of AP. Interestingly, if one compares patients with traditional CAP to those with AP there are no significant increases in risk of aspiration in the former group.

## CLINICAL MANIFESTATIONS

The initial macroaspiration event is often unwitnessed so the magnitude of exposure is often unknown. Even in situations such as aspiration during anesthesia neither clinical nor chest radiographic features develop in 64% of patients.

The clinical presentation may range from mild to severe and occasionally respiratory failure may occur. The findings are dependent to some extent on the nature of the aspirate; for example, bacteria versus noninfectious material such as gastric acid, blood, tube feeding, or a foreign body.

Cases of CP usually require macroaspiration of large-volume, low-pH (<2.5) gastric content. Initially described in the setting of obstetrical anesthesia, it is now uncommon in such circumstances. Lung damage in CP is due to the inflammatory response to the acidic material and involves various chemokines, pro-inflammatory cytokines, and neutrophils.

Following the acute inciting event symptoms may appear within hours to days. These can include cough, wheezing, shortness of breath, tachycardia, and fever. In cases of a solid foreign body aspiration distal airway obstruction may result which could lead to postobstructive pneumonia.

Findings on physical examination may include cyanosis, use of accessory muscles of respiration, wheezing, and rales. Chest radiographic findings may initially be normal but early on or subsequently are more likely to demonstrate bronchopneumonia rather than lobar pneumonia (68% vs. 15%) and posterior infiltrates are common. As a result of airway anatomy right-sided clinical and radiographic findings are more frequent. In cases of aspiration of blood or tube feedings neither CP nor AP usually develops because of the relatively high pH and lack of bacteria.

## DIAGNOSIS

AP is distinguished from other forms of pneumonia by the history (witnessed macroaspiration), the presence of risk factors, and the finding of new radiographic infiltrates in gravity-dependent lung segments (Fig. 29.2). Although findings in the posterior upper lobe or superior segment of the lower lobes are characteristic, as are basal infiltrates (right more than left) early on, the radiograph may appear clear, even though computed tomography (CT) scanning may demonstrate an infiltrate. In patients with a new infiltrate following general anesthesia, the differential diagnosis also includes negative pressure pulmonary edema from forceful inhalation against a closed airway leading to bilateral symmetric infiltrates.

The diagnosis is usually clinical, but quantitative bronchoscopic and non-bronchoscopic lung lavage may distinguish chemical and bland (blood, food material) aspiration from bacterial infection. Biomarkers are unreliable for the diagnosis of aspiration, including procalcitonin, which has been unable to distinguish chemical from bacterial pneumonitis in patients with aspiration risks. In ventilated patients, airway secretions had elevated levels of alpha—amylase from salivary and pancreatic sources at a frequency reflecting the number of patient aspiration risk factors present. This approach may not be relevant for diagnosing AP and CP in nonventilated patients.

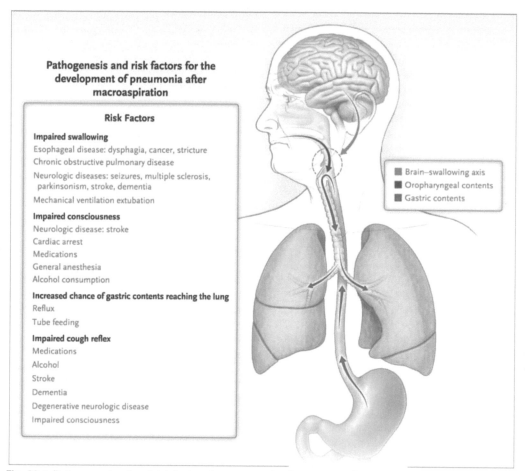

**Pathogenesis and risk factors for the development of pneumonia after macroaspiration**

### Risk Factors

**Impaired swallowing**
Esophageal disease: dysphagia, cancer, stricture
Chronic obstructive pulmonary disease
Neurologic diseases: seizures, multiple sclerosis, parkinsonism, stroke, dementia
Mechanical ventilation extubation

**Impaired consciousness**
Neurologic disease: stroke
Cardiac arrest
Medications
General anesthesia
Alcohol consumption

**Increased chance of gastric contents reaching the lung**
Reflux
Tube feeding

**Impaired cough reflex**
Medications
Alcohol
Stroke
Dementia
Degenerative neurologic disease
Impaired consciousness

■ Brain–swallowing axis
■ Oropharyngeal contents
■ Gastric contents

**Fig. 29.1** Pathogenesis of and risk factors for pneumonia after macroaspiration. Macroaspiration can occur as a result of abnormalities in the swallowing mechanism or altered swallowing due to dysfunction of the central nervous system. In patients with these disorders, oropharyngeal or gastric contents can enter the lung. An impaired cough reflex increases the likelihood that aspirated material will reach the lung. Shown are the disease processes that serve as risk factors for macroaspiration by impairing consciousness, swallowing, and cough and by increasing the chance that gastric contents will reach the lung. (Reprinted with permission from Mandell LA, Niederman MS. Aspiration pneumonia. *N Engl J Med* 2019;380(7):651-663, Massachusetts Medical Society. Reprinted with permission from Massachusetts Medical Society. Figure 1.)

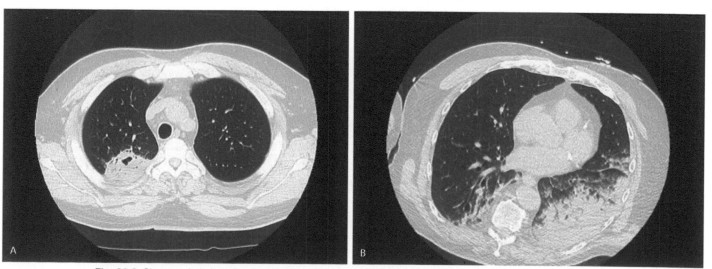

**Fig. 29.2** Characteristic imaging findings in patients with aspiration pneumonia. (A) A 56-year-old male with cough after tooth extraction under local anesthesia. He drinks 4 beers per day. Bronchoscopic cultures revealed *Klebsiella pneumoniae*. Note the cavitary infiltrate in the right upper lobe posteriorly. (B) A 79-year-old male with dyspnea and new bilateral infiltrates following upper endoscopy complicated by vomiting. Note that the infiltrates are located in posterior, gravity-dependent lung segments. (Reprinted with permission from Mandell LA, Niederman MS. Aspiration pneumonia. *N Engl J Med* 2019;380(7):651-663, Massachusetts Medical Society. Reprinted with permission from Massachusetts Medical Society. Figure 2CD.)

# TREATMENT

## Aspiration Pneumonia

Cases of AP may be divided into those acquired in the community, hospital, or LTCF setting and these are further considered based on an initial clear or abnormal chest radiograph (Fig. 29.3). When anaerobes were felt to be the major pathogens initial treatment was aimed at these organisms. With the shift in pathogens the treatment regimens have been modified as well.

Antibiotics should be used only when needed and ideally with an appropriate antibacterial spectrum of activity. Aside from consideration of adverse events, selection of a drug must also take into account the site of acquisition of the infection; for example, community, hospital, LTCF, and risks of infection with multidrug resistant pathogens. These include treatment with a broad-spectrum antibiotic within the past 90 days and hospitalization for at least 5 days.

For community-acquired cases, if dental health is reasonable, a variety of drugs with antiaerobic activity may be used and some also have antianaerobic activity. Generally, we only consider addition of clindamycin when the risk of predominantly anaerobic organisms is high as in cases with severe periodontal disease, necrotizing pneumonia, or lung abscess.

If resistance is a concern then drugs with broader activity may be used either singly or in combination (see Fig. 29.3). If multidrug resistance is a possibility one may have to resort to an aminoglycoside or colistin with or without vancomycin or linezolid if there has been documented nasal or respiratory colonization with methicillin-resistant *S. aureus*.

In the absence of extrapulmonary infection and with a good clinical response, duration of treatment should be 5 to 7 days. Longer treatment is required in cases of necrotizing pneumonia, lung abscess, or empyema. With abscess or empyema, drainage is important for diagnostic and therapeutic purposes.

## Chemical Pneumonitis

Treatment of CP involves management of the airways including bronchospasm as well as mitigation of tissue damage. Depending upon the situation and the clinical presentation management may also include suctioning, bronchoscopy, intubation, and mechanical ventilation.

Given the nature of the tissue injury in CP antibiotics are usually unnecessary unless bacteria are present as when gastric pH is elevated secondary to the use of acid-suppressing medication or in the presence of small bowel obstruction.

We suggest that in mild to moderate cases, even if a chest radiograph shows a new infiltrate, that antibiotics be withheld and the patient reassessed after 48 hours. In more serious cases, however, empiric therapy may be started initially and reassessed after 48 to 72 hours.

Supportive therapy including fluids, maintenance of blood pressure, airways, and ventilatory status are of critical importance. Adjuvant measures such as glucocorticoids, however, have not been shown to play a role in either AP or CP.

# PREVENTION

Prevention efforts are based on an understanding of disease pathogenesis (Box 29.1). Certain strategies should be used in the appropriate clinical setting while others are not as well established but should be considered. Some approaches are still too undeveloped to be recommended.

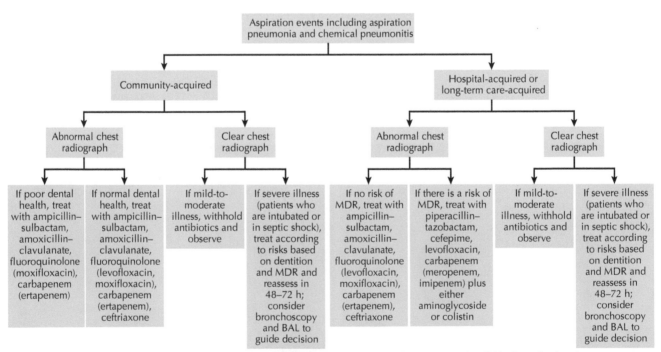

**Fig. 29.3** An algorithmic approach to antibiotic therapy for aspiration pneumonia. *BAL*, Bronchoalveolar lavage; *MDR*, multidrug resistant. (Reprinted with permission from Mandell LA, Niederman MS. Aspiration pneumonia. *N Engl J Med* 2019;380(7):651-663, Massachusetts Medical Society. Reprinted with permission from Massachusetts Medical Society. Figure 3.) If considering outpatient therapy, the following drugs are available orally: ampicillin-sulbactam, moxifloxacin, levofloxacin, clindamycin. Consider adding clindamycin in cases of necrotizing pneumonia or lung abscess. In a setting of CP, consider withholding antibiotics even if chest x-ray is abnormal in patients with mild-moderate illness and reassess in 48 hours. Severe refers to patients who are intubated or in septic shock. Treat according to risks based on dentition and MDR.

## Recommended in the Appropriate Setting

Prior to elective surgery, patients should be fasting for at least 8 hours after a meal and at least 2 hours after clear liquids. Medications that can interfere with swallowing and/or promote aspiration should be avoided if possible, including sedatives, antipsychotics, and antihistamines. In patients requiring emergent intubation, especially after cardiac arrest or with coma from neurologic injury, 24 hours of antibiotic therapy directed at pathogens likely to have been colonizing the oropharynx at the time of intubation has been shown to prevent very early onset pneumonia. The antibiotics studied include cefuroxime, ceftriaxone, and ertapenem.

## To Be Considered in the Appropriate Setting

These strategies are directed at patients with stroke and impaired swallowing, those with poor oral hygiene in the setting of aspiration risks, and patients recently extubated following mechanical ventilation.

Swallowing disorders are common after stroke and a full speech and swallowing evaluation may allow oral feeding rather than enteral tube feeding, using a mechanical soft diet with thickened liquids in place of pureed food or thin liquids. Feeding should be done in a semi-recumbent rather than supine position. Patients with oropharyngeal dysphagia should be fed with the chin down and head turned to one side, with small volume feedings and a cough after each swallow. Post pyloric tube feeding and checking for post feeding residual volume may not reduce aspiration risk. Enteral feeding tubes (particularly percutaneous placement) are controversial and not recommended for patients with dementia as a long-term approach to aspiration risk.

---

**BOX 29.1 Prevention of Aspiration Pneumonia**

**Recommended in the appropriate clinical setting:**
- 24 h of antibiotics in comatose patients following emergent intubation
- Hold general anesthesia for elective surgery: at least 8 h after a meal and 2 h after clear liquids

**To be considered in the appropriate clinical setting:**
- Swallowing evaluation after stroke
- Preference for ACE inhibitors to control blood pressure after stroke
- Oral care with brushing, removal of poorly maintained teeth

**More data needed before recommending**
- Swallowing exercises for those with dysphagia after stroke
- Oral chlorhexidine in those at risk for aspiration

*ACE,* Angiotensin converting enzyme.

---

Many stroke patients require hypertension therapy, and the use of angiotensin converting enzyme inhibitors may reduce aspiration risk, particularly in Asian patients, by elevating substance P levels, which promotes cough and improves the swallowing reflex. Cilostazol, an antiplatelet agent, may also be beneficial through a similar mechanism.

Oral hygiene maintenance has not been consistently beneficial. In nursing home patients, a program of comprehensive oral care including manual tooth and gum brushing, chlorhexidine mouth washes, and upright positioning when feeding was not effective over 1 year in preventing pneumonia. In patients undergoing surgery for esophageal cancer, lack of preoperative oral care including tooth scaling, mechanical cleaning, and tooth extraction if necessary are important risk factors for post-operative pneumonia.

## Not Recommended Currently, More Data Needed

Swallowing exercises and early mobilization for post-stroke dysphagia is a promising approach to prevent recurrent aspiration. Oral chlorhexidine has been studied in nonventilated patients at risk for aspiration, and when oral care with this agent was combined with mechanical cleaning, was effective in preventing pneumonia. However, chlorhexidine is controversial as it has been associated with increased mortality when used to prevent pneumonia in ventilated patients, possibly due to toxicity when aspirated into the lung.

## ADDITIONAL RESOURCES

Dickson RP, Erb-Downward JR, Martinez FJ et al. The microbiome and the respiratory tract. *Annu Rev Physiol* 2016;78:481-504. *An overview of the lung microbiome and its role in the lung in healthy and disease states.*

Macht M, White D, Moss M. Swallowing dysfunction after critical illness. *Chest* 2014;146:1681-89. *A comprehensive review of swallowing disorders in the critically ill, particularly after extubation from mechanical ventilation with an approach to diagnosis and prevention.*

Mandell LA, Niederman MS. Aspiration Pneumonia. *N Engl J Med* 2019;380(7):651-663. *A review of the microbiology, pathogenesis, diagnosis, management, and prevention of aspiration pneumonia.*

Shinohara Y, Origasa H. Post-stroke pneumonia prevention by angiotensin-converting enzyme inhibitors: results of a meta-analysis of five studies in Asians. *Adv Ther* 2012;29(1):900-912. *A meta-analysis of 8693 stroke patients showing that angiotensin converting enzyme inhibitors, but not other antihypertensives, reduced the risk of post-stroke pneumonia by nearly 40% with an even greater benefit in Japanese patients.*

Vallés J, Peredo R, Burgueño MJ, et al. Efficacy of single-dose antibiotic against early-onset pneumonia in comatose patients who are ventilated. *Chest* 2013;143(5):1219-1225. *A prospective cohort study of comatose patients (25% post cardiac arrest) which showed that a single dose of ceftriaxone or ertapenem, within 4 hours after intubation, could prevent early onset pneumonia.*

# 30

# Viral Respiratory Infections

*Michael J. Tan*

## ABSTRACT

Viruses are an important but underrecognized cause of pneumonias. The exact percentage of all pneumonias caused by viral infection is unknown, but viruses are probably responsible for at least 25% of cases. This figure will likely change as a result of utilization of molecular diagnostic panels. The most common respiratory viruses are influenza, rhinovirus, coronavirus, respiratory syncytial virus (RSV), parainfluenza virus, human metapneumovirus, and adenovirus. Multiplex polymerase chain reaction (PCR) detection enables relatively simple detection. Patients with severe chronic lung disease, chronic medical conditions, immunosuppression, and the elderly are groups most susceptible to viral pneumonia. With the exception of influenza, treatment is largely supportive.

## CLINICAL VIGNETTE

A 68-year-old woman with chronic obstructive pulmonary disease comes to the emergency department in early July with fever, cough, and wheezing for 3 days. She has young grandchildren that she was watching for her family. Her vital signs were stable, and her physical examination was unremarkable other than fatigued appearance and scattered expiratory wheezes. She had no leukocytosis and her procalcitonin was less than 0.1. Chest radiograph demonstrated chronic changes associated with her lung disease. Legionella and pneumococcal urine antigens were not detected. Nasopharyngeal swab sent for respiratory PCR reveals parainfluenza virus. The patient is discharged with supportive care measures, no antimicrobial agents, and she recovers uneventfully.

## INFLUENZA VIRUS

Influenza viruses are enveloped, single-stranded ribonucleic acid (RNA) viruses of the family Orthomyxoviridae. The viruses are classified as type A, B, or C and subtyped based on differences in the surface hemagglutinin (H) and neuraminidase (N) glycoproteins. Influenza A is the leading cause of influenza in adults in the United States and is responsible for 90% of all epidemic influenza. Prevention, diagnosis, and treatment are important, as secondary bacterial pneumonia can be severe and is not uncommon.

In the United States, influenza virus has no geographic predilection. It is spread by respiratory secretions from individuals who are actively shedding the virus. Incubation is approximately 1 to 5 days. Epidemics occur annually during the winter months. It is associated with 10,000 to 40,000 excess deaths. For seasonal influenza, 80% of these deaths are in patients older than 65. Patients with chronic lung diseases such as chronic obstructive pulmonary disease and emphysema, congestive heart failure, hemoglobinopathies, and immunosuppression are at risk for severe disease (Fig. 30.1 and Box 30.1).

Clinical manifestations include an acute febrile respiratory illness with cough, sore throat, headache, malaise, and myalgias. Symptoms are usually self-limited, with the major symptoms improving after 3 to 5 days. Complications can include secondary bacterial infections caused by *Streptococcus pneumoniae*, *Staphylococcus aureus*, *Haemophilus influenzae*, or gram-negative pathogens. These secondary infections are often suggested by initial improvement followed by clinical worsening (Fig. 30.2). Diagnosis of influenza is suggested by the symptoms listed earlier, usually with the presence of influenza in the community. Confirmation can be made with several available rapid diagnostic tests that detect viral nucleoproteins or viral neuraminidase, although RT-PCR that detects viral RNA is the preferred diagnostic modality. Rapid influenza diagnostic tests have high specificity (>90%) but have low to moderate sensitivity (20% to 70%) compared with other influenza tests and PCR detection. These and other virus and viral antigen detection methods may also be helpful in nonepidemic months (Table 30.1).

Treatment is largely supportive with management of symptoms with antipyretics and analgesics. The neuraminidase inhibitors oseltamivir (oral) and zanamivir (inhaled) are traditionally effective against influenza A and B. Peramivir offers an intravenous option, and baloxavir, a cap-dependent endonuclease inhibitor (CEN), offers a single oral dose. The tricyclic amines, amantadine and rimantadine, are classically active against only influenza A. Widespread resistance of influenza A (H3N2) and 2009 pandemic H1N1 to amantadine and rimantadine has been seen. Presently, therefore, only neuraminidase inhibitors or baloxavir should be used if influenza is suspected. The recent Infectious Diseases Society of America (IDSA) guideline emphasizes that during times of influenza activity, a clinical diagnosis is adequate for the decision to initiate antiviral therapy without the need for testing. The guideline also suggests immediate initiation of antiviral treatment for adults and children with documented or suspected influenza, regardless of vaccination history if they are (1) hospitalized with influenza regardless of duration prior to hospitalization, (2) outpatients with severe or progressive illness, regardless of duration of illness, (3) outpatients at high risk of complications from influenza, including those with chronic medical conditions or immunocompromising states, (4) patients younger than 2 years or 65 years or older, and (5) pregnant patients and those within 2 weeks postpartum. Treatment consideration can be given to those outpatients with uncomplicated disease with onset 2 days or less before presentation, and symptomatic outpatients who are household contacts of those who are at high risk for influenza complications. Resistance to neuraminidase inhibitors of predominant strains is rare (<1%). The neuraminidase inhibitors are also effective for chemoprophylaxis. The current antiviral recommendations are readily available from the Centers for Disease Control and Prevention website on seasonal influenza, www.cdc.gov/flu.

The most effective means of prevention is with annual influenza vaccination. All persons older than 6 months of age are recommended to have annual seasonal influenza vaccination. Several vaccine formulations

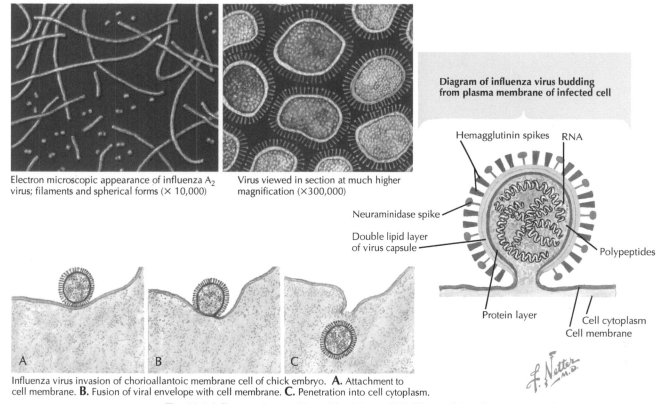

Electron microscopic appearance of influenza A₂ virus; filaments and spherical forms (× 10,000)

Virus viewed in section at much higher magnification (×300,000)

Diagram of influenza virus budding from plasma membrane of infected cell

Hemagglutinin spikes     RNA

Neuraminidase spike

Double lipid layer of virus capsule

Polypeptides

Protein layer

Cell cytoplasm
Cell membrane

Influenza virus invasion of chorioallantoic membrane cell of chick embryo. **A.** Attachment to cell membrane. **B.** Fusion of viral envelope with cell membrane. **C.** Penetration into cell cytoplasm.

**Fig. 30.1** Influenza virus and its epidemiology. *RNA*, Ribonucleic acid.

are available with the considerations being quadrivalent versus trivalent (most are quadrivalent now), intranasal live attenuated versus intramuscular recombinant or inactivated, high dose versus standard dose. These vaccines contain the three or four virus strains that are projected to be responsible for the annual epidemic. Vaccination is usually effective from about 2 weeks to 4 to 6 months postvaccination. However, data suggest vaccination efficacy is maintained late in the season and into the summer; vaccination should be given at earliest availability and should not be delayed until later in the season. Current influenza vaccine details are available at www.cdc.gov/flu.

## AVIAN INFLUENZA AND PANDEMIC INFLUENZA

Two influenza entities that have come to be of significant public concern in recent years are the avian and pandemic influenza. Avian influenza is caused by influenza viruses that occur naturally in wild birds. Typically they are transmitted between birds, and occasionally from birds to humans. Person-to-person transmission is currently rare and

uncommon. Low pathogenic avian influenza is common in birds and not a significant mortality risk, but highly pathogenic influenza, specifically influenza A viruses H5N1 and H7N9, can be deadly to domesticated birds. Because there is little human immunity and little human vaccine availability, these strains can also be deadly to humans. Patients at risk for avian influenza usually have significant contact with infected poultry. H5N1 has been resistant to tricyclic amines. Neuraminidase inhibitors have been suggested for prophylaxis and treatment if outbreaks occur; however, emerging resistance has the potential to limit efficacy.

*Pandemic flu* refers to a global influenza outbreak resulting from a highly virulent influenza to which humans have little natural immunity. This may result from introduction of an immunologically specific virus or development of an antigenically different mutation to which human hosts have little defense (see Fig. 8.1). It occurs rarely rather than seasonally. This type of influenza is spread easily from person to person and may result in rapid global spread, a lack of treatment or medical supplies, overwhelmed healthcare systems, and economic and social disruption. (H1N1)pdm09 was responsible for the last pandemic, and this strain of influenza continues to circulate as a seasonal influenza virus. Notably, the most adversely affected demographic differed from seasonal influenza, in that the greatest percentage of complications were observed in adolescents and young adults. For both H5N1 and pandemic influenza, isolation and quarantine may be needed to limit the spread of the infection. In response to concerns regarding pandemics, governments have been stockpiling antiviral medications for treatment and prophylaxis; resistance patterns, however, cannot be guaranteed.

## RESPIRATORY SYNCYTIAL VIRUS

RSV is an enveloped, single-stranded RNA virus of the Pneumoviridae family. RSV is the most common cause of lower respiratory tract

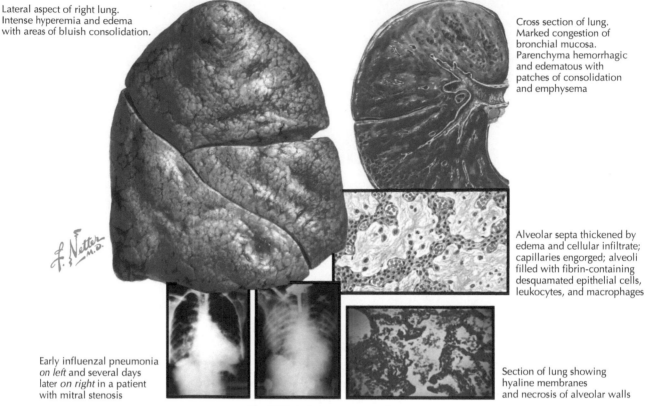

Lateral aspect of right lung. Intense hyperemia and edema with areas of bluish consolidation.

Cross section of lung. Marked congestion of bronchial mucosa. Parenchyma hemorrhagic and edematous with patches of consolidation and emphysema

Alveolar septa thickened by edema and cellular infiltrate; capillaries engorged; alveoli filled with fibrin-containing desquamated epithelial cells, leukocytes, and macrophages

Early influenzal pneumonia *on left* and several days later *on right* in a patient with mitral stenosis

Section of lung showing hyaline membranes and necrosis of alveolar walls

**Fig. 30.2** Influenza pneumonia.

| TABLE 30.1 | Diagnostic Tests for Viral Infections |
|---|---|
| Respiratory syncytial virus infection | Tracheal aspirate or bronchial alveolar lavage for viral culture, antigen testing by ELISA and fluorescein conjugate monoclonal or polyclonal antibody, RT-PCR |
| Parainfluenza | Nasal and bronchial secretions for viral culture and immunofluorescent assays, RT-PCR Serum for complement fixation and hemagglutination |
| Influenza | Respiratory secretions for viral cultures and immunofluorescent and ELISA assays, RT-PCR |
| Adenovirus infection | Respiratory secretions for viral culture, complement fixation, hemagglutination inhibition, and neutralization, PCR |
| Rhinovirus/Enterovirus | Respiratory secretions for viral culture, PCR |
| Coronavirus | Respiratory secretions for viral culture, PCR |

*ELISA,* Enzyme-linked immunosorbent assay; *RT-PCR,* reverse transcription–polymerase chain reaction.

infections in infants and children, although infection in adults is possible. It is responsible for 25% of hospitalizations of children with pneumonia and 40% to 50% of hospitalizations of children with bronchiolitis. Premature infants and children with bronchopulmonary dysplasia, congenital heart disease, and immunodeficiency are at greatest risk. There is no geographic predilection for infection. With the presence of pneumonia, the death rate is reported to range from 11% to 78%.

Infection occurs by inoculation of ocular, nasal, or oral mucosa after contact with fomites or infected secretions. The virus may persist on environmental surfaces for several hours. Good hand hygiene and contact precautions are the most effective means of prevention.

Asymptomatic infections with RSV are uncommon. Nasal congestion, sinusitis, otitis media, coryza, and pharyngitis are typical presentations in children. Lower respiratory tract infections may also be seen and include tracheobronchitis, bronchiolitis, and pneumonia. Infections in the elderly and immunocompromised more often tend to involve the lower respiratory tract. Patients often have fever, nonproductive cough, anorexia, and dyspnea. Physical examination may reveal wheezing and crackles. In adults, radiographs may exhibit bilateral interstitial or patchy infiltrates with occasional lobar consolidation. White blood cell (WBC) count is usually less than 10,000. Radiographs in children may demonstrate bronchial wall thickening and hyperinflation.

Diagnosis is made by examination of respiratory secretions. Viral culture is considered the gold standard, but results may require several days. Identification of RSV antigen by immunofluorescence is faster and fairly sensitive. PCR assays are available on upper and lower respiratory tract specimens and have the advantage of rapid results, and such results may help reduce the use of non-necessary antimicrobials. Acute and convalescent serologies may also be performed, but these are usually helpful only for retrospective diagnosis (see Table 30.1).

Ribavirin is the only effective antiviral agent for RSV, but its use in children is controversial; and the benefit of ribavirin therapy for healthy or immunocompromised adults has not been established. Ribavirin has not been shown to be effective prophylaxis for RSV. Currently there is no vaccination available.

# OTHER RESPIRATORY VIRUSES

Other respiratory viruses including parainfluenza virus, rhinovirus/enterovirus, coronavirus, human metapneumovirus, and adenovirus can be difficult to distinguish. Common symptoms of respiratory viruses include fever, cough, rhinorrhea, and sore throat. Shortness of breath, sneezing, myalgias, and headache are also not uncommon. Parainfluenza viruses, in children, can cause croup, bronchitis, pharyngitis, and pneumonia. Adults generally get mild upper respiratory infections. With these respiratory viruses, immunocompromised hosts and those with cardiopulmonary disease may be more likely to have more severe bronchitic and lower respiratory symptoms including pneumonia. Patients with adenovirus infection may also have conjunctivitis and lymphadenopathy. Two variants of coronavirus, Middle East respiratory syndrome (MERS-CoV) and severe acute respiratory syndrome (SARS-CoV), may cause severe respiratory symptoms with high mortality. SARS-CoV was last reported in 2004, although MERS-CoV still has transmission on the Arabian Peninsula. Suspicion for MERS should be reported to local health authorities. Chest radiographs do not generally demonstrate specific characteristic changes. Another novel variant, SARS-CoV2, was identified in 2019 and is responsible for a worldwide pandemic. Chest radiography frequently demonstrates nodular ground-glass infiltrates. Common respiratory viruses traditionally can be isolated by cell culture of respiratory samples and immunofluorescence, but currently available PCR testing on upper and lower respiratory samples can be more sensitive and can distinguish the different virus (Table 30.1). Such testing is also more rapid than traditional means and may help reduce unnecessary antimicrobial use by identifying specific viruses that may be causing a monomicrobial viral respiratory condition. For these viruses, treatment is generally supportive. Ribavirin has been used for parainfluenza virus and adenovirus, but there are scarce data to support its use. There are no currently available vaccines. Good hand hygiene and respiratory precautions are the best prevention.

## ADDITIONAL RESOURCES

Centers for Disease Control and Prevention. Influenza (Flu). Available at: http://www.cdc.gov/flu. Accessed August 21, 2019. *Gives up-to-date information regarding the current patterns of seasonal influenza, resistance patterns, and methods of prevention. This site also contains links to nonseasonal influenza sites including pandemic influenza.*

Metlay JP, Waterer G, Long AC, et al. Diagnosis and treatment of adults with community-acquired pneumonia. An official clinical practice guideline of the American Thoracic Society and Infectious Diseases Society of America. *Am J Respir Crit Care Med* 2019;200(7):e45-467. *The most recent guideline for diagnosis and treatment of community-acquired pneumonia.*

Uyeki TM, Bernstein HH, Bradley JS, et al. Clinical practice guidelines by the Infectious Diseases Society of America: 2018 update on diagnosis, treatment, chemoprophylaxis, and institutional outbreak management of seasonal influenza. *Clin Infect Dis* 2018;68:e1-e27. *Guideline summarizes modalities for treatment and prevention of seasonal influenza.*

# 31

# Sinus Infections

*Anthony W. Chow*

## ABSTRACT

*Sinusitis* is defined as an inflammation of the mucosal lining of the paranasal sinuses and can be caused by various factors including allergy, environmental irritants, and infection by viruses, bacteria, or fungi. It is also commonly referred to as *rhinosinusitis* because there is almost always coexisting inflammation in the nasal mucosa. Sinusitis can be classified as acute, subacute, and chronic. Acute sinusitis lasts up to 4 weeks and is usually caused by a viral or bacterial infection. Chronic sinusitis lasts more than 12 weeks and may result from a wide range of allergic and nonallergic causes. Subacute sinusitis lasts 4 to 12 weeks and usually represents a transition between acute and chronic sinusitis. Other patterns include recurrent acute sinusitis, defined as the occurrence of four or more episodes of acute sinusitis within 1 year, each lasting at least 7 days; and acute exacerbation of chronic sinusitis, defined as the presence of signs and symptoms of chronic sinusitis that worsen but return to baseline after treatment.

Sinusitis can also be categorized according to the mode of infection and underlying conditions, such as nosocomial sinusitis associated with nasotracheal intubation, odontogenic sinusitis, and sinusitis in severely immunocompromised hosts. From a clinical and management standpoint, the most important goal is to distinguish a bacterial infection from viral or allergic causes in acute sinusitis and to identify structural or fungal causes in chronic sinusitis. Distinguishing bacterial infection is critical for appropriate antimicrobial therapy, whereas structural or fungal causes may necessitate surgical intervention for diagnosis and treatment.

## CLINICAL VIGNETTE

A 27-year old woman presented to the Emergency Room with a history of an acute onset of nasal discharge, a frontal headache, and a temperature of 39.5°C. Her temperature normalized within 2 days, but after 12 days she developed bothersome nasal congestion and purulent postnasal drip that did not improve. Physical examination revealed facial tenderness over the right maxillary sinus, and purulent secretion in the nasal passages and posterior pharynx. How should this patient be managed? Does she require antimicrobial therapy?

COMMENT: Several features in this patient suggest a bacterial etiology of acute sinusitis. Firstly, worsening of symptoms after initial improvement (double-sickening) is consistent with a bacterial superinfection following initial acute viral sinusitis. Secondly, purulent discharge in the nasal passages and posterior nasopharynx is highly predictive of a bacterial sinusitis. In light of her persistent symptoms, "watchful waiting" with symptomatic management is not warranted. A computed tomography (CT) of the paranasal sinusitis is recommended to confirm the diagnosis and exclude the possibility of complications. If fluid level or opacification in the sinuses is demonstrated, empirical antibiotic therapy with standard dose amoxicillin-clavulanate for 5 days is recommended. If there is no clinical improvement after 5 days, endoscopically directed middle meatal culture should be obtained for semi-quantitative culture and antibiotic susceptibility testing to confirm the diagnosis and guide further antimicrobial therapy.

## ANATOMIC CONSIDERATIONS

The paranasal sinuses (maxillary, ethmoid, frontal, and sphenoid) are air-filled cavities lined by pseudostratified, ciliated columnar epithelium. They are interconnected through small tubular openings, the sinus ostia, which drain into different regions of the nasal cavity (Fig. 31.1). The frontal, anterior ethmoid, and maxillary sinuses open into the middle meatus, whereas the posterior ethmoid and sphenoid sinuses open into the superior meatus. The osteomeatal complex, an area between the middle and inferior nasal turbinates representing the confluence of drainage from the paranasal sinuses, is a particularly important anatomic site because of its potential for mucosal thickening and impaired drainage leading to sinus infection even without mechanical obstruction of the ostia.

The maxillary sinuses, either alone or in combination with the ethmoid or frontal sinuses, are the most frequent site of infection. The ostium of the maxillary sinus lies at an obtuse angle toward the roof (see Fig. 31.1), so the maxillary sinus does not empty well in the erect posture but drains best when the patient is lying on the side opposite the affected sinus. The floor of the maxillary sinus directly adjoins the maxillary bone in which the apices of the first, second, and third molar teeth reside; hence, extraction or root infection of these teeth is a frequent cause of maxillary (odontogenic) sinusitis. Furthermore, because the superior alveolar nerves (branches of the maxillary nerve) supply both the molar teeth and the mucous membranes of the sinus, maxillary sinusitis may frequently manifest as a toothache.

The frontal sinus is not a frequent site of infection but may be a focus for spread of infection into the orbit or the brain (Fig. 31.2). The frontal sinus is supplied by the supraorbital branch of the ophthalmic division of the trigeminal nerve. Thus headache is a prominent symptom of frontal sinusitis.

The ethmoid sinuses are composed of multiple air cells that are separated by thin bony partitions, and each air cell drains by an independent ostium. The ethmoid sinuses are separated from the orbit by a paper-thin orbital plate. Perforation of the plate allows direct spread of infection into the retro-orbital space. Ethmoid sinusitis can also spread to the superior sagittal vein or the cavernous venous sinus (see Fig. 31.2).

The sphenoid sinus occupies the body of the sphenoid bone in proximity to the pituitary gland above; the optic nerve and optic chiasm in front; and the internal carotids, the cavernous sinuses, and the temporal lobes of the brain on each side (see Fig. 31.2). Therefore sphenoid sinusitis can spread locally to cause cavernous sinus thrombosis, meningitis, temporal lobe abscess, and orbital fissure syndromes. The superior orbital fissure syndrome, characterized by orbital pain, exophthalmos, and ophthalmoplegia, is caused by involvement of the abducens, oculomotor, and trochlear nerves and the ophthalmic division of the trigeminal nerve as they pass through the orbital fissure.

**Paranasal Sinuses**

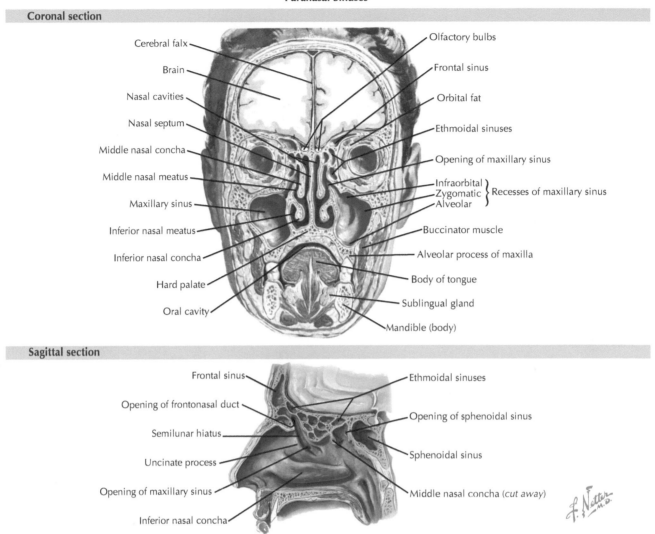

**Coronal section**

- Cerebral falx
- Brain
- Nasal cavities
- Nasal septum
- Middle nasal concha
- Middle nasal meatus
- Maxillary sinus
- Inferior nasal meatus
- Inferior nasal concha
- Hard palate
- Oral cavity

- Olfactory bulbs
- Frontal sinus
- Orbital fat
- Ethmoidal sinuses
- Opening of maxillary sinus
- Infraorbital
- Zygomatic } Recesses of maxillary sinus
- Alveolar
- Buccinator muscle
- Alveolar process of maxilla
- Body of tongue
- Sublingual gland
- Mandible (body)

**Sagittal section**

- Frontal sinus
- Opening of frontonasal duct
- Semilunar hiatus
- Uncinate process
- Opening of maxillary sinus
- Inferior nasal concha

- Ethmoidal sinuses
- Opening of sphenoidal sinus
- Sphenoidal sinus
- Middle nasal concha (cut away)

**Fig. 31.1** Anatomic relationships of the paranasal sinuses.

A patent osteomeatal complex and normal mucociliary clearance function are the key defense mechanisms of the paranasal sinuses.

## EPIDEMIOLOGY AND RISK FACTORS

According to the 2017 National Health Interview Survey, approximately 30 million cases of sinusitis are diagnosed each year, representing 12% of all adults 18 years of age or older. Children younger than age 15 and adults 25 to 64 years old are the most frequently affected. It is the fifth leading cause for antimicrobial prescriptions in office practice. Total direct costs for treating sinusitis are estimated at $6 billion per year, not to mention significant indirect costs such as days lost from work, decreased productivity, and impaired quality of life.

Several risk factors predispose to acute or chronic sinusitis (Box 31.1). The most common cause of acute sinusitis is a viral upper respiratory infection or the common cold. Adults typically develop two to three colds per year, and children have six to eight episodes per year. However, only up to 2% of adults and 6% to 13% of children with viral upper respiratory infections develop a secondary bacterial infection of the sinuses. Nose blowing that generates excessive positive intranasal pressures and propels contaminated fluid from the nasal cavity into the paranasal sinuses may be an important predisposing mechanism.

A strong association exists among allergic rhinitis, asthma, and recurrent sinusitis in both children and adults. It has been suggested that allergic rhinitis may be a predisposing factor for acute maxillary sinusitis in 25% to 30% of patients, and as many as 60% to 80% of patients with chronic sinusitis. The observation that asthma, allergic rhinitis, and rhinosinusitis frequently coexist raises the possibility that these conditions are manifestations of an inflammatory process within an integrated and contiguous upper and lower airway (the integrated airway hypothesis).

Mild and selective immune deficiencies have been frequently demonstrated in children and adults with recurrent or persistent symptoms of sinus disease. One study found that 52% of patients had selective immunoglobulin (Ig) deficiencies (IgG2, IgG4 subclass, or IgA) or poor responsiveness to some polysaccharide antigens. In addition, chronic or recurrent sinusitis is an important source of morbidity in patients with cystic fibrosis and patients infected with human immunodeficiency virus (HIV).

Certain actions, such as sustaining head trauma, swimming, diving, cocaine sniffing, and nasotracheal or nasogastric intubation, may repeatedly traumatize the nasal mucosa and facilitate microbial invasion of the paranasal sinuses. Dental extraction and periapical infections of the maxillary molar teeth are particularly important causes of odontogenic sinusitis.

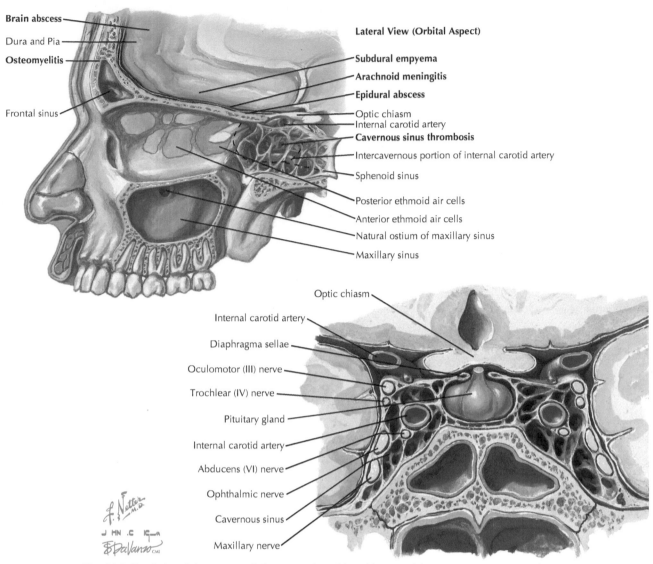

**Fig. 31.2** Proximity of the paranasal sinuses to the orbit and intracranial structures; cavernous sinuses.

## MICROBIOLOGY

Sinus puncture and aspiration is the gold standard for establishing the microbial cause of sinusitis. Although acute sinusitis is primarily caused by respiratory viruses such as rhinoviruses, influenza A, parainfluenza, respiratory syncytial virus, and adenoviruses, direct isolation of these viruses from antral aspirates has been relatively uncommon. In acute bacterial sinusitis, *Streptococcus pneumoniae* and nonencapsulated *Haemophilus influenzae* are the causative agents in 70% of adults, whereas the addition of *Moraxella catarrhalis* accounts for 80% of cases in children (Table 31.1). *Streptococcus aureus* together with viridans streptococci and *S. pneumoniae* are the predominant isolates in acute sphenoid sinusitis. Anaerobes are uncommonly isolated in acute sinusitis but are the predominant flora in chronic sinusitis. The isolation of anaerobes during acute sinusitis suggests an odontogenic source. Nosocomial sinusitis secondary to prolonged nasotracheal intubation is commonly a polymicrobial infection caused by gram-negative bacteria, *S. aureus,* and anaerobes.

Up to 40% of *H. influenzae,* 80% of *M. catarrhalis,* and 30% of respiratory tract anaerobes produce β-lactamase. Over 25% of *S. pneumoniae* is now resistant to trimethoprim-sulfamethoxazole (TMP-SMX).

Macrolide-resistant *H. influenzae* and *S. pneumoniae* are also isolated with increasing frequency.

The microbiology of chronic sinusitis is more closely linked to underlying comorbid diseases and differs considerably from that of acute sinusitis. In patients with cystic fibrosis, *Pseudomonas aeruginosa* and nontypable *H. influenzae* are the most frequent pathogens. Sinusitis in patients with HIV infection is often caused by gram-negative bacilli and unusual pathogens such as *Aspergillus* species and cytomegalovirus. In debilitated and severely immunocompromised hosts, such as those with uncontrolled diabetes mellitus, advanced HIV infection, or chemotherapy-induced neutropenia, invasive fungal infection with *Aspergillus, Mucor, Pseudallescheria boydii, Fusarium,* and other saprophytic fungi may occur. In chronic sinusitis associated with nasal polyposis or allergic fungal rhinosinusitis, hypersensitivity to and colonization by *Aspergillus* and other saprophytic fungi in the paranasal sinuses can often be demonstrated. In addition, there is an increased prevalence of nasal colonization by enterotoxin-producing *S. aureus* in patients with chronic sinusitis associated with nasal polyposis. Such patients typically demonstrate local production of enterotoxin-specific IgE antibodies and coexisting aspirin sensitivity or asthma.

## BOX 31.1   Predisposing Factors for Sinusitis

**Impaired Mucociliary Function**
- Viral upper respiratory tract infection
- Allergic rhinitis
- Irritants from cold or dry air
- Chemicals or drugs (rhinitis medicamentosa)
- Human immunodeficiency virus infection
- Cystic fibrosis
- Ciliary dysmotility syndrome

**Obstruction of Sinus Ostia**
- Viral upper respiratory tract infection
- Allergic rhinitis
- Anatomic abnormalities (e.g., nasal polyps, deviated nasal septum, choanal atresia, foreign body, tumors)

**Immune Defects**
- Common variable immunodeficiency
- Selective immunoglobulin deficiency (immunoglobulin A [IgA], IgG subclasses)
- Acquired immunodeficiency syndrome
- Wegener granulomatosis
- Diabetes mellitus

**Direct Microbial Invasion of the Sinuses**
- Odontogenic infection
- Nasotracheal or nasogastric intubation
- Head trauma
- Swimming or diving
- Cocaine sniffing

## CLINICAL FEATURES

### Acute Sinusitis

Acute sinusitis is often difficult to distinguish from the common cold or allergic (vasomotor) rhinitis. The presence of at least two major symptoms or one major and two or more minor symptoms may distinguish acute sinusitis (whether viral or bacterial) from rhinitis (Table 31.2). Three hallmarks that suggest a bacterial sinusitis rather than a viral infection are: (1) persistence (i.e., more than 10 days); (2) severity; and (3) worsening of respiratory symptoms after initial improvement (double-sickening). The probability of identifying a bacterial infection by sinus aspiration is approximately 60% for patients with symptoms persisting beyond 10 days. The presence of purulent postnasal discharge, maxillary toothache, facial pain, or unilateral maxillary sinus tenderness further increases the likelihood of a bacterial infection. Hyposmia, jaw pain with mastication, nasal congestion, and a recent history of upper respiratory tract infection are other manifestations. The combination of these clinical findings greatly enhances the diagnostic probability. In children, the most common manifestations of bacterial sinusitis are cough (80%), nasal discharge (76%), and fever (63%). Parents of preschoolers often report malodorous breath. Headache, facial pain, and swelling are rare.

In ethmoid sinusitis, edema of the eyelids and excessive tearing may be a prominent feature. Retro-orbital pain and proptosis indicate extension of infection into the orbit. Severe intractable headache is dominant in sphenoid sinusitis and can mimic ophthalmic migraine or trigeminal neuralgia. Neurologic deficit with hypoesthesia or hyperesthesia of the ophthalmic or maxillary dermatomes of the trigeminal nerve may be detected in one third of the patients.

In nosocomial sinusitis secondary to prolonged nasotracheal or nasogastric intubation, the clinical features may be relatively silent apart from unexplained fever. The presence of purulent rhinorrhea or a middle ear effusion may be the only physical finding. A high index of suspicion is required for early diagnosis.

### Subacute and Chronic Sinusitis

Chronic sinusitis may mimic asthma, allergic rhinitis, or chronic bronchitis. Pain or tenderness on palpation may be present over the affected sinuses. Fever is uncommon. Physical findings may be subtle. Fatigue, general malaise, and an ill-defined feeling of unwellness and irritability can be more prominent than local symptoms of nasal congestion, facial pain, or postnasal drip. Four cardinal findings in chronic sinusitis include: (1) anterior and/or posterior mucopurulent drainage; (2) nasal obstruction; (3) facial pain, pressure, or fullness; and (4) decreased sense of smell (hyposmia or anosmia). The presence of at least two of these signs or symptoms together with objective signs of mucosal inflammation is required to make a firm diagnosis. In addition, chronic sinusitis may manifest in three distinctive clinical syndromes: (1) chronic sinusitis with nasal polyposis (20% to 33% of cases); (2) allergic fungal rhinosinusitis (6% to 12%); and (3) chronic sinusitis without nasal polyposis (60% to 65%).

Chronic sinusitis without nasal polyposis typically shows abundant eosinophils and neutrophils but lacks evidence of fungal hyphae. Chronic sinusitis with polyposis is clinically more distinctive because of the presence of bilateral nasal polyps with eosinophilic infiltration in the middle meatus or the sinus cavities. There is a high association with aspirin sensitivity and asthma, and colonization with enterotoxin-producing *S. aureus* with evidence of IgE-mediated hypersensitivity to these superantigens. Allergic fungal sinusitis is characterized by abundant "allergic mucin" associated with sinus opacification, the presence of degranulating eosinophils, and fungal hyphae. IgE-mediated hypersensitivity to one or more colonizing fungi can usually be demonstrated.

## DIAGNOSTIC APPROACH

In primary care the diagnosis of acute bacterial sinusitis is mainly based on the history and physical examination. Sinus puncture is seldom performed owing to its invasiveness and poor patient acceptance. However, the ability to distinguish bacterial from viral sinusitis based on history or physical findings alone is limited.

### Imaging Studies

Routine imaging is unnecessary for uncomplicated sinusitis. Imaging studies are essential for patients with suspected orbital or intracranial complications, in recurrent or chronic sinusitis, and for those conditions which do not improve despite appropriate medical therapy. Plain sinus radiographs are now largely superseded by CT for evaluating the sinonasal cavity. A single Waters (occipitomental) view should suffice to visualize the maxillary and frontal sinuses. The Caldwell (occipitofrontal) and lateral views are used for evaluating the ethmoid and sphenoid sinuses, respectively. Radiographic findings in acute sinusitis include thickened mucosa (>6 mm), air-fluid level, or complete opacification of the involved sinuses. They have low sensitivity (60%) for bacterial infection as determined by sinus puncture, but are more helpful for excluding sinus disease when clinical manifestations are unclear (specificity 80%).

A coronal CT scan is the most cost-efficient imaging technique for the diagnosis of acute sinusitis. Compared with plain radiographs, CT provides greater definition of the sinus cavity and its

**TABLE 31.1   Microbial Causes of Acute and Chronic Sinusitis Determined by Antral Sinus Aspirate or Sinus Surgery Specimens**

| Microbial Agent | PREVALENCE MEAN (RANGE) | |
| --- | --- | --- |
| | Adults (%) | Children (%) |
| **Acute Sinusitis** | | |
| *Streptococcus pneumoniae* | 20–43 | 36–37 |
| *Haemophilus influenzae* | 6–35 | 23–25 |
| *Moraxella catarrhalis* | 2–10 | 19–25 |
| *Streptococcus pyogenes* | 1–7 | 2 |
| *Staphylococcus aureus* | 0–8 | 8–10 |
| Gram-negative bacilli (includes Enterobacteriaceae species, *Pseudomonas aeruginosa*) | 0–24 | 2 |
| Anaerobes *(Bacteroides, Fusobacterium, Peptostreptococcus, Veillonella)* | 0–12 | 0–4 |
| Respiratory viruses (rhinovirus, influenza, parainfluenza, adenovirus) | 3–15 | 0–2 |
| **Chronic Sinusitis** | | |
| Aerobes | 29–43 | 20 |
| *Streptococcus* species | 9–14 | 6 |
| *Staphylococcus* species | 5–14 | 6 |
| *Haemophilus influenzae* | 1–6 | 3–5 |
| Anaerobes | 57–88 | 80 |
| *Peptostreptococcus* species | 25–38 | 23–73 |
| *Bacteroides* species | 14–27 | 27–29 |
| *Fusobacterium* species | 3–4 | 5–16 |

Data from Chow AW: Acute sinusitis: current status of etiologies, diagnosis, and treatment, *Curr Clin Top Infect Dis* 2001;21:31-63; Gwaltney JM Jr: Acute community-acquired sinusitis, *Clin Infect Dis* 1996;23:1209-1225; Wald ER: Microbiology of acute and chronic sinusitis in children, *J Allergy Clin Immunol* 1992;90:452-460; Noye KA, Brodovsky D, Coyle S, et al: Classification, diagnosis and treatment of sinusitis: evidence-based clinical practice guidelines, *Can J Infect Dis* 1998;9(suppl B):3B-24B; Brooke I: The role anaerobic bacteria in sinusitis, *Anaerobe* 2006;12:5-12; and Brook I, Foote PA, Hausfeld JN: Increase in the frequency of recovery of methicillin-resistant *Staphylococcus aureus* in acute and chronic maxillary sinusitis, *J Med Microbiol* 2008;57:1015-1017.

**TABLE 31.2   Diagnostic Criteria of Sinusitis[a]**

| Major Symptoms | Minor Symptoms |
| --- | --- |
| Purulent anterior nasal discharge | Headache |
| Purulent or discolored posterior nasal discharge | Ear pain, pressure, or fullness |
| | Halitosis |
| Nasal congestion or obstruction | Dental pain |
| Facial congestion or fullness | Cough |
| Facial pain or pressure | Fever (for subacute or chronic sinusitis) |
| Hyposmia or anosmia | |
| Fever (for acute sinusitis only) | Fatigue |

[a]A diagnosis of sinusitis is probable in the presence of at least two major symptoms or one major and two or more minor symptoms. Modified from Meltzer EO, Hamilos DL, Hadley JA, et al: Rhinosinusitis: establishing definitions for clinical research and patient care, *J Allergy Clin Immunol* 2004;114:S155-S212.

contents and offers better visualization of the ethmoid and sphenoid sinuses. Contrast-enhanced CT is invaluable for assessing orbital or intracranial complications (Fig. 31.3) and can clarify anatomic variations that may play an important role in recurrent or chronic sinusitis. However, although CT is very sensitive for detecting sinus abnormalities, it lacks specificity for bacterial infection, because abnormalities can be demonstrated in over 80% of patients with the common cold.

Magnetic resonance imaging (MRI) is used rarely because of its cost and limitations in assessing cortical bone. Although MRI provides better visualization of soft tissues than CT, it is best reserved for the investigation of intracranial suppurative complications, vascular involvement, and for better delineation between the intraorbital and extraorbital soft tissue compartments.

Ultrasonography is technically difficult to perform in young children, and its ability to demonstrate retained fluid or thickened mucosa of affected sinuses is highly operator dependent. It may be useful for individuals in whom radiography or CT scanning is not feasible (such as in obtunded and critically ill patients with suspected nosocomial sinusitis). More recently, point-of-care sinus ultrasound has been found useful in excluding the diagnosis of bacterial sinusitis in patients who do not have sinus fluid collection, thus avoiding antibiotic over-prescription.

## Sinus Cultures

Surface cultures of the nasal vestibule or the nasopharynx are unreliable for microbiologic diagnosis of sinusitis owing to regular contamination by the resident microflora. Although quantitative culture of aspirates from sinus puncture remains the gold standard, this procedure is poorly accepted by patients and impractical in most primary care situations. An alternative is to obtain endoscopically directed middle meatal cultures in which the accuracy has been reported to range from 76% to 78% in both adults and children. If the analysis is restricted to isolation of the major pathogenic bacteria in acute bacterial sinusitis (i.e., *S. pneumoniae, H. influenzae,* and *M. catarrhalis*), the

**Contrast-Enhanced CT of the Orbits**

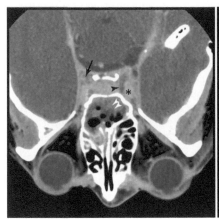 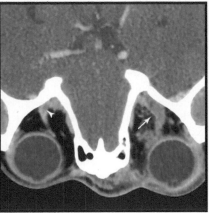 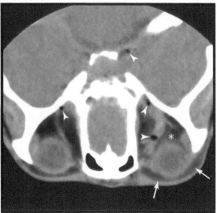

Inspissated material with gas bubbles in the sphenoid sinus (*white arrowheads*), fat stranding reflecting right intraorbital edema (*curved arrow*), lack of normal enhancement in the right cavernous sinus (*arrowhead*) around the right cavernous carotid artery (*), and heterogeneous enhancement in the left cavernous sinus caused by partial thrombosis (*arrow*).

Expansion of thrombosed right superior ophthalmic vein (*arrow*) with a nonocclusive filling defect representing partial thrombosis of the left superior ophthalmic vein (*arrowhead*).

Repeat post-contrast CT several days later shows increased edema in the right preseptal (*arrows*) and retro-orbital fat (*). Gas is visible now in the right and left superior ophthalmic veins and in the right cavernous sinus (*arrowheads*).

**Fig. 31.3** Computed tomographic findings in cavernous sinus thrombosis secondary to acute sphenoid sinusitis. *CT,* Computed tomography. (Reproduced with permission from Hurley MC, Heran MKS: Imaging studies for head and neck infections, *Infect Dis Clin North Am* 2007;21:305-353.)

accuracy rate improves to 87%. Thus endoscopically directed middle meatal cultures is an acceptable method for the microbiologic diagnosis of acute bacterial sinusitis.

## Other Investigations

Mucosal biopsy and histopathologic studies may be indicated in patients with recurrent or chronic sinusitis to rule out malignancy or invasive fungal infections. Skin testing with environmental or seasonal allergens such as pollens, dust mites, animal danders, and saprophytic fungi is warranted to rule out allergy. Serum immunoassays for allergen-specific IgE antibodies may be useful in some cases, particularly *Aspergillus*-precipitating antibodies for patients with suspected allergic fungal sinusitis. Overall, the sensitivity of immunoassays ranges from 70% to 75% compared with skin testing. Studies for immune function in patients with chronic or recurrent sinusitis may include total quantitative immunoglobulins (IgG, IgA, IgM, and IgG subclasses), preimmunization and postimmunization specific antibody responses to tetanus toxoid or pneumococcal vaccine, and T-cell subset analysis.

## CLINICAL MANAGEMENT AND DRUG TREATMENT

The goals of management of sinusitis are (1) to optimize symptomatic relief and facilitate drainage of congested sinuses, (2) to eradicate infection and restore sinus function, (3) to prevent recurrence and complications, and (4) to reduce antibiotic overuse in ill-defined upper respiratory tract infections and minimize the development of antibiotic resistance. Various management options exist (Box 31.2). Because acute sinusitis is most commonly caused by respiratory viruses that spontaneously resolve after 7 to 10 days, supportive and symptomatic management should suffice. Antimicrobial therapy is reserved for those with severe, persistent, or worsening symptoms despite 7 to 10 days of "watchful waiting" with symptomatic management. Surgical treatment is considered when medical management has failed or complications are suspected.

---

**BOX 31.2  Management Options for Sinusitis**

**Supportive and symptomatic**
- Intranasal glucocorticosteroids
- Sinus irrigation
- Humidification and hydration
- Analgesics and antipyretics
- Decongestants (topical and systemic)
- Mucolytic agents

**Antimicrobial (oral or parenteral)**

**Surgical (functional endoscopic sinus surgery)**
- Correct intranasal, ostial, or other abnormalities
  - Turbinectomy
  - Septal surgery
  - Polypectomy
  - Adenoidectomy
  - Tonsillectomy
- Promote drainage
  - Intranasal antrostomy
  - Ethmoidotomy
  - Frontal sinus trephination
- Remove diseased tissue
  - Caldwell-Luc operation
  - Ethmoidectomy
  - Frontal sinus obliteration
  - Sphenoidectomy

---

## Supportive and Symptomatic Management
### Intranasal and Systemic Corticosteroids

Topical corticosteroid nasal sprays reduce inflammation and edema in the nasal mucosa and may be beneficial as adjunctive therapy for acute bacterial sinusitis in both adults and children. Several double-blind, placebo-controlled, randomized trials found that the use of intranasal glucocorticoids (mometasone, fluticasone, flunisolide, or beclomethasone) alone or as adjuvant therapy to antibiotics increased the rate of symptom response compared with placebo. Adverse effects were minimal. Intranasal steroids may also benefit patients with chronic or recurrent sinusitis.

Few randomized controlled trials of systemic corticosteroids for acute sinusitis have been published. A recent systematic review found that monotherapy with systemic corticosteroids was ineffective. Used in combination with antibiotics, only minimal benefit with short-term symptomatic relief was realized. Although adverse effects were minimal, systemic corticosteroids are not currently recommended for the management of acute sinusitis.

## Sinus Irrigation

Saline irrigation of the nasal cavity has been found in several randomized controlled trials to improve clinical symptom scores and radiographic findings in patients with chronic sinusitis. Hypertonic and physiological saline appear equally effective in children and adults.

## Decongestants and Mucolytic Agents

Nasal and oral decongestants are α-adrenergic agonists that shrink the erectile vascular tissue of the nasal turbinates and theoretically may help to relieve osteomeatal and nasal obstruction. However, there is conflicting evidence as to whether these agents functionally improve aeration of the sinuses. Furthermore, prolonged use of nasal decongestants beyond 3 or 4 days can cause rebound vasodilatation (rhinitis medicamentosa). Mucolytic agents (e.g., guaifenesin) thin nasal secretions and theoretically may promote drainage. There is currently insufficient evidence to recommend either nasal decongestants or mucolytic agents as adjunctive therapy in acute or chronic sinusitis.

## Antihistamines

The role of oral antihistamines as adjunctive treatment of acute sinusitis without coexisting allergic rhinitis remains to be determined. There are theoretic concerns that antihistamines might impede mucociliary clearance by drying the mucous membrane and thickening nasal secretions.

## Miscellaneous Measures

The symptomatic value of hydration, warm facial packs, steam baths, antipyretics, and analgesics (acetaminophen or nonsteroidal antiinflammatories) for comfort and pain relief is self-evident. However, the potential benefit of heated humidification, zinc lozenges, or *Echinacea* preparations remains unclear.

## Antimicrobial Therapy

Even though only 0.5% to 2% of patients with acute sinusitis develop a bacterial infection, over 80% are prescribed an antibiotic. This misuse of antimicrobial therapy not only exposes patients to unnecessary adverse effects with escalating healthcare costs, it is also the major driving force for the emergence of antibiotic resistance among respiratory pathogens. The appropriateness and efficacy of antimicrobial therapy in acute bacterial sinusitis have been critically evaluated in various double-blind, placebo-controlled clinical trials. However, until more reliable methods for the diagnosis of acute bacterial sinusitis become available, the efficacy of antimicrobial therapy will likely be under-valued due to dilution by the presence of acute viral sinusitis misdiagnosed as acute bacterial sinusitis.

### Meta-Analyses of Placebo-Controlled Randomized Trials

A recent systematic review examined the benefit of antimicrobial therapy for acute bacterial sinusitis in 3057 patients enrolled in 15 double-blind, placebo-controlled, randomized trials. Main findings were: (1) without antibiotics, 46% of patients were cured after 1 week and 64% after 2 weeks; (2) antibiotics can shorten the duration of symptoms, but only 5 to 11 more people per 100 will be cured faster if they received antibiotics instead of a placebo or no treatment, while 13 more per 100 will experience side effects; (3) cure rates with antibiotics were higher if purulent secretions were present in nasal passages, or if fluid level or total opacification in any sinus was found on CT; and (4) 5 fewer people per 100 will experience clinical failure if they received antibiotics instead of placebo or no treatment. The high rate of spontaneous resolution (≈70%) in the control patients supports the strategy of "watchful waiting" with symptomatic management in patients with acute sinusitis. Only severely ill patients (high fever with temperature ≥ 39°C, periorbital edema, purulent nasal discharge, facial pain, or headache), those with suspected orbital or intracranial complications, and those whose symptoms persist or worsen after 7 to 10 days of symptomatic management should be considered for empirical antimicrobial therapy.

### Empirical Antimicrobial Regimens

Antibiotic therapy is primarily directed against *H. influenzae* and *S. pneumoniae*, whereas in children, *M. catarrhalis* should also be covered. Amoxicillin-clavulanate is preferred over amoxicillin alone based on the increasing prevalence of β-lactamase–producing respiratory pathogens, particularly *H. influenzae* (25% to 35%) and *M. catarrhalis* (90%). Standard dose amoxicillin-clavulanate should suffice *S. pneumoniae* infections since the prevalence of penicillin-nonsusceptible *S. pneumoniae* has sharply declined since the widespread use of pneumococcal vaccines. A 5 to 7 day course of therapy is recommended. Doxycycline is recommended for adults with type-1 penicillin hypersensitivity. In children, levofloxacin (for type-1 penicillin hypersensitivity) or an oral cephalosporin (cefuroxime or cefpodoxime) (for non–type-1 penicillin hypersensitivity) may be used. TMP-SMX is no longer recommended because of high rates of resistance among both *S. pneumoniae* and *H. influenzae* (30% to 40%). Similarly, newer macrolides (azithromycin or clarithromycin) are no longer recommended because of high rates of resistance among *S. pneumoniae* (30%). High-dose amoxicillin-clavulanate is recommended for patients from geographic regions with high endemic rates of penicillin-nonsusceptible *S. pneumoniae*. In patients with odontogenic sinusitis, treatment should be directed at mixed anaerobes and streptococci. The respiratory fluoroquinolones (levofloxacin or moxifloxacin) or a parenteral third-generation cephalosporin (cefotaxime or ceftriaxone) should be reserved for patients who have severe symptoms, who are immunocompromised, or whose condition has not responded clinically despite 3 to 5 days of empirical therapy with first-line agents (Table 31.3).

### Treatment Failure

Patients in whom symptoms persist or worsen despite a 3 to 5 day course of first-line antibiotic therapy should receive an additional course of high-dose amoxicillin-clavulanate or with respiratory fluoroquinolones. If this regimen fails, imaging studies should be performed and a sinus aspirate or endoscopically directed middle meatal culture obtained, and further antimicrobial therapy should be guided by culture and susceptibility data. Patients with subacute or recurrent symptoms that fail to respond to the earlier-described approach should be investigated for structural abnormalities by sinus endoscopy and for comorbid conditions such as cystic fibrosis, Churg-Strauss vasculitis, Wegener granulomatosis, or immunodeficiency syndromes. Repeated antral lavage in addition to antibiotics may be required before consideration of a surgical approach (Fig. 31.4).

**TABLE 31.3 Empirical Antimicrobial Regimens for Acute and Chronic Bacterial Sinusitis**

| Stratification | First Line (Daily Dose)[a,b] | Second Line (Daily Dose)[a,b] |
|---|---|---|
| **Acute Sinusitis in Adults** | | |
| Initial empirical therapy | Amoxicillin-clavulanate (500 mg/125 mg tid, or 875 mg/125 mg q12h) | Doxycycline (200 mg qd on day 1, then 100 mg qd) |
| β-Lactam allergy | Doxycycline (200 mg qd on day 1, then 100 mg qd) | Levofloxacin (500 mg qd) or Moxifloxacin (400 mg qd) |
| Failed initial therapy | Amoxicillin-clavulanate (2000 mg/125 mg bid) Levofloxacin (500 mg qd) Moxifloxacin (400 mg qd) | Ceftriaxone (1–2 g IV q24h) or Cefotaxime (2 g IV q6h) |
| Hospitalized patients | Ceftriaxone (1–2 g IV q12-24h) Cefotaxime (2 g IV q6h) | Levofloxacin (500 mg qd) or Moxifloxacin (400 mg qd) |
| **Chronic Sinusitis in Adults** | | |
| Empiric therapy | Amoxicillin-clavulanate (500 mg/125 mg bid × 3 weeks) plus clindamycin (450 mg tid × 3 weeks) | Levofloxacin (500 mg qd × 3 weeks) Moxifloxacin (400 mg qd × 3 weeks) |
| **Acute Sinusitis in Children** | | |
| Initial empirical therapy | Amoxicillin-clavulanate (45 mg/kg/day bid) | Cefuroxime (30 mg/kg/day bid) or cefpodoxime (10 mg/kg/day bid) |
| β-Lactam allergy | Levofloxacin (10–20 mg/kg/day q12–24h) (type I hypersensitivity) | Cefuroxime (8 mg/kg/day bid) or cefpodoxime (10 mg/kg/day bid) (non-type I hypersensitivity) |
| Failed initial therapy | Amoxicillin-clavulanate (90 mg/kg/day bid) | Cefixime (8 mg/kg/day bid) or cefpodoxime (10 mg/kg/day bid), each plus clindamycin (30–40 mg/kg/day tid) |
| Hospitalized patients | Ceftriaxone (50 mg/kg/day IV q12h) or Cefotaxime (100–200 mg/kg/day IV q6h) | Levofloxacin (10–20 mg/kg/day IV q12-24h) |
| **Chronic Sinusitis in Children** | | |
| Empiric therapy | Amoxicillin-clavulanate (45 mg/kg/day bid × 3 weeks) plus clindamycin (30 mg/kg/day qid × 3 weeks) | Levofloxacin (10–20 mg/kg/day qd × 3 weeks) |

[a]Oral dose, unless specified otherwise.
[b]Duration of therapy usually 5 to 7 days, unless specified otherwise.
*IV,* Intravenously.

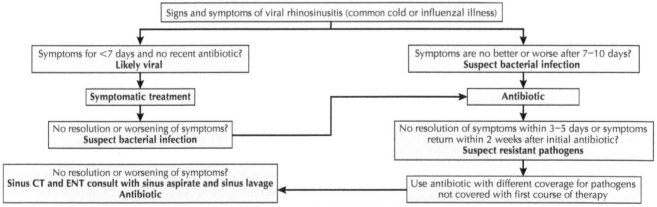

**Fig. 31.4** Algorithm for the management of acute sinusitis. (Modified from Brooks I, Gooch WM 3rd, Jenkins SG, et al: Medical management of acute bacterial sinusitis: recommendations of a clinical advisory committee on pediatric and adult sinusitis, *Ann Otol Rhinol Laryngol* 2000;182[suppl]:2-20.)

## Fungal Sinusitis

Antimicrobial therapy for fungal sinusitis is required only if the disease is invasive or the patient is severely immunocompromised and the risk of progression is high. Noninvasive disease usually responds to surgical debridement alone. Most immunocompromised patients with invasive fungal infection will require a combination of surgery and high-dose intravenous amphotericin B. Voriconazole is more effective than amphotericin B in invasive aspergillosis and *P. boydii* infections.

## Subacute and Chronic Sinusitis

Antibiotic therapy alone is of questionable value in chronic sinusitis. Medical treatment options should begin with topical intranasal steroids and saline irrigations. Surgical procedures to correct sinus abnormalities, relieve obstruction of the osteomeatal complex, and improve drainage are often required. In choosing an antibiotic for chronic sinusitis, the initial coverage should include *S. aureus* and β-lactamase–producing organisms, including anaerobic species. Clindamycin may be added if anaerobic organisms are suspected. Subsequent antimicrobial selection (including antifungal agents) should be guided by endoscopically directed middle meatus or sinus puncture culture results. Many of the second-line antibiotics used for acute bacterial sinusitis are also effective in chronic sinusitis, but the course of treatment is generally prolonged to 4 to 6 weeks.

## Surgical Management

The indications for surgical treatment include refractory sinusitis caused by osteomeatal obstruction, invasive fungal infections, and orbital or intracranial complications. Traditional open sinus procedures have been largely replaced by functional endoscopic sinus surgery (FESS), which facilitates drainage by removing any soft tissue causing obstruction of the ostia and allows improved access to the ethmoid and sphenoid sinuses. Approximately 80% to 90% of patients undergoing FESS experience significant improvement in sinusitis symptoms and function, and complication rates are low (<1%).

## COMPLICATIONS

Suppurative complications of acute and chronic sinusitis are relatively rare in the postantibiotic era. Extension of infection from the maxillary or ethmoid sinuses into the adjacent structures may result in osteomyelitis of the facial bones, including prolapse of the orbital antral wall with retro-orbital cellulitis, proptosis, and ophthalmoplegia. Direct intracranial extension from the maxillary sinus is rare, except in rhinocerebral mucormycosis. Frontal sinusitis may lead to osteomyelitis of the frontal bones (Pott's puffy tumor) or thrombosis of the superior sagittal sinus. Intracranial extension of infection from the ethmoid or sphenoid sinusitis may result in cavernous sinus thrombosis, epidural or subdural empyema, meningitis, and brain abscess (see Fig. 31.2). The diagnosis and management of these life-threatening complications require an aggressive and multidisciplinary approach.

## PREVENTION AND CONTROL

Apart from pneumococcal and *H. influenzae* vaccines, there are currently no effective preventive measures for acute or chronic sinusitis. Efforts should be directed to early and aggressive treatment of acute sinusitis, surgical correction of anatomic deformities of the sinus ostia, promotion of good dental hygiene, and effective control of underlying allergic manifestations.

## EVIDENCE

Autio TJ, Koskenkorva T, Narkio M, et al. Diagnostic accuracy of history and physical examination in bacterial acute rhinosinusitis. *Laryngoscope* 2015;125(7):1541-1546. *This prospective study evaluated symptom progression as well as physical findings among military recruits with sinus-puncture confirmed acute bacterial sinusitis. The presence of symptoms or their change had little value in identifying acute bacterial sinusitis. In contrast, physical findings at 9 or 10 days after onset, particularly the presence of purulent secretions in the nasal passages or posterior pharynx, predicted acute bacterial sinusitis accurately.*

Karageorgopoulos DE, Giannopoulou KP, Grammatikos AP, et al. Fluoroquinolones compared with β-lactam antibiotics for the treatment of acute bacterial sinusitis: a meta-analysis of randomized controlled trials, *CMAJ* 2008;178:845-854. *This meta-analysis evaluated the effectiveness and safety of newer fluoroquinolones compared with β-lactams for the treatment of acute bacterial sinusitis from eight randomized controlled trials. The authors concluded that the newer fluoroquinolones cannot be endorsed as first-line therapy because they conferred no benefit over β-lactam antibiotics.*

Lemiengre MB, Van Driel ML, Merenstein D, et al. Antibiotics for acute rhinosinusitis in adults. *Cochrane Database Syst Rev* 2018;9:CD006089. *This comprehensive review concluded that the potentially beneficial therapeutic effect of antibiotic therapy in clinically diagnosed acute sinusitis in adults is small and must be balanced by the potentially harmful adverse effects. Only 5 more people per 100 would be cured faster if they received antibiotics instead of a placebo.*

Venekamp RP, Thompson MJ, Hayward G, et al. Systemic corticosteroids for acute sinusitis. *Cochrane Database Syst Rev* 2014;3:CD008115. *This meta-analysis concluded that monotherapy with systemic corticosteroids was ineffective, while adjuvant therapy in conjunction with antibiotics was modestly beneficial in short-term symptom relief, with number needed to treat for additional benefit in 7 patients.*

Zalmanovici A, Yaphe J. Steroids for acute sinusitis. *Cochrane Database Syst Rev* 2007;4:CD005149. *This Cochrane meta-analysis assessed the effectiveness and safety of topical steroids for the treatment of acute sinusitis in four double-blind, placebo-controlled, randomized trials. The authors concluded that intranasal corticosteroids are useful as monotherapy or adjuvant therapy with antibiotics for acute sinusitis with modest but clinically important benefits and relatively minor adverse effects.*

## ADDITIONAL RESOURCES

Benninger MS, Payne SC, Ferguson BJ, et al. Endoscopically directed middle meatal cultures versus maxillary sinus taps in acute bacterial maxillary rhinosinusitis: a meta-analysis. *Otolaryngol Head Neck Surg* 2006;134:3-9. *Endoscopically directed middle meatal cultures are highly sensitive (81%) and specific (91%), with a positive predictive value of 83%, negative predictive value of 89%, and overall accuracy of 87%.*

Chow AW, Benninger MS, Brook I, et al. IDSA clinical practice guideline for acute bacterial rhinosinusitis in children and adults. *Clinical Infectious Diseases* 2012;54(8):e72-e112. *This clinical practice guideline provides recommendations for empirical antimicrobial therapy of acute bacterial sinusitis based upon current status of antimicrobial resistance patterns, and a shortened duration of therapy. An algorithm for diagnosis and treatment is presented.*

Ebell MH, McKay B, Dale A, et al. Accuracy of signs and symptoms for the diagnosis of acute rhinosinusitis and acute bacterial rhinosinusitis. *Ann Fam Med* 2019;17(2):164-172. *This systematic review found that only one-third of patients with clinically diagnosed acute sinusitis have bacterial sinusitis. The overall clinical impression, fetid odor in the breath, and pain in the teeth are the most reliable findings pointing towards acute bacterial sinusitis.*

Harris AM, Hicks LA, Qaseem A. Appropriate antibiotic use for acute respiratory tract infection in adults. *Ann Intern Med* 2016;165(9):674. *This article from the American College of Physicians and the US Center for Disease Control provides five high-value care advices for appropriate antimicrobial use in patients with respiratory tract infections.*

Rosenfeld RM, Piccirillo JF, Chandrasekhar SS, et al. D. Clinical practice guideline (update): adult sinusitis. *Otolaryngol Head Neck Surg* 2015;152(2):S1-S39. *This updated guideline from the American Academy of Otolaryngology-Head and Neck Surgery Foundation summarizes the rationale, purpose, and key recommendations that address diagnostic accuracy for adult rhinosinusitis, appropriate use of ancillary tests to confirm diagnosis, and the judicious use of systemic and topical therapy.*

# 32

# Acute Otitis Media

*Blaise L. Congeni*

## ABSTRACT

Acute otitis media (AOM) is the most common bacterial infection seen in pediatric patients, and treatment of AOM is the most common reason children receive antibiotics. Physician visits and antibiotic use for otitis media, however, have decreased in the last decade. AOM follows eustachian tube dysfunction, which is most often seen with a viral upper respiratory infection. Consequently, those organisms that are part of the normal flora of the nasopharynx are the major pathogens responsible for AOM. Although the cause has remained relatively constant over the last few decades, changes secondary to immunization and antibiotic pressure may now be occurring. Accurate diagnosis is essential for appropriate management. Of all the features associated with AOM, establishing appropriate therapy has recently undergone the greatest change; current guidelines now advocate that a substantial proportion of patients might appropriately be observed and not treated with antibiotics. These guidelines also now emphasize that more stringent criteria be applied for diagnosis, more attention must be paid to pain control, and patients must have access to antibiotics within 48 to 72 h if the patient fails to improve. This option is now recommended though only in the context of shared decision making and only in certain patients meeting very specific clinical criteria supported by the literature.

## CLINICAL VIGNETTE

An 18-month-old white male was seen in the office of his primary care physician. He had recently had three episodes of AOM within the last 3 months. In his most recent episode he was treated with high-dose amoxicillin and he completed the amoxicillin one week prior to this current visit. He presented with irritability, low-grade fever of 100.8, and poor oral intake. Since his last visit he had lost 5 ounces. Physical exam revealed that both tympanic membranes were erythematous and mildly bulging. He was started on amoxicillin-clavulanate at 90 mg/kg/day in two divided doses. Close follow-up was recommended and 72 hours later he returned to the office without much improvement. He continued to sleep poorly and eat poorly. Physical exam revealed continued bulging of both tympanic membranes. The patient was then treated for 3 consecutive days with ceftriaxone, 50 mg/kg/day in a single daily dose. This was administered intramuscularly with lidocaine. At follow-up 1 week later the patient was markedly improved. The tympanic membranes were now normal and the patient was again gaining weight. Following completion of therapy quantitative immunoglobulins were obtained which were normal, the child's parents were now able to have in-home childcare with only two other children, and referral was also made to otolaryngology for discussion of other options.

## MAGNITUDE OF THE PROBLEM, RISK FACTORS, PATHOGENESIS, AND ETIOLOGY

Virtually all children experience at least one middle ear infection during the first decade of life. AOM, however, is not limited to pediatric patients, and the disease seen in adults is similar to that seen in pediatrics with regard to pathogenesis, etiology, and treatment. In addition, at least in children, physician visits and antibiotic use for AOM have seen a decline in the last decade. This decline may be in part related to widespread use of pneumococcal conjugate vaccine (PCV) 7 and PCV 13 along with use of the influenza vaccine. The more stringent criteria advocated for the diagnosis of otitis media may also have contributed to this decline. However, AOM remains the most common reason children visit a physician, and it is also the second most common reason for a surgical procedure in the pediatric population, behind only circumcision. The persistent effusion seen in the middle ear after an episode of AOM is responsible for significant hearing loss and delay in development of language skills.

Recurrent and persistent disease is also more commonly seen now. There are several risk factors for acute and recurrent otitis media (Box 32.1). By age 3, approximately one-third of children will have been identified as otitis prone, one-third as having no trouble with otitis media, and one-third as occasionally infected. Those with recurrent disease contribute greatly to antibiotic usage and development of antibiotic resistance.

Anatomic, physiologic, and immunologic factors all contribute to this epidemiology. Furthermore, the eustachian tube is shorter and more horizontally positioned in children (Fig. 32.1). This makes it easier for nasopharyngeal flora to gain access to the middle ear.

Eustachian tube malfunction is generally the final step that leads to nasopharyngeal organisms gaining access to the middle ear. Mucosal edema, which can lead to obstruction of the eustachian tube, impairs drainage of secretions of the middle ear. Bacteria that then gain access to the middle ear can readily multiply. Respiratory syncytial virus, parainfluenza virus, and influenza virus are the viruses most commonly responsible for causing impairment of eustachian tube function. Less commonly, other factors play a pivotal role in the development of AOM, including allergy, genetic factors (Native American or Eskimos), or immunologic factors. The role of daycare centers in the epidemiology of AOM and recurrent disease is a significant issue as well. Daycare attendance in the United States has increased substantially, and currently half of US children attend a daycare center regularly. The larger the number of children in attendance at the center, the greater the likelihood of exposure to a wide variety of viral pathogens. This can obviously lead to the increase in diagnosis of AOM seen in such children (Fig. 32.2). Besides daycare attendance, several other factors appear to be associated with recurrent episodes of otitis media, otitis proneness. These associations include male sex, family history, atopic disease, age at first diagnosis (the younger the age the greater the risk), and non-Hispanic white race.

The cause of AOM naturally reflects the pathophysiology. *Streptococcus pneumoniae*, nontypable *Haemophilus influenzae,* and *Moraxella catarrhalis* are the leading pathogens. Whereas viruses play a

## BOX 32.1   Major Risk Factors for Acute and Recurrent Otitis Media

- Onset of otitis media in infancy
- Male gender
- Sibling with recurrent acute otitis media
- Bottle feeding only
- Parents who smoke at home
- Daycare attendance
- Craniofacial anomalies

pivotal role in predisposing the host, middle ear taps of children with AOM rarely yield a viral pathogen alone.

Since 2000, universal immunization of all infants with conjugated PCV has been recommended. Since the introduction of this vaccination, several changes in the epidemiology of AOM have been seen. There have been several benefits, although the reduction in AOM overall has been modest compared with the reduction in invasive disease caused by S. pneumoniae. Introduction of the 13 Valent conjugated pneumococcal vaccine (PCV-13) occurred in 2010. Subsequent to introduction of these vaccines, there appears to be some decline in the occurrence of otitis media due to vaccine serotypes. Vaccine serotypes continue to more likely be drug resistant. Replacement of these vaccine types with infections with nontypable H. influenzae or nonvaccine strains of S. pneumoniae has been reported.

## CLINICAL FEATURES AND DIAGNOSIS

Ear pain is the most specific finding associated with AOM. This finding occurs in approximately two-thirds of patients; however, it is less often seen in younger patients. This may relate to an inability to adequately verbalize this manifestation. In general, symptoms in younger patients tend to be less specific, as is the case in other pediatric infections. Patients younger than 1 year old are asymptomatic about half of the time. Likewise, otorrhea is a highly specific but infrequently occurring finding and is often associated with sudden relief of pain. Less specific symptoms include coryza, irritability, poor feeding, sleep disturbance, fever, hearing loss, "tugging" at the ears, and difficulty with balance.

The initial presentation is helpful in predicting the causative organism. Disease caused by S. pneumoniae or S. pyogenes is more likely to manifest suddenly, with high fever and severe otalgia. Spontaneous resolution is also less likely, and complications are more likely to occur with infection with these organisms. On the other hand, M. catarrhalis is associated with the greatest likelihood of spontaneous resolution and the mildest disease. Disease caused by H. influenzae falls between these two extremes. Recent studies have not consistently demonstrated a correlation between symptoms with the initial diagnosis of AOM. Ear pain, otalgia, is frequently associated with AOM. American Academy of Pediatrics (AAP) guidelines continue to emphasize the need to inquire about the presence of pain and consider options for symptomatic treatment of otalgia, including topical therapies.

Making the diagnosis of AOM can be challenging. The diagnosis is almost invariably confirmed in clinical practice by physical examination alone (Fig. 32.3). Although the symptoms are often nonspecific, the patient may be very uncooperative, and adequately restraining such a patient is essential. Even with adequate restraint, additional difficulties with examination may arise. The canals may be occluded with cerumen, and removal is difficult. Moreover, the practitioner must have appropriate equipment, which means adequate illumination and a proper seal with the speculum on the otoscope. Disposable specula are not as satisfactory in producing an adequate seal. Only when all these conditions are satisfied is an adequate inspection and attempt at insufflation (pneumatic otoscopy) possible.

In 2019, the AAP adopted more stringent criteria for the diagnosis of AOM. More emphasis is placed on the physical exam especially with regard to the presence of a bulging tympanic membrane. The presence of moderate or severe bulging of the TM or the new onset of otorrhea not due to external otitis confirms the diagnosis of AOM. Mild bulging of the TM, and recent onset of ear pain or intense erythema of the tympanic membrane is a second method recommended for confirmation of the diagnosis.

Much of the difficulty associated with establishing the diagnosis of AOM results from the fact that a middle ear effusion that is not purulent can be seen as part of a viral upper respiratory infection. Such an effusion is referred to as otitis media with effusion (OME). This can also occur before AOM or after therapy for AOM. Antibiotic therapy is of no benefit for the patient with OME, and distinguishing OME from AOM can be especially challenging.

## CLINICAL MANAGEMENT AND DRUG TREATMENT

Several important concepts need to be considered when deciding on the appropriate management of the patient with AOM. The most important concept that drives the entire discussion is that AOM is so often self-limited. With or without appropriate antibiotic treatment, patients generally improve. This is referred to as the "Pollyanna phenomenon." This greatly complicates the interpretation of studies that look to compare different antibiotic regimens. When a clinical endpoint is used (e.g., physical examination of the ear), significant differences are not seen even when a placebo-treated group is included, unless very large numbers of subjects are included. Demonstrating sterilization of the middle ear by using a repeat ear tap is therefore a preferred method of studying treatment options for patients with otitis media. These issues need to be kept in mind when studies are evaluated.

The other major issue that must be conceded early on is that it is likely that many of the studies regarding AOM may be flawed. Studies that rely only on physical examination findings for establishing the diagnosis must begin with a stringent definition of AOM. Confirmation of the diagnosis, skill on the part of the enrolling physician, and the presence of sufficient patients at risk for failure (such as those younger than 2 years of age) are also essential. Developing recommendations from flawed studies is particularly problematic with AOM because spontaneous resolution is so common. Finally, the goal or goals of antibiotic therapy must be clearly understood. Complications such as mastoiditis or cholesteatoma are rare, even when antibiotic therapy is initially withheld, as long as patients are followed closely and appropriate therapy is offered if symptoms do not improve (i.e., "watchful waiting"). Younger patients, however, being at greater risk for complications and failure, will generally demonstrate a greater benefit from antibiotic therapy. Other outcomes, such as duration of pain, must also be considered when considering withholding of antibiotics. In other words, the assumption that antibiotic therapy results in more rapid resolution of pain, as some studies suggest, might shift the equation toward antibiotic therapy for some physicians.

Short- and long-term prognoses are good with or without antibiotic therapy. Having said that, studies that have included a placebo group consistently demonstrate lower failure rates in the group treated with antibiotics. In most patients treated with antibiotics, recurrent disease is caused by new pathogens or new serotypes of the same pathogen. The major severe complications include mastoiditis, facial palsy, otorrhea, brain abscess, cholesteatoma, and epidural abscess. All of these complications have dramatically decreased in the antibiotic era, and no convincing data have emerged to suggest that that has changed even with more children having antibiotics withheld ("watchful waiting").

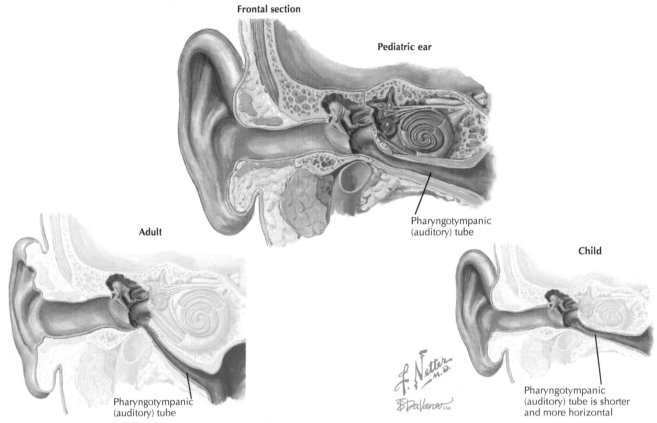

**Fig. 32.1** Eustachian tube.

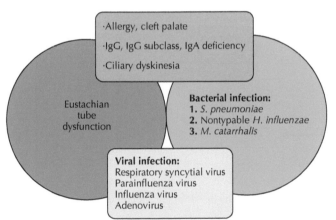

**Fig. 32.2** Pathogenesis of otitis media. *IgA,* Immunoglobulin A; *IgG,* immunoglobulin G.

## Antibiotic Therapy

In short, the 2019 recommendations for diagnosis and management of AOM recommend appropriate antibiotic therapy be given to those patients who are more likely to benefit from therapy. Those are generally patients with severe disease and/or those that are younger in age. Amoxicillin (80 to 90 mg/kg/day in two divided doses; maximum dose 2 to 3 grams per day) has generally been the mainstay of treatment for patients with AOM. Advantages include low cost, acceptable taste, and the safety of a β-lactam. Currently in the United States approximately 10% or more of *S. pneumoniae* isolates are resistant to amoxicillin owing to alteration of penicillin-binding proteins. One-third to one-half of the *H. influenzae* and virtually all of the *M. catarrhalis* strains are resistant because of production

of β-lactamase. In patients with severe disease, concurrent conjunctivitis, or those treated with amoxicillin in the last 30 days should instead be started on amoxicillin with clavulanate, 90 mg/kg/day in 2 divided doses.

Consequently, if, as noted previously, *H. influenzae* is in fact becoming more prevalent with widespread use of the PCV, earlier use of a β-lactamase stable agent such as amoxicillin-clavulanate might make good sense. Patients who fail initial therapy with high-dose amoxicillin at 48 to 72 hours should then be treated either with amoxicillin with clavulanate 90 mg/kg/day in 2 divided doses or ceftriaxone 50 mg/kg/day, I M, for 3 consecutive days (maximum dose 2 grams).

For patients with non–type 1 allergies to penicillins, cephalosporins including cefuroxime, cefpodoxime, or cefdinir may be used. Injectable ceftriaxone may also be useful for patients in whom these regimens fail. For patients with a type 1 history of allergy to a penicillin, the recommended approach is somewhat unsettled. Cross-reactivity between cephalosporins especially second- and third-generation cephalosporins and penicillins is negligible. 2019 guidelines recommend a cephalosporin might be considered in those patients not having a severe or recent penicillin allergy reaction. Because it appears that perhaps as many as 90% of patients with a penicillin or ampicillin allergy can in fact be treated safely with a penicillin such as amoxicillin many hospitals, including ours, have become more aggressive with regard to developing a much more extensive and accurate history. Following that, skin testing or oral challenge are now being utilized (Fig. 32.4). Many of the drugs frequently used do not have a realistic likelihood of being effective. Macrolides should not be used and clindamycin lacks activity against *H. influenzae*. Using regimens that provide appropriate pharmacokinetic and pharmacodynamic parameters is greatly preferred. This means selecting appropriate agents and using appropriate doses for the offending pathogen. β-Lactam antibiotics that achieve levels greater than the minimum inhibitory concentration for over

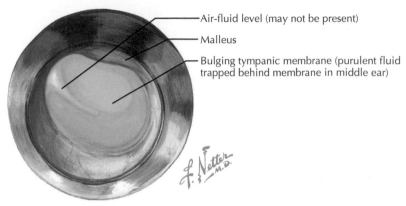

Air-fluid level (may not be present)

Malleus

Bulging tympanic membrane (purulent fluid trapped behind membrane in middle ear)

Otoscopic view demonstrating clinical appearance of otitis media

**Fig. 32.3** Tympanic membranes.

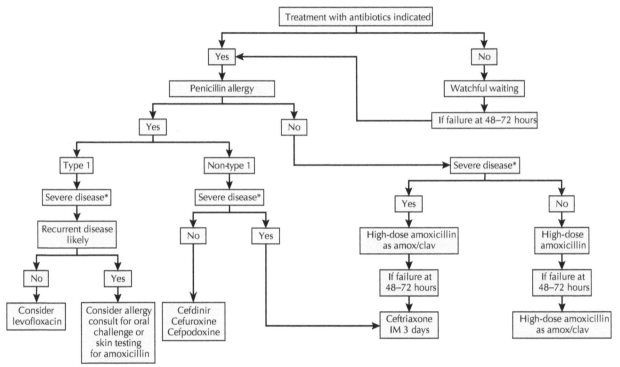

*Severe disease is defined by temperature≥39°C and/or severe otalgia.

**Fig. 32.4** Acute otitis media treatment algorithm. (Data from American Academy of Pediatrics Subcommittee on Management of Acute Otitis Media: Diagnosis and management of acute otitis media, *Pediatrics* 2004;113:1451-1465.)

40% of the dosage interval are more likely associated with a bacteriologic cure. Virtually all patients who experience an early bacteriologic cure improve more rapidly and are much less likely to experience a relapse.

Most patients with AOM are treated with a 10-day course of treatment, excluding parenteral therapy. Although compelling evidence is not available, most experts recommend the standard 10-day course for patients younger than 2 years of age, patients with severe disease, and patients with immunologic disorders. Children 2 to 5 years of age with mild to moderate disease can be treated with 7 days of therapy. Children 6 years of age and older again with mild to moderate disease may be treated with a 5- to 7-day course of treatment.

## Observation Without the Use of Antimicrobial Agents

Because the benefit of antibiotic therapy appears to be modest, recent recommendations have offered options and strategies that do not

initially employ antibiotic therapy. The guidelines developed by the AAP and American Academy of Family Practice look at three characteristics of patients with AOM that are useful in establishing a strategy for treatment. These include age of the patient and severity of the disease. It is important to understand that a consensus does not yet exist as to how robust the evidence is for using this approach. A consensus also does not exist relative to what the stated goal is in treating patients with AOM. If the goal is to reduce antibiotic use in the general population, some would argue that this can best be accomplished by reducing antibiotic use in patients with viral disease. If modest clinical improvement with regard to pain is the goal of the practitioner, the benefit of antibiotic therapy might be seen in a different light, and withholding of antibiotics would be a less attractive option. These recommendations make it clear that medications to reduce pain (analgesics) should be offered if pain is present. It is also clear in the recommendations that the observation or "watchful waiting" option is appropriate only when follow-up can be ensured so that antibiotics can be offered if symptoms persist or worsen beyond 48 to 72 hours.

The decision to treat or to withhold antibiotic therapy and observe may be considered in children 6 to 23 months of age with unilateral disease, ill for less than 48 hours, temperature less than 39 degrees without severe signs and symptoms provided that follow-up and a mechanism for starting antibiotic therapy if symptoms persist at 48 hours is ensured. Children who are 24 months of age and older, with bilateral or unilateral disease and mild symptomatology as noted above may also be managed with antibiotic therapy initially or with observation. Appropriate therapy for pain should be used in all these children. The recent guidelines have emphasized that shared decision-making should be in place during these discussions.

## Special Considerations

Because the conjugate PCV has been widely used since 2000, some investigators have reported an increase in AOM caused by *H. influenzae* along with a decrease in otitis media secondary to *S. pneumoniae*. This may lead some to consider use of a β-lactamase stable drug as an earlier option. In addition, since 2000, concern for emergence of *S. pneumoniae* serotypes not included in the vaccine has increased. Recently, serotype 19A, a multidrug-resistant strain of *S. pneumoniae*, has emerged. Serotype 19A should be considered as a potential pathogen in patients in whom conventional therapy has repeatedly failed. Tympanocentesis to confirm 19A as the cause should be considered. Currently no licensed drug is available for use in pediatric patients, including injectable ceftriaxone, that would reliably treat them. Hence, confirmation that serotype 19A is, in fact, the pathogen must be undertaken.

## PREVENTION

Several strategies should be considered in the pediatric patient with recurrent AOM. The first step is developing reliable data with regard to the frequency of the problem. If intervention is necessary, modifying environmental or host risk factors is generally the appropriate starting place. Encouraging breastfeeding, reducing pacifier use, and reducing parental smoking in the home may all provide some benefit. However, the most significant intervention probably will occur if out-of-home daycare or childcare is used. Studies have consistently demonstrated benefit to moving the child to a care center with fewer children. When parents cannot make alternative arrangements, they should be advised that some benefits may actually occur as a result of attendance at an out-of-home center for childcare.

After environmental risk factors have been evaluated, several medical interventions should be considered in the otitis-prone child. Appropriate immunization has minimal risk and the potential for significant benefit. Currently all children 6 months to 18 years of age should be immunized with an influenza vaccine annually. This has been shown to reduce subsequent episodes of AOM. Universal immunization with the conjugate PCV is also expected to provide some benefit to the otitis-prone patient. Chemoprophylaxis has been shown to provide only very modest benefit and any benefit may be associated with the potential for side effects. In addition any benefit does not appear to persist beyond the use of the preventative agent. A comprehensive discussion of the role of surgical options for such patients is beyond the scope of this presentation. Placement of tympanostomy tubes has the greatest benefit in patients with recurrent AOM associated with persistent middle ear effusion. Adenoidectomy has also been of modest benefit. The rationale revolves around the idea that swollen adenoids may harbor bacteria that can gain access to the middle ear from that location.

## EVIDENCE

Craig WA, Andes D. Pharmacokinetics and pharmacodynamics of antibiotics in otitis media. *Pediatr Infect Dis J* 1996;15:255-259. *This review presents the evidence and basis for making decisions regarding antibiotic selection for AOM. The pharmacokinetic-pharmacodynamic model is used to establish the basis for drug selection.*

Pichichero ME, Casey JR. Emergence of a multiresistant serotype 19A pneumococcal strain not included in the 7-valent conjugate vaccine as an otopathogen in children. *JAMA* 2007;298:1772-1778. *These authors review the presentation and approach to the patient with AOM secondary to S. pneumoniae serotype 19A.*

## ADDITIONAL RESOURCES

American Academy of Pediatrics Subcommittee on Management of Acute Otitis Media. Diagnosis and management of acute otitis media. *Pediatrics* 2004;113:1451-1465. *This is the definitive review for recommendations regarding AOM. The committee presents the evidence for the views expressed.*

American Academy of Pediatrics Subcommittee on Management of Acute Otitis Media. *Pediatrics* 2019;131:e965-e999.

Wald ER. Acute otitis media: more trouble with the evidence. *Pediatr Infect Dis J* 2003;22:103-104. *This is an editorial detailing problems with the evidence that led to the recommendations of the AAP and AAFP.*

Wald ER. To treat or not to treat. *Pediatrics* 2005;115:1087-1089. *This is an editorial that details an alternative view on reasons to treat or withhold antibiotics in patients with AOM.*

# 33

# Pharyngitis

*John R. Bower*

 **ABSTRACT**

"Sore throat" or pharyngitis is one of the most frequent complaints of patients in the acute care setting, accounting for nearly 7 million pediatric and 6 million adult visits each year. On the surface, pharyngitis would appear to pose few challenges to the clinician; the site of infection is both visible and accessible for inspection and culture, and most pharyngeal pathogens are self-limiting respiratory viruses. Yet the diagnosis and management of acute pharyngitis is complicated by bacterial pathogens that occur in 10% to 30% of cases and pose potential risks for serious sequelae such as acute rheumatic fever (ARF) and deep neck infection. Moreover, culture and rapid antigen testing are alone unable to distinguish bacterial infection from normal colonization. This combination of diagnostic uncertainty and the risks for sequelae from bacterial pharyngitis have contributed to the widespread practice of empiric antimicrobial therapy. The consequences of antimicrobial overuse, however—measured by cost, adverse events, and bacterial resistance—have refocused attention on the need for targeted therapy based on an understanding of the epidemiology and diverse clinical presentations of acute pharyngitis.

 **CLINICAL VIGNETTE**

An 11-year-old female presents with a worsening sore throat that began 2 days earlier. She complains of worsening pain and difficulty swallowing but is tolerating oral fluids reasonably well. Her parents report a fever of 101.5°F and no cough or nasal drainage. On physical exam she is alert and in no distress. Her conjunctivae and tympanic membranes are normal. Her posterior pharynx reveals bilateral tonsillar enlargement with a white exudate and surrounding erythema. No palatal petechiae are noted. There is mild bilateral anterior cervical lymphadenopathy and the abdomen is soft and nontender. Her rapid test for group A β hemolytic streptococcus (GAS) is negative. Based on this exam and the rapid antigen test, she is discharged from the office with only symptomatic care. A throat culture for GAS obtained at the time was subsequently negative, and her symptoms resolved after 5 days.

## EPIDEMIOLOGY

### Viral Causes of Pharyngitis

Between 70% and 90% of acute episodes of pharyngitis are viral, involving a wide array of common agents that vary with the season and patient's age (Table 33.1). The leading viral cause of pharyngitis is rhinovirus, the common cold agent. Outbreaks of rhinovirus begin in September, with the start of school, and pass efficiently from child to adult at home. A second peak of rhinovirus activity appears in the spring.

Adenovirus is another significant cause of viral pharyngitis in both children and adults and involves multiple serotypes, including 1, 2, 3, 5, 6, and 7. During late winter and early spring, approximately 20% of pharyngitis cases are attributable to adenovirus, especially in children younger than 5 years of age. Young adults, including military recruits, constitute another high-incidence group. A unique form of adenoviral infection is pharyngoconjunctival fever, which commonly involves children, especially in outbreaks arising from common-source exposures such as contaminated swimming pools.

In the summer and fall months, the enteroviruses (including enterovirus, echovirus, and coxsackievirus groups) frequently cause a nonexudative pharyngitis. These viruses, particularly the coxsackieviruses, can also trigger discrete ulcerative lesions in the oropharynx known as herpangina.

Pharyngitis has long been recognized as an important feature of mononucleosis, caused by the Epstein-Barr virus (EBV) in adolescents and young adults. A lesser known cause of exudative pharyngitis in this age group is herpes simplex virus (HSV). Among college students complaining of sore throat, HSV has been reported as the cause in 6% of cases, with nearly one-third presenting as exudative pharyngitis.

### Bacterial Causes of Pharyngitis

Group A β hemolytic streptococcus (GAS) is not only the most common bacterial pathogen but also the most likely to result in complications of pharyngitis when untreated. In children 5 years of age and older who present with pharyngitis, GAS is identified in up to 30% of the cases, whereas among children younger than 5 years of age the incidence is lower. For adults, the incidence of GAS pharyngitis is 5% to 10%. Asymptomatic carriage of GAS also occurs in 10% of school-age children, a number that likely increases during the winter.

Although less common than GAS, other bacterial causes of pharyngitis may play a role. Up to 7% of throat cultures from symptomatic children and young adults grow groups C or G streptococci; however, the significance of these organisms is unclear, since they are also frequent colonizers. Importantly, there is no evidence that pharyngitis due to organisms other than GAS poses a risk for acute rheumatic fever (ARF). Rare reports linking group C streptococcus to poststreptococcal glomerulonephritis have not been validated.

Among other aerobic nonstreptococcal bacteria causing pharyngitis, *Arcanobacterium haemolyticum* occurs in nearly 3% of cases in adolescents and young adults; it may appear similar to GAS, including a scarlatiniform-like rash. *Neisseria gonorrhoeae* should be considered in individuals with pharyngitis and a history of orogenital sex with a partner at risk for gonorrhea. Rates of *N. gonorrhoeae* pharyngitis vary according to a patient's risk factors and community prevalence but are generally highest in adolescents and young adults. The isolation of *N. gonorrhoeae* from a prepubescent child should always raise suspicion for sexual abuse; however, care should be taken to verify *N. gonorrhoeae*, since nonpathogenic *Neisseria* species are frequently present as normal flora in the oropharynx.

The anaerobe *Fusobacterium necrophorum* has been increasingly associated with a variety of head and neck infections and potentially

## TABLE 33.1 Microbial Causes of Pharyngitis by Type of Pathogen

| | |
|---|---|
| Bacterial | Group A streptococci |
| | Groups C, G streptococci |
| | *Neisseria gonorrheae* |
| | *Arcanobacterium haemolyticum* |
| | *Fusobacterium necrophorum* |
| | *Corynebacterium diphtheriae* |
| Viral | Rhinovirus |
| | Coronavirus |
| | Adenovirus |
| | Parainfluenza virus types 1, 2, and 3 |
| | Influenza A and B viruses |
| | Coxsackievirus |
| | Herpes simplex virus |
| | Epstein-Barr virus |
| | Cytomegalovirus |
| | Human immunodeficiency virus |
| Atypical agents | *Mycoplasma pneumonia* |
| | *Chlamydophila pneumoniae* |

Data from Bisno AL, Gerber MA, Gwaltney JM Jr, et al: Practice guidelines for the diagnosis and management of group A streptococcal pharyngitis, *Clin Infect Dis* 2002;35:113-125.

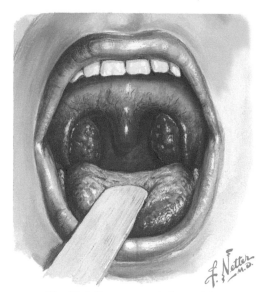

**Fig. 33.1** Streptococcal pharyngitis.

## TABLE 33.2 Clinical and Epidemiologic Characteristics of Group A Beta-Hemolytic Streptococcal Pharyngitis

| | |
|---|---|
| Features suggestive of group A *Streptococcus* as causative agent | Sudden onset |
| | Sore throat |
| | Fever |
| | Headache |
| | Nausea, vomiting, and abdominal pain |
| | Inflammation of pharynx and tonsils |
| | Patchy discrete tonsils |
| | Tender, enlarged anterior cervical nodes |
| | Patient aged 5–15 years |
| | Presentation in winter or early spring |
| | History of exposure |
| Features suggestive of viral cause | Conjunctivitis |
| | Coryza |
| | Cough |
| | Diarrhea |

Data from Bisno AL, Gerber MA, Gwaltney JM Jr, et al: Practice guidelines for the diagnosis and management of group A streptococcal pharyngitis, *Clin Infect Dis* 2002;35:113-125.

life-threatening septic complications. In patients with pharyngitis, *Fusobacterium* has been most commonly identified in 14- to 20-year-olds, where the reported incidence is 13% and possibly higher. By comparison, fewer than 2% of anaerobic throat cultures from younger children grow the organism.

The most common atypical agent associated with pharyngitis is *Mycoplasma pneumoniae*. The organism has been implicated as a cause of pharyngitis in 3% to 23% of cases; among patients diagnosed with *Mycoplasma* pneumonia, nearly half complain of sore throat.

### Noninfectious Causes of Pharyngitis

Noninfectious processes may present as inflammation involving the posterior pharynx, including Stevens-Johnson syndrome, mucositis, toxic shock syndrome, Kawasaki disease, Behçet syndrome, aphthous stomatitis, and PFAPA (periodic fever, aphthous stomatitis, pharyngitis, and adenopathy) syndrome. The latter is frequently misdiagnosed as recurrent tonsillitis in children younger than 5 years of age.

## DIAGNOSIS

### Clinical Manifestations

Pharyngitis is broadly defined as mucous membrane inflammation either localized to the posterior pharynx or contiguous with the adjacent membranes of the posterior nares or larynx. The principal challenge in managing pharyngitis, however, is distinguishing GAS infection from all other infectious causes, particularly viral. Patients with GAS most often present with a sudden onset of sore throat, marked pain with swallowing, and fever (Fig. 33.1). Despite these straightforward signs and symptoms, GAS can be predicted clinically with only 50% accuracy. Yet the extent of pharyngeal inflammation and the presence or absence of other accompanying signs and symptoms remain helpful in distinguishing GAS from non-GAS etiologies (Table 33.2).

Signs and symptoms often associated with GAS include headache, lymphadenopathy, palatal petechiae, nausea, vomiting, and the absence of cough and coryza. Whereas findings that favor viral pathogens include cough, nasal congestion, punctate ulcerative tonsillar lesions, and stomatitis (Fig. 33.2). Certain strains of GAS elicit a scarlatiniform rash by elaborating an erythrogenic toxin. Starting over the neck, the bright fine maculopapular rash extends to the trunk and extremities and is particularly prominent over flexural creases and the perineum. After several days the rash fades, followed by a fine desquamation reminiscent of a sunburn. The lingual papillae become quite prominent, producing the characteristic "strawberry tongue" appearance.

Confusion in the diagnosis of GAS commonly arises when patients without evidence of pharyngeal inflammation are inappropriately tested and have a positive culture or rapid test for GAS. This circumstance most often reflects a GAS carrier state, which does not require treatment unless there is a history of ARF or glomerulonephritis. The significance of GAS in patients with mild pharyngitis is uncertain, although it appears that most of these patients are at lower risk for ARF.

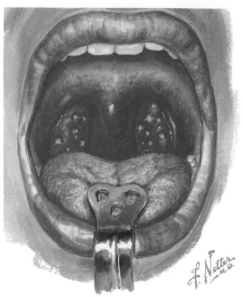

Fig. 33.2 Viral pharyngitis. More discrete punctuate pattern versus the strep pharyngitis.

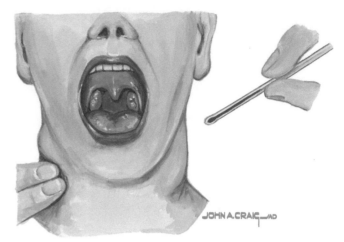

Fig. 33.3 Epstein-Barr virus pharyngitis.

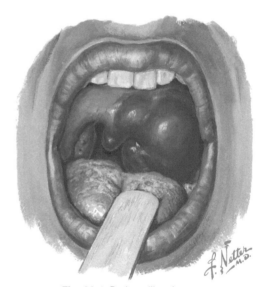

Fig. 33.4 Peritonsillar abscess.

Groups C and G streptococci, *Fusobacterium*, and *A. haemolyticum* can mimic GAS infection, causing exudative pharyngitis, adenopathy, and fever.

Viral infections frequently share the same clinical features as GAS pharyngitis. Adenoviral pharyngitis presents with a tonsillar exudate, adenopathy, and fever over half of the time, mostly in children. EBV infection often presents acutely as sore throat, fever, and cervical lymphadenopathy, and in one half of cases the pharyngitis is exudative (Fig. 33.3). Palatal petechiae may also appear, making EBV even more difficult to differentiate from GAS pharyngitis. Distinguishing clinical features of EBV infection include splenomegaly (present in 50% of patients), generalized lymphadenopathy, and periorbital edema. HSV pharyngitis may manifest with fever, exudative pharyngitis, and cervical lymphadenopathy. This is distinct from the herpetic gingivostomatitis common in young children, which typically involves the lips, gingiva, and tongue.

Enteroviral infections frequently include mild pharyngitis as part of a nonspecific febrile illness. A characteristic presentation of enteroviruses is herpangina, which manifests with fever and discrete vesicular lesions involving the soft palate, tonsils, and surrounding posterior pharynx. Likewise, influenza is often characterized by the abrupt onset of fever and sore throat as well as myalgia, malaise, and dry cough. The rapid onset of cough and constitutional symptoms, coinciding with known influenza activity in the community, is usually sufficient to suggest influenza.

## Complications

Peritonsillar abscess, or quinsy, is the most frequent suppurative complication associated with pharyngitis (Fig. 33.4). It occurs most often in adolescents and young adults and is rare in younger children. Organisms associated with peritonsillar abscesses include GAS and anaerobes, sometimes presenting as a mixed infection. Patients appear unwell, with dysphagia, muffled "hot-potato voice," and fetid breath. The soft palate is displaced on the affected side, with accompanying deviation of the uvula. Lateral pharyngeal abscesses may arise secondary to pharyngitis or peritonsillar abscess and are characterized by swelling

over the lateral aspect of the neck and restriction in neck movement. Asymmetry of the posterior pharyngeal wall may be present, but its absence does not exclude the diagnosis. Pharyngeal infections with *F. necrophorum* may extend locally to cause deep neck infection, septic thrombophlebitis, and potentially life-threatening septic embolization (Lemierre syndrome).

Serious nonsuppurative complications are most always a result of GAS and include ARF and poststreptococcal glomerulonephritis (Fig. 33.5). ARF occurs most commonly in children 5 to 14 years of age and is rare in those under 3 and in adults over 40 years of age. Despite a marked decline in the prevalence of ARF, sporadic outbreaks continue to appear. Poststreptococcal acute glomerulonephritis (PSAGN) occurs most often in children; it typically appears 10 days after infection and is associated with the emergence of nephritogenic strains of GAS. Gross hematuria or hypertension in children should raise concern for PSAGN.

Another complication of GAS pharyngitis is scarlet fever, a toxin mediated process characterized by a distinctive fine papular erythematous "sandpaper" rash. The rash typically begins in major flexural areas, such as the axillae and groin, spreading diffusely to the trunk and extremities. Large flexural creases may develop linear arrays of petechia termed "pastia lines." Another common feature of scarlet fever is a "strawberry" tongue, where the lingual papillae appear red and prominent. As the rash fades, a fine desquamation develops that is often more pronounced over the palms and soles.

## Laboratory Diagnosis

The primary objective in caring for patients with pharyngitis is to avoid the nonsuppurative complication of ARF caused by GAS. Reducing the duration of localized acute pharyngitis remains a secondary concern, since the great majority of affected patients follow a self-limiting course. Key to proper management, therefore, is distinguishing GAS from other infectious agents. The problem, however, is that GAS pharyngitis cannot be reliably diagnosed by clinical evidence alone. Even with exudative pharyngitis, experienced physicians are only 50% accurate in diagnosing GAS. Thus reliance on clinical scoring tools alone for initiating antimicrobial therapy can cause up to 40% of patients to receive unnecessary treatment.

Despite its limitations, clinical judgment remains pivotal in the diagnosis and management of pharyngitis. Clinically based scoring tools, such as the modified Centor score, have proven useful in identifying which patients should be further screened for GAS (Table 33.3). In this approach, patients with clinical scores of 0 to 1 are judged at low risk for GAS and require no testing or antimicrobial treatment, whereas those with scores equal to or greater than 2 require laboratory evaluation for GAS. To aid clinicians in the diagnosis and treatment of GAS, national guidelines have been formulated that incorporate both clinical and laboratory criteria for treatment.

Evaluating a patient with pharyngitis begins by assessing the risk for GAS based on history and physical examination (Fig. 33.6). This risk assessment relies on either clinical judgment alone or clinical judgment in combination with a scoring tool (see Tables 33.2 and 33.3). Patients who are judged to be at low risk for GAS can be managed with symptomatic care and require no further evaluation. Because of their low risk for developing ARF, children with acute pharyngitis who are younger than 3 years of age ordinarily do not require evaluation for GAS.

Patients judged to be at risk for GAS require either a throat culture or a rapid antigen detection test (RADT)—in some cases both—to confirm or exclude the presence of GAS (Fig. 33.7). Because of the high specificity of RADTs, culture and antigen detection are considered equivalent in confirming GAS. Less certain is the role of RADTs in excluding GAS infection, where the reported sensitivity varies from 70% to 95%. Consequently it is recommended that negative RADT results be confirmed by culture. Decisions to waive culture confirmation of a negative RADT should be based on an individual laboratory's internal validation of RADT sensitivity. More recently, studies using nucleic acid amplification testing (NAAT) suggest that this technique may be equivalent to culture in sensitivity.

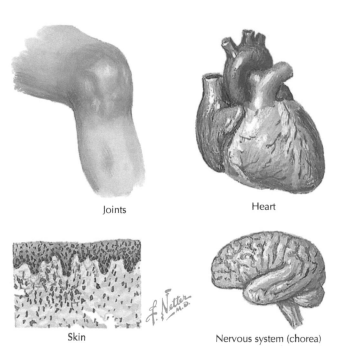

Joints

Heart

Skin

Nervous system (chorea)

**Fig. 33.5** Acute rheumatic fever.

| TABLE 33.3 | Modified Centor Score | |
| --- | --- | --- |
| **Criteria** | | **Points** |
| Temperature >38°C | | 1 |
| Absence of cough | | 1 |
| Swollen, tender anterior cervical nodes | | 1 |
| Tonsillar swelling or exudate | | 1 |
| Age | | |
| 3–14 years | | 1 |
| 15–44 years | | 0 |
| 45 years or older | | −1 |

Adapted from McIsaac WJ, Kellner JD, Aufricht P, et al: Empirical validation of guidelines for the management of pharyngitis in children and adults, *JAMA* 2004;291:1587-1595.

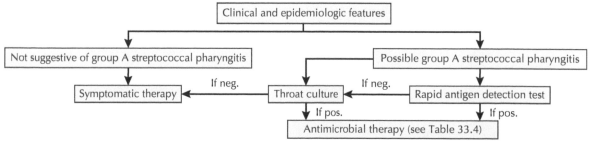

**Fig. 33.6** Risk assessment for group A streptococci. (From Bisno AL, Gerber MA, Gwaltney JM Jr, et al: Practice guidelines for the diagnosis and management of group A streptococcal pharyngitis, *Clin Infect Dis* 2002;35:113-125.)

Equally important to a test's sensitivity is the adequacy of the sample. Obtaining a proper throat swab, especially in children, is frequently challenging and time consuming. An adequate specimen for testing requires that the swab contact both the tonsillar surface and the posterior pharyngeal wall. Results obtained by a buccal swab or blind pass have no value when they prove negative for GAS.

Testing is rarely required for pathogens other than GAS in uncomplicated acute pharyngitis. Isolating non–group A streptococci from patients with uncomplicated pharyngitis is unhelpful, as it may represent normal flora. And routine aerobic throat cultures are insufficient to recover *F. necrophorum* or routinely identify *A. haemolyticum*. If indicated, specimens obtained for virus isolation should be transported in the appropriate viral medium at 4°C. Any concern for *C. diphtheriae* should be addressed to an infectious disease specialist and local health officials.

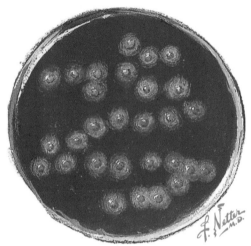

**Fig. 33.7** Example of group A streptococcus on plate with beta-hemolysis.

## TREATMENT

Symptomatic patients with laboratory confirmation of GAS should receive antimicrobial therapy (Table 33.4). The principal benefit of treatment is prevention of ARF, and starting appropriate antimicrobial therapy within 9 days of pharyngitis onset is sufficient to achieve this goal. Early treatment of GAS pharyngitis may shorten the course of illness; however, the benefit of antimicrobials in avoiding suppurative complications is less clear.

Penicillin remains the drug of choice for treatment of GAS pharyngitis because of its narrow spectrum of activity, proven record of efficacy, safety, and cost. Oral penicillin is effective given either twice daily or three times daily for 10 days. Although injectable benzathine penicillin offers single-dose convenience, its use affords no advantage except to aid compliance, and this concern may be addressed by once-a-day oral dosing with amoxicillin.

Amoxicillin is equally effective in the treatment of GAS; for children requiring an oral suspension, it is often preferred due to its improved palatability over penicillin V. Amoxicillin is administered orally as a single daily dose or in divided doses for 10 days. Shorter treatment courses are not recommended, since studies confirming their effectiveness in preventing ARF are lacking.

Cephalosporins have been suggested as first-line therapy for GAS pharyngitis based on studies showing their potential superiority over penicillin in eradicating GAS from the oropharynx. Differences over the interpretation of these studies, together with concerns of cost and drug resistance, have led both the Infectious Diseases Society of America (IDSA) and the American Academy of Pediatrics (AAP) to continue recommending penicillin and amoxicillin as first-line therapies, leaving cephalosporins to be considered as an alternative for patients reporting a moderate penicillin allergy.

Because of increasing reports of macrolide resistance among strains of GAS in the United States and worldwide, macrolides such as azithromycin should be limited to individuals with a history of severe allergy to penicillin. For such patients, clindamycin is also acceptable.

| TABLE 33.4 | **Antimicrobial Therapy for Group A Beta-Hemolytic Streptococcal Pharyngitis** | |
|---|---|---|
| **Drug** | **Dose** | **Duration** |
| **Oral** | | |
| Penicillin V | 250 mg twice or three times daily for children <27 kg<br>500 mg twice or three times daily for children ≥27 kg<br>500 mg twice or three times daily for adolescents and adults | 10 days |
| **Intramuscular** | | |
| Penicillin G: benzathine | 600,000 units for patients <27 kg<br>1.2 million units for patients ≥27 kg | 1 dose |
| Penicillin G: benzathine and procaine mixtures[a] | Varies with formulation | 1 dose |
| **Patients Allergic to Penicillin** | | |
| Azithromycin | 12 mg/kg/day on day 1 (maximum dose 500 mg), then 6 mg/kg/day (maximum dose 250 mg) daily for days 2–5. | 5 days |
| Cephalexin (mild allergy with nonspecific rash)[b]<br>Cefuroxime or cefdinir (mild allergy with urticarial rash)[b] | Varies with formulation | 10 days |

[a]Dose should be based on benzathine component.
[b]Avoid in patients with a severe penicillin allergy.

# EVIDENCE

Alcaide ML, Bison AL. Pharyngitis and epiglottitis. *Infect Dis Clin North Am* 2007;21:449-469. *A standard overview of pharyngitis with emphasis on viral and bacterial pathogens. Well referenced.*

Casey JR, Pichicero ME. The evidence base for cephalosporin superiority over penicillin in streptococcal pharyngitis. *Diagn Microbiol Infect Dis* 2007;57(3 suppl) 39S-45S. *A meta-analysis of 35 studies in support of cephalosporin as the drug of choice for GAS pharyngitis.*

Luo R, Sickler J, Vahidnia F, et al. Diagnosis and management of group a streptococcal pharyngitis in the United States, 2011-2015. *BMC Infect Dis* 2019;19:193. *This retrospective analysis reviewed the management of 11 million patients with respect to antibiotic prescribing, provider type, place of service, and GAS pharyngitis diagnostic testing methods.*

McIsaac WJ, White D, Tannenbaum D, Low DE. A clinical score to reduce unnecessary antibiotic use in patients with sore throat. *CMAJ* 1998; 158:75-83. *The authors provide a useful tool designed for general practitioners that incorporates a modification of the Centor score by including patient age in calculating the score.*

McMillan JA, Weiner LB, Higgins AM, Lamparella VJ. Pharyngitis associated with herpes simplex virus in college students. *Pediatr Infect Dis J* 1993; 12:280-284. *An example of the many studies evaluating the role of common pathogens in pharyngitis. Although focusing on HSV, it also provides a look at pharyngitis pathogens in general.*

Shulman ST, Bisno AL, Clegg HW et al. Clinical practice guideline for the diagnosis and management of group A streptococcal pharyngitis: 2012 update by the Infectious Diseases Society of America. *Clin Infect Dis* 2012; 55:e86-e102. *These are the 2012 IDSA guidelines for pharyngitis and provide a comprehensive review of diagnostics and therapeutics.*

Shulman ST, Gerber MA, Tanz RR, Markowitz M. Streptococcal pharyngitis: the case for penicillin therapy. *Pediatr Infect Dis J* 1994;13:1-7. *A useful review in support of the continued recommendation for penicillin as the drug of choice for GAS pharyngitis.*

Stewart, EH, Davis B, Clemens-Taylor BL et al. Rapid antigen group A Streptococcus test to diagnose pharyngitis: a systematic review and meta-analysis. *PLOS One* 2014;9(11):1-10. *An examination of the sensitivity and specificity for rapid group A streptococcal testing in adults and children.*

# ADDITIONAL RESOURCE

American Academy of Pediatrics. 2018-2021 report of the Committee on Infectious Diseases, 31st edition. The IDSA guidelines for the diagnosis and management of GAS pharyngitis are available at www.idsociety.org, in the guidelines section.

# 34

# Acute Exacerbations of Chronic Obstructive Pulmonary Disease

*Sanjay Sethi*

## ABSTRACT

Exacerbations are episodes of increased respiratory and sometimes systemic symptoms in patients with underlying chronic obstructive pulmonary disease (COPD). They reflect increased airway inflammation usually induced by bacterial and/or viral tracheobronchial infection. These episodes are a major contributor to the morbidity and, in advanced disease, to the mortality associated with COPD. A careful clinical evaluation with selected application of diagnostic tests should be followed by individualized management including supportive care, bronchodilators, antibiotics, and corticosteroids. Exacerbations can be partially prevented with current treatment of COPD. Our understanding of exacerbation mechanisms has evolved and led to better outcomes and greater use of effective preventative measures.

## CLINICAL VIGNETTE

A 72-year-old female who was diagnosed with COPD a year earlier presents for an acute care office visit. She has not smoked for the last 10 years but did smoke a pack of cigarettes per day for 40 years. She had spirometry performed a year ago because of gradually increasing dyspnea on exertion demonstrating moderate airflow obstruction. She was started on an inhaled long-acting muscarinic antagonist to be used once daily and a short-acting beta-agonist inhaler to be taken as needed. She states that she was doing well until the past 5 days, during which she has had an increase in her dyspnea accompanied by cough and sputum that is yellow-green in color. She feels more fatigued and has not been able to sleep well but denies fever, pleuritic chest pain, or hemoptysis. Her other medical conditions include hypertension, a myocardial infarction, and congestive heart failure.

On examination she is not in acute distress and her vital signs are stable. Oxygen saturation by pulse oximetry on room air is 91%. Lung examination demonstrates decreased air entry and scattered rales and rhonchi in both lower lobes. Cardiac exam is unremarkable, and there is no evidence of decompensation of her congestive heart failure.

COMMENT: The clinical presentation is compatible with an exacerbation of COPD. A careful clinical evaluation was conducted to exclude other reasons for the worsening respiratory symptoms, such as congestive heart failure, pneumonia, and pulmonary embolism. No further diagnostic testing was required, and there was no indication for hospitalization. Purulent sputum indicated a need for a short course of oral antibiotics. Short-acting bronchodilator therapy was recommended at increased frequency, as well attention to adequate hydration and nutrition. The patient was instructed to seek additional care if there was no improvement within 48 hours or there was worsening at any point.

## DISEASE BURDEN

COPD is a universal disease related primarily to tobacco smoking but also to other noxious smoke and fume exposures. The current estimate of prevalence in adults is about 10%, and COPD is the fourth leading cause of death worldwide. Almost all patients with COPD experience repeated episodes of worsening respiratory symptoms and lung function, termed *exacerbations*. Exacerbation incidence increases with worsening airflow obstruction. Exacerbations are major reasons for health care usage in patients with COPD and, in advanced disease, major causes of hospitalization and mortality. They are associated with worsening health status and airflow obstruction. Consequently adequate clinical management and the prevention of exacerbations have become important parts of managing this disease.

## RISK FACTORS

The risk factors for the development of COPD are well defined; however, there is wide variation in the frequency of exacerbations in patients with COPD, which is only partially understood. Frequency of exacerbations does increase with worsening lung function. Other clinical susceptibility factors for exacerbation are gastroesophageal reflux, vitamin D deficiency, and comorbid bronchiectasis and asthma. Infection with bacteria and/or viruses is the underlying cause of the majority of exacerbations. COPD does increase susceptibility to these infections in the lower respiratory tract. The normal lung has a multi-faceted defense system to maintain a very sparse and transitory microbiome in the lower airways in spite of repeated exposure to infectious organisms by inhalation or microaspiration. Impairment of specific innate lung defense mechanisms, such as macrophage function and localized reduction in secretory IgA, have been associated with more frequent exacerbations.

## CLINICAL FEATURES

Exacerbations are defined by increased respiratory symptoms, the cardinal ones being dyspnea, sputum production, and sputum purulence (Fig. 34.1). Additional respiratory symptoms include cough, chest discomfort, and wheezing. Systemic symptoms are common and include fatigue and sleep disturbance, with fever seen in only 20% of episodes. Symptom intensity has to exceed day-to-day variability, and symptoms are usually present for 2 to 5 days before presentation. The severity of an exacerbation depends on the underlying lung disease as well as the magnitude of the acute episode. Findings on examination can therefore range from minimal findings to respiratory failure. Chest examination usually reveals diminished breath sounds, with wheezing and localized rales seen in a few patients. Findings consistent with lung

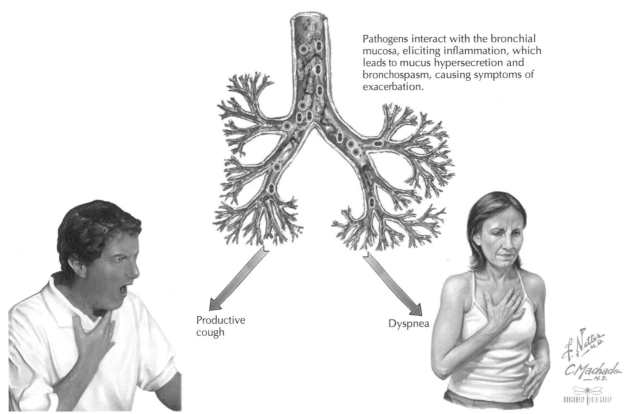

Pathogens interact with the bronchial mucosa, eliciting inflammation, which leads to mucus hypersecretion and bronchospasm, causing symptoms of exacerbation.

Productive cough

Dyspnea

**Fig. 34.1** Cardinal manifestations of exacerbation of chronic obstructive pulmonary disease.

consolidation or pleural effusion should prompt a search for alternative diagnoses as discussed later. Exacerbations with severe underlying COPD can manifest as respiratory distress and alteration of mental status because of hypercapnia and consequent acute respiratory acidosis.

## DIAGNOSTIC APPROACH

Tracheobronchial infection with bacteria and/or a virus is responsible for up to 80% of exacerbations, with eosinophilic inflammation related to environmental stimuli accounting for the rest. Such infection induces airway inflammation, hypersecretion of mucus, and bronchial narrowing; these, in turn, result in the classic clinical manifestations. However, symptoms that resemble an exacerbation can occur in patients with COPD because of other clinical entities. Most important among these differential diagnoses are pneumonia and congestive heart failure. Other important but less common clinical entities include pulmonary embolism, other systemic infections, and noncompliance with COPD maintenance medications.

Because of wide variations in severity and clinical findings, the diagnostic approach must be tailored specifically to each exacerbation. In mild to moderate exacerbations, when the expectation is that outpatient treatment will suffice and when the other possible diagnoses are unlikely based on clinical evaluation, no further diagnostic testing is necessary. Chest x-ray examination is indicated if there is clinical suspicion for pneumonia or congestive heart failure (Fig. 34.2). A spiral computed tomography angiogram for pulmonary embolism should be considered if the exacerbation has an atypical presentation, including sudden onset, absence of bronchitic symptoms, pleuritic chest pain, and hemoptysis.

Sputum Gram stain and culture have limited utility in exacerbations because of their limited sensitivity and specificity and long turnaround times. Clinical situations in which they may be useful

are in patients with early relapse or nonresponse to treatment and in patients in whom infection with *Pseudomonas aeruginosa* is a possibility, such as those with very severe airflow obstruction and comorbid bronchiectasis. In severe exacerbations, assessment of gas exchange with pulse oximetry and arterial blood gases is essential to detect respiratory failure.

## MANAGEMENT

A multifaceted approach to treatment of exacerbations is usually indicated. Intensification of bronchodilators, attention to nutrition and hydration, and symptomatic treatment of cough and expectoration are often implemented. In patients with hypoxemia, oxygen supplementation and, in the presence of respiratory failure, ventilatory support—preferably with noninvasive ventilation—are indicated. Specific treatment of exacerbation includes systemic antibiotics and corticosteroids (Fig. 34.3).

Several placebo-controlled trials support the use of antibiotics in exacerbations. Concerns about adverse effects, emergence of antibiotic-resistant pathogens, and appropriate use are best addressed by adopting a stratification approach to antibiotic choice in exacerbations so as to maximize the probability that a bacterial infection is responsible and that an improvement in clinical outcomes is likely. The presence of bronchitic symptoms and the absence of wheezing make a bacterial cause more likely. Among these patients, older age, a history of frequent exacerbations, severe airflow obstruction, and comorbid cardiac disease are indications for broader-spectrum antibiotics.

Systemic corticosteroids have also been found to be useful in m exacerbations in placebo-controlled trials. A blood eosinophil count that is high normal or above normal has been associated with benefit with systemic steroids. The currently recommended dose is 40 mg of prednisone (or equivalent) daily for 5 days.

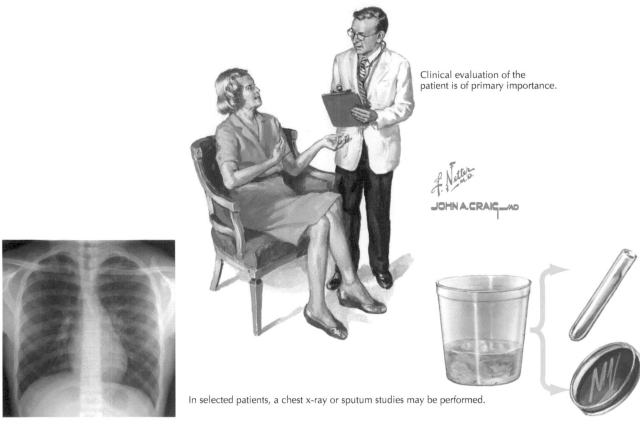

Clinical evaluation of the patient is of primary importance.

In selected patients, a chest x-ray or sputum studies may be performed.

**Fig. 34.2** Clinical evaluation of exacerbation.

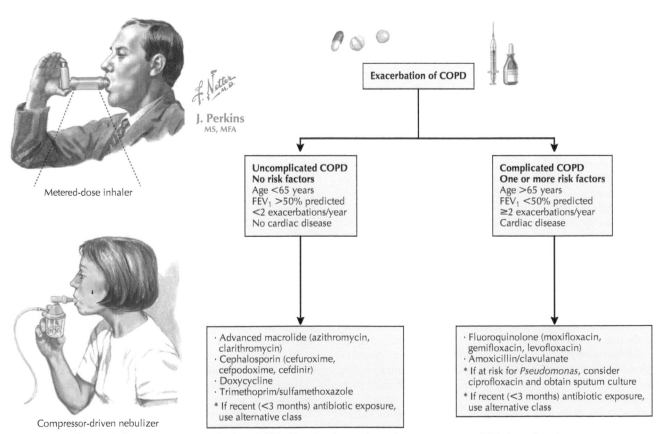

Metered-dose inhaler

Compressor-driven nebulizer

**Exacerbation of COPD**

**Uncomplicated COPD**
**No risk factors**
Age <65 years
FEV₁ >50% predicted
<2 exacerbations/year
No cardiac disease

**Complicated COPD**
**One or more risk factors**
Age >65 years
FEV₁ <50% predicted
≥2 exacerbations/year
Cardiac disease

· Advanced macrolide (azithromycin, clarithromycin)
· Cephalosporin (cefuroxime, cefpodoxime, cefdinir)
· Doxycycline
· Trimethoprim/sulfamethoxazole
* If recent (<3 months) antibiotic exposure, use alternative class

· Fluoroquinolone (moxifloxacin, gemifloxacin, levofloxacin)
· Amoxicillin/clavulanate
* If at risk for *Pseudomonas*, consider ciprofloxacin and obtain sputum culture
* If recent (<3 months) antibiotic exposure, use alternative class

**Fig. 34.3** Treatment of exacerbation. *COPD,* Chronic obstructive pulmonary disease; *FEV₁,* forced expiratory volume in 1 second.

## PROGNOSIS

Exacerbations are associated with considerable morbidity and mortality. Among patients admitted to intensive care, a 90-day mortality of 20% in those requiring noninvasive ventilation and 35% in those requiring invasive ventilation is seen. Patients hospitalized with exacerbation but not in intensive care have a 9% mortality rate. In outpatient exacerbations, a 25% to 33% rate of treatment failure and/or relapse is seen; this can be improved with early appropriate intervention.

## PREVENTION

Prevention of exacerbations has now become one of the major goals of COPD management. With current therapeutic approaches, a 40% to 50% relative reduction in exacerbations has been seen. Yearly influenza vaccination and a single pneumococcal vaccination are indicated. Maintenance treatment of the underlying COPD with inhaled long-acting anticholinergic and β-agonist bronchodilators, as well as with inhaled corticosteroids, has been shown to decrease exacerbation frequency in COPD. In individuals who continue to exacerbate frequently with standard maintenance treatment, the use of an oral phosphodiesterase inhibitor and chronic antibiotics such as a macrolide are indicated. Home noninvasive ventilation, surgical or bronchscopic lung volume reduction, and pulmonary rehabilitation also reduce exacerbation-related health care usage in selected individuals.

## EVIDENCE

Anthonisen NR, Manfreda J, Warren CPW, et al. Antibiotic therapy in exacerbations of chronic obstructive pulmonary disease. *Ann Intern Med* 1987;106:196-204. *An older study that still remains the largest and best placebo-controlled trial of antibiotics in exacerbations of COPD.*

Calverley PM, Anderson JA, Celli B, et al. Salmeterol and fluticasone propionate and survival in chronic obstructive pulmonary disease. *N Engl J Med* 2007;356:775-789. *A large study demonstrating that a long-acting β-agonist (salmeterol) and inhaled corticosteroid (fluticasone) are effective in reducing exacerbations of COPD.*

Leuppi JD, Schuetz P, Bingisser R, et al. Short-term vs conventional glucocorticoid therapy in acute exacerbations of chronic obstructive pulmonary disease. *JAMA* 2013;309(21):2223-2231. *A well-conducted controlled trial demonstrating the benefit of short-term systemic corticosteroids in exacerbations of COPD.*

Sethi S, Evans N, Grant BJB, Murphy TF. Acquisition of a new bacterial strain and occurrence of exacerbations of chronic obstructive pulmonary disease. *N Engl J Med* 2002;347:465-471. *A prospective longitudinal cohort study of bacterial infection in COPD, which was the first one to show that acquisition of new strains of bacteria is clearly related to exacerbation occurrence.*

Sethi S, Murphy TF. Infection in the pathogenesis and course of chronic obstructive pulmonary disease. *N Engl J Med* 2008;359:2355-2365. *A review outlining current evidence that infections cause exacerbations of COPD and containing a suggested algorithm for rational use of antibiotics to treat exacerbations.*

Tashkin DP, Celli B, Senn S, et al. A 4-year trial of tiotropium in chronic obstructive pulmonary disease. *N Engl J Med* 2008;359:1543-1554. *A large study demonstrating that a long-acting inhaled anticholinergic (tiotropium) is effective in reducing exacerbations of COPD.*

Vollenweider DJ, Frei A, Steurer–Stey CA, et al. Antibiotics for exacerbations of chronic obstructive pulmonary disease. *Cochrane Database Syst Rev* 2018;10(10):CD010257. *A systematic analysis of placebo-controlled trials of antibiotics in exacerbations of COPD.*

Walters JAE, Tan DJ, White CJ, et al. Different durations of corticosteroid therapy for exacerbations of chronic obstructive pulmonary disease. *Cochrane Database Syst Rev* 2018;3(3):CD006897. *A systematic analysis of different durations of corticosteroids in exacerbations of COPD.*

## ADDITIONAL RESOURCES

American Thoracic Society official documents regarding COPD. Available at: https://www.thoracic.org/statements/copd.php.

Global Initiative for Chronic Obstructive Lung Disease. Available at: www.goldcopd.com. *Global guidelines for COPD.*

National Heart, Lung, and Blood Institute, National Institutes of Health: COPD. Available at: www.nhlbi.nih.gov/health/public/lung/copd. Patient information for COPD.

# Systemic Infections

*Dennis L. Stevens*

# 35

# Introduction to Systemic Infections

*Dennis L. Stevens*

Systemic infections comprise a broad range of clinical entities including endocarditis, meningitis, osteomyelitis, nonbacterial infections such as disseminated tuberculosis, and fungal infections such as coccidiomycosis and histoplasmosis. In all these scenarios, infection develops in organs normally protected by innate and adaptive immune processes. In the last decade, the systemic manifestations of microbial infections have been categorized in terms of severity. For example, the terms *systemic inflammatory response syndrome* (SIRS), *sepsis*, and *septic shock* all describe increasingly severe inflammatory responses and/or systemic infections. One caveat to this stratification system is that an infection may result in profound systemic signs of inflammation without causing a systemic infection per se. For example, localized pneumococcal infection of the sinuses, middle ear, or lung may result in systemic signs of inflammation. However, only when the infection disseminates—resulting in bacteremia and meningitis—does a systemic infection ensue.

Systemic infection invariably results when the innate and adaptive immune systems fail to contain a primary infection. Thus complement deficiency and hypogammaglobulinemia are risk factors for systemic pneumococcal and meningococcal infections. Clearly the additive effects of type-specific immunoglobulin G (IgG), coupled with complement-enhanced opsonophagocytosis, prevent systemic complications of infection with *Haemophilus influenzae*, *Streptococcus pneumoniae*, or *Neisseria meningitides*. In fact, in populations immunized against *H. influenzae* and *S. pneumoniae*, systemic infections caused by these microbes have become rare. Splenic sequestration and clearance of blood-borne pathogens are also crucially important in preventing systemic infections, and clearly asplenia is commonly associated with fulminant infections caused by pyogenic microbes.

In recent years, systemic and disseminated infections have been more commonly recognized and reported in immunocompromised patients because of the epidemic of acquired immunodeficiency syndrome (AIDS), the increased frequency of transplantation, and the introduction of potent immunomodulatory therapies directed against tumor necrosis factor (TNF) and interleukin-1 (IL-1). These reports substantiate the important role of the normal host response to infection.

There is evidence that some bacterial systemic infections have specific relationships to age, whereas others are associated with gender or race. For example, group B streptococcal meningitis is related to the neonatal period, and *H. influenzae* infection is most common in children aged 6 months to 6 years. Systemic mycoses such as coccidiomycosis show a clear increase in the frequency of dissemination among patients of Filipino or African descent. In terms of gender, there is also a clear predilection for disseminated *Coccidioides immitis* infection in females, although this is entirely related to pregnancy. These relationships are of sufficient strength that pregnant women, particularly those of Filipino or African descent, should not be employed in clinical or research laboratories that perform studies on *C. immitis*.

Besides host factors, specific microbial traits are clearly associated with the ability to cause systemic infection. For example, endocarditis is most commonly caused by gram-positive cocci such as viridans streptococci, enterococcal species, *Streptococcus bovis*, *Staphylococcus aureus*, and *Staphylococcus epidermidis*. This remarkable molecular epidemiology is supported by simple laboratory studies demonstrating that these gram-positive bacteria adhere to heart valve endothelium with far greater avidity than gram-negative microbes.

Hematogenous osteomyelitis is clearly a systemic infection because bacteremia seeds proximal and distal long bones or paravertebral plexuses, resulting in acute bone infection and destruction. Here there is an age relationship, because proximal and distal osteomyelitis usually occurs in younger individuals before the epiphyseal plate closes and at a time when blood flow to this plate is maximal. With increasing age beyond adolescence, osteomyelitis of the proximal and distal long bones is a rare event. In sharp contrast, vertebral osteomyelitis is predominately related to increasing age, and the source of *Staphylococcus* may be either arterial or venous. The venous channels include the Batson plexus, which drains the lower urinary tract and provides retrograde flow to the paravertebral plexus. Although *S. aureus*, including methicillin-resistant *S. aureus* (MRSA), is most commonly implicated in hematogenous osteomyelitis, the mechanism causing the transient bacteremia and "homing" to proximal long bone or vertebrae is entirely unknown. Interestingly, the likelihood of cure of hematogenous osteomyelitis of the long bones is excellent, particularly in preadolescent children, probably a result of the augmented blood flow to the epiphyseal centers and enhanced delivery of complement, polymorphonuclear leukocytes, and antibiotics to the site in reasonable concentrations.

In contrast, osteomyelitis of the midshaft of long bones occurs in people of any age group and is more common in younger individuals participating in activities conducive to fractures. Closed fractures rarely become infected. However, compound fractures are more prone to infection because (1) the injury site is exposed to skin microflora as well as exogenous microbes from clothing, soil, water, and so on, and (2) injury induced by the penetration of bone fragments through the skin, fascia, and muscle frequently also destroys the blood supply. Thus this type of osteomyelitis is not really a systemic infection. Management of midshaft long bone osteomyelitis is complex because of poor bone repair, attenuated inflammatory reaction, and diminished delivery of antibiotics to the site of infection. Postadolescent osteomyelitis of this type is even more difficult in patients with intrinsic deficits of vascular integrity, such as individuals with diabetes. An aggressive approach including surgical debridement, pathogen-directed antibiotic treatment for 6 weeks, and long-term suppression with oral agents provides the best results. In complex cases, myocutaneous transplantation combined with the approach just described offers some hope.

In summary, systemic infections are a diverse group of clinical entities caused by bacteria, fungi, and other pathogens. Clinical and epidemiologic studies have clearly implicated host factors including age, gender, and specific deficiencies of adaptive and innate immunity as prerequisites or risk factors for poor outcomes. Microbial virulence factors are clearly important for the development of specific systemic infections, and innate immune responses contribute to the systemic manifestations. Thus the initial encounter between the innate immune system and the invading microbe determines whether infection will develop and what the signs, symptoms, and morbidity will be. It is rather ironic that early attenuation of the innate immune response predisposes to the establishment of infection, yet after infection a robust response may contribute to morbidity. Clearly, continued basic scientific and clinical research is necessary to broaden our knowledge of bacterial, viral, and fungal pathogenesis and provide new tools to improve outcomes.

# 36

# Endocarditis

*Sky R. Blue, Casi M. Wyatt, Birgitt L. Dau*

## ABSTRACT

Infective endocarditis (IE) is a serious infection involving the interior of the heart, most commonly the heart valves. It is predominantly bacterial, most commonly caused by *Staphylococcus aureus* and streptococcal species. This chapter reviews the pathogenesis, clinical features, diagnosis, treatment, and prevention of this condition for both native and prosthetic valves. IE has always been a diagnosis that many physicians approach with anxiety and trepidation. The diagnosis can be elusive, the complications severe, and the treatment arduous. This has not changed significantly since the time of William Osler, who said that "Few diseases present greater difficulties in the way of diagnosis than malignant endocarditis, difficulties which in many cases are practically insurmountable" (from the Gulstonian Lectures on Malignant Endocarditis, 1885). Endocarditis and its many protean manifestations have long been known; since the mid-1700s they have been eloquently described by notable pioneers of medicine, including Morgagni, Virchow, Wilks, Janeway, Roth, and Bowman. The modern era has seen an increase not only in the risk of endocarditis but also in the ability to diagnose IE. Endocarditis is now likely to manifest much differently—more acutely and without the widespread stigmata observed in the past. However, despite great strides in the microbiologic isolation of causative organisms, echographic diagnosis, and antimicrobial treatment, the mortality remains high, contributing to the constant anxiety associated with this age-old disease.

## CLINICAL VIGNETTE

A 28-year-old woman presented with fatigue and associated symptoms of arthralgias, malaise, and subjective low-grade fevers that had begun 2 weeks earlier. Initially a viral illness was suspected; further evaluation noted leukocytosis, mild thrombocytopenia, slightly elevated creatinine, microscopic hematuria, and an elevated sedimentation rate. Her rheumatoid factor was also elevated, prompting referral to rheumatology. Over the next few days, however, she noticed increased dyspnea with mild exercise and nonproductive cough. She presented to the emergency department with chills, fever and dyspnea. She had not injected drugs but had had a routine dental cleaning the month before the onset of her symptoms. On examination she was found to have a detectable systolic murmur, dependent rales on lung exam, and multiple small petechiae on her hands. Blood cultures grew *Streptococcus mutans* with a penicillin minimal inhibitory concentration (MIC) of 0.12 mcg/mL. A transthoracic echocardiogram noted a previously unknown bicuspid aortic valve with some aortic regurgitation. A subsequent transesophageal study revealed a mobile, oscillating 7-mm mass on the aortic valve leaflet. Chest x-ray showed pulmonary edema, and she had an elevated brain natriuretic peptide. Her neurologic examination was normal, but magnetic resonance imaging of her brain noted several lesions compatible with very small cortical and subcortical ischemic lesions without evidence of hemorrhage. After cardiovascular surgical consultation, early surgery for aortic valve replacement was performed on her fifth hospital day. A mechanical valve was placed successfully and, postoperatively, she was treated with 4 weeks of intravenous penicillin. Six months later, her exercise tolerance was back to baseline, she was anticoagulated on warfarin, and advised to have antibiotic prophylaxis before dental procedures.

COMMENT: This case demonstrates the nonspecific protean symptoms that may be early indications of endocarditis. The subacute presentation, although initially nonspecific, was eventually associated with many of the classic findings. *S. mutans* is a common agent of dental caries and a common cause of endocarditis after transient bacteremia. At the time of her eventual diagnosis, the patient was found to have congestive heart failure (CHF), which was an indication for early surgery. Appropriately, the magnetic resonance imaging (MRI) findings did not delay her surgery. Endocarditis associated with bicuspid aortic valves tends to occur in younger individuals who have a paucity of comorbid findings.

## PATHOGENESIS

The pathogenesis of IE usually starts with an area of endocardial injury leading to platelet/fibrin deposition. Then, based on predisposing factors, when certain microorganisms enter the bloodstream, adherence occurs to the area of injury, as was demonstrated in a rabbit model of endocarditis by Durack and colleagues in 1972. Injury and infection most commonly occur on the valve leaflets but can also occur on or near congenital defects, chordae, chamber walls, prosthetic material, pacemaker leads, or any other endocardial location where favorable conditions exist. Microorganism adherence factors facilitate the initial colonization of the injured valve surface. *Staphylococcus* and *Streptococcus* species produce the majority of the human cases of IE because of their ability to adhere to damaged tissues of the heart. Conversely, *Escherichia coli* can be a common cause of bacteremia from urinary or gastrointestinal sources but is a rare cause of endocarditis because it lacks these adherence factors. The causes of native valve endocarditis (NVE) and their frequencies, as found by the International Collaboration on Endocarditis, are as follows: *S. aureus*, 31%; *viridans* group streptococci, 17%; coagulase-negative staphylococci, 11%; enterococci, 10%; *Streptococcus gallolyticus* (formerly *S. bovis*), 6%; other streptococci, 6%; HACEK*, 2%; fungi/yeast, 2%; polymicrobial, 1%; other, 4%; and culture negative, 10%. The most common cause of culture-negative endocarditis is *Coxiella burnetii*, the etiologic agent of Q fever.

In 1978 Drs. Sheld, Valone, and Sande described the role of dextran, platelets, and fibrin in the adherence of streptococci to damaged endocardial tissue. Staphylococci use a variety of surface-bound adhesion

---

*\*Haemophilus parainfluenza, Aggregatibacter spp, Cardiobacterium hominis, Eikenella corrodens, Kignella spp.*

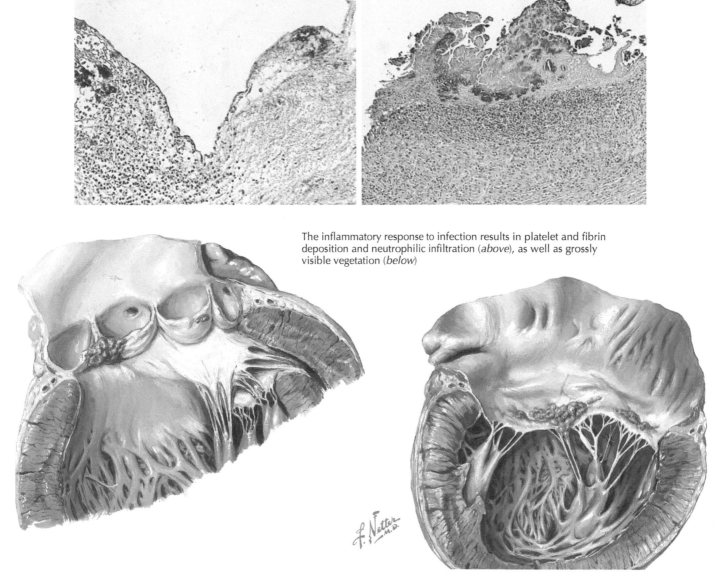

The inflammatory response to infection results in platelet and fibrin deposition and neutrophilic infiltration (*above*), as well as grossly visible vegetation (*below*)

**Fig. 36.1** Bacterial endocarditis: early lesions.

components to bind to fibrinogen for colonization and to fibronectin for invasion. Certain coagulase-negative staphylococci such as *Staphylococcus lugdunensis* and *Staphylococcus schleiferi* possess a clumping factor that also binds fibrinogen and fibronectin as virulence factors. *Enterococcus faecalis* utilizes pili and collagen adhesion to facilitate adherence and aggregation. Once attached to the platelet-fibrin nidus on injured cardiac tissue, bacteria begin to multiply, increasing activation of coagulation cascade, attracting leukocytes, and grow in a further inflammation-promoting vegetation (Fig. 36.1). This, in effect, buries bacteria deep within the mature vegetation, creating a biofilm environment. The unique behavior of microorganisms within this biofilm and the difficulty of antimicrobial penetration contributes to the treatment challenge of IE.

## RISK FACTORS

Patient factors that predispose to IE are those that create damage to the valvular surface, increase the incidence of bacteremia, or slow the immune response to infection. Endocardial and valvular pathology that can be seen with congenital processes such as bicuspid aortic valves or septal defects predispose to injury. Acquired valvular damage then may occur from atherosclerotic changes and rheumatic fever. Prosthetic heart valves of any type, prosthetic material used to repair congenital heart disease, and implantable cardiac devices are considered significant risk factors. Injection drug use leads to both valvular injury and frequently introduces bacteria into the bloodstream and therefore is a synergistic risk for IE. Hemodialysis likewise creates opportunity for frequent episodes of transient bacteremia owing to regular vascular access. Increasingly, exposure to health care procedures has become recognized as a risk factor. Diabetes mellitus and other diseases leading to phagocytic cell dysfunction are predisposing factors in some cases. Men tend to be predominant in most series, with ratios ranging from 3:2 to 9:1. The elderly have nearly five times the risk of younger individuals. This is likely because of the increased prevalence of degenerative valve disease, increased use of invasive procedures, implanted medical devices, atrophy of skin, and senescence of the immune system. Children with IE almost always have underlying congenital heart disease as the main risk. In both children and adults, a previous episode of IE is a notable risk factor for a subsequent infection.

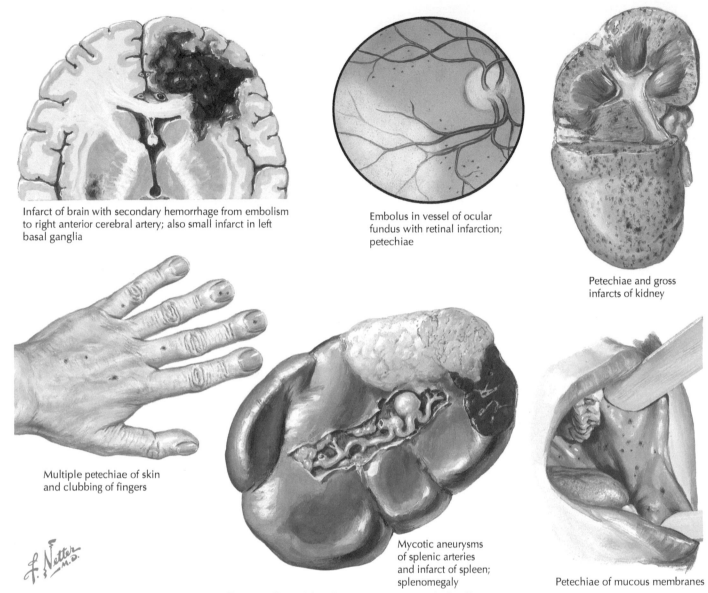

Infarct of brain with secondary hemorrhage from embolism to right anterior cerebral artery; also small infarct in left basal ganglia

Embolus in vessel of ocular fundus with retinal infarction; petechiae

Petechiae and gross infarcts of kidney

Multiple petechiae of skin and clubbing of fingers

Mycotic aneurysms of splenic arteries and infarct of spleen; splenomegaly

Petechiae of mucous membranes

**Fig. 36.2** Bacterial endocarditis: remote embolic effects.

## CLINICAL PRESENTATION

Historically, endocarditis was a syndrome recognized by the chronic and diverse clinical findings depicted in the Netter illustrations. Now the rapid detection of bacteria with automated continuously monitored blood culture systems allow for much earlier diagnosis. Often, the first presenting sign of an eventual diagnosis of IE is a positive blood culture obtained in the evaluation of fever. Ninety-six percent of IE patients present with fever, and about 85% have an audible heart murmur, either new, worsening, or preexisting. Some patients still present with the classic manifestations of endocarditis after weeks of constitutional symptoms, depending on the virulence of the infecting organism. This subacute presentation is more common with *viridans* group streptococci, HACEK organisms, and *Enterococcus*. Endocarditis is often classified based on the acuity of the presentation and the type of valve: acute or subacute; native valve or prosthetic. The diagnosis should be suspected in any patient with risk factors, evidence of chronic inflammation (e.g., fever, chills, or night sweats) and a corresponding laboratory result of an elevated erythrocyte sedimentation rate, C-reactive protein, or rheumatoid factor with findings such as leukocytosis,

anemia, and thrombocytosis. The distal physical findings of subungual splinter hemorrhages (8%), Janeway lesions (5%), Osler nodes (3%), and Roth spots (2%) occur rarely. More common clinical findings are worsening valvular dysfunction with resultant early CHF, conduction abnormalities including atrioventricular block, mycotic aneurysms, vascular embolic lesions (Fig. 36.2) including embolic stroke, intracerebral hemorrhage, splenic or renal infarcts, septic pulmonary embolic from right-sided IE, petechiae of the conjunctivae, oral mucosa, or extremities, metastatic abscesses, septic arthritis, and vertebral osteomyelitis.

## DIAGNOSTIC APPROACH

Part of the challenge of IE is that the diagnosis is usually made based on a constellation of clinical and laboratory findings without direct pathologic evidence of endocardial infection. In general, the diagnosis rests on the presence of an intravascular infection and evidence of endocardial involvement. The former is most commonly the growth of a typical organism on several blood cultures. Endocardial involvement is inferred by a new cardiac murmur, echocardiographic evidence of an

## BOX 36.1  Definition of Terms Used in the Modified Duke Criteria

**Major Criteria**
- Positive blood cultures
  - Microorganism typical of endocarditis from two separate blood cultures *or*
  - Persistently positive blood cultures drawn more than 12 hours apart or three or more positive cultures drawn over more than 1 hour *or*
  - Single positive blood culture for *Coxiella burnetii* or antiphase I IgG titer of greater than 1:800
- Evidence of endocardial involvement
- Positive echocardiogram for endocarditis
- New valvular regurgitation (change in preexisting murmur is not sufficient for major criteria)

**Minor Criteria**
- Predisposing heart condition or injection drug use
- Fever (38°C or 100.4°F)
- Vascular phenomena
  - Major arterial emboli, septic pulmonary infarcts, mycotic aneurysms, intracranial hemorrhage, conjunctival hemorrhages, Janeway lesions, and so on
- Immunologic phenomena
  - Glomerulonephritis, Osler nodes, Roth spots, positive rheumatoid factor
- Microbiologic evidence not included previously
  - Positive blood culture or cultures not meeting major criteria (not including single positive culture for coagulase-negative staphylococci not typically causing endocarditis) or serologic evidence of active infection with organism capable of causing endocarditis
- Echocardiographic minor criteria eliminated

Adapted from Li JS, Sexton DJ, Mick N, et al: Proposed modifications to the Duke Criteria for the diagnosis of infective endocarditis, *Clin Infect Dis* 2000;30:633.

## BOX 36.2  Definition of Infective Endocarditis According to the Modified Duke Criteria

**Definite Endocarditis**
- Pathologic Criteria
  - Microorganisms identified by culture or histology in a vegetation, embolic lesion, or intracardiac abscess *or*
  - Pathology confirmation of active endocarditis by histology of vegetation or intracardiac abscess
- Clinical Criteria
  - Two major criteria *or*
  - One major and three minor criteria *or*
  - Five minor criteria

**Possible Endocarditis**
- Clinical Criteria
  - One major and one minor criterion *or*
  - Three minor criteria

**Rejected Endocarditis**
- Clinical Criteria
  - Firm alternate diagnosis for symptoms thought to be manifestations of endocarditis *or*
  - Resolution of symptoms thought to be manifestations of endocarditis with antibiotic therapy for 4 days or less *or*
  - Patient does not meet the criteria for definite or possible endocarditis
- Pathologic Criteria
  - No pathologic evidence of endocarditis on surgical or autopsy specimen after 4 days (or less) of antibiotic therapy

Adapted from Li JS, Sexton DJ, Mick N, et al: Proposed modifications to the Duke Criteria for the diagnosis of infective endocarditis, *Clin Infect Dis* 2000;30:633.

endocardial vegetation, or the pathologic evaluation of a surgical specimen. Because there is no single definitive test result that is pathognomonic, syndromic diagnostic criteria have been used for many years. The first criteria were proposed in 1977 by Pelletier and Petersdorf. The most recently accepted criteria are often referred to as the "Duke criteria" and have been used in their modified state since 2000 (Boxes 36.1 and 36.2). These criteria are more suited to left-sided NVE than to right-sided or prosthetic valve endocarditis (PVE) and must be utilized with significant clinical judgment, as their sensitivity and specificity for diagnosing IE are only around 80%.

Culture-negative endocarditis presents a unique diagnostic challenge. It is defined as a clinical syndrome of endocarditis in which three separate blood cultures are negative after 5 days of incubation. Culture-negative IE may result from antibiotic therapy given prior to blood cultures being drawn or from bacteria that are fastidious and difficult to culture. Common causes of culture-negative endocarditis include *C. burnetii*, *Bartonella* spp., *Brucella* spp., *Tropheryma whipplei*, and fungal organisms. The highest-yield diagnostic testing is serology for *C. burnetii* and *Bartonella* spp., and broad-range polymerase chain reaction (PCR) testing of valve tissue obtained at the time of surgery. Noninfective endocarditis must also remain on the differential diagnosis of culture-negative endocarditis. Noninfective endocarditis may be secondary to marantic endocarditis in the setting of metastatic cancer or a rheumatologic illness, most commonly lupus erythematosus.

Attributed to Hertz and Edler in the 1950s, the advent of two-dimensional and Doppler echocardiography in the 1970s and 1980s changed the diagnostic process for IE. Instead of relying on the pathologic and

microbiologic examination of a surgical or autopsy specimen, clinicians could now obtain evidence of a characteristic, pathologic structural abnormality (oscillating intracardiac mass, perivalvular abscess, leaflet perforation, or partial dehiscence of a prosthetic valve) to help establish a definitive diagnosis. However, an absence of echocardiographic findings on transthoracic echocardiography (TTE) will not rule out endocarditis in a patient with high clinical risk. Various patient and valve factors can contribute to the lack of sensitivity, such as obesity, hyperinflated lungs, narrow costal interspaces, presence of a prosthetic heart valve, and functional or structural valvular abnormalities. Given these constraints, transesophageal echocardiography (TEE) is more sensitive, and because of its improved image quality, it can more easily demonstrate endocardial complications of endocarditis. Studies comparing the two approaches have found sensitivity to differ significantly: 90% or better for TEE compared with 40% to 60% for TTE, whereas specificity was 95% for both approaches. TTE is inadequate to evaluate prosthetic valve endocarditis; TEE should be selected in such a case. TEE may, however, fail to identify very early lesions. In patients with a high clinical probability of endocarditis—for example those with prolonged bacteremia with *S. aureus* lasting longer than 72 hours—repeat TEE is indicated 5 days after the initial image to improve sensitivity. The echocardiographic appearance of lesions can help to diagnose complications, estimate the risk of embolic events, determine the need for surgery, and evaluate the treatment response. Newer imaging studies such as electrocardiography (ECG)-gated computed tomography (CT) angiography or MRI, positron emission tomography (PET), or radionuclide-labeled white blood cell scintigraphy have proved useful in diagnostically difficult cases.

**TABLE 36.1    Treatment of Endocarditis Caused by _Streptococcus_ Species Including Viridans Group and _S. gallolyticus_**

| Organism | NVE | PVE | Comments |
|---|---|---|---|
| PCN susceptible (MIC ≤0.12 mcg/mL) | Penicillin G 12–18 million units/day IV divided in 6 doses for 4 weeks _or_ Ceftriaxone 2 g IV daily for 4 weeks _or_ Vancomycin 30 mg/kg IV divided in 2 daily doses for 4 weeks | Penicillin G 24 million units/day IV divided in 6 doses for 6 weeks _or_ Ceftriaxone 2 g IV daily for 6 weeks _or_ Vancomycin 30 mg/kg IV divided in 2 daily doses for 6 weeks | Infectious disease consultation recommended<br><br>Use vancomycin only for patients unable to tolerate β-lactams |
| Relative PCN susceptible (MIC 0.12-0.5 mcg/mL) | Penicillin G 24 million units/day IV divided in 6 doses for 4 weeks _plus_ gentamicin 3 mg/kg IV daily for first 2 weeks _or_ Ceftriaxone 2 g IV daily for 4 weeks _or_ Vancomycin 30 mg/kg IV divided in 2 daily doses for 4 weeks | Penicillin G 24 million units/day IV divided in 6 doses for 6 weeks _plus_ gentamicin 3 mg/kg IV daily for 6 weeks _or_ Ceftriaxone 2 g IV daily for 6 weeks _plus_ gentamicin 3 mg/kg IV daily for 6 weeks _or_ Vancomycin 30 mg/kg IV divided in 2 daily doses for 6 weeks | Dose gentamicin or vancomycin with consultation from clinical pharmacist |
| Relative PCN resistance (MIC >0.5 mcg/mL) streptococci including _Abiotrophia_ sp., _Gemella_ sp. or _Granulicatella_ sp. | Penicillin G 24 million units/day IV divided in 6 doses for 4 weeks _plus_ gentamicin 1 mg/kg IV q8h for 4 weeks _or_ Ampicillin 12 g/day IV divided in 6 doses for 4 weeks _plus_ gentamicin 1 mg/kg IV q8h for 4 weeks _or_ Vancomycin 30 mg/kg IV divided in 2 daily doses for 4 weeks | Penicillin G 24 million units/day IV divided in 6 doses for 6 weeks _plus_ gentamicin 1 mg/kg IV q8h for 6 weeks _or_ Ampicillin 12 g/day IV divided in 6 doses for 6 weeks _plus_ gentamicin 1 mg/kg IV q8h for 6 weeks _or_ Vancomycin 30 mg/kg IV divided in 2 daily doses for 6 weeks | |

_IV_, Intravenous; _MIC_, minimal inhibitory concentration; _NVE_, native valve endocarditis; _PCN_, penicillin; _PVE_, prosthetic valve endocarditis.

# TREATMENT

Since the beginning of the era of effective antimicrobial therapy, there has been an expected and significant decline in mortality. But despite modern diagnostic techniques, advances in surgical interventions, and timely treatment, individuals still die from this infection. The general principles that guide treatment have evolved over the years but have their roots in Netter's time. Optimal treatment now includes the involvement of a multidisciplinary team including an infectious diseases specialist, cardiologist, cardiovascular surgeon, microbiologist, and clinical pharmacist. However, several principles of treatment still remain valid. Because of the relative paucity of leukocytes and high numbers of bacteria in a vegetation, bactericidal rather than bacteriostatic therapy is needed. Because organisms reside deep within the vegetation and may be in a state of reduced metabolism within the biofilm, a prolonged course and, in some cases, combination therapy may be required. The selection of appropriate antimicrobial therapy depends on quickly and accurately identifying an etiologic agent. See Tables 36.1, 36.2, and 36.3 for representative antimicrobial therapy regimens for common etiologies.

## Streptococci

As a group, streptococci account for many cases of IE. Penicillin has long been the standard therapy for these organisms. Today the choice, dose, and duration of antimicrobial treatment still depend on the MIC to penicillin. Early studies looked at 6 and 14 days of therapy and found that although many patients were cured, mortality and relapse rates were still unacceptably high. For susceptible streptococcal species

(MIC <0.12), a study of 4 weeks of therapy with penicillin demonstrated no deaths and no relapses. Therefore penicillin or ceftriaxone for 4 weeks is a reasonable treatment. For organisms with intermediate MIC (0.12 to 0.5), a combination of a β-lactam, either penicillin or ceftriaxone, for 4 weeks and an aminoglycoside for 2 weeks is recommended. Isolates with a MIC below 0.5 need to have individual therapy optimized based on full susceptibility results and be treated for 6 weeks.

## Staphylococci

The treatment of staphylococcal IE is determined on the basis of methicillin susceptibility. Methicillin-susceptible _S. aureus_ (MSSA) are best treated with semisynthetic penicillins such as nafcillin or oxacillin or cefazolin. Methicillin-resistant _S. aureus_ (MRSA) has historically been treated with vancomycin, but daptomycin has been shown to be noninferior to standard therapy and is considered a reasonable alternative. Coagulase-negative staphylococci are treated similarly.

## Enterococci

Because of intrinsic resistance in enterococci to cephalosporins and potential for resistance to penicillins, vancomycin, and aminoglycosides, treatment needs to be individualized based on susceptibility patterns. The combination treatment with ampicillin or penicillin with an aminoglycoside for 6 weeks has been considered the standard for susceptible strains. Because of the toxicity from aminoglycosides, combination treatment is often truncated and therapy concluded with a β-lactam alone. Increasing experience with the combination of

## TABLE 36.2 Treatment of Endocarditis Caused by Staphylococci

| Organism | NVE | PVE | Comment |
|---|---|---|---|
| Methicillin-susceptible *Staphylococcus aureus* (MSSA) and methicillin-susceptible coagulase-negative staphylococcus sp. | Nafcillin or oxacillin 12 g/day IV continuously or divided in 6 doses for 6 weeks *or* Cefazolin 6 g/day divided in 3 doses for 6 weeks *or* Daptomycin 8–12 mg/kg IV daily for 6 weeks *or* Vancomycin 15–20 mg/kg IV divided in 2 daily doses for 6 weeks | Nafcillin or oxacillin 12 g/day IV continuously or divided in 6 doses for 6–8 weeks *plus* Gentamicin 1 mg/kg IV q8h for first 2 weeks *plus* Rifampin 900 mg/day PO/IV divided in 2–3 doses for 6–8 weeks | Infectious disease consultation recommended Dose gentamicin or vancomycin with consultation from clinical pharmacist |
| Methicillin-resistant *S. aureus* (MRSA) and methicillin-resistant coagulase-negative staphylococcus sp. | Vancomycin 15–20 mg/kg IV divided in 2 daily doses for 6 weeks *or* Daptomycin 8–12 mg/kg IV daily for 6 weeks | Vancomycin 15–20 mg/kg IV divided in 2 daily doses for 6–8 weeks *plus* Gentamicin 1 mg/kg IV q8h for first 2 weeks *plus* Rifampin 900 mg/day PO/IV divided in 2–3 doses for 6–8 weeks | Use daptomycin or vancomycin in MSSA only for patients unable to tolerate β-lactams Vancomycin trough goal is 15–20 mcg/mL |

*IV,* Intravenous; *NVE,* native valve endocarditis; *PVE,* prosthetic valve endocarditis.

## TABLE 36.3 Treatment of Endocarditis Caused by Enterococci

| Organism | NVE | PVE | Comment |
|---|---|---|---|
| *Enterococcus* sp. susceptible to penicillin, gentamicin and vancomycin | Ampicillin 12 g/day IV divided in 6 doses plus ceftriaxone 2 g q12h for 6 weeks *or* Penicillin G 24 million units per day IV divided in 6 doses plus gentamicin 1 mg/kg IV q8h for 4–6 weeks *or* Ampicillin 12 g/day IV divided in 6 doses plus gentamicin 1 mg/kg IV q8h for 4–6 weeks *or* Vancomycin 30 mg/kg IV divided in 2 daily doses plus gentamicin 1 mg/kg IV q8h for 4–6 weeks | Ampicillin 12 g/day IV divided in 6 doses plus ceftriaxone 2 g q12h for 6 weeks *or* Penicillin G 24 million units per day IV divided in 6 doses plus gentamicin 1 mg/kg IV q8h for 6 weeks *or* Ampicillin 12 g/day IV divided in 6 doses plus gentamicin 1 mg/kg IV q8h for 6 weeks *or* Vancomycin 30 mg/kg IV divided in 2 daily doses plus gentamicin 1 mg/kg IV q8h for 6 weeks | Infectious disease consultation recommended Dose gentamicin or vancomycin with consultation from clinical pharmacist |
| *Enterococcus* sp. β-lactamase–producing and penicillin-resistant but susceptible to gentamicin and vancomycin | *β-lactamase producing strains:* Ampicillin-sulbactam 12 g/day divided in 4 doses plus gentamicin 1 mg/kg IV q8h for 6 weeks *Intrinsic penicillin resistance or intolerance to β-lactams:* Vancomycin 30 mg/kg IV divided in 2 daily doses plus gentamicin 1 mg/kg IV q8h for 6 weeks | Same as NVE | Use vancomycin only for patients unable to tolerate β-lactams Vancomycin trough goal is 15–20 mcg/mL |
| Multiply-resistant *Enterococcus* including vancomycin-resistant *E. faecium* | Treatment with infectious disease consultation *Options include:* Daptomycin 8–12 mg/kg IV daily Linezolid 1200 mg/day IV/PO divided in 2 doses Duration is often 8 weeks and longer | Same as NVE | Consider surgical valve replacement Combination therapy preferred |

*IV,* Intravenous; *NVE,* native valve endocarditis; *PVE,* prosthetic valve endocarditis.

ampicillin and ceftriaxone has now made this a reasonable choice in NVE. Other combinations including newer agents have been evaluated in individual cases or experimental series and should be considered only with expert guidance.

## Surgical Therapy

The decision to proceed with surgical therapy for the treatment of IE is complex and requires a multidisciplinary team. Surgery is indicated in approximately 50% of IE cases, and the risk of reinfection after valve replacement is 3% to 5%. The majority of the guidelines for the surgical management of IE are based on retrospective and prospective cohort studies, and inherent bias limits these studies. Without large controlled trials that provide clear, unambiguous data regarding the timing of surgery, clinicians must continue to utilize expert guidelines and specialty collaboration on an individualized patient basis to decide if surgical therapy is indicated.

The most recent guidelines from the American Heart Association (AHA) for IE address surgical recommendations. Acknowledging the paucity of data defining optimal timing of surgery, the guidelines are based on available evidence in conjunction with expert opinion. Early surgical therapy, defined as "during initial hospitalization before completion of a full course of antibiotics," is recommended for IE in the following situations: valve dysfunction leading to CHF; infections from specific organisms such as *S. aureus*, multidrug-resistant organisms (MDROs) or fungal infections; penetrating infection causing abscess, fistula or conduction disturbances such as heart block; persistent infection despite maximal medical therapy; recurrent emboli and enlarging vegetation despite maximal medical therapy; severe valvular regurgitation; mobile vegetation greater than 10 mm in size, especially when involving the anterior leaflet of the mitral valve; relapsing infection after completing a recommended antibiotics course; or when cardiac devices are involved (Box 36.3).

The patient's hemodynamic stability is the most critical component in determining the need for surgical therapy because it has the greatest impact on prognosis and is the most common cause of death from IE. Moderate to severe CHF is an independent risk factor for 6-month mortality. Aortic valve endocarditis is more likely to result in hemodynamic instability than mitral or tricuspid valve endocarditis. Heart failure can occur suddenly from perforation of a valve leaflet, ruptured mitral chordae, valve obstruction by bulky vegetation, or development of an intracardiac shunt from a fistula. Echocardiography can confirm the clinical diagnosis of heart failure (HF). If signs of HF are present, an urgent evaluation by a cardiac surgeon should be performed. Patients who develop HF may have higher operative mortality rates, but overall in-hospital mortality decreases from about 45% to 20% in these patients when surgery is performed in conjunction with medical therapy, thus improving 1-year mortality to 30% versus 59% with medical therapy alone. During medical management, clinical vigilance must be maintained, because HF can develop slowly from progressive valve dysfunction even during the course of appropriate medical therapy leading to a surgical indication.

IE can progress to perivalvular abscess and extensive tissue destruction. Aortic valve involvement is most likely to result in extension to the conduction system due to the proximity of the AV node and can cause heart block, leading to sudden death. TEE can fail to identify an intracardiac abscess up to 60% of the time. Surgery is required to treat these cases, as mortality is greater than 40% with medical therapy alone. The 5-year survival rate after surgery is as high as 80%, which correlates most closely with the surgeon's ability to remove all of the infected tissue. Medical therapy alone can be considered for perivalvular abscesses smaller than 1 cm without heart block, echocardiographic progression of abscess, or valve dehiscence or insufficiency.

---

### BOX 36.3  Indications for Surgery in Infective Endocarditis

**Class I: Benefit Is Much Greater Than Risk: Surgery Should Be Performed**
- Surgery is recommended by multidisciplinary infective endocarditis (IE) team
- Valve dysfunction resulting in congestive heart failure
- Left-sided IE from *Staphylococcus aureus*, fungal, multidrug-resistant organisms
- Heart block, annular or aortic abscess
- Persistent bacteremia for more than 5 to 7 days while on appropriate antibiotics and no other metastatic foci
- If a cardiac device is associated, remove all leads and devices
- Relapsing prosthetic valve endocarditis after clearance of bacteremia and completing appropriate antimicrobials

**Class IIa: Benefit Greater Than Risk: It Is Reasonable to Perform Surgery**
- Remove all devices and leads in IE from *S. aureus*, fungal or multidrug-resistant organisms even if not confirmed to be involved
- Recurrent emboli or persistent vegetation despite appropriate antibiotics
- Remove pacemaker and leads empirically in all patients undergoing valve surgery for IE

**Class IIb: Benefit Equal to or Greater Than Risk: Surgery Should Be Considered**
- Mobile vegetation larger than 10 mm
- Embolic stroke without hemorrhage or extensive neurologic damage

Adapted from Nishimura RA, Otto CM, et al: 2017 AHA/ACC focused update of the 2014 AHA/ACC guideline for the management of patients with valvular heart disease: a report of the American College of Cardiology/American Heart Association task force on clinical practice guidelines. *J Am Coll Cardiol* 2017;70:252-289.

---

One of the most common causes of complications of IE are embolic events. Early surgery may prevent embolization, which is greatest in the first 1 to 2 weeks of infection. Anterior mitral valve vegetations, especially those larger than 10 mm, pose the highest risk of embolization. Embolization to the central nervous system causing stroke complicates 20% to 50% of left-sided endocarditis cases and is the second leading cause of death in IE after CHF. Ruptured mycotic aneurysms can also cause stroke. MRI or other appropriate cerebral imaging should be considered in all cases of left-sided IE. When present, stroke further complicates the decision regarding surgery because of the anticoagulation needed during cardiopulmonary bypass. When stroke has been identified either clinically or by imaging, neurology consultation is indicated and the neurologist becomes part of the multidisciplinary endocarditis team. Intracranial mycotic aneurysms need to be treated prior to cardiac surgery for IE. Increasing data have shown that early surgical intervention in certain stroke cases does not increase risk for hemorrhagic conversion or neurologic decompensation and is reflected in current guidelines. In cases when surgical therapy is needed for IE, valve surgery can be performed without delay in patients with subclinical cerebral emboli or stroke provided that intracranial hemorrhage has been excluded or neurologic damage is not severe. If the stroke has resulted in major ischemia, severe neurologic sequelae, or intracranial hemorrhage, valve surgery should be delayed at least 4 weeks.

In preparation for surgery, other metastatic foci of infection should be addressed first to decrease the risk for infecting the new prosthetic valve. Dental extractions prior to valve surgery have not been shown to decrease the future risk of prosthetic valve

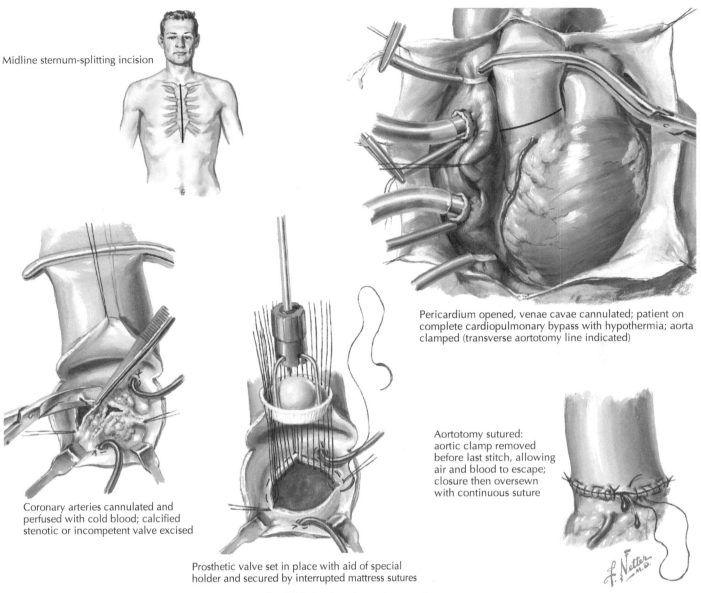

Midline sternum-splitting incision

Pericardium opened, venae cavae cannulated; patient on complete cardiopulmonary bypass with hypothermia; aorta clamped (transverse aortotomy line indicated)

Coronary arteries cannulated and perfused with cold blood; calcified stenotic or incompetent valve excised

Aortotomy sutured: aortic clamp removed before last stitch, allowing air and blood to escape; closure then oversewn with continuous suture

Prosthetic valve set in place with aid of special holder and secured by interrupted mattress sutures

**Fig. 36.3** Aortic valve replacement.

endocarditis. Surgical repair consists of valve repair versus replacement with either a mechanical or bioprosthetic valve. Repairing the valve rather than performing a full replacement decreases the amount of prosthetic material in place, which may decrease the risk of future infection. Valve repair is possible in only a limited number of cases and usually consists of vegetectomy with pericardial patch for small leaflet perforations. Valve repair may be preferable to replacement in young patients and in patients with a history of intravenous drug use and may require only chordal repair or annular support. Valve replacement is necessary if severe valve dysfunction is present. If the aortic valve is involved, surgery may include placing an aortic homograft, reconstructing the aorta, and replacing the aortic valve with a mechanical valve (Fig. 36.3). Right-sided endocarditis is more common in patients using intravenous drugs and can often be managed medically, but surgery may occasionally be indicated. When surgery is needed for right-sided IE, options include valvectomy or vegetectomy with valvuloplasty. Tricuspid valve replacement is rarely performed; even though the patient may later develop right-sided HF, this can often be managed medically.

## PROSTHETIC VALVE ENDOCARDITIS

PVE complicates up to 6% of all prosthetic valves and accounts for approximately 15% to 30% of all cases of IE. Men are affected more frequently than women, and the mean age of presentation is 65. Risk factors for PVE include health care–associated infections, central venous catheters, and hemodialysis. The in-hospital mortality rate of PVE is about 23%, primarily due to HF, arrhythmias, multiorgan system failure from sepsis, or major neurologic complications from emboli.

The pathogenesis of PVE begins at the suture line between valve and native tissue rather than on a leaflet. Infection can occur at the time of surgery or at any time postoperatively from hematogenous spread. As with native valves, prosthetic surfaces are coated with host proteins, especially fibrin and fibrinogen. These proteins facilitate adhesion by certain organisms, especially *S. aureus*, the most common cause of PVE. Because of the location of the infection, perivalvular abscess is more common than in NVE and can result in dehiscence of the valve or other complication such as periprosthetic leak or intracardiac shunting. With time, the prosthetic valve endothelializes and

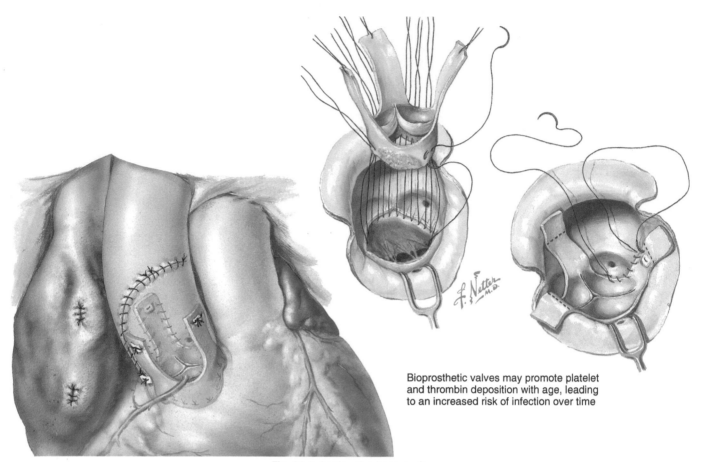

Bioprosthetic valves may promote platelet and thrombin deposition with age, leading to an increased risk of infection over time

**Fig. 36.4** Aortic valve homograft.

| TABLE 36.4 | **Common Causes of Early- and Late-Onset Prosthetic Valve Endocarditis** | |
|---|---|---|
| **Incidence** | **Early-Onset PVE** | **Late-Onset PVE** |
| Highest | *Staphylococcus aureus* | *Staphylococcus aureus* |
| | Coagulase-negative *Staphylococcus* | Coagulase-negative *Staphylococcus* |
| | Culture-negative organisms | *Streptococcus* including the *viridans* group |
| | *Enterococcus* | *Enterococcus* |
| | Gram-negative organisms | Culture negative |
| Lowest | *Streptococcus* | Fungi |

*PVE,* Prosthetic valve endocarditis.

functions more like native tissue, making it more difficult for organisms to adhere and cause infection. A bioprosthetic valve may promote platelet and thrombin deposition with age and theoretically may increase the risk for infection (Fig. 36.4).

Although diagnosis and treatment are similar to those of NVE, the timing of infection in relation to the surgery affects the pathogenesis of infection, the clinical course, and often the pathogen involved. Infection is categorized as early-onset PVE (EO-PVE) and late-onset PVE (LO-PVE) from the time of surgery until development of infection. The greatest risk of developing infection is within 2 months after surgery. *S aureus* and coagulase-negative staphylococci are the most common pathogens in both EO-PVE and LO-PVE, but the *viridans* group of streptococci and enterococci are more likely to occur later (Table 36.4).

Treatment of PVE should be a collaborative effort among cardiac surgeons, cardiologists, and infectious disease experts. See Tables 36.1, 36.2, and 36.3 for representative antimicrobial therapy. Culture-negative PVE should include antibiotics targeted to the most common pathogens for either early or late PVE. Surgery is generally not indicated for an uncomplicated infection with a susceptible organism. Otherwise indications for early surgery are similar to those for NVE.

## PREVENTION

In 2007 the AHA revised guidelines regarding the prophylactic use of antibiotics before dental, genitourinary, and gastrointestinal procedures, and these remain authoritative. Historically, antibiotics were given before these procedures to prevent IE caused by the *viridans* group streptococci—which are normal skin, oral, and gastrointestinal tract flora—because they cause up to 50% of community-acquired NVE. It is well known that certain dental procedures can cause bacteremia, but it is also known that normal daily activities such as

## TABLE 36.5  Antibiotic Prophylaxis With Dental Procedures

| | Antibiotic | Dose |
|---|---|---|
| Oral | Amoxicillin | 2 g |
| Unable to take oral | Ampicillin | 2 g IM or IV |
| | or | |
| | Cefazolin or ceftriaxone | 1 g IM or IV |
| | or | |
| | Clindamycin | 600 mg IM or IV |
| Allergy to penicillin | Cephalexin | 2 g |
| | or | |
| | Clindamycin | 600 mg |
| | or | |
| | Azithromycin | 500 mg |

All are given as a single dose 1 hour before the procedure.
*IM,* Intramuscular; *IV,* intravenous.

### BOX 36.4  Antibiotic Prophylaxis Before Dental Procedures

Such prophylaxis before dental procedures is recommended in the following situations:
- Patients with prosthetic heart valves or prosthetic material from previous repair
- Patients with previous infective endocarditis
- Patients with congenital heart disease limited to
  - Unrepaired cyanotic heart disease
  - Completely repaired cyanotic heart disease in the first 6 months after the procedure
  - Repaired congenital heart disease with residual defect near the site of the prosthetic material
- Cardiac transplantation patients who develop valvulopathy

Adapted from Wilson W, Taubert KA, Gewitz M, et al: Prevention of infective endocarditis: guidelines from the American Heart Association: a guideline from the American Heart Association Rheumatic Fever, Endocarditis, and Kawasaki Disease Committee, Council on Clinical Cardiology, Council on Cardiovascular Surgery and Anesthesia, and the Quality of Care and Outcomes Research and Interdisciplinary Working Group, *Circulation* 2007;116:1736–1754.

brushing of the teeth and flossing, chewing, and using a toothpick can also result in bacteremia of a similar magnitude as that of dental procedures. Promoting overall good oral health may have a greater impact on decreasing rates of IE than routine prophylaxis before dental procedures.

The AHA recommends the use of antibiotic prophylaxis (Table 36.5) in conditions with a high risk of complications from IE (Box 36.4) prior to dental procedures that involve the "manipulation of gingival tissue or the periapical region of teeth or perforation of oral mucosa," including routine dental cleaning. Guidelines no longer recommend antibiotic prophylaxis for patients with an increased risk of endocarditis but who are not at high risk of complications, such as those with mitral valve prolapse or stable rheumatic heart disease.

When a patient with one of the underlying conditions listed (see Box 36.4) undergoes an invasive procedure of the respiratory tract that involves an incision or biopsy, including tonsillectomy, then prophylactic antibiotics should also be used. It is no longer recommended to use antibiotic prophylaxis solely to prevent IE with invasive genitourinary or gastrointestinal procedures.

## EVIDENCE

Chu VH, Park LP, Athan E, et al. Association between surgical indications, operative risk, and clinical outcome in infective endocarditis: a prospective study from the International Collaboration on Endocarditis. *Circulation* 2015;131:131-140. *A prospective cohort study reviewing indications for surgery and showing a positive association with outcome in the surgical group at 6 months.*

Durack DT, Lukes AS, Bright DK. New criteria for diagnosis of infective endocarditis: utilization of specific echocardiographic findings. Duke Endocarditis Service. *Am J Med* 1994;96:200-209. *Now somewhat of a historical reference but useful in understanding the use of echocardiography and criteria-based diagnosis for endocarditis.*

Fernandez-Hildalgo N, Almirante B, Galvadà J, et al. Ampicillin plus ceftriaxone is as effective as ampicillin plus gentamicin for treating Enterococcus faecalis infective endocarditis. *Clin Infect Dis* 2013;56:1261-1268. *Observational, nonrandomized review showing efficacy of a nonaminoglycoside regimen for Enterococcus endocarditis.*

Leisman RM, Pritt BS, Maleszewski JJ, Patel R. Laboratory diagnosis of infective endocarditis. *J Clin Microbiol* 2017;55:2599-2608. *A recent review of the culture-based and molecular methods to identify etiology in endocarditis.*

Li JS, Sexton DJ, Mick N, et al. Proposed modifications to the Duke Criteria for the diagnosis of infective endocarditis. *Clin Infect Dis* 2000;30:633. *This paper outlines the "modified Duke criteria" currently accepted as the diagnostic standard.*

Murdoch DR, Corey GR, Hoen B, et al. Clinical presentation, etiology, and outcome of infective endocarditis in the 21st century: The International Collaboration on Endocarditis-Prospective Cohort Study. *Arch Intern Med* 2009;169:463-473. *The authors describe the presentation and bacterial etiology of infective endocarditis with a modern prospective cohort study.*

Vikram HR, Buenconsejo J, Hasbun R, Quagliarello VJ. Impact of valve surgery on 6-month mortality in adults with complicated, left-sided native valve endocarditis: a propensity analysis. *JAMA* 2003;290:3207-3314. *This was an important paper demonstrating a reduced mortality in patients treated with valve surgery for complicated left-sided NVE.*

Wang A, Athan E, Pappas PA, et al. Contemporary clinical profile and outcome of prosthetic valve endocarditis. *JAMA* 2007;297:1354-1361. *A good description of prosthetic valve endocarditis in a prospective cohort from the International Collaboration on Endocarditis.*

## ADDITIONAL RESOURCES

Baddour LM, Wilson WR, Bayer AS, et al. Infective endocarditis: diagnosis, antimicrobial therapy, and management of complications: a scientific statement for healthcare professionals from the American Heart Association. Endorsed by the Infectious Diseases Society of America. *Circulation*

2015;132:1435-1486. *This is considered to be the most authoritative guideline for treatment currently available.*

Bennett JE, Dolin R, Blaser MJ (eds). *Mandell, Douglas and Bennett's Principles and Practice of Infectious Diseases.* 8th edition, 2015 Elsevier/Saunders, Philadelphia, PA. *Complete resource on diagnosis and treatment of endocarditis.*

Nishimura RA, Otto CM, et al. 2017 AHA/ACC Focused update of the 2014 AHA/ACC guideline for the management of patients with valvular heart disease: a report of the American College of Cardiology/American Heart Association task force on clinical practice guidelines. *J Am Coll Cardiol* 2017;70: 252-289. *This guideline updates the recommendations for surgery in cases of endocarditis.*

Silverman ME, Upshaw CB Jr. Extracardiac manifestations of infective endocarditis and their historical descriptions. *Am J Cardiol* 2007;100:1802-1807.

*A review of the rich history and descriptions of the extracardiac manifestations of IE.*

Wilson W, Taubert KA, Gewitz M, et al. Prevention of infective endocarditis: guidelines from the American Heart Association: a guideline from the American Heart Association Rheumatic Fever, Endocarditis, and Kawasaki Disease Committee, Council on Clinical Cardiology, Council on Cardiovascular Surgery and Anesthesia, and the Quality of Care and Outcomes Research and Interdisciplinary Working Group. *Circulation* 2007;116: 1736-1754. *Guidelines for the prevention of infective endocarditis from a panel of experts and review of the current evidence on prevention are presented here. This paper represents a major change in recommendations from the previous version in 1997.*

# 37

# Meningitis

*Peter P. McKellar*

## ABSTRACT

Meningitis is a relatively rare inflammation of the membranes surrounding the brain and spinal cord. This inflammation results from locally produced cytokines (primarily interleukin [IL]-1, IL-6, and tumor necrosis factor [TNF]) and is most commonly caused by infectious agents such as viruses and bacteria (Box 37.1). The antibiotic-induced lysis of bacteria produces additional inflammatory cytokines that complicate therapy. Meningeal inflammation may also result from parameningeal infective foci. Noninfectious causes of meningeal inflammation include reactions to medications, autoimmune diseases, vasculitis, and tumors. Untreated bacterial meningitis is usually fatal. Early antibiotic therapy has reduced but not eliminated the mortality and morbidity. Viral meningitis is often self-limited. Vaccines that reduce meningitis risks are available and in development.

## CLINICAL VIGNETTE

A 50-year-old farmer is admitted with an unsteady gait, diplopia, headache, nausea, and vomiting. He had been in excellent health until fever, unsteady gait, and neck soreness started 7 days before admission and progressed to diplopia, dysarthria, dysphagia, headache, and a vague numbness in his right cheek. Nausea and vomiting started 3 days before admission. He lived on a farm in Vermont with his family, who were all healthy. He had many animal exposures but none that were unusual. He denied drinking unpasteurized milk, eating rare meat, or drinking excessive alcohol. In the past his medical health had been excellent. On physical exam his temperature was 37.5°C. He was markedly irritable and had diminished light touch sensation over his right cheek as well as a left sixth nerve palsy with right-eye nystagmus when he looked to the right. He was ataxic with a broad-based gait and no muscle weakness. Nuchal rigidity was absent and plantar reflexes were normal. Hemoglobin was 15.3 g/dL; white blood cells (WBCs) 11,400 with 72% polymorphonuclear leukocytes (PMNs), 11% bands, and 1 eosinophil. A comprehensive metabolic panel was normal. Lumbar puncture (LP) showed an opening pressure of 130 mm of fluid, 103 red blood cells (RBCs), 78 WBCs (4% PMNs, 90% lymphocytes, 6% monocytes), protein 91 mg/dL, glucose 59 mg/dL (140 mg/dL in blood). Gram smear was negative. Cryptococcal antigen (cerebrospinal fluid [CSF] and blood), syphilis, and HIV serologies were negative. A chest x-ray and cranial computed tomography (CT) scan were normal. On the third day his temperature was 38.5°C. Blood cultures were repeated. His laboratory data were unchanged. A second LP revealed clear fluid with 10 RBCs and 250 WBCs (22% PMNs, 53% lymphocytes, 25% monocytes). CSF glucose was now 40 mg/dL (120 mg/dL in blood), CSF protein had increased to 106 mg/dL. Gram stain was negative. Culture was sterile on no antibiotic. He became progressively obtunded. Four-drug tuberculosis (TB) therapy was started. On the fifth day his temperature was normal, he was more alert, and his headache had diminished. One of two blood cultures from day 3 was growing a small gram-positive bacillus identified as *Listeria monocytogenes*.

His antituberculous therapy was switched to ampicillin 12 g/day for 3 weeks. At the end of therapy his mental status was normal, his ataxia and gaze palsies had resolved, and he had no neurologic sequelae.

This patient had rhombencephalic listeriosis. He lacked the more classic, abrupt onset of meningeal changes and demonstrated the value of repeat LPs and blood cultures. The rifampin in his TB regimen probably produced the initial improvement. Fortunately the blood culture result established the true cause. The lack of risk factors for TB and the farm animal exposures in this healthy 50-year-old were clues. A chronic meningitis presentation (usually more than 7 days) should be treated as TB initially, but a specific diagnosis must be established.

## ANATOMIC AND PHYSIOLOGIC CONSIDERATIONS

The meninges consist of three layers (Fig. 37.1). The outer layer (dura mater) is a tough white fibrous connective tissue that forms the inner periosteum of the cranium and the inner meningeal layer protecting the brain. The middle layer (arachnoid) is a thin layer with numerous threadlike strands attaching it to the innermost layer, the pia mater. The pia mater is a thin, delicate membrane tightly bound to the surface of the brain and spinal cord. The subarachnoid space between the arachnoid and the pia is filled with CSF and traversed by the blood vessels of the brain.

Meningitis is an inflammation in the subarachnoid and ventricular spaces; it involves the adjacent meninges, the traversing blood vessels, and the brain structures. This inflammation occurs within a closed anatomic space that is devoid of significant defense mechanisms. Many presentations of meningitis include altered mentation due to brain inflammation and can be described as meningoencephalitis. Infection of the subdural space (empyema) usually stems from a sinus, middle ear, facial, or scalp infection. Except in infants, meningitis is a very rare cause of subdural empyema. The differential diagnosis of subdural infection is similar to that of meningitis, meningoencephalitis, brain abscess, epidural infection, subdural hematoma, and even thrombophlebitis of cerebral vessels, since headache, fever, neck stiffness, altered mental status, seizures, and focal neurologic signs can be seen in all of these entities. An epidural abscess from a contiguous infection or hematogenous spread may have a more gradual onset and feature local spine pain, myalgias, and minimal fever. Although brain abscess, subdural and epidural abscess, and encephalitis are often in the differential diagnosis of meningitis, the primary focus of this chapter is on meningitis.

CSF is produced by the choroid plexus in the lateral, third, and fourth ventricles. The choroid plexus consists of projections of vessels and pia mater into the ventricular cavities. Normally about 500 mL of CSF is produced per day by both filtration and active transport at a rate of about 20 mL/h. CSF circulates from the lateral ventricles into the third and fourth ventricles and then into the subarachnoid space over the surfaces of the brain and down the spinal cord. CSF is reabsorbed back into the bloodstream via the arachnoid villi located along the superior sagittal and intracranial venous sinuses and around the spinal nerve roots. The movement of CSF and cellular components across arachnoid villi occurs via transport within giant vesicles and functions as a one-way valve from CSF to peripheral blood (see Fig. 37.1). This description of the transport mechanism for CSF is overly simplistic, since brain lymphatic pathways have recently been discovered.

Two interfaces exist between the blood and the brain. The larger and better known is the blood-brain barrier (BBB). It functions to protect the interstitial fluid of the brain from changes in the blood levels of ions, amino acids, peptides, and other substances. Lipid-soluble molecules—including those of oxygen, carbon dioxide, anesthetics, and alcohol—can pass through the BBB and gain access to all parts of the brain depending on the rate of blood flow. The permeability of the BBB is primarily dependent on lipid solubility, molecular size, and protein binding of drugs. Drugs that are bound to plasma protein must be in a free form (unbound) to adequately penetrate the BBB. An example is ceftriaxone, a beta lactam, which is highly protein bound, hydrophilic rather than lipid soluble, and penetrates relatively poorly across the BBB. Fortunately the high antibacterial potency of ceftriaxone against *Streptococcus pneumoniae, Neisseria meningitidis,* and *Haemophilus influenzae* makes it a drug of choice in bacterial meningitis. (See resistance concerns further on regarding organisms with increasing minimal inhibitory concentrations [MICs] to beta-lactams.) The smaller, less direct interface between blood and brain is between the blood and the CSF (BcsfB). This interface controls the composition

of the CSF, and its permeability is primarily dependent on meningeal inflammation and secretion in the choroid plexus. Both barrier interfaces have distinct pharmacokinetics and are important for the regulation of CSF hydrodynamics and the exchange of substances, including antibiotics between the blood and the CNS compartments.

Normal CSF is clear and colorless. CSF sampled from the lumbar spine area in adults contains 15 to 45 mg/dL of protein and 50 to 80 mg/dL of glucose, corresponding to two-thirds of the simultaneous blood glucose value. It may also contain up to five mononuclear cells and five RBCs per microliter. More than three PMNs per microliter is considered abnormal. Trauma, spinal anesthesia, stroke, and CNS malignancy may increase the PMN count in the absence of infection. The volume of CSF sampled is usually about 15 mL divided into four tubes. Higher amounts do not seem to add to the post-LP headache risk, but elderly patients with cerebral atrophy may have a higher risk of subdural bleed.

The normal pressure measured at the lumbar space by LP is 100 to 200 mm $H_2O$ with the patient in the lateral recumbent position. The reliability of LP pressure measurements when the patient is sitting is uncertain. During the pressure reading, the patient's legs should be slightly straightened to avoid compression of the intra-abdominal cavity and artificially increasing cerebral venous blood pressure. Children may have normal CSF opening pressures to 280 mm $H_2O$. High CSF pressure with normal CSF content can indicate cerebral vein or dural sinus thrombosis. Cerebral venous thrombosis can be acute, subacute, or chronic in onset and often presents with new severe headache that worsens with recumbency. The headache is not a "thunderclap" type headache but can mimic a subarachnoid hemorrhage. Cerebral venous thrombosis is rare and more likely seen in young women with increased coagulation risks, such those who are pregnant or taking oral contraceptives. Specific diagnosis requires brain MRI and MR venography. Additional manifestations of intracranial hypertension include focal neurologic deficits, seizures, and encephalopathy.

Accidental breach of a small vessel in the lumbar region at the time of spinal tap may allow peripheral blood to mix with CSF, thus increasing the number of both WBCs and RBCs. The color of CSF from a traumatic tap may vary from pink to red and often shows progressive clearing in sequentially gathered collection tubes. Correction of the WBC count when CSF is contaminated with peripheral blood is done by subtracting one WBC from the CSF count for every 1000 RBCs seen, provided that the peripheral WBC count is within the normal range.

The first and third tubes usually go for microscopy to distinguish cellular constituents. A culture and potentially molecular testing—for example, via the polymerase chain reaction (PCR)—is done on the third tube of CSF. Fungal and mycobacterial cultures require larger CSF volumes. *Mycobacterium tuberculosis* may take 4 weeks to grow. There are nucleic acid amplification assays for *M. tuberculosis,* but in the United States they are approved only for sputum specimens. The second CSF tube is sent for protein and glucose. The fourth CSF tube can be held for later testing (e.g., CSF antibodies, syphilis serology, and fungal serology, especially a cryptococcal antigen). The detection of CSF autoimmune disorders requires looking for anti–N-methyl-D-aspartate (NMDA) antibodies in some encephalitis settings and occasionally a CSF protein 14-3-3 biomarker of neurodegenerative disorders such as Creutzfeldt-Jakob disease.

Xanthochromia, a yellow to amber color of the CSF, results from the breakdown of RBCs and the release of hemoglobin components. Xanthochromia requires about 4 hours to develop and suggests a subarachnoid or intracerebral bleed. However, high CSF protein concentrations or high bilirubin levels can also produce xanthochromia. Subarachnoid hemorrhage as opposed to a traumatic bloody tap would produce many RBCs, which would not reduce in number from

**A. Circulation of cerebrospinal fluid**

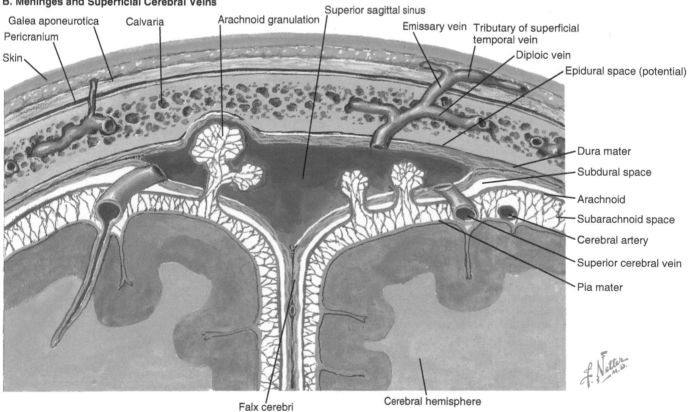

Choroid plexus of lateral ventricle (*phantom*)
Dura mater
Arachnoid
Cistern of corpus callosum
Choroid plexus of third ventricle
Interventricular foramen (of Monro)

Superior sagittal sinus
Subarachnoid space
Arachnoid granulations

Chiasmatic cistern
Interpeduncular cistern
Cerebral aqueduct (of Sylvius)
Prepontine cistern
Lateral aperture (foramen of Luschka)
Choroid plexus of 4th ventricle
Dura mater
Arachnoid
Subarachnoid space
Central canal of spinal cord

Quadrigeminal cistern (of great cerebral vein)

(Posterior) cerebellomedullary cistern
Median aperture (foramen of Magendie)

**B. Meninges and Superficial Cerebral Veins**

Galea aponeurotica
Pericranium
Skin
Calvaria
Arachnoid granulation
Superior sagittal sinus
Emissary vein
Tributary of superficial temporal vein
Diploic vein
Epidural space (potential)
Dura mater
Subdural space
Arachnoid
Subarachnoid space
Cerebral artery
Superior cerebral vein
Pia mater

Falx cerebri
Cerebral hemisphere

**Fig. 37.1** (A) Circulation of cerebrospinal fluid and (B) meninges and superficial cerebral veins.

## TABLE 37.1    Organisms That Cause Bacterial Meningitis

| Organism | Age Range/Frequency | Pathogenesis | Risk Factors | Note |
|---|---|---|---|---|
| *Streptococcus pneumoniae* | Most common cause, highest mortality in all ages. | Hematogenous from nasopharyngeal colonization or contiguous otitis, sinusitis. | Infants or elderly, HIV, head trauma, splenectomy, IgG deficiencies, immunosuppression. | Decreased prevalence since introduction of pneumococcal vaccines. |
| *Neisseria meningitidis* | Endemic worldwide, epidemic in sub-Saharan Africa, increased prevalence in youth. | Hematogenous from nasopharyngeal colonization. | Close contacts, HIV, terminal complement deficiencies. | Fluoroquinolone resistance increasing. Prophylaxis and carriage eradication needed. Droplet isolation day 1. |
| *Haemophilus influenzae* | Any age. *H. influenzae* non–type b since Hib vaccine used. | Hematogenous from nasopharyngeal colonization. | Nonimmunized and Native American Indian population at increased risk. | Vaccines since 1980s changed disease to nontypeable or non–type b. |
| *Listeria monocytogenes* | All ages but highest in newborns and elderly. Third or fourth most common meningitis. | Gastrointestinal tract, food outbreaks, placenta. | Elderly, pregnancy, immunosuppression, but about 30% occur in normal hosts. | Soft cheese, milk, processed meats as sources. *Listeria* is neurotropic. Resistant to cephalosporins. |
| Coagulase-negative *Staphylococcus* | All ages. | Dermal sinus or CNS foreign body. | Surgery, ventricular shunt (LPs done without mask). | Difficult to diagnose. |
| Group B *Streptococcus* (GBS) | Meningitis, early onset (within 6 days). Late onset (7–90 days). | Acquired in utero or on passage through the birth canal. | Vaginal colonization, premature membrane rupture, chorioamnionitis, lack of prenatal screening. | Vaginal screening and therapy in pregnancy reduces early GBS but not late GBS meningitis. Universal GBS vaccine is needed. |
| *Staphylococcus aureus* | All ages. Rare cause of community-onset meningitis. | Bacteremia or contiguous infective focus. | Neurosurgery, ventricular shunt, endocarditis. | MRSA complicates therapy. |
| Gram-negative rods | Most common in neonates, infants, or after neurosurgery. | Bacteremia. | Advanced age, severe comorbidities, neurosurgery. | Multidrug resistance is a problem. |

*CNS,* Central nervous system; *Hib, H. influenzae* type b; *HIV,* human immunodeficiency virus; *LP,* lumbar puncture; *MRSA,* methicillin-resistant *Staphylococcus aureus.*

tube one to tube three. The classic "thunderclap" headache onset is seen in many subarachnoid hemorrhages, most of which are due to leaking or ruptured cerebral aneurysms. A prodromal sentinel headache is present in as many as 40% of patients with aneurysmal subarachnoid hemorrhages. A sudden, severe headache with nausea and vomiting would not be typical of meningitis. Brain CT is more reliable than xanthochromia for diagnosing subarachnoid hemorrhage in the emergency department.

LP should always be done using sterile technique. Meningitis produced by a nonsterile LP has been well documented. Hematoma and post-LP headache are potential complications. There are concerns for brain herniation as well. When should brain imaging precede an LP to reduce the risk of intracranial hypertension and possible brain herniation? There are no studies to guide a decision here; only case reports. Herniation may occur due to increased intracranial pressure produced by meningeal inflammation unrelated to the decreased CSF pressure resulting from an LP. The presence of a brain abscess is a major concern but is more commonly associated with focal neurologic findings and a longer symptom duration than seen in meningitis. Cranial CT scans do not rule out the possibility of a very high intracranial pressure. A retrospective study of the Swedish quality registry for acute community-acquired bacterial meningitis from 2005 to 2012 suggests that LP without imaging is safe in the setting of altered mental status and leads to earlier antibiotic treatment.

Skin or deep tissue infection at the site of LP or a significant coagulation problem is a contraindication to an LP. Some guiding principles are as follows: Never delay antibiotic therapy in suspected bacterial meningitis. If brain imaging (CT scan) is imperative or a difficult LP needs to be done by interventional radiology, blood cultures should immediately be collected and appropriate antibiotic therapy started beforehand. Concomitant dexamethasone therapy is usually indicated. Generally accepted reasons for brain imaging before LP include the following: altered mental status with strong suspicion of intracranial pressure, papilledema, focal neurologic deficits, new-onset seizures, a history of stroke or CNS mass, especially in persons over 60 years of age, and an immunocompromised host. Increased intracranial pressure heightens the risk of cerebral herniation and requires emergency measures, including neurosurgical assistance (Table 37.1).

## EPIDEMIOLOGY

The incidence of meningitis in the United States and the developed world has changed greatly over the past 30 years, in large part due to the development of vaccines for *H. influenzae* type b, *S. pneumoniae*, and *N. meningitidis*. The incidence of bacterial meningitis in the United States is about 0.8/100,000 persons per year (approximately 2700 bacterial meningitis cases per year). Most community-acquired bacterial meningitis is caused by *S. pneumoniae, N. meningitidis, H. influenzae,* and *L. monocytogenes*.

In the population older than 50 years of age, *Pneumococcus* is the predominant pathogen. In neonates, group B streptococcus (GBS) (*Streptococcus agalactiae*) and *Escherichia coli* predominate. GBS meningitis can manifest within the first 6 days of life or have a late onset within 90 days. Prenatal rectovaginal screening cultures for GBS at 35 to 37 weeks of gestation and, in the setting of premature onset of labor or rupture of membranes at less than 37 weeks of gestation, the use of targeted intrapartum antibiotic prophylaxis decreased the early post-partum incidence of neonatal infection. Unfortunately, the incidence of late-onset invasive GBS infection in infants has not changed. A mul-tivalent vaccine to prevent vaginal colonization is in development. Of note, the incidence of invasive GBS disease among nonpregnant adults in the United States has increased, abetted by chronic diseases, diabetes, and obesity. Most of this GBS increase in nonpregnant adults does not involve CNS infection.

Before the introduction of *Hemophilus influenzae* type b (Hib) vaccines in the late 1980s, an average of 25,000 children developed Hib yearly in the United States, amounting to about 45% of all cases of bacterial meningitis. The conjugate Hib vaccine (1990) resulted in decreased nasopharyngeal colonization and a herd immunity in the unvaccinated population. However, the lack of vaccines for other serotypes or nontypeable *H. influenzae* has allowed the non–type b strains to increase, especially among the elderly. Pneumococcal and meningococcal meningitis are still much more common than meningitis due to *H. influenzae*.

The meningitis mortality rate decreased by 70% from 1980 (1.36 deaths per 100,000) to 2014 (0.41 deaths per 100,000). Mortality is driven by host defenses, organism virulence factors, and access to care. Faster diagnosis and treatment along with immunization programs reduced mortality rates, but there is still a significant geographic variation in the United States. *S. pneumoniae* accounts for the most common bacterial meningitis in the United States and has the highest mortality and morbidity. Increasing age and comorbidities are the major risks. Pneumococcal vaccines—pneumococcal polysaccharide vaccine (Pneumovax 23) in the 1980s and pneumococcal conjugate vaccines (Prevnar 7 and PCV13) in 2000 and 2010—have caused a significant reduction in pneumococcal invasive disease in most age groups, especially under the age of 2 years. However, a reduction in pneumococcal meningitis specifically due to vaccines is difficult to demonstrate. Any reduction in invasive disease is serotype specific and includes unvaccinated populations by "herd immunity." Of note, the incidence of serotype 3 invasive pneumococcal disease in unvaccinated children has not significantly changed with PCV13 (containing serotype 3) infant immunization programs. An increase in mucosal carriage of non-vaccine serotypes (serotype replacement) with lower virulence has been noted.

Since 1980, antimicrobial resistance among *S. pneumoniae* isolates has dramatically increased worldwide. Beta-lactam resistance is associated with reduced enzyme binding site affinity and can be overcome with higher doses. However, the CNS is an area that is both difficult to penetrate with antibiotics and relatively lacking in immunologic defenses. Consequently antibiotic resistance definitions (MICs) of beta-lactams defining resistance are much lower for CNS infections than for other anatomic locations. A relatively low dose of a beta-lactam will treat a pneumonia or skin/soft tissue infection, whereas meningitis will call for much higher dosing.

*N. meningitidis* can cause epidemics; the most devastating occur in the sub-Saharan "meningitis belt." In the United States, *N. meningitidis* is the second most common cause of meningitis, with serogroups B, C, and Y accounting for about two-thirds of the cases. The Centers for Disease Control and Prevention (CDC) reported a total of 350 meningococcal meningitis cases in 2017; 92% were sporadic, unrelated to an outbreak. Risk factors include close contact among cases (within households, in military barracks or college dorms), men who have sex with men, travelers to hyperendemic areas (e.g., the Hajj pilgrimage), and complement component deficiencies. Quadrivalent meningococcal polysaccharide conjugate vaccines (MenACWY-DT) in 2005 and (MenACWY-CRM) in 2010 reduced the risk, but serogroup B increased. Serogroup B recombinant protein meningococcal vaccines

(MenB) were introduced in 2015. Ongoing studies suggest that they have benefits in reducing outbreaks and nasopharyngeal carriage, but herd immunity seems lacking. Mortality from meningococcal meningitis is still around 15%, and at least 20% of survivors develop major clinical sequelae.

*Listeria monocytogenes* meningitis is the third or fourth most common cause of bacterial meningitis. Approximately 30% of *Listeria* meningitis occurs in normal hosts; 60% is seen in compromised hosts, including alcoholics, diabetics, pregnant women, and the extremes of age. Outbreaks are often associated with foodborne sources. Rhombencephalitis or brain-stem involvement is rare and can present over days as headache, fever, and gastrointestinal complaints followed by cranial nerve deficits, cerebellar signs of tremor or ataxia, seizures, and obtundation. Brain-stem involvement may mimic herpes simplex encephalitis and a variety of other infective and noninfective brain-stem diseases. *Listeria* under the microscope is a tiny gram-positive rod that may be difficult to see on a CSF smear and may be misidentified as a diphtheroid or possible contaminant on culture. Of note, *Listeria* is intrinsically resistant to cephalosporins, and adjuvant steroid use in *Listeria* meningitis may be detrimental.

Nosocomial meningitis usually occurs as a complication of neurosurgery or head trauma. Postoperative CNS infection can manifest as meningitis, brain abscess, or subdural/epidural empyema. Depending on the virulence of the involved organisms, infection becomes clinically apparent within hours to days or even weeks after invasive procedures on the head or spine. Unlike community-acquired meningitis, the causative organisms differ significantly, with *S. aureus*, coagulase-negative staphylococci, *Cutibacterium* spp. (*Propionibacterium* spp.), and gram-negative rods predominating. The incidence of nosocomial meningitis is approximately 1% following craniotomy procedures. In addition to recent neurosurgery, risk factors for infection include placement of CSF drainage tubes, head trauma within 1 month, and CSF leaks, which may cause recurrent meningitis.

Recurrent meningitis can arise in both the community and nosocomial settings and may have an incidence as high as 6% in the combined settings. Recurrent meningitis is also associated with deficiencies of one or more terminal complement components in the case of recurrent meningococcal infections or an immunoglobulin deficiency, a splenectomy, or a chronic CSF leak in the case of recurrent pneumococcal meningitis. Immunizations seem justified but are often overlooked. Of note, a case control study (2008 to 2017) from Kaiser Permanente Northern California found that a history of prior head injury or spine surgery well beyond the 30-day postoperative period was a potential risk for pneumococcal meningitis among adults.

The term "aseptic meningitis" has been applied to patients with clinical and laboratory signs of meningeal inflammation who have negative bacterial or fungal cultures. Viruses produce most cases of aseptic meningitis, which is much more common than bacterial meningitis. Other infective causes include partially treated bacterial meningitis, parameningeal infections, infections with mycobacteria, fungi, spirochetes (syphilis, neuroborreliosis, or leptospirosis), and parasites. Hypersensitivity reactions to medication (e.g., nonsteroidal antiinflammatory drugs [NSAIDs], sulfa drugs), malignancy, and immunologic disorders such as anti-NMDA-receptor encephalitis can produce meningeal inflammation. The extent of CSF inflammation and the cellular composition can vary widely and may be of limited diagnostic value. Predominantly lymphocytic CSF pleocytosis and mild chemical abnormalities are suggestive of viral meningitis, whereas other causes can result in mixed cellular inflammations (e.g., tuberculous meningitis or predominantly granulocytic inflammation resembling bacterial meningitis). Unfortunately the clinical and laboratory findings are not sensitive or specific. The clinical history (time of year, location, exposures, etc.) and the physical findings (rash, encephalopathy, seizures, etc.) are of some differential value but lack both sensitivity and specificity.

Most viral meningitis in adults and children is caused by enteroviruses and typically occurs from May through October in the Western Hemisphere. All members of the herpesvirus family can produce an aseptic meningitis syndrome year round. PCR testing has helped to define the cause of aseptic meningitis. Herpes simplex virus 2 (HSV-2) may be the second most common cause of aseptic meningitis in adults. A retrospective review of adults with HSV-2 in CSF found that most did not have concomitant genital lesions. It also found recurrent viral meningitis uncommon, although HSV-2 has been said to be a common cause of Mollaret syndrome, a recurrent benign lymphocytic meningitis. Other common causes of viral meningitis include mumps, varicella zoster virus, arthropod-borne viruses, lymphocytic choriomeningitis, and several adenovirus serotypes. A subset of patients with acute human immunodeficiency virus (HIV) infections will develop meningitis or meningoencephalitis manifested by headache and confusion as well as occasional focal neurologic deficits.

## PATHOGENESIS AND PATHOPHYSIOLOGY

The pathogenesis of community-acquired bacterial meningitis involves initial mucosal colonization followed by hematogenous spread to the subarachnoid space. Colonization of the nasopharynx requires successful mucosal attachment and evasion of the mucosal defenses. In some cases meningitis evolves from a contiguous local infection, such as dental, otitis media, mastoiditis, or sinusitis, and bacteria gain direct access to the subarachnoid space by venous or lymphangitic spread. Occasionally bacteremia originates from endovascular foci such as endocarditis. Endocarditis can mimic meningitis even in the absence of bacterial seeding of the CSF.

The CSF is an "immunologic desert," since WBCs, complement, immunoglobulins, and other defense mechanisms are relatively sparse. CSF inflammation is produced by the interaction of bacterial components, with host cells equipped to recognize pathogen-associated molecular patterns—that is, bacterial cell wall fragments, lipopolysaccharides (endotoxin), and others. Proinflammatory cytokines (e.g., IL-1, IL-6, and TNF-$\alpha$) are essential mediators of the CSF inflammation and contribute to ensuing pathophysiologic changes such as brain edema. Antibiotic therapy contributes to the cytokine cascade by lysing bacterial cells and releasing inflammatory mediators. Cerebral blood flow is impaired by increased intracranial pressure, loss of cerebral blood flow autoregulation, vasospasms, and thrombosis of cerebral vessels. The clinical CNS effects include headache, altered mental status, coma, focal sensory and motor deficits, and seizures. Analogous to bacterial meningitis, various inflammatory cytokine elevations have been detected in the CSF of patients with viral meningitis. The clinical distinction between bacterial and viral meningitis is difficult. Apart from actual identification of an infecting agent there are no sensitive or specific laboratory findings to distinguish bacterial from viral infection. Bacterial meningitis is clearly more severe than viral, but viral meningitis can also be lethal.

## CLINICAL FEATURES

The cardinal signs and symptoms of bacterial meningitis comprise fever, headache, nuchal rigidity, and altered mental status. Few patients manifest all four, but at least two of the four are present in more than 90% of adult patients with bacterial meningitis. Other common signs and symptoms of bacterial meningitis include photophobia, nausea and vomiting, seizures, and sensory or motor deficits. Patients with meningococcal infection may develop a petechial rash that progresses to frank ecchymosis. Some viral infections or rickettsial infections (e.g., Rocky Mountain spotted fever) may produce a petechial rash. Symptoms of meningococcal inflammation are usually present for less than a day in typical bacterial meningitis. The varicella zoster virus rash may be so distinct as not to require proof by CSF PCR when ganglionitis promotes a meningitis. TB and fungal meningitis (cryptococcosis or coccidioidomycosis) may have a more subtle, chronic presentation. Rarely, a psychiatric presentation is most prominent, with fever, headache, and focal neurologic changes obscured. The extremes of age and the severely immunocompromised hosts may have a delayed presentation. HIV-positive patients with cryptococcal meningitis may initially have normal CSF findings but a positive cryptococcal antigen test (CRAG) in both serum and CSF.

A typical presentation in older children and adults is fever, headache, neck stiffness, and confusion. Since neck stiffness may be subtle in elderly patients and may be noninfective in origin (cerebrovascular accident, Parkinson disease, or cervical arthritis) or not apparent in the comatose patient, it has less diagnostic import. Elderly patients can demonstrate confusion or generalized malaise without prominent fever, headaches, or other characteristic meningeal signs, and their diagnosis may be obscured by comorbidities. Nuchal rigidity in *Listeria* meningitis is reported to be less common and movement disorders more common than in other bacterial meningitides. The severity of headaches and fever is similar in viral and bacterial, but mental status changes are typically mild or absent with viral meningitis unless encephalitis is present. Rarely, fever, headache, meningismus, and elevated levels of inflammatory markers may be seen with arthritis involving the atlantoaxial (C1/C2) joint, surrounding the odontoid process (crowned dens syndrome) in the absence of infection.

Neonatal meningitis presentations are often nonspecific, with fever, irritability, excessive crying or sleepiness, and difficulty with feeding. Seizures or a bulging fontanelle may suggest CNS infection. The clinical spectrum of illness varies depending on age and immune status. Viral causes are much more common than bacterial, but bacterial meningitis has a very high morbidity and mortality. Age, host defenses, and organism virulence are drivers of illness. Enteroviral meningitis may be quite mild in adults, but in neonates it may present as very severe meningoencephalitis with rapidly progressive sepsis, organ failure, and a very high morbidity and mortality rate. Severe hepatitis and/or myocarditis may be more prominent than meningitis. Since antibody production is an important enteroviral clearance mechanism, patients with agammaglobulinemia are more severely affected and may develop chronic meningoencephalitis. The infecting serotype of enterovirus is a potential virulence factor, particularly if an infected neonate lacks maternal antibody to that serotype.

Tuberculous meningitis causes a subacute febrile illness with headache and lassitude, often with a gradual progression over several weeks. Over time, complaints intensify and are associated with nausea, vomiting, confusion, and cranial nerve deficits. Two fungal infections (cryptococcosis and coccidioidomycosis) may have a delayed meningitis presentation that mimics tuberculous meningitis. Immunocompromised patients with suspected CNS infection should always have blood and CSF testing for cryptococcal antigen. The epidemiology of cryptococcal disease is changing. In a retrospective cohort study (2002 to 2017) 39% of non-HIV or non-transplant-related hosts with cryptococcosis appeared to be immunocompetent.

## TABLE 37.2 Cerebrospinal Fluid Analysis in the Diagnosis of Meningitis

| Parameters | WBC Count (cells per microliter) | Differential | Glucose (mg/dL) | Protein (mg/dL) | Stains | Serology or Molecular Studies |
|---|---|---|---|---|---|---|
| Normal values | 0-5/µL | Mononuclear. | >60% of serum glucose | <40 mg/dL | Negative. | |
| Bacterial, community acquired | >1000 | >70% PMNs, but lymphocytes seen in early infection. Differential useful only when total WBCs >1000. | <40 mg/dL but abnormal if <60% of blood glucose | 100–2000 (very high when CSF obstructed) | CSF Gram stain positive >60% but dependent on bacterial numbers in CSF. | PCR for partially treated disease or negative cultures. |
| Listeria | 100–1200 | Usually PMNs greater than lymphocytes. | <40 mg/dL | 100-2000 | Tiny gram-positive diphtheroid-like. | PCR may prove helpful. |
| Tuberculous | 50–500 | Mononuclear. PMNs seen in early and treated disease. | <45 mg/dL | 100-500 | Rarely seen unless a large volume is spun down. | Molecular tests on sputa only. Xpert MTB/RIF Ultra good, but off label for CSF testing. |
| Cryptococcus HIV negative. (In HIV-positive patient cell count may be normal, but Cryptococcus antigen is often positive.) | 20–500 | 50%–80% mononuclear. | <45 mg/dL | >50 | Antigen test better than India ink staining. | Crypto antigen very sensitive in serum and CSF. Multiplex PCR assay similar accuracy but expensive. |
| Viral, general | 50–1000 (>1180 suggests a bacterial diagnosis) | Most children have PMNs persisting >48 h. PMNs not useful to distinguish bacterial from viral. | >45 mg/dL | 50–250 | Negative. | Viral PCR testing has become standard. Antibody test for West Nile preferred. Newer molecular tests coming. |

CSF, Cerebrospinal fluid; GNB, gram-negative bacilli; HIV, human immunodeficiency virus; HSV, herpes simplex virus; IgG, immunoglobulin G; IgM, immunoglobulin M; PCR, polymerase chain reaction; PMN, polymorphonuclear leukocyte; WBC, white blood cell.

## DIAGNOSIS

History is essential. Meningitis should be considered in the differential diagnosis of all febrile patients with significant headache and neurologic dysfunction. Almost any microbial agent can cause meningitis, but viruses and bacteria are most common. When a CNS infection is suspected, lab testing (especially blood cultures) should be expeditiously initiated while the history and physical exam are under way. Time is critical in starting meningitis therapy. The history should focus on the onset and evolution of symptoms, risk factors, recent illnesses, travel, contacts, medications, and allergies. Past health issues are important. An expedited but reasonably complete physical examination with special attention to neurologic deficits and skin rashes is essential. The laboratory data must include two separately drawn blood cultures before antibiotic therapy proceeds. A complete blood count (CBC) with WBC differential and a comprehensive metabolic panel are needed. Inflammatory markers such as C-reactive protein and serum procalcitonin levels are not useful in discerning the type of meningitis. CSF should be promptly examined (usually within 30 minutes of consideration) for WBC with differential count, glucose, protein, and Gram stain. An elevated CSF WBC count documents an inflammatory process in the subarachnoid space. Bacterial meningitis must be the primary consideration in all patients with compatible clinical findings. Acute bacterial meningitis is typically associated with a marked predominance of PMNs over mononuclear cells as well as elevated opening pressure (>300 mm on a water manometer), increased CSF protein concentration, and reduced

glucose concentration relative to the concomitant serum glucose level. The CSF Gram smear may rapidly identify a bacterium in many cases but is dependent on the number of bacteria in the CSF. Blood and CSF cultures are essential, since definitive diagnosis is often dependent on isolation of the causative microorganism (Table 37.2).

Viral meningitis can produce elevated CSF WBC counts with a predominance of granulocytes early in the course, shifting to lymphocytic cells by 48 to 96 hours. A predominance of PMNs in the CSF is of less diagnostic value if the total CSF WBC count is less than 1000. A low CSF glucose (compared with a simultaneous serum glucose) can occasionally be seen with viral meningitis but is more common with bacterial and fungal meningitis. A normal CSF glucose does not rule out a bacterial meningitis. Mixed lymphocytic and granulocytic CSF inflammation is typically seen with meningitis caused by M. tuberculosis and L. monocytogenes. The finding of eosinophils in the CSF is suggestive of coccidioidomycosis, parasitic (Angiostrongylus cantonensis), or allergic causes but is not diagnostic and may occur with bacterial or viral infections. Travel history may support a specific diagnosis. Rarely, in early bacterial meningitis, microorganisms may be seen on CSF Gram smear before the recruitment of WBCs occurs.

The diagnosis of meningitis in the postneurosurgical patient is often difficult, especially if the patient has had a ventricular or posterior fossa procedure. The triad of fever, neck stiffness, and change in neurologic status is nonspecific after neurosurgery. CSF analysis may be difficult to interpret, as many of the usual markers may be abnormal from underlying brain pathology and manipulation or bleeding. The neurosurgical literature suggests the use of total CSF WBC cellularity, glucose, and lactate levels to diagnose postoperative meningitis, but definitive proof depends on the documentation of a pathogen. β–D-glucan and galactomannan levels in CSF may be useful in the diagnosis of fungal ventriculitis and meningitis. In the setting of hydrocephalus, a basal meningitis may rarely produce a normal ventricular CSF when an external ventricular drain is placed but a very abnormal lumbar CSF when an LP is done. TB, cryptococcosis, coccidioidomycosis, and neurocysticercosis classically present as basilar meningitis. Outbreaks of fungal meningitis have been described following epidural injections of methylprednisolone for low back pain. A positive CSF culture is essential. Empiric therapy is warranted in clinically suggestive cases. The Practice Guidelines for Healthcare-Associated Ventriculitis and Meningitis published in 2017 are helpful.

Despite advances in microbiologic culture and molecular methods, no organism is identified in 10% to 30% of the CSF specimens from suspected bacterial meningitis. Although some organisms are difficult or impossible to culture, prior antibiotic therapy accounts for some of the negative cultures. Multiplex CSF nucleic acid testing panels that detect the genetic material of pathogens are sensitive, specific and may yield results within several hours. The potential for contamination exists, and costs are significant when compared with the past gold standard of culture and Gram smear. Molecular testing is of value when prior antibiotic or comorbidities obfuscate a diagnosis. Enteroviral molecular testing takes less than 2 hours. Syndromic molecular testing needs to confirm clinical relevance, sensitivity, and specificity, and costs and will likely remain complementary to classic microbiologic methods in the near future. Molecular sequencing techniques on CSF may be helpful in the investigation of outbreaks.

Serologic investigation for cryptococcus, *A. cantonensis*, *Mycoplasma pneumoniae* or *hominis*, *Bartonella*, *Treponema pallidum*, *Leptospira*, and neurologic Lyme disease should be guided by history and physical examination. The compromised host has virus, bacteria (especially *Nocardia* spp. and *Listeria*), TB, fungi (especially *Cryptococcus* spp.), neurocysticercosis, toxoplasmosis, primary CNS lymphoma, and vasculitis in the differential diagnoses. CSF findings of pleocytosis, high protein, and low glucose are nonspecific at best and may be normal. Brain or leptomeningeal biopsy may be needed for some diagnoses. When molecular testing is requested, a specialist in clinical microbiology should be consulted.

## MANAGEMENT OF PATIENTS WITH SUSPECTED MENINGITIS

The prognosis of patients with bacterial meningitis is critically dependent on the rapid initiation of effective therapy. Host deficiencies and organism virulence are factors that cannot be immediately changed. Appropriate antibiotic therapy along with dexamethasone are the initial treatments for most community-acquired bacterial meningitis. There are no randomized controlled studies comparing "cidal" to "static" antibiotics in the therapy of bacterial meningitis. Since the CSF lacks real immunologic defenses, including WBCs, bactericidal antibiotics are likely to be more effective than bacteriostatic agents provided that pharmacokinetics and dosing are optimal. It seems logical to assume that bactericidal antibiotics are warranted in CNS infection, much as they are in endocarditis and in neutropenic hosts. One must keep in mind that bacteriostatic antibiotics kill bacteria in vitro, as do bactericidal agents, but they need a much higher concentration compared with their MICs. Recall that host defenses—similar to WBCs, complement, and immunoglobulins—are very important. Unfortunately, part of the ongoing inflammatory response within the CNS is due to antibiotic lysis of microorganisms and the release of proinflammatory products. Ideally, we need a rapidly bactericidal, non–cell-wall-lysing antibiotic that readily penetrates the CSF—one that has yet to be developed.

## Empiric Antibiotic Therapy

Several guidelines, including those from the Infectious Diseases Society of America (IDSA) in 2004, cover in detail the recommendations regarding empiric and guided antimicrobial therapy of bacterial meningitis. Specific antibiotic resistance is rarely known initially so therapy is empiric and directed at common but possibly resistant organisms. Once bacterial susceptibility is known, therapy can be focused. Age and immunosuppression are helpful clues to specific bacterial causes of meningitis; but in adults, *S. pneumoniae*, *N. meningitidis*, and nontypeable *H. influenzae* must be appropriately treated initially. Immunosuppression (including pregnancy), bowel changes, or an obvious foodborne outbreak makes the addition of *Listeria* coverage important. Neither cephalosporins nor vancomycin will treat *Listeria*. *Listeria* meningitis warrants high-dose ampicillin (or trimethoprim-sulfamethoxazole if serious beta-lactam allergy exists). The potential adverse effect of concomitant dexamethasone has already been mentioned.

Pneumococci are typically treated with ceftriaxone 2 g IV q12h over 30 minutes, followed by vancomycin (20 mg/kg IV q12h to obtain a serum trough concentration of 15 to 20 micrograms/mL) pending beta-lactam susceptibility test results. Vancomycin should be stopped if the MIC of the *S. pneumoniae* is less than 1.0 mcg/mL to ceftriaxone. Note that even when ceftriaxone resistance is present, ceftriaxone should continue along with vancomycin. Since both antibiotics are hydrophilic and pass the BBB better when inflammation is present, there is concern that dexamethasone will reduce CSF antibiotic levels. Animal data including data from humans suggest that this concern is real; thus the need for high-dose vancomycin. The IDSA guidelines call for daily vancomycin doses of 30 to 45 mg/kg in adults and 60 mg/kg in children. European guidelines suggest 60 mg/kg/day in adults. Toxicities of vancomycin must be weighed with concurrent drugs, severe underlying infection, preexisting renal impairment, and many other factors. The potential vancomycin toxicity with high doses is lessened by the usual brevity of therapy. When ceftriaxone resistance is present and continued high-dose vancomycin is needed, clinical pharmacy assistance should be sought.

Ceftriaxone is also effective against meningococci and *H. influenzae* and can be infused quickly, so it should be the first antibiotic given. Using the MIC breakpoints based on meningitis criteria for *S. pneumoniae*, there was about 6% ceftriaxone resistance, with some geographic variation, in 2018. Due to the relatively poor penetration across an inflamed BBB and achievable CSF levels, ceftriaxone may cover only about 80% of *S. pneumoniae*; thus the need for high-dose vancomycin pending susceptibility results. In severe beta-lactam allergy (history of anaphylaxis, giant urticarial hives), a fluoroquinolone such as moxifloxacin 400 mg/day can be used with vancomycin. Once a specific organism has been isolated and susceptibility data are available, the antibiotic regimen should be narrowed.

Neonates should be empirically treated with ampicillin plus cefotaxime or ampicillin plus an aminoglycoside to include *Listeria*, group B streptococci, and gram-negative rods in the coverage. Doses for neonates, infants, and children can be found in the *Red Book 2018 from the Committee on Infectious Diseases*; American Academy of Pediatrics; Kimberlin, DW; Brady, MT; Jackson, MA; et al., or in *UpToDate* (https://www.uptodate.com).

For infants and children, the choice of empiric therapy (ceftriaxone plus vancomycin) is not different from that for adults younger than 50 years of age without immunocompromising conditions. For postneurosurgery meningitis, staphylococci and gram-negative rods have to be treated empirically. The combination of vancomycin with a broad-spectrum beta-lactam (usually cefepime or meropenem) is recommended. The cited IDSA guidelines for meningitis also specify the most appropriate therapy once the infecting organism has been identified and antibiotic sensitivities are known. The guidelines recommend a duration of antibiotic therapy depending on the isolated pathogen; 7 days for *N. meningitidis* or *H. influenzae*, 10 to 14 days for *S. pneumoniae*, and up to 21 days for *Listeria* meningitis. Intravenous dexamethasone 0.15 mg/kg IV q6h for 2 days should start immediately before antibiotic therapy begins. Improved clinical outcomes have been noted using dexamethasone in the treatment of acute bacterial meningitis (*S. pneumoniae* especially) and in adults with tuberculous meningitis presumably by reducing proinflammatory cytokines. However, dexamethasone does not improve outcomes in HIV-related cryptococcal meningitis, where it appears that proinflammatory immune responses assist fungal clearance from the CSF. *Listeria* meningitis may be another setting where a concomitant corticosteroid is not indicated.

If they were not being treated with ceftriaxone, patients with meningococcal meningitis require rifampin 600 mg orally every 12 hours for 2 days or ciprofloxacin 500 mg orally once to eradicate nasopharyngeal carriage of *N. meningitidis*. Meningococcal disease also requires droplet isolation for the first 24 hours of therapy. Close contacts (family members living in the same household) should receive rifampin or ciprofloxacin prophylaxis, as already described for the eradication of nasopharyngeal carriage. The risk of meningococcal disease in close contacts is highest within the first 10 days. The value of meningococcal vaccination in close contacts is unclear.

Sometimes the possibility of herpes simplex encephalitis is entertained in patients with suspected bacterial meningitis and mental status alterations. New-onset seizures along with rapidly progressing mental obtundation and focal neurologic findings are suggestive of encephalitis. In these patients empiric coverage with intravenous acyclovir (15 mg/kg q8h) is indicated until the cause of the disease has been clarified. Empiric acyclovir for HSV should commence immediately along with antibacterial therapy. Among the common viral causes of encephalitis, effective therapy is available only for HSV.

## PROGNOSIS

Mortality rates remain high in meningitis caused by pneumococci and *Listeria* (20% to 30%); they are lower with meningococcal disease (5% to 15%). When carefully examined, up to 50% of patients who survive pneumococcal meningitis may show signs of permanent neuronal impairment. Early antibiotic therapy is essential. Viral meningitis (excluding encephalitis) has a much more favorable prognosis, with negligible rates of mortality and long-term sequelae, even though many patients take weeks or months to recover fully.

## CONCLUSIONS

Viral meningitis is much more common than bacterial but has fewer sequelae. Bacterial meningitis is a relatively rare infection most commonly caused by *Pneumococcus*, *Meningococcus*, nontypeable *H. influenzae*, or *Listeria*. Vaccines have reduced the incidence of bacterial meningitis. Considerable mortality and morbidity still occur. Rapid diagnosis and therapy initiation are crucial. Neuroimaging should never delay therapy; in fact, antibiotics and dexamethasone should start while an LP is under way. Antibiotic resistance is always a concern. There are clinical clues in the specific diagnosis of some types of meningitis.

Until a specific diagnosis is established, culture-negative meningitis should be treated as TB if the patient has risk factors and CSF findings of low glucose and high protein. Molecular techniques for diagnosis are available but should be used selectively. Keeping in mind that "common things happen commonly" and that bacterial meningitis is

not common, it is still useful to broaden one's differential diagnosis, particularly in culture-negative meningitis. Medications, autoimmune disorders, vasculitis, malignancies, and subtle genetic deficiencies can cause meningitis. An underlying systemic infection such as endocarditis, Rocky Mountain spotted fever, syphilis, or Lyme disease may occasionally be the source.

The host is the primary driver of infective causes. The microorganisms contribute virulence and resistance properties. The immunologic desert of the CSF makes bactericidal antibiotic therapy essential. Speed in starting appropriate therapy is crucial. When complications occur, one must look for cerebral vasculitis and swelling causing increased intracranial pressure and subsequent coma, bradycardia, respiratory depression, hypertension, or cranial nerve palsies. Seizures occur in up to 30% of patients with bacterial meningitis. On occasion, status epilepticus must be ruled out with an electroencephalogram. Bacterial meningitis in the setting of otitis or sinusitis is more likely to be complicated by the rare occurrence of subdural empyema. The new onset of focal neurologic deficits, seizures, and worsening mentation along with neuroimaging changes warrants a neurosurgical consultation, since subdural empyema has a very poor prognosis. Hydrocephalus formation is slightly more common (5% in bacterial meningitis) and has a poor prognosis. Early repeat LP and/or radiologic imaging is essential when the clinical condition fails to improve. There is no reason to do an LP as a test of cure. Antibiotic prophylaxis and droplet isolation are recommended with meningococcal disease.

## ACKNOWLEDGMENT

The author would like to acknowledge the first edition authors Thomas A. Kurrus[†] and Martin G. Tauber.

## EVIDENCE

Bard JD, Albyc K. Point-counterpoint: meningitis/encephalitis syndromic testing in the clinical laboratory. *J Clin Microbiol* 2018;56:1-10. *Rapid multiplex PCR testing exists for multiple pathogens simultaneously, but is the need, cost, and specificity worth it? Syndromic molecular testing will need ordering and interpretation guidelines as they become more common in point-of-care settings. Gram smears are rapid and still very useful.*

Brouwer MC, van de Beek D. Epidemiology of community-acquired bacterial meningitis. *Curr Opin Infect Dis* 2018;31:78-84. *A very cogent discussion of the incidence of bacterial meningitis in the United States, Europe, and Africa, the impact of vaccines, and the concerns for antibiotic resistance.*

Charlier C, Perrodeau E, Leclercq A, et al. Clinical features and prognostic factors of listeriosis: the MONALISA national prospective cohort study. *Lancet Infect Dis* 2017;17:510-519. *Large French observational cohort study of all forms of invasive listeriosis with 212 patients manifesting meningoencephalitis. More than 30% had no known risk factors. Adjunctive steroid use significantly reduced survival. Long-term sequelae were noted in 44% of survivors.*

Cooper R, Dudley J, Farmakiotis D. Man with headache, fever, and neck stiffness; crowned dens syndrome. *JAMA* 2019;321:1624-25. *Rarely, fever, headache, and meningismus may be seen with arthritis involving the atlantoaxial (C1/C2) joint, surrounding the odontoid process in the absence of infection. Associated with osteoarthritis, rheumatoid arthritis, and crystal-forming disorders such as hyperparathyroidism and hemochromatosis.*

Glimåker M, Sjölin J. Lumbar puncture is safe in bacterial meningitis: impaired mental status alone does not motivate cranial computed tomography before lumbar puncture. *Clin Infect Dis* 2019;68:168. *A letter to the editor with references pointing out that lumbar puncture rarely (much less than 1%) causes clinical deterioration in bacterial meningitis, that cerebral CT is a poor predictor of high intracranial pressure and risk of herniation, and that early antibiotic use improves outcomes.*

†Deceased.

Khan, SF, Macauley T, Tong SYC, et al. ventricular cerebrospinal fluid assessment misleads: basal meningitis and the importance of lumbar puncture sampling. *Open Forum Infect Dis* 2019;6:ofz324. *Four very instructive basilar meningitis cases that emphasize the potential discordance between ventricular CSF chemistries, cultures, and serologies compared to lumbar CSF findings in the setting of hydrocephalus. Normal ventricular CSF does not always equate to normal lumbar CSF.*

Tunkel AR, Hasbun R, Bhimra A, et al. 2017 Infectious Diseases Society of America's Clinical Practice Guidelines for Healthcare-Associated Ventriculitis and Meningitis. *Clin Infect Dis* 2017;64:e34-e65. *These guidelines are primarily based on expert opinion of best practice rather than randomized controlled trials but are nevertheless practical.*

## ADDITIONAL RESOURCES

Langereis JD, de Jonge MI: Invasive disease caused by nontypeable Haemophilus influenzae. *Emerg Infect Dis* 2015;21(10):1711-18. *Since H. influenzae type b vaccines, invasive, nontypeable H. influenzae has increased, especially among the elderly. Most were blood isolates and not causing meningitis. Reasons for this shift are unclear but suggested.*

McCarthy M, Rosengart A, Schuetz AN, et al. Mold infections of the central nervous system. *N Engl J Med* 2014;371:150-160. *A review of the epidemiology, clinical characteristics, and treatment of CNS mold infections. Early diagnosis, therapy including host immune status improvement, along with neurosurgical evaluation are emphasized. Although uncommon and more likely seen in immunocompromised hosts, CNS mold infections can occur in immunocompetent hosts following trauma, intravenous drug use, and contaminated medical supplies.*

McGill F, Griffiths MJ, Solomon T. Viral meningitis: current issues in diagnosis and treatment. *Curr Opin Infect Dis* 2017;30:248-256. *An overview of viral meningitis emphasizing how common it is, how difficult a distinction from bacterial meningitis may be, how seldom a proven etiology is identified, and how limited our therapy is today. Molecular testing is rapid and may help to reduce unnecessary antibiotic use.*

Scheld WM, Whitley RJ, Marra CM. *Infections of the Central Nervous System.* 4th Edition. 2014; Lippincott Williams & Wilkins, Philadelphia. *At more than 950 pages and 90 authors, this text is a definitive resource.*

Straus SE, Thorpe KE, Holroyd-Leduc J. How do I perform a lumbar puncture and analyze the results to diagnose bacterial meningitis? *JAMA* 2006;296:2012-22. *An excellent review of the techniques, adverse effects, and contraindications for LPs in the diagnosis of bacterial meningitis.*

Swartz MN. Bacterial meningitis—a view of the past 90 years. *N Engl J Med* 2004;351:1826-1828. *A concise historical perspective of the epidemiology of bacterial meningitis.*

Tunkel AR, Hartman BJ, Kaplan SL, et al. Practice guidelines for the management of bacterial meningitis. *Clin Infect Dis* 2004;39:1267-1284. *Review of major recommendations for initial management, diagnostic testing, and specific antimicrobial agents for the treatment of patients with suspected or proven bacterial meningitis. These guidelines are dated but still relevant.*

van de Beek D, Cabellos C, Dzupova O, et al. ESCMID guideline: diagnosis and treatment of acute bacterial meningitis. *Clin Microbiol Infect* 2016;22:S37-S62. *This is a comprehensive, practical guideline from the European Society for Clinical Microbiology and Infectious Diseases.*

Venkatesan A, Michael BD, Probasco JC, et al. Acute encephalitis in immunocompetent adults. *Lancet* 2019;393:702-16. *A comprehensive review of infective and autoimmune encephalitis. Viral causes predominate, with other infective causes account for about 50%. The postinfectious syndrome of acute disseminated encephalomyelitis (ADEM) and autoimmune encephalitis account for another 20% to 30%. The clinical features and diagnostic strategies overlap those seen in meningitis.*

# 38

# Osteomyelitis

*Russell W. Steele*

 **ABSTRACT**

The three distinct presentations of osteomyelitis are defined by the mechanism whereby infectious agents are introduced into bone: (1) hematogenous infection from bacteremia; (2) local spread from contiguous foci such as abscesses, insect bites, or infected exanthematous lesions; and (3) direct inoculation after trauma, invasive procedures, or surgery. There is no particular geographic distribution, and the incidence in the United States is 8 per 100,000 population per year, with cases diagnosed by most primary care physicians at least yearly. It is 2.5 times more prevalent in males than in females, and approximately 40% of cases occur in patients younger than 20 years of age.

*Staphylococcus aureus* is the most common pathogen in all age groups, with methicillin-resistant *S. aureus* (MRSA) accounting for 25% to 40% of these organisms. This prevalence necessitates the addition of vancomycin to the initial empiric therapy, since currently all MRSA isolates ate resistant to cefazolin and 25% to 50% are resistant to clindamycin. Abundant data now support deescalating to oral therapy after 2 to 5 days of intravenous antibiotics for all organisms except perhaps MRSA, since there are currently few published studies examining oral antibiotics for this pathogen.

## CLINICAL VIGNETTE

An 11-year-old female presented with bone pain, her third episode of a gradual onset, this time involving her proximal left femur and pelvis. She had had two similar episodes over the previous year, one involving the right tibia and mandible and the other the right humerus. Each time the pain was moderate, interrupting her sleep and worsening with physical activity such as playing volleyball. Fever was present intermittently. Each time the magnetic resonance imaging (MRI) scan was interpreted as osteomyelitis, but cultures of the tibia and humerus were negative. Each time the patient was treated for 3 days with intravenous vancomycin followed by oral cephalexin to complete 21 days of therapy.

She was otherwise healthy, with no weight loss and no respiratory, gastrointestinal, or urinary tract symptoms. She had not traveled out of the country and did not have any pets.

Physical exam was positive only for pain on palpation of the left femur and iliac crest.

Pertinent lab findings included a white blood cell (WBC) count of 18,200/mm³ (normal 4500–13,500), C-reactive protein (CRP) of 12.3 (0–8.2), and procalcitonin of 0.6 (<0.25). A chest x-ray was normal.

The most likely diagnosis was chronic recurrent multifocal osteomyelitis (CRMO), an inflammatory bone condition thought to be an autoimmune process, since it is often associated with psoriasis or inflammatory bowel disease or as part of the Majeed syndrome or deficiency of the IL-1 receptor antagonist

(DIRA). Signs and symptoms of these include recurrent episodes of pain and joint swelling with or without fever. Symptoms typically begin in childhood, usually occurring alone. For most children, CRMO resolves after many years without lasting effects. However, it can also cause slow growth and permanent bone deformity.

- **Tuberculosis of bone** (Pott disease) usually involves the spine and is seen with advanced disease. A normal chest x-ray, absence of a travel or exposure history, and the recurrent nature of the disease in multiple bones rules out this etiology.
- **Cat scratch disease** can cause multifocal osteomyelitis but is an acute process that resolves with proper treatment. Our patient had no exposure to cats, but if the etiology is in question, as it would have been with the first episode, serologic studies for IgM and IgG antibody to *Bartonella henselae* would be most useful in eliminating this etiology.
- **Kingella kingae** is increasingly recognized as a cause of culture negative osteomyelitis and more commonly septic arthritis in children <5 years of age. The reason is the difficulty in culturing this fastidious organism so a PCR is often required to confirm the etiology. This patient was 10 years old with the first episode and the recurrent disease involving multiple bones rules out this etiology.
- **Ewing sarcoma** and other bone tumors are often in the differential for bone pain with or without fever; a biopsy would be needed to rule out this etiology. An elevated CRP with a low procalcitonin (<2.0) along with a negative bacterial culture should raise suspicion for this diagnosis.

## CLINICAL VIGNETTE

A 3-week-old neonate is brought in because he has not moved his right arm for the past 24 hours. He is afebrile. An x-ray demonstrates a lytic lesion of the proximal humerus with periosteal elevation. A complete blood count (CBC) included a WBC count of 24,400/mm³ with 62% PMNs and 7% bands. The CRP and procalcitonin were both markedly elevated. The Gram stain of an aspirate and biopsy showed gram-positive cocci.

This is the classic presentation for group B streptococcus (GBS) late-onset disease involving bone, with the right humerus being the most common site. It is hypothesized that the reason for this location is that the right humerus lies over the mother's sacral promontory during labor, producing trauma to the bone. Late-onset GBS neonatal disease is defined as an infection appearing after the first 7 days of life.

It is not uncommon for neonates with osteomyelitis to be afebrile. They may also be afebrile with bacteremia and even meningitis, manifesting only poor feeding and lethargy. Therefore even very subtle signs of illness warrant a full sepsis workup in a neonate (0–28 days of life). With evidence of osteomyelitis in this neonate, a full sepsis evaluation (including cultures of blood, urine, and cerebrospinal fluid [CSF]) should be undertaken. If the neonate also has meningitis, the dosage of antibiotics is usually higher than for osteomyelitis alone, and the duration of intravenous antibiotics is longer. Recommended antibiotics for GBS include penicillin or ampicillin plus gentamicin. Intravenous ceftriaxone is an option for treating GBS and offers the advantage of once-a-day therapy. However, because ceftriaxone has the property of binding to bilirubin protein-binding sites, it should not be used in a jaundiced neonate or young infant because of the theoretic possibility of increasing the likelihood of kernicterus.

*S. aureus* osteomyelitis is certainly a possibility in this case, warranting initial empiric antibiotics to cover MRSA as well as methicillin-sensitive *S. aureus* (MSSA). This would require the addition of vancomycin. The increasing resistance of MRSA to clindamycin (25%–40%) would make this a poor choice for initial empiric therapy.

It would be helpful to review the Gram stain to see whether these organisms are primarily in chains (streptococci) or clusters (staphylococci), but it is often difficult to classify the morphology. Therefore an antibiotic regimen to cover both groups of organisms must be continued until cultures are complete.

# HEMATOGENOUS OSTEOMYELITIS

Hematogenous osteomyelitis originates in the metaphysis of tubular long bones adjacent to the epiphyseal growth plate. Pediatric patients almost exclusively have this mechanism of infection. Thrombosis of the low-velocity sinusoidal vessels from trauma or embolization is considered the focus for bacterial seeding in this process. This avascular environment allows invading organisms to proliferate while avoiding the influx of phagocytes, the presence of serum antibody and complement, interaction with tissue macrophages, and other host defense mechanisms. The proliferation of organisms, the release of organism enzymes and by-products, and the fixed-volume environment contribute to progressive bone necrosis (Fig. 38.1).

## Clinical Features

The signs, symptoms, and pathologic progression vary by age (Fig. 38.2 and Table 38.1). Tubular long bones are primarily involved, especially of the lower extremities (Table 38.2).

## Pathogenesis

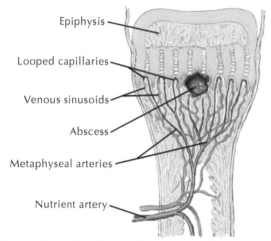

Terminal branches of metaphyseal arteries form loops at growth plate and enter irregular afferent venous sinusoids. Blood flow is slowed and turbulent, predisposing to bacterial seeding. In addition, lining cells have little or no phagocytic activity. Area is catch basin for bacteria, and abscess may form.

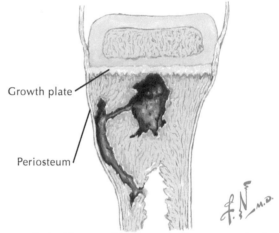

Abscess, limited by growth plate, spreads transversely along Volkmann canals and elevates periosteum; extends subperiosteally and may invade shaft. In infants under 1 year of age, some metaphyseal arterial branches pass through growth plate, and infection may invade epiphysis and joint.

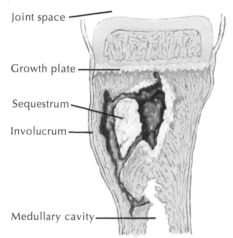

As abscess spreads, segment of devitalized bone (sequestrum) remains within it. Elevated periosteum may also lay down bone to form encasing shell (involucrum). Occasionally, abscess is walled off by fibrosis and bone sclerosis to form Brodie abscess.

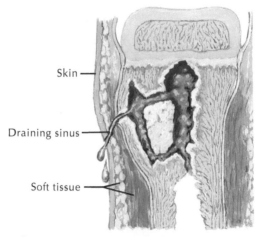

Infectious process may erode periosteum and form sinus through soft tissues and skin to drain externally. Process is influenced by virulence of organism, resistance of host, administration of antibiotics, and fibrotic and sclerotic responses.

**Fig. 38.1** Pathogenesis of hematogenous osteomyelitis.

## Clinical Manifestations

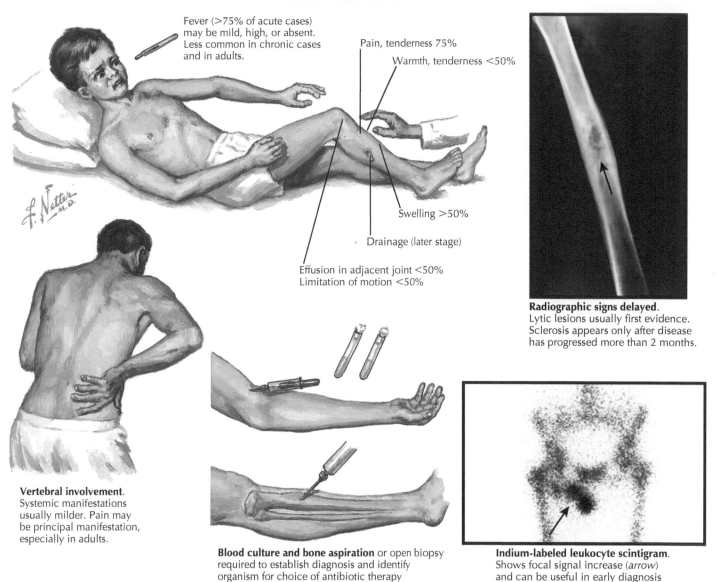

Fever (>75% of acute cases) may be mild, high, or absent. Less common in chronic cases and in adults.

Pain, tenderness 75%

Warmth, tenderness <50%

Swelling >50%

Drainage (later stage)

Effusion in adjacent joint <50%
Limitation of motion <50%

**Radiographic signs delayed.** Lytic lesions usually first evidence. Sclerosis appears only after disease has progressed more than 2 months.

**Vertebral involvement.** Systemic manifestations usually milder. Pain may be principal manifestation, especially in adults.

**Blood culture and bone aspiration** or open biopsy required to establish diagnosis and identify organism for choice of antibiotic therapy

**Indium-labeled leukocyte scintigram.** Shows focal signal increase (*arrow*) and can be useful in early diagnosis

**Fig. 38.2** Clinical manifestations of hematogenous osteomyelitis.

| TABLE 38.1 | Hematogenous Osteomyelitis: Signs and Symptoms | | |
|---|---|---|---|
| | **Newborn** | **Older Infant and Young Child (2 wk to 4 yr)** | **Older Child, Adolescent, and Adult (>4 yr)** |
| Systemic symptoms[a] | Clinical sepsis; irritable, especially to touch; pseudoparalysis | Pain; limp; refusal to use affected limb | Focal signs and symptoms; less restriction of movement; local pain; mild limp; fever; malaise |
| Signs | Red, swollen, discolored local site; massive swelling | Marked focality; point tenderness; well-localized pain | Focal signs; point tenderness very localized |
| Pathology | Thin cortex; dissects into surrounding tissue | Cortex thicker; periosteum dense | Metaphyseal cortex thick; periosteum fibrous and dense |
| Progression | Nidus (purulent) progresses rapidly[b]; subperiosteal purulence spreads; secondary septic arthritis | Subperiosteal abscess and edema; metaphyseal involvement | Cortical rupture rare |
| Radiograph | Useful early—periosteal and bony changes | Later findings confirmatory; early changes—deep soft-tissue swelling | Bony changes apparent only after 7–10 days of involvement |

[a]May be subclinical; constitutional symptoms (fever, malaise, anorexia, irritability) are no different among the different age groups; also, there is no correlation with the severity of constitutional symptoms and ultimate severity of subsequent osteomyelitis.
[b]Residual effects may be anticipated in up to 25% of newborns.

| TABLE 38.2 | Site of Bone Involvement |
|---|---|
| **Site** | **Frequency (%)** |
| Femur | 23–36 |
| Tibia | 19–26 |
| Humerus | 5–13 |
| Fibula | 4–10 |
| Radius | 1–4 |
| Calcaneus | 4–11 |
| Ilium | 3–14 |

| TABLE 38.3 | Etiology of Hematogenous Osteomyelitis | | |
|---|---|---|---|
| **Neonates** | | **Infants, Children, and Adults** | |
| Staphylococcus aureus | 40% | S. aureus (MSSA and MRSA) MRSA | 80% 30%–50% |
| Group B streptococci | 30% | Group A streptococci | 7% |
| | | Kingella kingae | 3% |
| Coliforms | 10% | Salmonella species | 3% |
| Others | 20% | Others | 7% |
| Neisseria gonorrhoeae Pseudomonas aeruginosa Candida species | | Coliforms Streptococcus pneumoniae Candida species Anaerobes | |

MRSA, Methicillin-resistant Staphylococcus aureus; MSSA, methicillin-sensitive Staphylococcus aureus.

| TABLE 38.4 | Specific Causes of Osteomyelitis |
|---|---|
| **Clinical Circumstances** | **Probable Cause** |
| **Common Causes** | |
| Human bite | Eikenella corrodens, Staphylococcus |
| Dog or cat bite | Pasteurella multocida, Streptococcus, Staphylococcus |
| Puncture wound of foot | Pseudomonas aeruginosa |
| Sickle cell disease | Salmonella species |
| Rheumatoid arthritis | Staphylococcus aureus (from joint) Pasteurella multocida |
| Diabetes mellitus | Fungi |
| Newborns | Group B streptococci Escherichia coli Salmonella species |
| **Uncommon Causes** | |
| Facial and cervical area; in the jaw; sinus drainage; lytic bone changes with "eggshell" areas of new bone | Actinomyces species |
| Vertebral body or long bone abscesses; systemic signs and symptoms | Brucella Salmonella |
| Regional distribution; systemic findings; vertebral body, skull, long bone involvement | Coccidioides |
| Skin lesion; pulmonary involvement; skull and vertebral bodies most common, but long bone involvement is reported | Blastomyces |
| Very distinct, slowly progressive bony lesions can occur | Cryptococcus |
| Exposure to cats, fever of unknown origin, liver granulomas, chronic adenitis | Bartonella henselae (cat scratch disease) |

## Etiology

The bacterial and fungal causes of hematogenous osteomyelitis demonstrate an age-specific pattern (Table 38.3). Other epidemiologic factors, predisposing chronic diseases, and exposure history may suggest unusual pathogens (Table 38.4).

## Differential Diagnosis

The differential diagnosis of hematogenous osteomyelitis includes diskitis, pyomyositis, cellulitis, toxic synovitis, septic arthritis, thrombophlebitis, or, in a patient with sickle cell disease, a bone infarction.

# CHRONIC OSTEOMYELITIS

By definition, continuation of bone infection for more than 30 days is termed *chronic osteomyelitis*. This can occur with both hematogenous- and contiguous-focus disease, but it is most commonly seen with contaminated open fractures or placement of orthopedic stabilizing devices such as metal rods and screws. Persistent drainage and development of sinus tracts are common. Bone loss may be extensive (Fig. 38.3). Removal of the foreign body and any necrotic bone is usually necessary to clear the infection, and antibiotics must be continued for months until there is clear clinical and radiographic evidence of resolution.

## Diagnostic Approach

The diagnosis of osteomyelitis is confirmed with the isolation of organisms from bone, subperiosteal exudate, or contiguous joint fluid.

Needle aspiration through normal skin over involved bone at a subperiosteal site or at the metaphyseal area combined with aspiration from a potentially involved joint should be performed by an orthopedic surgeon. Aspirates of involved focal areas yield positive cultures in only 50% to 60% of cases that have not been pretreated with antibiotics. However, because early institution of antimicrobial therapy is so common, even fewer suspected cases are culture positive.

Blood cultures have been reported to be positive in as many as 50% of cases, so samples should be obtained routinely before the initiation of antimicrobial therapy.

The most sensitive radiographic procedure for establishing a diagnosis of osteomyelitis, localizing disease, and determining the need for surgical intervention is MRI. Plain radiographs may be considered because they are readily available and may identify other causes of bone pain. The diagnosis of osteomyelitis on routine radiographs can range from subtle to obvious (Table 38.5), depending on the duration of disease, and they are adequate for differentiating patients with trauma, including physical abuse. Radiographs are also fairly sensitive for identifying leukemic infiltrates, which represent one important cause of bone pain. For osteomyelitis, changes are not apparent until approximately 14 days after infection begins.

Diagnosis of bone infection is enhanced with the use of MRI scans, which can be completed in minutes, or technetium scanning, which requires 2 hours to complete. One of these studies should be obtained for any patient who has obvious evidence of focal bone pathology, fever

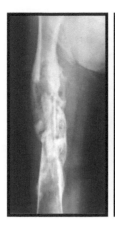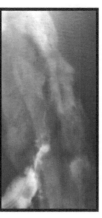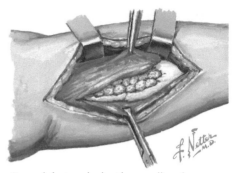

Fracture of femur treated with intramedullary rod. Postoperative infection developed and rod was removed. Persistent drainage and sinus tract developed. Radiograph *at left* shows sequestra and radiolucent lesions consistent with chronic osteomyelitis. Sinogram *at right* reveals sinus tract extending to site of fracture.

Sequestra surgically removed, sinus tract excised, and all dead and infected tissue thoroughly débrided.

Bone defect packed with cancellous bone autograft from ilium overlaid with muscle flap from vastus lateralis muscle. Patient remained free of pain and drainage.

**Fig. 38.3** Chronic osteomyelitis.

| TABLE 38.5 | **Radiographic Diagnosis of Hematogenous Osteomyelitis** |
|---|---|
| **Day** | **Changes** |
| 0–3 | Local, deep soft tissue swelling; near metaphysical region or with localized findings |
| 3–7 | Deep soft tissue swelling; obscured translucent fat lines (spread of edema fluid) |
| 10–21 | Variable—bone specific; bone destruction, periosteal new bone formation |

of undetermined cause with bone tenderness on physical examination, and an elevated CRP or procalcitonin, or suggestive findings on routine radiographs. Both imaging procedures can also aid in directing aspirate procedures for diagnosis and culture.

In situations where osteomyelitis is suspected on physical examination, MRI should be performed. In selected cases—including pelvic, vertebral, and small bone (hands or feet) osteomyelitis—MRI or computed tomography (CT) can be useful in establishing a diagnosis or directing surgical intervention. A technetium- or indium-labeled WBC scan using tagged autologous leukocytes requires 24 hours of imaging for completion and has only limited usefulness. Gallium scanning requires 24 to 48 hours, may be difficult to read (midline scans) owing to uptake in the bowel, and has been replaced by MRI, CT, and WBC scans.

## NONHEMATOGENOUS OSTEOMYELITIS

Bone involvement arises through spread from a contiguous focus of infection or direct inoculation. Common predisposing factors are trauma, burns, and nail puncture wounds of the foot. Infections in deep sites, such as retropharyngeal and renal abscesses, may also spread to bone (Fig. 38.4). The following are the more common types of nonhematogenous osteomyelitis.

## *PSEUDOMONAS* OSTEOCHONDRITIS

The predilection of *Pseudomonas* to involve cartilaginous tissue and the relative amount of cartilage in children's tarsal-metatarsal region explain why this infection is classified as an "osteochondritis." The classic history is a nail puncture through a tennis shoe. This entity is seen as early as 2 days postinjury but frequently requires up to 21 days to manifest clinically. The proper initial management is vigorous

irrigation and cleansing of the puncture wound in conjunction with tetanus prophylaxis. On diagnosis of *Pseudomonas* osteochondritis from wound drainage culture or surgical curettage culture, intravenous antibiotics should be initiated and guided by antibiotic sensitivity testing. The crucial factor in successful therapy is complete evacuation of all necrotic infected bone and cartilage. If this is accomplished, only 7 to 10 days of parenteral antibiotics will be needed (depending on soft tissue healing and appearance) for completion of therapy. *Pseudomonas* osteochondritis of the vertebrae or pelvis should lead one to suspect intravenous drug abuse as a cause.

## PATELLAR OSTEOCHONDRITIS

Patellar osteochondritis is seen most commonly in children 5 to 15 years of age, when the patella has significant vascular integrity. Direct inoculation via a puncture wound will yield symptoms within 1 week to 10 days. Constitutional symptoms are uncommon. *S. aureus* is the most common cause. Radiographs may take 2 to 3 weeks to show bone sclerosis or destruction.

## CONTIGUOUS OSTEOCHONDRITIS

In contrast to other forms of bone and cartilage infection, contiguous osteochondritis is more common in adults. It is associated with nosocomially infected burns or penetrating wounds. The clinical course characteristically includes 2 to 4 weeks of local pain, skin erosion, ulceration, or sinus drainage. Multiple organisms are common, and draining sinus cultures correlate well with bone aspirate or biopsy cultures. *S. aureus*, streptococci, anaerobes, and nosocomial gram-negative enterics are the causative organisms; the peripheral leukocyte count or erythrocyte sedimentation rate is usually normal. It is important to be aware of predisposing conditions (Table 38.6).

## PELVIC OSTEOMYELITIS

Bones involved, in order of frequency (highest to lowest), are the ilium, ischium, pubis, and sacroiliac areas. The tendency for multifocal involvement is high in the pelvis compared with other sites. The symptoms may be poorly localized with vague onset; hip and buttock pain with a limp are frequently the only findings. Tenderness to palpation in the buttocks or the sciatic notch or positive sacroiliac joint findings suggest this diagnosis. The differential diagnosis includes mesenteric lymphadenitis, urinary tract infection, and acute appendicitis. Patients with inflammatory bowel disease have an increased risk for the development of pelvic osteomyelitis.

## Secondary to Contiguous Focus of Infection

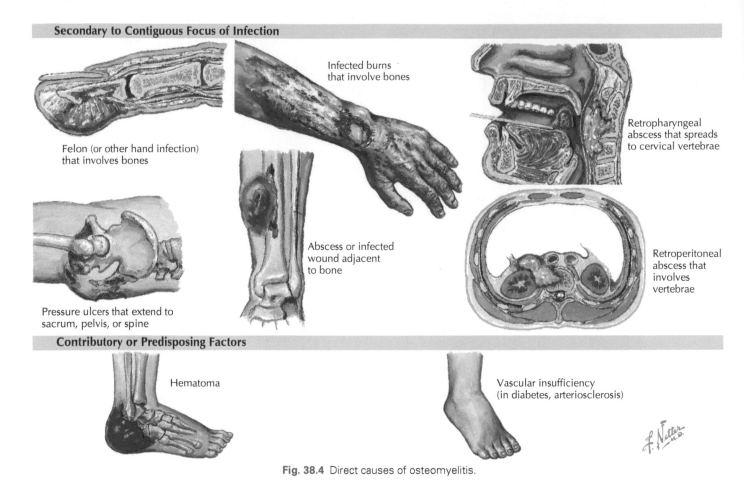

Felon (or other hand infection) that involves bones

Infected burns that involve bones

Retropharyngeal abscess that spreads to cervical vertebrae

Pressure ulcers that extend to sacrum, pelvis, or spine

Abscess or infected wound adjacent to bone

Retroperitoneal abscess that involves vertebrae

## Contributory or Predisposing Factors

Hematoma

Vascular insufficiency (in diabetes, arteriosclerosis)

**Fig. 38.4** Direct causes of osteomyelitis.

| TABLE 38.6 | **Predisposing Conditions for Contiguous Osteochondritis** |
| --- | --- |
| Closed fractures | Osteomyelitis one to several weeks postfracture; after postfracture pain subsides, the pain recurs with progression; local erythema, warmth, and fluctuation; fever common; osteomyelitis applies to this circumstance |
| Open fractures | Thorough debridement and wound cleansing paramount; lower infection rates have been reported in patients receiving prophylactic first-generation cephalosporin for open fractures; the consequence of infection can be significant; staphylococci, streptococci, anaerobes, and *Clostridium* species or gram-negative enterics, depending on the environment related to the trauma, should be considered; tetanus prophylaxis vital |
| Hemodialysis | Increased risk because of multiple procedures with intravascular cannulas; ribs, thoracic spine, and bones adjacent to indwelling catheters; *Staphylococcus aureus* and *Staphylococcus epidermidis* commonly found |

## VERTEBRAL OSTEOMYELITIS

The vertebral venous system is valveless, with a low-velocity bidirectional flow, likely the predisposing factors for vertebral body osteomyelitis. It is common for two adjacent vertebrae to be involved while sparing the intervertebral disk. Spread to the internal venous system (epidural abscess) or the external venous system (paraspinous abscess) are complications of this infection.

The symptoms of vertebral osteomyelitis include constant back pain (usually dull), low-grade fever, and pain on exertion; the signs may include paraspinous muscle spasm, tenderness on palpation or percussion of the spinal dorsal processes, and limitation of motion. The symptoms can be present for 3 to 4 months without overt toxicity or signs of sepsis. Radiographs show rarefaction in one vertebral edge as early as several days and progress to marked destruction, usually anteriorly, followed by changes in the adjacent vertebrae and new bone formation.

### Diskitis

Infection of the intervertebral disk most commonly occurs in infants and young children with the lower lumbar disks most frequently affected. Diskitis typically presents with the gradual onset of irritability and back pain, limp, or difficulty crawling or walking, without systemic signs or symptoms; fever is often absent. The pathogenesis is incompletely understood. *S. aureus* is recovered from as many as 60% of biopsied disks. But despite the frequency of positive cultures, children recover without antibiotic therapy.

## CHRONIC RECURRENT MULTIFOCAL OSTEOMYELITIS

CRMO, also called chronic nonbacterial osteomyelitis (CNO), is an inflammatory bone condition thought to be an autoimmune process, since it is often associated with psoriasis or inflammatory bowel disease, or as part of Majeed syndrome or DIRA. Signs and symptoms

## TABLE 38.7 Simplified Management of Osteomyelitis

| Phase | Management |
|---|---|
| *Initial:* day 0–3, inpatient | Obtain CBC and CRP<br>Begin intravenous antibiotics (e.g., clindamycin)<br>Repeat CBC and CRP when patient is afebrile and clinical response observed<br>CRP <3 mg/dL or returning to normal, proceed to next phase |
| *Continued therapy:* day 4–21, outpatient | Oral antibiotics (e.g., linezolid [or cephalexin if susceptible]), at two to three times the usual dose |
| *Completion:* day 21 (ESR <30 mm/h) | Obtain ESR; if <30 mm/h, stop antibiotics |
| *Delayed response:* day 21–42 (ESR >30 mm/h) | ESR >30 mm/h<br>Obtain MRI scan<br>Surgical debridement if bone inflammation and destruction identified<br>Continue oral antibiotics<br>Repeat ESR at 42 days<br><30 mm/h: stop antibiotics<br>>30 mm/h: repeat MRI scan, consider continued surgical and/or medical management at 6 weeks |

*CBC,* Complete blood count; *CRP,* C-reactive protein; *ESR,* erythrocyte sedimentation rate; *IV,* intravenous; *MRI,* magnetic resonance imaging.

## TABLE 38.8 Antimicrobial Therapy for Osteomyelitis

| Infection | Antimicrobial Agents |
|---|---|
| **Empiric Therapy for Osteomyelitis and Septic Arthritis** | |
| Neonate (0–28 days) | Vancomycin *plus* cefotaxime or ceftriaxone |
| Infants and children | Clindamycin, vancomycin, or linezolid |
| Puncture wound to foot | Gentamicin,[a] tobramycin, or amikacin *plus* ceftazidime, ticarcillin, or meropenem *plus* clindamycin, vancomycin, or linezolid[b] |
| **Specific Therapy[c]** | |
| Methicillin-sensitive *Staphylococcus aureus* (MSSA) | Cefazolin, clindamycin, oxacillin, nafcillin |
| Methicillin-resistant *S. aureus* (MRSA) | Clindamycin, vancomycin, or linezolid |
| Group B streptococci | Penicillin |
| Group A streptococci | Penicillin |
| *Streptococcus pneumoniae* | Penicillin sensitive—penicillin<br>Penicillin resistant—ceftriaxone, cefotaxime<br>Penicillin or cephalosporin resistant—clindamycin, vancomycin, or linezolid |
| Enterobacteriaceae | Meropenem, cefepime<br>Alternative: aminoglycoside or third-generation cephalosporins, depending on sensitivities |
| *Neisseria gonorrhoeae* | Ceftriaxone |
| *Pseudomonas aeruginosa* | Aminoglycoside *plus* cefepime, ceftazidime, or ticarcillin |
| *Salmonella* species | Third-generation cephalosporins |
| *Candida albicans* | Amphotericin B ± 5-flucytosine or liposomal AmpB or fluconazole |
| Anaerobes | Penicillin, clindamycin, or metronidazole |
| **Continuation: Oral Therapy[d]** | |
| *Staphylococcus aureus* | Clindamycin or linezolid<br>Cephalexin if susceptible |
| Streptococci (group A) | Penicillin or amoxicillin |
| *S. pneumoniae* | Penicillin or amoxicillin; third-generation cephalosporins; clindamycin |
| Enterobacteriaceae | Ampicillin or trimethoprim-sulfamethoxazole (TMP-SMX) |
| *Neisseria gonorrhoeae* | Cefixime |
| *Pseudomonas aeruginosa* | Ciprofloxacin or other quinolones |
| *Salmonella* species | Amoxicillin, TMP-SMX or third-generation cephalosporins |
| *Candida albicans* | Fluconazole |
| Anaerobes | Penicillin, metronidazole, or clindamycin |
| Culture negative | Coverage for *S. aureus* (see earlier) |

[a]Aminoglycoside choice guided by *Pseudomonas* sensitivities in your hospital.
[b]Concomitant wound infection, Gram stain positive.
[c]Inpatient or home intravenous therapy.
[d]Oral continuation (modified by sensitivity testing).

include recurrent episodes of pain and joint swelling with or without fever. Symptoms typically begin in childhood, usually occurring alone. For most children, CRMO resolves after many years without lasting effects. However, it can cause slow growth and permanent bone deformity.

## Treatment

Staphylococci are the infecting organisms in 80% to 90% of cases with gram-negative enterics (associated with urinary tract infection), *Pseudomonas* species (intravenous drug abusers), and a small percentage of miscellaneous organisms (see Table 38.3). Because approximately half of *S. aureus* strains are methicillin resistant (MRSA), parenteral antibiotic therapy for presumed staphylococcal involvement including MRSA (e.g., vancomycin or linezolid) should be initiated, with consideration given to a needle aspirate or bone biopsy for culture and sensitivity testing first. Treatment should also include surgical drainage (especially if cord compression is present) and immobilization (bed rest versus casting).

Most cases of hematogenous osteomyelitis are culture negative after 21 days of therapy (Table 38.7). Exceptions are vertebral osteomyelitis, which should be treated for 6 weeks, and puncture wound osteochondritis caused by *Pseudomonas,* which requires only 7 to 10 days of antimicrobial therapy after adequate surgical debridement (Table 38.8).

## ADDITIONAL RESOURCES

Floyed RL, Steele RW. Culture-negative osteomyelitis. *Pediatr Infect Dis J* 2003; 22:731-735. *This paper describes the complexities of diagnosing and treating patients with negative blood and bone cultures.*

Kaplan SL. Challenges in the evaluation and management of bone and joint infections and the role of new antibiotics for gram positive infections. *Adv Exp Med Biol* 2009;634:111-120. *This review article discusses old and new treatments for bone and joint infections in the era of methicillin-resistant Staphylococcus aureus infections.*

Li H, Rombach I, Zambellas R, et al. Oral versus intravenous antibiotics for bone and joint infection. *N Engl J Med* 2019;380:425-436. *A large collaborative trial at 26 UK centers of adults treated for bone and joint infection supporting oral antibiotic therapy for osteomyelitis.*

Peltola H, Paakkonen M. Acute osteomyelitis in children. *N Engl J Med* 2014; 370:352-360. *A comprehensive review of the diagnosis and treatment of hematogenous osteomyelitis in children.*

Sawyer JR, Kapoor M. The limping child: a systematic approach to diagnosis. *Am Fam Physician* 2009;79:215-224. *A clinically useful article that discusses the differential diagnosis of children with painful bone and joint symptoms and recommends a diagnostic and therapeutic approach.*

Weichert S, Sharland M, Clarke NM, Faust SN. Acute haematogenous osteomyelitis in children: is there any evidence for how long we should treat? *Curr Opin Infect Dis* 2008;21:258-262. *An in-depth review article that discusses the type and duration of treatment of hematogenous osteomyelitis in children.*

# 39

# Urinary Tract Infections

*Dimitri Drekonja, James R. Johnson*

 **ABSTRACT**

Urinary tract infection (UTI), an acute bacterial infection of the urinary bladder, kidney, or collecting system, is among the most commonly diagnosed infectious diseases. The spectrum of disease is broad, ranging from asymptomatic bacteri-uria (ABU) and simple cystitis to septic shock. Highly active and bioavailable oral antimicrobials have made therapy for UTI convenient and inexpensive. However, widespread use (including overuse) of these drugs has promoted the emergence of antimicrobial resistance, so clinicians increasingly find themselves without reliably active oral options for empiric UTI therapy. Strategies for optimizing care and pro-longing the utility of currently available drugs include (1) following evidence-based practice guidelines, (2) not treating patients with ABU unless they are pregnant or about to undergo an invasive urologic procedure, and (3) using fluoroquinolone (FQ)-sparing therapy in most patients.

## ✷ CLINICAL VIGNETTE

An 82-year-old woman was brought to the emergency department from her nursing home because she had manifested confusion, decreased responsiveness, and decreased oral intake over the previous 2 days without localizing symptoms or fever. She had a history of dementia, urinary incontinence, extensive antibiotic exposure related to recurrent urinary tract infections (RUTIs), past *Clostridioides difficile* disease, and allergies to multiple antibiotics. Physical examination showed mild tachycardia, flat neck veins, dry mucous membranes, and impaired responsiveness without focal neurologic findings or nuchal rigidity. Blood chemistries suggested mild prerenal azotemia. The complete blood count and differential were normal, as was a chest roentgenogram. Urinalysis (UA) showed a high specific gravity, hyaline casts, bacteriuria, and pyuria, leading to a presumptive initial diagnosis of UTI. Blood cultures were drawn. Before receiving any antimicrobial therapy, the patient was admitted to the observation ward. The receiving hospitalist deferred antimicrobial therapy and instead administered 1.5 L normal saline overnight. The next morning the patient was more alert and responsive, with resolved tachycardia. Blood cultures yielded no growth. Urine culture (UC) grew more than 100,000 cfu/mL of multidrug-resistant *Escherichia coli*, which was not treated but prompted the institution of contact precautions. The patient was returned to her nursing home in baseline condition, with a final diagnosis of metabolic encephalopathy from dehydration, cause unknown, and ABU due to a multidrug-resistant organism.

## RISK FACTORS

Known risk factors for UTI include female gender, a history of previous UTI, and, among adolescent or adult women, sexual intercourse and use of spermicide-based contraception. The risk of UTI also increases with age and the presence of certain underlying conditions, including medical illnesses and anatomic or functional abnormalities of the urinary tract, which also lead to a more diverse range of causative microorganisms and decrease the likelihood of treatment success. The presence of such so-called complicating conditions is classically considered to define a UTI episode as "complicated" versus "uncomplicated." Examples of complicating conditions include urinary obstruction (anatomic or functional), other urinary tract abnormalities, intermittent or indwelling use of a urinary catheter, nephrolithiasis, chronic kidney disease, and diabetes mellitus (Fig. 39.1). Increasingly, a severe initial presentation (e.g., pyelonephritis or febrile UTI: see further on) is also sometimes regarded as qualifying a UTI episode as complicated, regardless of underlying host status; therefore caution is needed with the use of this label to avoid misunderstandings.

## CLINICAL FEATURES

The clinical presentation of UTI varies with the anatomic site of infection. This chapter presents UTI as two main syndromes, cystitis and pyelonephritis, but it also touches on entities such as febrile UTI, ABU, and catheter-associated UTI (CAUTI). Cystitis denotes symptomatic infection or inflammation of the bladder, whereas pyelonephritis (Fig. 39.2) denotes infection of the renal pelvis and parenchyma. Although the patient's clinical presentation may not accurately reflect the actual anatomic localization of UTI and inflammation, precise localization is not needed for effective management because this can be guided adequately by clinical presentation alone. The implications of complicating conditions are discussed as appropriate.

### Cystitis (Lower Urinary Tract Infection)

The typical presentation of cystitis is the acute onset of irritative voiding symptoms, including dysuria, frequency, and urgency. Gross hematuria and a change in urine odor may occur. Suprapubic discomfort and tenderness are sometimes present. Historical features that increase the likelihood of cystitis in a woman with some combination of these manifestations include absence of vaginal discharge, personal or family history of UTI, recent sexual activity, and spermicide-based contraception.

Risk factors for UTI caused by an antimicrobial-resistant organism include recent antimicrobial use and, in the United States, recent travel to a developing country. Possible alternative diagnoses in patients with symptoms suggesting cystitis include urethritis (caused by *Chlamydia trachomatis*, *Neisseria gonorrhoeae*, herpes simplex virus, *Ureaplasma*, and other organisms) and vaginitis (caused by *Candida* or *Trichomonas* species). Other considerations include nephrolithiasis, irritant or atrophic vaginitis or urethritis, prostatism (in elderly men), and diabetes mellitus (especially if the main symptom is urinary frequency).

The microbiology of uncomplicated cystitis is well described. *E. coli* is the predominant organism, causing 80% to 90% of cases. *Staphylococcus saprophyticus* is responsible for 5% to 10% of cases in women, with the remainder being caused by non–*E. coli* gram-negative bacilli or, especially in men, enterococci.

**Predisposing Factors in Urinary Tract Infections**

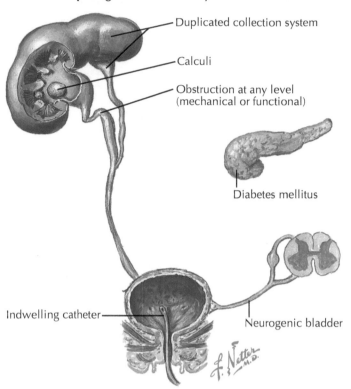

— Duplicated collection system

— Calculi

— Obstruction at any level (mechanical or functional)

Diabetes mellitus

Indwelling catheter —

— Neurogenic bladder

*f. Netter M.D.*

Fig. 39.1 Possible complicating conditions predisposing to urinary tract infections that are more frequent, are more difficult to eradicate, and are caused by a more diverse list of microorganisms than uncomplicated urinary tract infection.

In the presence of a complicating condition, the microbiology of cystitis is more variable, although relevant data are scarce. *E. coli* remains the most commonly isolated pathogen but is less predominant than in uncomplicated cystitis, whereas other gram-negative bacilli and gram-positive cocci—such as enterococci, *Streptococcus* species, *Staphylococcus aureus*, and coagulase-negative staphylococci—occur relatively more frequently.

## Pyelonephritis (Upper Urinary Tract Infection)

The presentation of pyelonephritis is varied, ranging from mild flank pain and malaise to urosepsis requiring aggressive hemodynamic support and other intensive care modalities. Classic features include flank or back pain and fever, often accompanied by nausea and vomiting. Relevant historical information is similar to that for cystitis, although gastrointestinal symptoms may be more important to elicit in order to assess for both systemic toxicity and the feasibility of oral therapy.

The pathognomonic physical finding of pyelonephritis is tenderness over the costovertebral angle, usually accompanied by fever; abdominal tenderness, tachycardia, and/or hypotension also may occur. The differential diagnosis includes diverticulitis, appendicitis, ectopic pregnancy, pelvic inflammatory disease, endocarditis, and nephrolithiasis. Perinephric and intrarenal abscess should also be considered, the latter particularly if *S. aureus* is isolated from blood cultures. For both cystitis and pyelonephritis, underlying complicating conditions should be sought because their presence may influence the diagnostic and therapeutic approach and the risk of subsequent infections.

The spectrum of causative organisms in uncomplicated pyelonephritis is largely similar to that in uncomplicated cystitis, with *E. coli* being by far the most common pathogen. Minor differences include that *S. saprophyticus* is less common, whereas non–*E. coli* gram-negative bacilli are more common. In patients with a complicating condition, the range of potential microorganisms is wider, which confounds accurate prediction, although *E. coli* is usually still the most common organism.

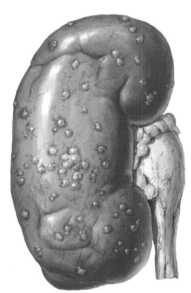

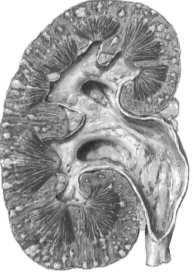

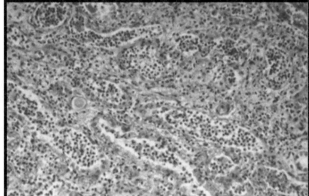

Surface aspect of kidney: multiple minute abscesses (surface may appear relatively normal in some cases)

Cut section: radiating yellowish gray streaks in pyramids and abscesses in cortex; moderate hydronephrosis with infection; blunting of calyces (ascending infection)

Acute pyelonephritis with exudate chiefly of polymorphonuclear leukocytes in interstitium and collecting tubules

*f. Netter M.D.*

Fig. 39.2 Pyelonephritis, with numerous small abscesses visible on the serosal surface and within the renal parenchyma and pelvis. The image at right shows renal cortex with the typical neutrophilic infiltrate seen in acute pyelonephritis.

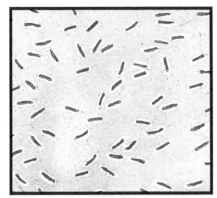

Urine smear stained to demonstrate Gram-negative bacilli, typical for either cystitis or pyelonephritis

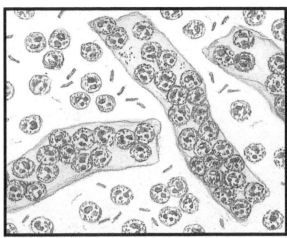

Pyuria with white blood cell casts, typically seen only in pyelonephritis or noninfectious renal inflammatory processes

**Fig. 39.3** Microscopy findings of urinary tract infection.

## Other Urinary Tract Infection Syndromes

*Febrile UTI* and *urosepsis* are terms sometimes applied to a syndrome characterized by fever presumably caused by UTI, with or without irritative voiding symptoms suggesting cystitis but without localizing evidence of kidney involvement (i.e., flank pain or tenderness). The likelihood of a positive blood culture increases in relation to the patient's age, illness severity, and degree of host compromise. ABU, which is infrequent in otherwise healthy individuals, also increases in frequency with age and the presence of complicating conditions. Screening for ABU is discouraged except during pregnancy and before urologic surgery.

CAUTI, a subset of complicated UTI, denotes UTI occurring in the presence of an indwelling urinary catheter. This term is often mistakenly applied whenever a positive UC is encountered in a patient with an indwelling catheter irrespective of the presence or absence of clinical manifestations attributable to UTI (e.g., suprapubic or urethral discomfort or otherwise unexplained indicators of systemic inflammation); it is best reserved for when such manifestations are present. Presence of bacteriuria (or pyuria) in a catheterized patient who lacks such symptoms or findings should be interpreted not as CAUTI but as catheter-associated ABU, which does not require treatment.

Acute prostatitis, a relatively uncommon condition, is characterized by fever, perineal pain, dysuria, and extreme prostate tenderness. In contrast, chronic prostatitis, which is much more common, rarely causes significant prostate tenderness. Men with a flare of chronic prostatitis often have only mild irritative voiding symptoms and therefore may be diagnosed initially as having cystitis. Chronic prostatitis should be considered as a possible underlying persisting internal nidus of infection in a man who has repeated episodes of cystitis, particularly if the same organism is consistently isolated from UCs. Unapparent prostatic involvement is common in men with febrile UTI, as reflected in an acutely elevated prostate-specific antigen concentration that returns to baseline after resolution of the UTI episode.

## DIAGNOSTIC APPROACH

### Cystitis

For patients with a first episode of cystitis, a history and physical examination directed toward the symptoms and findings discussed earlier in combination with UA are usually sufficient to make the diagnosis of cystitis with high certainty. The typical microscopic UA findings of cystitis are pyuria and bacteriuria (Fig. 39.3). If a rapid test (e.g., urine dipstick) is used, surrogate markers (leukocyte esterase for pyuria,

nitrites for bacteriuria) should be present. Positive urine microscopy or a rapid test in a patient with symptoms consistent with cystitis support the initiation of antimicrobial therapy, which should be chosen based on local susceptibility data.

UC is not generally recommended for reproductive-age women with a first episode of cystitis, although routine avoidance of UCs in such patients can lead to biased cumulative susceptibility data because of the resulting relative oversampling of patients with more severe or recurrent (previously treated) infections. In contrast, a pretherapy UC should be performed for suspected UTI occurring in older women, men, and patients with complicating conditions because of the less predictable urine microbiology in these patient groups.

For patients with suspected cystitis who lack complicating conditions and have had a prior episode of laboratory-confirmed cystitis, the typical clinical constellation is sufficient to diagnose cystitis without a need for urine testing. Indeed, management over the telephone or by using self-initiated therapy without the patient coming for in-person evaluation is safe, convenient, and cost effective. However, it may be advisable for a patient who in the past 6 months has been treated for cystitis or who has received antimicrobial therapy for any other indication to have a pretherapy UC done as a part of the evaluation for suspected cystitis, even if the diagnosis is not in question, because of the increased risk of a resistant urine organism.

### Pyelonephritis

The recommended diagnostic approach to pyelonephritis is similar to that for cystitis with the exception that a pretreatment UC is highly advisable, both for early detection of possible resistance to the empiric antimicrobial regimen and to guide a step down to narrower-spectrum oral therapy when possible. Aside from the UC and a serum creatinine check for adjustment of antimicrobial dosage, very little laboratory testing or imaging is needed. If the diagnosis is uncertain, the imaging modality of choice (to demonstrate pyelonephritis and its complications and to screen for other possible causes) is contrast-enhanced computed tomography.

### Complicated Cystitis and Pyelonephritis

Patients with cystitis who have an underlying complicating condition should have a pretreatment UC obtained because of the more varied microbiology than in uncomplicated cystitis. In contrast, with pyelonephritis, the presence of a complicating condition per se need not influence the diagnostic approach for most patients because the UC is already mandatory. However, a suspicion of possible urinary obstruction (e.g., based on a history of past obstruction, underlying conditions

**Sepsis Associated with Indwelling Urinary Catheters**

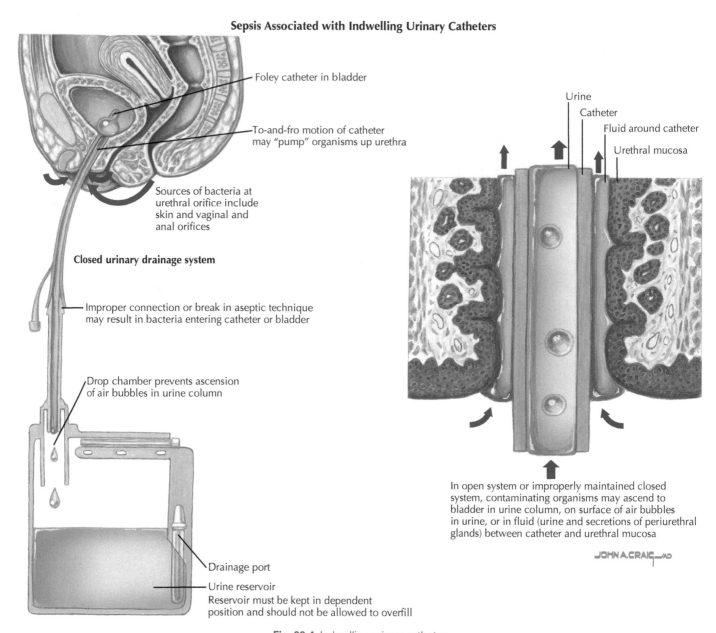

Foley catheter in bladder

To-and-fro motion of catheter may "pump" organisms up urethra

Sources of bacteria at urethral orifice include skin and vaginal and anal orifices

**Closed urinary drainage system**

Improper connection or break in aseptic technique may result in bacteria entering catheter or bladder

Drop chamber prevents ascension of air bubbles in urine column

Drainage port

Urine reservoir
Reservoir must be kept in dependent position and should not be allowed to overfill

Urine
Catheter
Fluid around catheter
Urethral mucosa

In open system or improperly maintained closed system, contaminating organisms may ascend to bladder in urine column, on surface of air bubbles in urine, or in fluid (urine and secretions of periurethral glands) between catheter and urethral mucosa

JOHN A. CRAIG—AD

**Fig. 39.4** Indwelling urinary catheter.

predisposing to obstruction, an acutely elevated serum creatinine level, or suggestive symptoms) should lower the threshold for imaging to exclude obstruction. Emphysematous pyelonephritis, a rare but severe form of pyelonephritis in which gas from microbial metabolism accumulates within the renal parenchyma, should be considered if a patient with pyelonephritis and poorly controlled diabetes mellitus has severe abdominal or flank pain, septic shock, or rapid clinical deterioration.

## Other Urinary Tract Infection Syndromes

Patients with febrile UTI should be evaluated in the same manner as those with pyelonephritis, including a pretreatment UC and, usually, blood cultures. Patients suspected of having symptomatic CAUTI should also have a UC obtained, preferably from the sampling port on the drainage tubing (not the collecting bag), using aseptic technique (Fig. 39.4). For patients who lack symptoms or other manifestations suggesting infection, especially those referable to the urinary tract, UTI should neither be considered in the differential diagnosis nor tested for except during pregnancy and before urologic surgery, in which settings ABU should be routinely screened for and treated if found.

Significant prostatic tenderness suggests acute prostatitis. The causative organism can usually be isolated by standard UC, although testing for *N. gonorrhoeae* and other sexually transmitted pathogens should be considered in sexually active men. Chronic prostatitis is suggested by repeated isolation of the same organism in serial UCs despite the absence of prostate-localizing symptoms or tenderness. In men with febrile UTI, prostate-specific antigen testing may be useful to detect prostatic involvement and assess response to therapy.

## MANAGEMENT AND THERAPY

### Cystitis

Therapy for cystitis in an otherwise healthy woman should be of short duration; for example, 3 days of trimethoprim-sulfamethoxazole (TMP-SMX) if local susceptibility data are favorable, 5 days of nitrofurantoin (or pivmecillinam, where available), or a single dose of fosfomycin. Guidelines from the Infectious Diseases Society of America (IDSA) list these regimens as first-line therapy for cystitis and specifically recommend that FQs be reserved as alternative agents, given their side-effect profile and utility for more serious infections. Amoxicillin-clavulanate and oral extended-spectrum cephalosporins are additional (but little studied) alternatives if antimicrobial resistance or patient factors preclude the use of the previously listed agents.

The optimal duration of therapy for cystitis in men and in women with complicating conditions is poorly defined. Cystitis in men with no complicating conditions may respond to short-course (3-day) therapy, although experience is limited. Some authorities consider male sex as a moderately complicating condition and would thus extend therapy to 7 days, which is also a frequently recommended (albeit poorly studied) duration of therapy for complicated cystitis in women. Cystitis in men with known underlying complicating conditions is commonly treated with longer courses of therapy (e.g., for 10 to 14 days), although better evidence is needed here also.

## Recurrent Cystitis

Recurrent episodes of cystitis are a significant burden for both patients and providers because repeated exposure to antimicrobials selects for resistant microorganisms, increases costs, and imposes the inconvenience of repeatedly accessing the medical care system. Preventive options for women with recurrent UTI include continuous (e.g., daily or thrice weekly) prophylactic antimicrobial therapy, postcoital therapy (if UTI episodes typically occur after sexual intercourse), and behavior change (increased fluid intake, urination after sex, and avoidance of spermicide-based contraception).

Self-initiated antimicrobial therapy is an effective way for patients to terminate recurrent UTI episodes quickly once they are noticed, without the inconvenience, cost, and delays associated with a visit to a clinic or urgent care center or the extensive antimicrobial exposure associated with continuous prophylaxis. Daily consumption of cranberry juice or extract is a nonpharmacologic method for preventing UTI that may be effective for motivated patients, although the supporting evidence from clinical trials is limited and inconsistent. Probiotic therapy—for example, with *Lactobacillus* preparations and inhibition of bacterial attachment using oral D-mannose—has even less supporting clinical evidence, notwithstanding their biologic plausibility and intuitive appeal. Anti-UTI vaccines, bacteriophage therapy, and fecal microbiota transplantation (as a way to eliminate an intestinal reservoir of recurrent UTI-causing pathogens) are intriguing potential preventive options that are currently under development. For older men with recurrent UTI episodes that presumably involve the same organism (according to species and susceptibility profile), longer-duration therapy to eradicate a possible persisting prostatic focus is a potentially useful strategy.

## Pyelonephritis

For patients with pyelonephritis, the preferred initial route of therapy and treatment setting are influenced by the severity of illness and presence of comorbidities. Oral outpatient therapy is appropriate for otherwise healthy patients who are tolerating oral intake, lack hypotension and tachycardia, and have a stable psychosocial situation. By contrast, hospital admission for parenteral therapy and close observation is advisable for patients with severe underlying medical conditions who exhibit hemodynamic instability or vomiting or who have an unstable psychosocial situation.

However, some patients (for instance, an otherwise healthy woman with suspected pyelonephritis who is normotensive but tachycardic) do not fit neatly into either group. Such a patient can be managed in the emergency department with intravenous fluids and an initial dose of a parenteral antimicrobial. If the patient responds adequately to this therapy, discharge on an oral antimicrobial agent is appropriate; if not, hospital admission is advisable.

Regarding drug choice, the increasing prevalence of TMP-SMX resistance in *E. coli*, combined with the serious consequences of inadequately treated pyelonephritis, has made TMP-SMX monotherapy unsuitable for empiric pyelonephritis therapy in many locales. However, if susceptibility of the urine organism to TMP-SMX is confirmed, TMP-SMX remains an excellent drug for completing therapy.

Likewise, FQ resistance is as common or more so as TMP-SMZ resistance in many locales, which complicates selection of an empiric oral regimen that is reliably active. If the causative organism is susceptible, ciprofloxacin 250 mg bid for 7 days and levofloxacin 750 mg qd for 5 days are both effective and proven regimens, with ciprofloxacin having an inexpensive generic formulation. By contrast, nitrofurantoin and fosfomycin, although both typically active against *E. coli* and other gram-negative bacilli, lack adequate blood and tissue levels and are therefore inappropriate for the treatment of pyelonephritis.

Parenteral options for pyelonephritis include ceftriaxone (and other extended-spectrum cephalosporins), piperacillin-tazobactam, carbapenems (including imipenem, ertapenem, and meropenem), aztreonam, aminoglycosides, and FQs. Local susceptibility data, toxicity, ease of administration, possible drug-drug interactions, and cost all influence drug selection. The deescalation of broad-spectrum empirical therapy should occur as soon as susceptibility data are available, preferably to an oral antimicrobial if the patient's condition has improved sufficiently.

High-dose levofloxacin is the only regimen with clinical trial data supporting a 5-day course of therapy; for all others, a treatment period of 7 to 14 days is recommended. Longer treatment courses are unnecessary and increase the risk of adverse events and selection for resistance. Initial use of an agent with antienterococcal or antistaphylococcal activity can be considered if suspicion for these organisms is high based on recent culture, urine Gram stain, or patient-specific risk factors.

Uncomplicated pyelonephritis usually responds rapidly to an active antimicrobial regimen, with fever and other manifestations generally improving if not resolving within the first 48 hours. If the response is insufficient or delayed, the patient should be reevaluated, with consideration given to the possibility of a resistant organism, a renal abscess, emphysematous pyelonephritis, obstruction, or an incorrect initial diagnosis. Repeat UC, an empiric change in antimicrobial therapy, and evaluation for a drainable focus of infection or other cause of the patient's symptoms should all be considered.

The optimal duration of therapy for pyelonephritis in men and in any patient with underlying complicating conditions is unclear. Seven days was slightly but statistically significantly inferior to 14 days in a subgroup analysis of male subjects in a trial of febrile UTI that included patients with pyelonephritis, although subgroup analyses should typically be considered to be hypothesis generating.

## Other Types of Urinary Tract Infection

As with pyelonephritis occurring in men and patients with complicating conditions, the optimal duration of therapy for febrile UTI and CAUTI is unknown. Febrile UTI is customarily treated similarly to pyelonephritis, with treatment duration adjusted for gender (i.e., possibly longer for men) and presence of complicating conditions. For CAUTI, 7 to 14 days of therapy is reasonable, depending on the rapidity of response. Catheter exchange during antimicrobial therapy may reduce the risk of relapse caused by persisting organisms within a catheter-adherent biofilm. ABU should not be treated outside of specifically defined situations (pregnancy and urologic surgery), as discussed earlier.

Preferred therapy for prostatitis consists of FQ or TMP-SMX, assuming a susceptible organism, because of their superior prostatic penetration compared with other antimicrobials. FQ therapy for 2 weeks usually suffices for acute prostatitis or febrile UTI in men (which often involves the prostate), whereas extended therapy (i.e., up to 4 weeks of an FQ or 6 to 12 weeks of TMP-SMX) is sometimes used for chronic prostatitis, particularly after the failure of shorter-duration therapy.

## SUMMARY

UTI is a commonly encountered clinical problem, making an efficient, cost-effective approach to diagnosis and treatment essential. Awareness of relevant host factors, local susceptibility data, and common pitfalls in UTI management (including failure to differentiate ABU from symptomatic UTI) is necessary for effective and safe management. A major challenge in the field is the rising prevalence of antimicrobial resistance among uropathogens. Widespread adoption of evidence-based treatment approaches may help to slow this worrisome trend, although the difficult scenario of a patient with an infection amenable only to parenteral therapy is likely to become increasingly common. Efforts should be made to establish a specific syndrome diagnosis, select empiric therapy appropriate for the syndrome, determine local susceptibility data and patient characteristics, deescalate to narrow-spectrum therapy as soon as possible, and treat only as long as necessary based on patient-specific factors.

## EVIDENCE

Gupta K, Hooton TM, Roberts PL, Stamm WE. Short-course nitrofurantoin for the treatment of acute uncomplicated cystitis in women. *Arch Intern Med* 2007;167:2207-2212. *An open-label trial evaluating TMP-SMX for 3 days versus nitrofurantoin for 5 days. Clinical and microbiologic outcomes were not significantly different, suggesting that nitrofurantoin may be a viable FQ-sparing regimen for cystitis treatment.*

Harding GKM, Zhanel GG, Nicolle LE, et al. Antimicrobial treatment in diabetic women with asymptomatic bacteriuria. *N Engl J Med* 2002;347:1576-1583. *A trial randomizing female subjects with ABU to antimicrobial treatment versus no treatment. No benefit was observed in the treatment group, whereas there was significant evidence of harm from adverse drug events.*

Hooton TM, Winter C, Tiu F, Stamm WE. Randomized comparative trial and cost analysis of 3-day antimicrobial regimens for treatment of acute cystitis in women. *JAMA* 1995;273:41-45. *A study examining the efficacy, safety, and cost of four different three-day regimens for uncomplicated cystitis. TMP-SMX was superior to the three comparators; however, none of the comparators was a quinolone or FQ.*

Iravani A, Tice AD, McCarty J, et al. Short-course ciprofloxacin treatment of acute uncomplicated urinary tract infection in women: the minimum effective dose. The Urinary Tract Infection Study Group. *Arch Intern Med* 1995;155:485-494. *A multicenter study demonstrating that ciprofloxacin given for 3 days was an effective treatment for uncomplicated cystitis.*

Johnson JR, Russo TA. Acute pyelonephritis in adults. *New Engl J Med* 2018;378:48-59. *This recent review summarizes relevant treatment trials and outlines a patient-centered algorithmic management approach.*

Talan DA, Stamm WE, Hooton TM, et al. Comparison of ciprofloxacin (7 days) and trimethoprim-sulfamethoxazole (14 days) for acute uncomplicated pyelonephritis in women: a randomized trial. *JAMA* 2000;283:1583-1590. *A multicenter trial evaluating therapy for pyelonephritis in the ambulatory setting. Ciprofloxacin was associated with an increased proportion achieving clinical and microbiologic cure, with the TMP-SMX failures occurring among subjects with an isolate that was resistant to TMP-SMX.*

Thänert R, Reske KA, Hink T, et al. Comparative genomics of antibiotic-resistant uropathogens implicates three routes for recurrence of urinary tract infections. *mBio* 2019;10:e01977-19. *A comparative genomic analysis of longitudinally collected urine and fecal isolates from patients with recurrent and nonrecurrent UTI supported three sources for urine isolates: reintroduction from the gut reservoir, reemergence from a persisting urinary tract reservoir, and de novo acquisition from the external environment.*

van Nieuwkoop C, van der Starre WE, Stalenhoef JE, et al. Treatment duration of febrile urinary tract infection: a pragmatic randomized, doubleblind, placebo-controlled non-inferiority trial in men and women. *BMC Medicine* 2017;15:70. *Outcomes of 7- vs. 14-day ciprofloxacin therapy for adults with febrile UTI (including pyelonephritis) varied by sex. Among women, both durations were similarly effective; among men, 7 days was statistically significantly (but only slightly) inferior to 14 days. Whether the marginal efficacy increase would justify doubling the treatment duration is unclear.*

## ADDITIONAL RESOURCES

Infectious Disease Society of America (IDSA). IDSA website. Available at: www.idsociety.org. Accessed September 18, 2019. *A site maintained by IDSA. Practice guidelines are available for cystitis, ABU, and catheter-associated UTI.*

UpToDate. UpToDate website. Available at: uptodateonline.com. Accessed September 18, 2019 (personal or institutional subscription required). *An online resource supplying topic-specific reviews, which are written and frequently updated by experts in the field.*

# 40

# Systemic Fungal Infections

*Carol A. Kauffman*

## ABSTRACT

The endemic mycoses are geographically restricted pathogens that exist as molds in specific environmental niches, where humans encounter them. In the United States, the major endemic mycoses are histoplasmosis, blastomycosis, and coccidioidomycosis. The extent of disease manifested by a given patient depends on both the size of the inoculum of the organism and the ability of the host to mount an effective immune response. The route of infection is almost always by inhalation into the alveoli of infectious conidia produced by the mold in the environment; therefore the major clinical manifestations are pulmonary. In addition, all of the endemic mycoses have the potential to disseminate hematogenously, and disease manifestations, especially in immunosuppressed patients, can reflect this widespread dissemination. In general, severe infection is initially treated with an intravenous amphotericin B formulation, which can later be changed to an oral azole agent. Mild to moderate infection is treated with an oral azole agent. With appropriate antifungal therapy, the prognosis is excellent for most such infections.

This chapter focuses on infection with the major endemic mycoses: histoplasmosis, blastomycosis, and coccidioidomycosis. Other systemic fungal infections—such as cryptococcosis, aspergillosis, and candidiasis, which primarily cause infection in immunosuppressed hosts—are not discussed here. The fungi that cause the endemic mycoses are all dimorphic; they exist as molds in the environment and as yeasts, or in the case of coccidioidomycosis, as spherules in tissues. Each organism occupies a different environmental niche and occurs in specific geographic areas (Fig. 40.1). Infection occurs when a person is exposed to the conidia (sometimes called *spores*) produced by the organism as it grows in the environment as a mold. These fungi cause infection in normal hosts as well as those who have defects in their immune response. A spectrum of disease manifestations can occur after inhalation of the conidia depending on the number of conidia inhaled and the immune status of the host. Hematogenous dissemination is common for all of these organisms, and it is possible for reactivation of a quiescent focus to occur years later.

## EPIDEMIOLOGY

Histoplasmosis is the most common endemic mycosis, likely infecting hundreds of thousands of persons yearly. Most persons in the highly endemic area are infected before adulthood. *Histoplasma capsulatum* thrives in highly nitrogenous soils in the Mississippi and Ohio River valleys, in multiple locations in Central and South America, and in focal areas of the eastern United States, Southeast Asia, and the Mediterranean basin. The organism grows to high concentrations in caves rich in bat guano. Activities that disperse the conidia include landscaping, demolition of old buildings, cleaning debris from attics or barns, and spelunking. Although most cases are sporadic, outbreaks of varying sizes frequently occur. The largest outbreak occurred during an urban demolition project in Indianapolis and caused infection in as many as a hundred thousand people.

Blastomycosis occurs most frequently in the south central and north central United States, the midwestern Canadian provinces, and areas bordering the St. Lawrence seaway, but *Blastomyces dermatitidis* is also found in the Middle East and Africa. The environmental niche for *B. dermatitidis* is thought to be soil and decaying wood. In many reports, middle-aged men account for most cases, and a well-described scenario is blastomycosis occurring in both a hunter and his dog.

There are two species of *Coccidioides*. *Coccidioides immitis* includes isolates from California, and *Coccidioides posadasii* includes isolates from all other areas. The endemic area for *Coccidioides* species includes the southwestern US desert regions known as the *lower Sonoran life zone* and areas of Central and South America that have the same type of ecosystem. Environmental cycles of rain and drought in the desert are important in the growth and dispersal of *Coccidioides* species. Dust storms and activities that involve the disruption of desert soil can lead to widespread dispersal of the conidia and increased risk for persons in the area. The exodus to the sun belt has contributed to an increase in the number of cases of coccidioidomycosis in older adults who are experiencing their first exposure to this organism.

## CLINICAL FEATURES

Most patients infected with one of the endemic mycoses are asymptomatic or have such mild symptoms that they are thought to have a self-limited viral illness. Therefore the discussion of symptoms and signs that follows concentrates on fewer than 5% of persons exposed to these organisms. When symptoms do occur, the predominant manifestations are pulmonary, which is not surprising given that the portal of entry for these fungi is the lungs. For all of these fungi, dissemination during the early stages of infection is common; this occurs before the host establishes cellular immunity to the organism and is able to contain the infection. In most cases, dissemination is silent and not associated with clinical manifestations. However, patients who are immunosuppressed or exposed to a high inoculum of the organism can become acutely ill with disseminated infection. Patients also can be seen at a later time point with focal infection at a site to which the organism had spread hematogenously. Although the pulmonary manifestations are often similar among these three endemic mycoses, the clinical manifestations of disseminated infection are somewhat different for each organism.

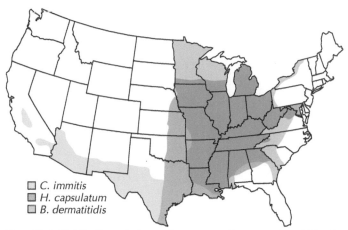

**Fig. 40.1** Distribution of the major geographic areas in which the endemic mycoses are found in the United States.

☐ *C. immitis*
☐ *H. capsulatum*
☐ *B. dermatitidis*

## Histoplasmosis

> ### ✸ CLINICAL VIGNETTE
>
> A 66-year-old man with ankylosing spondylitis who had been on adalimumab for 10 months presented with fever and night sweats of 10 days' duration and a dry cough and dyspnea that had lasted for 5 days. On examination, his temperature was 103°F and he appeared quite ill. Crackles were heard at both bases and hepatosplenomegaly was present. Computed tomography (CT) of the thorax showed diffuse reticulonodular lesions. His white blood cell count (WBC) was 3200/μL, hemoglobin 9.6 g/dL, platelets 86,000/μL. Liver enzymes were abnormal: AST 145 units, ALT 210 units, alkaline phosphatase 494 units, bilirubin 2.9 mg/dL. The laboratory called to say that they saw organisms in several neutrophils in the peripheral blood smear. *Histoplasma* antigen in urine was 19 ng/mL and in serum 11 ng/mL. Blood cultures yielded *H. capsulatum* 7 days later. Treatment with liposomal amphotericin B was begun, and the patient's temperature rapidly returned to normal; within a week, he felt much better. After 2 weeks, therapy was changed to oral itraconazole, 200 mg twice daily for a total treatment course of 12 months. CT of the thorax showed complete resolution, pancytopenia resolved, and liver enzymes and bilirubin returned to normal.
>
> COMMENT: Patients treated with tumor necrosis factor (TNF) antagonists, such as adalimumab, are at risk for developing widespread disseminated infection with *H. capsulatum*. Blood cultures are frequently positive, and the yeast form can be seen in neutrophils and monocytes in the peripheral blood smear. Involvement of liver, spleen, bone marrow, lungs, and adrenal glands is common. The diagnosis must be made quickly, allowing amphotericin B therapy to begin promptly if a fatal outcome is to be avoided.

Patients who have symptoms associated with acute pulmonary histoplasmosis usually manifest fever, nonproductive cough, anterior chest discomfort, myalgias, and fatigue. Patchy nodular infiltrates are noted on chest radiographs. Most patients are initially treated with antibiotics for community-acquired bacterial pneumonia. Only when the infection persists is the possibility of a fungal infection considered, and then appropriate diagnostic studies are undertaken.

Severe pneumonia develops in a minority of patients and manifests as high fever, dyspnea, nonproductive cough, and hypoxemia (Fig. 40.2). Diffuse infiltrates are noted on the chest radiograph and may progress to acute respiratory distress syndrome (ARDS). Immunosuppressed patients are more likely to develop severe pulmonary infection.

Chronic cavitary pulmonary histoplasmosis, which mimics tuberculosis, occurs mostly in older adults who have chronic obstructive pulmonary disease (COPD). Patients usually manifest fever, fatigue, anorexia, weight loss, cough productive of purulent sputum, and hemoptysis. Chest radiographs show upper lobe cavitary lesions, and fibrosis is seen in the lower lung fields (see Fig. 40.2).

The complications of pulmonary histoplasmosis include mediastinal granuloma and mediastinal fibrosis. Patients with mediastinal granuloma have persistent mediastinal and hilar lymphadenopathy; symptoms related to the enlarged nodes include dysphagia, chest pain, and nonproductive cough when the involved nodes impinge on mediastinal structures. Most patients have a gradual resolution of symptoms as the lymphadenopathy regresses. Mediastinal fibrosis is a rare and, in many patients, fatal complication in which excessive fibrosis occurs in response to mediastinal histoplasmosis; this can entrap the great vessels and bronchi. The symptoms are progressive and include dyspnea, cough, wheezing, and hemoptysis. Heart failure, pulmonary emboli, and superior vena cava syndrome can occur.

Acute disseminated histoplasmosis is seen mostly in patients who are immunosuppressed, including those who have acquired immunodeficiency syndrome (AIDS), have received a transplant, or have been treated with corticosteroids or TNF antagonists. Young infants are also at risk for disseminated histoplasmosis. Symptoms include chills, fever, fatigue, anorexia, and weight loss; sepsis syndrome with ARDS and disseminated intravascular coagulation can occur. Hepatosplenomegaly and skin and mucous membrane lesions are frequently present, and pancytopenia and elevated liver enzymes are common.

Chronic progressive disseminated histoplasmosis occurs mostly in middle-aged to elderly patients with no known immunosuppression. These patients often have fever of unknown origin with weeks to months of fever, night sweats, weight loss, and fatigue. Hepatosplenomegaly and mucosal ulcerations (Fig. 40.3) and symptoms and signs of adrenal insufficiency are often noted.

## Blastomycosis

> ### ✸ CLINICAL VIGNETTE
>
> A 36-year-old healthy woman from upper Michigan was seen with complaints of fevers, malaise, and left-sided chest pain of 2 weeks' duration. She was given levofloxacin, 250 mg daily, but her symptoms progressed over the next week. A chest radiograph showed a left-lower-lobe pneumonia, and the levofloxacin was increased to 750 mg daily. One week later she was admitted to hospital with consolidation of most of the left lung and increased dyspnea. Further history revealed that in the preceding 2 months she had done most of the work of cleaning out a vacant cottage near the lakeshore that her family had purchased. Her temperature was 102°F, pulse 117 beats/min, oxygen saturation 93% on 3 L of oxygen. The WBC count was 21,600/μL. Treatment was changed to vancomycin and piperacillin-tazobactam, resulting in no improvement in her fever or dyspnea. On the fourth hospital day, she was intubated and bronchoscopy was performed. Bronchoalveolar lavage (BAL) fluid showed broad-based budding yeasts typical of *B. dermatitidis*. Culture later yielded *B. dermatitidis*. Treatment with liposomal amphotericin B, 5 mg/kg daily was begun; 2 weeks later, therapy was changed to oral itraconazole, 200 mg twice daily. She was treated for a total of 12 months and recovered completely.
>
> COMMENT: This woman likely had exposure to a large number of conidia of *B. dermatitidis* in the course of cleaning out a vacant cabin in a highly endemic area for this fungus. Identification of the typical appearance of the yeast form of *B. dermatitidis* in the BAL fluid allowed treatment to begin before the culture revealed the organism.

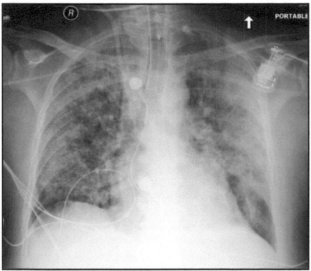 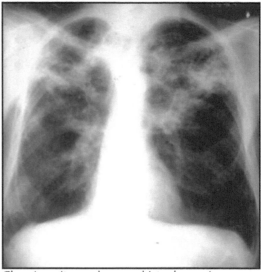

Severe acute pulmonary histoplasmosis in a young man with a heavy exposure to H. capsulatum

Chronic cavitary pulmonary histoplasmosis in 60-year-old man with emphysema

**Fig. 40.2** Pulmonary histoplasmosis.

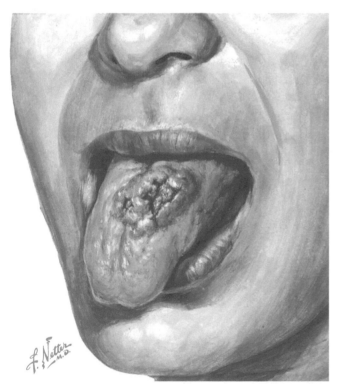

Ulcerating lesion of tongue due to histoplasmosis. Lesion is identical in appearance to carcinoma of tongue

**Fig. 40.3** Mucosal ulceration.

The symptoms and signs of acute pulmonary blastomycosis are fever, cough, and myalgias; a localized pulmonary infiltrate is seen on the chest radiograph. The diagnosis is usually community-acquired bacterial pneumonia, and most patients are treated with antibiotics. Only when the infection persists is the possibility of a fungal infection considered, at which point appropriate diagnostic studies are undertaken. A minority of patients have severe pneumonia that progresses quickly to severe hypoxemia and ARDS (Fig. 40.4).

Chronic pulmonary blastomycosis can be mistaken for tuberculosis or lung cancer. The symptoms include fever, night sweats, weight loss, fatigue, dyspnea, cough, sputum production, and hemoptysis. The chest radiograph shows upper-lobe cavitary infiltrates, mass-like lesions, or multiple nodular lesions (see Fig. 40.4). Hilar and mediastinal lymphadenopathy occur less often than with histoplasmosis.

Cutaneous lesions are the most common manifestation of disseminated blastomycosis. The lesions are nonpainful well-circumscribed nodules or plaques that are verrucous and have punctate draining areas in their centers (Fig. 40.5). The lesions can also manifest as painful ulcerations or pustular nodules. The skin lesions of blastomycosis can be mistaken for those caused by nontuberculous mycobacteria, bromide use, and pyoderma gangrenosum. Patients with skin lesions may or may not have pulmonary manifestations; in many, the pneumonia has resolved by the time the skin lesions appear. Osteoarticular blastomycosis may be contiguous to skin lesions or may occur at distant sites. The genitourinary tract is another frequent site of disseminated infection, and the prostate is the usual target organ. Prostatic nodules, with or without symptoms of prostatism, are found.

Blastomycosis is more severe in immunosuppressed patients, although it is seen much less commonly than histoplasmosis and coccidioidomycosis in patients with AIDS, transplant recipients, and patients receiving TNF antagonists. In these patients, the disease is usually disseminated to multiple organs; severe pulmonary manifestations also are more common.

## Coccidioidomycosis

### ✳ CLINICAL VIGNETTE

A 71-year-old retired lawyer from Michigan who spends the winters in Phoenix developed fever and a dry cough in February. On examination, he was febrile to 102°F and was coughing but did not look ill. He was thought to have community-acquired pneumonia and was treated with azithromycin for 5 days. After his return to Michigan in April, he complained of continued fatigue and trouble concentrating; his wife noted that he had hesitant speech and word-finding problems. On examination, he was afebrile; the only finding was that he was slow to respond to questions but he was oriented to time, place, and person. A CT scan of the brain showed no abnormalities. Lumbar puncture was performed and CSF revealed a

**CLINICAL VIGNETTE—CONT'D**

WBC count of 693/μL, all mononuclear cells; glucose 19 mg/dL; and protein 170 mg/dL. Cryptococcal antigen in serum and CSF was negative; complement fixation (CF) antibody to *Coccidioides* was 1:8 in serum and 1:2 in CSF. The diagnosis of coccidioidal meningitis was made and treatment with 800 mg fluconazole daily was begun with a plan to treat with this agent for life. CSF values returned to normal. The patient's cognitive function improved but did not return to preillness levels.

COMMENT: Coccidioidal meningitis can cause classic meningeal symptoms and signs, including headache and nuchal rigidity, or be quite subtle, as in this case. This older man likely had primary coccidioidal pneumonia in Arizona, cleared that infection, and then developed coccidioidal meningitis as a complication. This form of coccidioidomycosis requires lifelong azole therapy to prevent recurrence.

Acute pulmonary coccidioidomycosis usually manifests with fever, fatigue, cough, dyspnea, myalgias, and arthralgias. Erythema nodosum is not uncommon during acute pulmonary coccidioidomycosis. Chest radiographs show a patchy pneumonia; hilar lymphadenopathy can also occur (Fig. 40.6). Symptoms may persist for weeks to months but gradually resolve in almost all persons. Acute pulmonary coccidioidomycosis is often diagnosed as community-acquired bacterial pneumonia and antibiotics are prescribed. The diagnosis is thought of only later when there is no response to antibiotic therapy. In immunosuppressed hosts or when exposure to the organism is extensive, severe pneumonia, sometimes progressing to ARDS, can occur (see Fig. 40.6). This is a common manifestation of coccidioidomycosis in AIDS patients with low CD4 counts.

Several pulmonary complications are more common with coccidioidomycosis than with the other endemic fungi. Solitary thin-walled cavities may persist for months to years. Hemoptysis is frequent, and a cavity can rupture into the pleural space. Older adults and those who have COPD or diabetes mellitus are more likely to develop chronic progressive cavitary coccidioidomycosis.

Very few patients infected with *Coccidioides* species go on to develop symptomatic disseminated infection. This is more common in African Americans and people of other dark-skinned ethnicities; those who are immunosuppressed, especially AIDS patients; and in some women during the second or third trimester of pregnancy. The sites involved

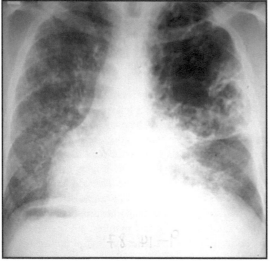

Severe acute pulmonary blastomycosis with ARDS in a 56-year-old man.

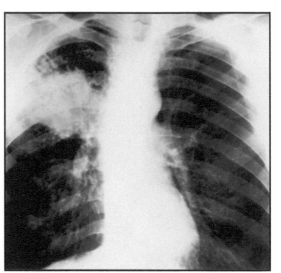

Chronic pulmonary blastomycosis showing lesion in upper lobe of right lung. Radiographic pattern may, however, be very diverse.

**Fig. 40.4** Pulmonary blastomycosis. *ARDS*, Acute respiratory distress syndrome.

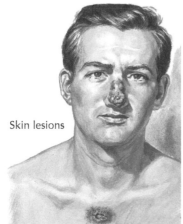

Skin lesions

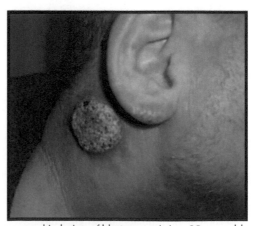

Verrucous skin lesion of blastomycosis in a 30-year-old man

**Fig. 40.5** Skin lesions in blastomycosis.

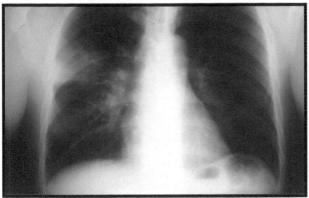 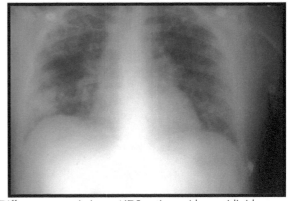

Acute pulmonary coccidioidomycosis in a healthy 40-year-old man    Diffuse pneumonia in an AIDS patient with coccidioidomycosis

**Fig. 40.6** Pulmonary coccidioidomycosis.

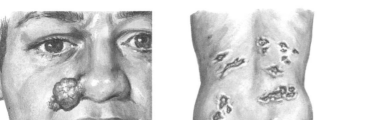

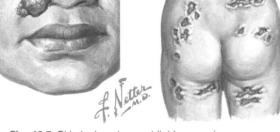

**Fig. 40.7** Skin lesions in coccidioidomycosis.

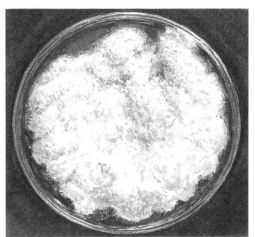

**Fig. 40.8** Endemic fungi grow as molds at room temperature.

most often are skin, subcutaneous tissues, osteoarticular structures, and meninges (Fig. 40.7). The skin lesions can be papular, pustular, or plaquelike and can ulcerate and drain. Subcutaneous abscesses often form sinus tracts with purulent drainage. Osteoarticular infection is common and may occur contiguous to subcutaneous abscesses and skin lesions or at distant sites. Vertebral involvement is common and extensive destruction is often seen.

Meningitis is the most feared complication of disseminated coccidioidomycosis. The patient may have only isolated chronic meningitis or meningitis can be one manifestation of disseminated infection. Symptoms include headache, confusion, behavioral changes, cranial nerve palsies, and signs of increased intracranial pressure. Cord involvement at any level can also occur, causing back pain, weakness, and bowel and bladder dysfunction.

## DIAGNOSTIC APPROACH

In general, the diagnosis of infection with one of the endemic mycoses can be made most expeditiously by histopathologic demonstration of the organisms in biopsy specimens from involved organs. However, the definitive diagnosis still requires growth of the organism in culture. Serology and antigen detection are variably helpful in certain clinical circumstances for each of the endemic fungi.

Growth of *B. dermatitidis* and *H. capsulatum* in culture takes several weeks; *Coccidioides* species usually grow in several days to a week. *H. capsulatum*, but rarely the other fungi, can be grown from blood using the lysis-centrifugation (isolator tube) system. The endemic

fungi grow as molds at room temperature (Fig. 40.8). Commercially available deoxyribonucleic acid (DNA) probes that are specific for *B. dermatitidis* and *H. capsulatum* are used to confirm the identification of the mold as soon as growth occurs. Because *Coccidioides* species are classified as potential bioterrorism agents, clinical laboratories are obligated to send the mold to a reference laboratory for identification.

Tissue obtained for biopsy should be stained with methenamine silver or periodic acid–Schiff (PAS) stain to visualize the yeast forms of *H. capsulatum* and *B. dermatitidis*. *Coccidioides* spherules are large and readily seen with hematoxylin and eosin stain, but fungal stains better define these structures. *H. capsulatum* appears as uniform, 2- to 4-μm oval budding yeasts (Fig. 40.9). For patients with disseminated disease, biopsy samples of bone marrow, liver, lymph nodes, or lesions on mucous membranes or skin reveal the organisms. *B. dermatitidis* can be seen in respiratory secretions and lung tissue; the calcofluor fluorescent stain or tissue fungal stains show the distinctive 8- to 15-μm broad-based budding yeasts (Fig. 40.10). Biopsy of skin lesions characteristically shows pseudoepitheliomatous hyperplasia, and fungal stains reveal the thick-walled yeasts with a single broad-based bud. Identification of the large 20- to 80-μm spherules that contain numerous endospores in tissue biopsies, respiratory secretions, or purulent material from abscesses establishes the diagnosis of coccidioidomycosis (Fig. 40.11).

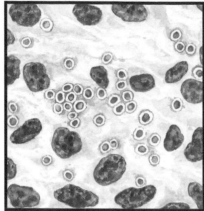

H. capsulatum in tissue

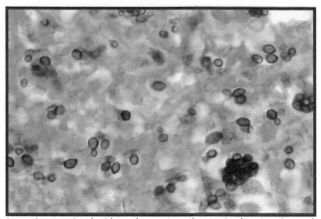

Lung tissue stained with methenamine silver stain showing the oval, narrow-based budding yeasts of H. capsulatum

**Fig. 40.9** *Histoplasma capsulatum.*

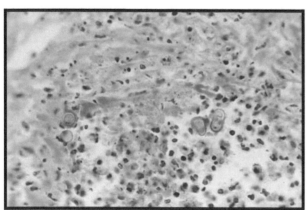

Lung tissue stained with periodic acid Schiff stain showing the large, thick-walled, broad-based budding yeasts of B. dermatitidis

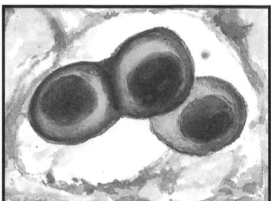

Very high-power view of a budding and nonbudding organism

**Fig. 40.10** *Blastomyces dermatitidis.*

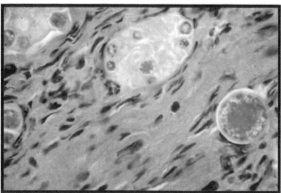

Lung tissue stained with hematoxylin and eosin showing several large spherules, some of which contain endospores, typical of *Coccidioides* species

**Fig. 40.11** *Coccidioides.*

Serology plays an important role in the diagnosis of coccidioidomycosis and certain forms of histoplasmosis but is less useful for blastomycosis. Serology is not as useful in immunosuppressed patients, who may not mount an antibody response. For histoplasmosis, CF and immunodiffusion tests are available. Patients with chronic cavitary and acute pulmonary histoplasmosis and those with disseminated histoplasmosis almost always have positive results with both the CF and immunodiffusion assays. The diagnosis of acute coccidioidomycosis is aided by the detection of IgM antibodies, which are usually measured by immunodiffusion. An enzyme immunoassay (EIA) that measures IgM antibodies is also available, but false-positive results have been reported with this test. IgG antibodies measured by CF assay appear later and persist longer. Rising CF titers may reflect worsening of infection and dissemination. A positive CF antibody test for *Coccidioides* in CSF is diagnostic of coccidioidal meningitis. CF and immunodiffusion assays for blastomycosis are neither sensitive nor specific; however, a new EIA may prove useful for diagnosis but requires further evaluation.

Antigen assays using EIA techniques to detect cell wall galactomannans are available for all three major endemic mycoses. The greatest experience has been with the EIA for *H. capsulatum*, which has proved extremely useful for the diagnosis of disseminated histoplasmosis and is assuming a useful role in the diagnosis of acute pulmonary histoplasmosis. As experience increases with the antigen assays for *B. dermatitidis* and *Coccidioides* species, it is likely that these tests will assume a greater role in diagnosis. For all patients suspected of having one of the endemic mycoses, testing for both antibody and antigen enhances the diagnostic approach.

## TABLE 40.1  Recommended Treatment Regimens for Histoplasmosis

| Type of Infection | Recommended Therapy |
| --- | --- |
| **Acute Pulmonary** | |
| Mild, moderate | Less than 4 weeks of symptoms, therapy not recommended; more than 4 weeks of symptoms, itraconazole 200 mg once or twice daily for 6–12 weeks. |
| Severe | Lipid AmB 3–5 mg/kg daily until improved clinically, then itraconazole 200 mg twice daily; length of therapy depends on the host's immune status and response to therapy and is usually 3–6 months. |
| **Chronic Cavitary Pulmonary** | Itraconazole 200 mg twice daily for at least 12 months. |
| **Mediastinal Granuloma** | Antifungal therapy not usually recommended, but itraconazole 200 mg twice daily can be tried for 6–12 weeks. |
| **Mediastinal Fibrosis** | Antifungal therapy not effective; vascular stents may be useful. |
| **Disseminated** | |
| Mild, moderate | Itraconazole 200 mg twice daily for 12 months. |
| Severe | Liposomal AmB (AmBisome) 3 mg/kg daily until improved clinically, then itraconazole 200 mg twice daily for 12 months total. Long-term suppressive therapy with itraconazole may be required for immunosuppressed patients. |

For patients who cannot tolerate itraconazole, voriconazole 200 mg twice daily *or* posaconazole, 300 mg daily can be prescribed.
*AmB,* Amphotericin B.

## TREATMENT

For all of the endemic mycoses, treatment of mild to moderate infection should be with an azole and treatment of moderately severe to severe infection should be with an amphotericin B formulation, followed by an azole after the patient has shown a clinical response. Use of a lipid amphotericin B formulation decreases the risk of nephrotoxicity and has been shown to improve outcomes for disseminated histoplasmosis. Itraconazole is the azole of choice for histoplasmosis and blastomycosis, and coccidioidomycosis can be treated with either itraconazole or fluconazole. Itraconazole capsules require both food and acid for absorption; agents that decrease gastric acid must be avoided. Itraconazole suspension is given on an empty stomach and does not require acid for absorption. With either formulation, a loading dose of 200 mg three times daily for 3 days should be given, followed by 200 mg twice daily thereafter. Itraconazole serum concentrations should be obtained to help ensure efficacy. Absorption of fluconazole is not problematic and serum concentrations are not required. There is increasing use of voriconazole and posaconazole for patients who cannot tolerate therapy with itraconazole. For all azoles, drug-drug interactions are extremely common and must be carefully sought and dealt with.

### Histoplasmosis

The following recommendations are modified from the Infectious Diseases Society of America (IDSA) guidelines for the management of histoplasmosis (Table 40.1). Patients who have acute pulmonary histoplasmosis usually do not require treatment. However, if the patient has symptoms that last more than 4 weeks, therapy with itraconazole 200 mg daily for 6 to 12 weeks is recommended. All patients who have a chronic cavitary pulmonary infection require antifungal therapy. Itraconazole is recommended at a dosage of 200 mg twice daily for at least 12 months. Patients who develop severe pulmonary infection should be treated initially with a lipid formulation of amphotericin B, 3 to 5 mg/kg daily. After a favorable response is noted, therapy can be changed to oral itraconazole, 200 mg twice daily; the length of treatment depends on the patient's immune status and response to therapy and can vary from 3 to 6 months.

Mediastinal granuloma is often treated with a course of itraconazole, 200 mg once or twice daily, for 6 to 12 weeks, but benefit is not always found. Mediastinal fibrosis does not respond to antifungal therapy. The most effective treatment appears to be selective placement of stents in obstructed great vessels.

All patients who have disseminated histoplasmosis should be treated with an antifungal agent. Mild to moderate disease can be treated with itraconazole, 200 mg twice daily; severe infection and infection in immunosuppressed patients should be treated initially with a lipid formulation of amphotericin B. Liposomal amphotericin B (AmBisome) 3 mg/kg daily is recommended based on the results of a blinded randomized treatment trial that showed this agent to be superior to amphotericin B deoxycholate for initial therapy in AIDS patients with severe disseminated histoplasmosis. Therapy can be changed to itraconazole, 200 mg twice daily, after clinical improvement is noted. Immunosuppressed patients may require prolonged suppressive therapy with itraconazole.

There are an increasing number of reports on the use of voriconazole and posaconazole for histoplasmosis. These azoles can be used when itraconazole is not tolerated and are preferable to fluconazole as second-line therapy. The echinocandins do not have activity against *H. capsulatum* and should not be used.

### Blastomycosis

The following recommendations are modified from the IDSA guidelines for the management of blastomycosis (Table 40.2). With the exception of acute pulmonary blastomycosis, in which all symptoms and signs have already resolved before diagnosis, all patients who have blastomycosis—even mild pneumonia or a single cutaneous lesion—should be treated with an antifungal agent. For immunosuppressed patients and those who have severe pulmonary disease, central nervous system involvement, or widespread visceral involvement, lipid formulation amphotericin B, 3 to 5 mg/kg daily, should be used as initial therapy. After the patient has shown clinical improvement, it is appropriate to change to oral itraconazole, 200 mg twice daily. Patients with mild to moderate pulmonary blastomycosis and those who have skin lesions, osteoarticular involvement, or other focal infection can be treated with itraconazole, 200 mg twice daily. Treatment is generally given for a total of 6 to 12 months; patients who have moderately severe to severe infection and those with osteoarticular involvement should receive therapy for at least 12 months. Immunosuppressed patients for whom the immunosuppression cannot be reversed may require long-term suppressive azole therapy.

| TABLE 40.2 | Recommended Treatment Regimens for Blastomycosis |
|---|---|
| **Type of Infection** | **Recommended Therapy** |
| **Pulmonary** | |
| Mild to moderate | Itraconazole 200 mg twice daily for 6–12 months. |
| Severe | Lipid AmB 3–5 mg/kg daily until improved clinically, then itraconazole 200 mg twice daily for 6–12 months. |
| **Disseminated** | |
| Mild to moderate | Itraconazole 200 mg twice daily for 6–12 months. |
| Severe | Lipid AmB 3–5 mg/kg daily until improved clinically, then itraconazole 200 mg twice daily for 6–12 months. |
| Immunosuppressed host | Lipid AmB 3–5 mg/kg daily until improved clinically, then itraconazole 200 mg twice daily for 12 months. Long-term suppressive therapy with itraconazole may be required for immunosuppressed patients. |

For patients who cannot tolerate itraconazole, voriconazole, 200 mg twice daily, *or* posaconazole 300 mg daily can be prescribed.
*AmB,* Amphotericin B.

| TABLE 40.3 | Recommended Treatment Regimens for Coccidioidomycosis |
|---|---|
| **Type of Infection** | **Recommended Therapy** |
| **Acute Pulmonary** | |
| Mild to moderate | Antifungal therapy usually not needed. Itraconazole 200 mg twice daily *or* fluconazole 400 mg daily for 3–6 months for those who remain symptomatic for more than 4 weeks. |
| Severe | Lipid AmB 3–5 mg/kg daily until improved clinically, then itraconazole 200 mg twice daily, *or* fluconazole 400 mg daily for at least 12 months. |
| **Chronic Pulmonary** | Itraconazole 200 mg twice daily, *or* fluconazole 400 mg daily, for at least 12 months. |
| **Disseminated** | |
| Mild, moderate | Itraconazole 200 mg twice daily, *or* fluconazole 400 mg daily, for at least 12 months. |
| Severe | Lipid AmB 3–5 mg/kg daily until improved clinically, then itraconazole 200 mg twice daily, *or* fluconazole 400 mg daily for at least 12 months. Long-term suppressive therapy with an azole may be required for immunosuppressed patients. |
| **Central Nervous System** | Fluconazole 800 mg daily; intrathecal AmB-d may be needed for refractory cases; azole treatment must be lifelong. |

For patients who cannot tolerate itraconazole, voriconazole, 200 mg twice daily *or* posaconazole 300 mg daily can be prescribed.
*AmB,* Amphotericin B; *AmB-d,* amphotericin B deoxycholate.

Voriconazole and posaconazole, rather than fluconazole, are increasingly used as second-line therapy for blastomycosis; for central nervous system involvement, voriconazole has become the preferred azole agent following initial amphotericin B treatment. The echinocandins are not active against *B. dermatitidis* and should not be used.

## Coccidioidomycosis

The following recommendations are modified from the IDSA guidelines for the management of coccidioidomycosis (Table 40.3). Most patients with acute pulmonary coccidioidomycosis do not require therapy with an antifungal agent. However, if the patient has symptoms for longer than 4 weeks with no improvement, is immunosuppressed, or is at high risk for dissemination (African American, other dark-skinned ethnic groups), treatment with either fluconazole, 400 mg daily, or itraconazole, 200 mg twice daily for 3 to 6 months should be given. Severe diffuse coccidioidal pneumonia should be treated initially with a lipid formulation of amphotericin B; after the patient has had a clinical response, therapy can be changed to an oral azole. Patients with chronic pulmonary coccidioidomycosis also require antifungal therapy, generally with an azole. Except for acute pulmonary infection, treatment should continue for at least 12 months.

Patients who have disseminated coccidioidomycosis should always be treated. Those who are seriously ill and immunosuppressed patients should receive amphotericin B until improvement is noted. Patients who have mild to moderate disease and are not immunosuppressed can be treated with either itraconazole or fluconazole. For osteoarticular coccidioidomycosis, itraconazole has proved modestly superior to fluconazole. Treatment is generally given for at least 12 months. Immunosuppressed patients often require lifelong azole therapy to prevent relapse.

There are increasing reports of success in treating coccidioidomycosis with voriconazole and posaconazole. The echinocandins appear to have no activity against *Coccidioides* species and should not be used.

Coccidioidal meningitis is difficult to treat. Fluconazole is the agent of choice, in part because of its superior penetration into the CSF. The initial studies used 400 mg daily, but most physicians in the endemic area now use at least 800 mg daily. Voriconazole also achieves high CSF concentrations and has been used successfully in patients with meningitis. For patients in whom remission is not achieved with azoles, amphotericin B deoxycholate can be given as an intrathecal injection into the cistern or ventricle. Patients who have *Coccidioides* meningitis must be treated with suppressive azole therapy for life to avoid relapse.

## PROGNOSIS

Most patients who have pulmonary or disseminated histoplasmosis, even those with AIDS, respond quickly to antifungal agents, and treatment success rates are greater than 90%. One exception is those patients who have chronic cavitary pulmonary histoplasmosis; they frequently have progressive respiratory insufficiency in spite of treatment with antifungal agents. More than 90% of patients with blastomycosis respond to antifungal therapy. However, the mortality rate for those who have severe pulmonary blastomycosis and ARDS remains over 50%. Coccidioidomycosis is the least likely endemic mycosis to respond to antifungal therapy. Success rates with azole agents are 70% for soft tissue infections and 50% to 60% for chronic pulmonary infections. Patients who have *Coccidioides* meningitis must be treated with suppressive azole therapy for life to avoid relapse. For all the endemic mycoses, if immunosuppression cannot be reversed, long-term suppressive azole therapy is recommended.

## EVIDENCE

Galgiani JN, Catanzaro A, Cloud GA, et al. Comparison of oral fluconazole and itraconazole for progressive, nonmeningeal coccidioidomycosis: a randomized, double-blind trial. Mycoses Study Group. *Ann Intern Med* 2000;133:676-686. *Randomized, blinded, controlled trial comparing fluconazole with itraconazole for the treatment of various forms of coccidioidomycosis that showed that either azole was effective for the treatment of coccidioidomycosis.*

Johnson PC, Wheat LJ, Cloud GA, et al. Safety and efficacy of liposomal amphotericin B compared with conventional amphotericin B for induction therapy of histoplasmosis in patients with AIDS. *Ann Intern Med* 2002; 137:105-109. *Randomized, blinded, controlled trial comparing liposomal amphotericin B with amphotericin B deoxycholate for the initial treatment of severe disseminated histoplasmosis in patients with AIDS that established the superiority of the liposomal formulation.*

## ADDITIONAL RESOURCES

Azar MM, Hage CA. Laboratory diagnostics for histoplasmosis. *J Clin Microbiol* 2017;55:1612-1620. *Excellent overview of the availability and usefulness of various diagnostic tests for histoplasmosis.*

Chapman SW, Dismukes WE, Proia LA, et al. Clinical practice guidelines for the management of blastomycosis: 2008 update by the Infectious Diseases Society of America. *Clin Infect Dis* 2008;46:1801-1812. *Latest IDSA guidelines for the management of blastomycosis.*

Galgiani JN, Ampel NM, Blair JE, et al. Infectious Diseases Society of America (IDSA) clinical practice guidelines for the treatment of coccidioidomycosis. *Clin Infect Dis* 2016;63:717-729. *Latest IDSA guidelines for the management of coccidioidomycosis.*

Kauffman CA. Treatment of the Midwestern endemic mycoses, blastomycosis and histoplasmosis. *Curr Fungal Infect Rep.* 2017;11:67-74. *Updated review of treatment aspects of histoplasmosis and blastomycosis.*

Nguyen C, Barker BM, Hoover S, et al. Recent advances in our understanding of the environmental, epidemiological, immunological, and clinical dimensions of coccidioidomycosis. *Clin Microbiol Rev* 2013;26:505-525. *Recent review of various nuances of coccidioidomycosis.*

Pappagianis D. Current status of serologic studies in coccidioidomycosis. *Curr Fungal Infect Rep* 2007;1:129-134. *Older but still excellent review that simplifies the often confusing area of serologic testing for coccidioidomycosis.*

Smith JA, Gauthier G. New developments in blastomycosis. *Sem Respir Crit Care Med* 2015;36:715-728. *Important update on the pathogenesis and clinical aspects of blastomycosis.*

Wheat LJ, Freifeld AG, Kleiman MB, et al. Clinical practice guidelines for the management of patients with histoplasmosis: 2007 update by the Infectious Diseases Society of America. *Clin Infect Dis* 2007;45:807-817. *Latest IDSA guidelines for the management of all forms of histoplasmosis.*

# SECTION V

# Surgical Infections

*E. Patchen Dellinger*

# 41

# Surgical Infections: Introduction and Overview

*E. Patchen Dellinger*

Surgical infections comprise a broad range of infections, many of which are not obviously similar to one another, do not occur in the same organ system or anatomic location, or do not even necessarily share common pathogenic flora. Many think of surgical infections as those that follow a surgical procedure, and this can certainly be true for an incisional surgical site infection (SSI) or a postoperative intra-abdominal abscess (organ space SSI). However, the common thread of a surgical infection is that it stems from—or causes in its evolution—an anatomic or physical condition that must be corrected or ameliorated for the infection to resolve. Common features of many surgical infections include transgression of an epithelial barrier (either skin or gastrointestinal, respiratory, or urologic epithelium) by a surgeon; trauma; tumor; ischemia; obstruction of a hollow organ, as may occur with an appendicolith; a bowel obstruction; a common bile duct stone; or a ureteral stone. Some surgical infections, such as subcutaneous abscesses or superficial SSIs, may resolve with this correction alone and not require any antibiotic treatment. Others need a combination of anatomic manipulation and antimicrobial treatment. The physical treatment of a surgical infection has been termed "source control," which may involve extensive surgical treatment such as debridement of a necrotizing soft-tissue infection or something as simple as removing sutures from a recent surgical incision to let the skin fall open and allow the site to drain. Source control may require a surgeon to remove an appendix, close a perforated ulcer, or remove a segment of colon involved in diverticulitis; it may be managed by a gastroenterologist who removes a common duct stone in a person with cholangitis and places a stent across the ampulla; or it may involve an interventional radiologist who inserts a percutaneous drain into an intra-abdominal abscess. Prosthetic infections usually require removal of the prosthesis for resolution. In each case a mechanical or anatomic source-control maneuver is essential to the resolution of the surgical infection, and in most cases antimicrobial treatment is needed as well.

Although some surgical infections, such as a superficial abscess or superficial SSI, may allow diagnosis in a straightforward manner, others, such as appendicitis, diverticulitis, and cholangitis, require a knowledge of the relevant anatomy and physiology and the natural history of the infection for accurate and timely diagnosis. Some surgical infections, such as necrotizing fasciitis, masquerade as "medical" infections, initially resembling a simple cellulitis. Some, such as chronic foreign body infections involving a vascular prosthesis or prosthetic joint, can be occult, with minimally virulent pathogens. In contrast to many medical infections that are caused by a single pathogen possessing a virulence characteristic that allows it to evade host defenses, many surgical infections are caused by a mixture of normal endogenous flora residing on mucosal surfaces that cause disease only after the epithelial barrier has been violated. Thus, many surgical infections are caused by an anatomic injury, just as they subsequently require an anatomic correction for resolution. The authors of the chapters that follow illustrate these facets of the individual surgical infections under discussion.

# 42

# Acute Appendicitis

*Jasmina Ehab, Daniel A. Anaya*

 **ABSTRACT**

Acute appendicitis is the most common surgical emergency and appendectomy is the most common emergency operation, with more than 250,000 procedures reported annually in the United States. Acute appendicitis results from appendiceal endoluminal obstruction, typically caused by a fecalith. Although no specific risk factors have been identified, appendicitis is slightly more common in males, the young, and the elderly, with more advanced disease on presentation in the geriatric population. Clinical findings in combination with basic laboratory tests are often enough to establish the diagnosis, and imaging studies, such as abdominal ultrasound or computed tomography (CT), can be helpful to confirm the diagnosis and rule out other potential pathologies in selected individuals. The treatment for acute appendicitis is appendectomy and the laparoscopic approach is preferred, given its association with better postoperative outcomes and recovery. In later stages, appendicitis may be complicated by phlegmon or intra-abdominal abscess. For these patients, aggressive medical treatment with broad-spectrum antibiotics and percutaneous drainage when indicated is the initial treatment of choice; the operative approach is reserved for when this treatment fails and in the setting of peritonitis. Interval appendectomy after an episode of appendicitis treated with antibiotics, although still controversial, must be considered to minimize the risk of recurrent inflammation, which is associated with worse outcomes. Better outcomes are generally seen in patients diagnosed and treated early, and this should be the main goal in approaching patients with suspected acute appendicitis.

## ✳ CLINICAL VIGNETTE

A 51-year-old male arrives at the emergency department complaining of abdominal pain as well as nausea and vomiting. He started having dull periumbilical pain several hours earlier, but now the pain is more severe and localized to the right lower abdominal quadrant. His temperature is 101°F; blood pressure, 124/78 mm Hg; and respirations are 14 per minute. On physical exam, the patient keeps his hips flexed and there is intense tenderness one-third of the distance from the anterior superior iliac spine to the umbilicus on the right (McBurney point). The patient exhibits abdominal guarding and has referred pain to the lower right quadrant upon deep palpation of the lower left quadrant (Rovsing sign). Laboratory studies show a leukocyte count of 17,000/mm³.

The patient is taken to the operating room with a diagnosis of acute appendicitis. Prophylactic antibiotics are given and a laparoscopic appendectomy is performed, revealing an inflamed nonperforated appendix. The patient has an uneventful recovery and is discharged home on postoperative day 2. The final pathology report says, "acute edematous appendicitis, with no signs of perforation or gangrene."

Acute appendicitis is the most common surgical emergency in the United States and worldwide. It is characterized by acute inflammation of the appendix and is generally caused by proximal endoluminal obstruction. No specific geographic or endemic distribution has been reported. The cumulative lifetime risk of appendicitis ranges from 7% to 15%. Appendectomy is the most common emergency operation performed worldwide, with approximately 300,000 operations per year in the United States and a median cost close to $11,000 per patient; the estimated overall cost in the United States is more than $2.5 billion. In 2007, it was determined that $7.8 billion per year is spent on hospitalizations for appendectomy in the United States. However, with new data emerging supporting the nonsurgical treatment of nonperforated acute appendicitis, it is predicted that the costs will decrease as surgical hospitalizations decrease. A randomized controlled trial published in 2017 predicted that—if a nonsurgical approach is ultimately proven to have equivalent outcomes and is implemented routinely for selected patients with increased provider education on the effectiveness of antibiotic treatment of nonperforated appendicitis—the US financial burden should gradually decrease.

## RISK FACTORS AND PATHOPHYSIOLOGY

Endoluminal obstruction is the most common cause of appendicitis, with fecaliths being found in 40% of cases of simple appendicitis, 65% of gangrenous appendicitis, and 90% of ruptured appendicitis. Less common causes of obstruction include lymphoid hypertrophy—usually caused by a viral infection in children (mesenteric adenitis)—tumors, seeds, parasites, and inspissated barium. Obstruction leads to appendiceal dilatation with subsequent transmural congestion, venous obstruction, and ischemia. Delayed presentation can result in gangrenous appendicitis with associated perforation and subsequent abscess formation or peritonitis (Fig. 42.1). Gender is associated with different incidence of appendicitis, with a male/female ratio of 1.3 to 1. Similarly, certain age groups are associated with an increased incidence of appendicitis at its different stages. Appendicitis (nonperforated) is more common among adolescents and young adults, with a peak incidence in the second and third decades, whereas perforated appendicitis is more common in children and the elderly.

## CLINICAL PRESENTATION

An accurate history and physical examination are of utmost importance and provide the main clues to an early and accurate diagnosis of appendicitis. The classic presentation is characterized by early periumbilical colicky abdominal pain, which subsequently migrates to the right lower quadrant. This was first described by Murphy and is present in at least 50% of all patients. The periumbilical early pain (present within 1 to 12 hours of the onset of symptoms) is referred from

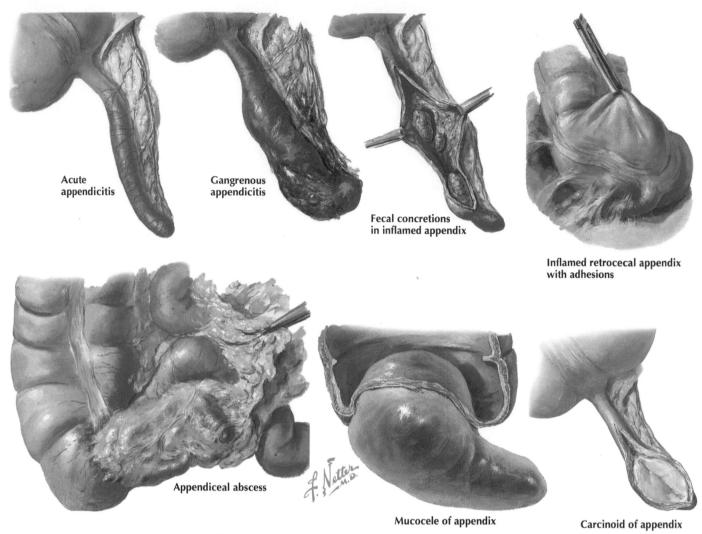

Acute
appendicitis

Gangrenous
appendicitis

Fecal concretions
in inflamed appendix

Inflamed retrocecal appendix
with adhesions

Appendiceal abscess

Mucocele of appendix

Carcinoid of appendix

**Fig. 42.1** Diseases of the appendix: inflammation, mucocele, tumors.

the visceral innervation of the midgut. With progression of the local inflammatory process, the parietal peritoneum becomes inflamed, which results in migration of the pain to the right lower quadrant. Occasionally pain is described as originating at this site during the early phase of appendicitis. Nausea and vomiting occur in 75% of patients, although it is not prominent or prolonged. The sequence of symptoms helps to confirm the diagnosis; in more than 95% of cases of acute appendicitis, anorexia is the first symptom, followed by abdominal pain and then vomiting. A different sequence of events is usually associated with other pathologies included in the differential diagnosis. A recent meta-analysis evaluating signs and symptoms associated with acute appendicitis was unable to identify any one specific isolated diagnostic finding to help confirm acute appendicitis but identified migrating pain as a sequence of events strongly associated with this diagnosis.

Physical examination is extremely helpful when a patient is being evaluated for appendicitis. Although vital signs in early nonperforated appendicitis are relatively unchanged, a temperature elevation of 1°C and a mildly elevated heart rate are common. High fever and tachycardia with hypotension are more common in advanced presentations and should raise suspicion of other diagnoses when present during the early course. Tenderness on palpation over the McBurney point is the classic finding on physical examination, particularly when the appendix is located in an anterior or posterior retrocecal but intraperitoneal position (Fig. 42.2). Other findings can be present with locations in

the retroperitoneal or pelvic areas. Specifically, a retrocecal retroperitoneal appendix typically causes flank or back tenderness, whereas a pelvic appendix may cause rectal and/or suprapubic pain (see Fig. 42.2). Other important findings include rebound tenderness as well as referred pain in the right lower quadrant when pressure is exerted in the left lower quadrant (Rovsing sign). More uncommon but important signs associated with the specific location of the inflamed appendix include the psoas and the obturator signs. A psoas sign, or pain while extending the psoas muscle, is present in cases of a retroperitoneal inflammation, and an obturator sign, or pain on passive internal rotation of the right thigh, suggests a pelvic location.

## DIAGNOSTIC APPROACH

Once a thorough history and physical examination have been completed, laboratory findings and the integration of all data can help to confirm the diagnosis. Laboratory tests should include a complete blood count, urinalysis, basic metabolic panel, and—for all female patients of childbearing age—a pregnancy test. Leukocytosis with cell numbers ranging from 10,000 to 18,000/mm³ along with an elevated absolute neutrophil count is common in patients with simple early appendicitis. White blood cell counts greater than 18,0000/mL³ correlate with perforated appendicitis or appendiceal abscess. A urinalysis helps to rule out a urinary tract infection

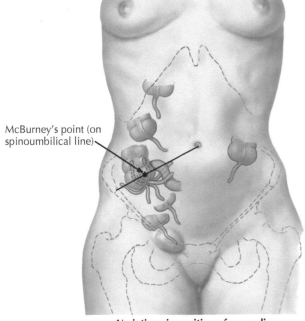

McBurney's point (on spinoumbilical line)

**Variations in position of appendix**

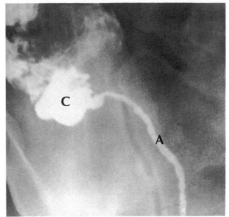

Barium radiograph of unusually long appendix (*A*, Appendix; *C*, cecum)

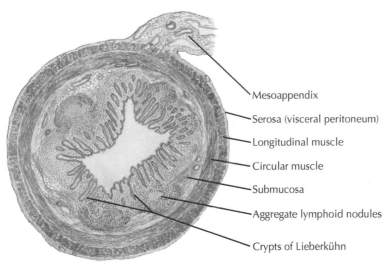

Mesoappendix

Serosa (visceral peritoneum)

Longitudinal muscle

Circular muscle

Submucosa

Aggregate lymphoid nodules

Crypts of Lieberkühn

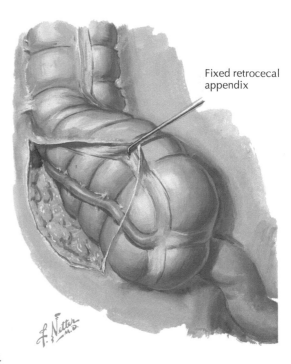

Fixed retrocecal appendix

**Fig. 42.2** Vermiform appendix.

because bacteriuria is not typically seen in patients with appendicitis. A basic metabolic panel may reveal electrolyte abnormalities derived from anorexia, vomiting, and secondary dehydration. A pregnancy test in all females of childbearing age is important to rule out an ectopic pregnancy.

The combination of findings derived from the history and physical examination and results from the initial laboratory tests can usually be enough to confirm the diagnosis or suspicion of acute appendicitis, particularly in the most common and typical cases. Less commonly, late presentations progress to diffuse peritonitis characterized by diffuse abdominal pain and a more severe systemic inflammatory response.

Other tools can be used to help confirm the diagnosis. Findings from history, physical examination, and laboratory tests can also be used to calculate the Alvarado score. This score is based on eight data points and was developed to help confirm the diagnosis of acute appendicitis using clinical and initial laboratory findings (Table 42.1). Higher scores are directly associated with a higher likelihood of appendicitis; scores lower than 5 are unlikely to represent appendicitis,

whereas those higher than 6 or a score of 10 are considered consistent and highly consistent with appendicitis, respectively.

Special consideration should be given to the elderly, young children, and pregnant women. Both the elderly and children usually present later, with a longer duration of vague prehospital symptoms. In addition, findings in these populations are somewhat atypical. In older patients, periumbilical migratory pain is almost always absent, and pain located in the right lower quadrant is reported in only 80% of cases. The accuracy of the Alvarado score also declines in this population, with less than 50% of patients having scores higher than 7. This uncommon presentation is associated with more advanced disease, including higher rates of perforations and abscess formation, which in combination with more associated comorbidities result in worse overall outcomes, including longer lengths of stay and higher rates of postoperative complications and death. Children younger than 5 years old are generally unable to give an accurate history. This often limits an earlier diagnosis and can result in similar patterns of diagnosis, treatment, and outcomes as those observed in the elderly. In the pregnant patient, as the gravid uterus enlarges, the appendix is displaced

| TABLE 42.1 | Alvarado Score |
|---|---|
| **Findings** | **Score Points** |
| **Symptoms** | |
| Pain migration | 1 |
| Anorexia | 1 |
| Nausea and vomiting | 1 |
| **Signs** | |
| Right-lower-quadrant pain | 2 |
| Rebound tenderness | 1 |
| Temperature ≥37.3°C | 1 |
| **Laboratory Results** | |
| Leukocytosis | 2 |
| Left shift | 1 |
| **Total score** | **10** |

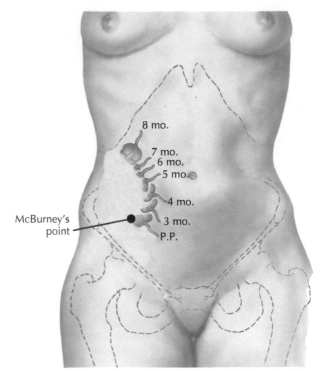

Fig. 42.3 Appendiceal migration at different stages of pregnancy.

cephalad (Fig. 42.3). This is an important consideration that changes the location of pain as well as the surgical approach when the diagnosis is confirmed.

Finally, other differential diagnoses must be considered when patients with right-lower-quadrant abdominal pain are evaluated, including urinary tract infections, ureteral/kidney stones, diverticulitis, perforated ulcer, ectopic pregnancy, ruptured ovarian cyst, ovarian torsion, pelvic inflammatory disease, testicular torsion, inguinal hernia, Meckel diverticulum, Crohn enteritis, gastroenteritis, and complications derived from a tumor of the colon or small bowel.

When the described approach is used to work up patients with suspected acute appendicitis, a false-positive diagnosis can still result. This leads to a negative appendectomy, which—in both clinical and population-level analyses—has been reported to occur in 15% of patients. The negative appendectomy rate is higher for females (22% vs. 9% for males) and even higher for women of reproductive age (up to 26%). A negative appendectomy rate of 10% to 15% is generally accepted, considering that a low threshold for operation can avoid the complications stemming from a delayed diagnosis.

A more thorough workup including the use of imaging studies (CT or ultrasound) has been advocated with the primary goal of improving the accuracy of the diagnosis. Two prospective studies have shown a decrease in the number of unnecessary admissions and appendectomies with CT. However, a longitudinal population-level study suggested that despite the introduction of ultrasound and CT, the rates of negative appendectomies have remained unchanged over time, arguing against the routine use of these diagnostic strategies. Imaging studies should be considered selectively, as when the diagnosis is unclear and the patient is at higher risk for a negative appendectomy. CT is the preferred diagnostic test, although ultrasound is a good alternative, particularly in children and thin patients, as well as in pregnant women or women of childbearing age, to delineate uterine and/or ovarian pathology.

## TREATMENT

The treatment for acute appendicitis has traditionally been surgical, or an appendectomy. When treatment is being planned, it is important to know whether the appendix is perforated, as then treatment will be handled differently. The differentiation between a perforated and nonperforated appendix can be made by CT on hospital admission. Early

acute appendicitis is generally managed with surgery and prophylactic antibiotics to minimize the risk of surgical site infection (SSI). If no perforation or focal peritonitis is encountered, there is generally no need for the continuation of antibiotics after surgery because the main infectious source has been removed (i.e., by source control). Numerous studies have emerged in recent years regarding the treatment of acute appendicitis. As previously described, the mainstay of treatment for acute nonperforated appendicitis has historically been surgery. However, there have been numerous randomized clinical trials, systematic reviews, and meta-analysis showing that nonsurgical treatment for acute nonperforated appendicitis is also a viable option. In this case, patients are treated with antibiotics and intravenous fluids. Antibiotic combinations in these studies have included amoxicillin plus clavulanic acid, cefotaxime and metronidazole, and ciprofloxacin and metronidazole, among others. Most regimens included the in-hospital intravenous administration of antibiotics until fever and leukocytosis have resolved, followed by a 10-day oral course after discharge. In these studies, approximately 90% of patients treated with intravenous fluids and antibiotics were able to avoid surgery during their admission. Approximately 70% of patients who were treated with an antibiotic regimen were not readmitted within the first year for acute appendicitis. One study conducted a 5-year observational follow-up and found that the collective incidence of recurring appendicitis was approximately 27%, and those patients were treated with surgery (appendectomy). Antibiotic therapy alone is only an option for those who are diagnosed with a nonperforated appendicitis. Contraindications to sole antibiotic therapy include patients who have medical comorbidities that would exclude them from a 10-day antibiotic regimen, those who are immunocompromised, and the elderly.

For perforated appendicitis, appendectomy should be performed and systemic antibiotics continued for 5 to 7 days or until fever and leukocytosis have resolved. *Escherichia coli* and *Bacteroides fragilis* are the main organisms isolated in acute simple and perforated appendicitis. However, both anaerobic bacteria and other gram-negative organisms may be present, and polymicrobial infections are the most common type. The Surgical Infection Society (SIS) has recommended different single- and multiple-agent regimens based on the best available evidence generally derived from randomized controlled trials. Single-agent regimens include cefoxitin, cefotetan, and ticarcillin-clavulanic acid; multiple-agent therapies include a third-generation cephalosporin, monobactam, or aminoglycoside combined with antianaerobic coverage with agents such as metronidazole or clindamycin.

Delayed presentation, usually defined as presentation after 5 days or longer since onset of symptoms, is associated with abscess formation plus or minus phlegmon or diffuse peritonitis. Outcomes in these patients are worse than in those with early presentation. For these patients a CT scan is generally recommended. If a phlegmon is identified, the patient is admitted and treated with systemic antibiotics, bowel rest, and physiologic support. If the CT scan reveals an abscess, it should be drained via a percutaneous approach whenever feasible and medical treatment as described previously should be initiated. Nonsurgical treatment has been reported to fail in approximately 7% to 10% of patients. Failure of medical treatment is defined as worsening abdominal pain, continuous fever, leukocytosis, and/or progression to focal or diffuse peritonitis. In these cases the patient should be taken promptly to the operating room for surgical management including drainage and resection of the involved tissues, which often requires a cecectomy or hemicolectomy with drain placement.

Last, there has been some debate regarding the best surgical approach for appendectomy: laparoscopic versus open. Results from multiple randomized controlled trials have been reported, and a recent review of the literature favored the laparoscopic approach because of better postoperative outcomes including lower rates of SSI, shorter length of stay, and faster return to work. Given the reported benefits after surgery, the laparoscopic approach is currently the standard of care.

## PROGNOSIS

The outcomes after appropriate treatment for acute appendicitis have improved significantly over the last two decades, with current low rates of morbidity and mortality. However, complications and mortality increase significantly among the elderly and pregnant females. In the latter, fetal mortality ranges from 0% to 1.5% in cases of simple appendicitis to 20% to 35% in cases of perforation. The most common complication after appendectomy is SSI, typically affecting the superficial wound. The risk of SSI ranges from 1% to 20% and increases in perforated appendicitis and in the elderly population. Most commonly, superficial SSI can be treated with local wound care and antibiotics if surrounding cellulitis is present. Deeper SSI (intra-abdominal abscess) is less common but does occur, particularly in those treated for perforated appendicitis. The diagnosis is made with CT or ultrasonography in patients with worsening abdominal pain during the postoperative period or those with signs and/or symptoms of sepsis. In the vast majority of cases, these abscesses can be treated with percutaneous drainage. Surgical management is seldom required, although it must be considered when large multiloculated abscesses are present and in patients with multiple diffuse abscesses not amenable to the percutaneous approach.

## EVIDENCE

Alvarado A. A practical score for the early diagnosis of acute appendicitis. *Ann Emerg Med* 1986;15:557-564. *This study describes the Alvarado score and its predictive ability for acute appendicitis.*

Andersen BR, Kallehave FL, Andersen HK. Antibiotics versus placebo for prevention of postoperative infection after appendicectomy. *Cochrane Database Syst Rev* 2005;3:CD001439. *This systematic review evaluates the role and benefits of prophylactic antibiotics for appendectomies.*

Andersson R. Meta-analysis of the clinical and laboratory diagnosis of appendicitis. *Br J Surg* 2004;91:28-37. *This meta-analysis evaluates the predictive value of different variables to help establish the diagnosis of acute appendicitis.*

Andersson RE, Petzold MG. Nonsurgical treatment of appendiceal abscess or phlegmon: a systematic review and metaanalysis. *Ann Surg* 2007;246:741-748. *This is a good review of the risks and benefits of a nonsurgical approach to advanced, complicated acute appendicitis using evidence-based analysis.*

Campbell MR, Johnston SL 3rd, Marshburn T, et al. Nonoperative treatment of suspected appendicitis in remote medical care environments: implications for future spaceflight medical care. *J Am Coll Surg* 2004;198:822-830. *This analysis evaluates the role of medical management of acute appendicitis.*

Davidson GH, et al. Comparison of Outcomes of antibiotic Drugs and Appendectomy (CODA) trial: a protocol for the pragmatic randomised study of appendicitis treatment. *BMJ Open* 2017;7:e016117. *A study discussing the future of the cost of treating appendicitis, with new data supporting the use of antibiotics for nonperforated appendicitis.*

Flum DR, Koepsell T. The clinical and economic correlates of misdiagnosed appendicitis: nationwide analysis. *Arch Surg* 2002;137:799. *This article explores the population-level rate of misdiagnosis of acute appendicitis, its impact, and the role and effect of imaging modalities as diagnostic adjuncts on the final outcome in patients with acute appendicitis.*

Flum DR, McClure TD, Morris A, Koepsell T. Misdiagnosis of appendicitis and the use of diagnostic imaging. *J Am Coll Surg* 2005;201:933. *This article explores the population-level rate of misdiagnosis of acute appendicitis, its impact, and the role and effect of imaging modalities as diagnostic adjuncts on the final outcome in patients with acute appendicitis.*

Hansson J, Körner U, Khorram-Manesh A, et al. Randomized clinical trial of antibiotic therapy versus appendicectomy as primary treatment of acute appendicitis in unselected patients. *Br J Surg* 2009;96:473. *A randomized controlled trial comparing antibiotic therapy vs surgery for the treatment of nonperforated appendicitis.*

Salminen P, Tuominen R, Paajanen H, Rautio T, Nordström P, Aarnio M, Rantanen T, Hurme S, Mecklin JP, Sand J, Virtanen J, Jartti A, Grönroos JM. Five-year follow-up of antibiotic therapy for uncomplicated acute appendicitis in the APPAC randomized clinical trial. *JAMA* 2018;320(12):1259. *This study looks at the 5-year follow-up of patients who received antibiotic therapy for nonperforated appendicitis.*

Sheu BF, Chiu TE, Chen JC, et al. Risk factors associated with perforated appendicitis in elderly patients presenting with signs and symptoms of acute appendicitis. *ANZ J Surg* 2007;77:662. *This study identifies specific age-related characteristics in elderly patients predictive of complicated (perforated) appendicitis.*

Styrud J, Eriksson S, Nilsson I, et al. Appendectomy versus antibiotic treatment in acute appendicitis: a prospective multicenter randomized controlled trial. *World J Surg* 2006;30:1033. *This is a reasonably good randomized controlled trial evaluating the role of medical management of acute appendicitis (antibiotics treatment) as compared with surgical management in a well-selected group of patients.*

Vons C, Barry C, Maitre S, et al. Amoxicillin plus clavulanic acid versus appendicectomy for treatment of acute uncomplicated appendicitis: an open-label, non-inferiority, randomised controlled trial. *Lancet* 2011;377:1573. *This study looks at amoxicillin plus clavulanic acid for treatment of non-perforated appendicitis.*

## ADDITIONAL RESOURCES

Corfield L. Interval appendectomy after appendiceal mass or abscess in adult: what is the best practice? *Surg Today* 2007;37:1-4. *This article discusses the still controversial issue of interval appendectomy and its risks and benefits in patients initially treated nonoperatively for acute appendicitis.*

Lawson RS. Murphy's triad. *Br Med J* 1971;1.5745:401.

Meeks DW, Kao LS. Controversies in appendicitis. *Surg Infect* 2008;9:553-558. *This is a good evidence-based review of the most controversial issues of the management of acute appendicitis.*

Solomkin JS, Mazuski JE, Bradley JS, et al. Diagnosis and management of complicated intra abdominal infection in adults and children: guidelines by the Surgical Infection Society and the Infectious Diseases Society of America. *Surg Infect (Larchmt)* 2010;11(1):79-109. *This is a good evidence-driven guideline summarizing the consensus from the SIS and the ISDA regarding the management of complicated intra-abdominal infection applicable to selected patients with advanced, complicated acute appendicitis.*

# 43

# Acute Ascending Cholangitis and Suppurative Toxic Cholangitis

*Kevin P. Labadie, David W. Miranda, Patrick S. Wolf, James O. Park*

## ABSTRACT

Acute ascending cholangitis is a biliary tract infection resulting from bile duct obstruction. It carries potential for significant morbidity and mortality and requires prompt diagnosis and treatment. Clinical presentation can range from a mild, self-limited course to a serious, life-threatening condition requiring emergent intervention. Patients typically have some combination of fever, jaundice, and abdominal pain. A methodical patient history and imaging workup can usually elucidate the cause of the biliary obstruction. Therapy should be initiated swiftly; it relies on aggressive resuscitation, appropriate antibiotics, and prompt biliary decompression. Outcome largely depends on the underlying cause of biliary obstruction and timely recognition and treatment.

## CLINICAL VIGNETTE

A 58-year-old woman with type 2 diabetes, obesity, and hypertension presented to the emergency department (ED) with a chief complaint of abdominal pain and fevers. She reported having experienced this pain before but never of this severity. The pain has worsened over the past 12 hours and she began to experience subjective chills and rigors, prompting her to present to the ED for evaluation.

Upon arrival at the ED, the patient's vital signs were as follows: temperature 103.6°F, heart rate 112 beats/min, blood pressure 110/45 mm Hg, and respiratory rate 22 breaths/min. On physical examination, her skin was moist and warm; eye exam revealed scleral icterus. Her cardiopulmonary exam was normal. Her abdomen was obese, soft without guarding but with moderate tenderness to deep palpation in the right upper quadrant. Murphy sign was negative. There was no evidence of ascites or lower extremity edema. Serum chemistry and hematologic panel were remarkable for a leukocytosis with a white blood cell (WBC) count of 19,000/µL; glucose, 245 mg/dL; blood urea nitrogen, 35 mg/dL; creatinine, 1.2 mg/dL; potassium, 2.7 mEq/L; and total bilirubin, 7.6 mg/dL. Blood cultures were positive for *Escherichia coli*. Right-upper-quadrant ultrasound demonstrated gallstones and a dilated common bile duct.

A diagnosis of acute ascending cholangitis secondary to choledocholithiasis was suspected. She was started on broad antibiotics and fluid resuscitated. Endoscopic retrograde cholangiography demonstrated a stone completely obstructing the common bile duct. The stone was removed with graspers, suppurative cholangitis was confirmed, and the biliary obstruction was relieved. The patient improved on antibiotics and was eventually discharged.

## GEOGRAPHIC DISTRIBUTION AND MAGNITUDE OF DISEASE BURDEN

An estimated 10% to 15% of the more than 1 million patients diagnosed with gallstones annually in the United States will have stones in the common bile duct. Of these patients, over half will develop symptoms of acute cholangitis—a conservative estimate of 50,000 to 75,000 cases per year. Although the disease is prevalent worldwide, in Western countries over 85% of bile duct stones originate from the gallbladder, whereas in East Asian countries, where liver fluke infections are endemic, primary brown pigment stones are a significant source of recurrent pyogenic cholangitis.

## PATHOGENESIS

Ascending cholangitis is an infection of the intrahepatic and extrahepatic biliary system that occurs as a consequence of stagnant bile. It is well established that both biliary obstruction and bacterobilia are necessary for clinical disease manifestation (Fig. 43.1). Diminished biliary outflow, most commonly caused by stone disease, predisposes to infection of the normally sterile, free-flowing bile. The consequent bacterial proliferation, together with disrupted hepatocellular tight junctions from increased intraductal pressures, results in the release of systemic mediators and even overt bacteremia through translocation into the hepatic veins or perihepatic lymphatics. Patients with indwelling biliary stents for malignant strictures and those who have undergone surgical biliary-enteric reconstruction have colonized biliary systems and can develop cholangitis even with low-grade obstruction. Other causes of biliary obstruction include primary sclerosing cholangitis, parasitic infestation *(Clonorchis, Opisthorcis)*, and acquired immunodeficiency syndrome (AIDS) cholangiopathy.

## RISK FACTORS

Because the majority of cases in the United States result from choledocholithiasis, the risk factors for ascending cholangitis parallel those of gallstone disease: advanced age, female gender, obesity, metabolic syndrome, rapid weight loss, gallbladder stasis, cirrhosis, and Crohn ileitis. Also, given that bacterobilia is a prerequisite for the development of cholangitis, biliary interventions that introduce bacteria into the bile duct place the patient at risk of developing cholangitis. This is evident in patients with malignant strictures. Although it is rare for such patients to develop cholangitis de novo, the risk of cholangitis is significantly increased after biliary manipulation, especially with inadequate drainage.

## CLINICAL FEATURES

The classic presentation of acute ascending cholangitis as described by Jean-Martin Charcot in 1877 consists of the triad of fever, jaundice, and right-upper-quadrant pain (Fig. 43.2). However, only 56% to 70% of all patients seeking medical attention for cholangitis exhibit all three elements. Fever is the most common presenting symptom, occurring in upwards of 90% of patients. Jaundice is variable and can be absent,

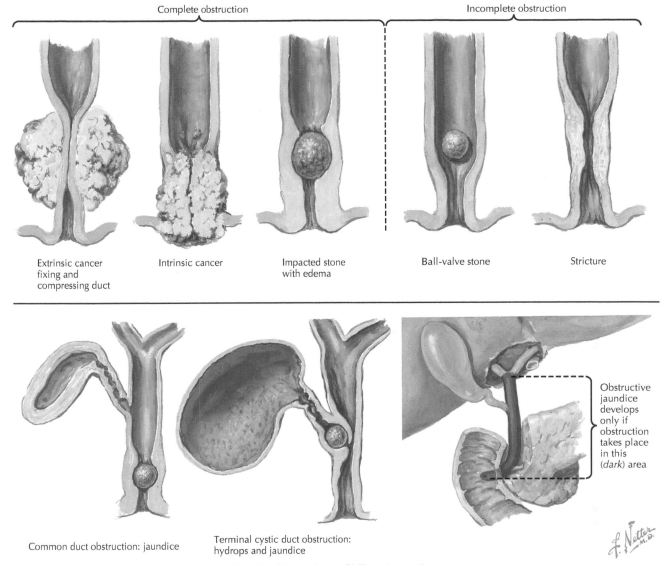

**Fig. 43.1** Mechanisms of biliary obstruction.

especially in patients with a partially occluded biliary stent as the cause of cholangitis. Abdominal pain, although common (70%), is typically mild in nature; severe pain or tenderness on examination should raise the suspicion of an alternative diagnosis, such as acute pancreatitis, cholecystitis, or perforated viscus. A more advanced form of disease, manifested by signs and symptoms of septic shock including hemodynamic instability and altered mental status (Reynolds pentad), is identified in a small subset (5% to 7%) of critically ill patients and requires prompt recognition and intervention for optimal outcome, given the higher risk of multiorgan failure.

## DIAGNOSTIC APPROACH

Cholangitis should be suspected in any patient with the triad of fever, jaundice, and right-upper-quadrant abdominal pain. Whereas the differential diagnosis remains broad for patients with more ambiguous complaints, a careful patient history may provide clues to narrow the diagnostic focus and elucidate the mechanism causing biliary obstruction. Patient-related factors that should raise the suspicion of mechanical biliary obstruction predisposing to cholangitis include a known history of gallstones, biliary colic, and pancreatitis. A history of back pain, weight loss, and acholic stools may indicate a biliary or pancreatic malignancy. Last, any history of surgical biliary reconstruction and endoscopic or percutaneous bile duct manipulation must be sought. Physical findings in patients with cholangitis are often nonspecific. Rigors are frequently described but variably detected. Abdominal tenderness is common and typically localized to the right upper quadrant, but overt peritoneal signs are unusual. Persistent tachycardia and hypotension herald advanced disease and warrant expedient resuscitation and treatment.

Laboratory evaluation typically demonstrates conjugated hyperbilirubinemia and elevation of alkaline phosphatase and gamma-glutamyltranspeptidase, confirming biliary obstruction and injury. Aminotransferases are generally normal; however, they can be elevated in advanced disease secondary to acute hepatic necrosis from pyogenic abscess formation. Leukocytosis with neutrophil predominance is common in immunocompetent patients and may be quite marked, indicating the systemic nature of the disease. Hyperamylasemia is present in roughly one-third of patients and indicates more distal common duct obstruction with or without concomitant pancreatitis. Bile and blood cultures, which are positive in more than 80% and 20% to 70%, respectively, may help tailor antibiotic selection. Enteric gram-negative

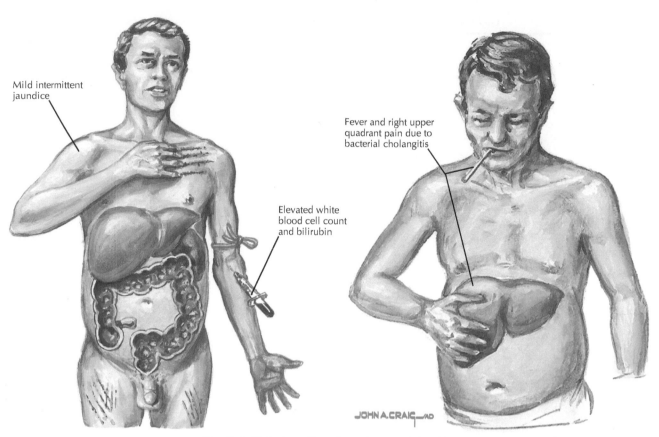

**Fig. 43.2** Clinical manifestations of acute cholangitis.

organisms such as *E. coli*, *Klebsiella*, and *Enterobacter* and gram-positive organisms such as *Enterococcus* are the most common isolates. Antibiotic-resistant strains (methicillin-resistant *Staphylococcus aureus*, vancomycin-resistant *Enterococcus*), anaerobic organisms (*Clostridium, Bacteroides)*, and fungi are more commonly seen in patients with indwelling stents, prior biliary-enteric surgery, or severe infections.

Imaging is an important adjunct in the workup of patients suspected of having cholangitis; it can confirm biliary dilatation, demonstrate the cause of cholangitis (e.g., choledocholithiasis or pancreatic mass), and rule out other causes for the patient's symptoms. Right-upper-quadrant ultrasound is an excellent modality to evaluate biliary pathology as it is a quick, noninvasive, and cost-effective study that is sensitive in detecting biliary ductal dilatation and gallstones, although the sensitivity to common bile duct stones is relatively low and is operator dependent. Computed tomography (CT) effectively detects biliary dilatation and has the added benefit of a more detailed assessment of coexistent pathology such as liver abscesses, although it is not effective in diagnosing choledocholithiasis. Magnetic resonance cholangiography is costly and time-consuming; it should not be the first test of choice in evaluating a patient with suspected cholangitis. However, it can be reserved for cases in whom more accurate delineation of intraductal pathology (e.g., cholangiocarcinoma) is necessary. Direct cholangiography via either an endoscopic or percutaneous transhepatic approach is generally required for patients diagnosed with cholangitis, as these interventions clearly delineate biliary ductal anatomy and can be used to relieve the biliary obstruction. Endoscopic ultrasound is an alternative option for identifying choledocholithiasis and has similar sensitivity to endoscopic retrograde cholangiopancreatography (ERCP) while avoiding invasive bile duct exploration.

The Tokyo Guidelines for the diagnosis of acute cholangitis—first published in 2007 and updated in 2013 and 2018 to include severity-based therapy—detail four criteria for the diagnosis of acute cholangitis: history of biliary disease, clinical presentation indicating cholestasis, laboratory data suggesting evidence of inflammation and biliary obstruction, and diagnostic imaging indicating biliary dilatation.

## CLINICAL MANAGEMENT AND TREATMENT

Once the diagnosis of cholangitis has been established, effective treatment relies on two principles: antibiotic therapy and biliary ductal decompression. Although most patients are clinically stable, with normal hemodynamics, it must be stressed that those who have advanced disease require intensive care unit monitoring and aggressive resuscitation consisting of intravenous hydration and correction of coagulopathy. Treatment delay in these critically ill patients frequently portends a poor prognosis.

Early administration of antibiotic therapy is of paramount importance in all patients with cholangitis. In the absence of microbiologic data on a given patient, broad-spectrum, empiric antibiotic therapy covering enteric gram-negative and gram-positive aerobes and anaerobes should be initiated. Antibiotic selection should favor drugs that achieve high biliary concentrations. Fluoroquinolones, extended-spectrum penicillins, carbapenems, and aminoglycosides are all excellent choices for empiric therapy while awaiting blood or bile culture results. Treatment duration depends in part on the presence of bacteremia. Patients found to have positive blood culture results require 10 to 14 days of antibiotics. In the absence of bacteremia, length of antibiotic therapy depends on adequate treatment response as manifest by resolution of leukocytosis and fever as well as success of biliary decompression.

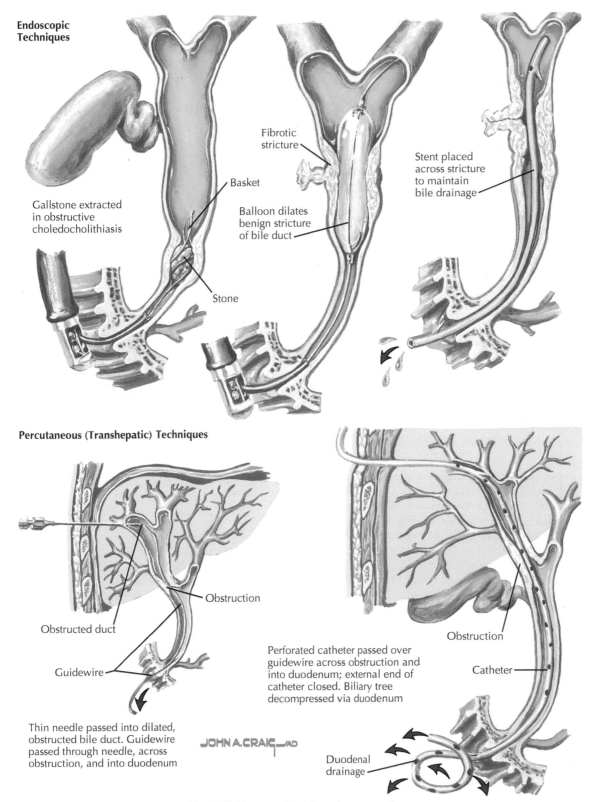

**Endoscopic Techniques**

Gallstone extracted in obstructive choledocholithiasis

Basket

Stone

Fibrotic stricture

Balloon dilates benign stricture of bile duct

Stent placed across stricture to maintain bile drainage

**Percutaneous (Transhepatic) Techniques**

Obstructed duct

Obstruction

Guidewire

Thin needle passed into dilated, obstructed bile duct. Guidewire passed through needle, across obstruction, and into duodenum

JOHN A. CRAIG___MD

Perforated catheter passed over guidewire across obstruction and into duodenum; external end of catheter closed. Biliary tree decompressed via duodenum

Obstruction

Catheter

Duodenal drainage

Fig. 43.3 Nonoperative biliary decompression.

Biliary decompression is the next step in the management of acute cholangitis. Endoscopic and percutaneous transhepatic modalities are well established and effective at achieving successful biliary drainage and are preferred over a surgical approach because they have lower morbidity and provide better outcomes in prospective randomized trials (Fig. 43.3). ERCP is the first-line procedure of choice; therefore consultation with an experienced gastroenterologist early in the course of therapy is essential. ERCP is effective in decompressing the biliary tract in 90% of cases, and culture or pathologic specimens can be obtained concurrently. Patients in whom endoscopic biliary decompression fails or those who cannot undergo the procedure secondary to anatomic restrictions (e.g., Roux-en-Y hepaticojejunostomy or gastric

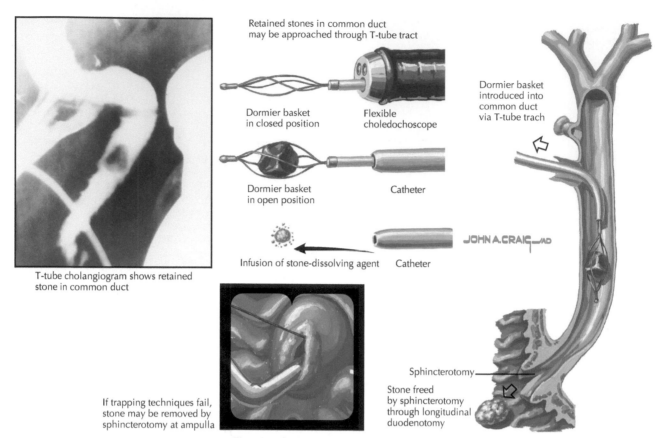

T-tube cholangiogram shows retained
stone in common duct

Retained stones in common duct
may be approached through T-tube tract

Dormier basket
in closed position

Flexible
choledochoscope

Dormier basket
in open position

Catheter

Infusion of stone-dissolving agent    Catheter

Dormier basket
introduced into
common duct
via T-tube trach

JOHN A.CRAIG─MD

Sphincterotomy

Stone freed
by sphincterotomy
through longitudinal
duodenotomy

If trapping techniques fail,
stone may be removed by
sphincterotomy at ampulla

**Fig. 43.4**  Operative biliary decompression.

bypass) should be referred for percutaneous transhepatic cholangiography (PTC) with external biliary drain placement. A small subset of patients will not achieve decompression with either ERCP or PTC and should be referred to a surgeon for discussion of surgical options (Fig. 43.4). Choledochotomy and T-tube placement are advocated over common bile duct (CBD) exploration and cholecystectomy to reduce morbidity in these cases. Mortality rates are historically high for this group of patients, so all efforts should be made to achieve decompression via less invasive approaches.

Timing of biliary decompression depends on the patient's response to intravenous hydration and antibiotics. Biliary drainage can be performed on an elective basis before discharge in 80% of cases when the cholangitis responds well to antibiotic therapy. Emergent biliary decompression is necessary in severe cases, allowing for better biliary penetration of antibiotics. This small subset of patients who continue to manifest signs of sepsis after 12 to 24 hours of antibiotics should proceed emergently to decompression. Although beyond the scope of this discussion, once the acute episode has been controlled, definitive therapy should be targeted at the underlying cause of biliary obstruction. Depending on the cause, this frequently involves a multidisciplinary approach comprising a surgeon, gastroenterologist, and radiologist.

## PROGNOSIS

Prompt recognition and initiation of treatment are of paramount importance in reducing morbidity from acute cholangitis. In general, outcomes are worse for malignant causes of obstruction and in patients with significant comorbidities. Most series cite an overall mortality of 5% for all patients with ascending cholangitis. However, mortality figures remain high (50%) in patients with severe cholangitis resulting in multiorgan dysfunction. Predictors of worse outcome include advanced age, renal failure, coexistent liver abscess, and malignant stricture.

## PREVENTION

The prevention of cholangitis and its recurrence is focused on minimizing biliary stasis. In patients with cholelithiasis and biliary colic, this means recommended elective cholecystectomy. In patients with benign biliary stricture after bile duct injury, endoscopic or surgical repair can minimize risk of developing stasis-related complications such as stones and cholangitis. In patients with malignant stenosis, endoscopic stent placement as part of a broader oncologic treatment plan is important for avoiding cholangitis.

## EVIDENCE

Amouyal P, Amouyal G, Lévy P, et al. Diagnosis of choledocholithiasis by endoscopic ultrasonography. *Gastroenterology* 1994;106:1062-1067. *Prospective study comparing endoscopic ultrasonography, transabdominal ultrasonography, and CT for diagnosing choledocholithiasis. Endoscopic ultrasound has improved diagnostic accuracy but carries with it the risks of an invasive procedure.*

Boey JH, Way LW. Acute cholangitis. *Ann Surg* 1980;191:264-270. *Case series describing the clinical presentations and outcomes of a heterogeneous patient population with acute cholangitis.*

Gigot JF, Leese T, Dereme T, et al. Acute cholangitis: multivariate analysis of risk factors. *Ann Surg* 1989;209:435-438. *Large retrospective study that used multivariate analysis to determine risk factors predictive of mortality in patients with acute cholangitis.*

Hui CK, Liu CL, Lai KC, et al. Outcome of emergency ERCP for acute cholangitis in patients 90 years of age and older. *Aliment Pharmacol Ther* 2004;19:1153-1158. *Describes the safety of emergent ERCP in an elderly patient population treated for acute cholangitis.*

Jacobsson B, Kjellgander J, Rosengren B. Cholangiovenous reflux: an experimental study. *Acta Chir Scand* 1962;123:316-321. *Describes the basis for systemic sepsis secondary to cholangitis.*

Kiriyama S, Kozaka K, Takada T, et al. Tokyo Guidelines 2018: diagnostic criteria and severity grading of acute cholangitis. *J Hepatobiliary Pancreat Surg* 2018;25:17-30. *Updated criteria and severity assessment developed by a consensus conference of experts after reviewing the best available literature.*

Lai EC, Mok FP, Tan ES, et al. Endoscopic biliary drainage for severe acute cholangitis. *N Engl J Med* 1992;326:1582-1586. *Randomized trial comparing surgical versus endoscopic biliary drainage in patients with acute cholangitis. Mortality was lower in the endoscopic treatment group.*

Lai EC, Tam PC, Paterson IA, et al. Emergency surgery for severe acute cholangitis: the high risk patients. *Ann Surg* 1990;211:55-59. *Retrospective study that developed five clinical risk factors predictive of increased morbidity and mortality after surgery for acute cholangitis.*

Lee WJ, Chang KJ, Lee CS, Chen KM. Surgery in cholangitis: bacteriology and choice of antibiotic. *Hepatogastroenterology* 1992;39:347-349. *Case series examining the outcomes of patients with hepatolithiasis based on microbiologic biliary culture results and choice of antibiotics.*

Muller EL, Pitt HA, Thompson JE, et al. Antibiotics in infections of the biliary tract. *Surg Gynecol Obstet* 1987;165:285-292. *Prospective clinical study comparing outcomes of three different antibiotic regimens in the treatment of cholecystitis or cholangitis.*

O'Connor MJ, Schwartz ML, McQuarrie DG, Sumer HW. Acute bacterial cholangitis: an analysis of clinical manifestation. *Arch Surg* 1982;117:437-441. *Retrospective case series of patients with acute bacterial cholangitis.*

Pitt HA, Postier RG, Cameron JL. Consequences of preoperative cholangitis and its treatment on the outcome of operation for choledocholithiasis. *Surgery* 1983;94:447-452. *Retrospective study describing the complications of treating cholangitis with aminoglycosides.*

Reynolds BM, Dargan EL. Acute obstructive cholangitis: a distinct clinical syndrome. *Ann Surg* 1959;150:299-303. *Description of the pathogenesis and clinic syndrome associated with fulminant cholangitis.*

Sugiyama M, Atomi Y. Endoscopic ultrasonography for diagnosing choledocholithiasis: a prospective comparative study with ultrasonography and computed tomography. *Gastrointest Endosc* 1997;45:143-146. *Prospective study concluded that endoscopic ultrasound is more accurate than conventional transabdominal ultrasound and CT imaging in diagnosing choledocholithiasis.*

Thompson J, Bennion RS, Pitt HA. An analysis of infectious failures in acute cholangitis. *HPB Surg* 1994;8:139-144. *Describes clinical factors predictive of treatment failure in patients with acute cholangitis, including malignancy, high levels of hyperbilirubinemia, and bacteremia.*

Thompson JE, Pitt HA, Doty JE, et al. Broad spectrum penicillin as an adequate therapy for acute cholangitis. *Surg Gynecol Obstet* 1990;171:275-282. *Prospective study concluded that single-agent penicillin therapy is as efficacious as penicillin plus an aminoglycoside in the treatment of acute cholangitis.*

Wada K, Takada T, Kawarada Y, et al. Diagnostic criteria and severity assessment of acute cholangitis: Tokyo Guidelines. *J Hepatobiliary Pancreat Surg* 2007;14:52-58. *Diagnostic criteria and severity assessment developed by a consensus conference of experts after reviewing the best available literature.*

## ADDITIONAL RESOURCES

Attasaranya S, Fogel EL, Lehman GA. Choledocholithiasis, ascending cholangitis, and gallstone pancreatitis. *Med Clin North Am* 2008;92:925-960. *Review of current practice in diagnosis and management of complicated biliary calculus disease.*

Leung JW, Yu AS. Hepatolithiasis and biliary parasites. *Baillieres Clin Gastroenterol* 1997;11:681-706. *Review of the pathogenesis of intrahepatic biliary stones secondary to parasitic infection and the role of various treatment modalities.*

Lipsett PA, Pitt HA. Acute cholangitis. *Surg Clin North Am* 1990;70:1297-1312. *Review of the clinical management of acute cholangitis with a focus on antibiotic selection and the use of nonsurgical catheter-based treatment options.*

National Institutes of Health Consensus Development Conference statement on gallstones and laparoscopic cholecystectomy. *Am J Surg* 1993;165:390-398. *Early consensus statement on the role of laparoscopic cholecystectomy in the treatment of gallstone disease.*

Van Erpecum KJ, Venneman NG, Portincasa P, et al. Review article: agents affecting gallbladder motility—role in treatment and prevention of gallstones. *Aliment Pharmacol Ther* 2000;14:66-70. *Review of nonsurgical treatment options in cholelithiasis.*

# 44

# Acute Diverticulitis

*Seth I. Felder, Daniel A. Anaya*

## ABSTRACT

Diverticulitis of the colon is an extremely common disease, accounting for close to 30,000 annual hospital admissions in the United States. Asymptomatic diverticulosis is widely pervasive among older individuals of Westernized countries; however, the rate of progression to acute diverticulitis is likely less than 10%. Although the precise pathogenesis of progression from diverticulosis to diverticulitis is not clear, herniation of bowel mucosa through the colon wall results in the clinical manifestation of diverticular disease. The presentation and clinical course can range from mild to severe, including an acute abdomen in cases of free perforation. Computed tomography (CT) is considered the standard diagnostic test, allowing stratification of a clinical presentation into "uncomplicated" or "complicated" categories as well as guiding and informing subsequent treatment decisions. Because diverticulitis represents a wide spectrum of acuity, treatment must be individualized to each patient's clinical presentation and may consist of antibiotic treatment alone, percutaneous drainage, or possible urgent surgical intervention for source control or sequelae secondary to complications related to diverticular disease (e.g., fistula, obstruction). With better understanding of the disease process and its natural history, the necessity for antibiotics as well as surgical resection has been increasingly studied. Management of patients initially treated without surgery continues to remain an unsettled debate. Currently, the American Society of Colon and Rectal Surgeons (ASCRS) recommends an individualized approach for each patient based on age, comorbidities, frequency and intensity of diverticular episodes.

## CLINICAL VIGNETTE

A 68-year-old woman presented with new-onset pneumaturia and dysuria. She denied associated abdominal pain, fever, or fecaluria. She had experienced two prior episodes of CT–proven uncomplicated diverticulitis managed successfully with antibiotics, each requiring a short hospitalization. These episodes were not associated with a diverticular abscess, and both attacks occurred within the previous 12 months. She had undergone colonoscopy following clinical resolution of her most recent diverticulitis episode, which was notable only for sigmoid diverticulosis. She had a benign abdominal examination eliciting no tenderness to palpation. A urinalysis contained bacteria and leukocyte esterase, consistent with a urinary tract infection. A CT scan of the abdomen and pelvis demonstrated air within the bladder and a redundant sigmoid colon loop with wall thickening opposing the dome of the bladder, consistent with a diverticular colovesical fistula.

COMMENT: In a surgically fit patient, a diverticular colovesical fistula should call for surgery. Colovesical fistula, colovaginal fistula, and diverticular stricture are often secondary sequelae of prior diverticular episodes and are classified as "complicated" diverticulitis.

There are two main types of diverticular disease. True diverticula are congenital, contain all layers of the bowel wall, and are more common in the Eastern Hemisphere. These are not the subject of this review. Acquired (false) diverticula are more common and form when the mucosa and submucosa of the bowel herniate through the muscular layer of the bowel wall. Between the longitudinal muscular layer of the colon (which forms the taenia coli), intrinsic weakness within the colon wall exists where the vasa recta penetrate. These vessels penetrate the mesenteric side of the taenia and continue submucosally and symmetrically around the bowel wall. Diverticular herniation occurs at these weak points due to concomitant localized, elevated sigmoid intraluminal pressure (Fig. 44.1). Although acquired (false) diverticula can affect any segment of the colon, they are, in the industrialized western world, predominantly concentrated in the left side of the colon. The sigmoid colon is disproportionally affected. This is attributed to the higher intraluminal pressure generated within the segment when stool is being propelled antegrade. Diverticula may extend proximally within the colon and rarely throughout the entire colon (pandiverticulosis). However, they do not involve the rectum (no taenia).

Diverticula themselves are asymptomatic but can become symptomatic due to inflammation (diverticulitis) or bleeding. It is difficult to assess the overall prevalence of diverticulosis, but autopsy studies have suggested that the prevalence increases with age, with less than 10% of individuals younger than 40 years of age having diverticula compared with over 80% by age 80. Historically, it has been estimated that 10% to 25% of persons with diverticula progress to acute diverticulitis. However, more contemporary studies suggest that this rate is likely less than 10%. Within this proportion of patients progressing to diverticulitis, 75% of cases are "uncomplicated" and the remaining 25% are "complicated," associated with free perforation, abscess, obstruction, or fistula.

Several studies have reported a higher incidence of diverticula and diverticular disease in Western countries. Diverticular disease is rare in Asia and Africa, although the incidence appears to be rising as these areas become more industrialized. This finding has supported the hypothesis that environmental factors, specifically diet, may play a key role in development of the disease.

A recent review using the National Hospital Discharge Survey estimated that diverticulitis accounts for 300,000 admissions and 1.5 million days of inpatient care annually in the United States. One study reported that over 50% of all sigmoid resections and over 30% of all colostomies performed in the state of Washington were done for diverticulitis.

## RISK FACTORS AND PATHOPHYSIOLOGY

The risk of diverticular disease increases with age. The mean age of affected patients is 62 years, with an increasing prevalence among older adults. Because the US population older than 75 years of age is growing rapidly (a 33% increase from 1998 to 2005), this cohort is an important subset who will be encountered with diverticulitis.

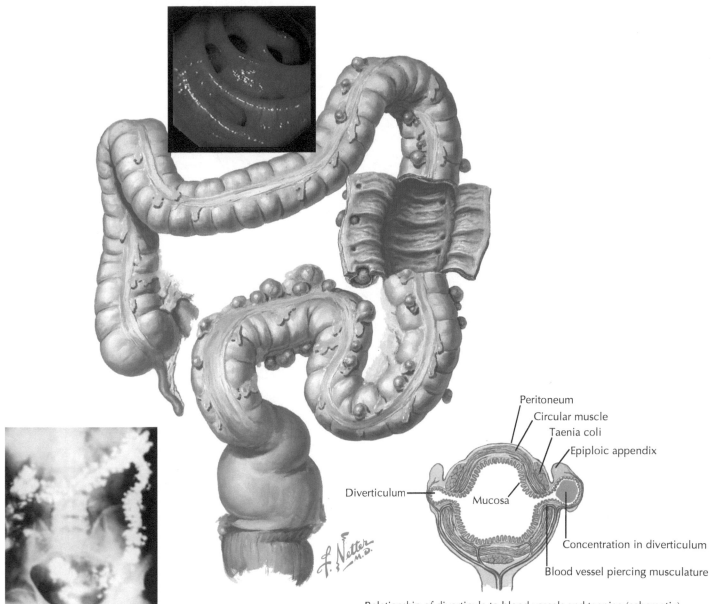

Fig. 44.1 Diverticulosis.

Peritoneum
Circular muscle
Taenia coli
Epiploic appendix
Diverticulum
Mucosa
Concentration in diverticulum
Blood vessel piercing musculature

Relationship of diverticula to blood vessels and taeniae (schematic)

The male-to-female ratio of prevalence is estimated to be between 2:1 and 3:1. Diverticulitis has been linked with several environmental factors including diet, obesity, and smoking. It is theorized that a low-fiber diet leads to small hard stools that lead to increased intra-luminal colonic pressures, muscular hypertrophy, and segmentation of the colon, which leads to radially directed pressure and subsequent perforation. Support for this theory comes from several studies including an autopsy review, which found that Japanese immigrants exposed to a Westernized diet had a 52% incidence of colonic diverticula versus 1% of time-matched native Japanese. Links have been sought between obesity and smoking and diverticular disease, but studies have yielded conflicting results and the confounder of diet complicates interpretation.

Although patients who present with uncomplicated disease as their first episode are unlikely to have another attack (complicated or uncomplicated), patients who are admitted for a recurrence of uncomplicated diverticulitis are at increasing risk of having further recurrent episodes. Patients who have had one episode of diverticulitis

are at increased risk for subsequent attacks. A retrospective review of a statewide hospital discharge database showed that 19% of patients who underwent initial nonoperative treatment for diverticulitis had a subsequent admission for a recurrent episode.

## CLINICAL PRESENTATION

Diverticulitis presents with symptoms ranging from mild intermittent abdominal pain to an acute abdomen with peritonitis, sepsis, and multiorgan failure. Suspected acute early or mild diverticulitis (contained microperforation) causes abdominal pain localizing to the left lower quadrant as peritoneal irritation develops. Patients may develop nausea, fever, and malaise. Additional key symptoms to consider are fecaluria, pneumaturia, and vaginal drainage, which may suggest a fistulous communication between the colon and the bladder or vagina (colovesical or colovaginal fistula). Symptoms may be nonspecific, and alternative diagnoses to consider when patients present with suspected diverticulitis include enteritis, irritable bowel syndrome, inflammatory

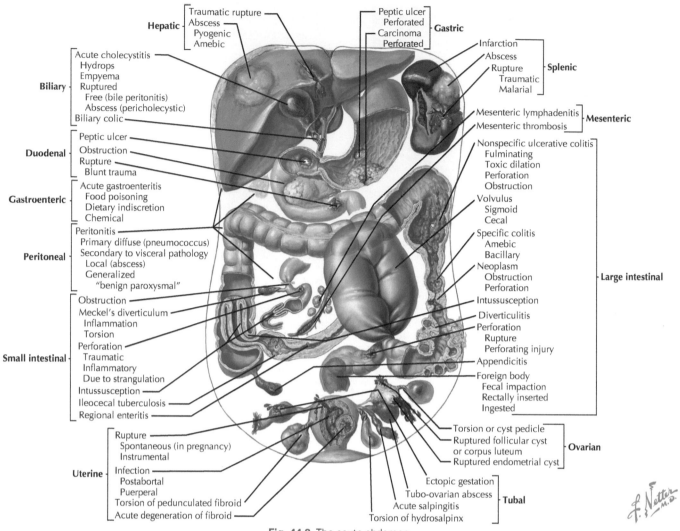

**Fig. 44.2** The acute abdomen.

bowel disease, urinary tract infections or obstruction (nephrolithiasis), appendicitis, neoplasia, gynecologic disease, and bowel obstruction (Fig. 44.2). If the inflammatory process extends beyond a contained perforation, the patient usually experiences more severe abdominal pain, nausea, vomiting, ileus, fever, and systemic signs of sepsis. This may progress to generalized peritonitis, resulting in an acute abdomen, which can potentially evolve into multiorgan system failure. Aggressive resuscitation and emergent abdominal exploration for the control of intra-abdominal sepsis are indicated for patients presenting in extremis.

## DIAGNOSTIC APPROACH

The evaluation of a patient with suspected acute diverticulitis includes a problem-specific history, focused physical examination, and laboratory as well as selective radiographic evaluation. Prior episodes of diverticulitis should be carefully explored, including any associated hospitalizations, imaging, and treatments (e.g., antibiotics and duration, image-guided percutaneous drainage). If the patient has undergone a colonoscopy, the indication and result of the procedure should be reviewed, with particular attention to any diverticulosis and/or other colonic diseases (e.g., neoplasms, colitis). The initial basic laboratory workup for abdominal pain should include a complete blood count, electrolytes, and urinalysis to narrow the differential diagnosis. There is limited evidence to support additional laboratory testing—such as

C-reactive protein, procalcitonin, or fecal calprotectin—when acute diverticulitis is highly suspected, since the role of these potential biomarkers in stratifying disease severity and prognosis remains unclear. Coagulation studies and electrocardiograms should be obtained in patients requiring surgical intervention.

Plain abdominal radiographs are often obtained initially and can be useful to efficiently identify pneumoperitoneum in cases of free colonic perforation. Historically, contrast enema was used for diagnosis. CT imaging is currently the standard diagnostic study for suspected diverticular disease. A complete history and physical in combination with CT scan can classify diverticulitis into two main categories: "uncomplicated" or "complicated" diverticulitis. Complicated diverticulitis refers to diverticulitis associated with free perforation with a systemic inflammatory response, fistula, abscess, stricture or obstruction. Uncomplicated diverticulitis is characterized by left-lower-quadrant pain without signs of systemic involvement and absence of the aforementioned features.

CT can often determine the severity and extraluminal extent of disease, assess other possible intra-abdominal sources mimicking diverticular disease, and guide therapeutic percutaneous drainage of an abscess when indicated (Fig. 44.3). CT imaging commonly depicts colonic wall thickening, mesenteric fat stranding and edema, phlegmon, abscess, fistula, extraluminal air, free peritoneal fluid, obstruction, and/or ileus. CT findings may then be applied to stratify diverticulitis patients using

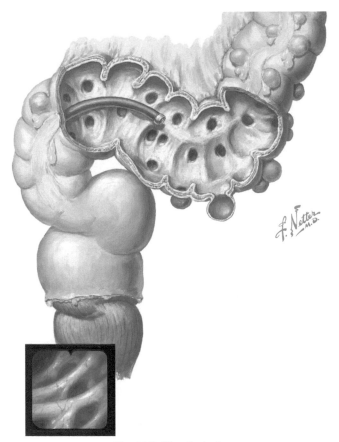

**Fig. 44.3** Diverticulosis.

the Hinchey system, providing practical information to guide therapy. Hinchey stage I denotes a pericolic (mesenteric) abscess or phlegmon; stage II a retroperitoneal, intra-abdominal or pelvic abscess; stage III generalized purulent peritonitis; and stage IV feculent peritonitis.

For patients nonsurgically managed following an episode of acute diverticulitis, a colonoscopy is recommended (if a colonoscopy has not been recently performed) after symptom resolution to confirm the diagnosis and absence of an occult neoplasm (Fig. 44.4). In general, the colonoscopic examination should be completed at least 6 to 8 weeks after resolution of the acute episode to minimize the potential risk of perforation. The risk of an occult malignancy following an episode of complicated diverticulitis is up to 10%; however, it is less than 1% in patients with an episode of uncomplicated diverticulitis. Patients with uncomplicated diverticulitis diagnosed on CT scan are considered to be at similar risk of having colorectal cancer or advanced polyps as compared with the general population.

## TREATMENT

In general, principles for the management of diverticulitis include appropriate antimicrobial therapy with or without additional source control (i.e., percutaneous or operative drainage and resection when indicated), bowel rest, and physiologic support. Two recently published randomized controlled trials found no significant differences in outcomes (complications, recurrence, time to recovery) in patients with CT-proven uncomplicated diverticulitis treated with or without antibiotics, challenging the long-held tenet regarding the necessity of routine antibiotic treatment. Although these data, along with several meta-analyses, suggest that antibiotics in uncomplicated diverticulitis may be superfluous in some patients, ideal patient selection remains a significant limiting factor when the appropriateness of withholding

antibiotics is being determined. Currently the majority of patients with uncomplicated diverticulitis should be treated with antibiotics covering colonic microorganisms (including anaerobes, gram-negative rods, and gram-positive rods). Despite the broad application of antibiotic treatment for all degrees of diverticulitis severity, optimal antibiotic dosing and duration remain largely unstudied. Patients who demonstrate systemic signs of illness or are unable to tolerate oral intake require hospitalization and parenteral antibiotics initially until symptoms resolve. Once clinically improved, patients can then be transitioned to an oral antibiotic regimen. Recommended antibiotic regimens are listed in Table 44.1.

Percutaneous drainage (PD) of a pericolic or pelvic abscess in a stable patient is usually recommended, yet it remains unclear what size of abscess warrants drainage versus antibiotic treatment alone. Diverticular abscesses occur in up to 25% of patients with complicated diverticulitis, and PD is highly effective in obtaining source control and resolution of the acute episode. Although up to 80% of patients can be successfully managed nonoperatively with initial PD with or without antibiotics, up to 61% of patients may experience a recurrent complicated diverticular episode and require an urgent colectomy. In addition, some patients may not be suitable candidates for PD due to unfavorable anatomy; consequently they require consideration for surgical abscess drainage.

Although a proportion of patients may not experience recurrence of diverticulitis, interval elective surgical resection rather than expectant observation following successful treatment of a diverticular abscess is generally recommended. Likewise, patients who have recovered from an episode of complicated diverticulitis—including fistula, obstruction, or stricture—are generally recommended to undergo interval resection. These patients are significantly more likely to experience a complicated diverticulitis recurrence following medical treatment of the initial complicated episode.

Urgent sigmoid colectomy is indicated for patients with diffuse peritonitis or those in whom initial nonoperative management of acute diverticulitis has failed. Although less than 10% of patients present with purulent or feculent peritonitis, these episodes almost always occur as the index attack of diverticulitis. Patients requiring an emergent operation have significant mortality and morbidity, with 30-day mortality reported to range between 5% to over 50%, depending on underlying comorbidities. Treatment for free perforation includes immediate operation with drainage and resection of the involved colon. However, a proportion of patients presenting with free perforation may have only limited pneumoperitoneum, along with physiologic stability and localized peritonitis. For these highly selected patients, initial nonoperative management may be considered in order to convert a nonelective operation into an elective colectomy, attempting to diminish the high morbidity and mortality associated with immediate surgery.

In the past, a three-stage operative approach was standard, with initial drainage and a diverting colostomy, followed by resection of the involved colon with a Hartmann procedure (blind rectal stump) and a third operation for colostomy closure with a colorectal anastomosis. By the 1980s, operative treatment had evolved to a two-stage approach, a Hartmann procedure with formation of a temporary end colostomy. Advantages of this operation include adequate source control, relative ease, and safety, since no anastomosis is constructed in the setting of intra-abdominal contamination. However, this approach requires a second operation to take down the colostomy. Because a significant proportion of patients never undergo colostomy closure and the operation is of moderate risk, some have questioned whether similar outcomes can be obtained with a modified two-stage operation of resection with primary colorectal anastomosis and diverting loop ileostomy. This approach allows for a technically easier and

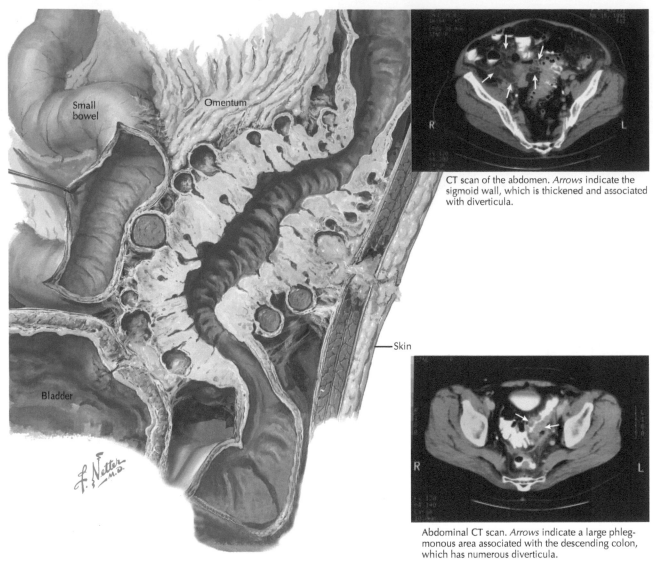

CT scan of the abdomen. *Arrows* indicate the sigmoid wall, which is thickened and associated with diverticula.

Small bowel

Omentum

Skin

Bladder

*f. Netter M.D.*

Abdominal CT scan. *Arrows* indicate a large phlegmonous area associated with the descending colon, which has numerous diverticula.

**Fig. 44.4** Diverticulitis, with CT scan showing thickened wall and diverticula.

**TABLE 44.1    Initial Intravenous Antibiotic Treatment for Intra-abdominal Infections Per the Surgical Infection Society Guidelines (2010)**

| Antimicrobial Agent | Intravenous Dose | Duration |
|---|---|---|
| Ticarcillin clavulanate | 3.1 g q6h | 4–7 days after adequate source control obtained |
| Cefoxitin | 2 g q6h | |
| Ertapenem | 1 g q24h | |
| Moxifloxacin | 400 g q24h | |
| Tigecycline | 100 mg once then 50 mg q12h | |
| Cefazolin (+ metronidazole) | 2 g q6h | |
| Cefuroxime (+ metronidazole) | 1.5 g q8h | |
| Ceftriaxone (+ metronidazole) | 1–2 g q12–24h | |
| Cefotaxime (+ metronidazole) | 1–2 g q6–8h | |
| Levofloxacin (+ metronidazole) | 750 mg q24h | |
| Ciprofloxacin (+ metronidazole) | 400 mg q12h | |

Data from Solomkin JS, Mazuski JE, Bradley JS, et al. Diagnosis and management of complicated intraabdominal infection in adults and children: guidelines by the Surgical Infection Society and the Infectious Diseases Society of America. *Clin Infect Dis* 2010;50:133-164.

less morbid subsequent second operation (loop ileostomy closure) to restore intestinal continuity. Despite two recent randomized trials supporting a benefit for primary anastomosis with proximal ileostomy compared with the Hartmann procedure in emergent Hinchey III and IV patients, adoption rates of primary anastomosis with proximal diversion remain low (<5%). As expected, the benefits reported in these studies were largely attributable to a reduction in long-term stoma rates and stoma reversal surgical complications. Although the data are compelling, primary anastomosis with diverting ileostomy remains an area of controversy, and most would agree that the decision to perform a primary colorectal anastomosis must be individualized and determined on a case-by-case basis, weighing the risks associated with anastomotic failure carefully.

In 2000, the ASCRS recommended that elective colectomy should be performed in all patients who experienced two episodes of uncomplicated diverticulitis. This recommendation was based in part on the belief that patients having an episode of diverticulitis were at much higher risk for eventual recurrent episodes and emergent surgery. Recent studies have challenged this assumption, since the overwhelming majority of the 8% to 30% of patients experiencing a recurrent episode will present again with an uncomplicated attack. A Markov model constructed to evaluate the mortality risk, risk of colectomy, and cost of treating diverticular disease nonsurgically after an acute episode of diverticulitis has resolved with medical management showed that nonoperative treatment is associated with lower mortality rates and is cost saving. Additionally, up to 90% of emergency resections are performed during the index diverticulitis presentation. The minority of resolved uncomplicated episodes recurs as complicated disease or need emergency resection requiring a colostomy. Recommending an elective colectomy to prevent a future recurrence is not supported by the literature, such that interval elective surgery for patients recovering from an uncomplicated episode of diverticulitis should be individualized. The indications for elective colon resection continue to evolve, such that the neither the number of uncomplicated diverticulitis episodes nor the age of patient should determine the appropriateness of surgery. The ASCRS recommendations were revised in 2014, and it is now recommended that elective colectomy after resolved episodes of acute diverticulitis should be reviewed individually based on medical condition, frequency and severity of attacks, and presence of persistent symptoms.

## PROGNOSIS

After an episode of diverticulitis treated surgically, most patients do not develop recurrence. The appropriate resection does not require that all colonic diverticula be removed; only that the proximally transected colon is soft and pliable and the distal transection margin includes the upper rectum, such that a colorectal, not colosigmoid, anastomosis can be constructed.

Although patients who recover from an uncomplicated diverticular episode as their index presentation are unlikely to experience recurrence (complicated or uncomplicated), those patients who do experience a recurrent uncomplicated diverticular episode are then at an increased risk for future recurrent attacks. A large population review reported a 9% risk of a second diverticular-related hospital admission following recovery from a medically treated initial uncomplicated episode; a 23% risk of a third diverticular-related hospital admission if hospitalized twice; and after three hospitalizations, a 36% risk of requiring future hospitalizations for diverticulitis. As the risk of recurrence increases after each recurrence, patients may prefer resection over repeated medical therapy and elect to pursue colectomy.

Consideration of elective colectomy versus antibiotic treatment after repeated episodes of uncomplicated diverticulitis should then be individualized with the patient, factoring in frequency, severity, persistence of symptoms, quality of life, and surgical risks. Although the literature suggests a higher risk of recurrence in younger patients managed successfully for uncomplicated diverticulitis, routine elective resection should not be based solely on age. Younger patients as compared with older ones may have a higher rate of recurrence; however, they likely have a comparable risk of developing a future complicated episode requiring nonelective surgery.

## PREVENTION AND CONTROL

There is little evidence supporting effective methods to either prevent diverticulitis or reduce the risk of recurrent diverticulitis. Since a low-fiber diet has been linked to a higher incidence of diverticulosis at a population level, it follows that a high-fiber diet may reduce the incidence of diverticulum formation (Fig. 44.5). Fiber increases stool weight, decreases gut transit time, and lowers colonic intraluminal pressure. A high-fiber diet and fiber supplements have been reported to decrease the risk of acute diverticulitis, yet their effectiveness in the secondary prevention of recurrent diverticulitis remains unknown. Due to the perception about an association between nuts, popcorn, and seeds with development of complications from diverticular disease, it has been a traditional recommendation to avoid this type of product once one has been diagnosed with diverticulosis in an effort to reduce the risk of diverticulitis or diverticular bleeding. However, this recommendation is not based on any scientific data; in fact, a recent study that prospectively followed over 45,000 men with recording of diet and incident health events found an inverse association between the ingestion of nuts and popcorn and diverticulitis. The authors concluded that this traditional recommendation should be reconsidered. At this point we do not recommend avoidance of these products in relation to diverticulitis prevention. Additional low-quality evidence has shown an increased risk of diverticulitis among tobacco smokers as well as obese patients. Mesalamine-like compounds (e.g., 5-aminosalicylic acid, sulfasalazine), immunomodulatory agents that are believed to reduce a proinflammatory state, have been studied in symptomatic uncomplicated diverticulitis; however, trials have produced conflicting results regarding decreasing the risk of recurrent diverticulitis. Similarly, both rifaximin treatment and probiotics for prevention of recurrent diverticulitis have not provided convincing efficacy. Based on these data, it is reasonable to recommend a high-fiber diet, smoking cessation, and weight loss as risk-reducing strategies. Routine use of additional therapies (e.g., mesalamine, rifaximin, probotics) for secondary diverticular prevention is not recommended at this time. It is important to educate patients regarding the risk of recurrent episodes of acute diverticulitis so that subsequent episodes can be identified early and managed appropriately.

## EVIDENCE

Anaya DA, Flum DR. Risk of emergency colectomy and colostomy in patients with diverticular disease. *Arch Surg* 2005;140:681-685. *This is a large population-level study that evaluates the risk of adverse outcomes in patients who have experienced a first episode of diverticulitis after medical (nonoperative) management. It is the foundation of many current recommendations regarding the role of elective colectomy.*

Dahl C, Crichton M. Evidence for dietary fibre modification in the recovery and prevention of reoccurrence of acute, uncomplicated diverticulitis: a systematic literature review. *Nutrients* 2018;20:137. *This systematic review included eight studies that provided "very low" quality evidence for restricting patients to a high- or low-fiber diet during an acute medically managed*

In diverticular disease, parts of the colon wall are weak and puff out as small sacs (diverticula). Diverticulitis is the disorder when diverticula become inflamed and infected and cause symptoms. It can be a minor inflammation or serious infection.

Symptoms are cramping and pain in the abdomen, usually in the left lower part. Pain is usually severe and starts suddenly. Other symptoms are fever, chills, constipation or diarrhea, and loss of appetite and nausea.

Rest, stool softeners, liquid diet, and oral antibiotics are used for treatment. For severe or complicated cases, or frequent diverticulitis, colon surgery is possible.

To avoid constipation, eat a high-fiber, low-salt, low-fat diet between attacks. Drink lots of fluids. But don't use laxatives.

Maintain a healthy weight and exercise daily.

Call your doctor if you have progressive weight loss, bowel movements with blood in them; dark, tarry bowel movements; fever; or abdominal pain.

Your doctor will make a diagnosis from your medical history, physical examination, blood tests, x-rays, and CT.

**Fig. 44.5** Managing diverticulosis.

diverticulitis episode as well using a high-fiber diet compared with standard or low dietary fiber following resolution of an acute episode to improve reoccurrence risk.

Desai M, Fathallah J, Nutalapati V, Saligram S. Antibiotics versus no antibiotics for acute uncomplicated diverticulitis: a systematic review and meta-analysis. *Dis Colon Rectum* 2019;62(8):1005-1012. *This meta-analysis of recent literature shows that select patients with uncomplicated diverticulitis can be monitored off antibiotics; however, it includes only two randomized controlled trials.*

Etzioni DA, Mack TM, Beart RW Jr, Kaiser AM. Diverticulitis in the United States: 1998-2005. *Ann Surg* 2009;249:210-217. *This population-level study evaluates critical epidemiologic trends of diverticular disease and its management over almost a decade in the United States. It brings attention to important health care burden issues as well as to the increasing incidence of diverticular disease in the young.*

Gregersen R, Mortensen LQ, Burcharth J, Pommergaard HC, Rosenberg J. Treatment of patients with acute colonic diverticulitis complicated by abscess formation: a systematic review. *Int J Surg* 2016;35:201-208. *This systematic review presents evidence on patient outcomes following treatment of diverticular abscess, reporting that regardless of the nonoperative management strategy used (antibiotics or percutaneous drainage), recurrence is moderately high.*

Li D, Baxter NN, McLeod RS, Moineddin R, Nathens AB. The decline of elective colectomy following diverticulitis: a population-based analysis. *Dis Colon Rectum* 2016;59(4):332-339. *This population-based retrospective cohort study evaluated the temporal trends in the use of elective colectomy following nonoperatively managed diverticulitis, reporting a decrease in the odds of elective surgery by 0.93 per annum from 2002 to 2011, consistent with evolving practice guidelines.*

Parks T. Natural history of diverticular disease of the colon. A review of 521 cases. *Br Med J* 1969;4:639-645. *This is a classic article that focused on evaluating the natural history of diverticular disease and represents the basis for most of the recently changed indications for elective colectomy.*

Salem L, Veenstra DL, Sullivan SD, Flum DR. The timing of elective colectomy in diverticulitis: a decision analysis. *J Am Coll Surg* 2004;199:904-912. *This is one of the first large studies evaluating the timing of and current indications for elective colectomy after acute diverticulitis.*

Strate LL, Liu YL, Syngal S, Aldoori WH, Giovannucci EL. Nut, corn, and popcorn consumption and the incidence of diverticular disease. *JAMA* 2008;300(8):907-914. *This is a prospective cohort study evaluating the association of nut/popcorn intake and diverticular disease, which found an inverse relation calling for strongly reconsidering the recommendation of avoiding these products as a strategy to prevent acute attacks.*

## ADDITIONAL RESOURCES

Acuna SA, Wood T, Chesney TR, Dossa F, Wexner SD, Quereshy FA, Chadi SA, Baxter NN. Operative strategies for perforated diverticulitis: a systematic review and meta analysis. *Dis Colon Rectum* 2018;61(12):1442-1453. *This study reviewed operative strategies for treating perforated diverticulitis and effects on postoperative mortality, morbidity, and future stoma reversal rates.*

Commane DM, Arasaradnam RP, Mills S, et al. Diet, aging, and genetic factors in the pathogenesis of diverticular disease. *World J Gastroenterol* 2009; 15:2479-2488. *This article goes over the most important recognized risk factors for diverticular disease and highlights some of the evidence supporting each one.*

Feingold D, Steele SR, Lee S, et al. Practice parameters for sigmoid diverticulitis. *Dis Colon Rectum* 2014;57:284-294. *The most recent guidelines and standards for the management of diverticulitis as recommended by the American Society of Colon and Rectal Surgeons are presented.*

Hall JF, Roberts PL, Ricciardi R, Read T, Scheirey C, Wald C, Marcello PW, Schoetz DJ. Long-term follow-up after an initial episode of diverticulitis: what are the predictors of recurrence? *Dis Colon Rectum* 2014;54(3):283-288. *This retrospective study analyzed clinical predictors of recurrent disease following a first episode of diverticulitis managed nonoperatively, citing a family history of diverticulitis, long segment of involved colon, and/or retroperitoneal abscess as high-risk factors for recurrence.*

Lee JM, Bai PCJ, El Hechi M, Kongkaewpaisan N, Bonde A, Mendoza AE, Saillant NN, Fagenholz PJ, Velmahos G, Kaafarani HM. Hartmann's procedure vs primary anastomosis with diverting loop ileostomy for acute diverticulitis: nationwide analysis of 2,729 emergency surgery patients. *J Am Coll Surg* 2019;229(1):48-55. *This American College of Surgeons National Surgical Quality Improvement Program (NSQIP)study showed that when controlling for patient population differences, primary colorectal anastomosis with diverting loop ileostomy appears to be at least a safe alternative to a Hartmann's procedure with end colostomy for select patient populations needing emergent surgical management of acute diverticulitis.*

Li D, de Mestral C, Baxter NN, McLeod RS, Moineddin R, Wilton AS, Nathens AB. Risk of readmission and emergency surgery following nonoperative management of colonic diverticulitis: a population-based analysis. *Ann Surg* 2014;260(3):423-430; discussion 430-431. *This article retrospectively analyzed the risk of subsequent emergency surgery following nonoperative management of diverticulitis, reporting a risk of emergent surgery of 4.3% for patients with a prior complicated episode and to 1.4% in patients with a prior uncomplicated episode at a median of 3.9 years' follow-up.*

Martel J, Raskin JB. History, incidence, and epidemiology of diverticulosis. *J Clin Gastroenterol* 2008;42:1125-1127. *This review focuses on the epidemiology of diverticular disease and the association to important factors such as age, sex, race, and geography.*

Mazuski JE, Tessier JM. The Surgical Infection Society revised guidelines on the management of intra-abdominal infection. *Surg Infect (Larchmt)* 2017;18(1):1-76. *This article is a recent summary of the most current recommendations for the antimicrobial treatment of complicated intra-abdominal infections and should help to guide antibiotic selection for acute diverticulitis*

Tadlock MD, Karamanos E, Skiada D, Inaba K, Talving P, Senagore A, Demetriades D. Emergency surgery for acute diverticulitis: which operation? A National Surgical Quality Improvement Program study. *J Trauma Acute Care Surg* 2013;74(6):1385-1391. *This article used the American College of Surgeons National Surgical Quality Improvement Program (NSQIP) database to analyze whether varying types of surgery (Hartmann's procedure with end colostomy; primary colorectal anastomosis, no ostomy; primary colorectal anastomosis with diverting ileostomy) in patients requiring emergency surgery for acute diverticulitis affected morbidity and mortality.*

# 45

# Hydatid Cyst Disease (Echinococcosis)

*Kevin P. Labadie, David W. Miranda, James O. Park, Paul S. Pottinger*

## ABSTRACT

Hydatid cyst disease or echinococcosis is a parasitic infection caused by the *Echinococcus* tapeworm larvae. Of the seven recognized species, *Echinococcus granulosus* and *Echinococcus multilocularis*, which cause cystic echinococcosis (CE) and alveolar echinococcosis (AE), respectively, pose clinically significant and potentially lethal public health risks. Human infection occurs after incidental ingestion of the parasite eggs from the stool of infected animals. Most of these cysts are solitary and occur in either the liver or the lungs. Although the cysts displace healthy tissue, the primary infection is typically asymptomatic unless rupture or mass effect occurs. Definitive diagnosis for most cases is by imaging with ultrasound or computed tomography (CT), although immunodiagnostic assays can be a useful adjunct in both primary diagnosis and follow-up. Surgery has the potential to remove the cysts and lead to complete cure; however, successful eradication requires the entire elimination of the parasite without intraoperative contamination or compromise of affected organ systems. Alternatively, the puncture, aspiration, injection, and reaspiration (PAIR) procedure has emerged as a less invasive treatment option in patients with CE who present high surgical risks because of underlying pathophysiology or who are remote from surgical care facilities.

## CLINICAL VIGNETTE

A 58-year-old woman presents with gradual worsening of abdominal pain and distention. She lives on a sheep ranch in Turkey, in the home in which she grew up. For approximately a year, she has noticed increased abdominal girth and early satiety. In recent months, her abdomen has begun to ache around the clock, without any discernible relation to meals or time of day. Her bowel movement pattern is unchanged.

On examination, she is found to be a short woman in no acute distress. Her exam is unremarkable except for her abdomen, which is protuberant, especially in the right upper quadrant, where she reports tenderness on palpation without a rebound phenomenon. By percussion, her liver seems to be markedly enlarged. Routine bloodwork is unremarkable.

Ultrasonography of the right upper quadrant reveals a large, well-circumscribed, complex cystic mass involving most of the right lobe of the liver. Innumerable spheres are present within this mass, all of which have an echodensity suggestive of water. CT imaging confirms this finding (Fig. 45.1A) and demonstrates that the mass abuts and impinges on the inferior vena cava. The diagnosis of multicystic echinococcosis (World Health Organization [WHO] class CE2) is made. The patient confirms that there are always sheepdogs on her ranch and that they are fed the uncooked entrails of sheep after they have been harvested. She is told that she probably acquired this infection years earlier from consuming food soiled with the feces of her sheepdogs.

Treatment is indicated because she is symptomatic. PAIR is not appropriate because the cyst contains many daughter cysts. Complete surgical resection is not possible because of the size and location of the lesion. She is treated with antiparasitic medications (praziquantel and albendazole) for 4 weeks and then undergoes palliative surgery: the mother cyst is carefully entered via a small incision surrounded by sponges soaked in hypertonic 15% saline (see Fig. 45.1B). Liquid and solid cyst contents are removed, the cyst is instilled with hypertonic saline for 10 min that is then removed, and the case is concluded. No hemodynamic instability concerning for anaphylaxis is noted. Direct microscopic examination of cyst debris ("sand") reveals hooklets and viable protoscolices of *Echinococcus granulosus* (see Fig.45.1C). Postoperatively, the patient reports feeling substantial relief from abdominal bloating and pain. She continues to receive antiparasitic medications for the next 4 months, which she tolerates well except for mild leukopenia. She remains symptomatically improved at 2 years of follow-up. She states that her sheepdogs have all been dewormed, and they are no longer fed offal on her ranch.

A common feature of all strains of *Echinococcus* is the use of dogs and other canids (e.g., wolves and coyotes for *E. granulosus* and foxes for *E. multilocularis*) as definitive hosts, which are infected during ingestion of raw visceral organs from intermediate hosts (e.g., sheep refuse) that contain hydatid cysts with viable protoscolices. After ingestion, the protoscolices attach to the canid's intestinal mucosa, where they mature into adult tapeworms, each several millimeters long. After 4 to 5 weeks, the released eggs are shed into the feces; this is when these eggs are accidentally ingested by a human host, initiating infection. The larvae are released from the eggs and penetrate the intestinal epithelium. Subsequently, larvae are passively transported through blood or lymph to target organs, where they may develop into a hydatid cyst.

The host ultimately forms a pericyst, a capsule of connective tissue, in an effort to isolate the parasite, which forms two inner layers: a nucleated germinal layer and an acellular laminated layer (Fig. 45.2). The exact time for the development of protoscolices within cysts is unknown, although it is believed to be more than 10 months after exposure. The liver (60%) and/or lungs (20%) are infected in the majority of cases, and in *E. granulosus* infection a solitary lesion typically develops; however, some patients develop multiple cysts. In endemic areas, the kidney is the third most common organ involved, constituting 2% to 3% of cases. While involvement of the heart and mediastinum is extremely rare, one cases series reported an incidence of 0.5% in patients who underwent operative intervention for thoracic hydatidosis (Figs. 45.3 and 45.4). Classically, an intact hydatid cyst has been classified as a simple cyst. A perforated cyst, with or without infection, has been referred to as "complicated" if it has ruptured into neighboring areas.

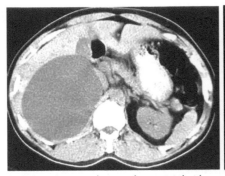

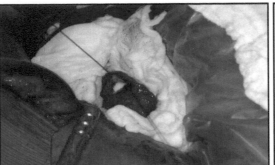

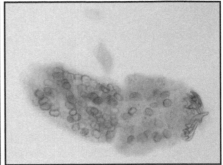

A. CT imaging reveals a very large cyst that has virtually replaced the right lobe of the liver, containing innumerable smaller spherical "daughter" cysts.

B. Gauze sponges soaked in hypertonic saline surround the cyst incision, to guard against spillage of debris into the peritoneal cavity, which could trigger anaphylaxis or metastatic spread of germinal tissue.

C. Intact protoscolex of E.granulosus seen in microscopic examination of hydatid debris ("sand").

Fig. 45.1 Multicystic echinococcosis. (Courtesy of Paul S. Pottinger, MD)

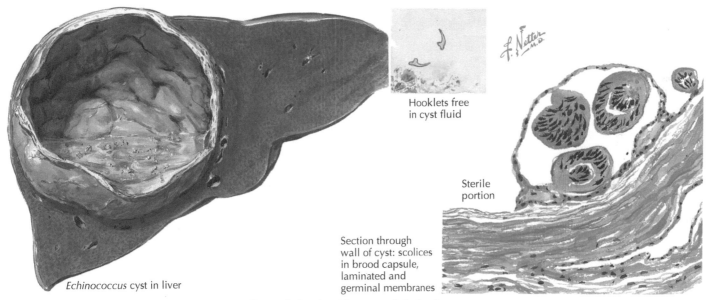

Hooklets free in cyst fluid

Sterile portion

Section through wall of cyst: scolices in brood capsule, laminated and germinal membranes

*Echinococcus* cyst in liver

Fig. 45.2 Cystic echinococcosis in the liver.

## GEOGRAPHIC DISTRIBUTION AND MAGNITUDE OF DISEASE BURDEN

Hydatid cyst disease is seen worldwide but is endemic in the Mediterranean region, Asia, South America, western China, and the former Soviet Union, where the definitive and intermediate hosts live in close contact in herding environments. The prevalence is underestimated in most series because population surveys are not performed in endemic areas. However, several studies have demonstrated by ultrasound that 2% to 9% of endemic populations have detectable CE cysts. CE occurs in all age groups; however, in areas of endemic infection, most symptomatic cases occur in individuals from 4 to 40 years of age (Fig. 45.5).

## RISK FACTORS

In endemic areas, transmission of human echinococcosis is significantly related to certain occupations such as farm laborers and workers in animal husbandry. Communities involved in sheep herding demonstrate the highest rates of infection.

## CLINICAL FEATURES

The initial phase of infection is asymptomatic. Small, well-encapsulated or calcified cysts typically do not induce major pathology; thus patients may remain asymptomatic for years or permanently. There are no pathognomonic features, and clinical symptoms occur only after a highly variable incubation period. However, hepatic hydatid cysts can cause significant upper abdominal pain, hepatomegaly, cholestasis, biliary cirrhosis, portal hypertension, and ascites. Furthermore, the cysts may rupture into the biliary tract or peritoneal and pleural cavities, possibly causing anaphylaxis and seeding secondary infections.

## DIAGNOSTIC APPROACH

The primary diagnosis of hepatic CE disease is typically based on radiologic identification of cystic structures, the clinical evolution of the disease, and appropriate history revealing risk factors for hydatid cysts. Ultrasound demonstrates well-defined, circumscribed, anechoic lesions without infiltration of the surrounding tissues in uncomplicated cases and can also demonstrate complications such as intrabiliary rupture and infection. In addition, in questionable cases of CE, specimens obtained by ultrasound-guided aspiration can be examined for protoscolices, rostellar hooks, and *Echinococcus* antigens (see Fig. 45.2). Albendazole prophylaxis is recommended for 4 days before aspiration and should be continued for at least 1 month after puncture of a lesion ultimately diagnosed and resected as *E. granulosus*. CT, if available, is considered the diagnostic modality of choice, given its superior accuracy, and magnetic resonance imaging (MRI) may help define changes in the hepatic venous system or identify atypical presentations. The WHO has developed a standard classification system for the diagnosis and management of CE based on ultrasonographic and CT imaging.

There are three groups consisting of the active (CE1 and 2), transitional (CE3), and inactive (CE4 and 5) classes.

Immunodiagnosis via serum antibody detection with indirect hemagglutination, enzyme-linked immunosorbent assay (ELISA), or latex agglutination with the hydatid cyst fluid antigens is a useful adjunct in the primary diagnosis and in the follow-up after surgical or pharmacologic treatment. However, many of these tests are not specific for CE and confirmatory tests are often needed. Furthermore, many of these diagnostic assays are unavailable in the nations where disease is endemic.

## CLINICAL MANAGEMENT AND DRUG TREATMENT

There are several options for treatment, including surgery, PAIR, and chemotherapy. For asymptomatic individuals, as previously mentioned, an observational approach can be attempted with appropriate supervision provided that the cysts are considered to be at relatively low risk for rupture based on size, location, and patient activities. Entire surgical removal of the cyst cures the patient; however, there are both temporary and permanent contraindications to surgery based on the difficulty of reaching the lesion, the patient's advanced age or comorbidities, pregnancy, small or calcified cysts, and potential lack of adequate medical care in certain endemic areas. In these individuals, albendazole is the drug of choice, although—given the recurrence rates after discontinuation—it is suspected not to be parasiticidal.

Although the technical procedure of choice is still debated, given the lack of controlled trials, the accepted objective is the entire elimination of the parasite without intraoperative spillage or compromise of healthy tissue. In light of the concern for recurrent disease, debate exists regarding the importance of removing the pericyst and hepatic tissue (radical resection) versus conservatively evacuating the cyst alone. The level of evidence is inadequate to inform the correct level of aggressiveness; however, there is support for the safety of a laparoscopic approach and the use of omentoplasty to prevent abscess formation. Because spilled cyst fluid may contain viable protoscolices that could implant in the peritoneal cavity during surgery or cause anaphylaxis, protection of the operating field is imperative before emptying or resecting the cyst with either the radical or the conservative approach. The peritoneal and/or pleural cavities should be isolated with dry gauze

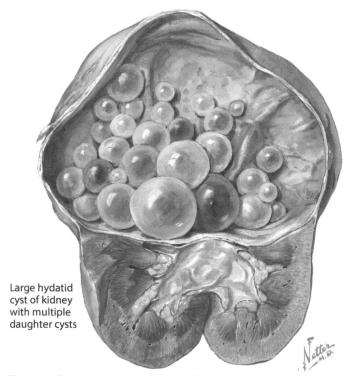

Large hydatid cyst of kidney with multiple daughter cysts

**Fig. 45.3** Cystic echinococcosis in the kidney.

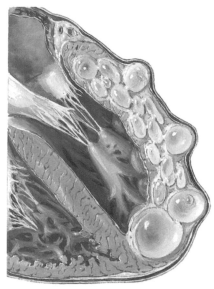

Multiple myocardial cysts intramurally located

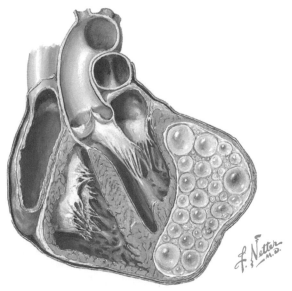

More common form: multiple cysts within unilocular sac in heart wall

**Fig. 45.4** Cystic echinococcosis in the heart.

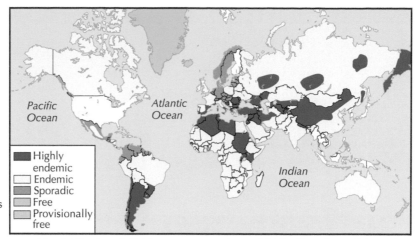

**Fig. 45.5** Approximate geographic distribution of *Echinococcus granulosus* (1999). (Data from Eckert J, Schantz PM, Gasser RB, et al. Geographic distribution and prevalence of Echinococcus granulosus. In *WHO/ OIE manual on echinococcosis in humans and animals: a public health problem of global concern*, Paris, France, 2001, World Organization for Animal Health and World Health Organization.)

or gauze soaked with parasiticidal solution or 20% hypertonic saline. After access is established and control of the cyst wall verified, the cyst is punctured and evacuated using a large-caliber suction device; resection is then performed in either fashion. If there is any question of possible spillage of cyst contents during the case, patients are offered postsurgical treatment with albendazole, such as 10 to 15 mg/kg/day divided into two doses per day for approximately 3 months.

An alternative to surgical intervention is the PAIR technique, in which the cyst is punctured transcutaneously under ultrasound guidance and the parasite is killed through repeated aspiration and injection of a scolicidal agent such as 20% hypertonic saline. For simple hydatid liver cysts that do not abut the liver capsule, this appears to be a safe and attractive option, especially in endemic areas without the option for more aggressive intervention. Many infectious disease specialists recommend treatment with antiparasitic drugs such as albendazole for up to 8 weeks postprocedure.

## PROGNOSIS

The mortality rate for CE is estimated to be 0.2 per 100,000 population, with a case-fatality rate of 2.2%. In contrast, the mortality rate for untreated AE caused by *E. multilocularis* has generally been accepted to be 100% at 15 years from diagnosis. However, survival has improved dramatically among those with alveolar disease who are treated with albendazole.

## PREVENTION AND CONTROL

The risk of developing CE is directly related to contact with definitive hosts like dogs and other canids. Effective CE prevention involves minimizing contact with stray dogs and especially dogs that can be in contact with intermediate hosts like sheep. Preventive measures include improved education and sanitation (e.g., thoroughly cooking food, washing hands vigorously, and not feeding raw sheep offal to work dogs), contact avoidance of dog and fox waste, and deworming treatment of dogs with praziquantel. Vaccination of farm animals with the EG95 vaccine, containing antigens cloned from the parasite oncosphere, has shown promise in the

prevention of transmission among intermediate hosts such as sheep and cattle. Efforts to interrupt the life cycle of the parasite in the definitive canine hosts have not been as successful but would ideally complement the effect of vaccinating the intermediate hosts.

## ACKNOWLEDGMENTS

The authors would like to acknowledge Dr. Austin L. Spitzer for his contribution to this chapter in the previous edition.

## ADDITIONAL RESOURCES

Akbulut S, Senol A, Sezgin A, et al. Radical vs conservative surgery for hydatid liver cysts: experience from a single center. *World J Gastroenterol* 2010;16:953-959. *A review of outcomes in a group of 59 well-characterized patients from an endemic area who had undergone radical or conservative surgical procedures for liver hydatid disease from 2004 to 2009. Postoperative recurrence was lower after radical surgery.*

Dziri C, Haouet K, Fingerhut A. Treatment of hydatid cyst of the liver: where is the evidence? *World J Surg* 2004;28:731-736. *A systematic review of published literature on different modalities of treatment for hydatid cyst of the liver, leading to evidence-based recommendations based on cyst classification. The level of evidence is low regarding treatment of complicated cysts.*

Eckert J, Deplazes P. Biological, epidemiological, and clinical aspects of echinococcosis, a zoonosis of increasing concern. *Clin Microbiol Rev* 2004;17:107-135. *A comprehensive review of the epidemiology and clinical aspects of echinococcosis featuring life-cycle illustrations as well as radiographic and clinical images from human cases.*

Frider B, Larrieu E. Treatment of liver hydatidosis: how to treat an asymptomatic carrier? *World J Gastroenterol* 2010;16:4123-4129. *This article considers alternatives in the treatment of an asymptomatic carrier—surgery, albendazole, PAIR, or wait and watch—with a review of the natural history of CE and the evolution of treatment modalities.*

Pawlowski ZS, Eckert J, Vuitton D, et al. Echinococcosis in humans: clinical aspects, diagnosis and treatment. In Eckert J, Gemmell MA, Meslin FX, Pawlowski ZS, eds: WHO/OIE manual on echinococcosis in humans and animals: a public health problem of global concern, Paris, France, 2001, World Organization for Animal Health and World Health Organization, pp 20-71. *An essential reference on the epidemiology, clinical aspects, diagnosis, and treatment of human echinococcosis.*

# 46

# Intraabdominal Abscess

Derek T. Tessman, Graham W. McLaren, Robert G. Sawyer

## ABSTRACT

Intraabdominal abscesses (IAAs) can serve as significant causes of morbidity if not promptly diagnosed and treated. IAAs occur when a previously sterile site, such as a solid intraabdominal organ or the peritoneal fluid, becomes inoculated via natural or iatrogenic means and the infection is walled off by a local inflammatory response. In reaching a diagnosis, risk factors to consider include recent intraabdominal procedures, penetrating trauma, or inflammatory conditions involving one or more abdominal viscera. Taking these into account, an experienced clinician can make the diagnosis by combining the clinical examination with one of several imaging modalities. Once diagnosed, treatment involves correcting the resulting physiologic derangements, if septic, while ultimately achieving source control and initiating early, judicious use of appropriate antibiotics. If managed appropriately, the prognosis is favorable. Prevention of IAAs remains a topic of active investigation.

## CLINICAL VIGNETTE

A 65-year-old female presented to the emergency department with 3 days of progressively worsening left lower quadrant pain. She had never experienced similar pain before. Associated symptoms included a low-grade fever to 38°C and loose bowel movements. Her past medical history was remarkable for hypertension alone. She had no history of prior abdominal operations, and she had yet to undergo a screening colonoscopy as recommended by her primary care provider (PCP). On examination, she was tachycardic to 105 and normotensive with a blood pressure of 135/75. Her abdominal exam demonstrated mild left lower quadrant tenderness without significant guarding or peritoneal signs. Initial lab work revealed a leukocytosis to 15,000 white blood cells/dL but no other significant abnormalities. Computed tomography (CT) imaging subsequently identified a pericolonic abscess measuring 6×4 cm with associated inflammatory changes about the sigmoid colon with numerous diverticula. This patient with diverticulitis complicated by pericolonic abscess subsequently was started on broad-spectrum antibiotic therapy with ceftriaxone and metronidazole prior to undergoing CT-guided percutaneous drain placement later that same day. Over the course of a 2-day hospital stay, the output of her Interventional Radiology (IR) drain decreased significantly, and her pain and appetite both improved. Her examination improved as the drain output decreased, and she was discharged home with oral antibiotics to complete a 4-day course following source control with the IR drain placement. Follow-up imaging demonstrated resolution of the pericolonic abscess, and the drain was subsequently removed. Colonoscopy completed 6 weeks later confirmed the presence of diverticulosis with no underlying malignancy.

## ETIOLOGY AND GEOGRAPHIC DISTRIBUTION

All IAAs result from the inoculation of a normally sterile site within the abdominal cavity. These sites include the nonluminal aspects of various organs, as well as the small quantity of peritoneal fluid that naturally resides within the abdomen. When healthy, this fluid amounts to roughly 50 mL and travels in a typical pattern determined by the various peritoneal reflections and potential spaces. These potential spaces within the abdomen include the pelvis, the lesser sac, the subdiaphragmatic spaces, both right and left paracolic gutters, and between "loops" or folds of the peritoneum, mesentery, or omentum (Fig. 46.1). The volume of this fluid is maintained via absorption in two ways: It either is continuously reabsorbed as it traverses the peritoneal cavity, or it gets reabsorbed in the pleural spaces after it passes through tiny pores in the diaphragm. Increased amounts of fluid, such as occurs with ascites or after irrigation during an operation, will continue to flow along typical routes unless altered by the surgical procedure itself. For example, after gastric resection and anastomosis, a common site for fluid to collect is in the retrogastric space made as a result of operative dissection. Additional mechanisms that can alter the flow of peritoneal fluid may arise in association with intraabdominal infections (IAIs). The inflammatory process associated with these infections often results in localized adhesions that may result in the isolation of this fluid. This leads to an increased risk of infection as fluid stasis allows for either inoculation by direct spread from the inflammatory process or from indirect spread from the bloodstream. Solid organs themselves may also become inoculated with pathogens via the bloodstream, which results in a localized infection. This includes hepatic and splenic abscesses. Similarly, hollow organs may also develop abscesses within their walls from microperforations or macroperforations, as seen with periappendiceal and pericolonic abscesses.

It is unclear how the prevalence of IAAs differs around the world. Studies from Europe, North America, and Japan seem to imply a relatively similar incidence and pathophysiology, including the organisms most commonly cultured. It is obvious, however, that a large number of abscesses are iatrogenic in nature following abdominal surgery or arising from device-associated bloodstream infection. Because of this, non-iatrogenic infections, such as periappendiceal or peridiverticular abscesses, most likely compose a higher percentage of cases in geographic regions where high-technology medical care is unavailable.

## RISK FACTORS

The most common risk factors for IAA are diseases that lead to the perforation of a hollow viscus. These conditions can be categorized as either naturally occurring or iatrogenic. The most common naturally occurring conditions associated with IAA include congenital gastrointestinal, urologic, or gynecologic anomalies; inflammatory conditions of the intraabdominal organs (such as appendicitis and diverticulitis); penetrating abdominal trauma; and causes of intestinal obstruction (such as hernias or adhesions) that can secondarily lead to perforation and abscess formation. The most common medical interventions that are associated with subsequent IAA are gastrointestinal endoscopy and

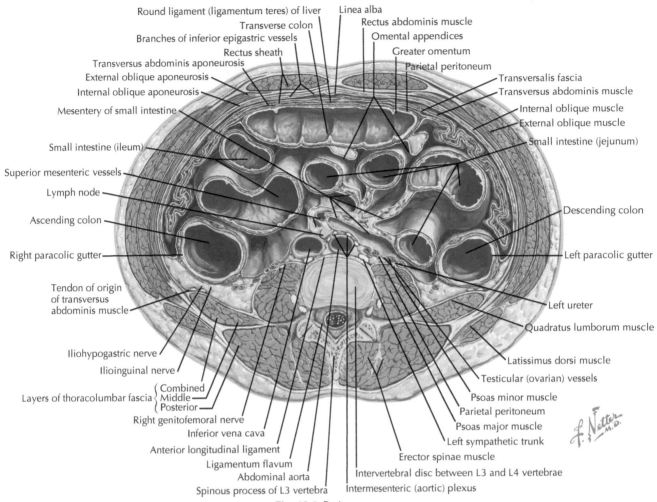

Round ligament (ligamentum teres) of liver
Transverse colon
Branches of inferior epigastric vessels
Rectus sheath
Transversus abdominis aponeurosis
External oblique aponeurosis
Internal oblique aponeurosis
Mesentery of small intestine
Small intestine (ileum)
Superior mesenteric vessels
Lymph node
Ascending colon
Right paracolic gutter
Tendon of origin of transversus abdominis muscle
Iliohypogastric nerve
Ilioinguinal nerve
Layers of thoracolumbar fascia { Combined  Middle  Posterior
Right genitofemoral nerve
Inferior vena cava
Anterior longitudinal ligament
Ligamentum flavum
Abdominal aorta
Spinous process of L3 vertebra

Linea alba
Rectus abdominis muscle
Omental appendices
Greater omentum
Parietal peritoneum
Transversalis fascia
Transversus abdominis muscle
Internal oblique muscle
External oblique muscle
Small intestine (jejunum)
Descending colon
Left paracolic gutter
Left ureter
Quadratus lumborum muscle
Latissimus dorsi muscle
Testicular (ovarian) vessels
Psoas minor muscle
Parietal peritoneum
Psoas major muscle
Left sympathetic trunk
Erector spinae muscle
Intervertebral disc between L3 and L4 vertebrae
Intermesenteric (aortic) plexus

**Fig. 46.1** Peritoneum.

surgical procedures on the gastrointestinal, urologic, or gynecologic organs. In the absence of any connection to a hollow viscus, a postoperative IAA would be more specifically categorized as an organ space surgical site infection, although the management of such an abscess is more similar to a de novo IAA than an incisional surgical site infection.

## CLINICAL PRESENTATION

In the same way that many different processes can result in an IAA (Fig. 46.2), the symptoms that manifest as a result are also quite variable. If occurring in conjunction with a primary inflammatory process, such as diverticulitis, pancreatitis, or appendicitis, abscesses often result in vague abdominal complaints, possibly localized to the site of inflammation, sometimes with accompanying fever and leukocytosis. In a similar fashion, patients with either hepatic or splenic abscesses may have indistinct symptoms, including fever and malaise, with possible focal abdominal pain. In the postoperative state, patients frequently have a blunted examination secondary to a prolonged hospital stay and/or various pain medications, both narcotic and anti-inflammatory. Because of this, the signs and symptoms of a developing abscess may be incorrectly attributed to incisional pain and tenderness, such as the patient who develops a pelvic abscess following a colorectal surgical procedure. Alternatively, rather than having focal tenderness on examination, postoperative patients may exhibit only either tachycardia and/or a prolonged ileus as the outward manifestations of an underlying abscess. While postoperative patients are diagnostically

challenging, those with an immunosuppressed state are even more so. Because of the iatrogenic immunosuppression associated with the treatment of inflammatory bowel disease and ongoing functioning solid organ transplants, those patients may have no specific, localizing symptoms at all. As can be seen, at all times, there must be a high index of suspicion in order to make the diagnosis expeditiously, given the deep anatomic location of the underlying process.

## DIAGNOSTIC EVALUATION

Each evaluation begins with a thorough history and physical examination, which may reveal some of the signs and symptoms mentioned above. If concerned about an intraabdominal inflammatory process such as an abscess, CT is the diagnostic modality of choice. If not contraindicated due to allergy or renal insufficiency, intravenous (IV) contrast is administered in order to discriminate between noninflammatory fluid collections and abscesses. The latter will often have a hyperenhancing wall, meaning that the IV contrast causes the wall of the abscess to appear lighter/brighter than its contents. In the instance where IV contrast cannot be used, other inflammatory changes such as stranding of surrounding tissues or extraluminal air may suggest the presence of an abscess. However, if the patient is in the early postoperative state (generally <5 to 7 days after an operation), those findings are not unexpected, and the images may still be considered indeterminate. If rapid diagnosis is critical, percutaneous aspiration may be performed in order to sample the fluid for pathogens and confirm

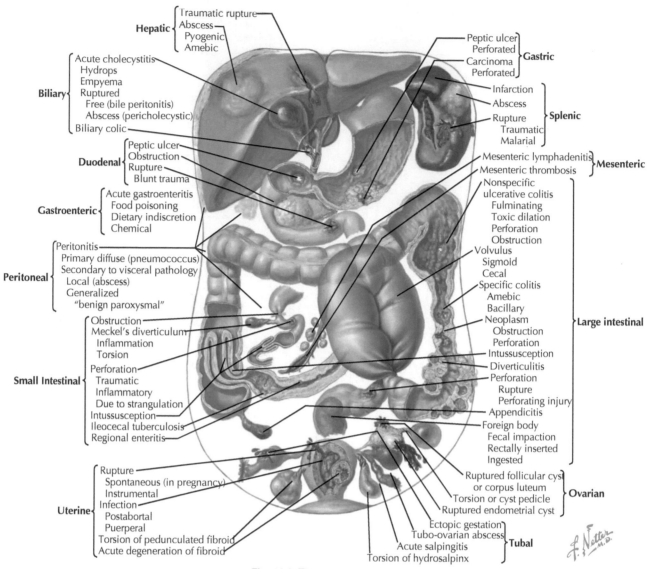

**Hepatic**
- Traumatic rupture
- Abscess
  - Pyogenic
  - Amebic

**Biliary**
- Acute cholecystitis
- Hydrops
- Empyema
- Ruptured
  - Free (bile peritonitis)
  - Abscess (pericholecystic)
- Biliary colic

**Duodenal**
- Peptic ulcer
- Obstruction
- Rupture
- Blunt trauma

**Gastroenteric**
- Acute gastroenteritis
- Food poisoning
- Dietary indiscretion
- Chemical

**Peritoneal**
- Peritonitis
- Primary diffuse (pneumococcus)
- Secondary to visceral pathology
- Local (abscess)
- Generalized
- "benign paroxysmal"

**Small Intestinal**
- Obstruction
- Meckel's diverticulum
  - Inflammation
  - Torsion
  - Perforation
- Traumatic
- Inflammatory
- Due to strangulation
- Intussusception
- Ileocecal tuberculosis
- Regional enteritis

**Uterine**
- Rupture
  - Spontaneous (in pregnancy)
  - Instrumental
- Infection
  - Postabortal
  - Puerperal
- Torsion of pedunculated fibroid
- Acute degeneration of fibroid

**Gastric**
- Peptic ulcer
  - Perforated
- Carcinoma
  - Perforated

**Splenic**
- Infarction
- Abscess
- Rupture
  - Traumatic
  - Malarial

**Mesenteric**
- Mesenteric lymphadenitis
- Mesenteric thrombosis

**Large intestinal**
- Nonspecific ulcerative colitis
  - Fulminating
  - Toxic dilation
  - Perforation
  - Obstruction
- Volvulus
  - Sigmoid
  - Cecal
- Specific colitis
  - Amebic
  - Bacillary
- Neoplasm
  - Obstruction
  - Perforation
- Intussusception
- Diverticulitis
  - Perforation
  - Rupture
  - Perforating injury
- Appendicitis
- Foreign body
  - Fecal impaction
  - Rectally inserted
  - Ingested

**Ovarian**
- Ruptured follicular cyst or corpus luteum
- Torsion or cyst pedicle
- Ruptured endometrial cyst

**Tubal**
- Ectopic gestation
- Tubo-ovarian abscess
- Acute salpingitis
- Torsion of hydrosalpinx

**Fig. 46.2** The acute abdomen.

or exclude active infection. If CT imaging itself is contraindicated, focused abdominal ultrasound or magnetic resonance imaging can be used as alternative imaging modalities. Once an abscess is confirmed or suspected based on radiologic findings, aspiration of the fluid of interest (with or without placement of a drain), if feasible, should be performed under CT or ultrasound guidance. At the time of aspiration and/or drain placement, the fluid should be sent for various laboratory studies to assist with ongoing management. These studies include a Gram stain; white blood cell count; aerobic, anaerobic, and fungal cultures; and occasional chemistries, if doing so would clarify a diagnosis. For example, fluid amylase or creatinine can be used to confirm the suspected origin if the fluid of a pancreatic or urologic process is suspected, respectively.

While CT imaging remains the diagnostic modality of choice, other adjuncts can be used to complete a more thorough evaluation, depending on the anatomic location of the IAA (Fig. 46.3). For those abscesses associated with penetrating abdominal trauma, inflammatory bowel disease, or any intraabdominal operation, evaluation of intestinal continuity should be performed as soon as possible. Frequently the diagnosis of a gastrointestinal leak can be made simply by examining drainage material when it is noted to be consistent with succus or bile.

Alternatively, a more definitive diagnosis can be achieved by obtaining a triple-contrast CT scan, which is one that uses IV contrast as well as water-soluble contrast material administered by both the mouth and rectum. Violation of the bowel is then confirmed if extravasation of contrast is seen outside of the bowel wall. Another method of identifying a leak from a hollow viscus is to perform a sinogram through a previously placed drain, looking for a possible fistulous connection between the abscess cavity and nearby hollow viscus.

Such a fistula tract can be observed in patients with a peridiverticular abscess—one of the more common IAAs, given the prevalence of diverticular disease in the Western world. CT imaging as mentioned previously has greatly aided in the diagnosis of complicated diverticulitis, but unfortunately, imaging findings alone cannot differentiate a benign peridiverticular abscess from an abscess associated with localized colon cancer. Because of this, an underlying malignancy must be ruled out either by operative pathology (if emergency operation is required) or by follow-up colonoscopy in 6 to 8 weeks following successful nonresectional management.

Apart from bowel-associated IAAs, pancreatic abscesses present a unique diagnostic dilemma because of the overlap in presentation with two associated though noninfectious processes: pancreatic pseudocyst

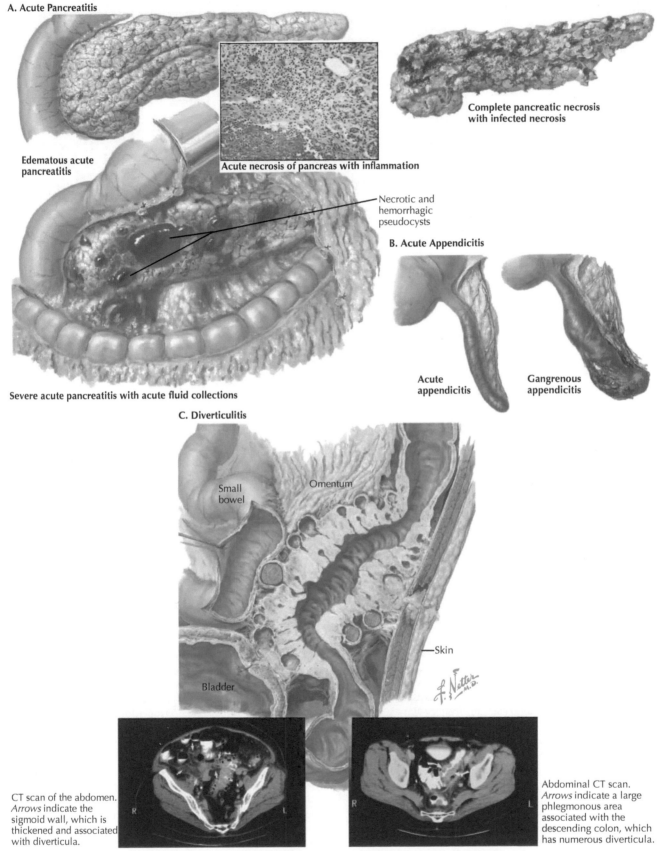

**A. Acute Pancreatitis**

Edematous acute pancreatitis

Acute necrosis of pancreas with inflammation

Complete pancreatic necrosis with infected necrosis

Necrotic and hemorrhagic pseudocysts

Severe acute pancreatitis with acute fluid collections

**B. Acute Appendicitis**

Acute appendicitis

Gangrenous appendicitis

**C. Diverticulitis**

Small bowel

Omentum

Skin

Bladder

CT scan of the abdomen. *Arrows* indicate the sigmoid wall, which is thickened and associated with diverticula.

Abdominal CT scan. *Arrows* indicate a large phlegmonous area associated with the descending colon, which has numerous diverticula.

**Fig. 46.3** (A) Acute pancreatitis, (B) acute appendicitis, and (C) diverticulitis.

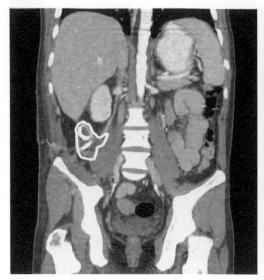 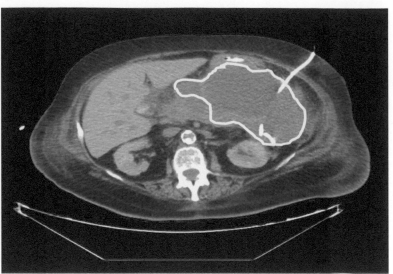

Right pericecal abscess (*outlined in yellow*) with drain inserted

Pancreatic abscess (*outlined in yellow*) with drain inserted

**Fig. 46.4** Abscess drainage.

and pancreatic necrosis. All three are frequently a result of pancreatitis, and all three can cause abdominal pain, tenderness, and evidence of the systemic inflammatory response syndrome (SIRS). Furthermore, a pseudocyst or necrosis can progress to a pancreatic abscess should either become secondarily infected through local invasion, translocation from the gastrointestinal tract, or seeding from the bloodstream. If active infection is suspected and its diagnosis would lead to a change in management, aspiration of peripancreatic fluid is indicated with or without placement of a drain. In the setting of a pseudocyst, this quite frequently is done via percutaneous or transgastric open or endoscopic drainage. In the setting of known pancreatic necrosis, simple needle aspiration is performed, because drain placement does not appear to improve the outcome from uninfected necrosis.

## CLINICAL MANAGEMENT AND DRUG TREATMENT

After the diagnosis of an IAA is made, attention should first be directed at appropriate goal-directed resuscitation of a patient with an intraabdominal abscess who presents with signs of abdominal sepsis or septic shock, although a patient with an abscess is much less likely to present with shock than one with a free intraabdominal perforation and peritonitis. The Surviving Sepsis Campaign, first published in 2008 and revised again in 2012 and 2016, outlines a goal-directed management strategy for appropriately and successfully resuscitating a patient presenting in septic shock. Besides the hemodynamic and cardiovascular end points of resuscitation outlined in this campaign, the strategy also emphasizes the prompt administration of empiric antibiotic therapy in conjunction with early source control. Source control means removing or draining the nidus or focus of infection as quickly as possible and ideally with the least disruption to normal anatomic and physiologic function. This end point of source control in patients presenting with IAA can often be achieved initially with percutaneous drainage and catheter placement (Fig. 46.4), but there may be instances that require open surgical intervention if the IAA is in an area that cannot be accessed safely by image-guided percutaneous approaches. More recently, it has become apparent that some IAAs, particularly those that are less than 5 cm in diameter, multiple in nature, or in inaccessible locations, can be successfully treated with antibiotics alone, with operative intervention reserved for those where repeat imaging demonstrates a lack of resolution.

With regard to management strategies specific to different IAAs, organ space surgical site infections without an associated fistula can almost always be managed with a percutaneous drain and a brief course of antimicrobials lasting no more than 4 days beyond source control. On the contrary, postoperative infections that are associated with a connection to a hollow viscus require more complicated management. Occasionally, relatively long-term percutaneous drainage in a clinically stable patient combined with a brief course of antimicrobials will result in the closure of a small fistula, and the patient will need no other intervention. Many patients, however, will require an additional operation to attain adequate source control, ranging from diversion to resection of a leaking anastomosis to both resection and diversion combined.

Pancreatic abscesses also present a unique and challenging problem. An isolated abscess without necrosis can be treated as other IAAs with urgent aspiration and catheter drainage combined with antibiotics. If surrounding necrosis is present, though, the trend in management over the past two decades has diverged from that of most IAAs. As long as the patient does not clinically deteriorate, a strategy of antibiotics alone in conjunction with physiologic support followed by delayed debridement of pancreatic necrosis has been shown to have improved patient outcomes compared to early intervention/source control. Following the same "less is more" approach, the optimal interventional approach at the delayed time interval has shifted from open surgical necrosectomy to much less invasive approaches, including laparoscopic, endoscopic, or retroperitoneal debridement. However, if the patient demonstrates signs of clinical deterioration while receiving appropriate antimicrobial therapy and physiologic support, urgent debridement of necrotic tissue must be undertaken in spite of the associated increased risk of morbidity and mortality.

With regard to other inflammatory processes that can lead to IAAs, a similar conservative approach has been adopted with much success. As with the organ space surgical site infections mentioned previously, diverticular abscesses are often amenable to catheter-based drainage and no more than 4 days of appropriate antimicrobial therapy. Similarly, abscesses associated with appendicitis can be successfully treated by catheter-based drainage, if technically feasible, and appropriate antibiotic therapy. For those with appendicitis, interval appendectomy performed 4 to 6 weeks following the initial presentation has been demonstrated to be both safe and effective in both the adult and pediatric populations. However, beyond ensuring that follow-up colonoscopy is performed to rule out an underlying malignancy as the cause of the abscess for those with diverticulitis, recent data suggests that a strategy of watchful waiting rather than elective resection of the involved segment of colon can be pursued without an increased risk of requiring an emergent operation or experiencing life-threatening complications.

A common theme throughout the management strategies for each type of IAA is the appropriate and judicious use of antibiotic therapy. Empiric therapy is dictated by the likely organism arising from the site of suspected perforation, the individual patient's history of resistant pathogens, and local antibiograms. Many different antibiotic regimens are available and appropriate, as outlined in various evidence-based guidelines, and none have been found to be clearly superior to others. Most recently updated in 2017 by the Surgical Infection Society, the newest guidelines for antimicrobial management of IAIs, including IAAs, contain additional specific recommendations regarding the duration of antibiotic therapy. Specifically, a fixed four-day course of appropriate antibiotic therapy following adequate source control is sufficient treatment for complicated IAIs such as IAAs. The duration of therapy for infections with poor source control, such as those with abscesses associated with pancreatic necrosis, is less well-defined but may be sufficient if administered for 5 to 7 days.

## PROGNOSIS

If IAA is treated appropriately and early, the prognosis is favorable for most patients with IAAs. Many IAAs resolve quickly, drains can be removed after days to weeks, and the patient can be followed clinically for signs of recurrence. These good results are probably related to the fact that abscess formation itself indicates that a robust immune response has occurred and that the infection has already been naturally contained. For patients managed with antibiotics alone, repeat imaging should be performed 1 to 2 weeks later to check for abscess resolution. If an IAA does recur, which can happen up to 20% to 25% of the time, the recurrence can often be managed in a similar fashion to the initial presentation—typically with percutaneous drainage of the recurrent collection and a new short course of antimicrobials or placement of a drain in a previously undrained collection. It should be mentioned that repeat cultures are beneficial with any recurrence because the likelihood of a resistant pathogen being present increases.

## PREVENTION AND CONTROL

Prevention must be targeted at changing underlying risk factors for IAA. Naturally occurring risk factors can probably be controlled only in a population-based manner, such as through dietary interventions to decrease the risk of diverticular disease or societal efforts to reduce violence and prevent traumatic bowel injuries. It is worth noting that for patients experiencing a hollow viscus injury following penetrating abdominal trauma, a randomized controlled trial demonstrated that 24 hours of perioperative broad-spectrum antibiotics was sufficient for the prevention of postoperative infections. In terms of preventing IAA after other nonelective procedures, perioperative antibiotics have been shown to be similarly effective in this regard, as demonstrated in a recent Cochrane review of antibiotic use for appendectomy in the setting of appendicitis. To further reduce the occurrence of iatrogenic IAA, proper and precise sterile surgical technique serves a valuable role but cannot prevent them completely. Numerous studies have evaluated adjunctive intraoperative measures employed in the hopes of reducing IAAs. One randomized trial evaluating intraoperative drainage in patients undergoing a pancreaticoduodenectomy clearly

demonstrated a mortality benefit to continuing the practice, as the study had to be halted due to the increased frequency of complications, including IAAs in the group who did not have an intraoperative drain placed. However, similar studies have evaluated the use of intraoperative drainage in cholecystectomy, appendectomy, and distal pancreatectomy, and all have failed to demonstrate a reduced rate of IAA in the group receiving drainage. In a similar fashion, intraoperative irrigation or lavage has been traditionally used to mitigate the development of IAA based on anecdotal evidence, but a recent meta-analysis of patients undergoing laparoscopic appendectomy showed no benefit to irrigation in comparison to suction alone when looking at the rate of postoperative IAA.

## ACKNOWLEDGMENTS

The authors would like to acknowledge Dr. Christopher M. Watson for his work on the previous edition chapter.

## EVIDENCE

Andersen BR, Kallehave FL, Andersen HK. Antibiotics versus placebo for prevention of postoperative infection after appendicectomy. *Cochrane Database Syst Rev* 2005;3:CD001439. *A Cochrane review of 45 studies including 9576 patients found that the use of antibiotics is superior to placebo for preventing wound infection and IAA, regardless of whether the appendicitis was simple or complicated.*

Bugiantella W, Rondelli F, Boni M, et al. Necrotizing pancreatitis: a review of the interventions. *International Journal of Surgery* 2016;28:S163-S171. *A review of the management strategies employed when treating necrotizing pancreatitis and how the approach has changed dramatically over the past 20 years.*

Garfinkle R, Kugler A, Pelsser, V, et al. Diverticular abscess managed with long-term definitive nonoperative intent is safe. *Dis Colon Rectum* 2016;59:648-655. *A single-center retrospective review of 73 patients managed expectantly following nonoperative treatment of diverticular abscess found the practice to be safe.*

Hajibandeh S, Hajibandeh S, Kelly A, et al. Irrigation versus suction alone in laparoscopic appendectomy: is dilution the solution to pollution? A systematic review and meta-analysis. *Surgical Innovation* 2018;25:174-182. *A meta-analysis of 2511 patients involved in three randomized controlled trials and two retrospective observational studies found no difference in the rate of intraabdominal abscess between those undergoing peritoneal irrigation versus suction alone following laparoscopic appendectomy.*

Kirton OC, O'Neill PA, Kestner M, et al. Perioperative antibiotic use in high-risk penetrating hollow viscus injury: a prospective randomized, double-blind, placebo-control trial of 24 hours versus 5 days. *J Trauma* 2000;49(5):822-832. *A multicenter randomized controlled trial demonstrated the effectiveness of 24 hours alone of perioperative antibiotics in the prevention of intraabdominal abscess following penetrating hollow viscus injury.*

Li Z, Zhao L, Cheng Y, et al. Abdominal drainage to prevent intra-peritoneal abscess after open appendectomy for complicated appendicitis. *Cochrane Database Syst Rev* 2018;5:CD010168. *A review of six randomized controlled trials evaluating 521 participants found no evidence supporting the use of intraabdominal drain placement in preventing intraabdominal abscess following open appendectomy for complicated appendicitis.*

Paul JS, Ridolfi TJ. A case study in intra-abdominal sepsis. *Surg Clin North Am* 2012;92(6):1661-1677. *An overview of the management of a septic patient presenting with diverticular abscess, including a detailed discussion of the Surviving Sepsis Campaign guidelines.*

Sawyer RG, Claridge JA, Nathens AB, et al. Trial of short-course antimicrobial therapy for intraabdominal infection. *N Engl J Med* 2015;372:1996-2005. *A multicenter randomized controlled trial demonstrating the effectiveness of a fixed-duration 4-day course of antibiotic therapy following source control for the management of intraabdominal infection.*

Van Buren G 2nd, Bloomston M, Hughes SJ, et al. A randomized prospective multicenter trial of pancreaticoduodenectomy with and without routine intraperitoneal drainage. *Ann Surg 2014;259:605-612. The elimination of intraperitoneal drainage in all cases of the Whipple procedure increased the frequency and severity of complications, including intraabdominal abscess formation, in this multicenter randomized controlled trial of 137 patients.*

Van Buren G 2nd, Bloomston M, Schmidt CR, et al: A prospective randomized multicenter trial of distal pancreatectomy with and without routine intraperitoneal drainage. *Ann Surg 2017;266:421-431. Similar to the above study, this multicenter randomized controlled trial of 344 patients undergoing a distal pancreatectomy instead demonstrated the feasibility of eliminating routine intraabdominal drainage in this subset of patients with no increase in postoperative intraabdominal abscess formation.*

## ADDITIONAL RESOURCES

Kambadakone A, Mueller PR. Abdominal abscess. *Textbook of Gastrointestinal Radiology 2015;72:1254-1278. Thorough overview of the various imaging modalities used to diagnose abdominal abscesses along with descriptions of various approaches to image-guided drainage of the same.*

Mazuski JE, Tessier JM, May AK, et al. The Surgical Infection Society revised guidelines on the management of intra-abdominal infection. *Surgical Infections 2017;18(1):1-76. An exhaustive and recently updated review and series of recommendations for the management of intraabdominal infections, including evidence-based recommendations for antimicrobial therapy for adult and pediatric populations.*

# 47

# Liver Abscess: Pyogenic and Amebic Hepatic Abscess

*David W. Miranda, Kevin P. Labadie, Patrick S. Wolf, James O. Park*

## ABSTRACT

Liver abscess is comprised of two main types: pyogenic and amebic. The most common type is pyogenic, which usually results from biliary and intra-abdominal infections. On the other hand, amebic liver abscess occurs far less frequently and is a complication of *Entamoeba histolytica* infection. Both abscess types present in a similar manner, often manifesting vague and nonspecific symptoms, necessitating a high index of suspicion for prompt diagnosis and treatment. Ultrasonography and computed tomography, along with culture and serology, correlated to clinical signs and symptoms, aid in establishing the diagnosis. Treatment of pyogenic liver abscess requires antimicrobial therapy and percutaneous drainage, with surgical drainage reserved for treatment failures or for patients undergoing simultaneous treatment of an intra-abdominal source. In contrast, amebic liver abscess responds well to amebicidal treatment and rarely requires drainage. Outcomes depend on the severity of illness at presentation, presence of underlying malignancy, and need for surgical drainage.

## CLINICAL VIGNETTE

A 42-year-old man presents with gradually worsening right-sided abdominal pain, fatigue, and anorexia. On further history, he is noted to have several weeks of inter-mittent fevers, general malaise, and occasional severe right upper quadrant pain after eating. He is obese but without other medical problems. He is a heterosexual male without drug use or significant travel history.

On physical examination, he has a heart rate of 109, blood pressure 130/84, respiratory rate of 20 breaths per minute, temperature of 37.5°C, and 97% oxygen saturation on room air. He is alert, oriented, tired appearing, and oth-erwise unremarkable, except for an obese abdomen, which is tender to deep palpation throughout the right upper quadrant without rebound tenderness. He has no hepatomegaly on percussion or palpation.

Significant laboratory values include a white blood cell count of 13,000/μL, alkaline transaminase (ALT) 109 μ/L, aspartate transaminase (AST) 95 μ/L, and bilirubin 0.9 mg/dL.

A right upper quadrant ultrasound demonstrated a right lobe predominant hyperechoic; a loculated lesion in the liver as well as cholelithiasis, with-out evidence of cholecystitis; and a positive Murphy sign. As a result of the abscess findings, a computed tomography (CT) scan was performed of the chest, abdomen, and pelvis, which revealed a small simple right pleural effu-sion and a rim enhancing loculated cystic lesion in the right lobe of the liver.

Treatment was initiated with piperacillin/tazobactam, and percutaneous drainage was performed by interventional radiology with the placement of a pigtail drain. The patient was admitted to the acute care surgery service, and the following day, cultures revealed *Klebsiella pneumoniae* and *Escherichia coli*. He was discharged with 4 weeks of ciprofloxacin and drain removal in a clinic.

## GEOGRAPHIC DISTRIBUTION AND MAGNITUDE OF DISEASE BURDEN

Pyogenic liver abscess is the predominant (>80%) form of liver abscess and accounts for the majority of visceral abscesses. Its inci-dence has risen in the last three decades, currently estimated between 2.9 and 3.6 per 100,000 population years in the United States. It is more prevalent in Southeast Asia, and a primary invasive liver abscess syndrome caused by *Klebsiella pneumoniae* has been described most commonly in this region. Approximately 10% of the world's popu-lation (50 million annually) is infected with *Entamoeba histolytica*, with the majority of cases occurring in developing countries. Amebic liver abscess is the most common form of extraintestinal manifesta-tion of amebiasis.

## PATHOGENESIS

Identifying the underlying source of the liver abscess is important for prompt recognition of this challenging diagnosis, aiding in accurate prognostication and complete treatment of the disease. For pyogenic abscess, etiologies include biliary disease, portal venous seeding from gastrointestinal infections, bacteremia from external sites, direct extension from either right upper quadrant abscesses, and liver trauma (Fig. 47.1). The cause of pyogenic liver abscess often results from ascending infection due to biliary disease or from portal pyemia due to peritoneal infection. Moreover, procedures like endoscopic retrograde cholangiopancreatography (ERCP), percutaneous transhepatic chol-angiography (PTC), or surgical biliary reconstruction can introduce microorganisms into the liver and predispose to hepatic abscess for-mation. Intra-abdominal infections such as appendicitis and divertic-ulitis are well-known sources of liver abscess via portal venous seeding. Hematogenous arterial seeding can occur from extra-abdominal sites such as in infective endocarditis or intravenous drug use. Suppurative cholecystitis and perforated peptic ulcers may lead to hepatic abscess via direct extension. Pyogenic liver abscess formation through direct inoculation of microbes into devitalized liver tissue from blunt or penetrating liver trauma or following thermal ablation using either radiofrequency ablation (RFA) or microwave ablation (MWA) of liver tumors is an increasingly recognized etiologic mechanism. A small subset of patients will have cryptogenic hepatic abscess without an identifiable source.

Amebic liver abscess is an extraintestinal complication of *E. histo-lytica* dysentery, which is transmitted via a fecal-oral route. Amebiasis results from the ingestion of cysts of the protozoan *E. histolytica*, which liberate the trophozoite form of the parasite in the intestine (Fig. 47.2). In complicated cases, intestinal wall invasion and subsequent seeding of the liver via the portal vein occur. In the liver, the trophozoites cause

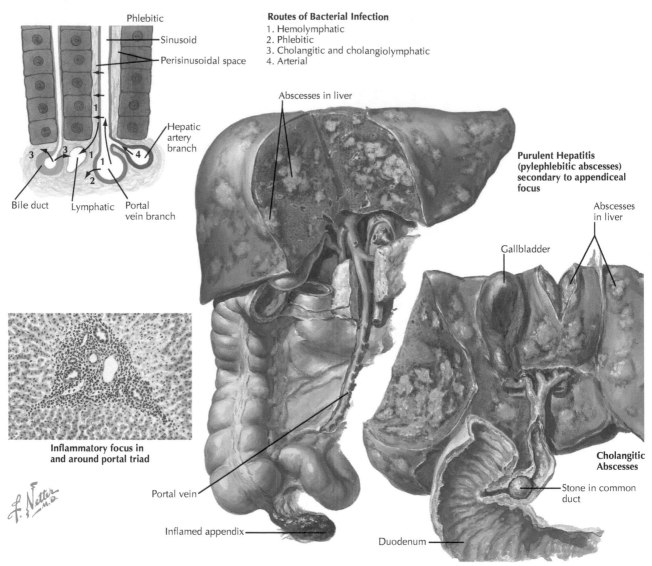

Routes of Bacterial Infection
1. Hemolymphatic
2. Phlebitic
3. Cholangitic and cholangiolymphatic
4. Arterial

**Fig. 47.1** Etiology of pyogenic liver abscess.

an acute inflammatory response that results in granuloma formation and liver necrosis, leading to the classic "anchovy paste" amebic liver abscess.

## RISK FACTORS

Diabetes mellitus, underlying hepatobiliary or pancreatic malignancy, and previous biliary manipulation or reconstruction are risk factors for the development of pyogenic liver abscess. The risk of developing amebic liver abscess relates to travel history in endemic areas and exposure to unclean water or poor sanitation conditions. It is also more prevalent in adult men and immunocompromised hosts.

## CLINICAL FEATURES

The clinical presentation of pyogenic and amebic liver abscess can be quite variable and nonspecific, making accurate and prompt diagnosis a challenge. Although fever and abdominal pain are frequently observed (75%–90%), patients often have a wide range of other signs and symptoms, including anorexia, lethargy, jaundice, and weight loss (Fig. 47.3). Peritonitis is an infrequent (≈5%) finding but may occur

with free rupture of the abscess. A small subset of patients may have overt sepsis manifested by high fevers and cardiovascular collapse. The differential diagnosis of a patient with these nonspecific symptoms is extremely broad, and additional history to narrow the diagnostic focus is necessary. Inquiries regarding a history of biliary disease, intra-abdominal infection, trauma, or endocarditis should be sought. Amebic abscess should be considered if a history of travel to an endemic area is discovered. The majority of patients with amebic abscess develop signs and symptoms of illness within 3 to 5 months of travel to the endemic region.

## DIAGNOSTIC APPROACH

As history and physical examination findings are generally nondiagnostic of pyogenic or amebic liver abscess, laboratory and radiographic studies are necessary. Leukocytosis is a common finding, although the expected eosinophilia is not commonly seen with amebiasis. Mild elevation of the transaminases is also common. Hyperbilirubinemia and alkaline phosphatase elevation may occur and may indicate an obstructive biliary source of disease. Blood cultures demonstrate gram-negative and anaerobic bacteremia in the setting of a lower gastrointestinal

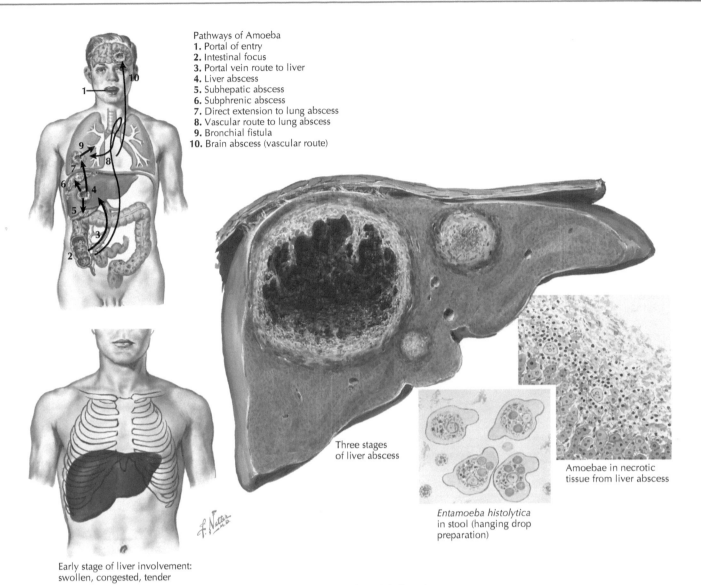

Pathways of Amoeba
1. Portal of entry
2. Intestinal focus
3. Portal vein route to liver
4. Liver abscess
5. Subhepatic abscess
6. Subphrenic abscess
7. Direct extension to lung abscess
8. Vascular route to lung abscess
9. Bronchial fistula
10. Brain abscess (vascular route)

Three stages
of liver abscess

Amoebae in necrotic
tissue from liver abscess

*Entamoeba histolytica*
in stool (hanging drop
preparation)

Early stage of liver involvement:
swollen, congested, tender

**Fig. 47.2** Pathogenesis of amebic liver abscess.

source of disease, whereas gram-positive cultures are more frequent in arterial sources of pyogenic liver abscess, such as endocarditis. If amebic abscess is suspected, serologic testing for antibodies to *E. histolytica*, present in 95% of cases, is useful. Although fecal microscopy and culture can be performed, these are usually low yield or difficult to perform.

Imaging is a key component in the diagnostic workup of liver abscesses. Ultrasound is frequently used as an initial study, as it is rapid, noninvasive, avoids radiation exposure, and evaluates concomitant biliary pathology. It demonstrates greater than 80% sensitivity but poor specificity (due to similar appearing benign cysts) for the detection of abscesses. Small abscesses and those located near the dome of the liver may be more difficult to detect, and ultrasound is less likely to diagnose extrabiliary, intra-abdominal source for the liver abscess. Therefore, CT is a superior modality to diagnose liver abscess and provides higher specificity than ultrasound. CT imaging can also detect the primary cause of liver abscess in a majority of cases. Magnetic resonance imaging (MRI) is time-consuming, expensive, and does not generally provide a diagnostic advantage over CT. Neither CT nor MRI is able to differentiate between pyogenic and amebic liver abscess. Occasionally a gallium scan can be used to differentiate the two entities, if necessary.

## CLINICAL MANAGEMENT AND TREATMENT

As with any abscess, the primary principle of treatment of pyogenic liver abscess is drainage. Adjunctive measures include appropriate antibiotic therapy and a thorough search for and treatment of the precipitating cause. Empiric broad-spectrum antibiotics should be promptly instituted in any patient diagnosed with a hepatic abscess. Therapy should ideally be instituted before drainage and should be based on the probable source of infection and local antibiotic resistance data. Antibiotic therapy can subsequently be tailored once culture and susceptibility results become available. Length of therapy should be individualized based on clinical response and the complexity of the abscess and its source.

Abscess drainage can be accomplished via various modalities, including percutaneous, laparoscopic, or open surgical approaches. Although percutaneous drainage of liver abscess was uncommon until the 1980s, improvements in imaging technology have made this approach the initial treatment modality of choice in managing pyogenic liver abscesses, with success rates reported in upward of 90% of cases (Fig. 47.4). Both percutaneous aspiration and drain placement have been used to treat abscesses. Simple, one-time aspiration has published success rates of around 60%~90%. Factors that predict success

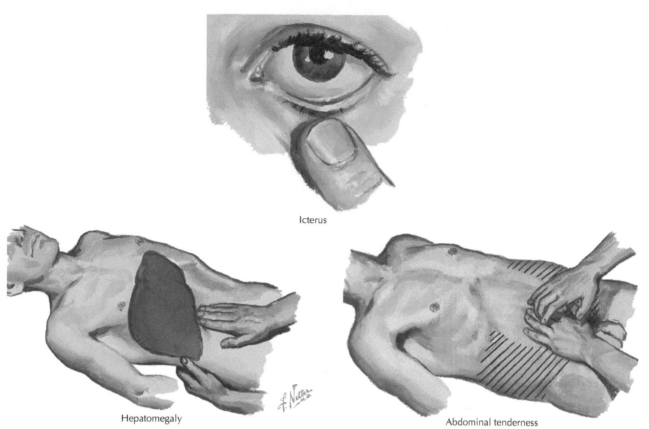

Icterus

Hepatomegaly

Abdominal tenderness

**Fig. 47.3** Physical examination findings suggestive of hepatic abscess.

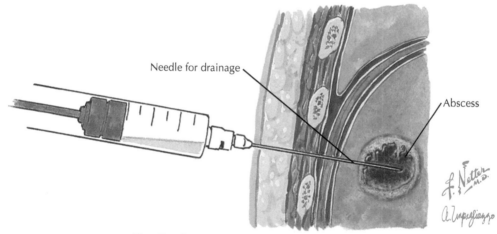

Needle for drainage

Abscess

**Fig. 47.4** Percutaneous drainage of liver abscess.

with aspiration include solitary, unilocular, small (<5 cm) abscesses with thin walls and nonviscous content. If aspiration fails, drainage catheter placement is indicated. Although drain placement, in general, has been demonstrated to be more effective than simple aspiration, it typically results in longer hospital stays and is more prone to complications such as bleeding or iatrogenic injury to the biliary system. Successful nonoperative management of pyogenic liver abscess is predicated on an aggressive approach to catheter management, with repeat imaging and catheter manipulation and/or upsizing should drainage be suboptimal.

Open surgical drainage of pyogenic liver abscess is reserved for cases in which percutaneous drainage attempts have failed, or when an acute concomitant intra-abdominal source of an abscess requires surgical

therapy, such as perforated diverticulitis. Frank rupture of an abscess into the intraperitoneal space is also an indication for primary operative therapy. The operative approach to liver abscess involves accurate localization of the abscess cavity within the liver parenchyma, complete evacuation of pus, and thorough debridement of necrotic liver tissue and loculations. Drain placement is commonly used. In rare instances, formal liver resection may be necessary for adequate treatment. Circumstances prone to require resection include multiple small abscesses confined to a specific anatomic location of the liver, hepatic atrophy caused by biliary obstruction and abscess of the affected segment(s), abscess that causes significant parenchymal destruction, and hepatolithiasis.

In contrast to pyogenic liver abscess, amebic liver abscesses are generally treatable with antibiotics alone. Treatment consists of a tissue agent and a luminal agent (which targets intraluminal cysts). Metronidazole is the treatment of choice, as it is effective in treating both liver abscess and intestinal amebiasis. More than 90% of patients achieve a cure with a 10-day course of therapy. Chloroquine is used as a second-line agent if patients do not respond to metronidazole therapy. Drainage of amebic liver abscess is rarely necessary and is generally performed only if medical management fails or if there is suspicion of a superimposed pyogenic abscess.

## PROGNOSIS

Overall mortality from hepatic abscess has declined over the past several decades. This is attributed to precise imaging, effective antimicrobial therapy, and improved availability of intensive care, and has paralleled the shift from primary operative therapy to more conservative percutaneous treatment approaches. In general, expected mortality is less than 20% for both pyogenic and amebic sources of disease. Outcome is becoming increasingly dependent on the underlying cause of the hepatic abscess and comorbidities, rather than on the liver abscess itself. Risk factors predicting a graver prognosis include more severe underlying illness, as predicted by Acute Physiology and Chronic Health Evaluation (APACHE) score, abscess rupture, and multiple abscesses.

## EVIDENCE

Barakate MS, Stephen MS, Waugh RC, et al. Pyogenic liver abscess: a review of 10 years' experience in management. *Aust N Z J Surg* 1999;69:205-209. *Retrospective case series of patients with pyogenic liver abscess. Factors associated with failed nonoperative management are outlined.*

Ch Yu S, Hg Lo R, Kan PS, Metreweli C. Pyogenic liver abscess: treatment with needle aspiration. *Clin Radiol* 1997;52:912-916. *Case series describing a 96% success rate in the treatment of pyogenic liver abscess with percutaneous needle aspiration.*

Chou FF, Sheen-Chen SM, Chen YS, Chen MC. Single and multiple pyogenic liver abscesses: clinical course, etiology, and results of treatment. *World J Surg* 1997;21:384-388. *Large series of over 400 patients diagnosed with a hepatic abscess at a single institution.*

Giorgio A, Tarantino L, Mariniello N, et al. Pyogenic liver abscesses: 13 years of experience in percutaneous needle aspiration with US guidance. *Radiology* 1995;195:122-124. *Case series demonstrating effective treatment of pyogenic liver abscess with percutaneous needle aspiration and adjunctive antibiotic therapy.*

Huang CJ, Pitt HA, Lipsett PA, et al. Pyogenic hepatic abscess. Changing trends over 42 years. *Ann Surg* 1996;223:600-607. *Retrospective analysis of over 200 patients treated with pyogenic liver abscess of 42 years. Documents changing etiology but lack of improved outcomes over time.*

Rajak CL, Gupta S, Jain S, et al. Percutaneous treatment of liver abscesses: needle aspiration versus catheter drainage. *AJR Am J Roentgenol* 1998;170:1035-1039. *Randomized trial comparing needle aspiration and catheter drainage of liver abscess found that catheter drainage had a higher rate of abscess resolution.*

## ADDITIONAL RESOURCES

Braiteh F, Golden MP. Cryptogenic invasive Klebsiella pneumoniae liver abscess syndrome. *Int J Infect Dis* 2007;11:16-22. *Review of Klebsiella pneumoniae liver abscess as a distinct clinical syndrome.*

Branum GD, Tyson GS, Branum MA, Meyers WC. Hepatic abscess. changes in etiology, diagnosis and management. *Ann Surg* 1990;12:655-662. *Retrospective review emphasizing the importance of the underlying cause of hepatic abscess on outcomes.*

Chou FF, Sheen-Chen SM, Lee TY. Rupture of pyogenic liver abscess. *Am J Gastroenterol* 1995;90:767-770. *Retrospective review of treatment and outcomes of a series of patients with ruptured hepatic abscess.*

Hughes MA, Petri WA. Amebic liver abscess. *Infect Dis Clin North Am* 2000;14:565-582. *Review article of the clinical pathogenesis, diagnosis, treatment, and outcomes of amebic liver abscess.*

Mischinger HJ, Hauser H, Rabl H, et al. Pyogenic liver abscess: studies of therapy and analysis of risk factors. *World J Surg* 1994;18:852-857. *Retrospective review of nonoperative versus operative therapy for pyogenic liver abscess. Adverse outcomes were related more to underlying causes than to mode of therapy.*

Ng FH, Wong WM, Wong BC, et al. Sequential intravenous/oral antibiotic vs continuous intravenous antibiotic in the treatment of pyogenic liver abscess. *Aliment Pharmacol Ther* 2002;16:1083-1090. *Retrospective review of antimicrobial therapy in the treatment of liver abscess. Sequential intravenous/oral antibiotic therapy is a safe and effective treatment.*

Saini S. Imaging of the hepatobiliary tract. *N Engl J Med* 1997;336:1889-1894. *Review of advances in hepatic imaging in the diagnosis of liver pathology.*

Salles JM, Salles MJ, Moraes LA, Silva MC. Invasive amebiasis: an update on diagnosis and management. *Expert Rev Anti Infect Ther* 2007;5:893-901. *Clinical review of amebic liver abscess with a focus on diagnosis and management.*

Seeto RK, Rockey D. Amebic liver abscess: epidemiology, clinical features, and outcome. *West J Med* 1999;170:104-109. *Retrospective review of patients diagnosed with amebic liver abscess at two large institutions in the United States.*

Seeto RK, Rockey D. Pyogenic liver abscess: changes in etiology, management, and outcome. *Medicine* 1996;75:99-113. *Comprehensive review of the presentation, diagnosis, and management options of pyogenic liver abscess.*

# 48

# Necrotizing Soft-Tissue Infections

*Sarah S. Zhu, E. Patchen Dellinger, Daniel A. Anaya*

##  ABSTRACT

Necrotizing soft-tissue infections (NSTIs) are highly fatal infections that share features, including the presence of necrotic tissue and the need for surgical debridement (among other therapies). NSTIs are under the category of complicated skin and soft-tissue infections (SSTIs). The nomenclature of these infections has typically been complicated and confusing because different terms are used to describe specific types of NSTI based on location, causative organisms, and other features, which may ultimately delay diagnosis and/or surgical intervention. The use of the term *necrotizing soft-tissue infection* is advocated because it groups all of the categories together and helps establish a common pathway to diagnosis and management. Advanced NSTIs are relatively easy to recognize based on characteristic local findings, severe systemic derangement, and are associated with a high mortality rate. Early diagnosis is key to improved clinical outcomes, but it is not always straightforward. A high index of suspicion coupled with biochemical and radiological studies can help confirm or rule out the diagnosis. Surgical exploration is the ultimate diagnostic (and therapeutic) strategy. Management of NSTIs should include early debridement, broad-spectrum antimicrobial therapy, and supportive care. Prognostic factors have been identified and prognostic scores developed; these tools may help in selecting patients who may benefit from a more aggressive surgical strategy and/or novel treatments.

## ✳ CLINICAL VIGNETTE

A 40-year-old male with a history of type 2 diabetes mellitus presents with a 2-day history of severe left arm pain to the ED. He states the left upper extremity pain is associated with redness. The patient explained that the area has been getting progressively "more red" and painful throughout the day. Social history is significant for heavy alcohol use and IV drug use on the weekends.

Vitals BP 160/82, HR 112, $O_2$ stat 97% on room air, RR 18, Temperature 38°C.

Physical exam within normal limits, except for mild distress on initial examination. 5×3 cm area of erythema on left antecubital fossa—without sharp margins, tender to palpation, edema involving the whole extremity. Weakened strength in left upper extremity, but normal ROM.

Labs reveal an elevated WBC 16 k/uL, hemoglobin 12.1 g/dL, hematocrit 23%, sodium 134 mmol/L, potassium 4.0 mmol/L, glucose 305 mg/dL, creatinine 1.0 mg/dL, C reactive protein (CRP) 15 mg/dL, lactate 0.8 mmol/L.

**Differential diagnosis:** NSTI versus cellulitis. Other: abscess versus phlebitis versus hidradenitis suppurativa

The Laboratory Risk Indicator for Necrotizing Fasciitis (LRINEC) score = 7. The patient was started on meropenem, vancomycin, and clindamycin, and general surgery completed an emergent surgical debridement for complete diagnosis and treatment. Intraoperative findings revealed necrotic pale appearing tissues, thrombosed vessels, easy dissection, thin exudate without clear purulence, and dull gray appearance of fascia. Blood culture was positive for CA-MRSA. Procalcitonin (PCT) POD 1= 12.9 ng/mL, PCT POD 2= 11.5 ng/mL; PCT ratio= 1.12. Based on these results and persistent erythema and pain, the patient was taken back for serial debridements.

## ✳ CLINICAL VIGNETTE

A 63-year-old female with a history of a strangulated ventral hernia is status post laparoscopic converted to open exploratory laparotomy, lysis of adhesions, and bowel resection. The patient started to develop worsening abdominal pain at the surgical site on POD 2. The patient was evaluated and found to have purulent drainage when staples were removed at the surgical site. The following day, the patient developed induration and worsening erythema at the surgical site associated with fever and tachycardia.

**Vitals:** BP 155/62, HR 104, $O_2$ stat 95% on room air, RR 19, Temperature 39°C

The physical exam was within normal limits, except for mild distress on initial examination. Erythema and edema along laparotomy and laparoscopic incisions on abdomen, mild purulent drainage, soft, tender to palpation, bowel sounds × 4, distension difficult to assess due to body habitus.

Labs revealed WBC 17.29 k/uL, hemoglobin 9.3 g/dL, hematocrit 35%, sodium 135 mmol/L, potassium 3.8 mmol/L, glucose 201 mg/dL, creatinine 1.3 mg/dL, CRP 19.6 mg/dL, lactate 0.7 mmol/L.

**Differential diagnosis:** postoperative wound infection versus NSTI

A computed tomography (CT) of the abdomen and pelvis was performed and revealed subfascial fluid collection and areas of inflammation and subcutaneous emphysema at the areas of port sites. The LRINEC score= 6. GI surgery removed the remaining staples and examined the wound. Residual purulent fluid from the surgical sites was evacuated and the fascia was found to be necrotic with dehiscence. The patient was started on meropenem and clindamycin, and general surgery completed an emergent surgical debridement for complete diagnosis and treatment. Intraoperative findings revealed residual purulent drainage associated with necrotic fascia and infection extending through the tissue planes. The tissues had foul-smelling drainage that appeared thin and serosanguinous, and there was woody induration of the tissue. PCT POD 1 = 10.8 ng/mL, PCT POD 2 = 9.7 ng/mL; PCT ratio: 1.11. The patient had to be taken for serial debridement × 3 and ultimately needed a negative pressure wound therapy dressing placed. Blood culture × 2 negative. Fluid culture revealed *Clostridium perfringens*.

NSTIs include a wide range of skin and soft-tissue infections (Fig. 48.1) characterized by the presence of necrotic tissue and the need for debridement. Common terms include necrotizing cellulitis/fasciitis, "flesh-eating bacteria," gas gangrene, and Fournier gangrene. There is no known geographic distribution for NSTIs, except those related to specific risk factors.

Given the rarity of NSTIs, it is hard to estimate accurate disease burden. Population-level studies using administrative and insurance-based databases have estimated an incidence of 4 per 100,000 person-years and an estimated 500 to 1500 new cases diagnosed yearly in the United States. As a consequence of the increasing number of *Staphylococcus*-related SSTIs, the incidence of NSTIs appears to have increased over time.

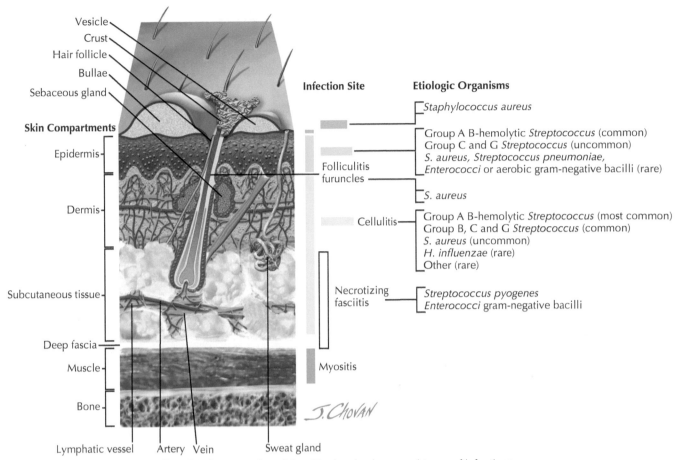

Fig. 48.1 Cross-section of the skin showing layers and types of infections.

Despite a lack of evidence-derived data on the impact on health and cost of NSTIs, it is well known that NSTIs are infections with high risk for fatal outcome or limb loss, and when successfully treated, they are associated with significant effects on quality of life, including prolonged hospitalizations, disfiguring procedures, and the need for long-term rehabilitation. Furthermore, these effects carry substantial healthcare costs.

## RISK FACTORS

Few studies have compared patient and other baseline characteristics of NSTI with those of nonnecrotizing SSTI populations. Two such studies were able to identify intravenous drug use, as well as diabetes mellitus, as conditions more commonly associated with NSTI. Intravenous drug users with muscle or subcutaneous injections are particularly susceptible to developing NSTI. NSTI outbreaks have been reported in patients with drug use associated with contaminated heroin lots (e.g., "black tar heroin"). Other series evaluating NSTI populations, although not methodologically able to identify risk factors, have found associations with older age, chronic comorbidities, obesity, alcohol and drug use, malnutrition, immune suppression, and specific medications (steroids and nonsteroidal antiinflammatory drugs [NSAIDs], among others). These conditions do appear to be commonly associated with patients presenting with NSTI; however, to date, there is no evidence suggesting their predictive, risk-related, or causative association. Although adequate epidemiologic studies have identified an association between NSTIs and NSAID use, it is more likely that patients with NSTIs take NSAIDs during the initial phase of their infection than NSAID use predisposes individuals to developing NSTI. It is important to remember that NSTIs can occur in all populations.

## CLINICAL PRESENTATION

Clinical manifestations of NSTI vary significantly based on the timing of presentation and the extent of the infectious process. The necrotizing component can involve any layer of the skin and soft tissues, including the skin, subcutaneous tissue, superficial/deep fascia, and/or muscle (see Fig. 48.1). Different anatomic areas can be involved, including, in order of frequency, the extremities (lower more than upper), perineum, trunk, and head and neck.

The vast majority of patients with early NSTI start with innocuous signs and symptoms. A preceding event localized to the involved area can be recalled, although in up to 20% of cases, no precipitating event is identified. Initial symptoms are difficult to differentiate from those of nonnecrotizing SSTIs (e.g., cellulitis, erysipelas) and include warmth, erythema, and pain with or without fever and tachycardia. Progression of the infection is variable and can be prolonged but usually occurs over 2 to 5 days. Once the necrotizing component starts to spread, the more ominous findings ensue and progress rapidly, leading to an overwhelming systemic infection. It is essential to identify these NSTIs early in their course, in order to prevent uncontrolled spread and systemic involvement. Progression from the time of the inciting event can be extremely rapid. Severe and advanced NSTIs are characterized by tense edema extending beyond erythema, ecchymosis, bullae or blisters, erythema without sharp margins, pain out of proportion, and crepitus. At these advanced stages, systemic manifestations, including fever, tachycardia, hypotension, multisystem organ dysfunction, and shock, are the *sine qua non* of NSTI (Fig. 48.2).

Clostridial and group A *Streptococcal* infections (GAS), as well as those caused by mucormycosis and *Vibrio* species, are characteristically

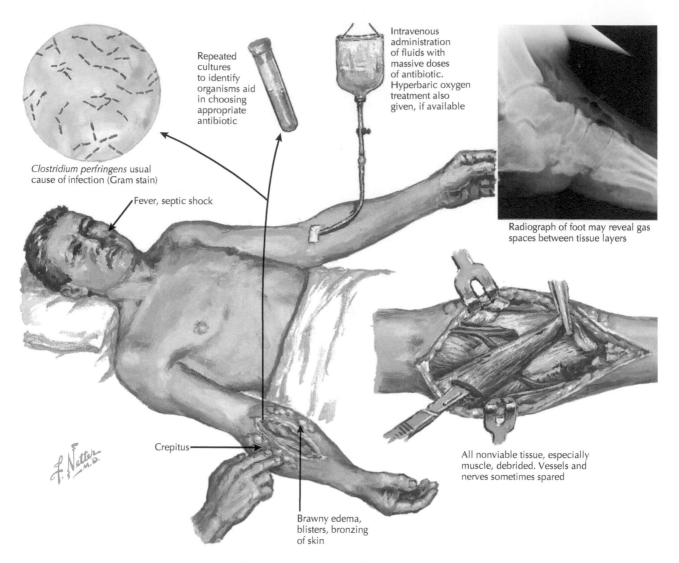

Repeated cultures to identify organisms aid in choosing appropriate antibiotic

Intravenous administration of fluids with massive doses of antibiotic. Hyperbaric oxygen treatment also given, if available

*Clostridium perfringens* usual cause of infection (Gram stain)

Radiograph of foot may reveal gas spaces between tissue layers

Fever, septic shock

Crepitus

All nonviable tissue, especially muscle, debrided. Vessels and nerves sometimes spared

Brawny edema, blisters, bronzing of skin

**Fig. 48.2** Necrotizing soft-tissue infections.

aggressive, with rapid local progression and severe systemic derangement such as signs of systemic inflammation response or sepsis features, and are associated with the highest mortality rates. Clostridial infections are usually characterized by thin serosanguinous fluid and woody induration of the involved tissue. Upon exploration, there is typically thin, "dishwater" appearing fluid.

## DIAGNOSTIC APPROACH

One of the most challenging aspects of managing patients with NSTI is the ability to diagnose it early. Early diagnosis with adequate surgical debridement is the most important therapy. The diagnosis of NSTI is purely clinical, and when in doubt, it should be confirmed or ruled out with surgical exploration. This is a crucial concept that allows early identification with timely debridement, maximizing the opportunity to control the infection and its systemic manifestations and leading to better overall outcomes.

Multiple studies have focused on different tools that may help identify patients with NSTIs early. The most important step in achieving this is to have a high index of suspicion based on risk factors and clinical presentation. Once NSTI is considered as a potential diagnosis, biochemical and radiological studies can be used. In a study by Wall et al. (2000), the authors found that a white blood cell count (WBC) greater

than 15,400 and serum sodium level less than 135 mmol/L were associated with NSTI. Although the positive predictive value (PPV) was 26%, the negative predictive value (NPV) was 99%, making these values useful for ruling out NSTI when neither of these criteria is present. Wong and colleagues (2004) identified a series of factors associated with NSTI and designed a diagnostic score based on their presence: WBC, hemoglobin, serum sodium, serum glucose, serum creatinine, and C-reactive protein. The score is referred to as the LRINEC. The PPV and NPV of this score were 92% and 96%, respectively. Although useful for confirming the index of suspicion, the LRINEC score and other laboratory tests have limited sensitivity. One should be careful not to use them as standalone tests to rule out NSTIs. Lab results such as left shift, acidosis, and coagulopathy may be more useful for specific circumstances. It may be beneficial to obtain creatinine kinase and aspartate transaminase levels when there is concern for muscle or fascial involvement.

Radiological tests can also help establish the diagnosis of NSTI and are useful in patients with equivocal findings and no evidence of sepsis or shock. Radiographs, ultrasound, CT, and magnetic resonance imaging (MRI) have all been studied. In general, evidence of subcutaneous gas is pathognomonic of NSTI and should prompt emergency surgical debridement; however, this is a late finding and is less frequent than previously thought, present in only 24% of patients. Additional radiological findings include thickening of the underlying soft tissues and decreased enhancement of the deep fascial layers. These findings are not specific for NSTI but can increase the suspicion and direct further diagnostic efforts or surgical exploration for diagnosis.

Other studies have focused on evaluating the transcutaneous oxygen saturation with good PPV and NPV. However, this test is limited to patients with adequate underlying vasculature and hence is less useful for those with peripheral vascular disease. Finally, frozen section biopsy has been used to confirm the diagnosis of NSTI before formal surgical debridement. Typical microscopic findings include leukocyte infiltration, thrombosis of small arteries and veins, and necrosis. This test is limited by the subjective interpretation and availability of the pathologist, and when there is sufficient clinical suspicion to perform biopsy, the diagnosis is usually evident to the naked eye.

The most important strategy in diagnosing NSTI is a high index of suspicion followed by immediate surgical exploration. This cannot be emphasized enough and constitutes the difference between early infection control versus extensive delayed debridement in the setting of septic shock, multisystem organ failure, and a significantly higher risk of death. Intraoperative findings supporting the diagnosis of NSTI include the presence of necrotic tissue, easy finger dissection through normally fixed planes, foul-smelling "dishwater" purulence, thrombosed vessels, and lack of bleeding during transection of involved tissues.

## TREATMENT

The management of NSTI follows the principles for any other surgical infection: source control, antimicrobial therapy, and supportive care; however, more than with any other infection, source control is paramount and needs to be instituted as early as possible. Early and complete debridement of necrotic and involved tissues is associated with lower mortality rates. Scheduled re-explorations for a thorough examination and re-excision of newly involved areas are necessary and should be performed anytime the initial process is extensive or with worsening infection. It is worth considering amputation early, rather than serial attempts at limb salvage for certain cases—particularly those with severe systemic manifestations and extensive local disease, unable to be cleared.

Nutritional support is paramount for these patients, who experience a long period of a hypercatabolic state. ICU care is essential and has contributed significantly to decreasing early mortality rates. A high proportion of these patients will develop multisystem organ failure, requiring aggressive physiologic support, including ventilatory, cardiac, renal support, and so forth.

### Antimicrobial Treatment

NSTIs are divided into three subgroups: polymicrobial (type 1), monomicrobial (type 2), and monomicrobial secondary to a variety of pathogenic bacilli (type 3; e.g., gas gangrene from clostridial myonecrosis). NSTIs are polymicrobial in the majority of cases, and antimicrobial therapy should be instituted at the time of diagnosis, with broad-spectrum antibiotics to cover gram-positive, gram-negative, and anaerobic bacteria. Monomicrobial infections, although less common, are becoming more frequent. Methicillin-resistant *Staphylococcus aureus* (MRSA) and GAS represent the two most frequently identified monomicrobial infections, with the *Clostridium* species not falling too far behind. For MRSA NSTI, the progression of the infection is not as fast or as extensive as with other monomicrobial infections, and specific antibiotics covering MRSA (e.g., daptomycin, linezolid, ceftaroline, vancomycin, or tigecycline) should be empirically started.

The causative agent is often difficult to determine before culture results; however, aggressive, rapidly progressing infections should raise the index of suspicion, and surgical exploration must be rapidly performed until source control is achieved. Initial empirical broad-spectrum antimicrobials that cover these organisms should be used, and subsequent narrowing of the antibiotics when sensitivities are reported is encouraged. The empirical antibiotic regimen should include daptomycin or linezolid for Gram-positive and MRSA coverage and piperacillin-tazobactam or carbapenems for Gram-negative coverage. Piperacillin-tazobactam should be used in low prevalence of extended-spectrum beta-lactamase *Enterobacteriaceae* (ESBL), and carbapenem should be used in high prevalence of ESBL.

Clindamycin at high doses or linezolid should be used as a protein synthesis inhibitor to reduce toxin production in patients with toxic shock syndrome, in which the rapid local and systemic events are exotoxin mediated. Other studies have evaluated intravenous immunoglobulin (IVIg) and hyperbaric oxygen as adjunctive strategies in managing these patients. At the current time, the use of these treatments is experimental, and although some studies have reported adequate outcomes, there are no high-level data supporting their routine use.

Antibiotic therapy should be given until no further debridement is indicated and there is marked clinical improvement and resolution of fever for 48 to 72 hours. Currently, there are no studies that evaluate the optimal duration of antibiotic therapy, but several controlled clinical studies have evaluated the potential of PCT as a guide to antimicrobial discontinuation. Friederichs (2013) performed a study to evaluate serum levels of PCT and CRP in 38 postoperative patients who were treated with clinical signs of sepsis due to NSTI. All patients received surgical debridement, and the PCT ratio of day 1 to day 2 was studied. The ratio was significantly higher in patients with successful surgical debridements (1.665 vs. 0.9, $P < .001$). The established cutoff was 1.14. The PPV was 75.8% and the NPV was 80.0%, which supports that the PCT ratio can be a useful clinical tool for indicating successful surgical intervention, source control, and clinical recovery of the patient.

### Novel Agents

There is a new agent found to regulate the inflammation in patients with NSTIs. AB103 (Reltecimod) is a synthetic CD28 mimetic octapeptide that selectively inhibits the direct binding of superantigen exotoxins to the CD28 costimulatory receptor on T-helper 1 lymphocytes. It is intended that AB103 will minimize the inflammatory immune response to bacterial toxins, therefore limiting the capability of the toxins to cause organ failure. A prospective, randomized, placebo-controlled, double-blind study was conducted in 6 academic medical centers in the United States. The Sequential Organ Failure Assessment (SOFA) score improved from baseline in both low and high dose treatment groups when compared to placebo group at 14 days with a similar number of surgical debridements. No drug-related adverse events were seen. Further studies are still needed to evaluate AB103's efficacy, though it appears to be a promising adjunct to current treatment.

## PROGNOSIS

A recent pooled analysis of published data from NSTI series revealed an overall mortality rate of 34%. Multiple studies have focused on identifying specific prognostic factors in different NSTIs, including age, comorbidities, diabetes mellitus, obesity, injection drug use, clostridial and GAS infections, mucormycosis, leukocytosis, renal failure, acidosis, Acute Physiology and Chronic Health Evaluation II (APACHE II) score, quick SOFA score, and delayed surgical treatment, among others. The findings from these series are not generalizable owing to the inherent selection of patients in these single-institution reports. Most of these factors develop during the management of these patients, hence limiting their utility in stratifying patients by prognosis early.

To overcome some limitations, Anaya and colleagues (2009) evaluated a wide range of patient and physiologic characteristics at the time of emergency room evaluation in 2 large referral centers and identified 6 independent prognostic factors: heart rate >110, temperature <36°C, creatinine >1.5 mg/dL, age >50, WBC >40,000, and hematocrit >50. A prognostic score able to stratify patients by risk of mortality at the time of first assessment was developed. When present, the first three variables add 1 point each, and the last three add 3 points each. A score of 6 was associated with a mortality rate of over 88%, scores of 3 to 5 with 24%, and scores of 0 to 2 with 6%. Although this scoring system warrants external validation, it is useful in early stratification and identification of high-risk patients who may benefit from more aggressive surgical debridement or alternative treatments. These findings are valid when early and aggressive surgical debridement is performed, as was done by these experienced centers—this is probably a reason why the overall mortality rate in their population was below 20%. It is clear that delayed surgical treatment is the most important prognostic factor and has been universally correlated with the highest mortality rates and complications.

## CONTROL AND PREVENTION

It is important to consider a multidisciplinary approach to treatment and recovery after debridement. Patients will require extensive psychiatric, nutrition, and physical therapies. Negative pressure devices can promote and enhance vascularity of wound bed and eliminate daily dressing changes. 25 kcal/kg/day should be the initial nutritional goal for the first week of recovery and may be advanced to 30 to 35 kcal/kg/day. There has not been adequate data on early enteral nutrition, but parenteral nutrition should be reserved for patients when the enteral route is contraindicated or unlikely to meet nutritional requirements within a week.

There are few data on how to prevent NSTIs. An important concept is the observation that a proportion of patients with nonnecrotizing infections, when left untreated, can progress to develop NSTI. Without a clear understanding of the pathophysiology of NSTI at present time, it should be emphasized that any type of SSTI has the potential for developing into a NSTI, and early diagnosis and treatment are the most important steps in avoiding such progression and/or improving outcomes when it has occurred.

## EVIDENCE

Anaya DA, Bulger EM, Kwong YS, et al. Predicting mortality in necrotizing soft tissue infections: a clinical score. *Surg Infect (Larchmt)* 2009;10:517-522. *This is the largest multi-institutional study evaluating predictors of survival in NSTI. A clinical score was developed to stratify patients based on the risk of death at the time of initial evaluation.*

Anaya DA, McMahon K, Nathens AB, et al. Predictors of mortality and limb loss in necrotizing soft tissue infections. *Arch Surg* 2005;140:151-157; discussion 158. *This is one of the largest studies evaluating prognostic factors and determinants of adverse outcomes (mortality and/or limb amputation) in patients with NSTI. It also highlights the more severe course typically seen in patients with clostridial infections.*

Bulger EM, Maier RV, Sperry J, et al. A novel drug for treatment of necrotizing soft-tissue infections: a randomized clinical trial. *JAMA Surg* 2014;149(6):528-536. *This randomized clinical trial evaluates the safety and clinical parameters of AB103 when used in patients with NSTI.*

Friederichs J, Hutter M, Hierholzer C, et al. Procalcitonin ratio as a predictor of successful surgical treatment of severe necrotizing soft tissue infections. *Am J Surg* 2013;206:368-373. *This study was to develop a PCT ratio indicating successful surgical intervention in patients with sepsis caused by NSTI.*

McHenry CR, Piotrowski JJ, Petrinic D, Malangoni MA. Determinants of mortality for necrotizing soft-tissue infections. *Ann Surg* 1995;221:558-563; discussion 563-565. *This is a large study evaluating specific prognostic factors and determinants of mortality in patients with NSTI.*

Miller LG, Perdreau-Remington F, Rieg G, et al. Necrotizing fasciitis caused by community-associated methicillin-resistant Staphylococcus aureus in Los Angeles. *N Engl J Med* 2005;352:1445-1453. *This study brings attention to the trend of increasing CA-MRSA infections, including those causing NSTI.*

Singh G, Ray P, Sinha SK, et al. Bacteriology of necrotizing infections of soft tissues. *Aust N Z J Surg* 1996;66:747-750. *This study focuses on describing the nature (polymicrobial or monomicrobial) of and the most common microorganisms leading to NSTI.*

Wall DB, Klein SR, Black S, De Virgilio C. A simple model to help distinguish necrotizing fasciitis from nonnecrotizing soft tissue infection. *J Am Coll Surg* 2000;191:227-231. *This article identifies predictors of NSTI versus nonnecrotizing infections and develops a simple score to help establish the diagnosis.*

Wong CH, Khin LW, Heng KS, et al. The LRINEC (Laboratory Risk Indicator for Necrotizing Fasciitis) score: a tool for distinguishing necrotizing fasciitis from other soft tissue infections. *Crit Care Med* 2004;32:1535-1541. *This is a large study that evaluates multiple potential predictors of NSTI and develops a much more elaborate and accurate score to help confirm (or rule out) the diagnosis of NSTI.*

## ADDITIONAL RESOURCES

Anaya DA, Dellinger EP. Necrotizing soft-tissue infection: diagnosis and management. *Clin Infect Dis* 2007;44:705-710. *This review encompasses all aspects of NSTI.*

Ebright JR, Pieper B. Skin and soft tissue infections in injection drug users. *Infect Dis Clin North Am* 2002;16:697-712. *This article reviews the increased risk and microbiologic characteristics of NSTI in intravenous drug users.*

Green RJ, Dafoe DC, Raffin TA. Necrotizing fasciitis. *Chest* 1996;110:219-229. *This is a comprehensive review of NSTI.*

Hakkarainen TW, Kopari NM, Pham TN, et al. Necrotizing soft tissue infections: review and current concepts in treatment, systems of care, and outcomes. *Curr Probl Surg* 2014;51(8):344-362. *This article is an up-to-date comprehensive review of NSTI.*

Pallin DJ, Egan DJ, Pelletier AJ, et al. Increased U.S. emergency department visits for skin and soft tissue infections, and changes in antibiotic choices, during the emergence of community associated methicillin-resistant Staphylococcus aureus. *Ann Emerg Med* 2008;51:291-298. *This article goes over important epidemiologic and healthcare-burden aspects of soft-tissue infections within the United States.*

Sartelli M, Guirao X, Hardcastle TC, et al. 2018 WSES/SIS-E consensus conference: recommendations for the management of skin and soft-tissue infections. *World J Emerg Surg* 2018;13-58. *This article is an up to date review on clinical practice recommendations for skin and soft-tissue infections.*

Weiss KA, Laverdiere M. Group A Streptococcus invasive infections: a review. *Can J Surg* 1997;40:18-25. *This review goes over the available data on epidemiology, pathophysiology, clinical presentation, treatment, and prognosis of group A Streptococcus infections, including those in the skin and soft tissues.*

# 49

# Anorectal Abscess and Fistula in Ano

*Arden M. Morris, Natalie Kirilcuk*

## ABSTRACT

Anal abscess and fistula in ano are not uncommon and have a high recurrence rate. Both are easy to identify on clinical examination. Identification of the internal opening or discerning the pathway of a fistula may be challenging and may require radiologic imaging. In most cases, surgical drainage of the abscess or fistula infection is the primary treatment. Some fistulas will require staged treatment to avoid anal sphincter injury that could threaten continence. In spite of initial enthusiasm for fibrin glue, it no longer has a place in the management of anal fistulas. Cumulative data show that collagen plugs fail in more than 40% of cases. The Ligation of Intersphincteric Fistula Tract (LIFT) procedure and anal advancement flaps continue to have a place in the management of persistent or complex fistulas.

## CLINICAL VIGNETTE

A 65-year-old man who experienced anorectal pain and itching was diagnosed by his primary care provider with a thrombosed hemorrhoid, which he stated later became infected. He was treated with three antibiotics sequentially, which did not resolve his pain or drainage. One month after the initial symptoms, a seton was placed, and 2 months later, a combination collagen plug placement and anal advancement flap were performed. Within a month of the anal advancement flap, his wound separated resulting in constant anal drainage and severe pain exacerbated by bowel movements. Clinical exam revealed macerated perianal skin, bilateral external fistula openings consistent with a transsphincteric horseshoe fistula, and internal anal sphincter spasm. He was treated with zinc oxide on the perianal skin and taken to the operating theater for an exam under anesthesia with a possible fistulotomy versus seton placement and a staged procedure. Operative findings included a posterior midline intersphincteric fistula and a second deeper posterior midline fistula tract into the deep post-anal space. Both were opened with fistulotomy, flushed with hydrogen peroxide, and curetted. The wound was marsupialized with a running locked absorbable suture. The patient reported complete pain resolution within 2 weeks and intact anal continence to gas and stool, but minor serosanguinous drainage that persisted for 3 months.

## INTRODUCTION

Considering the frequent exposure to a large bacterial load and high pressure, infection of the anorectal crypts with a resultant abscess or fistula is relatively uncommon. Limited epidemiologic data indicate an annual incidence of about 9 cases per 100,000 in the population, most commonly affecting people aged 30 to 50 years. Approximately 30% to 50% of patients with an initially treated abscess will experience a subsequent abscess recurrence or frank anorectal fistula formation. Male gender, smoking, diabetes, and inflammatory bowel disease (IBD) are risk factors for the initial anorectal abscess formation. However, recurrence is not associated with gender, smoking, human immunodeficiency virus (HIV) status, sedentary lifestyle, or perioperative antibiotic use according to the majority of studies. Instead, the single most important predictor of recurrence or fistula formation is age under 40 years. Other possible predictors of recurrence, supported by studies in widely divergent settings, include infection with *Escherichia coli* and the absence of diabetes mellitus.

## ETIOLOGY AND CLINICAL FEATURES

In more than 90% of cases, the pathogenesis of anorectal abscess formation is thought to be anal crypt obstruction by inspissated mucus or stool (Fig. 49.1). As trapped bacteria proliferate and mucus and pus accumulate, an abscess forms and erodes through adjacent tissue planes, resulting in classic signs of tenderness, redness, swelling, and heat. Thus most patients become aware of their symptoms later in the stages of abscess formation. Patients will generally complain of throbbing or dully aching pain that is aggravated by walking, sitting, straining, coughing, and sneezing.

Abscess progression can proceed in any direction. If the abscess is eroding superiorly or in an intersphincteric manner, a swollen mass may not be obvious. Urinary retention, fever, or even septicemia may accompany the anorectal pain and is an urgent and important clue to the presence of an abscess in cases of cephalad or otherwise obscure infectious erosion. Fortunately, in most cases, the abscess will erode toward the perianal margin. Without intervention, most abscesses eventually will rupture through the anal margin skin. Formation of this external opening provides tremendous relief of pain, as well as some anxiety to the patient. The initial cryptoglandular insult results in abscess, but it is the persistent internal opening of the initial crypt which results in fistula in ano.

## HISTORY AND PHYSICAL EXAMINATION

During the initial consultation, a thorough history is paramount and should include documentation of any previous similar events; previous anorectal surgery or other trauma; previous obstetric injury; previous history of sexual assault; personal or family history of IBD or colorectal cancer; symptoms consistent with potentially undiagnosed IBD such as unintentional weight loss, chronic diarrhea, or abdominal pain; and symptoms consistent with lymphoma, leukemia, or HIV infection such as weight loss, night sweats, lymphadenopathy, or unexplained fevers. Obtaining and documenting a thorough history will provide guidance for an appropriately aggressive treatment plan and potential use of medical as well as surgical therapies.

Careful documentation of bowel habits is also prudent and should include frequency of defecation; fecal urgency or incontinence to gas, liquid, or solids; presence of pain or bleeding with defecation; and sexual dysfunction. If there are any concerns regarding the risk of

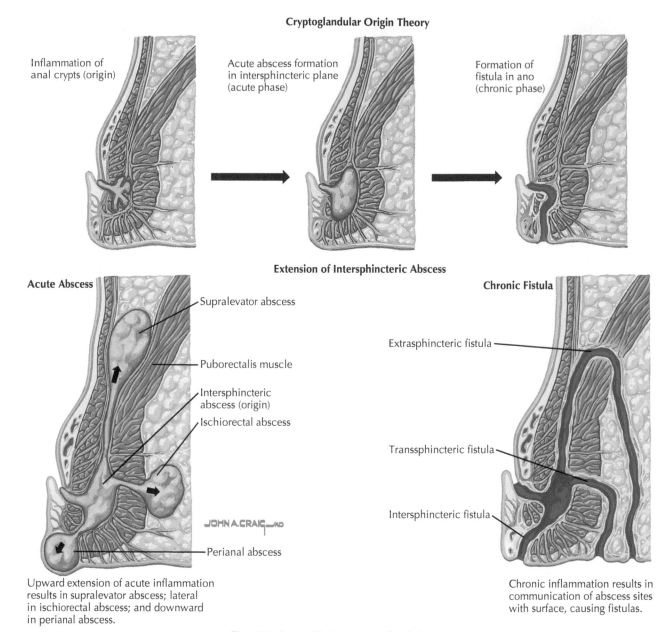

**Cryptoglandular Origin Theory**

Inflammation of anal crypts (origin)

Acute abscess formation in intersphincteric plane (acute phase)

Formation of fistula in ano (chronic phase)

**Extension of Intersphincteric Abscess**

**Acute Abscess**

Supralevator abscess

Puborectalis muscle

Intersphincteric abscess (origin)

Ischiorectal abscess

JOHN A. CRAIG—AD

Perianal abscess

Upward extension of acute inflammation results in supralevator abscess; lateral in ischiorectal abscess; and downward in perianal abscess.

**Chronic Fistula**

Extrasphincteric fistula

Transsphincteric fistula

Intersphincteric fistula

Chronic inflammation results in communication of abscess sites with surface, causing fistulas.

**Fig. 49.1** Anorectal abscess and fistula in ano.

colorectal cancer, such as bleeding per anus, or changes in stool caliber, the patient should be referred for colonoscopic evaluation after the anorectal pain and infection have been addressed.

A comprehensive physical examination must be conducted to rule out underlying or concomitant diseases. The examination of the perineum, often referred to in the medical record as the "rectal exam," should be preceded by ensuring patient privacy and respectful treatment, and should include an additional medical staff person in the examination room. The most important part of the physical examination is alerting the patient before physically touching the perineum.

Documentation of the physical examination should include the appearance of the perineum, specifically the condition of the skin; presence of erythema; presence of abnormal pigmentation, papular lesions, or masses; and potential perianal soiling, which can indicate compromised continence. It is important to examine the perineum anterior to the anus including the intertriginous folds between the perineum and thighs. In the presence of induration, an inflamed mass, or an external opening, the location should be documented as "posterior,"

"anterior," "right," and/or "left." Describing lesions using a clock face can be very confusing in subsequent examinations, during which the patient may be in prone, supine, or lateral positions.

After visualization, the anus and perineum should be tested for neuromuscular function *if the patient is not excessively tender*. The presence of an intact sacral spinous pathway is documented by the presence of an "anal wink" with light touch (after warning the patient). Previously noted erythema, induration, or external fistula opening should be palpated for the presence of a firm cord of tissue that can help define a fistulous tract. The gluteal muscles should be distracted to examine for the presence of an anterior or posterior midline fissure. This maneuver also helps identify whether the patient is too tender to tolerate the insertion of a finger into the anal canal or digitation. If the patient is unable to tolerate digitation, 2% viscous lidocaine can be applied or the examination should be conducted in a setting that permits sedation or general anesthesia. If the patient is able to tolerate digitation, a finger should be gently and slowly inserted into the anal canal after verbally warning the patient that his will happen. After noting baseline

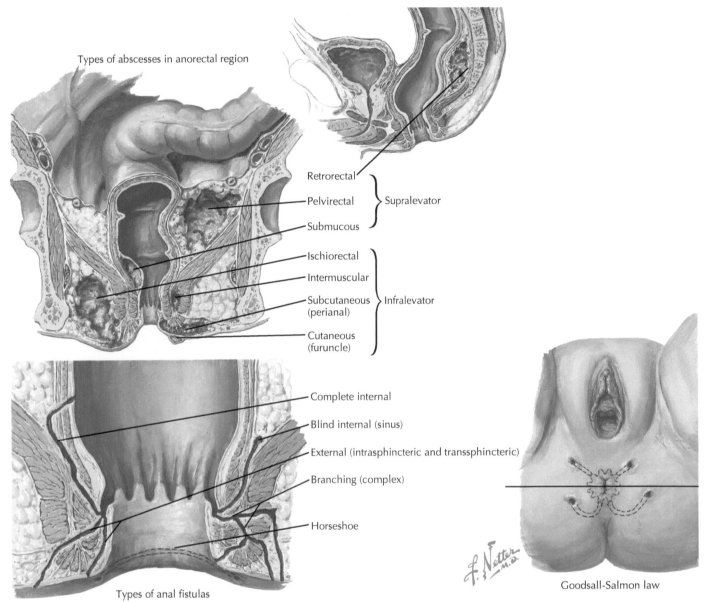

Types of abscesses in anorectal region

Retrorectal ⎫
Pelvirectal ⎬ Supralevator
Submucous ⎭

Ischiorectal ⎫
Intermuscular ⎪
Subcutaneous ⎬ Infralevator
(perianal) ⎪
Cutaneous ⎭
(furuncle)

Complete internal

Blind internal (sinus)

External (intrasphincteric and transsphincteric)

Branching (complex)

Horseshoe

Types of anal fistulas

Goodsall-Salmon law

**Fig. 49.2** Proctologic conditions: anorectal abscess and fistula.

anal sphincter tone, the examiner should request that the patient squeeze then relax the anal sphincter. This permits the assessment of anal sphincter function; it is particularly important to note a diminished squeeze in the medical record. The patient should also be asked to bear down and then relax. This request may be confusing and it can be helpful to repeat and explain. The puborectalis muscle will feel like a thick band posterior and just cephalad to the anal sphincter. The puborectalis muscle should relax when the patient bears down. If the puborectalis muscle tightens instead, the patient may have paradoxic puborectalis function, thus substantially increasing anal canal pressures during defecation and mechanically promoting cryptoglandular infection. Finally, the distal rectum should be palpated to search for evidence of a supralevator fluctuance or tenderness that would require intraoperative drainage.

## ABSCESS LOCATION AND MANAGEMENT

The anal glands often extend at least into the space between the internal and external sphincter muscles, and therefore abscesses commonly

originate in the intersphincteric space. As the infection spreads along fascial planes and potential spaces, abscesses and their management are defined by their location. In order of frequency, these abscesses are perianal, ischioanal, intersphincteric, and supralevator (Fig. 49.2). Anorectal abscesses that have not spontaneously ruptured should be incised and drained as soon as possible. Antibiotics are rarely indicated except in the presence of cellulitis or immune compromise. If widespread induration or crepitance is present, the patient will require urgent operative debridement. A plain radiograph can help clarify the presence of gas in the tissue planes. Necrotizing perineal fasciitis or Fournier gangrene is a surgical emergency that will be addressed more fully in the chapter describing necrotizing fasciitis.

Perianal abscesses are usually small and often can be drained with the patient under local anesthesia in the clinic or emergency room (Fig. 49.3). An abscess can be treated with an elliptical incision, a cruciate incision, or a Pezzer drain through a small skin opening. A linear incision alone is avoided to prevent premature healing of the external opening. The cavity should be curetted and irrigated. Packing of a perianal or perirectal abscess should be avoided, as it is painful and difficult

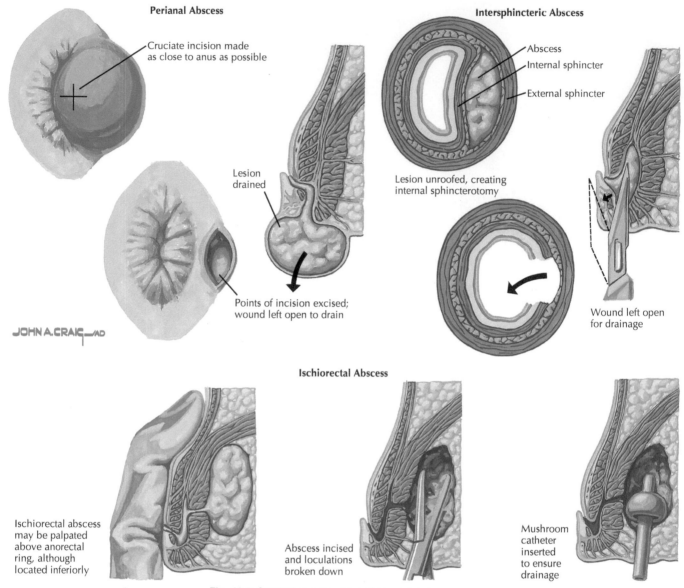

**Fig. 49.3** Surgical management of anorectal abscess.

for the patient to do at home and no better than sitz baths for wound care. If the elliptic incision appears to close with dependent positioning, it can be marsupialized with a running locked baseball stitch circumferentially using absorbable suture. Postoperative care includes pain control, warm sitz baths, a fiber supplement, and avoidance of constipation.

Unilateral ischioanal abscesses, located in the ischioanal fossa, should be incised and drained similarly to perianal abscesses. Bilateral ischioanal or "horseshoe abscesses" originate in the deep postanal space with extension to the bilateral ischioanal spaces. A horseshoe abscess should be drained through the deep postanal space by incising the skin longitudinally between the tip of the coccyx and the anus to expose the anococcygeal ligament (Fig. 49.4). The anococcygeal ligament is incised longitudinally, and the deep postanal space is opened. After the abscess cavity is drained, a counterincision is made on one or both limbs of the ischioanal space, and a loose seton (preferably of soft Silastic tubing) can be placed.

Intersphincteric abscesses differ from perianal and ischioanal abscesses in several important ways. Visualization is less useful because evidence of induration or swelling is absent. Anorectal pain is often

so severe that digitation of the anal canal in an awake patient is not feasible. The patient will experience deep tenderness most notably in the posterior deep anal space, and an indurated mass can be palpated cephalad to the dentate line. Intersphincteric abscesses must be drained by incising the internal sphincter muscle. Incision of 25% to 50% of the internal anal sphincter muscle rarely results in changes to bowel control. Among male and many younger female patients, even incision of 100% of the internal anal sphincter is unlikely to result in noticeable changes.

Supralevator abscesses are uncommon and originate in several ways—via cephalad extension of an intersphincteric abscess, cephalad extension of an ischioanal abscess, or caudad extension of an intra-abdominal process such as Crohn's disease. The origin of the abscess dictates the treatment plan. Incision and drainage of the abscess into the incorrect space may result in the formation of an iatrogenic fistula. Therefore an appropriate workup is key, beginning with digitation that reveals a tender mass above the level of the anorectal ring and followed by imaging with computed tomography, magnetic resonance imaging (MRI), or less commonly endorectal ultrasound. A supralevator abscess that is secondary to upward extension of an intersphincteric

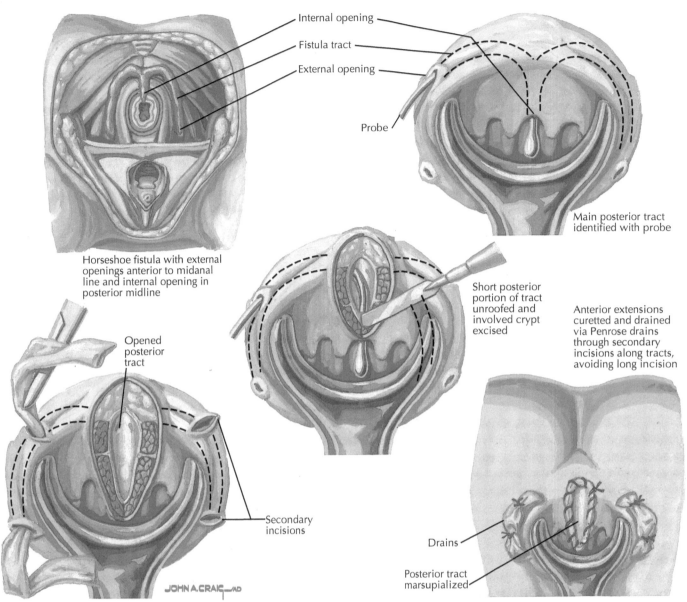

Internal opening

Fistula tract

External opening

Probe

Main posterior tract
identified with probe

Horseshoe fistula with external
openings anterior to midanal
line and internal opening in
posterior midline

Short posterior
portion of tract
unroofed and
involved crypt
excised

Anterior extensions
curetted and drained
via Penrose drains
through secondary
incisions along tracts,
avoiding long incision

Opened
posterior
tract

Secondary
incisions

Drains

Posterior tract
marsupialized

JOHN A. CRAIG—MD

**Fig. 49.4** Anorectal disorders: surgical management of horseshoe fistula.

abscess should be drained into the rectum. An abscess that is an upward extension of an ischiorectal abscess should be drained through the ischioanal fossa. If the abscess is secondary to an intra-abdominal disease such as neoplastic or Crohn's disease, the supralevator abscess is drained into the rectum, through the ischioanal fossa, or through the abdominal wall with concurrent treatment of the primary disease.

## FISTULA IN ANO ETIOLOGY AND EVALUATION

In multiple studies conducted in separate settings, anorectal abscesses recur or evolve into fistulas in approximately 30% to 50% of cases. By inference, therefore, the internal opening representing the original source of the problem spontaneously heals in at least 50% of cases. When the internal opening does not heal, either the abscess drainage wound will not resolve or a new abscess will develop after apparent complete healing and an asymptomatic interval. Most frustrating to patients is cyclical wound healing, reaccumulation of the pus, and reopening of the wound with new drainage.

Appropriate evaluation and treatment of a fistula in ano are predicated on correct identification of the internal opening and the fistula tract. Evaluation begins with a careful history, as described earlier. Several disorders may mimic fistula in ano and must be considered in the differential diagnosis. Hidradenitis suppurativa is differentiated by the presence of multiple perianal skin openings with surrounding leatherlike skin or with no obvious association with the anal canal. A careful history may reveal previous intertriginous infections in the groin folds or axilla. A pilonidal sinus with perianal extension can be identified by prior history or by the presence of midline pilonidal pits in the gluteal cleft cephalad to the point of the coccyx. IBD, specifically Crohn's disease, should be considered and ruled out as noted previously. Diverticulitis of the sigmoid colon with fistulization to the perineum is exceedingly rare. Aggressive anal condylomata and low rectal and anal canal carcinomas also may manifest as fistula in ano.

On physical examination, in addition to the maneuvers described for assessing anorectal abscess, the provider may note that the external opening of the fistula is an erythematous area of granulation tissue with purulence expressed during compression. The track may be palpable as an indurated cord. Identification of the internal opening during physical examination can be challenging. A useful but not foolproof rule of thumb is Goodsall's rule that external openings anterior to the anus will track radially to an internal opening at the dentate line, whereas external openings posterior to the anus will track to the posterior midline at the dentate line. In the clinic setting and if the patient can tolerate it, anoscopy may help clarify the location. In the operating room, a fistula probe can be inserted into the external opening and passed along the tract. It is not always possible to pass the probe, and *it is paramount to avoid creating a false passage or an artificial internal opening during this process.* The suspected internal opening should be confirmed with an injection of hydrogen peroxide through an angiocatheter into the external opening. Concurrent endorectal ultrasound can be extremely effective in demonstrating the tract and internal opening. If the internal opening remains elusive in spite of these maneuvers, the external opening should be extended, vigorously curettaged, and reflushed with peroxide. The external opening heals quickly and well in most patients, so opening it further generally should not present a problem, except in the case of underlying Crohn's disease, leukemia, or lymphoma. If the internal opening still is not apparent, the preferential imaging study is MRI.

Similar to anorectal abscess, the four main forms of fistula in ano are defined by the relation of the fistula tract to the sphincter muscles (see Fig. 49.2). An intersphincteric fistula tract is in the intersphincteric plane. The external opening usually is in the perianal skin close to the anal verge. A transsphincteric fistula tract crosses both the internal and the external sphincter and opens externally from the ischioanal fossa. A horseshoe fistula is a specific form of transsphincteric abscess, which starts in the posterior midline anal canal, traverses into the deep postanal space and from there travels through one or both ischiorectal fossae, and erodes toward the skin uni- or bilaterally. Suprasphincteric fistulas start in the intersphincteric plane proceeding cephalad laterally and superiorly to the puborectalis muscle, then caudad between the puborectalis and levator ani muscles into the ischioanal fossa. Extrasphincteric fistula tracts can traverse in either direction from the perineal skin through the ischioanal fossa and levator ani muscle to penetrate the rectal wall.

## FISTULA IN ANO MANAGEMENT

Management of fistula in ano is based on eliminating the source of infection (the internal opening) and establishing drainage without compromising anal continence. When the fistula does not much of the external sphincter, the simplest means to accomplish this is fistulotomy, wherein a fistula probe is placed through the tract and electrocautery is used to open the overlying tissues down to the level of the probe. Epithelial granulation tissue can be curetted, and if necessary, the unroofed fistula can be marsupialized. Failure to open the entire track may lead to fistula recurrence. Fistulectomy or excision of the entire fistulous track confers no healing benefit over fistulotomy, results in greater pain and higher rates of anal incontinence, and should be relegated to historical accounts only unless a biopsy of suspicious tissue is intended.

When the fistula extends more deeply than the subcutaneous external anal sphincter, a simple fistulotomy may increase the risk to anal continence. Although transection of the posterior external sphincter muscles can be performed, older patients and especially women with anterior fistulas may be prone to reduced anal continence postoperatively. In lieu of transecting the external anal sphincter, drainage can be accomplished with the placement of a cord of nonreactive material or seton in the fistula tract (Fig. 49.5). The seton will facilitate drainage of the infected space by stenting the external opening. Over a 6- to 8-week interval, the seton permits the resolution of infection and narrowing of the tract. Historically, tight or cutting setons were used to slowly erode through the sphincter and create fibrosis. Cutting setons are associated with substantial pain and loss of anal continence. At this time, their use is strongly discouraged.

After adequate drainage with the use of a seton and resolution of the infection, if fistulotomy is not an option, then the internal opening may be closed with absorbable suture and a collagen plug may be placed in the fistula tract. Although a collagen plug fails in more than 40% of cases by 6 months postoperatively, it has several distinct advantages over other treatment options. It is associated with few to no side effects, can be applied multiple times, and can be used in patients with complex fistulas and few other options.

In patients in whom less invasive therapy fails, other accepted surgical options include the Ligation of Intersphincteric Tract (LIFT) procedure or anal advancement flap. The LIFT procedure, first described in 2007, functions to close the internal opening and remove the fistula tract. With a probe in the fistula tract, the surgeon creates a transverse incision at the intersphincteric groove, identifies the intersphincteric tract, ligates the tract with absorbable suture close to the internal opening (after removing the probe), divides the tract, and curettes the remaining intersphincteric tract. The defect at the external sphincter muscle is loosely closed with an absorbable suture. Consistent with many fistula solutions, the reported success of the LIFT procedure has declined with time from its first description, but studies still cite a 40% to 80% success rate.

An anal advancement flap is a slightly more complex interventional option and can be combined with the placement of a plug. With the patient in an appropriate position, the fistula tract is identified. A U-shaped flap of mucosa is created that includes and extends just beyond the internal fistula opening. The distal strip of the flap is excised, removing the internal opening. The flap is then advanced to cover the tract and sutured in place. Owing to the disruption of tissue planes, an anal advancement flap has the best chance of success with its first attempt.

## ANAL FISTULA ASSOCIATED WITH CROHN'S DISEASE

The presence of Crohn's disease should be suspected in cases of multiple fistula recurrence, extremely delayed perineal healing, or the presence of multiple tracts originating in the upper anal canal, lower rectum, or even the abdomen. Aggressive medical therapy is encouraged. Metronidazole and ciprofloxacin may be useful. Infliximab, a monoclonal antibody directed against tumor necrosis factor, has shown some efficacy in reducing the fistula burden. Aggressive surgical intervention is discouraged because of extremely poor healing and risk to continence. The most judicious intervention is local curettage, drains, and setons, laying open only superficial tracts, and opening abscess cavities. In some hands, anorectal advancement flaps have met with success and should be considered during Crohn's quiescence. Long-term indwelling setons are often the best choice for this patient population.

### Rectovaginal Fistula

A rectovaginal fistula is a communication between the anterior wall of the anal canal or rectum and the posterior wall of the vagina. Rectovaginal fistulas are classified as low if a perineal approach to repair is possible and high if a repair can be accomplished only transabdominally. For low or small fistulas, the most common complaint is the passage of gas per the vagina. On digital examination of the anal

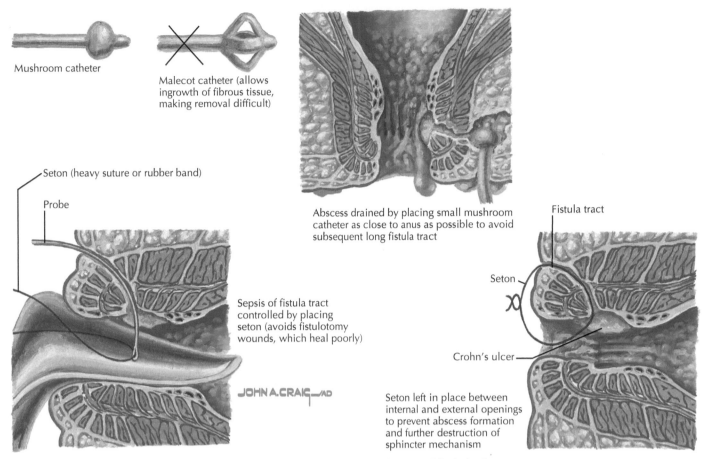

Mushroom catheter

Malecot catheter (allows ingrowth of fibrous tissue, making removal difficult)

Abscess drained by placing small mushroom catheter as close to anus as possible to avoid subsequent long fistula tract

Seton (heavy suture or rubber band)

Probe

Sepsis of fistula tract controlled by placing seton (avoids fistulotomy wounds, which heal poorly)

JOHN A. CRAIG—AD

Fistula tract

Seton

Crohn's ulcer

Seton left in place between internal and external openings to prevent abscess formation and further destruction of sphincter mechanism

Fig. 49.5 Appearance and management of anorectal Crohn's disease.

canal, an anterior scar or defect may be present. Bimanual examination of the rectum and the vagina can also detect the internal opening in the anal canal. Obstetric injury formerly accounted for most cases, but this has declined in recent years. For a low, simple fistula that does not heal spontaneously, an anorectal advancement flap yields the best result after the inflammation has subsided—generally 3 to 6 months postpartum. If the rectovaginal fistula is associated with anal incontinence, an anal sphincteroplasty is also performed. In experienced hands, endorectal advancement flap treatment for simple rectovaginal fistula has shown 83% primary healing. For recurrent rectovaginal fistulas, a tissue interposition graft may facilitate complete healing.

For large fistulas, the most common complaint is vaginal discharge with fecal odor and painful vaginitis. Patients with middle or high fistulas should be assessed with proctoscopy. If a fistula is suspected but cannot be demonstrated, a tampon can be placed in the vaginal canal and 100 mL of diluted methylene blue instilled into the anorectum. The tampon is then removed and checked for evidence of blue staining. Barium enema is rarely helpful but may be indicated in patients with IBD or previous radiation to the pelvis. For midlevel or high rectovaginal fistulas, simple fistulas with healthy adjacent tissue are repaired by transabdominal mobilization of the rectovaginal septum, division of the fistula, and layer closure of the rectal defect. If the local tissues have been damaged by irradiation, infection, or inflammatory diseases, an extended low anterior resection with coloanal anastomosis should be considered.

Although a diverting colostomy is unnecessary for most simple rectovaginal fistulas, a preliminary colostomy should be considered for complex rectovaginal fistulas. For elderly or unfit patients,

radiation-induced fistulas, and rectovaginal fistulas associated with Crohn's disease, a permanent colostomy may be the procedure of choice.

## SUMMARY AND FUTURE EFFORTS

In summary, anorectal abscess and fistula disease generally originate with anal cryptoglandular disease. With appropriate drainage, most abscesses are cured. However, a substantial minority will recur or redevelop as fistulas, indicating that the original insult or internal opening never resolved. Treatment is focused on drainage, obliteration of the internal opening when feasible, and avoidance of anal sphincter compromise. Understanding the relationship of the fistula tract to the pelvic anatomy is crucial to appropriate treatment planning. Stepwise increase in the aggressiveness of operative treatment is prudent, especially in patients with Crohn's disease. After all, as noted by J. Alexander-Williams (1976), "fecal incontinence … is the result of aggressive surgeons and not of progressive disease."

## EVIDENCE

Anan M, Emile SH, Elgendy H, Shalaby M, Elshobaky A, Abdel-Razik MA, Elbaz SA, Farid M. Fistulotomy with or without marsupialisation of wound edges in treatment of simple anal fistula: a randomised controlled trial. *Ann R Coll Surg Engl* 2019;101(7):472-478. *This randomized trial compared healing rates among patients with uncomplicated anal fistula treated with simple fistulotomy versus fistulotomy with marsupialization. The latter group achieved significantly more rapid complete healing (5.1 vs. 6.7 weeks; P < 0.0001).*

Buchanan GN, Halligan S, Bartram CI, et al. Clinical examination, endosonography, and MR imaging in preoperative assessment of fistula in ano: comparison with outcome-based reference standard. *Radiology* 2004;233:674-681. *This study compared the effectiveness of fistula tract and internal opening identification using digital examination (61% correct), endoanal ultrasound (81% correct), and MRI (90% correct).*

Eitan A, Koliada M, Bickel A. The use of the loose seton technique as a definitive treatment for recurrent and persistent high trans-sphincteric anal fistulas: a long-term outcome. *J Gastrointest Surg* 2009;13:1116-1119. *This study examined 97 Israeli patients with fistulas and followed for a mean duration of 5.1 years. Those treated with the "loose seton technique" experienced superior healing over time and minimal incontinence compared with those who underwent incision of the external anal sphincter.*

Emile SH, Khan SM, Adejumo A, Koroye O. Ligation of intersphincteric fistula tract (LIFT) in treatment of anal fistula: an updated systematic review, meta-analysis, and meta-regression of the predictors of failure. *Surgery* 2020;167(2):484-492. *This review combined subjects from 26 studies and identified a mean success rate for the LIFT procedure of 76% over mean follow-up of 16 months. Failure was associated with horseshoe fistulas, Crohn's disease, and previous fistula surgery.*

Hamadani A, Haigh PI, Liu IL, Abbas MA. Who is at risk for developing chronic anal fistula or recurrent anal sepsis after initial perianal abscess? *Dis Colon Rectum* 2009;52:217-221. *This retrospective review examined patients with an incident anorectal abscess treated in a US managed care organization and then followed for a 12-year period to identify risk factors associated with recurrence.*

Han JG, Wang ZJ, Zheng Y, Chen CW, Wang XQ, Che XM, Song WL, Cui JJ. Ligation of intersphincteric fistula tract vs ligation of the intersphincteric fistula tract plus a bioprosthetic anal fistula plug procedure in patients with transsphincteric anal fistula: early results of a multicenter prospective randomized trial. *Ann Surg* 2016;264(6):917-922. *This multicenter prospective study in China randomized patients with transsphincteric fistula to a LIFT procedure versus LIFT procedure with a collagen plug. The LIFT-plug group experienced significantly more rapid healing but no other differences. Neither group reported incontinence or recurrence over a 6-month period.*

Lowry AC, Thorson AG, Rothenberger DA, Goldberg SM. Repair of simple rectovaginal fistulas. Influence of previous repairs. *Dis Colon Rectum* 1988;31:676-678. *This retrospective cohort study examined the relative success of simple anorectal advancement flaps among first-time and recurrent fistula patients. The rate of success was 88% for first-time repairs, 85% for second repairs, and 55% for third-time repairs. The authors recommend a muscle interposition graft for third-time anorectal fistula repairs with an advancement flap.*

Ortiz H, Marzo J, Ciga MA, et al. Randomized clinical trial of anal fistula plug versus endorectal advancement flap for the treatment of high cryptoglandular fistula in ano. *Br J Surg* 2009;96:608-612. *This study randomized 31 patients with fistula plug versus endorectal advancement flap. At 1-year follow-up, 80% of the patients treated with a plug had experienced recurrence, whereas only 13% of patients treated with a flap had done so.*

Present D, Rutgeerts P, Targan S, et al. Infliximab for the treatment of fistulas in patients with Crohn's disease. *N Engl J Med* 1999;340:1398-1405. *This important trial randomized patients with Crohn's disease and anorectal fistulas to receive placebo, low dose infliximab, or high dose infliximab at 0, 2, and 6 weeks. At a median of 3 months follow-up, 68% of patients on low-dose infliximab and 56% of those on high-dose infliximab had achieved a reduction of >50% in the number of draining fistulas, compared with 26% of patients in the placebo group (P = .002 and P = .02). Moreover, 55% of patients on low-dose infliximab and 38% of those on high-dose infliximab had closure of all fistulas, compared with 13% of the patients on placebo (P = .001 and P = .04).*

Rojanasakul A, Pattanaarun J, Sahakitrungruang C, et al. Total anal sphincter saving technique for fistula-in-ano; the ligation of intersphincteric fistula tract. *J Med Assoc Thai* 2007;90:581-586. *This was the first published description of the LIFT procedure among 17 patients who reported 94% successful healing over 4 weeks. Although the study was not rigorous, the novel procedure which has held up with time makes it an important contribution.*

Sonoda T, Hull T, Piedmonte MR, Fazio VW. Outcomes of primary repair of anorectal and rectovaginal fistulas using the endorectal advancement flap. *Dis Colon Rectum* 2002;45:1622-1628. *This retrospective cohort of more than 100 patients who underwent anorectal advancement flap as a fistula repair sought risk factors for flap failure after following for a median of 17 months. The two factors statistically significantly associated with failure were a fistula associated with Crohn's disease (P = .027) and the presence of a rectovaginal fistula (P = .002).*

## ADDITIONAL RESOURCES

Alexander-Williams J. Fistula-in-ano: management of Crohn's fistula. *Dis Colon Rectum* 1976;19:518-519. *This is a classic how-to text describing judgment and technical issues in anorectal fistula repair among Crohn's disease patients.*

Hanley PH, Ray JE, Pennington EE, Grablowsky OM. Fistula-in-ano: a ten-year follow-up study of horseshoe-abscess fistula-in-ano. *Dis Colon Rectum* 1976;19:507-515. *This article described 10-year follow-up of a then-novel method for treating horseshoe fistulas with setons rather than incision, thus preserving anal continence.*

Ritchie RD, Sackier JM, Hodde JP. Incontinence rates after cutting seton treatment for anal fistula. *Colorectal Dis* 2009;11:564-571. *This review of the literature identified an average anal incontinence rate of 12% after the use of a cutting seton for anal fistulas.*

# 50

# Peritonitis

*Graham W. McLaren, Derek T. Tessman, Robert G. Sawyer*

 **ABSTRACT**

The peritoneum is the thin lining separating the intra-abdominal and extra-abdominal spaces. Inflammation of this lining is termed *peritonitis*. This inflammation alone is nonspecific and can be caused by many different etiologies. Peritonitis can be subdivided into three main categories: primary, secondary, and tertiary. Primary peritonitis is the inflammation of the peritoneum caused by the translocation of bacteria into the abdominal cavity. This condition is also known as spontaneous bacterial peritonitis. Primary peritonitis is typically a nonsurgical process related to ascites or an indwelling peritoneal dialysis catheter and will not be explored deeply in this chapter. Secondary peritonitis is peritoneal inflammation that is able to be directly attributed to a specific source. Common examples include diverticulitis, appendicitis, Crohn's disease, anastomotic leak, and penetrating abdominal trauma. Tertiary peritonitis, also known as recurrent peritonitis, is the persistence of peritonitis greater than 48 hours after adequate, definitive surgical treatment of the initial insult.

 **CLINICAL VIGNETTE**

The patient is a 53-year-old Caucasian male with a past medical history of osteoarthritis for which he takes ibuprofen regularly. He presented to the emergency department with complaints of 48 hours of upper abdominal pain, nausea, and vomiting. The patient reports that he does have some coffee-ground emesis as well as darker stools. Vitals were taken in the emergency department and his blood pressure was 95/63, heart rate of 110, respiratory rate of 12, temperature of 101°F, and saturating and 98% on room air. The patient was bolused with 1 L of lactated Ringer's, and his pressures improved to 110/74 and his heart rate came down to 95. A computed tomography (CT) scan of the abdomen and pelvis revealed free air under the diaphragm, gastric wall thickening, and some free fluid around a suspected gastric perforation. General surgery was consulted. The patient was started on broad-spectrum antibiotics and antifungals with ceftriaxone, metronidazole, and fluconazole. IV fluids were continued, and the patient was taken to the operating room for a diagnostic laparoscopy. Intraoperatively, the patient was found to have a perforated gastric ulcer. Intra-abdominal contamination was limited to the left upper quadrant, and all gross contamination was removed from the abdomen. The ulcer edges were biopsied and sent to pathology—eventual results revealed ischemic necrosis along the edge of the ulcer and were negative for *Helicobacter pylori* and cancer. The ulcer was closed primarily and then covered with an omental patch. The remainder of the abdomen was inspected, and no other findings were noted. A nasogastric (NG) tube was placed, and positioning was confirmed intraoperatively.

Postoperatively, the patient was started on a proton pump inhibitor (PPI) and continued on antibiotics. The NG tube was kept to suction, and the patient maintained n.p.o. status. On postoperative day 2, an upper GI contrast study was performed, which showed no leak at the site of the ulcer. The patient's NG tube was removed, and the patient's diet was slowly advanced. The patient had return of bowel function, and his antibiotics were discontinued after 4 days. On postoperative day 4, the patient was discharged home with continued PPI therapy and instructions to avoid nonsteroidal antiinflammatory drugs (NSAIDs). The patient was scheduled for a follow-up esophagogastroduodenoscopy (EGD) in 4 to 6 weeks.

## GEOGRAPHIC DISTRIBUTION AND DISEASE BURDEN

### Secondary Peritonitis

Secondary peritonitis (an intra-abdominal infection caused by a breach in the gastrointestinal tract, another hollow viscus, or the abdominal wall) is one of the most commonly treated conditions by general surgeons. The overall prevalence of secondary bacterial peritonitis and intra-abdominal infection in an inpatient surgical service may be as high as 75%. There is little reason to believe that there is a significant geographic difference in the distribution of secondary peritonitis across the world. Reports from Europe, North America, and Asia appear to describe similar origins and pathogens associated with the pathology of this state. Some areas do have region-specific considerations such as bowel perforations related to typhoid or parasitic infections in areas where those diseases are prevalent.

### Tertiary Peritonitis

Tertiary peritonitis, defined loosely as a recurrent or persistent intra-abdominal infection after initial treatment, has become more common as modern technology allows the support of increasingly ill patients and is often characterized by a unique, healthcare-associated microbiology. In a study from 1998, the rate of tertiary peritonitis after secondary peritonitis was 74%, but more recently, depending on the definition and study type, 20% to 25% of patients with secondary peritonitis go on to develop tertiary peritonitis. Unlike secondary peritonitis, the geographic distribution of tertiary peritonitis is not even. Because tertiary peritonitis by definition results from the treatment of secondary peritonitis, it generally occurs in areas where the resources required for high-intensity medical care are found. In fact, tertiary peritonitis must be considered one of the surprise sequelae of the rise of critical care and is fairly uniquely a disease found in wealthy countries.

## RISK FACTORS

### Secondary Peritonitis

Since secondary peritonitis is, by definition, secondary to another disease process, the risk factors mirror those of the inciting process. Causes of secondary bacterial peritonitis include any inflammatory condition in the peritoneal cavity. This can range from the extension of localized inflammatory conditions (e.g., appendicitis, pancreatitis, cholecystitis, pelvic inflammatory disease, etc.) to perforation of hollow visceral organs such as the stomach, small bowel, or colon. It can also arise after blunt or penetrating abdominal trauma in which organ damage is sustained or the integrity of the abdominal cavity is violated. Additionally, inflammatory bowel diseases such as ulcerative colitis and Crohn's disease can lead to peritonitis. Finally, peritonitis can develop in any patient after undergoing an abdominal operation resulting in a subsequent organ space surgical site infection or anastomotic leak. There are many factors that can increase a patient's postoperative peritonitis risk, including older age, longer surgery, emergency/trauma surgery, additional procedures, comorbidities, blood loss, tumor location/burden, obesity, and hypoalbuminemia. In a study evaluating only abdominal trauma patients, multivariate logistic regression analysis implicated an abdominal trauma index score greater than 24, contamination, and admission to an intensive care unit (ICU) as independent predictors for the development of organ space surgical site infections. Fig. 50.1 illustrates many different causes of the acute abdomen.

## Tertiary Peritonitis

The major risk factors for the development of tertiary peritonitis include malnutrition, a high Acute Physiology and Chronic Health Evaluation (APACHE) II score, the presence of organisms resistant to antimicrobial therapy, organ system failure, pancreatic/small bowel source, drainage alone at initial intervention, and Gram-positive/fungal pathogens. Obviously, previous secondary peritonitis is a prerequisite for tertiary peritonitis and therefore cannot be truly considered a risk factor.

## CLINICAL FEATURES

### Secondary Peritonitis

A complex interplay between the patient's systemic response to contamination and the peritoneum's ability to contain the contamination ultimately determines the presentation. For example, the immunocompetent, healthy patient may rapidly contain contamination and manifest minimal systemic illness, resulting in a small, localized abscess, as is common with peri-appendiceal abscesses or diverticular abscesses. Another patient who is immunocompromised may display no systemic signs of disease and be found to have diffuse contamination with no apparent natural containment of the source, resulting in a rapid decline without intervention. Between these two extremes lies the classic patient with initial contamination resulting in minimal systemic manifestations that progress to diffuse peritonitis and systemic toxicity. Complete source control (the repair, resection, bypass or diversion of perforation, and removal of infected fluid and necrotic tissue), combined with appropriate antimicrobial therapy, will resolve the peritonitis. However, if source control is incomplete, this may result in abscess formation or tertiary peritonitis (Fig. 50.2).

In general, patients will report symptoms of progressive abdominal pain, often beginning acutely in a focal area and progressing to diffuse involvement. This sequence is different from that seen for appendicitis without perforation, in which abdominal discomfort begins throughout the abdomen and then localizes to the right lower quadrant. The disparity can be explained by the fact that appendicitis is not considered secondary peritonitis until rupture, at which time diffuse peritonitis and diffuse abdominal pain does occur. On examination, the immunocompetent patient with peritonitis will have exquisite tenderness with minimal pressure and possibly rebound tenderness and rigidity of the abdominal wall (often referred to as *involuntary guarding*). If the patient is examined early, the tenderness, similar to subjective pain, may be localized to a small area or region, but as the disease worsens and more peritoneum becomes inflamed, the tenderness will become more diffuse. In addition to the typical systemic signs of infection, patients may exhibit respiratory insufficiency, hepatic insufficiency, oliguria, ileus, or obstipation.

## Tertiary Peritonitis

Due to the typical critical nature of a patient with tertiary peritonitis, getting a complete history and physical exam can be challenging. These patients can often be in the ICU and on sedating medications. A possible need for mechanical ventilation can also complicate the information gathering process. Also, immunosuppression, which can be variably present in critical illness, can contribute to the lack of clinical findings. When signs and symptoms are present, they are the same as those of secondary peritonitis.

## DIAGNOSTIC APPROACH

### Secondary Peritonitis

Once peritonitis has been diagnosed, the exact cause must be further investigated and treated. A patient with peritonitis who is unstable or in shock and *does not* respond to resuscitation should go directly to the operating room for exploratory laparotomy or laparoscopy as a diagnostic study, as well as to achieve definitive source control. A similar patient that *does* respond to resuscitation may undergo specific studies to evaluate the source of peritonitis. This may allow for nonoperative management in certain limited situations.

Specific pain distributions both temporally and spatially can direct the experienced clinician toward the correct differential diagnosis. In general, however, the best initial radiographic examination is CT of the abdomen and pelvis. Whether or not intravenous, oral, and/or rectal contrast agents are used is directed by the initial examination and the differential diagnosis, and this decision can be aided by consultation with the radiologist. Intravenous contrast agents aid in the visualization of specific inflammatory conditions, such as an abscess or appendicitis. In addition, other signs of inflammation include stranding or localized inflammation of the mesentery, intramesenteric fat, or retroperitoneal fat. Fluid collections can be seen without contrast, but in the absence of findings such as air-fluid levels or layering, the nature of the collection may be difficult to determine. Oral and rectal contrast allow the clinician to evaluate the integrity of hollow viscera. Obstruction will be seen as an abrupt cutoff or transition of oral or rectal contrast, and perforation may show extravasation of contrast into the peritoneum.

An alternative examination is abdominal ultrasound. This may be the initial imaging study of choice in pregnant patients, young children, or patients with suspected gallbladder pathology. It is important, however, to have sufficiently narrowed the differential diagnosis prior to ultrasound in order to allow the technician to focus on a particular region, because generalized examinations of the abdomen frequently lack sensitivity. Gaseous bowel distention may prevent useful ultrasound visualization.

Recently, diagnostic laparoscopy has been evaluated as an initial diagnostic study in peritonitis. A Turkish study showed that laparoscopy rendered a definitive diagnosis in 93% of patients with peritonitis and was used to treat the underlying disease in 86% (2008). An unnecessary laparotomy was avoided in 17%, and the conversion rate was 14%. Many more studies have shown benefits for laparoscopy in the evaluation and treatment of mild peritonitis, but in most series, patients with severe hypotension or generalized peritonitis were excluded. In the ICU, bedside laparoscopy for the evaluation of peritonitis has been determined to be safe and feasible with high accuracy, though it is still rarely used. Finally, it has been theorized that open abdominal exploration may be the "second hit," leading to worse organ failure postoperatively in many patients. Animal studies appear to support this theory, but human studies are lacking. In the end, it is the

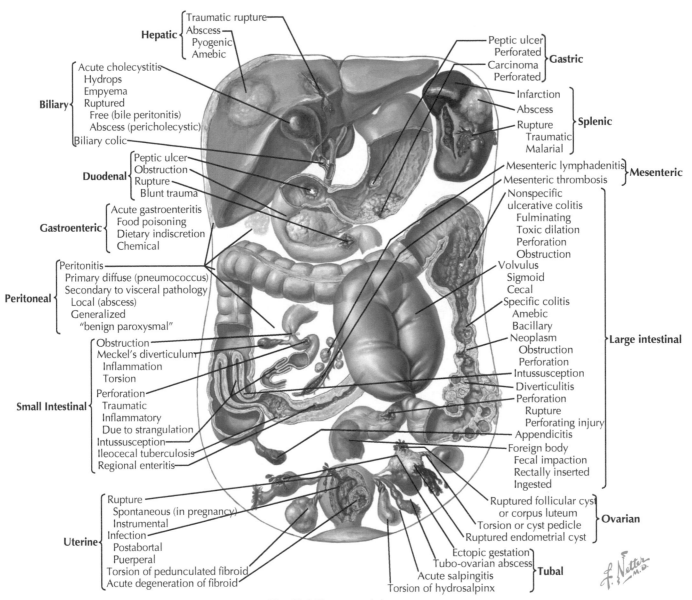

**Fig. 50.1** The acute abdomen.

stability of the patient and the experience of the surgeon that will dictate whether this is an option for the diagnosis and treatment of peritonitis.

Peritonitis stemming from most perforations will be caused by a fairly consistent set of pathogens, including aerobic Gram-negative and anaerobic organisms. Cultures should generally be obtained at the time of exploration or percutaneous drainage, but definitely in the setting of healthcare-associated infections. Isolation of yeast, particularly *Candida* species, is common, and certain risk factors have been associated with an increased likelihood of this, including upper gastrointestinal perforation, biliary or pancreatic perforation or necrosis, an immunosuppressed state, female gender, previous antimicrobial therapy for longer than 48 hours, and intraoperative cardiovascular failure.

### Tertiary Peritonitis

The clinical features of tertiary peritonitis are often subtle because many patients are in the ICU and sedated; therefore a high index of suspicion is necessary. Diagnosis is further clouded by concurrent

nosocomial infections, often in different stages of therapy. Although careful clinical examination is important, CT scanning is quite frequently necessary to make the diagnosis of tertiary peritonitis in these patients who, by definition, have already had at least one abdominal intervention for infection.

## CLINICAL MANAGEMENT AND DRUG TREATMENT

For all types of peritonitis, the management scheme is the same: fluid resuscitation and blood pressure support; appropriate, broad-spectrum antibiotic administration; early source control; and de-escalation of antibiotics after culture results and antibiotic sensitivities are known. Appropriate empirical therapy is usually based on the specific disease state or organ of origin (e.g., colonic source infections are initially treated with antimicrobials active against aerobic Gram-negative rods and anaerobes). Modifications of this principle should be based on local antibiograms and take into account local resistance rates as well as patient-specific factors, such as previous antibiotic usage, recent hospitalization or institutionalization, and previous resistance carrier

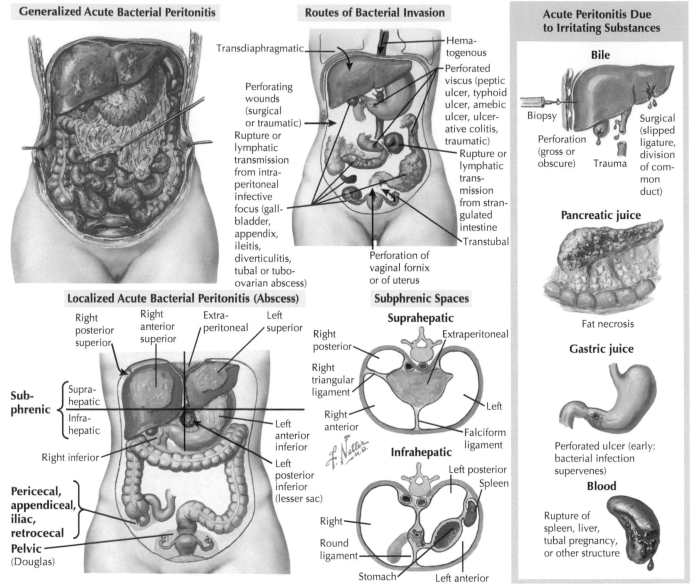

**Fig. 50.2** Acute peritonitis.

status. Newer studies are supporting the de-escalation and discontinuation of antibiotics shortly after definitive source control. One study from 2015 found that shorter duration antibiotic courses had similar outcomes to antibiotics tailored to the resolution of systemic symptoms (e.g., fever, leukocytosis, or ileus).

## Secondary Bacterial Peritonitis

Source control should be sought concurrently with antibacterial administration and consists of repair, resection, bypass or diversion of perforation, and removal of infected fluid and necrotic tissue. Blood cultures should be obtained in severely septic patients and empirical antibiotics started before surgical intervention. While surgical source control is being achieved, specimens should be obtained to guide specific antibiotic therapy. Empirical antibiotics should be based on the most likely organism, determined by the site of perforation, the patient's history of resistant pathogens, and local antibiograms. Of note, most cases of secondary peritonitis should be considered community-acquired intra-abdominal infections if the patent has received antimicrobials for less than 48 hours and did not have a prolonged stay in the hospital or other healthcare facility before the onset of secondary peritonitis. Tertiary peritonitis is always a healthcare-associated intra-abdominal infection and will be considered in the next section. For community-acquired infections, the location of the gastrointestinal perforation (stomach, duodenum, jejunum, ileum, appendix, or colon) guides the clinician toward the infecting flora. Established infection beyond the proximal small bowel is usually caused by facultative and aerobic Gram-negative organisms; infections beyond the proximal ileum frequently are caused by a variety of aerobic and anaerobic microorganisms. Many antibiotics are available and appropriate for empirical therapy, although none has been found to be superior to others. Guidelines from the Infectious Diseases Society of America (IDSA), the Surgical Infection Society, the American Society for Microbiology, and the Society of Infectious Diseases Pharmacists contain evidence-based recommendations for selection of antimicrobial therapy for adult patients with complicated intra-abdominal infections. Recommendations are based on whether infections are community acquired or healthcare associated and include options for monotherapy and combination therapy.

In the case of localized peritonitis that does not extend beyond the wall of the organ of origin, such as nonperforated appendicitis or cholecystitis with only focal spillage and containment, antibiotic treatment is basically prophylaxis for the surgical incision and is not needed after the operation. Likewise, a patient with gastric, duodenal, or proximal jejunal perforations without associated malignancy or acid suppression therapy and rapid attainment of source control (within 24 hours of the onset of symptoms) requires only prophylactic antibiotics. Also, if a traumatic bowel perforation is repaired within 12 hours of injury, then a 24-hour course of antibiotics is appropriate. In the 2015 STOP-IT trial, fixed duration (4 ± 1 day) course of antibiotics after source control was found to be comparable to antibiotic duration tailored to the resolution of fever, leukocytosis, or ileus and is becoming standard of care since it reduces the potential side effects of unnecessary antimicrobial exposure.

## Tertiary Peritonitis

Tertiary peritonitis is by definition a nosocomial or healthcare-associated infection and thus should be treated more aggressively with broader-spectrum antibiotics when antibiotics are required. Unfortunately, many times the presentation is insidious and often masked by other critical illness and underlying organ failure. Patients should be started on antibiotics that cover resistant pathogens common locally but also cover previously treated pathogens. In addition, reevaluation for missed sources of infection, such as another bowel leak or intra-abdominal abscess, should be performed. Again, the Guidelines from the IDSA, the Surgical Infection Society, the American Society for Microbiology, and the Society of Infectious Diseases Pharmacists should be reviewed for specific empirical choices, which are beyond the scope of this publication.

## PROGNOSIS

### Secondary Bacterial Peritonitis

Prognosis is tied directly to underlying disease process, but overall mortality is estimated to be 10% to 20%. Postoperative deep and organ or organ space infection showed an odds ratio (OR) of death of 2.6 to 4.5 in different studies. The key to achieving good outcomes, however, is most dependent on the time it takes to obtain adequate source control, which may require multiple procedures and modalities—for example, serial surgical procedures or surgical procedures in combination with percutaneous drainage.

### Tertiary Peritonitis

The mortality of tertiary peritonitis is approximately 30%—twice that of secondary peritonitis. Independent predictors of mortality in this population include increasing age (OR 1.06), increasing APACHE II score (OR 1.18), and four comorbidities: cerebrovascular disease (OR 4.3), malignant disease (OR 2.9), hemodialysis dependency (OR 3.8), and liver disease (OR 4.2). Compared with secondary peritonitis, however, tertiary peritonitis has not been found to be an independent predictor of mortality but rather a marker of overall disease severity.

## ACKNOWLEDGMENTS

The authors would like to acknowledge Dr. Christopher M. Watson for his work on the previous edition chapter.

## EVIDENCE

Ates M, Coban S, Sevli S, Terzi A. The efficacy of laparoscopic surgery in patients with peritonitis. *Surg Laparosc Endosc Percutan Tech* 2008;18:453-456. *A case series of 147 patients with acute abdomen who were explored laparoscopically; 86% were successfully managed without laparotomy.*

Bisclone FM, Cuoto RC, Pedrosa TM, Neto MC. Factors influencing the risk of surgical site infection following diagnostic exploration of the abdominal cavity. *J Infect* 2007;55:317-323. *A retrospective review of multiple institutions over a 13-year period found that laparoscopic surgery was less likely to be complicated by surgical site and organ or organ space infections.*

Blumetti J, Luu M, Sarosi G, et al. Surgical site infections after colorectal surgery: do risk factors vary depending on the type of infection considered? *Surgery* 2007;142:704-711. *A retrospective review of colon and rectum surgeries over a 4-year period found an overall incidence of surgical site infection to be 25%, with body mass index and creation, revision, or reversal of an ostomy to be risk factors for incisional infection and perioperative transfusion and with previous abdominal surgery to be risk factors for organ and organ space infections.*

Delgado-Rodríguez M, Gómez-Ortega A, Liorca J, et al. Nosocomial infection, indices of intrinsic infection risk, and in-hospital mortality in general surgery. *J Hosp Infect* 1999;41:203-211. *A single center looked prospectively at 1483 general surgery patients and found that both the Study on the Efficacy of Nosocomial Infection Control (SENIC) index and the National Nosocomial Infection Surveillance (NNIS) index are independent predictors of several sites of nosocomial infection and in-hospital death.*

DuPont H, Bourichon A, Paugam-Burtz C, et al. Can yeast isolation in peritoneal fluid be predicted in intensive care unit patients with peritonitis? *Crit Care Med* 2003;31:752-757. *On analysis of 278 ICU patients, the following four risk factors were found to be associated with the isolation of yeast in peritonitis: female gender, upper gastrointestinal origin, intraoperative cardiovascular failure, and receipt of antimicrobial agents for at least 48 hours before intervention.*

Evans HL, Raymond DP, Pelletier SJ, et al. Diagnosis of intra-abdominal infection in the critically ill patient. *Curr Opin Crit Care* 2001;7:117-121. *A discussion of peritonitis in critically ill patients, emphasizing the importance of radiologic and careful microbiological diagnosis.*

Evans HL, Raymond DP, Pelletier SJ, et al. Tertiary peritonitis (recurrent diffuse or localized disease) is not an independent predictor of mortality in surgical patients with intraabdominal infection, *Surg Infect (Larchmt)* 2001;2:255-263. *A comparison of 473 patients with secondary peritonitis with 129 patients with tertiary peritonitis demonstrating that after controlling for severity of illness, mortalities are similar.*

Garrouste Orgeas MG, Timsit JF, Soufir L, et al. Impact of adverse events on outcomes in intensive care unit patients. *Crit Care Med* 2008;36:2041-2047. *A prospective review of a database from 12 medical or surgical ICUs exploring the effect of adverse events on mortality.*

Haridas M, Malangoni MA. Predictive factors for surgical site infection in general surgery. *Surgery* 2008;144:496-501. *A single center's review of 10,253 general and vascular surgeries over a 6-year period found previous operation and hypoalbuminemia to be independent predictors of intra-abdominal infections.*

Jaramillo EJ, Trevino JM, Berghoff KR, Franklin ME Jr. Bedside diagnostic laparoscopy in the intensive care unit: a 13-year experience. *JSLS* 2006;10:155-159. *Describes 13 critically ill patients who underwent bedside diagnostic laparoscopy in the ICU; 30% had negative examination findings, whereas a diagnosis was confirmed in the rest, without complications.*

Morales CH, Villegas MI, Villavicencio R, et al. Intra-abdominal infection in patients with abdominal trauma. *Arch Surg* 2004;139:1278-1285. *A prospective evaluation of 768 blunt and penetrating trauma patients showed that abdominal trauma index score greater than 24, contamination of the abdominal cavity, and admission to the ICU were independent risk factors for the development of organ or organ space surgical site infection.*

Nathens AB, Rotstein OD, Marshall JC. Tertiary peritonitis: clinical features of a complex nosocomial infection. *World J Surg* 1998;22:158-163. *The first large study of tertiary peritonitis, noting the longer ICU length of stay and higher ICU mortality in patients with tertiary peritonitis compared with secondary peritonitis.*

Sawyer RG, Claridge JA, Nathens AB, Rotstein OD, Duane TM, Evans HL, Cook CH, O'Neill PJ, Mazuski JE, Askari R, Wilson MA, et al. Trial of short-course antimicrobial therapy for intraabdominal infection. *N Engl J Med* 2015;372(21):1996-2005. *A randomized control trial of 518 patients showing similar outcomes in patients with fixed, short-term antibiotic course (4 ± 1day) and variable longer antibiotic duration (8 days).*

## ADDITIONAL RESOURCES

Malangoni MA. Evaluation and management of tertiary peritonitis. *Am Surg* 2000;66:1571-1561. *A review of the risk factors, diagnosis, and management of tertiary peritonitis.*

Solomkin JS, Mazuski JE, Bradley JS, et al. Diagnosis and management of complicated intra-abdominal infection in adults and children: guidelines by the Surgical Infection Society and the Infectious Diseases Society of America. *Clin Infect Dis* 2010;50:1331-1364. *An exhaustive and recently updated review and series of recommendations for the management of intra-abdominal infections, including evidence-based recommendations for antimicrobial therapy.*

# 51

# Pyomyositis (Pyomyositis Tropicans)

*Alan F. Utria, E. Patchen Dellinger, James O. Park*

## ABSTRACT

Pyomyositis, a disease historically seen in tropical climates, is characterized by primary abscess formation in the skeletal musculature. It is increasing in incidence in temperate climates, especially in immunocompromised hosts. In all cases, *Staphylococcus aureus* is the most commonly implicated organism. Clinical presentation is nonspecific, with muscle pain, tenderness, and swelling, accompanied by leukocytosis and fever in cases with bacteremia. Computed tomography (CT) and magnetic resonance imaging (MRI) are the diagnostic modalities of choice. Treatment most often involves appropriate antimicrobial therapy coupled with abscess drainage.

## CLINICAL VIGNETTE

A 17-year-old male with no past medical history presents to the emergency department with 1 week of left thigh pain. His pain initially began 24 hours after he had completed a 10-mile run. He went to an urgent care 3 days ago, where he was diagnosed with a muscle strain and instructed to take nonsteroidal antiinflammatory medications, apply ice, and rest the leg. His pain has continued to worsen and he noted that the skin overlying his anterior thigh appears red. He denies any trauma to the thigh, fevers, or recent illnesses. On physical examination, his temperature was 37°C (98.6°F), pulse 105 beats/min, respirations 18/min, and blood pressure 120/70 mmHg. The skin overlying his anterior left thigh was erythematous and tender to palpation. There was no appreciable fluctuance. Point-of-care ultrasound showed a complex fluid collection measuring 5.5 × 4.2 × 2.1 cm in the rectus femoris muscle. He was found to have a slightly elevated white blood cell (WBC) count of 13,500/mm³, an elevated ESR of 22, and an elevated C-reactive protein (CRP) of 9. Blood cultures were sent and ultimately returned as negative. A surgical consult was placed and he was started on vancomycin for broad coverage of gram-positive organisms. The surgical team performed an ultrasound-guided aspiration of the collection and were able to obtain 9 mL of purulent fluid which was sent for culture. An MRI demonstrated hyperintensity on T2 in the right rectus femoris. On T1, a small rim enhancing area within the rectus femoris was present, which was concerning for residual abscess. The patient was admitted to the surgical team for observation. On hospital day 2, the patient's pain had improved and the erythema appeared to be resolving. Aspirate cultures returned as positive for methicillin-sensitive staphylococcus, and the patient was transitioned to oral trimethoprim-sulfamethoxazole and discharged to home with clinic follow-up scheduled in 2 weeks. In the clinic, the patient was doing well with no lingering sequelae. ESR, CRP, and WBC had all returned to normal. Antibiotics were discontinued after a 2 weeks total course.

## INTRODUCTION AND EPIDEMIOLOGY

Pyomyositis, also known as *tropical pyomyositis* because of its proclivity for warm climates, is an uncommon disease characterized by primary abscess formation in the skeletal musculature. First described in 1885, it is presumed to arise not from contiguous infections but by hematogenous seeding. Tropical pyomyositis can affect patients of all ages, with a predominance in children and young adults. It has been widely reported in Asia, Africa, and the Caribbean, and accounts for 1%~4% of all admissions in some tropical countries. In temperate climates, it occurs most commonly in children, as well as in patients with an immunodeficiency. As many as 75% of reported cases are in the immunocompromised, with the incidence of pyomyositis in those with human immunodeficiency virus (HIV) infection as high as 31%. However, reports of pyomyositis in immunocompetent hosts are emerging.

Predisposing factors include immunodeficiency, trauma, injection drug use, concurrent infection, and malnutrition. Many of these risk factors weaken host defenses, possibly because of underlying muscle damage and impaired local immunity.

## MICROBIOLOGY

Causative organisms of pyomyositis are most commonly *S. aureus* or group A streptococci, but other streptococcal groups, *Haemophilus influenzae*, *Aeromonas hydrophila*, *Bartonella* species, *Fusobacterium* species, anaerobes, and other enteric flora have also been implicated. *S. aureus* as the offending organism affecting a single muscle group is the most common presentation, though cases of disseminated infection have been reported. Recently, methicillin-resistant *S. aureus* (MRSA) isolates have been recognized as a cause of tropical pyomyositis. *S. aureus* produces virulence factors including enterotoxins, exfoliative toxins, and extracellular proteins that act against host defense mechanisms. Certain strains of MRSA, particularly strains in the community setting, have acquired the Panton-Valentine leukocidin virulence factor and demonstrate a proclivity for aggressive, disseminated infections.

## CLINICAL FEATURES

While pyomyositis typically presents in a single muscle, 12%~40% of patients present with multiple affected muscles. Larger muscle groups are more commonly affected with a predilection for the lower extremities. However, various case reports indicate that any skeletal muscle can be affected. Clinical features on presentation include muscle pain, stiffness, swelling, and tenderness. Fever can also be present, often in a cyclic pattern owing to bacteremia (Fig. 51.1). The combination of an often insidious onset and vague symptoms can make the diagnosis elusive.

Three stages of pyomyositis are described. After transient bacteremia, the first or *invasive* stage develops, with edematous and painful musculature caused by bacterial seeding. Aspiration is

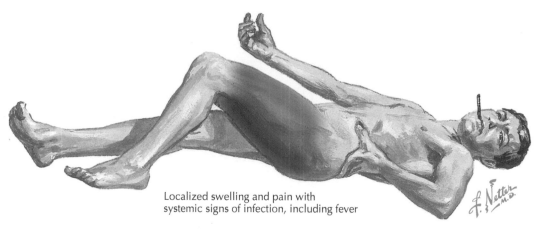

Localized swelling and pain with
systemic signs of infection, including fever

**Fig. 51.1** Clinical presentation of pyomyositis.

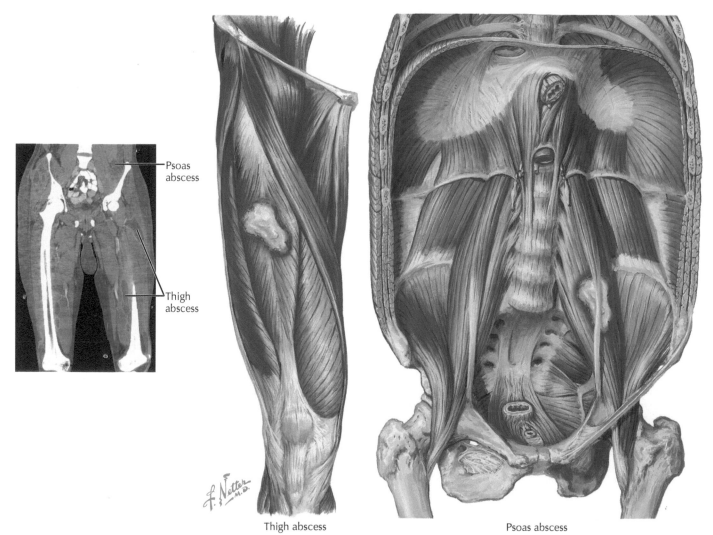

Thigh abscess                              Psoas abscess

**Fig. 51.2** Possible anatomic locations of abscesses in pyomyositis.

nondiagnostic, as no abscess has formed. The second or *suppurative* stage is characterized by abscess formation, and the majority of patients are in this stage at presentation (Fig. 51.2). In the final or *late* stage, septicemia ensues, with disseminated abscesses and resultant multiorgan dysfunction.

## DIAGNOSIS

Laboratory findings are usually nonspecific. Markers of inflammation such as CRP and erythrocyte sedimentation rate (ESR) may be elevated. Leukocytosis with a left shift can be seen in immunocompetent patients; however, this finding is not consistent in the first stage of the disease. Similarly, early on patients may only be transiently bacteremic and accordingly blood cultures are negative in 70%~90% of patients. When possible, abscess aspirate should be cultured prior to the initiation of antibiotics in order to determine a causative pathogen and guide antimicrobial therapy. For patients that have already been treated with antibiotics or where abscess cultures are negative, there is emerging data on the use of broad-range polymerase chain reaction (PCR) targeting 16S ribosomal DNA for bacterial identification.

Imaging plays an important role in the diagnosis and can guide treatment. Plain radiographs can be used to rule out other pathologies but will often miss the diagnosis of pyomyositis. Ultrasonography is a cost-effective screening tool that can expedite the diagnosis and treatment. It is of more use during the purulent stage where hypoechoic areas in the muscle and increased muscle bulk can be seen. MRI is the gold standard for detecting early pyomyositis and evaluating the extent of infection due to the higher signal intensity of involved muscles. If MRI is unavailable CT can be performed, however, it may fail to differentiate an abscess from inflamed muscle.

## TREATMENT

Though early-stage pyomyositis can be treated with antibiotics alone, later stages require treatment with appropriate long-term antibiotic therapy, coupled with surgical drainage of large abscesses. If feasible, image-guided percutaneous drainage should be the first approach, though surgical intervention may be required in cases with deep infection or extensive muscle involvement. Aspiration or drainage cultures should guide antimicrobial therapy, but empirical therapy should be initiated based on patient characteristics. In immunocompetent patients, empirical therapy against staphylococci, including methicillin-resistant *Staphylococcus aureus* (MRSA) and streptococci should be initiated. Broad gram-positive, gram-negative, and anaerobic coverage should be initiated in immunocompromised hosts. There is no consensus on the duration of antibiotic treatment, with reports describing both long-term parenteral therapy and early oral therapy with adequate drainage. However, it is generally accepted that the duration of antimicrobial therapy should be directed by clinical and radiographic improvement. Assessment of other sources for bacteremia should also be undertaken, and antimicrobial therapy should be modified accordingly. Persistent symptomatology warrants a renewed search and drainage of abscesses using interventional or surgical approaches.

With appropriate antimicrobial therapy and drainage of abscesses, most patients do very well after treatment for pyomyositis. There can be little residual deformity and minimal loss of function, even in cases with extensive muscle damage. Physical therapy and rehabilitation may be required in severe disseminated cases, though most function is fully recovered. Mortality rates from 1%~10% have been reported, with rare recurrence of infection, usually in immunocompromised hosts.

## ACKNOWLEDGMENTS

The authors would like to acknowledge Dr. Joseph F. Woodward for his contribution to this chapter in the previous edition.

## EVIDENCE

Agarwal V, Chauhan S, Gupta RK. Pyomyositis. *Neuroimaging Clinics* 2011;21(4):975-983. *A concise review of pyomyositis.*

Arizono T, Saito T, Matsuda S, et al. Pyomyositis in adults without any predisposing factors in a non-tropical region. *Orthopedics* 2005;28:324-326. *Case reports of two immunocompetent adults with pyomyositis.*

Chiedozi LC. Pyomyositis: review of 205 cases in 112 patients. *Am J Surg* 1979;137:255-259. *A study of pyomyositis, a pyogenic infection of the muscle, in 112 patients is presented, along with the clinical findings; criteria for diagnosis are suggested.*

Dumitrescu O, Tristan A, Meugnier H, et al. Polymorphism of the Staphylococcus aureus Panton-Valentine leukocidin genes and its possible link with the fitness of community-associated methicillin-resistant *S. aureus*. *J Infect Dis* 2008;198:792-794. *Virulence of community-acquired MRSA (CA-MRSA) associated with PVL gene.*

Evans JA, Ewald MB. Pyomyositis: a fatal case in a healthy teenager. *Pediatr Emerg Care* 2005;21:375-377. *Pyomyositis is a common disease in the tropics that is reported with increasing frequency in the United States. An unusually fulminant, fatal case in a previously healthy adolescent male is described, illustrates the clinical progression of pyomyositis from localized muscle infection to disseminated disease, and highlights the importance of considering this rare diagnosis in any stage of occult sepsis.*

Fowler A, Mackay A. Community-acquired methicillin-resistant Staphylococcus aureus pyomyositis in an intravenous drug user. *J Med Microbiol* 2006;55(Pt 1):123-125. *A case of CA-MRSA pyomyositis in an IDU is described; to the authors' knowledge, this is the first reported case of CA-MRSA pyomyositis in the United Kingdom.*

Gabas T, Podglajen I, Cheminet G, Gagnard J-C, Wyplosz B. Diagnostic accuracy of 16S rDNA PCR in bacterial pyomyositis: a prospective study. *J Infect* 2019;79(5):462-470. *A letter to the editor describing the ability to determine the causative pathogen in pyomyositis using 16s rDNA PCR.*

Gibson RK, Rosenthal SJ, Lukert BP. Pyomyositis: increasing recognition in temperate climates. *Am J Med* 1984;77:768-772. *Pyomyositis is common in the tropics yet is rarely reported in temperate climates. A woman in whom pyomyositis developed in a temperate climate is presented, and the 31 cases reported in the United States are reviewed.*

Hossain A, Reis ED, Soundararajan K, et al. Nontropical pyomyositis: analysis of eight patients in an urban center. *Am Surg* 2000;66:1064-1066. *Nontropical pyomyositis is rare and usually associated with HIV infection. Retrospective chart review assessing the manifestations and response to treatment of nontropical pyomyositis in an area with a high prevalence of HIV seropositivity.*

Lin MY, Rezai K, Schwartz DN. Septic pulmonary emboli and bacteremia associated with deep tissue infections caused by community-acquired methicillin-resistant Staphylococcus aureus. *J Clin Microbiol* 2008;46:1553-1555. *Case reports of four adult patients with septic pulmonary emboli and CA-MRSA bacteremia associated with deep tissue infections, such as pyomyositis, osteomyelitis, and prostatic abscess.*

Pannaraj PS, Hulten KG, Gonzalez BE, et al. Infective pyomyositis and myositis in children in the era of community-acquired, methicillin-resistant Staphylococcus aureus infection. *Clin Infect Dis* 2006;43:953-960. *Retrospective chart review of 45 children with bacterial pyomyositis. The number of cases increased between 2000 and 2005, primarily as a result of an increase in the prevalence of CA-MRSA. CA-MRSA is an increasing cause of pyomyositis and myositis in children.*

Ruiz ME, Yohannes S, Wladyka CG. Pyomyositis caused by methicillin-resistant Staphylococcus aureus. *N Engl J Med* 2005;352:1488-1489. *Case reports of four patients with pyomyositis caused by MRSA.*

Scriba J. Beitragzuraetiologie der Myositis acuta. *Dtsch Z Chir* 1885;22:497-502. *First description of pyomyositis in the literature.*

Wang CM, Chuang CH, Chiu CH. Community-acquired disseminated methicillin-resistant Staphylococcus aureus infection: case report and clinical implications. *Ann Trop Paediatr* 2005;25:53-57. *A 6-year-old girl with community-acquired disseminated infection caused by MRSA is described.*

Woodward JF, Sengupta DJ, Cookson BT, et al. Disseminated community-acquired USA300 methicillin-resistant Staphylococcus aureus pyomyositis and septic pulmonary emboli in an immunocompetent adult, *Surg Infect (Larchmt)* 2010;11:59-63. *Case report of an immunocompetent adult with CA-MRSA disseminated pyomyositis, illustrating the importance of aggressive surgical intervention.*

## ADDITIONAL RESOURCES

Bickels J, Ben-Sira L, Kessler A, Wientroub S. Primary pyomyositis. *J Bone Joint Surg Am* 2002;84:2277-2286. *Review of primary pyomyositis and stages of progression, diagnosis, and treatment.*

Chauhan S, Jain S, Varma S, Chauhan SS. Tropical pyomyositis (myositis tropicans): current perspective. *Postgrad Med J* 2004;80:267-270. *Overview of the history, diagnosis, and treatment of pyomyositis.*

Crum NF. Bacterial pyomyositis in the United States. *Am J Med* 2004;117:420-428. *The incidence of reported bacterial pyomyositis is increasing in the United States, especially among immunocompromised persons. This review summarizes all reported cases of pyomyositis among HIV-infected persons worldwide and HIV-negative persons in the United States since 1981.*

Roberts S, Chambers S. Diagnosis and management of *Staphylococcus aureus* infections of the skin and soft tissue. *Intern Med J* 2005;35(suppl 2):S97-S105. *Overview of S. aureus skin and soft-tissue infections.*

Small LN, Ross JJ. Tropical and temperate pyomyositis. *Infect Dis Clin North Am* 2005;19:981-989, x-xi. *This article discusses the pathogenesis, clinical presentation, diagnosis, and management of pyomyositis in the tropical and temperate settings.*

Struk DW, Munk PL, Lee MJ, et al. Imaging of soft tissue infections. *Radiol Clin North Am* 2001;39:277-303. *Description of common imaging techniques used to diagnose soft-tissue infections.*

# 52

# Surgical Site Infections

*E. Patchen Dellinger*

## ABSTRACT

Surgical site infections (SSIs), previously known as *wound infections*, remain one of the most common adverse events that occur with hospitalized surgical patients or after outpatient surgical procedures, despite many advances in preventive techniques. A 2015 report from the National Healthcare Safety Network (NHSN) of the Centers for Disease Control and Prevention (CDC) documents SSI as accounting for 16% of all healthcare-associated infections (HAIs) among hospitalized patients. Of 100,000 HAIs reported in 1 year, deaths followed SSI in 8000 cases. The incidence of SSI after a surgical procedure is highly variable, depending on the type of operation being done and the underlying risk factors of the patient, but the average across the United States is estimated to be 2% to 3% of all procedures. Length of stay and associated costs are dramatically increased when an SSI develops.

## CLINICAL VIGNETTE

A 54-year-old woman, otherwise healthy, underwent a low anterior resection through a lower midline incision with a stapled, end to end anastomosis (EEA) colorectal anastomosis for a suspicious polyp at a prior rectal cancer site. She had preoperative bowel prep with oral antibiotics and appropriate parenteral prophylactic antibiotics. Everything went well, and she had a smooth recovery and was discharged on the third postoperative day, tolerating a diet and with an intact incision.

Eleven days postop she called with concerns for feeling feverish and noting redness around the lower portion of her incision for 1½ inches. She was seen in the office the next day and found to have inflamed skin with some swelling around the lower portion of her incision, which was tender to palpation. Her temperature was 38°, with a heart rate of 90. Probing the wound revealed a 3 cm deep cavity, which extended 6 cm cephalad. The overlying skin was opened and dressed with gauze, and she was sent home with instructions for dressing changes.

Eleven days later, she was seen with no fever, wound clean with granulation tissue, and fascia intact.

Two weeks after the last visit her wound was clean and skin was approximated with Steri Strips.

A month later, she was seen with a well-healed incision without any inflammation.

## DEFINITIONS

SSSIs are divided into three categories according to the anatomic extent of the infection at the time of diagnosis: superficial incisional SSI, deep incisional SSI, or organ or organ space SSI (Table 52.1). A superficial incisional SSI occurs within 30 days after the operative procedure *and* involves only skin and subcutaneous tissue of the incision. In addition, the patient has at least one of the following:

- Purulent drainage from the superficial incision
- Organisms isolated from an aseptically obtained culture of fluid or tissue from the superficial incision

- At least one of the following signs or symptoms of infection: pain or tenderness, localized swelling, redness, or heat, *and* the superficial incision is deliberately opened by the surgeon and is culture positive if cultured *or* is not cultured (a culture-negative finding does not meet the definition for SSI)
- A diagnosis of superficial incisional infection that has been made by the surgeon or attending physician

There are two specific types of superficial SSI. A primary superficial SSI is one that occurs in the primary incision in a patient who has had an operation with one or more incisions. A secondary superficial SSI is one that occurs in the secondary incision in a patient who has had an operation with more than one incision. An example would be the donor site in the leg for a patient who has had a coronary artery bypass with a vein graft taken from the leg.

A deep incisional SSI is one that develops within 30 days of the operative procedure if no implant was left in place during the operation or within 90 days if an implant was left and the infection appears to be related to the operative procedure. By definition, a deep incisional SSI involves deep soft tissues (e.g., fascia or muscle layers of the incision) *and* the patient has purulent drainage *or* the deep incision spontaneously dehisces or is deliberately opened by the surgeon and is cultured or, if not cultured, the patient has fever (temperature >38°C) or localized pain or tenderness (a culture-negative finding does not meet the definition for SSI). As with superficial incisional SSI, deep incisional SSI can be either primary or secondary.

An organ or organ space SSI involves any part of the body, excluding the skin incision, fascia, or muscle layers, that is opened or manipulated during the operative procedure. When an organ or organ space SSI is reported, the specific anatomic site involved is also reported. Thus an intra-abdominal abscess, an empyema, a mediastinal infection, or a joint space infection after an operation at one of those sites would be reported as an organ or organ space SSI in the abdomen, chest, mediastinum, or joint space. By definition, an organ or organ space SSI occurs within 30 days of the operation if no implant was left or within 90 days if an implant was involved. Most organ or organ space SSIs are covered in other chapters of this book and so will not be dealt with further here.

## RISK FACTORS

The risk that an SSI will follow a surgical procedure depends on several characteristics of the operative procedure and has been well established for the degree of bacterial contamination expected during the procedure and for the duration of the procedure. Operations are divided into four categories of increasing risk of intraoperative contamination and thus the risk of subsequent SSI. Class 1 or clean wounds are incisions made through uninfected tissues without any evidence of inflammation and that do not enter any portion of the respiratory, alimentary,

| TABLE 52.1 | Summary of Surgical Site Infections | | |
|---|---|---|---|
| **Category** | **Onset** | **Characteristics** | **Criteria** |
| Superficial incisional SSI | Within 30 days of operation | Involves skin and subcutaneous tissue of the incision only | 1. Purulent drainage from the superficial incision *or*<br>2. Organisms isolated from an aseptically obtained culture of fluid or tissue from the superficial incision *or*<br>3. At least one of the following signs or symptoms of infection<br>• Pain or tenderness<br>• Localized swelling<br>• Redness<br>• Heat<br>*and*<br>The superficial incision is deliberately opened by the surgeon and is culture positive if cultured or is not cultured (a culture-negative finding does not meet the definition for SSI) *or*<br>4. A diagnosis of superficial incisional infection is made by the surgeon or attending physician |
| Deep incisional SSI | Within 30 days of operation if no implant left in place<br>Within 90 days of operation if implant left in place | Involves deep soft tissues (e.g., fascia or muscle layers of the incision) | 1. Purulent wound drainage *or*<br>2. Spontaneous wound dehiscence *or*<br>3. Surgeon deliberately opens the wound and obtains a positive culture sample *or*<br>4. Localized pain and tenderness |
| Organ or organ space SSI | Within 30 days of operation if no implant left in place<br>Within 90 days of operation if implant left in place | Involves any part of the body, excluding the skin incision, fascia, or muscle layers, that is opened or manipulated during the operative procedure | See chapters in this Surgical Infections Section for organ or organ space SSIs involving specific anatomic sites. |

*SSI,* Surgical site infections.

genital, or urinary tract. A clean wound must be closed primarily at the time of the operation.

A class 2 or clean-contaminated wound is one in which the respiratory, alimentary, genital, or urinary tract is entered under controlled conditions and without unusual contamination. This includes any operation involving the biliary tract, appendix, vagina, lungs, or oropharynx.

A class 3 or contaminated wound includes open, fresh, traumatic wounds; any operation that encounters acute, nonpurulent inflammation; operations involving gross spill from the gastrointestinal tract; or major breaks in sterile technique.

A class 4 or dirty wound is one that involves wounds with established infection, perforated viscera, or old traumatic wounds with retained devitalized tissue. This class implies that any postoperative infection is caused by organisms present at the time of the original operation.

Data from many sources demonstrate that operations that take longer to perform have a higher risk of infection. Whether this is because having the incision open longer increases risk or because operations that take longer to perform are intrinsically different from those that can be done more rapidly is not known. It is likely that both factors contribute. An operation may take longer because the tumor was larger, the adhesions were more troublesome, or bleeding was more difficult to stop, all of which could increase infection risk. The NHSN has traditionally divided cases by determining the 75th percentile for the duration of a particular type of procedure, and they assign an extra risk point for any operation that exceeds the 75th percentile. Data also demonstrate that risk is increased for patients with underlying illnesses that cause the anesthesiologist to assign a higher risk score. This score was developed by the American Society of Anesthesiologists (ASA) and is divided into five categories. ASA 1 refers to a normal healthy patient, ASA 2 indicates a patient with mild systemic disease, ASA 3 indicates a patient with severe systemic disease, ASA 4 indicates a patient with

severe systemic disease that is a constant threat to life, and ASA 5 indicates a moribund patient who is not expected to survive without the operation. NHSN assigns an additional risk point for a patient with ASA score of 3 or above. Currently, NHSN reports SSI rates at participating hospitals in terms of Standardized Infection Ratios (SIR) that express the rates as a ratio of the observed divided by the expected SSI rate. To calculate the SIR NHSN looks at data for specific procedure categories and applies between 1 and 11 specific risk factors.

Many other factors specific to the individual patient affect risk of SSI. These include advancing age, diabetes mellitus, high perioperative blood glucose levels independent of a diagnosis of diabetes, hypothermia during the operation, increased blood loss, perioperative transfusion, obesity, malnutrition, cigarette smoking, and preexisting distant site infections, among others. Data regarding the concentration of oxygen ($FiO_2$) administered during and immediately after the operative procedure have been conflicting, although animal data and observational data in limited numbers of humans have demonstrated a relationship between low levels of oxygen tension in the fresh incision and an increased risk for SSI. These levels are influenced not only by $FiO_2$ but also by blood volume, cardiac output, patient temperature, and vasoconstrictive medications.

Preparation for and the conduct of the operation also affect the risk for SSI. The use of good skin preparation with effective antiseptic and sterile instruments is basic. Glove perforations on the surgical team are associated with an increased risk for SSI and may be reduced by wearing double gloves. Surgical technique is widely regarded as an important risk factor, but there is no agreement on how to measure it. Shaving the operative site the day before the operation has been shown to increase risk. The administration of appropriate prophylactic antibiotics during the hour before the incision is well documented to reduce the risk of SSI in the range of 30% to 60%, and efforts to keep the patient warm and to prevent hyperglycemia also have demonstrated efficacy in reducing risk.

# DIAGNOSTIC APPROACH

The majority of SSIs become evident within 2 weeks of the operation, although a minority may be evident much later. It is rare for an SSI to be clinically evident during the first 3 to 5 days. An exception is SSI caused by beta-hemolytic streptococci, which can manifest clinically within 24 to 36 hours. However, these infections are extremely rare. Very obese patients may take considerably longer to manifest local signs of infection. It is common for a patient who has had a major abdominal or chest operation to manifest fever during the first few postoperative days. The majority of these fevers are not associated with any diagnosed infection. Early fever should be followed by an examination of the patient to rule out any obvious early infection. However, empirical antibiotic administration is not indicated for an otherwise well postoperative patient with a fever, and the majority of these fevers resolve without a specific diagnosis. Fever that occurs or persists after the fourth or fifth postoperative day has a much higher likelihood of representing a true infection and should trigger a more vigorous effort at diagnosis.

The physical appearance of the incision provides the most helpful information regarding a possible SSI. Local signs of erythema, swelling, and tenderness are usually present. Purulent drainage may occur spontaneously or only after the wound is opened. Flat erythema around an incision without induration or increased tenderness can occur during the first week and does not usually represent infection. If observed, this will usually resolve spontaneously without the administration of antibiotics. The erythema may have a local cause such as tape sensitivity or other noninfectious trauma. Considerable evidence demonstrates that antibiotics begun after a surgical procedure and continued for long periods after the procedure do not prevent or treat SSI but do increase the risk of *C. difficile* infection.

# TREATMENT

When the diagnosis of SSI is made, the incision must be opened for the evacuation of the infected material. This is the single most important element of treatment. Although it is common to administer an antimicrobial agent when an SSI is drained, there are actually no data supporting the necessity of this practice for superficial SSIs. Studies of the management of subcutaneous abscesses found no additional benefit over incision and drainage from the addition of antimicrobial agents, and the only study of SSI management that looked at this issue found no benefit to the addition of antibiotics. If evidence for an invasive aspect of the infection is absent with local erythema and induration less than 5 cm from the wound edges, and if the patient does not exhibit signs of a major systemic response (temperature <38.5°C, pulse <100, white blood cells [WBCs] <12,000), then it is quite reasonable to omit any antibiotics. If there is a marked local or systemic reaction, then the administration of antibiotics for a few days is reasonable, although unstudied. For clean procedures, the most common infecting organisms are skin flora, and a first-generation cephalosporin or penicillinase-resistant penicillin would be reasonable, as long as there is not a significant methicillin-resistant *Staphylococcus aureus* (MRSA) problem in the local institution and the patient has no risk factors for MRSA (recent hospitalization, previous MRSA, or recent antibiotics). If the operation was clean-contaminated or contaminated or involved incision in the axilla or perineum, then a drug choice that provides activity against coliforms and anaerobes would be indicated.

When the incision is opened it should be inspected for necrotic tissue and foreign bodies, which should be debrided or removed. Small bits of infected tissue will separate over time with dressing changes and do not need aggressive debridement. When an SSI is drained, it should

be examined shortly afterward by a surgeon familiar with the original operative procedure. If the original operation entered the abdominal or thoracic cavity, the integrity of the abdominal or chest wall closure should be confirmed. If these are involved, it is a deep incisional SSI, and particular care must be exercised to prevent evisceration or damage to underlying viscera. Typically a wound will be managed with a gauze dressing changed two or three times per day initially and then less often as the wound cleans up. Most SSIs are left open to heal by secondary intention. The dressings should not be packed tightly into the incision but should be placed so that all surfaces of the wound are in contact with the dressing. Dressings placed forcefully increase pain and retard wound healing.

# EVIDENCE

Bowater RJ, Stirling SA, Lilford RJ. Is antibiotic prophylaxis in surgery a generally effective intervention? Testing a generic hypothesis over a set of meta-analyses. *Ann Surg* 2009;249:551–556. *This meta-analysis analyzes the evidence that prophylactic antibiotics, when used appropriately, result in a reduction of the incidence of SSI for essentially all surgical procedures. Whether or not to use antibiotics before a specific procedure depends on an analysis of the relative cost and risk of SSI compared with the administration of an antibiotic rather than on an assessment of whether or not the antibiotic would lower the risk of SSI.*

Rajendran PM, Young D, Maurer T, et al. Randomized, double-blind, placebo-controlled trial of cephalexin for treatment of uncomplicated skin abscesses in a population at risk for community methicillin-resistant *Staphylococcus aureus* infection. *Antimicrob Agents Chemother* 2007;51:4044. *This nicely done, prospective, randomized, double-blind trial established that for subcutaneous abscesses that are, for all practical purposes, very similar to superficial incisional SSI, there is no added benefit to treating with antibiotics if adequate incision and drainage have been done.*

# ADDITIONAL RESOURCES

Bratzler DW, Dellinger EP, Olsen KM, Perl TM, Auwaerter PG, Bolon MK, et al. Clinical practice guidelines for antimicrobial prophylaxis in surgery. *Surgical Infections* 2013;14(1):73–156. *This article contains the joint recommendations of the Infectious Diseases Society of America, the Society for Healthcare Epidemiology of America, the American Society of Health-System Pharmacists, and the Surgical Infection Society for use of prophylactic antibiotics for surgical procedures and the recommended antibiotic choices.*

Berrios-Torres SI, Umscheid CA, Bratzler DW, Leas B, Stone EC, Kelz RR, et al. Centers for Disease Control and Prevention guideline for the prevention of surgical site infection, 2017. *JAMA Surg* 2017;152(10):784–91. *This very extensive guideline prepared by the Healthcare Infection Control Practices Advisory Committee (HICPAC) of the CDC contains a wealth of information regarding the prevention of SSI, along with detailed definitions and extensive references.*

Dellinger EP. Approach to the patient with postoperative fever. In Gorbach SL, Bartlett JG, Blacklow NR, eds: *Infectious diseases*, Philadelphia, 2004, Lippincott Williams and Wilkins, pp 817–823. *This is an extensive discussion of the role and significance of postoperative fever.*

National Healthcare Safety Network (NHSN): Surgical site infection (SSI) event. Procedure-associated Module.SSI, January 2019. Available at: www.cdc.gov/nhsn/PDFs/pscManual/9pscSSIcurrent.pdf. *This web resource for NHSN provides detailed definitions and procedures for surveillance and categorization of SSI and has links to many other resources regarding SSI and other HAIs.*

Stevens DL, Bisno AL, Chambers HF, Dellinger EP, Goldstein EJ, Gorbach SL, et al. Executive summary: practice guidelines for the diagnosis and management of skin and soft tissue infections: 2014 update by the Infectious Diseases Society of America. *Clin Infect Dis* 2014;59(2):147–59. *This article contains the recommendations of the Infectious Diseases Society of America for the diagnosis and management of all skin and soft-tissue infections, including SSIs.*

# Sexually Transmitted Infections

*Jeanne M. Marrazzo*

# 53

# Introduction to Sexually Transmitted Infections

*Jeanne M. Marrazzo*

## ABSTRACT

Sexually transmitted infections (STIs) are currently at historic high rates in the United States and several other countries. These pathogens, including the bacteria *Treponema pallidum* (syphilis), *Chlamydia trachomatis* and *Neisseria gonorrhoeae*, the viruses herpes simplex and HIV-1, and protozoan *Trichomonas vaginalis*, exert untold adverse effects on global sexual and reproductive health. Advances in the last decade have been made in progress toward mitigating the carcinogenic effects of the human papillomavirus (HPV) and in development of new antibiotics to treat gonorrhea.

It has been more than 20 years since the Institute of Medicine published the landmark report entitled *The Hidden Epidemic: Confronting Sexually Transmitted Diseases*. In that document, the committee articulated a vision statement to guide its deliberations:

> *"An effective system of services and information that supports individuals, families, and communities in preventing STDs, including HIV infection, and ensures comprehensive, high-quality STD-related health services for all persons."*

The committee concluded that four major strategies should form the public and private sector response to what was then recognized as the growing (yet still hidden) epidemic of STIs: (1) overcome barriers to adoption of healthy sexual behaviors; (2) develop strong leadership, strengthen investment, and improve information systems for STI prevention; (3) design and implement essential STI-related services in innovative ways for adolescents and underserved populations; and (4) ensure access to and quality of essential clinical services for STIs.

Since those ambitious strategies were published, how has the approach to managing and preventing STIs changed? How has the epidemiology of these important infections changed? And most important for readers of this section, has the fourth strategy been addressed? Ensuring the quality of essential clinical services directly touches on the skills of providers of clinical care to people concerned about, at risk for, or manifesting STI-related clinical syndromes.

The need for effective prevention and management of STIs, including human immunodeficiency virus (HIV) infection, remains an exceedingly high priority, both internationally and domestically. The United Nations office on acquired immunodeficiency syndrome (UNAIDS) reported that although new HIV infections among women aged 15 to 24 years were reduced by 25% between 2010 and 2018, more than 6000 women in this age group still become infected every week. The US Centers for Disease Control and Prevention (CDC) estimated that in 2017, 38,739 people received a new HIV diagnosis in the United States. Unfortunately, a large proportion of new HIV infections continue to be diagnosed in late stages of the disease. This fact highlights the need for clinicians to remain familiar with the recognition and management of the common opportunistic infections that define clinical AIDS.

In the United States in 2017, most new HIV infections occurred in men who have sex with men (MSM), a population that also continues to sustain the highest incidence of syphilis—an infection many physicians had experience with primarily in the pre-AIDS era and one completely new to many young physicians. As summarized in the chapter on syphilis, this resurgence has highlighted that clinical recognition of this protean disease—called "the great pretender" by Sir William Osler—continues to present diagnostic and management challenges to those who care for patients at risk, particularly when co-infection with HIV is involved. Persons with HIV are more likely to have atypical manifestations of the genital ulcerations caused by *Treponema pallidum* and atypical results of diagnostic serology tests, and very probably have an increased risk of neuroinvasive disease because of this pathogen.

Apart from HIV, rates of other reportable STIs either have not declined or have actually increased in the last decade. In 2018, more than 1.7 million diagnoses of *Chlamydia trachomatis* were reported to the CDC. Despite this, interventions to detect this common infection in populations most at risk are infrequently performed. For example, rates of routine annual screening for genital chlamydial infections in young women, especially adolescents, remain suboptimal, and many women at low risk (primarily those over age 30 years without other indications) are tested unnecessarily. Moreover, recommendations to routinely retest infected persons 4 to 6 months after treatment (a practice termed *repeat testing*, which is distinct from test of cure) are not frequently adhered to, despite the fact that this approach detects repeat infection in approximately 15% to 40% of those tested.

In 2017 to 2018, the rate of reported cases of gonorrhea in the United States increased 5% from the prior year and 82.6% from the historical low in 2009. Moreover, the relentless evolution of antimicrobial resistance in *Neisseria gonorrhoeae* continues to present a major challenge. Both fluoroquinolones, which offered a new class of effective single-dose oral therapy for this organism in the 1990s, and macrolides (azithromycin) are no longer reliably effective owing to increasing resistance, a trend especially notable in MSM. Providers are now effectively left with only a single class of antibiotics—the cephalosporins—that reliably treat this infection. Concern for nascent development of resistance to this class—a phenomenon that has already been reported—has prompted the study of several new alternatives, with at least two hopeful candidates moving into late-phase clinical trials.

As noted previously, syphilis is enjoying a global resurgence in MSM, especially those co-infected with HIV, but rates of syphilis in the last several years have increased for all races and ethnicities (with the exception of Asians/Pacific Islanders). Even more alarming is the historic high reached for cases of congenital syphilis in 2018: 1306 in the United States alone. Finally, sexual transmission of hepatitis C has been increasingly recognized in MSM who report sexual practices involving exposure to blood or even minimal trauma to the rectal mucosa and, most recently, in injection drug users.

These worrisome trends emphasize the need for physicians to be aware of emerging STI-related challenges and of the availability of guidelines and tools to help manage their patients. The CDC STD treatment guidelines are currently being updated, with new guidance expected by 2021. In the years succeeding the previous version, there have been several developments that clinical providers should be aware of. These include very positive developments, such as further evidence of the significant protective benefits of immunization against several common genital HPV types for women and men. Antiretroviral-based preexposure prophylaxis (PrEP) offers a potent prevention tool against acquisition of HIV, and along with definitive proof that undetectable plasma viral load confers protection from transmitting HIV ("*Undetectable equals untransmissible,*" or U = U), its deployment in some metropolitan settings has already effected a marked reduction in HIV incidence.

The excellent chapters that make up this section cover all of these developments and more. Against the backdrop of providing key epidemiologic trends for each disease, the authors have emphasized that clinical recognition and diagnosis of these infections are not always straightforward. Moreover, therapeutic management of some STIs can be complicated by limited diagnostic capability, co-infections, and immune compromise resulting from HIV infection. In addition to biomedical management of the individual patient who is affected by STIs, clinicians must remember that the prevention of these infections requires combinations of biomedical, behavioral, and structural interventions. Among the most promising approaches is expedited partner management—a strategy that allows for the treatment of sex partners without in-person evaluation. This approach has been widely adopted and should provide a major tool to control the stubborn epidemics of chlamydia and gonorrhea, in particular. Finally, new technologies have opened up our routes for communicating important information to patients, whether that be diagnostic tests results or health promotion messaging through apps, texts, or other electronic platforms.

Although the Institute of Medicine's vision for controlling the hidden epidemic of STIs has not been fully realized, the healthcare field is in an exciting period of renewed hope for advances in diagnosis, therapy, and prevention of these stubborn infections. The state-of-the-art information in the chapters that follow will undoubtedly assist clinicians in contributing to the overall goal of improving the community's sexual health through recognition, management, and prevention of STIs.

## EVIDENCE

Baeten JM, et al. Antiretroviral prophylaxis for HIV prevention in heterosexual men and women. *New Engl J Med* 2012;367(5):399-410. *The investigators randomized the HIV-uninfected partner in HIV-serodiscordant couple in Sub-Saharan Africa to daily oral tenofovir-based PrEP versus placebo, and demonstrated a high degree of protection in those who received active study product.*

Barbee LA, et al. Increases in Neisseria gonorrhoeae with reduced susceptibility to azithromycin among men who have sex with men (MSM), in Seattle, King County, Washington: 2012–2016. *Clin Infect Dis* 2018;66(5):712-8. *The oropharynx in MSM yielded a relatively high rate of azithromycin resistance for gonorrhea, prompting the authors to recommend cessation of this antibiotic's use in the treatment of gonorrhea.*

Cohen MS, et al. Antiretroviral therapy for the prevention of HIV-1 transmission. *New Engl J Med* 2016;375(9):830-9. *This study showed definitively that people whose HIV viral load was suppressed below detectable levels in plasma by antiretroviral therapy did not transmit their infection to susceptible sex partners, giving rise to the now widely accepted maxim that "Undetectable equals untransmissible" (U = U).*

Molina JM, et al. Efficacy, safety, and effect on sexual behaviour of on-demand pre-exposure prophylaxis for HIV in men who have sex with men: an observational cohort study. *Lancet HIV* 2017;4(9): e402-e410. *This trial studied peri-sexual (episodic, on-demand) use of four doses of oral tenofovir-emtricitabine in MSM, and showed that the strategy significantly and markedly reduced rates of HIV acquisition.*

Unemo M, Lahra MM, Cole M, et al. World Health Organization Global Gonococcal Antimicrobial Surveillance Program (WHO GASP): review of new data and evidence to inform international collaborative actions and research efforts. *Sex Health* 2019;16(5):412-425. *This is a comprehensive overview of global patterns of gonococcal antimicrobial resistance, providing a sobering picture of the relentless advance of this pathogen's ability to evade modern antibiotics.*

## ADDITIONAL RESOURCES

Centers for Disease Control and Prevention (CDC). *HIV/AIDS surveillance report, 2017, vol. 29.* Atlanta, 2018, U.S. Department of Health and Human Services, CDC. Available at: https://www.cdc.gov/hiv/library/reports/hiv-surveillance.html.

Eng TR, Butler WT, eds. *Committee on Prevention and Control of Sexually Transmitted Diseases, Institute of Medicine: The hidden epidemic: confronting sexually transmitted diseases.* Washington DC, 1996, National Academies Press. *This is the most recent comprehensive overview of the state of STI control in the United States. It is currently being updated by a new panel, with recommendations expected in the coming year.*

Fauci AS, Redfield RR, Sigounas G. Ending the HIV epidemic: a plan for the United States. *JAMA* 2019;321(9):844-5. *The federal government has laid out an ambitious plan to end the HIV epidemic by 2030, as summarized in this overview.*

Meites E, Szilagyi PG, Chesson HW, et al. Human papillomavirus vaccination for adults: updated recommendations of the Advisory Committee on Immunization Practices (ACIP). *MMWR Recomm Rep* 2019;68(32):698-702. *The available HPV vaccines have received strong endorsement from numerous committees and should continue to favorably modify the epidemiologic trajectory of associated cancers across the globe.*

Sexually transmitted disease treatment guidelines, 2015. *MMWR Morb Mortal Wkly Rep* 2015;64(3). *These guidelines, derived from comprehensive evidence-based and expert review, are an invaluable resource in assisting clinicians in the appropriate diagnosis and management of STIs.*

UNAIDS. UNAIDS Data 2019. www.unaids.org. *This is the most up-to-date source for relevant global statistics on the HIV/AIDS epidemic.*

Unemo M, et al. Sexually transmitted infections: challenges ahead. *Lancet Infect Dis* 2017;17(8):e235-279.

# 54

# Trichomonas vaginalis

Patricia Kissinger, Christina A. Muzny

 **ABSTRACT**

*Trichomonas vaginalis* is estimated to be the most common treatable sexually transmitted infection (STI) worldwide. It is an important contributor to perinatal morbidity, cervical cancer, poor sperm quality, and HIV acquisition and transmission. Less is known about the importance of asymptomatic infection. *T. vaginalis* shares many of the same risk factors as other STIs but tends to be more prevalent among older women. Risk factors for *T. vaginalis* include female sex, older age, lower educational level, lower socioeconomic status, history of incarceration, and having had more than one sexual partner in the previous year. General screening for *T. vaginalis* in the United States has not been recommended by the Centers for Disease Control and Prevention (CDC) except in HIV-infected women. Recent evidence from two randomized trials and a meta-analysis has demonstrated that single-dose metronidazole (MTZ) is not effective among women and a 7-day

course of 500 mg twice daily should be used. Less is known about the most effective treatment in men and more investigation is needed. Given the high rate of *T. vaginalis* among male sex partners of women with *T. vaginalis*, presumptive partner treatment is recommended by the CDC. There has been a proliferation of *T. vaginalis* diagnostic tests, including nucleic acid amplification tests as well as point-of-care (POC) tests that are far more sensitive than the traditionally used wet-mount microscopy. Treatment of persistent infection may require higher doses of medication and hypersensitivity reactions to 5-nitroimidazole medications (i.e., MTZ) may require consultation with an allergy specialist. Condoms and male circumcision can prevent *T. vaginalis* infection. A better understanding of the influence of the vaginal microbiota on *T. vaginalis* pathogenicity and treatment outcomes is needed.

 **CLINICAL VIGNETTE**

A 28-year-old woman with a past medical history significant for bacterial vaginosis (BV) and chlamydial infection presents to the clinic with complaints of a 3-day history of a foul-smelling, frothy vaginal discharge and genital pruritus. She reports unprotected sex with two casual male sexual partners within the past 2 weeks. She denies a history of HIV infection. She is currently taking no medications and is not on birth control. She smokes one pack of cigarettes daily, drinks alcohol socially, and occasionally uses crack cocaine. On pelvic examination, she has no vaginal sores or other lesions. There is mild bilateral inguinal lymphadenopathy. She has a profuse, thin, white/yellow vaginal discharge. No cervical discharge is noted. There is no uterine or adnexal tenderness on bimanual exam. Vaginal pH is high at 7.0. Wet mount of vaginal secretions shows more than 20 clue cells per high power field (hpf) in addition to motile trichomonads. No white blood cells or red blood cells are noted. The whiff test is positive. KOH examination shows no yeast. A vaginal specimen is sent for chlamydia, gonorrhea, and trichomonas nucleic acid amplification testing (NAAT).

The patient is diagnosed with BV and trichomoniasis. She is treated with oral MTZ 500 mg bid for 7 days and counseled to have all sexual partners within the past 60 days present to the clinic to be treated for trichomoniasis as contacts. She is also counseled to avoid any sexual activity while on treatment. Her NAAT results come back positive only for trichomoniasis.

Approximately 2 weeks later, the patient returns to the clinic with complaints of persistent vaginal discharge. Repeat wet mount again shows motile trichomonads. She states that both of her male partners were treated and denies any sexual activity for the preceding 2 weeks. She is prescribed a second course of oral MTZ 500 mg bid for 7 days. She continues to be symptomatic with a persistently positive wet mount over the next several weeks. MTZ drug resistance is suspected. A trichomonas culture is obtained and sent to the CDC for drug resistance testing. While awaiting the results, the patient is prescribed MTZ 2 g PO bid for 7 days, which is also unsuccessful in clearing her infection. Her trichomonas isolate is found to have low-level MTZ resistance with a minimum lethal concentration (MLC) of 75 µg/mL. She is subsequently treated with high-dose oral tinidazole (TDZ) 2 g qd for 7 days. Her symptoms resolve and her wet mount is negative at a follow-up visit in 14 days.

COMMENT: This case represents difficult-to-treat, persistent trichomonas infection in the absence of sexual reexposure. Low-level MTZ resistance (i.e., MLC 50–100 µg/mL) is more common than moderate (MLC 200 µg/mL) or high-level resistance (MLC >400 µg/mL) and can be overcome with high-dose TDZ, as occurred with this patient.

## EPIDEMIOLOGY

*Trichomonas vaginalis* infection is likely the most common nonviral STI in the world. Although it is not a reportable disease, global estimates indicate that there are 143 million new cases per year; among women, *T. vaginalis* is more prevalent than *Chlamydia trachomatis*, *Neisseria gonorrhea*, and syphilis combined. The global prevalence of trichomoniasis is 5.0% in women and 0.6% in men. In the United States, the prevalence in a recent population-based study was 1.8% among women and 0.5% among men; however, African Americans

in both sexes had more than a fivefold higher rate than the general population.

Additional risk factors for *T. vaginalis* include more than one sex partner in the previous year, having less than a high school education, living below the poverty level, having a history of incarceration, and HIV infection. The prevalence of *T. vaginalis* infection, unlike that of many other STIs, is higher among older women. In contrast, data from the National Health and Nutrition Examination Survey (NHANES), a population-based study, found the *T. vaginalis* prevalence among men to be far lower than that among women (<1%), and the rate did not vary by age. Whether this sex difference is real or due to a detection bias as a result of the lack of general screening, biologic reasons, or other factors is unknown. *T. vaginalis* is uncommon in men who have sex with men.

## ETIOLOGY/PATHOPHYSIOLOGY

*T. vaginalis* is a unicellular flagellate parasite that primarily infects the squamous epithelium of the genital tract by adhering to the epithelial cells and obtaining nutrients by lysing the cells. *T. vaginalis* can harbor two *Mycoplasma* species (*M. hominis* and *M. girerdii*), which may be linked to persistent infection. It can also harbor a *T. vaginalis* virus (TVV). This is a double-stranded RNA (dsRNA) virus, of which there are four types (TVV1–TVV4). The presence of TVV2 has been linked to more severe genital symptoms in some studies but not others. *T. vaginalis* infects the female lower genital tract (vagina, urethra, and endocervix) and the male urethra and prostate, where it replicates by binary fission and causes inflammation. Infection with *T. vaginalis* has been linked to a greater likelihood of both the transmission and acquisition of HIV, largely due to inflammation and recruitment of target CD4 cells.

*T. vaginalis* is transmitted among humans, its only known host, primarily by sexual intercourse, including digital-vaginal manipulation and mutual masturbation where there is exchange of genital fluids. *T. vaginalis* has been found in the oral cavity and in the rectum. Although transmission from these sites is possible, there has been less research on these modes of extragenital transmission.

The parasite does not appear to have a cyst form and does not survive well in the external environment, but it can survive outside the human body in a wet environment (e.g., wet wash clothes) for more than 3 hours and perhaps even up to 24 hours. Possible transmission of *T. vaginalis* within a family in Ghana through the shared use of a bathing towel and sponge has also been reported. A case of iatrogenic transmission of *T. vaginalis* by a traditional healer to a patient following genital manipulation has been reported from The Gambia. There may be a pseudocyst form, more virulent in animals, that could have relevance for humans. Infection in women may persist for long periods of time, possibly months or even years. Less is known about the natural history in men, although spontaneous resolution has been documented as early as 15 to 19 days after initial diagnosis.

The majority of women (50%) and men (77%) with *T. vaginalis* infection are asymptomatic. One-third of asymptomatic women may become symptomatic within 6 months. Symptomatic women can have vaginal erythema, dyspareunia, dysuria, vaginal discharge (which is often diffuse, malodorous, and yellow-green), as well as pruritus in the genital region. The normal vaginal pH is 4.4; with *T. vaginalis* infection, this increases markedly, often to more than 5. However, *T. vaginalis* infection can also be seen in the presence of a normal vaginal pH. *Colpitis macularis,* or strawberry cervix, is seen in about 5% of infected women; with colposcopy it can be seen in nearly 50%. In men, *T. vaginalis* is associated with urethritis, epididymitis, prostatitis, decreased sperm motility, and altered sperm morphology (Fig. 54.1). It is worth noting that most of the studies of sequelae from *T. vaginalis* have been done in symptomatic women. Less is known about asymptomatic women and men.

*T. vaginalis* has been associated with poor birth outcomes, such as low birth weight, preterm delivery, and premature rupture of membranes. Although rare, *T. vaginalis* infection can be transmitted perinatally and cause vaginal and respiratory infections in neonates. In a study of HIV-infected women, those with *T. vaginalis* had a significantly higher risk of pelvic inflammatory disease (PID) than those without this infection. These data are corroborated from results recently published in the Prenatal Environment And Child Health (PEACH) Study. This study found that the odds of having endometritis were twice as high among women with trichomoniasis as compared with those without. Persistent endometritis was also more common, although nonsignificantly, among women with *T. vaginalis*. A meta-analysis of 17 published studies found that women with *T. vaginalis* were at twofold higher risk for cervical cancer, but meta-analyses of *T. vaginalis* and prostate cancer have failed to find a statistically significant association. *T. vaginalis* has been associated with HIV acquisition in women and has been shown to increase genital shedding of HIV among women who are coinfected.

*T. vaginalis* has also been associated with increased rates of high-risk human papillomavirus (HPV). Coinfection with other STIs appears to be common and may vary by demographic and geographic factors. *T. vaginalis*, BV, and *Candida* spp. are often seen together and are the three most common causes of vaginitis, but they present differently (Fig. 54.2). Up to 50% of women with *T. vaginalis* also have BV. Whether *T. vaginalis* is causal or coincidental to BV has not been established. More cohort studies are needed to determine whether causality exists.

Certain components in the vaginal microbiome have been associated with higher prevalence of *T. vaginalis* and less successful MTZ treatment. Women with a community state type (CST) dominated by BV–associated bacteria (i.e., CST-IV) were eight times more likely to have *T. vaginalis* compared with those with a *Lactobacillus*

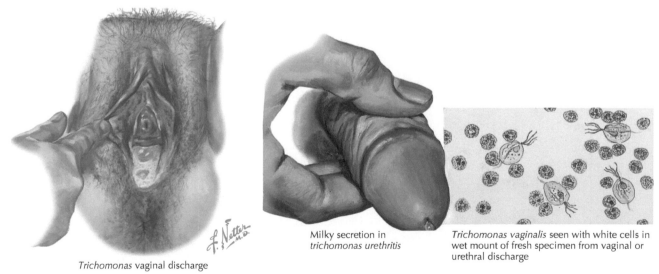

*Trichomonas* vaginal discharge

Milky secretion in *trichomonas urethritis*

*Trichomonas vaginalis* seen with white cells in wet mount of fresh specimen from vaginal or urethral discharge

**Fig. 54.1** Trichomoniasis discharge in females and males.

*crispatus*–dominated (CST-I). One study also found that *T. vaginalis* reduced colonization by *Lactobacillus* spp. but not by BV-associated bacteria, suggesting that *T. vaginalis* itself may alter the vaginal milieu. Other studies have found that although most women with *T. vaginalis* had a BV-like vaginal microbiota, other *T. vaginalis*–infected women had two unique vaginal microbiomes: one dominated by *M. hominis* and the other by *M. girerdii*. Women with the latter vaginal microbiome showed evidence of intense vaginal inflammation (i.e., vaginal mucosal erythema) compared with other *T. vaginalis*–infected women, suggesting the importance of unidentified co-infections in *T. vaginalis* infection. One small cohort study (*n* = 50) of women in Mombasa found *Sneathia sanguinegens* more common in women who acquired *T. vaginalis* compared with those who did not. More studies of the influence of the genital microbiome on *T. vaginalis* acquisition are needed.

## DIAGNOSTIC APPROACH

The diagnosis of *T. vaginalis* is becoming more precise, and more tests have become available in the last decade. In choosing which diagnostic test to use, the need for rapid diagnosis should be balanced with the sensitivity of the test its cost (Tables 54.1 and 54.2). In addition,

some of the tests have not been approved by the US Food and Drug Administration (FDA) for use in men, and laboratories should undergo internal validation testing before use.

Wet-mount microscopy has been used for many decades to diagnose *T. vaginalis* in women. The test is inexpensive and can be easily performed in clinical settings as a POC test, but it is less sensitive than many of the newer diagnostics. Sensitivities range from 44% to 68% depending on the expertise of the reader and should be read within 10 minutes of collection, as sensitivity decreases quickly over time, up to 20% within 1 hour after collection. Although *T. vaginalis* culture has better sensitivity than wet mount in women, it is more expensive and time consuming; it also demonstrates poor sensitivity in men. The Inpouch culture system (Biomed, White City, OR) is a Clinical Laboratory Improvement Amendments (CLIA)-waived diagnostic test that requires a maximum of three readings over a 5- to 7-day period, and the specimen must be incubated during that time. The sensitivity of culture has been found to range between 75% and 96% in longitudinal studies of *T. vaginalis* treatment in women. Studies of HIV-infected and uninfected women found that *T. vaginalis* infection posttreatment was undetectable for months via culture; it then reappeared in the absence of reported sexual exposure, underscoring the need for more sensitive testing. Cultures in men are less sensitive with urine than with

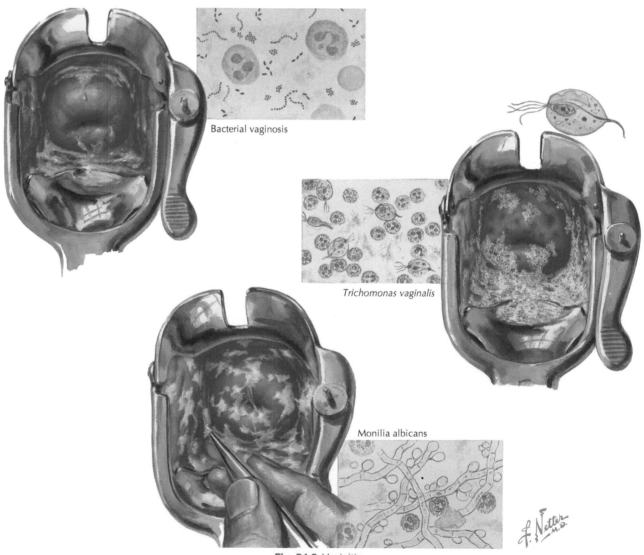

Bacterial vaginosis

*Trichomonas vaginalis*

Monilia albicans

**Fig. 54.2** Vaginitis.

### TABLE 54.1   Point-of-Care Diagnostic Tests for *Trichomonas vaginalis*

| Test | Sample Type | Sensitivity/Specificity Compared With NAAT | Complexity/Time | Cost |
|---|---|---|---|---|
| Wet mount preparation | Vaginal swab | Women only. Sensitivity: 44%–68% Specificity: 100% | CLIA waved. Microscope with 40× objective required. Results in 5 min. | $ |
| OSOM | Vaginal swab | Symptomatic women only. Sensitivity: 83%–92% Specificity: 99%–100% | CLIA waved. Detects antigen. No instrumentation needed. Results in 10 min. | $$ |
| Solana | Vaginal swab or urine from women | Asymptomatic or symptomatic women. Sensitivity: 90%–99% Specificity: 97%–99% | Not CLIA waved. Requires some instrumentation. Results in less than 1 h. | $$ |
| Cepheid GeneXpert | Endocervical or vaginal swabs and urine from women; Urine for men | In women: Sensitivity: 95%–100% Specificity: 98%–99% | CLIA waved. Requires instrumentation. Results in approximately 40 min. | $$$ |
| Amplivue | Vaginal swab | Asymptomatic or symptomatic women. Sensitivity: 97%–100% Specificity: 97%–99% | Not CLIA waved. Requires some instrumentation. Results in approximately 45 min. | $$ |

*NAAT,* Nucleic acid amplification testing.

### TABLE 54.2   Non–Point-of-Care Diagnostic Tests for *Trichomonas vaginalis*

| Test | Sample Type | Sensitivity/Specificity Compared With NAAT | Complexity/Time | Cost |
|---|---|---|---|---|
| Culture | Women: Vaginal swab Men: Urethral or penile-metal swabs or semen | Women: Sensitivity: 75%–96% Specificity: 100% Men: Sensitivity: 50%–80% Specificity: 100% | Requires incubator and microscope with 40× objective. Results in 5–7 days. | $$ |
| Liquid Pap | Women: Endocervical swab | Women: Sensitivity: 60%–96% Specificity: 98%–100% | Special brushes and instrumentation required. Results can take up to 1 week. | $$ |
| Hologic APTIMA *T. vaginalis* Assay | Women: Vaginal/endocervical swab or urine Men: Urethral or urine specimens (if test is internally validated) | Women: Sensitivity: 88%–100% Specificity: 98%–100% | Panther, Viper, or Tigris instrumentation needed. Results in hours to weeks. | $$$ |
| BD Probe Tec TV NAAT | Women: Vaginal/endocervical swab or urine Men: Urethral or urine specimens (if test is internally validated) | Women: Sensitivity: 98%–100% Specificity: 98%–100% | CLIA high complexity. Requires VIPER system. Results available in less than 8 h. | $$$ |

*NAAT,* Nucleic acid amplification testing.

urethral or metal swabs and have the highest sensitivity when multiple specimens are tested.

*T. vaginalis* NAATs are the most sensitive tests; they are moderately priced and generally take less time than culture but require specific instruments and facilities. In addition, except for the Cepheid TV NAAT, these tests are not considered POC. The APTIMA *T. vaginalis* Assay (Hologic Gen-Probe, San Diego, CA) was cleared by the FDA in 2011 for use with urine, endocervical and vaginal swabs, and endocervical specimens collected in the Hologic PreserveCyt (ThinPrep) solution from asymptomatic and symptomatic women. This assay has not been approved by the FDA for use in men and must be internally validated prior to use. The BD Probe Tec TV Qx Amplified DNA Assay (Becton Dickinson, Franklin Lakes, NJ) is FDA-cleared for detection of *T. vaginalis* from vaginal patient- or clinician-collected swabs, endocervical swabs, and urine specimens from

women. Similar to the Aptima *T. vaginalis* assay, this test is FDA-approved for use only in women and must be internally validated prior to use in men. Becton Dickinson also has a BD Max GCCTTV assay 2 that is expected to become available in 2020 and be used to detect *T. vaginalis* DNA in patient- or clinician-collected vaginal specimens (in a clinical setting) from asymptomatic and symptomatic patients as well as male and female urine specimens (https://www.accessdata.fda.gov/cdrh_docs/reviews/K182692.pdf). Although it is feasible to perform these NAAT tests on the same genital specimens used for chlamydia and gonorrhea testing, testing for these STIs is primarily recommended for sexually active women 25 years of age and younger. By contrast, trichomoniasis is common in women older than 25 years, and this demographic should not be overlooked.

In addition to wet-mount microscopy, additional POC FDA-approved diagnostic tests for *T. vaginalis* among women are the OSOM lateral flow test (Sekisui Diagnostics, Bedford, MA) (to be used only with clinician-obtained vaginal specimens in symptomatic women; this test has poor sensitivity in male specimens), the Isothermal Helicase-Dependent AmpliVue test (Quidel, San Diego, CA; can be used on vaginal specimens from both asymptomatic and symptomatic women), and the Solana TV-Assay (Quidel, San Diego, CA; can be used on vaginal and urine specimens from both asymptomatic and symptomatic women). The Cepheid GeneXpert TV (Cepheid, Sunnyvale, CA) is a moderately complex FDA-approved rapid POC NAAT test that can be performed in less than 1 hour for use on female urine samples, endocervical swabs, and patient- and clinician-collected vaginal specimens as well as male urine specimens.

A repeat *T. vaginalis* NAAT too soon after treatment can result in the detection of remnant trichomonad DNA, thus producing a false positive. The results of three studies indicate that retesting for *T. vaginalis* using NAAT can result in a false positive if done less than 3 weeks after completion of treatment.

# TREATMENT

For nearly four decades, single-dose (2 g) MTZ has been the recommended treatment regimen for women with trichomoniasis, with single-dose (2 g) TDZ as an alternative. Both drugs belong to the 5-nitroimidazole drug class. Recommendations from the World Health Organization (WHO) and the US CDC STD Treatment Guidelines for the treatment of *T. vaginalis* currently include as the preferred regimens MTZ or TDZ 2 g PO as a single dose and MTZ 400 to 500 mg PO twice daily for 7 days as the alternative regimen. However, two multicenter randomized trials (one in HIV-infected women and one in HIV-uninfected women) and a recent meta-analysis has demonstrated that single-dose therapy leads to nearly twice as many treatment failures as the 7-day 500 mg bid MTZ regimen. Because of this, it is likely that preferred treatment recommendations will change to multidose MTZ over single-dose treatment for all women. Head-to-head comparisons of different MTZ dosages in men using sensitive *T. vaginalis* testing have not been done; thus single-dose MTZ treatment will likely continue to be preferred for men.

There are several possible explanations for the superior effectiveness of multi- over single-dose MTZ in women. One plausible explanation is competition for MTZ by other micro-organisms present in the vaginal microbiota that are sponge-organisms known to absorb MTZ. Such examples are *Escherichia coli*, *Enterococcus faecalis*, *Proteus* spp., and *Klebsiella* spp. This process of MTZ inactivation may result in drug concentrations at the site of infection that are lower than those required to be trichomonacidal. Treatment success may depend on reducing these organisms through scheduled, successive doses of MTZ (i.e., multidose therapy). Another potential explanation is inadequate accumulation of active MTZ metabolites (i.e., hydroxyl-metronidazole) during single-dose therapy, potentially decreasing the drug's therapeutic contribution. A significant accumulation of these active metabolites may occur with repeat dosing in the multidose MTZ regimen. Future pharmacokinetic and pharmacodynamic studies with MTZ should be performed to investigate these hypotheses.

MTZ is often preferred over TDZ because it is considerably less expensive. An additional 5-nitroimidazole, secnidazole (SEC), which was recently approved by the FDA for the treatment of BV in women, is currently under investigation for the treatment of *T. vaginalis*–infected women in a phase 3 multicenter prospective randomized, placebo-controlled delayed-treatment double-blind study (NCT03935217). SEC is given as a single oral dose (2 g of granules), which must be taken with unsweetened applesauce, pudding, or yogurt. Its primary advantage is its ability to be taken as a single dose. However, similar to TDZ, it is more expensive than MTZ.

If a patient has failed one round of MTZ treatment and there is no history of sexual re-exposure, higher doses of medication can be considered (i.e., MTZ or TDZ 2 g PO qd for 5 to 7 days). If a patient remains positive after two rounds of treatment, a consultation for medication resistance testing should be done. Consultation and *T. vaginalis* susceptibility testing is available in the United States from the CDC (telephone: 404-718-4141; website: http://www.cdc.gov/std).

## Treatment of Pregnant and Lactating Women

MTZ is a class B drug, and several meta-analyses have found it to be safe in pregnant women in all stages of pregnancy. TDZ has not been evaluated in pregnant women and remains a class C drug. Treatment with 2 g MTZ is recommended by CDC at any time during pregnancy, whereas the WHO does not recommend treatment in the first trimester unless it is indicated for the prevention of an adverse birth outcome.

In lactating women who are administered MTZ, withholding breastfeeding during treatment and for 12 to 24 hours after the last dose will reduce the exposure of the infant to MTZ. For women treated with TDZ, interruption of breastfeeding is recommended during treatment and for 3 days after the last dose.

## Treatment of Persistent *T. vaginalis* or in the Setting of Severe Hypersensitivity Reactions to 5-Nitroimidazoles

Persistent *T. vaginalis* is usually treated with multidose MTZ or TDZ. The most common reactions reported from MTZ are urticaria and facial edema, whereas other adverse reactions include flushing, fever, and anaphylactic shock from immediate-type hypersensitivity. Consultation with an allergist for potential desensitization should be done in these cases. If *T. vaginalis* remains persistent or desensitization is contraindicated, other intravaginal treatments studied include acetarsol, boric acid, furazolidone, and paromomycin. Nitrazoxanide has also been examined as an alternative oral agent for MTZ-resistant *T. vaginalis* but was not found to be very effective. Some plant extracts have shown anti-*T. vaginalis* activity, but these have not yet been tested in clinical trials.

## Treatment of HIV-Infected Women

In a randomized clinical trial (RCT) among HIV-infected women with *T. vaginalis*, multidose MTZ was found to be superior to single-dose treatment. Further analysis revealed that the superiority occurred only in the presence of BV. Studies have also found that antiretroviral therapy (ART) may interfere with the efficacy of MTZ in HIV-infected women. Given the high rate of BV in *T. vaginalis*/HIV coinfected women, the CDC currently recommends that all HIV-infected women with *T. vaginalis* be given MTZ 500 mg bid for 7 days.

Several review papers and meta-analyses have demonstrated an association between *T. vaginalis* and HIV acquisition. Because of this, the CDC currently recommends that all HIV-infected women be screened for *T. vaginalis* at entry to care and then annually. One study found that if these recommendations were upheld in the United States, the lifetime cost of new HIV infections prevented annually would approximate $159,264,000.

## Repeat or Persistent Infections

Repeat *T. vaginalis* infections are common, ranging from 5% to 31%, and share similar sequelae to primary infections. The source of repeat infection is often thought to be reinfection from an untreated sex partner. However, data from two RCTs and one meta-analysis demonstrated that many of the repeat infections actually represent treatment failures from using single-dose MTZ. Resistance appears to play only a minor role in explaining repeat infections. Because of the high rate of repeat infections, the CDC recommends rescreening women with *T. vaginalis* 3 months after treatment.

## Partner Management

One study found that 70% of male sex partners of women infected with *T. vaginalis* were also infected; thus the sex partners of patients with *T. vaginalis* should be treated. Commonly, patients are told by their providers to tell their partners to seek testing and treatment. Providers should consider presumptively treating sexual partners of positive patients. One method of presumptive partner treatment is called expedited partner therapy (EPT). This is the clinical practice of treating the sex partners of patients diagnosed with an STI by providing prescriptions or medications to the patient to take to the partner without the health care provider first examining the partner. State laws may or may not allow such a practice (https://www.cdc.gov/std/ept/legal/default.htm).

One RCT demonstrated that partner treatment with 2 g TDZ resulted in a greater than fourfold reduction in repeat infections in *T. vaginalis*–infected women. Two other studies using 2 g MTZ for male partners of *T. vaginalis*–infected women found either no effect of EPT or a borderline effect. Although it is possible that the two studies using MTZ were either underpowered or did not use a correct control arm, it is also possible that TDZ is a better treatment for men.

## Antimicrobial Resistance

Currently approved drugs for the treatment of trichomoniasis are from the 5-nitroimidazole class of medications (i.e., MTZ and TDZ). Resistance rates in US populations are relatively low for both MTZ (4.3%) and TDZ (1%). However, these data are older (2009–2010), and newer data are needed. Low-level MTZ resistance (i.e., MLC 50 to 100 µg/mL) is more common than moderate (MLC 200 µg/mL) or high-level resistance (MLC >400 µg/mL) and can be overcome with high-dose TDZ, as described in the clinical vignette.

## PREVENTION AND CONTROL

Condoms serve as a mechanical barrier; when used consistently and correctly, they can prevent the transmission of *T. vaginalis*. Male circumcision has also been shown to prevent infection in both men and women. Although there is no vaccine against *T. vaginalis* approved by the FDA and long-term immunity from a prior infection does not occur, an existing vaccine against *T. foetus* may serve as a model for vaccine development. Given the high percentage of persons with *T. vaginalis* who are asymptomatic, screening is an important control measure; however, general screening of women other than those who are infected with HIV is not currently recommended by the CDC. Targeted screening (e.g., pregnant women, incarcerated persons, and/or persons in an area of high *T. vaginalis* prevalence) should be considered.

## FUTURE DIRECTIONS

Future studies are needed to examine the effect of asymptomatic *T. vaginalis* on reproductive outcomes. More research is needed to understand the importance of the vaginal microbiota on *T. vaginalis* pathogenicity and treatment. In addition, a head-to-head comparison of the 2-g oral dose of MTZ and MTZ 500 mg PO twice daily for 7 days should be conducted in men, as *T. vaginalis* treatment regimens for women and men are likely going to differ moving forward.

## EVIDENCE

Howe K, Kissinger PJ. Single-dose compared with multidose metronidazole for the treatment of trichomoniasis in women: a meta-analysis. *Sex Transm Dis* 2017;44(1):29-34. *In this meta-analysis, women receiving 2 g MTZ were two times more likely to retest positive after treatment compared with those receiving multidose MTZ.*

Kissinger P, Mena L, Levison J, et al. A randomized treatment trial: single versus 7-day dose of metronidazole for the treatment of Trichomonas vaginalis among HIV-infected women. *JAIDS* 2010;55(5):565-571. *In this RCT, women who were infected with HIV and receiving 2 g dose of metronidazole were two times more likely to retest positive 1 month after completion of treatment compared with those receiving MTZ 500 mg twice daily for 7 days.*

Kissinger P, Muzny CA, Mena LA, et al. Single-dose versus 7-day-dose metronidazole for the treatment of trichomoniasis in women: an open-label, randomised controlled trial. *Lancet Infect Dis* 2018;18(11):P1251-1259. *In this RCT, women who were uninfected with HIV and receiving a 2-g dose of metronidazole were two times more likely to retest positive 1 month after completion of treatment than women receiving MTZ 500 mg twice daily for 7 days.*

## ADDITIONAL RESOURCES

Committee on Practice B-G. Vaginitis in Nonpregnant Patients: ACOG Practice Bulletin, Number 215. *Obst Gynecol* 2020;135:e1-e17.

Gaydos CA, Klausner JD, Pai NP, et al. Rapid and point-of-care tests for the diagnosis of Trichomonas vaginalis in women and men. *Sex Trans Infect* 2017;93(S4):S31-S35. *This paper discusses POC testing for T. vaginalis.*

Newman L, Rowley J, Hoorn SV, et al. Global estimates of the prevalence and incidence of four curable sexually transmitted infections in 2012 based on systematic review and global reporting. *PLoS ONE* 2015;10(12). *Given that T. vaginalis is not a reportable disease, this paper models the magnitude of the infection globally.*

Patel EU, Gaydos CA, Packman ZR, et al. Prevalence and correlates of trichomonas vaginalis infection among men and women in the United States. *Clin Infect Dis* 2018;67(2):211-217. *This paper describes population-based epidemiology of T. vaginalis.*

Silver BJ, Guy RJ, Kaldor JM, et al. Trichomonas vaginalis as a cause of perinatal morbidity: a systematic review and meta-analysis. *Sex Transm Dis* 2014;41(6):369-76. *This meta-analysis evaluates the association of T. vaginalis and perinatal outcomes.*

Van der Pol B. Clinical and laboratory testing for trichomonas vaginalis infection. *Journal of Clinical Microbiology* 2016;54(1):7-12. *This paper compares many of the new tests for T. vaginalis.*

Workowski KA, Bolan GA. Sexually transmitted diseases treatment guidelines, 2015. *MMWR Recomm Rep* 2015;64(RR03):1-137. *These are the most recent CDC guidelines for treatment. New recommendations are targeted to be released in 2021.*

# 55

# Herpes Simplex Virus Genital Infection

*Abir Hussein, Nicholas J. Moss, Anna Wald*

 **ABSTRACT**

Genital herpes is a globally endemic sexually transmitted infection (STI) and the most common cause of genital ulcer disease. Classically, genital herpes manifests as a cluster of painful vesicular or ulcerative mucocutaneous lesions; however, such presentations account for a minority of cases, as the clinical manifestations vary widely. Genital herpes is caused by herpes simplex virus type 1 (HSV-1) or type 2 (HSV-2), two closely related but genetically distinct viruses. HSV-2 causes the greater burden of genital disease worldwide, especially in resource-poor settings, and risk factors for HSV-2 acquisition are similar to those of other STIs. Both HSV-1 and HSV-2 establish latent infection in sensory nerve root ganglia and can reactivate to cause mucosal or skin recurrences throughout the life of the patient. Most infected patients have mild symptoms or are asymptomatic. Asymptomatic persons can still shed virus in the genital secretions and transmit it to their sex partners. The clinical diagnosis of genital herpes is unreliable, and laboratory testing is necessary for definitive diagnosis. Complications of genital herpes infection include aseptic meningitis and, rarely, disseminated herpes simplex infection in which multiple organ systems can be affected. Life-threatening neonatal herpes infection is the most severe consequence of genital herpes infection in women of childbearing age. Genital herpes shedding and recurrences are more frequent in human immunodeficiency virus (HIV)–infected patients and other immunocompromised individuals. There is no cure, but effective antiviral therapy is available for the treatment of active lesions and suppression of recurrences; daily use also reduces the risk of sexual transmission. Counseling patients about the disease should be a part of any management strategy. Although behavioral measures such as condom use provide partial protection against infection, no broadly effective prophylactic or therapeutic vaccine exists at this time.

## CLINICAL VIGNETTE

A 25-year-old woman presented to an urgent care center complaining of 1 week of burning and pain in her left labium. She stated that about 8 months earlier she had a similar presentation but at that time she had pain bilaterally and recalled feeling as if she had the flu, with fever, body aches, and headaches. During that visit, a swab of the area was taken; she was told that she had genital herpes but did not recall being told the specific type. She reported taking medication for 7 days and that her symptoms then resolved. She had not had symptoms since then. She had had one partner for the previous 3 months and they had been monogamous. He had not reported any symptoms, although the patient admitted that she had never told him that she had been diagnosed with genital herpes in the past. They had not been using condoms as she had not had any other symptoms or lesions and she did not think she could transmit HSV when she was asymptomatic.

On examination, she appeared well and her vital signs were normal. On genital exam, she had a shallow ulceration on an erythematous base on her left labium (Fig. 55.1). The remainder of her exam was normal. The lesion was swabbed and sent for HSV polymerase chain reaction (PCR) examination.

Blood tests were also done and sent for HIV and syphilis serology, and additional genital samples were taken for chlamydia and gonorrhea.

The patient was given valacyclovir 500 mg twice a day for 3 days, leading to the resolution of her symptoms. She was counseled on the persistence of HSV in the body and that, despite not having active lesions or symptoms, she could still be shedding the virus and transmitting it to her partner. Her risk for recurrence will depend on whether she has HSV-1 or HSV-2. Genital herpes caused by HSV-1 recurs rarely, especially after the first year. Genital herpes caused by HSV-2 recurs a median of four times in the first year. She was educated on the possibility of recurrence. Suppressive therapy was offered as an option if she was concerned about future recurrences or wanted to take antiviral therapy to reduce the risk for transmission. Disclosure to her partner was strongly recommended. She was encouraged to use a condom to reduce his risk of acquiring the infection and to avoid sex if she had active lesions. Serologic testing of the partner may be appropriate in some settings.

## EPIDEMIOLOGY

Genital herpes infections occur throughout the world in all settings, including developed and developing nations as well as rural and urban areas. HSV-2 causes the majority of genital herpes infections, with an estimated 417 million people between 15 and 49 years of age infected worldwide. HSV-1 has been reported as an increasing cause of first episodes of genital ulcer disease in sexually active patients in developed countries, where childhood orolabial infection with HSV-1 has been decreasing. Transmission typically occurs during sexual intercourse or other intimate contact between an infected source partner who is shedding virus from a mucosal site or genital skin and an uninfected partner. Infection requires direct contact of virus-containing secretions with mucosal surfaces or breaks in the skin; a clinically visible lesion is not required for transmission. *Primary infection* refers to the first infection with either HSV-1 or HSV-2 in an immunologically naïve host. Subsequent infection by the other virus is often called *nonprimary initial infection*. For example, a person with *primary infection* by HSV-1 is still at risk for *nonprimary initial infection* caused by sexual acquisition of HSV-2.

Because HSV-infected persons are so frequently asymptomatic, most large surveys of HSV epidemiology rely on assays that detect antibodies to HSV-1 and HSV-2 in sera. In the United States, data from the National Health and Nutrition Examination Surveys (NHANES) for 2015 through 2016 suggest an HSV-2 seroprevalence of 11.9% in persons aged 14 to 49 years, with prevalence increasing with age. This represents a decrease from the HSV-2 seroprevalence of 16.2% reported in NHANES for 2005 through 2008 and 21.9% for 1988 through 1994. Consistently, only approximately 20% of persons with HSV-2 antibodies report having been diagnosed with genital herpes. In contrast, between 1999 and 2004, seroprevalence of HSV-1 was 57.7%, and 1.8% of those persons had a history of a

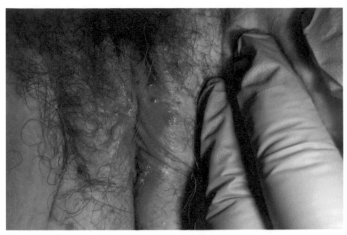

**Fig. 55.1** Shallow ulceration on the erythematous base on the left labium.

genital herpes diagnosis. More recent surveys note a decline in HSV-1 prevalence, especially in younger age groups, reporting a prevalence of 47.8%.

HSV-2 seroprevalence varies widely across different populations. Women are more susceptible to genital HSV-2 infection and bear a greater burden of disease. In the United States, non-Hispanic black persons have the highest HSV-2 seroprevalence. Because infection persists in the host, seroprevalence also increases with age. The highest HSV-2 seroprevalences, reaching almost 100%, have been reported in persons with HIV infection and in female commercial sex workers, especially in sub-Saharan Africa.

Risk factors for genital herpes acquisition, as with other STIs, include a higher number of lifetime sex partners and a history of unprotected sex. It is interesting to note that risk factors prevalent in the community from which one chooses sex partners are more influential than individual sexual behavior in estimating the risk of HSV-2 infection. For example, African American women with few sexual partners remain at increased risk of acquiring HSV-2 because of the prevalence of the infection among African American men. Condom use reduces the risk for HSV-2 acquisition, especially among women, and prior infection with HSV-2 appears to protect against subsequent infection with HSV-1 but not vice versa.

## CLINICAL FEATURES

Genital infections with both types of HSV have similar presentations. A single episode of genital herpes cannot be attributed to HSV-2 or HSV-1 by history or physical examination alone. The pattern of disease recurrence provides important information, though, as HSV-2 recurs more frequently than HSV-1, especially after the first year of infection. With both viruses, true primary episodes are the most severe, followed by nonprimary initial episodes. Recurrent episodes are the least severe. However, substantial overlap exists in the severity of all types of episodes. Symptoms of the primary episode usually occur within 2 to 12 days of inoculation. Serologic studies have shown that some patients may have a delayed presentation and recognize clinical symptoms only months or years after the infection is established. Such a presentation is termed the *first recognized episode*, as serologic testing can document the presence of a mature antibody profile. Many patients who have clinical disease report atypical or mild manifestations that are not recognized as genital herpes by the patient or the healthcare provider.

## Primary Episode

Genital herpes manifests classically as clusters of painful vesicles on an erythematous base. The primary episode can last 2 to 3 weeks. During this time the lesions progress to form pustules and then shallow ulcers. New lesions form while older ones coalesce and crust over; therefore lesions in various stages are found at the time of presentation. Mucosal lesions are typically ulcerative without a detectable vesicular stage. Lesions can occur on the external genitalia of either sex as well as on the upper thighs, buttocks, and in the perianal region (Fig. 55.2). Of note, primary lesions are often bilateral, although recurrent lesions may be bilateral as well. Local tender lymphadenopathy and cervicitis may be present. Primary herpes proctitis can also occur in patients engaging in receptive anal intercourse, and ulcers may be seen with anoscopy or sigmoidoscopy.

Atypical lesions can have an appearance ranging from papules to macular lesions, fissures, or excoriations. A spectrum of symptoms— including itching, burning, dysuria, and urethral discharge—can occur, and lesions may not be obvious on visual inspection. Patients usually report headache, fevers, malaise, and myalgias accompanying the primary genital infection. Nonprimary initial infections with HSV-2 in patients already infected with HSV-1 tend to be milder.

## Recurrent Episodes

Recurrences are more frequent with genital HSV-2 infection than with HSV-1, thus underscoring the need to differentiate the infecting virus. Most patients with genital HSV-2 infection and around 50% of those with genital HSV-1 infection experience a recurrence within 1 year of the first episode. Among patients who experience recurrences, those with genital HSV-1 typically have a median of one recurrence in the first year of infection and only a few afterward. In contrast, patients with genital HSV-2 have a median of four recurrences per year, and 20% have more than 10 recurrences in the first year after a primary episode.

With reactivation of either virus, patients frequently report a local prodrome that consists of itching, tingling, or pain before the development of a frank lesion. Recurrent episodes tend to be milder and shorter than the primary episode, lasting 4 to 7 days on average. Compared with the primary episode, patients have fewer lesions, usually in a unilateral distribution, and typically lack systemic symptoms (see Fig. 55.2). Recurrent lesions can occur on or near the genitalia, and HSV infection should always be considered during evaluation of lower abdominal, lower back, thigh, or buttock sores. In men, recurrences are typically seen on the shaft of the penis, not on the glans or in the urethra. In women, recurrences usually occur on the vulva. Recurrent lesions can also affect the perianal area, even in patients without a history of receptive anal intercourse, because of the shared nerve supply with other genital sites of primary infection. As with primary infection, atypical presentations are common. Patients and clinicians often confuse genital herpes sores with minor superficial trauma (e.g., penis caught in the zipper or trauma from intercourse), tinea cruris, vulvovaginal candidiasis, or other irritating skin abnormalities.

The triggers for recurrent episodes are incompletely understood, and there is usually no identifiable predisposing event. Psychological stress has been reported as a trigger, and animal data support the concept of physiologic stress resulting in HSV reactivation. The frequency and severity of recurrent genital herpes is significantly increased in immunocompromised persons such as HIV-infected and transplant patients, highlighting the role of cellular immunity in containing HSV infection.

Periodic asymptomatic reactivation and shedding of virus are universal features of HSV-1– and HSV-2–seropositive individuals. Asymptomatic shedding of HSV has been detected to occur as frequently as 1 day in 4 in some studies. Recent research has established

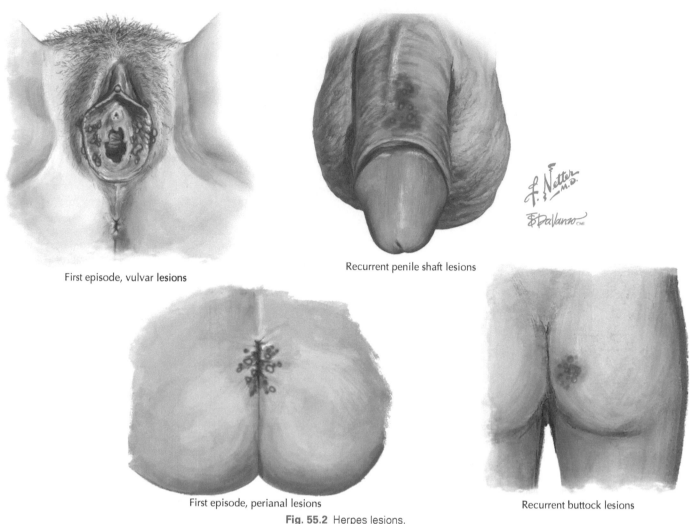

First episode, vulvar lesions

Recurrent penile shaft lesions

First episode, perianal lesions

Recurrent buttock lesions

**Fig. 55.2** Herpes lesions.

that most shedding episodes are short, with about half lasting less than 12 hours. Shedding episodes of longer duration and with higher viral loads are more likely to produce clinical disease.

## Differential Diagnosis

HSV-2 is by far the most common cause of genital ulcer disease, and genital herpes should always be considered in persons with a compatible presentation. The differential diagnosis of genital sores includes genital herpes, primary syphilis, chancroid, and lymphogranuloma venereum (LGV). Some genital herpes lesions are occasionally confused with herpes zoster, especially if they occur at a nongenital site. Laboratory testing is required for definitive diagnosis. It is important to note that genital herpes can manifest with only mild discomfort and dysuria, as can genital *Neisseria gonorrhoeae* and *Chlamydia trachomatis* infections. Given the high prevalence of these infections in patients seeking care for STIs, screening for them is appropriate in all patients in whom genital herpes is a consideration. Rarely, lesions caused by human papillomavirus or scabies may be confused with genital herpes infection. Primary Epstein-Barr virus (EBV) infection can also occasionally cause genital ulcerations. Finally, malignancies such as squamous cell carcinoma and inflammatory conditions—including fixed drug eruption, Behçet disease, and mucocutaneous manifestations of Crohn's disease—should be considered in patients with persistent ulcerative lesions. If the initial evaluation of a persistent genital lesion is negative, a biopsy should be considered.

## Complications

Important complications of genital herpes include aseptic meningitis and disseminated infection. Some degree of central nervous system involvement occurs in many primary HSV infections, although the need for hospitalization is uncommon. Aseptic meningitis as a complication of genital herpes in adults is seen more often with HSV-2 than HSV-1 and is characterized by fever, severe headache, stiff neck, photophobia, and vomiting. Onset usually occurs within 10 to 14 days of the primary genital lesion. The cerebrospinal fluid (CSF) shows a lymphocytic pleocytosis with mild elevation of protein; the CSF glucose is often normal. Diagnosis should be confirmed with deoxyribonucleic acid (DNA) amplification testing for HSV on the spinal fluid. Central nervous system involvement in genital herpes infections typically has a benign prognosis, although it can recur and is an indication for long-term suppressive antiviral therapy. Mollaret meningitis, or idiopathic benign recurrent lymphocytic meningitis, is now considered to be caused by recurrent HSV-2 infection. Other possible neurologic sequelae of genital herpes include hyperesthesia and anesthesia in affected sacral dermatomes, urinary retention, and transverse myelitis. Herpes encephalitis is a life-threatening infection requiring hospitalization and intravenous acyclovir but mostly arises as a complication of orolabial HSV-1 infection.

Disseminated herpes simplex infection is also life-threatening and can occur as sequela of genital infections. It happens most frequently in immunocompromised patients but has also been described in primary infections of normal hosts. Fulminant hepatitis is common with visceral dissemination. Pregnancy is a recognized risk factor for HSV dissemination and hepatitis. The diagnosis of HSV hepatitis is often missed, and the subsequent mortality is high. Cutaneous vesicular lesions may be seen with dissemination but are not necessarily the dominant feature. Patients with disseminated herpes infection are very ill and require emergent hospitalization, even in the absence of a definitive diagnosis, with intravenous antiviral therapy. Such patients should be managed in consultations with experts in infectious diseases.

Although patients are often concerned about cancer risk, HSV infections are not associated with the development of malignancy.

## NEONATAL HERPES

Neonatal herpes is an infrequent but devastating consequence of maternal genital herpes infection. Most cases result from neonatal exposure to HSV-1 or HSV-2 in the birth canal at delivery. The highest risk to the neonate occurs when the mother acquires primary or nonprimary initial genital herpes infection late in pregnancy. In such cases, the risk arises from the absence of circulating maternal antibodies and from a high maternal mucosal viral load in the early stages of infection. Viral transmission to the neonate most commonly occurs during asymptomatic shedding. Reactivation of genital HSV acquired before the third trimester can also lead to neonatal infection, but this occurs relatively infrequently, given the prevalence of genital HSV and the frequency of asymptomatic shedding. Other factors that contribute to neonatal risk include prolonged time with ruptured membranes and mucocutaneous injury to the neonate at birth; both provide a portal of entry for the virus. Prevention strategies for neonatal herpes are discussed later.

## GENITAL HERPES AND HUMAN IMMUNODEFICIENCY VIRUS

As noted earlier, HIV-infected patients have very high rates of HSV-2 infection. This is true partially because of behaviors associated with acquisition of both infections but also because genital HSV-2 increases the risk of HIV acquisition, and HIV infection may increase the risk of HSV-2 acquisition. Genital ulcerations facilitate HIV exposure, and the inflammatory infiltrate of herpetic lesions is rich in HIV target cells. The increased risk of HIV acquisition in HSV-2 seropositive persons persists, even in persons using preexposure antiretroviral therapy (PrEP). However, large randomized controlled trials have failed to show a benefit of suppressive therapy with acyclovir in reducing either the acquisition or transmission of HIV.

Patients coinfected with HIV and HSV-2 have more frequent clinical recurrences of genital herpes and substantially increased rates of asymptomatic shedding of HSV-2. In addition, HSV-2 reactivation leads to increased plasma HIV RNA and increased HIV mucosal shedding. Given the burden of HSV-2 in HIV-infected persons, it is reasonable to test them for HSV infection and consider long-term suppressive therapy even in the absence of a recurrence history. Long-term therapy with antiretrovirals reduces the clinical symptoms and shedding in persons with HIV. However, as part of the immune reconstitution syndrome, both lesions and shedding tend to be more frequent in the first several months following initiation of antiretrovirals.

## PATHOPHYSIOLOGY

Both types of HSV have tropisms for epithelial cells and neurons. New infection occurs with viral entry via mucosal surfaces or skin breaks and penetration into the epithelial layers. Viral replication

in epidermal and dermal cells leads to cytolysis and release of large quantities of new virus that subsequently spread locally to neighboring epithelial cells as well as to sensory and autonomic nerve endings. Epithelial cell death and detachment caused by this process lead to lesion formation. Vesicular lesions contain cell debris, inflammatory cells, and free virus. The histology is notable for epithelial cell necrosis, with basal ulceration or vesicle formation, multinucleated giant cells with intranuclear viral inclusions, and an inflammatory infiltrate of CD4- and CD8-receptor–positive T cells as well as neutrophils.

After infecting peripheral nerve endings, virus migrates up the axons and replicates in nerve cell bodies in the lumbosacral nerve root ganglia but is not thought to cause nerve cell death. In primary infection, virus spreads within the ganglia and down the axons of newly infected nerve cells to cause lesions at other mucocutaneous sites. Thus the lesions in primary infection are distributed across the territory of multiple nerves, and recurrent lesions may occur at sites different from the original site of inoculation.

After genital infection, HSV establishes latency in the lumbosacral nerve root ganglia, where it persists throughout the life of the host. Successful evasion of the immune response results in periodic viral reactivation, and reactivated virus migrates down the axon to mucocutaneous sites at or near the original site of infection. Asymptomatic shedding of virus at these peripheral sites is frequent. Clinical recurrences are associated with an increased quantity of virus and viral spread within the epithelium to a degree sufficient to cause noticeable lesions. Autonomic nerves are also infected by HSV, but for unclear reasons viral reactivation from autonomic nerves seems to be rare.

The mechanisms of latency and the biologic determinants of recurrent episodes of genital herpes versus asymptomatic shedding are incompletely understood, as noted earlier. Only one HSV DNA transcript is expressed in latently infected nerve cells, and it produces no protein but rather seems to regulate or suppress transcription of other viral genes. The factors that lead to viral reactivation and influence the quantity of virus released are unknown. When clinical lesions do form, CD4-receptor–positive T cells and neutrophils are found early in the local inflammatory infiltrate. Research has shown that infiltration of HSV-2–specific CD8-receptor–positive T cells into lesions is associated with local viral clearance. These cells persist for weeks after lesion healing; tissue-resident memory cells may persist indefinitely. Notably, several viral gene products have been shown to have immunomodulatory activity that may contribute to latency and recurrent disease.

## LABORATORY TESTING

Clinical diagnosis of genital herpes has low sensitivity and specificity; therefore confirmatory laboratory testing is essential. Patients with active genital lesions or a history of recurrences should be tested to make the diagnosis of HSV disease and, if needed, to exclude other causes of genital ulcers. Testing should be performed with a methodology that will distinguish HSV-1 and HSV-2 ("type-specific" testing), as this will inform the prognosis for recurrences. Several testing options are available (Table 55.1). If the clinical presentation is strongly suggestive of genital herpes, treatment should begin before the test results return or if laboratory testing is not readily available.

Serologic testing for HSV-1 and HSV-2 provides evidence of past infection and is commercially available, yet it has limitations. Type-specific serologic testing identifies immunoglobulin G (IgG) antibodies to viral glycoprotein G, which differs between HSV-1 and HSV-2, allowing providers to distinguish these infections. In acute HSV infection, results of routine serologic testing for HSV-1 and HSV-2 will be negative because the IgG antibody response is delayed and seroconversion takes 2 to 12 weeks. Accordingly, a negative antibody test result

| TABLE 55.1 | **Diagnostic Tests for Genital Herpes** | |
|---|---|---|
| Method[a] | Clinical Utility | Limitations |
| Serology[b] | Positive HSV-2 serology indicates genital herpes infection. Isolated positive HSV-1 serology may indicate genital infection.[c] | Takes 2–12 weeks to turn positive following acquisition; negative during acute infection. Does not test lesions directly. Commercial tests have limited accuracy. |
| PCR | Most sensitive test to diagnose lesions. Swab vesicle fluid or ulcer base to obtain specimen. | |
| Culture | Use to diagnose lesions. Specimen collection as with PCR. | Low sensitivity, especially in old or recurrent lesions. |

[a]All methods shown can differentiate HSV-1 and HSV-2. Such typing should be ordered routinely.
[b]Serology = type-specific HSV glycoprotein G serology.
[c]A positive HSV-1 serology may be challenging to interpret in a person who lacks a history of genital or oral herpes.
*HSV,* Herpes simplex virus; *PCR,* polymerase chain reaction.

in early infection does not rule out the diagnosis of HSV, and repeat testing may be indicated if there is a high suspicion for genital herpes. In addition to laboratory-based serologic testing, point-of-care tests have been developed for HSV-2. Unfortunately, in practice the commercially available HSV serologic assays appear to fall short of the accuracy shown in preclinical evaluations. These assays yield false-positive results, especially at lower index values. While the manufacturer's cutoff for positivity is 1.1, only about 50% of results with index values of 1.1 to 3.5 can be confirmed with the University of Washington western blot assay (see further on). For HSV-1, the commercial antibody assays lack sensitivity; thus HSV-1 infections may be missed by these tests. HSV-2 serology results with low index values (1.1 to 3.5) should be confirmed with either the gold standard for HSV antibody testing, the University of Washington western blot, or the Biokit, which is sometimes more easily accessible.

No reliable HSV IgM antibody testing is commercially available to diagnose acute or recent infection, and IgM can be detected in recurrent infection, precluding its use to define new acquisition of HSV.

HSV-1 seropositivity is frequently a result of primary oral infection, but serologic testing does not distinguish the site of infection. Serologic evidence of HSV-2, however, is virtually always indicative of genital infection. The combination of compatible genital ulcer disease and serologic evidence of HSV-2 infection is sufficient to make the diagnosis; however, other etiologies including syphilis and LGV should still be ruled out. Persons with HSV-2 antibodies but without any current symptoms are still considered to have genital herpes; they are at risk to transmit the infection to sexual partners and, for childbearing persons, to neonates.

Because of the inherent limitations of serologic testing, additional laboratory workup may be indicated in some symptomatic individuals. Direct testing of the lesion can be done with DNA amplification testing or viral culture. The most sensitive tests are based on the PCR amplification of viral DNA obtained with a swab from a lesion. HSV PCR is available in many laboratories and is routinely used for CSF testing, but it is also increasingly used for the diagnosis of genital ulcer disease. One caveat is that HSV-infected patients can have asymptomatic shedding of the virus even with genital lesions from another cause, although this situation is unusual. If the clinical presentation is highly suggestive of an alternative diagnosis, this should not necessarily be excluded based on a positive HSV PCR test alone. Viral culture from affected lesions is very specific but is substantially less sensitive than PCR, especially from healing lesions, and many laboratories no longer perform viral cultures. Type-specific fluorescent antibody (FA) tests can be used to test cellular material from HSV lesions, but they have largely been eclipsed by other testing methods. A Tzanck smear from

the base of a herpes lesion may show viral inclusion bodies, but these are nonspecific and the test has limited clinical utility compared with newer methods.

If genital herpes is suspected, providers can collect samples for these tests by unroofing a vesicle and rubbing its base, where the virus is replicating in epithelial cells, with a swab. Intact vesicles may not be present, but ulcer bases can still harbor virus. With a compatible lesion, if the PCR assay result or culture is positive, clinicians can be confident in the diagnosis of genital herpes. In this setting, a negative serologic test result with positive PCR assay result or culture suggests primary genital herpes infection. Some laboratories require additional requests to perform HSV typing by culture or PCR; typing should be ordered routinely, as it is essential for discussing the prognosis.

## MANAGEMENT

As with other chronic diseases and STIs, management of genital herpes involves counseling patients in addition to providing appropriate medical therapy to treat episodes and suppress recurrences. Prevention of infection in sexually active patients, reproductive-aged women, and the partners of infected persons is an important aspect of the management of genital herpes.

### Counseling

With any diagnosis of genital herpes, providers should counsel their patients about the meaning of test results, the natural history of HSV infection, transmission risks, and treatment options. Counseling may need to take place at a visit subsequent to initial diagnosis, as patients often cannot comprehend additional information beyond the diagnosis of genital herpes at that time. Patients will want to know about the risk and severity of recurrences as well as possible complications. They are frequently concerned about transmitting the infection to their sex partners and should be informed that transmission can occur even in the absence of active lesions. Sex partners should be aware that they might already be infected even if they have no symptoms. It is appropriate to recommend disclosure of genital herpes diagnosis to sex partners, avoidance of sex if active lesions or prodrome symptoms are present, use of suppressive antiviral therapy when one is in a relationship with a susceptible partner, and condom use. These measures substantially decrease but do not eliminate the risk of transmission.

The possibility of neonatal herpes should be discussed with all patients, including men. The risk of neonatal herpes is greatest for infants born to women newly infected late in pregnancy. Because genital herpes can have implications for management at delivery, the diagnosis should be conveyed to the obstetrician and the pediatrician

## TABLE 55.2    Oral Treatment Options for Genital Herpes

|  | Acyclovir | Valacyclovir | Famciclovir |
|---|---|---|---|
| First episode, all hosts | 400 mg three times a day for 7–10 days[a]<br>200 mg five times a day for 7–10 days | 1000 mg twice a day for 7–10 days | 250 mg three times a day for 7–10 days |
| **Recurrent Episodes**<br>Immunocompetent host | 400 mg three times a day for 5 days<br>800 mg twice a day for 5 days<br>800 mg three times a day for 2 days | 500 mg twice a day for 3 days[b]<br>1000 mg daily for 5 days | 125 mg twice a day for 5 days<br>1000 mg twice a day for 1 day |
| Immunocompromised | 400 mg three times a day up to 800 mg<br>    five times a day for 5–10 days | 1000 mg twice a day up to 1000 mg<br>    three times a day for 5–10 days | 500 mg twice a day up to 750 mg<br>    three times a day for 5–10 days |
| **Suppression**<br>Immunocompetent host | 400 mg twice a day | 500–1000 mg once a day | 250 mg twice a day |
| Immunocompromised | 400–800 mg two to three times a day | 500 mg twice a day | 500 mg twice a day |

[a]All durations may be extended for persistent disease.
[b]May be less effective than other regimens in patients with ≥10 recurrences a year.
Adapted from Centers for Disease Control and Prevention (CDC), Workowski KA, Berman SM. Sexually transmitted diseases treatment guidelines, 2015. *MMWR Recomm Rep* 2015;64:1-137.

of the newborn. Reproductive ability is not compromised by genital herpes infection.

## Medical Therapy

There are several treatment options for primary and recurrent genital herpes as well as for viral suppression (Table 55.2). Antiviral therapy can be administered for an individual episode of HSV, or daily to abrogate most subsequent recurrences during the therapy. In patients with a clinical syndrome compatible with a primary or nonprimary first episode of genital herpes, antiviral therapy should be administered even before diagnostic testing is completed, as it reduces the severity of primary infection and prevents neurologic complications.

The antiviral medications acyclovir, valacyclovir, and famciclovir have all been shown to reduce the severity and duration of genital herpes symptoms. These agents form nucleoside analogues in infected cells that impair the function of the viral DNA polymerase, thereby halting viral replication. Human polymerases are unaffected. Valacyclovir is the prodrug of acyclovir, and famciclovir is the prodrug of penciclovir. The prodrugs are available only in oral form but are more efficiently absorbed in the digestive tract than acyclovir or penciclovir, thus allowing less frequent administration. Acyclovir is available in intravenous, oral, and topical formulations. Initial intravenous acyclovir therapy may be necessary for very severe primary genital herpes cases, immunocompromised patients, and patients with complications of genital infection. Topical therapies for genital herpes, such as topical acyclovir, have not been shown to have substantial clinical utility, and their use is discouraged.

Acyclovir, valacyclovir, and famciclovir are typically well tolerated. Rarely, renal dysfunction caused by crystallization of drug in the renal tubules can be seen in patients receiving intravenous acyclovir, usually for other indications and in patients with comorbid conditions. Type 1 allergic responses to acyclovir have also been reported, and desensitization has been used successfully.

Acyclovir-resistant HSV-2 is infrequent and almost always occurs in severely immunocompromised patients who have received prior antiviral therapy. Resistance is not a concern in immunocompetent hosts, even when they are on long-term suppressive therapy. Resistance testing is indicated in cases of clinical failure of antiviral therapy. In immunocompromised patients with documented or suspected acyclovir resistance, intravenous foscarnet could be administered. Such

patients should be managed in consultation with an expert. If intravenous foscarnet is contraindicated because of renal failure, anecdotal reports suggest that topical foscarnet, cidofovir, or imiquimod may be of benefit.

Primary genital herpes outbreaks should be treated for 7 to 10 days, but a longer course may be used if lesions persist, new lesions appear, or the patient is immunocompromised. Recurrent episodes can be treated with a shorter course. The standard approach is 2 to 5 days, with a variety of dosing regimens available. Antiviral therapy does not eradicate HSV from the body and will not prevent the virus from establishing latency, nor will it abrogate future recurrences once therapy is stopped. Patients opting for episodic therapy of recurrent episodes should be encouraged to have medication on hand and to begin therapy as soon as they notice prodromal symptoms. Prompt initiation of antiviral medication results in reduction in the duration of recurrences by 1 to 2 days.

Suppressive therapy with daily oral acyclovir has been shown to reduce the frequency of asymptomatic shedding, and, when given to patients with a history of genital herpes, it reduces the risk of transmission of HSV-2 to sexual partners in heterosexual couples by half. Long-term oral acyclovir usage in the doses used for suppression of genital herpes is safe, and no laboratory monitoring is needed. Fewer long-term data are available for valacyclovir and famciclovir, although clinically people often use valacyclovir because it can be dosed once daily. Asymptomatic source patients and same-sex couples were not included in this transmission study, but therapy should be considered for persons in these situations if there is a susceptible partner. Other characteristics that support the use of suppressive therapy include frequent recurrences, psychologically or physically bothersome recurrences, and patients with recurrent aseptic meningitis symptoms. The need for continued suppressive therapy should be assessed in a discussion with the patient annually, but it is not necessary to interrupt therapy in a patient who desires to continue daily antiviral treatment.

Novel antivirals are in development but have not been licensed for clinical use.

## PREVENTION AND CONTROL

In the United States, a prevention strategy for genital herpes has not been designed and may not be possible without an effective prophylactic vaccine. Therefore prevention can be considered only in the setting of patient management rather than as a public health intervention. In sexually active individuals, certain practices can decrease the risk of infection. As noted earlier, condom use has been shown to decrease transmission, especially transmission from men to women. In serodiscordant couples, the risk of transmission is decreased but not eliminated when the partner with genital herpes is treated with daily antiviral therapy. Long-term suppressive therapy should be offered to such couples in the hope of mitigating the risk of transmission. Suppressive antiviral therapy may also have a benefit in reducing transmission from infected patients who have multiple sex partners, although this has not been evaluated.

The risk of neonatal herpes can be decreased with various strategies. Susceptible pregnant women should avoid sexual contact with infected partners during the third trimester. Suppressive therapy for the partner may reduce the risk of infection, but pregnant women may have heightened susceptibility, and data on the efficacy of this approach are lacking. The risk of neonatal transmission can be decreased with cesarean section before membrane rupture in women who are actively shedding virus; current management strategies do not identify such women. Suppressive therapy starting at week 36 of gestation has been shown to reduce the risk of cesarean sections done because it reduces the frequency of HSV lesions at delivery. Whether suppressive therapy mitigates the risk of neonatal herpes is unknown, and failures to prevent neonatal herpes have been reported among women who received oral antivirals toward the end of pregnancy. Acyclovir and probably valacyclovir can be used safely during pregnancy.

Vaccination is likely the key to genital herpes prevention. An HSV vaccine must reduce the incidence of infection and viral shedding in breakthrough infections to have a public health impact. Currently there is no commercially available vaccine for either HSV-1 or HSV-2; prior candidate vaccines have failed in clinical trials.

## SUMMARY

Genital herpes infections are widespread and cause patients significant distress. The spectrum of disease is broad, ranging from latent infection with only asymptomatic viral shedding to significant recurrent and painful ulcerations. Many patients have atypical symptoms, and genital herpes should be considered in any comprehensive evaluation of skin lesions occurring on or near the external genitalia and in any workup for STDs. Several laboratory methods are available to make an accurate diagnosis of genital herpes, and these methods should be used to confirm clinically consistent lesions. Effective antiviral therapies are available to treat primary and recurrent disease, and both medical and behavioral strategies can modify the risk of sexual transmission of the infection. There is an urgent need to develop commercial, type-specific antibody assays with high sensitivity and specificity for asymptomatic infection and a therapeutic and prophylactic vaccine.

## EVIDENCE

Benedetti J, Corey L, Ashley R. Recurrence rates in genital herpes after symptomatic first-episode infection. *Ann Intern Med* 1994;121:847-854. *The authors report on rates and correlates of symptomatic genital herpes recurrence after a documented first episode in 457 patients with new HSV-2 infection. Median follow-up time was 391 days.*

Benedetti J, Zeh J, Corey L. Clinical reactivation of genital herpes simplex virus infection decreases in frequency over time. *Ann Intern Med* 1999;131:14-20. *The authors report on changes in recurrence rates over time in 664 patients with genital herpes caused by either HSV-1 or HSV-2. The study included patients taking suppressive therapy and patients not taking it. Some patients were followed for more than 9 years.*

Brown ZA, Selke S, Zeh J, et al. The acquisition of herpes simplex virus during pregnancy. *N Engl J Med* 1997;337:509-515. *In this prospective study, in 7046 pregnant women at risk for HSV acquisition the risk of vertical transmission of HSV to the neonate was highest for women infected late in pregnancy.*

Brown ZA, Wald A, Morrow RA, et al. Effect of serologic status and cesarean delivery on transmission rates of herpes simplex virus from mother to infant. *JAMA* 2003;289:203-209. *The authors report that in 202 pregnant women from a large cohort study found to be shedding genital HSV at delivery, cesarean section greatly reduced risk of subsequent neonatal herpes. Also, risk was much higher for offspring of women shedding HSV-2 but without HSV-2 serum antibodies at labor than for infected women with serum antibodies.*

Corey L, Adams HG, Brown ZA, Holmes KK. Genital herpes simplex virus infections: clinical manifestations, course, and complications. *Ann Intern Med* 1983;98:958-972. *The authors describe the symptoms and disease course of 648 patients with first and recurrent episodes of genital herpes.*

Corey L, Wald A, Patel R, et al. Once-daily valacyclovir to reduce the risk of transmission of genital herpes. *N Engl J Med* 2004;350:11-20. *The authors report on a double-blind randomized controlled trial of suppressive therapy with valacyclovir for genital HSV-2 infection in monogamous couples in which one partner was infected and one was not. Suppressive therapy for the infected partner reduced transmission of HSV-2 to the susceptible partner by half compared with placebo.*

Douglas JM, Critchlow C, Benedetti J, et al. A double-blind study of oral acyclovir for suppression of recurrences of genital herpes simplex virus infection. *N Engl J Med* 1984;310:1551-1556. *A double-blind randomized controlled trial that established the efficacy of acyclovir in reducing recurrences of genital herpes.*

Langenberg AG, Corey L, Ashley RL, et al. A prospective study of new infections with herpes simplex virus type 1 and type 2. *N Engl J Med* 1999;341:1432-1438. *In this prospective cohort study, the authors report incidence, risk factors, and clinical presentations of HSV-1 and HSV-2 infections in 2393 initially HSV-2 seronegative subjects followed for more than 3000 cumulative person-years. Many genital infections were asymptomatic, and new HSV-1 infections were as likely to be genital as oral.*

Looker KJ, et al. Global estimates of prevalent and incident herpes simplex virus type 2 infections in 2012. *PLoS ONE* 2015;10(1):e114989. *A study of the prevalence and incidence of HSV-2 world-wide that estimated around 417 million between ages 15 to 49 were living with the infection.*

Magaret AS, Mujugira A, Hughes JP, et al. Effect of condom use on per-act HSV-2 transmission risk in HIV-1, HSV-2-discordant couples. *Clin Infect Dis* 2016;62(4):456-461. doi:10.1093/cid/civ908. *These authors studied the transmission of HSV-2 in HIV positive serodiscordant couples with and without condoms and noted significant decrease of transmission with condom use.*

Mark KE, Wald A, Magaret AS, et al. Rapidly cleared episodes of herpes simplex virus reactivation in immunocompetent adults. *J Infect Dis* 2008;198:1141-1149. *In this study the authors detected frequent, short bursts of asymptomatic oral and anogenital viral shedding in 43 HSV-1 or HSV-2 infected patients using PCR DNA amplification techniques.*

Martin ET, Krantz E, Gottlieb SL, et al. A pooled analysis of the effect of condoms in preventing HSV-2 acquisition. *Arch Intern Med* 2009;169:1233-1240. *In this meta-analysis, prospective data on condom use from 5384 initially HSV-2 seronegative subjects from six studies were analyzed. Subjects who used condoms 100% of the time had a 30% lower risk of HSV-2 acquisition than those who never used condoms.*

Posavad CM, et al. Frequent reactivation of herpes simplex virus among HIV–1–infected patients treated with highly active antiretroviral therapy. *J Infect Dis* 2004;190(4):693-696. doi:10.1086/422755. *The authors discuss their findings from 28 HAART treated and 49 untreated subjects with HIV-1 and HSV-2 in regard to herpes reactivation.*

Xu F, Sternberg MR, Kottiri BJ, et al. Trends in herpes simplex virus type 1 and type 2 seroprevalence in the United States. *JAMA* 2006;296:964-973. *The authors report on HSV-1 and HSV-2 seroprevalence rates in different demographic groups in NHANES, a large study designed to evaluate a representative sample of the US population. Using these data, they estimate the national prevalence of both viruses.*

Zhu J, Hladik F, Woodward A, et al. Persistence of HIV-1 receptor–positive cells after HSV-2 reactivation is a potential mechanism for increased HIV-1 acquisition. *Nat Med* 2009;15:886-892. *The authors report on the composition and long duration of the inflammatory infiltrate at HSV-2 genital lesion sites, even after lesion healing and in the setting of antiviral therapy. HSV-2–specific inflammatory cells and cell types with HIV receptor targets are present in the milieu.*

## ADDITIONAL RESOURCES

American Social Health Association (ASHA): Herpes resource center. Available at: www.ashastd.org/herpes/herpes_overview.cfm. Accessed August 5, 2010. *An online resource for patients seeking information about genital herpes.*

Centers for Disease Control and Prevention (CDC), Workowski KA, Berman SM. Sexually transmitted diseases treatment guidelines, 2015. *MMWR Recomm Rep* 2015;64:1-137. *Evidence-based treatment guidelines for genital herpes and other STIs.*

Fleming T, et al. Herpes simplex virus type 2 in the United States, 1976 to 1994. *N Engl J Med* 1997;337:1105-1111. *A study that reviewed the data collected by the National Health and Nutrition Examination questionnaire on seroprevalence on HSV-1 and HSV-2 between 1976 and 1994.*

Gupta R, Warren T, Wald A. Genital herpes. *Lancet* 2007;370:2127-2137. *A detailed review of the pathophysiology, natural history, and diagnosis of genital herpes.*

McQuillan G, Kruszon-Moran D, Flagg EW, Paulose-Ram R. Prevalence of herpes simplex virus type 1 and type 2 in persons aged 14–49: United States, 2015–2016. NCHS Data Brief, no 304. Hyattsville, MD: National Center for Health Statistics; 2018. *The authors reviewed the National Health and Nutrition Examination Survey in regard to HSV-1 and HSV-2 seroprevalence in the United States between 2015 and 2016.*

Schiffer JT, Corey L. Herpes simplex virus. *Mandell, Douglas, and Bennett's Principles and Practice of Infectious Diseases,* 8th ed., Saunders, 2015, pp. 1713-1730. *A comprehensive review of sexually transmitted HSV infections in humans.*

# 56

# Human Immunodeficiency Virus and Acquired Immunodeficiency Syndrome

*Eric A. Meyerowitz, Shireesha Dhanireddy*

## ABSTRACT

Human immunodeficiency virus (HIV) is a sexually transmitted infection that causes progressive immune dysfunction, putting affected individuals at risk for opportunistic infections and certain cancers. It was first recognized in the early 1980s, when men who had sex with men (MSM) in several cities in the United States presented with an unusual fungal pneumonia caused by *Pneumocystis carinii* (now called *Pneumocystis jiroveci*) and others with a rare cancer called Kaposi sarcoma. Over the decades, HIV has been identified in all countries, although it continues to disproportionately affect certain groups. After acquiring the infection, most untreated individuals remain asymptomatic for years, eventually progressing to acquired immunodeficiency syndrome (AIDS) and developing clinical symptoms from opportunistic infections. Treatment for HIV has improved dramatically over the past three decades. Today HIV can be managed as a chronic disease, and most virally suppressed people are predicted to have a near normal life expectancy. With ongoing prevention efforts and new treatment innovations, the field is continually advancing toward ending the HIV epidemic, but gaps remain in treating all infected individuals and preventing new infections.

## CLINICAL VIGNETTE #1

A 26-year-old man who has sex with men presented to the sexual health clinic with rectal pain with defection as well as a new rash that included his palms and soles. He was afebrile with a blood pressure of 125/78 mmHg, pulse 78 beats/min, and oxygen saturation 99% on ambient air. He had a normal neurologic examination, including no visual symptoms or tinnitus. He had erythematous rectal mucosa with some bleeding and purulent exudate. Testing revealed that the patient was HIV negative by the fourth-generation HIV 1/2 Ab/Ag test, treponemal Ab positive with a rapid plasma reagin (RPR) titer of 1:128, and positive for *Chlamydia trachomatis* on nucleic acid amplification testing of the rectal exudate. The patient was treated for syphilis and chlamydial proctitis and started on tenofovir disoproxil fumarate with emtricitabine for preexposure prophylaxis (PrEP) to prevent HIV acquisition.

## CLINICAL VIGNETTE #2

A 31-year-old woman who injects drugs presented to the emergency department with fevers and altered mental status, requiring intubation for airway protection. Her initial vital signs showed that she was febrile, with a temperature of 102.9°F and tachycardic, with a pulse of 115 beats/min. Her blood pressure was 110/76 mmHg and her oxygen saturation 96% on ambient air. Blood cultures were drawn, and she was started on broad-spectrum antimicrobials. Computed tomography (CT) of the head revealed no acute process and no space-occupying lesions. A cerebrospinal fluid (CSF) examination by lumbar puncture showed 84 nucleated cells (90% lymphocytes) in tube 4, with normal glucose and elevated protein. Blood and CSF cultures proved negative. Polymerase chain reaction (PCR) testing of the CSF

proved negative for herpes simplex virus (HSV). The patient's HIV 1/2 Ab/Ag test returned positive, with a differentiation assay negative for HIV-1 and HIV-2 antibodies. An HIV-1 RNA showed a viral load greater than 1 million copies per milliliter. The patient was started on antiretroviral therapy (ART) with dolutegravir and tenofovir alafenamide with emtricitabine. Thereafter her mental status normalized rapidly. Two months later she was feeling well, with an undetectable viral load.

## ORIGINS AND GEOGRAPHIC DISTRIBUTION OF HUMAN IMMUNODEFICIENCY VIRUS INFECTION

HIV infection, caused by the HIV-1 and HIV-2 viruses, originated after the cross-species transmission of simian immunodeficiency viruses from African primates to humans. HIV-1 is responsible for the vast majority of the global HIV disease burden, with a predictable illness course marked by progressive immunosuppression and, eventually, death. HIV-2, which is mostly found in West Africa, is less transmissible and progresses more slowly than HIV-1.

The latest World Health Organization (WHO) data indicate that at the end of 2017, nearly 37 million people worldwide were living with HIV, with 1.8 million new infections and 940,000 deaths occurring during 2017. The WHO data show that overall global incidence has been steadily declining since its peak in 1996, when there were greater than 3 million new infections annually. Significant geographic variation exists in the burden of HIV, as nearly 70% people living with the infection and nearly 70% of new infections are concentrated in Africa. In recent years incidence has been either stable or decreasing in all regions of the world.

In the United States, HIV incidence declined steadily until around 2013, when it plateaued. Since then there has been a significant rise in new infections among young African American and Latino MSM. High rates of new infection have been identified in transgender individuals as well. Preventive strategies have worked in some groups, but significant disparities remain. Although approximately 1.1 million Americans are likely to benefit from preventive strategies like PrEP (see later for a discussion of PrEP), only about 100,000 people were actually prescribed PrEP in 2017, and high-risk men of color were less likely to be prescribed PrEP as compared with white MSM.

## MODES OF TRANSMISSION

HIV is a sexually transmitted and blood-borne virus. Sexual transmission is the primary mode of transmission of HIV-1 infection worldwide. Sexual acts have differential risk rates for transmission, with the highest risk from receptive anal intercourse, which has a transmission rate of about 1% per sex act (138 transmissions per 10,000 exposures).

Receptive vaginal intercourse has a much lower rate of transmission of around 0.1% per sex act. Higher risk of transmission is seen in individuals with higher viral loads, as is the case during acute infection and advanced AIDS, and for people with active anogenital ulcerations as from HSV or in patients with bacterial vaginosis. Circumcised males have a lower risk of HIV acquisition compared with those who are uncircumcised.

Blood-borne exposure is another potential mode of HIV transmission. Before blood transfusion guidelines were updated, infected blood products posed a major risk for transmission, and this was the predominant mechanism for HIV acquisition among individuals with hemophilia in the early 1980s. Since intensive screening tests of the blood supply were implemented, this is no longer a risk factor for HIV transmission in the United States. Intravenous drug use continues to be an important means of transmission. Sharing hollow-bore needles for the intravenous use of recreational drugs carries an estimated risk of transmission of around 0.5 to 1% per act, again with higher rates from sources with very high viral loads. In recent years, several clusters of acute HIV infection among people who inject drugs have been seen in the United States, including in Indiana and Massachusetts.

HIV transmission in an occupational setting—as from percutaneous needlestick injuries or (even more unusual) mucous membrane exposure—is another possible mode of transmission, although it has become extremely rare in the last 20 years (with only one confirmed case since 1999). Percutaneous injuries from community needlestick exposures (such as from stepping on a discarded needle on a sidewalk) are a theoretical risk of transmission, since the virus can live outside the body for some time, but there has never been a documented case of transmission by this route.

Vertical transmission, in which the virus is passed from mother to child during pregnancy or in the peripartum period, is another important mechanism for spread of the virus. Without effective ART, perinatal transmission occurs in about 25% of pregnant women with HIV and up to about 40% with prolonged breastfeeding. The risk of transmission is highest after 36 weeks of gestation and during delivery. In recent years rates of vertical transmission have decreased significantly. With appropriate antenatal care, universal ART for mothers and the availability of chemoprophylaxis for newborns, mother-to-child transmission can be eliminated. ART interruption during the third trimester increases the risk of vertical transmission by nearly 50-fold, likely due to viral rebound.

## PREVENTION OF INFECTION

Pre- and postexposure prophylaxis is an effective way to decrease the chance of transmission of HIV. Postexposure prophylaxis (PEP) is indicated if a person has a substantial risk of HIV acquisition based on the nature of an exposure in either an occupational or nonoccupational setting. Transmission is considered possible when a potentially HIV-containing body fluid (i.e. blood, semen, vaginal secretions, breast milk, or any bodily fluid contaminated with blood) comes into contact with a mucous membrane or broken skin. PEP is a 28-day course of three active antiretroviral drugs that must be started within 72 hours of the exposure, ideally as soon as possible. If the source is HIV positive, PEP is usually recommended. If the HIV status of the source is unknown, PEP is recommended on a case-by-case basis. The Centers for Disease Control and Prevention (CDC) guidelines are helpful in counseling patients about risk of HIV acquisition based on their exposure and whether to recommend PEP. The University of California San Francisco also has a PEP consultation hotline where clinicians can get expert recommendations for managing healthcare worker exposures (available at http://nccc.ucsf.edu/clinician-consultation/

pep-post-exposure-prophylaxis/; accessed February 2020) for more details. Although there are no randomized controlled trials assessing the protective benefit of PEP, animal studies and retrospective data show that it likely has at least 90% efficacy in preventing transmission, with the greatest benefit seen when it is started early after exposure.

PrEP is approved for certain patients at elevated risk for HIV acquisition. PrEP currently includes two active antiretroviral agents (most commonly tenofovir disoproxil fumarate and emtricitabine) and, when taken regularly, has been shown in several large studies to prevent transmission more than 90% of the time. This reduction in risk is best substantiated for MSM and for heterosexual couples in serodiscordant relationships; data on women are less robust, but high levels of adherence should yield adequate protection. Screening for acute HIV, by either symptom screen or HIV RNA, should be performed for all patients who will initiate PrEP. Cases of transmission of resistant virus have rarely been reported, despite protective serum levels of prophylactic antiviral drugs, most commonly if PrEP was started during acute HIV infection that was missed on baseline testing. Additional agents, including long-acting injectables, are emerging options for PrEP.

## CLINICAL FEATURES, PATHOGENESIS, AND PROGRESSION OF HUMAN IMMUNODEFICIENCY VIRUS-1 INFECTION

HIV virus gains entry to cells by binding to the CD4 receptor and a chemokine coreceptor (most frequently CCR5 or, less frequently, CXCR4). In sexual transmission, infection is established most often by a single "founder virus" (>80% of the time), which quickly arrives at regional lymph nodes, where rapid viral replication takes place. Some individuals, perhaps around 1%, mount an effective early immune response to the infection. In the vast majority, however, immunologic response is less robust. After the initial high viremia of acute infection, a viral set point is established, with a steady decline of CD4 cells at a rate of 50 to 100 cells per year; without treatment, progression to AIDS ensues by around 8 to 10 years after initial infection. Of note, certain HLA types are associated with slower or more rapid progression to AIDS. Some individuals remain HIV negative despite repeated exposures to HIV; these include individuals with the CCR5Δ32 mutation, who lack the CCR5 chemokine coreceptor.

### Acute Retroviral Syndrome

Early infection manifests as an acute retroviral syndrome (ARS) in 40% to 90% of people who acquire HIV, which typically includes nonspecific symptoms such as fever, fatigue, rash, headache, lymphadenopathy, and pharyngitis. These symptoms typically present within days to 2 weeks after initial HIV exposure. Aseptic meningitis or meningoencephalitis can be seen in about 25% of patients with ARS. Acute HIV is notable for extremely high viral loads and a transient drop in the CD4 cell count, sometimes to below 200/mL (Fig. 56.1). Occasionally opportunistic infections can be seen during acute HIV infection, also known as primary infection. Individuals with symptomatic acute HIV lasting longer than 14 days have a faster rate of progression to AIDS than those with acute infectious symptoms of shorter duration.

### Neurologic Manifestations

Patients with HIV can have neurologic manifestations at any stage of the infection. HIV invades the central nervous system (CNS) within days of systemic infection. Neurologic symptoms are seen in 10% to 50% of patients with ARS; they include cranial and peripheral neuropathies, meningoencephalitis, acute disseminated encephalomyelitis, and/or acute inflammatory demyelinating polyneuropathy. Most of these symptoms remit completely with ART.

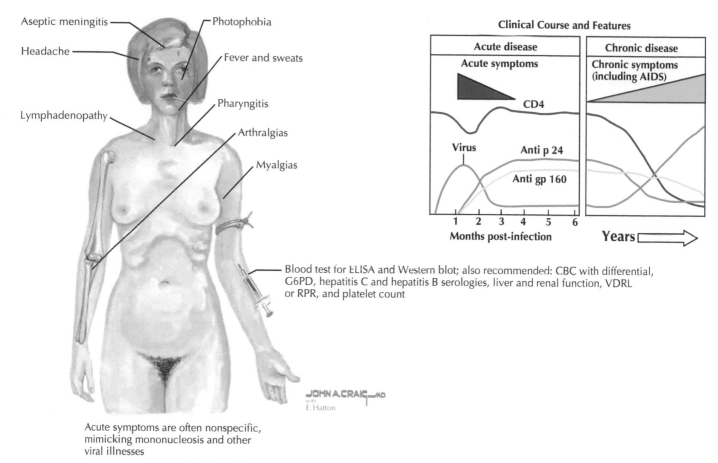

Fig. 56.1 Clinical course and features of the human immunodeficiency virus.

Neurologic manifestations of advanced HIV/AIDS include HIV-related dementia, opportunistic infections, and neoplasms. Cerebral toxoplasmosis is caused by reactivation of the parasite *Toxoplasma gondii*, typically presenting in patients with sustained CD4 counts below 100/mL, causing multiple lesions (often in the brain stem) and often presenting with seizures (Fig. 56.2). Fungal meningitis from *Cryptococcus neoformans*, also most commonly seen in patients with CD4 cell counts below 100/mL, usually presents subacutely with severe headaches and elevated intracranial pressure. In severely immunocompromised patients, typically with CD4 cell counts below 50/mL, progressive multifocal leukoencephalopathy (PML), caused by reactivation of the John Cunningham (JC) virus, can develop. This is a severe demyelinating disease that often presents with multiple neurologic symptoms, including hemiparesis, ataxia, and visual symptoms. Before ART, it was nearly universally fatal. Despite prompt ART and immune reconstitution, affected patients are often left with severe neurologic deficits.

CNS HIV escape occurs when an individual has suppressed virus in the peripheral blood but develops symptoms associated with detectable CSF levels of HIV RNA. There is a broad spectrum of symptoms including headache, behavioral changes, meningoencephalitis, psychosis, and coma. Patients on certain ART regimens, protease inhibitors (PIs) in particular, and with certain mutations (including the M184V combined with thymidine analog mutations) are more likely to develop CNS escape. The syndrome can be treated by changing the ART regimen to a more CNS-active combination, often with resolution of symptoms.

## Oral Manifestations

Oral candidiasis is a common manifestation of HIV, typically occurring in patients with CD4 cell counts below 350/mL (Fig. 56.3).

Angular cheilitis presents as painful fissured lesions at the angles of the mouth and is also caused by *Candida* species. Oral hairy leukoplakia (OHL) is an Epstein-Barr virus–associated disease, which presents as white plaques on the side of the tongue; these resolve with ART. HIV can also cause unilateral or bilateral parotid swelling, particularly during early infection, and is caused by the deposition of CD8-rich lymphocytes in parotid glands.

## Dermatologic Manifestations

There is a wide range of dermatologic disease seen in individuals with HIV. Some represent limited cutaneous diseases whereas others represent manifestations of an underlying systemic process. Superficial fungal infections, such as seborrheic dermatitis, are more common in people living with HIV (PLWH). Reactivation of viral infections from HSV with oral and genital ulcerations and varicella zoster virus (VZV) causing a dermatomal rash (at any CD4 count) or a disseminated vesicular/papular rash (at lower CD4 counts) are common. Viral warts from human papillomavirus (HPV) are also common and can progress to fungating condyloma acuminata or undergo malignant transformation to squamous cell carcinoma. Molluscum contagiosum, caused by a pox virus, manifests as small umbilicated cutaneous lesions that can look similar to HPV-related warts (Fig. 56.4).

Eosinophilic folliculitis (EF) is a noninfectious cutaneous condition that is seen in patients with low CD4 cell counts; it is characterized by pruritic papules on the face, scalp, and trunk. This condition can flare after starting ART, during immunologic reconstitution.

Kaposi sarcoma (KS) is a human herpesvirus 8–associated malignancy that most commonly presents as nonblanching violaceous papules or nodules on the skin, but it can also involve the lymph nodes,

## Toxoplasmosis

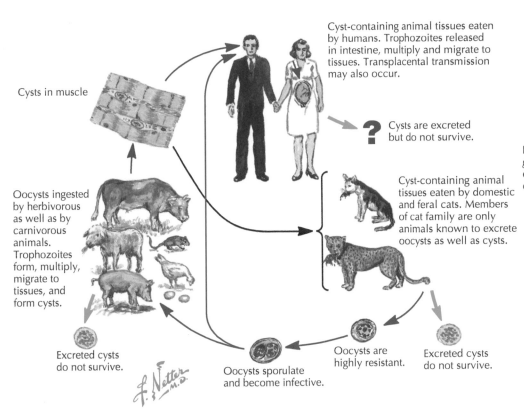

Cyst-containing animal tissues eaten by humans. Trophozoites released in intestine, multiply and migrate to tissues. Transplacental transmission may also occur.

Cysts in muscle

**?** Cysts are excreted but do not survive.

Cyst-containing animal tissues eaten by domestic and feral cats. Members of cat family are only animals known to excrete oocysts as well as cysts.

Oocysts ingested by herbivorous as well as by carnivorous animals. Trophozoites form, multiply, migrate to tissues, and form cysts.

Excreted cysts do not survive.

Oocysts sporulate and become infective.

Oocysts are highly resistant.

Excreted cysts do not survive.

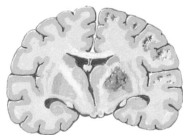

Brain section with nodule of *Toxoplasma gondii* in basal ganglia and necrotizing encephalitis in left frontal and temporal corticomedullary zones

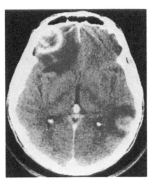

CT scan showing enhancing lesions of toxoplasmosis in right frontal and left temporal lobes of immunocompromised patient

**Fig. 56.2** Toxoplasmosis.

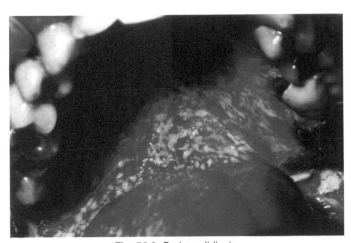

**Fig. 56.3** Oral candidiasis.

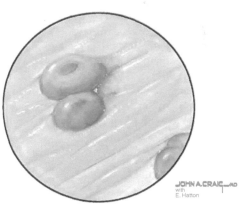

Magnified view showing typical umbilicated lesion

**Fig. 56.4** Molluscum contagiosum.

lungs, and gastrointestinal tract (known as "visceral KS"). Although steroids can be helpful in calming an overly robust immune response after the initiation of ART in patients with other opportunistic infections (i.e., disseminated *Mycobacterium avium* complex [MAC], *Pneumocystis jiroveci* pneumonia [PJP]), steroids cause KS to flare and progress and should therefore be avoided. Cutaneous violaceous lesions in a PLWH with a low CD4 cell count may alternatively be due to bacillary angiomatosis, which is a disseminated form of *Bartonella* species infection.

## Pulmonary Manifestations

Bacterial pneumonia affects PLWH at any CD4 count, but higher risk is associated with lower CD4 cell counts. The polyvalent pneumococcal vaccine series helps protect against *Streptococcus pneumoniae* pneumonia. *P. jiroveci* (formerly *P. carinii*) pneumonia is a fungal pneumonia that occurs in PLWH with a CD4 cell count below 200/mL and is commonly accompanied by oral thrush. Patients are typically profoundly short of breath and hypoxemic, with an elevated alveolar-arterial (AA) oxygen gradient. Diffuse lung opacities without effusions are characteristic radiographic findings, although normal,

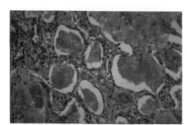

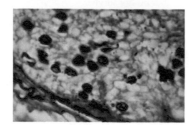

Interstitial lymphocyte and plasma cell infiltration with foamy exudate in alveoli

Methenamine AgNO₃ stain showing *Pneumocystis* organisms in lung (black spots)

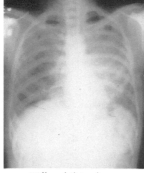

Diffuse bilateral pulmonary infiltrates

**Fig. 56.5** *Pneumocystis jiroveci* pneumonia.

nodular, or lobar findings are also possible (Fig. 56.5). Prophylaxis with trimethoprim-sulfamethoxazole (preferred agent) or alternative agents can protect against this infection and is typically recommended for patients with poorly controlled HIV and CD4 cell counts below 200/mL. *Mycobacterium tuberculosis* (TB) coinfection occurs in about one-third of people living with HIV worldwide. PLWH are at elevated risk for pulmonary TB at any CD4 cell count and more likely to have disseminated or extrapulmonary findings with increased amounts of immune dysfunction. In certain areas endemic pathogens, such as *Histoplasma capsulatum*, are important causes of infection, often causing a spectrum of disease similar to TB.

Noninfectious manifestations of pulmonary disease can also be seen in PLWH. Pulmonary hypertension occurs more commonly in PLWH and can lead to cor pulmonale and death. Lymphocytic interstitial pneumonia (LIP) presents with cough, dyspnea, and pulmonary infiltrates and is the result of lymphocytic infiltration of the lung. Nonspecific interstitial pneumonitis (NSIP) is quite common in individuals with low CD4 cell counts and presents with dyspnea and interstitial infiltrates that may be treated with steroids or may improve on its own.

## Gastrointestinal Manifestations

Gastrointestinal manifestations of HIV may be directly related to HIV or due to an opportunistic pathogen. Diarrhea is a common manifestation of acute HIV. In fact, during early infection, massive, irreversible CD4 T-cell depletion occurs in the gut in association with lymphoid tissue. With untreated infection and resulting immunodeficiency, infectious esophagitis is commonly seen, most often caused by *Candida* species (usually accompanied by oral thrush). Esophagitis may also be caused by cytomegalovirus (CMV) or HSV. Diarrhea is a common manifestation of lower gastrointestinal tract disease and can be due to nontyphoidal *Salmonella* species or other

enteric bacterial pathogens, CMV disease, or parasitic diseases such as *Cryptosporidium parvum*, *Isospora*, or *Microsporidia*. Diarrhea is also commonly seen in patients with disseminated MAC infection, which can also cause fever, lymphadenopathy, and pancytopenia with bone marrow infiltration.

## Malignancy and Human Immunodeficiency Virus

PLWH are at higher risk for AIDS-related and non–AIDS related malignancies. AIDS-related cancers include non-Hodgkin lymphoma, Kaposi sarcoma (HHV8 related), primary CNS lymphoma (EBV driven) and cervical and anal cancer (HPV-related). Individuals with HIV are at higher risk not only for AIDS-related malignancies but also for non–HIV-associated malignancies such as liver and lung cancer. In the era of effective ART and with an aging population of PLWH, the incidence of AIDS-related cancers has decreased while incidence of non–AIDS related cancers continues to increase.

## DIAGNOSTIC APPROACH

The CDC currently recommends one-time HIV screening for all adults and adolescents between the ages of 13 and 75, with more frequent testing for individuals at potential risk for ongoing acquisition. Individuals with symptoms or laboratory features consistent with HIV should be tested, as should all those with possible exposure to HIV. Repeat testing is recommended for high-risk individuals, including MSM and people who inject drugs. Patients on PrEP are recommended to have HIV testing every 3 months.

The testing algorithm currently includes an HIV1/2 antigen/antibody combination test that detects antibodies to HIV-1 and HIV-2 as well as HIV-1 p24 antigen. A positive result is confirmed with a differentiation antibody assay that distinguishes HIV-1 infection from HIV-2 infection. In acute infection, the p24 antigen may be positive and the antibodies may be negative. In these cases, the HIV RNA test is confirmatory. Viral RNA is detectable in the serum around 10 days after initial infection. The p24 antigen is detected approximately 14 days after exposure. The earliest antibody detection is around 21 days. The "window period," during which a person could have acute HIV but a nonreactive fourth-generation HIV 1/2 Ab/Ag test, is generally 2 to 3 weeks after exposure.

Once a new diagnosis has been confirmed, typical baseline testing includes blood counts, kidney function, liver tests, hepatitis B and C serologies, a baseline CD4 cell count, and a baseline genotypic resistance test, which looks for possible transmitted drug resistance in the protease and reverse transcriptase genomes. Baseline integrase resistance testing is not routinely recommended. Additional testing depends on the degree of immunosuppression and may include *Toxoplasma* IgG and CMV IgG (if the number of CD4 cells is less than 100/mL), HLA-B5701 testing (if abacavir is being considered for treatment), and G6PD testing (if the patient is allergic to sulfa antibiotics and needs an alternative for *Pneumocystis* pneumonia prophylaxis).

## CLINICAL MANAGEMENT AND TREATMENT

ART has been available since zidovudine was introduced in 1987. Combination therapy became the standard of care in 1996. There are multiple classes of ART that target essential HIV enzymes (reverse transcriptase, protease, integrase) or entry mechanisms of the virus into cells (fusion inhibitors, CCR5 antagonists, CD4 post-attachment inhibitors) (Fig. 56.6). There are now more than 30 agents approved as ART for HIV in the United States. Most regimens contain three active drugs in two different drug classes; however, more recently certain two-drug regimens have been shown to be noninferior to traditional

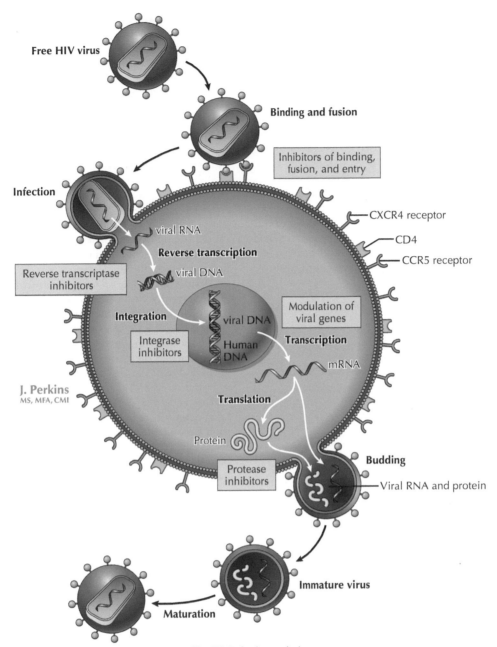

**Fig. 56.6** Antiretrovirals.

three-drug regimens after achieving virologic suppression with standard therapy for certain ART-naïve patients.

The four primary classes of ART are the nucleoside reverse transcriptase inhibitors (NRTIs), the nonnucleoside reverse transcriptase inhibitors (NNRTIs), the PIs, and the integrase strand transfer inhibitors (INSTIs). Currently emtricitabine (FTC), lamivudine (3TC), abacavir (ABC), tenofovir disoproxil fumarate (TDF), and tenofovir alafenamide (TAF) are the most commonly used NRTIs. Efavirenz (EFV), rilpivirine (RPV), etravirine (ETR), and doravirine (DOR) are the most commonly used NNRTIs. The PIs include atazanavir (ATV) and darunavir (DRV), which must be pharmacologically boosted with ritonavir or cobicistat to attain high enough trough levels to prevent the development of resistance. The INSTIs include elvitegravir (EVG), which requires boosting with cobicistat, and raltegravir (RAL) as well as dolutegravir (DTG) and bictegravir (BIC). Resistance to older HIV regimens emerges quickly in the setting of poor or intermittent adherence. Resistance to regimens with darunavir, dolutegravir, or bictegravir serving as the anchor drug is extremely unusual, as these drugs are extremely potent and have a high genetic barrier to resistance. Previously, PWLH had to take numerous tablets several times throughout the day, but now there are many single-tablet fixed-dose combinations so that most patients, even those with extensive prior resistance, can remain suppressed with one to three tablets each day. Current first-line regimens in the United States are all INSTI-based regimens (Box 56.1).

Early ART regimens had many significant side effects, including lipodystrophy (older PIs), pancreatitis (didanosine and stavudine), and Stevens-Johnson syndrome (nevirapine). Newer regimens are much better tolerated, with fewer long-term toxicities. Tenofovir can cause a Fanconi tubulopathy in the kidneys; however, TAF is considered less nephrotoxic than TDF. Abacavir can cause a severe hypersensitivity reaction in more than 50% of people with HLA B5701 and, in some studies, has been associated with increased cardiovascular risk. Finally, multiple observational studies have suggested that dolutegravir-based therapy may lead to more weight gain than other regimens.

National and international guidelines recommend universal ART for patients regardless of CD4 cell count. Prior to the Strategic Timing of AntiRetroviral Therapy (START) trial, published in 2015, HIV treatment was initiated based on CD4 cell count. This landmark trial randomized over 4000 PLWH to start ART at CD4 cell counts above 500/mL or to wait until their counts declined to less than 350/mL. AIDS and pooled non-AIDS events, including other cancers and cardiovascular events, were significantly lower in the immediate treatment group. Since then, many other studies have supported the benefit of early ART for individual health. In addition, treatment has also been shown to be an extremely important part of prevention efforts. The landmark study in this area was the HPTN-052 study, which studied serodiscordant partners and showed a 96% reduction in transmission to the uninfected partner when ART was started early, regardless of CD4 cell count. Multiple other studies have shown similar results. Analysis of all existing data has led to the "undetectable = untransmissible" campaign, which is supported by overwhelming scientific evidence and says that if a PLWH is virologically suppressed for at least 6 months and remains adherent to ART, he or she cannot transmit the virus.

In addition to ART, patients with lower CD4 cell counts should receive immunizations against vaccine-preventable diseases and prophylaxis against opportunistic infections (Tables 56.1 and 56.2). For a complete list of pathogens and recommended prophylaxis, guidelines for the prevention and treatment of opportunistic infections published by the CDC are readily available.

## PROGNOSIS AND FUTURE DIRECTIONS

The prognosis for PLWH has improved over the last 20 years, since the introduction of combination ART. Patients who are diagnosed early in the course of their infection and remain virally suppressed on ART have roughly the same life expectancy as the general population.

Despite incredible advances in treatment, no cure exists for HIV. Currently most patients need to take their daily combination ART lifelong, without interruptions. A small minority of patients (<5%), called long-term nonprogressors, are able to maintain high CD4 cell counts and low or undetectable HIV RNA levels for years. Some of these patients will eventually develop rapidly progressive disease, so they continue to need close monitoring and/or initiation of ART.

New and emerging therapeutic options are becoming available. Long-acting intramuscular injections of combination ART will soon be available. A potent drug in a novel class of ART called a nucleoside reverse transcriptase translocation inhibitor may soon be available as a long-term implant for PrEP and as part of a combination ART regimen for treatment. Finally, novel work on broadly neutralizing antibodies (bNAbs) is under way and may eventually be available for treatment (with intermittent infusions) and prevention.

---

**BOX 56.1    Department of Health and Human Services, October 2018: Preferred Initial Combination Therapy Regimens for Antiretroviral Therapy–Naïve Adults**

- Bictegravir/tenofovir alafenamide/emtricitabine
- Dolutegravir/abacavir/lamivudine (for patients who are HLA-B*5701-negative)
- Dolutegravir plus tenofovir/emtricitabine
- Raltegravir plus tenofovir/emtricitabine

Adapted from the Department of Health and Human Services (DHHS) panel recommendations on "Guidelines for the Use of Antiretroviral Agents in Adults and Adolescents with HIV." Available at: https://aidsinfo.nih.gov/guidelines/html/1/adult-and-adolescent-arv/11/what-to-start.

---

**TABLE 56.1    Immunization Recommendations for People Living With Human Immunodeficiency Virus**

| Vaccine | Regimen | Notes |
| --- | --- | --- |
| Influenza vaccine (inactivated) | Annual | Recommended for all people with HIV. |
| Polyvalent pneumococcal vaccines | PCV13 followed by PPSV23 8 weeks later | Recommended for all people with HIV. |
| Hepatitis B vaccine | 3 dose series (0, 1, and 6 months) | Recommended for all people with HIV without immunity to HBV. Note that recombinant, adjuvanted vaccine (Heplisav-B) is currently under study for people with HIV. |
| Hepatitis A virus | 2 dose series (0 and 6 months) | Recommended for many people with HIV without immunity to HAV, including MSM, homeless individuals, and people who inject drugs. |
| Human papillomavirus vaccine | 3 dose series (0,1, and 6 months) | Recommended for all people younger than 26 years of age. |
| Tetanus, diphtheria, pertussis vaccine | Tdap ×1, then Td booster every 10 years | Recommended for all people. |
| *Neisseria meningitis* vaccine for serogroups A, C, W, Y | Two-dose series (0 and 8 weeks) followed by booster every 5 years | Recommended for all people with HIV. |
| Recombinant adjuvant zoster vaccine | Two-dose series (0 and 2–6 months) | Currently being studied for patients with HIV. Many providers currently recommend this for their patients with CD4 cell counts greater than 200/mL. |

*HIV,* Human immunodeficiency virus; *MSM,* men who had sex with men.
Based on recommendations from Department of Health and Human Services (DHHS) panel on "Guidelines for the Prevention and Treatment of Opportunistic Infections in Adults and Adolescents with HIV." Available at https://aidsinfo.nih.gov/guidelines/html/4/adult-and-adolescent-opportunistic-infection/365/figure--immunization.

## TABLE 56.2    Opportunistic Infection Prophylaxis

| Organism | Risk | Notes About Prophylaxis |
|---|---|---|
| *Pneumocystis jiroveci* | CD4 cells <200/mL or CD4 <14% or thrush | Chemoprophylaxis indicated. Note: can discontinue prophylaxis if CD4 cells >100/mL and virally suppressed for >3–6 months. |
| *Toxoplasma gondii* | CD4 cells <100/mL and Toxo IgG positive | Chemoprophylaxis indicated. Note: if IgG negative, should be counseled about reducing risk of exposure. |
| *Mycobacterium avium* complex | CD4 cells <50/mL | No prophylaxis recommended. Start antiretroviral therapy. |
| Cytomegalovirus | CD4 cells <50/mL and CMV IgG positive | Baseline ophthalmologic examination recommended. |

Based on information from the Department of Health and Human Services (DHHS) panel on "Guidelines for the Prevention and Treatment of Opportunistic Infections in Adults and Adolescents with HIV." Available at: https://aidsinfo.nih.gov/guidelines/html/4/adult-and-adolescent-opportunistic-infection/354/primary-prophylaxis.

## EVIDENCE

Antiretroviral Therapy Cohort Collaboration. Survival of HIV-positive patients starting antiretroviral therapy between 1996 and 2013: a collaborative analysis of cohort studies. *Lancet* 2017;4:e349-356. *This study showed continuing improvement in mortality for patients on ART. People starting ART between 2008 and 2010 who survived the first year after treatment initiation could expect to have an essentially normal life expectancy.*

Gray RH, Wawer MJ, Brookmeyer R, et al. Probability of HIV-1 transmission per coital act in monogamous, heterosexual, HIV-1–discordant couples in Rakai, Uganda. *Lancet* 2001;357:1149-1153. *This study demonstrated that higher viral loads and genital ulceration are associated with increased risk of sexual transmission of HIV.*

Kahn JO, Walker BD. Acute human immunodeficiency virus type 1 infection. *N Engl J Med* 1998;339:33-39. *Review of the symptoms associated with acute HIV as well as the early pathophysiology of HIV.*

Riddell J, Rivet Amico K. Mayer KH. HIV preexposure prophylaxis: a review. *JAMA* 2018;319(12):1261-1268. *This is a comprehensive review of all of the studies to date on preexposure prophylaxis and their findings. The pooled findings suggest that TDF/FTC is safe and effective for prevention of HIV transmission and that with good adherence it has greater than 90% efficacy in preventing HIV sexual transmission.*

Shiels MS, Engels EA. Evolving epidemiology of HIV-associated malignancies. *Curr Opin HIV AIDS* 2017;12(1):6-11. *In this excellent review, the authors note that rates of AIDS-related malignancies have declined dramatically in people with HIV in ART (though still remain elevated above the general population) and that with an aging HIV-positive population there has been an increasing incidence of non-AIDS-defining cancers in HIV-positive individuals.*

Siegfried N, Muller M, Deeks JJ, Volmink J. Male circumcision for prevention of heterosexual acquisition of HIV in men. *Cochrane Database Syst Rev* 2009;2:CD003362, 2009. *This systematic review of randomized controlled trials of male circumcision for HIV prevention found that male circumcision decreased the acquisition of HIV by heterosexual men by 38% to 66% over a 24-month period.*

The INSIGHT START Study Group, Lundgren D, Babiker AG, et al. Initiation of antiretroviral therapy in early asymptomatic HIV infection. *N Engl J Med* 2015;373(9):795-807. *This landmark study showed benefit of starting ART early with decreased AIDS and non-AIDS related mortality and led to recommendations for universal ART for all people with HIV, regardless of CD4 count.*

Walker AS, Ford D, Gilks CF, et al. Daily co-trimoxazole prophylaxis in severely immunosuppressed HIV-infected adults in Africa started on combination antiretroviral therapy: an observational analysis of the DART cohort. *Lancet* 2010;375:1278-1286. *This observational analysis of a large cohort in Africa found that co-trimoxazole prophylaxis improved mortality and reduced frequency of malaria.*

## ADDITIONAL RESOURCES

Centers for Disease Control and Prevention (CDC). National Center for HIV/AIDS, Viral Hepatitis, STD, and TB Prevention (NCHHSTP) website. Available at: www.cdc.gov/nchhstp/. *The NCHHSTP is the branch of the CDC responsible for public health surveillance, prevention research, and programs to prevent and control HIV and AIDS, other STDs, viral hepatitis, and TB.*

National HIV Curriculum. The AIDS Education and Training Centers National HIV Curriculum is a free educational website led by the University of Washington. Available at: https://www.hiv.uw.edu/. *This website offers a variety of educational material and interactive case-based tutorials about clinically relevant HIV-related topics.*

National Network of STD/HIV Prevention Training Centers (NNPTC). NNPTC website. Available at: https://nnptc.org/. *The NNPTC is a CDC-funded group of regional centers created in partnership with health departments and universities. The NNPTC provides health professionals with state-of-the-art educational opportunities with an emphasis on prevention, with the goal of increasing the knowledge and skills of health professionals in the areas of sexual and reproductive health.*

U.S. Department of Health and Human Services. AIDSinfo. Available at: www.aidsinfo.nih.gov. *This link provides information about current treatment and prevention guidelines as well as information about medications and clinical trials.*

World Health Organization (WHO). WHO website. Available at: www.who.int. *This site offers information about the latest worldwide epidemiology of HIV and international guidelines.*

# 57

# Human Papillomavirus

*Nancy McClung, Eileen F. Dunne, Lauri E. Markowitz*

## ABSTRACT

Genital human papillomavirus (HPV) infection is estimated to be the most common sexually transmitted infection (STI) in the United States and worldwide. Oncogenic, or "high-risk" HPV types, such as HPV types 16 and 18, can cause cervical and other anogenital precancers, anogenital cancers, and oropharyngeal cancers. Other HPV types, such as HPV types 6 and 11, can cause anogenital warts as well as recurrent respiratory papillomatosis and benign or low-grade cervical cell changes. Most infections are asymptomatic and do not result in clinical disease. However, persistent oncogenic HPV infection over time can lead to precancers and cancers. Approximately 35,000 people are diagnosed with HPV-attributable cancers each year in the United States, and over 600,000 people worldwide. In the United States, oropharyngeal cancers are the most common HPV-attributable cancer in men; cervical cancers are the most common in women. Globally, cervical cancer accounts for 84% of all HPV-attributable cancers. Between 70% and 90% of cervical cancers and a substantial number of other HPV-attributable cancers potentially could be prevented by prophylactic HPV vaccines.

## CLINICAL VIGNETTE

A 22-year-old female presented with a new onset of "bumps." She noticed the bumps about 2 months earlier and reports that they are located on her vulva; she has no genital discharge, discomfort, itching, or pain. She began a relationship with a new male sex partner about 6 months earlier but reports no other notable history. She has never had a Papanicolaou (Pap) smear and has not received HPV vaccine. On physical examination she had small (<2 mm) painless grouped papules on the vulva. There was no ulceration, umbilication, or vesiculation. External genital and pelvic examinations revealed no other external lesions; there was no vaginal discharge or odor, and the cervix was mobile, nontender, and nonfriable. Serum venereal disease research laboratory (VDRL), human immunodeficiency virus (HIV), chlamydia, gonorrhea, and Pap tests were done. She returned for follow-up 2 weeks later, by which time testing had proven negative. She was given a dose of HPV vaccine, with scheduled follow-up doses in 1 to 2 and 6 months.

The differential of the anogenital papules included molluscum contagiosum, syphilis (condylomata lata), and anogenital warts. On the differential but rare is lichen planus or intraepithelial neoplasia. She had no evidence of umbilicated papules, and syphilis testing was negative, ruling out syphilis and molluscum contagiosum. Because she was a young, otherwise healthy woman, intraepithelial neoplasia was unlikely.

Several questions arise with this scenario:

- What should be recommended regarding treatment for the anogenital warts? The patient should be counseled that if left untreated, genital warts may go away, stay the same, or increase in size or number. The types of HPV that cause genital warts are different from the types that can cause cancer. If the patient wants treatment, one of the recommended treatment options can be offered. There is no evidence that any recommended treatment is superior to another.
- What are the recommendations for the testing/evaluation of other STIs? Testing for HIV should occur at any clinical visit with a diagnosis of an STI. The Centers for Disease Control and Prevention (CDC) recommends that everyone between the ages of 13 and 64 get tested for HIV at least once as part of routine health care. Even though she is not symptomatic, this patient should have chlamydia and gonorrhea testing as she is younger than 24 years of age. Chlamydia and gonorrhea are often asymptomatic.
- What are the recommendations for cervical cancer screening in the setting of anogenital warts? Cervical cancer screening recommendations are the same regardless of anogenital warts. Screening for cervical cancer is recommended in women age 21 to 65 years with cytology (Pap smear) every 3 years; for women age 30 to 65 years who want to lengthen the screening interval, screening with a combination of cytology and HPV testing every 5 years or an HPV test alone every 5 years can be recommended.
- What are the recommendations for HPV vaccination? A 22-year-old woman who has not been immunized should get HPV vaccine even if she already has or has had genital warts. The vaccine will not protect her from disease due to infections she has already acquired, and because of her history of genital warts, she has likely already acquired nononcogenic types HPV 6 and/or 11. However, the vaccine may have benefits for protecting against HPV types she has not yet acquired. There are a number of oncogenic HPV types that can cause cervical cancer; seven of these are prevented by the 9-valent HPV vaccine. Three doses are recommended because this patient will have started the series after her 15th birthday. Two doses are recommended for persons who start the series before age 15.

## BACKGROUND

Papillomaviruses are a family of deoxyribonucleic acid (DNA) viruses that infect the epithelium and have a double-stranded, closed, circular genome of approximately 8 kb and a nonenveloped icosahedral capsid. There are over 150 HPV types, which are further characterized as mucosal or cutaneous depending on the epithelium they primarily infect. Cutaneous types cause common skin warts; mucosal types have primary affinity for the genital and oral mucosa and can cause a variety of diseases and cancers including anogenital warts, recurrent respiratory papillomatosis, cervical and other anogenital precancers and cancers, and oropharyngeal cancers. Most infections are asymptomatic, cause no disease, and become undetectable within 2 years of acquisition. However, persistent HPV infection with oncogenic types is the most important risk factor for the development of serious diseases, including cancers. This chapter focuses on HPV infection and the genital diseases that can result from infection, including anogenital warts, cervical and other anogenital precancers, and cancers.

## GEOGRAPHIC DISTRIBUTION AND MAGNITUDE OF INFECTION AND DISEASE BURDEN

Infection with genital HPV is ubiquitous and occurs commonly throughout the world. In the United States, an estimated 79 million people, most in their late teens and early 20s, are infected with HPV; each year, about 14 million people are newly infected. The prevalence of genital HPV among women varies in different world regions, from 8% to 9% in Western/Central Asia and Europe to more than 20% in Africa, North America, South/Central America, and Oceania. There is some variation in prevalence of HPV types by world region, but HPV-16 is frequently the most common oncogenic type. In many settings, HPV prevalence is highest in adolescent females and young women and then decreases; in other settings, there is a U-shaped curve, with the highest prevalence in the early twenties and a second smaller peak in older females. Some of the differences in HPV prevalence throughout the world may reflect differences in sexual debut and sexual behavior among females and their partners. There are less data on genital HPV prevalence in males, but the available data show a slightly higher prevalence among males than females and with the highest prevalence in the late 20s, which, in contrast to females, does not decline with age. There are fewer data on HPV prevalence at other anatomic sites. Oral HPV prevalence is much lower than genital HPV prevalence and also slightly higher among males; in the United States, the estimated oral HPV prevalence in the general population is approximately 7%.

Anogenital warts result from infection with non-high-risk HPV types (HPV type 6 or 11 most commonly). In the United States, prior to the introduction of HPV vaccination, an estimated 340,000 people were diagnosed with anogenital warts annually, with the highest prevalence in women aged 20 to 24 years and men aged 25 to 29 years. Global estimates of anogenital wart incidence vary from 160 to 289 per 100,000 person-years.

HPV causes nearly all cervical cancers and a substantial number of other anogenital (vulvar, penile, anal) and oropharyngeal cancers. In the United States, an estimated 35,000 HPV-attributable cancers occur each year, 21,000 in females and 14,000 in males. HPV-16 and HPV-18 account for the majority of HPV-attributable cancers worldwide, including 70% of cervical cancers; HPV-31, HPV-33, HPV-45, HPV-52, and HPV-58 account for an additional 20% of cervical cancers. Cervical cancer is the most common HPV-attributable cancer worldwide, accounting for 84% of all HPV-attributable cancers; each year, over 500,000 new cases occur and over 300,000 women die of the disease. In countries with effective cervical cancer screening programs, the incidence of cervical cancer is low compared with low-resource settings in which limited or no screening exists. For example, in the United States, there has been a reduction of over 70% in cervical cancer cases due to decades of cervical cancer screening. Cervical cancer is the 12th most common cancer among women in the United States, with approximately 13,000 cases and 4000 deaths annually. Globally, cervical cancer is the fourth most common type of cancer in women and a leading cause of death. In the United States and other western countries, oropharyngeal cancers have been increasing, particularly among men. Because cervical cancer incidence is decreasing due to screening, the highest burden of cancer due to HPV in the United States is oropharyngeal cancer.

Cervical precancers can be detected through cervical cancer screening. In the United States, prior to the introduction of HPV vaccination, an estimated 216,000 women per year were diagnosed with high-grade cervical precancers. The highest rate of disease was among women aged 20 to 24 years, and HPV-16 or HPV-18 was detected in over 50% of lesions. Routine screening programs are not recommended for other HPV-associated cancers due to a lack of evidence for the safety and effectiveness of screening and treatment to prevent cancer. No precancerous lesion for oropharyngeal cancer has been identified.

## RISK FACTORS

Genital HPV infection is primarily transmitted by genital contact, usually through sexual intercourse. The risk for genital HPV infection increases with an increasing lifetime number of sex partners; however, a substantial percentage of persons acquire infection even with one sex partner. More than 80% of women and men are estimated to acquire HPV by age 45 years. Transmission can also occur by sexual contact other than sexual intercourse; one study of young women who reported no previous sexual intercourse and had other sexual activity such as genital-genital or genital-oral contact found that the 24-month cumulative incidence of infection was 7.9% (95% confidence intervals [CIs] 3.5 and 17.1). HPV infection is acquired soon after sexual debut; in one study, the 1-year cumulative incidence of first HPV infection was 28.5%, and the incidence increased to almost 50% by 3 years in young women with one sex partner. In virtually all studies of HPV prevalence and incidence, the most consistent predictors of infection have been various measures of sexual activity (i.e., the number of sex partners). Risk factors for anogenital warts are similar to those for HPV infection and include increasing numbers of sex partners and young age.

Both men and women with compromised immune function (including HIV infection and posttransplant) are at higher risk for HPV infection and HPV-associated outcomes. For example, prevalence of HPV-16 has been shown to be significantly higher in HIV-infected compared to HIV-uninfected men, both among men who have sex with women (prevalence ratio [PR] 3.5; 95% CI, 1.6–7.7) and men who have sex with men (PR 2.1; 95% CI, 1.8–2.5). People living with HIV have also been found to have significantly higher rates of HPV-associated cancers compared with the general population, including a threefold increase in cervical cancer among women.

An important risk factor for cervical cancer is not having been screened and/or not having received appropriate follow-up after screening. In the United States, rates of cervical cancer are higher among black and Hispanic women than non-Hispanic white women; rates also vary geographically, with higher rates reported in the South. The median age for cervical cancers is 49 years. Other HPV-associated cancers are typically diagnosed at later ages. In women, the median age of diagnosis is 62 years for anal and oropharyngeal cancer, 66 years for vulvar cancer, and 67 years for vaginal cancer; in men, the median age of diagnosis is 59 years for anal cancer and 69 years for oropharyngeal cancer.

## CLINICAL FEATURES

Infection with genital HPV usually causes no signs or symptoms, and infections clear or become undetectable within 2 years. Asymptomatic infection is typically not diagnosed or treated. A common question posed to clinical providers is whether partners of patients with anogenital warts or cervical disease caused by HPV should be tested for HPV infection. There is no reason to test partners routinely because HPV infection is common and often clears on its own. Most sex partners share HPV types and readily transmit infection to one another. The most important risk factor for the development of precancers and cancers is persistent oncogenic HPV infection.

One of the most common clinical manifestations of HPV is anogenital warts; more than 90% of anogenital warts are caused by nononcogenic HPV-6 or HPV-11. Warts typically occur 4 to 6 months after the incident HPV infection. In one study, 12% of young women with incident HPV-6 or HPV-11 infection developed anogenital warts over a 4-year follow-up. Warts appear as small papules or flat, smooth, or pedunculated lesions. Sometimes they can be soft, pink, or white "cauliflower-like" sessile growths on moist mucosal surfaces (condylomata acuminata), or keratotic lesions on squamous epithelium of the skin with a thick, horny layer (Fig. 57.1).

If anogenital warts are detected, experts recommend that clinicians conduct a complete anogenital examination and screen for other STIs. Detection of anogenital warts is not a reason to recommend more frequent cervical cancer screening because the HPV types that cause anogenital warts are different from those that cause cancer.

Cervical precancers caused by HPV are typically detected through cervical screening using Pap tests (either conventional or liquid based), or Pap tests combined with HPV tests, or with HPV test alone (Fig. 57.2). Cervical lesions can be caused by either nononcogenic or oncogenic HPV types. Nononcogenic types that infect the cervix can cause low-grade cervical abnormalities (termed low-grade squamous intraepithelial lesions [LSILs]). Oncogenic types, when they persist, can lead over time to high-grade cervical abnormalities (termed high-grade squamous intraepithelial lesions [HSILs]) and cervical cancers. Because cervical precancers and cervical cancers take years to develop, cervical cancer screening is often able to detect lesions early, when lesions are small and easier to treat.

## DIAGNOSTIC APPROACH

Cervical precancers and cancers are diagnosed during routine cervical cancer screening (https://www.cdc.gov/cancer/cervical/basic_info/screening.htm). Guidelines for cervical cancer screening are made by several organizations and differ slightly; the US Preventive Services Task Force recommends routine cervical cancer screening starting at age 21. Conventional or liquid-based cytologic tests (Pap tests) are recommended every 3 years in women aged 21 to 29 years. Women aged 30 to 65 years can receive a Pap test every 3 years, a Pap test with an HPV test (cotesting) every 5 years, or an HPV test alone every 5 years. Recommendations for screening intervals and follow-up with colposcopy depend on the type of screening used and the screening results. All women who are sexually active, including women who have sex only with women, are at risk for cervical dysplasia and cancer. Biopsies for histologic diagnosis of the cervix are directed to cervical changes noted by colposcopy. As of 2020, cervical cancer screening recommendations are not different for unvaccinated women and vaccinated women in the United States.

Testing for HPV infection should not be done to screen for HPV infection outside the specific indications in the setting of cervical cancer screening; as emphasized earlier, most HPV infections become undetectable and do not result in clinical disease. The tests are not recommended for screening sex partners, screening men, screening adolescents, diagnosing anogenital warts, or as a "sexually transmitted infection (STI) test."

Anogenital warts are diagnosed by physical examination. Topical acetic acid is sometimes used as a diagnostic aid but is not specific for anogenital warts and may result in unnecessary interventions. Biopsy of suspicious lesions may be necessary in some settings (e.g., if the diagnosis is uncertain; the lesions do not respond to standard therapy; the disease worsens during therapy; the patient is immunocompromised; or warts are pigmented, indurated, fixed, bleeding, or ulcerated).

Routine screening tests are not recommended for other anogenital and oropharyngeal cancers. In populations at high risk for anal cancer, such as men and women who are living with HIV, some specialists recommend anal cytologic screening or high-resolution anoscopy. An annual digital anal examination may be useful to detect masses that could be anal cancer in high-risk populations.

## CLINICAL MANAGEMENT AND DRUG TREATMENT

Treatment is directed at the clinical manifestations of HPV infection but not the infection itself. The treatment options differ depending on the condition. Treatments for anogenital warts can include patient-applied and provider-administered therapies. Available therapies for anogenital warts might reduce but probably do not eradicate HPV infectivity. Patients with warts that are located on the rectum or cervix, those with extensive anogenital warts, and patients whose condition does not respond to a standard course of therapy for anogenital warts should be managed by a specialist.

Anogenital warts may remain the same, grow in size and/or number, or regress without treatment. Therefore some patients elect to wait and see whether their anogenital warts will regress on their own. Treatment is directed at removing warts. Different treatments are available, and no single treatment is ideal for all patients or all lesions. Patients and providers may decide on a treatment option based on convenience, cost, availability, the methods of administering therapy, or other factors (Table 57.1). Treatment can induce wart-free periods, but the underlying viral infection can persist and may result in recurrence (in about 30% of cases). No data suggest that treatment modalities for external anogenital warts should be different in the setting of HIV infection; however, squamous cell carcinomas resembling anogenital warts occur more frequently among immunocompromised individuals, and biopsy should be considered in such cases.

Patient-applied therapies should be used only when the warts can be identified and accessed for treatment and there is the likelihood of high compliance. A follow-up appointment several weeks into therapy to determine appropriateness of medication use and response to treatment may be useful. Patient-applied therapies include podofilox 0.5% solution or gel, imiquimod 5% cream, and sinecatechins 15% ointment. Patients should apply podofilox solution with a cotton swab or podofilox gel with a finger to visible anogenital warts twice a day for 3 days, followed by 4 days of no therapy. This cycle may be repeated as necessary for up to four cycles. Patients should apply imiquimod cream once daily at bedtime, three times a week for up to 16 weeks. The treatment area should be washed with soap and water 6 to 10 hours after the application. Sinecatechins 15% ointment should be applied three times a day for up to 16 weeks.

Provider-applied therapies include cryotherapy with liquid nitrogen or cryoprobe, trichloroacetic acid (TCA) or bichloroacetic acid (BCA) 80% to 90%, and surgical removal. There are also other provider-applied therapies with fewer data available and/or more reported side effects. Cryotherapy should be performed every 1 to 2 weeks. Both TCA and BCA should be applied sparingly and allowed to dry before the patient sits or stands. If pain is intense, the acid can be neutralized with soap or sodium bicarbonate. A white "frosting" will develop on the wart after the TCA or BCA dries. TCA or BCA treatments can be repeated weekly. Anogenital warts may be surgically removed by tangential scissor excision, tangential shave excision, curettage, electrosurgery, or other methods.

Recommendations for the management and treatment of cervical lesions detected during cervical cancer screening are updated regularly (https://www.asccp.org/guidelines). LSIL, such as cervical intraepithelial neoplasia (CIN) 1, are managed by follow-up rather than treatment, as these lesions often regress. Treatments for HSIL (CIN 2, CIN 3, AIS) are tailored and may include loop electrosurgery excision procedure (LEEP), conization, other surgical options, or ablation (cryotherapy, laser ablation, and thermoablation). Treatments for cervical cancer depend on the stage of the cancer, the size of the tumor, the patient's desire to have children, and the patient's age; they may include chemotherapy, radiation therapy, surgery, or other therapies.

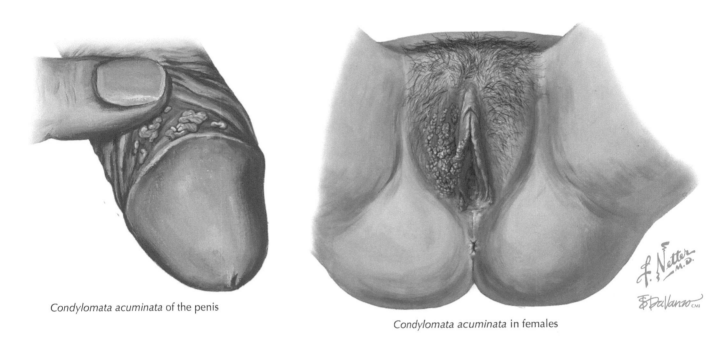

Condylomata acuminata of the penis

Condylomata acuminata in females

Fig. 57.1 Condylomata acuminata.

## PROGNOSIS

Anogenital warts are generally a benign condition that resolves with time; most resolve within 6 months (with or without treatment). However, a large proportion of anogenital warts recur, which can result in frequent office visits, sometimes costly and debilitating treatment, and a psychosocial burden. There are rare cases of giant condylomata of Buschke and Lowenstein, a slow-growing, highly destructive lesion caused by HPV-6 or HPV-11 infection, which may have a focus of squamous cell carcinoma. This tumor does not metastasize but causes severe local destruction. This condition most commonly occurs in immunocompromised individuals, including patients with HIV infection or those who have undergone transplantation.

The prognosis for women with LSIL is also good; CIN 1 usually clears spontaneously (60% of cases) and rarely progresses to cancer (1%). Lower percentages of high-grade lesions including CIN 2 and 3 spontaneously clear (40% for CIN 2 and 33% for CIN 3), and a substantial number progress to cancer if not treated (5% for CIN 2 and >12% for CIN 3). Treatment of these lesions prevents the progression to invasive cancer. The prognosis for cervical cancers depends on the stage at diagnosis, size of the tumor, and age of the woman.

## PREVENTION AND CONTROL

Three prophylactic HPV vaccines have been licensed and recommended for use: a quadrivalent HPV vaccine (Gardasil, Merck & Co., Kenilworth, NJ) that targets HPV types 6, 11, 16, and 18; a bivalent HPV vaccine (Cervarix, GlaxoSmithKline, Rixensart, Belgium) that targets HPV types 16 and 18; and a 9-valent HPV vaccine (Gardasil 9, Merck & Co., Kenilworth, NJ) that targets HPV types 6, 11, 16, 18, 31, 33, 45, 52, and 58. The quadrivalent HPV vaccine was licensed for use in the United States in June 2006, the

bivalent HPV vaccine in October 2009, and the 9-valent vaccine in December 2014. Almost all HPV vaccine used in the United States before 2015 was quadrivalent vaccine; after the end of 2016, only 9-valent vaccine has been available in the United States. All three vaccines are used in other countries. In the United States, routine vaccination is recommended for all adolescents at age 11 or 12 years and can be started at age 9. Catch-up vaccination is recommended through age 26 years for persons not adequately vaccinated already. For persons aged 27 through 45 years, shared clinical decision making is recommended regarding potential HPV vaccination in this age group. Either two or three doses are recommended, depending on the age at initiation of the vaccination series. Two doses are recommended for persons who start the series before their 15th birthday. Three doses are recommended for persons who start the series at older ages and for persons with immunocompromising conditions regardless of age at vaccination initiation.

Although the HPV vaccines have high prophylactic efficacy, they have no therapeutic effect; HPV vaccines do not prevent progression of infection to disease or enhance regression of existing lesions. Vaccination has maximum benefit when given before exposure to HPV infection (i.e., before sexual debut). HPV vaccination has decreased vaccine-type HPV prevalence, anogenital warts, and cervical precancers in countries where HPV immunization programs have been introduced. In the United States, vaccination coverage has been increasing gradually. Routine vaccination was first recommended for girls in 2006 and for boys in 2011. In 2019, 71.5% of 13- to 17-year-old adolescents had received at least one vaccine dose and 54.2% had completed the recommended series. Even though vaccination coverage is lower than target goals in the United States, vaccine impact has been demonstrated through a reduction in the prevalence of HPV types targeted by the quadrivalent vaccine among females in the vaccine era compared with the prevaccine era: an 86% decline among 14- to 19-year-olds and

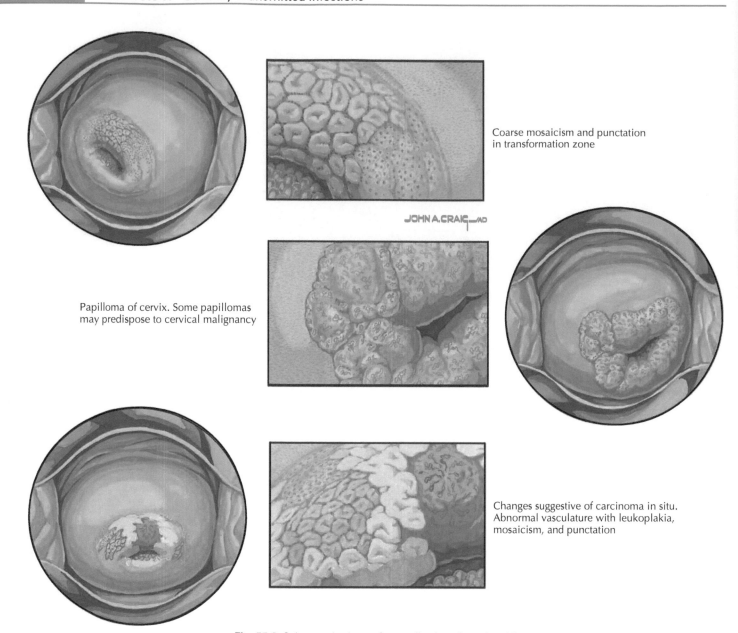

Coarse mosaicism and punctation in transformation zone

Papilloma of cervix. Some papillomas may predispose to cervical malignancy

Changes suggestive of carcinoma in situ. Abnormal vasculature with leukoplakia, mosaicism, and punctation

**Fig. 57.2** Colposcopic views after application of acetic acid.

a 71% decline among 20- to 24-year-olds. Data are also available from other countries. In a systematic review of impact of HPV vaccination programs in 14 high-income countries, the prevalence of HPV-16 and HPV-18 decreased over 80% among girls aged 13 to 19 years and over 60% in women aged 20 to 24 years within 5 to 8 years of vaccine introduction. In vaccination programs using vaccines that target HPV 6 and 11, the prevalence of anogenital warts decreased over 60% among girls aged 15 to 19 years and over 50% in women aged 20 to 24 years within 5 to 8 years of vaccine introduction; declines have also been observed among males in female-only vaccination programs, demonstrating herd protection.

Treatment of existing lesions (anogenital warts or cervical lesions) may reduce but does not eliminate HPV infection. Condoms have been shown to reduce the risk for genital warts, cervical cell changes, and HPV infection; however, infection or disease can occur on sites not covered or protected by a condom. Vaccination is the best way to prevent HPV infection and attributable disease. The impact of HPV vaccines on infection and disease, as well as tools for cervical cancer screening and treatment, has raised interest in cervical cancer elimination goals.

## DISCLAIMER

The findings and conclusions in this report are those of the authors and do not necessarily represent the official position of the Centers for Disease Control and Prevention.

**TABLE 57.1 Recommended Treatment Regimens for Anogenital Warts Based on Anatomic Location**

| Anatomic Location | PATIENT-APPLIED THERAPY | | | PROVIDER-ADMINISTERED THERAPY | | | |
| --- | --- | --- | --- | --- | --- | --- | --- |
| | Podofilox[a] | Imiquimod[a] | Sinecatechins[a] | Cryotherapy | TCA or BCA | Surgical Removal | Other |
| External anogenital[b] | X | X | X | X | X | X | |
| Meatus | | | | X | | X | |
| Vagina | | | | X | X | X | [c] |
| Cervical | | | | X | X | X | Biopsy, consult a specialist |
| Intra-anal | | | | X | X | X | Consult a specialist |

[a]Podofilox, podophyllin, imiquimod, and sinecatechins are not recommended during pregnancy.
[b]Including penis, groin, scrotum, vulva, perineum, external anus, and perianus.
[c]Some experts recommend the use of podofilox and imiquimod, but data are limited.
*BCA,* Bichloroacetic acid; *TCA,* trichloroacetic acid.
Adapted from Centers for Disease Control and Prevention (CDC): Sexually Transmitted Diseases Treatment Guidelines, 2015, MMWR.

# EVIDENCE

Chesson WH, Dunne EF, Hariri S, Markowitz LE. The estimated lifetime probability of acquiring human papillomavirus in the United States. *Sex Transm Dis* 2014:41(11):660-4. *An estimated 80% of women and men with at least one opposite sex partner will acquire HPV by age 45 years.*

Curry SJ, Krist AH, Owens DK, et al. Screening for cervical cancer: US Preventive Services Task Force recommendation statement. *JAMA* 2018; 320:674-86. *This statement provides national recommendations for cervical cancer screening.*

de Martel, C, Plummer M, Vignat J, et al. Worldwide burden of cancer attributable to HPV by site, country and HPV type. *Int J Cancer* 2017;141: 664-70. *Over 600,000 HPV-attributable cancers occur each year worldwide.*

Drolet M, Benard E, Perez N, Brisson, M. Population-level impact and herd effects following the introduction of human papillomavirus vaccination programmes: updated systematic review and meta-analysis. *Lancet* 2019;394:497-509. *This systematic review and meta-analysis summarizes the evidence of population-level impact of HPV vaccination programs on HPV infections, anogenital wart diagnoses, and high-grade cervical lesions; 65 articles from 14 high-income countries were included.*

FUTURE II Study Group. Quadrivalent vaccine against human papillomavirus to prevent high-grade cervical lesions. *N Engl J Med* 2007;356:1915-1927. *This randomized controlled clinical trial of the quadrivalent HPV vaccine found high efficacy for prevention of CIN 2/3.*

Gargano JW, Park IU, Griffin MR, et al. Trends in high-grade cervical lesions and cervical cancer screening in five states—United States, 2008–2014. *Clin Infect Dis* 2019;68(8):1282-1291. *In this report, the incidence of high-grade cervical lesions has declined in young women since HPV vaccine introduction in the United States.*

Garland SM, Steben M, Sings HL, et al. Natural history of genital warts: analysis of the placebo arm of the 2 randomized phase III trials of a quadrivalent human papillomavirus (types 6, 11, 16, and 18). *J Infect Dis* 2009;199:804-14. *This follow-up evaluation of the placebo arm of a clinical trial of HPV vaccination found that 3.4% of women without incident HPV infection developed genital warts over a 4-year period.*

Gillison ML, Broutian T, Pickard RK, et al. Prevalence of oral HPV infection in the United States, 2009-2010. *JAMA* 2012;307:693-703. *Prevalence of oral HPV infection is 6.9% in the United States and slightly higher among males than females.*

Hernández-Ramírez RU, Shiels MS, Dubrow R, Engels EA. Cancer risk in HIV-infected people in the USA from 1996 to 2012: a population-based, registry-linkage study. *Lancet HIV* 2017;4(11):e495-e504. *Persons living with HIV have a higher risk of cancers caused by HPV than persons without HIV infection.*

Hoy T, Singhal PK, Willey VJ, Insinga RP. Assessing incidence and economic burden of genital warts with data from a US commercially insured population. *Curr Med Res* 2009;25(10):2349-2351. *Based on medical claims data, approximately 340,000 people in the United States were diagnosed with anogenital warts in 2004.*

Insinga RP, Glass AG, Rush BB. Diagnoses and outcomes in cervical cancer screening: a population-based study. *Am J Obstet Gynecol* 2004;191:105-113. *This study examined routine cervical cancer screening diagnoses and outcomes on an age-specific basis in one managed care organization.*

Joura EA, Giuliano AR, Iversen OE, et al. for the Broad Spectrum HPV Vaccine Study. A 9-valent HPV vaccine against infection and intraepithelial neoplasia in women. *N Engl J Med* 2015;372:711-723. *This randomized controlled clinical trial of 9-valent HPV vaccine demonstrated high efficacy.*

Lewis RM, Markowitz LE, Gargano JW, et al. Prevalence of genital human papillomavirus among sexually experienced males and females aged 14–59 years, United States, 2013–2014. *J Infect Dis* 2018;217(6): 869-877. *HPV prevalence is higher among males than females, and patterns of HPV prevalence by age group vary by sex.*

Marra E, Lin C, Clifford GM. Type-specific anal human papillomavirus prevalence among men, according to sexual preference and HIV status: a systematic literature review and meta-analysis. *J Infect Dis* 2019;219(4):590-598. *HPV prevalence is higher among men living with HIV than HIV-uninfected men, regardless of sexual identity.*

McClung NM, Lewis RM, Gargano JW, et al. Declines in vaccine-type human papillomavirus prevalence in females across racial/ethnic groups: data from a national survey. *J Adol Health* 2019;65(6):715-722. *This study reports a declining trend in HPV vaccine-type prevalence in young women in all reported racial/ethnic groups since the HPV vaccine was introduced.*

Meites E, Szilagyi PG, Chesson HW, et al. Human papillomavirus vaccination for adults: updated recommendations of the Advisory Committee on Immunization Practices. *MMWR Morb Mortal Wkly Rep* 2019;68:698-702. *This report includes the most recent HPV vaccination recommendations in the United States, 2019.*

Ostor AG. Natural history of cervical intraepithelial neoplasia: a critical review. *Int J Gynecol Pathol* 1993;12:186-192. *An evaluation of the natural history of CIN demonstrates that most low-grade lesions regress.*

Patel H, Wagner M, Singhal P, Kothari S. Systematic review of the incidence and prevalence of genital warts. *BMC* 2013;13:39. *Global estimates of annual incidence of anogenital warts ranges from 160 to 289 per 100,000 person-years.*

Satterwhite CL, Torrone E, Meites E, et al. Sexually transmitted infections among U.S. women and men: prevalence and incidence estimates, 2008. *Sex Transm Dis* 2013;40:187-193. *This analysis estimated that 79 million people, most in their late teens and early 20s, are infected with HPV; each year, about 14 million people are newly infected.*

Senkomago V, Henley SJ, Thomas CC, Mix JM, Markowitz LE, Saraiya M. Human papillomavirus-attributable cancers—United States, 2012-2016. *MMWR* 2019;68(33):724-728. *This surveillance report describes the incidence of cancers caused by HPV in the United States between 2012 and 2016, by cancer type and state of diagnosis.*

Winer RL, Feng Q, Hughes JP, et al. Risk of female human papillomavirus acquisition associated with first male sex partner. *J Infect Dis* 2008;197:279-282. *This evaluation of the acquisition of HPV in young women demonstrated a high incidence of infection with first male sex partners.*

## ADDITIONAL RESOURCES

American College of Obstetricians and Gynecologists. Available at: https://www.acog.org/Clinical-Guidance-and-Publications/Search-Clinical-Guidance. *The American Congress of Obstetricians and Gynecologists is an organization supporting providers of care in the field of obstetrics and gynecology.*

ASCCP. Consensus guidelines. Available at: http://www.asccp.org/guidelines. *This organization seeks to educate practitioners about the appropriate screening and management of lower genital tract diseases.*

Centers for Disease Control and Prevention (CDC). Cancer prevention and control. Available at: https://www.cdc.gov/cancer/hpv/index.htm. *The Division of Cancer Prevention and Control of the CDC conducts monitoring of, research on, and evaluation of cancers, including HPV-associated cancers. The Division of Cancer Prevention and Control is responsible for the Breast and Cervical Cancer Early Detection Program (NBCCEDP) and the National Program of Cancer Registries (NPCR).*

Centers for Disease Control and Prevention (CDC). Human papillomavirus. Available at: www.cdc.gov/hpv. *The CDC HPV web portal provides patient, provider, and general audience information on HPV, cervical cancer, and HPV vaccines.*

Centers for Disease Control and Prevention (CDC). Sexually transmitted diseases treatment guidelines, 2015, *MMWR Recomm Rep. The STD treatment guidelines are the primary reference for clinicians evaluating and treating patients for STDs in the United States. Available at:https://www.cdc.gov/std/tg2015/default.htm.*

Elam-Evans LD, Yankey D, Singleton JA, et al. National, Regional, State, and Selected Local Area Vaccination Coverage Among Adolescents Aged 13–17 Years—United States, 2019. *MMWR Morb Mortal Wkly Rep* 2020;69:1109-1116. DOI: *http://dx.doi.org/10.15585/mmwr.mm6933a1.*

International Agency for Research on Cancer. Cancer Today. Available at: http://gco.iarc.fr/today/home. *Cancer Today is a data visualization tool for exploring the global burden of cancer, including incidence, mortality, and prevalence of 36 specific cancer types in 185 countries or territories.*

# 58

# Infections Caused by *Chlamydia trachomatis*, Including Lymphogranuloma Venereum

Stephen J. Jordan, William M. Geisler

 **ABSTRACT**

*Chlamydia trachomatis* is an obligate intracellular bacterium that infects mucosal surfaces of humans, including oropharyngeal, anogenital, and conjunctival surfaces. *C. trachomatis* can be classified through molecular typing into strains causing ocular infections (trachoma), nonulcerative oropharyngeal and/or anogenital infections (chlamydia), and a distinct ulcerative syndrome called *lymphogranuloma venereum* (LGV). *C. trachomatis* infections are highly prevalent worldwide, especially in adolescents and young adults, and can cause substantial reproductive tract morbidity. The majority of persons with chlamydia are asymptomatic, with normal genital examination findings; diagnosis of chlamydia usually relies on testing for the bacterium. Despite highly sensitive tests and highly effective therapy, the number of reported chlamydial infections continues to rise. As a chlamydial vaccine is not yet available, better prevention and control efforts are needed. This chapter reviews the epidemiology, clinical features, diagnosis, treatment, and prevention of chlamydia and LGV.

 **CLINICAL VIGNETTE**

A 19-year-old female presents to the emergency department (ED) complaining of lower abdominal pain, nausea, and fever. Her abdominal pain began one day prior and rapidly progressed, causing her to present to the ED. She describes the pain as severe, constant abdominal cramping, 8/10 in severity, and located in her lower abdomen. One week prior, she had experienced a vaginal discharge and some mild dysuria. She has no known medical problems. Her only medication is an oral contraceptive medication; she has no known drug allergies. She denies tobacco, alcohol, and drug use. She is sexually active with a single male partner and does not use barrier protection; she has had six lifetime sexual partners. She has never had a sexually transmitted infection (STI) before. She was last sexually active 2 days ago and reports mild pain during sexual intercourse as well as postcoital bleeding. A review of systems is positive for nausea and two episodes of vomiting within the last 12 hours.

Her vital signs include an oral temperature of 101.5°F, pulse of 110 beats/min, respirations 12 breaths/min, and an oxygen saturation of 100% on room air. She has no evidence of conjunctivitis and her oropharynx is not erythematous. On physical examination, she is noted to have mild tenderness on palpation of her lower abdomen. A pelvic examination by speculum reveals a white vaginal discharge and a normal-appearing cervical os without cervical discharge. Bimanual examination reveals cervical motion tenderness and fundal tenderness but no adnexal tenderness. A complete blood count reveals a mild leukocytosis; a pregnancy test is negative. A cervical swab is obtained and sent for a nucleic acid amplification test (NAAT) for gonorrhea and chlamydia.

The patient is diagnosed with pelvic inflammatory disease (PID) and admitted to the hospital, given the concern that she will not tolerate oral outpatient antibiotics. Intravenous antibiotics are administered and, 24 hours later, she is markedly improved, with resolution of her nausea and improvement in her abdominal pain. Her gonorrhea test is negative, but her chlamydia test is positive. Given her clinical improvement, her intravenous antibiotics are stopped and oral antibiotics are started. She is discharged home with instructions to avoid sexual activity until she has completed her antibiotics, her symptoms have resolved, and her partner has been treated. Seventy-two hours later, she reports tolerating her antibiotics well, near complete resolution of her symptoms, and confirms that her partner has been treated. Three months later, she undergoes repeat chlamydia testing, which is negative.

## GEOGRAPHIC DISTRIBUTION AND MAGNITUDE OF DISEASE BURDEN

### Chlamydia

Chlamydia is highly prevalent worldwide in both industrialized and developing countries. The World Health Organization (WHO) estimates that approximately 127 million new cases of chlamydia occur worldwide each year. Chlamydia is a reportable infection in the United States and remains the most commonly reported bacterial STI. Despite declining prevalence in some geographic areas with successful screening and treatment programs, the total number of chlamydia cases reported in the United States continues to increase each year and now exceeds 1.7 million annually. The majority of chlamydia cases in the United States remain undiagnosed, and it is estimated that the true prevalence may be up to 3 million new chlamydial infections each year. Nationally representative surveys of chlamydia prevalence in adolescents and young adults have revealed a prevalence of up to 5% in females and up to 4% in males. The actual prevalence of chlamydia may be much higher, often greater than 10%, in some venues such as clinics aimed at STI care and screening and in correctional facilities. In the United States, the region with the highest burden of chlamydia is the Southeast.

Females may experience significant health consequences from chlamydial complications. One important complication in women is PID, in which the infection spreads to the upper genital tract and can involve the uterus (endometritis), fallopian tubes (salpingitis), ovaries, and/or peritoneum. In the United States, it is estimated that there are over 1 million new cases of PID each year and that at least 10% of women with PID will become infertile (with repeated PID episodes resulting in higher infertility rates). In the United States, the estimated annual cost attributable to chlamydia and its complications is several billion dollars.

### Lymphogranuloma Venereum

There are no reliable estimates of the worldwide burden of LGV because the diagnosis is often missed or is made on clinical findings rather than the results of diagnostic testing. In general, LGV was previously thought to be rare and sporadic in industrialized countries. However, following an LGV outbreak in the Netherlands in 2013, LGV infections have become more frequent in Europe, North America, and Australia, primarily among men who have sex with men (MSM) and mostly occurring at the anogenital site. In developing countries, LGV may be endemic and is a cause of genital ulcers. Hundreds to thousands of LGV cases occur annually in parts of East and West Africa, Southeast Asia, India, South America, and the Caribbean.

## RISK FACTORS

### Chlamydia

Among many risk factors associated with chlamydia, young age is the most prominent. Surveillance studies from the US Centers for Disease Control and Prevention (CDC) have demonstrated that the highest rates are in persons younger than 25 years of age. The chlamydia rate declines significantly in persons 30 years of age and older but still may be substantial in those who have other relevant risk factors, including unprotected sexual intercourse with a new partner or multiple sexual partners and engaging in transactional sex (trading sex for money or drugs). A prior chlamydial infection is a major risk factor, primarily because of high rates of reinfection from untreated partners. Use of human immunodeficiency virus (HIV) preexposure prophylaxis (PrEP) (taking HIV medications to prevent HIV acquisition) is associated with an increased risk for multiple STIs, including chlamydia; most chlamydia infections in this setting are likely due to non-LGV serovars, but some anorectal LGV infections also likely occur. Although chlamydia is common in all racial and ethnic groups in the United States, African Americans have the highest rate; Hispanics and American Indian/Alaska Natives are also disproportionately affected compared with Caucasians.

In addition to risk factors related primarily to higher-risk sexual behaviors or socioeconomic status (which may be a proxy for limited access to healthcare and availability of educational or prevention measures), biologic factors can also increase chlamydia risk. Cervical ectopy (expansion of intracervical columnar epithelial cells across the surface of the cervix) likely increases the risk of chlamydia acquisition by providing a greater surface area of susceptible cells for chlamydial infection. Cervical ectopy is more common in younger females, especially those on hormonal contraceptive therapy. Bacterial vaginosis (BV) has also been associated with a twofold increased risk of having chlamydia. The bacterial species present in BV may alter vaginal levels of specific metabolites (increased indole, decreased D-lactic acid) that favor *C. trachomatis* acquisition and survival. Finally, host immune responses and host genetic determinants (e.g., human leukocyte antigen [HLA] types and immune gene polymorphisms) likely play a role in susceptibility to or protection against chlamydia. Limited data suggest that interferon gamma–producing CD4[+] T cells protect against chlamydia incidence and reinfection. HLA-DQB1*06 is a risk for chlamydia reinfection and complications, but more research is needed on the role of immunogenetic factors in chlamydia and its complications.

### Lymphogranuloma Venereum

Studies of LGV outbreaks in industrialized countries have revealed that most LGV cases were anorectal infections in MSM who practiced unprotected receptive and/or insertive anal intercourse. Other identified LGV risk factors included having multiple sexual partners, engaging in anonymous or group sex, having sex under the influence of drugs ("chemsex"), and being infected with HIV. Although risk factors for endemic genital LGV in developing countries are less well understood, they likely resemble traditional chlamydia risk factors except

| TABLE 58.1  Clinical Syndromes Caused by *Chlamydia* | | |
|---|---|---|
| **Women** | **Men** | **Neonates**[a] |
| Urethritis | Urethritis | Conjunctivitis |
| Cervicitis | Epididymitis | Pharyngitis |
| Pelvic inflammatory disease | Proctitis or proctocolitis | Respiratory infection |
| Proctitis or proctocolitis | Conjunctivitis[a] | |
| Conjunctivitis[a] | Pharyngitis | |
| Bartholinitis | Prostatitis | |
| Pharyngitis | Reactive arthritis | |
| Reactive arthritis | Trachoma[a] | |
| Trachoma[a] | | |

[a]Not discussed in this chapter.

that, unlike with chlamydia, LGV can also be transmitted by skin-to-skin genital contact.

## CLINICAL FEATURES

### Chlamydia

As discussed in more detail in the following paragraphs, chlamydia causes a diverse spectrum of clinical syndromes (Table 58.1). The incubation period for chlamydia is estimated to be between 7 and 21 days.

#### Genital Infection in Females

In the female genital tract, *C. trachomatis* primarily infects the endocervix, but it may also infect the urethra and Bartholin glands. The majority (>75%) of women with endocervical chlamydia are asymptomatic; even when symptoms are present, they are frequently nonspecific and overlap with those caused by other vaginal infections (e.g., trichomoniasis or BV) or endocervical infections (e.g., gonorrhea). Symptoms may include new or increased vaginal discharge, intermenstrual bleeding, or pain during intercourse (dyspareunia). Even in the absence of symptoms, 10% or more of women with chlamydia will have signs of infection on pelvic examination. Endocervical discharge (purulent, cloudy, or bloody, Fig. 58.1), easily induced endocervical bleeding ("friability") on insertion of an endocervical swab, and edematous ectopy are the examination findings most suggestive of endocervical chlamydia, yet they are nonspecific. The presence of one or more of these findings supports the clinical diagnosis of cervicitis. Abnormal vaginal discharge originating from the endocervix may also be present. Although cervical ectopy may predispose to chlamydia, the presence of cervical ectopy without edema or congestion is not indicative of chlamydia; ectopy is very common in female adolescents and young adults.

The majority of women with endocervical chlamydia will have concomitant urethral infection, and a small proportion of these urethral infections may cause painful urination and/or increased urinary frequency. In young sexually active women, dysuria due to urethral chlamydia is often misdiagnosed as a urinary tract infection (UTI) and treated with antibiotics that are not effective against chlamydia (e.g., trimethoprim-sulfamethoxazole). Consideration should be given to screening sexually active young women with urinary symptoms for chlamydia. Concomitant infection of the Bartholin glands may rarely occur and manifests as ductal erythema and swelling, typically unilateral and often with purulent ductal exudate.

Symptoms of PID include fever, nausea, and pelvic or lower abdominal pain (especially with intercourse). Pelvic examination findings of PID may include cervical motion tenderness and/or tenderness of the uterus, fallopian tubes, ovaries, and/or lower abdomen.

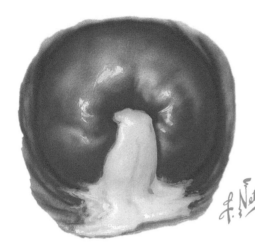

Mucopurulent endocervical discharge may be present on examination in a small proportion of females with chlamydial endocervical infection, but most chlamydia-infected females have a normal-appearing cervix on examination

**Fig. 58.1** Mucopurulent cervicitis.

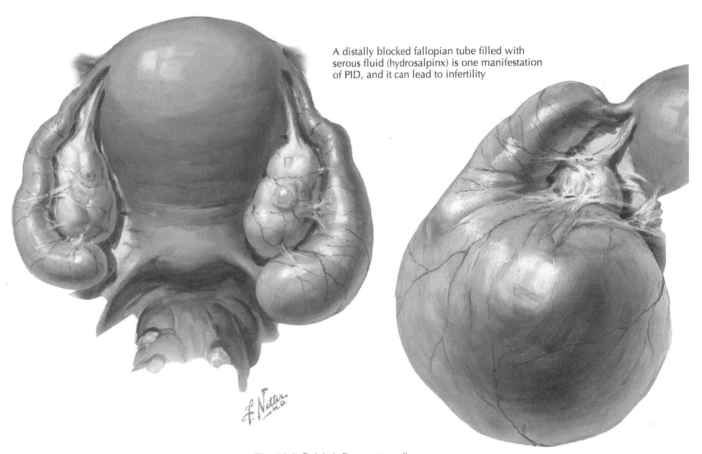

A distally blocked fallopian tube filled with serous fluid (hydrosalpinx) is one manifestation of PID, and it can lead to infertility

**Fig. 58.2** Pelvic inflammatory disease.

A distally blocked fallopian tube can fill with serous fluid and become substantially dilated, a condition called *hydrosalpinx* (Fig. 58.2). The majority of subjects with PID will have either no symptoms or mild pelvic complaints. Short-term complications of PID include tubal and/or ovarian abscesses; long-term complications include chronic pelvic pain, ectopic pregnancy, and infertility. Chlamydia is a leading preventable cause of ectopic pregnancy and infertility worldwide. Overall, the most common clinical presentation of genital chlamydia in females is the absence of symptoms with normal pelvic examination findings (i.e., no cervical signs).

## Genital Infection in Males

In the male genital tract, *C. trachomatis* primarily infects the urethra. Studies employing a universal screening approach have demonstrated that over 50% of males with urethral chlamydia are asymptomatic. Symptoms of urethral chlamydia in males may include painful urination, urinary frequency, meatal itching or discomfort, and urethral discharge. A urethral discharge may be apparent on examination and is typically clear or cloudy, often with mucus strands (Fig. 58.3). Meatal erythema and/or swelling may accompany the discharge. If no spontaneous urethral discharge is noted, the urethra should be stripped and

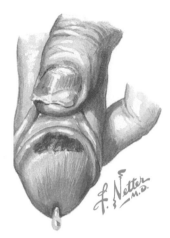

Cloudy or purulent urethral discharge may be present on examination in males with urethral chlamydial infection

**Fig. 58.3** Urethritis.

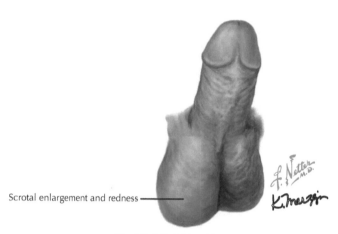

Scrotal enlargement and redness ——————

**Fig. 58.4** Epididymitis.

examined again. Abundant purulent discharge, as seen in gonorrhea, is uncommon in chlamydia.

Males with urethral symptoms but normal urethral examination findings are sometimes misdiagnosed as having a UTI and not tested for chlamydia or treated with antibiotics effective against chlamydia. However, UTIs caused by enteric bacteria are rare in young men with anatomically and functionally normal urinary and renal collecting systems. Therefore such males with urethral symptoms should be presumed to have a STI and undergo appropriate diagnostic testing and treatment. Inguinal lymph node swelling is uncommon in urethral chlamydia (non-LGV) but can occur and is typically mild.

Chlamydia may spread to the upper genital tract in males and cause infection of the epididymis, termed *epididymitis* (Fig. 58.4). The findings in chlamydial epididymitis are usually unilateral. Symptoms include swelling or pain in the scrotum, testicle, and/or epididymis and, infrequently, fever. On examination, males with epididymitis may have redness, warmth, and/or swelling of the overlying scrotum. Swelling and tenderness of the involved epididymis is usually present and may be accompanied by testicular tenderness. Short-term complications of epididymitis include testicular abscesses; long-term complications include chronic testicular pain, chronic epididymitis, and infertility. The most important differential diagnosis of epididymitis is testicular torsion (pain resulting from compromise of testicular arterial blood flow), which is a urologic emergency. Males with severe acute scrotal pain, especially if associated with recent testicular trauma

or prior recurring episodes of testicular pain, should be sent immediately for a testicular ultrasound to rule out torsion.

Chlamydia has also been reported as a cause of prostatitis, but the epidemiology, clinical features, and course of this clinical presentation are poorly understood. Overall, the majority of urethral chlamydial infections in males have no urethral symptoms and a normal genital examination.

### Anorectal Infection (Non-Lymphogranuloma Venereum Chlamydia)

Similar to anorectal LGV (discussed later), non-LGV *C. trachomatis* strains can cause anorectal infection in men or women who practice receptive anorectal intercourse. Anorectal chlamydia may also be diagnosed in women who deny having anorectal intercourse, possibly due to cross-contamination of the anus with vaginal chlamydial organisms or orally acquired chlamydial organisms that survive transit through the gastrointestinal (GI) tract and infect the anorectum; this latter mechanism has been shown to occur in animals but has not been proven to occur in humans. The majority of men and women with anorectal chlamydia are asymptomatic. Symptoms that may occur include anorectal pain, itching, mucus, discharge, bleeding, or diarrhea. Anorectal chlamydia can be complicated by strictures, which may affect the frequency of bowel movements and the character of the stools. Specific complaints may include an inability to defecate, spasms and cramping. Patients with anorectal symptoms should ideally be evaluated by anoscopy. Anoscopy findings suggestive of anorectal chlamydia include anorectal erythema, discharge (mucous, cloudy, purulent, or bloody), or occasionally ulceration (Fig. 58.5). Anorectal chlamydia may be misdiagnosed as proctitis caused by a non-STI etiology (e.g., Crohn's disease, ulcerative colitis, etc.). However, any patient with proctitis who reports anorectal sexual activity should receive testing for chlamydia and other STIs.

### Oropharyngeal Infection

*C. trachomatis* has been detected in oropharyngeal specimens and saliva, yet it remains unclear whether chlamydial infection of the oropharynx causes clinically significant disease. Sparse studies in MSM suggest that oropharyngeal chlamydia may be transmissible to the urethral site. One reason for the limited knowledge regarding oropharyngeal chlamydia is that traditional chlamydia tests, such as culture, perform poorly on oropharyngeal specimens. However, with the availability of a *C. trachomatis* NAAT, the clinical presentation and course of oropharyngeal chlamydia should become better understood.

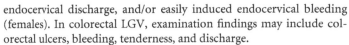

Limited clinical experience to date suggests that chlamydial infection of the oropharynx is mostly asymptomatic, with normal oropharyngeal examination findings or minimal pharyngeal erythema. Until the clinical significance of oropharyngeal chlamydial infections is better understood, routine chlamydia testing of the oropharynx is not recommended.

### Reactive Arthritis

Reactive arthritis (previously called *Reiter syndrome*) may occur during or shortly after a genital chlamydial infection and is characterized primarily by conjunctivitis and an aseptic rheumatoid factor–negative asymmetric polyarthritis. Other clinical findings may include oral ulcers, uveitis, rashes, inflammation of the sacroiliac joints, and cardiac and neurologic complications. Reactive arthritis occurs more commonly in men and is often linked to the HLA-B27 genotype; persons with HLA-B27 may have a more aggressive clinical course.

### Lymphogranuloma Venereum

The clinical features of LGV can be divided into three stages. The primary stage involves development of a primary lesion at the inoculation site approximately 3 to 30 days after inoculation. The primary lesion is usually a small asymptomatic papular lesion and/or ulcers, that frequently goes unnoticed by the patient and heals spontaneously. Other symptoms of urethral, endocervical, or rectal LGV may resemble those described for chlamydia. With genital LGV, examination findings may include genital papules and/or ulcers, urethral discharge (males),

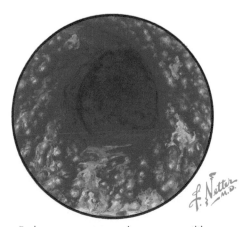

Endoscopy appearance in severe proctitis
**Fig. 58.5** Proctocolitis.

endocervical discharge, and/or easily induced endocervical bleeding (females). In colorectal LGV, examination findings may include colorectal ulcers, bleeding, tenderness, and discharge.

Approximately 2 to 6 weeks after the primary stage, the secondary stage begins and involves the lymph nodes that drain the inoculation site. The involved lymph nodes rapidly increase in size, sometimes coalescing with adjacent involved nodes, and are often associated with pain and erythema. The enlarged lymph nodes become fluctuant and are termed *buboes* (Fig. 58.6). Up to one-third of buboes may spontaneously rupture and drain purulent material, which often relieves the pain. The draining buboes may form fistulas or sinus tracts. An examination finding of enlargement of both inguinal and femoral nodes has been termed the *groove sign* and is highly characteristic of LGV. The secondary stage may be accompanied by constitutional symptoms including fever, chills, headache, myalgias, and fatigue.

The tertiary stage involves chronic or late complications of LGV. Genital LGV may be complicated by chronic extensive ulceration of the external genitalia (which can be destructive), and lymph node destruction may obstruct lymphatic drainage, resulting in genital elephantiasis. Colorectal LGV may be complicated by strictures, fistulas, and/or perirectal abscesses.

## DIAGNOSTIC APPROACH

### Chlamydia

Because most chlamydial infections in men and women are asymptomatic and still cause substantial morbidity, population-based detection of *C. trachomatis* involves selective screening of those at highest risk. Chlamydia screening facilitates the identification and treatment of infection, and prevention of complications as well as patient education, sexual partner treatment, and communicable disease reporting. The CDC recommends annual chlamydia screening for all sexually active women 24 years of age or younger as well as women older than age 24 with relevant risk factors (e.g., multiple sexual partners, a new sexual partner). Women of any age with genital complaints or signs of cervicitis or PID should undergo diagnostic testing as well. Although early evidence suggested that screening may lower both the prevalence and complication rates of chlamydia, subsequent studies found that screening has resulted only in a decrease in complication rates. Pregnant women should be screened for chlamydia at their first prenatal visit; pregnant women younger than 25 years of age or with risk factors should be tested again for chlamydia during the third trimester. Although annual chlamydia screening of sexually active males is not recommended, males with genital symptoms and/or chlamydia risk

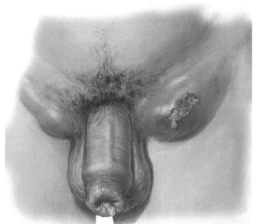

Lymph nodes in LGV infection can become enlarged, erythematous, and fluctuant, termed *buboes*. Buboes may spontaneously rupture and drain purulent material, which can provide pain relief.

**Fig. 58.6** Buboes.

factors should be tested. Also, males in venues with a high chlamydia prevalence (e.g., clinics that provide STI care, correctional facilities, etc.) should be tested for chlamydia if resources permit and if testing does not hinder screening efforts in women. Among sexually active MSM, annual screening for chlamydia should be performed at exposed genital and anorectal sites; more frequent testing should be determined by relevant risk behavior (new sexual partners), particularly in those taking HIV PrEP.

Over the last 30 years there have been considerable advances in the development of diagnostic assays for detecting *C. trachomatis*. *Chlamydia* culture, traditionally considered the gold standard, was the first chlamydia test available and is still used in some centers for research. However, because culture was technically demanding, there was a shift toward nonculture chlamydia tests. *C. trachomatis* serologic assays are available but are not useful in the diagnosis of active chlamydial infection because they do not distinguish past from current chlamydia and may be negative in early or acute chlamydia.

Around the mid-1990s, NAATs became available for genital chlamydia testing and offered two distinct advantages over earlier tests: (1) higher test sensitivity resulting from amplification of *C. trachomatis* nucleic acids and (2) greater patient convenience and/or satisfaction in that, as an alternative to the usual genital swab specimens collected by examination, NAATs can also be performed with similar accuracy on noninvasively or minimally invasively collected specimens, including urine and self-collected vaginal, meatal/urethral, and rectal swabs. The NAAT is now the test of choice for diagnosing genital chlamydia. The recommended specimen for the urogenital *C. trachomatis* NAAT in men is first-catch urine (i.e., the first portion of urine); in women it is a vaginal swab. Alternate urogenital specimen options that can be tested in men include a urethral swab and in women a cervical swab or urine; NAAT sensitivity is lower on urine from women than on vaginal or cervical swabs. Individuals undergoing testing for chlamydia should generally also receive testing for gonorrhea, because coinfection with chlamydia and gonorrhea is common and their recommended treatments differ; most NAATs include testing for both infections.

Patients who report engaging in receptive anorectal intercourse should undergo anorectal chlamydia testing. NAATs are superior to culture for detecting *C. trachomatis* in anorectal specimens. Recently some NAATs have been approved by the US Food and Drug Administration (FDA) for testing on anorectal specimens. Non-FDA-approved NAATs can also be used with anorectal specimens, but laboratories (including commercial laboratories such as LabCorp, QUEST, and others) must first perform in-house validation in order to adhere to Clinical Laboratory Improvement Amendments (CLIA) regulations.

Whether to perform oropharyngeal chlamydia testing on persons reporting receptive oropharyngeal sex is debatable, because it is unclear if *C. trachomatis* causes clinical disease of the oropharynx or has subsequent clinical consequences. As with using NAATs on anorectal specimens, validation is necessary for laboratories performing NAATs on oropharyngeal specimens.

## Lymphogranuloma Venereum

In developing countries, a presumed diagnosis of LGV is usually made based on clinical findings. In industrialized countries, LGV diagnosis is still difficult owing to limitations in commercially available tests and is typically either presumed based on clinical findings or is made through specialized LGV testing at the CDC or other specialized laboratories. Serologic tests for *C. trachomatis*, including complement fixation (CF) and microimmunofluorescence (MIF) tests, are not standardized and do not reliably differentiate between LGV and non-LGV *C. trachomatis* strains; higher antibody titers (CF titers >1:64 and MIF titers >1:256) are strongly suggestive of LGV when accompanied by compatible

clinical findings but are not confirmatory. Detection of *C. trachomatis* by culture or NAATs from anogenital specimens also does not differentiate LGV from non-LGV strains. Therefore the only means to confirm LGV is to have the CDC or a specialized laboratory perform OmpA typing on the *C. trachomatis* strain to detect the presence of an LGV OmpA genotype. If buboes are present, *C. trachomatis* may be detected in a bubo aspirate, which would be highly suggestive of LGV in this clinical context; however, confirmation by OmpA typing should still be performed if possible. Treatment of LGV should not be delayed while awaiting the results of LGV testing.

## CLINICAL MANAGEMENT AND DRUG TREATMENT

To ensure eradication of infection and prevention of reinfection, management of chlamydia and LGV requires a multifaceted approach (Box 58.1).

### Antibiotic Therapy
#### Uncomplicated Chlamydia

The 2015 CDC STD Guidelines recommend either azithromycin 1 g PO as a single dose or doxycycline 100 mg PO bid for 7 days for uncomplicated genital or anorectal chlamydia in nonpregnant individuals (Table 58.2). These regimens are equally efficacious for urogenital chlamydia, but increasing evidence suggests that azithromycin may be an inferior treatment for non-LGV anorectal chlamydia; many experts now recommend doxycycline for anorectal chlamydia. Azithromycin and doxycycline are well tolerated but should be taken with food to minimize GI side effects. Doxycycline can cause photosensitivity of the skin, and patients should be educated to take appropriate precautions to prevent sunburn. Treatment adherence is higher with azithromycin. Erythromycin base 500 mg PO four times a day for 7 days is an alternative regimen that has a low rate of completion. Two fluoroquinolone regimens, levofloxacin 500 mg PO daily for 7 days or ofloxacin 300 mg PO twice a day for 7 days, are other alternative regimens for chlamydia that do not offer an efficacy advantage over the recommended regimens and are costlier. Ciprofloxacin has poor in vitro activity against *C. trachomatis* and should not be used. Moxifloxacin has in vitro activity against *C. trachomatis*, but its use is not recommended due to limited data on

---

**BOX 58.1    Multifaceted Approach to Management of Chlamydia or Lymphogranuloma Venereum**

- Centers for Disease Control and Prevention (CDC)–recommended antibiotic therapy for infected patients and their partners
- Abstinence from sexual activity for patients and their partners until treatment is completed
- Test of cure for pregnant patients at 3–4 weeks after completing treatment
- Repeat testing approximately 3 months after completing treatment

---

**TABLE 58.2    Centers for Disease Control and Prevention (CDC)–Recommended Antibiotic Regimens for Uncomplicated Chlamydia**

| Nonpregnant | Pregnant |
|---|---|
| Azithromycin 1 g PO in a single dose<br>*or*<br>Doxycycline 100 mg PO twice a day for 7 days | Azithromycin 1 g PO in a single dose |

Note: See the text for alternative treatment regimens for chlamydia.

its clinical efficacy. Although there has been only limited clinical experience with azithromycin and doxycycline regimens for treating oropharyngeal chlamydia, these regimens are likely efficacious; the efficacy of alterative regimens for treating oropharyngeal chlamydia is unknown. Antimicrobial resistance has not been a major concern to date. There are no data to suggest that the efficacy of CDC-recommended regimens is altered in HIV-positive individuals, in whom the same regimens are recommended.

Doxycycline, ofloxacin, and levofloxacin are contraindicated in pregnancy. The CDC recommends azithromycin 1 g PO as a single dose for uncomplicated chlamydia in pregnant women (see Table 58.2). Amoxicillin 500 mg PO three times a day for 7 days is now an alternative chlamydia treatment regimen in pregnant women due to the potential concern that treating chlamydia with amoxicillin may induce persistence of *C. trachomatis*, which has been demonstrated in vitro after exposure to penicillin. Limited clinical studies of amoxicillin for chlamydia in pregnancy have had study design limitations and have not been able to sufficiently address this concern. Azithromycin is advantageous in terms of compliance, and anecdotal experience suggests that azithromycin is far more widely used than amoxicillin. Erythromycin base 500 mg PO four times a day for 7 days is an alternative regimen for chlamydia in pregnancy; but again, it is a difficult regimen to complete. All three antibiotics are pregnancy category B drugs. Considering its efficacy, tolerability, and compliance advantage over other regimens, azithromycin is the overall best option for the treatment of chlamydia in pregnancy.

## Complications of Chlamydia

An initial consideration in managing PID or epididymitis is whether hospitalization is necessary. Patients with PID should be hospitalized if they are pregnant, unable to tolerate oral outpatient antibiotics, not clinically improved within 72 hours of outpatient treatment, or there is concern about an alternative diagnosis or a complication. Hospitalization for epididymitis is rare and usually occurs because of concerns about a complication, an alternative diagnosis, or failure of outpatient treatment.

Antimicrobial treatment for PID or epididymitis caused by chlamydia is usually started empirically before chlamydia test results are available and is broad enough to cover at least chlamydia and gonorrhea. The CDC recommends one of the following empiric combination regimens for inpatient treatment of PID: (1) cefotetan 2 g intravenously (IV) every 12 hours *or* cefoxitin 2 g IV every 6 hours *plus* doxycycline 100 mg PO or IV every 12 hours, or (2) clindamycin 900 mg IV every 8 hours *plus* gentamicin intramuscularly (IM) or IV with a 2-mg/kg loading dose followed by 1.5 mg/kg every 8 hours (once-daily dosage regimens are an alternative). After 24 hours of clinical improvement, patients may be switched to an oral regimen depending on the initial regimen: either doxycycline 100 mg twice a day or clindamycin 450 mg four times a day for a total PID treatment duration of 14 days.

The CDC-recommended outpatient treatment for PID is doxycycline 100 mg PO twice a day for 14 days plus a single dose of ceftriaxone 250 mg IM with or without metronidazole 500 mg PO twice a day for 14 days. There are limited studies on the use of azithromycin for PID (given as 1 g PO weekly for 2 weeks), but azithromycin is not currently recommended as a first-line treatment for PID. Recommended outpatient treatment for epididymitis is doxycycline 100 mg PO twice a day for 10 days plus a single dose of ceftriaxone 250 mg IM. Adjunctive measures for pain relief in epididymitis include analgesics and scrotal support (e.g., wearing briefs rather than boxers and placing a rolled-up towel or blanket under the scrotum for support while sitting or lying). It is recommended that subjects with PID or epididymitis undergo repeat genital examination in about 72 hours to evaluate the clinical response to treatment; lack of treatment response or worsening pain may suggest treatment failure, a complication (e.g., an abscess), or another cause for the clinical presentation.

Reactive arthritis associated with chlamydia is initially treated with antichlamydial antibiotics (necessary duration of treatment is unknown) and nonsteroidal anti-inflammatory agents. More aggressive reactive arthritis may require disease-modifying antirheumatic drugs (e.g., sulfasalazine and methotrexate). Management of reactive arthritis often requires consultation with a rheumatologist.

## Lymphogranuloma Venereum

The CDC-recommended treatment for LGV is doxycycline 100 mg PO twice a day for 21 days. An alternative regimen is erythromycin base 500 mg four times a day for 21 days, and azithromycin 1 g PO once weekly for 3 weeks is probably an effective alternative; however, data supporting the use of these nontetracycline alternative regimens for LGV are limited. As with chlamydia, LGV treatment recommendations are not different for HIV-infected patients. In individuals with tender, swollen inguinal lymph nodes, symptomatic relief can be achieved by aspiration or incision and drainage of the affected lymph node or nodes and can prevent cutaneous fistula formation. For pregnant women with LGV, the erythromycin or azithromycin regimens should be used because doxycycline is contraindicated in pregnancy.

## Sexual Activity

To prevent transmission to uninfected individuals or reinfection of the patient, it is important to instruct patients with chlamydia or LGV to abstain from sexual activity until both the patient and his or her partner or partners have completed a CDC-recommended treatment regimen and symptoms (if present) have resolved. Completion of the single-dose azithromycin regimen is considered to be 7 days after the single dose is administered. For those persons unlikely or unwilling to abstain from sexual activity during this period, strict compliance with condom use should be reinforced.

## Treatment of Sexual Partners

Most chlamydia (and presumably many LGV) treatment failures are a result of reinfection from an untreated sexual partner or a new infection. Treatment of sexual partners of patients with chlamydia or LGV is important, both for preventing reinfection of the patient and for preventing further transmission to other susceptible individuals. Sex partners should be treated if they had sexual contact with the patient during the 60 days preceding symptom onset or the diagnosis. The most recent sex partner should be treated even if the time of the last sexual contact was more than 60 days before symptom onset or diagnosis. Partners of patients with chlamydia should receive either the azithromycin or doxycycline regimen recommended by the CDC for chlamydia treatment. The CDC recommends that sexual partners of patients with LGV receive CDC-recommended chlamydia treatment regimens rather than CDC-recommended LGV regimens if the partner does not have clinical evidence of LGV disease (the rationale for this may be that subjects exposed to LGV without clinical evidence of infection likely have less invasive disease and may not need the 3 weeks of treatment normally recommended for LGV). There are two main strategies for getting sexual partners treated: partner referral and expedited partner therapy (EPT).

## Partner Referral

Because many clinical providers do not have the necessary resources, experience, or willingness to perform "partner notification" for chlamydia or LGV, the standard method to get partners of patients with chlamydia or LGV treated is partner referral, whereby patients notify their partners that they have been exposed to the infection and should themselves be seen by a clinical provider for testing and treatment. However, repeat chlamydia rates are high in patients instructed to refer their partners, suggesting that partner referral is sometimes ineffective in ensuring partner treatment (especially for male partners of female patients).

### Expedited Partner Therapy

A promising approach to ensuring that more partners of patients with chlamydia are treated is EPT, whereby a partner receives the treatment in an expedited manner without seeing a clinical provider. The primary way that EPT is used is for a patient to deliver the medication directly to the partner, which is termed "patient-delivered partner therapy" (PDPT). An alternative strategy is for the patient to deliver a prescription to the partner. Azithromycin is the chlamydia treatment primarily used in PDPT (rather than doxycycline). EPT is highly acceptable to patients and their partners. Several studies, both observational and clinical trials, have demonstrated a trend toward lower repeat chlamydia rates in patients with chlamydia who use PDPT versus partner referral. The major advantage of EPT is that more sexual partners will get treated, which should decrease rates of repeat chlamydia and lead to an overall decrease in chlamydia prevalence in a given community. Additional information can be found at www.cdc.gov/std/ept.

### Follow-Up and Repeat Testing

Nonpregnant patients with uncomplicated chlamydia or LGV do not need a test of cure after completion of therapy unless symptoms or signs of infection persist or recur. However, a test of cure (by NAAT) at 3 to 4 weeks after completion of treatment for uncomplicated chlamydia or LGV infections should be performed for pregnant women, a population in which treatment failures could lead to both maternal and neonatal complications. Repeat chlamydia NAAT earlier than 3 weeks after completion of treatment could yield a false-positive test result because of residual DNA or RNA from nonviable *Chlamydia*.

Chlamydia recurrence is common in males and females, occurring in up to 10% to 20% within 6 months of treatment. All chlamydia-infected persons should be retested for chlamydia at approximately 3 months after treatment. Some experts recommend repeat testing for women with chlamydial PID as early as 6 weeks after therapy. Some researchers in the United States (and several in Europe) are evaluating the feasibility of repeat chlamydia testing by home self-collection and mailing in specimens, but this practice is still under investigation in the United States. The rate of LGV recurrence in unknown, but it is likely high enough to also warrant repeat testing at approximately 3 months after treatment.

## PROGNOSIS

The CDC-recommended chlamydia and LGV antibiotic regimens have high cure rates when the full course of treatment is completed (the exception may be using azithromycin to treat anorectal chlamydia). If not treated, chlamydia tends to persist for weeks to months and perhaps for more than a year in a small proportion of persons. Delays in chlamydia treatment can lead to acute (e.g., upper genital tract disease) and long-term (e.g., infertility, chronic pelvic pain) chlamydia complications. Studies have shown that about 2% to 5% of females who have a positive chlamydia screening test but are not treated at the time of screening (due to no clinical indication for treatment) will develop

PID in the interval between screening and returning for treatment. Delays in treatment may increase the likelihood that exposed partners will also acquire infection. Even though patients may complete treatment, there is a high likelihood that they will be reinfected within a few months after treatment if their partners are not treated; irrespective of their symptoms, sexual partners should be treated. Because the majority of chlamydial infections are asymptomatic, many patients with chlamydia never see providers for testing and treatment; as a result, they are at significant risk for "silent" chlamydial complications such as infertility. Genital chlamydia can also increase the risk of acquisition and transmission of HIV.

## PREVENTION AND CONTROL

Until an effective chlamydia vaccine is developed, prevention and control of chlamydia will rely on a comprehensive approach, including STI education, chlamydia screening, timely treatment of patients and their sexual partners, abstinence until treatment completion, and repeat chlamydia testing at 3 months after treatment. Unfortunately, provider adherence rates with CDC-recommended chlamydia screening are often less than 50% in women in whom chlamydia screening is recommended and likely also low in other select populations, including MSM and persons taking PrEP. Many barriers to screening—including patients seeking chlamydia screening, patient access to healthcare providers, and some providers not performing chlamydia testing—exist and need to be addressed.

Providers need STI education regarding taking a sexual history, performing STI testing, providing CDC-recommended therapy, and educating patients about STIs. Patients need education about risk factors, barrier prevention methods (used properly, condoms are highly effective in preventing chlamydia), symptoms of STIs, and available STI screening tests and treatments. In order to prevent recurrent chlamydia, patient and partner adherence with treatment and abstinence until treatment is complete should be stressed. In order to help prevent chlamydia complications, efforts should be put into place for prompt notification of chlamydia test results and expediting treatment. As efforts for providing treatment to sexual partners are not always effective, repeat chlamydia testing at approximately 3 months after treatment completion to rule out reinfection should be stressed.

## ADDITIONAL RESOURCES

Centers for Disease Control and Prevention (CDC): Expedited partner therapy in the management of sexually transmitted diseases, Atlanta, 2006, U.S. Department of Health and Human Services. Available at: www.cdc.gov/std/treatment/EPTFinalReport2006.pdf. *This CDC report summarizes the available literature on EPT for the management of the partners of persons with chlamydia and discusses implementation of EPT.*

Centers for Disease Control and Prevention (CDC): Male chlamydia screening consultation. Available at: www.cdc.gov/std/chlamydia/ChlamydiaScreening-males.pdf. *This CDC report provides guidance for performing chlamydia screening in men based on available scientific data.*

Centers for Disease Control and Prevention (CDC): Recommendations for the laboratory-based detection of *Chlamydia trachomatis* and *Neisseria gonorrhoeae*—2014. *MMWR* 2014;63 (No. RR-2). Available at: https://www.cdc.gov/std/laboratory/2014labrec/default.htm. *This CDC report provides current recommendations for diagnostic tests and patient sample types to be used for diagnosing chlamydial infections.*

Centers for Disease Control and Prevention (CDC). Sexually transmitted disease surveillance 2018. Atlanta, 2019, U.S. Department of Health and Human Services. Available at: https://www.cdc.gov/std/stats18/default.htm. *This CDC report presents statistics and trends for sexually transmitted diseases (STDs), including chlamydia, in the United States through 2018.*

Centers for Disease Control and Prevention (CDC), Workowski KA, Bolan GA: Sexually transmitted diseases treatment guidelines. 2015. *MMWR Recomm Rep* 2015;64 (No. RR-03):1-137. Available at: https://www.cdc.gov/std/tg2015/default.htm. *This CDC report provides evidence-based guidelines for diagnosis and management of STDs, including chlamydial infections.*

Schachter J, Moncada J, Liska S, et al: Nucleic acid amplification tests in the diagnosis of chlamydial and gonococcal infections of the oropharynx and rectum in men who have sex with men. *Sex Transm Dis* 2008;35:637-642. *This article provides evidence that nucleic acid amplification tests perform better than culture for the detection of oropharyngeal and anorectal chlamydial infections.*

# 59

# Infections With *Neisseria gonorrhoeae*

*Stephanie N. Taylor, Lori M. Newman, Kimberly A. Workowski*

## ABSTRACT

Diagnosis of infection with *Neisseria gonorrhoeae* (NG), commonly known as *gonorrhea*, is important because the sequelae of untreated gonorrhea can include pelvic inflammatory disease (PID), perihepatitis, ectopic pregnancy, infertility, and chronic pelvic pain in women and epididymitis or infertility in men. Infection with gonorrhea also increases risk of both acquiring and transmitting human immunodeficiency virus (HIV). Gonorrhea is the second most common notifiable condition in the United States and disproportionately affects minorities, persons of lower socioeconomic status, and those with poor access to medical care. Although gonorrhea can be treated with a single dose of antibiotics, increasing spread of antimicrobial resistance has limited optimal treatment to a single class of drugs, systemic cephalosporins. Diagnosis of gonorrhea in asymptomatic individuals, appropriate treatment, partner treatment, and prevention of gonorrhea are all important strategies for the control of gonorrhea and gonococcal antimicrobial resistance.

## CLINICAL VIGNETTE

A 24-year-old woman presented with fever, nausea, right-upper-quadrant abdominal pain, and a white blood cell (WBC) count of 16,400/μL (3400–10,800 μL). She also reported right wrist and left dorsal foot pain along with a few upper extremity pustules on an erythematous base. Her partner was treated for gonococcal urethritis at a sexual health clinic 3 weeks prior, but her NG and *Chlamydia trachomatis* (CT) results were negative at that time by nucleic acid amplification test (NAAT) performed at her physician's office. Due to her negative results, she was not given prophylactic treatment as recommended.

Pelvic exam revealed minimal purulent cervical discharge. Ultrasound of the liver revealed thickening of the gallbladder wall but no gallstones. A computed tomography scan of the abdomen and pelvis revealed hepatic capsular enhancement. Vaginal and pharyngeal swabs were both positive for NG and negative for CT.

The patient was diagnosed with disseminated gonococcal infection (DGI) with clinical findings of left foot tenosynovitis, right wrist arthritis, and perihepatitis (Fitz-Hugh-Curtis syndrome). She responded well to ceftriaxone 1 g IV qd. After 3 days, she had resolution of all clinical and radiographic findings and was discharged to complete the 7-day course of therapy with cefixime 400 mg once daily. Although her partner was asymptomatic at the time of this patient's diagnosis, he was retreated for reexposure to gonorrhea.

## EPIDEMIOLOGY

The World Health Organization (WHO) estimates that there are approximately 87 million new gonorrhea infections globally per year. In the United States, gonorrhea is the second most common notifiable condition, second only to chlamydial infection. 583,405 cases of gonorrhea were reported in 2018, and this number represents less than half of all gonorrhea infections because many cases go undiagnosed and unreported. The national gonorrhea rate was 179 cases per 100,000 population, representing an 82% increase since a historic low of 99 cases per 100,000 population in 2009. In addition, when last published, the total direct medical cost of gonorrhea in the United States was $162 million among all ages adjusted to 2010 dollars.

Gonorrhea disproportionately affects adolescents and young adults, African Americans, men who have sex with men (MSM), and individuals living in the South, urban areas, or low-income communities. Racial disparities are greater for gonorrhea than for any other notifiable condition, with reported infection rates among African Americans 7.7 times higher than among whites in 2018. In the same year, 19-year-old women had the highest rate of gonorrhea, at 877 cases per 100,000 population. For 20- to 24-year-old women, the case rate was 703 cases per 100,000 population, representing an increase of 1.5% from 2017 to 2018 but a 32% increase from 2014 to 2018. Last, in men, the gonorrhea rates increased from 119 cases per 100,000 population in 2014 to 213 cases per 100,000 population in 2018, representing a 79% increase.

## CLINICAL FEATURES

### Urogenital Infection

The clinical presentation of urogenital infection with NG varies for women and men (Fig. 59.1). Gonorrhea is frequently asymptomatic in women, with only approximately 50% of women reporting symptoms such as vaginal discharge, pain or spotting with intercourse, burning with urination, or lower abdominal pain. Although men are more likely to have symptoms, approximately 10% of men infected with gonorrhea are thought to be asymptomatic. Common symptoms for men include urethral discharge, dysuria, and testicular tenderness. Because some strains of gonorrhea are less likely to cause symptoms than others, the proportion of patients whose presentation is symptomatic will vary by population.

Clinical examination findings of urogenital infection with gonorrhea are generally quite similar to those of other sexually transmitted infections. For women, examination findings may include a purulent yellow or green-tinged cervical discharge, inflammation of the cervix, cervical motion tenderness, or Bartholin gland abscess. Clinical examination findings in men include purulent urethral discharge, unilateral testicular tenderness consistent with epididymitis and lymphedema of the penile shaft and/or foreskin. Recent reports have reemerged that describe clusters of patients with urethritis, but with the etiology identified as *Neisseria meningitidis*.

### Rectal and Pharyngeal Infection

Infection with NG can also cause infections outside of the urogenital tract (Fig. 59.2). In both men and women, NG can cause proctitis with symptoms such as rectal discharge, tenesmus and pain, or without symptoms. In MSM, it is usually the result of unprotected receptive anal

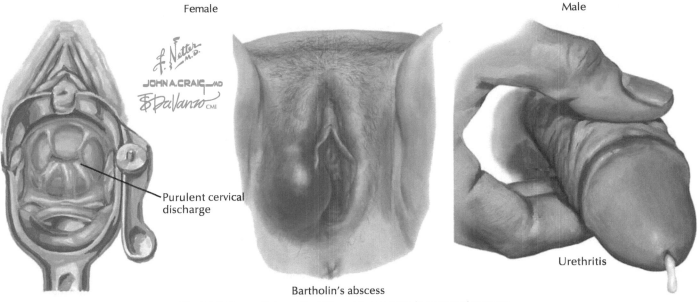

Female

Male

Purulent cervical discharge

Bartholin's abscess

Urethritis

**Fig. 59.1** Urogenital manifestations of gonorrhea in men and women.

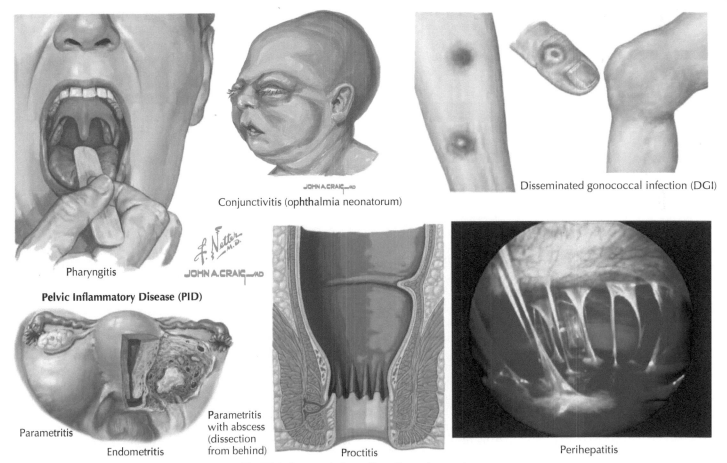

Pharyngitis

Conjunctivitis (ophthalmia neonatorum)

Disseminated gonococcal infection (DGI)

**Pelvic Inflammatory Disease (PID)**

Parametritis

Endometritis

Parametritis with abscess (dissection from behind)

Proctitis

Perihepatitis

**Fig. 59.2** Extragenital manifestations of gonorrhea.

intercourse. However, rectal gonococcal infection in men and women can also be acquired through perineal spread from the cervicovaginal or genital areas or in women from receptive anal intercourse as well.

Pharyngeal infection with NG is generally asymptomatic, although some persons may have sore throat or pain with swallowing. Although the prevalence of gonococcal proctitis and pharyngitis varies by population, in MSM who were attending sexually transmitted infection (STI) clinics in 2018, median urogenital positivity was 7.9%, rectal positivity was 14.8%, and pharyngeal positivity was 12.9%. Women had a median rectal positivity rate of 1.9% and pharyngeal positivity rate of 2.1%. Heterosexual men had a median positivity rate of 3.4% rectal and 1.6% pharyngeal.

## Ophthalmic Infection

Ophthalmic infection with NG can occur in both adults and infants, causing a severe conjunctivitis that can result in corneal damage and blindness (see Fig. 59.2). In the infant, symptoms appear approximately 2 to 5 days after delivery, generally with a profuse ophthalmic discharge and lid edema. Symptoms may occur later, however, especially if the infant received ophthalmic prophylaxis at birth. Neonatal ophthalmia can also be caused by infection with *C. trachomatis, Haemophilus influenzae, Streptococcus pneumoniae,* and *Klebsiella pneumoniae.* Although gonococcal ophthalmic infection in the newborn is generally through perinatal transmission, gonococcal ophthalmia in an adult is usually through autoinoculation from an individual infected at a urogenital site or via direct contact with infected genitalia or urine. Adult infection may also cause subtler symptoms of conjunctivitis (injected sclera, mild discharge) than are commonly seen in the newborn.

## Pelvic Inflammatory Disease

Complicated infection in women most commonly manifests as PID and occurs in approximately 10% to 20% of women with acute gonococcal infection. PID occurs due to infection of the endometrium, the fallopian tubes, or the peritoneum. Women with PID may be asymptomatic or may have lower abdominal pain, fever, and pain or spotting with intercourse. Even if treated appropriately, PID may result in intra-abdominal scarring and adhesions that increase the risk of infertility, ectopic pregnancy, and chronic pelvic pain. PID may be caused by single or polymicrobial infections, including CT, *Mycoplasma genitalium,* and various anaerobes.

## Testicular Infection

Complicated infection in men can cause penile edema or epididymitis, with characteristic symptoms of unilateral testicular pain, swelling, dysuria, and fever. Because symptomatic men generally seek treatment, very few men progress to develop urethral strictures. Gonococcal infections in men may play a role in male infertility.

## Disseminated Gonococcal Infection

DGI occurs as the result of gonococcal bacteremia. Risk factors for dissemination have included female sex, menstruation, pregnancy, and terminal complement deficiency; however, there are increasing reports in men. Persons receiving monoclonal antibodies that inhibit terminal complement activation are also at higher risk of dissemination. DGI classically manifests as a dermatitis-arthritis syndrome. The classic presentation of gonococcal skin lesions is a necrotic pustule on an erythematous base, but the lesions may also appear as macules, hemorrhagic lesions, or papules (see Fig. 59.2). The arthritis associated with gonococcal infection generally affects wrists, knees, and ankles, although any joint may be affected. DGI may also cause nonspecific signs and symptoms of sepsis such as fever and low blood pressure. Although rare, DGI can also cause endocarditis, meningitis, or perihepatitis (Fitz-Hugh–Curtis syndrome) (see Fig. 59.2). The differential diagnosis for DGI includes a broad range of organisms responsible for bacteremia, endocarditis, and meningitis; in particular, special care must be taken to distinguish systemic infection with NG from infection with *N. meningitidis.*

## DIAGNOSTIC APPROACH

### History and Physical Examination

Given the importance of early treatment of gonorrhea to avoid continued sexual transmission of disease, diagnosis of gonorrhea in a symptomatic individual relies on a careful history and physical examination. However, a large proportion of both women and men infected with

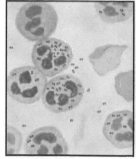
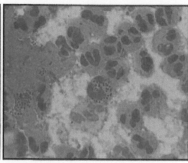

Gonococcal infection (Gram stain)

Gonococcal infection (Methylene blue/ Gentian violet stain; Courtesy of Stephanie N. Taylor, MD)

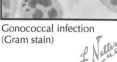

**Fig. 59.3** Gram and methylene blue/gentian violet stains.

gonorrhea are asymptomatic. A history should include detailed questions about age of patient; number and sex of sex partners; unprotected penetration of the pharynx, vagina, penis, and rectum; symptoms (e.g., sore throat, genital or rectal discharge, pain on urination, pain on intercourse); STI history (especially previous episodes of gonorrhea); associated high-risk behaviors (e.g., exchange of sex, illicit drug use); pregnancy status; HIV status; concomitant medical conditions; travel history; and recent antibiotic exposure. Physical examination should focus on the skin, pharynx, lower abdomen, external genitalia, vagina, cervix, and rectum.

### Laboratory Testing

Laboratory testing should be conducted for all patients with suspected gonococcal infection, even if history or physical examination findings are strongly suggestive of gonorrhea, because management of the patient and the patient's partner or partners is facilitated by having a definitive diagnosis. Given the poor sensitivity and specificity of serologic tests for NG, serologic tests should not be used for the clinical diagnosis of gonorrhea.

### Diagnosis of Urogenital Infection

The presence of gram-negative intracellular diplococci on a Gram or methylene blue/gentian violet stain of a urethral exudate is both highly sensitive (90% to 95%) and highly specific (95% to 100%) for symptomatic males and can be considered sufficiently diagnostic without further testing for infection with NG (Fig. 59.3). In asymptomatic women, Gram stain of urethral or cervical exudate is less sensitive (50% to 70%) but still highly specific (95% to 100%) and should be considered only an adjunct to more sensitive molecular testing.

In general, NAATs of first-catch urine from men and vaginal swabs from women are the optimal test types for the diagnosis of urogenital gonococcal infection. A NAAT for gonococcal infections is usually performed jointly with a NAAT for chlamydia. NAATs have been found to be more sensitive than culture and can be performed on a wider variety of specimen types (urine and vaginal, endocervical, or urethral swabs) than can culture or nonamplified deoxyribonucleic acid (DNA) probe tests. However, gonococcal culture is still critical for management to provide antimicrobial susceptibility test results. Currently several molecular resistance gene assays have been developed and are under evaluation. These assays detect genes that code for specific mechanisms of resistance that are associated with specific organisms (i.e., DNA gyrase coding for NG ciprofloxacin resistance). They do not have the capacity to detect susceptibility to multiple antibiotics simultaneously, as in traditional methods.

## Diagnosis of Extragenital Infection

NAATs also have a higher sensitivity and specificity compared with culture for the diagnosis of rectal and pharyngeal NG infections. Certain NAATs have been cleared by the US Food and Drug Administration (FDA) for use at non-urogenital sites. Therefore individual laboratory validation to establish performance specifications to satisfy Clinical Laboratory Improvement Amendments (CLIA) compliance standards before reporting results for patient management are no longer necessary. Culture of the rectum and pharynx can also be used for diagnosis at these sites and has the additional benefit of allowing for antimicrobial susceptibility testing.

## Screening for Infection With *Neisseria gonorrhoeae*

The prevalence of gonorrhea varies widely by population. The United States Preventive Services Task Force recommends annual screening for NG in all sexually active women younger than 25 years of age and for older women at increased risk for infection (e.g., those who have a new sex partner, more than one sex partner, a sex partner with concurrent partners, or a sex partner who has an STI). Additional risk factors for gonorrhea include inconsistent condom use among persons who are not in mutually monogamous relationships, previous or coexisting sexually transmitted infections, and exchanging sex for money or drugs.

## CLINICAL MANAGEMENT AND DRUG TREATMENT

Antibiotic therapy for gonorrhea should be safe, highly effective, single dose, and affordable whenever possible. Both efficacy at the site of infection and antimicrobial resistance considerations are important when treatment for NG infections is being selected. The Centers for Disease Control and Prevention (CDC) currently recommends an antibiotic for the treatment of gonorrhea if the efficacy in summed clinical trials is ≥95%, with a lower 95% confidence interval of ≥95%, and considers antibiotics as alternative therapies if the efficacy is ≥95% with a lower 95% confidence interval of ≥90%. In addition to efficacy considerations, however, both recommended and alternative antibiotics are advised by both the CDC and the WHO only if fewer than 5% of strains tested demonstrate antibiotic resistance.

Unfortunately, selection of antimicrobial therapy for gonorrhea is severely limited due to the rapidity with which NG develops resistance to antibiotics. Over the years, NG has developed widespread resistance to nearly every class of antibiotics routinely used for treatment, including sulfonamides, penicillins, tetracyclines, fluoroquinolones, and oral cephalosporins. As the number of gonorrhea cases has increased, so has the number of isolates with decreased susceptibility and resistance to currently recommended antibiotic regimens.

There have been some reports of treatment failures and isolates with decreased susceptibility and resistance to oral cephalosporins. The CDC recommends ceftriaxone 500 mg IM for treatment of gonococcal infection (Table 59.1). Treatment for chlamydia, doxycycline 100 mg bid, should occur concurrently if chlamydia infection has been identified or has not been excluded.

Cephalosporins are the last class of antibiotics remaining effective for the treatment of gonorrhea, although recent failures have been documented in patients treated with injectable cephalosporins. Due to these circumstances, NG is now considered a public health threat as a multidrug-resistant organism, and the CDC and WHO have made calls for the development of new antibiotics to treat gonorrhea. Due to unprecedented collaboration between public health organizations and industry, two new antimicrobials are currently under evaluation. Both gepotidacin and zoliflodacin show promise in the treatment of uncomplicated gonorrhea and have entered phase 3 trials.

Patients with a history of severe reaction to penicillin or cephalosporins (e.g., anaphylaxis, Stevens-Johnson syndrome, or toxic epidermal necrolysis) should be treated with single doses of gentamicin 240 mg IM plus azithromycin 2 g PO. Additionally, if a patient is asymptomatic and the treating facility is able to perform gyrA testing to identify ciprofloxacin susceptibility (wild type), ciprofloxacin 500 mg orally in a single dose can be given. Data from the Gonococcal Isolate Surveillance Program (GISP) demonstrate that azithromycin resistance has increased steadily from 0.3% in 2012 to 4.6% in 2018, and treatment failures have been reported. Therefore azithromycin 2 g in a single oral dose is no longer recommended.

Pregnant women and persons with HIV infection should receive the same treatment as other persons with the exception of pregnant women with severe penicillin or cephalosporin allergy. Pregnant women should not be treated with aminoglycosides or quinolones. When cephalosporin allergy or other considerations preclude treatment with the standard regimen, consultation with an infectious disease specialist is recommended.

Neonatal ophthalmia caused by infection with NG should be treated with ceftriaxone 25 to 50 mg/kg IV or IM in a single dose, and the dose should not exceed 250 mg.

Because patterns of resistance of NG are constantly changing, it is important for clinicians and health departments to remain alert for changing recommendations for gonorrhea therapy. Updates on the treatment of gonococcal infections can be found on the CDC website, at www.cdc.gov/std/treatment.

### Partner Management

Treatment of recent sexual contacts is necessary to avoid reinfection and prevent ongoing disease transmission. Patients should be instructed to contact any partners with whom they have had sexual intercourse within the last 60 days to inform them of the possibility of infection and encourage them to seek treatment for gonorrhea and chlamydia. Patients should be instructed to avoid sexual intercourse until they and their partners have received treatment and no longer have symptoms.

If it is unlikely that the partner will seek care, expedited partner therapy (EPT) is an option in most states (https://www.cdc.gov/std/ept/gc-guidance.htm). EPT is a partner treatment approach where sex partners of patients with gonorrhea are provided treatment without previous medical evaluation. Because of EPT's effectiveness in reducing gonorrhea reinfection rates, the CDC has recommended its use since 2006 for the heterosexual partners of patients diagnosed with gonorrhea and/or chlamydia if it seemed unlikely that the partners would seek timely evaluation and treatment. Depending on the state, a prescription or the medication for the treatment and educational materials are provided to the patient to give to his or her partner or partners. Educational materials should encourage the partners to seek professional evaluation, especially if they are symptomatic or allergic to antibiotics. There are limited data on EPT in MSM and the potential for inadequate treatment of bacterial STI in partners; therefore shared clinical decision making regarding EPT is recommended.

Because gonorrhea is a notifiable condition in all states, providers should be aware of both the importance and the utility of notifying the state or local health department of all laboratory-confirmed gonococcal infections. Some state or local health departments will provide assistance in notifying and counseling partners.

### Suspected Treatment Failure

Treatment failures with oral and injectable cephalosporins have been reported, as well as gonococcal isolates with decreased susceptibility

**TABLE 59.1    Centers for Disease Control and Prevention—Recommended Antibiotic Regimens for the Treatment of *Neisseria gonorrhoeae* Infections**

| Type of Gonococcal Infection | Adult Treatment Regimen | Commentary |
|---|---|---|
| Uncomplicated infection (All sites) | | |
| Cervix, urethra, rectum, pharyngeal | Ceftriaxone 500* mg IM in a single dose | *For persons weighing ≥150 kg (~300 lbs.), 1 g ceftriaxone should be administered.<br>Single dose of ceftizoxime (500 mg IM), or cefoxitin (2 g IM with probenecid 1 g PO), or cefotaxime (500 mg IM) may be used in place of ceftriaxone.<br>If chlamydial infection has been identified or has not been excluded, providers should treat for chlamydia with doxycycline 100 mg orally twice a day for 7 days.<br>None of these injectable cephalosporins offer any advantage over ceftriaxone for urogenital infection, and efficacy for pharyngeal infection is less certain.<br>There are no reliable alternative treatments for pharyngeal gonorrhea at this time. |
| | Gentamicin 240 mg IM in a single dose PLUS azithromycin 2 g PO in a single dose<br><br>*or* Cefixime 800 mg PO in a single dose | Alternatives only if ceftriaxone is not available. |
| Conjunctiva | Ceftriaxone 1 g IM in a single dose* | *If chlamydial infection has not been excluded, providers should treat for chlamydia with doxycycline 100 mg orally twice a day for 7 days. |
| Severe cephalosporin allergy | Gentamicin 240 mg IM in a single dose PLUS azithromycin 2 g PO in a single dose | If patient is pregnant, give azithromycin 2 g PO in a single dose without gentamicin.<br>If a patient is asymptomatic and the treating facility is able to perform gyrase A testing to identify ciprofloxacin susceptibility (wild type), then ciprofloxacin 500 mg orally in a single dose can be given. |
| Disseminated gonococcal infection | Ceftriaxone 1 g IM or IV q24h<br>*or* Cefotaxime 1 g IV q8h<br>*or* Ceftizoxime 1 g IV q8h | If chlamydial infection has been identified or has not been excluded, providers should treat for chlamydia with doxycycline 100 mg orally twice a day for 7 days.<br>Treatment can be switched to an oral agent guided by antimicrobial susceptibility testing 24–48 h after substantial clinical improvement for a total treatment course of at least 7 days. |
| Meningitis | Ceftriaxone 1–2 g IV q24h | Continue for 10–14 days and consult specialist. |
| Endocarditis | Ceftriaxone 1–2 g IV q24h | Continue for at least 4 weeks and consult specialist. |

*IM,* Intramuscular; *IV,* intravenous; *PO,* oral.

and resistance to cephalosporins. Most suspected treatment failures in the United States are likely to be reinfections rather than actual treatment failures. However, before retreatment in cases where reinfection is unlikely and treatment failure is suspected, clinical specimens should be obtained for culture (preferably with simultaneous NAAT) and antimicrobial susceptibility testing and partner treatment should be confirmed. All isolates of suspected treatment failures should be sent to the CDC for antimicrobial susceptibility testing by agar dilution and local laboratories should store isolates for possible further testing if needed. Suspected treatment failures first should be retreated routinely with the recommended regimen (ceftriaxone 500 mg IM with or without an antichlamydial agent). An alternative therapeutic option includes a single dose of intramuscular gentamicin 240 mg plus oral azithromycin 2 g. A test of cure at relevant clinical sites should be obtained 7 to 14 days after retreatment; culture is the recommended test, preferably with simultaneous NAAT and antimicrobial susceptibility testing of

*N gonorrhoeae* if isolated. If this regimen fails or if isolates with decreased susceptibility or resistance are recovered, advice from an infectious diseases specialist should be sought and state or local health authorities contacted.

## Follow-Up and Repeat Testing

The CDC recommends that all patients diagnosed with gonorrhea undergo a repeat test in 3 months at the anatomic site of infection, as reinfection with gonorrhea is quite common. Repeat testing for gonorrhea is distinct from a test of cure (testing shortly after treatment to determine if treatment was effective). A test of cure is not needed for persons who receive a diagnosis of uncomplicated urogenital or rectal gonorrhea who are treated with any of the recommended or alternative regimens; however, any person with pharyngeal gonorrhea should return between 7 and 14 days after initial treatment for a test of cure using either culture or NAAT.

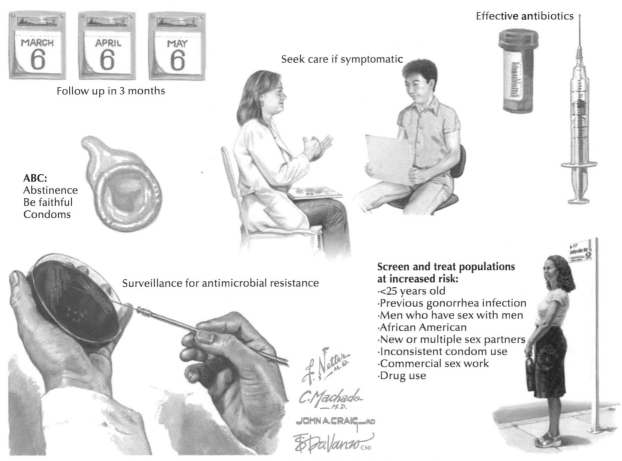

MARCH 6  APRIL 6  MAY 6

Follow up in 3 months

ABC:
Abstinence
Be faithful
Condoms

Seek care if symptomatic

Effective antibiotics

Surveillance for antimicrobial resistance

Screen and treat populations at increased risk:
·<25 years old
·Previous gonorrhea infection
·Men who have sex with men
·African American
·New or multiple sex partners
·Inconsistent condom use
·Commercial sex work
·Drug use

Fig. 59.4 Prevention and control strategies for gonorrhea.

## PROGNOSIS

Persons who undergo a CDC-recommended or alternative regimen have a high probability of resolution of infection with *NG*. Approximately 10% to 20% of patients are reinfected within several months of initial infection, highlighting the importance of partner management in the treatment of gonorrhea. PID, either asymptomatic or symptomatic, may occur in approximately 10% to 20% of women with gonorrhea. Even with appropriate treatment, however, the inflammation and scarring from PID can result in infertility, ectopic pregnancy, and chronic pelvic pain. However, prompt evaluation and treatment may decrease the duration of inflammation and reduce the risk of the sequelae of PID. In men, postinflammatory urethral strictures are rare after antibiotic treatment, and it is unclear how the treatment of gonorrhea affects future fertility.

## PREVENTION AND CONTROL

### General Prevention and Control Strategies

The prevention and control of gonorrhea require multiple strategies (Fig. 59.4). On an individual basis, as with other STIs, the most effective ways to avoid transmission are abstinence, long-term mutual monogamy, or correct and consistent use of latex condoms. Patients must be aware of the importance of recognizing symptoms early and seeking care promptly. Clinicians must educate patients regarding the importance of informing partners of their infection and encouraging partners to seek care. Clinicians must also ensure that they provide effective antibiotics as well as appropriate counseling and education and that they encourage repeat testing in 3 months to rule out reinfection and ongoing transmission. Health departments should educate communities about STI symptom recognition, provide information on access to STI care, provide support for partner management strategies, ensure that clinicians are up to date on the latest treatment recommendations, screen and treat populations at increased risk of infection, and conduct routine surveillance for antimicrobial resistance to recommended treatment regimens.

### Ophthalmia Neonatorum Prophylaxis

Diagnosis and treatment of gonococcal and chlamydial infections in pregnancy are the best method for preventing neonatal gonococcal and chlamydial disease. However, not all women receive prenatal care, and not all women are tested for gonorrhea and chlamydia in pregnancy. Ophthalmia neonatorum prophylaxis for all newborn infants is important because it is a preventable cause of blindness and is safe, easy to administer, and inexpensive. Therefore all newborns should have erythromycin 0.5% ophthalmic ointment applied once into both eyes as soon as possible after delivery. Infants born to mothers with known untreated gonococcal infection are at high risk of infection; such infants should receive ceftriaxone 25 to 50 mg/kg IV or IM once, not to exceed 250 mg.

## EVIDENCE

Antibiotic Resistance Threats in the United States 2019. Available at: https://www.cdc.gov/drugresistance/biggest-threats.html. Accessed November 20, 2019. *This CDC report presents data supporting the recognition of N. gonorrhoeae as one of the top five antibiotic-resistant organisms considered urgent public health threats.*

Centers for Disease Control and Prevention. Sexually transmitted disease surveillance 2018. Atlanta: U.S. Department of Health and Human Services; 2019. DOI: 10.15620/cdc.79370. *Annual publication that presents statistics and trends for STDs in the United States through 2018.*

Kidd S, Moore PC, Kirkcaldy RD, et al. Comparison of antimicrobial susceptibility of urogenital Neisseria gonorrhoeae isolates obtained from women and men. *Sex Transm Dis* 2015;42:434-439. *Presents N. gonorrhoeae antimicrobial susceptibility data.*

Kirkcaldy RD, Weston E, Segurado AC, Hughes G. Epidemiology of gonorrhoea: a global perspective. *Sexual Health (Online)* 2019;16(5):401-411. DOI:10.1071/SH1906. *Current epidemiology of gonorrhoea is reviewed through an international lens and with a focus on selected populations.*

Owusu-Edusei K Jr, Chesson HW, Gift TL, et al. The estimated direct medical cost of selected sexually transmitted infections in the United States, 2008. *Sex Transm Dis* 2013;40(3):197-201. *A synthesis of existing literature to estimate the lifetime medical cost per case of eight major STDs among youths: HIV infection, human papillomavirus (HPV) infection, genital herpes simplex virus type 2 infection, hepatitis B, chlamydia, gonorrhea, trichomoniasis, and syphilis.*

Unemo M, Shafer WM. Antimicrobial resistance in Neisseria gonorrhoeae in the 21st century: Past, evolution, and future. *Clin Microbiol Rev* 2014;27:587-613. *Describes mechanisms of N. gonorrhoeae resistance.*

World Health Organization. Report on global sexually transmitted infection surveillance 2019. Available at: https://www.who.int/news-room/detail/06-06-2019-more-than-1-million-new-curable-sexually-transmitted-infections-every-day. Accessed December 4, 2019. *Estimates the global burden of syphilis, gonorrhea, chlamydia, and trichomoniasis 2016.*

## ADDITIONAL RESOURCES

Centers for Disease Control and Prevention (CDC): Sexually transmitted diseases. Available at: www.cdc.gov/std. Accessed October 20, 2019. *Provides links to surveillance data, treatment updates, fact sheets for patients, control strategies, and more.*

Centers for Disease Control and Prevention. Sexually transmitted diseases treatment guidelines, 2020. *MMWR 2020* (in press). *Describes current gonorrhea treatment regimens recommended by the CDC.*

Chlamydia and Gonorrhea: Screening. U.S. Preventive Services Task Force. September 2016. (Updated guidelines anticipated in 2020.) Available at: https://www.uspreventiveservicestaskforce.org/Page/Document/UpdateSummaryFinal/chlamydia-and-gonorrhea-screening. Accessed November 2, 2019. *Summarizes the evidence base and the USPSTF recommendations for gonorrhea screening.*

Hook EW, Handsfield HH: Gonococcal infections in the adult. In Holmes KK, Sparling PF, Stamm WE, et al, eds: *Sexually transmitted diseases*, ed 4, New York, 2008, McGraw Hill. *Reviews the epidemiology, pathology, clinical manifestation, diagnosis, and treatment of uncomplicated and complicated gonococcal infection.*

Hook EW III, Kirkcaldy RD. A brief history of evolving diagnostics and therapy for gonorrhea: Lessons learned. *Clin Infect Dis* 2018;67:1294-1299. *Provides historical review of diagnostics and therapy for gonorrhea.*

World Health Organization. Sexually Transmitted and Reproductive Tract Infections. Available at: https://www.who.int/reproductivehealth/publications/rtis/monitoring/en/. Accessed October 1, 2019. *Provides clinical guides, policy and program guidance, advocacy, journal articles, and monitoring and evaluation data.*

# 60

# Syphilis *(Treponema pallidum)*

*Paul C. Adamson, Jeffrey D. Klausner*

## ABSTRACT

Syphilis is a complex systemic disease caused by the spirochete *Treponema pallidum* subspecies *pallidum*. Syphilis is transmitted sexually or congenitally and can involve nearly every organ system. Its clinical progression involves several well-characterized stages: (1) an incubation period of 1 week to 3 months; (2) a primary stage characterized by a chancre (an indurated, nontender ulcer at the site of exposure); (3) a secondary stage, usually several weeks after the resolution of the chancre, associated with a diffuse rash, mucocutaneous lesions, and lymphadenopathy; (4) a latent stage of subclinical infection detected by reactive serologic tests; and (5) a late or tertiary stage involving end-organ damage, including neurologic, cardiovascular, and gummatous (or late benign) syphilis (Fig. 60.1). Penicillin is highly effective against syphilis and remains the treatment of choice. This chapter reviews the etiology, epidemiology, clinical features, diagnostic approach, treatment, and prevention of syphilis.

## CLINICAL VIGNETTE

A 34-year-old man with no significant past medical history presented with a 3-week history of progressive blurry vision as well as a rash on his chest and the palm of his hand. He works as a special education teacher and lives alone. He is sexually active with both men and women and reports having had five sex partners in the previous 3 months. He reports occasional use of alcohol and methamphetamine.

On exam, he was afebrile and well appearing. He had patchy alopecia on his scalp. His pupils were normal and his right eye exhibited conjunctival injection. He had no signs of meningismus. He had lymphadenopathy of his anterior cervical nodes. Genital and rectal examinations were unremarkable. His skin examination was notable for a diffuse, erythematous maculopapular rash over his upper chest and neck and erythematous papules on the palm of his right hand. An ophthalmologic examination was notable for decreased acuity in his right eye (20/80), as well as increased intraocular pressure and bilateral optic disc edema. There was no evidence of retinitis. Laboratory evaluation revealed a reactive rapid plasma reagin (RPR) titer of 1:1024. He tested negative for antibodies to human immunodeficiency virus. A lumbar puncture was notable for an elevated opening pressure and a reactive Venereal Disease Reference Laboratory (VDRL) test.

He was diagnosed with secondary syphilis complicated by ocular syphilis and neurosyphilis and was started on intravenous aqueous penicillin G 4 million units q4h. Six hours following his first infusion of penicillin, he developed a fever to 38.5°C, rigors, and myalgias, consistent with a Jarisch-Herxheimer reaction, which was managed with nonsteroidal anti-inflammatory medications. Following 24 to 48 h of penicillin, his vision and rash began to improve. He was continued on intravenous aqueous penicillin G for a total of 14 days. Upon discharge he was given benzathine penicillin G 2.4 million units IM and scheduled for follow-up with an infectious diseases specialist.

## ETIOLOGY AND PATHOGENESIS

Syphilis is caused by infection with the spirochetal bacterium *T. pallidum* subspecies *pallidum*, a highly motile organism with tapering ends presenting 6 to 14 spirals. Of uniform cylindrical shape, the bacteria measure approximately 6 to 15 μm in length and 0.25 μm in width. *T. pallidum* is a slowly metabolizing organism with an average multiplying time of approximately 30 hours. Humans are the only host. Most cases of syphilis are transmitted by sexual contact (vaginal, anogenital, and orogenital), but it can also be spread congenitally (in utero or, less commonly, during passage through the birth canal). Rare cases of acquisition through blood products have also been reported. On skin-to-skin contact, the motile spirochetes from an infected person enter a new host through areas of microtrauma in the skin or mucosa, multiplying locally, with resultant systemic dissemination in less than 24 hours. The phospholipid-rich outer membrane of the spirochete contains few surface-exposed proteins; this may help it to evade the host's immune system. The primary pathologic lesion, found at all stages of the disease, is an obliterative endarteritis, which leads to many of the clinical manifestations of syphilis. Histologic examination of a chancre is characterized by an intense infiltrate of plasma cells with scattered macrophages and lymphocytes. A granulomatous reaction can also occur.

## EPIDEMIOLOGY

The epidemiology of syphilis follows two patterns that are distinct between higher-income and lower-income countries. Syphilis incidence in high-income nations declined dramatically after the introduction of mass screening programs and the advent of penicillin therapy. In 2000, primary and secondary syphilis rates were at historic lows in the United States, with only 2.1 cases per 100,000, compared with 70.9 cases per 100,000 in 1946. Similar declines were observed across Europe and Australia. However, those declines have been followed by a resurgence in syphilis cases, most concentrated in men who have sex with men (MSM). In the United States, approximately half of MSM diagnosed with primary and secondary syphilis in 2016 were also living with HIV infection. Reversal in the control of syphilis in low-socioeconomic, black, heterosexual subpopulations and in those experiencing the opiate epidemic has also been observed across the United States. The rise in syphilis seen among women has been accompanied by a dramatic rise in congenital syphilis.

In contrast to the changing epidemiology of syphilis in high-income countries, syphilis has remained endemic in many low- and middle-income countries. Although high-quality surveillance data from low- and middle-income countries are limited, syphilis cases among the general population appear to be declining, but increases remain in certain high-risk populations. Maternal and congenital syphilis remain

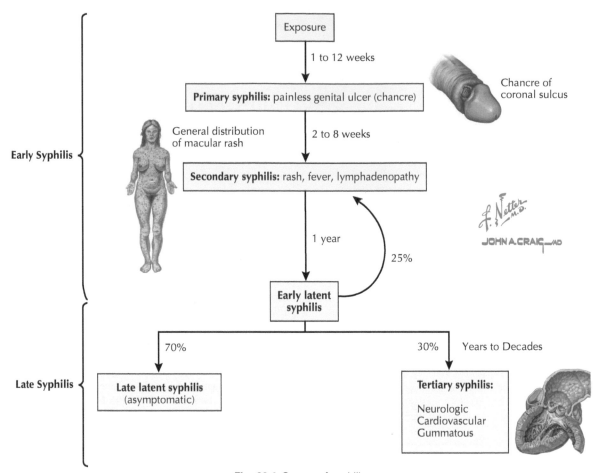

**Fig. 60.1** Stages of syphilis.

major public health problems in many low- and middle-income countries. In 2016, the World Health Organization (WHO) estimated there were approximately 1 million maternal syphilis cases and 661,000 cases of congenital syphilis worldwide. The WHO's African and Eastern Mediterranean regions accounted for approximately three-quarters of congenital syphilis cases in 2016.

## CLINICAL FEATURES

### Primary Syphilis

Primary syphilis manifests 1 week to 3 months (median 21 days) after exposure with a painless lesion, called a chancre, at the site of inoculation along with nontender regional lymphadenopathy. The lesion starts as a papule and rapidly forms a typically nonexudative ulcer with a clean base. Primary lesions are most commonly found on the external genitalia but can develop on any site of exposure including the perineum, cervix, anus, rectum, lips, and oropharynx (Fig. 60.2). Multiple chancres can occur and are more common in patients with HIV infection. Without treatment, the chancre usually heals on its own within 1 to 3 weeks. Primary syphilis must be differentiated from other causes of genital ulcer disease including other infectious causes (herpes simplex virus [HSV], chancroid, lymphogranuloma venereum, and pyogenic ulcers), as well as noninfectious causes (including trauma, neoplasia, and fixed drug eruptions). Herpetic ulcers, unlike chancres, are usually superficial, vesicular, nonindurated, and painful. Chancroid, caused by *Haemophilus ducreyi*, is uncommon in the United States and is typically nonindurated, painful, and exudative with a necrotic base (Fig. 60.3).

### Secondary Syphilis

The timing of onset of the secondary stage of syphilis is highly variable. It typically occurs 2 to 8 weeks after the disappearance of a chancre. In some cases, however, the primary chancre may still be present, and secondary syphilis has been described up to 4 years after initial exposure. The overlap in clinical manifestations of primary and secondary syphilis is more common in patients coinfected with HIV. Many patients do not recall a history of a primary lesion. Secondary syphilis typically manifests with rash, fever, headache, pharyngitis, and lymphadenopathy but has a wide range of possible systemic manifestations including hepatitis, uveitis, meningitis, glomerulonephritis, periostitis, aortitis, and cerebrovascular accidents.

The cutaneous manifestations of secondary syphilis are quite diverse. The classic exanthem of secondary syphilis is a diffuse maculopapular rash that often involves the palms and soles (Fig. 60.4). However, the rash can also be papular, annular, or pustular and can have a fine overlying scale. Other mucocutaneous manifestations include (1) *condylomata lata*, moist, heaped-up broad plaques found in intertriginous areas such as the perianal area, vulva, and inner thighs; (2) *mucous patches*, gray, superficial erosions or plaques on the buccal mucosa and tongue, under the prepuce, and on the inner labia; (3) *split papules*, fissured, nodular lesions at the angle of the lips and in the nasolabial folds; and (4) *patchy alopecia*, or "moth-eaten" thinning of hair, eyebrows, and beard from syphilitic involvement of the hair follicle (Fig. 60.5). The cutaneous lesions of syphilis, particularly the nonkeratinized mucocutaneous lesions (condylomata lata and mucous patches), contain large concentrations of spirochetes and are highly infectious.

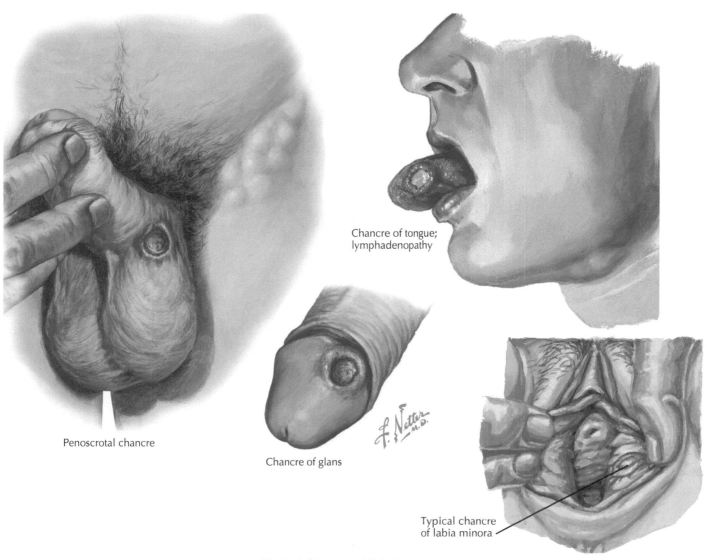

Penoscrotal chancre

Chancre of tongue; lymphadenopathy

Chancre of glans

Typical chancre of labia minora

**Fig. 60.2** Primary syphilitic chancres.

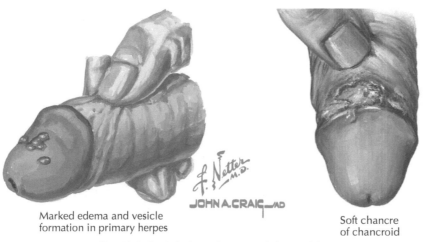

Marked edema and vesicle formation in primary herpes

Soft chancre of chancroid

**Fig. 60.3** Genital ulcers: herpes and chancroid.

The diverse manifestations of secondary syphilis earn it the name "the great imitator." Other diseases that should be considered in the differential diagnosis of fever, rash, pharyngitis, and lymphadenopathy include mononucleosis (acute Epstein-Barr virus or cytomegalovirus infection), acute HIV infection, and other viral syndromes. The condylomata lata of secondary syphilis should be distinguished from *condylomata acuminata*—multiple small, raised genital warts caused by human papillomaviruses (Fig. 60.6). Mucous patches can be mistaken for oral candidiasis. Other infections that cause a rash involving the palms and soles include Rocky Mountain spotted fever, meningococcemia, measles, and certain coxsackievirus infections (hand, foot, and mouth disease).

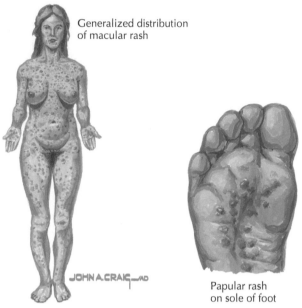

Generalized distribution of macular rash

Papular rash on sole of foot

**Fig. 60.4** Secondary syphilis: diffuse maculopapular rash.

## Early Neurosyphilis

Invasion of the central nervous system (CNS) can occur during secondary syphilis and may manifest as an aseptic meningitis, with headache, neck stiffness, and a lymphocytic pleocytosis of cerebrospinal fluid (CSF). The meningeal inflammation is often basilar, leading to unilateral or bilateral cranial nerve abnormalities, particularly of cranial nerves II, III, VI, VII, and VIII.

Patients with syphilis who exhibit cochleovestibular signs or symptoms without an alternative diagnosis are considered to have otologic syphilis. Those symptoms include vertigo, tinnitus, and hearing loss, which is typically sensorineural and bilateral. The diagnosis of otologic syphilis is typically presumptive, as CSF analysis is normal in most cases. Otologic syphilis can occur at any stage of disease, although most cases occur in early syphilis. Treatment is more likely to restore hearing if this condition is detected early in the disease course.

Ocular syphilis can occur at any stage of disease and can involve nearly all structures of the eye, but posterior uveitis, panuveitis, and retinitis are the most common. Symptoms are most commonly eye redness, decreased acuity, and/or eye pain. Outcomes for ocular syphilis can be severe, including blindness, and longer duration of symptoms is associated with poor prognosis. Ocular syphilis is generally classified as neurosyphilis, and up to 70% of patients with ocular symptoms will have CSF abnormalities. Therefore any visual signs or symptoms in a patient with syphilis should prompt an evaluation for neurosyphilis.

It should be noted that neurosyphilis can occur at any stage of syphilis, including the primary and secondary stages. Likewise, ocular and otosyphilis can also occur during any stage of the disease.

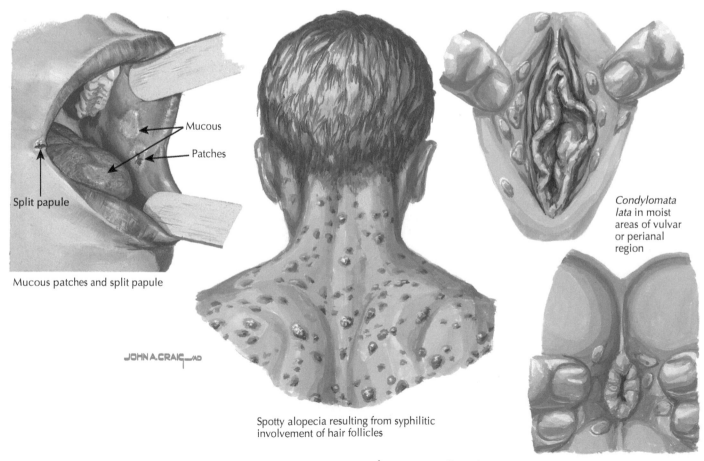

Mucous

Patches

Split papule

Mucous patches and split papule

*Condylomata lata* in moist areas of vulvar or perianal region

Spotty alopecia resulting from syphilitic involvement of hair follicles

**Fig. 60.5** Secondary syphilis: mucocutaneous manifestations.

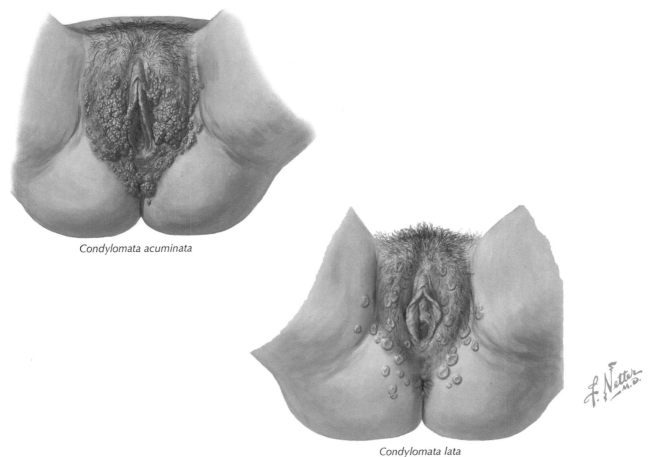

*Condylomata acuminata*

*Condylomata lata*

**Fig. 60.6** Condylomata lata versus condylomata acuminata.

## Latent Syphilis

Without treatment, the manifestations of secondary syphilis generally resolve within a few weeks. The disease then enters a latent phase, characterized by a lack of clinical signs of syphilis but positive serologic test results. Observational studies have shown that recrudescent secondary syphilis symptoms can occur in untreated patients up to 5 years after their initial presentation, but generally these relapses occur within the first year. Early latency therefore has been defined as the first year after initial syphilis infection. An asymptomatic patient with a newly reactive serologic test result who had a nonreactive serologic test result during the previous year is also designated as having early latent syphilis. Late latency is the asymptomatic phase longer than 1 year after syphilis infection. Late latent syphilis, unlike early latent syphilis, is not thought to be infectious except in pregnant women and requires a longer duration of treatment compared with early latent syphilis (see later section on treatment).

## Late Syphilis

*Late* syphilis, or tertiary syphilis, occurs in up to one-third of untreated patients with latent syphilis and is characterized by end-organ damage. It has become very uncommon in the antibiotic era. In tertiary syphilis, endarteritis leads to cellular necrosis, fibrosis, sclerosis, scarring, and loss of normal tissue parenchyma. The three main types of tertiary disease are neurologic, cardiovascular, and gummatous (or late benign) syphilis.

### Late Neurosyphilis

As described previously, acute syphilitic meningitis can occur early in syphilis infection and is a well-described feature of secondary syphilis. Late neurologic complications of syphilis, which manifest after long periods of latency, are caused by meningovascular disease or parenchymal damage or both and have a range of manifestations. Vascular involvement leading to focal ischemia can cause a myriad of neurologic deficits including hemiparesis, aphasia, and focal or generalized seizures. Classic late neurosyphilis syndromes attributed to parenchymal damage include general paresis and tabes dorsalis.

General paresis, also known as *general paralysis of the insane*, is a meningoencephalitis with direct invasion of the cerebrum by *T. pallidum* (Fig. 60.7). The encephalitis is chronic and usually manifests in middle to late adulthood after a 15- to 25-year incubation period. A wide range of manifestations includes progressive dementia with changes in personality, affect, sensorium, intellect, and speech. Defects in judgment, emotional lability, grandiose delusions, megalomania, depression, catatonia, amnesia, and hyperreflexia have been described. The Argyll-Robertson pupil—a small, often irregularly shaped pupil that constricts on accommodation but not to light—is an uncommon feature of general paresis and is more commonly seen in tabes dorsalis.

Tabes dorsalis, syphilitic involvement of the posterior columns of the spinal cord, affected about one-third of patients with neurosyphilis in the preantibiotic era (see Fig. 60.7). Currently it is a very rare condition. As with general paresis, the incubation period ranges from 15 to 25 years. Clinical symptoms include lightning pains shooting down the posterior legs, paresthesias, decreased reflexes, abnormalities in peripheral sensation, difficulty walking, and bladder and bowel dysfunction. Patients often have a positive Romberg sign. A classic description of tabes dorsalis includes patients who walk with their heels landing hard on the floor, knees

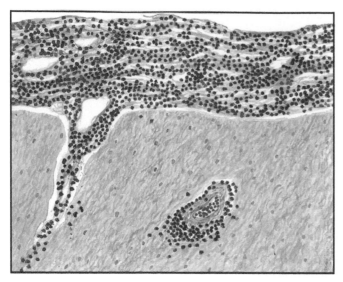

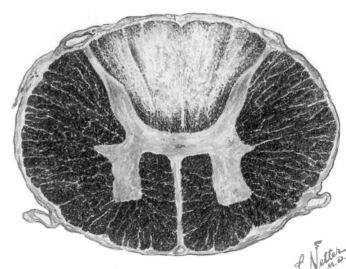

Syphilitic meningoencephalitis with perivascular infiltration

Section of thoracic spinal cord in tabes dorsalis

**Fig. 60.7** Tertiary syphilis: late neurosyphilis.

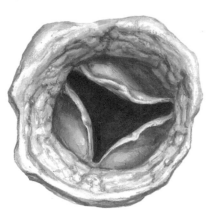

Incompetent aortic valve with taut, separated cusps viewed from above

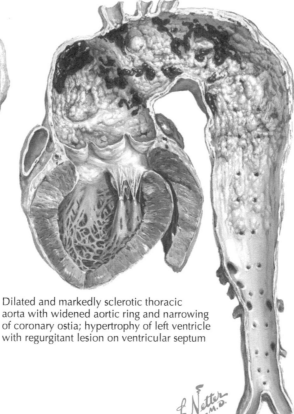

Dilated and markedly sclerotic thoracic aorta with widened aortic ring and narrowing of coronary ostia; hypertrophy of left ventricle with regurgitant lesion on ventricular septum

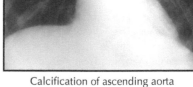

Calcification of ascending aorta and dilatation of thoracic aorta

**Fig. 60.8** Tertiary syphilis: cardiovascular complications.

positioned outward, with the feet slapping. Pupillary findings, including the Argyll-Robertson pupil, are more commonly found in patients with tabes dorsalis.

Although syphilis should be considered in the differential diagnosis of nearly any psychiatric or neurologic presentation—including dementia, late-onset psychosis, and neuropathy—late neurologic manifestations of syphilis are rare in the antibiotic era.

### Cardiovascular Syphilis

Endarteritis of the vasorum of the aorta leads to aortitis and aneurysm formation. This usually involves the ascending aorta, which in turn can cause dilation of the aortic ring and aortic regurgitation (Fig. 60.8). After the ascending aorta, the transverse aorta and then the descending arch are the next most common sites involved. Chronic inflammation of the coronary arteries can lead

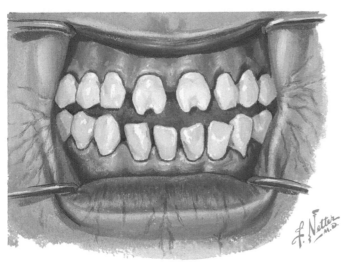

Hutchinson's teeth, scars of healed rhagades (congenital syphilis)
**Fig. 60.9** Congenital syphilis: Hutchinson teeth.

**Fig. 60.10** Spirochetes in darkfield examination.

to narrowing and stenosis of the coronary ostia, which can ultimately lead to myocardial ischemia, infarction, and congestive heart failure.

## Gummatous (Late Benign) Syphilis

Gummatous disease is extremely uncommon and is characterized by indolent destructive lesions of the skin, soft tissue, and bony structures. Although the lesions are destructive, they respond rapidly to treatment. Visceral organs, bones, and the CNS can also be involved. The differential diagnosis of lesions of the skin and mucous membranes is broad and will depend on the local epidemiology of other infectious diseases and neoplasms. Conditions to consider in the differential diagnosis of gummatous-appearing skin lesions include Hodgkin disease, mycosis fungoides, tuberculosis, systemic lupus erythematosus, fungal infections, sarcoid, and granuloma annulare.

## Congenital Syphilis

The manifestations of congenital syphilis are variable and include asymptomatic disease, spontaneous abortion, intrauterine growth restriction, neonatal disease, and neonatal death. The fetus is usually infected in utero transplacentally, and congenital infection rarely occurs before the fourth month of gestation. Congenital infection is most likely to be acquired in the setting of maternal early syphilis; however, it has been documented at any stage of syphilis. Some of the classic features of neonatal disease include rhinitis (snuffles), which typically occurs early in the course of the disease, as well as rash, hepatitis, splenomegaly, and perichondritis or periostitis. Untreated neonates who survive neonatal syphilis enter a latent period. The perichondritis and periostitis can lead to deformities of the nose (saddle nose) and of the metaphyses of the lower extremities (saber shin). Other late manifestations of congenital syphilis include peg-shaped central incisors (Hutchinson teeth), frontal bossing, and recurrent arthropathy (Fig. 60.9).

The prevention and early detection of congenital syphilis depend on routine screening of pregnant women for syphilis. All pregnant women should be screened at the first prenatal visit. Women who are at high risk for syphilis infection or who live in an area with a high syphilis prevalence should be screened again in the third trimester or at delivery (see Diagnostic Approach, next).

## DIAGNOSTIC APPROACH

Historically, *T. pallidum* has been difficult to culture in vitro, although new laboratory techniques have recently demonstrated the ability to grow *T. pallidum* in culture. The organism is too slender to be observed by light microscopy and fails to take up traditional Gram stains. It can be visualized using darkfield microscopy, which uses refracted light on a darkened background to identify the spirochete in clinical specimens (Fig. 60.10). Whereas polymerase chain reaction (PCR) technology has been used to amplify genetic elements of *T. pallidum* in clinical specimens, there are currently no US Food and Drug Administration (FDA)–cleared PCR assays. The clinical diagnosis of syphilis is based on the characteristic findings of the skin and mucous membranes and is confirmed with serologic assays measuring antibodies to nontreponemal (RPR or VDRL tests) and treponemal antigens (*T. pallidum* particle agglutination [TPPA], fluorescent treponemal antibody absorption [FTA-ABS], and enzyme immunoassays [EIAs]).More recently, rapid lateral flow assays (LFAs) for syphilis have been developed for use at the point of care. Although most of these tests detect treponemal antibodies and cannot differentiate between active and prior infections, they can be useful in screening, especially in populations at risk for loss to follow-up, at high-risk for infection, and in low-resource settings, where laboratory capacity is limited.

Nontreponemal tests use a laboratory-prepared lecithin-cholesterol antigen to detect treponemal-directed antibody in the patient's serum specimen. Nontreponemal tests have a sensitivity of approximately 86% in primary syphilis and 100% in secondary syphilis. Nontreponemal tests are 98% specific, with false-positive results associated with older age, autoimmune disease (e.g., lupus), other infections (e.g., bacterial endocarditis, rickettsial infection, HSV), chronic liver disease, intravenous drug use, and recent vaccination. Nontreponemal tests can be performed quantitatively, and response to treatment can be demonstrated by declining nontreponemal titers over time.

Treponemal-specific tests are used to confirm the diagnosis of syphilis and to rule out false-positive nontreponemal test results. The treponemal-specific tests currently available in the United States—the FTA-ABS, TPPA, and *T. pallidum* hemagglutination assay (TPHA)—use true treponemal antigens as key reagents. Unlike nontreponemal antibody tests, which decline in titer with treatment, treponemal-specific tests remain reactive for the remainder of the life of the individual, irrespective of the success of treatment. Similar to nontreponemal tests, their sensitivity is lower in primary disease, although they may become reactive before nontreponemal tests in the earliest stages of primary infection. They are 100% sensitive and 99% specific in secondary disease. Reactive laboratory tests for syphilis (treponemal and nontreponemal) are reportable to public health authorities.

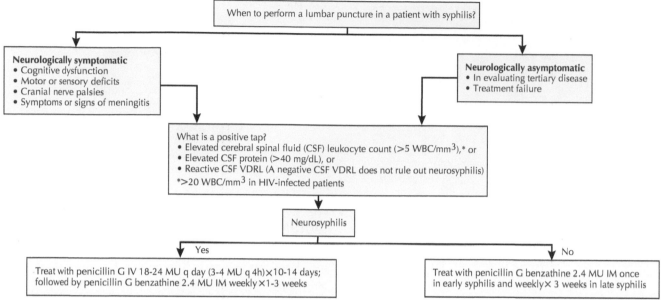

Fig. 60.11 Syphilis management: indications for lumbar puncture and cerebrospinal fluid interpretation.

Some clinical laboratories and blood banks have begun to use a "reverse algorithm" as a screening strategy for syphilis antibodies, in which a treponemal EIA is used in place of a nontreponemal assay as an initial screening test (Fig. 60.12). A positive treponemal EIA result will identify persons with a history of treated syphilis and those with untreated or incompletely treated syphilis. If the treponemal EIA result is positive, a nontreponemal test should be performed to confirm infection and determine the titer for monitoring response to treatment. If the nontreponemal test is nonreactive, a second treponemal-specific antibody test (TPPA or FTA-ABS) should be performed, as the initial EIA result could be falsely positive. If both treponemal-specific antibody test results are positive and the nontreponemal test is nonreactive, this could represent latent infection or a previously treated case. In that situation, providers should attempt to document prior treatment; sexually transmitted disease control programs within local health departments can often assist in this effort. If that is not possible, treatment for latent syphilis should be considered.

## Primary Syphilis

Evaluation of a patient with a genital ulcer should include (1) sexual, medical, travel, and medication history; (2) examination of the oral cavity, skin (trunk, upper and lower extremities, palms and soles, scrotum), and genital and anal areas; (3) examination by darkfield microscopy, if available, of serous exudate from a chancre for the presence of spirochetes; (4) serum RPR or VDRL and TPPA, TP-EIA or FTA-ABS (because treponemal-specific tests may be more sensitive in early infection); (5) HSV type-specific PCR or viral culture from swab of ulcer; and (6) antibody test for HIV infection (particularly essential if syphilis is diagnosed). Diagnostic testing is recommended for definitive diagnosis of primary syphilis, as the appearance of the chancre is often difficult to distinguish from that of other genital ulcers, such as genital herpes and chancroid.

## Secondary Syphilis

A rash of any type in a sexually active individual should be considered as potential syphilis until proven otherwise, particularly if it is bilaterally symmetrical. The typical rash of secondary syphilis does not yield moist specimens for darkfield examination; however, if condylomata lata is present and darkfield microscopy is available, these can be

swabbed and examined directly for spirochetes. Darkfield microscopy is limited in the evaluation of oral mucous patches, as the presence of nonpathogenic spirochetes of the oral flora decreases specificity. Patients with secondary syphilis should be evaluated for signs of complicated infection, including visual and hearing symptoms, which, if present, should prompt evaluation for neurosyphilis. Nontreponemal tests are highly sensitive in secondary syphilis. A prozone phenomenon can occur when the cardiolipin antibody titer is so high that antigen-antibody complexes cannot form, leading to the inability to visualize the characteristic agglutination reaction. When the clinical suspicion for secondary syphilis is high and the nontreponemal test result is negative, the test should be repeated with a 10-fold dilution of serum.

## Tertiary Syphilis

Syphilis should be considered in patients with ascending aortic aneurysms or aortic regurgitation. Serologic tests are usually reactive in tertiary syphilis; titers of nontreponemal tests can range from low to very high but are usually lower than in early syphilis. In patients with neurologic findings suggestive of late neurosyphilis and positive serum nontreponemal and treponemal antibody test results, the CSF should be examined. A positive CSF VDRL result establishes the diagnosis of neurosyphilis.

## Cerebrospinal Fluid Analysis: Indications and Interpretation

As discussed earlier, syphilis can involve the CNS at any stage of disease. The CSF should therefore be examined in any patient with syphilis and any neurologic symptoms or signs (cognitive dysfunction, motor or sensory deficits, cranial nerve palsies, or meningismus). The CSF should also be examined in a patient diagnosed with any other form of tertiary syphilis (cardiovascular or gummatous) and in patients who fail to respond to therapy with an appropriate decline in nontreponemal antibody titers (Fig. 60.11).

Lymphocytic pleocytosis (>5 white blood cells [WBCs] per cubic millimeter) and elevated CSF total protein are characteristic of the acute, syphilitic meningitis seen in early syphilis. Fewer cells are seen in the CSF in late neurosyphilis, including syphilitic cerebrovascular disease, general paresis, and tabes dorsalis. In the setting of a reactive serum nontreponemal and treponemal antibody test, a reactive CSF VDRL confirms the diagnosis of neurosyphilis. However, the CSF

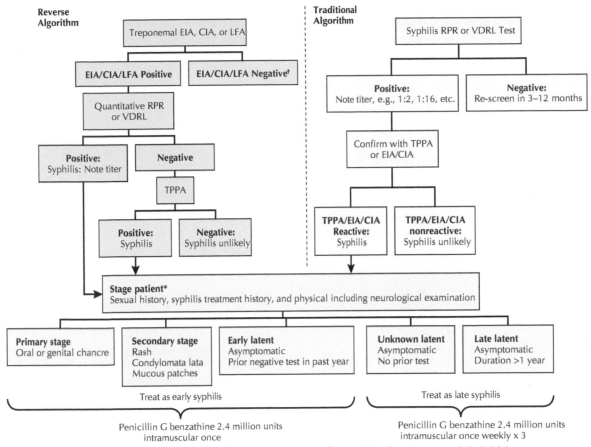

**Fig. 60.12** Summary of syphilis diagnosis and management. †If suspicion for primary syphilis is high, treat as early syphilis. *RPR or VDRL titer ≥1:32 suggests infection in the prior year. *CIA*, Chemiluminescence immunoassay; *EIA*, enzyme immunoassay; *LFA*, lateral flow assay; *RPR*, rapid plasma reagin; *TPPA*, *Treponema pallidum* particle agglutination; *VDRL*, Venereal Disease Reference Laboratory.

VDRL, particularly in early syphilitic meningitis, is not highly sensitive. The role of other serologic tests in the CSF is uncertain. The CSF FTA-ABS has a high false-positive frequency but is more sensitive than the CSF VDRL. Because it is highly specific, the CSF FTA-ABS can be used to exclude neurosyphilis in at-risk patients with abnormal CSF and a negative CSF VDRL result.

## Congenital Syphilis

The diagnosis of congenital syphilis rests on the identification of syphilis in the mother and a combination of clinical, radiologic, and laboratory findings in the infant. All infants born to mothers with reactive nontreponemal and treponemal test results should be screened for congenital syphilis by performing a quantitative nontreponemal antibody test on infant serum (not umbilical cord blood, which can become contaminated with maternal blood). The infant should be examined carefully for signs and symptoms of syphilis. If clinically indicated, the workup may include long-bone radiographs, a chest radiograph, liver function tests, cranial ultrasound, ophthalmologic examination, auditory examination, and CSF analysis for VDRL, cell count, and protein. The evaluation and management of congenital syphilis should be performed in consultation with a pediatric infectious disease specialist.

## TREATMENT

Penicillin G remains the treatment of choice for all stages of syphilis. The treatment regimen (route of administration and duration) depends on the stage of disease and site of infection. Early syphilis (primary, secondary, and early latent) can be treated with a single injection of penicillin G benzathine 2.4 million units IM. A nontreponemal antibody test should be performed on the day of treatment to establish a baseline titer for monitoring the response to therapy. Late syphilis (late latent syphilis and syphilis of unknown duration) is treated with penicillin G benzathine 2.4 million units IM weekly for a total of three injections. A lapse of more than 14 days between doses requires restarting treatment. Neurosyphilis is treated with aqueous penicillin G, 3 million to 4 million units IV q4h for 10 to 14 days, followed by penicillin G benzathine, 2.4 million units intramuscularly weekly for 1 to 3 weeks. Patients with otologic and ocular syphilis should be treated according to the recommendations for neurosyphilis; patients with ocular syphilis should be managed in consultation with an ophthalmologist.

Although penicillin G remains the treatment of choice, other antibiotics have efficacy and can be used, if necessary, in the setting of penicillin allergy. Doxycycline is an effective alternative for early-stage syphilis (100 mg PO bid for 14 days) and for late syphilis (100 mg PO twice daily for 28 days). Ceftriaxone can also be used as an alternative to penicillin, although the doses and duration of treatment have not been standardized. Given documented cases of azithromycin treatment failures and evidence of *T. pallidum* resistance to azithromycin in the United States, there is a limited role for azithromycin in the treatment of syphilis (Fig. 60.12 and Table 60.1).

## TABLE 60.1 Recommended Treatment

| | Recommended Therapy | Alternative Therapy | Comment |
|---|---|---|---|
| Primary, secondary, and early latent syphilis | Penicillin G benzathine, 2.4 million units IM as a single dose | Doxycycline 100 mg PO bid for 14 days <br> *or* <br> Ceftriaxone, 1–2 g IV or IM qd for 10–14 days <br> *or* <br> Tetracycline 500 mg PO qid for 14 days | |
| Latent syphilis of unknown duration and late latent syphilis | Penicillin G benzathine, 2.4 million units IM once weekly for 3 weeks | Doxycycline 100 mg PO bid for 28 days <br> *or* <br> Tetracycline 500 mg PO qid for 28 days | |
| Neurosyphilis | Penicillin G aqueous, 18–24 million units IV qd (3–4 million units q4h or by continuous infusion) for 10–14 days | Procaine penicillin, 2.4 million units IM qd *plus* probenecid 500 mg PO qid, both for 10–14 days <br> *or* <br> Ceftriaxone, 2 g IV or IM qd for 10–14 days | Follow-up treatment with one to three additional weekly injections of penicillin G benzathine, 2.4 million units IM |
| Tertiary syphilis (not neurosyphilis) | Penicillin G benzathine, 2.4 million units IM once weekly for 3 weeks | | CSF evaluation should be performed before therapy to rule out neurosyphilis |

*bid,* Twice daily; *CSF,* cerebrospinal fluid; *IM,* intramuscular; *IV,* intravenous; *PO,* by mouth, *qd,* daily; *qid,* four times daily.

## Follow-Up

Patients with early-stage syphilis, particularly those with high-titer secondary syphilis, should be counseled about the possibility that they may experience a Jarisch-Herxheimer reaction after treatment. This immune-mediated process occurs within 6 to 24 hours of receiving penicillin G and is characterized by the acute onset of fever, headache, and myalgias. Peripheral leukocytosis and transaminitis can also occur. This reaction is usually self-limited and can be managed with antipyretics and nonsteroidal anti-inflammatory medications.

The response to treatment should be assessed by the resolution of clinical manifestations and by the decline in nontreponemal antibody titers over time. Successful therapy is determined by a fourfold decline in the nontreponemal antibody test result (e.g., 1:32–1:8). When response to treatment is monitored, the same nontreponemal test (either RPR or VDRL), ideally performed at the same laboratory, should be followed serially owing to between-test and between-laboratory variability in nontreponemal antibody titer. In early-stage syphilis, a fourfold decline should occur within 12 months of treatment. In late syphilis, this decline should occur within 24 months. For example, if a patient presented with a 1-week history of a rash, an RPR of 1:64, and was treated for early syphilis, then at the 12-month follow-up visit, the RPR should be 1:16 or less. HIV-uninfected patients should have follow-up titers done at 6 and 12 months for early syphilis and at 6, 12, and 24 months for late syphilis; patients with HIV infection should have follow-up titers done at 3, 6, 9, 12, and 24 months for early syphilis and at 6, 12, 18, and 24 months posttreatment for late syphilis. A nontreponemal antibody titer that fails to fall by fourfold over the expected period may represent reinfection or, less likely, treatment failure. By following serial nontreponemal antibody titers using the same quantitative assay (for example, RPR or VDRL), it is easier to distinguish reinfection from treatment failure. If there is concern regarding treatment failure, a CSF analysis should be performed to rule out neurosyphilis. If the CSF is abnormal, the patient should be treated for neurosyphilis. If the CSF is normal, treatment should be reinitiated with intramuscular penicillin G benzathine 2.4 million units weekly for 3 weeks.

## Pregnancy and Congenital Syphilis

Pregnant women with syphilis who are allergic to penicillin should be desensitized and treated with penicillin according to the guidelines listed earlier. In pregnant women the Jarisch-Herxheimer reaction can

## BOX 60.1 Syphilis Screening Recommendations From the US Centers for Disease Control and Prevention

**Pregnancy**
- First antenatal care visit
- Repeat at 28–32 weeks or at delivery in areas with high syphilis prevalence or for women at high risk

**Men Who Have Sex With Men**
- Annual screening for all sexually active men who have sex with men
- Screening every 3–6 months for those at increased risk for syphilis, including
- More than 1 sex partner in the past 12 months
- Substance use (e.g., methamphetamine)
- Meeting sex partners online or in sex venues
- Having sex partners who participate in the activities already listed

**Persons Living With HIV Infection**
- Screening at first HIV clinical evaluation
- Annual screening
- More frequent screening as needed based on local epidemiology and individual risk behaviors

precipitate a miscarriage; therefore pregnant women should be treated in a monitored setting where available. Treatment of neonates with proven or probable congenital syphilis should be done in consultation with a pediatric infectious disease specialist.

## PREVENTION AND CONTROL

Syphilis screening is important to the treatment and prevention of disease. The CDC recommends syphilis screening for certain populations (Box 60.1). It is recommended that all pregnant women be screened at the first antenatal visit and again later in pregnancy, depending on risk factors. For MSM, annual screening is recommended for all those who are sexually active or every 3 to 6 months for those with specific risk factors. Among persons living with HIV infection, screening for syphilis should be done at the first clinical evaluation and then annually, but more frequent screening might be indicated depending on risk factors.

Providers can work together with local health departments to prevent the spread of syphilis. Presumptive and confirmed cases of syphilis should be reported within 1 working day of diagnosis. Staff in public health departments are then able to contact and notify sex partners and provide testing and treatment as appropriate. For patients exposed to early syphilis within the previous 3 months, the proper management includes examination, nontreponemal testing (stat if available) and treatment with intramuscular penicillin G benzathine regardless of serologic test results. For patients exposed to early syphilis who are beyond the 3-month incubation period, treatment depends on clinical examination findings and serologic test results. Presumptive treatment of contacts based on exposure history is essential to prevent reinfection and control the spread of disease.

## FUTURE DIRECTIONS

There is a growing body of evidence showing that chemoprophylaxis is an effective way to reduce syphilis acquisition. An open-label randomized control trial of doxycycline (100 mg PO qd) in men living with HIV infection showed a 73% reduction in the incidence of any bacterial STI over 1 year (syphilis, gonorrhea, or chlamydia). Another open-label randomized controlled trial using doxycycline 200 mg once after each sexual contact also showed a 73% reduction in risk of syphilis incidence over 1 year. There are several ongoing clinical trials aiming to provide further evidence of the acceptability, safety, as well as efficacy of syphilis chemoprophylaxis.

## EVIDENCE

Dunaway SB, Maxwell CL, Tantalo LC, Sahi SK, Marra CM. Neurosyphilis treatment outcomes after intravenous penicillin G versus intramuscular procaine penicillin plus oral probenecid. *Clinical Infectious Diseases* 2019. *A retrospective review of neurosyphilis cases that found no difference in treatment outcomes between treatment with intravenous penicillin G compared to intramuscular procaine penicillin with oral probenecid.*

Ghanem KG, Erbelding EJ, Cheng WW, Rompalo AM: Doxycycline compared with benzathine penicillin for the treatment of early syphilis. *Clin Infect Dis* 2006;42:e45-e49. *A retrospective chart review demonstrating the efficacy of doxycycline for the treatment of early syphilis.*

Grant JS, Stafylis C, Celum C, Grennan T, Haire B, Kaldor J, Luetkemeyer AF, Saunders JM, Molina JM, and Klausner JD. Doxycycline prophylaxis for bacterial sexually transmitted infections. *Clinical Infectious Diseases* 2019. https://doi.org/10.1093/cid/ciz866. *A review of the current evidence around the use of doxycycline prophylaxis to prevent syphilis and other sexually transmitted infections.*

Liu HY, Han Y, Chen XS, Bai L, Guo SP, Li L, Wu P, Yin YP. Comparison of efficacy of treatments for early syphilis: A systematic review and network meta-analysis of randomized controlled trials and observational studies. *PLoS One.* 2017;12(6):e0180001. *A systematic review and meta-analysis comparing efficacy of treatments for early syphilis and finding that ceftriaxone is as effective as penicillin in treating early syphilis.*

Long CM, Klausner JD, Leon S, et al. Syphilis treatment and HIV infection in a population based study of persons at high risk for sexually transmitted disease/HIV infection in Lima, Peru. *Sex Transm Dis* 2006;33:151-155. *Describes the epidemiology of syphilis in a high-risk population in Peru and illustrates efficacy of treatment with penicillin or doxycycline for early syphilis in HIV-infected and HIV-uninfected patients.*

Lukehart S, Godornes C, Molini B, et al. Macrolide resistance in Treponema pallidum in the United States and Ireland. *N Engl J Med* 2004;351:154-158. *Describes treatment failures associated with azithromycin-resistant syphilis, along with epidemiology and molecular characteristics.*

Rolfs R, Joesoef M, Hendershot E, et al. A randomized trial of enhanced therapy for early syphilis in patients with and without human immunodeficiency virus infection. *N Engl J Med* 1997;337:307-314. *Randomized, controlled trial demonstrating that penicillin G benzathine is equal to enhanced therapy in the treatment of syphilis in HIV-infected patients.*

## ADDITIONAL RESOURCES

Augenbraun M. Syphilis. In Klausner JD, Hook EW III, eds: *Current diagnosis and treatment: sexually transmitted diseases.* New York, 2007, McGraw-Hill, pp 119-129. *Brief and thorough review of the presentation, diagnosis, and management of syphilis.*

Janier M, Hegyi V, Dupin N, Unemo M, Tiplica G, Potočnik M, French P, Patel R. 2014 European guideline on the management of syphilis. *J Eur Acad Dermatol Venereol* 2014;28:1581-1593. *European guidelines for the management of syphilis, published by the International Union Against Sexually Transmitted Infections.*

Osman C, Clark TW. Tabes dorsalis and Argyll Robertson pupils. *New England Journal of Medicine* 2016;375(20):e40. *A clinical video showing the physical examination findings of late neurosyphilis.*

US Preventive Services Task Force. Screening for syphilis infection in pregnant women: US Preventive Services Task Force reaffirmation recommendation statement. *JAMA* 2018;320(9):911-917. *Summary of the US Preventive Services Task Force recommendations on screening for syphilis in pregnant women and the supporting evidence.*

US Preventive Services Task Force (USPSTF). Screening for syphilis infection in nonpregnant adults and adolescents: US Preventive Services Task Force recommendation statement. *JAMA* 2016;315(21):2321-2327. *A review of recent evidence on the benefits and harms of screening for syphilis in pregnancy and the harms of treatment with penicillin.*

Workowski KA, Bolan GA, and Centers for Disease Control and Prevention. Sexually transmitted diseases treatment guidelines, 2015. *MMWR* 2015;64(RR3):1-137. *The CDC's evidence-based guidelines on diagnosis and management of sexually transmitted diseases.*

# 61

# Bacterial Vaginosis, Related Syndromes, and Less Common Sexually Transmitted Diseases

*Sandra G. Gompf, John F. Toney*

## ABSTRACT

Bacterial vaginosis (BV), the most common cause of vaginal discharge in the United States, is not traditionally described as a sexually transmitted infection (STI) but is well documented to be sexually associated. In the United States, two STIs—chancroid and granuloma inguinale—are uncommon diseases associated with genital ulcerations. These two infections may be encountered outside the United States and are significant risk factors for human immuno-deficiency virus (HIV) infection.

## CLINICAL VIGNETTE

A 27-year-old female presents with a 1-week history of thin, white vaginal discharge. She states that the discharge began 2 days after her last sexual intercourse when she also noted the occurrence of an unpleasant "fishy" odor. She denies any fevers, chills, pain with urination, pelvic pain, pain during sex, or vaginal itching. In the last 3 months she engaged in vaginal sexual intercourse with one new male partner; however, she has had three male sex partners during the past 12 months and she inconsistently uses condoms. She has no other medical problems and currently takes no medications. To her knowledge, she has never had routine STI screening since she infrequently has medical issues and does not have consistent healthcare. A speculum exam of the vagina showed a thin white discharge adherent to the vaginal wall. The vaginal discharge had a fishy odor and tested positive on a card test for a pH of greater than 4.5 and amines. Vaginal swabs were sent for chlamydia and gonococcal nucleic acid amplification (NAAT) testing. A blood sample was taken for syphilis, HSV-2 and HIV screening. The patient was given the diagnosis of BV and prescribed a regimen of oral metronidazole for 7 days. She was told to abstain from vaginal intercourse during treatment and to return for additional therapy if the symptoms recurred.

## BACTERIAL VAGINOSIS

### Geographic Distribution and Magnitude of Disease Burden

BV is the most common cause of vaginal discharge in the United States and has a worldwide presence. The prevalence of BV varies by the population studied; reports of the occurrence of BV in college students of 5% to 25% compared with 12% to 35% among STI clinic patients have been published. Estimates from the Centers for Disease Control and Prevention (CDC) 2001–2004 National Health and Nutrition Survey indicate that as many as one in three women of reproductive age are affected at any given time. African American and Hispanic women had significantly higher prevalence (33.2% and 30.7%, respectively) than Caucasian or Asian groups (22.7% and 11.1%, respectively). The occurrence of BV in our global population can range from 23% to 29%.

### Risk Factors

BV is more common in women with a new sexual partner or many sexual partners, in those who have frequent intercourse, when there is a lack of condom use, and in those who practice douching. Studies of women who have sex with women (WSW) have found concordance between BV diagnosis and the prevalence of BV-associated bacteria (BVAB); digital-vaginal and digital-anal sex practices significantly increased the incidence of BV. BV has not been considered an STI, but recent studies support the idea that BVAB exchange between sexual partners is related to the occurrence of BV. Additionally, male circumcision also can reduce the risk of BV in women. These data suggest that BV is sexually transmitted; however, the definite etiology remains unclear. The use of an intrauterine device (IUD)—particularly copper IUDs (thought to increase BVAB colonization)—has been associated with an increased risk of BV; interestingly, women using hormonal contraception had no BV prevalence change. In pregnancy, BV has been associated with preterm delivery and premature rupture of membranes. BV can increase susceptibility to STIs, including *Chlamydia trachomatis*, *Neisseria gonorrhoeae*, *Trichomonas vaginalis*, *Mycoplasma genitalium*, human papillomavirus (HPV), and herpes simplex virus type-2. Several of these bacteria may lead to the development of pelvic inflammatory disease (PID), postoperative infections after gynecologic procedures, and HIV infection—the latter is thought to be associated with vaginal dysbiosis, even without BV, which may increase susceptibility to HIV.

### Clinical Features

In BV, vaginal dysbiosis results from replacement of healthy, hydrogen peroxide and lactic acid-producing *Lactobacillus* spp. with high concentrations of facultative and strict anaerobic bacteria (i.e., *Gardnerella vaginalis*, *Prevotella* spp., *Atopobium vaginae*, and many others) which alters the vaginal pH and is associated with the occurrence of BV. That said, lactobacilli may not be the predominant species in all women without symptoms. A growing spectrum of species form the "normal" microbiome in women of different ethnicities. In addition, different periods of the menstrual cycle with hormonal status and immune variations may play a role.

Women with apparent microbiologic dysbiosis may or may not be symptomatic. BV is occasionally characterized by mucosal inflammation that may produce irritation, itching, or burning, but the most common symptom is a vaginal discharge that may be malodorous (often described as "fishy") and acquired more commonly after unprotected sexual intercourse or during their menses. On vaginal speculum examination, a homogeneous, milky discharge that is adherent to the walls of the vagina is commonly present

(Fig. 61.1). BV may occur concomitantly with other causes of vaginitis, including trichomoniasis (discussed in Chapter 54) and vulvovaginal candidiasis.

## Diagnostic Approach

Evaluating patients with suspected BV is dependent on physical examination of the vagina, microscopic examination of the discharge, and pH determination. The amount, consistency, and location of the discharge within the vagina should be noted.

The diagnosis of BV is accomplished by using clinical criteria (Amsel criteria) or a Gram stain of the vaginal discharge. Clinical criteria diagnosis requires three of the following symptoms or signs: (1) a homogeneous, thin, white discharge that smoothly coats the vaginal walls; (2) a fishy odor from the vaginal discharge before or after addition of 10% potassium hydroxide (KOH; i.e., the amine or "whiff" test); (3) a pH of vaginal discharge fluid greater than 4.5; or (4) the presence of clue cells (vaginal epithelial cells with adherent coccobacilli best recognized at the edge of the cell accounting for at least 20% of the epithelial cells on microscopic examination). A comparison of these criteria with characteristics of other common causes of vaginitis is shown in Table 61.1.

A Gram stain of a vaginal discharge sample for Nugent scoring is considered the gold standard for BV laboratory diagnosis because it shows the relative concentrations of bacteria in the vaginal ecosystem—including the reduction of lactobacilli and the increase of bacteria associated with BV, including gram-negative and gram-variable rods and cocci morphotypes. Performing a vaginal Gram stain for Nugent score is time consuming and usually not a point-of-care (POC) test. White blood cells are not usually present in vaginal fluid from a patient with BV; the presence of neutrophils in vaginal fluid suggests the potential of a co-infection at either the cervical or vaginal sites. Cervical Papanicolaou (Pap) testing is not recommended for the diagnosis of BV because of low sensitivity.

Saline wet mount, OSOM BV Blue test (which detects vaginal sialidase activity), FemExam card (detects metabolic by-products of *G. vaginalis*), and the Affirm VP assay (oligonucleotide probe test detecting high concentrations of *G. vaginalis*) are POC tests with varying sensitivities and specificities; no new POC tests for BV have emerged in the past few years. There are currently five commercially available NAATs for BV diagnosis in symptomatic women; two are US Food and Drug Administration (FDA) approved and the other three are laboratory-developed tests (LDTs), which need to be internally validated prior to their use. These three NAATs have very good sensitivity and specificity similar to FDA-approved assays, and include LabCorp NuSwab VG, MDL OneSwab BV Panel PCR w/ Lactobacillus Profiling by qPCR, and Quest Diagnostics SureSwab BV. FDA-approved assays include the BD MAX Vaginal Panel and Hologic Aptima BV; both NAATs perform nearly as well as Nugent scoring with the BD MAX also reporting Candida ssp. and *T. vaginalis* detection results. Advantages of NAAT tests over POC tests for BV diagnosis are that they do not require microscopy, are objective, are able to detect fastidious bacteria, provide quantitation, and are ideal for self-collected vaginal swabs. However, Amsel criteria and the Nugent score still remain useful for the diagnosis of symptomatic BV with their lower cost and ability to provide a rapid diagnosis.

## Clinical Management and Treatment

In nonpregnant women, BV treatment is intended to relieve vaginal signs and symptoms of infection and reduce the potential for infectious complications associated with abortion or hysterectomy as well as the potential reduction in the risk of acquiring HIV and other STIs. Because of the increased risk of BV postoperative infectious complications, some providers screen for and treat women with BV before performing elective surgical abortion or hysterectomy. All women with symptomatic disease should be offered treatment and screened for STIs at appropriate intervals depending on their sexual risk behaviors.

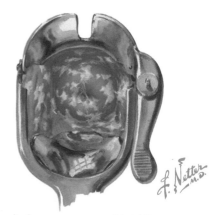

The discharge appears grayish-white, homogenous, and adheres to the vaginal wall

**Fig. 61.1** Discharge of bacterial vaginosis.

| TABLE 61.1 | Differentiating Bacterial Vaginosis, Candidiasis, and Trichomoniasis | | | |
|---|---|---|---|---|
| | **Normal** | **Bacterial Vaginosis** | **Candidiasis** | **Trichomoniasis** |
| Symptoms or presentation | | Odor, discharge, itch | Itch, discomfort, dysuria, thick discharge | Itch, discharge, 50% asymptomatic |
| Vaginal discharge | Clear to white | Homogenous, adherent, thin, milky-white; malodorous "fishy" | Thick, clumpy, white "cottage cheese" | Frothy, gray, or yellow-green; malodorous |
| Clinical findings | | | Inflammation and erythema | Cervical petechiae, "strawberry cervix" |
| Vaginal pH | 3.8–4.2 | >4.5 | Usually ≤4.5 | >4.5 |
| KOH "whiff test" | Negative | Positive | Negative | Often positive |
| NaCl wet mount | Lactobacilli | Clue cells (≥20%), no or few WBCs | Few WBCs | Motile flagellated protozoa, many WBCs |
| KOH wet mount | | | Pseudohyphae or spores if non-*albicans* species | |

*KOH*, Potassium hydroxide; *NaCl*, sodium chloride, *WBC*, white blood cell.

The 2015 CDC-recommended treatment for BV (Box 61.1) includes using oral metronidazole, 500 mg twice a day for 7 days; 0.75% metronidazole gel, one full applicator (5 g) intravaginally once a day for 5 days; or 2% clindamycin phosphate cream, one full applicator (5 g) intravaginally at bedtime for 7 days. Three additional FDA-approved BV treatments are available in the United States, all using a single dose: secnidazole 2 g oral granules, metronidazole 1.3% vaginal gel, and clindamycin phosphate 2% vaginal cream (a sustained-release formulation). Research involving biofilm-disrupting agents is underway; studies evaluating the clinical and microbiologic efficacy of using intravaginal suppositories of a human-derived strain of *Lactobacillus* (*Lactobacillus crispatus* strain CTV-05) given as adjunctive therapy in managing BV were inconclusive. There are no data supporting the use of douching for treatment or relief of BV symptoms, and the use of vaginal probiotics is currently not recommended. Prior recommendations suggested patients should avoid alcohol ingestion during metronidazole treatment and for 24 hours after treatment; however, a 2014 systematic review of the literature revealed no in vitro studies, animal models, reports of adverse effects, or clinical studies provided any evidence of a disulfiram-like interaction between alcohol and metronidazole. Clindamycin cream is oil-based and might weaken latex condoms and diaphragms for 5 days after use.

Adverse outcomes associated with BV during pregnancy include premature rupture of the membranes, preterm labor, preterm birth, intra-amniotic infection, and postpartum endometritis. Past studies indicate that treatment of pregnant women with BV at high risk for preterm delivery (women who previously delivered a premature infant) might reduce their risk for additional prematurity. Clinicians should consider the evaluation and treatment of high-risk pregnant women with asymptomatic BV. A meta-analysis of metronidazole use during pregnancy failed to identify any associated adverse events. The

treatment of asymptomatic BV in women with low- or average-risk pregnancies is controversial and not currently recommended.

## Prognosis and Recurrence

When recommended therapies are used, cure rates for BV are 70% to 80%. Recurrences with BV are common; up to 66% of women experience a recurrence within 12 months of treatment. A different treatment regimen from the original management may be considered to control recurrent disease; however, optimal management strategies for women with persistent or recurrent BV is unclear. In one trial, after completion of a recommended BV regimen, metronidazole gel 0.75% used topically twice per week for 6 months as suppressive management was effective in maintaining a clinical cure for 6 months. Additional studies using high-dose intravaginal metronidazole 750 mg in conjunction with 200 mg miconazole as well as the compound astodrimer 1% vaginal gel, a dendrimer-based microbicide, looks promising. The utility of yogurt therapy, acidifying agents, or exogenous oral *Lactobacillus* treatment is disappointing.

There are no data suggesting that BV should be treated differently in HIV-infected women. Additionally, treatment of male sexual partners of women with BV remains not recommended based on data from prior male partner treatment trials. There are currently no data on whether treatment of female sexual partners of women with BV improves outcomes.

## CHANCROID

### Geographic Distribution and Magnitude of Disease Burden

Chancroid, an infection caused by bacteria of the gram-negative species *Haemophilus ducreyi*, is prevalent in many areas of the world, including Africa, Asia, Latin America, parts of the United States, and the Caribbean. Chancroid is difficult to culture and requires molecular assays for detection. Given the prevalence, likelihood of co-infections, and poverty of resources, syndromic therapy for syphilis and chancroid has been adopted widely, based on clinical diagnosis alone, for genital ulceration syndrome. Given that both chancroid and herpes simplex virus (HSV) are common causes of painful genital ulcers, this approach has limited both reporting and validation of clinical diagnoses. It is thus difficult to determine the true incidence of chancroid worldwide. Similarly, it is difficult to ascertain whether declining reports of chancroid reflect a true decline in global incidence. In the United States, chancroid appears to be very uncommon and is likely on the decline. Accurate reporting is similarly hindered by lack of molecular assays approved by the FDA as of this writing, and diagnosis remains clinical in most cases. Chancroid usually occurs in discrete outbreaks in the United States, although the disease is endemic in some areas, principally among migrant farm workers and poor inner-city residents. Previous US endemic and epidemic chancroid outbreaks have occurred in New York City, New Orleans, Florida, and Texas.

Chancroid is a cofactor for HIV transmission; high rates of HIV infection in patients who have chancroid occur in the United States and other countries. Approximately 10% of US individuals who have chancroid are co-infected with *Treponema pallidum* or HSV; this percentage is higher in persons who acquired chancroid outside the United States.

### Risk Factors

Chancroid occurs more frequently in men than in women. Infected men are less likely to have used condoms and more likely to report a history of contact with female commercial sex workers or multiple

---

### BOX 61.1    Recommended Treatment Regimens for Bacterial Vaginosis

Metronidazole 500 mg orally twice a day for 7 days
or
Metronidazole gel, 0.75%, one full applicator (5 g) intravaginally once a day for 5 days
or
Clindamycin cream, 2%, one full applicator (5 g) intravaginally at bedtime for 7 days

**Alternative Regimens**
Clindamycin 300 mg orally twice a day for 7 days
or
Clindamycin ovules 100 mg intravaginally once at bedtime for 3 days
or
Tinidazole 2 g orally once a day for 2 days
or
Tinidazole 1 g orally once a day for 7 days

**Recommended Regimens for Pregnant Women**
Metronidazole 500 mg orally twice a day for 7 days
or
Metronidazole 250 mg orally three times a day for 7 days
or
Clindamycin 300 mg orally twice a day for 7 days

Data from Centers for Disease Control and Prevention (CDC), Workowski KA, Bolan GA: Sexually transmitted diseases treatment guidelines, 2015, *MMWR Recomm Rep* 2015;64(3) 1-135.

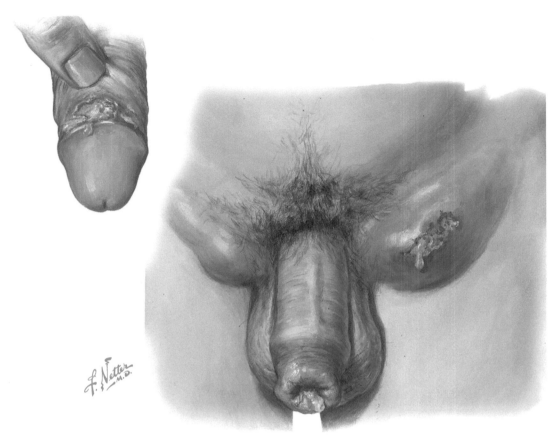

Note the ulcer on the glans and shaft of the penis with a "dirty" base consisting of inflammatory debris and the significant suppurative inguinal lymphadenitis or "bubo"

**Fig. 61.2** Male chancroid.

sexual partners in the preceding 3 months. Oral sex has occasionally been implicated in the transmission of chancroid.

## Clinical Features

Chancroid is often referred to as a "soft chancre" because the lesions are usually not indurated. After an incubation period of 3 to 7 days, the patient develops painful erythematous papules at the site of contact. The chancroid papules become pustular and then rupture to form shallow painful ulcers with purulent exudates and granulomatous bases. The ulcer edge is typically ragged and undermined.

Men usually have chancroid symptoms directly related to the painful genital lesions or inguinal tenderness (Fig. 61.2). In men, lesions typically occur on the prepuce and frenulum. Most affected women are asymptomatic but may have lesions, commonly occurring on the labia majora and less frequently on the vulva, cervix, and perianal area (Fig. 61.3). Women have less obvious signs and symptoms, such as dysuria, dyspareunia, vaginal discharge, pain on defecation, or rectal bleeding. Constitutional symptoms of chancroid, such as malaise and low-grade fevers, may be present. Painful, tender inguinal lymphadenitis typically occurs in up to 50% of cases, and the lymph nodes may develop into fluctuant, suppurative nodes termed *buboes*. The lymphadenopathy is usually unilateral and tends to be more prevalent in men. If not aspirated or drained through incision, fluctuant buboes can rupture spontaneously. Complications of chancroid in men include phimosis, balanoposthitis, and rupture of buboes with fistula formation and scarring.

## Diagnostic Approach

The diagnosis of chancroid based exclusively on the ulcer's appearance is accurate in only 30% to 50% of cases, even in areas where this disease is common and where physicians are experienced in the management of genital ulcer disease (GUD). Significant overlap exists among the major causes of GUD—HSV, syphilis, and chancroid; often, coinfection occurs with two diseases at the same time. Understanding that no cause can be found in 25% to 50% of all cases of GUD is important.

A probable chancroid diagnosis can be made if all the following criteria are met:

The patient has one or more painful genital ulcers.

The patient has no evidence of *T. pallidum* infection on darkfield examination of ulcer exudate or on serologic testing for syphilis performed at least 7 days after the onset of ulcers.

The clinical presentation, the appearance of genital ulcers, and, if present, regional lymphadenopathy are typical for chancroid.

Results of tests for HSV performed on the ulcer exudate are negative.

Gram staining of an ulcer specimen may show gram-negative coccobacilli singly, in clusters, or in various morphologic forms described as "schools of fish," "railroad tracks," or "fingerprints" (Fig. 61.4). Gram-stained ulcer material should not be routinely examined as a tool to diagnose chancroid owing to poor sensitivity and specificity of this test.

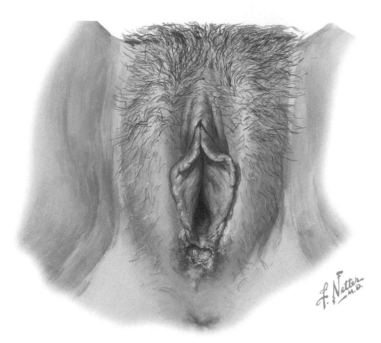

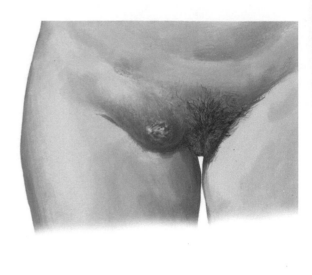

Note the ulcer at the inferior introitis and the significant right inguinal bubo

**Fig. 61.3** Female chancroid.

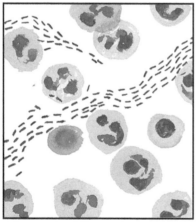

**Fig. 61.4** Gram stain of *Haemophilus ducreyi* illustrating the "school of fish" bacterial forms.

*H. ducreyi* is a fastidious bacterium requiring a selective nutritive medium for growth and is an extremely difficult organism to culture from clinical specimens. Culture is the current accepted standard for chancroid diagnosis in most areas, but even in an experienced laboratory, it is only 60% to 80% sensitive. Patients' specimens must either be plated out directly on an appropriate culture medium or sent to the microbiology laboratory for culture as soon as possible for optimal results.

Polymerase chain reaction (PCR) amplification is replacing culture as the diagnostic test of choice in some major medical centers. PCR amplification using a variety of primers may provide a useful alternative to culture for the detection of *H. ducreyi*. Although PCR assays perform well on samples prepared from *H. ducreyi* cultures, they are less sensitive when used to test genital ulcer specimens. A multiplex PCR (M-PCR) assay has been developed for the simultaneous amplification of DNA targets from *H. ducreyi*, *T. pallidum*, and HSV types 1 and 2;

**BOX 61.2   Recommended Chancroid Regimens**

Azithromycin 1 g orally in a single dose
or
Ceftriaxone 250 mg intramuscularly (IM) in a single dose
or
Ciprofloxacin 500 mg orally twice a day for 3 days
or
Erythromycin base 500 mg orally three times a day for 7 days

Data from Centers for Disease Control and Prevention (CDC), Workowski KA, Bolan GA: Sexually transmitted diseases treatment guidelines, 2015, *MMWR Recomm Rep* 2015;64(3) 1-135.

it appears more sensitive than standard diagnostic tests for the detection of these causative agents in genital ulcer specimens. The sensitivity of *H. ducreyi* culture relative to the M-PCR assay has been shown to be approximately 95% to 98.4% and specificity of 99.6% to 100% in studies that have used genital ulcer–derived swabs; unfortunately, the M-PCR assay is not commercially available. No FDA-cleared PCR test for *H. ducreyi* is available in the United States as of this writing, but the CDC suggests that clinical laboratories that have developed their own PCR test and conducted a Clinical Laboratory Improvement Amendments (CLIA) verification study can perform testing.

## Clinical Management and Treatment

CDC-recommended chancroid regimens include azithromycin, 1 g orally in a single dose; ceftriaxone, 250 mg given intramuscularly (IM) in a single dose; alternatively ciprofloxacin, 500 mg orally twice a day for 3 days; or erythromycin base 500 mg orally three times a day for 7 days can also be used (Box 61.2). Single-dose oral azithromycin or intramuscular ceftriaxone regimens offer advantages in terms of improved patient compliance; also, there have been worldwide reports

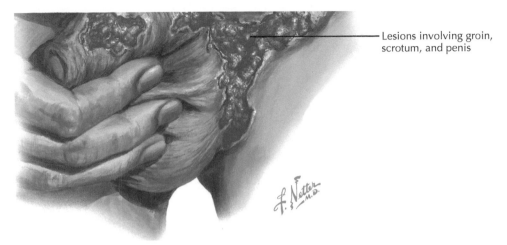

Lesions involving groin, scrotum, and penis

Fig. 61.5 Male donovanosis.

of several isolates with intermediate resistance to either ciprofloxacin or erythromycin.

Fluctuant buboes should be aspirated to provide symptomatic pain relief for the patient and to avoid the further complication of spontaneous rupture. Chancroid ulcers treated with the appropriate antibiotic agent generally resolve within 7 to 14 days; the time to complete healing depends on the size of the ulcer (larger ulcers may require more than 2 weeks). Chancroid relapses may occur in as many as 5% of patients after antibiotic therapy, and relapses are more common in patients who are uncircumcised or are infected with HIV. If the patient is not HIV infected, repeating the original therapy is usually effective.

## Prognosis and Recurrence

The prognosis is excellent if chancroid is treated properly and HIV co-infection is not present. No adverse effects of chancroid on pregnancy outcome have been reported. Chancroid-infected patients who have HIV should be monitored closely because they are more likely to experience treatment failure and have ulcers that heal slowly.

## Prevention and Control

Sex partners of patients who have chancroid should be examined and treated, regardless of whether symptoms of the disease are present, if they had sexual contact with the patient during the 10 days preceding the patient's onset of symptoms. The patient should be strongly advised to avoid sexual contacts while the ulcers are open because they are highly infectious. Patients should be advised to avoid commercial sex workers, properly and consistently use condoms, and avoid having multiple partners. Chancroid cases should be reported to the local STI program in states where reporting is mandated.

## GRANULOMA INGUINALE (DONOVANOSIS)

### Geographic Distribution and Magnitude of Disease Burden

Granuloma inguinale, or donovanosis, is a chronic, slowly destructive, ulcerative disease of skin and subcutaneous tissues caused by *Klebsiella* (formerly *Calymmatobacterium*) *granulomatis*. Rarely identified in developed countries, donovanosis has been endemic in adolescents and adults in some tropical and developing regions, including Papua New Guinea, central Australia, South Africa, and areas of India and Brazil, with sporadic cases reported in the West Indies, South America, and other areas of southern Africa. Accurate current data for most endemic areas are limited; however, in recent years several endemic regions have reported substantial declines in prevalence. The published literature on donovanosis represents few geographic locations, reflects limited microbiologic testing, and relies on syndromic GUD surveillance in areas where donovanosis is thought to be most common.

### Risk Factors

The sexual transmission of donovanosis has been controversial, but there is substantial evidence that *K. granulomatis* is transmitted sexually. The proportion of steady sexual partners of people diagnosed with donovanosis who develop the disease is 12% to 52%. Although rare, vertical transmission of donovanosis has been reported. There also is evidence that transmission may occur through fecal contamination and autoinoculation. Children rarely are diagnosed with donovanosis; cases in children have been attributed to sitting on the laps of infected adults rather than sexual abuse.

### Clinical Features

The incubation period of donovanosis is uncertain; however, experimental lesions have appeared in humans 50 days after inoculation. Donovanosis lesions involve the genitalia (typically the prepuce or glans in men and the vagina or labia minora in women) in 80% to 90% of cases but also can involve the inguinal and anal regions (Figs. 61.5 and 61.6). Beginning as a small, single papule or multiple papules at the site of inoculation, donovanosis typically causes painless, easily bleeding ulcers or vegetative lesions. Ulcerative lesions slowly expand and become clean, shallow, well-demarcated ulcer(s) with a beefy red granular base. The clinical presentation also may include hypertrophic, necrotic, or sclerotic variants. In most cases clinical findings are suggestive of donovanosis but are not highly specific.

Untreated lesions can cause extensive local tissue damage, including pelvic and perianal fistulas, urethral obstruction, and lymphedema. Although uncommon, lesions can develop secondary bacterial infection and cellulitis. Systemic infections can cause fever, weight loss, and anemia. Involvement of the bone, joint, and liver may occur infrequently and is thought to be more common in pregnant women. Involvement of the head and neck also has been described. The prominent inguinal swellings seen in patients with donovanosis have been called "pseudobuboes" because they are subcutaneous granulomas that occur superficially in the area of inguinal lymph nodes.

The differential diagnosis of donovanosis is broad and includes syphilis, lymphogranuloma venereum, chancroid, lymphoma, carcinoma, amebiasis, tuberculosis, blastomycosis, and other granulomatous

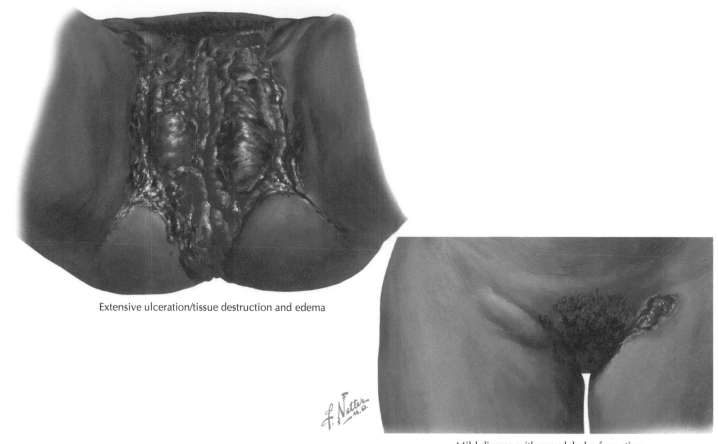

Extensive ulceration/tissue destruction and edema

Mild disease with pseudobubo formation

**Fig. 61.6** Female donovanosis.

diseases. People suspected of having donovanosis also should be tested for syphilis and other STIs, including HIV, which may coexist with donovanosis. Donovanosis is thought to increase risk of HIV transmission similar to other GUDs, such as syphilis and HSV infection.

## Diagnostic Approach

*K. granulomatis* is a pleomorphic, intracellular, gram-negative bacterium. Because it is difficult to grow on artificial media, tissue "crush" or biopsy specimens can also be used to identify *K. granulomatis*. For tissue crush specimens, granulation tissue is collected from a lesion and crushed between two microscopic slides. "Donovan bodies," safety pin–appearing and clustered in macrophages when stained with Wright or Giemsa stain, are diagnostic for *K. granulomatis* (Fig. 61.7). Although not FDA approved for clinical use, a PCR test for *K. granulomatis* is available for use in some research settings.

## Clinical Management and Treatment

A limited number of studies on donovanosis treatment have been published. Treatment halts progression of lesions, although prolonged therapy is usually required to permit granulation and re-epithelialization of the ulcers. Healing typically proceeds inward from the ulcer margins. Several antimicrobial regimens have been effective but results from only a limited number of controlled trials have been published. The recommended first-line therapy for adults with donovanosis according to the CDC's STI treatment guidelines is azithromycin 1 g orally once per week or 500 mg daily for at least 3 weeks until all lesions have completely healed. Alternative regimens include doxycycline 100 mg orally twice a day, ciprofloxacin 500 mg orally twice a day, erythromycin base 500 mg

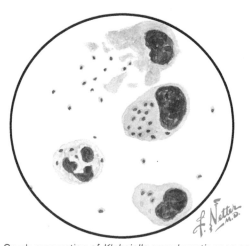

**Fig. 61.7** Crush preparation of *Klebsiella granulomatis* seen as Donovan bodies intracellularly in mononuclear cells.

orally four times a day, or trimethoprim-sulfamethoxazole one double-strength (160 mg/800 mg) tablet orally twice a day for at least 3 weeks and until all lesions have fully healed (Box 61.3). Pregnancy is a relative contraindication to the use of sulfonamides. Pregnant and lactating women should be treated with the erythromycin regimen; azithromycin might prove useful for treating granuloma inguinale during pregnancy, but published data are lacking. Additionally, doxycycline and ciprofloxacin are contraindicated in pregnant women. Persons with both granuloma inguinale and HIV infection should receive the same

---

**BOX 61.3  Recommended Regimens for Treatment of Granuloma Inguinale**

Azithromycin 1 g orally once per week or 500 mg daily for at least 3 weeks and until all lesions have completely healed

**Alternative Regimens**

Doxycycline 100 mg orally twice a day for at least 3 weeks and until all lesions have completely healed

or

Ciprofloxacin 750 mg orally twice a day for at least 3 weeks and until all lesions have completely healed

or

Erythromycin base 500 mg orally four times a day for at least 3 weeks and until all lesions have completely healed

or

Trimethoprim-sulfamethoxazole one double-strength (160 mg/800 mg) tablet orally twice a day for at least 3 weeks and until all lesions have completely healed

---

Data from Centers for Disease Control and Prevention (CDC), Workowski KA, Bolan GA: Sexually transmitted diseases treatment guidelines, 2015, *MMWR Recomm Rep* 2015;64(3) 1-135.

regimens as those who are HIV negative. With both of these groups, consideration should be given to the addition of a parenteral aminoglycoside (e.g., gentamicin) to these regimens if improvement is not evident within the first few days of therapy.

## Prognosis and Recurrence

Patients should be followed clinically until signs and symptoms have resolved, as relapse has been noted to occur 6 to 18 months after apparently effective therapy. The risk of complications, such as complete genital erosion and urethral obstruction, can be minimized with early therapy. Carcinoma is the most serious complication but is relatively rare. Practitioners should consider giving a monitored trial of antibiotic treatment for granuloma inguinale, as the histologic distinction between squamous cell carcinoma and granuloma inguinale may sometimes be difficult to differentiate. Surgical intervention may be required for advanced disease resulting in tissue destruction.

## Prevention and Control

Persons who have had sexual contact with a patient who has granuloma inguinale within the 60 days before onset of the patient's symptoms should be examined and offered therapy. However, the value of empirical therapy in the absence of clinical signs and symptoms has not been established.

## EVIDENCE

Fethers KA, Fairley CK, Hocking JS, et al. Sexual risk factors and bacterial vaginosis: a systematic review and meta-analysis. *Clin Infect Dis* 2008;47:1426-1435. *This article is the first to summarize available observational data for BV. It shows that BV is significantly associated with sexual contact with new and multiple male and female partners and that decreasing the number of unprotected sexual encounters may reduce incident and recurrent infection.*

Muzny CA, Schwebke JR. Pathogenesis of bacterial vaginosis: Discussion of current hypotheses. *J Infect Dis* 2016;214(S1):S1-5. *An in-depth review of the current epidemiology and pathogenesis of BV with several hypotheses on pathogenesis discussed.*

Nygren P, Fu R, Freeman M, et al. Evidence on the benefits and harms of screening and treating pregnant women who are asymptomatic for bacterial vaginosis: an update review for the U.S. Preventive Services Task Force. *Ann Intern Med* 2008;148:220-233. *There has been continued debate about the value of screening and treating asymptomatic pregnant women for BV; this review found no benefit in treating women with low- or average-risk pregnancies for asymptomatic BV.*

Oakley BB, Tina L, Fiedler TL, et al. Diversity of human vaginal bacterial communities and associations with clinically defined bacterial vaginosis. *Appl Environ Microbiol* 2008;74:4898-4909. *The compositions of vaginal bacterial communities differ dramatically between subjects with and without BV. These data describe a previously unrecognized diversity in the vaginal ecosystem and, in particular, in BV-associated bacteria.*

Onderdonk AB, Delaney ML, Fichorova RN. The human microbiome during bacterial vaginosis. *Clin Microbiol Rev* 2016;29:223-238. *Great overview of the vaginal microbiome with a focus on BV epidemiology, detection methods, individual microbial species involved in BV, and their role in immunity and immune alterations.*

## ADDITIONAL RESOURCES

Amsel R, Totten PA, Spiegel CA, et al. Nonspecific vaginitis. Diagnostic criteria and microbial and epidemiologic associations. *Am J Med* 1983;74:14-22. *This landmark paper provided clearer diagnostic criteria for the diagnosis of BV and remains the standard clinical assessment tool.*

Centers for Disease Control and Prevention (CDC), Workowski KA, Bolan GA. Sexually transmitted diseases treatment guidelines, 2015. *MMWR Recomm Rep* 2015;64(3):1-135. Available at: https:// www.cdc.gov/std/ tg2015/default.htm. *This document contains the CDC's recommendations for the management of the wide variety of STIs. Updates to this document can be found at www.cdc.gov/std as well as the link to download the free CDC 2015 STD Treatment Guide smartphone app.*

Evans AL, Scally AJ, Wellard SJ, Wilson JD. Prevalence of bacterial vaginosis in lesbians and heterosexual women in a community setting. *Sex Transm Infect* 2007;83:470-475. (Also see related editorial for this paper: Marrazzo JM. Elusive etiology of bacterial vaginosis—do lesbians have a clue? *Sex Transm Infect* 2007;83:424-425.) *This article provides insights into the interesting finding of higher BV prevalence in women who report sexual activity with other women. The accompanying editorial is insightful and discusses the limitations and complexities of studies into sexual orientation and disease investigation.*

Hart G. Donovanosis. *Clin Infect Dis* 1997;25:24-30. *Although an older review, this paper provides significant insight regarding the transmission, clinical features, related laboratory testing, and clinical management of granuloma inguinale.*

Lewis DA. Chancroid: clinical manifestations, diagnosis, and management. *Sex Transm Infect* 2003;79:68-71. *A concise review of the diagnosis and management of chancroid, including color plates of clinical manifestations.*

Mohammed TT, Olumide YM. Chancroid and human immunodeficiency virus infection—a review. *Int J Dermatol* 2008;47:1-8. *An informative review of the difficulties of chancroid management in individuals with concomitant HIV infection.*

World Health Organization (WHO). Guidelines for the management of sexually transmitted infections, 2003. Geneva, Switzerland: WHO. Available at: https://www.who.int/hiv/pub/sti/pub6/en/. *Recommendations for the diagnosis and management of STIs outside of the United States.*

# Infections Associated With International Travel and Outdoor Activities

*Elaine C. Jong*

# 62

# Introduction to Infections Associated With International Travel and Outdoor Activities

*Elaine C. Jong*

 **ABSTRACT**

*Travel* in the broadest sense means going from one place to another, especially journeys to distant or unfamiliar places, and the theme of travel being associated with exposure to exotic infectious diseases (IDs) has been a recurring one throughout history. This section covers a variety of IDs that may be encountered during international travel and through participation in outdoor activities: these two kinds of exposures often intersect in pursuit of wilderness adventure travel. No matter what the destination, duration, or primary purpose of the trip, travelers more likely than not are going to spend some time outdoors enjoying the scenery; mingling with the local population while shopping, sightseeing, or exploring; and eating food prepared by others—thus exposing themselves to multiple modes of disease transmission. Familiarity with the geographic distribution of IDs and modes of transmission that are commonly associated with travel and outdoor activities (whether domestic or international) allows clinicians to offer appropriate advice and prophylaxis before patients embark on travel or participate in outdoor activities; it also aids clinicians in the diagnosis and treatment of returned travelers who seek consultation for illnesses associated with travel, outdoor activities, or both. In providing pre- and posttravel medical advice, the "travel medicine triad" is a useful risk-assessment tool, considering the traveler, the trip risks, and the possible interventions (Fig. 62.1).

## ILLNESS IN RETURNED TRAVELERS

GeoSentinel is a network of 53 tropical or travel disease units in 24 countries that collaboratively contribute data to a central database on travel-associated illnesses seen in patients at their units. In a published report by Leder et al. the GeoSentinel database, records of 42,173 ill returned travelers seen between 2007 and 2011 were reviewed. Typical diseases in returned travelers were categorized according to region, travel reason and patient demographic characteristics; the pattern of low-frequency travel-associated diseases was also described. The most common regions where illnesses were acquired were Asia (32.6%) and sub-Saharan Africa (26.7%). Gastrointestinal illness accounted for about 34% of travel-related illnesses, followed by febrile illness (23.3%), dermatologic diagnoses (19.5%), and respiratory illness (10.9%). The proportion of major syndromic groupings for gastrointestinal, febrile, dermatologic, and respiratory illnesses among ill returned travelers is shown in Fig. 62.2.

There were regional differences in the frequencies of a given diagnosis: in Asia, dengue fever, typhoid fever, and bacterial pneumonia (in descending order) were the leading causes of illness, followed by hepatitis A and malaria; in the Pacific, hepatitis A was the leading diagnosis, followed by malaria and then bacterial pneumonia; in Africa, malaria was the most frequent diagnosis, followed by bacterial pneumonia and typhoid fever; in the Middle East and Latin America, respectively,

typhoid fever and malaria were reported. When the GeoSentinel data were analyzed by mode of transmission, more than 35% of the diseases were vectorborne, almost 25% were respiratory, and approximately 23% were foodborne and/or waterborne.

Returned travelers presenting with acute febrile illnesses warrant prompt medical attention: malaria (especially when caused by *Plasmodium falciparum*) can rapidly progress to a life-threatening condition. The early clinical signs and symptoms of several serious febrile illnesses among returned travelers are similar—however, malaria should be at the top of the list if a patient's travel included a malaria-endemic region.

Travelers' diarrhea is a well-known scourge of international travelers, and commonly episodes are experienced during the trip rather than after return home. Travelers to regions with suboptimal sanitation and sewage treatment systems may experience acute onset of loose watery stools sometimes accompanied by abdominal cramps, anorexia, nausea, and general malaise: this symptom complex is called travelers' diarrhea (TD). This is usually a self-limited disease that runs its course over 5 to 7 days, but it can seriously affect the enjoyment of a short vacation trip, threaten the success of a business trip, and impair the performance of participants in competitive sports. In some cases, episodes of travel-associated diarrhea may cause lingering symptoms that prompt the traveler to seek a full medical workup after returning home. TD is dreaded by most travelers, but other gastrointestinal (GI) pathogens can cause serious, even life-threatening illnesses: cholera, typhoid, paratyphoid, and viral hepatitis fall into this category.

Diseases acquired by inadvertent contact with animals, insects, and the outdoor environment must be considered in the case of rural travelers and sometimes urban travelers as well. This category of infections includes yellow fever and other arboviruses (e.g., dengue, chikungunya tick-borne encephalitis, West Nile, Japanese encephalitis), rabies, Lyme disease, leptospirosis, and primary amebic meningoencephalitis.

## "WHERE ARE YOU GOING…WHERE HAVE YOU BEEN?"

In the 21st century, advances in telecommunications have resulted in unprecedented access to news of disease outbreaks all over the world. New disease-mapping technologies allow for sophisticated epidemiologic analysis, and advances in molecular biology yield new avenues for rapid diagnosis and novel treatments. Despite all this technologic progress, however, the prevention of diseases in outbound travelers and the detection of imported IDs in returned travelers depend on the acumen of individual healthcare providers, who should always ask, "Where are you going?" and "Where have you been?" as part of the standard medical history. The geography of possible exposures to IDs must be added to the traditional considerations of mode of transmission, incubation

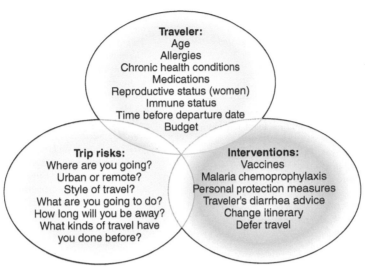

**Fig. 62.1** The travel medicine triad. (Reused with permission from Sanford CA, Jong EC, Pottinger PS. *The Travel and Tropical Medicine Manual*, 5th edition, Elsevier, 2017, Fig. 1.1.)

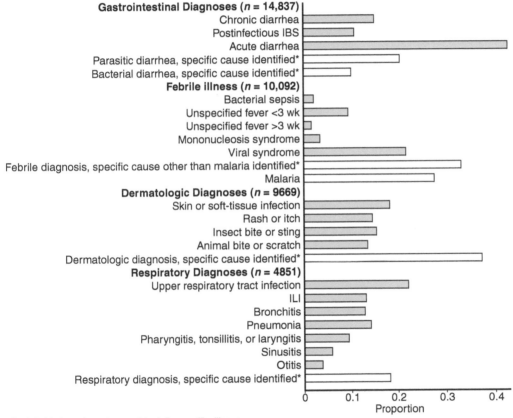

IBS = irritable bowel syndrome; ILI = influenza-like illness.
* Each bar represents a mutually exclusive classification. Green bars depict the proportion of each diagnostic category with the given syndromic grouping. White bars depict the proportion with the given specific cause.

**Fig. 62.2** Proportion of major syndromic groupings for gastrointestinal, febrile, dermatologic, and respiratory illness among ill returned travelers. (Reused with permission from Leder K, Torresi J, Libman MD, et al. Geo-Sentinel surveillance of illness in returned travelers, 2007-2011. *Ann Intern Med*. 2013;158(6):456-468. Fig. 1. doi:10.7326/0003-4819-158-6-201303190-00005.)

time, clinical presentation, and pathophysiology of disease when one is evaluating a returned traveler who is ill.

This section presents a selection of IDs that may pose significant health risks to the international traveler and to others whose recreational or occupational activities place them at risk for certain infections transmitted in geographic environments away from home.

## EVIDENCE

Leder K, Torresi J, Libman MD et al. GeoSentinel surveillance of illness in returned travelers, 2007-2011. *Ann Intern Med* 2013;158:456-468. Doi: 10.7326/0003-4819-158-6-201303190-00005. *Analysis of 5 years of sentinel surveillance data that identifies many travel-associated illnesses that may have been preventable with appropriate advice, chemoprophylaxis, or vaccination. Diagnoses among ill returned travelers suggest common and low-frequency imported diseases according to region, purpose of travel, and patient demographic characteristics.*

## ADDITIONAL RESOURCES

Centers for Disease Control and Prevention (CDC). *Health Information for International Travel 2020.* Atlanta: U.S. Department of Health and Human Services, Public Health Service; 2019. Available at: http://wwwnc.cdc.gov/travel/content/yellowbook/home-2020.aspx. *This publication is known as the "Yellow Book" and is the authoritative source of US government recommendations for immunizations and prophylaxis for foreign travel. Both communicable and noncommunicable diseases associated with travel are covered, and the country-by-country and region-specific descriptions of endemic diseases and prevailing public health make this an invaluable resource for guiding medical advice to travelers.*

Jong EC. Approach to travel medicine and contents of a personal travel medicine kit. In: Sanford CA, Pottinger PS, Jong EC: *The Travel and Tropical Medicine Manual.* 5th ed. Philadelphia: Elsevier; 2017, pp 1-16. *The manual is a practical guide for clinicians who provide care to travelers, immigrants, and refugees. Covers diseases and conditions, diagnosis, treatments, and prevention for many travel and tropical medicine issues that are commonly seen in primary care settings.*

World Health Organization (WHO). *International Travel Health 2020.* Geneva, Switzerland, WHO, 2020. Available at: http://www.who.int/ith/en/. *This publication is known as the "Green Book" and is a compendium of WHO regulations and recommendations for immunizations and prophylaxis for foreign travel, and is a resource for information on drugs and regimens available outside of the United States. This webpage provides links to the latest updates on WHO information and guidance on current global disease outbreaks and disasters.*

# 63

# Malaria

Winnie W. Ooi, Elaine C. Jong

 **ABSTRACT**

Malaria is a bloodborne protozoan parasite mostly transmitted from person to person by bites of infected female *Anopheles* mosquitoes in tropical and subtropical regions. Despite a substantial reduction in the malaria burden observed since 2010, largely attributed to effective control measures and treatment regimens, the World Health Organization (WHO) still reported 228 million infections and 405,000 deaths in 2018. The majority of deaths from malaria occur among children in sub-Saharan Africa, where over 85% of the world's malaria deaths occur. Travelers going to malaria-endemic countries are at risk for contracting the disease, and almost all of the approximately 1700 cases per year of malaria in the United States are imported. Despite taking antimalarial chemoprophylaxis, travelers may become infected and return home with asymptomatic infections, becoming ill weeks to months after exposure, when parasites that had been incubating in liver cells are released into the circulation, invade red blood cells (RBCs), and replicate to produce succeeding generations of parasites.

Four main species of parasites cause human-only malaria: *Plasmodium falciparum (PF)*, *Plasmodium vivax (PV)*, *Plasmodium ovale (PO)*, and *Plasmodium malariae (PM)*. Human infections with the zoonotic parasite *Plasmodium knowlesi (PK)* occur throughout Southeast Asia in all countries where both the major macaque monkey reservoir and anopheles mosquito vector are present. The greatest number of cases have been reported from Malaysia, where *PK* has become the most common cause of malaria. A second simian species, *Plasmodium cynomolgi*, has been shown also to cause human infection in a traveler to Southeast Asia. *PF* and *PV* infections occur most commonly, but *PF* causes the most severe disease and is the target of major prevention programs, especially in Africa. If not promptly diagnosed and treated with efficacious drugs, *PF* infections may result in death within a few days after the onset of fever because of the high levels of parasitemia (particularly in young children and nonimmune adults such as travelers to hyperendemic areas). *PV* is also not always benign and has increasingly been described to cause severe disease, including acute respiratory distress syndrome and death.

The emergence of resistance to chloroquine, sulfadoxine-pyrimethamine, mefloquine, and increasingly to artemisinin drugs among *PF* strains has prompted the search for and development of new antimalarial drugs and regimens for both preventing disease (e.g., tafenoquine [TQ]) and treating diagnosed infections (e.g., three-drug artemisinin combination therapy for *PF* malaria). Research on the development of vaccines against malaria has produced several candidate vaccines that are being studied in clinical trials among children in endemic areas.

## ✳ CLINICAL VIGNETTE

A 35-year-old male lawyer was seen for 7 days of fever to 105°F, headache, chills, and myalgias. He was born in the United States but for several years had traveled to Sierra Leone multiple times for trips lasting several months. He reports taking antimalarial chemoprophylaxis irregularly and had been treated for at least four different episodes of malaria, the last of which was about 7 months ago.

For the present episode he had already been admitted to an outside hospital, where he was diagnosed with malaria; he was discharged the next day on chloroquine and primaquine (PQ).

He followed up with his primary care provider who started him also on doxycycline which he took for 2 days without improvement. He was then admitted to our hospital with a hemoglobin of 7.2, platelets of 30,000/mL$^3$, and 8% parasitemia. His malarial smear was consistent with *P. falciparum* (see Fig. 63.4). Despite this the patient was alert and normotensive; he was tachycardic but otherwise stable. He received quinidine gluconate intravenously and was transfused with 2 units of packed RBCs, which produced a persistent drop in his hematocrit. He subsequently received intravenous artesunate and responded with a rapid decline in his parasitemia to 3% and felt much better.

COMMENT: This was a semi-immune healthy young patient who, despite his high parasite count, was relatively stable in the face of severe anemia and thrombocytopenia due to his disease. Inappropriate treatment with chloroquine despite known travel to a highly endemic area for chloroquine-resistant *Plasmodium falciparum* (CR*PF*) led to a higher blood parasite count, which was quickly brought down with intravenous artesunate, a rapidly acting drug.

## EPIDEMIOLOGY

Malaria transmission occurs in Africa, Asia (including Southeast Asia, South Asia, and the Middle East), eastern Europe, the South Pacific, and Central and South America. The heaviest burden of infection is in sub-Saharan Africa, where malaria is the leading cause of childhood mortality (Fig. 63.1).

Humans become infected with malaria from the bites of infected female *Anopheles* mosquitoes, which are nighttime feeders biting from dusk to dawn; they generally inhabit altitudes of less than 1500 m. Mosquitoes acquire the infection when biting and taking blood meals from individuals with untreated malaria, during which RBCs containing male and female malaria gametocytes are ingested. After the parasite goes through several developmental stages in the mosquito gut, sporozoites (the stage of the parasite infective for humans) reach the

**A. Malaria-endemic countries in the Eastern Hemisphere[1]**

**B. Malaria-endemic countries in the Western Hemisphere[1]**

■ Endemic countries
☐ Non-endemic countries

Boundary representation is not necessarily authoritative.
[1]In this map, countries with areas endemic for malaria are shaded completely even if transmission occurs only in a small part of the country.

Boundary representation is not necessarily authoritative.
[1]In this map, countries with areas endemic for malaria are shaded completely even if transmission occurs only in a small part of the country.

**Fig. 63.1** Geographic distribution of malaria. (Reused with permission from Centers for Disease Control and Prevention [CDC]: *CDC health information for international travel 2020*, https://wwwnc.cdc.gov/travel/yellowbook/2020/travel-related-infectious-diseases/malaria. Atlanta, 2019, U.S. Department of Health and Human Services, Public Health Service. Accessed January 2020.)

mosquito's salivary glands and are subsequently injected into a capillary during the bite of the infected mosquito, leading to new human infections (Fig. 63.2). The intensity of transmission in a given area is dependent on the presence of untreated malaria infections in the human population, the prevalent species of *Anopheles* mosquitoes, their host feeding preferences, and environmental conditions—such as rainfall, temperature, and humidity—that favor mosquito breeding habitats.

The majority of human malarial infections are transmitted by mosquito vectors, but transplacental infections and blood transfusion and organ transplant–acquired infections can occur. These modes of transmission are special concerns when they occur outside of endemic areas because healthcare providers may be unfamiliar with them and unsuspecting of the diagnosis.

## LIFE CYCLE AND CLINICAL FEATURES

The minimum incubation period for malaria is 7 days from the first exposure (bite by an infected mosquito) to the development of fever, although an incubation period of 14 to 30 days is more typical (Table 63.1). The incubation period may be longer if malarial chemoprophylaxis has been taken and there is suboptimal parasite suppression. Initially the clinical onset of a malarial attack is nonspecific, with fever, fatigue, headache, muscle aches, and chills resembling a flulike illness. As the illness progresses, cycles of high fever and chills begin, which correspond to parasite replication cycles in the RBCs.

After entering the human body through an infected mosquito bite, the malarial sporozoites reach the liver, where they infect hepatocytes, incubate, and multiply (see Fig. 63.2). After the parasites multiply, each infected hepatocyte ruptures, releasing hundreds to thousands of malarial merozoites into the bloodstream. In the case of *PV* and *PO*

infections, some parasites may persist as dormant hypnozoites within hepatocytes for periods of several months to 4 years, before reactivation and multiplication to initiate a new episode of clinical disease.

The merozoites released by the hepatocytes infect circulating RBCs, subsequently developing into early trophozoites, which appear as ring-shaped forms within RBCs on Giemsa-stained peripheral blood smears. The intraerythrocytic trophozoites subsequently progress through the schizont phase to ultimately produce merozoites that are released on rupture of the infected RBCs. After release, these second-generation merozoites rapidly infect other circulating RBCs and the infection is amplified, with the subsequent completion of new erythrocytic cycles resulting in the release of new generations of infective merozoites.

Each cycle ending in rupture of infected RBCs is accompanied by the onset of chills and fever. The symptomatic response is thought to be mediated through cytokines and other factors produced by macrophages and inflammatory cells triggered by hemozoin pigment and other toxic wastes produced during intraerythrocytic parasite development that are released with each new crop of merozoites when the infected RBCs rupture. In addition to chills and fever, the onset of clinical malaria may be accompanied by headache, myalgias, nausea, and vomiting. In nonimmune hosts the fever pattern during the initial days of clinical illness may be irregular, particularly in *PF* infections. In established infections and in semi-immune patients, each species has a characteristic periodic fever spike separated by an afebrile interval corresponding to the incubation time of the new generation of parasites in the erythrocytic cycle (Fig. 63.3). *PF*, *PV*, and *PO* have a 48-hour periodicity (tertian malaria, meaning fever every third day), and *PM* has a 72-hour periodicity (quartan malaria, with a fever spike every fourth day), although *PM* tends to cause chronic, asymptomatic infections with low parasitemia. *PK* infections have a 24-hour periodicity; patients with this infection may exhibit a daily fever spike.

Malaria is an infection caused by a parasite carried by a mosquito found in tropical and subtropical areas. It can be very serious in travelers and young children, who don't have good protection against it.

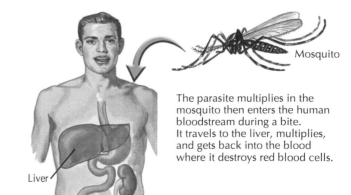

Mosquito

The parasite multiplies in the mosquito then enters the human bloodstream during a bite. It travels to the liver, multiplies, and gets back into the blood where it destroys red blood cells.

Liver

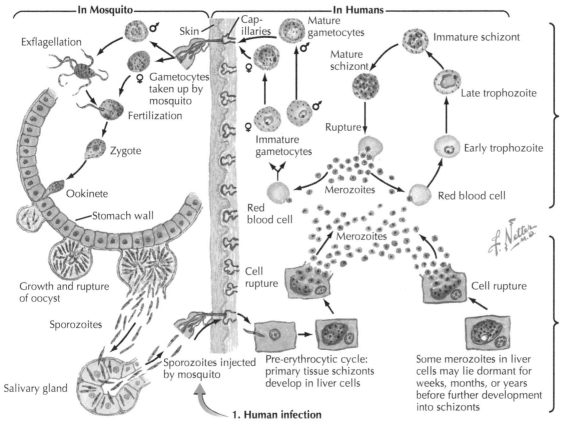

**In Mosquito**

Exflagellation

Skin

Gametocytes taken up by mosquito

Fertilization

Zygote

Ookinete

Stomach wall

Growth and rupture of oocyst

Sporozoites

Salivary gland

Sporozoites injected by mosquito

**1. Human infection**

Pre-erythrocytic cycle: primary tissue schizonts develop in liver cells

**In Humans**

Cap-illaries

Mature gametocytes

Mature schizont

Immature schizont

Late trophozoite

Rupture

Immature gametocytes

Early trophozoite

Merozoites

Red blood cell

Red blood cell

Merozoites

Cell rupture

Cell rupture

Some merozoites in liver cells may lie dormant for weeks, months, or years before further development into schizonts

**3. Erythrocytic cycle:**
Malaria merozoites from ruptured liver cells infect and multiply in red blood cells leading to cyclical red cell rupture, and release new crops of merozoites, which infect other red cells. Some red cells infected by merozoites may develop into male and female malaria gametocytes.

**2. Hepatic cycle:**
Malaria sporozoites injected by an infected mosquito during a bite incubate silently and multiply in liver cells. Liver cells eventually rupture and release merozoites into the bloodstream.

**Fig. 63.2** Malaria life cycle and transmission to humans.

Severe systemic complications may develop during *PF* replication within the RBCs. The merozoites can potentially infect any circulating RBC, leading to high levels of parasitemia (>5%). Parasitized RBCs tend to stick to nonparasitized red cells (called *rosetting*) and also to the capillary endothelium in many end organs. Impairment of the microcirculation by the RBC rosettes and the sludging and sequestration of parasitized RBCs in the end-organ capillaries are thought to affect the microcirculation of vital organs such as the brain and lead to multiple end-organ failure.

During the erythrocytic cycles, some of the merozoites in RBCs may develop into male and female gametocytes. Gametocytes cannot directly cause malarial infections in other humans but, if ingested by female *Anopheles* mosquitoes, can undergo excystation, fertilization, and reproductive stages in the gut of the mosquito host and vector, promoting the continuation of mosquito-borne transmission (see Fig. 63.2).

Malarial gametocytes may be found on peripheral blood smears, even after appropriate treatment has been given, with apparently successful clearance of RBC asexual parasite forms. The gametocytes do not have pathologic implications for the individual patient. However, malaria patients should be protected from mosquito bites after treatment whenever possible, both to prevent reinfection from new bites by infected mosquitoes and to prevent new mosquito infections from gametocytes ingested during the bite of a treated patient. One review of the duration of *PF* gametocyte carriage in East Africa showed that the duration of gametocyte carriage after artemisinin-based combination therapy (ACT) was an average of 13 days, compared with an average of 55 days after non-ACT treatment. The addition of PQ to ACT (recommended by the WHO in 2012 for all patients posttreatment to decrease malarial transmission) resulted in a further reduction in this duration.

## DIAGNOSTIC APPROACH

### Thick and Thin Blood Smears

The gold standard of malaria diagnosis is a parasitic diagnosis based on finding diagnostic morphologic forms by microscopic examination

| TABLE 63.1 | Characteristics of Malaria Species Infecting Humans | | | | |
|---|---|---|---|---|---|
| Feature | Plasmodium falciparum | Plasmodium vivax | Plasmodium ovale | Plasmodium malariae | Plasmodium knowlesi |
| Reservoir of infection | Humans | Humans | Humans | Humans | Macaque monkeys |
| Predominant geographic area | Sub-Saharan Africa, Asia, South America | Asia, Southeast Asia, Oceania, South America, Africa | Sub-Saharan Africa | South America, Asia, Africa | Southeast Asia |
| Incubation period[a] | 7–14 days | 8–14 days | 8–14 days | 7–30 days | 9–12 days |
| Dormant liver-stage parasites (hypnozoite stage)— relapses may occur | Absent | Present (relapses may occur several months up to 4 years after initial exposure) | Present (relapses may occur several months up to 4 years after initial exposure) | Absent (occult low-grade parasitemia may persist for years) | Absent |
| RBC preference | Circulating RBCs | Reticulocytes[b] | Reticulocytes | Older RBCs | Circulating RBCs |
| Typical percent parasitemia | May be >1.0%; severe malaria >5.0%; consider exchange transfusion for >10% | <0.5%[b] | <0.5% | <0.5% | <0.5%; in some cases may be >1.0% |
| Rapid diagnostic test (RDT) (United States) | BinaxNOW | BinaxNOW | NA | NA | NA |

*RBC,* Red blood cell.

[a]The incubation period tends to be shorter, as short as 4 days after infection by direct inoculation (blood transfusion, needlestick, transplacental). The incubation period tends to be longer when there is suboptimal suppression of malaria parasites because of prophylactic drugs that have been taken.
[b]In pregnancy, the reticulocyte count may be increased and there may be a correspondingly higher level of parasitemia.

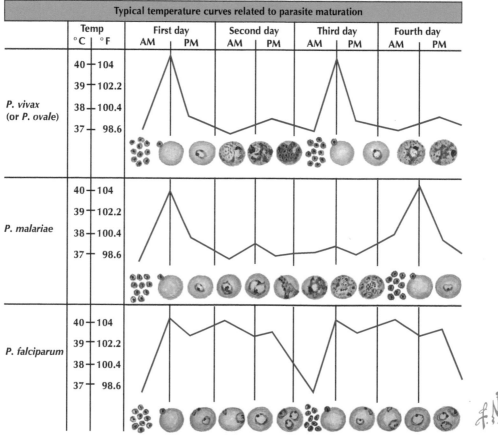

**Fig. 63.3** Typical temperature curves related to parasite development in red blood cells.

of thick and thin blood smears. The severity of the malarial infection is expressed as percent parasitemia (percentage of RBCs with parasites). The morphology of the ring trophozoites, mature trophozoites, schizonts, and gametocytes is used to differentiate the various malaria species when blood smears are examined microscopically, and up to three blood samples taken 6 to 8 hours apart should be examined by an experienced microscopist (Fig. 63.4). The gametocytes of PF have a distinctive banana-shaped appearance on blood smears and may be contained within RBCs or lie outside of the cells.

On blood smears from infected humans, early ring trophozoite forms of PK may resemble those of PF, whereas mature forms of PK trophozoites, schizonts, and gametocytes can be mistaken for those of PM. An atypical appearance of the *Plasmodium* species seen in the blood smears and a higher-than-expected level of parasitemia than is usual for PM should raise suspicions of PK if the patient has traveled in Asia. A definitive diagnosis may not be possible from light microscopy (especially PK, which is microscopically indistinguishable from PM), and samples of the patient's blood may have to be sent to a reference laboratory for species confirmation by molecular testing, such as polymerase chain reaction (PCR) amplification and DNA sequencing.

### Rapid Diagnostic Tests

Microscopy is both time consuming and expensive and requires trained microscopists. The human and laboratory resources necessary for high-quality parasitic diagnosis by microscopy may not be available in some malaria-endemic areas and in clinics in nonendemic areas that are outside of regional medical centers. Therefore there has been extensive adaptation of rapid diagnostic tests (RDTs) for malaria in such situations. The RDTs are meant to aid in the diagnosis of malaria, especially in healthcare settings, and are not recommended for use by lay travelers for self-diagnosis. There are several RDTs that demonstrate consistent detection of malaria at low parasite densities ($200/\mu L$) as well as high parasite densities ($2000–5000/\mu L$); they have low false-positive rates, are relatively easy to use, and can detect PF or PV infection or both. If positive, the test result may facilitate a more rapid initiation of empiric therapy for the patient with signs and symptoms of malaria and a history of possible exposure: this may be lifesaving in the case of PF.

Standardized evaluations of RDTs have been conducted by the WHO in collaboration with the Centers for Disease Control and Prevention (CDC). Only one RDT is available in the United States and approved by the US Food and Drug Administration (FDA): the BinaxNOW malaria test. This RDT can detect PF and PV antigens using a whole blood sample drawn from a vein or a finger stick. Performance data showed that for PF, the sensitivity and specificity of this test were 99.7% and 94.2%, respectively; for PV, the sensitivity and specificity were 93.5% and 99.8%, respectively.

The test consists of antimalarial antibodies immobilized at one end of a test strip that target the histidine-rich protein II (HRPII) antigen specific to PF as well as a pan-malarial antigen common to the four major malaria species. When a patient's blood sample is applied to the designated spot on the test strip, any malarial antigens present bind to the antibodies, forming antigen-antibody complexes. The results must be confirmed by microscopy of thick and thin blood smears, because clinical performance of the test has not been adequately established for PO, PM, and PK. In some countries, increasing levels of histidine-rich protein 2 and 3 (hrp2/3) gene deletions causing false-negative RDT tests threaten the ability of health providers to diagnose and appropriately treat people infected with PF malaria. Although the prevalence of hrp2/3 gene deletions in most countries with high malaria transmission remains low, the WHO is monitoring this situation closely.

It is important to note that testing for malarial antibodies does not play a role in the clinical management of acute malaria.

## CLINICAL MANAGEMENT AND DRUG TREATMENT

Patients with malaria can be divided into two categories: those who have uncomplicated malaria and others with severe malaria. Patients with uncomplicated malaria have low parasitemia, are fully conscious, and do not have preexisting comorbidities that might complicate the course of illness or its treatment. Such patients can often be treated with oral drug regimens and do not require hospitalization (if the patient is judged to be reliable, home conditions are supportive, and outpatient follow-up is feasible) (Tables 63.2 and 63.3). The use of the same or related drugs that have been taken for prophylaxis is not recommended to treat malaria. For example, lumefantrine-artemether may be used as a treatment medication by travelers who were taking atovaquone-proguanil for prophylaxis. After initial treatment of infections caused by PV or PO to eradicate the blood-borne parasites, a course of PQ phosphate is administered to eradicate the latent hypnozoite forms that may be present in the liver and to prevent relapses.

Severe malaria is most often caused by PF infection, when a parasitemia of greater than 5% develops or when a person with a blood smear–confirmed malarial infection (or a history of recent possible exposure and no other recognized disease cause) develops one or more of the following clinical problems: impaired consciousness or coma, severe normocytic anemia, renal failure, pulmonary edema, acute respiratory distress syndrome, circulatory shock, disseminated intravascular coagulation, spontaneous bleeding, acidosis, hemoglobinuria, jaundice, and/or repeated generalized convulsions.

Malaria in pregnancy can cause severe maternal and fetal complications. Infection with PF can result in miscarriage, stillbirth, premature delivery, and fetal growth restriction. Chloroquine and mefloquine are considered safe to use during pregnancy, but scant data exists for the newer drugs. To date, no studies exist that have convincingly reported increased adverse events to the mother or infant following the use of atovaquone-proguanil prophylaxis for malaria. Most studies have lacked sufficient statistical power to detect associations between AP exposure and adverse fetal outcomes.

ACTs are composed of two components: an artemisinin derivative and a partner drug. The artemisinin derivative is rapidly effective, killing a large proportion of parasites; but it is rapidly excreted, leaving residual parasites to be killed by a more slowly eliminated partner drug. ACTs are well tolerated and also reduce transmission due to their gametocidal ability.

The ACTs were introduced in the mid-1990s, when there was an imminent prospect of untreatable malaria in Southeast Asia, where resistance to all available antimalarial drugs had developed. In 2005, the WHO recommended that ACTs be used as first-line treatments for uncomplicated PF malaria in all countries where malaria was endemic.

Artemisinin resistance in Southeast Asia is now well characterized, recognized clinically by delays in the clearance of parasites after treatment with artemisinin-based regimens, molecularly by mutations in the propeller domain of the K13 gene (*K13PD*), and parasitologically by decreased clearance in the ring survival assay. Resistance has since been documented in Brazil, eastern India, isolated parts of Africa and China; it is likely to spread to other areas over time. The delayed clearance phenotype is associated with frequent treatment failures when resistance is also seen in its partner drug as seen in dihydroartemisinin-piperaquine (DHA-PP) treatment failures in Western Cambodia. Several strategies to overcome this rapidly evolving problem include triple drug ACTs, newer drugs, and longer administration of current drugs. Two promising clinical trials that are examining the efficacy of triple ACTs in Southeast Asia are ongoing.

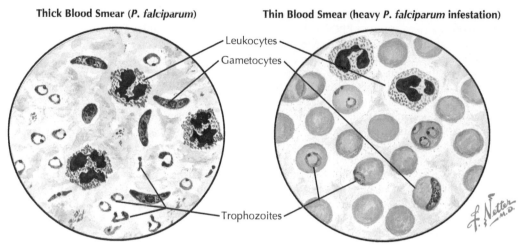

**Thick Blood Smear (*P. falciparum*)**    **Thin Blood Smear (heavy *P. falciparum* infestation)**

Leukocytes

Gametocytes

Trophozoites

**Fig. 63.4** Diagnosis of *Plasmodium falciparum* malaria by thick and thin blood smears stained with Giemsa.

**TABLE 63.2    Drugs for the Treatment of Uncomplicated Malaria in Travelers Returning to Nonendemic Countries: Malaria Caused by Chloroquine-Sensitive *Plasmodium falciparum*, *Plasmodium vivax*, *Plasmodium ovale*, *Plasmodium malariae*, or *Plasmodium knowlesi***

| Drug-Drug Combination | Adult Treatment Regimen | Pediatric Treatment Regimen[a] | Main Contraindications and Precautions[b] |
|---|---|---|---|
| Chloroquine phosphate (Aralen and generics) | First dose: 600 mg base (1000 mg salt) PO, followed by 300 mg base (500 mg salt) PO at 6, 24, and 48 h<br>Total dose: 1500 base (2500 mg salt) | First dose: 10 mg base/kg PO, followed by 5 mg base/kg PO at 6, 24, and 48 h<br>Total dose: 25 mg base/kg | Hypersensitivity to chloroquine; history of epilepsy; psoriasis |
| Hydroxychloroquine (Plaquenil and generics) | First dose: 620 mg base (800 mg salt) PO, followed by 310 mg base (400 mg salt) PO at 6, 24, and 48 h<br>Total dose: 1550 mg base (2000 mg salt) | First dose: 10 mg base/kg PO, followed by 5 mg base/kg PO at 6, 24, and 48 h<br>Total dose: 25 mg base/kg | Hypersensitivity to chloroquine; history of epilepsy; psoriasis |
| Primaquine phosphate (used to decrease the risk of relapses of *P. vivax* and *P. ovale* infections) | 30 mg base PO qd for 14 days. Give the first dose of primaquine phosphate with the last dose of chloroquine phosphate or hydroxychloroquine used for treatment of *P. vivax* or *P. ovale* | 0.5 mg base/kg PO qd for 14 days: Give the first dose of primaquine phosphate with the last dose of chloroquine phosphate or hydroxychloroquine used for treatment of *P. vivax* or *P. ovale* | Quantitative G6PD screening must be done before starting primaquine (value has to be more than 70%), because primaquine can cause hemolytic anemia in G6PD-deficient persons |
| Tafenoquine (similar use as Primaquine) | 300 mg once with the last dose of chloroquine phosphate or hydroxy-chloroquine used for treatment of *P. vivax* or *P. ovale* | Not for children under the age of 16 years | Quantitative G6PD screening must be done before starting tafenoquine (value has to be more than 70%), because tafenoquine can cause hemolytic anemia in G6PD-deficient persons |

*G6PD*, Glucose-6-phosphate dehydrogenase.
[a]Pediatric doses should never exceed the adult dose.
[b]See package inserts for full information on contraindications and precautions.
Data from World Health Organization (WHO): *International travel and health 2010.* Available at: http://www.who.int/ith/en/index.html and from Centers for Disease Control and Prevention (CDC): *Health information for international travel 2020.* Available at: https://www.cdc.gov/malaria/resources/pdf/Malaria_Treatment_Table_120419.

The treatment of severe complicated *P. falciparum* malaria is beyond the scope of this chapter. Healthcare providers may consult CDC malaria experts through the CDC Malaria Hotline (1-770-488-7788, 8 AM–4:30 PM, EST; after hours, call 1-770-488-7100 and ask to speak to a CDC Malaria Branch clinician). Intravenous quinidine gluconate, an effective treatment for severe malaria, is no longer available in the United States. However intravenous artesunate from Amivas (Frederick, MD), another highly effective treatment for severe malaria, was approved by the United States FDA (Food and Drug Administration) in 2020. During the interim period while widespread availability through hospital supply chains is being established, healthcare providers should contact the CDC for emergency access to intravenous artesunate from CDC quarantine stations. Intravenous artesunate clears malarial parasitemia more rapidly than quinine/quinidine, reducing fatalities by about 20%, and it is the treatment of choice for severe malaria. It is safe, well tolerated, and can be administered to infants and children as well as to pregnant women in their second and third trimesters; it can also be given to lactating women. In the first trimester of pregnancy, the benefits of intravenous artesunate treatment outweigh the risk of death and poor outcomes due to severe malaria.

**TABLE 63.3  Drugs for the Treatment of Uncomplicated Chloroquine-Resistant *Plasmodium falciparum* Malaria in Travelers Returning to Nonendemic Countries**

| Drug-Drug Combination | Adult Treatment Regimen | Pediatric Treatment Regimen[a] | Main Contraindications and Precautions[b] |
|---|---|---|---|
| Atovaquone-proguanil combination tablet (Malarone) | Adult tablet: 250 mg atovaquone and 100 mg proguanil<br>Adult dose: 4 adult tablets as a single oral dose daily for 3 consecutive days<br>*Note:* Must be taken with fatty foods to enhance absorption. | Pediatric tablet: 62.5 mg atovaquone and 25 mg proguanil (¼ the adult tablet)<br>One dose daily (based on weight) for 3 consecutive days:<br>5–8 kg: 2 pediatric tablets daily<br>9–10 kg: 3 pediatric tablets daily<br>11–20 kg: 1 adult tablet daily<br>21–30 kg: 2 adult tablets daily<br>31–40 kg: 3 adult tablets daily<br>>40 kg: 4 adult tablets daily | Hypersensitivity to atovaquone and/or proguanil; severe renal insufficiency (creatinine clearance <30 mL/min)<br>Plasma concentrations of atovaquone are reduced when the drug is coadministered with rifampicin, rifabutin, metoclopramide, or tetracycline |
| Artemether-lumefantrine combination tablet (Coartem) | One tablet: 20 mg artemether plus 120 mg lumefantrine<br>3-day course of six doses total taken at 0, 8, 24, 36, 48, and 60 h<br>Adult dose ≥35 kg: four tablets per dose | One tablet: 20 mg artemether plus 120 mg lumefantrine<br>3-day course of six doses total taken at 0, 8, 24, 36, 48, and 60 h<br>Pediatric weight-based dose:<br>5 to <15 kg: 1 tablet per dose<br>15 to <25 kg: 2 tablets per dose<br>25 to <35 kg: 3 tablets per dose<br>*Note:* Coartem Dispersible is a new formulation that dissolves into a sweetened suspension for easy administration to children. | Hypersensitivity to artemether and/or lumefantrine |
| Quinine sulfate *plus* doxycycline *or* tetracycline *or* clindamycin | Quinine sulfate: 542 mg base (650 mg salt)<br>8 mg base/kg tid for 3 days (or 7 days for infections acquired in Southeast Asia)<br><br>Doxycycline: 100-mg tablet<br>Adult >50 kg: 1 tablet PO bid for 7 days<br>Tetracycline: 250-mg tablet<br>Adult: 250-mg tablet PO qid for 7 days<br><br>Clindamycin: 300-mg base tablet<br>Adult >60 kg: 20-mg base/kg/day divided into 3 doses/day for 7 days | Quinine sulfate: 8.3-mg base/kg (10 mg salt/kg) PO tid for 3 days (or 7 days for infections acquired in Southeast Asia)<br><br>Children 8 years of age and older: doxycycline 2.2 mg/kg PO bid for 7 days<br>Children 8 years of age and older: tetracycline 25 mg/kg/day divided into 4 doses/day for 7 days<br><br>Children <60 kg: clindamycin 20-mg base/kg/day by mouth divided into 3 doses/day for 7 days | Hypersensitivity to quinine or quinidine; caution in persons with G6PD deficiency, cardiac dysrhythmias, and conduction abnormalities<br>Hypersensitivity to tetracyclines; liver dysfunction<br>Hypersensitivity to tetracyclines; liver dysfunction<br><br>Hypersensitivity to clindamycin or lincomycin; history of colitis; severe kidney or liver impairment |
| Mefloquine (Lariam and generics) | One tablet: 228 mg base (250 mg salt)<br>Adult: 684 mg base (750 mg salt) PO as the initial dose followed by 456 mg base (500 mg salt) PO given 6–12 hours after the initial dose<br>Total dose: Five (250 mg salt) tablets | One tablet: 228-mg base (250 mg salt)<br>Pediatric weight-based dose:<br>13.7 mg base/per kilogram (15 mg salt/kg) PO as the initial dose followed by 9.1 mg base/kg (10 mg salt/kg) PO given 6–12 h after the initial dose<br>Total dose = 25 mg salt/kg | Hypersensitivity to mefloquine; history of neuropsychiatric disease; treatment with mefloquine in previous 4 weeks; concomitant halofantrine treatment<br>Do not give mefloquine within 12 h of last dose of quinine treatment. Caution: Many drug-drug interactions |
| Dihydroartemisinin-piperaquine (Eurartesim) | One dose daily for 3 consecutive days<br>Adults >50 kg: 3 tablets daily for 3 days<br>Target dose: 4 mg of dihydroartemisinin per kilogram per day and piperaquine 18 mg/kg qd<br>*Note:* Undergoing regulatory review by the European Medicines Agency.<br>*Note:* A similar drug combination, Artekin, is being studied in clinical investigational trials in Asia. | *Note:* Clinical trials to establish safety and efficacy in children are ongoing. | Hypersensitivity to dihydroartemisinin and/or piperaquine |

*G6PD,* Glucose-6-phosphate dehydrogenase.
[a]Pediatric doses should never exceed the adult dose.
[b]See package inserts for full information on contraindications and precautions.
Data from World Health Organization (WHO): *International travel and health 2020.* Available at: http://www.who.int/ith/en/index.html; and from Centers for Disease Control and Prevention (CDC): *Health information for international travel 2020.* Available at: https://www.cdc.gov/malaria/resources/pdf/Malaria_Treatment_Table_120419.pdf.

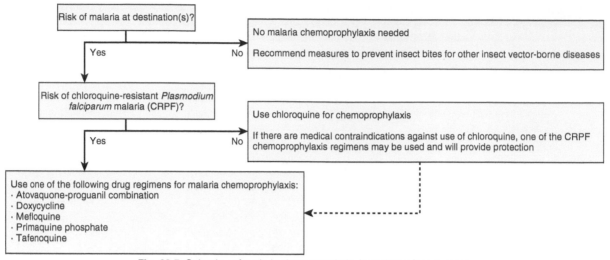

**Fig. 63.5** Selection of malaria chemoprophylaxis regimen for travelers.

Current criteria for CDC approval for intravenous artesunate include a positive blood smear AND severe malaria as defined by one or more of the following: hyperparasitemia (>5%), impaired consciousness, seizures, shock, acute respiratory distress syndrome, acidosis, acute renal failure, abnormal bleeding or disseminated intravascular coagulation, jaundice, or severe anemia (Hb <7 g/dL) or inability to take oral medications regardless of severity of malaria.

Expert consultation should also be obtained for cases of malaria in patients with underlying health conditions, including pregnancy, splenectomy, infection with human immunodeficiency virus (HIV), and immunocompromising conditions or treatments. There is no evidence that exchange transfusion (ET), in addition to antimalarial drugs, reduces mortality in patients with severe malaria. This practice has largely been abandoned.

## PREVENTION AND CONTROL

Measures to prevent and control malaria fall into two categories: environmental interventions and human interventions. Environmental interventions include draining swampy areas and collections of standing fresh water where mosquitoes breed; introducing fish into freshwater ponds to eat mosquito larvae; applying larvicidal chemicals into ponds, lakes, and slow-moving streams; and spraying the indoors of human dwellings with residual insecticides such as dichlorodiphenyl-trichloroethane (DDT). When possible, screens should be installed on window and door openings of dwellings.

Human interventions include preventing exposure to mosquito bites, taking antimalarial drugs to prevent disease (chemoprophylaxis), and using effective antimalarial drugs to treat human infections, thus interrupting the life-cycle stage whereby the mosquitoes become infected while feeding on humans with circulating gametocytes. Exposure to mosquitoes can be lessened by not going outdoors between dusk and dawn and sleeping under a bed net. When it is necessary to go outdoors, applying an insect repellent, such as N,N-diethyl-meta-toluamide (DEET), to all exposed skin areas and wearing external clothing that has been treated with a knock-down insecticide, such as permethrin, can greatly decrease exposure to mosquito bites. In several published reports, field studies have shown that sleeping every night under insecticide-treated nets (ITNs) is a very effective measure to decrease transmission of malaria in highly endemic areas in Africa.

Recommendations for drugs to be taken for malaria chemoprophylaxis are based on the type of malaria that is likely to be transmitted at a given destination. The recommendations are usually based on whether exposure to CRPF malaria is anticipated. In scenarios where both CRPF and chloroquine-sensitive malaria (PV, PM, PO, PK) are present, chemoprophylaxis regimens used against CRPF will cover the chloroquine-sensitive species as well. Fig. 63.5 presents an algorithm that may be used to select appropriate malarial chemoprophylaxis. Table 63.4 lists the drugs and regimens that are recommended by the CDC for malaria chemoprophylaxis in adults.

Chemoprophylaxis is highly effective against malaria when taken as prescribed, beginning before travel as well as during travel and used for an interval after travel in a malaria-endemic area. Table 63.5 illustrates that the dosage schedule and total duration of taking malaria chemoprophylaxis varies depending on the length of the trip and the regimen selected. The most common mistake among travelers is skipping or forgetting doses. Therefore, after determining what antimalarial drugs are appropriate for the malaria risk at a given destination, considering daily versus weekly administration schedules may contribute to the final selection. TQ is an aminoquinoline drug related to PQ that has been FDA approved for chemoprophylaxis of malaria in adults and for radical cure of PV in persons greater than 16 years old. However, TQ like PQ only should be used in individuals shown not to have a deficiency of glucose-6-phosphate dehydrogenase (G6PD). Quantitative and not just qualitative G6PD testing must be checked prior to TQ or PQ administration because these drugs can cause severe hemolytic anemia in persons who are G6PD deficient (value <70%). TQ's long half-life of approximately 2 weeks allows for fewer doses and a shorter duration of use in chemoprophylaxis regimens. Used in treatment regimens, one dose of TQ effectively kills the dormant liver stage of PV responsible for relapses of malaria (see further on).

Although some chemoprophylaxis regimens have been taken with apparent safety for periods of up to 6 years (e.g., weekly chloroquine phosphate), safety of drug regimens currently used to protect against CRPF has been studied only for periods of up to 1 to 2 years of continuous use. For long-stay visitors and others making trips to CRPF areas where there is no access to medical care within 24 hours after onset of a febrile illness, standby emergency therapy (SBET) for malaria may be considered for certain knowledgeable and capable travelers. For SBET, travelers are instructed on how to decrease the risk of exposure to mosquito bites and how to recognize the clinical signs and symptoms of malaria; they are prescribed a treatment course of one of the oral malarial treatment regimens to be used as an emergency measure while seeking professional medical assistance for a presumed clinical attack of malaria.

## TABLE 63.4  Drugs Used for Malaria Chemoprophylaxis (Adult)[a]

| Species | Malaria Chemoprophylaxis Drugs (Adult Doses) |
|---|---|
| Plasmodium falciparum, chloroquine resistant | Atovaquone-proguanil (Malarone), atovaquone 250 mg plus proguanil 100 mg per tablet: 1 dose daily<br>Doxycycline, 100-mg tablet: 1 dose daily<br>Mefloquine (Lariam), 250-mg tablet: 1 dose weekly<br>Primaquine phosphate,[b] 30-mg base tablet: one dose daily<br>Tafenoquine 200 mg daily for 3 days pretravel, weekly during travel (start 7 days after last pretravel dose) and once after travel (7 days after last travel dose) (Maximal duration of 6 months) |
| Plasmodium vivax | Chloroquine phosphate (Aralen and generics), 500-mg tablet (300-mg chloroquine base): 1 dose weekly<br>Hydroxychloroquine (Plaquenil and generics), 400 mg salt (310-mg hydroxychloroquine base): 1 dose daily<br>Tafenoquine: same as P. falciparum |
| Plasmodium malariae | Same as P. vivax |
| Plasmodium ovale | Same as P. vivax |
| Plasmodium knowlesi | Same as P. vivax except no known data for tafenoquine |
| P. vivax, Chesson strain (chloroquine-resistant P. vivax) | Primaquine phosphate[b] |

[a]Consult www.cdc.gov/travel for recommended pediatric doses and regimens and for contraindications and precautions for use in pregnant and breastfeeding women.
[b]Glucose-6-phosphate dehydrogenase (G6PD) screening must be done before starting primaquine, because primaquine can cause hemolytic anemia in G6PD-deficient persons.

## TABLE 63.5  Duration of Therapy for Malaria Chemoprophylaxis Regimens

| Drug | Chloroquine-Sensitive Malaria | Chloroquine-Resistant Malaria | Pretravel Administration: Time Before First Potential Exposure to Begin Medication | Posttravel Administration: Time After Last Known Exposure to Continue Medication |
|---|---|---|---|---|
| Atovaquone-proguanil (daily dose) | Yes | Yes | 1–2 days | 7 days |
| Tafenoquine (daily for 3 days pre exposure) then once a week till one week after exposure | Yes | Yes | 3 days | 1 week |
| Primaquine (daily dose) | Yes | Yes | 1–2 days | 7 days |
| Doxycycline (daily dose) | Yes | Yes | 1–2 days | 4 weeks |
| Chloroquine (weekly dose) | Yes | No | 1 week | 4 weeks |
| Mefloquine (weekly dose) | Yes | Yes | 1–3 weeks | 4 weeks |

Either *atovaquone-proguanil* (Malarone) or artemether-lumefantrine (Coartem) can be used for SBET. Artemether-lumefantrine is thought to be safe in the first trimester of pregnancy but has been shown to have higher treatment failure rates in the nonimmune traveler. Travelers should stop taking the drug being used for antimalarial chemoprophylaxis while taking the SBET drug. If atovaquone-proguanil was being used as chemoprophylaxis, artemether-lumefantrine should be used for SBET; then the atovaquone-proguanil chemoprophylaxis regimen should be resumed immediately upon completion of SBET. If another antimalarial is being used for chemoprophylaxis, atovaquone-proguanil may be used for SBET, then the drug being used for malaria chemoprophylaxis can be restarted 1 week after *initiating* SBET.

Prolonged chemoprophylaxis is neither feasible nor recommended for the most vulnerable victims of malaria—children living in sub-Saharan Africa. They must rely on environmental interventions and personal protection measures to prevent mosquito bites, treatment of individual infections, and periodic mass treatment programs until a safe and efficacious malaria vaccine is developed and distributed. RTS,S recombinant protein-based malaria vaccine (Mosquirix, GlaxoSmithKline, UK) is approved by European regulators and is the first and only malaria vaccine to show a protective effect against *P. falciparum* malaria among young children in a phase 3 trial.

RTS,S is the scientific name given to this malaria vaccine candidate and represents its composition. The "R" stands for the central repeat region of *Plasmodium (P.) falciparum* circumsporozoite protein (CSP); the "T" for the T-cell epitopes of the CSP; and the "S" for hepatitis B surface antigen (HBsAg). These are combined in a single fusion protein ("RTS") and co-expressed in yeast cells with free HBsAg. The "RTS" fusion protein and free "S" protein spontaneously assemble in "RTS,S" particles. RTS,S also contains the AS01 adjuvant system and in scientific papers is usually referred to as "RTS,S/AS01." Among children aged 5 to 17 months who received four doses of the vaccine, it prevented approximately 4 in 10 (39%) cases of malaria over 4 years of follow-up and about 3 in 10 (29%) cases of severe malaria, with significant reductions also seen in overall hospital admissions as well as in admissions due to malaria or severe anemia. The vaccine also reduced the need for blood transfusions required to treat life-threatening malarial anemia. Beginning in 2019, it was the first malarial vaccine provided to young children through routine immunization programs in three sub-Saharan African countries as part of a pilot implementation program recommended by the WHO.

## PREVENTION OF RELAPSE

Frequency of relapse and the latent period before relapse of *PV* malaria vary between geographic zones. The zone comprising Southeast Asia, Papua New Guinea, and Melanesia has the highest predicted incidence of relapse and the fastest mean time to relapse (approximately 45 days). The zone comprising northern Asia and Europe has the lowest incidence of relapse, which may occur several years after exposure. A relapse can occur even if the person harboring hypnozoites of *PV* or *PO* had no symptoms of malaria during or after initial infection because parasites in the bloodstream would have been completely suppressed by the chemoprophylactic drug.

TQ and PQ can both be given to prevent relapsing malaria after the traveler has left the endemic area in persons who have not taken these two drugs for primary chemoprophylaxis. Presumptive antirelapse therapy (PART) is generally indicated for persons who have had more than 6 months exposure in malarious areas where *PV* or *PO* transmission is high or where exposure is intense (e.g., Indonesia, Papua New Guinea, Timor-Leste) regardless of duration. TQ has the advantage of only requiring only a single dose for radical cure.

According to the American Association of Blood Banks' guidelines, persons who have taken malaria chemoprophylaxis or who have had treatment for diagnosed malaria are deferred from donating blood for 3 years after taking antimalarial drugs. Under these guidelines, solid organ donation is similarly deferred. However, if there is a compelling need for donation of blood or organs from someone who has recently lived or traveled in a malaria-endemic area and has taken malaria chemoprophylaxis, the risk of malaria transmission can be addressed by consulting an infectious diseases or tropical medicine specialist to consider chemotherapy for eradication of possible relapsing or latent malaria. As mentioned previously in the treatment section, the CDC Malaria Branch has telephone consultation with CDC malaria experts available to healthcare providers to discuss difficult or complex cases.

## EVIDENCE

Bousema T, Okell L, Shekalaghe S, et al. Revisiting the circulation time of Plasmodium falciparum gametocytes: molecular detection methods to estimate the duration of gametocyte carriage and the effect of gametocytocidal drugs. *Malar J* 2010;9:136. *New data on the gametocyte carriage in treated malaria patients, an essential consideration for developing malaria control and elimination programs.*

de Laval F, Oliver M, Rapp C, et al. The challenge of diagnosing Plasmodium ovale in travelers: report of clustered cases in French soldiers returning from West Africa. *Malar J* 2010;9:358. *Review of 62 cases of P. ovale with analysis of clinical symptoms, laboratory findings, diagnosis, and response to treatment.*

Efficacy and safety of RTS,S/AS01 malaria vaccine with or without a booster dose in infants and children in Africa: final results of a phase 3, individually randomised, controlled trial. *Lancet* 2015;386(9988):31-45.

Grigg MJ, William T, Barber BE, et al. Age-related clinical spectrum of plasmodium knowlesi malaria and predictors of severity. *Clin Infect Dis*

2018;67(3):350-9. *Comparison of 481 P. knowlesi, 172 P. vivax, and 96 P. falciparum cases in Sabah over a 3.5-year period and the predictors of severity.*

Hassett MR, Roepe PD. Origin and spread of evolving artemisinin-resistant Plasmodium falciparum malarial parasites in Southeast Asia. *Am J Trop Med Hyg* 2019;101(6):1204-1211.

McGready R, White NJ, Nosten F. Parasitological efficacy of antimalarials in the treatment and prevention of falciparum malaria in pregnancy 1998-2009: a systematic review. *BJOG* 2011;118:123-135. *Review of therapeutic efficacy of antimalarials used for treatment and intermittent preventive treatment in pregnancy. Artemisinin combination therapy (ACT) provided lower parasitological failure and gametocyte carriage rates, but many of the other treatments used were associated with lower cure rates.*

Sinclair D, Zani B, Donegan S, et al. Artemisinin-based combination therapy for treating uncomplicated malaria. *Cochrane Database Syst Rev* 2009;3:CD007483. *Evidence-based review of the relative benefits and harms of the available treatment options using artemisinin-based combination therapy.*

World Health Organization (WHO). WHO Special Program for Research and Training in Tropical Diseases. Malaria rapid diagnostic test performance: results of WHO product testing of malaria RDTs—round 6 (2014-2015). Geneva: WHO; 2015. Available at: https://www.who.int/malaria/publications/atoz/9789241510035/en/. *Results of the sixth round of product testing of malaria antigen-detecting RDTs completed in 2015. Results of round 6, and details on product performance and on the interpretation of results are given.*

## ADDITIONAL RESOURCES

Centers for Disease Control and Prevention (CDC). *Health Information for International Travel 2020.* Atlanta: U.S. Department of Health and Human Services, Public Health Service; 2019. Available at: https://wwwnc.cdc.gov/travel/page/yellowbook-home. *The chapter on malaria in this reference is essential reading for healthcare providers with patients going to destinations where malaria is a risk. This publication, known as the "Yellow Book," is the authoritative source of US government recommendations for malaria prophylaxis.*

Centers for Disease Control and Prevention (CDC). Malaria diagnosis (U.S.)—Rapid diagnostic test. Available at: http://www.cdc.gov/malaria/diagnosis_treatment/rdt.html. *Information and instructions on the use of the RDT approved in the United States.*

World Health Organization. *Compendium of WHO Malaria Guidance: Prevention, Diagnosis, Treatment, Surveillance and Elimination,* 2019. Available at: https://apps.who.int/iris/bitstream/handle/10665/312082/WHO-CDS-GMP-2019.03-eng.pdf?ua=1. *This publication provides, for the first time, a complete list of all formal WHO policy recommendations on malaria in a single resource. The document also serves as a catalogue of all WHO publications on malaria prevention, diagnosis, treatment, surveillance, and elimination.*

World Health Organization (WHO). Guidelines for the treatment of malaria. 3rd ed. Geneva: WHO; 2015. Available at: www.who.int/malaria/publications/atoz/9789241549127/en/. *Guidelines for malaria case management and treatment recommendations based on updated evidence. Includes discussion of some treatment options that are not yet prequalified by WHO or registered by any stringent medical regulatory authority.*

World Health Organization (WHO). *International Travel and Health 2019.* Geneva: WHO; 2019. Available at: http://www.who.int/ith/en. *The chapter on malaria in this reference presents recommendations for malaria prophylaxis that differ in some cases from those in the CDC's Health Information for International Travel 2020. Drugs not licensed in the United States are discussed, as well as expanded indications for SBET for presumed malaria in travelers.*

# 64

# Yellow Fever

*Mark D. Gershman, J. Erin Staples*

## ABSTRACT

Yellow fever (YF) is a vector-borne disease resulting from the transmission of the YF virus to a human from the bite of an infected mosquito. It is endemic to sub-Saharan Africa and tropical South America. Infection in humans can cause mild, undifferentiated febrile illness to severe disease with jaundice and hemorrhagic manifestations. All travelers to YF-endemic countries should be advised of the risks of acquiring YF disease and available preventive techniques, including personal protective measures and vaccination.

## CLINICAL VIGNETTE

A 56-year-old previously healthy American male presented to the hospital emergency department with a 3-day history of fever, headache, myalgia, and abdominal pain after returning from Brazil. Symptoms began 4 days after he completed a 5-day hiking trip in the Amazon rainforest. He had not obtained any travel-related information or vaccinations before departure.

On assessment, he had a fever of 39.2°C and appeared mildly dehydrated. He was treated with intravenous fluids; malaria smears were negative. He was prescribed oral doxycycline for possible rickettsial disease and discharged home. He returned 3 days later, complaining of persistent fever, headache, vomiting, and watery diarrhea; he appeared jaundiced. Blood tests revealed a white blood cell (WBC) count of 2500/mm³, platelets of 42,000/mm³, INR of 2.8, creatinine of 4.9 mg/dL, alanine aminotransferase (ALT) of 4672 U/L, aspartate aminotransferase (AST) of 5826 U/L, and total bilirubin of 3.7 mg/dL. He was admitted to the intensive care unit and started on broad-spectrum antibiotics. The following day he developed shock, excessive bleeding at venipuncture sites, epistaxis, and melena. Tests of blood and cerebrospinal fluid for selected viral, bacterial, and parasitic agents were performed. Three days after admission the patient developed ventricular fibrillation and died despite attempts at resuscitation.

A serum specimen collected 7 days after illness onset was positive for YF virus ribonucleic acid (RNA) by reverse transcription–polymerase chain reaction (RT-PCR). Postmortem histopathologic examination demonstrated fulminant hepatitis with diffuse hepatocellular necrosis, marked microvesicular steatosis, and apoptosis. Wild-type YF virus RNA was detected by RT-PCR, and YF virus antigen was demonstrated by immunohistochemistry in hepatocytes and Kupffer cells.

## ETIOLOGY AND TRANSMISSION

YF virus is an RNA virus belonging to the genus *Flavivirus*. It is transmitted to humans primarily through the bite of an infected mosquito, primarily *Aedes* or *Haemagogus* species.

There are three transmission cycles for YF virus: sylvatic (jungle), intermediate (savannah), and urban (Fig. 64.1). The sylvatic cycle involves transmission of YF virus between nonhuman primates and tree hole–breeding mosquito species in the forest canopy. The virus is transmitted via mosquitoes from monkeys to humans when humans enter the jungle during occupational or recreational activities. In Africa, there is also an intermediate cycle that involves transmission of YF virus by tree hole–breeding *Aedes* species to humans living or working in jungle border areas. In this cycle, mosquitoes transmit YF virus from monkey to human or from human to human. The urban transmission cycle involves anthroponotic transmission of the virus between humans and urban mosquitoes, primarily *Ae. aegypti*.

Humans infected with YF virus have high levels of viremia and are infectious to mosquitoes shortly before fever onset and for 3 to 5 days thereafter. This high level of viremia theoretically could cause blood-borne transmission through transfusions or needlestick injuries. One case of perinatal transmission of wild-type YF virus has been documented from a woman who developed her initial symptoms of YF 3 days prior to delivery.

## GEOGRAPHIC DISTRIBUTION

YF occurs in sub-Saharan Africa and tropical South America, where it is endemic and intermittently epidemic (Fig. 64.2). The World Health Organization (WHO) estimates that 200,000 cases of YF occur annually, with 30,000 deaths. However, only a small percentage of cases are identified because of lack of clinical recognition and underreporting. Unvaccinated travelers to YF-endemic countries are at risk of acquiring YF.

## RISK FACTORS

A traveler's risk of acquiring YF is determined by various factors, including immunization status, use of personal protection measures against mosquito bites, local virus transmission rates, and travel itinerary, such as destination, duration, and occupational and recreational activities. In West Africa, YF virus transmission is seasonal, with an increased risk from July to October, the end of the rainy season through the beginning of the dry season. However, YF virus may be transmitted by *Ae. aegypti* during the rest of the year (dry season) in both rural and densely settled urban areas. In South America, the risk of infection is highest from January to May, the rainy season. Reported cases of human disease are the principal, but imperfect, indicator of disease risk. Case reports may be absent because of a low level of virus transmission but also because of a high level of immunity in the population (e.g., because of vaccination) and insufficient local surveillance to detect cases. Because "epidemiologic silence" does not mean absence of risk, travelers should not enter endemic areas without taking protective measures.

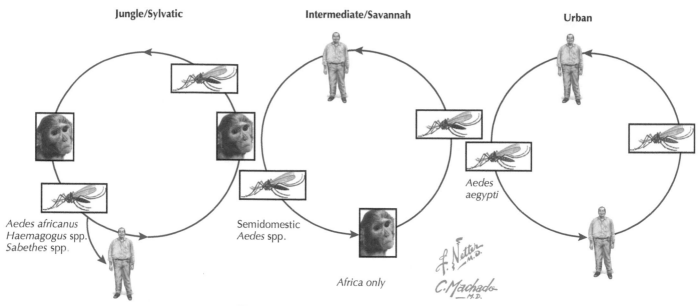

**Fig. 64.1** Yellow fever virus transmission cycles.

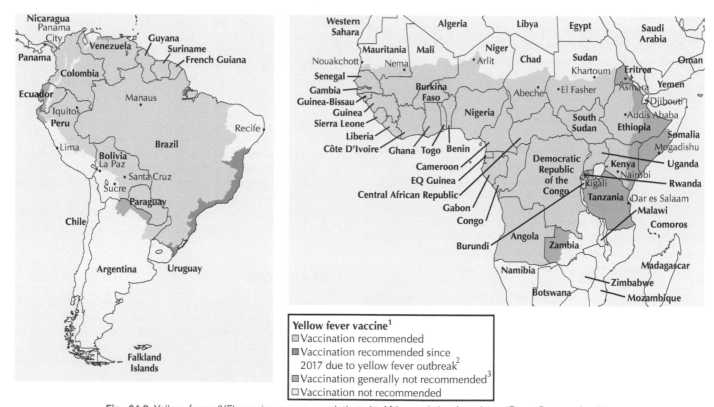

**Yellow fever vaccine[1]**
☐ Vaccination recommended
■ Vaccination recommended since 2017 due to yellow fever outbreak[2]
☐ Vaccination generally not recommended[3]
☐ Vaccination not recommended

**Fig. 64.2** Yellow fever (YF) vaccine recommendations in Africa and the Americas. (From Centers for Disease Control and Prevention. *CDC Yellow Book 2020: Health Information for International Travel.* New York: Oxford University Press; 2017.) [1]This map is an updated version of the 2010 map created by the Informal WHO Working Group on the Geographic Risk of Yellow Fever. [2]In 2017, CDC expanded YF vaccination recommendations for travelers to Brazil because of a large outbreak of YF in multiple states in that country. For the most current map, please consult www.cdc.gov/travel. [3]YF vaccination is generally not recommended in areas where there is low potential for YF virus exposure. However, vaccination might be considered for a small subset of travelers to these areas who are at increased risk for exposure to YF virus because of prolonged travel, heavy exposure to mosquitoes, or inability to avoid mosquito bites. Consideration for vaccination of any traveler must take into account the traveler's risk of being infected with YF virus, country entry requirements, and individual risk factors for serious vaccine-associated adverse events (e.g., age, immune status).

## CLINICAL FEATURES

Most persons infected with YF virus experience asymptomatic or clinically inapparent infection. Symptomatic infection can range from a mild, undifferentiated febrile illness to severe hemorrhagic disease with jaundice, resulting in death (Fig. 64.3). The incubation period is usually 3 to 6 days.

In its mildest symptomatic form, YF is a self-limited illness characterized by sudden onset of fever and headache. Some patients experience abrupt onset of high fever (up to 40°C), chills, severe headache, myalgia, lumbosacral pain, anorexia, nausea, vomiting, and dizziness. These patients appear acutely ill, and examination may demonstrate relative bradycardia in association with the elevated body temperature (Faget's sign). The patient is usually viremic during this period, which lasts for approximately 3 days before the symptoms abate (the "period of remission"). Many patients have an uneventful recovery at this stage.

Approximately 12% of infected persons experience a reappearance of the illness in more severe form within 48 hours of remission.

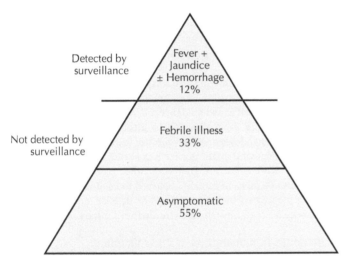

**Fig. 64.3** Clinical presentation of yellow fever. (Data from Johansson et al. *Trans R Soc Trop Med Hyg* 2014;108:482-487.)

Clinical findings consist of fever, nausea, vomiting, epigastric pain, jaundice, renal insufficiency, and cardiovascular instability. Viremia might be absent during this phase. Coagulopathy may occur, causing hematemesis, melena, petechiae, epistaxis, and bleeding from gingiva and needle-puncture sites (Fig. 64.4). Physical findings include scleral and dermal icterus, hemorrhage, and epigastric tenderness without hepatomegaly.

Patients with YF have multiple laboratory abnormalities that vary based on the severity and stage of illness. Leukopenia may occur early, and leukocytosis later. Bleeding dyscrasias associated with decreased platelet count, elevated prothrombin and partial thromboplastin times, and elevated fibrin split products may occur. Hyperbilirubinemia usually peaks toward the end of the first week of illness but can occur as early as day 3. Serum transaminase levels may be elevated in severe disease and remain so for up to 2 months.

## DIAGNOSTIC APPROACH

Preliminary diagnosis is based on a patient's clinical features, YF vaccination status, and travel information. Mild YF cannot be distinguished clinically from a variety of other infections. The clinician must differentiate cases of YF with jaundice from viral hepatitis, malaria, leptospirosis, Crimean-Congo hemorrhagic fever, Rift Valley fever, typhoid, Q fever, and typhus, as well as surgical, drug-induced, and toxic causes of jaundice. Other viral hemorrhagic fevers, such as dengue hemorrhagic fever; Lassa fever; Marburg and Ebola virus diseases; and Bolivian, Argentinean, and Venezuelan hemorrhagic fevers usually manifest without jaundice.

Laboratory diagnosis involves using serologic assays to detect virus-specific immunoglobulin M (IgM) and G (IgG) antibodies or using RT-PCR testing to detect viral RNA. Because serologic cross-reactions occur with other flaviviruses, such as Zika, West Nile, and dengue viruses, providers should confirm positive test results with a more specific test, such as the plaque reduction neutralization test. During the first week of illness, YF viral RNA can often be detected in the serum. However, by the time more overt symptoms are recognized, viral RNA might be undetectable. Therefore providers should not use a negative RT-PCR result to rule out a diagnosis of YF. Immunohistochemical staining of formalin-fixed material may detect YF viral antigen in histopathologic specimens.

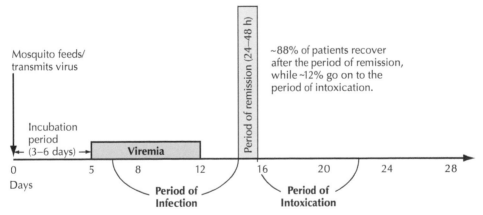

**Fig. 64.4** Timeline of yellow fever disease.

Healthcare providers should contact their state or local health department and the Centers for Disease Control and Prevention (CDC) (970-221-6400) for assistance with diagnostic testing for YF infections, including assessing antibody responses to wild-type virus or the vaccine.

## CLINICAL MANAGEMENT

To date, no drugs have shown specific benefit to treat YF disease. Management is supportive and based on symptoms. Rest, fluids, and acetaminophen (paracetamol) may relieve symptoms of fever and myalgias. Avoid prescribing aspirin or other nonsteroidal antiinflammatory drugs (e.g., ibuprofen or naproxen) because they may increase the risk of bleeding. Protect infected persons from further mosquito exposure (staying indoors and/or under a mosquito net) during the first few days of illness so they do not contribute to the transmission cycle.

## PROGNOSIS

Most persons with mild illness recover without long-term sequelae. For those with severe disease, the length of illness is variable, and the case fatality ratio is 30% to 60%. For those who survive, convalescence is often prolonged, lasting several weeks. Rarely, death can occur at the end of convalescence or even weeks after complete recovery from acute illness, possibly because of myocardial damage and cardiac arrhythmia. Secondary bacterial infections, such as pneumonia, can complicate recovery. Jaundice has been observed for up to 3 months after recovery.

## PREVENTION AND CONTROL

Advise all travelers to YF-endemic countries of the risk of disease and available methods to prevent it, including personal protective measures and vaccination.

### Personal Protective Measures

Personal protective measures against mosquito bites are the best way to prevent mosquito-borne diseases and, in the case of YF, are the only way to prevent the disease in persons who cannot be vaccinated. Advise travelers to use insect repellent containing compounds recommended by the Environmental Protection Agency on exposed skin and to follow the label directions. Additional protective measures include wearing long sleeves, long pants, hats, and socks; treating clothes with permethrin; and staying in well-screened or air-conditioned accommodations (Box 64.1). More detailed information can be obtained at https://wwwnc.cdc.gov/travel/yellowbook/2020/noninfectious-health-risks/mosquitoes-ticks-and-other-arthropods.

### Vaccine

YF can be prevented by a live-attenuated viral vaccine. YF-VAX (Sanofi-Pasteur, Swiftwater, Pennsylvania) is the only vaccine licensed for use in the United States. A single injection of 0.5 mL of reconstituted vaccine is administered subcutaneously. Because studies suggest that there is no difference in the efficacy of licensed vaccines produced outside the United States, persons who receive YF vaccines in other countries should be considered protected against YF.

In the United States, periodic shortages of YF-VAX have occurred, including one that started in late 2015. This shortage was addressed by the importation and distribution of Stamaril, another YF vaccine produced by Sanofi Pasteur in France, under an expanded-access investigational new drug protocol.

---

> **BOX 64.1** **Prevention Measures for Yellow Fever**
>
> - Undergo vaccination.
> - Use insect repellent on exposed skin.
>   - DEET (chemical name: *N,N*-diethyl-*m*-toluamide or *N,N*-diethly-3-methyl-benzamide)
>   - Picaridin
>   - Oil of lemon eucalyptus
>   - IR3535 (chemical name: 3-[*N*-butyl-*N*-acetyl]-aminopropionic acid, ethyl ester)
>   - 2-undecanone (chemical name: methyl nonyl ketone)
> - Wear long-sleeved shirts, long pants, hats, socks.
> - Treat clothes with permethrin.
> - Stay in well-screened or air-conditioned accommodations.

### Indication

The Advisory Committee on Immunization Practices (ACIP) recommends that the vaccine be administered to persons aged 9 months or older who are traveling to or living in areas with risk of YF virus transmission in South America and Africa. Furthermore, certain countries require proof of vaccination for entry (see International Health Regulations [IHR] section below for more detail). YF vaccine needs to be given 10 days before entering a country that requires proof of vaccination. Although revaccination was previously recommended at 10-year intervals, in 2015 ACIP issued guidance that a single dose of YF vaccine provides long-lasting protection and is adequate for most travelers. However, additional doses of YF vaccine are recommended for certain groups of travelers who might not have as strong or durable immune response to YF vaccine compared with other recipients. Additional doses may also be given to travelers believed to be at increased risk for YF disease because of their itineraries. For detailed information see: https://www.cdc.gov/mmwr/preview/mmwrhtml/mm6423a5.htm.

### Simultaneous Administration of Other Vaccines

Inactivated vaccines can be administered either simultaneously or at any time before or after YF vaccination. To minimize the potential risk of immune interference, give YF vaccine either at the same time as—or at least 30 days apart from—other injectable live viral vaccines. Oral Ty21a typhoid vaccine, a live bacterial vaccine, can be administered simultaneously or at any interval before or after YF vaccine. There are no data on the immune response to live-attenuated oral cholera vaccine (Vaxchora) or nasally administered live-attenuated influenza vaccine administered simultaneously with YF vaccine.

### Vaccine Adverse Events

Most reactions to YF vaccine are not serious and include injection site reactions. Mild systemic symptoms have been reported in 10% to 30% of vaccinees. Reported events typically include low-grade fever, headache, and myalgias that begin within days of vaccination and last 5 to 10 days. Serious adverse events, although rare, do occur. Immediate hypersensitivity reactions characterized by rash, urticaria, or bronchospasm can occur. Anaphylaxis after YF vaccine is reported to occur at a rate of 1.3 cases per 100,000 doses.

### Yellow Fever Vaccine–Associated Neurologic Disease

YF vaccine–associated neurologic disease (YEL-AND) is a serious, but rarely fatal, adverse event. Historically, YEL-AND was reported primarily among infants and manifested as encephalitis. Consequently, a contraindication to use of YF vaccine in infants aged less than 6 months was instituted in the 1960s. More recently, YEL-AND has been reported among persons of all ages and with a variety of clinical presentations. The illness onset for documented cases in the United States is 2 to 56 days after vaccination, and almost all cases have been reported in first-time vaccine recipients.

YEL-AND can manifest in several distinct clinical syndromes, including aseptic meningitis, meningoencephalitis, Guillain-Barré syndrome (GBS), acute disseminated encephalomyelitis (ADEM), bulbar palsy, and Bell palsy. Aseptic meningitis and meningoencephalitis occur as a result of direct vaccine virus infection of the meninges and/or the brain (neurotropic disease). The other neurologic syndromes represent autoimmune manifestations in which antibodies produced in response to the vaccine cause either central or peripheral demyelination. The overall incidence of YEL-AND in the United States is 0.8 per 100,000 doses. The rate is higher in people 60 years of age or older, at 2.2 per 100,000 doses.

Suspect YEL-AND in any patient who develops neurologic symptoms, with or without fever, following YF vaccination. Various tests may be helpful in diagnosing the specific neurologic syndrome, including studies of cerebrospinal fluid (culture and RT-PCR for vaccine-strain YF virus, or YF-specific IgM antibodies), neuroimaging, electroencephalography, electromyelography, and nerve conduction studies.

Treatment for neurologic disease related to direct viral invasion of the central nervous system (e.g., meningitis or encephalitis) is supportive. For autoimmune neurologic manifestations associated with YF vaccine, definitive treatments include various combinations of intravenous immune globulin, plasmapheresis, and corticosteroids.

### Yellow Fever Vaccine–Associated Viscerotropic Disease

YF vaccine–associated viscerotropic disease (YEL-AVD) was first reported in 2001, although retrospective studies indicate probable cases occurring decades prior. YEL-AVD is a severe illness similar to wild-type disease, with vaccine virus proliferating in multiple organs, causing multiorgan dysfunction or failure and sometimes death. To date, more than 100 cases have been reported globally. The illness onset for YEL-AVD cases reported in the United States occurs at a median of 4 days (range 1 to 18 days) after vaccination, and the case fatality ratio is approximately 48%. Laboratory-confirmed YEL-AVD has only been reported to occur after the first dose of YF vaccine. The overall incidence of YEL-AVD in the United States is 0.3 cases per 100,000 doses. The rate is higher for people 60 years of age or older, with a rate of 1.2 per 100,000 doses. Consider YEL-AVD in patients who develop fever and other systemic symptoms such as nausea, vomiting, or myalgias typically within 10 days of vaccination, especially if symptoms persist or worsen. Patients suspected to have YEL-AVD require close observation and often hospitalization, including intensive care. In addition to routine hematology, chemistry, and coagulation studies, other helpful laboratory tests include creatine phosphokinase (for rhabdomyolysis) and tests for disseminated intravascular coagulation. Draw blood for YF virus testing, particularly for RT-PCR and sequencing to detect YF vaccine–strain virus. In fatal cases, if autopsy is performed, collect both fixed and frozen tissue samples for immunohistochemical staining and RT-PCR testing. There is no specific therapy for YEL-AVD; treatment is supportive, once other diseases have been excluded.

### Contraindications to Yellow Fever Vaccine

YF vaccine is contraindicated in infants younger than 6 months of age because of the increased risk of vaccine-associated encephalitis. It is also contraindicated in anyone with a history of acute hypersensitivity reaction to any vaccine components, including eggs, chicken proteins, and gelatin. Because YF vaccine contains live virus, it is contraindicated

---

**BOX 64.2   Yellow Fever Vaccine Contraindications and Precautions**

**Contraindications**
- Age <6 months of age
- History of hypersensitivity to any vaccine component
  - Eggs
  - Chicken proteins
  - Gelatin
- Primary immunodeficiencies
- Malignant neoplasms
- Symptomatic human immunodeficiency virus (HIV) infection or CD4+ T lymphocytes <200/mm$^3$ (or <15% of total in children age <6 years)
- Transplantation
- Immunosuppressive or immunomodulatory therapies[a]
- Thymus disorder associated with abnormal immune function

**Precautions**
- Adults ≥60 years of age
- Infants 6–8 months of age[b]
- Asymptomatic HIV and CD4+ T-lymphocytes 200–499/mm$^3$ (or 15%–24% of total in children age <6 years)
- Pregnancy
- Breastfeeding

[a] Corticosteroids, alkylating drugs, antimetabolites, tumor necrosis factor (TNF)-α inhibitors, interleukin (IL)-blocking agents, and other monoclonal antibodies targeting immune cells; current or recent radiation therapy (see https://wwwnc.cdc.gov/travel/yellowbook/2020/travelers-with-additional-considerations/immunocompromised-travelers).
[b] In special circumstances, infants aged 6–8 months can be considered for vaccination.

---

in persons with immunocompromising conditions or those receiving immunosuppressive or immunomodulatory therapy (Box 64.2).

### Precautions to Use of Yellow Fever Vaccine

YF vaccine should not be routinely given to infants 6 to 8 months of age. However, there are situations in which vaccination might be considered, such as residence in or travel to an area experiencing a YF outbreak. Studies indicate that persons aged 60 years or older may be at increased risk of serious adverse events after YF vaccination, with an even higher risk for persons 70 years of age or older. When discussing YF vaccination with patients, weigh risks and benefits of vaccination against the destination-specific risk for exposure to YF virus.

The safety of YF vaccination during pregnancy or breastfeeding has not been studied thoroughly. Ideally, neither pregnant nor nursing women should be vaccinated with YF vaccine unless travel to a YF-endemic area is unavoidable and the risk of acquiring the disease outweighs the risk of vaccination. Infants born to women vaccinated during pregnancy should be monitored for evidence of congenital infection or other possible vaccine-associated adverse effects. Consider vaccinating persons with asymptomatic human immunodeficiency virus (HIV) infection and CD4+ T lymphocytes 200 to 499/mm$^3$ (or 15% to 24% of total lymphocytes in children age <6 years) who cannot avoid potential exposure to YF virus (see Box 64.2). Because both pregnant patients and patients with asymptomatic HIV infection can exhibit reduced seroconversion rates to YF vaccine, additional doses of YF vaccine might be necessary for continued or repeat exposure.

### International Health Regulations

The IHR (2005) allow countries to require proof of YF vaccination from travelers arriving from and transiting through certain countries to prevent importation and autochthonous YF virus transmission. Proof of vaccination must be provided on an International Certificate of Vaccination or Prophylaxis (ICVP), issued by a medical provider authorized to give YF vaccine and validated with the provider's signature and official YF vaccination center stamp. Travelers without a valid ICVP, who arrive in a country with a YF vaccination entry requirement, may be quarantined up to 6 days or denied entry.

In the United States, YF vaccination stamps are issued to medical providers by state health departments. A completed ICVP becomes valid beginning 10 days after vaccination and remains valid for the lifetime of the person vaccinated. Per the 2014 amendment to the IHR, a booster dose of YF vaccine cannot be required as a condition of entry into any country. A medical provider may issue a waiver to a person for whom the YF vaccine is medically contraindicated. In such a case the provider should fill out and sign the "Medical Contraindications to Vaccination" section of the ICVP and give the traveler a signed and dated exemption letter on letterhead stationery, clearly stating the contraindications to vaccination and containing the validation stamp used by the YF vaccination center. The provider should advise the traveler that issuance of a waiver does not guarantee its acceptance by the destination country and inform them of any increased risk of YF virus infection associated with lack of vaccination and how to minimize this risk by using mosquito bite prevention measures.

For more information about the location of YF vaccination centers, the ICVP, and medical waivers of YF vaccination, refer to the YF section in the *CDC Yellow Book 2020: Health Information for International Travel* (available at https://wwwnc.cdc.gov/travel/yellowbook/2020/travel-related-infectious-diseases/yellow-fever).

Country entry requirements for proof of YF vaccination under the IHR are different from the CDC's recommendations. YF vaccine recommendations are public health advice given by CDC to prevent YF virus infections among travelers and are subject to change at any time based on disease activity. Therefore both healthcare providers and travelers should check for relevant information on the CDC Travelers' Health website (www.cdc.gov/travel) well in advance of international travel.

## EVIDENCE

Gershman MD, Staples JE, Bentsi-Enchill AD, et al. Viscerotropic disease: case definition and guidelines for collection, analysis, and presentation of immunization safety data. *Vaccine* 2012;30:5038-5058. *The authors detail a standardized case definition of viscerotropic disease as an adverse event following immunization and include separate YF vaccine causality criteria.*

McMahon AW, Eidex EB, Marfin AA, et al. Neurologic disease associated with 17D-204 yellow fever vaccination: a report of 15 cases. *Vaccine* 2007; 25:1727-1734. *The authors summarize 15 cases of YEL-AND and describe the clinical features of the three most common types of neurologic disease associated with YF vaccine (encephalitis, GBS, and ADEM).*

## ADDITIONAL RESOURCES

Gershman MD, Staples JE. Yellow fever. In Centers for Disease Control and Prevention. *CDC Yellow Book 2020: Health Information for International Travel*. New York. Oxford University Press; 2017. *A practical and concise overview of YF epidemiology, YF vaccine and vaccine safety, the IHR, and the ICVP as they pertain to travelers' health.*

Monath TP, Cetron MS. Prevention of yellow fever in persons traveling to the tropics. *Clin Infect Dis* 2003;34:1369-1378. *The authors discuss risk of YF for travelers to endemic countries, summarize cases of imported YF in travelers, and present general considerations regarding YF vaccine for travel.*

Staples JE, Bocchini JA Jr, Rubin L, Fischer M. Yellow fever vaccine booster doses: recommendations of the Advisory Committee on Immunization Practices, 2015. *MMWR Recomm Rep* 2015;64:647-650. *Current guidelines for YF vaccine boosters in the United States.*

Staples JE, Gershman M, Fischer M, Centers for Disease Control and Prevention (CDC). Yellow fever vaccine: recommendations of the Advisory Committee on Immunization Practices (ACIP). *MMWR Recomm Rep* 2010;59:1-27. *The current guidelines for YF vaccine use in the United States.*

Staples JE, Monath TP, Gershman MD, Barrett ADT. Yellow fever vaccine. In Plotkin SA, Orenstein WA, Offit PA, Edwards KM, eds: *Vaccines*, ed 7. Philadelphia: Elsevier; 2018, pp 1181-1265. *Comprehensive review of YF virus and YF vaccine, including historical, epidemiologic, clinical, immunologic, and virologic aspects.*

# 65

# Travelers' Diarrhea

*Mark S. Riddle, Bradley A. Connor*

## ABSTRACT

Acute diarrhea associated with travel is referred to as travelers' diarrhea (TD) and is the most frequent illness among travelers. TD is defined as an acute illness associated with an increase in frequency and change in stool form (loose or liquid) compared with normal in an individual from one region who has traveled to another. It is frequently associated with other gastrointestinal symptoms that may include nausea, distention, urgency, and abdominal cramps as well as systemic symptoms such as fever, muscle aches, joint aches, and malaise. On average untreated bacterial diarrhea lasts 3 to 7 days. When accompanied by bloody stools and high fever, it is often termed dysentery. Persistent (2 to 4 weeks) or chronic (>30 days) diarrhea can also manifest, but it is a less frequent occurrence.

Although TD occurs more frequently in travelers visiting resource-poor settings, it can also occur among travelers worldwide. Higher risk regions include Central America and northern South America, Africa, the Middle East, and Asia (not including Japan); some Caribbean islands, China, and eastern European countries are intermediate-risk destinations; and North America, northern and western Europe, Australia, New Zealand, Japan, and a number of other Caribbean islands are low-risk destinations. Average attack rates for a 2-week itinerary are estimated to be 10% to 40%, although higher rates can occur based on location and itinerary and season of travel.

Risk factors for developing TD beyond location include age, itinerary type, duration of travel, source of meals, comorbidities, and medication. Younger travelers are generally at higher risk, and travel itineraries where food and water are consumed frequently from unhygienic sources pose greater risk. There is a greater incidence of TD in younger, more adventurous travelers. Bacteria are the predominant cause of acute TD and viruses (mainly norovirus) can account for 10% to 15%. Onset is usually during travel, as common infectious causes have relatively short incubation periods; however, illness frequently occurs on the transit home or shortly after return as well. Parasites are relatively rare and usually occur with more austere itineraries, in adventurous travels, and on longer trips; symptoms may not appear until well after a traveler returns home, and they can be prolonged.

Due to ubiquitous exposure and low inoculation doses for some pathogens, prevention of TD is a challenge. TD is best managed by the traveler through pretravel counseling, with a heavy emphasis on appropriate hydration/rehydration, provision of antibiotics and loperamide to take when ill, and recommendations on appropriate follow-up and testing in case of self-treatment failure, prolonged illness, or severe symptoms.

## CLINICAL VIGNETTE

A 21-year-old man traveled to a resort in Mexico for his spring break holiday. He had traveled to Mexico before, and each time he was there he developed diarrhea. This time, he asked his doctor what he could do to avoid getting sick. His doctor reviewed food and water precautions, telling him to eat only freshly cooked foods served hot, to avoid the local water and anything washed in water, and to avoid eating food prepared by street vendors. During his trip he decided to take a probiotic on a daily basis, beginning a day before his trip and continuing for the duration. He was fine during the week he was in Mexico, carefully avoiding salads at the buffet and drinking only carbonated bottled beverages, including beer. On the plane ride home, he developed a sudden onset of cramps and diarrhea. He continued to have symptoms after arriving at home. He then returned to see his doctor and reported that he had been very careful about his diet. He took his probiotics faithfully but may have had ice in a margarita 2 days before he got sick.

The patient's physical examination was unremarkable, and there was no blood or mucus in the stool. The doctor suggested a single dose of azithromycin 1000 mg, which the patient took. By the end of the day, he felt better. He continued to have loose stool for some 24 hours, after which his symptoms resolved completely.

COMMENT: This vignette highlights a few important points. Younger individuals are more likely to get sick with travelers' diarrhea. Whether this is related to more adventurous food consumption, a lack of acquired immunity, or other factors is unclear. Despite the patient's attention to food and water precautions, it is very difficult on a vacation to adhere to these recommendations fully. It is possible that the ice cubes in his margarita were responsible for his illness, although often the source of contamination is unknown. In addition, even with the most careful food and water precautions, one can never completely eliminate the risk of travelers' diarrhea at an at-risk destination. Many factors are beyond the traveler's control. Information regarding the symptoms of travelers' diarrhea and possible mitigation strategies (e.g., attention to hydration, self-treatment either with nonantibiotic remedies such as loperamide or bismuth subsalicylate or even a single dose or short course of an appropriate antibiotic) can be usefully offered to individuals planning to travel. The patient's clinical presentation is most consistent with a bacterial diarrhea with a fairly sudden onset of uncomfortable diarrhea. Diarrhea in the first week of travel is almost always bacterial, although viral pathogens may present with a fairly short incubation period; however, they are generally associated with vomiting. Most enteric parasitic pathogens have an incubation period of at least 7 to 10 days.

## AGENTS OF INFECTION

Several bacterial, viral, and parasitic agents cause TD; they have varied incidence and differential features.

### Bacterial

Enterotoxigenic and enteroaggregative *Escherichia coli* are the most common bacterial causes of TD, accounting for 30% to 50% or more of cases and typically causing a watery (secretory) diarrhea. Less frequent but often more serious (invasive) bacteria include *Campylobacter*, *Shigella*, and invasive *Salmonella* species. Their ability to invade the colonic mucosa, leading to dysentery, can cause more severe, debilitating illness. Less common bacteria of concern are *Aeromonas* species, *Plesiomonas shigelloides*, and, particularly in the Far East, *Vibrio parahaemolyticus*. *Vibrio cholerae*, the agent of cholera, very rarely causes diarrhea among travelers with the exception of humanitarian response workers involved in cholera outbreaks. (See Chapter 66.) Bacterial infections have an incubation period of generally some 24 to 72 hours.

### Food Intoxications

Occasionally foodborne bacterial intoxications occur (so called "toxic gastroenteritis" or traditional "food poisoning"), usually with an incubation periods of only a few hours. Toxigenic agents include *Clostridium* species, *Staphylococcus aureus*, and *Bacillus cereus* as well as seafood-associated intoxications such as ciguatera toxin (caused by dinoflagellates) and scombroid fish poisoning (caused by a variety of bacteria). The heating and freezing resistance of these preformed toxins make them difficult to avoid where food has become contaminated through harvesting, handling, processing, and preparation. The illnesses associated with food intoxications are generally short lived, although prolonged effects can occur. Ciguatera toxin illnesses—which most often occur from the consumption of large reef fish (e.g., barracuda, grouper, red snapper, amberjack, Spanish mackerel) in areas of the Caribbean, Hawaii, and coastal Central America—can have prolonged effects in the gastrointestinal, cardiovascular, and nervous systems for weeks to months.

### Viral

The most common viral causes, comprising up to 10% to 15% of TD cases, are noroviruses. Sapovirus and astrovirus are also occasionally identified; rotavirus infections occur less frequently and predominately in the older traveler. Norovirus infections often occur as outbreaks, particularly on cruise ships and in other closed-living-group situations. Vomiting may be a prominent feature of norovirus infection, but illness can also present as a predominately diarrheal syndrome. Most viral cases are short lived, but they can be particularly morbid in very young and elderly travelers who are less able to handle the rapid fluid and electrolyte losses associated with these infections.

### Parasitic

Intestinal protozoan parasites make up about less than 5% of TD cases. Because of relatively longer incubation periods, symptoms may not develop until the traveler returns home. Protozoa with shorter incubation periods include *Cryptosporidium*, with an incubation period of 3 to 8 days. *Cryptosporidium* infections are usually self-limited, with a duration of about 2 weeks of symptoms in travelers with normal immune systems, but they may cause a more prolonged illness in persons with compromised immunity. *Cyclospora* has a 2- to 11-day incubation period and can cause troublesome symptoms unless it is recognized and treated; it is generally self-limited after 6 to 8 weeks. *Giardia lamblia* is the most common pathogenic protozoan infection in travelers. It has an incubation period of 12 to 14 days. Travelers presenting with TD during a 1- to 2-week trip are not likely to have

*Giardia* as the cause unless it was preexisting. Infections with *Entamoeba histolytica* may have a shorter incubation period, perhaps less than 7 days, but usually the onset of symptoms is later. *Dientamoeba fragilis* and *Cystisospora belli* are potential intestinal protozoan causes of TD, but their incubations periods are not known. The intestinal helminth most associated with persistent abdominal symptoms is *Strongyloides stercoralis*, with an incubation period of at least 2 weeks before intestinal symptoms commence. (See Chapters 77, 78, and 79.)

## DIAGNOSIS

Given the conflation of symptoms and presentations across various pathogen etiologies, it is difficult to clinically diagnose specific etiologies of infection. However, there are some rules of thumb that can be considered with an eye toward guiding appropriate empiric therapy in the traveler setting. Acute dysentery, particularly associated with high fever and constitutional symptoms, is most likely attributed to invasive pathogens such as *Campylobacter jejuni*, *Shigella* spp., or nontyphoidal *Salmonella* spp. A vomiting-predominant gastroenteritis illness picture in the setting of an outbreak on a cruise ship or other mass-gathering travel event could be considered likely attributable to a viral etiology. Most sporadic acute watery diarrhea is associated with one or more bacterial pathogens based on historical and recent epidemiologic studies.

Routinely, studies to examine the cause of TD are not necessary given the location in which illness usually occurs (e.g., while traveling), and empiric treatment for acute diarrhea with antibiotics is usually effective against the broad variety of bacteria commonly encountered. However, there are certain scenarios where stool examinations for etiology may be important to aid in targeted anti-infectious therapy or identify noninfectious cause of severe acute, persistent or chronic abdominal symptoms associated with travel. Postinfectious irritable bowel syndrome (PI-IBS) is described to occur in 4% to 8% of those experiencing TD and risk increases with severe and prolonged illness. The pathogenesis of PI-IBS is not completely understood. Fortunately, most patients' symptoms resolve within a year; however, symptoms may persist for several years in some. In a returning traveler with persistent abdominal symptoms, common infectious and noninfectious etiologies should be worked up before a diagnosis of PI-IBS is made.

Current guidelines recommend microbiologic testing in travelers who return with severe or persistent symptoms or who fail a trial of appropriate antibiotic empiric therapy. Travelers who present with severe diarrhea or dysentery and have had an antibiotic exposure within the past 2 to 3 months should be evaluated specifically for infection with *Clostridium difficile*. In travelers whose symptoms have persisted for 2 weeks or more, stool testing targeting the identification of protozoal etiologies should be performed. Because of increased sensitivity, shorter duration until results, and the availability of multipathogen panels, nucleic acid amplification–based testing has grown in use. However, due to the high frequency of copathogen infections identified, clinical and epidemiologic considerations are important in order to interpret the results of such tests and guide treatment. It is important that reflexive cultures be performed in the context of suspected bacillary dysentery or failed acute TD antibiotic therapy so that antimicrobial sensitivities may be performed to add in appropriate antibiotic management.

## TREATMENT

### Hydration and Rehydration

An otherwise healthy individual with TD is unlikely to develop dehydration and can replace lost fluids and electrolytes with safe water and low-sugar noncaffeinated beverages along with some form of starch,

Traveler's diarrhea (TD) is diarrhea in people who are traveling or recently returned from traveling. It's the most common illness in travelers.

The usual cause is bacteria, but viruses and parasites can also cause TD. It's usually related to travel in developing countries or places with a contaminated water supply.

Symptoms, which start suddenly, include diarrhea, abdominal cramps and tenderness, and sometimes nausea or vomiting and fever.

Before traveling you should see your health care provider to talk about the risks of TD and what to do if you get sick

Prevention is best. Drink bottled water and avoid ice cubes when you travel. Drinking enough clear fluids is important, especially for babies and young children and older adults, because diarrhea can lead to dehydration.

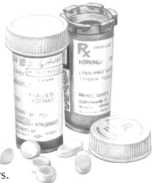

For symptoms which interfere with your ability to carry out travel activities, medications provided to you prior to travel can be used to rapidly improve your symptoms. Take these medications as directed. Often a single dose of antibiotics will lead to cure in less than 24 hours.

When you travel, wash all fruits and vegetables carefully in uncontaminated water. Be careful of tap water, ice, unpasteurized milk, dairy products, undercooked meat, and seafood.

Be sure to wash your hands every time you use the toilet.

**Fig. 65.1** Counseling your patients on traveler's diarrhea.

such as salted crackers (Fig. 65.1). Severe comorbidities due to dehydration are more of a threat for young children, elderly travelers, and those on medications such as diuretics. For these individuals, packets of oral rehydration salts are available commercially and can be carried by high-risk travelers or acquired in local pharmacies in the country of travel.

For mild diarrhea, an illness that does not affect the traveler's function in performing planned daily activities, this may be the only treatment required, because the diarrheal illness will be of short duration and self-limited.

## Antimotility and Nonspecific Agents

Loperamide (Imodium) is widely used to treat TD. It can be accompanied by an antimicrobial agent for moderate and severe watery diarrhea (and avoided if dysentery or high fever is present). Loperamide

liquid and caplets are available as over-the-counter products; these products should not be administered to children younger than 2 years of age. If symptoms persist over 48 hours or blood or mucus appears in the stool, loperamide should be discontinued. Another antimotility agent, diphenoxylate (Lomotil) can also be used. Diphenoxylate contains atropine and should not be used in older men, as it may cause urinary retention. It has been replaced by loperamide as the drug of choice for treatment of TD.

The antisecretory agent bismuth subsalicylate (Pepto-Bismol) taken as a liquid (1 oz every 30 minutes until eight doses have been taken) works more slowly than antimotility agents. In the treatment of mild to moderate TD, bismuth subsalicylate decreases the number of stools passed and duration of diarrhea by 50% as compared with no treatment. This product may produce darkening of the stool and

| TABLE 65.1 | Drugs for the Treatment of Travelers' Diarrhea | |
|---|---|---|
| **Drug** | **Adult Dosage** | **Children's Dosage** |
| Azithromycin | 1000 mg once (1-day divided)[a] OR 500 mg qd for 3 days | 10 mg/kg/day on day 1, 5 mg/kg/day on days 2 and 3 |
| Levofloxacin | 500 mg single dose[a] OR 500 mg qd for 3 days | Not recommended |
| Ciprofloxacin | 750 mg single dose[a] OR 500 mg bid for 3 days | Not recommended |
| Rifaximin[b] | 200 mg tid for 3 days | Not recommended below age 12 years |
| Rifamycin-SV MMX[b] | 388 mg bid for 3 days | Not recommended below age 18 years |

[a]If symptoms are not resolved after 24 hours, continue daily dosing for up to 3 days.
[b]Should not be used in infections associated with fever or blood or when invasive pathogens such as *Campylobacter, Salmonella,* or enteroinvasive *Escherichia coli* are suspected.
For more information on intestinal protozoa, see Chapters 77 and 78.

tongue from the bismuth component, but that is harmless. Other agents, such as kaolin-pectin and probiotics, have been found ineffective in clinical trials.

## Antimicrobial Treatment

Antibiotics are the most important agents in treating moderate (significantly impacting activities) and severe,(preventing activities) diarrhea, including dysentery. However, the use of antibiotics in mild diarrhea should be avoided in most circumstances. The current recommended first choice in the treatment of TD is azithromycin (Table 65.1). Fluoroquinolones are also commonly used; however, impacts on the microbiome, off-target systemic effects (e.g., tendonopathies), a purportedly increased risk of *C. difficile* infection, and the development of small intestinal bacterial overgrowth (SIBO) as well as a growing resistance in common bacterial infections make these antibiotics secondary choices, particularly in areas where *Campylobacter* spp. are encountered (e.g., South and Southeast Asia). Azithromycin is also the antibiotic of choice in pregnant women, where fluoroquinolones are contraindicated. Azithromycin is also preferable for children, as fluoroquinolones are not approved for use in children owing to potential drug toxicity (see Table 65.1).

Two new nonabsorbable antibiotic products, rifaximin and rifamycin-SV MMX, have been found to be useful in treating diarrheagenic *E. coli*, such as enterotoxigenic *E. coli* (ETEC) and enteroaggregative *E. coli* (EAEC), but it has not proven effective against more serious invasive organisms such as *Shigella* and *Campylobacter*.

The hydroxyquinoline class of drugs (e.g., Entero-Vioform, Mexaform) should never be used for treating diarrhea because they have never been proven to be effective for this purpose and their use has been associated with blindness and a paralytic syndrome. Although these products have been removed from the US market, they may still be available over the counter in some foreign countries.

Recent studies have identified that there is a risk of acquiring multidrug-resistant organisms (MDROs) associated with travel, a risk that increases among those traveling to Asia who develop diarrhea and take antibiotics for any reason during travel. Rates of carriage have been reported to be as high as 75% in travelers returning from the Indian subcontinent. Although colonization with these MDROs does not appear to affect the average traveler, recent reports have identified travel as an important risk factor in MDRO urinary tract infections in community- and hospital-based settings as well as following transrectal prostate biopsy infection in those who have traveled internationally during the previous 3 months Beyond the potential for individual impact in susceptible travelers, carriage appears to be transient, but it has been described to persist in approximately 10% at a year posttravel

and can be transmitted to close household contacts. Providers should have discussions with their patients about the benefits of antibiotic therapy for moderate and severe TD but also the risks of MDRO acquisition. Travelers who are at risk for recurrent urinary tract infections or who plan to undergo surgery upon return from travel should be informed of these potential risks.

## Combination Therapy

Multiple clinical trials have demonstrated that when loperamide is combined with an effective antibiotic (single-dose or multiday dosing), health outcome measures such as clinical cure at 24 hours or time to last unformed stool after starting therapy are much improved compared with antibiotics alone. Therefore if quick return to function is a priority, combination therapy is recommended.

## Antiparasitic Drugs

Diarrhea and other gastrointestinal symptoms that develop after return from travel may indicate a parasitic infection with a longer incubation period or a bacterial infection contracted late on the trip. Persistent symptoms that recur intermittently suggest a parasitic infection. *G. lamblia* is the most common pathogen causing diarrhea. Other parasites include *E. histolytica, Cryptosporidium, D. fragilis,* and *Cyclospora*. These parasites should always be considered in a returned traveler with continued symptoms such as abdominal pain, loose stools, foul gas or belching, intestinal rumbles, distention, fatigue, or weight loss. Descriptions of stool examinations for these parasites and recommendations for treatment are thoroughly described in Chapters 77 and 78.

## PREVENTION

Fecally contaminated food, water, and beverages are the most common vehicles of transmission in infectious diarrhea (see Fig. 65.1). During travel in areas of low sanitation, TD may still occur even with diligent attempts to avoid ingesting untreated tap water and products made with it as well as raw vegetables and fruits, which are often cultivated, rinsed, and freshened with contaminated water. Frequent oversights include brushing one's teeth with contaminated water, having ice cubes in drinks, drinking diluted fruit juices and mixed drinks, and eating leafy green salads. Other potentially hazardous foods include dairy products, raw or undercooked seafood and meat, buffet meals set out in warm climates, and food served by street vendors; these food and beverage items should be avoided when possible. A traveler should consume only hot, well-cooked food, fruits that can be peeled, carbonated beverages (weak carbonic acid inhibits bacterial growth), coffee, tea, and reliably purified water. Beer and wine are also safe beverages when consumed in moderation.

Bottled water sold at some high-risk travel destinations may not be as labeled and may actually be tap water. If possible, sparkling bottled water ("with gas") should be selected instead of still water ("without gas"). Tap water or water from local water supplies in areas of low sanitation should be boiled for 3 minutes, which will kill all dangerous organisms, including hepatitis viruses, and will compensate for thermal barriers (filth) and the lower boiling point of water at high altitude. Alternatively, water can be purified with iodine water-purification tablets such as Potable Aqua. If water is particularly cold, contact time for the tablets should be increased; if the water is turbid, the concentration of tablets should be increased according to package directions. Many portable filters on the market claim to provide safe drinking water. The most effective are those with iodide-impregnated resins, a micropore-type filter, and a carbon filter to improve taste.

Meticulous attention to food and beverage selection and preparation can decrease the likelihood of developing TD, but this is admittedly difficult to achieve. In addition to appropriate food, beverage, and water hygiene, other preventive measures against TD include hand washing, the use of certain nonantimicrobial medications, and the use of prophylactic antibiotics (reserved for special situations posing a very high risk). Currently there are no licensed vaccines in the United States against the main causative agents that cause TD, including enterotoxigenic *E. coli*, *Shigella*, and *Campylobacter*. Vaccines against these pathogens are, however, being developed and may be available in the future. Oral cholera vaccines are licensed in many countries including the United States. Dukoral (a two-dose nonliving vaccine) and Vaxchora (a single-dose live vaccine) may be indicated in adults (18 to 64 years) traveling to cholera-affected area if maximum protection is desired. (See Chapter 66.) However, cholera in adult travelers is very rare and is easily treated by empiric antibiotics, which should be provided to travelers to these regions to self-treat for TD. A nonantimicrobial agent found helpful in preventing TD is bismuth subsalicylate. A dose of two tablets four times daily taken by adults appears to be a safe and effective means of reducing the occurrence of TD by about 65% in persons at risk for periods of use up to 3 weeks. Salicylate absorption from bismuth subsalicylate may be enough to cause toxicity in those already taking aspirin-containing compounds and may alter anticoagulant control in patients taking Coumadin. High-dose bismuth subsalicylate may cause blackening of the tongue or stool. Bismuth subsalicylate should not be used concurrently with doxycycline used for antimalarial prophylaxis and may not be readily available outside the United States.

A number of antibiotics have been shown to prevent TD when taken prophylactically. However, because the risk of adverse side effects may outweigh the benefits in many situations, most experts advise against the routine use of prophylactic antibiotics by travelers. Some high-risk travelers with underlying medical conditions that could be significantly worsened by diarrhea may consult with their physicians and, once the risks and benefits are clearly understood, elect to use prophylactic antibiotic agents for only short periods of time. Probiotics, prebiotics, and passive immunoprophylactics have not yet been demonstrated to be consistently effective in clinical trials in the TD setting. If a traveler wishes to use them, he or she should not be complacent regarding proper food and water hygiene.

## EVIDENCE

Armand-Lefèvre L, Andremont A, Ruppé E. Travel and acquisition of multidrug-resistant Enterobacteriaceae. *Med Mal Infect* 2018;48(7):431-441. *Well-referenced and thorough review article on the epidemiology and impacts of travel on acquisition of multidrug-resistant bacteria, a growing concern in the context of antimicrobial resistance and spread worldwide.*

Duplessis CA, Gutierrez RL, Porter CK. Review: chronic and persistent diarrhea with a focus in the returning traveler. *Trop Dis Travel Med Vaccines* 2017;3:9. *A heavily referenced systematic review on the epidemiology of chronic and persistent diarrhea in the returning traveler.*

Riddle MS, Connor BA, Beeching NJ, et al. Guidelines for the prevention and treatment of travelers' diarrhea: a graded expert panel report. *J Travel Med* 2017;24(suppl_1):S57-S74. *High-quality graded expert panel–developed guidelines on definitions, prophylaxis, therapy, and diagnosis in travelers' diarrhea.*

Riddle MS, Connor P, Fraser J, et al; TrEAT TD Study Team. Trial Evaluating Ambulatory Therapy of Travelers' Diarrhea (TrEAT TD) study: a randomized controlled trial comparing 3 single-dose antibiotic regimens with loperamide. *Clin Infect Dis* 2017;65(12):2008-2017. *Recently completed large randomized controlled trial from three geographic regions evaluating single-dose antibiotic therapies (with loperamide) for watery diarrhea.*

Schwille-Kiuntke J, Mazurak N, Enck P. Systematic review with meta-analysis: post-infectious irritable bowel syndrome after travellers' diarrhoea. *Aliment Pharmacol Ther* 2015;41(11):1029-37. *A high-quality systematic review that describes the increased risk of irritable bowel syndrome after travelers' diarrhea.*

## ADDITIONAL RESOURCES

Connor BA. Travelers' diarrhea, Chapter 2: Preparing International Travelers. Centers for Disease Control and Prevention. *CDC Yellow Book 2020: Health Information for International Travel.* New York: Oxford University Press. *Accessible and up-to-date online source on all aspects of TD.*

DuPont HL. Persistent diarrhea: a clinical review. *JAMA* 2016;315(24):2712-23. *Succinct summary, well-referenced, narrative review article on persistent diarrhea.*

Steffen R, Hill DR, DuPont HL. Traveler's diarrhea: a clinical review. *JAMA* 2015;313(1):71-80. *A relatively recent, high-quality narrative review article on travelers' diarrhea.* Idi tem. Oviduci quodit entur, cuptae ni aut

# 66

# Cholera

*Ana A. Weil, Edward T. Ryan*

## ABSTRACT

*Vibrio cholerae* is a noninvasive small intestinal pathogen that produces cholera toxin, resulting in severe acute watery diarrhea in humans. *Vibrio* are gram-negative bacteria that live in marine and freshwater environments, and the O1 and O139 serogroups of *V. cholerae* can cause both endemic and epidemic cholera. Transmission occurs via the fecal-oral route, and humans lacking access to clean water are most likely to be affected. Cholera can be diagnosed using stool culture or rapid antigen testing; however, diagnostic testing is usually not conducted in most resource-limited settings, where ongoing cholera cases occur. The mainstay of treatment is rapid rehydration to avoid the consequences of hypovolemia from fluid losses; with prompt treatment, death is rare. Antibiotic treatment also decreases the duration and severity of symptoms. Cholera causes disease primarily in impoverished areas of Asia and Africa, and large outbreaks have recently occurred in Haiti and Yemen. Prevention and control should be a coordinated effort including provision of safe water, sanitation and hygiene (WASH) and vaccination. Several cholera vaccines are available, and cholera vaccines are currently maintained in a global stockpile to assist with outbreak prevention and control.

## CLINICAL VIGNETTE

A 15-year-old girl presented to a cholera treatment center during an outbreak in Dhaka, Bangladesh. Upon arrival, she was not responding to questions and was carried by her father, who reported that she had been vomiting three to four times per hour for 8 hours and had had several watery bowel movements per hour for 6 hours. The household water source was reported to be a shared tap with water stored in household containers, and the patient's neighbor reported recent diarrheal illness. Upon examination, the patient was awake but lethargic and had dull skin with sunken eyes. She was found to have a faint radial pulse. Two large-bore antecubital intravenous catheters were immediately placed for rehydration. She was placed on a cot with a hole in the middle to capture ongoing stool losses; these ongoing stool volumes and vomitus volumes were carefully measured in buckets. A green-tinged stool sample with white flecks of mucus was obtained for stool culture. As intravenous rehydration commenced, her mental status rapidly improved and she was offered oral rehydration solution (ORS) to drink. Once she was tolerating ORS, 700 mg of azithromycin was given orally based on her weight of 35 kg. After 5 L of intravenous fluids had been administered, the patient was alert and requesting food. The next day, the patient had urinated, appeared fully recovered, and was discharged from the treatment center. The day after admission, her stool culture resulted as positive for *V. cholerae* serogroup O1.

COMMENT: Vomiting can precede the profuse watery diarrhea of cholera, and severe disease progresses rapidly over a period of hours. Dehydration can occur quickly, and patients presenting with altered mental status or other signs of severe disease require prompt intravenous rehydration. In the setting of an outbreak, diagnostic and laboratory testing is rarely needed unless there are complications of disease, such as pneumonia, fluid overload, or comorbid malnutrition.

## ETIOLOGY AND PATHOPHYSIOLOGY

The causative agent of cholera is *V. cholerae*, a gram-negative comma-shaped human-restricted pathogen. *V. cholerae* naturally inhabits brackish and saltwater environments in association with copepods and is transmitted among humans via the ingestion of contaminated water and food. After ingestion, organisms that survive the acidic environment of the stomach reach the small intestine and use flagellar motion to reach the epithelial surface. Microcolonies form in small intestinal crypts, where *V. cholerae* secrete cholera toxin, an ADP-ribosylating enzyme that causes intestinal epithelial cells to secrete $Cl^-$, resulting in the movement of $Na^+$ and $H_2O$ out of cells and thus the watery diarrhea that typifies the disease. *V. cholerae* are then flushed from the intestines, classically in a "rice water" stool with a fishy odor. These bowel movements may have an extremely high density (100 million/mL) of *V. cholerae*. These organisms have a transient hyperinfectious phenotype, which can significantly contribute to transmission and result in explosive outbreaks. The inoculum usually needed to cause infection in humans is high ($10 \times 6$ to 8 colony-forming units [CFU]), although low acid states of the gastric environment can significantly reduce the infectious dose. The life cycle of *V. cholerae* is illustrated in Fig. 66.1.

Based on the lipopolysaccharide (LPS) O antigen, *V. cholerae* can be classified into more than 200 serogroups, of which the O1 and O139 serogroups can cause outbreaks of disease. The O1 serogroup is divided into El Tor and classic biotypes. Both biotypes can be further described as Inaba and Ogawa serotypes, which differ by one methyl group on the surface LPS. The serogroup O1 El Tor biotype *V. cholerae* is the current pandemic strain; the rarely found O139 serogroup was previously endemic only to certain areas of Asia. The pathogenic O1 and O139 serogroups have two critical virulence factors responsible for causing disease. The first is the toxin coregulated pilus (TCP), which facilitates microcolony formation at the intestinal surface. Upon attachment to the epithelial surface, the second virulence factor, cholera toxin, is secreted. Cholera toxin is composed of six subunits: a single copy of the A subunit and five copies of the B subunit. The cholera toxin B (CTB) subunit binds to GM1 gangliosides on the surface of the intestinal epithelial cells and enables the translocation of the enzymatically active cholera toxin A subunit into the epithelial cell. Other virulence determinants include biofilm formation, quorum sensing, and antibiotic resistance factors that are acquired through horizontal transfer of the encoding genes.

## EPIDEMIOLOGY

Cholera has occurred in pandemic waves over several centuries and first emerged from South Asia in the early 19th century. The current circulating strain is part of the 7th cholera pandemic, which has been ongoing for more than 50 years. Cholera is currently endemic

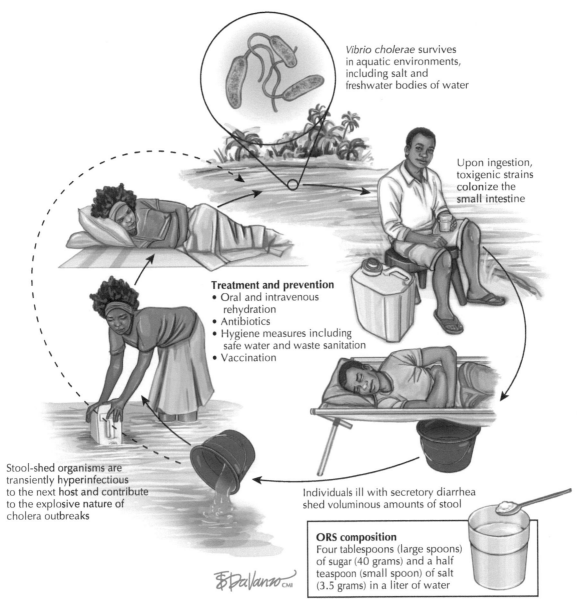

Vibrio cholerae survives in aquatic environments, including salt and freshwater bodies of water

Upon ingestion, toxigenic strains colonize the small intestine

**Treatment and prevention**
- Oral and intravenous rehydration
- Antibiotics
- Hygiene measures including safe water and waste sanitation
- Vaccination

Stool-shed organisms are transiently hyperinfectious to the next host and contribute to the explosive nature of cholera outbreaks

Individuals ill with secretory diarrhea shed voluminous amounts of stool

**ORS composition**
Four tablespoons (large spoons) of sugar (40 grams) and a half teaspoon (small spoon) of salt (3.5 grams) in a liter of water

**Fig. 66.1** Cholera life cycle. *ORS,* Oral rehydration solution.

in approximately 50 countries, where it survives in aquatic reservoirs, primarily in Asia and Africa (Fig. 66.2). To cause disease, the environment must be hospitable to fecal-oral spread of the pathogen, and *V. cholerae* must have the opportunity to be repeatedly introduced from an environmental source. In these regions, regular sporadic or seasonal cholera outbreaks are recorded, with intermittent periods of low or no incidence. In a nonendemic region, an outbreak is defined as at least one confirmed case of cholera accompanied by evidence of local transmission.

Over 3 to 5 million cases of cholera occur each year, resulting in tens of thousands deaths annually. Most cases occur in areas of endemic spread, with sporadic, year-round transmission. Several countries—including Yemen, Democratic Republic of Congo, Somalia, Haiti, Sudan, and Zambia—have had recent large outbreaks of cholera. Since cholera transmission is facilitated by poor sanitation and lack of clean water, the disease primarily occurs in countries without improved sanitation. Crowded urban slums and makeshift areas housing large numbers of people in areas of disrupted sanitation (such as refugee camps) are at particularly high risk of outbreaks. In addition, humanitarian

crises that disrupt the normal supply of drinking water can increase the risk of cholera transmission. Over the last several decades, cholera has emerged and reemerged in areas affected by displacement, natural disasters, and ongoing lack of clean drinking water. Approaches to modeling cholera outbreaks have recently led to an increased understanding of disease distribution, showing that small areas are often severely affected. The current case fatality rate for cholera varies widely depending on the resources and level of health care preparedness for an outbreak and case management; it remains highest when cholera newly emerges in a complex humanitarian emergency.

## CLINICAL PRESENTATION

The primary symptom of cholera is diarrhea often with vomiting, which usually occurs within a few hours to 5 days following ingestion of *V. cholerae*, depending on the inoculum size and host susceptibility. In severe cases, known as cholera gravis, the patient excretes copious amounts of watery rice-water stool (Fig. 66.3) with mucous specks, named for the appearance of water after cooking rice. In an

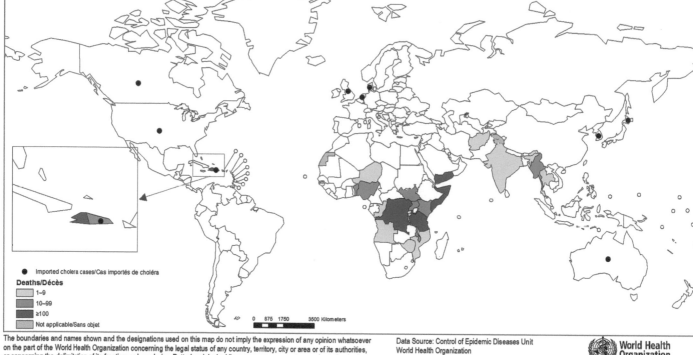

The boundaries and names shown and the designations used on this map do not imply the expression of any opinion whatsoever on the part of the World Health Organization concerning the legal status of any country, territory, city or area or of its authorities, or concerning the delimitation of its frontiers or boundaries. Dotted and dashed lines on maps represent approximate border lines for which there may not yet be full agreement.
Les appellations employées dans la présente publication et la présentation des données qui y figurent n'impliquent de la part de l'Organisation mondiale de la Santé aucune prise de position quant au statut juridique des pays, territoires, villes ou zones, ou de leurs autorités, ni quant au tracé de leurs frontières ou limites. Les lignes discontinues et en pointillé sur les cartes représententdes frontières approximatives dont le tracé peut ne pas avoir fait l'objet d'un accord définitif.

Data Source: Control of Epidemic Diseases Unit
World Health Organization

Map Production: Information Evidence and Research (IER)
World Health Organization

**Fig. 66.2** Global burden of cholera, based cases reported in 2016 by the World Health Organization. (Reused with permission from Control of Epidemic Diseases Unit, World Health Organization. http://gamapserver. who.int/mapLibrary/Files/Maps/Global_Cholera(WER)_2016.png)

untreated adult, the stool output can be as high as 10 to 20 L within 24 hours, resulting in dramatic fluid losses and electrolyte derangement. Cholera is an acute illness and the symptoms and possible complications of the disease are usually due to dehydration. Symptoms of severe dehydration include lethargy, sunken eyes, dull skin that is slow to retract, dry mucous membranes, and depressed mental status (Table 66.1). Loss of bicarbonate and acidosis due to diarrhea can cause a labored breathing pattern in severe cases. Other complications of extreme dehydration include acute kidney injury due to hypovolemia and seizures due to hypoglycemia, the latter which occurs more commonly in children. Fever is typically absent in cholera and should raise suspicion of a comorbid bacterial infection such as sepsis or pneumonia. Death is most likely to occur in children under 5 years of age, and pregnant women may suffer poor fetal outcomes depending on the level of dehydration that occurs. In children, pneumonia and comorbid malnutrition require close attention and are significant contributors to death after cholera.

Host characteristics also contribute to susceptibility to *V. cholerae* infection and the development of severe disease. Persons of younger age and who are immunologically naïve have increased susceptibility to disease. Individuals with blood group O are less likely to become infected with *V. cholerae* but if infected are more likely to have severe disease. In addition, single nucleotide polymorphisms in the gene encoding long palate, lung, and nasal epithelium clone 1, a secretory innate immunity protein, have been linked to susceptibility to *V. cholerae* infection. The gut microbiome has also recently been identified as a host factor that may impact susceptibility to disease.

## DIAGNOSTIC APPROACH

The definitive diagnosis of cholera depends on the successful isolation of *V. cholerae* from freshly passed watery "cholera" stool. Stool is cultured overnight on selective media such as thiosulfate citrate bile sucrose (TCBS) or taurocholate tellurite gelatin (TTG) agar. Darkfield microscopy is less sensitive but can provide a rapid diagnosis by observing a characteristic "shooting star" motility pattern of *Vibrio* in fresh rice-water stools at 400× magnification. Although specific for *V. cholerae*, this technique is operator dependent and not adequately sensitive for routine diagnosis. During a suspected outbreak, a rapid diagnosis can be made using commercially available dipsticks, such as Crystal VC (Span Diagnostics), which can detect the O1 or O139 antigen in stool samples. Although Crystal VC has a high sensitivity of 95%, its low specificity of 65% to 85% limits its diagnostic utility in nonendemic areas.

In endemic areas lacking clean water sources, a diagnosis of cholera is often made clinically based on the epidemiologic context. If watery diarrhea in a noncholera endemic area is accompanied by rapid fluid loss causing death or incapacitating disease in a healthy adult or child over 5 years of age without serious comorbid conditions, cholera should be suspected. Prompt identification of an outbreak can be essential for identifying an emerging epidemic, which requires rapid mobilization of resources for treatment and prevention. Other pathogens can also cause acute, severe watery diarrhea resulting in dehydration (such as rotavirus, *Cryptosporidium*, enterotoxic *Escherichia coli*, *Salmonella*, and *Campylobacter*, among others), and symptoms can be difficult to distinguish, especially in young children.

**A. Rice-water stool.**

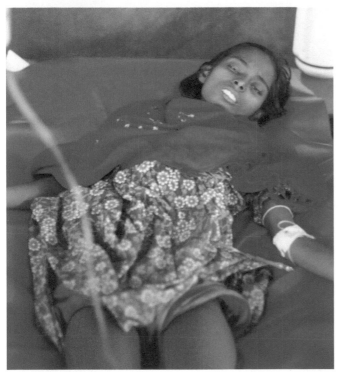

**B. A cholera patient with sunken eyes.**

**Fig. 66.3** (A) "Rice water" stool and (B) severe dehydration. (Reused with permission from Saha D, LaRocque RC. Cholera and other vibrios. In Ryan ET, Hill DR, Solomon T, et al., eds: *Hunter's tropical medicine and emerging infectious diseases*. 10th ed. Elsevier; 2020, Figures 43.3 and 43.4.)

## TREATMENT

The central tenet of cholera treatment is rehydration, and adequate and timely fluid replacement can reduce fatalities to less than 1%. The quantity of fluids administered is determined based on the volume depletion at the time of presentation and the ongoing fluid losses. Degree of dehydration is assessed clinically by evaluating mental status, eyes, mucous membranes, the patient's skin and pulse; it is then graded as none (<5% of body weight), moderate (5% to 10%), or severe (>10%), as per World Health Organization (WHO) guidelines (see Table 66.1). Continued volume loss is measured through ongoing stool and vomitus losses, assisted by a "cholera cot" (i.e. a bed/cot covered by a plastic sheet with a hole in the center that allows the stool output to collect in a calibrated bucket placed underneath the cot). If ongoing losses cannot be recorded, volume loss can be estimated as 10 to 20 mL/kg of body weight for each stool. For patients with mild or moderate fluid loss, an ORS made up of glucose with sodium and potassium salts is the mainstay of therapy. This treatment utilizes the glucose-mediated cotransport of water and sodium across the small intestinal mucosal surface, which remains intact during diarrheal illness. The WHO currently recommends a hypo-osmolar ORS consisting of 75 mEq/L sodium and 75 mM glucose. ORS can also be prepared by dissolving 6 teaspoons of sugar and ½ teaspoon of salt in 1 L of clean drinking water. In severe volume depletion, fluid replacement is performed in two phases: first, an intravenous rescue phase to replace deficits and, second, a maintenance phase consisting of ORS (if possible) or intravenous fluid administration to replenish ongoing losses (Fig. 66.4). (Note that fluid management in severely malnourished children is distinct.) Ringer lactate (RL) is the most commonly used intravenous rehydration solution, and isotonic solutions are recommended for rehydration. Antimicrobial treatment is also recommended as an adjunctive therapy in moderate or severe cholera once fluid losses are being corrected; this is effective in reducing the duration of both diarrhea and bacterial shedding in endemic areas. Doses and regimens of different antibiotics against cholera are summarized in Table 66.2. The choice of antibiotic is based on local antimicrobial resistance patterns and patient age. Antibiotic resistance to *V. cholerae* is common, especially in areas of antibiotic overuse. Mechanisms of resistance include plasmid gene transfer of drug-altering enzymes, efflux pumps, and other mechanisms. In Asia, many *V. cholerae* strains are now fluoroquinolone and tetracycline resistant, and extended-spectrum beta-lactamase (ESBL) and carbapenemase-expressing *V. cholerae* strains have been identified.

After severe diarrhea in young children, an important adjunctive treatment is adequate nutrition. Cholera patients are typically encouraged to begin taking in solids once the initial fluid deficit has been corrected; this practice assists in preventing ongoing electrolyte derangement. Breastfeeding should be continued for infants receiving breast milk, along with ORS. Acceptable additional fluids include coconut water, unsweetened juice, or broth. Medicinal teas, caffeine, and high-glucose drinks should be avoided due to their potential diuretic effects. Zinc and vitamin A supplementation during the recovery phase also reduces the duration and volume of stool output in children with cholera.

## Immune Response to Cholera

Based on field studies in cholera-endemic areas, persons who develop severe cholera are thereafter protected from the same serogroup infection for 3 to 10 years. The vibriocidal titer is the most commonly used indicator of immune responses to *V. cholerae* infection and is measured as a complement-mediated bactericidal antibody response from the peripheral blood. The primary target of the vibriocidal host immune response is the O-specific polysaccharide (OSP) antigen of the *V. cholerae* surface LPS. In cholera-endemic areas, vibriocidal titers increase with age and vacillates over short time periods, presumably based on low-level ongoing exposure to *V. cholerae*. Persons recovering from cholera and their household contacts in endemic areas have circulating memory B cells (MBCs) specific for anti-OSP responses, and these correlate with protection for at least 1 year after infection. Although CTB-specific circulating IgG and MBCs have also been detected in patients, these have not been associated with long-term protection. Since *V. cholerae* is a noninvasive pathogen, immunologic studies have largely focused on the adaptive antibody

**TABLE 66.1　Assessment and Treatment of Dehydration**

| Degree of Dehydration | None | Some (Two or More Criteria Are Present) | Severe (Two or More Criteria Are Present) |
|---|---|---|---|
| General appearance | Alert, well | Restless, irritable | Lethargic or unconscious |
| Eyes | Normal | Sunken | Very sunken |
| Thirst | Drinks normally or no thirst | Thirsty, drinks eagerly | Drinks poorly or unable to drink |
| Skin pinch | Retracts quickly | Retracts slowly (≥2 s) | Retracts very slowly (≥3 s) |
| Percent body fluid loss | <5% | 5%–10% | >10% |
| Estimated fluid deficit | <50 mL/kg | 50–100 mL/kg | >100 mL/kg |
| General treatment plan | Maintenance hydration: ORS to match stool volume | Hydration with ORS and observation | Rapid intravenous hydration and ORS (if possible) with close monitoring |

*ORS*, Oral rehydration solution.
Reused with permission from World Health Organization. http://whqlibdoc.who.int/publications/2005/9241593180.pdf.

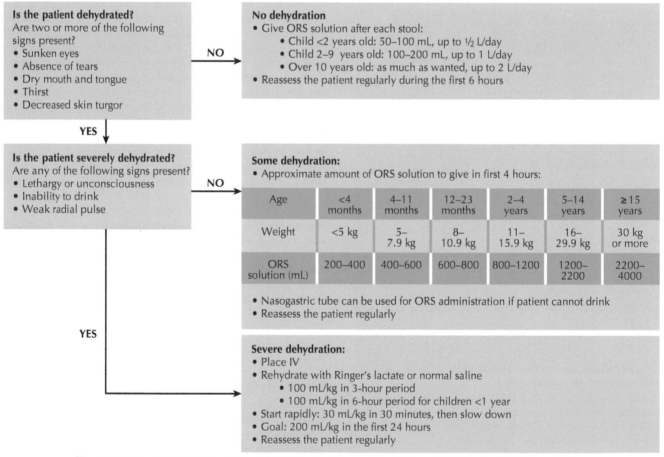

**Fig. 66.4** Approach to fluid management in patients with dehydration. *ORS*, Oral rehydration solution. (Reused with permission from Saha D, LaRocque RC. Cholera and other vibrios. In Ryan ET, Hill DR, Solomon T, et al., eds: *Hunter's tropical medicine and emerging infectious diseases.* 10th ed. Elsevier; 2020, Figure 43.5.)

response. However, recent evidence indicates that innate immune responses may also affect durable protective immunity. Blood and stool samples of patients with acute cholera have high levels of innate immune mediators (e.g., reactive oxygen species such as nitric oxide, bacteriocidal proteins such as myeloperoxidase and lactoferrin, and inflammatory cytokines such as TNFα and IL-1β). These innate responses and T-cell activation during acute disease correlate with the development of protective MBC responses.

## PREVENTION AND VACCINATION

Because cholera is a water-borne disease, prevention relies on a clean water supply and sanitation systems that prevent fecal contamination of drinking water. In situations where a clean water supply is lacking, boiling water or treating it with commercial chlorine tablets is effective to prevent disease. At present, three killed oral cholera vaccines (kOCVs) are WHO-prequalified and approved for use in endemic

| TABLE 66.2 | Antibiotic Treatment for Cholera | | |
|---|---|---|---|
| Antibiotic | Older Children and Adults | Child[a] | Comments |
| Tetracycline | 500 mg every 6 h for 3 days | 50 mg/kg/day divided every 6 h for 3 days | In endemic areas, resistance is common. Not recommended in pregnant women and children less than 8 years of age |
| Doxycycline | 300 mg single dose | 4–6 mg/kg in a single dose | Same as above |
| Erythromycin | 500 mg every 6 h for 3 days | 40 mg/kg/day divided every 6 h for 3 days | Widely used in the pediatric population |
| Azithromycin | 1 g in a single dose | 20 mg/kg in a single dose | Preferred therapy in areas of resistance to fluoroquinolones; reports of resistance are rare |
| Ciprofloxacin | 500 mg once daily for 3 days OR 1 g single dose | 20 mg/kg body weight in a single dose | Reduced susceptibility in Asia and Africa |

[a]Not to exceed adult dose.
Reused with permission from Saha D, LaRocque RC. Cholera and other vibrios. In Ryan ET, Hill DR, Solomon T, et al., eds: *Hunter's tropical medicine and emerging infectious diseases.* 10th ed. Elsevier; 2020, Table 43.2.)

regions. Euvichol Plus/Euvichol (Eubiologics, Seoul, Korea) and Shanchol (Shantha Biotechnics-Sanofi Pasteur, India) are bivalent vaccines with similar formulations containing killed whole-cell mixtures of O1 and O139 serogroup *V. cholerae*; Dukoral (Valneva, France) contains killed whole cell O1 serotype *V. cholerae* plus the CTB subunit. The efficacy of these kOCVs is higher in older children and adults compared to infants and children younger than 5 years (66% vs 38% in the first 2 years after vaccination, respectively). The bivalent vaccine has been tested after storage at ambient temperatures for up to several weeks, reducing the need for cold chain in vaccine administration. However, all three vaccine products require a two-dose administration schedule for optimal protection. The bivalent vaccines are the versions maintained in the cholera global vaccine stockpile. A live attenuated single-dose oral vaccine against the O1 serotype (Vaxchora, Emergent Biosolutions) has also been developed and is available in the United States for certain travelers to high-risk areas. In Canada and Europe, Dukoral is recommended for certain travelers to endemic areas.

As of August 2017, the WHO has recommended kOCVs for use in endemic areas, during humanitarian crises in areas with a high risk of cholera, and during outbreaks. Due to ongoing large outbreaks resulting in thousands of deaths during the last decade, a kOCV stockpile was established by the WHO in 2013 with assistance from global partners. This stockpile is meant for prompt deployment in the setting of rapidly spreading outbreaks or to prevent such outbreaks and was first used in South Sudan in 2014. Vaccination is now recognized as a critical tool in addition to other interventions focused on safe water and sanitation. Since 2014, the kOCV stockpile has been used repeatedly, and millions of doses have been administered in 15 countries. At this time, improved cholera vaccines are needed to provide longer-lasting protection, especially in young children who are most likely to die from cholera.

## EVIDENCE

Bi Q, Ferreras E, Pezzoli L, Legros D, Ivers LC, Date K, Qadri F, Digilio L, Sack DA, Ali M, Lessler J, Luquero FJ, Azman AS; Oral Cholera Vaccine Working Group of the Global Task Force on Cholera Control. Protection against cholera from killed whole-cell oral cholera vaccines: a systematic review and meta-analysis. *Lancet Infect Dis* 2017;17(10):1080-1088. *This review summarizes the studies of whole-cell oral cholera vaccines and the protection afforded by them in various populations and demographic groups.*

Deen J, Mengel MA, Clemens JD. Epidemiology of cholera. *Vaccine* 2019. https://doi.org/10.1016/j.vaccine.2019.07.078. *An up-to-date report of*

cholera epidemiology including spread of the seventh global pandemic and outbreaks of the last several years.

Dick MH, Guillerm M, Moussy F, Chaignat CL. Review of two decades of cholera diagnostics—how far have we really come? World Health Organization Global Task Force on Cholera Control. *Plos Negl Trop Dis* 2012;6(10):e1845. *This report discusses diagnostic tests developed for V. cholerae infection, highlighting test accuracy based on field compared with laboratory settings.*

Midani FS, Weil AA, Chowdhury F, Begum YA, Khan AI, Debela MD, Durand HK, Reese AT, Nimmagadda SN, Silverman JD, Ellis CN, Ryan ET, Calderwood SB, Harris JB, Qadri F, David LA, LaRocque RC. Human gut microbiota predicts susceptibility to *Vibrio cholerae* infection. *J Infect Dis* 2018;218(4):645-653. *In this study, authors define the gut microbiota at the time of V. cholerae exposure and discover that specific microbes are correlated with increased susceptibility to cholera.*

Saha D, Karim MM, Khan WA, Ahmed S, Salam MA, Bennish ML. Single dose azithromycin for the treatment of cholera in adults. *N Engl J Med* 2006;354:2452-2462. *This study is the basis for the most commonly used treatment for cholera in Asia and other areas of antibiotic resistant V. cholerae.*

## ADDITIONAL RESOURCES

Cholera Outbreak Training and Shigellosis (COTS). Free materials available in several languages for diarrheal outbreak treatment and prevention. http://www.cotsprogram.com. Accessed August 13, 2019. *A practical guide for treating cholera and severe diarrhea in the field with useful resources for health care providers at all levels of training.*

World Health Organization. Cholera vaccines: WHO position paper August 2017. *Wkly Epidemiol Rec* 2017;92;34:477-98. *The World Health Organization's statement on the rationale for adding cholera vaccines to the global armamentarium of tools for fighting cholera.*

World Health Organization. Deployments from the oral cholera vaccine stockpile, 2013-2017. *Wkly Epidemiol Rec* 93;32:437-42. *A review of the first 5 years of administration of oral cholera vaccines from the stockpile created for cholera outbreaks and emergencies.*

World Health Organization. Ending cholera—A global roadmap to 2030. 2017. http://www.who.int/cholera/publications/global-roadmap/en/. Accessed August 13, 2019. *This report details the World Health Organization's goals and strategy for decreasing cholera cases by 90% by the year 2030.*

World Health Organization. The treatment of diarrhea: A manual for physicians and other senior health workers. 2005. http://whqlibdoc.who.int/publications/2005/9241593180.pdf. Accessed August 13, 2019. *A complete guideline of the principles and practices for treatment of severe diarrhea.*

# 67

# Enteric Fever: Typhoid and Paratyphoid Fever

*Elaine C. Jong*

## ABSTRACT

Infection with the gram-negative bacteria *Salmonella enterica* serotype Typhi (*S.* Typhi) or *S. enterica* serotype Paratyphi A, B, or C (*S.* Paratyphi A, B, or C) may result in a serious febrile illness known as enteric fever. The illness is called typhoid fever or paratyphoid fever once the causative agent has been identified. Pathogenic *Salmonella* bacteria causing invasive disease are a particular problem in parts of the world with poor sanitation and hygiene, as the disease is transmitted by food and water contaminated with human waste. In endemic areas, the estimated number of individuals with enteric fever at any one time may exceed 10 per 100,000 population. Cases reported in the United States, Canada, Western Europe, Australia, Japan, and other industrialized countries are relatively rare—usually acquired by travelers abroad and diagnosed when symptoms become manifest after return home. Enteric fever can be a mild uncomplicated illness with fever, headache, and malaise, or it can develop into a more severe life-threatening illness manifested by prolonged high fever as high as 39°C to 40°C (102°F to 104°F), prostration, hepatosplenomegaly, intestinal hemorrhage or perforation, and altered mental status. Children and older adults tend to experience more serious illness with infection, although persons in all age groups are susceptible. Infections caused by *S.* Typhi and *S.* Paratyphi cannot be distinguished from each other based on clinical presentation alone (which may be similar to that of other infectious diseases—such as malaria, dengue, leptospirosis, brucellosis, and typhus—causing fever in endemic regions), although paratyphoid fever tends to be a milder typhoid-like illness. The diagnosis is confirmed by isolation of the organism from blood, stool, urine, or other clinical specimens including bone marrow. The Widal-Felix agglutination reaction is still used as a test for typhoid fever in certain endemic countries but has relatively low sensitivity and specificity and should not be performed before the second week of illness. Rapid diagnostic tests (RDTs) for detecting typhoid with relatively high sensitivity and specificity have become commercially available in several countries where enteric fever is endemic and yield results that are more timely and less labor intensive than culture. Multidrug-resistant (MDR) typhoid strains resistant to chloramphenicol, ampicillin, and cotrimoxazole first appeared in the 1980s. In the 1990s and 2000s *S.* Typhi and *S.* Paratyphi strains with decreased susceptibility to fluoroquinolone antibiotics and to some cephalosporins emerged on the Indian subcontinent and in Southeast Asia and Central Asia, making treatment of serious infections more challenging. Among untreated typhoid patients the fatality rate can be as high as 20%. Approximately 5% of those with acute disease become chronic carriers. Efficacious vaccines against typhoid fever are available for the protection of travelers and for mass immunization programs of children residing in highly endemic areas, but a safe and effective vaccine against paratyphoid fever is not currently available.

## CLINICAL VIGNETTE

A 29-year-old graduate student came to the university health center with complaints of fever, headache, stomachache, loss of appetite, and feeling weak over the previous 2 days. He and his wife had just returned a week earlier from a 2-week trip to Mumbai to visit his ailing father, who was in the hospital. The patient stated that he was generally healthy before the onset of his current illness and that the trip to India was unexpected, so neither he nor his wife received any pretravel health advice. His review of systems was remarkable for his report of general malaise, anorexia, diffuse abdominal discomfort, and no bowel movements in the previous 2 days. He denied shaking chills and sweats, stiff neck, photophobia, nausea and vomiting, or change in the color of his urine. Vital signs were blood pressure (BP) 116/78, temperature 103°F, pulse 96, and respiration 20. The patient was an ambulatory but ill-appearing Indian male accompanied by his wife. Physical exam was remarkable for a macular erythematous rash on the trunk that was barely visible against the patient's natural skin tone; bowel sounds were present and abdominal exam showed mild tenderness to deep palpation in all four quadrants but no enlargement of liver and spleen; rectal exam was unremarkable and the stool guaiac test was negative. Blood samples were drawn for a complete blood count, blood chemistries, malaria smear, and blood cultures. Additional blood was drawn to perform a RDT for malaria, and an extra tube was drawn as an acute serum sample. A urine sample was sent for a urinalysis (UA) and a stool sample was sent for culture of enteric pathogens. The patient and his wife waited in clinic for the initial stat lab results. The blood count and white blood cell differential were within normal limits (WNL), the malaria smear and RDT test were reported negative; glucose, blood urea nitrogen (BUN), bilirubin, and electrolytes were WNL; the UA was normal; and cultures of blood and stool were pending. A presumed diagnosis of typhoid fever was made on the basis of the history of travel to a highly endemic area for enteric fever, the clinical presentation, and the preliminary laboratory results. Another blood sample was sent for culture, and the patient was started on an oral antibiotic regimen of azithromycin 1 g by mouth once daily for 7 days at the advice of an infectious diseases consultant. The patient was discharged to home with arrangements for daily telephone follow-up and a clinic follow-up appointment in 1 week. Three days after his discharge from the clinic, the lab reported that two out of three blood cultures were positive for *S.* Typhi but the stool culture was negative. The patient reported having no fever and feeling much better by day 4 on treatment. He was instructed to complete his antibiotic course and to return for his follow-up appointment.

## GEOGRAPHIC DISTRIBUTION

Enteric fever is a global public health problem and tends to be highly endemic in countries where inadequate sanitation and hygiene allows for the ready contamination of food and water with human waste. Epidemics of typhoid fever and high endemic disease rates have been reported in India and countries in South Asia, the Middle East, Central Africa, and South America. According to World Health Organization (WHO) estimates, typhoid fever accounts for up to 21 million cases and 161,000 deaths annually worldwide. In industrialized nations, most reported cases occur among returned international travelers, with local outbreaks resulting from contact with asymptomatic carriers. Approximately 350 culture-confirmed cases of typhoid fever and 90 cases of paratyphoid fever caused by *S.* Paratyphi A per year were

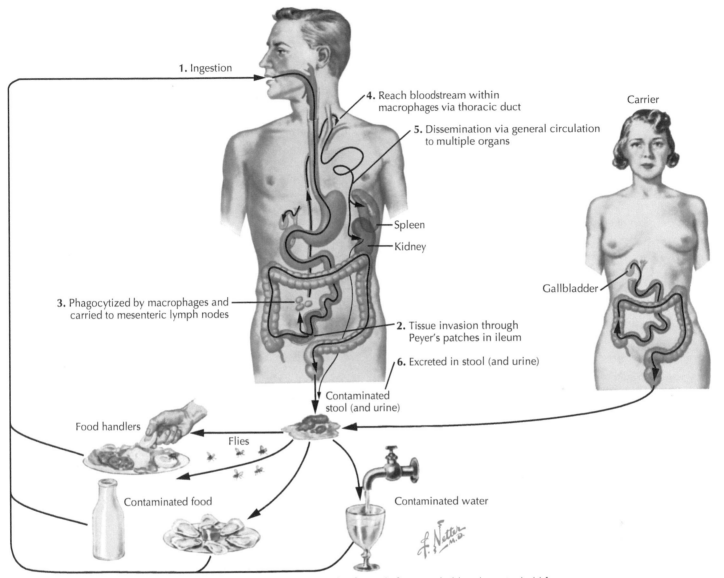

**Fig. 67.1** Transmission and pathogenesis of enteric fever: typhoid and paratyphoid fever.

reported during 2008–2015 to the Centers for Disease Control and Prevention (CDC) in the United States.

S. Paratyphi A accounts for a substantial number of enteric fever cases among returned travelers from India and countries in South Asia, the Middle East, and East Africa. In India, Pakistan, and Indonesia, an analysis of enteric fever cases published in 2005 showed that S. Paratyphi A had emerged as the causative agent in a growing proportion of enteric fever cases; in China it exceeded S. Typhi as the cause of enteric fever.

## TRANSMISSION AND PATHOGENESIS

Humans are the reservoir for S. Typhi and S. Paratyphi infections, with human-to-human transmission taking place through the fecal-oral route. Drinking water and ingesting raw vegetables and fruits, undercooked shellfish, ice cream, and other food products contaminated by impure water sources or the dirty hands of infected food handlers are the usual sources of infection, although flies landing on infected feces and then on food also contribute to food contamination (Fig. 67.1). Airborne transmission through aerosols during large outbreaks has also been considered a possibility.

Inoculum size influences the severity of an infection, with a large inoculum tending to cause more severe illness after a relatively short incubation period. Host factors associated with higher morbidity include being very young or very old, having lower gastric acidity due to medications (e.g., histamine-2 [$H_2$] blockers, proton pump inhibitors, antacids), or anatomy (partial gastrectomy); immune suppression caused by human immunodeficiency virus infection, cancer chemotherapy, and other conditions or treatments; coinfection with *Helicobacter pylori*; and human leukocyte antigen (HLA) tissue type.

After S. Typhi or S. Paratyphi bacteria are ingested in contaminated foods and beverages, the bacteria that survive passage through gastric acidity in the stomach pass in the fecal stream to the small intestine. During invasive disease the bacteria invade the microfold cells (M cells) that overlie the lymphoid tissue (Peyer patches) in the ileum and pass through them into the underlying tissue (lamina propria), where they are phagocytized by macrophages and carried to the mesenteric lymph nodes, where they continue to multiply intracellularly. The bacteria within infected macrophages disseminate in the bloodstream to multiple end organs including skin, spleen, liver, biliary tract, kidneys, and bone, causing multiple foci of infection and inflammation. Fever

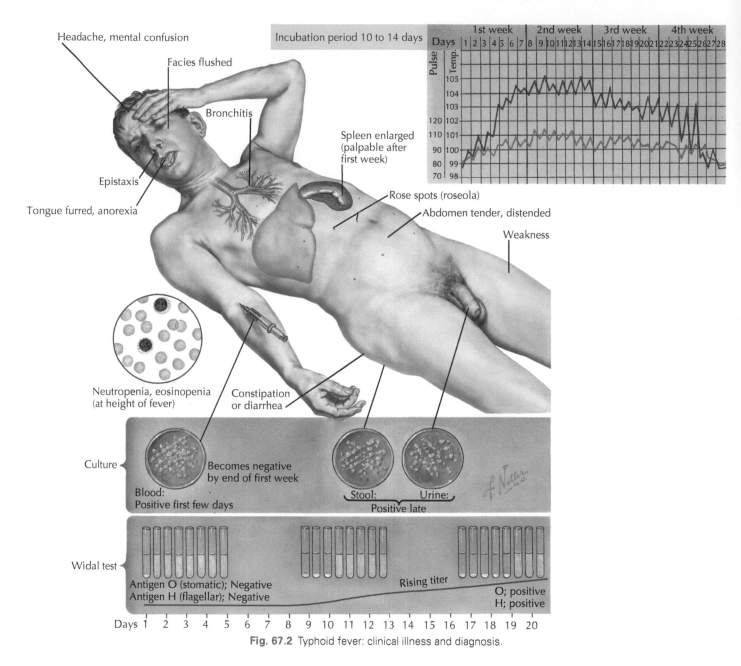

**Fig. 67.2** Typhoid fever: clinical illness and diagnosis.

and prostration are caused by the release of bacterial lipopolysaccharides (LPSs) and other cytotoxic inflammatory mediators (Fig. 67.2).

## CLINICAL FEATURES

The incubation period is commonly 7 to 14 days (with a range of 3 to 60 days). Onset of symptoms corresponds with bacteremia. The acute illness is characterized by fever, headache, malaise, and disturbed bowel function, with constipation more likely to be reported in adults and diarrhea more likely to be reported in children. Some patients develop a macular erythematous rash on the trunk (rose spots) (see Fig. 67.2). Without appropriate antibiotic treatment, the uncomplicated illness may last for 3 to 4 weeks with gradual recovery, although there is still a risk of late-stage relapse or complications. The clinical illness may be accompanied by a sustained but low level of secondary bacteremia. Although paratyphoid fever may have similar signs and symptoms, it is generally a milder disease than typhoid. In some cases, however, it may also cause severe life-threatening disease.

In severe illness there is prolonged fever as high as 39°C to 40°C (102°F to 104°F), with stepwise temperature elevation and a relative bradycardia (dissociation of the fever-pulse curve). Progressive abdominal discomfort, abdominal distention, and altered mental status (delirium, obtundation, and stupor) may develop over the second week of illness. Complications may include anemia, bacterial hepatitis, melena, intestinal perforation, peritonitis, myocarditis, pneumonia, disseminated intravascular coagulation, thrombocytopenia, hemolytic uremic syndrome, and ultimately death. Despite treatment with appropriate antibiotics and life support measures, typhoid death rates can range from 12% to 30%.

## DIAGNOSTIC APPROACH

A history of recent travel to an area of known endemicity justifies inclusion of typhoid and paratyphoid fever in the differential diagnosis. In a returned traveler from the tropics with fever, the differential diagnosis would also include malaria, dengue fever, and leptospirosis as well as more ubiquitous infections such as influenza, pneumonia, meningitis, and pyelonephritis.

A *confirmed case* of typhoid fever is when the patient has had a fever of 38°C (100.4°F) and higher for longer than 3 days and has a laboratory-confirmed positive culture of *S.* Typhi. Isolation of *S.* Typhi bacteria in cultures of blood, stool, urine, or other clinical specimens including bone marrow is the gold standard for diagnosis. Blood cultures are likely to be positive during the first 7 to 10 days of illness. Bone marrow aspiration cultures are useful if the diagnosis is considered after a prolonged illness, especially if antibiotics have previously been administered. If the diagnosis has been confirmed by a positive culture, serologic testing is unnecessary.

A *probable case* of typhoid fever is defined when the febrile patient has a positive serodiagnosis or antigen detection test for *S.* Typhi without isolation of the bacteria in culture. The Felix-Widal test (Widal test) measures serum agglutinating antibody levels against the somatic O (LPS) and flagellar H antigens of *S.* Typhi bacteria. Two serum specimens, acute and convalescent, must be drawn from the patient at least 5 days apart. A fourfold rise between the acute and convalescent antibody levels is diagnostic (see Fig. 67.2). However, this test has a false-negative rate of up to 30% in culture-proven cases, and cross-reacting antigens in patients with malaria, typhus, and bacteremia from other enteric bacteria may also confuse the interpretation of test results.

Newer diagnostic tests including polymerase chain reaction (PCR) assays are not universally available. The sensitivity, specificity, positive predictive value, and negative predictive value of rapid serologic tests for typhoid fever may offer superior sensitivity and specificity as well as convenience under field conditions compared with culture and the Widal test, but the rapid diagnostic tests (RDTs) are not yet widely used outside of endemic areas. The RDTs include IDL Tubex-TF (specific immunoglobulin M [IgM] antibodies to *S.* Typhi–specific O9 antigen factor); Typhidot (specific IgM and IgG antibodies to an *S.* Typhi 50-kD antigen); Typhidot-M (specific IgM-only antibodies to an *S.* Typhi 50-kD antigen); and the IgM dipstick test (specific IgM antibodies to *S.* Typhi LPS antigen).

Diagnosis of paratyphoid fever is based on isolation of *S.* Paratyphi in bacterial culture from specimens, usually blood or stool, followed by serologic tests for somatic (O) and flagellar (H) antigens to confirm serotype. A new rapid test for paratyphoid fever (Tubex-PA) has been developed based on technology similar to that of the Tubex-TF test. The paratyphoid test relies on the detection of *S.* Paratyphi A–specific $O_2$ antibodies in patient sera and is still under evaluation.

## CLINICAL MANAGEMENT AND DRUG TREATMENT

Patients with mild illness can be managed as outpatients on oral antibiotics, antipyretics, oral hydration, and oral nutrition. Patients with more severe illness manifested by persistent vomiting, severe diarrhea, abdominal discomfort, and/or altered mental status require in-hospital monitoring with parenteral administration of antibiotics, fluids, and ancillary drugs.

The optimal treatment for typhoid fever is not established. Prompt treatment with an appropriate antibiotic is generally recommended. Corticosteroid (dexamethasone) treatment is sometimes considered in severe typhoid fever with neurologic disorders (seizures, hallucinations, altered mental status). Antibiotic treatment should be guided by antimicrobial susceptibility testing of culture isolates; however, it is often necessary to initiate presumptive antibiotic therapy while awaiting the results of bacterial cultures and antibiotic susceptibility tests. Regional epidemiologic data on antibiotic susceptibility patterns

among *S.* Typhi strains should guide the choice of a presumptive treatment regimen. The approach to the management and treatment of paratyphoid fever follows similar lines.

The choice of antibiotics is complicated by the emergence in Asia in the late 1980s of *S.* Typhi strains that were multidrug-resistant (MDR) to the traditional first-line antibiotics ampicillin, chloramphenicol, and trimethoprim-sulfamethoxazole; subsequently the MDR strains spread globally. Fluoroquinolones were then widely used as the antibiotic treatment of choice, and *S.* Typhi strains with decreased susceptibility to fluoroquinolones emerged in the 1990s and 2000s. After strains fully resistant to fluoroquinolones were recognized in South Asia and spread to sub-Saharan Africa, other antibiotics such as extended-spectrum cephalosporins (cefixime and ceftriaxone) and azithromycin became the antibiotics of choice in the affected regions. Over the past decade, there have been increasing reports of *S.* Typhi resistant to the extended-spectrum cephalosporins in Asia and Africa, and there have been sporadic reports of resistance to azithromycin. Knowledge and understanding of local resistance patterns is essential for the selection of appropriate antibiotics for typhoid fever cases. Antibiotic treatment guidelines are given in Tables 67.1 and 67.2.

As with *S.* Typhi, decreased susceptibility to fluoroquinolone antibiotics is widespread among the *S.* Paratyphi A isolates tested. In a CDC survey of 162 paratyphoid cases reported from April 1, 2005, through March 31, 2006, 92% were *S.* Paratyphi A infections associated with recent travel to Asia, with 87% of the isolates showing decreased susceptibility to ciprofloxacin.

## TYPHOID CARRIER

Approximately 1% to 5% of typhoid fever patients become chronic carriers, with carriage of *S.* Typhi in the gallbladder. A chronic carrier is defined as an asymptomatic person who remains culture positive in stool or rectal swab cultures a year after recovery from acute illness. Factors associated with chronic carriage of *S.* Typhi include being over 50 years of age, being female, having gallstones, or having chronic schistosomiasis. Carriage can be eradicated by prolonged administration of antibiotics (e.g., ciprofloxacin 750 mg by mouth twice daily for 28 days) along with definitive treatment of any comorbid condition. A small number of individuals have also been reported to carry the *S.* Paratyphi bacteria after recovery from illness.

## PREVENTION

Prevention of enteric fever, as with other common gastrointestinal infections that are spread from human to human, involves both human and environmental interventions. Human factors include personal measures, such as thorough hand washing after defecation and immediately before food preparation, sanitary disposal of human urine and feces, adequate cooking of food, covering of prepared food to protect it from flies, isolation of infected patients, identification and treatment of chronic carriers, and immunization of susceptible persons with efficacious vaccines. Environmental controls include community systems for pure water supplies, sanitary disposal and treatment of human waste products, insect control measures against flies, and mass immunization programs in highly endemic areas.

Three vaccines against typhoid fever are recommended by the WHO: a purified Vi capsular polysaccharide typhoid vaccine given by injection (ViCPS), a live-attenuated Ty21a typhoid vaccine (Ty21a) administered orally, and typhoid conjugate vaccine (TCV) given by injection (Table 67.3).

**TABLE 67.1    Antibiotics for the Treatment of Uncomplicated Typhoid Fever[a]**

| Antibiotic Susceptibility | Uncomplicated Typhoid Fever Cases Oral Antibiotic Drugs (7 Days) |
|---|---|
| Sensitive to ciprofloxacin or decreased sensitivity to ciprofloxacin[b] | **Ciprofloxacin PO**<br>Children: 15 mg/kg bid (max 1 g qd)<br>Adults: 500 mg bid<br>or<br>**Ofloxacin PO**<br>Children: 15 mg/kg/day in two divided doses (max 600 mg qd)<br>Adults: 300 mg bid |
| Full resistance to ciprofloxacin and other fluoroquinolones[c] | **Azithromycin PO**<br>Children: 10–20 mg/kg qd (max 1 g qd)<br>Adults: 1 g qd<br>or<br>**Cefixime PO**<br>Children: 10 mg/kg bid (max 400 mg qd)<br>Adults: 200 mg bid |
| Region with data on susceptibility to these antibiotics from recent drug susceptibility tests | **Chloramphenicol PO**<br>Children >1 year and <13 years: 25 mg/kg tid (max 3 g qd)<br>Children >13 years and adults: 1 g tid<br>or<br>**Amoxicillin PO**<br>Children: 30 mg/kg tid (max 3 g qd)<br>Adults: 1 g tid<br>or<br>**Sulfamethoxazole/trimethoprim sulfa** (cotrimoxazole) PO<br>Children: 20 mg SMX + 4 mg TMP/kg bid (max 1600 mg SMX + 320 mg TMP qd)<br>Adults: 800 mg SMX + 160 mg TMP bid |

[a]Use knowledge of local resistance patterns where the typhoid fever was transmitted to guide the selection of appropriate antibiotic regimen.
[b]Microorganisms resistant to other fluoroquinolones may be susceptible to ofloxacin.
[c]The optimum treatment for quinolone-resistant typhoid fever has not been determined. Azithromycin, the third-generation cephalosporins, or a 10- to 14-day course of high-dose fluoroquinolones *may be* effective. Combination antibiotic regimens are under evaluation.
Adapted from: World Health Organization (WHO). Typhoid vaccines: WHO position paper—March 2018. *Wkly Epidemiol Rec* 2018;93(13):153–172. Accessed February, 2020; and Médecins Sans Frontières. Clinical Guidelines/Chapter 7: Bacterial Diseases—Enteric (typhoid and paratyphoid) fevers. https://www.medicalguidelines.msf.org/viewport/CG/english/typhoid-fever/6689926.html. Accessed February, 2020.

**TABLE 67.2    Antibiotics for the Treatment of Severe Typhoid Fever**

| S. Typhi Antibiotic Susceptibility | Severe Cases of Typhoid Fever Parenteral Antibiotic Drugs (Switch to Oral Route as Soon as Possible to Complete 14 to 21 Days of Treatment) |
|---|---|
| Decreased ciprofloxacin susceptibility or full resistance | **Ceftriaxone IV**<br>Children: 50–100 mg/kg qd (max 4 g qd)<br>Adults: 2 g qd or bid |
| Region with data on susceptibility to these antibiotics from recent drug susceptibility tests | **Chloramphenicol IV**<br>Children >1 year and <13 years: 25 mg/kg q8h (max 3 g qd)<br>Children ≥13 years and adults: 1 g q8h<br>or<br>**Ampicillin IV**<br>Children: 50 mg/kg q6–8h (max 3 g qd)<br>Adults: 1 g q6–8h |

Adapted from: Médecins Sans Frontières. Clinical Guidelines/Chapter 7: Bacterial Diseases—Enteric (typhoid and paratyphoid) fevers. https://www.medicalguidelines.msf.org/viewport/CG/english/typhoid-fever/6689926.html. Accessed February 2020; and Crump JA, Mintz ED. Global trends in typhoid and paratyphoid fever. *Clin Infect Dis* 2010;50:241-246. Available at: https://www.ncbi.nim.nih.gov/pmc/articles/PMC2798017/pdf/nihms154999.pdf. Accessed February, 2020.

The live oral attenuated Ty21a typhoid vaccine may be used in persons older than 6 years of age and consists of an enteric-coated capsule given for four doses (United States), which are taken 2 days apart on an empty stomach. Protection is elicited 10 to 14 days after the last dose, and protective efficacy is estimated to be around 70%. Revaccination is recommended every 5 years for continued risk of exposure, although significant protective immunity has been shown among populations living in endemic areas up to 7 years after immunization. The live oral typhoid vaccine is licensed in Africa, Asia, Europe, and South America. Outside the United States, it is available in both enteric-coated capsule and liquid formulations, is licensed for persons older than 5 years of age, and is approved as a three-dose vaccine series. To ensure vaccine efficacy, antibiotics should not be used for at least 3 days before and after the immunization series, and most experts recommend waiting 3 days after completion of the immunization series before administration of the first dose of mefloquine or proguanil (if either drug is to be used for malaria chemoprophylaxis).

New typhoid Vi conjugate vaccines (TCVs) appear to have improved efficacy in eliciting strong, long-lived immune responses in both infants and children compared with the existing ViCPS and Ty21a vaccines already discussed and can be used in children younger than 3 years of age. Thus TCVs could be incorporated into routine childhood immunization programs in countries with the highest burden of typhoid disease. One vaccine, Typbar-TCV, is approved for use in children 6 months of age and older. Typbar-TCV is licensed in India, Nepal, Cambodia, and Nigeria; it was prequalified by the WHO in 2017 and gained WHO recommended status in 2018. Another TCV vaccine is PedaTyph, licensed in India for children 3 months of age and older but not yet submitted for WHO prequalification.

There is no vaccine against paratyphoid fever, and the typhoid vaccines in current use do not offer cross protection. Thus improved water, sanitation, and hygiene (WASH) are extremely important in preventing this disease as well as preventing the transmission of other gastrointestinal infections.

The Vi polysaccharide typhoid vaccine may be used in persons older than 2 years of age; it consists of a single dose administered by subcutaneous injection. Protection is elicited after 14 days, and protective efficacy is estimated to be around 70%. Revaccination is recommended every 2 years for continued risk of exposure.

| TABLE 67.3 **Currently Available Typhoid Vaccines** | | |
|---|---|---|
| **Vaccine Name** | **Oral, Live, Attenuated Ty21a Vaccine (Ty21a)** | **Vi Capsular Polysaccharide Vaccine (ViCPS)** | **Typhoid Conjugate Vaccine (TCV)** |
| Trade name(s) (Manufacturer) | Vivotif (PaxVax, Emergent BioSolutions, Gaithersburg, MD) | Typhim Vi (Sanofi Pasteur, Sanofi US, Bridgewater, NJ); Typherix (GlaxoSmith-Kline, US-GSK, Philadelphia, PA; UK-GSK, Brentford, Middlesex) | TypBar-TCV (Bharat Biotech, Hyderabad, India) |
| Administration | Oral capsules | Intramuscular injection | Intramuscular injection |
| Age | >6 years of age | >2 years of age | >6 months of age |
| Number of doses | Four doses with booster every 5 years (United States) | One dose with booster every 2 years (United States) | 1 dose |
| Duration of protection | 7 years | 2 years | >3 years (studies currently under way) |
| Effectiveness | 50%–80% | 50%–80% | 87% (estimated from an adult challenge study; effectiveness studies currently under way) |

Adapted from: Coalition against Typhoid: Typhoid Vaccines. https://www.coalitionagainsttyphoid.org/prevent-treat/typhoid vaccines. Accessed February, 2020; and CDC Health Information for International Travel 2020. https://www.cdc.gov/travel/yellowbook/2020/travel-related-infectious-diseases/typhoid-and-paratyphoid-fever. Accessed February, 2020.

# EVIDENCE

Effa EE, Lassi ZS, Critchley JA, et al. Fluoroquinolones for treating typhoid and paratyphoid fever (enteric) fever. *Cochrane Database Syst Rev* 2011;10:CD004530. https://doi:10.1002/14651858. *Evidence-based review of the use of fluoroquinolones for treating enteric fever in the face of emerging antimicrobial resistance patterns showing decreased sensitivity to fluoroquinolones. Review conclusions: fluoroquinolones (ciprofloxacin, ofloxacin) are generally effective in treatment of MDR typhoid, but clinicians need to consider local antibiotic resistance when selecting treatment options and clinical response must be monitored. Gatifloxacin, a relatively new fluoroquinolone, seems to remain effective in some regions where resistance to older fluoroquinolones has developed. There are insufficient data to support important differences with regard to clinical failures between the fluoroquinolones versus second-line options (ceftriaxone, cefalexin, and azithromycin) in populations where both multidrug-resistant and nalidixic acid–resistant enteric fever is being transmitted.*

Tiwaskar M. Cefixime-ofloxacin combination in the management of uncomplicated typhoid fever in the Indian community setting. *J Assoc Physicians India* 2019;67(3):75-80. *Background and summary of clinical studies in Indian patients diagnosed with typhoid fever using cefixime-ofloxacin instead of monotherapy. The combination was well tolerated, promoted quick symptomatic relief and high cure rates in clinical studies. This drug combination is approved by the Indian Regulatory Authority and recommended by the Association of Physicians of India as a treatment option for uncomplicated typhoid in the Indian community setting.*

Wijedoru L, Mallett S, Parry CM. Rapid diagnostic tests for typhoid and paratyphoid (enteric) fever. *Cochrane Database Syst Rev* 2017; 5:CD008892. *Evaluation of selected studies on commercially available rapid diagnostic tests and their prototypes (including TUBEX, Typhidot, Typhidot-M, Test-it Typhoid, and other tests) for detecting typhoid and paratyphoid fever in people living in highly endemic countries showed that the tests have moderate diagnostic accuracy (TUBEX sensitivity 78%, specificity 87%; Typhidot tests sensitivity 84%, specificity 79%; Test-It Typhoid and prototypes [KIT] sensitivity 69%, specificity 90%). The RDTs may be useful in community outbreak situations where laboratory facilities and trained staff are lacking; however, the RDTs are not yet accurate enough to replace the current WHO-recommended diagnostic test: culture of S. Typhi or S. Paratyphi bacteria from a patient's blood or bone marrow.*

# ADDITIONAL RESOURCES

Appiah GD, Hughes MJ, Chatham-Stephens K. Chapter 4: Travel-Related Infectious Diseases—Typhoid and Paratyphoid Fever. In: *Centers for Disease Control and Prevention (CDC) Health Information for International Travel 2020.* Available at: https://www.cdc.gov/travel/yellowbook/2020/travel-related-infectious-diseases/typhoid-and-paratyphoid-fever. Accessed February, 2020. *This chapter gives the CDC recommendations on the diagnosis, treatment, and prevention of typhoid fever among international travelers and discusses the administration of the two typhoid vaccines licensed in the United States.*

Crump JA, Mintz ED. Global trends in typhoid and paratyphoid fever. *Clin Infect Dis* 2010;50:241-246. Available at: https://www.ncbi.nim.nih.gov/pmc/articles/PMC2798017/pdf/nihms154999.pdf. Accessed February, 2020. *A well-organized review and perspective on global trends in typhoid and paratyphoid fever. The section on antimicrobial resistance and patient management is especially useful.*

Médecins Sans Frontières. Clinical Guidelines/Chapter 7: Bacterial Diseases—Enteric (typhoid and paratyphoid) fevers. https://www.medicalguidelines.msf.org/viewport/CG/english/typhoid-fever/6689926.html. Accessed February 2020. *A clearly written summary of clinical guidelines for the care of patients with enteric fever.*

World Health Organization (WHO). Typhoid vaccines: WHO position paper—March 2018. *Wkly Epidemiol Rec* 2018;93(13):153–172. Available at: http://www.who.int/wer. Accessed February, 2020. *This position paper replaces the 2008 position paper on typhoid vaccines. It reviews the current global status of typhoid fever and reemphasizes the importance of vaccination to control typhoid fever. It also presents the WHO recommendations on the use of a new generation of typhoid conjugate vaccine.*

# 68

# Hepatitis A, B, C, D, and E: Focus on Adult Infections

*Latha Rajan, Elaine C. Jong*

 **ABSTRACT**

Viral hepatitis is caused by infection with five distinctly different human viruses that cannot be distinguished from one another without serologic testing (Table 68.1). Hepatitis A virus (HAV), hepatitis B virus (HBV), hepatitis C virus (HCV), hepatitis delta virus (HDV), and hepatitis E virus (HEV) have distinct differences in their epidemiology, physical structure, pathobiology, and prognosis, yet all involve the liver as the primary target organ. The severity of liver disease that accompanies acute infection with these viruses generally distinguishes them from cytomegalovirus and Epstein-Barr virus, whose primary target is not the liver and which typically cause much milder liver dysfunction during primary infections. The major burden of disease caused by hepatitis virus infections stems from chronic liver damage with progression to cirrhosis and hepatocellular carcinoma (HCC), which occurs in individuals who develop persistent infection. Development of persistent infection is highly dependent on the infecting virus and host factors such as age, co-infections, comorbidities, and underlying immune status. In areas of the world with the highest incidence of HCC (Southeast Asia, sub-Saharan Africa), most cases are attributed to HBV infection. In areas of the world where an increase in the incidence of HCC (and cirrhosis) has been more recent (United States, Western Europe, Japan, Australia), the leading cause is HCV infection.

 **CLINICAL VIGNETTE**

A hospital nurse sustains a needlestick injury with a contaminated needle to her left thumb, through her protective glove, while caring for a seriously ill patient during the night shift. The patient is an intravenous drug user with recurrent endocarditis. She washes the wound vigorously with soap and water as soon as possible, and goes to the hospital's employee health clinic (EHC) in the morning when she gets off duty, before going home. The nurse is 29 years old and reports herself in good health, with no acute illness or chronic health conditions. She is single, has a boyfriend, and lives alone at present. She went on a volunteer medical mission to Indonesia last summer for a month. Her vaccine records show that she has received tetanus-diphtheria-acellular-pertussis vaccine, measles-mumps-rubella vaccine, varicella vaccine, poliovirus booster, hepatitis B vaccine (HBsAb response ≥10 mIU/mL), hepatitis A vaccine, typhoid vaccine, and the annual flu vaccine.

In the EHC, the needlestick injury is examined, an incident report form completed, and her blood is drawn for standard screening tests for bloodborne pathogens (BBPs): HBV, HCV, and human immunodeficiency virus (HIV). Three days after the incident, the nurse receives a call back from the EHC: her test results show that she is HCV antibody positive and HCV RNA positive. This is a change from when she was last tested 3 years ago, when her HCV antibody test was nonreactive. The current test results are compatible with active HCV infection, and she is referred to her primary-care clinician for further medical evaluation, counseling, specialty consultation, and treatment. Although the nurse had a needlestick injury associated with a high-risk patient in this reported incident, given the HCV incubation period range of 14 to 182 days it is most likely that her active HCV infection is the result of an earlier exposure. The majority of acute HCV cases are asymptomatic or have unrecognized mild symptoms, and the presence of chronic HCV infection is often discovered during later screening.

**CLINICAL VIGNETTE**

A 34-year-old homeless man is referred to the primary-care clinic for evaluation by the city's outreach team. Considering his history of IV drug abuse for several years, he is screened and found positive for HIV. In view of his homeless state and the risk of being lost to follow-up before appropriate management can be started, he is admitted and the recommended workup for a newly diagnosed HIV patient initiated. He gives a history of becoming homeless after losing his job in the construction industry 3 years ago. He received the hepatitis B vaccine upon induction into the army when he was 18 years old. His lab results show an unremarkable complete blood count and chemistry profile including liver function tests. Serology tests confirm that he is HIV-1 reactive by immunoassay, hepatitis B surface antibody positive, hepatitis B core antibody negative, and hepatitis C antibody positive. The patient's CD4+ T-cell count is 1100 per $mm^3$; his plasma HIV RNA and hepatitis C RNA are pending. Although he relapsed after a prior IDU treatment attempt, he realizes that his health is now on a downward spiral and desires treatment for his drug dependence and newly detected HIV and hepatitis C infections.

Risk factors associated with homelessness include unsafe injectable drug use, HIV infection, HCV infection, and incarceration. Community outbreaks of hepatitis A also have been associated with the poor sanitary conditions in homeless encampments. The US homeless population is highly impacted by HCV, with prevalence estimates ranging from 20 to >50%. Effective drug treatment regimens to manage HIV and hepatitis C infections are available but considering possible drug-drug interactions, a decision may be made to treat the two infections sequentially in this patient. Management and follow-up of this patient will be challenging: homeless people lead unstable lives, move around frequently, may not have good access to healthcare, and often have food and shelter priorities that take precedence over access and adherence to healthcare recommendations.

## HEPATITIS A

### Epidemiology

According to the World Health Organization (WHO), hepatitis A caused approximately 7134 deaths in 2016, which accounted for 0.5% of mortality due to viral hepatitis. Transmission being primarily fecal-oral, risk factors are lack of safe water and poor sanitation and hygiene (such as dirty hands). Worldwide, the endemicity of HAV infection differs markedly among and within countries depending on the age groups in which the majority of transmission occurs. In areas with a high endemic pattern of infection, represented by low-income countries in parts of Africa, Asia, and Central and South America, most persons are infected as young children, and essentially the entire population becomes infected and immune before reaching adolescence. At the other end of the spectrum are the higher-income countries in most areas of North America and Western Europe, in which the endemicity of HAV infection is low. Relatively fewer children are infected, and disease often occurs in the context of communitywide outbreaks, as well as in defined risk groups such as men who have sex with men (MSM),

## TABLE 68.1  Overview of Viral Hepatitis Agents

|  | HAV | HBV | HCV | HDV | HEV |
|---|---|---|---|---|---|
| Incubation period | 15–50 days (average 28 days) | 60–150 days (average 90 days) | 14–182 days (average range 14–84 days) | Similar to HBV | 15–60 days (average 40 days) |
| Source of virus | Feces | Blood (highest virus concentration), semen, other body fluids | Blood (highest virus concentration), semen, other body fluids | Blood (highest virus concentration), semen, other body fluids | Feces |
| Routes of transmission | Fecal-oral: close person-to-person contact, contaminated food and water, rarely through contaminated blood transfusion | Percutaneous, mucosal, or nonintact skin exposure: birth to an infected mother, sexual contact, sharing contaminated needles, syringes, or other injection drug equipment, needle-sticks or other sharp instrument injuries, unscreened blood transfusions or organ transplants, dialysis, shared use of personal items such as razors or toothbrushes of an infected person | Similar to HBV | Similar to HBV | Fecal-oral: through contaminated water, contact with zoonotic source (e.g., pigs) |
| Potential for persistent or chronic infection | No | Yes: 90% of infants infected at birth, 25%–50% of children infected at ages 1–5 years, 5% of people infected as adults | Yes: Chronic infection develops in over 50% of newly infected people | Yes | No |
| Vaccine available | Yes | Yes | No | No: prevent HDV co-infection by preventing HBV infection with HBV vaccine. | No |

*HAV,* Hepatitis A virus; *HBV,* hepatitis B virus; *HCV,* hepatitis C virus; *HDV,* hepatitis D virus; *HEV,* hepatitis E virus.
Adapted from: The ABCs of Hepatitis—for Health Professionals. www.cdc.gov/hepatitis, updated 2020. https://www.cdc.gov/hepatitis/resources/professionals/pdfs/abctable.pdf.

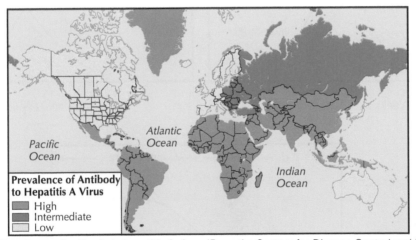

**Fig. 68.1** Prevalence of antibody to hepatitis A virus. (From the Centers for Disease Control and Prevention [CDC]: CDC health information for international travel 2010. Atlanta: U.S. Department of Health and Human Services, Public Health Service; 2009.)

injection drug users (IDUs), and travelers returning from areas with a high or intermediate endemicity of infection (Fig. 68.1).

HAV replicates in the liver, is excreted in bile, and is shed in stool. Peak infectivity occurs during the 2-week period before onset of jaundice or elevation of liver enzymes, when concentration of the virus in stool is highest. The concentration of the virus in stool declines after jaundice appears. Children can shed HAV for longer periods than adults, with shedding lasting up to 10 weeks after onset of clinical illness; infants infected as neonates in one nosocomial outbreak shed HAV for up to 6 months, but lifelong shedding of the virus does not occur.

In the United States, person-to-person transmission through the fecal-oral route is the primary means of HAV transmission. Transmission occurs most frequently among close contacts, especially in households, extended family settings, daycare facilities, correctional

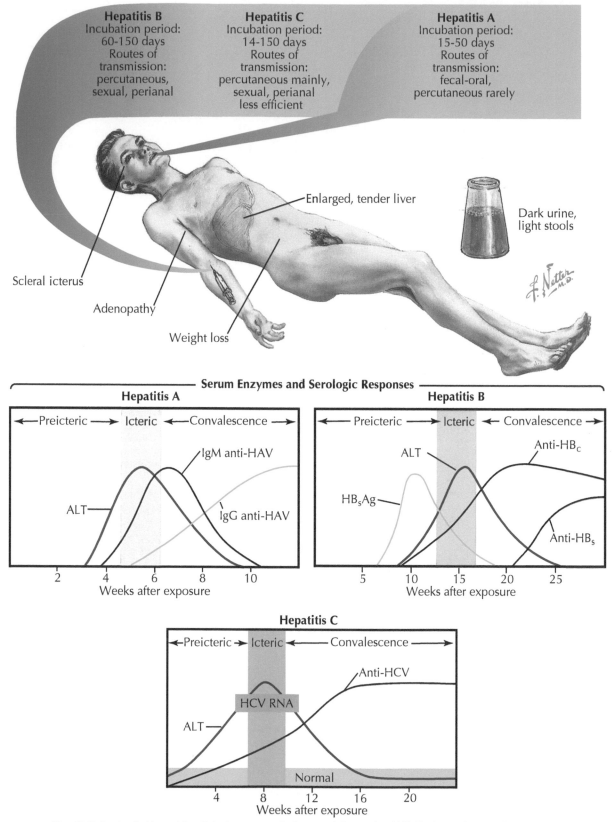

**Fig. 68.2** Acute viral hepatitis: clinical course, pathology, and diagnosis. *ALT,* Alanine aminotransferase; *HAV,* hepatitis A virus; *HBc,* hepatitis B core antigen; *HBsAG,* hepatitis B surface antigen; *HCV,* hepatitis C virus.

## TABLE 68.2 Vaccines to Prevent Hepatitis A Virus Infection

| Vaccine | Trade Name (Manufacturer) | Age | Dose | Route | Schedule | Booster |
|---|---|---|---|---|---|---|
| Hepatitis A, inactivated | Havrix (GlaxoSmithKline) | 1–18 years | 0.5 mL (720 ELU) | IM | 0, 6–12 months | None |
| | | ≥19 years | 1.0 mL (1440 ELU) | IM | 0, 6–18 months | None |
| Hepatitis A, inactivated | VAQTA (Merck & Co, Inc) | 1–18 years | 0.5 mL (25 U) | IM | 0, 6–18 months | None |
| | | ≥19 years | 1.0 mL (50 U) | IM | 0, 6–18 months | None |
| Hepatitis A+B, combined | Twinrix (GlaxoSmithKline) | ≥18 years | 1.0 mL (720 HAV ELU + 20 µg HBsAg) | IM | 0, 1, 6 months | None |

*ELU,* ELISA units of inactivated HAV; *HAV,* hepatitis A virus; *HBsAg,* hepatitis B surface antigen; *IM,* intramuscular; *U,* units HAV antigen.
From Health Information for International Travel, Table 4-02. Available at wwwnc.cdc.gov/travel/yellowbook/2020/travel-related-infectious-diseases/hepatitis-a.

facilities, and unsheltered populations. Common-source outbreaks and sporadic cases can occur from exposure to food or water contaminated at the source or by infected food handlers. Outbreaks in the context of floods or other natural disasters (e.g., hurricanes) have not been reported in the United States. On rare occasions, HAV infection has been transmitted by transfusion of blood or blood products collected from donors during the viremic phase of their infection. In the United States and other countries where the hepatitis A vaccine has been incorporated into standard childhood immunization programs, the burden of HAV infections has shifted to older unimmunized age groups who are more likely to be symptomatic (see Chapter 14).

### Clinical Illness

HAV, a 27-nm ribonucleic acid (RNA) agent classified as a picornavirus, can produce either asymptomatic or symptomatic infection in humans after an average incubation period of 28 days (range 15 to 50 days). Illness caused by HAV infection typically has an abrupt onset that can include fever, malaise, anorexia, nausea, abdominal discomfort, dark urine, and jaundice. The likelihood of having symptoms with HAV infection is related to age. In children younger than 6 years, 70% of infections are asymptomatic; if illness does occur, it is typically not accompanied by jaundice. Among older children and adults, infection is typically symptomatic, with jaundice occurring in more than 70% of patients. Signs and symptoms typically last less than 2 months, although 10% to 15% of symptomatic persons have prolonged or relapsing disease lasting up to 6 months (Fig. 68.2). However, persistent infections with HAV are not well documented and may never occur. The overall case-fatality ratio among reported cases in the United States is approximately 0.3% to 0.6% but reaches 1.8% among adults older than 50 years; persons with preexisting chronic liver disease are at increased risk of acute liver failure from HAV infection.

### Diagnosis

HAV cannot be differentiated from other types of viral hepatitis on the basis of clinical or epidemiologic features alone. Two serologic tests are licensed for the detection of antibodies to HAV: (1) immunoglobulin M antibody to HAV (IgM anti-HAV) and (2) total anti-HAV (i.e., IgM and IgG anti-HAV). Serologic testing to detect IgM anti-HAV is required to confirm a diagnosis of acute HAV. In the majority of persons, serum IgM anti-HAV becomes detectable 5 to 10 days before onset of symptoms and declines to undetectable levels less than 6 months after infection. IgG anti-HAV, which appears early in the course of infection, remains detectable for the person's lifetime and provides lifelong protection against the disease.

### Treatment

HAV does not cause persistent infection and rarely causes hepatic failure. Supportive care is the mainstay of treatment for HAV infections.

Individuals with acute HAV with altered mental status or severe dehydration from nausea and vomiting require hospitalization. Laboratory results revealing increased international normalized ratio (INR) and/or reduced albumin may indicate systemic dysfunction that requires consultation with a hepatologist. If hepatic encephalopathy develops, liver transplantation may be indicated. Patients with minimal symptoms can be managed on an outpatient basis.

### Prevention

Depending on conditions, HAV can be stable in the environment for months. Heating foods at temperatures greater than 185°F (85°C) for 1 minute or disinfecting surfaces with a 1:100 dilution of sodium hypochlorite (i.e., household bleach) in tap water is necessary to inactivate HAV. HAV can be prevented by (1) general measures of good personal hygiene, particularly hand washing, provision of safe drinking water, and proper disposal of sanitary waste; (2) preexposure or postexposure immunization with HAV vaccine; and (3) preexposure or postexposure immunization with immunoglobulin (see Chapter 14). In the United States, preexposure administration of HAV vaccine is recommended for all children 1 year of age and older, international travelers, MSM, injection and noninjection illicit drug users, and individuals with chronic liver disease. Inactivated HAV vaccines are safe, highly immunogenic, and well tolerated. Table 68.2 lists the HAV vaccines and schedules approved by the US Food and Drug Administration (FDA). Other HAV vaccine products, including attenuated live vaccines, may be available in other countries.

## HEPATITIS B

### Epidemiology

Approximately one-third of the global population has been infected with HBV. Around 5% of this population are chronic carriers and a quarter of these carriers develop serious liver diseases such as chronic hepatitis, cirrhosis, and hepatic carcinoma. Globally, hepatitis B prevalence is highest in the WHO Western Pacific Region and the WHO African Region, where 6.2% and 6.1% of the adult population is infected, respectively. In the general US population, the overall age-adjusted prevalence of HBV infection (including persons with chronic infection and those with previous infection) is 4.9%. The estimated prevalence of chronic HBV infection is 0.5%, corrected for the disproportionate contribution of foreign-born persons (particularly Asian/Pacific Islanders) who have emigrated from countries in which HBV is highly endemic. It is estimated that there were 862,000 persons with chronic HBV infection in 2016, and 22,000 new infections in 2017 nationwide.

HBV is a blood-borne and sexually transmitted virus. Transmission from infected mothers to their newborns at the time of birth is the most common source for HBV infection among infants.

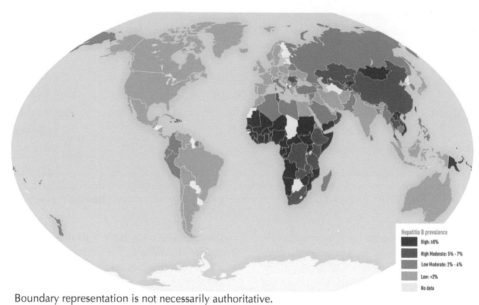

Boundary representation is not necessarily authoritative.

**Fig. 68.3** Prevalence of chronic infection with hepatitis B virus. (From the Centers for Disease Control and Prevention [CDC]: CDC health information for international travel 2020. Atlanta: U.S. Department of Health and Human Services, Public Health Service; 2019.)

For adults, the primary sources are sexual (male to female and male to male) and percutaneous (e.g., injection drug use) exposures to blood. For all age groups, unsafe therapeutic injections, unscreened blood transfusions, and long-term nonsexual household contact with chronically infected persons are risks for acquiring HBV infection.

Chronic infection (serum positivity for hepatitis B surface antigen [HBsAg]) is more likely to develop in infants infected at birth (90%) or young children infected at ages 1 to 5 years (25% to 50%), most of whom do not have symptoms. Rates of symptomatic acute disease are higher among older children and adults, who are less likely to develop chronic infection. Prevalence of chronic infection with HBV is high in countries or regions where HBV infections were historically acquired primarily by infants and young children, including countries in South America, Africa, the Middle East, Asia, Southeast Asia, Greenland, the First Nation populations of northern Canada, and Alaskan Native Americans (Fig. 68.3). In those countries or regions where universal childhood hepatitis B immunization programs have been implemented, early-age infections have been virtually eliminated and adults account for the remaining chronic infections (see Chapter 15).

## Clinical Presentation

HBV is a 42-nm DNA virus classified in the family Hepadnaviridae. HBV infection can produce either asymptomatic or symptomatic infection. The average incubation period is 90 days (range 60 to 150 days) from exposure to onset of jaundice, 60 days (range 40 to 90 days) from exposure to onset of abnormal serum alanine aminotransferase (ALT) levels, and 30 days (range 6 to 60 days) from exposure to detection of HBsAg. The onset of acute disease is usually insidious. Infants, young children, and immunosuppressed adults with newly acquired HBV infection are typically asymptomatic. When present, clinical symptoms and signs might include anorexia, malaise, nausea, vomiting, abdominal pain, and jaundice (see Fig. 68.2). Extrahepatic manifestations of disease include skin rashes, arthralgias, and arthritis. The case-fatality rate for acute HBV is 0.5% to 1%.

Most (≥95%) primary infections in adults with normal immune status are self-limited, with elimination of the virus from the blood and development of lasting immunity to reinfection. In contrast, primary infection develops into chronic infection in 30% of children younger than age 5 and in 80% to 90% of infants, with continuing viral replication in the liver and persistent viremia.

Although the consequences of acute HBV can be severe, most of the serious sequelae associated with the disease occur in chronically infected persons, who are at increased risk for developing cirrhosis, decompensated liver disease, and primary HCC. Persons who should be tested for chronic HBV infection include those born in geographic regions with HBsAg prevalence ≥2% (see Fig. 68.3), persons with unexplained elevations in liver enzymes, persons infected with HIV, and persons undergoing immunosuppressive therapy.

Host and viral risk factors associated with disease progression include older age (longer duration of infection), high levels of HBV deoxyribonucleic acid (DNA), habitual alcohol consumption, and concurrent infection with HCV, HDV, or HIV. Persons with chronic infection also serve as the reservoir for continued HBV transmission. Extrahepatic manifestations related to chronic HBV infection include immune complex–mediated diseases such as membranoproliferative glomerulonephritis.

## Diagnosis

Although there are eight serologic markers associated with HBV infection, only four of these are necessary for diagnosis and screening (Table 68.3). The presence of HBsAg is indicative of ongoing HBV infection, and all HBsAg-positive persons should be considered potentially infectious. In newly infected persons, HBsAg is the only serologic marker detected during the first 3 to 5 weeks after exposure, and it persists for variable periods. Antibody to hepatitis B core antigen (anti-HBc) develops in all HBV infections, appearing at the onset of symptoms or liver test abnormalities in acute HBV infection, rising rapidly to high levels, and persisting for life. Acute or recently acquired infection can be distinguished by the presence of the IgM class of anti-HBc, which is detected at the onset of acute HBV and persists for approximately 6 months.

In persons who recover from HBV infection, HBsAg is eliminated from the blood, usually in 2 to 3 months, and anti-HBs develops during convalescence, indicating immunity from HBV infection. Most persons who recover from natural infection will be positive for both anti-HBs and anti-HBc, whereas persons who are successfully

## TABLE 68.3 Interpretation of Serologic Test Results for Diagnosis of Hepatitis B Virus Infection

| HBsAg[a] | Total Anti-HBc[b] | IgM[c] Anti-HBc | Anti-HBs[d] | Interpretation |
|---|---|---|---|---|
| – | – | – | – | Susceptible; never infected |
| +[e] | – | – | – | Early acute infection |
| + | + | + | – | Acute infection |
| – | + | + | – | Acute resolving infection |
| – | + | – | + | Past infection; recovered and immune |
| + | + | – | – | Chronic infection |
| – | + | – | – | False positive (i.e., susceptible); past infection; "low-level" chronic infection[f]; passive transfer to infant born to HBV-infected mother |
| – | – | – | + | Immune if titer is ≥10 mIU/mL |

[a]Hepatitis B surface antigen.
[b]Antibody to hepatitis B core antigen.
[c]Immunoglobulin M.
[d]Antibody to hepatitis B surface antigen.
[e]To ensure that an HBsAg-positive test result is not a false positive, samples with repeatedly reactive HBsAg results should be tested with a licensed neutralizing confirmatory test.
[f]Persons positive for only anti-HBc are unlikely to be infectious except under unusual circumstances involving direct percutaneous exposure to large quantities of blood (e.g., blood transfusion).
*HBc,* Hepatitis B core; *HBsAg,* hepatitis B surface antigen; *HBV,* hepatitis B virus; *IgM,* immunoglobulin M; *mIU,* milli–international units per milliliter.

vaccinated against hepatitis B develop only anti-HBs. In persons who do not recover from HBV infection and who become chronically infected, HBsAg and anti-HBc persist, usually for life.

Other markers associated with HBV infection include hepatitis B e antigen (HBeAg) (and its corresponding antibody) and HBV DNA, both of which correlate with viral replication and high levels of virus and can be detected in the serum of persons with acute or chronic HBV infection. The results of testing for these markers are most useful for monitoring chronically infected patients who are treated with antiviral drugs.

## Treatment

The primary goal of HBV treatment is suppression of viral replication. Indications of response to treatment include return of liver enzymes to normal, suppression of HBV DNA levels, and serologic conversion of HBeAg to anti-HBe. Antiviral treatment algorithms change regularly owing to newly implemented agents, and consultation with a liver specialist (or a provider who is knowledgeable about hepatitis B) is recommended before initiation of treatment. Not every patient with chronic HBV needs to be treated: the approved drugs are most effective when there are signs of active liver disease.

Current treatments for chronic HBV fall into two classes: immune modulator drugs and antiviral drugs. Pegylated interferon therapy, possessing both antiviral and immunomodulatory properties, is commonly used for its suppression of viral replication. The treatment course is a once-weekly subcutaneous injection for 6 to 12 months. Antiviral drugs include the nucleoside and nucleotide analogues, and the main focus of this class is disruption of HBV DNA synthesis either by DNA chain termination or direct inhibition of DNA polymerase. FDA-approved drugs in this class are oral antivirals taken as a pill once a day for at least 1 year and usually longer. Tenofovir disoproxil, tenofovir alafenamide, and entecavir are considered first-line treatments; telbivudine and adefovir dipivoxil are considered second-line options; and lamivudine is generally not used in the United States because it is less potent than the newer drugs. Initial sustained-response rates range from 50% to 70% with antiviral therapy, but long-term response

is affected by adherence to the treatment course and development of resistance. Frequent relapses or reactivation of hepatitis B after cessation of antiviral treatment mean that the duration of use can be indefinite beyond a year. A "cure" or sustained loss of viral DNA and loss of surface antigen (HBsAg) in the blood off of drugs can be achieved only in a small number of people treated with the currently available drugs. Several investigational drugs for adults infected with HBV and HDV targeting novel HBV therapeutic targets are being tested in HBV Clinical Trials.

## Prevention

Immunization with hepatitis B vaccine is the most effective measure to prevent HBV infection and its consequences. Over the past two decades, the incidence of acute hepatitis B in the United States has declined by 70%. The most significant decline (94%) has occurred among children and adolescents, coincident with an increase in HBV vaccine coverage. Although acceptance of vaccination is high among adults offered vaccination, the rates of HBV vaccine coverage among most adults are low. The low coverage is attributed to missed opportunities by healthcare professionals to offer HBV vaccine to high-risk adults seeking care in general medical care settings.

Screening of individuals at risk of acquiring HBV (as well as those most likely to be infected) should occur in primary-care settings (e.g., physician's offices, community health centers, travel clinics, occupational health programs) and high-risk settings (sexually transmitted infections [STIs] clinics and drug treatment facilities). Information should be provided to all adults regarding risk factors for HBV transmission and persons for whom the vaccine is recommended (Table 68.4). Also, the HBV vaccine should be readily available to anyone requesting vaccination.

Several safe and efficacious recombinant HBV vaccines are FDA-approved and involve a series of two or more doses. In addition to the standard schedule for primary immunization, some HBV vaccines have the option for an accelerated dosing schedule in adults to elicit protective antibodies (HBsAb) in a shorter time frame than the standard primary dosing schedule. Use of the newer HBV vaccine recombinant with novel adjuvant (1018) or other HBV vaccines with an accelerated dosing schedule is ideal for susceptible persons with anticipated risk of exposure to HBV in less than the time needed for a standard schedule, for example, healthcare personnel, relief workers, international travelers, military deployment, patients likely to require blood transfusions, organ transplant, or hemodialysis.

---

**TABLE 68.4   Hepatitis B Vaccination Recommendations for Adults by Risk Category**

| | |
|---|---|
| Sexual exposure | Partners of HBsAg-positive partners<br>Sexually active with multiple partners<br>Homosexual males<br>Individuals seeking treatment for STDs |
| Percutaneous exposure | Injection drug users (past or present use)<br>Residents and staff of institutions for mentally disabled persons<br>Healthcare and public safety workers<br>Household contacts of HBsAg-positive individuals<br>Individuals with end-stage renal disease and hemodialysis patients |
| Other high-risk groups | Travelers to endemic regions<br>Individuals with chronic liver disease<br>HIV-positive individuals<br>Individuals interested in vaccination |

*HBsAg,* Hepatitis B surface antigen; *HIV,* human immunodeficiency virus; *STD,* sexually transmitted disease.
Source: Mast EE, Weinbaum CM, Fiore AE, et al. A comprehensive immunization strategy to eliminate transmission of hepatitis B virus infection in the United States: recommendations of the Advisory Committee on Immunization Practices (ACIP). Part II: immunization of adults, *MMWR Rep* 2006;55:1-33.

---

The widely available inactivated HBV recombinant vaccine (Engerix-B, Recombivax HB) is given on a standard schedule of three doses over a 6-month period for primary immunization. The vaccine has a 95% response rate in healthy persons younger than 40 years old; in older persons, rates may be lower. Postvaccination HBV antibody testing (HBsAb) 4 to 6 weeks after the last vaccine dose is recommended for individuals who need to know their immune status, such as healthcare workers. A serum titer of ≥10 mIU/mL HBsAb is considered protective. Nonresponders to primary vaccination with HBV recombinant vaccine may undergo a second three-dose regimen with the same vaccine (which may elicit immunity in up to 70%), or another HBV vaccine product may be used such as the HBV vaccine recombinant with novel adjuvant (1018), or the combined hepatitis A and B vaccine (Table 68.5).

Postexposure prophylaxis is achieved with hepatitis B immunoglobulin (HBIG) and hepatitis B vaccine. HBIG is effective in preventing infection, but vaccination is recommended for sustained immunity. The estimated window for postexposure prophylaxis is ≤7 days for needlestick injury and ≤14 days for sexual exposure.

## HEPATITIS C

### Epidemiology

HCV is primarily transmitted by direct percutaneous exposures to infectious blood, most commonly through injection drug use, transfusions from infected donors, and unsafe therapeutic injections. Although less efficient, transmission of HCV by perinatal and sexual exposures also occurs. Globally, the estimated prevalence of HCV infection is 71 million and the estimated incidence is 2 million annually. Key populations at risk for HCV acquisition include IDUs and HIV-positive MSM. Unsafe healthcare practices, including nonsterile injections, contribute a large proportion to new HCV infections in low- and middle-income countries (LMICs).

Worldwide, the prevalence of HCV antibody averages about 2% (Fig. 68.4). In the United States antibody prevalence is 1.7% in the

---

**TABLE 68.5   Vaccine Products and Schedules for Adult Hepatitis B Immunization**

| Vaccine | Trade Name (Manufacturer) | Age | Dose | Route | Schedule | Booster |
|---|---|---|---|---|---|---|
| Hepatitis B vaccine, recombinant with novel adjuvant (1018) | Heplisav-B (Dynavax Technologies) | ≥18 years | 0.5 mL (20 µg HBsAg and 3000 µg of 1018) | IM | 0, 1 months | None |
| Hepatitis B vaccine, recombinant | Engerix-B (GlaxoSmithKline) | 0–19 years (primary) | 0.5 mL (10 µg HBsAg) | IM | 0, 1, 6 months | None |
| | | 0–10 years (accelerated) | 0.5 mL (10 µg HBsAg) | IM | 0, 1, 2 months | 12 months |
| | | 11–19 years (accelerated) | 1.0 mL (20 µg HBsAg) | IM | 0, 1, 2 months | 12 months |
| | | ≥20 years (primary) | 1.0 mL (20 µg HBsAg) | IM | 0, 1, 6 months | None |
| | | ≥20 years (accelerated) | 1.0 mL (20 µg HBsAg) | IM | 0, 1, 2 months | 12 months |
| Hepatitis B vaccine, recombinant | Recombivax HB (Merck & Co., Inc.) | 0–19 years (primary) | 0.5 mL (5 µg HBsAg) | IM | 0, 1, 6 months | None |
| | | 11–15 years (adolescent accelerated) | 1.0 mL (10 µg HBsAg) | IM | 0, 4–6 months | None |
| | | ≥20 years (primary) | 1.0 mL (10 µg HBsAg) | IM | 0, 1, 6 months | None |
| Combined hepatitis A and B vaccine | Twinrix (GlaxoSmithKline) | ≥18 years (primary) | 1.0 mL (720 ELU) HAV + 20 µg HBsAg) | IM | 0, 1, 6 months | None |
| | | ≥18 years (accelerated) | 1.0 mL (720 ELU HAV + 20 µg HBsAg) | IM | 0, 7, 21–30 days | 12 months |

From CDC Health Information for International Travel 2020. Available at: wwwnc.cdc.gov/travel/yellowbook/2020/travel-reated-infectious-diseases/hepatitis-b.

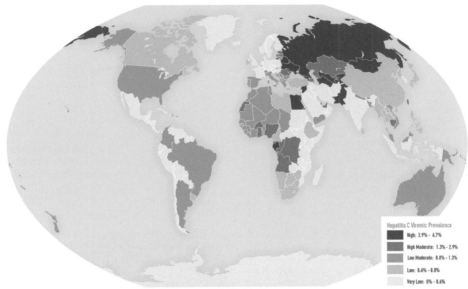

Boundary representation is not necessarily authoritative.

**Fig. 68.4** Global prevalence of hepatitis C virus infection. (From the Centers for Disease Control and Prevention [CDC]: CDC health information for international travel 2020. Atlanta: U.S. Department of Health and Human Services, Public Health Service; 2019.)

general population, corresponding to an estimated 4 million individuals ever infected with HCV, of which 2.4 million are chronically infected. The median anti-HCV prevalence in persons who use drugs is 54.2% (range 12.7% to 67.1%) according to studies that examined 2013–2016 National Health and Nutrition Examination Survey (NHANES) data for US adults. Approximately 44,700 new HCV infections were reported to the CDC in 2017, almost four times as many cases as were reported in 2010.

## Clinical Presentation

HCV, a single-stranded RNA agent, is a member of the Flaviviridae family. HCV strains show extraordinary genetic diversity. There are six major genotypes and more than 100 subtypes. The most common genotype in the United States is genotype 1 (70%), followed by genotypes 2 and 3 (30%). HCV genotype has no impact on disease progression; genotype 1 is associated with antiviral resistance.

HCV has a mean incubation period of 50 days (range 14 to 180 days). Although 70% to 80% of acute infections are asymptomatic, biochemical evidence of hepatitis, such as elevated ALT levels, is observed in most patients, and some patients demonstrate acute viral hepatitis–associated malaise, nausea, and right upper quadrant pain, followed by dark urine and jaundice. Such cases are indistinguishable from acute hepatitis A or B without the results of specific diagnostic tests (see the section on diagnosis). Chronic infection develops in about 70% to 85% of persons with HCV. They are at risk for developing clinically significant liver disease, including steatosis, progressive fibrosis, cirrhosis, and HCC (Fig. 68.5). Factors associated with more rapid disease progression include alcohol use, male gender, age older than 40 years at time of infection, and HIV co-infection. Extrahepatic manifestations associated with persistent HCV infection include cryoglobulinemia, glomerulonephritis, porphyria cutanea tarda, and possibly diabetes mellitus, lichen planus, and B-cell lymphoma.

## Diagnosis

Testing for HCV infection is performed for (1) clinical diagnosis of patients with signs, symptoms, or abnormal laboratory test results indicative of liver disease; (2) management of patients with chronic

HCV during therapy; and (3) screening of asymptomatic persons to identify those infected with HCV. Acute hepatitis C is usually asymptomatic and diagnosis of acute infection is uncommon. Most patients identified as HCV-positive have established chronic infection.

Serologic and virologic markers that are clinically useful include IgG antibody (anti-HCV), a marker of past or present HCV infection; HCV RNA, a direct indicator of ongoing HCV replication; and HCV genotype, for predicting drug resistance. In clinical practice the usual approach is to test initially for anti-HCV, and if results are positive, test for HCV RNA to document viremia. If HCV RNA test results are positive, active HCV infection is confirmed (Fig. 68.6). If HCV RNA test results are negative, additional testing with another HCV antibody assay can verify the anti-HCV result and determine the need for additional follow-up (Table 68.6). Although a confirmed anti-HCV–positive result does not distinguish between current or past infection, it does indicate the need for repeat HCV RNA testing and monitoring ALT activity if the person tested had HCV exposure within the past 6 months or has clinical evidence of HCV disease. Liver biopsy may be used for evaluation of chronic hepatitis and cirrhosis in confirmed cases but is not required for initiation of treatment. Persons diagnosed with HCV infection should be vaccinated against HAV and HBV if not previously immune as acute co-infections with these would negatively affect the course of the HCV infection.

## Treatment

Direct-acting antiviral (DAA) drugs in the treatment of HCV infection were introduced in 2011. The DAAs include protease inhibitors, nucleoside analog protease inhibitors, and nonstructural (NS5A) protein inhibitors, and are better tolerated and more effective than interferon-based regimens previously used. Many of the DAAs have antiviral activity against all genotypes. Approximately 90% of HCV-infected persons can be cured after completion of 8 to 12 weeks of therapy regardless of HCV genotype, prior treatment experience, fibrosis level or presence of cirrhosis. A cure (sustained virologic response [SVR]) is defined as the absence of detectable HCV RNA 12 weeks after completion of treatment.

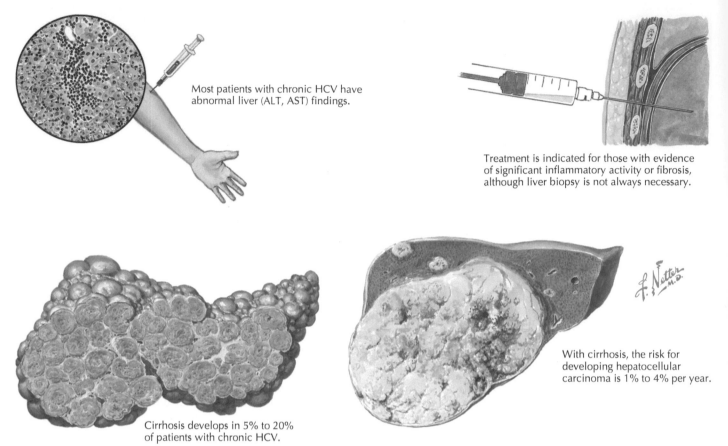

Most patients with chronic HCV have abnormal liver (ALT, AST) findings.

Treatment is indicated for those with evidence of significant inflammatory activity or fibrosis, although liver biopsy is not always necessary.

With cirrhosis, the risk for developing hepatocellular carcinoma is 1% to 4% per year.

Cirrhosis develops in 5% to 20% of patients with chronic HCV.

**Fig. 68.5** Progression of hepatitis C virus (HCV) disease. *ALT,* Alanine aminotransferase; *AST,* aspartate aminotransferase.

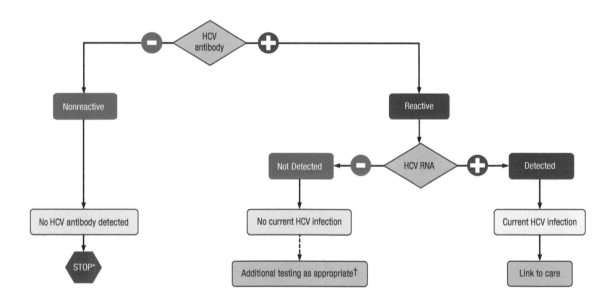

* For persons who might have been exposed to HCV within the past 6 months, testing for HCV RNA or follow-up testing for HCV antibody is recommended. For persons who are immunocompromised, testing for HCV RNA can be considered.

† To differentiate past, resolved HCV infection from biologic false positivity for HCV antibody, testing with another HCV antibody assay can be considered. Repeat HCV RNA testing if the person tested is suspected to have had HCV exposure within the past 6 months or has clinical evidence of HCV disease, or if there is concern regarding the handling or storage of the test specimen.

*Source: CDC. Testing for HCV infection: An update of guidance for clinicians and laboratorians. MMWR 2013;62(18).*

**Fig. 68.6** Recommended testing sequence for identifying current hepatitis C virus (HCV) infection. (Source: Centers for Disease Control and Prevention [CDC]. Testing for HCV infection: an update on guidance for clinicians and laboratories. *MMWR* 2013:62[18]:362-365.)

## TABLE 68.6 Interpretation of Serologic Test Results for Diagnosis of Hepatitis C Virus Infection

| Anti-HCV | HCV RNA | Diagnosis |
|---|---|---|
| Positive | Positive | Acute or chronic HCV infection |
| Positive | Negative | Resolution of HCV infection or false-positive antibody |
| Negative | Positive | Early acute HCV infection, HCV in immunocompromised patients, false-positive nucleic acid test result |
| Negative | Negative | Negative for HCV infection |

*HCV*, Hepatitis C virus.

Evaluation and treatment of patients with hepatitis C should be carried out in consultation with a hepatologist. Lecaprevir/pibrentasvir and sofosbuvir/velpatasvir are two current DAA HCV regimens used for treatment-naïve adults without cirrhosis or with compensated cirrhosis. Treatment with DAAs can lead to reactivation of hepatitis B virus, so testing HCV patients at higher risk of having chronic HBV co-infections (e.g., MSM and IDU) for HBV DNA along with hepatitis B core antibodies is recommended prior to DAA treatment. At the time of writing, DAAs are not yet approved for use in pregnancy or in infants and children aged <3 years. However, recommended HCV testing during pregnancy allows for medical management of pregnancy and delivery issues that may reduce HCV transmission to the infant, and identifies infected mothers and infants who should be monitored for HCV after delivery and treated when possible. Ledipasvir/sofosbuvir was approved for use in persons aged 12 to 17 years in 2017; glecaprevir/pibrentasvir was approved for persons aged ≥12 years, and ledipasvir/sofosbuvir became approved for children ages ≥3 years in 2019.

Co-infection with HIV causes rapid progression from primary HCV infection to end-stage liver disease. Combination antiviral therapy for HCV in HIV-positive individuals is recommended with early or mild disease. The decision to treat an HIV-positive patient with HCV is dependent on the current HIV regimen and stage of disease. Potential drug-drug interactions between medications used to treat HCV and HIV can interfere with optimal HIV therapy and the possibility needs to be assessed before starting HCV treatment.

### Prevention

Because there is no vaccine and no postexposure prophylaxis for HCV, the prevention of new infections worldwide requires procedures to ensure a safe blood supply, implementation of effective infection control and safe injection practices in healthcare and other settings, establishment of expanded harm-reduction and treatment programs for IDUs, and development of educational programs to prevent initiation of high-risk drug and sexual behaviors.

In countries with more developed economic, medical, and public health infrastructures, efforts to minimize morbidity and complications from chronic HCV need to start with the identification of persons already infected with HCV so they can be provided with appropriate counseling and medical management. In the United States, universal HCV screening is recommended at least once in a lifetime for all adults ≥18 years old and for all pregnant women during each pregnancy, as well as for all persons with recognized risk factors or exposures including persons with HIV, IDU, hemodialysis patients, recipients of blood transfusions or organ transplants, healthcare personnel and first responders, and children born to mothers with HCV infection (Box 68.1).

## BOX 68.1 Persons Recommended for Hepatitis C Testing

- **Universal hepatitis C screening:**
  - Hepatitis C screening at least once in a lifetime for all adults aged >18 years, except in settings where the prevalence of HCV infection (HCV RNA-positivity) is <0.1%
  - Hepatitis C screening for all pregnant women during each pregnancy, except in settings where the prevalence of HCV infection (HCV RNA-positivity) is <0.1%
- **One-time hepatitis C testing regardless of age or setting prevalence among persons with recognized risk factors or exposures:**
  - Persons with HIV
  - Persons who ever injected drugs and shared needles, syringes, or other drug preparation equipment, including those who injected once or a few times many years ago
  - Persons with selected medical conditions including persons who ever received maintenance hemodialysis and persons with persistently abnormal ALT levels
  - Prior recipients of transfusions or organ transplants, including persons who received clotting factor concentrates produced before 1987, persons who received a transfusion of blood or blood components before July 1992, persons who received an organ transplant before July 1992, and persons who were notified that they received blood from a donor who later tested positive for HCV infection
  - Healthcare, emergency medical, and public safety personnel after needlesticks, sharps, or mucosal exposures to HCV-positive blood
  - Children born to mothers with HCV infection
- **Routine periodic testing for persons with ongoing risk factors, while risk factors persist:**
  - Persons who currently inject drugs and share needles, syringes, or other drug preparation equipment
  - Persons with selected medical conditions, including persons who ever received maintenance hemodialysis
- **Any person who requests hepatitis C testing should receive it, regardless of disclosure risk, because many persons might be reluctant to disclose stigmatizing risks**

*HCV*, Hepatitis C virus.
From Schillie S, Wester C, Osborne M et al. CDC recommendations for hepatitis C screening among adults—United States, 2020. *MMWR Recomm Rep* 2020;69(No.RR-2):11.

# HEPATITIS D

## Epidemiology

Epidemiologic features of HDV infection strongly parallel those of HBV, as would be expected from its status as a hepatitis B–dependent virus. HDV shares with HBV similar reservoirs of infection and transmission pathways. Worldwide prevalence of HDV infection tends to parallel that of HBV, being highest in nations or subpopulations with highest HBV prevalence. In the United States, the most important routes of transmission for HDV are through injection drug use and sexual contact, although HDV is less efficiently transmitted than HBV in these settings. An estimated 5% of chronic HBV carriers worldwide are co-infected with HDV, which corresponds to an absolute number of 10 to 20 million patients infected with HDV. Co-infection with HBV and HDV is considered to be the most severe form of viral hepatitis, affecting 15 to 20 million individuals worldwide.

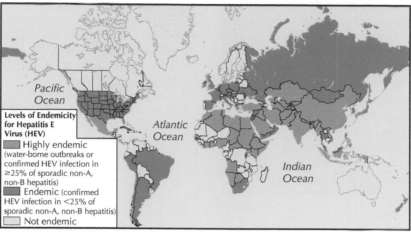

**Fig. 68.7** Global endemicity of hepatitis E virus. (From the Centers for Disease Control and Prevention [CDC]: CDC health information for international travel 2010. Atlanta: U.S. Department of Health and Human Services, Public Health Service; 2009.)

## Clinical Illness

HDV is a defective RNA virus that is dependent on the presence of HBV to cause infection and disease and greatly augments the severity of both acute and chronic liver disease in the HBV- and HDV-infected host. The incubation period and clinical presentation parallel those of HBV. HDV can cause infection in two ways: (1) by co-infection with HBV of an HBV-susceptible host and (2) by superinfection of an HBV chronic carrier. Co-infection with HDV can lead to accelerated fulminant hepatic failure, and the vast majority of HDV-superinfected carriers develop progressive hepatitis.

## Diagnosis

Accurate diagnosis of HDV infection in persons with acute hepatitis requires use of serologic tests for HBV (HBsAg and IgM anti-HBc) and HDV (anti-HDV and IgM anti-HDV). Acute HBV-HDV co-infection is diagnosed in persons with positive IgM anti-HBc (indicative of acute HBV infection) and a positive marker for HDV. Clinically, acute HDV superinfection often appears as a severe acute hepatitis that may run a fulminant course. It also may manifest as an exacerbation of preexisting HBV disease or as new hepatitis in a previously asymptomatic HBsAg carrier. The correct diagnosis is indicated by a negative result of a test for IgM anti-HBc and is confirmed by the detection of HDV markers in persons who are chronically infected with HBV (HBsAg and total anti-HBc positive but IgM anti-HBc negative).

## Treatment

HDV infection can be treated with interferon for 12 months, but therapy rarely leads to clearance of the infection. Lamivudine, a nucleoside analogue, has been used as monotherapy or in combination with interferon for treatment of HBV, but it does not suppress HDV infection.

## Prevention

The single most important tool in the prevention of HBV-HDV co-infection is immunization of HBV-susceptible individuals with the HBV vaccine to prevent HBV chronic infection. Sexually active young adults and new initiates of injection drug use are at highest risk, and efforts to vaccinate them need to be strengthened (see Table 68.4). The prevention of HDV superinfection must rely on awareness of HBV carrier status and counseling on modification of high-risk behaviors.

# HEPATITIS E

## Epidemiology

HEV is transmitted via the fecal-oral route, with contaminated water being the most common source of infection. HEV is the single most important cause of acute clinical hepatitis in adults throughout Central and Southeast Asia and the second most important cause, behind HBV, throughout the Middle East and North Africa. In contrast, HEV is responsible for a very small number of reported viral hepatitis cases in the United States and other industrialized countries; cases are usually associated with travelers returning from endemic areas. However seroprevalence of up to 20% has been found even in nonendemic countries such as the United States, where there are virtually no clinical cases (Fig. 68.7). The geographic disparities may be due to the HEV genotypes that are circulating in a given region.

Mammalian HEV strains typically infect pigs and humans and are grouped into four different genotypes, 1 through 4. Genotypes 1 and 2 are highly prevalent in developing countries, where they are responsible for widespread outbreaks of hepatitis E (from contaminated water). Infections with genotypes 1, 2, and 4 have high mortality rates in pregnant women, with development of fulminant liver failure and mortality rate up to 25%. Pigs commonly carry genotypes 3 and 4. Genotype 3 HEV strains are prevalent in industrialized countries, appear to be less virulent, and have been recovered from both swine and humans in the same regions. The virus can be transmitted by direct animal contact or by consumption of undercooked meat and may cause the occasional case of clinical HEV in these settings. This suggests that the relatively high prevalence of anti-HEV in industrialized countries may result from inapparent infections with attenuated strains of HEV derived from swine (or other domestic or wild animals) that rarely cause clinical disease. Unlike other enterically transmitted infections, HEV is rarely spread to household contacts, and infection rates are low in infants and young children. Cases of HEV genotype 3 infections have been related to consumption of shellfish, probably from contamination of shellfish harvesting in waters with agricultural waste runoffs. Avian strains of HEV that can cross-infect closely related species but have not been shown experimentally to infect primates are present around the world.

## Clinical Illness

HEV is a small nonenveloped RNA virus that causes disease indistinguishable from that caused by HAV without serologic testing. However,

HEV differs from HAV in a few respects: the incubation period of HEV is, on average, 10 days longer, with a mean of 40 days (range 14 to 60 days), and its overall mortality rate is higher (1% to 4%) but not age dependent. Similar to HAV, HEV causes an acute, self-limiting infection that may vary in severity from unapparent to fulminant and is not known to progress to chronicity, except in the rare event of infection of the immunologically compromised host.

A unique feature of HEV is an approximate 20% mortality rate among pregnant women, particularly in the third trimester. The exact cause of this phenomenon is uncertain, although hormonal and/or immunologic factors have been implicated. HEV infection during pregnancy is associated with low birth weight, prematurity, and increased perinatal mortality.

## Diagnosis

HEV can be diagnosed by the demonstration of IgM anti-HEV in the serum (generally present when the patient is first seen by a physician) or by detection of viral RNA in the serum or feces by nested or real-time PCR. Although the anti-HEV assay is commercially available, it is not FDA approved. Unfortunately, both serologic tests and molecular tests vary greatly in sensitivity, making diagnosis, and especially seroprevalence studies, less reliable than for the other human hepatitis viruses.

## Treatment

There is no specific treatment for acute HEV. The mainstay of treatment is supportive care with fluids, rest, and avoidance of medications and alcohol, which may exacerbate liver damage. There is limited experience with ribavirin being used to treat chronic hepatitis E in solid-organ transplant recipients.

## Prevention

In developing countries, preventive measures are aimed at proper treatment and disposal of human waste and purification of water in addition to improved personal hygiene. During epidemics, boiling and/or chlorination of water has been an effective means of prevention. Travelers to HEV endemic countries should observe protective measures against all enterically transmitted infections: drinking boiled or chlorinated water and eating only thoroughly cooked foods especially pork, offal, and seafood products. Serum immunoglobulin does not appear to protect against HEV. At least one vaccine has been shown to be safe and effective in protecting humans against HEV, but it is not in production. The potential market in industrialized countries is quite small, and such a vaccine may be cost-prohibitive for low-income countries.

## ACKNOWLEDGMENT

The authors acknowledge the work of Michael James Babineaux and Miriam J. Alter on this chapter in the previous edition.

## EVIDENCE

Alter MJ. Epidemiology of hepatitis C virus infection. *World J Gastroenterol* 2007;13(17):2436-2441. *Increase of HCV-related morbidity and mortality is the result of an unprecedented increase in the spread of HCV during the 20th century due to illicit drug use and injectable therapies.*

Bower WA, Nainan OV, Han X, Margolis HS. Duration of viremia in hepatitis A virus infection. *J Infect Dis* 2000;182:12-17. *Adults with HAV infection are viremic for as long as 30 days before the onset of symptoms and the average duration of viremia is 95 days.*

Dienstag JL. Hepatitis B virus infection. *N Engl J Med* 2008;359:1486-1500. *Review of epidemiology, virology, and treatment strategy for HBV.*

Goldstein ST, Alter MJ, Williams IT, et al. Incidence and risk factors for acute hepatitis B in the United States, 1982-1998: implications for vaccination programs. *J Infect Dis* 2002;185(6):713-719. *Acute HBV has dramatically declined in the United States, particularly in children, but most cases in adults represent missed opportunities for vaccination and may be prevented by HBV immunization in STD clinics and correctional systems.*

Hofmeister MG, Rosenthal EM, Barker LK, et al. Estimating prevalence of hepatitis C virus infection in the United States, 2013–2016. *Hepatology* 2019;69:1020-31. https://doi.org/10.1002/hep30297. *Analysis of 2013-2016 NHANES data for adults with anti-HCV and HCV RNA, and other data from the literature among populations not sampled by NHANES, to come up with national prevalence estimates.*

Nimgaonkar I, Ding Q, Schwartz RE, Ploss A. Hepatitis E virus: advances and challenges. *Nat Rev Gastroenterol* 2018;15:96-110. *A comprehensive review of hepatitis E including epidemiology, treatment, prevention, and molecular biology.*

Wasley A, Samandari T, Bell BP. Incidence of hepatitis A in the United States in the era of vaccination. *JAMA* 2005;294:194-201. *Implementation of routine hepatitis A vaccination results in decreased incidence of infection.*

## ADDITIONAL RESOURCES

American Association for the Study of Liver Diseases (AASLD); Infectious Diseases Society of America (IDSA). HCV guidance: recommendations for testing, managing and treating hepatitis C. Alexandria and Arlington, VA: AASLD and IDSA; 2019. https://www.hcvguidelines.org. *Updated standard of care guidelines for hepatitis C including "Simplified HCV Treatment Algorithm for Treatment-Naive Adults Without Cirrhosis" and "Simplified HCV Treatment Algorithm for Treatment-Naïve Adults With Compensated Cirrhosis".*

Centers for Disease Control and Prevention (CDC). CDC Health information for international travel 2020. Atlanta: U.S. Department of Health and Human Services, Public Health Service; 2019. Chapter 4: Travel-related infectious diseases, hepatitis A, B, C, D, E. Available at https://wwwnc.cdc.gov/travel/yellowbook/2020/travel-related-infectious-diseases/hepatitis. *Information on global prevalence of infectious hepatitis, risk of transmission factors, and personal prevention measures including vaccines against hepatitis A and B, and updated vaccine dosing schedules.*

Easterbrook PJ, Roberts T, Sands A, Peeling R. Special commentary—Diagnosis of viral hepatitis. *Curr Opin HIV AIDS* 2017;12(3): 302-314. *Testing and diagnosis open the path to accessing treatment and prevention services, furthering the world toward the declared goal of elimination of viral hepatitis as a public health threat by 2030.*

Schillie S, Wester C, Osborne M, et al. CDC recommendations for hepatitis C screening among adults—United States, 2020. *MMWR Recomm Rep* 2020;69(No.RR-2):1-17. *Updated and comprehensive recommendations from the CDC include universal opt-out hepatitis C screening for all adults at least once in a lifetime and all pregnant women during each pregnancy, review of epidemiology, summary of clinical management and treatment, and literature review.*

Wilkins T, Sams R, Carpenter M. Hepatitis B: screening, prevention, diagnosis, and treatment. *Am Fam Physician* 2019;99(5):314-323. https://www.aafp.org/afp/2019/0301/p314.html. *Comprehensive updated review on the medical management and treatment of HBV with a strong clinical perspective.*

# 69

# Rabies

*Elaine C. Jong*

## ⬤ ABSTRACT

Rabies has been recognized as a source of great human suffering and fear since ancient times. Characterized by a near 100% case fatality rate, it is among the deadliest infectious diseases known to humanity. The rabies virus (RABV) is present in the saliva of clinically ill mammals and is typically transmitted to humans through a bite. The incubation period is usually 1 to 3 months. After entering the central nervous system (CNS), the virus causes an acute, progressive encephalomyelitis. Although treatment options for rabies are currently limited, the disease is highly preventable by receipt of preexposure prophylaxis (PrEP) with rabies vaccine among persons at possible or known risk or by proper administration of rabies postexposure prophylaxis (PEP) after a possible high-risk rabies exposure.

## ✴ CLINICAL VIGNETTE[a]

A 15-year-old girl was brought by her mother to the emergency department (ED) of a community hospital with a one-day history of diplopia and feeling unsteady. Nausea and vomiting without fever developed on the day of her ED visit. The patient was a high school student who was active in sports and had been in excellent general health. Upon further questioning, she recalled that her illness probably began 3 days before, when she had experienced generalized fatigue and paresthesia of the left hand. "A neurologist noted partial bilateral sixth-nerve palsy and ataxia. The results of magnetic resonance imaging (MRI) and angiography of the brain were unremarkable. By the fourth day after the onset of symptoms, blurred vision, weakness of the left leg, and a gait abnormality were present. On the fifth day, fever (38.8°C), slurred speech, nystagmus, and tremors of the left arm developed." By this time additional history had been elicited from the patient's family: about a month before the onset of symptoms, the girl had rescued and released a bat that struck an interior window. She had sustained a 5-mm laceration to her left index finger from the bat and washed the wound with peroxide, but no medical attention was sought; thus no rabies PEP had been administered.

With the progression of symptoms, the patient was transferred to the university medical center hospital. "On the first hospital day, the patient was febrile (38.2°C) and semiobtunded but answered simple questions and complied with simple commands during diagnostic maneuvers. She had scanning speech, bilateral sixth-nerve palsies, decreased upward gaze, dysarthria, myoclonus, intention tremor of the left arm, and ataxia. Samples of serum, cerebrospinal fluid, nuchal skin, and saliva were submitted to the Centers for Disease Control and Prevention (CDC) for the diagnosis of rabies. Repeated MRI and angiography showed no abnormalities. The patient began salivating, with uncoordinated swallowing, and was intubated for airway protection. On the second hospital day, the presence of rabies virus-specific antibody in her CSF and serum was confirmed by the CDC. Attempts to isolate rabies virus,

detect viral antigen, and amplify viral nucleic acid from two skin biopsies and nine saliva samples were unsuccessful."

COMMENT: This case illustrates a typical incubation period of about a month between a high-risk rabies exposure and the onset of rabies disease symptoms. The patient's progression of symptoms, neurologic status, and diagnostic studies were well documented, as she was seen in a modern healthcare system where no effort was spared to secure the diagnosis and manage her care. Miraculously, she survived and was left with only minor neurologic impairments in the years following her discharge from the hospital and rehabilitation therapy.

[a] This clinical vignette is derived from the remarkable case report detailed in: Willoughby, Jr RE, Tieves KS, Hoffman GM, et al. Survival after treatment of rabies with induction of coma. *N Engl J Med* 2005;352:2508-2514; and personal communication with the patient.

## GEOGRAPHIC DISTRIBUTION

### Africa and Asia

The World Health Organization (WHO) estimates that rabies is responsible for 59,000 deaths every year, of which more than 99% occur as a result of dog bites in the developing countries of Africa and Asia. Exposure risk is highest in rural areas, where free-roaming dogs are commonplace. Approximately 40% of cases occur in children younger than 15 years of age owing to the high incidence of dog bites in this demographic group. In rabies-endemic regions, mass vaccination campaigns aimed at dogs attempt to interrupt RABV transmission between dogs, humans, and other mammals. Rabies and rabies-related viruses have also been isolated from African and Asian wildlife, including bats, mongooses, jackals, foxes, raccoons, skunks, and other species. Travelers to rabies-endemic countries who anticipate prolonged stays in rural areas and/or extensive outdoor activities should consider preexposure immunization before travel (Fig. 69.1).

### Latin America and the Caribbean

In many Latin American and Caribbean countries, human rabies has declined in recent years as canine vaccination rates and use of rabies PEP have risen. From 1993 to 2002, dogs were implicated in 65% of human cases reported in the region; in 2004, 22% were attributed to dogs. Despite this overall trend, canine rabies remains a concern in many places throughout the region.

A growing proportion of human cases are also mediated by hematophagous (vampire) bats. Notable outbreaks caused by these animals have occurred in Brazil, Peru, and Colombia. Rabid vampire bats are particularly a threat to human and cattle populations in remote tropical areas within the Amazon. Although rabies is also present in non-hematophagous bats in the region, a complete understanding of their contribution to human disease is lacking because of the absence of

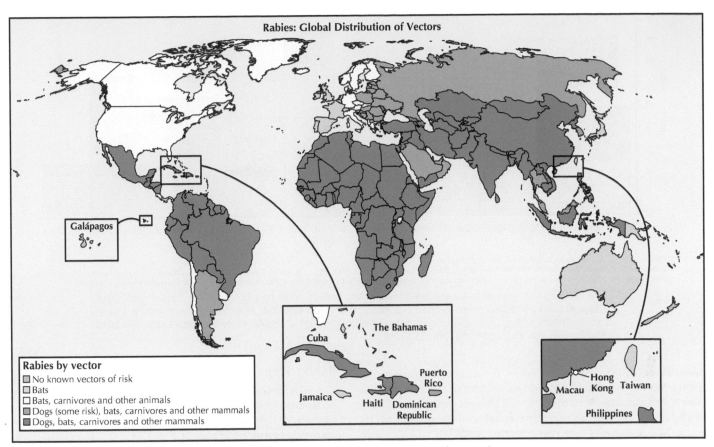

**Fig. 69.1** Rabies distribution and immunization recommendations. (Courtesy Nancy Gallagher and Kevin Liske, Division of Global Migration and Quarantine, Centers for Disease Control and Prevention.)

The following content is part of the figure:

**Rabies: Global Distribution of Vectors**

**Rabies by vector**
- No known vectors of risk
- Bats
- Bats, carnivores and other animals
- Dogs (some risk), bats, carnivores and other mammals
- Dogs, bats, carnivores and other mammals

**Recommendations for Pre-Exposure Immunization for Travelers**

| Exposure | Vaccine recommended for: |
|---|---|
| No known risk | No recommendation |
| • Bats | • Travelers with high occupational risks such as wildlife professionals, researchers, veterinarians |
| • Bats <br> • Carnivores and other mammals | • Adventure travelers visiting areas where vectors commonly found |
| • Bats <br> • Dogs (some risk) <br> • Carnivores and other mammals | All of the above, plus <br> • Long-term travelers <br> • Expatriates |
| • Bats <br> • Dogs <br> • Carnivores and other mammals <br> • High-risk activities explicitly identified | All of the above, plus <br> • Travelers spending a lot of time outdoors <br> • Travelers to rural areas <br> • Travelers involved in activities like bicycling, camping, hiking <br> • Children |

taxonomic specificity when bats are identified and reported as a source of exposure, in addition to other surveillance limitations. In multiple instances, human rabies has been linked to monkeys, skunks, foxes, raccoons, and livestock. Domestic cats have also served as an important source of infection, with 3% of reported human cases in 1993 to 2002 attributed to these animals. Travelers who come in contact with animals—particularly wild or stray animals—should be mindful of rabies risks and take steps to avoid bites and other exposures.

## Europe, Canada, and the United States

In Europe and temperate North America, human rabies is rare. From 1960 to 2018, there were 125 human rabies cases reported in the United States, an annual average of about two cases per year. Europe currently averages approximately nine reported cases a year, with most cases occurring in eastern Europe. The widespread availability of rabies vaccines and rabies immune globulin (RIG), a well-immunized dog population, and effective antistray programs are credited with the low human rabies incidence seen in most developed countries.

Wildlife has the highest burden of rabies in North America and Europe. Most human cases in the United States and Canada are associated with insectivorous bats, whereas in eastern Europe rabies transmission is largely driven by the red fox, with dogs playing an important role as victims of fox-associated spillover infections. Insectivorous bats in Europe also serve as important reservoirs of European bat

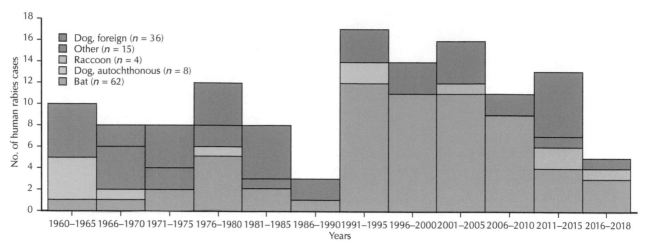

**Fig. 69.2** Rabies virus variants[a] associated with human rabies cases ($n$ = 125),[b] United States, 1960–2018. (Reused with permission from Pieracci EG, Pearson, CM, Wallace RM, et al. Vital signs: trends in human rabies deaths and exposures—United States, 1938-2018, Fig. 2. *MMWR Morb Mortal Wkly Rep* 2019; 68:524-528. doi: I0.15585/mmwr.mm6823e1.)

lyssaviruses (EBLs), which cause rabies in humans. Foreign-acquired rabies represents a significant portion of human cases reported in North America and western Europe, which are typically associated with dog exposures in rabies-endemic countries. During 1960 to 2018, some 28% of human rabies cases reported in the United States were acquired from dogs by people who were traveling abroad (Fig. 69.2).

More than 65% of animal rabies reported in the United States and Canada is found in terrestrial wildlife such as raccoons, skunks, and foxes. However, insectivorous bats are considered higher-risk vectors to humans because lesions inflicted by these mammals tend to be less conspicuous, less easily recognized, and/or taken less seriously and therefore are less likely to be treated than are bites and scratches from rabid carnivores. Among 125 reported human rabies cases in the United States from 1960 to 2018, a total of 62 were associated with bat rabies virus variants (see Fig. 69.2). A majority of bat-associated human cases in the United States and Canada have been attributed to so-called *cryptic bat exposures*—characterized by the absence of an elicited bite or scratch history—which most often involve rabies virus variants associated with the silver-haired bat (*Lasionycteris noctivagans*) and the eastern pipistrelle bat (*Pipistrellus subflavus*). Because of the risks associated with undetected bat bites, the US Advisory Committee on Immunization Practices (ACIP) recommends that any suspected contact with a bat be evaluated for possible RABV exposure if a bite cannot be reasonably excluded.

### Australia

In Australia, the emergence of Australian bat lyssaviruses (ABLs) in 1996 has elevated public health concerns in a country that has historically enjoyed "rabies free" status. These bat-associated viruses—like the EBLs seen in Europe and the Lagos bat and Duvenhage viruses seen in Africa—although phylogenetically distinct from the classic rabies virus seen in the New World and most of the Old, produce a fatal encephalomyelitis indistinguishable from that caused by the RABV. In the late 1990s, two people acquired rabies after incurring bites from ABL-infected bats. Variants of the virus have been isolated from both frugivorous and insectivorous bats. Exposures to these animals should therefore be regarded in the same way as they would be in countries where bat rabies is present.

### Rabies-Free Countries

The WHO may designate a country as "rabies free" if there have been no reports of indigenous cases in at least 2 consecutive years based on surveillance that is considered sufficiently sensitive. However, travelers should be aware that because surveillance for rabies often involves underreporting, an animal exposure may carry transmission risks even if the exposing animal is from a country considered to be rabies free.

## ETIOLOGY

The rabies and rabies-related viruses belong to the Rhabdoviridae family as members of the *Lyssavirus* genus. Lyssaviruses are neurotropic single-stranded ribonucleic acid (RNA) viruses characterized by a bullet-shaped morphology, a tightly coiled nucleocapsid, and five structural proteins. In keeping with other nonsegmented RNA viruses that have negative-sense polarity, genome replication and protein synthesis occur within the cytoplasm of infected cells under the direction of an RNA-dependent viral polymerase.

Eleven recognized lyssaviruses cause rabies, but only one species is formally called the *rabies virus*. Each species is further subdivided into phylogenetically distinct variants that are host adapted to the mammalian reservoirs in which they circulate. The phenomenon known as *spillover* occurs when a variant adapted to one host species (such as the dog) infects another species (e.g., the human) to which it is not adapted. Although disease in the newly infected host may result, spillover infrequently leads to sustained propagation of the variant in a new host population. Humans are poor conduits of disease transmission (naturally occurring human-to-human transmission has yet to be definitively established); thus they are considered dead-end hosts.

## PATHOPHYSIOLOGY

Rabies transmission usually occurs through the percutaneous bite of a rabid mammal shedding the virus in its saliva (Fig. 69.3). Nonbite exposures such as scratches and licks can also lead to rabies infection, although less frequently than bites. Under atypical conditions, transmission may also occur through the inhalation of highly concentrated aerosolized viral particles. Access to the nervous system is granted

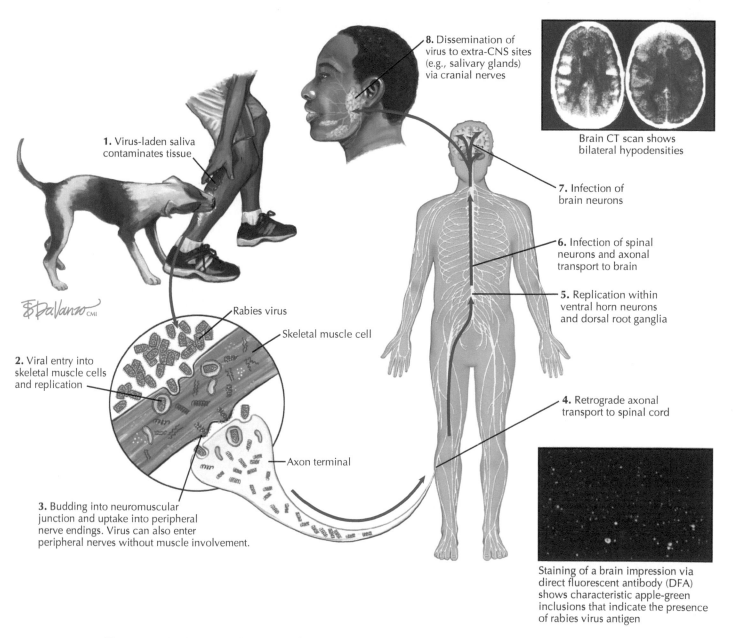

**8.** Dissemination of virus to extra-CNS sites (e.g., salivary glands) via cranial nerves

Brain CT scan shows bilateral hypodensities

**7.** Infection of brain neurons

**6.** Infection of spinal neurons and axonal transport to brain

**5.** Replication within ventral horn neurons and dorsal root ganglia

**4.** Retrograde axonal transport to spinal cord

**1.** Virus-laden saliva contaminates tissue

Rabies virus

Skeletal muscle cell

**2.** Viral entry into skeletal muscle cells and replication

Axon terminal

**3.** Budding into neuromuscular junction and uptake into peripheral nerve endings. Virus can also enter peripheral nerves without muscle involvement.

Staining of a brain impression via direct fluorescent antibody (DFA) shows characteristic apple-green inclusions that indicate the presence of rabies virus antigen

**Fig. 69.3** Pathophysiology of rabies. *CNS,* Central nervous system. (Computed tomography and direct fluorescent antibody courtesy Centers for Disease Control and Prevention.)

either through inoculation directly into peripheral nerves or via infection of surrounding tissue (e.g., muscle cells), with subsequent nerve entry at the neuromuscular junction (see Fig. 69.3).

After peripheral nerve invasion, the rabies virus reaches the CNS via retrograde axoplasmic transport. Once the virus has infected the ventral horn of the spinal cord and/or the dorsal root ganglia, viral amplification leads to rapid dissemination of the virus in the rostral gray matter of the spinal cord. Progression to the brain is achieved through axoplasmic transport within several ascending and descending fiber tracts, leading to early placement in the brain stem, followed by retrograde diffusion into the rest of the brain. Resulting neurologic signs are considered to be primarily a product of nerve cell dysfunction as opposed to necrosis or apoptosis; however, the exact functional impairment involved is unclear.

Viral migration from the brain into peripheral sites such as the salivary glands provides a gateway for virus particles to escape from the body and invade a new host. After infection of brain-stem nuclei, the facial and glossopharyngeal cranial nerves convey virus to the salivary glands via their associated ganglia. Subsequent infection of glandular epithelia results in considerable viral shedding into salivary secretions. Virions are also dispatched to ocular structures, such as the cornea and retina, and deposited in organs and tissues serviced by parasympathetic and sympathetic nerves, including the heart, kidneys, and liver. The iatrogenic implications of this latter phenomenon were demonstrated in 2004, when four people contracted rabies after receiving transplanted material from a then-undiagnosed rabies-infected organ donor. Rabies transmission via corneal transplants has also occurred on several occasions. Because the virus frequently accumulates in the free sensory nerve endings of nuchal tactile hair, a skin biopsy sample from this area is used as a standard diagnostic specimen.

## CLINICAL FEATURES

Onset of symptoms for most rabies patients occurs 3 to 12 weeks postexposure; however, incubation periods far outside of this time frame do occur. Multiple bite wounds, severe wounds, and bite wounds to the head, face, and neck are associated with shorter incubation times and should therefore be considered when the urgency of the need for PEP is being assessed. Incubation periods of a year or more have been described in a few cases of human infection; therefore rabies PEP should be reasonably considered regardless of the amount of time that has elapsed since an exposure occurred.

Rabies has two clinical manifestations: the encephalitic form (or furious rabies) and the paralytic form (or dumb rabies). Encephalitic rabies is the most common form overall, although the majority of patients infected by vampire bats exhibit the paralytic form for reasons that remain unknown. Prodromal signs and symptoms for both are nonspecific and include fever, chills, malaise, and headache. Paresthesias on the part of the body that received the bite and pain and/or pruritus at the site of the bite wound unrelated to the injury itself are also common features.

Encephalitic rabies is characterized by altered mental status, agitation, hyperreactivity to sensory stimuli, intermittent consciousness, myoclonus, and muscle tremors. Also featured are signs indicative of autonomic neuropathy, including hypersalivation, mydriasis, and excessive lacrimation. Dysphagia and hydrophobia are cardinal sequelae; affected patients often react fearfully when offered water and exhibit inspiratory laryngeal spasms. Aerophobia is also frequently observed, as reflected by an exaggerated response to air currents passing over the skin (commonly referred to as the "fan test"). Seizures may occur but are not typical. Paralysis leading to coma usually occurs within 10 to 14 days, with death ensuing shortly thereafter, frequently precipitated by multiple organ failure.

In paralytic rabies, patients initially develop ascending muscle weakness that rapidly progresses to flaccid paralysis. Mental status is often unremarkable at the onset, hydrophobic spasms are less likely to be present, and patients may be unable to speak because of laryngeal muscle weakness—an occurrence that can interfere with obtaining an animal bite history, thus complicating a clinical diagnosis. Peripheral neuropathy may be the cause of the muscle weakness seen in this form. Patients with the paralytic form overall tend to have longer periods of survival than patients exhibiting encephalitic rabies.

## DIFFERENTIAL DIAGNOSIS

Several viral infections causing encephalitis produce clinical signs and preliminary laboratory findings consistent with rabies. In particular, neuroinvasive arboviral diseases—including those caused by California serogroup equine encephalitis and West Nile viruses—are important differential diagnoses for rabies. Occurrence of these illnesses increases in late summer, which slightly resembles the seasonal pattern seen in bat-associated rabies in the United States. Serologic testing to rule out these diseases before testing for rabies (antemortem) is especially indicated if an encephalitic patient is older than age 50 years, lacks an animal exposure history, and resides in or has recently traveled to an area where West Nile disease and other mosquito-borne encephalitides are endemic.

Other acute neurologic diseases may be mistaken for rabies because of their epidemiologic link to animal exposures. Tetanus, which occurs secondary to animal bites and other contaminated wounds, causes sustained muscle rigidity that differs from the spasms expressed in encephalitic rabies and the flaccidity expressed in paralytic rabies. In addition, altered mental status is not a typical feature of tetanus. Herpes B infection is usually the result of bites or scratches from macaque monkeys, and skin vesicles at the bite wound are a common manifestation. Incubation periods for both tetanus and herpes B are generally shorter than for rabies. Severe forms of brucellosis, leptospirosis, and toxoplasmosis are other zoonoses associated with encephalopathy, but they usually produce systemic sequelae that rabies does not.

Another differential diagnosis is acute disseminated encephalomyelitis (ADEM), which is an immune-mediated disease triggered by exanthematous viral and bacterial infections including measles, varicella zoster, herpes simplex, and Rocky Mountain spotted fever. Less frequently, ADEM also occurs in association with certain vaccines, including those against rabies, smallpox, and measles. Signs of ADEM generally arise 1 to 20 days after a preceding illness or 1 to 3 weeks after vaccination is initiated. Rabies vaccines derived from neural tissue carry a higher risk for causing ADEM than do rabies cell-culture vaccines. MRI findings suggestive of diffuse or multifocal demyelination in the CNS along with cortical signs such as aphasia, cortical blindness, and seizures are more characteristic of ADEM than rabies.

A condition that clinically mirrors paralytic rabies is Guillain-Barré syndrome (GBS). Demyelination of peripheral nerves and axonal degeneration are features of both. As with ADEM, patients with GBS usually have a recent history of vaccination or febrile illness. Patients with postinfection GBS may exhibit signs associated with the precipitating infection (e.g., gastroenteritis), whereas localized pruritus and pain occurring in the prodromal phase favor a rabies diagnosis.

A psychosomatic condition termed *rabies hysteria* has been ascribed to individuals whose belief in having the disease leads them to exhibit rabies-like signs of aggression, swallowing difficulty, and other behavioral patterns. Objective evidence, such as fever and cerebrospinal fluid pleocytosis, that points to an encephalomyelitic infection is usually absent in these patients, as is the steady progression that exemplifies the clinical course of rabies disease.

## DIAGNOSTIC APPROACH

Rabies should be suspected in any person with an animal exposure history who has an unexplained encephalitis or myelitis. It should be noted that the lack of an elicited exposure history should not preclude suspicion, because absent exposure histories are common in rabies patients. Recent travel to or emigration from a rabies-endemic area should also elevate suspicion. Progressive worsening of neurologic signs over a period of days is an important positive indicator of disease.

Facilities capable of conducting human rabies testing are limited to a few reference laboratories. Most human rabies testing in the United States and Canada is conducted by the rabies laboratories at the CDC and the Canadian Food Inspection Agency, respectively. Once rabies is suspected, consultation with state or provincial health departments is advisable. Other, more likely causes should be ruled out before resources are expended in rabies testing. However, laboratory testing should be pursued soon after the disease is suspected to ensure that persons potentially exposed to infectious material can take appropriate actions and safeguards.

For antemortem diagnosis, specimens used to confirm rabies include serum, cerebrospinal fluid, saliva, and neck skin biopsy. A brain biopsy specimen may also be used for antemortem diagnosis; however, its collection is not recommended because its diagnostic value is outweighed by associated risks to the patient. Serum and cerebrospinal fluid are examined for the presence of rabies virus antibody via indirect fluorescent antibody and virus neutralization tests. Detection of viral antigen in a neck skin biopsy specimen is achieved through direct fluorescent antibody testing, and reverse transcription–polymerase chain reaction (RT-PCR) is used to detect the presence of viral RNA in both saliva and skin. Repeat testing may be necessary to rule out rabies if negative laboratory findings exist in the presence of strong clinical and epidemiologic evidence.

In deceased patients, brain tissue is the standard specimen for diagnosis. Direct fluorescent antibody testing is used to detect viral antigen in the brain stem, cerebellum, and hippocampus.

## TREATMENT

There is no standard treatment for rabies besides palliative support, which includes appropriately applied analgesia, sedation, and assisted ventilation. Given the poor prognosis, careful consideration should be given before pursuing aggressive treatment measures. Experimental therapeutic approaches have been used to treat human cases, including one patient in Wisconsin who successfully recovered from the disease. Treatment for this patient included antiviral therapy using ribavirin and coma induction using benzodiazepines and barbiturates, with ketamine and amantadine used to prevent excitotoxicity. Neither RIG nor vaccine was administered before or after illness. To date, this patient is the only documented survivor of rabies who had not received PEP or been previously vaccinated against rabies.

## PREVENTION

Rabies in humans and animals is highly preventable through vaccination and, when applicable, passive immunization with RIG. Currently two rabies vaccines approved by the US Food and Drug Administration (FDA) are available in North America: the human diploid cell vaccine (HDCV) and the purified chick embryo cell vaccine (PCECV). These biologics are used for both PrEP and PEP.

### Preexposure Prophylaxis

In areas where terrestrial animal rabies is present, occupational groups with frequent exposure to animals (e.g., veterinarians, wildlife workers, animal rehabilitators), or individuals engaged in activities that put them at risk of wildlife animal contact (e.g., ecotourists, cavers) should be vaccinated preventively, as should rabies laboratory workers who work closely with specimens from such individuals. Travelers who plan long-term travel to enzootic areas (e.g., expatriates and their young children) should also consider preexposure immunization. This recommendation also applies to travelers who are planning activities in settings where bats are abundant. Tiered recommendations for preexposure immunization for travelers are based on the global distribution of the principal reservoirs and vectors of rabies (see Fig. 69.1).

For preexposure immunization, a 1-mL dose of rabies vaccine (either HDCV or PCECV) should be injected intramuscularly (IM) in the deltoid (or outer thigh in children) on days 0, 7, and 21 (or 28). After primary immunization, boosters may later be indicated for individuals continuously or frequently at risk for inapparent rabies exposures, such as those encountered by rabies laboratory workers or bat handlers. For such occupational groups, antibody titers should be monitored using the rapid fluorescent focus inhibition test (RFFIT) every 6 months or 2 years, depending on the individual's risk category; a 1-mL booster is indicated if tested serum fails to exhibit complete virus neutralization at the 1:5 dilution. Periodic titer checks and booster shots are not recommended for individuals who are infrequently exposed to rabies and have a high likelihood of being aware of such exposures when they occur.

During the past decade, the supply of HDCV and PCEC has sometimes been limited owing to production constraints. During periods of limited supplies, preexposure rabies vaccination was restricted for most people in the United States and Canada, including overseas travelers. Exceptions included high-risk occupational groups such as animal control officers and rabies laboratorians, and the vaccine remained available for PEP for individuals possibly exposed to rabies.

### Postexposure Prophylaxis

To appropriately manage a potential rabies exposure, the risk of infection should be thoroughly assessed. The WHO has established three categories of RABV exposure, and the category of exposure determines the indicated PEP procedure (Box 69.1). Administration of rabies PEP is generally considered a medical urgency, not a medical emergency.

Any mammalian bite or scratch should receive prompt local first aid by thorough cleansing of the wound with copious amounts of soap, water, and a virucidal agent such as povidone iodine. Wound cleansing is considered an important component of rabies PEP as it reduces tissue contact with infectious material. In unvaccinated patients with severe wounds, suturing should be delayed to allow infiltration of the wound with RIG and to prevent further dissemination of the virus throughout the traumatized tissue. The recommended protocol for PEP in immunized and nonimmunized patients differs (Fig. 69.4).

---

**BOX 69.1  Risk of a Rabies Virus Exposure According to the Type of Contact With the Animal Suspected of Having Rabies**

- **Category I:** Touching or feeding animals, animal licks on skin (no exposure)
- **Category II:** Nibbling of uncovered skin, minor scratches or abrasions without bleeding (exposure)
- **Category III:** Single or multiple transdermal bites or scratches, contamination of mucous membrane or broken skin with saliva from animal licks, exposures due to direct contact with bats (severe exposure)

From WHO Weekly Epidemiological Record 2018, 93:203.

---

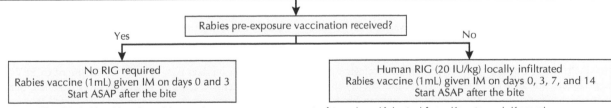

**What To Do If an Animal Bites During International Travel**

It is medically urgent to seek healthcare!

> Wash wound thoroughly with soap and water

> Get an immediate evaluation at a local medical facility. If they are unable to evaluate you, go to the closest modern medical facility as soon as possible. If no satisfactory evaluation is possible, then FLY TO THE NEAREST COUNTRY WITH ADVANCED MEDICAL FACILITIES as soon as possible.

> Rabies pre-exposure vaccination received?

Yes → No RIG required / Rabies vaccine (1mL) given IM on days 0 and 3 / Start ASAP after the bite

No → Human RIG (20 IU/kg) locally infiltrated / Rabies vaccine (1mL) given IM on days 0, 3, 7, and 14 / Start ASAP after the bite

**Fig. 69.4** Algorithm for rabies postexposure management of travelers. (Adapted from Keystone J, Kozarsky P, Freedman D, et al, eds. *Travel medicine.* 2nd ed. Philadelphia: Elsevier; 2008.)

When feasible, rabies transmission should be ruled out by having the exposing animal either euthanized and tested for rabies or—in the case of dogs, cats, and ferrets—confined and observed for any neurologic signs that appear within 10 days. In the absence of a negative animal rabies diagnosis, any patient who has been bitten by a wild terrestrial carnivore (e.g., raccoon, skunk, fox) should be suspected of rabies virus exposure and managed accordingly. Rabies infection in rodents is very uncommon and no human rabies cases due to rodent bites have been reported. State or local public health authorities can facilitate animal rabies testing and assist in conducting exposure risk assessments.

Travelers potentially exposed to rabies should contact local health authorities immediately for advice about the local availability of rabies PEP. Because RIG and/or rabies vaccine may not be available in the destination country, the individual should have a strategy in place before travel for responding to a possible exposure. This strategy may require the traveler to fly to a different country to obtain the appropriate care.

## Postexposure Prophylaxis for Previously Immunized Patients

PEP is indicated regardless of prior vaccination history. For patients who have previously received a full course of PrEP or PEP with PCECV, HDCV, or a comparable vaccine, PEP consists of two booster doses of rabies vaccine each given intramuscularly in the deltoid on days 0 and 3, in addition to wound cleansing. RIG should not be administered to previously vaccinated patients. Patients who were last vaccinated before the year 1980 (when lower-potency rabies vaccines were used) may not have adequate immunity against the virus to safely qualify for the two-dose course of PEP vaccination; in the absence of a documented history of an adequate antibody titer, it is advisable to manage this subset of patients identically to previously unvaccinated patients.

## Postexposure Prophylaxis for Nonimmunized Patients

In previously unvaccinated patients, recommended PEP consists of one dose of human RIG (HRIG) (20 IU/kg body weight) given on day 0 and a series of four 1-mL injections of rabies vaccine given intramuscularly on days 0, 3, 7, and 14 (see Fig. 69.4). HRIG should be infiltrated in and around the wound or wounds, with any remaining volume given intramuscularly at a site distant from the site of vaccine administration. The deltoid is the recommended injection site for vaccination; in children, the vastus lateralis is another acceptable location. Neither adults nor children should be given vaccine in the gluteus, and RIG and vaccine should not be administered in the same deltoid muscle site.

The window for administering RIG can be extended up to day 7 if not given when vaccination was initiated, but after that time RIG is not indicated owing to its likely interference with active immunity. Minor deviations of the vaccine schedule have not been shown to adversely influence the effectiveness of prophylaxis; however, adherence to the recommended schedule is advised whenever possible. If substantial schedule deviations have occurred, serologic testing using the RFFIT 7 to 14 days after the last dose is indicated to ensure that an adequate antibody titer has been reached.

Immunocompromised patients should receive, in addition to RIG and wound cleansing, a five-dose series of rabies vaccination on days 0, 3, 7, 14, and 28. Individuals who have HIV/AIDS or are recipients of chemotherapy, antimalarials, or other immunosuppressive medications are included in this group. Postvaccination serologic testing should be performed on these patients 7 to 14 days after the day 28 vaccine dose in order to verify an adequate antibody response. Placement of RIG and vaccine administration in these patients is the same as it is in healthy patients.

## Postexposure Prophylaxis Overseas

The first vaccines against rabies were derived from viruses extracted from the brains and spinal cords of infected animals. Some of these nerve tissue vaccines, also known as *Semple rabies vaccines*, are still in use in low resource countries because of their low production costs relative to modern purified cell culture and embryonated egg–based vaccines (CCEEVs), which are recommended by WHO for both PrEP and PEP. The WHO recommends that production and administration of nerve tissue vaccines be discontinued and replaced by CCEEVs. Nerve tissue vaccines are administered in daily injections over a period of 14 to 21 days in the subcutaneous tissue overlying the stomach or upper back. Doses may be relatively high in volume and somewhat uncomfortable to receive. These vaccines pose a greater risk of vaccine-associated adverse events and are of lower potency than cell culture vaccines. All attempts should be made to obtain modern cell culture vaccines before accepting prophylaxis with a nerve tissue vaccine.

Persons exposed to rabies while abroad may receive PEP with biologics that are not licensed and approved for use in the United States and Canada. Other cell culture–derived vaccines are available, including purified duck embryo and Vero cell vaccines, that meet the WHO recommended potency of at least 2.5 international units (IU) per dose for intramuscular injection. A photograph of the vaccine vial used or a copy of the package insert might simplify follow-up assessment once the traveler returns home. Purified equine RIG (ERIG) is frequently used in places where HRIG is unavailable. The frequency of reported adverse reactions associated with ERIG administration has been relatively low (0.8% to 6.0%), and most reactions reported are minor. Unpurified antirabies serum of equine origin may still be used in some countries where neither HRIG nor ERIG is available. More severe adverse reactions, including anaphylaxis, after antirabies serum administration have been reported.

If PEP is initiated with nonapproved biologics or regimens, a patient may require additional prophylaxis. State, provincial, or local health departments should be contacted for specific advice in such cases. Serologic testing using the RFFIT may be indicated to determine whether the patient's neutralizing antibody level precludes the need for additional vaccination.

Failure to prevent rabies in PEP recipients has not occurred in the United States since cell culture vaccines and RIG have been in routine use. However, failures have occurred abroad when biologics of low potency were used; when some deviation occurred from the recommended PEP protocol; or when RIG was not administered, was administered in insufficient amounts, or was improperly administered. Inadequate wound cleansing or administration of vaccine at incorrect anatomic sites (e.g., the gluteal area) may also be associated with ineffective PEP. In addition, substantial delays between exposure and PEP initiation increase the likelihood that disease will occur before an adequate immune response has developed.

In rural areas of Asia and Africa, where there is the greatest need for PEP among the local residents, modern rabies vaccines (CCEEVs) are frequently unavailable or unaffordable. The WHO has recently accepted a new postexposure IPC (Institut Pasteur du Cambodge) vaccine regimen consisting of a 0.1-mL intradermal (ID) injection at two sites on days 0, 3, and 7, based on preliminary serologic data, with the goal of extending the benefits of a costly resource in an economically feasible way. The IPC vaccine regimen is similar to the TRC (Thai Red Cross) 2-site intradermal vaccine regimen in use for 20 years, with doses given on days 0, 3, 7, and 28. The use of CCEEVs in ID injection regimens remains off label and investigational, but WHO consideration of novel vaccine regimens with regard to immunogenicity, practicability, and economy is ongoing in order to extend rabies PEP to the most vulnerable and underserved populations.

## Adverse Reactions

Patients receiving PrEP or PEP should be advised that they may experience local reactions after vaccination, such as pain, erythema, swelling, or itching at the injection site, or mild systemic reactions, such as headache, nausea, abdominal pain, muscle aches, and dizziness. Approximately 6% of persons receiving booster vaccinations with HDCV have reported an immune complex–like reaction characterized by urticaria, pruritus, and malaise. Fewer adverse events have been reported in association with PCECV. If exposure to the rabies virus is a valid concern, rabies PEP should not be interrupted or discontinued because of local or mild systemic reactions to rabies vaccine.

## Precautions and Contraindications

Pregnancy or age status are not contraindications for PEP. Known allergies to substances present in a particular vaccine (such as egg protein in the case of PCECV) may necessitate switching the vaccine to another type.

In immunocompromised individuals, rabies vaccination may fail to generate an adequate immune response. Such patients should postpone preexposure vaccinations and consider avoiding activities for which rabies PrEP is indicated. If PrEP and/or PEP must be administered to a person of poor immune status, serology is indicated to determine whether the patient obtained an adequate rabies virus neutralizing antibody titer. In the event that no acceptable antibody response is detected, the patient should be managed in consultation with his or her physician and appropriate public health officials.

## ACKNOWLEDGMENT

The author acknowledges the work of Kis Robertson, Nina Marano, and Katherine J. Johnson on this chapter in the previous edition.

## EVIDENCE

Angsuwatcharakon P, Khomvilai S, Limsuwun K, et al. Immunogenicity and safety of WHO-approved TRC-ID regimen with a chromatographically purified Vero cell rabies vaccine with or without rabies immunoglobulin in children. *Expert Rev Vaccines* 2018;17:185-188. *Using a chromatographically purified Vero-cell rabies vaccine (CPRV), the WHO-approved two-site ID modified Thai Red Cross regimen was administered to nonimmunized Thai children with possible or proven rabies exposure: two ID doses at both deltoid regions on days 0, 3, 7, and 28. WHO category III exposures also received RIG. Serum tests on days 14 and 90 for rabies neutralizing antibody titers showed an adequate immune response and no patients died of rabies infection.*

Hemachudha T, Wacharapluesadee S, Mitrabhakdi E, et al. Pathophysiology of human paralytic rabies. *J Neurovirol* 2005;11:93-100. *The authors discuss key clinical features and pathologic changes associated with paralytic rabies.*

Huynh W, Cordato DJ, Kehdi E, et al. Post-vaccination encephalomyelitis: literature review and illustrative case. *J Clin Neurosci* 2008;15:1315-1322. *The authors discuss etiologic associations, clinical features, and pathologic changes associated with postvaccination ADEM.*

Jackson AC. Pathogenesis. In Jackson AC, Wunner WH, eds. *Rabies.* 2nd ed. London: Academic Press; 2007, pp 341-381. *The author discusses current understanding of rabies pathophysiology.*

Pieracci EG, Pearson CM, Wallace RM, et al. Vital signs: trends in human rabies deaths and exposures—United States, 1938-2018. *MMWR Morb Mortal Wkly Rep* 2019;68:524-528. https://doi: l0.15585/mmwr.mm6823e1. Accessed September 17, 2019. *In the United States, wildlife rabies, especially in bats, continues to pose a risk to humans. International travel to regions where canine and wildlife rabies is present may result in high-risk rabies exposures in places where modern vaccines and biologics used for PEP are not readily accessible.*

Rupprecht CE, Briggs D, Brown CM, et al. Use of a reduced (4-dose) vaccine schedule for postexposure prophylaxis to prevent human rabies: recommendations of the Advisory Committee for Immunization Practices. *MMWR Recomm Rep* 2010;59(RR-2):1-9, 2010. *Describes ACIP guidelines and supporting evidence for a four-dose vaccination series in rabies PEP.*

Warrell MJ. Perspective Piece: simplification of rabies postexposure prophylaxis: a new 2-visit intradermal vaccine regimen. *Am J Trop Med* 2019;101:1199-1201. *Presents the rationale for investigating a new protocol consisting of a two-visit four-site rabies vaccine PEP regimen with or without a day 28 dose, especially for use in rural Asia and Africa, where optimizing use of an expensive, sometimes scarce vaccine and improved patient logistics might make possible the delivery of recommended PEP care to increased numbers of patients.*

Willoughby RE, Tieves KS, Hoffman GM, et al. Survival after treatment of rabies with induction of coma. *N Engl J Med* 2005;352:2508-2514. *The authors describe the approach used to treat a patient who survived rabies in the absence of any prior rabies vaccination or the administration of RIG.*

## ADDITIONAL RESOURCES

Centers for Disease Control and Prevention (CDC). Rabies. Available at: https://www.cdc.gov/rabies/. Accessed February 10, 2020. *Useful online reference for rabies information, including how to collect and submit patient samples to the Rabies Laboratory at the CDC for diagnosis; also gives updated information on the availability of rabies vaccine, rabies immunoglobulin, and reagents for rabies diagnostic tests.*

Wallace RM, Petersen BW, Shlim DR. Rabies. In *Centers for Disease Control and Prevention (CDC): Health information for international travel 2020.* Atlanta: U.S. Department of Health and Human Services, Public Health Service; 2019. https://www.cdc.gov/travel/yellowbook/2020/travel-related-infectious-diseases/rabies. *Describes risk factors, epidemiology, and management of rabies exposures in travelers.*

World Health Organization (WHO). Rabies vaccines: WHO position paper—April 2018. *WER* 2018;16(93):201-220. *Describes the global picture of rabies with respect to disease burden, prevention activities, and emerging research areas; also outlines World Health Organization–approved regimens for rabies PrEP and PEP.*

# 70

# Arboviruses of Medical Importance

*Johnnie A. Yates*

## ABSTRACT

Arboviruses are transmitted to humans primarily through the bites of arthropod vectors such as mosquitoes, ticks, and sandflies. Half of the world's population live in areas endemic for arboviruses, which cause significant morbidity and mortality in both developed and developing countries. More than 130 arboviruses are known to infect humans, but only a much smaller number are medically important. Since 2015, there have been major outbreaks of emerging (and reemerging) arboviruses such as dengue, Zika, chikungunya (CHIKV), and yellow fever (YF). The 2015–2016 Zika pandemic was declared a public health emergency of international concern by the World Health Organization (WHO) because of the association of Zika infection with microcephaly and Guillain-Barré syndrome (GBS). In 2019, more infections and deaths from eastern equine encephalitis virus (EEEV) were reported in the United States than in the previous 5 years combined. Arboviruses are also a major cause of illness in international travelers, and imported infections have resulted in localized outbreaks in nonendemic areas (e.g., dengue in Hawaii, Zika in Texas and Florida, and CHIKV in Italy and France). In addition to vector-associated infections, some arboviruses can be transmitted through blood transfusions from viremic donors, and Zika can be transmitted sexually. Climate change, demographic shifts, and increased international travel have resulted in a greater incidence of arboviral infections; it is therefore important for healthcare providers to be familiar with the distribution and clinical aspects of the more common arboviruses. Treatment is supportive, as there are no specific antiarboviral therapies. Personal protective measures such as the use of insect repellents and utilization of available vaccines can decrease the risk for acquiring arboviral infections.

## CLINICAL VIGNETTE

A 60-year-old man presented to the emergency department (ED) reporting 2 days of fever, chills, body aches, and a rash on his arms that appeared the day after he returned from a trip to Thailand to visit relatives. A complete blood count (CBC), liver transaminases, and malaria rapid diagnostic test and smear were normal; he was then diagnosed with a nonspecific viral infection. Four days later he presented to clinic because the rash had spread to his trunk and he continued to have fevers until the day before his clinic visit. Laboratory tests were repeated, and his CBC was notable for leukopenia, a 25% increase in his hematocrit, and a drop in platelets to 28,000 (from 218,000 4 days prior); he also had new liver transaminase elevations. His physical exam was notable for diffuse erythematous macules on his trunk and extremities, a positive tourniquet test, and right-upper-quadrant abdominal tenderness. He had been born and raised in Thailand and did not know whether he had had dengue as a child. He was diagnosed with presumptive dengue with warning signs of severe dengue and was hospitalized. During his hospital stay he remained afebrile, his laboratory abnormalities improved, and he was discharged after 2 days. Dengue serology obtained during his initial ED visit and a week later demonstrated IgM seroconversion from 0.09 to 1.61 (reference range <0.90); IgG was positive at 2.48 initially and increased to 8.34 on the follow-up specimen.

COMMENT: This patient was in the critical phase of dengue and had warning signs of severe dengue: his platelets dropped precipitously, he had a hemoconcentration suggestive of plasma leakage, and he developed liver enzyme abnormalities and abdominal tenderness. Because he was originally from Thailand, it was suspected that he had likely had a prior dengue infection (which was subsequently confirmed by the positive IgG) and was therefore at increased risk for severe dengue. An opportunity to perform polymerase chain reaction (PCR) testing for dengue (and Zika and CHIKV) was missed at his ED visit. A rapid rise in IgG during the first week of illness is commonly seen in patients with a secondary dengue infection.

## GEOGRAPHIC DISTRIBUTION AND RISK FACTORS

Arboviruses are endemic to every continent except Antarctica, and they are found in both tropical and temperate zones. The risk for exposure to arboviruses is dependent on the presence of competent vectors in addition to a variety of factors including the vector-host cycle, the ecologic niches of arthropods, temporal influences such as seasonality, and human factors (both biologic and behavioral). Thus the probability of acquiring an arboviral infection within a specific area is more nuanced than might be suggested from maps that highlight only endemic countries. Since arboviral infections provide strain-specific lifelong immunity, population susceptibility (and thus the epidemic potential of the virus) is influenced by the frequency of past outbreaks and the proportion of the population affected during each outbreak.

*Aedes aegypti*, the "yellow fever mosquito," is the vector for dengue, Zika, CHIKV, and YF. This anthropophilic mosquito is found throughout the tropics and subtropics, is abundant in urban and peridomestic environments, and can bite multiple people in a short period of time. Outbreaks triggered by *Ae. aegypti* are more common in the rainy season, during which mosquitoes lay eggs in small collections of standing water; these hatch and rapidly develop into more adult mosquitoes. *Ae. aegypti* is responsible for most dengue outbreaks and was also the vector implicated in the 2015–2016 Zika pandemic. *Aedes* mosquitoes are daytime feeders; consequently, the risk for exposure to dengue, Zika, CHIKV, and YF is greatest during daylight hours.

*Aedes albopictus* (the Asian tiger mosquito) is a secondary vector for dengue, Zika, and CHIKV. This mosquito has adapted to both temperate and tropical climates and has been found as far north as central Europe. The broad geographic distribution of *Ae. albopictus* was facilitated by the international transport of mosquito eggs in used tires and bamboo. Although *Ae. albopictus* is less efficient than *Ae. aegypti* in transmitting dengue and Zika, it was responsible for the 2005–2006 CHIK outbreak in Réunion Island, where over 250,000 residents were infected. A genetic mutation of an Asian CHIKV strain resulted in enhanced transmission by *Ae. albopictus*, the principal mosquito species in Réunion.

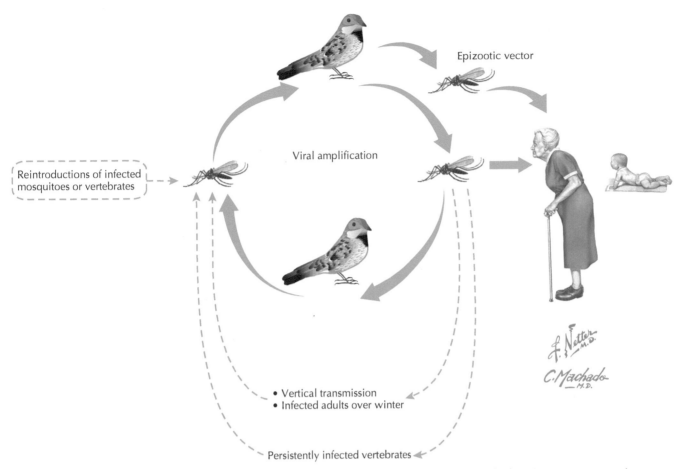

Epizootic vector

Viral amplification

Reintroductions of infected
mosquitoes or vertebrates

• Vertical transmission
• Infected adults over winter

Persistently infected vertebrates

In a typical arboviral transmission cycle, the virus is amplified among vertebrates, frequently birds, and infected enzootic vectors also
infect humans. In the transmission cycle of some viruses, an epizootic vector is required to bridge the amplification cycle to humans;
this circumstance applies when the enzootic vector does not feed on humans, necessitating other species with more catholic feeding
habits to acquire the infection from the vertebrate amplifying host and, on a subsequent feeding, to infect humans (e.g., eastern equine
encephalitis). A number of mechanisms allow viruses to survive periods when they are not actively amplified. Some viruses are transmitted
vertically in infected mosquito eggs; infected adults can survive in protected settings over extended periods, including winters in temperate
climates; persistently infected vertebrates (e.g., bats) can re-enter the virus amplification cycle locally; or the virus can be re-introduced
from a distance by windblown vectors or migratory viremic vertebrates.

**Fig. 70.1** Generic arbovirus transmission cycle.

Yellow fever is endemic in parts of sub-Saharan Africa and tropical South America. Sylvatic (jungle) and savannah cycles are maintained by *Haemagogus* and *Sabethes* mosquitoes in South America and *Aedes* mosquitoes in Africa, respectively. Nonhuman primates are the natural reservoirs. When YF is introduced into urban environments (the preferred habitat of *Ae. aegypti*) by viremic humans who acquired infection in the jungle or savannah, urban epidemics can occur. Estimates of the risk for YF in travelers spending 2 weeks in endemic areas are 50 per 100,000 for West Africa and 5 per 100,000 for South America. However, these estimates do not take into account the occurrence of YF outbreaks or the use of protective measures such as mosquito repellents. From 2016 to 2018, Brazil experienced the largest number of human YF cases in decades. The outbreak was confined to an epizootic cycle that spilled over to humans and did not become urbanized.

*Culex* mosquitoes, the vector of Japanese encephalitis virus (JEV) and West Nile virus (WNV), are present in tropical and temperate regions. JEV is endemic throughout Asia, where *C. tritaeniorhynchus* is the primary vector; water birds and pigs serve as amplifying hosts. JEV is transmitted year round in the tropics but is seasonal in more temperate climates, with peak transmission occurring during the rainy season. The risk for JEV illness in travelers to Asia has been estimated at 1 per

1 million. Long-term travel to rural areas (especially near pig farms or rice paddies) is a risk factor for infection. However, travelers visiting endemic areas for less than 2 weeks during periods of active transmission have contracted JEV. In contrast to *Ae. aegypti*, *Culex* mosquitoes feed during the evening and at night.

WNV is endemic in North America, Europe, Africa, and the Middle East. It was first detected in North America in 1999 and has since been found in every state in the continental United States. WNV is maintained by bird reservoirs; humans are dead-end hosts who do not develop sufficient viremia to sustain transmission.

The transmission cycle of EEEV involves *Culiseta melanura* mosquitoes and avian reservoirs in North America. Most cases of EEEV in the United States are sporadic and occur along the East and Gulf Coasts. *Aedes*, *Culex*, and *Coquillettidia* mosquitoes act as bridge vectors that infect horses and humans, who are dead-end hosts. Transmission of WNV and EEEV is seasonal, with peak transmission occurring in the summer and early fall. A typical arbovirus transmission cycle is shown in Fig. 70.1.

The flavivirus responsible for tick-borne encephalitis (TBE) is spread by *Ixodes* ticks in forested regions of Europe and Central Asia (see Chapter 72). In North America, *Ixodes* ticks transmit Powassan

**TABLE 70.1    Selected Medically Important Arboviruses**

| Clinical Syndrome[a] | Virus/Taxonomic Family | Geographic Distribution | Transmission[b] | Incubation Period |
|---|---|---|---|---|
| Febrile illness | Dengue 1-4/Flaviviridae | Tropical and subtropical Central and South America, Africa, Asia, Australia, Oceania | Mosquito–anthroponotic[c]: urban, peridomestic | 2–7 days |
| | Oropouche/Bunyaviridae | Central and South America | Mosquito and midge–vertebrates, anthroponotic: sylvatic, periurban | |
| Polyarthritis | Sindbis/Togaviridae | Europe, Africa, Asia, Australia | Mosquito–bird: sylvatic | 2–7 days |
| | Chikungunya/Togaviridae | Africa, Asia, Europe | Mosquito–anthroponotic: urban, peridomestic | 2–10 days |
| | Ross River/Togaviridae | Australia, Oceania | Mosquito–mammal: sylvatic, suburban | 3–21 days |
| Neurologic infection | Tick-borne encephalitis/Flaviviridae | Europe, Asia | Tick–mammal, bird: sylvatic Oral: milk products | 3–7 days |
| | West Nile/Flaviviridae | Cosmopolitan | Mosquito–bird: sylvatic, periurban | 3–10 days |
| | Japanese encephalitis/Flaviviridae | Asia, Australia, Oceania | Mosquito–vertebrate: rural | 4–14 days |
| | Toscana/Bunyaviridae | Europe, Africa | Sandfly–vertebrate: periurban | 2–7 days |
| Hemorrhagic fever | Yellow fever/Flaviviridae | Central and South America, Africa | Mosquito–mammal, anthroponotic: rural, urban | 3–6 days |
| | Rift Valley fever/Bunyaviridae | Middle East, Africa | Mosquito–vertebrate, anthroponotic: rural, periurban | 3–5 days |

[a]The diseases listed are displayed by their principal clinical presentation; other manifestations or complications may occur.
[b]Reports have included nosocomial transmission by needlestick, blood transfusion (dengue, West Nile viruses), or transplantation (West Nile virus); transmission by direct contact with infected animals or meat (tickborne encephalitis, Rift Valley fever viruses); transmission through breastfeeding (yellow fever virus); and vertical transmission to the fetus (dengue, Japanese encephalitis, West Nile, western equine encephalitis viruses, and possibly Colorado tick fever and Ross River viruses).
[c]Transmitted from human to human by the vector.

virus. Humans are exposed to Powassan virus through outdoor activities in the Great Lakes region of Canada and the United States from late spring until early fall, when the ticks are most active.

## CLINICAL PRESENTATION

A febrile illness with a potential exposure history (e.g., international travel or outdoor activity during the summer or early fall) should alert the clinician to the possibility of an arboviral infection. Online resources—such as the US Centers for Disease Control and Prevention (CDC), World Health Organization (WHO), and the Program for Monitoring Emerging Diseases (Pro-MED)—can help healthcare providers keep up to date about ongoing outbreaks. The incubation period of most arboviruses ranges from 2 to 14 days, so arboviral infections can usually be ruled out in patients who present with a fever that started more than 2 weeks after return from an at-risk area. Many arboviral infections—including dengue, Zika, CHIKV, WNV, and EEEV—are nationally notifiable diseases in the United States. Therefore healthcare providers should consult with state or local health departments as soon as an infection is suspected so that diagnostic assistance can be obtained and public health measures can be undertaken to reduce the threat of local transmission.

With the exception of CHIKV, the majority of arboviral infections are asymptomatic. Symptomatic infections present as febrile illnesses with or without rash or fever with arthritogenic, neurologic, or hemorrhagic manifestations. There is significant overlap of the clinical presentations of arboviral infections; shared symptoms include fever, headache, myalgias, and malaise. A maculopapular rash is commonly seen in dengue, Zika, and CHIKV. The features of selected arboviruses are described in the following text and summarized in Table 70.1.

These features, when combined with a detailed epidemiologic and travel history, can help narrow the differential diagnosis in a febrile patient with a potential arboviral infection.

### Dengue

Forty percent of the world's population is at risk for dengue and an estimated 50 to 100 million infections from this flavivirus occur each year. In 2019, the WHO named dengue as one of the top 10 threats to global health. Asia accounts for approximately 70% of the global burden of infection. Dengue is also the most common cause of febrile illnesses in travelers from Southeast Asia, Latin America, and the Caribbean. Symptomatic infections are notable for the abrupt onset of high fever, headache, retro-orbital pain, and intense myalgias and body aches (accounting for the colloquial name "breakbone fever"). A rash occurs in 50% to 75% of patients, but dermatologic findings can be subtle; they may consist only of flushing and may be difficult to discern in persons with darker skin. Petechiae can be seen and are the basis of the tourniquet test, a marker of capillary fragility (a sphygmomanometer is placed on the arm and inflated to halfway between the systolic and diastolic pressures; after 5 minutes, the number of petechiae in the antecubital fossa are counted greater than 10 petechiae per square inch constitutes a positive test). Minor hemorrhagic manifestations such as epistaxis, bleeding gums, and ecchymoses can also occur. In addition to the acute exanthem, a convalescent rash described as "islands of white on a sea of red" may be seen. Laboratory findings include leukopenia and thrombocytopenia on the CBC, as well as elevated liver transaminases. The febrile phase of dengue lasts 2 to 7 days.

Approximately 2% to 4% of dengue infections result in a severe, life-threatening illness characterized by vascular leakage, hemorrhage, and end-organ impairment. The risk for severe disease is increased in those with a previous dengue infection due to a phenomenon known as antibody-dependent enhancement. There are four dengue serotypes, and infection with one serotype confers lifelong type-specific immunity. Secondary infection with a heterologous serotype results in nonneutralizing antibodies that facilitate virus entry and replication in the target cells, which then triggers the inflammatory cascade responsible for the pathology of severe dengue. Infants are at increased risk for severe dengue due to transplacental transfer of antibodies from mothers with past dengue infections. The mortality rate from untreated severe dengue ranges from 10% to 20% but is less than 1% with appropriate medical care. After recovery from acute dengue, individuals may have fatigue that persists for several weeks.

## Zika

The 2015–2016 Zika pandemic mostly affected countries in the Western Hemisphere, but the virus is also present in Africa and Asia. The clinical manifestations of this flavivirus are similar to those of dengue, although the fever tends to be low grade and the illness is generally milder. Nonpurulent conjunctivitis occurs in approximately half of symptomatic Zika infections. Leukopenia may be seen, but thrombocytopenia or elevated transaminases are usually absent.

Zika infection during pregnancy is associated with microcephaly and other neurologic sequelae (called congenital Zika syndrome) in newborns. It is found in 5% to 14% of newborns exposed to the virus in utero, and the risk is greatest when the infection is acquired during the first trimester. Zika is unique among arboviruses in that it can be transmitted sexually; it is transmissible for up to 3 months in semen and 2 months in cervical secretions. Therefore current guidelines recommend that couples of childbearing age delay conception for 3 months after they return from an area with Zika risk (see Chapter 92).

GBS is another complication of Zika infection. The incidence of GBS during the 2013–2014 French Polynesia epidemic was estimated to be 2.4 cases per 10,000 Zika infections. Other flaviviruses can also cause GBS, but the incidence appears to be higher with Zika.

## Chikungunya

In contrast to dengue and Zika, CHIKV is an alphavirus and the majority of CHIKV infections are symptomatic. The primary distinguishing feature of CHIKV infection is severe joint pain. The arthralgias are symmetric and typically involve the hands, wrists, ankles, and knees. The debilitating nature of CHIKV arthralgias can lead to a contorted appearance, which accounts for the name *chikungunya*, which loosely translates to "that which bends" in the Makonde dialect. A CBC may show lymphopenia, but thrombocytopenia is less common than in dengue and hemoconcentration does not occur. Mortality is rare, with older adults being at greatest risk. The painful arthritis seen with CHIKV can persist for months or even years after an acute infection, and some patients eventually meet the criteria for seronegative rheumatoid arthritis. Risk factors for the development of chronic rheumatologic symptoms include age greater than 45 years, female gender, and preexisting arthritis.

## Yellow Fever

Yellow fever is a hemorrhagic flavivirus. As in the case of most arbovirus infections, a large proportion of YF infections are inapparent or minimally symptomatic. Initial symptoms include high fever, chills, headache, myalgias, and nausea/vomiting. In 15% of patients, a biphasic illness may herald progression to severe disease with jaundice (thus the name *yellow fever*), bleeding, and liver and kidney failure. Up to half of severe cases are fatal. During the recent YF outbreak in Brazil, there were 12 travel-associated infections and 4 deaths; none of the travelers had been vaccinated against YF (see Chapter 64).

## Japanese Encephalitis

This neurotropic flavivirus is the leading cause of epidemic encephalitis in Asia. Fewer than 1% of infections are symptomatic. Symptomatic infection results in fever, headache, nausea/vomiting, and weakness. Severe infections can present with altered mental status, focal neurologic signs, and seizures. Parkinsonian signs such as masked facies, cogwheel rigidity, and tremor are distinctive features of JEV. Up to one-third of symptomatic infections are fatal, and neurologic sequelae occur in half of the survivors.

## West Nile Virus

WNV is closely related to the JEV and is the most common arbovirus infection in North America. In 2018, over 2600 WNV infections were reported to the CDC—an increase of almost 25% compared with previous years. Approximately 80% of infections are asymptomatic. Of symptomatic patients, most develop "West Nile fever," characterized by fever, headache, myalgias, nausea, and weakness. Less than 1% develop neuroinvasive disease, with meningitis, encephalitis, or acute flaccid paralysis. The mortality rate is approximately 10% in patients with severe illness. The elderly, infants, and immune-compromised individuals are at highest risk for severe disease and death (see Chapter 71).

## Eastern Equine Encephalitis

EEEV is an alphavirus, and most infections are asymptomatic. Symptoms of acute infections are similar to those of JEV and WNV. Encephalitic manifestations include meningeal signs, vomiting, altered mental status, seizures, and coma. EEEV has the highest mortality rate (30% to 50%) of the neuroinvasive arboviruses. As with other arboviruses, individuals at the extremes of age and those with chronic medical conditions are at increased risk for death.

## DIAGNOSIS

Diagnosing arboviral infections can be challenging because many arboviruses circulate in the same regions and have overlapping clinical and laboratory findings. Nonetheless, accurate diagnosis is important, since the management of various arboviral infections differs. For example, patients with suspected dengue should not be treated with nonsteroidal antiinflammatory drugs (NSAIDs) due to the risk for bleeding, and they require close monitoring for severe dengue after defervescence. In contrast, NSAIDs are used in treating the arthralgias of CHIKV infections. Furthermore, a patient with possible Zika infection should be counseled about the risk for sexual transmission. Prompt diagnosis of arboviral infections can also help to reduce the risk for local transmission triggered by a viremic traveler returning to a nonendemic region with competent vectors.

Although some laboratory findings may be suggestive of specific arboviruses (e.g., leukopenia and thrombocytopenia in dengue), similar results can be seen in different arboviral infections as well as in infections due to non–vector borne viruses and bacteria (e.g., rickettsial infections, leptospirosis). Therefore definitive diagnosis relies on nucleic acid amplification tests (NAATs) or serology.

During the viremic period (typically the first 5 to 7 days of infection), NAATs are the preferred diagnostic tests. PCR testing of serum is available for dengue, Zika, CHIKV, and YF. Viral antigen detection (NS1 protein test) is another option for dengue, but it cannot determine serotype. Zika infection can also be diagnosed with urine PCR if performed within 2 weeks of illness onset.

Some arboviruses (e.g., JEV, WNV, EEE) do not produce a significant level of viremia in humans. In addition, it is common for patients to present for healthcare beyond the viremic period. Therefore virus-specific serology is often necessary for diagnosis. For most arboviral infections, IgM antibodies are detectable 3 to 5 days after illness onset and remain elevated for up to several months. IgG antibodies appear shortly after IgM and persist for life. A challenging aspect of arbovirus serology is cross-reactivity. Vaccines against flaviviruses can elicit cross-reacting antibodies; prior vaccination against YF, TBE, or JEV may result in falsely elevated values during antibody testing for a specific flavivirus. Therefore a history of prior vaccines received is relevant to the interpretation of serologic tests, and a single elevated IgM titer may not definitively diagnose an acute infection: documentation of a fourfold rise in titers may be necessary. When there is concern about cross-reactivity, plaque reduction neutralization tests (PRNTs) can confirm a recent infection.

Cerebrospinal fluid (CSF) testing can aid in the diagnosis of infections caused by the neuroinvasive arboviruses. Nonspecific CSF findings include pleocytosis, an elevated protein, and normal glucose. PCR and serology can be performed on CSF, although the latter is more sensitive.

Magnetic resonance imaging (MRI) of the brain in patients with arboviral encephalitis may show abnormalities in the thalamus, but these findings are nonspecific.

## TREATMENT

There are no approved antiviral or immunologic therapies against arboviruses. Therefore symptom management and supportive care are the mainstays of treatment. Antipyretics can help reduce fever, and analgesics can provide relief from headaches, myalgias, and arthralgias. Aspirin and NSAIDs should be avoided in dengue (or when dengue cannot be ruled out) because of thrombocytopenia and the risk for bleeding. However, they are useful in treating patients with joint pains due to CHIKV infection.

For patients with dengue, it is essential to monitor for warning signs of severe dengue such as bleeding, severe abdominal pain, fluid accumulation, hemoconcentration (as evidenced by a rise in hematocrit), or a rapid decrease in the platelet count. The 24- to 48-hour period between the febrile and convalescent phases of dengue is termed the critical phase. The CBC and liver enzymes should be closely monitored during the critical phase and patients with warning signs of severe dengue should be hospitalized. Careful attention to fluid status is critical in the management of severe dengue because redistribution of extravasated fluids from plasma leakage can result in fluid overload for patients who receive excessive intravenous hydration. Prophylactic blood or platelet transfusions are not recommended, and corticosteroids have not been shown to be of any benefit. The CDC has an excellent dengue clinical management course that is available free of charge at https://www.cdc.gov/dengue/training/cme.html.

Treatment of the chronic joint pains and arthritis associated with CHIKV can be challenging; if NSAIDs are ineffective, anti-rheumatic drugs may be necessary. Hydroxychloroquine has not been found to be effective, but methotrexate has shown promise in the treatment of CHIKV polyarthritis.

Guidelines for the management of pregnant women with Zika infection and infants with possible congenital Zika syndrome are available from the CDC (see Additional Resources, further on).

## PREVENTION

Strategies for the prevention of arbovirus infections include avoidance, the use of personal protective measures, and vaccination. Avoiding areas of ongoing outbreaks (e.g., pregnant women considering travel to areas with Zika outbreaks) or limiting outdoor activities during high-risk periods (e.g., evening and night in rural Asia) can help to reduce the risk for exposure to arbovirus-carrying vectors.

Effective insect repellents that are registered by the US Environmental Protection Agency (EPA) include N,N-diethyl-m-toluamide (DEET), picaridin, IR3535 and para-menthane-3,8-diol (oil of lemon eucalyptus), and 2-undecanone. For protection against *Aedes* mosquitoes, repellent should be applied during the day, when the mosquitoes are most active. In contrast, *Culex* mosquitoes are most active from dusk to dawn, so repellent use is most important during evenings and at night in order to prevent JEV and WNV infections.

Permethrin is an insecticide that repels and kills mosquitoes and ticks; it is used to treat clothes or mosquito nets. The combination of permethrin on clothing and an EPA-approved repellent on exposed skin can significantly reduce the risk for being bitten by arthropods. When traveling in tick-infested areas, pants should be rolled into socks and "tick checks" should be conducted after time spent outdoors.

Vaccines are available for YF, JEV, dengue and TBE. Routine childhood vaccination against YF and JEV occurs in many endemic countries. The YF vaccine is a highly effective live attenuated virus vaccine that is recommended for individuals 9 months of age and older who are traveling to (or living in) YF endemic areas. YF vaccine can be required for entry into some countries under the WHO International Health Regulations. Because it is a live virus vaccine, YF vaccine is contraindicated in immune-compromised individuals. The vaccine is associated with rare but serious adverse reactions including YF vaccine-associated neurologic disease (YEL-AND) and YF vaccine-associated viscerotropic disease (YEL-AVD). The risk for YEL-AVD (which has a fatality rate of 50%) is greater in people older than 60 years of age. Therefore a careful risk/benefit analysis should be conducted prior to vaccination against YF.

Several JEV vaccines are available worldwide but only the inactivated Vero cell–derived vaccine (Ixiaro) is available in North America. Two doses of the vaccine given at 0 and 28 days induce JEV neutralizing antibodies in over 95% of adults under 65 years of age. If the time before trip departure is less than 28 days, JE vaccine may be given on an accelerated schedule of two doses separated by 7 to 21 days, followed by a third dose given after a year to ensure long-lasting immunity. The JEV vaccine is recommended for travelers spending prolonged time in endemic areas (e.g. long-stay travelers and expatriates), those planning short-term travel to risk areas during high-transmission periods, and those who make frequent trips to risk areas.

A tetravalent dengue vaccine (Dengvaxia) was licensed in 2016, but use of the vaccine was suspended in some countries due to an increased risk for severe dengue if vaccinated individuals without a history of previous dengue subsequently acquired natural dengue infection. Currently, Dengvaxia is recommended only for those between 9 and 16 years of age who live in an endemic area and have a history of a past dengue infection confirmed by serology.

Inactivated vaccines for TBE are available in Europe but not in the United States or Canada.

## FUTURE DIRECTIONS

Efforts to reduce the global burden of arbovirus infections include improved surveillance, vaccine development, vector control, and climate change mitigation. Vector surveillance using geographic information systems (GIS) is becoming increasingly common in the detection and control of arbovirus outbreaks. On the vaccine front, a second dengue vaccine is undergoing phase 3 clinical trials and progress is being made in the development of Zika and CHIKV vaccines. A novel approach to vector control involves infecting *Ae. aegypti* mosquitoes with *Wolbachia* bacteria. *Wolbachia* inhibit dengue replication in *Ae. aegypti* and can also reduce *Aedes* egg hatching. This approach has been successfully trialed in Australia, Brazil, Singapore, and several other countries. In addition, climate change mitigation will help to limit the geographic expansion of arbovirus-carrying arthropods. It has been calculated that reaching a low carbon emission scenario (as outlined in the Paris Agreement) versus a high emission scenario would significantly reduce vector abundance by the end of the century.

## EVIDENCE

Adams LE, Martin SW, Lindsey NP, et al. Epidemiology of dengue, chikungunya, and Zika virus disease in U.S. states and territories, 2017. *Am J Trop Med Hyg* 2019;101(4):884-890. *Epidemiologic review of dengue, chikungunya, and Zika cases reported to ArboNET (the US national arbovirus surveillance system) in 2017. Approximately half of the cases were from the US territories of American Samoa and Puerto Rico.*

Goodyer LI, Croft AM, Frances SP, et al. Expert review of the evidence base for arthropod bite avoidance. *J Travel Med* 2010;17(3):182-192. *Summary of the evidence for mosquito and tick repellents.*

Halstead SB, Dans L. Dengue infection and advances in dengue vaccines for children. *Lancet Child Adolesc Health* 2019;3(10):734-741.

Halstead SB, Wilder-Smith A. Severe dengue in travelers: pathogenesis, risk and clinical management. *J Travel Med* 2019;26(7):taz062.

Leder K, Torresi J, Libman MD, et al. GeoSentinel surveillance of illness in returned travelers, 2007-2011. *Ann Int Med* 2013;158:456-468. *Analysis of over 42,000 ill returned travelers presenting to specialized travel and tropical medicine clinics. Dengue was the most common infection in travelers with systemic febrile illnesses returning from Southeast Asia, Latin America, and the Caribbean.*

Lee VJ, Chow A, Zheng X, et al. Simple clinical and laboratory predictors of chikungunya versus dengue infection in adults. *PLoSNegl Trop Dis* 2012;6(9):e1786. doi:10.1371/journal.pntd.0001786. *Study from Singapore identifying arthralgias in chikungunya patients and thrombocytopenia in dengue patients as distinguishing features between the two arboviruses.*

Lindsey NP, Martin SW, Staples JE, Fischer M. Notes from the field: multistate outbreak of eastern equine encephalitis virus—United States, 2019. *MMWR Morb Mortal Wkly Rep* 2020;69:50-51. DOI: http://dx.doi.org/10.15585/mmwr.mm6902a4external icon. *Describes an uptick in the number of cases reported to the CDC in 2019 of this relatively rare vector-borne infection in the United States.*

Ryan SJ, Carlson CJ, Mordecai EA, Johnson LR. Global expansion and redistribution of *Aedes*-borne virus transmission risk with climate change. *PLoSNegl Trop Dis* 2019;13(3):e0007213. https://doi.org/10.1371/journal.pntd.0007213. *Estimation of the risks of arbovirus transmission for Ae. aegypti and Ae. albopictus under varying climate change scenarios.*

Schilte C, Staikovsky F, Couderc T, et al. Chikungunya virus-associated long-term arthralgia: a 36-month prospective longitudinal study. *PLoSNegl Trop Dis* 2013;7(3): e2137. DOI:10.1371/journal.pntd.0002137. *Prospective analysis of a cohort of 180 CHIKV patients from Réunion; 60% had persistent arthralgias and over half had sleep difficulties at 3 years.*

## ADDITIONAL RESOURCES

Adebanjo T, Godfred-Cato S, Viens L, et al. Update: interim guidance for the diagnosis, evaluation, and management of infants with possible congenital Zika virus infection—United States, October 2017. *MMWR Morb Mortal Wkly Rep* 2017;66:1089-1099. *Recommendations from a multidisciplinary panel. Includes a summary of the clinical findings in congenital Zika syndrome and an evaluation/management algorithm.*

Centers for Disease Control and Prevention. *CDC Yellow Book 2020: Health Information for International Travel.* New York: Oxford University Press; 2019. *Definitive reference for travel-related risks and infections. Updated every 2 years. Online version available free of charge.*

Centers for Disease Control and Prevention Dengue Clinical Case Management Course. https://www.cdc.gov/dengue/training/cme.html. *Excellent online course on the diagnosis and management of dengue. Continuing medical education (CME) credits available.*

Halstead SB. Chikungunya (Ch 175D). In Cherry JD, Harrison GJ, Kaplan SL, Steinbach WJ, Hotez PJ, eds. *Textbook of Pediatric Infectious Diseases,* 8th ed. Philadelphia: Elsevier Saunders; 2019, pp 1364-1368.

Halstead SB. Dengue, dengue hemorrhagic fever and severe dengue (Ch. 176D). In Cherry JD, Harrison GJ, Kaplan SL, Steinbach WJ, Hotez PJ, eds. *Textbook of Pediatric Infectious Diseases,* 8th ed. Philadelphia: Elsevier Saunders; 2019, pp 1661-1670.

Hermance ME, Thangamani S. Powassan virus: an emerging arbovirus of public health concern in North America. *Vector Borne Zoonotic Dis* 2017;17(7):453-462. *Epidemiologic and clinical overview of Powassan virus, with suggestions for research priorities.*

Hills SL, Walter EB, Atmar RL, Fischer M. Japanese encephalitis vaccine: recommendations from the Advisory Committee on Immunization Practices. *MMWR Recomm Rep* 2019;68(2):1-33. *Updated ACIP recommendations regarding the Japanese encephalitis vaccine.*

Javelle E, Ribera A, Degasne I, et al. Specific management of post-chikungunya rheumatic disorders: a retrospective study of 159 cases in Réunion Island from 2006-2012. *PLoSNeglTrop Dis* 2015;9(3):e0003603. doi:10.1371/journal.pntd.0003603. *The experience of rheumatologists managing patients with CHIKV polyarthralgias after the 2006 Réunion epidemic.*

Liu-Helmersson J, Brannstrom A, Sewe MO, et al. estimating past, present, and future trends in the global distribution and abundance of the arbovirus vector *Aedes aegypti* under climate change scenarios. *Front Public Health* 2019;7:148. doi: 10.3389/fpubh.2019.00148. *Mathematical modeling of Aedes abundance under low, medium, and high carbon emissions scenarios.*

Morens DM, Folkers GK, Fauci AS. Eastern equine encephalitis virus—another emergent arbovirus in the United States. *N Engl J Med* 2019;381:1989-1992. *Commentary on EEEV and the unprecedented number of cases in the United States in 2019.*

Musso D, Ko AI, Baud D. Zika virus infection—after the pandemic. *N Engl J Med* 2019;381:1444-1457. *Review of the current knowledge of the epidemiology, transmission, and sequelae of Zika infection, with informative figures and tables.*

Oduyebo T, Polen KD, Walke HT, et al. Update: interim guidance for health care providers caring for pregnant women with possible Zika virus exposure—United States (including U.S. territories), July 2017. *MMWR Morb Mortal Wkly Rep* 2017;66:781-793. *Guidelines for the diagnosis of Zika in pregnant women and surveillance of pregnant women with confirmed Zika infection.*

Wilder-Smith A, Ooi E, Horstick O, Wills B. Dengue. *Lancet* 2019;393:350-363. *Comprehensive review of the epidemiology, pathophysiology, diagnosis, management, and prevention of dengue.*

# 71

# West Nile Virus Disease

*Nicole Lindsey, Emily McDonald, Marc Fischer, J. Erin Staples*

## ABSTRACT

Before 1999, West Nile virus (WNV) received little attention outside Africa, Asia, and Europe, where it caused an endemic, mosquito-borne febrile illness and sporadic encephalitis. After the dramatic emergence of WNV in New York City in 1999, the virus spread westward across the United States, resulting in the largest outbreaks of WNV disease ever reported. From 1999 through 2018, 50,830 cases of WNV disease were reported in the United States, including 24,657 infections affecting the central nervous system (neuroinvasive disease) and 2330 deaths. Over the past two decades, much has been learned about the virology, ecology, transmission, epidemiology, and clinical manifestations of WNV; these topics are reviewed in this chapter.

## CLINICAL VIGNETTE

A 62-year-old male with a history of hypertension presented to medical care in late August after he reportedly woke up confused and unable to get out of bed without assistance. Three days previously his wife noted that he was complaining of a fever, muscle aches, and diarrhea. In the previous 2 weeks, he was reportedly fixing fence on his farm in South Dakota.

On assessment, he had a temperature of 39.2°C, was tachycardic, and noted to be disoriented and dysarthric. Neurologic examination revealed bilateral asymmetric weakness with hyporeflexia of the upper and lower extremities. Laboratory tests showed a peripheral white blood cell (WBC) count of 5200/mm³ (75% segmented neutrophils, 15% lymphocytes, 7% monocytes, 3% bands), hemoglobin level of 13.3 g/dL, and platelet count of 125,000/mm³. His complete metabolic panel was normal except for a mildly elevated glucose (137 mg/dL). Computed tomography (CT) without contrast of the head was normal. Examination of a cerebrospinal fluid (CSF) sample revealed a WBC count of 157/mm³, with 82% lymphocytes and 18% granulocytes, red blood cell count of 15/mm³, elevated glucose of 84 mg/dL, and elevated protein of 107 mg/dL. He was admitted and started on broad-spectrum antibiotics. Over the next 24 hours, the patient became obtunded and was transferred to the intensive care unit (ICU), where he was intubated. Over the next week, the patient's level of consciousness improved, and he made attempts at communication. He was extubated 8 days following admission to the ICU and transitioned to the medical ward before being discharged to a rehabilitation center 3 weeks following his hospital admission. Three months following his acute illness, the patient returned home complaining of moderate fatigue and issues with memory.

The results of CSF and blood cultures were negative and polymerase chain reaction (PCR) for the detection of herpes simplex virus (HSV) in CSF was negative. Other infectious disease testing was negative except for a day 8 serum and CSF sample, both testing positive for WNV IgM antibodies by enzyme-linked immunosorbent assay (ELISA).

## ETIOLOGY AND TRANSMISSION

WNV is a ribonucleic acid (RNA) flavivirus that is related antigenically to St. Louis encephalitis and Japanese encephalitis viruses. WNV is transmitted to humans primarily through the bite of infected mosquitoes (Fig. 71.1). The predominant vectors worldwide are *Culex* mosquitoes, which feed primarily from dusk to dawn and breed mostly in peridomestic standing water or pools created by irrigation or rainfall. Mosquitoes become infected with WNV by feeding on a host that can sustain infectious levels of viremia, serving to amplify the virus. Birds are the most important amplifying hosts of WNV; they infect feeding mosquitoes, which then transmit the virus to humans and other mammals during subsequent feeding. Viremia usually lasts fewer than 7 days in immunocompetent persons, and WNV concentrations in human blood are generally too low to infect mosquitoes, making humans incidental or "dead end" hosts. However, person-to-person transmission can occur through blood transfusion and organ transplantation. Intrauterine transmission and probable transmission via human milk have also been described but appear to be uncommon, with most maternal infections having no noticeable impact (e.g., infection or clinical abnormalities) on the fetus or infant. Percutaneous and aerosol infections have occurred in laboratory workers, and an outbreak of WNV infection among turkey handlers also implicated aerosol transmission (Fig. 71.2).

## GEOGRAPHIC DISTRIBUTION

Prior to 1999, WNV activity was reported in Europe and the Middle East, Africa, India, parts of Asia, and Australia (in the form of Kunjin virus, a WNV subtype). WNV was first detected in North America in 1999, after which it rapidly spread across the United States and northward into Canada. The virus was subsequently detected in the Caribbean and Central and South America. In more recent years, WNV disease has reemerged in some areas of Europe.

In the United States, WNV activity in mosquitoes, birds, or other animals has been reported in all states except Alaska and Hawaii. The highest cumulative incidence of WNV neuroinvasive disease has occurred in central plains states (i.e., South Dakota, Nebraska, and North Dakota) (Fig. 71.3). The greatest burden of WNV disease occurs where areas of moderate to high incidence intersect metropolitan areas with high human population densities.

## PUBLIC HEALTH IMPORTANCE

WNV is the most common cause of locally acquired arthropod-borne viral (arboviral) disease in the continental United States and Canada. In the United States, WNV surveillance data are reported to the Centers for Disease Control and Prevention (CDC) through

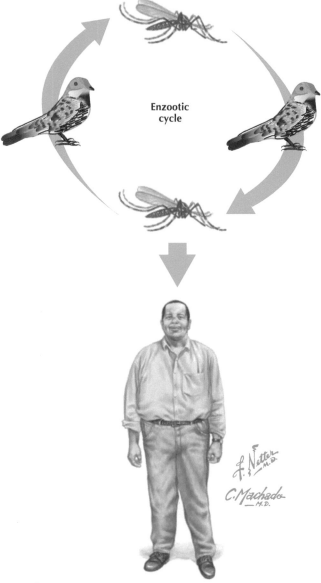

West Nile virus (WNV) is transmitted in an enzootic cycle between *Culex* mosquitoes and amplifying vertebrate hosts, primarily birds. WNV concentrations in human blood are generally too low to infect mosquitoes, making humans incidental or "dead-end" hosts.

**Fig. 71.1** West Nile virus transmission cycle.

ArboNET, an internet-based arbovirus surveillance system managed by the CDC and state health departments. From 1999 through 2018, 50,830 cases of WNV disease were reported to ArboNET, including 24,657 (49%) cases of neuroinvasive WNV disease and 26,173 (51%) cases of WNV nonneuroinvasive disease; there were 2330 (5%) associated deaths. Reports of neuroinvasive WNV disease are considered the most accurate indicator of WNV activity in humans, owing to the substantial associated morbidity and mortality and the presumed completeness of reporting by clinicians and laboratories. In contrast, WNV nonneuroinvasive disease (WNV fever) is likely underdiagnosed and underreported, as people with mild illness may not seek medical care or clinicians may not suspect or confirm WNV disease. Previous studies have estimated that between 30 and 70 nonneuroinvasive disease cases occur for every case of WNV neuroinvasive disease reported. Based on the number of neuroinvasive disease cases reported, an estimated 739,710 to 1,725,990 nonneuroinvasive disease cases of WNV occurred in the United States from 1999 through 2018 with roughly 1% to 3% of those reported.

## RISK FACTORS

In endemic areas of the Northern Hemisphere, the risk of WNV infection is higher during the warmer months, when mosquitoes are active and more abundant, typically July through October (Fig. 71.4). Although people of all age groups appear to be equally susceptible to WNV infection, the incidence of neuroinvasive WNV disease increases with age (see Fig. 71.4). In addition, among patients with neuroinvasive WNV disease, older adults are more likely to develop encephalitis or meningoencephalitis and have substantially higher case-fatality rates compared with children or younger adults (Table 71.1).

Solid organ transplant recipients also are at significantly higher risk of severe illness. With the exception of increased age and organ transplantation, risk factors for severe disease in persons infected with WNV have not been well defined. Severe WNV disease has been described in persons with malignancies, but the relative risk from these or other immunocompromising conditions remains unclear. Hypertension, cerebrovascular disease, chronic renal disease, alcohol abuse, and diabetes mellitus have also been identified as possible risk factors for severe WNV disease, but further research is warranted.

## CLINICAL FEATURES

Most human WNV infections are asymptomatic. The incubation period for WNV disease is typically 2 to 6 days but can be up to 21 days in immunocompromised people. Approximately 20% of infected people develop an acute systemic febrile illness, and less than 1% develop neuroinvasive WNV disease (i.e., encephalitis, meningitis, or acute flaccid paralysis) (Fig. 71.5). Most symptomatic patients have an acute systemic febrile illness that includes headache, myalgia, weakness, and often gastrointestinal symptoms and a transient maculopapular rash. Patients with neuroinvasive WNV disease may develop signs and symptoms of aseptic meningitis, encephalitis, and/or acute flaccid paralysis. WNV meningitis is generally indistinguishable from viral meningitis due to other causes and typically manifests with fever, headache, and neck stiffness. WNV encephalitis is a more severe clinical syndrome characterized by fever and altered mental status, seizures, focal neurologic deficits, or movement disorders such as tremor or parkinsonism. Acute flaccid paralysis associated with WNV infection has been attributed to WNV poliomyelitis, Guillain-Barré syndrome, and radiculitis. WNV poliomyelitis typically manifests as isolated limb paresis or paralysis and can occur without fever or apparent viral prodrome. It is clinically and pathologically identical to poliomyelitis caused by poliovirus, with damage of anterior horn cells, and may progress to respiratory muscle paralysis, necessitating mechanical ventilation. Rarely, cardiac dysrhythmias, myocarditis, rhabdomyolysis, optic neuritis, uveitis, chorioretinitis, orchitis, pancreatitis, and hepatitis have been described in patients with WNV disease.

## DIAGNOSTIC APPROACH

WNV disease should be considered in any person with a febrile or acute neurologic illness who has had recent exposure to mosquitoes, blood transfusion, or organ transplantation, especially during the summer months in areas where WNV activity has been reported in birds, animals, mosquitoes, or humans. The diagnosis of WNV disease should also be considered in any infant born to a mother infected with WNV during pregnancy or while breastfeeding. Guidelines for the evaluation of fetal and neonatal WNV infections are available at https://www.cdc.gov/mmwr/preview/mmwrhtml/mm5307a4.htm (accessed April

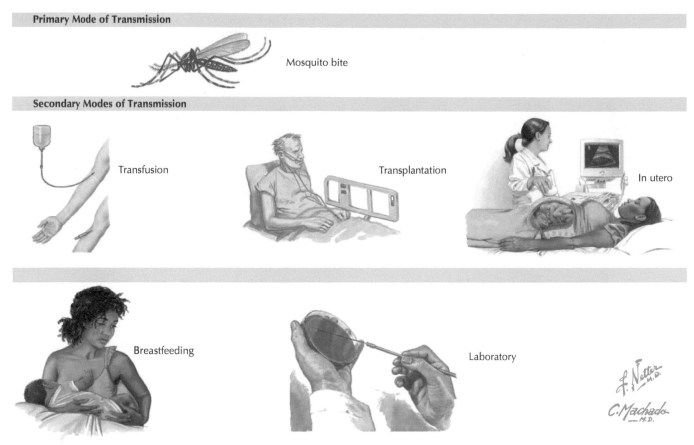

**Primary Mode of Transmission**

Mosquito bite

**Secondary Modes of Transmission**

Transfusion

Transplantation

In utero

Breastfeeding

Laboratory

Fig. 71.2 West Nile virus transmission to humans.

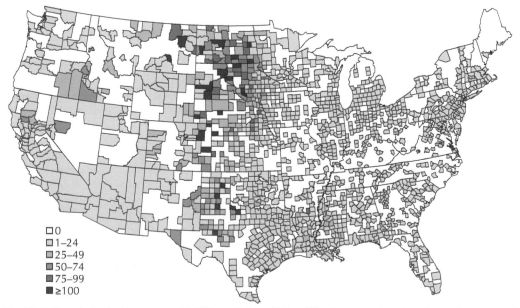

- □ 0
- ☐ 1–24
- ▨ 25–49
- ▨ 50–74
- ▨ 75–99
- ■ ≥100

Fig. 71.3 Cumulative incidence per 100,000 population of West Nile virus neuroinvasive disease by county of residence, United States, 2009–2018. (Adapted from McDonald E, Lindsey N, Staples JE, Martin S, Fischer M. Surveillance for human West Nile virus disease—United States, 2009-2018. Morbidity and Mortality Weekly Report (MMWR) Surveillance Summaries.)

2020). In a patient with suspected neuroinvasive WNV disease, other arboviruses (e.g., La Crosse, St. Louis encephalitis, eastern equine encephalitis, and Powassan viruses) also should be considered in the differential diagnosis where appropriate based on geographic distribution of these viruses.

Routine clinical laboratory studies are generally nonspecific in WNV disease. In patients with neuroinvasive WNV disease, CSF examination generally shows lymphocytic pleocytosis, but neutrophils may predominate early in the course of illness. Brain magnetic resonance imaging scans are frequently normal, but signal abnormalities in the

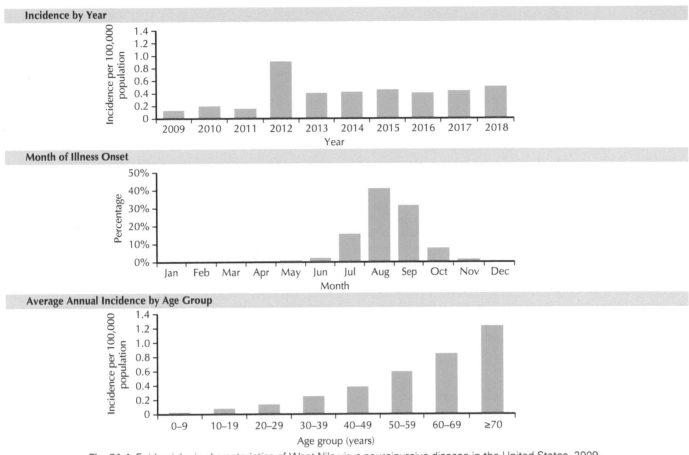

**Fig. 71.4** Epidemiologic characteristics of West Nile virus neuroinvasive disease in the United States, 2009–2018. (Adapted from McDonald E, Lindsey N, Staples JE, Martin S, Fischer M. Surveillance for human West Nile virus disease—United States, 2009-2018. Morbidity and Mortality Weekly Report (MMWR) Surveillance Summaries; pending publication.)

| TABLE 71.1 | Epidemiologic Features of West Nile Virus Neuroinvasive Disease, 2009–2018 | | | | | |
|---|---|---|---|---|---|---|
| | **ACUTE FLACCID PARALYSIS** | | **ENCEPHALITIS** | | **MENINGITIS** | |
| | *N* = 937 | | *N* = 6744 | | *N* = 4781 | |
| Median age in years (range) | 60 | (4–95) | 66 | (0.1–100) | 52 | (0.3–93) |
| Hospitalized (%) | 903 | (96) | 6505 | (96) | 4398 | (92) |
| Fatality (%) | 121 | (13) | 913 | (14) | 101 | (2) |

Adapted from McDonald E, Lindsey N, Staples JE, Martin S, Fischer M. Surveillance for human West Nile virus disease—United States, 2009-2018. Morbidity and Mortality Weekly Report (MMWR) Surveillance Summaries; pending publication.

basal ganglia, thalamus, and brain stem may be seen in patients with WNV encephalitis. Abnormalities may also be observed in the anterior spinal cord in patients with WNV poliomyelitis. Clinical features and electrodiagnostics can help differentiate WNV poliomyelitis from WNV-associated Guillain-Barré syndrome.

Serology continues to be the cornerstone of the laboratory diagnosis of WNV infection. Serum and, if indicated, CSF should be tested for WNV-specific immunoglobulin M (IgM) antibody. Enzyme immunoassays for WNV-specific IgM are currently available commercially and through local or state public health laboratories. If serum is collected within 8 days of illness onset, the absence of detectable WNV-specific IgM does not rule out the diagnosis of WNV infection, and the test may need to be repeated on a later sample. The presence of WNV-specific IgM in blood or CSF provides good evidence of recent WNV infection but may also result from cross-reactive antibodies after infection with other flaviviruses. Plaque-reduction neutralization tests performed in reference laboratories, including some state public health laboratories and the CDC, can help determine the specific infecting flavivirus and can confirm acute WNV infection by demonstrating a fourfold or greater change in WNV-specific neutralizing antibody titer between acute- and convalescent-phase serum samples collected 2 to 3 weeks apart. Because WNV-specific IgM can persist in some patients' serum for more than 1 year, the presence of WNV IgM may occasionally reflect past rather than recent WNV infection.

Viral cultures and nucleic-acid amplification tests (NAATs) for WNV RNA can be performed on serum, CSF, and tissue specimens that are collected early in the course of illness and, if results are positive,

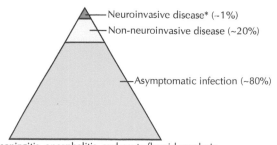

*Includes meningitis, encephalitis, and acute flaccid paralysis

**Fig. 71.5** Clinical spectrum of West Nile virus disease in humans. (Courtesy Centers for Disease Control and Prevention.)

can confirm WNV infection. However, the sensitivity of these tests to diagnose acute disease in immunocompetent individuals is low and negative results do not rule out infection. Because of the potentially prolonged period of viremia in immunocompromised individuals, NAAT is often more useful to make the diagnosis. Immunohistochemistry (IHC) can detect WNV antigen in formalin-fixed tissue particularly in fatal infections that might be missed if antemortem testing is not performed or is negative. Viral culture, NAAT, and IHC can be requested through state public health laboratories or the CDC.

## CLINICAL MANAGEMENT

No specific antiviral treatment for WNV disease exists; therefore treatment involves supportive care and management of complications. Patients with severe meningeal symptoms often require pain control for headaches and antiemetic therapy and rehydration for associated nausea and vomiting. Patients with WNV encephalitis require close monitoring for the development of elevated intracranial pressure and seizures, and patients with WNV encephalitis or poliomyelitis should be monitored for inability to protect their airway. Acute neuromuscular respiratory failure may develop rapidly, particularly in patients with prominent bulbar signs, and prolonged ventilatory support may be required. In patients hospitalized with WNV disease, standard infection control precautions are recommended. Although various drugs have been evaluated or empirically used for WNV disease, none has shown specific benefit to date. Information regarding ongoing clinical trials for treating WNV disease can be found at https://clinicaltrials.gov/ct2/home (accessed April 2020).

## PROGNOSIS

Most patients who develop WNV fever or meningitis recover completely, although some experience prolonged fatigue and other nonspecific symptoms. Among patients with neuroinvasive WNV disease, the overall case fatality rate is approximately 10%, but it is significantly higher in WNV encephalitis and poliomyelitis than in WNV meningitis or fever, in which case fatality rates are very low (see Table 71.1). Patients who recover from WNV encephalitis or poliomyelitis often have residual neurologic deficits that affect functional and cognitive performance.

## PREVENTION AND CONTROL

WNV infection can be prevented by avoiding exposure to WNV-infected mosquitoes and by systematic screening of blood donors for WNV infection. Coordinated mosquito-control programs in areas with enzootic WNV transmission can reduce the abundance of mosquito vectors. People who live in areas with WNV-infected mosquitoes

should apply insect repellent to skin and clothes and avoid being outdoors during peak mosquito-feeding times (usually dusk to dawn for WNV vectors). The CDC and the Environmental Protection Agency (EPA) recommend using an EPA-registered insect repellent. When used as directed, EPA-registered insect repellents are proven safe and effective, even for pregnant and breastfeeding women. The following active ingredients have been found to be effective: *N,N*-diethyl-*methyl*-toluamide (DEET), picaridin, IR3535, oil of lemon eucalyptus, para-methane-diol (PMD), and 2-undecanone. Products containing DEET or permethrin also can be applied to clothing. Oil of lemon eucalyptus or PMD should not be used on children younger than 3 years of age (https://www.cdc.gov/zika/prevention/prevent-mosquito-bites.html). According to the EPA (www.epa.gov/insect-repellents/deet), there is not a restriction on the percentage of DEET used for children versus adults.

In the United States, systematic screening of the blood supply for WNV infection was implemented in 2003. Nevertheless, healthcare providers should remain vigilant for possible WNV transmission through blood transfusion or organ transplantation. Any suspected WNV infections temporally associated with blood transfusion or organ transplantation should be reported promptly to public health authorities.

Pregnant women should take the aforementioned precautions to avoid mosquito bites. Products containing DEET can be used in pregnancy without adverse effects. Although WNV has probably been transmitted through human milk, such transmission appears rare, and no adverse effects on infants have been described.

Several vaccines against WNV are licensed for use in horses. Human WNV vaccines are not yet available, but several candidate vaccines are being evaluated.

## EVIDENCE

Lindsey NP, Hayes EB, Staples JE, Fischer M. West Nile virus disease in children, United States, 1999-2007. *Pediatrics* 2009;123:e1084-e1089. *The authors describe the epidemiologic features of pediatric WNV disease cases reported to the CDC from 1999 through 2007 and compared features of pediatric and adult neuroinvasive WNV disease. The clinical syndromes and severity of pediatric neuroinvasive WNV disease were similar to those reported for cases in young adults (aged 18–49 years). In contrast, a larger proportion of cases in older adults (aged 50 years or older) were classified as neuroinvasive WNV disease, cases of neuroinvasive WNV disease were more often reported as encephalitis or meningoencephalitis, and the case-fatality rate was substantially higher.*

Lindsey NP, Staples JE, Lehman JA, Fischer M. Medical risk factors for severe West Nile virus disease, United States, 2008-2010. *Am J Trop Med Hyg* 2012;87:179-184. *Describes medical risk factors for severe illness and neuroinvasive disease among WNV disease cases reported from selected states from 2008 to 2010. Chronic renal disease, history of cancer, history of alcohol abuse, diabetes, and hypertension were independently associated with hospitalization or encephalitis; only hypertension was associated with meningitis and immune suppression was independently associated with death.*

McDonald E, Lindsey N, Staples JE, Martin S, Fischer M. Surveillance for human West Nile virus disease—United States, 2009-2018, Morbidity and Mortality Weekly Report (MMWR) Surveillance Summaries (pending publication). *Summarizes the last 10 years of WNV disease cases reported through national surveillance. The report highlights both areas and populations at greatest risk for WNV in the United States.*

Nett RJ, Kuehnert MJ, Ison MG, Orlowski JP, Fischer M, Staples JE. Current practices and evaluation of screening solid organ donors for West Nile virus. *Transpl Infect Dis* 2012;14:268-277. *Provides a summary clusters of WNV transmission via solid organ transplantation and describes the current practices, concerns, and challenges related to screening organ donors for WNV in the United States.*

O'Leary DR, Kuhn S, Kniss KL, et al. Birth outcomes following West Nile virus infection of pregnant women in the United States: 2003-2004. *Pediatrics* 2006;117:e537-e545. *The authors report the results of birth outcomes among 77 women diagnosed with WNV infection during pregnancy. No adverse birth outcomes resulting from maternal WNV infection were conclusively demonstrated. Although larger studies would be useful, the results suggest that congenitally acquired WNV infections are uncommon and that birth outcomes in women infected during pregnancy are usually normal.*

Petersen LR, Brault AC, Nasci RS. West Nile virus: review of the literature. *JAMA* 2013;310:308-315. *Comprehensive review of the ecology, virology, epidemiology, clinical characteristics, diagnosis, prevention, and control of West Nile virus, with an emphasis on North America.*

Staples JE, Shankar MB, Sejvar JJ, Meltzer MI, Fischer M. Initial and long-term costs of patients hospitalized with West Nile virus disease. *Am J Trop Med Hyg* 2014;90:402-409. *The paper describes both short-term and long-term medical needs and estimates the costs for patients hospitalized with WNV disease in Colorado during 2003. Based on these costs and the number of cases reported to the CDC, the total cumulative costs of reported WNV hospitalized cases from 1999 through 2012 in the United States was estimated to be $778 million.*

## ADDITIONAL RESOURCES

American College of Physicians. Physicians' Information and Education Resource (PIER): West Nile virus disease. Available at: http://pier.acponline.org/physicians/public/d951/d951.html. Accessed February 24, 2010. *An authoritative and user-friendly online tool for clinicians, first published in 2004 and since updated periodically.*

Centers for Disease Control and Prevention (CDC). West Nile virus: information for healthcare providers. Available at: https://www.cdc.gov/westnile/healthcareproviders/index.html. Accessed December 19, 2019. *A concise online summary, including information on epidemiology, clinical features, diagnostic testing, prevention, surveillance, and public health reporting.*

United States Environmental Protection Agency (EPA). Using Repellent Products to Protect against Mosquito-Borne Illnesses. Available at: https://www.epa.gov/insect-repellents/using-repellent-products-protect-against-mosquito-borne-illnesses. Accessed December 19, 2019. *A brief summary of the current recommendations for repellent use in the prevention of mosquito-borne diseases; includes links to several additional useful resources.*

# 72

# Tick-Borne Encephalitis

*Martin Haditsch*

## ABSTRACT

For a long time, tick-borne encephalitis (TBE) was perceived as a risk only (if ever) for the local populations living in endemic areas. There was very little specific awareness and hardly any knowledge about this viral disease among doctors (not to mention lay people, including travelers) outside the endemic areas. Thus, despite high incidence rates in the local populations of some endemic areas, TBE has been and mostly still is a neglected travel-associated health risk.

World Tourism Organization statistics show that the European continent continues to be the main travel destination of international travelers. Considerable parts of this continent, especially those with the highest recent increases in international tourism (i.e., the Baltic States) as well as parts of Asia have proven to be TBE-endemic areas (Fig. 72.1). As shown in Fig. 72.2, there has been a substantial increase in nearly all European TBE-endemic countries (with the exception of Austria, where vaccination was introduced in 1981, and a general recommendation to vaccinate against TBE was implemented soon after). Thus it is increasingly important to provide the local population as well as travelers to these areas with current information on the risk of disease transmission and the possibilities of prophylaxis in general and vaccination in particular.

TBE is known by a variety of names, including spring-summer encephalitis, Central European encephalitis, Far-East Russian encephalitis, Taiga encephalitis, Russian spring-summer encephalitis, bi-undulating meningoencephalitis, diphasic milk fever, Kumlinge disease, Schneider disease, and, in German-speaking areas, Früh-Sommer-Meningo encephalitis (FSME) or early-summer meningoencephalitis—a description that might be misleading, as shown later.

In its most severe form, the infection affects the central nervous system (CNS), mostly as a meningoencephalitis leading to persistent sequelae in up to 58% of patients. Prophylaxis is focused on the avoidance of exposure and on vaccination. Because many of the TBE-endemic areas are increasingly popular tourist destinations, it seems of utmost importance to raise awareness among local people as well as travelers and their health advisors, primary care medicine practitioners, infectious disease specialists, and travel medicine and vaccination centers.

## CLINICAL VIGNETTE

A 32-year-old businessman went river rafting in the southern part of Austria. Two weeks later he developed a fever and subsequently neurologic symptoms, starting with headache and nausea followed by typical signs of meningitis and encephalitis: stiff neck, vomiting, and photophobia. He was the only member of his family (including his wife and 5-year-old son) who was not vaccinated against TBE, and he could not recall a tick bite. Based on the deterioration of his clinical status, he was brought to the emergency room of a nearby tertiary referral hospital.

Because this happened in a TBE-endemic area and the patient's history indicated a substantial likelihood of exposure, this disease was high on a short list of differential diagnoses. Thus, in addition to the usual examination of patients with symptoms of meningoencephalitis (including differential white blood cell count, lactate dehydrogenase [LDH], lumbar puncture, and examination of the cerebrospinal fluid [CSF]), a serologic test for TBE was performed (enzyme-linked immunosorbent assay [ELISA] only). All tests supported the likelihood of a viral infection involving the brain, and the ELISA test revealed positive IgM antibodies against TBE (IgG negative). A computed tomography (CT) scan of the head showed disseminated foci of inflammation and tissue destruction, which led to the most likely diagnosis of severe TBE (confirmed by a subsequent serologic test showing the classic immunoglobulin switch with TBE-IgG becoming highly positive and the TBE-IgM becoming negative). Although the patient was treated in the intensive care unit with maximal supportive care, the infection progressed and, after some days, led to the need for artificial ventilation. Subsequent CT scans showed further deterioration, which ultimately resulted in an apallic syndrome. After approximately 3 years of intensive care (owing to the need for artificial ventilation, the patient could not be discharged from the hospital), permission to withdraw all life support was obtained.

COMMENT: Although only a certain proportion of tick bites in TBE-endemic areas bears the risk of transmission and only a small percentage of infections result in classical disease, TBE can permanently impair previously healthy persons and poses the risk of death as a worst-case scenario. In contrast to other diseases, such as Lyme borreliosis, TBEV infection starts immediately after the attached tick starts its blood meal; therefore even timely removal of the tick may not suffice to prevent infection. Because there is no option of postexposure prophylaxis (administration of hyperimmune globulin was stopped due to the risk of viral enhancement and is no longer available in Western countries) and no causal treatment for this potentially devastating disease, strict adherence to safe behavior patterns as well as vaccination (for persons living in or going to or through high-risk areas) should be strongly recommended.

## THE PATHOGEN

The TBE virus (TBEV) is a single-stranded ribonucleic acid (RNA) virus belonging to the Flaviviridae family; it therefore shows some similarities with other viruses in this family (e.g., dengue virus [DENV], Powassan virus [POWV], West Nile virus [WNV], Japanese encephalitis virus [JEV], yellow fever virus [YFV], and Zika virus [ZV]). The TBEV is fairly homogeneous in endemic areas of Europe (European subtype). There are two additional subtypes within the same group with few genetic differences: the recently identified Siberian subtype (which genetically is quite closely linked to the European subtype) and the Far Eastern subtype.

■ Reported TBE cases and/or documented TBE virus isolation

**Fig. 72.1** Geographic distribution of tick-borne encephalitis. (Dobler G, Erber W, Schmitt HJ. Risk map of TBEV. Chapter 12c. In: Dobler G, Erber W, Bröker M, Schmitt HJ, eds. The TBE Book. 3rd ed. Singapore: Global Health Press;2020. http://doi:10.33442/26613980_12c.)

## VECTOR AND TRANSMISSION

TBE is transmitted by hard ticks and by the consumption of infected raw milk. In addition, there is an occupational health risk in laboratories for those working with material contaminated by or infected with TBEV. By far the most frequent mode of transmission to humans is by bites of certain species of hard ticks (i.e., *Ixodes ricinus* and *Ixodes persulcatus*) that have previously fed on infected reservoir hosts (Fig. 72.3). With some areas of overlap, *I. ricinus* (the common castor-bean tick) mostly inhabits the western part, and *I. persulcatus* (Taiga tick) is mainly found in the eastern part of the endemic area that roughly stretches from the Alsace region of France and eastern parts of the Netherlands (first identified in 2016) in the west to Hokkaido Island, Japan, in the east, and in Europe from Scandinavia in the north to Italy, Albania, and Greece in the south and from Russia in the north to southern China and South Korea in the south in Asia respectively. In 2019, the first autochthonous case was published in the United Kingdom.

According to old published data, TBEV prevalence in the tick population in endemic areas may be as low as 0.05% in Italy and 0.07% in Finland and as high as 26.6% in some regions of Latvia. Although replication in *Dermacentor marginatus* was already proven experimentally in 1985, TBEV infection linked with high probability to a bite by this different hard tick species was published only in 2017.

One phenomenon contributing to an increasing prevalence of TBEV-infected tick populations is the fact that a vertical or transovarian transmission (i.e., a TBEV-infected female tick infects her eggs before oviposition) takes place as well as from an infected animal to a tick during a bite (or even from tick to tick by cofeeding on the same blood pool). In contrast to other blood-feeding arthropod vectors, ticks do not directly puncture blood vessels but feed from a "feeding pool" produced by vasoactive mediators and coagulation inhibitors released via the tick's saliva. The blood meal of an adult female and

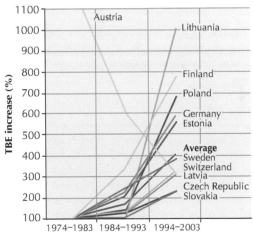

**Fig. 72.2** Tick-borne encephalitis cases reported in Europe, 1974 to 2003. (Data from Kunze U, Baumhackl U, Bretschneider R, et al: The golden agers and tick-borne encephalitis. Conference report and position paper of the International Scientific Working Group on Tick-Borne Encephalitis, *Wien Med Wochenschr* 155[11-12]:289-294, 2005.)

immature blood-sucking nymphs may last up to 5 days, resulting in an approximately 120-fold increase of the volume of an adult female tick. Male ticks do not feed on blood at all but may repeatedly feed on a small amount of tissue fluid during a relatively short feeding period, which nevertheless may be sufficient for the transmission of the TBEV.

In addition to tick bites, another mode of transmission is by the consumption of infected raw milk and products made from it. This mode of transmission used to occur quite frequently in the Baltic States (e.g., Lithuania) and in Slovakia. Between 2012 and 2016, there were at least 13 outbreaks with 141 cases linked to the consumption of unpasteurized dairy products made from sheep and goat milk. In 2008, some

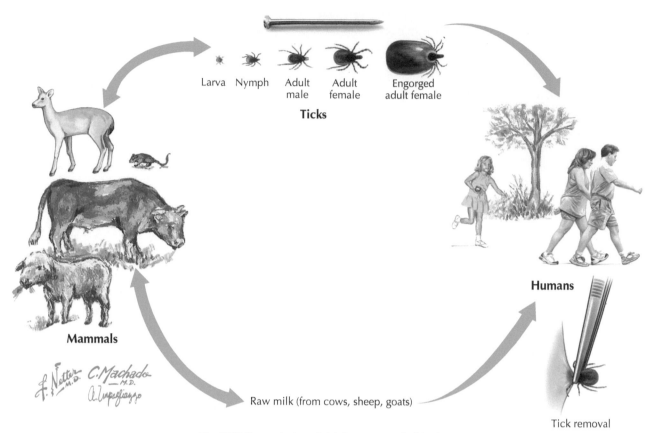

Larva　Nymph　Adult male　Adult female　Engorged adult female

**Ticks**

**Mammals**

Raw milk (from cows, sheep, goats)

**Humans**

Tick removal

**Fig. 72.3** Transmission of tick-borne encephalitis virus.

cases of alimentary infection were reported for the first time in Austria; since 2016, a few episodes were also reported from Germany.

Another risk of exposure involves laboratory workers who deal with this virus.

## EPIDEMIOLOGY

In the past, at least in central Europe, most cases were seen in early summer. Later on, two peaks of this disease were described, with the recognition of an additional peak in late summer. Tick-biting activity appears to be increased by higher humidity and temperature. Also, owing to recent climate change observed in the whole region, currently at least eight new phenomena have been observed:

- In addition to the steady growth of the geographic risk area at the same latitude accompanied by a confluence of scattered foci to larger endemic regions, the TBE endemic area boundary has moved northward, with the eruption of highly active foci in Scandinavia, for instance. In addition, the areas of risk include higher altitudes.
- In some areas the two peaks of reported clinical cases have merged into one broader peak occurring in July and August.
- In some areas, most likely as a result of warmer temperatures, the typical seasonal pattern with transmission-free intervals has vanished. Transmission seasons have expanded and, in some regions, year-round transmission is occurring.
- The tick populations in some areas have substantially increased, probably because of climate change, which has led to more rain, warmer winters, and moister springs. As a consequence, the tick-bite season starts earlier and the different stages develop more quickly. Ixodes tick activity starts with a soil temperature of about 7°C and a relative humidity of at least 92% necessary for these ticks' survival is provided by more frequent episodes of rainfall.

- *Dermacentor* ticks (e.g., *Dermacentor marginatus*) may transmit TBEV as well. The activity pattern of this tick species is different (e.g., the biting season lasts longer), which may also affect traditional seasonal transmission.
- In most TBE-endemic countries, people have achieved higher standards of living. They have more leisure time, which, at least during the warm season, contributes to more exposure-prone outdoor activities. Furthermore, many families have shifted from living in apartments to having their own homes and gardens, bushes, and trees, all of which are associated with an increased risk of tick bites.
- "Green thinking" results in some vaccination skepticism as well as addiction to "natural and healthy food," which may include unpasteurized dairy products.
- The emergence of the COVID-19 pandemic (SARS-CoV2) and the public health measures implemented to control transmission such as quarantine and travel restrictions had an unforeseen impact on the incidence of TBE subsequently experienced in some TBE-endemic countries. Weeks to months of quarantine at home mandates generated an increased demand among people for outdoor activities whenever possible. In addition, travel restrictions resulted in more people spending their holidays in their home country, which meant that nonimmune persons living in or going to TBE endemic areas like central Europe had increased exposure during their leisure time outdoor activities. This may have been one of the drivers for skyrocketing clinical TBE cases reported in Germany, Austria, and Switzerland in 2020. This should be considered in other epi-/pandemics to come, too.

These factors contribute to a higher risk of exposure to the TBEV. In most regions, this risk has shifted from being mainly occupational (forestry workers, farmers) to one affecting people anyone who spends leisure time outdoors (e.g., collecting mushrooms, playing golf, camping and picnicking, Nordic walking, hiking, trekking). This is particularly true for older people and those who are retired, who, if they acquire a TBE infection, tend to develop more severe symptoms.

As is the case with other infections, epidemiologic data reported officially depend on a variety of factors including awareness within the healthcare system; the availability of adequate diagnostic tools; the notification systems of regional, national, and international health authorities; and adherence among those who should report. The low awareness of TBE in nonendemic countries is likely to result in the underdiagnosis of clinical cases as well as in underreporting; these are the most probable explanations for the difference between the estimated attack rates among tourists and the numbers reported.

On a person-based calculation, the highest incidence published (98 per 100,000) was that among forestry workers in Austria in the prevaccination era; even today, however, case counts exceed 100 per 100,000 in high-risk areas (e.g., in the Baltic states) during high transmission seasons. As with other infectious diseases, not every infection results in specific clinical symptoms. The risk of symptomatic disease after a single tick bite in an endemic area varies from 1:200 to 1:1000. Actual epidemiologic data underline the importance of booster vaccinations of the local population as well as the vaccination of travelers going to endemic areas, as the risk of acquiring TBE in an endemic area is about the same as that for typhoid fever in nonvaccinated travelers going to India.

## CLINICAL FEATURES

TBE is typically a biphasic disease. About 1 week (2 to 28 days) after an infectious tick bite, most patients develop an influenza-like illness (ILI) for a few days (2 to 8 days) caused by the viremia (Fig. 72.4). For many patients, this remains the one and only symptomatic period. After an asymptomatic interval of about 1 more week (1 to 20 days), some patients develop a sudden rise in temperature, marking the beginning of the second stage. In this context, it seems noteworthy that TBE patients tend to have higher temperatures than those with other forms of viral meningitis or meningoencephalitis.

The second phase is that of CNS involvement. The symptoms are most frequently caused by meningoencephalitis: stiff neck, fever, headache, nausea/vomiting, photophobia, and impairment of CNS functions (Fig. 72.5). These result from either direct destruction of CNS tissue (irreversible) or from an inflammatory edema (reversible). The amount of CNS involved, the mass of destruction, and its location determine the severity of the clinical course. Some factors,

such as the presence of underlying diseases or conditions and older age, are associated with an increased severity of disease. In its most severe form, TBE may involve all parts of the CNS—that is, causing meningoencephalomyeloradiculitis.

Compared with some other vector-borne flaviviral infections, there is no impairment of the clotting system, and symptoms of a viral hemorrhagic fever have never been proven in a case of TBE in Western Europe. According to incidence figures, cases in travelers from nonendemic countries are likely to remain undiagnosed or misdiagnosed, showing that the clinicians responsible for the workup of these cases lack awareness of and experience in diagnosing this disease. In addition, underreporting must also be taken into account. The features of differential diagnosis between TBE and Lyme disease are given in Table 72.1.

## DIAGNOSTIC APPROACH

Typically, a patient will not seek medical care until the second stage of the disease—that is, with the onset of neurologic signs. As in other cases of meningoencephalitis, diagnostics usually include a neurologic examination, serologic tests (paired samples: blood and CSF), radiology (computed tomography [CT] and magnetic resonance imaging [MRI]), and general laboratory screening tests (including a differential blood count and general inflammation parameters). The typical results are as follows:

- Besides specific neurologic signs attributable to the cerebral region involved, patients show unspecific neurologic symptoms such as headache, stiff neck, photophobia, nausea with or without vomiting, and fever. The area most frequently affected in terms of paralysis is the shoulder girdle.
- The radiologic examination (CT or MRI) shows signs of meningitis and randomly (i.e., with an unpredictable pattern) the spread of foci in the CNS (encephalitis). Some of these abnormal signals are caused by destruction and others by perifocal edema.
- In the differential blood count at this stage there is a shift toward mononuclear cells with a usually decreased white blood cell (WBC) count. C-reactive protein (CRP) and procalcitonin (PCT) levels are low; LDH levels may be elevated (in accordance with CNS tissue destruction).
- CSF examination shows the typical picture of a viral meningitis—that is, elevated cell count (mononuclear cells) and protein, with a normal glucose level. The protein elevation is partly a result of unspecific inflammatory proteins and partly due to the autochthonous production of TBE-specific immunoglobulin (Ig).

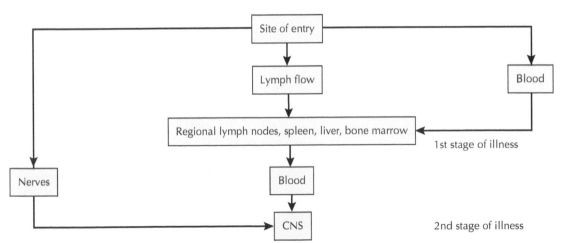

**Fig. 72.4** Spread of tick-borne encephalitis virus in the body. (Data from Kaiser R. Tick-borne encephalitis in Germany and clinical course of the disease. *Int J Med Microbiol* 2002;291[suppl 33]:58-61.)

- When paired samples are checked serologically, the first sign of a recent infection is the detection of specific anti-TBEV IgM antibodies in both the serum and CSF. These antibodies are often present at the time of hospital admission, which, as mentioned previously, is usually triggered by the appearance of neurologic signs (i.e., in the second phase). The tests routinely used are based on the ELISA technique. TBE antibodies may cross-react with other flaviviruses in these test systems, which is important, particularly in travelers returning from endemic areas of JEV and DENV or those previously vaccinated against JEV and/or yellow fever (YF). In case of inconclusive results, there are additional options such as the neutralization test (NT) or polymerase chain reaction (PCR) assay (which are usually limited to specialized laboratories).

## TREATMENT

As with most other viral diseases, no specific therapy is available. Treatment therefore is symptomatic only and focused on managing inflammation, pain, fever, and nausea as well as convulsions in serious infections. The most severe TBE cases may require intensive care, including parenteral nutrition and auxiliary ventilation. The availability of a high-quality standard of care may reduce the incidence and degree of neurologic sequelae and improve the survival rate. This should be recognized, especially by nonvaccinated travelers heading to remote areas within an endemic region or proceeding/returning to an area with limited/basic medical services, only.

## PROGNOSIS

The outcome of the disease depends on the size, amount, location, and function of the CNS structures destroyed. TBE viruses are known to have a particular predilection for anterior horn cells of the cervical spinal cord, thus leading to paresis of the upper limbs, the shoulder girdle, and the head levator muscles and having the potential to involve the respiratory muscles. The incidence rate of sequelae after TBE varies from 35% to 58%. The case-fatality rate (CFR) in central Europe is about 1% to 2%, whereas the CFR of the Siberian subtype (the Taiga strain, which is closely related to the Far Eastern form) is said to be as high as 20%, but there are doubts about whether this figure is correct; it might be biased by the fact that only people who are severely sick go to hospitals. On the other hand, there are investigations that show a higher virulence of the Siberian subtype in mice compared with the other Far Eastern strains. And there is scientific proof for a more serious course of the disease, a higher complication rate, and a significantly higher CFR among the elderly.

After an infection (whether symptomatic or asymptomatic), TBE always results in lifelong immunity.

## PREVENTION

There is no preexposure chemoprophylaxis against TBE, so the only methods of prevention are avoidance of exposure and vaccination. In contrast to widespread beliefs, ticks do not primarily live on trees, nor do they jump down on their victims. Their usual habitats are areas with grass, meadows, bushes, and small trees. In contrast to "hunting ticks" (which are common in the tropics), ticks transmitting TBEV (described as "questing ticks") increase their chances to attach to warm-blooded animals (and humans) by sitting at the tips of the plants, analyzing their surroundings with a very complex organ (the so-called *organ of Haller*, located at their first pair of legs). Although it is not understood in detail, temperature, $CO_2$, and butyric acid (which is also responsible for the smell of sweat) seem to attract ticks and to trigger the attachment of ticks to the skin, facilitated by a contact time of not more than 0.1 second.

Because ticks in TBE-endemic areas might also transmit other diseases (Lyme borreliosis being the most common), the prevention of all

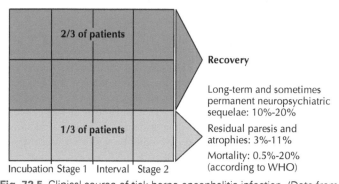

**Fig. 72.5** Clinical course of tick-borne encephalitis infection. (Data from Kaiser R. Tick-borne encephalitis in Germany and clinical course of the disease. *Int J Med Microbiol* 2002;291[suppl 33]:58-61.)

| TABLE 72.1 | Differential Diagnosis of Tick-Borne Encephalitis and Lyme Disease | |
|---|---|---|
| | **TICK-BORNE ENCEPHALITIS** | **LYME DISEASE** |
| Pathogen | Tick-Borne Encephalitis Virus | *Borrelia burgdorferi* (Bacterium) |
| **Clinical Picture** | | |
| Phase 1 | Temperature ≤38°C<br>Uncharacteristic influenza-like signs and symptoms | Red rash (usually round or oval, erythema chronicum migrans)<br>Uncharacteristic influenza-like signs and symptoms<br>Lymphadenosis benigna cutis (rare form of benign recurrent tumor) |
| Phase 2 | Fever (temperature ≥40°C)<br>Meningitis<br>Meningoencephalitis<br>Meningomyelitis<br>Meningoencephalomyelitis | Meningopolyneuritis (Garin-Bujadoux-Bannwarth)<br>Facial paresis<br>Cardiac arrhythmia<br>Myocarditis |
| Phase 3 | | Acrodermatitis chronica atrophicans<br>Arthritis |

## TABLE 72.2  General Preventive Measures

| Intervention | Measure | Comment |
|---|---|---|
| Behavior | Avoid tick-infested areas. | Whenever possible. |
| Clothing | Light-colored clothing that covers arms and legs (long-sleeved shirts, tight at the wrists; long pants, tight at the ankles and tucked into the socks), shoes covering the entire foot. | Dark clothing is more attractive for ticks (which in addition are more difficult to identify against a dark background). |
| Use of repellents | Apply adequate repellent (with proven activity against ticks) to exposed skin and to outer clothing. | DEET[a] in higher concentrations, picaridine 20%. and permethrin are proven to act against ticks; allow clothing to dry before wearing. |
| Early detection | Adults should check daily, children should be checked more frequently (i.e., after each episode of potential exposure [could result in two or three checks per day]). | The checks should especially focus on waistbands, sock tops, underarms, other moist areas (in children, check the head and behind the ears); even adults may need the help of a second person to check the whole body. |
| Early removal of ticks | Remove tick as soon as possible by using fine tipped tweezers or special cards (resembling carved credit cards); grasp the tick firmly as close to the skin as possible and simply tear it out without squeezing or rotating the tick. | Do not suffocate the tick (do not use oil, cream, nail polish, water); do not burn the tick; do not apply "traditional recipes"; do not wait for medical services when not promptly available—remove the tick yourself. |

[a]DEET, N,N-Diethyl-meta-toluamide.

tick exposure is crucial. This includes avoidance of running or walking through high grasses or on narrow paths with repeated unavoidable contact with grass and bushes. Equally important is the use of products qualified for repelling ticks on exposed skin surfaces as well as impregnating outer clothing. Products containing the chemical N,N-diethyl-meta-toluamide (DEET) (in higher concentrations, preferably >20%) and picaridine 20% are considered to be effective skin repellents against ticks (and other arachnids as well as insects). An insecticide containing permethrin is recommended for the impregnation of clothing. Several products are commercially available at sporting goods stores, to be used according to package directions. Another consideration is the fact that the color of the clothes has substantial impact on the attraction of ticks, light-colored clothing being significantly less attractive to ticks than dark-colored clothing. A summary of recommendations is provided in Table 72.2.

### First Aid

After an infected tick becomes attached to the skin, some time is required until the transmission of a specific pathogen takes place. The process involves the time it takes for the tick to select the location for its bite, penetration of the skin, production of a "feeding pool" in the case of the blood-sucking nymph stages and adult females (adult males feed on tissue fluid only), and time for the influx of a sufficient amount of tick saliva and/or regurgitated sucked blood to reach the minimal infectious dose in the human host. Therefore it makes sense to remove a tick as soon as possible to avoid an infection—even if the bug is already firmly attached to the skin.

To facilitate the removal of attached ticks in a timely fashion, the whole body should be examined after outdoor activities. A thorough inspection usually necessitates the help of a second person (e.g., before showering or bathing). The removal of an attached tick should be as atraumatic as possible. This can be managed by using a pair of fine-tipped tweezers. The recommended method for tick removal is to pick the part of the tick closest to the skin and to tear it off without rotation and without squeezing the body, which could result in an increased influx of pathogens (see Fig. 72.3). The tick should also not be drowned (e.g., by bathing, which—by the way—would take days to weeks), suffocated (by putting a drop of glue, nail polish, or oil on the tick), or burned (with a match or a cigarette), because an increased burden of infectious particles may be released into the bite wound while the tick is struggling to survive. Another misconception is that if a black dot remains in the wound, this might be the head; rather, it is probably

some part of the biting apparatus only. Because of the anatomy of the tick, the salivary glands containing the TBEV are removed by removing the body as described here; any remaining mouth parts make no significant contribution to the time span of possible virus transmission.

### Vaccination

In the past, postexposure prophylaxis (PEP) by using TBE hyperimmune globulin within 96 hours for passive immunization after a tick bite was practiced. However, TBE hyperimmune globulin is no longer available, and standard lots of commercially available human immune globulin do not qualify for PEP, as they are unlikely to contain a sufficient level of specific antibodies to protect against infection. In addition, giving a wrong quantity and/or the product too late might result in viral enhancement—that is, more serious symptoms and/or sequelae.

Immunization against TBE before exposure is advised for persons living in or travelers going to or traveling through TBE-endemic areas. The vaccine is currently available year 'round without any seasonal restrictions in many countries. It is not yet licensed in the United States and no longer available in Canada. In 2019 one Western brand was licensed in Israel.

The vaccine products that are registered in some western countries are branded as FSME-IMMUN and Ticovac (Pfizer; originally developed by the Immuno AG, Vienna, Austria) and Encepur (Valneva and Bavarian Nordic; originally developed by the Chiron Behring GmbH and Company KG, Marburg, Germany). Both vaccines contain very closely related inactivated TBEV strains (strain Neudörfl and strain Karlsruhe, respectively) as antigens that elicit protective antibodies against all known subtypes of the TBEV and are registered for intramuscular administration. Reported adverse events are mainly local side effects, with the likelihood of serious side effects such as neuritis being extremely low (neuritis occurs in less than 1:1,000,000 vaccinees). Neurologic disorders in general did not occur more often in the vaccinees than in the unvaccinated population. Standard and rapid immunization schedules are established; virtually all lead to a strong immune response (NB: in addition, locally produced vaccines are available in Russia and China).

The topic of the booster-dose interval is controversial. The intervals recommended by the companies are not always identical to those recommended by national vaccination boards or those listed on Internet websites (frequently used by travelers as a basic source of information). In case of a history of a probable TBE infection or irregular vaccination schemes, serologic testing to check or prove immunity should be stressed.

Data from 2004 show that about 60 million travelers visit TBE-endemic parts of Europe each year. Among these travelers, according to World Tourism Organization (WTO) data, there is an increasing trend to travel to those countries with an exceptionally high risk of TBE. Calculations show that for nonvaccinated travelers, the average risk of acquiring TBE in a highly endemic area is not less than the average risk of contracting typhoid fever in India (typhoid fever vaccination being recommended on a regular basis for travelers heading to this destination). As the TBE risk shows annual differences and seasonal changes during the high transmission season it may even exceed the risk of typhoid fever on the Indian subcontinent.

Local people as well as travelers tend to explore and enjoy their environment, which usually includes some outdoor exposure. Furthermore, green thinking in parts of the local population and the curiosity of travelers to sample authentic regional food in some areas might expose them to TBE via raw milk and dairy products. In addition to recreational travelers whose stated purpose for going to a TBE-endemic area is to participate in outdoor activities, education about the risks of TBE and prevention of transmission as well as discussion about the availability of vaccines to prevent disease should be offered by healthcare practitioners to families, scout groups, school classes, and students on exchange programs and those with occupational exposure, respectively. Since the decrease in the autochthonous population in some countries was accomplished by general vaccination programs only (vaccination of identified risk groups had limited effect on annual incidence rates), all travelers to endemic areas should at least be informed about the risk so that they can adapt their behavior accordingly.

## ACKNOWLEDGMENTS

I want to thank the companies of Baxter, Novartis, Pfizer, and GSK for providing information including data and graphics. In particular, Dr. Dieter Gniel was very helpful in sharing his vast experience on TBE.

## EVIDENCE

Centers for Disease Control and Prevention (CDC). Tick-borne encephalitis among U.S. travelers to Europe and Asia—2000-2009. *Morb Mortal Wkly Rep* 2010;26:59(11):335-338. *A review of all 2000–2009 laboratory records was conducted to identify cases of TBE among US travelers. Five cases were identified by IgM serum antibodies and confirmed as TBE by plaque-reduction neutralization tests: four patients had traveled to Europe or Russia and had a biphasic illness followed by nearly complete recovery. The fifth patient had traveled to China and had a monophasic illness with severe encephalitis and neurologic sequelae.*

Czupryna P, Moniuszko A, Pancewicz SA, et al. Tick-borne encephalitis in Poland in years 1993-2008—epidemiology and clinical presentation: a retrospective study of 687 patients. *Eur J Neurol* 2011;18(5):673-679. *The epidemiology and clinical features of TBE in this region of Europe were analyzed. In this group of patients, the initial disease presented with meningitis in 41%, meningoencephalitis in 51%, and meningoencephalomyelitis in 8%. Ataxia in 14% and pareses in 9% were the most common neurologic abnormalities that developed. Upon discharge, 23% had neurologic and 44% had psychiatric sequelae.*

Schoendorf I, Ternak G, Oroszlan G, et al. Tick-born encephalitis (TBE) vaccination in children: advantage of the rapid immunization schedule (i.e., days 0, 7, 21). *Hum Vaccin* 2007;4:42-47. *A clinical study involving 294 children 1 to 11 years of age who were vaccinated with a pediatric formulation of TBE vaccine (Encepur children) according to the conventional schedule on days 0, 28, and 300; the modified conventional schedule on days 0, 21, and 300; or the rapid schedule on days 0, 7, and 21. The rapid immunization schedule in children stimulated rapid protection and stable titers for at least 300 days after vaccination.*

Weinberger B, Keller M, Fischer KH, et al. Decreased antibody titers and booster responses in tick-borne encephalitis vaccinees aged 50-90 years. *Vaccine* 2010;28:3511-3515. *Cases of vaccine failures (clinical and serologic evidence of TBE infection despite adequate immunization) have been reported predominantly in older persons. The immune-responsiveness to TBE vaccinations for age groups 50 to 59, 60 to 69, and older than 69 years were compared to a control group aged below 30 years. The antibody titers and booster responses measured for each group suggest that responsiveness of the immune system to vaccination is already impaired at the age of 50 compared with the control group. Booster intervals of 3 years are currently recommended for persons at or older than 60 years of age in Austria, but this might be beneficially applied to persons at or older than 50 years of age.*

## ADDITIONAL RESOURCES

Centers for Disease Control and Prevention (CDC). Tick-borne encephalitis. Available at: www.cdc.gov/ncidod/dvrd/spb/mnpages/dispages/TBE.htm. *Good basic information on many aspects of this infectious disease.*

Centers for Disease Control and Prevention (CDC). Travelers' health. Available at: wwwnc.cdc.gov/travel/yellowbook/2020/travel-related-infectious-diseases/tickborne-encephalitis. *One of the reference sites on infectious diseases for US doctors.*

International Scientific Working Group on Tick-Borne Encephalitis. Prevention information. Available at: www.tbe-prevention.info. *Site providing information focused on tick bite and TBE prevention.*

International Scientific Working Group on Tick-Borne Encephalitis. Tick victims information. Available at: www.tick-victims.info. *Information on the disease focused on clinical aspects provided by tick victims as well as by people working with TBE patients (including the "self-help group of tick victims").*

International Scientific Working Group on Tick-Borne Encephalitis website. Available at: www.isw-tbe.info. *Information on TBE by the International Scientific Working Group on Tick-Borne Encephalitis.*

World Health Organization (WHO). The vector-borne human infections of Europe: their distribution and burden on public health. Available at: www.euro.who.int/__data/assets/pdf_file/0008/98765/e82481.pdf. *A WHO Internet site dedicated to European infectious diseases.*

World Health Organization (WHO). Tick-borne encephalitis. Available at: www.who.int/immunization/topics/tick_encephalitis/en/. *WHO site explaining prophylaxis and vaccination.*

World Health Organization (WHO). Tick-borne encephalitis vaccine. Available at: www.who.int/biologicals/areas/vaccines/tick_encephalitis/en/. *WHO site explaining prophylaxis and vaccination.*

Zuckerman JN, Jong EC, eds. *Travelers vaccines*, ed 2. Shelton, CT: Peoples Medical Publishing House; 2010. *This book includes a chapter on TBE, mostly focused on travel medicine aspects.*

# 73

# Lyme Disease

*Alexa R. Lindley, Christopher A. Sanford*

## ABSTRACT

Lyme disease, caused by infection with spirochetes of the *Borrelia burgdorferi* complex, is the most common arthropod-borne infection in the United States. The spirochetes are transmitted to humans by bites of infected ticks of the *Ixodes ricinus* complex, which are also known as *black-legged deer ticks*. As many as 90% of patients have a characteristic rash, erythema migrans (EM); other manifestations of early disease include lymphadenopathy, myalgias, and low-grade fever. If left untreated, the disease may progress to further systemic symptoms, including those involving the nervous and cardiac systems. Doxycycline appears to be a highly effective treatment for eradicating the infection when used in postexposure treatment of high-risk tick exposures and in cases of early localized (stage 1) and early disseminated (stage 2) Lyme disease, thus preventing the systemic complications of long-term infection. Reinfection of persons living in Lyme disease hyperendemic areas may complicate the assessment of treatment efficacy. Clinicians practicing in high-risk geographic areas should consider Lyme disease not only in patients with a history of tick exposure or a rash suggestive of EM but also in patients with a viral-like illness without a rash.

## CLINICAL VIGNETTE

A 32-year-old man presented to an urgent care clinic in suburban Pennsylvania for a rash on his leg. He stated that he noticed the lesion 5 days earlier and in that it had since increased in size. In addition, he complained of mild headache and malaise. Upon further questioning, he reported that 3 weeks prior he had taken an overnight backpacking trip with his brother, when they camped and hiked in nearby forested areas. He denied having traveled outside of the state.

On physical exam, his temperature was 37.4°C (99.2°F), blood pressure 121/72, pulse 89 respirations 16, and oxygen saturation 98% on room air. He was alert and oriented, demonstrating no photophobia or neck stiffness. His mucous membranes were moist. Examination of the heart, lungs, and abdomen was normal. His skin examination was notable for a 5- by 7-cm uniformly erythematous nontender oval patch located in the popliteal fossa area of his right leg.

Based on these clinical findings in a patient living in an endemic region, the lesion was identified as EM consistent with the diagnosis of early Lyme disease. He was prescribed a 21-day course of doxycycline 100 mg bid and provided with information for use during future outdoor activities as well as advice on tick avoidance, tick removal, and antibiotic prophylaxis against Lyme disease in case of a tick bite.

## GEOGRAPHIC DISTRIBUTION

Within the United States, Lyme disease is most commonly transmitted in the northeastern coastal region, although transmission in the north central and Pacific coastal regions also occurs (Fig. 73.1). In 2017, approximately 80% of the 42,743 cases of Lyme disease reported by state health departments occurred in the region from Maine through Virginia, 1.2% in Wisconsin and Minnesota, and 0.5% in Washington and Oregon. In the same year, Pennsylvania was the state with the highest number of Lyme disease cases, followed by New York and New Jersey. The incidence is highest in Maine, followed by Vermont and then Pennsylvania.

In general, the reported incidence has increased over the past 20 years, with the number of confirmed cases rising from 12,801 in 1997 to 29,513 in 2017. This increase has been seen in part because of improved surveillance systems and also due to the continued expansion of human settlement into deer habitats. Because early cases of Lyme disease are treated without laboratory testing and not all cases of Lyme disease are recognized by medical personnel, official numbers probably markedly underestimate the actual number of cases.

Outside of the United States, Lyme disease is transmitted in the temperate forested regions of Europe and Asia, but it is not known to be transmitted in the tropics.

## TICK LIFE CYCLE

*Ixodes* ticks are born as larval ticks in the summer; they feed only once, their preferred host being the white-footed mouse. In the following spring the larval ticks transform into nymphs and again feed only once, their preferred host again being the white-footed mouse. In the fall, the nymphs become adults and feed again, the preferred host being the white-tailed deer. Neither mice nor deer develop illness from *B. burgdorferi*. Only ticks that have fed on an infected animal can infect humans. Eighty-five percent of human infections are caused by nymphs in the spring and summer; 15% are caused by adult ticks in the fall.

In endemic areas, as many as 50% of ticks may be infected, but the risk of acquiring Lyme disease from a single tick bite is at most 3.5%, even in highly endemic areas. The time of year during which humans are at greatest risk of infection is midspring through late fall, which is when tick populations, particularly biting nymphs, are at their greatest levels; this is also when people in endemic areas have their highest level of outdoor exposure. A minimum of 36 to 48 hours of tick attachment is required for effective transmission to humans (Fig. 73.2).

## RISK FACTORS

People engaged in outdoor activities (e.g., hiking, gardening) are at highest risk for Lyme disease. Those with occupational exposure to brush, such as forest rangers and landscapers, are also at elevated risk. The age groups with the highest risk of infection range from 5 to 19 years and those older than 30 years. June and July are the peak months for infection in humans. There is no difference in risk of infection between the sexes.

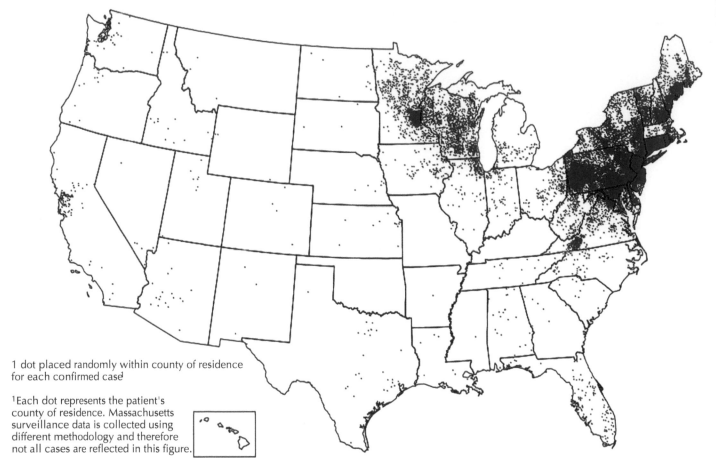

1 dot placed randomly within county of residence for each confirmed case[d]

[1]Each dot represents the patient's county of residence. Massachusetts surveillance data is collected using different methodology and therefore not all cases are reflected in this figure.

**Fig. 73.1** Reported cases of Lyme disease—United States, 2017. (Reused with permission from Centers for Disease Control and Prevention [CDC]: Reported Cases of Lyme Disease—United States, 2017. Available at https://www.cdc.gov/lyme/stats/maps.html. Accessed September 9, 2019.)

## CLINICAL PRESENTATION

For surveillance purposes, the clinical case definition of Lyme disease is used, which is as follows:

- EM of 5 cm (2 in) or greater or
- At least one late manifestation of neurologic, cardiovascular, or musculoskeletal disease and laboratory confirmation of infection with *B. burgdorferi*

Of patients with Lyme disease, 70% to 85% have early localized disease (stage 1) at presentation. The incubation period between the bite of an infected tick and the appearance of the characteristic rash ranges from a few days to 1 month. The EM rash (formerly termed *erythema chronicum migrans*) usually but not always occurs at the site of the tick bite. This large annular rash, which occurs in up to 90% of patients with Lyme disease, is the single best clinical indicator of the disease. Although it is sometimes described as a "bull's-eye" rash, it often consists of confluent erythema that expands, hence its name. Untreated, it persists for 2 to 3 weeks. Other manifestations of early localized disease include mild lymphadenopathy (23%), low-grade fever (19% to 39%), mild fatigue and malaise (54%), neck stiffness (35%), and mild arthralgias and myalgia (44%); headache is also a common stage 1 symptom.

In the absence of appropriate treatment, infection may progress to early disseminated disease (stage 2). Symptoms of stage 2 illness are dermatologic (disseminated EM), neurologic (meningitis, encephalitis), gastrointestinal (hepatitis, abdominal pain), cardiac (atrioventricular [AV] block, myopericarditis), and rheumatologic (monoarticular arthritis). Late disease (stage 3) manifestations are dermatologic (acrodermatitis chronica atrophicans), rheumatologic (arthralgias, oligoarthritis), and neurologic (cranial nerve palsy, ataxia, spastic paresis, encephalomyelitis). Unlike the case in syphilis, another disease caused by a spirochete infection, stage 3 illness may occur within 1 year of initial infection.

The Centers for Disease Control and Prevention (CDC) data collected from 2008 to 2017 show that EM was the most common presenting sign, with 71% of those with Lyme disease demonstrating this finding. Twenty-eight percent of patients with Lyme disease had arthritis, 9% had facial palsy, 4% had radiculopathy, 2% had meningitis or encephalitis, and 1% had carditis at presentation.

Clinical manifestations of Lyme disease differ between the United States and Europe. Those infected with *B. burgdorferi* in Europe have a higher rate of remaining asymptomatic relative to patients in the United States. Two dermatologic manifestations occur in some patients with Lyme disease in Europe but are apparently absent in patients in North America: borrelial lymphocytoma, a purplish nodular swelling usually occurring on the earlobe or nipple in stage 2 disease, and acrodermatitis chronica atrophicans, which occurs over extensor surfaces of the extremities. The clinical course of EM, arthritis, neuroborreliosis, and other manifestations may also differ between patients in the United States and those in Europe. The differences are probably a result of illness being caused by different strains of *Borrelia*. Isolates from the East Coast of the United States are termed *B. burgdorferi* sensustricto. The European isolate *Borrelia garinii* is associated with neurologic

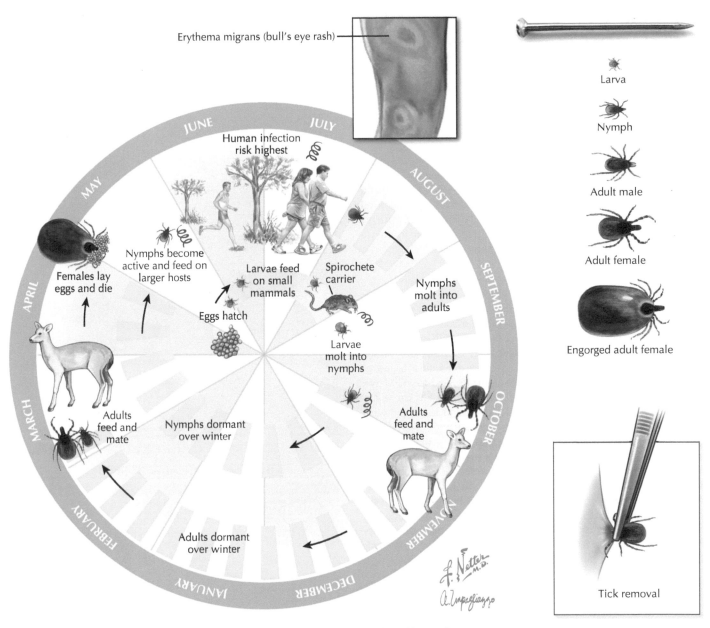

Erythema migrans (bull's eye rash)

JUNE  JULY

Human infection
risk highest

MAY

AUGUST

Nymphs become
active and feed on
larger hosts

APRIL

Females lay
eggs and die

Larvae feed
on small
mammals

Spirochete
carrier

Nymphs
molt into
adults

SEPTEMBER

Eggs hatch

Larvae
molt into
nymphs

MARCH

Adults
feed and
mate

Nymphs dormant
over winter

Adults
feed and
mate

OCTOBER

FEBRUARY

Adults dormant
over winter

NOVEMBER

JANUARY  DECEMBER

Larva

Nymph

Adult male

Adult female

Engorged adult female

Tick removal

**Fig. 73.2** Life cycle and transmission of Lyme disease.

symptoms, whereas *Borrelia afzelii,* also found in Europe, is associated with acrodermatitis chronica atrophicans.

## DIAGNOSTIC APPROACH

The diagnosis of early Lyme disease is based primarily on clinical findings. Any patient with a history of potential exposure to ticks who manifests EM should be presumed to have early localized disease and should be treated accordingly; resolution of symptoms with a course of antibiotics confirms the diagnosis.

When evidence of disseminated or late Lyme disease is present, laboratory investigation is indicated. Culturing the organism *B. burgdorferi* is limited to research settings. An enzyme-linked immunosorbent assay (ELISA) that tests for immunoglobulin M (IgM) and IgG antibodies to *B. burgdorferi* is available. However, in patients with Lyme disease, several weeks may elapse before an IgM response can be detected. Early antibiotic treatment may cause the antibody response to remain undetectable.

The IgM response, once positive, can remain positive for over a year; hence it indicates exposure but does not confirm recent infection. The IgG antibody response occurs 4 to 8 weeks after infection. A false-positive antibody response can be caused by several conditions and situations, including viral infections, rheumatoid arthritis, other autoimmune diseases, and healthy persons living in endemic areas (5%–10% of those living in Lyme disease–endemic regions have positive antibody results with no history of symptoms). Hence laboratory studies in and of themselves, in the absence of clinical findings and exposure history, are insufficient to confirm or exclude the diagnosis of Lyme disease.

A Western blot test is also available for Lyme disease. The CDC recommends a two-step sequence of testing, in which positive or equivocal ELISA findings are followed by confirmatory testing. The traditional confirmatory test has been Western blot testing for IgM and IgG bands specific to Lyme disease. The IgM Western blot has a high rate of false-positive results and should be performed only in patients with less than 1 month of symptoms. The test is read as positive if 2 of 3 Lyme disease–specific bands are present; the IgG test is considered positive if at least 5 of 10 Lyme disease–specific bands are positive. In 2019, the CDC approved several ELISA tests to be used as alternatives to Western blot testing in two-step testing.

Deoxyribonucleic acid (DNA) detection of *B. burgdorferi* by polymerase chain reaction (PCR) is problematic, with lack of standardization, poor sensitivity, and risk of contamination all contributing to limit its usefulness. One role of PCR may be in confirming active Lyme disease in synovial fluid.

The CDC warns against commercial laboratory tests for Lyme disease that use unproven techniques for diagnosis. These include immunofluorescent staining of blood for *B. burgdorferi*, DNA PCR tests on blood or urine, Western blot testing via unvalidated criteria and novel culture techniques. If in doubt, healthcare providers and patients should inquire whether or not these labs use tests approved by the US Food and Drug Administration (FDA) for Lyme disease.

## DIFFERENTIAL DIAGNOSIS

When EM is present in a patient with known or suspected exposure to ticks in an area endemic for Lyme disease, diagnosis is straightforward. EM may initially be mistakenly diagnosed as cellulitis, allergic dermatitis, or contact dermatitis in response to a plant such as poison oak or poison ivy. Manifestations other than EM have extensive differential diagnoses, and determining the presence of Lyme disease in the absence of EM may be difficult. The initial febrile illness occurring without EM can resemble influenza, viral gastroenteritis, infectious mononucleosis, and other acute febrile infections. Because the *Ixodes scapularis* and *Ixodes pacificus* ticks may also transmit the infective agents causing human granulocytic anaplasmosis (*Anaplasma phagocytophilum*, a small gram-negative bacterium) and babesiosis (*Babesia microti*, an intraerythrocytic protozoan parasite) in some of the same geographic areas as Lyme disease transmission, these tick-borne infections should also be considered in the differential diagnosis of a febrile illness without EM after a tick bite. The differential diagnosis may also include, depending on presenting symptoms, fibromyalgia, reactive arthritis, rheumatoid arthritis, systemic lupus erythematosus, and aseptic meningitis.

## TREATMENT

Lyme disease can usually be cured by an appropriate course of antibiotics. Doxycycline is the drug of choice in stage 1 disease; amoxicillin can be used in children younger than age 8 years and in pregnant or lactating women. Another option is oral cefuroxime. For stage 2 and 3 disease, treatment with intravenous antibiotics (ceftriaxone, cefotaxime, or penicillin G) is often indicated. Treatment regimens and drug doses are given in Table 73.1. With appropriate early treatment, prognosis is excellent. There is no demonstrated benefit to following serial serologies. Infection does not provide immunity from future infections.

Although most patients with Lyme disease can be cured by a 2- to 4-week course of antibiotics, some experience pain, fatigue, or difficulty thinking that can persist for over 6 months after finishing treatment. These symptoms are termed posttreatment Lyme disease syndrome (PTLDS). Although some practitioners use the term *chronic Lyme disease*, experts do not support its use. There is no evidence that prolonged courses of antibiotics are of benefit for PTLDS.

## POSTEXPOSURE PROPHYLAXIS

In Lyme-endemic regions, when an attached tick is large (suggesting a prolonged attachment time), a single dose of doxycycline 200 mg in

### TABLE 73.1   Antibiotic Treatment Regimens for Lyme Disease

| DRUG DOSES | | |
| --- | --- | --- |
| Medication | Adult Dosage | Pediatric Dosage |
| **Oral Regimens (Preferred)** | | |
| Doxycycline | 100 mg bid | 4.4 mg/kg divided bid (maximum 200 mg qd) |
| Amoxicillin | 500 mg tid | 50 mg/kg divided tid (maximum 500 mg per dose) |
| Cefuroxime axetil | 500 mg bid | 30 mg/kg divided bid (maximum 500 mg per dose) |
| **Alternative Oral Regimen** | | |
| Azithromycin | 500 mg qd | 10 mg/kg qd (maximum 500 mg per dose) |
| **Intravenous Regimen (Preferred)** | | |
| Ceftriaxone | 2000 mg qd | 50–75 mg/kg qd (maximum 2000 mg per dose) |
| **Alternative Intravenous Regimens** | | |
| Cefotaxime | 2000 mg tid | 150–200 mg/kg divided tid/qid (maximum 6000 mg qd) |
| Penicillin G | 18–24 million U divided q4h | 200,000–400,000 U/kg divided q4h (maximum 18–24 million units qd) |

| SUGGESTED TREATMENT REGIMENS | | |
| --- | --- | --- |
| Disease Manifestation | Medication | Duration |
| **Erythema Migrans** | Doxycycline | 10–21 days |
| | Beta lactams | 14–21 days |
| | Azithromycin | 7–10 days |
| **Neurologic Disease** | | |
| Cranial nerve palsy, radiculopathy, or meningitis (outpatient) | Doxycycline | 14–21 days |
| | Ceftriaxone | 14–21 days |
| Encephalitis, meningitis (inpatient) | Ceftriaxone | 14–28 days |
| **Carditis** | | |
| Outpatient | Doxycycline | 14–21 days |
| | Beta lactams | 14–21 days |
| Inpatient | Ceftriaxone | 14–21 days |
| **Arthritis** | | |
| Initial treatment | Doxycycline | 28 days |
| | Beta lactams | 28 days |
| Recurrent or refractory arthritis | Ceftriaxone | 14–28 days |

Adapted from Hu LT. Lyme disease. *Ann Intern Med* 2016;164(9). Lantos et al. Clinical practice guidelines by the Infectious Diseases Society of America (IDSA), American Academy of Neurology (AAN), and American College of Rheumatology (ACR): 2019 guidelines for the prevention, diagnosis and treatment of Lyme disease. 2019 Jun 26; [e-pub]; Sanchez E, Vannier E, Wormser GP, Hu LT. Diagnosis, treatment, and prevention of Lyme disease, human granulocytic anaplasmosis, and babesiosis: a review. *JAMA* 2016;315(16):1767-1777.

those older than the age of 8 years will prevent infection. This strategy appears to be less effective in Europe.

## PREVENTION

Deer do not become infected with the spirochete that causes Lyme disease; however, they do serve as the primary source of blood on which the adult ticks reproduce. Measures aimed at reducing deer populations have been shown to reduce the population of ticks. However, the amount of such reduction requisite to affect the transmission of Lyme disease has not been established.

Wearing long sleeves and long pants is of benefit, and tucking pants into boots or socks and tucking shirts into pants will help to keep ticks on the outside of clothing. Ticks are easier to see on light-colored clothing. Applying permethrin to clothing offers significant protection; applying *N,N*-diethyl-*meta*-toluamide (DEET) to exposed skin is less helpful. Avoiding thickly wooded and bushy areas and areas with high grass and large quantities of leaf litter may be of benefit, particularly during the months of May, June, and July, when the transmission of Lyme disease is most intense.

After outdoor activities, a head-to-toe tick check with the aid of a mirror or close friend will reduce risk. *Ixodes* ticks are small and may be difficult to detect. Ticks attached to the skin should be removed immediately with fine-tipped tweezers (see Fig. 73.2). If a tick has been attached for less than 24 hours, the risk of transmission of Lyme disease is extremely small.

An effective vaccine against Lyme disease (*B. burgdorferi* outer surface membrane protein A, or OspA) was available in the United States for a limited time but was withdrawn from the market by the manufacturer in 2002 owing to alleged adverse effects and low usage.

## OTHER TICK-BORNE DISEASES

Other tick-borne diseases include babesiosis, ehrlichiosis, Rocky Mountain spotted fever, anaplasmosis, southern tick-associated rash illness, tick-borne relapsing fever, tularemia, Colorado tick fever, Powassan disease, Q fever, *Borrelia mayonii*, *Borrelia miyamotoi*, Bourbon virus, Heartland virus, *Rickettsia parkeri rickettsiosis*, and 364D rickettsiosis. A bite by a Lone Star tick can transmit an allergy to red meat, including beef and pork (alpha-gal syndrome).

## EVIDENCE

Aucott J, Morrison C, Munoz B, et al. Diagnostic challenges of early Lyme disease: lessons from a community case series. *BMC Infect Dis* 2009;9:79-87. Available at: www.biomedcentral.com/1471-2334/9/79. Accessed February 17, 2010. *This community-based case series shows that primary care providers need to improve their clinical recognition of early Lyme disease, consider atypical presentations, and become familiar with appropriate antibiotic regimens.*

Centers for Disease Control and Prevention (CDC). Surveillance for Lyme Disease—United States, 1997-2017. Available at https://www.cdc.gov/lyme/datasurveillance/index.html. Accessed October 1, 2019. *Detailed summary of official surveillance data on Lyme disease cases reported in the United States.*

Kowalski TJ, Tata S, Berth W, et al. Antibiotic treatment duration and long-term outcomes of patients with early Lyme disease from a Lyme disease-hyperendemic area. *Clin Infect Dis* 2010;50:512-520. *This article discusses findings from a retrospective cohort study of 607 patients in a Lyme disease-hyperendemic area who had early localized disease or early disseminated*

*disease; it demonstrates that 10 days of doxycycline is as effective as treatment regimens of longer duration for the eradication of early Lyme disease.*

## ADDITIONAL RESOURCES

Baker CD, Charini WA, Duray PH, et al. Final report of the Lyme Disease Review Panel of the Infectious Diseases Society of America (IDSA). Available at: https://www.idsociety.org/globalassets/idsa/topics-of-interest/lyme/idsalymediseasefinalreport.pdf. Accessed October 1, 2019. *The review panel reaffirmed the recommendations of the 2006 guidelines and concluded that in the case of Lyme disease, there is no evidence to support long-term antibiotic therapy beyond 1 month. This report is a valuable resource to clinicians seeking evidence-based guidelines on Lyme disease and its treatment in the United States.*

Centers for Disease Control and Prevention (CDC). Concerns regarding a new culture method for Borrelia burgdorferi not approved for the diagnosis of Lyme disease. *MMWR Morb Mortal Wkly Rep* 2014;63(15):333-333. Available at https://www.cdc.gov/mmwr/preview/mmwrhtml/mm6315a4.htm?s_cid=mm6315a4_w. Accessed October 1, 2019. *This announcement stresses the importance of utilizing only FDA-approved diagnostic tests for Lyme disease.*

Feder HM Jr, Johnson BJ, O'Connell S, et al. A critical appraisal of "chronic Lyme disease". *N Engl J Med* 2007;357(14):1422-30. *This article evaluates and summarizes the evidence behind the clinical syndromes that have been labeled "chronic Lyme disease."*

Gilbert DN, Moellering RC, Eliopoulos GM, Chambers HF, et al. *The Sanford guide to antimicrobial therapy 2019.* Sperryville, VA; 2019. *The definitive source for recommended treatment regimens for infectious diseases.*

Hu LT. Lyme disease. *Ann Intern Med* 2016;164(9):ITC65-ITC80. *A practical overview of the prevention, diagnosis, and treatment of Lyme disease.*

Lantos P et al. Draft clinical practice guidelines by the Infectious Diseases Society of America (IDSA), American Academy of Neurology (AAN), and American College of Rheumatology (ACR): 2019 guidelines for the prevention, diagnosis and treatment of Lyme disease. 2019 Jun 26 [e-pub].

Mead P, Petersen J, Hinckley A. Updated CDC recommendation for serologic diagnosis of Lyme disease. *MMWR Morb Mortal Wkly Rep* 2019;68(32):703. Available at https://www.cdc.gov/mmwr/volumes/68/wr/mm6832a4.htm. Accessed October 1, 2019. *Recommendations for an FDA-approved enzyme immunoassay that may serve as an alternative to confirmatory Western blot.*

Moore A, Nelson C, Molins C, Mead P, Schriefer M. Current guidelines, common clinical pitfalls, and future directions for laboratory diagnosis of Lyme disease, United States. *Emerg Infect Dis* 2016;22(7):1169-1177. *Reviews the complexities of and recommendations for diagnostic testing of Lyme disease.*

Sanchez E, Vannier E, Wormser GP, Hu LT. Diagnosis, treatment, and prevention of Lyme disease, human granulocytic anaplasmosis, and babesiosis: a review. *JAMA* 2016;315(16):1767-77. *A review of the diagnosis and treatment of Lyme disease that features a table of recommended treatment regimens according to disease manifestation.*

Stafford KC. Tick management handbook, rev ed. New Haven: Connecticut Agricultural Experiment Station; 2007. Available at: www.ct.gov/caes. Accessed February 17, 2010. *A comprehensive handbook covering all aspects of tick-borne diseases including epidemiology, transmission, diagnosis, treatment, and prevention, as well as environmental controls. The handbook is illustrated with many photographs of tick vectors and habitats and integrates information from many sources. This is an essential reference for public health officers and clinical practitioners in tick-endemic areas.*

Wormser GP, Dattwyler RJ, Shapiro ED, et al. The clinical assessment, treatment, and prevention of Lyme disease, human granulocytic anaplasmosis, and babesiosis: clinical practice guidelines by the Infectious Diseases Society of America. *Clin Infect Dis* 2006;43:1089-1134. *A clinical practice guideline on Lyme disease developed by the Infectious Diseases Society of America. Published in 2006, the IDSA's 2006 Lyme disease guidelines underwent review by a special independent review panel. The review panel's final report, published on April 22, 2010, validated the 2006 Lyme disease guidelines and offered some additional advice.*

# 74

# Leptospirosis

*Vernon Ansdell*

## ABSTRACT

Leptospirosis is the most common zoonosis worldwide and has recently emerged as an important travel-related infection, particularly in adventure travelers to the tropics and subtropics. Characteristic symptoms may include fever, headache, myalgias, jaundice, and conjunctival suffusion, but this disease often manifests as a nonspecific febrile illness. Because it has a wide range of signs and symptoms, a high index of suspicion is necessary for clinicians to make an accurate diagnosis. Early treatment with an antibiotic—such as doxycycline, amoxicillin, penicillin, azithromycin, or ceftriaxone—before waiting for confirmation of the diagnosis is very important. Most cases are relatively mild and self-limited, but 5% to 10% of affected patients develop a severe, potentially life-threatening illness characterized by fever, jaundice, renal failure, bleeding, and/or severe pulmonary hemorrhage.

## CLINICAL VIGNETTE

A 42-year-old woman presented with a 3-day history of severe frontal headache, fever, chills, malaise, and intense muscle pains involving the calves, thighs, and lower back. Her headache had come on abruptly and she was unable to walk because the muscle pains were so severe.

Sixteen days earlier she had taken part in an adventure race in South Africa before returning to the United States. Just prior to the event there had been very heavy rainfall with flooding, and she remembered that the course was particularly muddy. As a result of the treacherous conditions she had developed multiple cuts and abrasions during the race. She remembered swallowing water and getting submerged frequently while swimming in rough water during the event.

Her temperature was 38.9°C. There was no rash, and careful examination did not reveal any eschars. She was mildly jaundiced and had conjunctival suffusion involving both eyes and a conjunctival hemorrhage in the left eye. The liver and spleen were mildly enlarged. There was no lymphadenopathy.

Laboratory results showed hemoglobin 11.8 g/dL, white blood cells (WBCs) 6400 with neutrophils 79%, bands 6%, platelets 115,200/μL, creatinine 2.1 mg/dL, bilirubin 4.2 mg/dL, alanine aminotransferase (ALT) and aspartate aminotransferase (AST) levels were mildly elevated, alkaline phosphatase levels were normal, and creatine kinase was 1700 international units (IU)/L. Urinalysis showed mild proteinuria, a few WBCs and red blood cells (RBCs), and granular casts. Chest x-ray was unremarkable.

Leptospirosis was considered the most likely diagnosis. However, in view of her recent history of travel to South Africa, the differential diagnosis included African tick typhus *(Rickettsia africae)*, viral hepatitis, and malaria. The malaria rapid diagnostic test and thick and thin films for malaria were negative. She was treated with intravenous ceftriaxone and oral doxycycline and given intravenous fluids. Leptospirosis real-time polymerase chain reaction (PCR) was positive. The patient made an uneventful recovery.

COMMENT: In leptospirosis, the differential diagnosis is often broad early in the course of the illness. This may be particularly common in travelers returned from the tropics and subtropics. In this case, leptospirosis was the most likely diagnosis, but it was important also to consider other possibilities such as African tick typhus, viral hepatitis, and malaria. Leptospirosis was more likely than viral hepatitis because of elevated creatine kinase (CK) levels, neutrophil leukocytosis with a left shift, and a relatively modest elevation of liver enzymes.

## TRANSMISSION AND RISK OF INFECTION

Leptospirosis *(Leptospira interrogans)* is a spirochete infection that occurs worldwide except in polar regions. The organism survives best in warm, humid conditions and is most common in the tropics and subtropics, with many wild and domestic animal reservoirs including rats, mice, dogs, pigs, and cattle (Fig. 74.1). After infection, animals may shed the organism in the urine for months or even years. Organisms proliferate in fresh water, damp soil, vegetation, and mud and can survive for weeks to months.

Infection in humans occurs after exposure to animal urine, either by direct contact or, more commonly, as the result of indirect exposure to contaminated mud or water in rivers, lakes, and streams. Infection is acquired through damaged skin or via exposed mucous membranes of the nose, mouth, or eyes. Contaminated drinking water may also be a significant source of infection.

Traditionally, leptospirosis was an occupational illness (e.g., farmers, abattoir workers, veterinarians, sewage workers, and miners). Recently, however, recreational exposure (e.g., of hikers, whitewater rafters, kayakers, and triathletes) has been recognized as an important exposure risk. Leptospirosis has recently been identified as an important emerging disease in adventure travelers to the tropics and subtropics, and it is an important cause of fever in returned travelers. The majority of reported cases have been linked with travel to Southeast Asia. Urbanization is increasingly recognized as a very important source of infection, and many outbreaks of leptospirosis have occurred after periods of heavy rainfall or flooding, as from hurricanes in more densely populated areas. With global climate change, extreme weather events such as hurricanes and flooding will likely increase in intensity and frequency, resulting in an increase in the number of cases of leptospirosis. Global warming will expand the geographic range for optimal survival and transmission of leptospires.

Because of the nonspecific presentation, the diagnosis of leptospirosis is often overlooked. In addition, underreporting is very common. Recent estimates by a World Health Organization expert group suggest that there may be 873,000 cases annually, with 48,600 deaths.

Dogs (and other domestic animals) Rats, mice   Livestock (cattle, pigs, etc.)

Soil and water are contaminated
by the urine of infected animals

Bacteria enter the body through
the skin and mucous membranes
of the eyes, nose, and mouth of
humans in contaminated aquatic
and muddy environments

White water rafters, triathletes and swimmers (freshwater recreation)    Farm workers (e.g., rice and taro farmers)

**Fig. 74.1** Transmission of leptospirosis.

## CLINICAL FEATURES

### Leptospirosis

The incubation period is usually 5 to 14 days (range 2 to 30 days). Some human cases are probably completely asymptomatic, and over 90% of symptomatic cases are relatively mild and self-limited. The remaining cases may be severe and potentially life threatening, however, and associated with jaundice, hemorrhage, renal failure, and myocarditis (Weil disease) or massive pulmonary hemorrhage.

The illness may be biphasic. The first or leptospiremic phase typically lasts 3 to 7 days and represents the period when organisms are present in the blood. The second or immune phase may be clinically silent or last for 4 to 30 days or longer. This phase coincides with the formation of circulating immunoglobulin M (IgM) antibodies. Aseptic meningitis (with or without symptoms) is characteristic of the immune phase and may occur in over 50% of cases (Fig. 74.2).

In the milder form of leptospirosis there may be a clinically apparent, symptom-free interval of 1 to 3 days between the first and second phases. In severe cases, however, the distinction between these two phases is usually not apparent.

The acute leptospiremic phase begins abruptly, with high fever (often >39°C), chills, and a severe frontal headache. Patients often report one of the worst headaches they have ever experienced. Severe muscle pain and tenderness are common, typically involving the muscles of the calves, thighs, and lower back. Conjunctival suffusion (dilatation of conjunctival vessels without inflammation or purulent discharge) is virtually pathognomonic of leptospirosis but must be distinguished clinically from conjunctivitis. It usually appears on the third or fourth day of illness and is probably very common, although it may be mild and easily overlooked. Subconjunctival hemorrhages are often present.

A wide range of other symptoms are common and may confuse the diagnosis. Gastrointestinal symptoms include abdominal pain, anorexia, nausea, vomiting, and diarrhea. Early respiratory symptoms may include sore throat, dry cough, dyspnea, and chest pains. The disease often manifests as a flulike illness without the typical upper respiratory symptoms associated with influenza or other respiratory viruses. A variety of rashes are present in up to 10% to 30% of patients during the first week of illness; they may be erythematous macular, maculopapular, urticarial, petechial, or purpuric. Less common physical signs during this phase include lymphadenopathy, hepatomegaly, and splenomegaly.

The second or immune phase is characterized by aseptic meningitis and symptoms such low-grade fever, headache, neck stiffness, nausea, vomiting, and photophobia. Other clinical features during this phase may include jaundice, renal insufficiency, cardiac arrhythmias, and various pulmonary symptoms. Unilateral or bilateral uveitis characterized by iritis, iridocyclitis, and chorioretinitis may develop up to 18 months after acute infection and persist for several years.

### Severe Leptospirosis

Approximately 10% of patients develop a severe, life-threatening form of the disease. Historically this was characterized by jaundice, hemorrhage, renal failure, and myocarditis; it was often referred to as *Weil disease*. More recently, however, another severe form of the disease characterized by massive pulmonary hemorrhage—representing

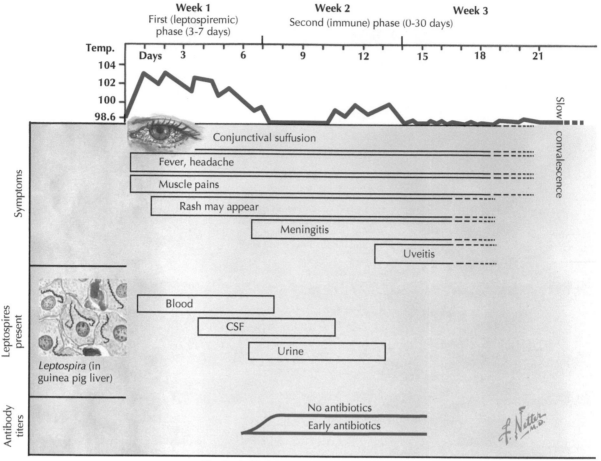

**Fig. 74.2** Clinical course of leptospirosis.

extensive alveolar hemorrhage—has been increasingly recognized and become known as severe pulmonary hemorrhage syndrome (SPHS). Frank hemoptysis may occur, but hemorrhage may not be apparent until after patients are intubated. It is therefore very important to suspect SPHS in patients with respiratory distress even if they do not have hemoptysis. Rapid progression to acute respiratory distress syndrome (ARDS) and high mortality (>50%) is not uncommon (Fig. 74.3).

The onset of severe illness is usually indistinguishable from the milder form of leptospirosis, but after 4 to 9 days there is progression to a severe, life-threatening form of the disease. Jaundice typically appears between the fifth and ninth days of illness and may last for several weeks. Bilirubin levels may be very high, but liver failure is extremely rare because severe hepatocellular damage is very unusual. Tender hepatomegaly and splenomegaly may be present. Renal insufficiency is evident within 3 to 4 days of onset. Important factors in the pathogenesis include hypovolemia, hypotension, and acute tubular necrosis. Oliguric or nonoliguric renal failure usually occurs during the second week of illness. Hemorrhagic manifestations are common in severe illness and are thought to be related to severe vasculitis with endothelial damage resulting in capillary injury. Thrombocytopenia and abnormal clotting factors serve to increase the risk of bleeding. Clinically there may be petechial and purpuric rashes, bleeding gums, epistaxis, hemoptysis, gastrointestinal hemorrhage, and, rarely, subarachnoid or adrenal hemorrhage. Cardiac involvement may result in myocarditis or pericarditis and arrhythmias such as atrial fibrillation, atrial flutter, and a variety of conduction disturbances. Congestive heart failure and myocarditis are common in fatal cases (Fig. 74.4).

## Recovery Phase

With appropriate supportive care, most patients with leptospirosis recover completely. There may be persistent mild impairment in renal function in some patients with acute renal failure who required dialysis. Ocular involvement in the form of uveitis may occur during the convalescent phase (see earlier). There is increasing evidence that many patients have chronic postleptospirosis symptoms. One recent study from the Netherlands showed that 30% of patients had symptoms such as fatigue, malaise, myalgia, headache, and weakness that often lasted for more than 2 years.

## DIAGNOSTIC APPROACH

Leptospirosis has protean manifestations and is often confused with other infectious diseases (Box 74.1). This is particularly common in returned travelers from the tropics. A high index of suspicion is often needed to make the diagnosis, and a number of diagnostic "red flags" may help to alert clinicians (Box 74.2).

The total WBC count is variable but is usually elevated in severe disease. A neutrophil leukocytosis with a shift to the left is common (in contrast to viral hepatitis). A mild to moderate thrombocytopenia (platelet counts 50,000 to 120,000/mm³) is common. Platelet counts of less than 50,000/mm³ are less common but may be seen in severe disease. Prothrombin time may be prolonged in Weil disease but can be corrected with vitamin K. The erythrocyte sedimentation rate (ESR) is very commonly elevated, often greater than 50 mm/h.

Liver function abnormalities include elevated bilirubin (up to 20 mg/dL or higher), with a relatively mild increase in transaminase

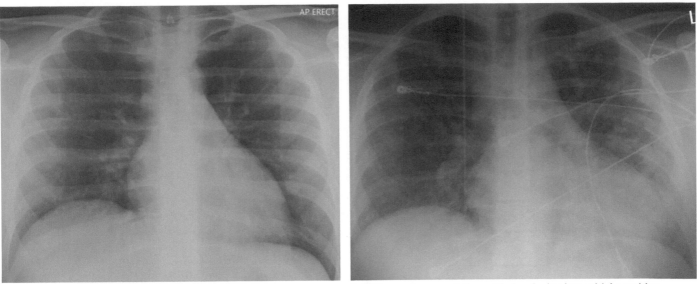

**A. Unremarkable CXR on initial presentation**

**B. CXR upon re-presentation with developing interstitial opacities**

**Fig. 74.3** Chest radiograph of leptospirosis. (Reused with permission from Schmalzle SA, Tabatabai A, Mazzeffi M, et al. Recreational 'mud fever': *Leptospira interrogans* induced diffuse alveolar hemorrhage and severe acute respiratory distress syndrome in a U.S. Navy seaman following 'mud-run' in Hawaii. *IDCases* 2019;15:e00529, Figure 1.)

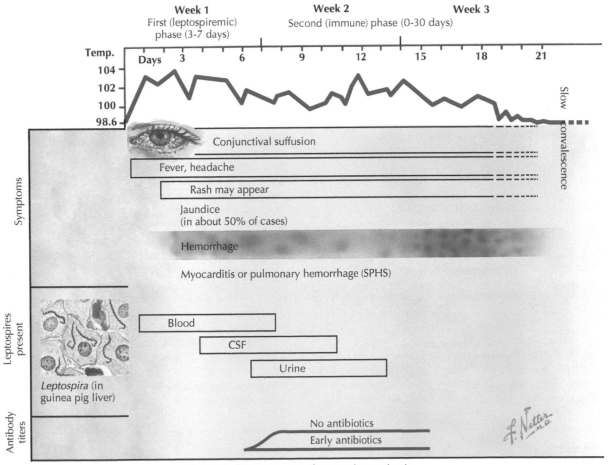

**Fig. 74.4** Clinical course of severe leptospirosis.

## BOX 74.1  Differential Diagnosis of Leptospirosis

- Influenza
- Aseptic meningitis
- Brucellosis
- Dengue fever
- Rickettsial diseases (e.g., typhus, Q fever)
- Streptococcal pharyngitis
- Acute human immunodeficiency virus (HIV) infection
- Toxoplasmosis
- Malaria
- Viral hemorrhagic fever
- Viral hepatitis
- Legionnaires' disease
- Hantavirus
- Typhoid fever
- Relapsing fevers
- Melioidosis
- Zika
- Chikungunya
- Toxoplasmosis

## BOX 74.2  Leptospirosis Diagnostic "Red Flags"

- History of contact with fresh water or mud
- History of contact with animals
- History of cuts or abrasions
- Abrupt onset of severe headache
- Severe myalgias (calves, thighs, lower back)
- Conjunctival suffusion
- Fever and new-onset atrial fibrillation
- Jaundice and relatively mild transaminase elevation
- Fever, jaundice, and thrombocytopenia
- Hepatitis and neutrophil leukocytosis with left shift
- Fever and elevated creatine kinase levels
- Fever and elevated amylase or lipase levels

## BOX 74.3  Antibiotics Used to Treat Leptospirosis (7–10 Days Administration in Adults)

- Doxycycline 100 mg PO bid
- Amoxicillin-clavulanate potassium 875–125 mg PO bid
- Azithromycin 1000 mg PO, then 500 mg × 3 days
- Penicillin G 1.5 MU IV q6h
- Amoxicillin 500 mg PO tid
- Erythromycin 500 mg PO qid
- Ceftriaxone 1–2 g IV or IM q24h

*IM,* Intramuscularly; *IV,* intravenously; *PO,* orally.

inoculated as soon as possible, using special media. It may take as long as 6 weeks or more for cultures to become positive.

### Immunodiagnosis

Antibodies usually appear in the second week of illness. The gold standard for immunodiagnosis remains the microscopic agglutination test (MAT), and paired sera should be obtained 14 to 28 days apart for testing. The diagnosis is usually based on demonstrating a fourfold rise in titer or a single MAT titer of at least 1 in 200. Rapid screening tests (e.g., enzyme-linked immunosorbent assay [ELISA], indirect hemagglutination assay [IHA]) may also be available. Early use of antibiotics may delay or blunt the appearance of antibodies (see Figs. 74.2 and 74.4).

### Molecular Diagnosis

Molecular tests such as real time PCR and loop-mediated isothermal amplification (LAMP) have become increasingly available for the diagnosis of leptospirosis. Advantages include early, rapid, and accurate diagnosis and, as compared with serology, there is no need for acute and convalescent specimens. Newer next-generation sequencing enables the sequencing of a much greater number of genes than conventional DNA tests. Leptospiral DNA may be present in blood, CSF, urine, and peritoneal fluid.

## CLINICAL MANAGEMENT AND DRUG TREATMENT

Early antibiotic treatment has been shown to reduce the duration and severity of illness, and antimicrobials should be started as soon as the diagnosis is suspected. Antibiotics are usually given for 7 to 10 days. The organism is sensitive to a wide range of antibiotics including penicillin, amoxicillin, doxycycline, erythromycin, macrolides (e.g., azithromycin and clarithromycin), third-generation cephalosporins (e.g., ceftriaxone and cefotaxime), and some fluoroquinolones (Box 74.3). The organism may be resistant to chloramphenicol, vancomycin, aminoglycosides, and first-generation cephalosporins. Jarisch-Herxheimer reactions after penicillin treatment occur less frequently than with other spirochetal infections. The diagnosis of leptospirosis may be uncertain in the early stages of illness; in that setting, antibiotic selection must cover other possible diagnoses. If, for example, a rickettsial disease such as typhus is in the differential diagnosis, doxycycline would be a good choice of antibiotic, as it would cover both infections. Supportive care, if necessary in an intensive care unit, is very important, and meticulous attention to fluid and electrolyte balance is essential. Renal failure is an important cause of death, and prompt initiation of hemodialysis or peritoneal dialysis helps to limit mortality.

## PROGNOSIS

Over 90% of cases of leptospirosis are mild and self-limited. In severe cases of Weil disease, however, mortality may be as high as 20%; in

and alkaline phosphatase levels. Elevated serum amylase and lipase levels may occur but are not necessarily associated with clinical evidence of pancreatitis. Creatine kinase levels are elevated in the majority of patients during the first week of illness, and this may help to differentiate leptospirosis from viral hepatitis. Hyponatremia is relatively common. Nonoliguric hypokalemia is an early feature of renal insufficiency. Urinalysis findings are frequently abnormal and may show proteinuria, pyuria, hyaline or granular casts, and hematuria.

During the second (immune) phase of illness, the cerebrospinal fluid (CSF) typically shows features of aseptic meningitis. The CSF cell count is usually less than 500/mm$^3$, with a lymphocytic pleocytosis. CSF protein is moderately elevated, but CSF glucose is normal.

Chest x-ray abnormalities may include small nodular densities, patchy alveolar infiltrates, areas of consolidation, pleural effusions, or the typical changes associated with ARDS (see Fig. 74.3).

### Culture

Cultures should be attempted whenever possible, and specimens should be obtained before antibiotics are started. Blood, CSF, urine, and peritoneal dialysis fluid may all be cultured. Specimens should be

SPHS mortality of greater than 50% has been reported. Mortality tends to increase with age and in association with underlying disease.

## PREVENTION AND CONTROL

Preventive measures include avoiding potentially contaminated freshwater, damp soil, or mud whenever possible, wearing protective clothing and covering cuts and abrasions with waterproof dressings. Submersion should be avoided, because the organism can enter via the mucous membranes of the eyes, nose, and mouth. Potentially contaminated drinking water should be boiled or treated with iodine or chlorine. Simple filtration may not provide adequate protection.

Chemoprophylaxis may be indicated in short-term, high-risk situations. Doxycycline, 200 mg once *weekly,* beginning before the first exposure and ending after the last possible exposure, appears to be effective. Travelers to malaria-endemic areas who are also at risk of leptospirosis may be protected against both infections by taking doxycycline, 100 mg *daily,* one of the chemoprophylaxis regimens recommended against chloroquine-resistant malaria (see Chapter 63). Azithromycin may also be effective for prevention of leptospirosis in situations where doxycycline is contraindicated.

Preventive measures are particularly important in situations that are associated with increased risk of infection—for example, whitewater rafting and adventure racing such as triathlons, particularly after heavy rainfall or flooding. Vaccines for human use are generally not available.

## EVIDENCE

de Vries SG, Bekedam MMJ, Visser BJ et al. Travel-related leptospirosis in the Netherlands 2009-2016: an epidemiological report and case series. *Travel Medicine and Infectious Diseases* 2018;24,44-50. *A very good review of travel related leptospirosis in the Netherlands.*

Frawley AA, Schafer IJ, Galloway R, Artus A, Ratard RC. Notes from the field: postflooding leptospirosis—Louisiana, 2016. *MMWR Morb Mortal Wkly Rep* 2017;66(42):1158-1159.

Goris GA, Kikken V et al. Towards the burden of human leptospirosis: duration of acute illness and occurrence of post-leptospirosis symptoms of patients in the Netherlands. *PLoS One* 2013;8(10):e76549. *An important review from the Netherlands highlighting the importance of ongoing complaints following acute leptospirosis.*

Haake DA, Levett PN. Leptospirosis in humans. *Curr Top Microbiol Immunol* 2015;387:65-97. *A valuable, detailed review of leptospirosis in humans by two recognized experts on the disease.*

Katz AR, Ansdell VE, Effler PV, et al. Assessment of the clinical presentation and treatment of 353 cases of leptospirosis in Hawaii, 1974-1998. *Clin Infect Dis* 2001;33:1834-1841. *Leptospirosis is more common in Hawaii than in any other region of the United States. This is one of the largest published reviews of the clinical features of leptospirosis.*

Lau CL, Smythe LD, Craig SB, Weinstein P. Climate change, flooding, urbanization and leptospirosis: fuelling the fire? *Trans R Soc Trop Med Hyg* 2010;104(10):631-638. *A detailed look at the impact of climate change, and urbanization on leptospirosis.*

Lau C, Smythe L, Weinstein P. Leptospirosis: An emerging disease in travellers. *Travel Med Infect Dis* 2010;8(1):33-39. *A review highlighting the importance of leptospirosis in travelers by experts on the subject.*

Panaphut T, Domrongkitchaiporn S, Vibhagool A, et al. Ceftriaxone compared with sodium penicillin G for treatment of severe leptospirosis. *Clin Infect Dis* 2003;36:1505-1513. *This study from Thailand showed that ceftriaxone and sodium penicillin G were equally effective for the treatment of severe leptospirosis.*

Schmalzle SA, Ali T, Mazzeffi M, et al. Recreational "mud fever": *Leptospira interrogans* induced diffuse alveolar hemorrhage and severe acute respiratory distress syndrome in a U.S. Navy seaman following a "mud-run" in Hawaii. *IDCases* 2019;15:e00529. *An example of severe leptospirosis with diffuse alveolar hemorrhage and severe acute respiratory distress syndrome. The patient was infected while participating in a mud-run in Hawaii.*

Takafuji ET, Kirkpatrick JW, Miller RN, et al. An efficacy trial of doxycycline chemoprophylaxis against leptospirosis. *N Engl J Med* 1984;310:497-500. *One of the few studies that has examined antibiotic prophylaxis against leptospirosis. Results suggested that weekly doxycycline helped to prevent leptospirosis in a large group of immune-naive US soldiers undergoing jungle training in Panama.*

Trevejo RT, Rigua-Pérez JG, Ashford DA, et al. Epidemic leptospirosis, associated with pulmonary hemorrhage—Nicaragua, 1995. *J Infect Dis* 1998;178:1457-1463. *One of the first papers to highlight the importance of severe pulmonary hemorrhage in leptospirosis. Since then, pulmonary hemorrhage has become increasingly recognized as an important cause of severe illness and death in leptospirosis.*

## ADDITIONAL RESOURCES

Bharti AR, Nally JE, Ricaldi JN, et al. Leptospirosis: a zoonotic disease of global importance. *Lancet Infect Dis* 2003;3:757-771. *This article provides a comprehensive review of several important aspects of leptospirosis by a very experienced group of authors.*

Sejvar J, Bancroft E, Winthrop K, et al. Leptospirosis in "Eco-Challenge" athletes, Malaysian Borneo. *Emerg Infect Dis* 2000;9:702-707. *This article emphasizes the importance of adventure travel and adventure racing in the epidemiology of leptospirosis.*

Victoriano AFB, Smythe LD, Gloriani-Barzaga N, et al. Leptospirosis in the Asia Pacific region. *BMC Infect Dis* 2009;9:147. *An excellent review of leptospirosis in the Asia Pacific region by an internationally recognized group of experts. The article describes current trends in the epidemiology of leptospirosis, existing surveillance programs, and some of the prevention control programs in the region.*

# 75

# Primary Amebic Meningoencephalitis

*Radhika Gharpure, Ibne Karim M. Ali, Jennifer R. Cope*

## ABSTRACT

Primary amebic meningoencephalitis (PAM) is an acute, progressive, and usually fatal condition caused by the free-living ameba *Naegleria fowleri*. Commonly referred to as the "brain-eating ameba," *N. fowleri* is naturally found in warm freshwater (e.g., lakes, rivers, and hot springs), and soil. Infection with *N. fowleri* can occur when water containing the ameba enters the body through the nose. The ameba migrates to the brain along the olfactory nerve, where it causes PAM, an acute disease of the central nervous system (CNS). Infection typically occurs when individuals swim or dive in warm freshwater, like lakes and rivers. Additionally, *N. fowleri* infections may occur when water from other sources (such as inadequately chlorinated swimming pool water or contaminated tap water) enters the nose, such as through sinus irrigation for therapeutic or ritual purposes. The rapid progression of PAM and the clinical similarity of signs and symptoms to bacterial meningitis pose challenges for diagnosis and early initiation of treatment.

## CLINICAL VIGNETTE

A previously healthy 18-year-old woman presented to the emergency department with a 3-day history of headache, fever, and lethargy. She had been seen by her primary care physician, who diagnosed possible sinusitis and prescribed amoxicillin. On presentation to the emergency department, she had altered mental status and responded only to noxious stimuli.

Lumbar puncture was performed, revealing an opening pressure of 36 cm $H_2O$. Cerebrospinal fluid (CSF) analysis showed a white blood cell count of 3808 cells/μL with predominantly neutrophils, a red blood cell count of 516 cells/μL, protein 410 mg/dL, and glucose less than 10 mg/dL. A computed tomography (CT) scan of the brain was interpreted as normal. However, over the next 36 hours, the patient rapidly declined and became obtunded, requiring intubation and critical care management.

A repeat CT scan of the brain showed interval increased effacement of cerebral sulci and decreased ventricular size compatible with diffuse cerebral edema. A right frontal external ventricular drain was placed, revealing an intracranial pressure (ICP) of 90 cm $H_2O$. Despite multiple aggressive measures, the patient's ICP remained greater than 50 cm $H_2O$, and she had a sudden change in hemodynamics and an acute drop in ICP suspicious for brain death. Cardiac death occurred approximately 48 hours after presentation.

Real-time PCR performed on CSF was positive for *N. fowleri*. The patient's family and friends were asked about the patient's freshwater exposures in the 2 weeks prior to her illness onset. Six days prior to illness onset, the patient participated in rafting on an artificial whitewater river, during which she fell out of the raft and was submerged under the water.

## ETIOLOGY/PATHOGENESIS

*N. fowleri* is a mitochondria-bearing aerobic protist that normally completes its life cycle in the environment as a free-living organism. Occasionally, however, the ameba invades the CNS of humans and animals, survives within the brain tissue, and causes an acute and fulminant infection called PAM.

*N. fowleri* is found worldwide and has been isolated from freshwater, thermal discharge from power plants, improperly disinfected swimming pools, hot springs, hydrotherapy pools, aquaria, sewage, public water systems, and even the nasal passages and throats of healthy individuals. The ameba has three life-cycle stages: a feeding and reproducing trophozoite, a transitory flagellate, and a resistant cyst. The only infective stage is the trophozoite. Trophozoites can transform into a temporary, non-feeding flagellated stage when stimulated by adverse environmental changes. These flagellates revert to the trophic stage when favorable conditions return. Additionally, during adverse conditions (e.g., in cold temperatures, when the food supply becomes scarce, or the environmental niche dries up), trophozoites can transform into resistant cysts.

Typically, cases of PAM occur in the hot summer months when large numbers of people visit lakes, rivers, and other warm freshwater bodies that may harbor *N. fowleri* (Fig. 75.1). Persons participating in recreational water activities, including swimming, diving, and water skiing, can have nasal exposure to the ameba in water, resulting in infection. Additionally, individuals using tap water for nasal irrigation (e.g., neti pots) or ritual ablution (e.g., religious practices in Yogic, Ayurvedic, and Islamic traditions) can also have nasal exposure to *N. fowleri* and develop PAM.

The portal of entry into the CNS is the olfactory neuroepithelium. *N. fowleri* enters the nasal passages and passes through the sieve-like cribriform plate of the ethmoid bone, penetrating the subarachnoid space and entering the brain parenchyma. The incubation period from exposure to disease may range from 1 to 12 days (median 5 days). The disease progresses rapidly and leads to death, often within a week of symptom onset. PAM has also been diagnosed in animal species including a South American tapir, domestic cattle, sheep and a black rhinoceros.

From 1962 to 2019, 148 human PAM cases were reported in the United States (Fig. 75.2). An additional 11 cases dating back to 1937 were identified through a retrospective examination of autopsy samples in Virginia. The median age in the 148 cases was 12 years (range 8 months to 66 years). Additionally, among the 148 cases, 112 (76%) occurred in male patients. Among the 137 cases for which the month of exposure was known, 117 (85%) occurred during July through September. Exposure occurred primarily from recreational water activities in warm freshwater lakes or rivers, most commonly in warm-weather southern-tier states (Fig. 75.3); however, two cases were exposed in the northern state of Minnesota in 2010 and 2012, suggesting an expanding geographic range for *N. fowleri*. Most patients were described as engaging in water-related activities such as diving or jumping into the water, swimming, or other water sports before illness onset. Additionally, three patients were exposed via nasal irrigation or ritual ablution using tap water in Louisiana (n = 2 cases) and the US Virgin Islands (n = 1).

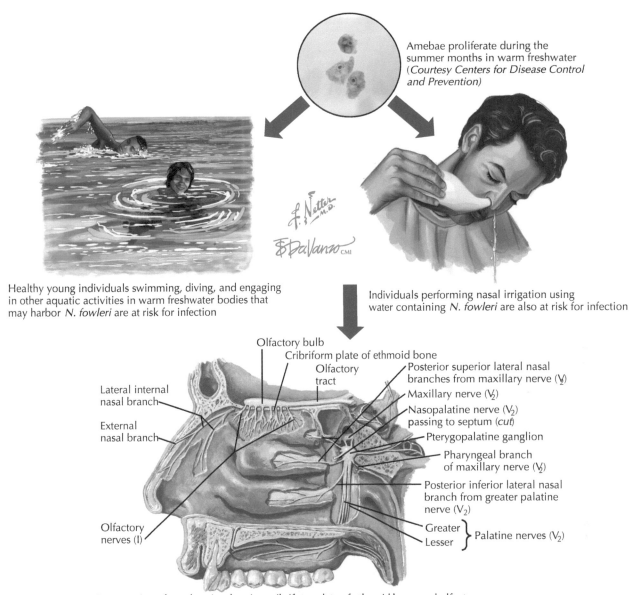

Amebae proliferate during the summer months in warm freshwater (*Courtesy Centers for Disease Control and Prevention*)

Healthy young individuals swimming, diving, and engaging in other aquatic activities in warm freshwater bodies that may harbor *N. fowleri* are at risk for infection

Individuals performing nasal irrigation using water containing *N. fowleri* are also at risk for infection

Olfactory bulb
Cribriform plate of ethmoid bone
Olfactory tract
Posterior superior lateral nasal branches from maxillary nerve (V₂)
Maxillary nerve (V₂)
Nasopalatine nerve (V₂) passing to septum (*cut*)
Pterygopalatine ganglion
Pharyngeal branch of maxillary nerve (V₂)
Posterior inferior lateral nasal branch from greater palatine nerve (V₂)
Greater } Palatine nerves (V₂)
Lesser
Lateral internal nasal branch
External nasal branch
Olfactory nerves (I)

Cross section of nasal cavity showing cribriform plate of ethmoid bone and olfactory nerves. Amebae enter the brain parenchyma through the cribriform plate

**Fig. 75.1** Exposures leading to infection with *Naegleria fowleri*.

## CLINICAL PRESENTATION

Symptoms and/or clinical features of PAM include headache, fever, nuchal rigidity, nausea, and vomiting, which resemble meningoencephalitis caused by bacteria or viruses. As the infection progresses, other signs and symptoms such as photophobia, lethargy, seizures, altered mental status, or coma may occur. Symptoms begin 1 to 12 days (median: 5 days) after swimming or other nasal exposure to water containing *N. fowleri*. PAM is usually fatal; of 148 reported PAM patients in the United States, only four (2.7%) have survived. Death from PAM occurs 1 to 18 days (median: 5 days) after symptom onset.

## DIAGNOSTIC APPROACH

Because the clinical presentation is nonspecific and similar to bacterial or viral meningitis, it can be difficult to differentiate PAM from other types of meningitis. PAM should be suspected in individuals with acute onset of symptoms following recent exposure to warm freshwater via recreational activity or nasal irrigation.

The CSF of patients with PAM is often characterized by low to normal glucose, elevated protein, and elevated opening pressure. The CSF is usually pleocytotic with mostly polymorphonuclear leukocytes and no bacteria. Microscopic examination of CSF wet smears may reveal the presence of actively moving amebae. Giemsa or trichrome staining of CSF smears can demonstrate the presence of amebae with a characteristically large nucleolus within the nucleus of the amebae, thus facilitating differentiation of amebae from neutrophils (Fig. 75.4). Amebae in CSF can be confirmed as *N. fowleri* using polymerase chain reaction (PCR).

Microscopically, the purulent leptomeningeal exudate generally consists of predominantly polymorphonuclear neutrophils (PMNs), few eosinophils, few macrophages, and some lymphocytes.

In the brain tissue of PAM patients, large numbers of *N. fowleri* trophozoites can frequently be seen under the microscope, usually in pockets within edematous and necrotic neural tissue. Amebic trophozoites ranging in size from 8 to 12 μm are also seen deep in Virchow-Robin spaces, usually around blood vessels with no inflammatory response. Scanning electron microscopic images have shown the presence of sucker-like structures, amebostomes, on the surface of the trophozoites, and it is believed

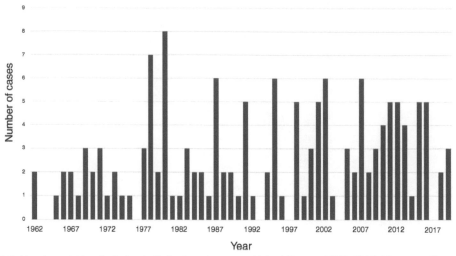

Fig. 75.2 Number of *Naegleria fowleri* infections by year—United States, 1962–2019. (Courtesy Centers for Disease Control and Prevention.)

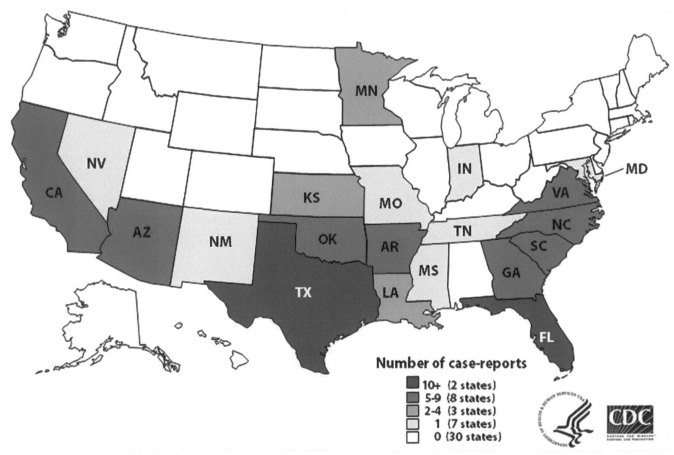

**Number of case-reports**

- 10+ (2 states)
- 5-9 (8 states)
- 2-4 (3 states)
- 1 (7 states)
- 0 (30 states)

Fig. 75.3 Number of cases of primary amebic meningoencephalitis caused by *Naegleria fowleri* by state of exposure—United States, 1962–2019. (Courtesy Centers for Disease Control and Prevention.)

that the amebostomes nibble away bits and pieces of the brain tissue by a process known as trogocytosis. The trophozoites in brain tissue can be specifically identified as *N. fowleri* by PCR or by immunofluorescence staining using polyclonal or monoclonal antibodies.

CT scan imaging of the brain is often normal early in the course of illness but later will usually show cerebral edema and evidence of brain herniation. At autopsy, the olfactory bulbs often demonstrate severe hemorrhagic necrosis and are usually surrounded by purulent exudate.

Cerebral hemispheres are usually soft, markedly swollen, edematous, and severely congested. The leptomeninges (arachnoid and pia mater) also are severely congested, hyperemic, and opaque, with limited purulent exudate within sulci, the base of the brain, the brainstem, and the cerebellum. Numerous superficial hemorrhagic areas are also seen in the cortex. Lesions are also found in and around the base of the orbitofrontal and temporal lobes, base of the brain, hypothalamus, midbrain, pons, medulla oblongata, and upper portion of the spinal cord.

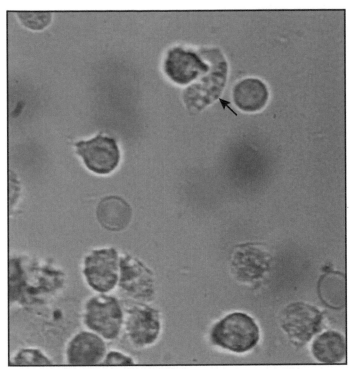

CSF smear showing an ameba *(arrow)*

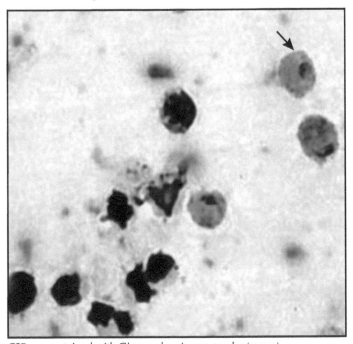

CSF smear stained with Giemsa showing an ameba *(arrow)*

**Fig. 75.4** *Naegleria fowleri* in the cerebrospinal fluid. (Courtesy Centers for Disease Control and Prevention.)

*N. fowleri* can be cultured from samples of CSF or from brain tissue obtained postmortem, by placing macerated brain tissue onto a non-nutrient agar plate coated with bacteria. *N. fowleri* culture can also be initiated by inoculating one or two drops of CSF or a small piece of the brain tissue directly into axenic growth medium or onto tissue culture monolayers such as E6 or HLF. Molecular techniques such as PCR and nested PCR assays have been developed for the specific identification of *N. fowleri* in clinical specimens, or cultured amebae from the CSF and brain tissue of patients and environmental samples. In addition, specific genotypes can be identified by sequencing the 5.8S ribosomal ribonucleic acid (rRNA) gene and the internal transcribed spacers 1 and 2 (ITS1 and ITS2) of *N. fowleri*.

A real-time multiplex PCR assay developed at the Centers for Disease Control and Prevention (CDC) can identify deoxyribonucleic acid (DNA) of *N. fowleri* in the CSF in about 3 hours upon receipt of the specimen, facilitating the rapid antemortem detection of *N. fowleri* so that treatment can be instituted. Furthermore, metagenomic deep sequencing (MDS) can also identify *N. fowleri* in clinical and environmental samples.

## TREATMENT

Of 148 reported PAM patients in the United States since 1962, only four survivors have been described. Three additional survivors with *N. fowleri* infection confirmed by PCR testing have been reported in Australia, Mexico, and Pakistan. Based on the treatment regimens used in survivors, a combination of drugs including amphotericin B (AMB), azithromycin, fluconazole, rifampin, miltefosine, and dexamethasone is recommended for treatment of PAM (Table 75.1). Recent mouse model and in vitro studies suggest that posaconazole may have greater efficacy against *N. fowleri* than fluconazole, and may be used instead of fluconazole in treating PAM.

Conventional AMB is preferred to liposomal AMB. When both formulations were experimentally compared, the minimum inhibitory concentration (MIC) against *N. fowleri* for conventional AMB was 0.1 µg/mL, while that of liposomal AMB was 10 times higher (1 µg/mL). Both liposomal AMB and AMB methyl ester were found to be less effective in the mouse model than conventional AMB, and liposomal AMB was also found to be less effective in in vitro testing than conventional AMB. Despite the more severe side effect profile of conventional AMB as compared to the liposomal formulation, the extremely poor prognosis of *N. fowleri* infection may warrant consideration of aggressive treatment.

Factors contributing to patient survival likely include early identification and treatment, use of the aforementioned combination of antimicrobials, and application of traumatic brain injury principles for management of elevated intracranial pressure.

## PREVENTION AND CONTROL

The only certain way to prevent *N. fowleri* infection from natural water bodies is for individuals to refrain from water-related activities in bodies of untreated freshwater. However, additional but unproven measures that might reduce risk by limiting the chance of water going up the nose include (1) holding the nose shut or using nose clips when taking part in water-related activities in bodies of warm fresh water such as lakes, rivers, and hot springs; (2) avoiding submersion of the head under the water in hot springs and other untreated thermal waters; (3) avoiding water-related activities in warm freshwater during periods of high water temperature; and (4) avoiding digging in or stirring up the sediment while taking part in water-related activities in shallow, warm freshwater areas. Swimmers and other recreational water users should assume that there is always a low level of risk whenever they enter warm freshwater lakes, rivers, and hot springs (for example, when swimming, diving, or waterskiing).

Chlorination (at a concentration of one part per million) can kill both *N. fowleri* trophozoites and cysts, so proliferation of this ameba can be prevented by adequate maintenance and disinfection of swimming pools and drinking water supplies. Inadequate chlorination may allow the growth of *N. fowleri*, resulting in colonization of pools and drinking water systems, proliferation of the ameba, and increased potential for infection and death of swimmers or people with nasal exposure to water containing the ameba. In former Czechoslovakia, 16 deaths from PAM occurred over a 2-year period (1963–1965) and were traced to a swimming pool with a low free chlorine concentration.

*N. fowleri* has caused deaths associated with using disinfected public drinking water supplies in Australia and Pakistan, an untreated, geothermal

| TABLE 75.1 | Recommended Treatment for Primary Amebic Meningoencephalitis | | | | |
|---|---|---|---|---|---|
| **Drug** | **Dose** | **Route** | **Maximum Dose** | **Duration** | **Comments** |
| Amphotericin B | 1.5 mg/kg/day in 2 divided doses | IV | 1.5 mg/kg/day | 3 days | |
| then | 1 mg/kg/day once daily | IV | | 11 days | 14-day course |
| Amphotericin B | 1.5 mg once daily | Intrathecal | 1.5 mg/day | 2 days | |
| then | 1 mg/day every other day | Intrathecal | | 8 days | 10-day course |
| Azithromycin | 10 mg/kg/day once daily | IV/PO | 500 mg/day | 28 days | |
| Fluconazole | 10 mg/kg/day once daily | IV/PO | 600 mg/day | 28 days | |
| Rifampin | 10 mg/kg/day once daily | IV/PO | 600 mg/day | 28 days | |
| Miltefosine | Weight <45 kg 50 mg BID<br>Weight >45 kg 50 mg TID | PO | 2.5 mg/kg/day | 28 days | 50 mg tablets |
| Dexamethasone | 0.6 mg/kg/day in 4 divided doses | IV | 0.6 mg/kg/day | 4 days | |

well-supplied drinking water system in Arizona, and a disinfected public drinking water system in Louisiana. While the act of drinking water containing the *N. fowleri* does not result in infection, individuals using tap water for nasal irrigation or ritual ablution should take precautions to boil, filter, or disinfect water prior to use, or use distilled or sterile water.

## FUTURE DIRECTIONS

Although *N. fowleri* infection is rare, it is of major concern due to the severity of the disease, its high fatality rate, and the disproportionate impact on children and adolescents. Further research may help to explain why PAM occurs primarily in these younger age groups and in males. These demographic groups may be more likely to engage in water activities that may result in high-risk *N. fowleri* exposure or may be predisposed by sex-linked hormones. Though recreational exposure to freshwater containing *N. fowleri* is relatively widespread, cases are rare; genetics or other host factors may contribute to increased susceptibility to PAM.

Because signs and symptoms of PAM are clinically similar to those of bacterial or viral meningitis, clinicians might not have a high index of suspicion, leading to delays in appropriate diagnostic testing and initiation of treatment. It is likely that many cases occur that go undiagnosed and unreported on a global scale. Improved clinician awareness may allow for earlier detection, diagnosis, and treatment, allowing for more favorable outcomes among patients. Additionally, the development of point-of-care tests for *N. fowleri*, allowing for rapid detection and diagnosis, may enhance recognition of PAM in clinical settings.

Because *N. fowleri* is a thermophilic ameba, it can proliferate in water when the ambient temperature increases. Although the impact of global climate change on this organism is unclear, rising temperatures may lead to cases of *N. fowleri* infection in an increasingly expanded geographic range. Further understanding the natural ecology of *N. fowleri* in the environment may allow for increased risk prediction and prevention strategies for PAM.

## EVIDENCE

Capewell LG, Harris AM, Yoder JS, et al. Diagnosis, clinical course, and treatment of primary amoebic meningoencephalitis in the United States, 1937–2013. *J Pediatric Infect Dis Soc* 2015;4(4):e68-75. *Clinical case series of PAM in the United States, providing a detailed description of the clinical picture, laboratory findings, and treatment of patients.*

Centers for Disease Control and Prevention. Notes from the field: primary amebic meningoencephalitis associated with ritual nasal rinsing—St. Thomas, U.S. Virgin Islands, 2012. *MMWR Morb Mortal Wkly Rep* 2013; 62(45):903. *Describes the investigation of a PAM case in the US Virgin Islands associated with nasal ablution for religious purposes.*

Cope JR, Conrad DA, Cohen N, et al. Use of the novel therapeutic agent miltefosine for the treatment of primary amebic meningoencephalitis: report of 1 fatal and 1 surviving case. *Clin Infect Dis* 2015;62(6):774-6. *Describes the use of miltefosine to treat PAM and provides a tabular description of the treatment of previous known survivors.*

Cope JR, Ratard RC, Hill VR, et al. The first association of a primary amebic meningoencephalitis death with culturable *Naegleria fowleri* in tap water from a US treated public drinking water system. *Clin Infect Dis* 2015;60(8):e36-42. *Presents a case report and environmental investigation findings of the first reported PAM death associated with culturable* Naegleria fowleri *in tap water from a US drinking water system.*

Linam WM, Ahmed M, Cope JR, et al. Successful treatment of an adolescent with Naegleria fowleri primary amebic meningoencephalitis. *Pediatrics* 2015;135(3):e744-8. *Detailed description of the clinical workup and treatment of the third well-documented PAM survivor in North America.*

Yoder JS, Eddy BA, Visvesvara GS, Capewell L, Beach MJ. The epidemiology of primary amoebic meningoencephalitis in the USA, 1962–2008. *Epidemiol Infect.* 2010;138(7):968-75. *Comprehensive report describing infections caused by* Naegleria fowleri *in the United States.*

Yoder JS, Straif-Bourgeois S, Roy SL, et al. Primary amebic meningoencephalitis deaths associated with sinus irrigation using contaminated tap water. *Clin Infect Dis* 2012;55(9):e79-85. *Describes the investigation of two PAM cases in Louisiana acquired via nasal irrigation using tap water.*

## ADDITIONAL RESOURCES

Colon BL, Rice CA, Guy RK, et al. Phenotypic screens reveal posaconazole as a rapidly acting amebicidal combination partner for treatment of primary amoebic meningoencephalitis. *J Infect Dis* 2018;219(7):1095-103. *Presents laboratory evidence supporting the efficacy of posaconazole for treatment of PAM.*

Cope JR, Ali IK. Primary amebic meningoencephalitis: what have we learned in the last 5 years? *Curr Infect Dis Rep* 2016;18(10):31. *Provides an overview of recent changes in the epidemiology and diagnosis of PAM.*

Kemble SK, Lynfield R, DeVries AS, et al. Fatal Naegleria fowleri infection acquired in Minnesota: possible expanded range of a deadly thermophilic organism. *Clin Infect Dis* 2012;54:805-9. *Discusses PAM acquired in Minnesota, north of the previously known geographic range for PAM in the United States.*

Matanock A, Mehal JM, Liu L, et al. Estimation of undiagnosed *Naegleria fowleri* primary amebic meningoencephalitis, United States. *Emerg Infect Dis* 2018;24(1):162. *Presents an estimate for the annual number of deaths consistent with PAM in the United States, suggesting that PAM might be underdiagnosed.*

Visvesvara GS, Moura H, Schuster FL. Pathogenic and opportunistic free-living amoebae: *Acanthamoeba* spp., *Balamuthia mandrillaris, Naegleria fowleri,* and *Sappinia diploidea.* *FEMS Immunol Med Microbiol* 2007;50(1):1-26. *A detailed review highlighting the morphology of the amebae (*Acanthamoeba, Balamuthia mandrillaris, Naegleria fowleri *and* Sappinia*), their taxonomic status, ecology, clinical manifestations of diseases caused by the amebae, pathologic features, epidemiology, and treatment modalities.*

# Parasitic Diseases

*Vernon Ansdell*

# 76

# Introduction to Parasitic Diseases

*Vernon Ansdell*

Parasitic diseases in humans are responsible for substantial morbidity and mortality in large parts of the world. At least 90 species of relatively common protozoal and helminthic infections occur in humans, and a small proportion of them cause some of the most important infectious diseases in the world. They are particularly common in rural areas of low-resource tropical countries, where access to health care is often limited and hygiene and sanitation are inadequate.

There is very good evidence that humans have been infected by parasites since prehistoric times. Examples exist confirming that almost all human nematode and protozoal parasite infections were present in ancient times. One of the very earliest examples appears to have been the discovery of *Enterobius vermicularis* ova in 10,000-year-old human coprolites (fossilized feces) from caves in Utah. Most of the evidence for parasitic infections has been found at burial sites, in ancient latrines or cesspits, in coprolites, and in mummified human remains. Some mummies were artificially prepared—for example, in ancient Egypt—but many have been naturally preserved by favorable environmental conditions. Examples of well-preserved naturally mummified human remains have been found in widely diverse environments such as deserts, peat bogs, calcareous caves, and glaciers in many parts of the world, including Europe, Africa, and the Americas. One of the earliest examples of parasitic disease in mummified remains was in an Egyptian mummy from 3200 BC that contained calcified *Schistosoma* ova.

Some parasites are better preserved than others. *Ascaris lumbricoides* ova have a thick chitinous shell and are particularly likely to be preserved. Multiple examples of ascaris infection have been found around the world. They include specimens in Peru from around 2300 BC; Brazil, 1600 BC; and Egypt, 1600 BC. In 2013 the body of Richard III, a medieval king, was found buried in Leicester, England. He was killed in the Battle of Bosworth Field in 1485. Examination of the burial site included soil from the sacral area where the intestines would have been during life and showed numerous *Ascaris lumbricoides* ova.

Discovery of certain parasitic ova may provide fascinating clues to the type of diet eaten in ancient times. For example, finding *Taenia* sp. and *Diphyllobothrium* ova in the salt mines of Central Europe from over 2000 years ago suggests that it was common, then, for miners in that part of Europe to eat raw or undercooked meat and fish.

For various reasons, eradication of parasitic diseases has proved very challenging. *Dracunculus medinesis*, or Guinea worm infection, is an important helminthic infection that was first mentioned in 1500 BC in the Ebers papyrus. The disease was clearly described in the Bible and calcified adult female worms have been found in in Egyptian mummies. There are encouraging signs that this disease, which is still responsible for considerable morbidity in some parts of the world, may be close to eradication, although significant obstacles remain.

New human parasitic diseases continue to be discovered. An important example is *Angiostrongylus cantonensis*, or rat lungworm

disease, which was recognized in animals in 1935; the first human case was identified in 1944. There is every reason to believe that other new parasitic diseases will be discovered in the future.

In terms of morbidity and mortality, malaria is probably the most important parasitic disease. Despite recent improvements in malaria control, the World Malaria Report from 2019 announced a staggering 228 million cases worldwide in 2018, with 405,000 deaths. Sixty-seven percent of deaths were in children under the age of 5 years. Although acute illness as a result of diseases such as malaria presents the greatest immediate risk to human life, chronic disease from parasitic infections creates the greatest impact as a result of malnutrition, chronic anemia, and lassitude, contributing to retardation of growth and loss of schooling in children and to inability to work and premature mortality among adults.

The greatest burden of parasitic diseases clearly falls on persons living in low-resource tropical countries. However, returned international travelers, international adoptees, immigrants, migrant workers, and residents of rural agricultural communities in high-resource temperate zone countries may harbor parasitic infections that, under the right circumstances, could become more widely transmissible. Vulnerable immunocompromised hosts in such countries may be at particularly increased risk of serious and life-threatening complications from imported parasitic infections.

The impact of climate change is unclear, but it is likely that there will be increased prevalence of many important parasitic diseases such as lymphatic filariasis and onchocerciasis due to an increase in insect vectors in certain parts of the world. At the same time there may be decreased prevalence of diseases such schistosomiasis in Africa due to factors such as shrinking snail host habitats.

The development of new drugs and the increased use of mass drug administration—together with new, improved diagnostic tests, particularly molecular testing—are encouraging signs for the control of parasitic diseases in the future.

Unfortunately there are no safe and uniformly effective vaccines against human parasitic infection. For various reasons the prospects of developing effective vaccines against parasitic infections are limited. Ultimately, the development of vaccines against the most serious parasitic infections will be a major step forward, but, unfortunately, we still remain a long way from achieving this goal.

## ADDITIONAL RESOURCES

Blum AJ, Hotez Peter J. Global "worming". Climate change and its projected general impact on human helminth infections. *PLOS Neglected Tropical Diseases* https//doi.org/10.1371/journal.pntnd July 18, 2018. *An editorial review of the potential impact of climate change on human helminth infections.*

Bouchet F, Guidon N, Dittmar K, et al. Parasite remains in archaeological sites. *Mem Inst Oswaldo Cruz Rio de Janeiro* 2003;98(Suppl 1):47-52. *An extensive review of paleoparasitology including details of the recovery and processing of specimens.*

Cook G. *History of Parasitology in Principles and Practice of Clinical Parasitology*, edited by S. Gillespie & Richard D Pearson 2001 John Wiley & Sons Ltd. *A fascinating and comprehensive review of the history of parasitology that covers all the main parasitic diseases. It is written by a clinician with extensive experience in clinical tropical medicine and a longstanding interest in the history of tropical diseases.*

Cox FEG. History of human parasitology. *Clin Microbiol Rev* 2002;595-212. *A very detailed review of the history of the most important helminth and protozoal diseases.*

Mitchell PD, Yeh HY, Appelby J, Buckley R. The intestinal parasites of King Richard III. *Lancet* 2013;382:888. *A fascinating account of the parasitological findings from the burial site of King Richard III.*

# 77

# Amebiasis

*Blaine A. Mathison, Bobbi S. Pritt*

## ABSTRACT

Infection with *Entamoeba histolytica* is a leading parasitic cause of morbidity and mortality in developing nations and is an important health risk to travelers. Amebiasis is transmitted by parasite cysts via fecal-oral contamination from infected individuals or from contaminated food or water. Although most infected individuals are asymptomatic, *E. histolytica* infections may cause varying degrees of symptoms through tissue invasion and dissemination outside of the intestinal tract.

## CLINICAL VIGNETTE

A 53-year-old entomologist who recently traveled to the jungles of Belize presented with a 2-week history of intermittent nausea, malaise, bloody diarrhea, weight loss, and abdominal pain. While in Belize, the patient had consumed local foods and beverages, including fresh salads and fruit juices, and had brushed his teeth using tap water. He was afebrile at the time of presentation, and physical examination was significant only for mild-to-moderate, nonfocal abdominal tenderness. Routine stool ova and parasite (O&P) examination revealed amebic cysts and trophozoites consistent with *E. histolytica* or one of the morphologically indistinguishable *Entamoeba* species (*E. dispar, E. bangladeshi, E. moshkovskii*). No ingested red blood cells were seen within the trophozoites. The stool specimen was also tested using a multiplex gastrointestinal nucleic acid amplification test (NAAT), which was positive for *E. histolytica*. The O&P examination was not reported to the species level despite the positive NAAT, as *E. histolytica* should only be reported at the species level by stool microscopy when ingested erythrocytes are observed in the cytoplasm of the trophozoites. Of note, some multiplex NAATs will give a false-positive result for *E. histolytica* when a high concentration of the nonpathogenic *E. dispar* is present; however, this NAAT result, in the context of the clinical findings and recent travel to an endemic setting with exposure to potentially contaminated food and water, is consistent with a diagnosis of intestinal amebiasis. The patient was started on the tissue amebicide, metronidazole, and made a rapid recovery. He was also given the luminal amebicide, paromomycin, to eliminate residual cysts in the bowel lumen. To prevent future infections with *E. histolytica* and other parasites transmitted through the fecal-oral route, the patient was cautioned against drinking inadequately treated water and against consuming raw, unpeeled fruits and vegetables in endemic, resource-limited settings.

## INTRODUCTION

The first description of *E. histolytica* was by Friedrich Lösch in 1875 from a case of acute dysentery. The cause of amebiasis is the protozoan parasite *E. histolytica*, which occurs in both cyst and motile trophozoite forms. There are several morphologically identical species in the *E. histolytica* complex, including *E. dispar, E. bangladeshi,* and *E. moshkovskii.* While *E. dispar* is considered a nonpathogenic parasite not requiring treatment, the pathogenic status of *E. bangladeshi* and *E. moshkovskii* is currently unresolved. These amebae cannot be differentiated through routine stool microscopy studies and, thus, must be differentiated from potentially pathogenic *E. histolytica* by means of a stool enzyme immunoassay (EIA) antigen test or NAATs. Another species, *Entamoeba hartmanni* (formerly known as "small-race" *E. histolytica*), is considered nonpathogenic. It is morphologically similar to *E. histolytica* and can best be distinguished by its size, which is consistently less than 10 μm.

## ETIOLOGY/PATHOPHYSIOLOGY

*E. histolytica* has trophozoite, precyst, and cyst stages (Fig. 77.1). The cyst is the infective stage and is ingested orally from contaminated food, water, or fingers. Ingested cysts pass through the stomach, and excystation occurs in the lower small bowel; four small metacystic trophozoites are formed, which mature to full size. Trophozoites generally measure 10 to 60 μm (invasive forms tend to be on the larger end of this range, often greater than 20 μm) and contain a single characteristic nucleus with a diameter of 2.8 to 4.5 μm that possesses a small, usually centrally located, karyosome and peripheral chromatin granules uniformly arranged along the nuclear membrane. Trophozoites are motile and move unidirectionally by rapidly thrusting out large, blunt, transparent pseudopodia. They pass along the lumen of the intestine until conditions favorable for colonization are found. This can occur anywhere in the large bowel but is more frequent in the cecal area. Multiplication is by rapid and repeated binary fission. Depending on various parasite and host factors, trophozoites may invade the tissue of the large intestine, primarily by lytic means, and may disseminate via the portal circulation to the liver and rarely to other extraintestinal sites. Invading trophozoites may contain ingested red blood cells. As the trophozoites are carried toward the rectum in the fecal stream, they eliminate food vacuoles and other cytoplasmic inclusions and become precysts. Precysts are rounded or oval with a cyst wall and contain a mass of glycogen vacuoles and large chromatoid bodies and a single nucleus. The precyst matures by two nuclear divisions to form a quadrinucleate cyst.

Mature cysts generally measure 10 to 20 μm and are round with a protective tough cyst wall. The four nuclei have the same characteristics as the nucleus of the trophozoite. Within the cyst are one or more chromatoid bodies with bluntly rounded ends, and early stage cysts contain glycogen. Cysts form only in the large intestinal lumen, and they exit the body in the feces. Cysts are relatively hardy and can survive in the environment long enough to be ingested and, thus, are the vehicles of transmission. However, cysts are quite sensitive to desiccation and to temperatures above 40°C or below −5°C. They can also be killed almost immediately by boiling. Cysts are relatively resistant to chlorine and are not destroyed by concentrations generally used for water purification. Motile trophozoites passed in the feces of infected persons with diarrhea or dysentery can survive for only a brief time outside the body and do not develop into cysts; therefore trophozoites do not play a role in transmission.

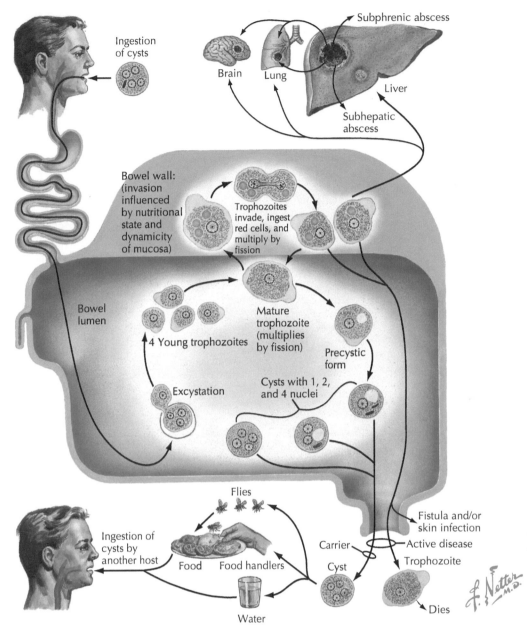

**Fig. 77.1** Amebiasis: life cycle of *Entamoeba histolytica*.

## EPIDEMIOLOGY

Amebiasis occurs worldwide but is most common in parts of the world with poor socioeconomic status and sanitation, especially Southeast Asia, Africa, and Latin America. In underdeveloped areas where drinking water is obtained from fecal-contaminated water sources and contaminated water or sewage is used to grow or wash vegetables, the incidence of amebiasis is high. While reported prevalence varies depending on the geographic region and method of diagnosis, *E. histolytica* infection is thought to cause up to 100 million symptomatic cases each year, with up to 100,000 associated deaths. Infection is rare in resource-rich countries such as the United States, Canada, and Europe, in which most cases are seen in travelers and immigrants.

Animal reservoirs of *E. histolytica* include nonhuman primates, dogs, and pigs, but these animals play a minor role in transmission in comparison to humans, who are the principal reservoir of infection. Infection usually occurs by either direct person-to-person transmission of cysts or by cysts contaminating food or water. Frequent transmission is recognized in institutionalized groups and daycare centers for young children. Men that have sex with men (MSM) are at increased risk of infection.

## CLINICAL PRESENTATION

*E. histolytica* is unique among the amebae parasitizing humans because of its ability to invade tissue. The fundamental pathology of *E. histolytica* is the trophozoite's lytic effect on the large bowel mucosa, leading to penetration of the host's tissue, necrosis of tissue cells, and formation of ulcers. Undermining of the ulcer margin and confluence of one or more ulcers lead to sloughing of the mucosa and development of broad ulcers with irregular outlines (Fig. 77.2). Host factors, including nutritional deficiencies and other variables associated with poor socioeconomic and environmental conditions, appear to promote the more invasive nature of amebiasis observed in the developing world.

There is a variable clinical response to *E. histolytica* infection including asymptomatic infection, amebic colitis with diarrhea, dysentery,

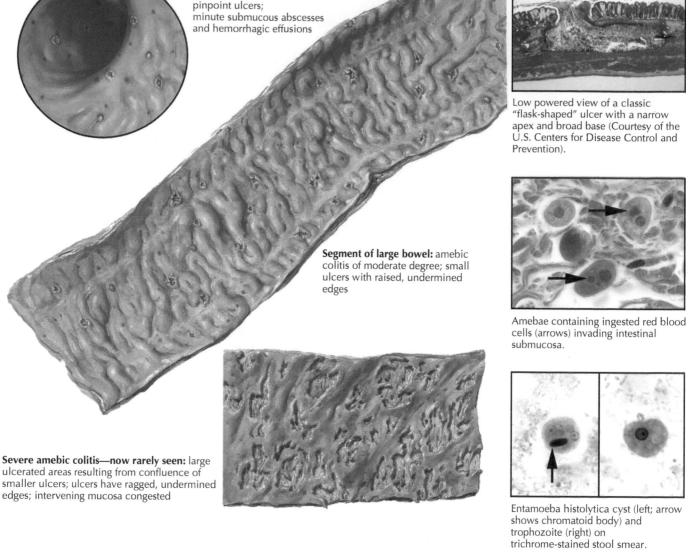

**Sigmoidoscopic view:** acute amebic colitis; pinpoint ulcers; minute submucous abscesses and hemorrhagic effusions

Low powered view of a classic "flask-shaped" ulcer with a narrow apex and broad base (Courtesy of the U.S. Centers for Disease Control and Prevention).

**Segment of large bowel:** amebic colitis of moderate degree; small ulcers with raised, undermined edges

Amebae containing ingested red blood cells (arrows) invading intestinal submucosa.

**Severe amebic colitis—now rarely seen:** large ulcerated areas resulting from confluence of smaller ulcers; ulcers have ragged, undermined edges; intervening mucosa congested

Entamoeba histolytica cyst (left; arrow shows chromatoid body) and trophozoite (right) on trichrome-stained stool smear.

**Fig. 77.2** Amebiasis: pathologic findings.

and even life-threatening fulminant disease, and rarely, extraintestinal spread to the liver and other organs. Most individuals (approximately 90%) are asymptomatic. Infections may remain asymptomatic or develop later into symptomatic and/or invasive infections. Importantly, asymptomatic individuals shed infectious cysts into the environment and thus unknowingly serve as a source of infection for others.

Patients with amebic colitis may have a wide range of symptoms. Mild-to-moderate, nonbloody diarrhea is the most common manifestation of amebic colitis, and may be accompanied by intermittent constipation, excessive intestinal distention and flatus, increased fatigue, anorexia, nausea, weight loss, and lower abdominal cramps (often localized over the cecum or sigmoid). The differential diagnosis includes giardiasis, *Dientamoeba fragilis* infection, strongyloidiasis, *Schistosoma mansoni* infection, low-level inflammatory bowel disease, diverticulitis, and irritable bowel syndrome.

Bloody diarrhea and dysentery are the more severe symptoms of disease, and may occur in approximately 15% to 33% of patients with amebic diarrhea. These manifestations may also develop in individuals with longstanding mild symptoms or who have been asymptomatic cyst passers. Symptom onset is typically subacute and

evolves over a period of 3 to 4 weeks. Signs and symptoms include marked abdominal cramps and severe diarrhea with blood and mucus. Surprisingly, fever is present in only a minority of patients. The white blood cell count may be elevated, with a polymorphonuclear leukocytosis. In very rare, severe fulminant cases, extensive colonic involvement may lead to massive destruction of the mucosa, hemorrhage and perforation, and peritonitis. These cases can be fatal, with fatality rates ranging from 40% to 89%. Risk factors for fulminant amebic colitis include corticosteroid use, alcoholism, diabetes, and chemotherapy for malignancy. The differential diagnosis of acute amebic colitis with dysentery includes infection with *Campylobacter* spp., *Shigella* spp., *Salmonella* spp., enteroinvasive *Escherichia coli*, enterohemorrhagic *E. coli*, *Clostridioides difficile*, *Strongyloides stercoralis* hyperinfection, and inflammatory bowel disease.

Uncommonly, intestinal involvement can result in extensive granulation tissue causing a mass lesion known as an ameboma. This mass may be palpable on exam and can grossly mimic colonic adenocarcinoma. Other uncommon intestinal manifestations include rectovaginal fistula formation and perianal cutaneous amebiasis.

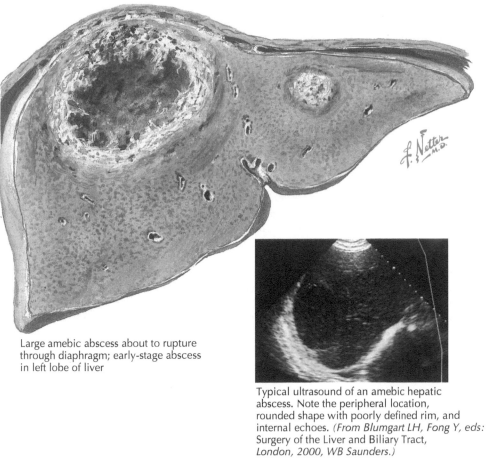

Large amebic abscess about to rupture through diaphragm; early-stage abscess in left lobe of liver

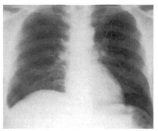

Chest x-ray of a 39-year-old man with a documented amebic liver abscess showing elevated right hemidiaphragm and a small right pleural effusion. *(From Mason RJ, Broaddus VC, Martin TR, et al: Murray and Nadel's Textbook of Respiratory Medicine, ed 5, Philadelphia, 2010, Saunders.)*

Typical ultrasound of an amebic hepatic abscess. Note the peripheral location, rounded shape with poorly defined rim, and internal echoes. *(From Blumgart LH, Fong Y, eds: Surgery of the Liver and Biliary Tract, London, 2000, WB Saunders.)*

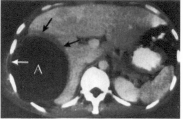

CT scan of amebic abscess. The lesion is peripherally located and round. The rim is nonenhancing but shows peripheral edema *(black arrows).* Note the extension into the intercostal space *(white arrow). (From Townsend CM, Beauchamp RD, Evers BM, et al: Sabiston Textbook of Surgery, ed 18, Philadelphia, 2008, Saunders.)*

**Fig. 77.3** Amebic liver abscess.

Extraintestinal spread is rare, occurring in less than 1% of cases. The liver is the most commonly involved organ. Dissemination of *E. histolytica* trophozoites through the portal circulation to the liver lead to formation of small lesions that then coalesce into an amebic liver "abscess," which is usually single. The right lobe is four times more likely to be involved than the left, as it receives the bulk of the intestinal venous drainage. Lesions consist primarily of necrotic hepatocytes and amebae, and are said to resemble "anchovy paste." Only a few inflammatory cells are present, thus, the term abscess is a misnomer. Instead, the preferred term for liver involvement is hepatic amebiasis. There is no entity of amebic hepatitis.

Amebic liver abscesses are 10 times more common in men than in women and are rare in children. They may occur in the presence or absence of intestinal symptoms and often develop after a latent period following an earlier diarrhea episode or other intestinal symptoms. Importantly, no more than 20% of patients with amebic liver abscess have *E. histolytica* organisms present on stool examinations. The clinical signs and symptoms may be variable, but typically there is a high fever, pain over the right lower area of the chest or right hypochondrium, marked tenderness over an enlarged liver, and a moderate leukocytosis. Chills and profuse sweats may be present. Jaundice occurs only with very large abscesses and is a poor prognostic sign. Other signs may include referred pain to the right shoulder, a visible mass with a large abscess, and a nonproductive cough. In many cases, abscesses extend upward to involve the diaphragm, leading to diaphragmatic elevation and immobility and compression of the right lower lobe of the lung. An abscess may rupture into the lung or peritoneum (Fig.

77.3). Differential diagnosis includes bacterial abscess of the liver, acute cholecystitis or cholangitis, infected hydatid cyst, acute hepatitis, malaria, subphrenic abscess, and carcinoma of the liver.

Pleuropulmonary amebiasis usually results from an extension of a hepatic abscess through the diaphragm, leading to pleural and pulmonary involvement. Pain occurs in the lower right area of the chest, and there may be a nonproductive cough.

Amebic pericarditis is an uncommon complication of an amebic liver abscess, usually resulting from the extension of a left lobe liver abscess through the diaphragm into the pericardium.

Finally, amebic brain abscess is a rare but generally fatal occurrence. It usually follows from amebic involvement of the liver and lungs. The most common symptoms are headache, fever, and convulsions.

## DIAGNOSTIC APPROACH

The preferred diagnostic method varies by the type of disease present, and more than one method is commonly required for definitive diagnosis. Intestinal amebiasis is usually diagnosed by demonstrating *E. histolytica* parasites, their antigens, or nucleic acid in the stool. Invasive trophozoites may also be observed on intestinal biopsies. In contrast, serology is the preferred diagnostic modality for amebic liver abscess in conjunction with clinical and radiologic features.

### Intestinal Disease

The traditional method for diagnosing intestinal amebiasis is by detection of morphologically compatible cysts and/or trophozoites in stool specimens (see Figs. 77.1 and 77.2). As *E. histolytica* is morphologically indistinguishable from *E. dispar*, *E. moshkovskii*, and *E. bangladeshi*, additional antigen or nucleic acid-based detection methods are required for definitive identification. *E. histolytica* must also be differentiated from other morphologically similar, mainly nonpathogenic, intestinal protozoa. *Specific identification of E. histolytica is recommended when possible, given its pathogenic potential.* *E. histolytica* and other protozoal cysts may remain viable for some time in unpreserved formed stools, but trophozoites that are seen in dysenteric and diarrheal stool specimens are labile and may disappear from the stool within 30 minutes of passage. Therefore specimens should preferably be collected with use of commercial stool collection kits containing a preservative. The traditional fixatives include a two-vial system consisting of 10% formalin and polyvinyl alcohol (PVA); however, there are now several single-vial systems available that use more environmentally friendly preservatives.

Because protozoal cysts and trophozoites may be passed intermittently, three stool specimens should be collected on consecutive days or over a 10-day period. A negative result with a single or even three specimens does not completely rule out infection. Stools should be processed per normal O&P procedures including both a concentrated wet mount to detect cysts and a permanent trichrome-stained slide to detect cysts and trophozoites. If a fresh, unfixed stool specimen is examined, the classically described unidirectional trophozoite motility may be observed. The presence of ingested erythrocytes in trophozoites observed on trichrome-stained stool smears is pathognomonic for *E. histolytica*; however, this phenomenon is rarely observed and the absence of ingested erythrocytes does not rule out true amebiasis. Antibiotic and antiparasitic drugs, antacids, kaolin products, enema products, oily laxatives, and barium can cause masking or disappearance of protozoa in the stool or interfere with their recognition.

Charcot-Leyden crystals may be found in the stools of patients with amebiasis but are not specific for *E. histolytica* infection as they can be seen with other parasitic and noninfectious conditions. Unlike stools in acute bacterial dysentery, in which large quantities of white blood cells are usually seen, stools in amebic dysentery seldom contain more than a few leukocytes.

Stool antigen tests are more sensitive than stool microscopy and may allow for differentiation of *E. histolytica* from other morphologically similar *Entamoeba* species. Several kits employing either EIA or lateral flow immunochromatographic (LF, or LFICA) formats are commercially available. Examples include the EIAs Techlab *E. histolytica* II (Abbott Laboratories, Chicago, IL), *Entamoeba* CELISA (Cellabs, Brookvale, Australia), and Remel ProSpecT *E. histolytica* (Thermo Fischer Scientific, Waltham, MA) and the LF assay Techlab *E. histolytica* QUIK CHEK (Abbot Laboratories). Different countries have different regulatory criteria; in the United States, only the Techlab EIA and LF assays are FDA-cleared for clinical use. Commercially available assays have varying degrees of sensitivity and specificity, and not all are capable of differentiating *E. histolytica* from *E. dispar*. It is therefore important to be familiar with the limitations of the assay used by the laboratory. The Techlab *E. histolytica* EIA and LF assays have comparable sensitivity and specificity (>95%) and do not cross-react with *E. dispar*.

NAATs are extremely sensitive and specific, and are considered the gold standard for diagnosis of intestinal amebiasis. Their increased sensitivity is particularly beneficial in patients with liver abscesses where few cysts are usually present in stool. Many laboratory-developed NAATS have been described; however, at the time of this writing there are three commercial assays FDA-cleared for clinical use in the United States: (1) FilmArray Gastrointestinal GI Panel (BioFire Diagnostics, Salt Lake City, UT), (2) BD MAX Enteric Parasite Panel (Becton Dickinson and Company, Franklin Lakes, NJ), and (3) xTag Gastrointestinal Pathogen Panel (Luminex, Austin, TX). Additional commercial NAATs are available outside of the United States. It is important to note that false-positive results may be noted with some NAATs in the presence of high concentrations of *E. dispar*.

When the differential diagnosis includes entities for which a tissue diagnosis is desired (e.g., ulcerative colitis), then proctoscopy or colonoscopy allows for visualization of typical ulcerated lesions and collection of scrapings or biopsies. Amebic ulcers are commonly "flask-shaped" as the trophozoites invade laterally into the intestinal submucosa, forming an ulcer with a narrow apex and broad base (see Fig. 77.2). Trophozoites of *E. histolytica* have the same morphologic characteristics in tissue biopsies that they have in stained stool specimens and are more likely to have ingested erythrocytes. Cysts are not seen in intestinal biopsy specimens. *E. histolytica* can be reported at the species level when organisms are observed in biopsy or extraintestinal specimens, as it is the only species that can invade the intestinal mucosa and disseminate to other organs.

Finally, serology may be useful in cases of invasive intestinal diseases, in which results may be positive in up to 90% of cases. It is insensitive for detecting cases of asymptomatic infection. Several formats are available, including ELISA and indirect hemagglutination-based assays. As with all serologic testing, positive results cannot distinguish between acute and past infection due to persistence of antibodies. Also, results may be negative in patients with an acute presentation of less than 7 days, but tests repeated 5 to 7 days later are generally positive.

## Extraintestinal Disease

Serology is the primary modality used to detect extraintestinal forms of disease such as amebic liver abscess. See the information above on serologic testing for *E. histolytica* infection.

Amebic liver abscess should be suspected in a patient with right upper quadrant pain, a tender liver, and a fever. Liver function test results are often normal except for elevated alkaline phosphatase. Demonstration of a filling defect in the liver on computed tomography, magnetic resonance imaging, or sonogram examination followed by a positive serum amebic antibody test result suggests definitive diagnosis (see Fig. 77.3). Where available, antigen tests and NAATs can be performed on aspirated liver lesions. Aspirates can also be examined using microscopy, although trophozoites are mostly present at the periphery of the lesion and may not be present in the material. Cysts are not seen in extraintestinal lesions.

## TREATMENT

The antiparasitic therapy of choice is based on the type and extent of infection. In general, both asymptomatic and symptomatic infection should be treated. The nitromidazoles, metronidazole, and the related tinidazole are tissue amebicides that are used to treat both invasive bowel and liver amebiasis. Paromomycin, iodoquinol, and diloxanide furoate are poorly absorbed luminal drugs that act primarily on the bowel lumen to eliminate cysts.

Patients with asymptomatic *E. histolytica* carriage need only be treated with a luminal agent. Treatment of the nonpathogenic *E. dispar* is not indicated. Although differences of opinion exist concerning the need to treat asymptomatic *E. histolytica* cyst passers, it is important to note that asymptomatic *E. histolytica* cyst passers are potential infectors of others, and long-term carriage of this parasite could lead to later active intestinal disease or amebic liver abscess.

For mild to severe amebic colitis, a tissue amebicide should be given, followed by a course of a luminal drug to prevent a later relapse (Table 77.1 lists treatment regimens). It is essential to attempt to differentiate amebic colitis from ulcerative colitis before administering corticosteroids because amebic infection may be worsened by corticosteroids.

## TABLE 77.1 Drug Therapy for Amebiasis

| Drug | Adult Dose | Pediatric Dose |
|---|---|---|
| **Asymptomatic Cyst Passer** | | |
| Paromomycin | 25–35 mg/kg/day by mouth in 3 divided doses × 7 days | 25–35 mg/kg/day by mouth in 3 divided doses × 7 days |
| Iodoquinol | 650 mg by mouth tid[a] × 20 days | 30–40 mg/kg/day (max 2 g) by mouth in 3 divided doses × 20 days |
| **Mild to Moderate Intestinal Disease** | | |
| Either metronidazole or tinidazole followed by either iodoquinol or paromomycin as described above for asymptomatic cyst passers | | |
| Metronidazole | 500–750 mg by mouth tid × 7–10 days | 35–50 mg/kg/day by mouth in 3 divided doses × 7–10 days |
| Tinidazole | 2 g once daily by mouth × 3 days | ≥3 years of age: 50 mg/kg/day (max 2 g) by mouth in 1 dose × 3 days |
| **Severe Intestinal Disease, Amebic Liver Abscess, and Other Extraintestinal Infection** | | |
| Either metronidazole or tinidazole followed by one of the luminal drugs used for asymptomatic cyst passers above | | |
| Metronidazole | 750 mg tid by mouth or IV × 7–10 days | 35–50 mg/kg/day by mouth in 3 divided doses × 7–10 days |
| Tinidazole | 2 g once daily by mouth × 5 days | ≥3 years of age: 50 mg/kg/day (max 2 g) by mouth in one dose × 5 days |

[a]*tid*, Three times a day.

Loperamide should also be avoided. Surgery may be necessary in patients with acute bowel perforation with localized abscess formation, perforation with peritonitis, or fulminating amebic colitis not responding to chemotherapy. Patients may also require intensive care support, broad-spectrum antimicrobial coverage for peritonitis, and aggressive fluid resuscitation.

With typical symptoms and signs of liver abscess, a scan or sonogram positive for a filling defect in the liver, and a positive serologic test for amebiasis, drug treatment is indicated. As with acute amebic dysentery, similar treatment with a nitroimidazole followed by a luminal drug should be administered. If satisfactory clinical improvement is not obtained after 3 days of the nitroimidazole treatment, there is a high risk of rupture (i.e., diameter of cavity >5 cm), so drainage of the abscess should be performed using either percutaneous aspiration or surgery. Consideration should also be given to drainage of a left lobe abscess due to the risk for rupture into the pericardium. Serial liver scans have shown that most amebic abscesses heal gradually over 2 to 4 months after successful treatment. Occasionally the resolution time may be longer.

After treatment of intestinal amebiasis, follow-up stool examinations and/or an amebic antigen test or NAAT should be performed about 4 weeks later to determine if it has been cured.

## PREVENTION

Preventive measures are similar to those for other enteric pathogens and should be directed toward education concerning means of transmission of amebiasis and methods of avoiding infection. Infected food handlers should be identified and treated, as should infected individuals in institutions and children in daycare centers. Contamination of food by flies may be prevented by screening and covering food items. Food handlers should wash their hands and have appropriate sanitary facilities. In endemic areas it is important to avoid raw, unpeeled, fruits and vegetables possibly infected from "night soil" (human feces used as fertilizer), sewage, and contaminated water. Community water sources should be protected from fecal contamination and made safe by filtration, sedimentation, and chlorination. Boiling of water destroys amebic cysts immediately.

Iodine water purification tablets or portable water filters using a filter and iodination are more effective than chlorine tablets at inactivating cysts. Only ice prepared from treated water should be used. Asymptomatic *E. histolytica* carriers should be treated to avoid possible transmission to others.

## ACKNOWLEDGMENTS

The authors would like to acknowledge the valuable contributions of the late Dr. Martin S. Wolfe to this chapter.

## EVIDENCE

Ali IL. Intestinal amebae. *Clin Lab Med* 35: 2015;393-422, 2015. *Updates on diagnosis of amebiasis and other intestinal amebae.*

Garcia LS, et al. Laboratory diagnosis of parasites from the gastrointestinal tract. *Clin Microbiol Rev* 2017;31(1):e00025-17. *Current updates on diagnostic procedures for intestinal parasites, including* E. histolytica.

Kirk MD, et al. World Health Organization estimates of the global and regional disease burden of 22 foodborne bacterial, protozoal, and viral diseases, 2010: a data synthesis. *PLOS Med* 2015;12(12):e1001921. *Global and regional estimates of illness, deaths, and disability adjusted life years (DALYs) due to amebiasis.*

Shirley DT, et al. A review of the global burden, new diagnostics, and current therapeutics for amebiasis. *Open Forum Infect Dis* 2018;5(7):ofy161. *Update on diagnosis and treatment of amebiasis.*

## ADDITIONAL RESOURCES

Ash LR, Orihel TC. *Atlas of human parasitology.* 5th ed. Chicago: ASM Press, 2007. *This is the "go-to" reference for the morphologic identification of intestinal amebae (and other parasites).*

Pritt BS (ed.). *Atlas of fundamental infectious diseases histopathology: a guide for daily use.* Northfield, IL: College of American Pathologists, 2018. *An atlas for the morphologic identification of parasites (and other microorganisms) in histopathologic specimens.*

*The Medical Letter: drugs for parasitic infections.* 3rd ed. New Rochelle, NY: The Medical Letter, Inc., 2013. *Treatment recommendations for parasitic infections.*

# 78

# Giardia, Cryptosporidium, Cyclospora, and Other Intestinal Protozoa

*Claire Panosian Dunavan*

 **ABSTRACT**

*Giardia* and *Cryptosporidium* are the two most common protozoan pathogens in the human intestine; in low-income countries, they also rank as leading causes of diarrhea in children under 5. *Giardia duodenalis* (previously *G. lamblia* or *G. intestinalis*), was first discovered by Antoine van Leeuwenhoek, a 17th-century lens-grinder who invented the microscope, then found *Giardia* in his own stool. For the next several centuries, *Giardia* was thought to be a harmless commensal. In the 1970s, that opinion changed after short-term travelers to the former Soviet Union developed acute, symptomatic illness. Today, the paradox continues. Although giardiasis is often silent in residents of low- and middle-income countries, it can also produce significant diarrhea, malaise, nausea, bloating, flatulence, and weight loss, both here and abroad. In its most severe form, giardiasis causes chronic malabsorption and, in youngsters, can hinder development. Although anti-parasitic treatment usually cures giardiasis, treatment-refractory infections are currently on the rise.

Compelling evidence for *Cryptosporidium's* pathogenicity also dates from the 1970s and 1980s. Risk factors for severe, even life-threatening diarrhea include HIV infection with a CD4 count below 100, organ transplantation, IgA deficiency, hypogammaglobulinemia, and the use of immunosuppressive drugs. Infection in immunocompetent hosts is self-limited. In the 1990s, *Cryptosporidium* was identified as a major waterborne pathogen. Today, like *Giardia*, *Cryptosporidium* infections also contribute to morbidity and mortality in children living in low-income countries, where it probably accounts for 20% of their diarrheal episodes.

*Cyclospora cayetanensis* was first identified and named in the early 1990s. *Cyclospora* has caused food- and waterborne outbreaks both in the United States and abroad (see Fig. 78.1). In the United States, sporadic foodborne outbreaks were solely linked to imported berries and produce until 2018, when mixed salads containing romaine grown in California were sold by a fast-food restaurant chain, causing roughly 500 human infections in 15 states. This event likely heralds future outbreaks stemming from domestic produce. On the other hand, despite rising cases of cyclosporiasis, the infection remains far less common than giardiasis or cryptosporidiosis. *Cyclospora* can be treated with antibiotics, and it is currently believed that humans are its only natural host.

 **CLINICAL VIGNETTE**

A 40-year-old male scientist camped and fished in a remote lake in Alaska. On his return, he developed diarrhea, cramps, and loose stools without blood or mucus in the absence of fever and was diagnosed with giardiasis.

A 3-year-old female living in the Florida Keys complained of intermittent stomachaches over a 2-month period. Her stools were variably loose. The patient was diagnosed with giardiasis, which led to examination of her mother, father, and brother, who were mildly symptomatic; all three were subsequently diagnosed with giardiasis. The child's only exposure was from swimming in a local community pool.

A 40-year-old from Mexico, who resided in Virginia and worked as a cook in a fast-food restaurant, was diagnosed with giardiasis. He denied any symptoms and was not allowed to prepare food. Treatment with metronidazole, nitazoxanide, and albendazole failed to eradicate the infection. He was successfully treated with a combination of paromomycin and metronidazole.

Reused from Nash TE: Unraveling how *Giardia* infections cause disease, *J Clin Invest* 23:2346-2347, 2013.

## ETIOLOGIC AGENTS

***Giardia duodenalis*** is a protozoan with two life-cycle stages: motile trophozoites and environmentally resistant cysts. The trophozoite containing two nuclei and four pairs of flagella is dorsally convex with a ventral sucking disk, which allows *Giardia* to attach to the surface of the small intestinal mucosa. Trophozoites divide by longitudinal binary fission, and cysts develop as feces slowly dehydrate while transiting the colon.

The cysts of *G. duodenalis* are oval with a tough hyaline cyst wall. Cysts readily survive in cool, damp conditions, sometimes for months. *Giardia* cysts can also withstand standard concentrations of chlorine used in water purification systems.

Giardiasis is contracted after humans ingest cysts present in contaminated water or food or through person-to-person contact, including sexual contact, with other infected people. After the cysts pass through the stomach unharmed, excystation occurs in the duodenum, and trophozoites emerging from mature cysts nestle among the fingerlike villi of the duodenum and upper jejunum. Trophozoites either attach to the intestinal microvillous surface by their sucking disks or move freely in the lumen. The trophozoites are most often found in liquid or soft stools; the hardier, infective cysts are found in firmer stools (Fig. 78.2).

The species *G. duodenalis* consists of eight genetic groups, also called assemblages. Assemblage A combines human and zoonotic isolates, assemblage B accounts for most human infections, and the other six assemblages exclusively infect animals.

***Cryptosporidium*** is caused by parasites that can complete their life cycle, which involves both sexual and asexual replication, in a single host. Infection begins with ingestion of oocysts shed from the stool of an infected animal or person (Fig. 78.3). Within an infected person, the round to oval organisms are typically found within superficial, bulging parasitophorous vacuoles in epithelial cells lining the intestine (Fig. 78.4) and, less commonly, in the biliary and respiratory tracts.

Roughly two dozen species of *Cryptosporidium* are now known to infect mammals, birds, reptiles, and fish, but the principal species infecting humans are *C. hominis* and *C. parvum*. Infection is acquired through the fecal-oral route and oocysts are immediately infectious after passage in stool. Waterborne transmission is especially common. The cyst stage of *Cryptosporidium* is relatively resistant to killing by chlorine and iodine.

**Vegetables and legumes**
Mesclun, salad,
coriander, basil, basil pesto,
garlic, watercress, leafy herbs,
sugar snap peas, and lettuce

**Fruit**
Raspberries, berry juice,
berry desserts,
and fruit salads

**Water**
Drinking water, irrigation water,
and recreational waters

**Fig. 78.1** Sources of cyclosporiasis. (Reused with permission from Giangaspero A, Gasser RB: Human cyclosporiasis. *Lancet Infect Dis* 19:e226-236, 2019, Figure 1.)

*Cyclospora* trophozoites and other asexual stages have rarely been observed in human tissue; the agent's spherical oocysts are roughly twice the size of *Cryptosporidium* and are not infectious until 10 to 12 days following their passage in stool. How they mature in the environment and whether their life cycle involves non-human animal hosts are currently unknown.

Within the human host, *Cyclospora* sporozoites attach to and penetrate small intestinal enterocytes, replicate asexually, spread to adjacent villi and crypts, destroy brush borders, and alter mucosal architecture.

## EPIDEMIOLOGY AND RISK FACTORS

Giardiasis occurs worldwide, but its prevalence is far higher in areas where sanitation is poor. Typical rates of infection are 2% to 7% in high-income countries and 2% to 30% in low-income countries, whose residents are almost universally infected in childhood. In addition, people who travel from high- to low-income settings—in particular, countries in South and Southeast Asia, North Africa, the

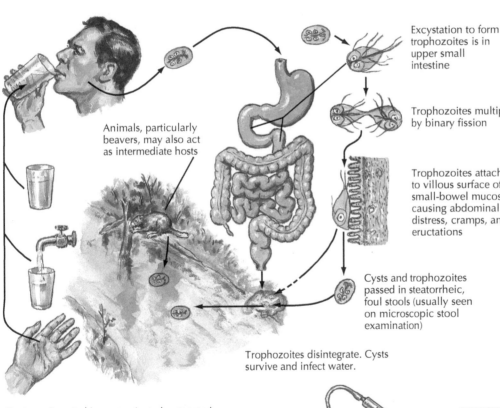

Excystation to form trophozoites is in upper small intestine

Trophozoites multiply by binary fission

Trophozoites attach to villous surface of small-bowel mucosa, causing abdominal distress, cramps, and eructations

Cysts and trophozoites passed in steatorrheic, foul stools (usually seen on microscopic stool examination)

Animals, particularly beavers, may also act as intermediate hosts

Trophozoites disintegrate. Cysts survive and infect water.

Cysts are ingested in contaminated, untreated stream water; in inadequately treated tap water; or via infected food handlers

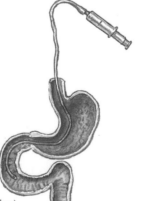

When infection is suspected but stool examination results are negative, duodenal or jejunal fluid (obtained by aspiration)

Cysts and trophozoite in stool

*Giardia* trophozoites in duodenal mucus

Jejunal biopsy specimen (obtained by suction or endoscopically) shows trophozoite on villous surface of mucosa

**Fig. 78.2** Giardiasis.

Caribbean, and South America—are at increased risk for acquiring the parasite.

Humans also contract giardiasis following exposure to rivers, lakes, and streams contaminated by animal hosts such as beavers since cysts

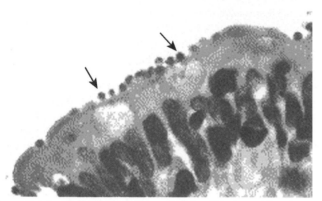

**Fig. 78.3** Intestinal biopsy specimen showing *Cryptosporidium* intracellular forms (trophozoites and merozoites) *(arrows)* inside the surface of intestinal epithelial cells. (Reused with permission from White AC: Cryptosporidiosis [*Cryptosporidium* Species]. In Bennett JE, Dolin R, Blaser MG, eds: *Principles and Practice of Infectious Diseases*, vol 2, Philadelphia, 2020, pp 3410-3420, Fig. 282-2.)

of *G. duodenalis* are especially long-lived in cold water. Food handlers and sexual partners can serve as sporadic sources of human infection. Outbreaks have occurred in custodial institutions as well as in childcare facilities where staff tend diapered infants. Dog ownership is associated with acquisition of *Giardia* assemblage A. Reinfection with *Giardia* can occur within family units or between close associates.

*Cryptosporidium* parasites have been found in every region of the world except Antarctica, and infections occur most commonly during warm or rainy months. *Cryptosporidium parvum* is generally linked to contact with animals, rural residence, and exposure to surface water, whereas *C. hominis* is more often contracted in densely populated, urban settings. Prospective studies following newborns suggest that virtually all children in South Asia are infected by age 2.

Hardy, chlorine-resistant oocysts can survive for as long as 12 months in cold water but are vulnerable to desiccation and killed by freezing, boiling, pasteurization, and microwave heating.

In recent years, numerous outbreaks of cryptosporidiosis have stemmed from exposure to contaminated drinking water as well as swimming pools and water parks. In 1993, oocysts in the public water supply serving roughly half of the local population of Milwaukee, Wisconsin, led to a massive *C. hominis* outbreak which affected an estimated 403,000 people. It was later determined that *Cryptosporidium* oocysts passed through the filtration system of a municipal water-treatment plant located 2 miles downstream from a sewage treatment station. Because *Cryptosporidium* has a vast zoonotic reservoir, contact

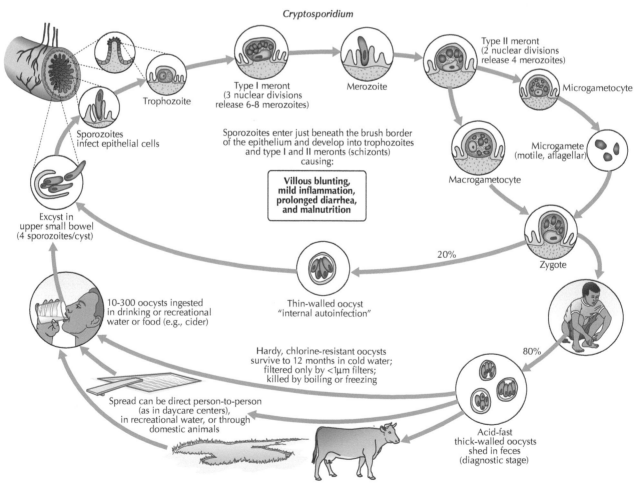

**Fig. 78.4** Life cycle of *Cryptosporidium*. (Reused with permission from White AC: Cryptosporidiosis [*Cryptosporidium* Species]. In Bennett JE, Dolin R, Blaser MG, eds: *Principles and Practice of Infectious Diseases*, vol 2, Philadelphia, 2020, pp 3410-3420, Fig. 282-1.)

with animals—especially calves, lambs, piglets, and other juvenile livestock—has also triggered outbreaks and individual cases, particularly among veterinary students.

Cyclosporiasis appears to have a worldwide geographic distribution but is more prevalent in tropical and subtropical regions. In endemic settings, fecal excretion predominates in children between 2 and 5 years of age; transmission is often seasonal. For example, in Nepal, higher rates of infection seem to occur during the summer and rainy seasons. In Peru, cases cluster during warm, dry summer months.

Because *Cyclospora* oocysts excreted in feces require days to weeks to become infectious, direct fecal-oral transmission from freshly passed stool does not occur.

In the United States, waterborne cyclosporiasis was first documented in the 1990s, when chlorinated water from a tank on the roof of a Chicago hospital dormitory infected 24 people. Five years later, 33 people contracted cyclosporiasis after drinking from a single water cooler at a golf course in New York. Since 2000, multiple outbreaks have been linked to imported raspberries from Guatemala as well as various items grown in Mexico including lettuce, mesclun, basil, cilantro, and mint. In 2018, a large, multi-state outbreak was traced to romaine lettuce grown in California, then served in salads sold by a national chain of fast-food restaurants. In addition to contaminated food and water, soil contaminated with human feces may play a role in transmission.

Along with *Giardia* and *Cryptosporidium, Cyclospora* should be included in the differential diagnosis of patients from high-income countries who develop acute traveler's diarrhea during or after overseas trips to destinations with poor sanitation and unclean water.

## CLINICAL FEATURES

The symptoms of giardiasis are variable. One to three weeks after ingesting as few as 10 cysts, roughly half of all infected people will have already cleared their infections; others who are clinically asymptomatic will continue to shed cysts; and the remainder will suffer symptoms that typically include acute, watery, diarrhea accompanied by abdominal cramps, bloating, foul-smelling flatus, sulfuric belching (also described as "eggy burps"), malaise, nausea, anorexia, and weight loss. In a few individuals, nausea, vomiting, and upper gastrointestinal symptoms predominate. At the onset of illness, diarrhea may precede the passage of cysts. Over time, the diarrhea becomes more intermittent. Malabsorption is relatively common, often resulting in stools that are yellow, frothy, and greasy. Chronic infection in children may cause failure to thrive. Risk factors for more severe clinical illness include cystic fibrosis and several immunodeficiency states, especially X-linked gammaglobulinemia, common variable immunodeficiency, and IgA deficiency. HIV infection does not make individuals more susceptible to giardiasis, but HIV-infected patients can prove difficult to treat.

Following an acute bout of giardiasis, secondary lactose intolerance—sometimes lasting for weeks to months—occurs in up to 40% of symptomatic patients. Post-*Giardia* irritable bowel syndrome is another common complication now increasingly recognized in non-endemic settings. Chronic giardiasis with frank malabsorption is associated with small intestinal histopathologic changes, which include shortened, atrophied villi and crypt hyperplasia accompanied by a modest degree of inflammation.

Important features which distinguish giardiasis from viral and bacterial diarrhea include its longer incubation, prolonged duration, waxing and waning symptoms, and associated weight loss. The presence of blood in stool should suggest another intestinal infection or illness.

In low-income areas of the world, *Giardia* is almost universal in children and is now felt to contribute, along with *Cryptosporidium*, to

malnutrition, stunting, and delays or deficits in cognitive development. But chronic exposure also induces partial immunity; in endemic areas, children under 10 have higher rates of giardiasis than older individuals.

Unlike *Giardia*, descriptions of *Cryptosporidium's* most dramatic manifestations preceded reports in otherwise healthy individuals. In your mind's eye, step back to the mid-1980s and picture emaciated patients with AIDS daily passing vast quantities of diarrheal stool. To maintain hydration, electrolyte balance, and nutrition in such individuals was a daunting prospect. It was not until effective anti-retroviral therapy became available that severely ill patients co-infected with HIV and *Cryptosporidium* truly recovered. Although chronic cryptosporidiosis is primarily a diarrheal illness, the parasite can also infect the biliary tract and present with acalculous cholecystitis, sclerosing cholangitis, or pancreatitis. Respiratory tract involvement can occur but is rarely symptomatic.

Today, cryptosporidiosis in immunocompetent adults and children is also better understood. Oocysts shed by infected hosts are immediately infectious. Although the pre-patent clinical period can range from 1 to 30 days, acute illness typically develops 1 week following ingestion of 10 or more thick-walled oocysts and presents with non-bloody diarrhea, abdominal cramps, nausea, vomiting, and fever. Symptoms normally last 1 to 2 weeks but sometimes recur after a 1- to 2-week hiatus. In elderly patients, intravenous fluids and electrolytes and inpatient care may be needed, but in large waterborne outbreaks, the majority of patients have not sought formal care. Chronic sequelae include irritable bowel syndrome, arthralgias, and fatigue.

In resource-poor countries, *Cryptosporidium* typically infects children before the age of 2 or 3 and causes persistent diarrhea in approximately one-third of cases. Malnutrition predisposes to more severe disease. Persistent diarrhea due to *Cryptosporidium* predisposes to weight loss and premature death. Long-term effects on physical fitness and cognitive development have also been observed in children who were initially infected during their first year of life.

Clinical illness due to *Cyclospora* is characterized by persistent bloating, flatulence, abdominal cramps, constipation, and fatigue, making it indistinguishable from infections due to other protozoal diarrheal agents including *Giardia, Cryptosporidium,* and other organisms reviewed in this chapter. Fever is present in 25% of cases. Asymptomatic infection also occurs in residents of developing countries, especially adults. This suggests that repeated exposure to *Cyclospora* can lead to partial immunity.

## DIAGNOSTIC APPROACH

The traditional microscopic ova and parasite (O&P) exam utilizing a wet mount or concentrated, preserved specimen of stool remains the classic method for detecting intestinal protozoa in many developing and developed countries. However, because some protozoa (*Giardia* in particular) are shed only intermittently, three separate specimens of stool are usually needed to diagnose 90% of infections. In addition, O&P exams are labor intensive, time-consuming, and require significant skill to interpret. To overcome these challenges, fecal antigen tests using enzyme-linked immunosorbent assays (ELISA) are now available as another method of diagnosing *Giardia* and *Cryptosporidium*, either singly or in combination. Excellent direct fluorescent diagnostics are also available to detect *Giardia* and *Cryptosporidium* in stool; the downside is the need for a fluorescent microscope. Molecular assays based on DNA detection represent a third-generation approach which offers even greater specificity and sensitivity along with ease of use and rapid turnaround, especially when polymerase chain reaction (PCR) technology is linked to automated DNA extraction. Such PCR-based tests are now combined in multiplex platforms designed to screen a single specimen of stool for 20 or more bacterial, viral, and protozoan pathogens including *G. duodenalis*, *C. parvum*, and *C. cayatenensis*. Recent research has shown that nested multiplex PCR can detect the same intestinal protozoa on fresh produce. Despite their improved sensitivity and specificity, however, the cost of nucleic acid–based tests remains a significant deterrent to their use in low-income settings.

Using the classic microscopic approach, *Giardia* infections are diagnosed by finding cysts or trophozoites in feces. Their characteristic "cartoon face" both in unstained and stained fecal specimens makes identification relatively easy. Although less commonly used today than in previous decades, duodenal aspirates obtained by a duodenal capsule or string test can also detect organisms. In cases where giardiasis is strongly suspected, a single fecal antigen test is often the easiest and most cost-effective method of diagnosis.

Unlike *Giardia*, PCR is now increasingly used to diagnose *Cryptosporidium* because the organism is small and does not stain particularly well. Commercially available immunofluorescent assays can also detect *Cryptosporidium* in fecal or tissue specimens. Nonetheless, modified acid-fast stains are still used by most US labs. Other possible techniques include Giemsa, Kinyoun's acid-fast, and iron-hematoxylin stains as well as flotation in Sheather's sugar/zinc sulfate solution. Prior to the 1980s, human cryptosporidiosis was sometimes diagnosed by small intestinal biopsies when the small, spherical life-cycle stages were found in microvillous regions of the intestinal mucosa. Because infection can be patchy, however, biopsies do not invariably detect *Cryptosporidium*.

When viewed in human stool, *Cyclospora* oocysts are spherical, smooth, thin-walled, and refractile and can be stained using Ziehl-Neelsen modified acid-fast, safranin, auramine, rhodamine, Kinyoun, Giemsa, and trichrome stains, among others. The modified Ziehl-Neelsen method, which renders oocysts pink to red, yields the highest sensitivity. Because of their similar appearance when acid-fast stained, the larger size of *Cyclospora* (which is roughly twice the diameter of *Cryptosporidium*) is one way to distinguish the two intestinal protozoa.

## CLINICAL MANAGEMENT AND DRUG TREATMENT

There is no standardized method to test *Giardia*'s susceptibility to antimicrobial agents, but recent clinical outcomes suggest a slow increase in drug resistance. In the United States, the current drug of choice for giardiasis is tinidazole, a nitroimidazole compound in the same family as metronidazole. Tinidazole's chief advantage over metronidazole is its longer half-life, which allows single-dose treatment both in children and adults. Alternate regimens include metronidazole in divided doses over 5 to 7 days or a 3-day course of nitazoxanide, which is FDA-approved for use in patients 1 year or older and eradicates infection in roughly 80% of *Giardia*-infected patients. Quinacrine, an older drug which is no longer manufactured in the United States, can still be obtained from compounding pharmacies. A 5- to 7-day course of quinacrine cures 90% of patients, but trade-offs include possible adverse effects ranging from nausea, vomiting, and abdominal cramps to yellow discoloration of skin, exfoliative dermatitis, and, rarely, psychosis. In pregnant women, the non-absorbable aminoglycoside paromomycin is safe to use and efficacious in roughly 60% to 70% of infected individuals.

Following an effective course of treatment, patients with giardiasis usually experience an improvement in their symptoms within 5 to 7 days with clearance of parasites from their stool. When patients recover this quickly, follow-up exams are unnecessary.

When symptoms do not resolve following standard single-drug treatments, practitioners must distinguish persistent carriage of *Giardia* from post-infectious syndromes such as secondary lactose intolerance

or irritable bowel syndrome. Patients with confirmed refractory, possibly drug-resistant, infections are sometimes treated with longer courses of their original drug, a second, different agent (nitazoxanide, for example, has demonstrated efficacy against nitroimidazole-resistant strains), or combination therapies such as metronidazole plus quinacrine. In a study conducted in Norway, metronidazole plus albendazole was effective in roughly 80% of patients who had previously failed metronidazole alone.

*Cryptosporidium* is far more difficult to treat. Despite studies of multiple agents and immunotherapies, no antiparasitic drug is reliably effective in eradicating *Cryptosporidium* from immunocompromised patients. In immunocompetent hosts, oral rehydration is the usual therapy for acute, symptomatic illness; in addition, nitazoxanide (the only FDA-approved treatment for cryptosporidiosis) shortens diarrhea and the duration of parasite excretion. Recent trials have also investigated combination therapies. Azithromycin and rifabutin, for example, may enhance the efficacy of nitazoxanide. In a pilot study of AIDS patients with chronic infection, azithromycin plus paromomycin also decreased parasite shedding. Nonetheless, the most important intervention in patients co-infected with HIV and *Cryptosporidium* is highly active antiretroviral therapy, which restores gastrointestinal host defense as it suppresses HIV replication, resulting in decreased diarrhea.

A randomized trial conducted in 40 adult expatriates in Nepal demonstrated that human infection with *C. cayetanensis* can be treated with a 7-day course of trimethoprim-sulfamethoxazole given as one double-strength (DS) 160 mg/800 mg tablet twice daily. Some patients may require longer courses or higher doses. An alternative treatment in adults who are sulfa-allergic is nitazoxanide 500 mg twice daily for 7 days. Ciprofloxacin has also been used to treat cyclosporiasis, but its efficacy is less reliable.

## OTHER INTESTINAL PROTOZOA

Other intestinal protozoa can produce human intestinal infection, including the following five species (Fig. 78.5).

### *Balantidium coli* (Balantidiasis)

*B. coli* is the largest infectious intestinal protozoan, measuring up 50 to 100 µm in length and 40 to 70 µm in width. Found in many mammals worldwide, its trophozoites have fine, visible cilia and a large macronucleus shaped like a kidney bean. In humans, balantidiasis can cause intermittent diarrhea, abdominal pain, and weight loss as well as rare cases of fulminant colitis and intestinal perforation, especially in malnourished or immunocompromised individuals. Most *B. coli* infections occur after individuals ingest animal-shed cysts (which can survive for weeks in moist conditions), particularly in unhygienic settings where pigs and humans are in close contact. Balantidiasis is typically diagnosed by finding large trophozoites in fresh or preserved stool. Tetracycline, iodoquinol, or metronidazole are the current treatments of choice.

### *Cystoisospora belli*

*C. belli* (formerly *Isospora belli*) is a worldwide protozoan pathogen exclusive to humans. It is commonly found in tropical or subtropical settings, especially in South America, Africa, and Southeast Asia. Infection follows ingestion of oocysts which become infectious roughly 48 hours after passage; oocysts can also remain viable in the environment, sometimes for months.

After oocysts enter the proximal small intestine, sporozoites reproduce and previously healthy patients experience non-inflammatory diarrhea lasting 2 to 3 weeks. *C. belli* can also produce prolonged,

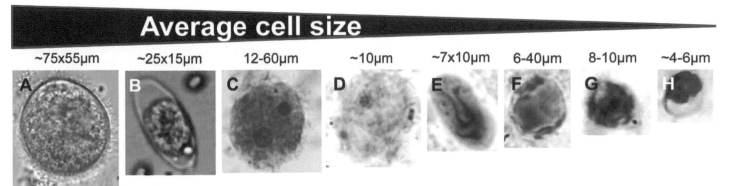

**Fig. 78.5** Key microscopic morphology of the enteropathogenic protozoa. Organisms are ordered from largest to smallest, based on average cell size. (A) *Balantidium coli* trophozoite unstained, wet mount. (B) *Cystoisospora belli* oocyst. (C, D, and F) Trophozoite forms are shown stained with trichrome for *Entamoeba histolytica* (C), *Dientamoeba fragilis* (D), and *Blastocystis hominis* (F). (E) Cyst form of *Giardia* stained with trichrome. (G and H) *Cyclospora cayetanensis* oocyst (G) and *Cryptosporidium spp.* oocyst (H) after modified acid-fast staining. (Reused with permission from McHardy IH, Wu M, Shimizu-Cohen R, et al: Detection of intestinal protozoa in the clinical laboratory. *J Clin Micro* 52:712-720, 2014, Figure 1; https://www.ncbi.nlm. nih.gov/pmc/articles/PMC3957779/table/T2/.)

debilitating diarrhea in immunosuppressed patients, including people with AIDS. In the latter case, diarrhea can persist for months or years, sometimes accompanied by peripheral eosinophilia. Hemorrhagic colitis, biliary tract involvement, and reactive arthritis due to *C. belli* have all been reported. Cystoisosporidiosis is usually diagnosed by finding oocysts in wet mounts of stool or characteristic pink-staining organisms with a single sporoblast in fecal smears treated with acid-fast stains. Villous atrophy, crypt hyperplasia, and inflammatory cells may be present in small bowel biopsies of infected persons. A 10-day course of trimethoprim-sulfamethoxazole DS tablets is the current treatment of choice for *C. belli* infection, although long-term suppressive therapy is often needed to prevent symptomatic relapse in HIV-infected patients. *C. belli* can also be treated with ciprofloxacin or nitazoxanide, among other agents.

## *Sarcocystis* Species (Sarcosporidiosis)

*Sarcocystis* species are zoonotic protozoa that infect more than 130 species of wild and domestic animals. In humans, intestinal sarcosporidiosis usually follows ingestion of raw or undercooked pork or beef containing tissue cysts of *S. hominis* or *S. suihominis* (freezing or heating to 60°F or higher renders meat non-infective). Other *Sarcocystis* species, once ingested, invade human muscles. A third route of transmission involves eating or drinking food or water contaminated with fecal sporocysts shed by animals or humans. Diagnosis of individual patients is difficult and there is no proven treatment, although corticosteroids provide symptomatic relief in acute myositis.

In areas endemic for sarcosporidiosis, particularly Southeast Asia and Malaysia, up to 20% of residents are seropositive. Outbreaks among travelers have also occurred. In 2011–2012, more than 100 travelers visiting Tioman island, Malaysia suffered muscular sarcosporidiosis. In this outbreak, myalgias were the clinical hallmark of infection, although fever and eosinophilia were also observed in some affected patients. A definitive source of infection was never identified.

## *Dientamoeba fragilis*

*D. fragilis* is an ameba-like flagellate typically found in preserved stool or permanently stained fecal smears; a cyst stage has also been reported. Molecular assays of stool can detect the organism but are not currently available in most clinical labs. *Dientamoeba's* principal transmission most likely involves fecal-oral transfer between humans

although animals such as pigs, rats, ruminants, and certain primates may serve as additional sources of human infection. Forty years ago, an intriguing hypothesis—that ova of *Enterobius vermicularis* (human pinworm) and *Ascaris lumbricoides* (human roundworm) could transmit *D. fragilis*—was proposed. At present, the evidence in support of this theory remains scant.

Because *D. fragilis* does not invade tissue, infected individuals' symptoms ranging from diarrhea, abdominal pain, flatulence, nausea, and vomiting to urticaria, biliary infections, pruritus, and irritable bowel most likely stem from irritation of large bowel mucosa. *D. fragilis* infections can also trigger mild eosinophilia. Treatments include paromomycin, iodoquinol, and tetracycline.

## *Blastocystis hominis*

For decades, there has been an ongoing debate about the pathogenicity of *B. hominis*. A common inhabitant of the human intestinal tract, *Blastocystis* was once considered a harmless yeast but is now classified as an anaerobic protozoan of a super-group including diatoms, water molds, and brown algae. Genetic studies indicate 17 subtypes, nine of which can infect humans. In some humans, symptoms accompanying *B. hominis* infection include mild diarrhea, nausea, anorexia, and fatigue. Immune compromise, travel to and immigration from developing countries, and exposure to contaminated food and water increase the risk of symptomatic disease. Small studies in which infected patients were treated for *Blastocystis* with trimethoprim-sulfamethoxazole, metronidazole, and nitazoxanide have demonstrated clinical improvement coincident with reduction or elimination of the agent; on the other hand, colonization with *Blastocystis* has also been associated with increased diversity of human gut microflora, a potentially beneficial state.

Today, the routine diagnosis of *Blastocystis* still relies on microscopy demonstrating parasites in wet mounts of stool or permanent stained fecal smears.

## PREVENTION AND CONTROL

Practical advice to limit infections due to *Giardia, Cryptosporidium,* and other enteric protozoa includes washing hands before preparing food and after using the lavatory; not sharing towels; not swallowing water while swimming in a community pool or water park; and maintaining excellent hygiene in custodial institutions, day care, play-groups, and other settings housing incontinent children or adults. In addition, when hiking or camping or traveling to countries with poor sanitation, it is recommended that drinking water be boiled in order to kill protozoal cysts. Adding chlorine and iodine disinfectants to water will also kill *Giardia* cysts depending on water temperature and clarity (contact time should be extended when treating cold water and halogen concentration increased when treating turbid water). *Cryptosporidium* oocysts are less susceptible to halogens than *Giardia*. To assure that public water is safe, coagulation, flocculation, sedimentation, and filtration followed by chlorination are also recommended. In recent years, both in the United States and Britain, better screening and treatment have led to a fall in *Cryptosporidium* outbreaks directly linked to public water.

Feces, soil, irrigation water, sewage, and dirty human hands can also contaminate food with intestinal protozoa. Reducing foodborne infections is a multi-stage process. Key steps in the farm-to-fork chain include working with suppliers who follow good agricultural practices and have appropriate facilities for washing and storing raw fruits and vegetables. Commercial producers should follow Hazard Analysis Critical Control Point (HACCP) systems designed to pinpoint and mitigate microbial hazards. Untreated animal or human manure should not be used to fertilize fruits and vegetables, nor should consumers add animal waste to household compost. Consumers should properly clean and wash fresh produce and fruit and refrain from drinking unpasteurized fruit juice and ciders, while persons working in commercial settings where food is prepared and served should be educated on their risk of infecting others. Food workers should also remain at home for at least 48 hours following their last diarrheal episode.

## EVIDENCE

Cama VA, Mathison BA. Infections by intestinal coccidia and *Giardia duodenalis*. *Clin Lab Med* 2015;35:423-444. *Excellent overview of Cryptosporidium spp, Cyclospora cayetanensis, Giardia duodenalis, and Cystisospora belli.*

Guerrant DI, Moore SR, Lima AA et al. Association of early childhood diarrhea and cryptosporidiosis with impaired physical fitness and cognitive function four-seven years later in a poor urban community in northeast Brazil. *Am J Trop Med Hyg* 1999;61:707-713. *A seminal paper showing that cryptosporidiosis in young children in a low-income setting affected their long-term physical and cognitive development.*

Hoge CW, Shlim DR, Ghimire M, et al. Placebo controlled trial co-trimoxazole for *Cyclospora* infections among travellers and foreign residents in Nepal. *Lancet* 1995;345:691-693. *The first study to show that a 7-day course of trimethoprim-sulfamethoxazole eradicates Cyclospora and improves clinical symptoms.*

Lalle M, Hanevik K. Treatment-refractory giardiasis: challenges and solutions. *Infection and Drug Resistance* 2018;11:1921-1933. *The authors of this paper propose a working definition of clinically drug-resistant giardiasis and also discuss its current management and future therapeutics.*

Mac Kenzie WR, Hoxie NJ, Proctor ME, et al. A massive outbreak in Milwaukee of *Cryptosporidium* infection transmitted through the public water supply. *New Engl J Med* 1994;331:161-167. *A detailed report of the largest outbreak to date of waterborne cryptosporidiosis in the United States.*

Magali C, Lokmer A, Segurel L. Gut protozoa: Friends or foes of the human gut microbiota? *Trends in Parasitology* 2017;33:925-934. *A provocative paper that explores whether fewer enteric parasites may contribute to decreased microbial diversity in the gastrointestinal tracts of residents of industrialized versus nonindustrialized countries.*

Ryan U, Paparini A, Oskam C. New technologies for detection of enteric parasites. *Trends in Parasitology* 2017;33:532-546. *Detailed discussion of commercially available PCR-based panels and emerging technologies that could lower cost and enhance detection of enteric protozoa in low-income countries.*

Shapiro K, Kim M, Rajal VB et al. Simultaneous detection of four protozoan parasites on leafy greens using a novel multiplex PCR assay. *Food Microbiology* 2019;84:103-252. *This pilot study supports the use of multiplex PCR as a relatively rapid, simple, and inexpensive technology for screening fresh produce for Cryptosporidium, Giardia, Cyclospora, and Toxoplasma gondii.*

Walzer PD, Wolfe MS, Schultz ME. Giardiasis in travelers. *J Infect Dis* 1971; 124:235-237. *A classic paper describing the clinical features of giardiasis in travelers recently returned from the former USSR compared with State Department cases with unknown duration of infection.*

## ADDITIONAL RESOURCES

Ash LR, Orihel C. *Human Parasitic Diseases—A Diagnostic Atlas*, 2020, ASCP Press. *The latest edition of a superb atlas containing more than 1500 images of parasites found in feces, blood, lab preps, and tissue sections.*

Fletcher SM, Stark D, Harkness J, Ellis J. Enteric protozoa in the developed world: A public health perspective. *Clin Micro Rev* 2012;25:420-449. *This richly referenced article outlines the many ways intestinal protozoa enter farm-to-fork chains, the complex interactions between humans and animal hosts, the emerging effects of climate change and urbanization, and modern measures that can aid in outbreak control.*

Giangaspero A, Gasser RB. Human cyclosporiasis. *Lancet Infect Dis* 2019; 19:e226-236. *A state-of-the-art review of the history, biology, epidemiology, and clinical features of Cyclospora cayetanensis.*

Nash TE, Bartelt LA. *Giardia lamblia*. In Bennett JE, Dolin R, Blaser MJ, eds: *Principles and Practice of Infectious Diseases*, vol 2. Philadelphia, 2020, Elsevier, pp 3388-3395. *Co-authored by a senior investigator whose research on Giardia spans 40 years, this scholarly chapter is a trove of scientific and clinical information.*

White AC. Cryptosporidiosis (*Cryptosporidium* Species). In Bennett JE, Dolin R, Blaser MJ, eds: *Principles and Practice of Infectious Diseases*, vol 2. Philadelphia, 2020, Elsevier, pp 3410-3420. *An authoritative, heavily referenced chapter covering every aspect of cryptosporidiosis.*

# 79

# Soil-Transmitted Helminths and Other Intestinal Roundworms

*Elaine C. Jong*

## ABSTRACT

Infections with soil-transmitted helminths (STHs) affect the health of around 1.5 billion people around the world. Individuals of all ages may be infected with the common roundworm (*Ascaris lumbricoides*), whipworm (*Trichuris trichiura*), hookworm (*Ancylostoma duodenale* and *Necator americanus*), and *Strongyloides stercoralis*, although school-aged children living in resource-poor endemic areas are more likely to be infected with heavy worm burdens that contribute to significant malnutrition, delayed physical growth, cognitive impairment, serious illness, and even death. Light worm infections are usually asymptomatic; however, when STHs are diagnosed among returning travelers and immigrants from endemic areas, there is usually a strong personal desire to be free from worms whether symptomatic or asymptomatic.

Chronic infections with hookworm and whipworm can be associated with the development of iron-deficiency anemia owing to daily blood loss in the stools. *Strongyloides* can cause chronic infections in humans that persist for decades because of an alternate parasite autoinfective cycle that can bypass obligatory developmental stages in the soil, and such infections may be associated with skin rashes and hypereosinophilia, as well as fatal hyperinfection syndrome in immune-compromised hosts.

Another nematode (roundworm) included in this chapter is pinworm (*Enterobius vermicularis*). Pinworm infections are a ubiquitous scourge among children and the households that they live in, usually causing perianal itching but occasionally associated with more serious pathology such as appendicitis.

There are many geographic areas where a high risk of STH transmission overlaps with high rates of HIV infections and acquired immunodeficiency syndrome (AIDS) among resident populations. Some studies have postulated that helminthic infections in persons co-infected with HIV may adversely affect HIV-1 progression, as measured by changes in CD4 count, viral load (measured by HIV-1 ribonucleic acid [RNA]), and/or clinical disease progression. Diagnosis of latent worm infections and appropriate treatment of HIV-1 co-infected persons and others with immunocompromised status are strongly recommended for those who live or have lived in high-risk geographic areas for STH transmission.

## CLINICAL VIGNETTE

A 3-year-old boy was brought to clinic by his parents, who complained that during the past week, he had developed a peculiar skin rash that seemed to enlarge each day. He was last seen in clinic 9 weeks ago for a general physical exam upon their return from a year-long missionary assignment in Nigeria. At his previous visit, he was found to be in the 70th percentile for height and weight for his age, and had a normal physical exam. Screening blood tests and urinalysis were normal. The complete blood count with white cell differential showed no elevation of the eosinophil count. Stool ova and parasite (O&P) exam showed ova of *Ascaris lumbricoides*, hookworm, and *Blastocystis hominis*. As the boy had normal stools and no abdominal symptoms, he was prescribed Albendazole, 400 mg once as a single oral dose, as treatment against the *Ascaris* and hookworm; no treatment was prescribed for the *Blastocystis*. A posttreatment stool specimen submitted 1 month after treatment was negative for O&P. At this visit, the physical exam was normal except for the finding of a serpiginous erythematous track-like rash on the skin of his left buttock, starting at the anus and extending about 7 cm. The parents reported that he frequently scratched that area. A blood sample was taken to submit for *Strongyloides* serological testing. Ivermectin 200 mcg/kg per day po × 2 days was prescribed for the presumed diagnosis of *Strongyloides stercoralis*. The parents reported that 1 week after treatment, the itching was completely gone and the rash was fading away. Three weeks following the clinic visit, the *Strongyloides* serology test was reported back as positive. A repeat *Strongyloides* serology test will be done 6 months after treatment for follow-up.

COMMENT: Children with only light worm burdens of STHs often show no signs or symptoms of morbidity from their infection. The absence of finding *S. stercoralis* larvae in the submitted stool specimens is common in light infections because of irregular and scanty output; peripheral blood eosinophilia may or may not be present. While the single dose of albendazole initially prescribed was efficacious against *Ascaris* and hookworm, a longer course of albendazole therapy at a higher daily dose (Table 79.3) would be recommended for intestinal *S. stercoralis* infections; the drug ivermectin is considered primary therapy for the treatment of intestinal strongyloidiasis.

## GEOGRAPHIC DISTRIBUTION

*Ascaris* is probably the most common helminthic infection, with a global prevalence of approximately 1.3 billion persons infected. The majority (more than 70%) of *Ascaris* infections occur in China, India, and Southeast Asia, followed by countries in Latin America and the Caribbean region (approximately 13%), and in sub-Saharan Africa (approximately 8%). It is estimated that whipworm and hookworm are each responsible for 500 to 900 million infections worldwide. Whipworm has a similar geographic distribution as *Ascaris*, whereas hookworm is highly prevalent in sub-Saharan Africa and South Asia. Transmission of STHs also occurs in developed countries and has been reported in persons living or working in resource-poor rural farming communities in the southern United States and southern Europe. The transmission of *Strongyloides* and pinworm occurs in urban as well as rural locales, because there is an auxiliary autoinfection cycle in addition to the life-cycle development in soil.

## RISK FACTORS

STHs are transmitted in human populations in tropical and temperate climates where poverty and poor sanitation result in fecal contamination of the environment. Parasite eggs of *Ascaris*, whipworm, and hookworm have an obligatory developmental period of several weeks in the soil before the larvae contained in the eggs become mature and

## TABLE 79.1   Summary of Parasite Life Cycles

| Parasite | Transmission | Incubation | Adult Habitat | Lifespan | Clinical Features |
|---|---|---|---|---|---|
| *Ascaris lumbricoides* (common roundworm) | Ingestion of eggs | 2–3 months | Small intestine | 1–2 years | Pulmonary larval migration (cough and eosinophilia) <br> Intestinal discomfort <br> Obstruction of a viscus, or intestinal perforation <br> Ova in stools <br> Spontaneous passage of adult worms per rectum, mouth, or nose |
| *Trichuris trichiuris* (whipworm) | Ingestion of eggs | 1–3 months | Large intestine in the cecum; gravid females migrate to the rectum | 3–8 years | Diarrhea, cramps <br> Blood in stools <br> Anemia <br> Tenesmus, rectal prolapse <br> Ova and occasional adults in stools |
| *Ancylostoma duodenale, Necator americanus* (hookworm) | Skin penetration by infective larvae after contact with contaminated soil | 2 or more weeks | Small intestine in the duodenum and upper jejunum | 1 year | Skin rash at site of infection ("ground itch") <br> Pulmonary larval migration (cough and eosinophilia) <br> Diarrhea, abdominal discomfort <br> Anemia <br> Hypoproteinemia <br> Occult blood and ova in stools |
| *Strongyloides stercoralis* | Skin penetration by infective larvae after contact with contaminated soil; autoinfection; skin-to-skin contact | 3 weeks | Small intestine | May persist up to 35 years through autoinfections | Skin rash at the site of infection <br> Pulmonary larval migration (cough and eosinophilia) <br> Diarrhea, abdominal discomfort <br> Persistent eosinophilia <br> Larvae in stools <br> Autoinfective cycle <br> Hyperinfection syndrome |
| *Enterobius vermicularis* (pinworm) | Ingestion of eggs | 2–4 weeks | Large intestine in the cecum | Gravid females, 3–6 weeks; males, 1–2 weeks | Anal and/or vulvar pruritus <br> Rare cause of appendicitis <br> Self-infection from fecal-oral contamination |

infective for humans. Humans usually acquire worm infections by fecal-oral transmission from contaminated fingers and food (*Ascaris*, whipworm, pinworm) or by direct skin contact with fecally contaminated soil (hookworm, *Strongyloides*). In addition, direct person-to-person transmission of *Strongyloides* and pinworm is possible among those having close personal contact with infected persons, and *Strongyloides* autoinfections are also possible.

## CLINICAL FEATURES

Clinical signs and symptoms reflect the life-cycle stages of each parasite within the human host (Table 79.1). Larval penetration of intact skin often elicits a pruritic skin rash (hookworm, *Strongyloides*). When immature larval parasite forms are migrating through the lungs and other host tissues during natural life-cycle stages, the elevation of peripheral blood eosinophils may occur. During larval migration of *Ascaris*, hookworm, and *Strongyloides* through the lungs as a part of their life cycle in the human host, a cough may develop and transient infiltrates may be seen on chest radiographs. During *Strongyloides* hyperinfection, larvae may be found in specimens of the blood-tinged sputum. Persons with light STH infections may have few specific signs or symptoms, and many are undiagnosed. Because worm infections do not elicit a protective immune response, persons (especially children) residing in areas of transmission experience repeated infections over

time and can acquire heavy worm burdens, eventually leading to serious manifestations of chronic infection.

## Common Roundworm: *Ascaris lumbricoides*

Infected persons may be asymptomatic or complain of vague abdominal symptoms. *Ascaris* worms become hyperactive when irritated by fever, starvation, or medications in the human host: a worm may ascend from normal residence in the lumen of the small intestine through the stomach and esophagus, exiting through the mouth or nose, or a worm may pass without symptoms per rectum, shocking the host who finds a spontaneously expelled gross specimen. Infection with only a single *Ascaris* worm can cause morbidity, owing to their relatively large size: a worm may migrate to ectopic locations such as the appendix or common bile duct, causing obstruction and inflammation. *Ascaris* is capable of perforating the intestines, resulting in fecal spillage and the development of peritonitis. In heavily infected children, small bowel obstruction may result from a bolus of worms and may necessitate emergency laparotomy. Taking all these possible scenarios into account, *Ascaris* infections should be treated when detected (Fig. 79.1).

## Whipworm: *Trichuris trichiura*

Whipworm is a parasitic infection with worldwide distribution, and although persons of any age may be infected, children account for the

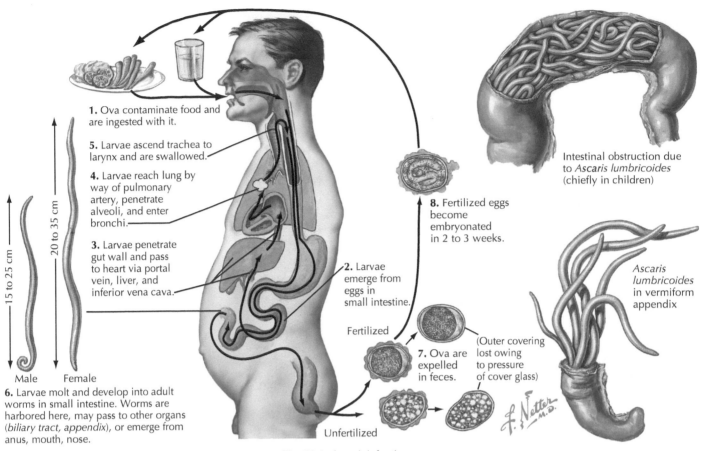

**1.** Ova contaminate food and are ingested with it.

**5.** Larvae ascend trachea to larynx and are swallowed.

**4.** Larvae reach lung by way of pulmonary artery, penetrate alveoli, and enter bronchi.

**3.** Larvae penetrate gut wall and pass to heart via portal vein, liver, and inferior vena cava.

**2.** Larvae emerge from eggs in small intestine.

Fertilized

Unfertilized

15 to 25 cm

20 to 35 cm

Male    Female

**6.** Larvae molt and develop into adult worms in small intestine. Worms are harbored here, may pass to other organs (*biliary tract, appendix*), or emerge from anus, mouth, nose.

**8.** Fertilized eggs become embryonated in 2 to 3 weeks.

**7.** Ova are expelled in feces.

(Outer covering lost owing to pressure of cover glass)

Intestinal obstruction due to *Ascaris lumbricoides* (chiefly in children)

*Ascaris lumbricoides* in vermiform appendix

**Fig. 79.1** *Ascaris* infection.

majority of reported cases. Whipworm infections are chronic and relatively silent, but moderate to severe infections (from around 200 to 1000 adult worms or more) are associated with iron-deficiency anemia, growth retardation, and chronic bloody mucoid diarrhea. The adult worms inhabit the human colon, from the cecum to the rectum, with the mouthpart of each worm firmly embedded in the bowel epithelium and the thicker posterior bodies of the worms moving freely in the bowel lumen. In heavy whipworm infections, rectal prolapse is thought to be associated with both physical factors and inflammatory changes caused by infection in the bowel wall: peristaltic contractions of the bowel push the worm bodies in the rectum toward the anus while the anterior ends remain firmly attached to the chronically inflamed bowel wall, and rectal prolapse may occur (Fig. 79.2).

## Hookworm: *Ancylostoma duodenale* and *Necator americanus*

*A. duodenale* and *N. americanus* were commonly known as "Old World" and "New World" hookworm, respectively. However, after recognition that infections with both species are transmitted in both the Eastern and Western Hemispheres, the geographic designations have decreased in usage. Infections with *A. duodenale* are potentially more harmful than infections with *N. americanus*; *A. duodenale* worms attach to the intestinal mucosa and suck blood at a rate of 0.15 to 0.26 mL/day per worm compared with *N. americanus* with a rate of 0.03 mL/day per worm. Thus blood loss is greater with *A. duodenale* for a comparable level of infection. Additional blood loss occurs from the multiple points of attachment and detachment of the adult hookworms in the duodenum and jejunum. When the worms bite into the mucosa to attach and feed, an anticoagulant is released into the local tissue,

and bleeding from these sites into the lumen of the small intestine persists after the worms detach and move on to fresh areas of mucosa. Although the two species can be distinguished by the morphology of the mouthparts and the copulatory bursae of the adult worms, the eggs of the two are indistinguishable. Once the diagnosis of hookworm is made, drug treatment is the same regardless of species (Fig. 79.3).

## Strongyloidiasis: *Strongyloides stercoralis*

In the soil-transmission cycle of *Strongyloides*, adult female worms developing in the submucosa of the small intestine lay eggs that mature within a few hours to produce rhabditiform larvae that enter the fecal stream in the lumen of the bowel. *Strongyloides* rhabditiform larvae exiting the body in feces that are deposited in moist soil develop into infective stage filariform larvae (through asexual or sexual free-living cycles). Filariform larvae are capable of penetrating intact human skin, and new infections occur when skin comes into direct contact with the contaminated soil.

Alternatively, the immature rhabditiform larvae in the fecal stream may rapidly develop into infective filariform larvae while still in the intestines. The filariform larvae in the fecal stream may then penetrate either the intestinal mucosa or the perianal skin and migrate to blood vessels, completing the life cycle without leaving the human host through this process of autoinfection. The pruritic, serpiginous erythematous skin rash on the buttocks elicited by this autoinfection is called *cutaneous larva currens*, because the track-like rash caused by the migrating subcutaneous larvae can extend at a rate of 5 to 10 cm an hour. Strongyloidiasis is a sexually transmitted infection when intimate skin-to-skin contact occurs while infective filariform larvae are present in the rectum and on the perianal skin.

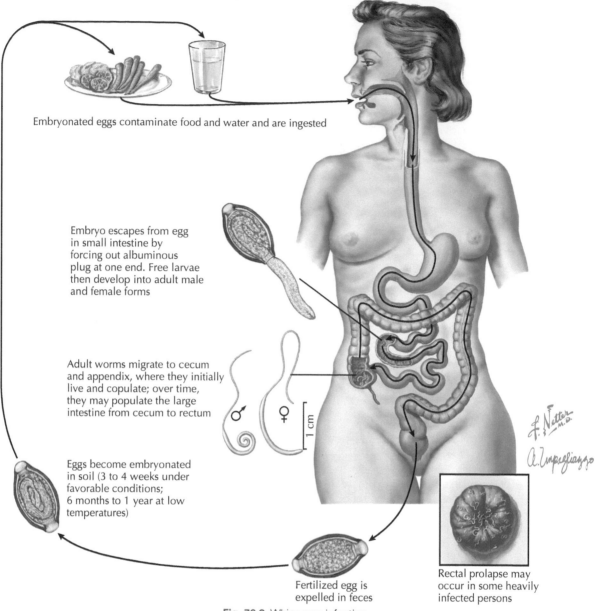

Embryonated eggs contaminate food and water and are ingested

Embryo escapes from egg in small intestine by forcing out albuminous plug at one end. Free larvae then develop into adult male and female forms

Adult worms migrate to cecum and appendix, where they initially live and copulate; over time, they may populate the large intestine from cecum to rectum

Eggs become embryonated in soil (3 to 4 weeks under favorable conditions; 6 months to 1 year at low temperatures)

Fertilized egg is expelled in feces

Rectal prolapse may occur in some heavily infected persons

**Fig. 79.2** Whipworm infection.

Chronic *Strongyloides* infections are often silent, although some patients complain of transient skin rashes and itching associated with the autoinfective cycle. Elevation of the peripheral blood eosinophils may be noted as an incidental finding on routine laboratory studies and may trigger a clinical investigation for occult parasite infection. Serious disease results if the infected host becomes immunocompromised; then, *Strongyloides* hyperinfection with dissemination of the parasites to all internal organs precipitates local inflammatory changes and severe enteritis, pneumonitis, and microabscesses, as well as other life-threatening secondary complications (Fig. 79.4).

### Pinworm: *Enterobius vermicularis*

Perianal itching in children is the hallmark of pinworm infections. However, there are rare reports of appendicitis, peritonitis, and salpingitis in which ectopic pinworms or ova were associated with inflammatory reactions in the tissue. Usually, adult pinworms inhabit the cecum, and gravid females migrate to the rectal area at night to deposit eggs on the perianal skin. The embryonated eggs mature after

4 to 6 hours of oxygenation outside the intestine. Fingers and fingernails touching or scratching the perianal area are easily contaminated and may reinfect the original host when the contaminated fingers or objects touched by the fingers are put in the mouth. Contamination of the household environment (e.g., blankets, sheets, clothing, dust) results from eggs shed from the skin, and infections are easily spread to other persons as a consequence of close household or personal contact (Fig. 79.5).

### DIAGNOSTIC APPROACH

Definitive diagnosis of helminthic infection depends on morphologic identification of the characteristic eggs (ova), larvae, and/or even adult forms in fecal samples, tissue biopsy specimens, or sputum. Identifying unique parasite ova and larval forms in submitted stool specimens by microscopic examination is the most common way of making the diagnosis. However, microscopic diagnosis and estimation of the worm burden by quantitative egg counts in the stool are

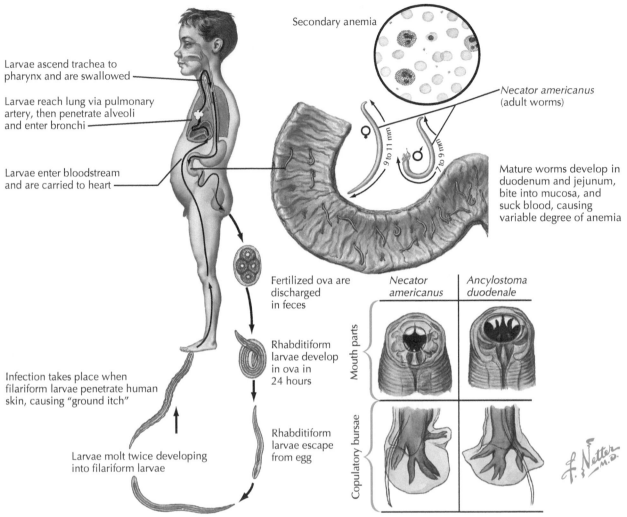

Secondary anemia

Larvae ascend trachea to pharynx and are swallowed

Larvae reach lung via pulmonary artery, then penetrate alveoli and enter bronchi

Larvae enter bloodstream and are carried to heart

*Necator americanus* (adult worms)

Mature worms develop in duodenum and jejunum, bite into mucosa, and suck blood, causing variable degree of anemia

Fertilized ova are discharged in feces

Rhabditiform larvae develop in ova in 24 hours

Infection takes place when filariform larvae penetrate human skin, causing "ground itch"

Rhabditiform larvae escape from egg

Larvae molt twice developing into filariform larvae

| *Necator americanus* | *Ancylostoma duodenale* |
|---|---|
| Mouth parts | |
| Copulatory bursae | |

**Fig. 79.3** Hookworm infection.

time-consuming, labor-intensive, and challenging, because parasite eggs may not be shed uniformly into the fecal stream on a daily basis and may be unevenly dispersed within a given stool specimen. When resources allow, the examination of three stool specimens from a given individual, each collected on a different day, yields a more comprehensive profile of potential parasite infections and allows a more accurate estimation of the parasite burden compared with an examination of a single stool specimen. There has been a lot of interest in using molecular diagnostics for STH diagnosis, particularly the quantitative polymerase chain reaction (qPCR). Studies comparing microscopy to qPCR show increased sensitivity and specificity of qPCR, but standardization of the tests is not yet completed and there are no US Food and Drug Administration (FDA)-approved tests for use in clinical laboratories at the time of writing.

Owing to their relatively large size, diagnosis of *Ascaris* infections also can be made by visual inspection of adult worms that are spontaneously passed through one of the body orifices (per rectum, mouth, or nose), are contained in surgical specimens, or are observed during radiologic imaging studies. Pinworm eggs can be recovered from suspected cases by pressing the sticky side of clear adhesive tape on the perianal skin first thing in the morning. *Strongyloides* eggs are rarely seen in stool specimens, and special laboratory techniques are usually required to visualize the larval forms. Serologic tests for diagnosis of *Strongyloides* are available from state public health and commercial reference laboratories.

## CLINICAL MANAGEMENT AND DRUG TREATMENT

Drug therapy is usually directed by the parasite diagnosis. The therapeutic goal of anthelmintic parasitic drug treatment is to eradicate or significantly lower the worm burden in infected individuals—except for *Strongyloides*-infected individuals, who should be treated until a total cure is achieved. The parasites have varying degrees of susceptibility to the anthelmintic drugs, and some of the drugs have a broad spectrum—a useful property for treating mixed infections. Single-dose drug treatment protocols (Tables 79.2 and 79.3) have been studied because of their utility in mass treatment programs. Published studies conducted in Africa, South America, and Asia have demonstrated that periodic mass drug administration (MDA) programs employing broad-spectrum anthelmintic drugs in school-aged children in endemic areas can result in catch-up and accelerated physical growth, as well as improved cognitive performance measurable in the months following treatment. However, the complex relationship between STH infection, STH morbidity, and the benefits of deworming over time continues to be evaluated. One of the critical issues in field studies has been the lack of standardized diagnostic tests that are less laborious and more sensitive and specific than microscopy for STH detection.

Individual drug regimens may feature a single-dose or a longer duration of treatment with a given drug to ensure optimal cure rates for a given parasite. **Pyrantel pamoate** is widely used and relatively inexpensive in developing countries. The drug is a tetrahydropyrimidine

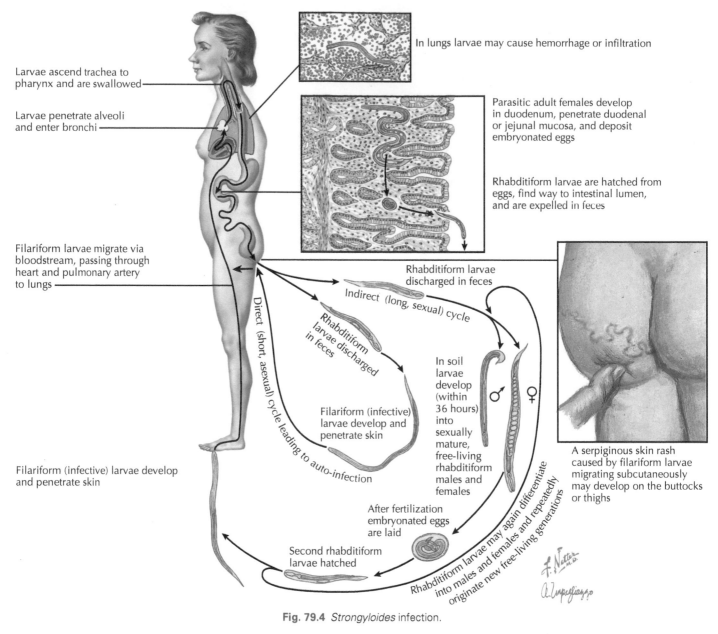

Larvae ascend trachea to pharynx and are swallowed

Larvae penetrate alveoli and enter bronchi

Filariform larvae migrate via bloodstream, passing through heart and pulmonary artery to lungs

Filariform (infective) larvae develop and penetrate skin

In lungs larvae may cause hemorrhage or infiltration

Parasitic adult females develop in duodenum, penetrate duodenal or jejunal mucosa, and deposit embryonated eggs

Rhabditiform larvae are hatched from eggs, find way to intestinal lumen, and are expelled in feces

Direct (short, asexual) cycle leading to auto-infection

Rhabditiform larvae discharged in feces

Indirect (long, sexual) cycle

Rhabditiform larvae discharged in feces

Filariform (infective) larvae develop and penetrate skin

In soil larvae develop (within 36 hours) into sexually mature, free-living rhabditiform males and females

After fertilization embryonated eggs are laid

Second rhabditiform larvae hatched

Rhabditiform larvae may again differentiate into males and females and repeatedly originate new free-living generations

A serpiginous skin rash caused by filariform larvae migrating subcutaneously may develop on the buttocks or thighs

**Fig. 79.4** *Strongyloides* infection.

derivative, which is thought to inhibit neuromuscular transmission in the helminth, causing spastic paralysis of the worm that promotes subsequent expulsion of the worm from the host's intestine. Pyrantel pamoate is poorly absorbed from the gastrointestinal tract, is generally well tolerated with few reported adverse side effects, and is not efficacious in the treatment of *Trichuris* and *Strongyloides*.

The anthelmintic drugs albendazole, mebendazole, and thiabendazole were developed as the result of research on the benzimidazole ring, an integral part of the chemical structure of vitamin $B_{12}$. Anthelmintic benzimidazole drugs are thought to preferentially bind with the cytoskeletal protein tubulin in parasite cells, impairing microtubule formation, and also appear to interfere with parasitic glucose uptake. The benzimidazole drugs are not efficiently absorbed from the gastrointestinal tract, although the amounts absorbed during oral treatment appear sufficient to affect some tissue-phase parasites.

**Thiabendazole** was discovered in 1961 and was the first anthelmintic benzimidazole drug introduced into clinical medicine. Although highly effective against several helminths, its usage has been limited by predictable unpleasant side effects (including anorexia, nausea,

vomiting, vertigo, and headache) and toxicity, notably erythema multiforme. Thiabendazole remains the drug of choice for the treatment of serious *Strongyloides* and *Trichinella* infections (Chapter 86).

**Mebendazole** became widely used in clinical medicine in the 1970s and is a highly efficacious drug against several intestinal parasite infections. In the United States, the drug is indicated for the treatment of *Ascaris*, whipworm, hookworm, and pinworm infections. Mebendazole has few adverse side effects (infrequently reported mild nausea, vomiting, abdominal discomfort) when used in the low-dose, short-term treatment schedules recommended for intestinal nematode infections.

**Albendazole** was introduced into clinical medicine in 1979, although it was not licensed in the United States until the mid-1990s. Albendazole's broad spectrum of activity and low profile of adverse reactions make it invaluable for the treatment of individuals, as well as a favored drug in mass treatment programs.

**Ivermectin**, a semisynthetic anthelmintic drug derived from the avermectins, antiparasitic agents isolated from the fermentation products of *Streptomyces avermitilis*, is considered the preferred treatment for uncomplicated (intestinal) *S. stercoralis* infections, given as a single

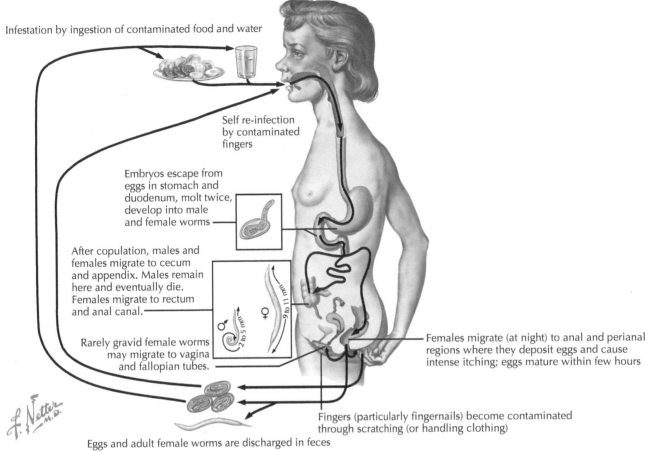

Infestation by ingestion of contaminated food and water

Self re-infection by contaminated fingers

Embryos escape from eggs in stomach and duodenum, molt twice, develop into male and female worms

After copulation, males and females migrate to cecum and appendix. Males remain here and eventually die. Females migrate to rectum and anal canal.

Rarely gravid female worms may migrate to vagina and fallopian tubes.

Females migrate (at night) to anal and perianal regions where they deposit eggs and cause intense itching; eggs mature within few hours

Fingers (particularly fingernails) become contaminated through scratching (or handling clothing)

Eggs and adult female worms are discharged in feces

**Fig. 79.5** Pinworm infection.

**TABLE 79.2   Summary of Overall Cure Rate (%) in Studies Reporting the Use of Single-Dose Oral Albendazole, Mebendazole, and Pyrantel Pamoate**

| | TREATMENT REGIMEN | | |
|---|---|---|---|
| Parasite | Albendazole, 400 mg Cure Rate (95% CI) | Mebendazole, 500 mg Cure Rate (95% CI) | Pyrantel Pamoate, 10 g/kg Cure Rate (95% CI) |
| Ascaris lumbricoides | 95.7% (93.2%–97.3%) | 96.2% (92.3%–98.1%) | 92.6% (85.6%–96.3%) |
| Trichuris trichiura | 30.7% (21.0%–42.5%) | 42.1% (25.9%–60.2%) | 20.2% (7.3%–44.7%) |
| Hookworm | 79.5% (71.5%–85.6%) | 32.5% (20.8%–46.9%) | 49.8% (29.5%–70%) |

Data from Moser W, Schindler C, Keiser J: Efficacy of recommended drugs against soil-transmitted helminths: Systematic review and network meta-analysis, *BMJ* 2017;358:j4307.

oral dose of 200 mcg/kg/day for 2 days in patients with body weight >15 kg. High dose albendazole (Table 79.3) is slightly less effective but is considered an alternative therapy. Ivermectin acts by binding selectively to glutamate-gated chloride ion channels present in nerve and muscle cells of the parasite. Subsequent hyperpolarization of these cells owing to chloride ion influx leads to paralysis and death of the parasite. The drug is active against both adults and larvae. The test of cure for intestinal *Strongyloides* infections is the absence of larvae in three or more follow-up stool samples collected over the period beginning 3 or 4 weeks after completion of therapy to 3 months afterward. Studies show that ivermectin (200 mcg/kg given once daily for 1 or 2 days) had a comparable cure rate with fewer reported side effects than thiabendazole (25 mg/kg twice a day for 3 days) against intestinal strongyloidiasis. Treatment with daily ivermectin doses (200 mcg/kg/day po) should be started immediately if the diagnosis of *Strongyloides* hyperinfection syndrome is being considered. The patient should be

treated for a minimum of 2 weeks or longer until parasite clearance is achieved. Non-oral (rectal, subcutaneous) administration of ivermectin and combination therapy with oral albendazole for hyperinfection syndrome may improve treatment efficacy.

## PREVENTION AND CONTROL

Prevention and control of STH require a multipronged approach involving public health measures, including health education and sewage treatment systems, personal hygiene, and drug treatment of infected persons. Improved levels of sanitation, especially the implementation of programs for collection and decontamination of human fecal wastes, are essential in regions with high rates of STH transmission, but such improvements require administrative infrastructure and resources over the course of years. MDA programs targeting school-aged children and other high-risk groups have been shown

| TABLE 79.3 | Drugs and Treatment Regimens for Selected Soil Transmitted Helminths | | | |
|---|---|---|---|---|
| | TREATMENT REGIMENS | | | |
| Parasite | Albendazole (200 mg Tablet; 100 mg/5 mL Oral Suspension) | Mebendazole (100 mg Chewable Tablet) | Pyrantel Pamoate (50 mg/mL Oral Suspension) | Thiabendazole (500 mg Chewable Tablet OR 500 mg/5 mL Oral Suspension) |
| *Ascaris lumbricoides* | 400 mg as a single dose for adults and children over 2 years of age; 200 mg as a single dose in children 1–2 years old | 100 mg twice daily × 3 days for adults and children over 2 years of age | 11 mg/kg in a single oral dose not to exceed a maximum dose of 1 g for adults and children over 2 years of age | Not recommended (drug toxicity concerns) |
| Whipworm (*Trichuris trichiura*) | 400 mg daily × 1 or 2 days | 100 mg twice daily × 3 or 4 days | Not recommended (low cure rates) | Not recommended (drug toxicity concerns) |
| Hookworm (*Ancylostoma duodenale* and *Necator americanus*) | 400 mg daily × 1 or 2 days | 100 mg twice daily × 3 or 4 days | 11 mg/kg in a single oral dose not to exceed a maximum dose of 1 g for adults and children over 2 years of age × 3 days | Not recommended (drug toxicity concerns) |
| Pinworm (*Enterobius vermicularis*) | 400 mg as a single dose for adults and children over 2 years of age; 200 mg as a single dose in children 1–2 years old; repeat the dose after 2 weeks | 100 mg as a single dose for adults and children over 2 years of age; repeat the dose after 2 weeks | 11 mg/kg in a single oral dose not to exceed a maximum dose of 1 g for adults and children over 2 years of age; repeat the dose after 2 weeks | Not recommended (drug toxicity concerns) |
| *Strongyloides stercoralis*— intestinal infection[a] | Ivermectin 200 mcg/kg/day po × 2 days for adults and children >15 kg; ALTERNATIVE: Albendazole 400 mg po bid × 7 days for adults and children >2 years of age | Not recommended (low cure rates) | Not recommended (low cure rates) | 25 mg/kg twice daily (not to exceed 1.5 g twice daily) × 2 or 3 days |
| *Strongyloides stercoralis*— hyperinfection syndrome[a] | Ivermectin 200 mcg/kg/day po, sc, or rectally × 14 days or longer for adults and children >15 kg PLUS Albendazole 400 mg po bid × 14 days or longer for adults and children >2 years of age | Not recommended (low cure rates) | Not recommended (low cure rates) | 25 mg/kg twice daily (not to exceed 1.5 g twice daily) × 10–15 days |

[a]See text for discussion of ivermectin drug therapy for the treatment of intestinal strongyloidiasis and hyperinfection syndrome.

to yield short-term improvements in affected populations, but these also depend on administrative infrastructure and continued availability of affordable, efficacious drugs. Personal prevention measures include wearing shoes and avoiding direct skin contact with moist ground in areas where there is known transmission of hookworm and *Strongyloides*; school-aged children should be taught good personal hygiene practices. Challenges to current control efforts include varying degrees of inherent parasite susceptibility to the commonly used anthelmintic drugs and the possibility of accelerated emergence of drug resistance as a consequence of repeated mass chemotherapy programs in endemic areas.

# EVIDENCE

Bogoch II, Speich B, Lo NC, et al. Clinical evaluation for morbidity associated with soil-transmitted helminth infection in school-age children on Pemba Island, Tanzania. *PLoS Negl Trop Dis* 2019;13:e0007581. https//doi.org/10.137/journal.pntd.0007581. *This report presents findings of a detailed longitudinal clinical evaluation in 434 school-age children to evaluate morbidity associated with soil-transmitted infection and responses to treatment. The study results demonstrate the challenges of measuring morbidity in the context of STH infection and treatment: the effects of infection were mainly subtle in this study.*

Igual-Adell R, Oltra-Alcaraz C, Soler-Company E, et al. Efficacy and safety of ivermectin and thiabendazole in the treatment of strongyloidiasis. *Expert Opin Pharamacother* 2004;5:2615-2619. *A retrospective review of 88 adult cases of chronic strongyloidiasis treated with either thiabendazole or ivermectin, from 1999 to 2002, in Valencia, Spain. Noncure after drug treatment was associated with continued eosinophilia.*

Kirwan P, Asaolu SO, Molloy SF, et al. Patterns of soil-transmitted helminth infection and impact of four-monthly albendazole treatments in preschool children from semi-urban communities in Nigeria: a double-blind placebo-controlled randomised trial. *BMC Infect Dis* 2009;9:20. *This placebo-controlled field study among Nigerian preschool children aged 1 to 4 years found that more than 50% of the preschool children were infected by one or more helminths. A. lumbricoides was the most prevalent infection (47.6%). Results of the study suggest that systematic treatment programs using a broad-spectrum anthelmintic drug are necessary to reduce the prevalence and intensity of STH infection among preschool children in a population characterized by moderate prevalence and low intensity.*

Pabalan N, Singian E, Tabangay L et al. Soil-transmitted helminth infection, loss of education and cognitive impairment in school-aged children: A systematic review and meta-analysis. *PLoS Negl Trop Dis* 2018;12:e0005523. https//doi.org/10.137/journal.pntd.0005523. *A total of 36 studies of 12,920 children were included in this analysis. Evidence of superior performance in five of six educational and cognitive domains was assessed for STH uninfected/dewormed versus STH infected/not-dewormed school-aged children from helminth endemic regions. This synthesis of data provides empirical support for the cognitive and educational benefits of deworming.*

Sasaki J, Seidel JS. Ascariasis mimicking an acute abdomen. *Ann Emerg Med* 1992;21:217-219. *Ascariasis is a common childhood infection worldwide. Whereas most Ascaris infections are benign, the two pediatric cases presented in this report illustrate that such infections are in the differential diagnosis of pediatric acute abdomen. Children at risk include immigrants and those with a history of travel to foreign countries, but cases of ascariasis have been reported in children who have not traveled outside the United States.*

Suputtamongol Y, Premasathian N, Bhumimuang K, et al. Efficacy and safety of single and double doses of ivermectin versus 7-day high dose albendazole for chronic strongyloidiasis. *PLoS Negl Trop Dis* 2011;5(5):e1044. doi:10.1371/journal pntd.0001044. *Comparison of ivermectin versus albendazole for treatment of Strongyloides infections.*

## ADDITIONAL RESOURCES

Boodman C, Chhonker YS, Murry DJ, et al. Case Report: Ivermectin and albendazole plasma concentrations in a patient with disseminated strongyloidiasis on extracorporeal membrane oxygen and continuous renal replacement therapy. *Am J Trop Med Hyg* 2018;99(5):1194-1197. *doi.10.4269/ajtmh.18-0487. Case report of an immunocompromised patient who succumbed to Strongyloides hyperinfection despite high plasma levels of ivermectin and albendazole plus intensive life support measures.*

Khurana S, Sethi S. Laboratory diagnosis of soil transmitted helminthiasis. *Trop Parasitol* 2017;7:86-91. https//doi.org/10.4103/tp.TP_29_17. *Recent review of conventional methods for STH diagnosis, including microscopy, culture, and egg counting compared with more rapid molecular techniques, particularly quantitative polymerase-chain reaction tests up to now, mainly used in research settings. The World Health Organization is considering adopting molecular tests in their STH elimination programs.*

Moser W, Schindler C, Keiser J. Efficacy of recommended drugs against soil transmitted helminths: systematic review and network meta-analysis. *BMJ* 2017;358:j4307. https//doi.org/10.1136/bmj.j4307: 10.1136/bmj.j4307. *This review and meta-analysis considered published studies from 1960 until December 31, 2016, that were at least level 1 randomized controlled trials. Four drugs currently recommended by the WHO for the treatment of STH (albendazole, metronidazole, pyrantel pamoate, and levamisole) were included in this review. Estimates for the years 1995 and 2015 showed significant reductions in efficacy of albendazole against T. trichiura. All drugs were highly efficacious against A. lumbricoides. Drug efficacy against hookworm was significantly higher with albendazole compared with mebendazole and pyrantel pamoate.*

Nutman TB. Human infection with Strongyloides stercoralis and other related Strongyloides species. *Parasitology* 2017;144:263-273. https//doi.org/10.1017/S0031182016000834. *A comprehensive review on human infection with Strongyloides, diagnosis, and treatment for this sometimes elusive infection with potentially serious implications for long-term health.*

# 80

# Intestinal Cestodes (Tapeworms)

*Douglas W. MacPherson*

## ABSTRACT

The relationship between humans and intestinal cestodes (tapeworms) goes well beyond the essential biologic interaction between host and parasite. Over time, tapeworms and tapeworm stories have contributed to human culture, literature, and dietary practices. In the late nineteenth century, part of the popular folklore was that a tapeworm infection could help make one thin, and tapeworm eggs were actually advertised commercially as a weight reduction aid. On the other hand, one of the well-known clinical manifestations of intestinal tapeworm infection is the passing of grossly visible ribbons or tapes (strobilae) of worm segments (proglottids), often alive and wriggling, from the anus of the human host, leading to considerable reaction in the infected person.

Human tapeworm infections usually result from food-borne transmission when raw, smoked, pickled, or undercooked pork, beef, or fish infected with larval stages of tapeworms are ingested. Pork *(Taenia solium)*, beef *(Taenia saginata)*, and fish *(Diphyllobothrium latum)* tapeworms account for the majority of human infections, but human infections with the rodent tapeworms *Hymenolepis nana* and *Hymenolepis diminuta* and the dog tapeworm *Dipylidium caninum* may occur after inadvertent ingestion of contaminated insect intermediate hosts. Invasive systemic diseases in humans caused by tapeworm infections such as cysticercosis (caused by *T. solium*) and cystic and multilocular echinococcosis (caused by *Echinococcus granulosus* and *Echinococcus multilocularis*, respectively) are addressed in separate chapters (see Chapter 45 and Chapter 85).

Intestinal tapeworm infections are rare in the general population of economically advanced countries but cause significant human morbidity and economic losses in the meat industry in regions where personal and environmental hygienic practices and regulatory controls are insufficient. Recently, however, with the emergence of a global food market, salmon aquaculture has been associated with the transmission of fish tapeworm infections thousands of miles beyond the region of original fish production. Similarly, it is possible that international meat markets could contribute to food-borne beef and pork tapeworm infections outside of known endemic regions, especially if breakdowns occur in regulatory infrastructure or meat inspection procedures.

## CLINICAL VIGNETTE

A colleague asks you about a 4-year-old child who has been treated many times for pinworms with over-the-counter medication by his mother, but she still sees worms in his stools and he complains of having an itchy backside. The mother is a working mom, and the child stays with his grandparents during the day and sometimes on weekends. The family lives in a poorer area of town where the housing is known for poor maintenance by the property owners. The mom has pictures on her phone of the worms she is seeing. The child is seen and is generally well nourished, with no evidence of chronic illness. The photographs show white glistening "things" not clearly discernable as worms and could just be mucus. Pinworm paddles are requested, and you ask the mom to catch some of the worms using the paddles, then place the paddles in a transport vial containing SAF (sodium acetate-acetic acid-formalin) ova and parasite fixative and specifically ask the laboratory to look for "worms." You consider treating with mebendazole for pinworms again but decide to wait for the parasitology results. Diagnosis: *Dipylidium caninum*. The mother was horrified when you described the life cycle and how her son probably got infected. Treatment with praziquantel was effective.

## CLINICAL PRESENTATION AND MANIFESTATIONS

Patients with the chief complaint of "passing a tapeworm" or with signs and symptoms of heavy worm burdens (vague abdominal discomfort, perianal irritation, anorexia, eosinophilia) are uncommon in most developed countries unless there is a history of specific risk, such as international travel, dietary exposures, migration, occupation, or certain circumstances of socioeconomic deprivation. Table 80.1 summarizes the geographic distribution and usual clinical features of the tapeworms infecting humans.

Humans serve as definitive hosts for beef and pork tapeworms; ingested larvae mature to sexually mature adult forms, and parasite eggs are produced in the small intestine. The eggs are excreted in the feces of the human host. Intermediate hosts ingest the eggs when human waste is deposited indiscriminately in the environment. On ingestion, the eggs develop into infective larvae in the tissues of the intermediate host. The reproductive life cycle is completed when humans ingest the infected intermediate hosts (Fig. 80.1; Fig. 81.3). In the case of *Diphyllobothrium* species (fish tapeworms), larval stages in two intermediate hosts are involved. The interdependence of the parasite on human definitive hosts and nonhuman intermediate hosts is complex (Fig. 80.2).

The rodent tapeworm infection caused by *H. nana* (Fig. 80.3) deserves special attention. It is one of very few helminthic (worm) infections of humans that can complete its entire life cycle in the human host without an obligate life stage in an intermediate host. Thus *H. nana* can cause auto-infections, resulting in a persistently infected state, heavy worm burdens in a given individual, and the possibility of direct human-to-human infections. Other important helminthic infections with a potential auto-infective cycle are the nematodes *Strongyloides stercoralis* (see Chapter 79), *Enterobius vermicularis* (see Chapter 79), and *Capillaria philippinensis*. *H. nana* infections are most commonly found in children living in dire socioeconomic conditions where rodent feces contaminate the environment and foodstuffs, leading to inadvertent oral exposure. The usual life cycle involves an infected rodent (definitive host) excreting eggs that are ingested by a beetle (intermediate host) where the larval forms develop; when a

| TABLE 80.1 | Distribution and Usual Clinical Significance of Tapeworms Affecting Humans | |
|---|---|---|
| Parasite Name | Endemic Geographic Parasite Distribution | Common Clinical Significance |
| **Tapeworms With Humans as the Definitive Host (Sexual Reproduction of the Parasite)** | | |
| *Taenia saginata* | The beef tapeworm is common in cattle-breeding regions worldwide. Humans are the definitive host and cattle the intermediate host. Areas with the highest (i.e., >10%) prevalence are central Asia, Near East Asia, and Central and Eastern Africa. Areas with low (i.e., 1%) prevalence are Southeast Asia, Europe, and Central and South America. *Prepatent period: 3–5 months* *Lifespan: up to 25 years* *Length of worms: 4–8 m* | Adult tapeworms live in the gastrointestinal tract of the human host. Eggs are excreted in the stools, and motile tapeworm segments can also be expelled from the bowels. The beef tapeworm does not cause invasive disease in humans but must be distinguished from the pork tapeworm, which does cause tissue infections in people, in regions where their distribution overlaps. People of all ages and races and both genders are susceptible to infection, which is acquired by eating larvae-infected undercooked beef meat. |
| *Taenia solium* | The pork tapeworm is endemic in Central and South America, Southeast Asia, India, the Philippines, Africa, Eastern Europe, and China, with humans being a definitive host and pigs the intermediate host. Areas of highest prevalence include Latin America and Africa. In some regions of Mexico, the prevalence of infection may reach 3.6% of the general population. *Prepatent period: 3–5 months* *Lifespan: up to 25 years* *Length of worms: 3–5 m* | Adult tapeworms live in the gastrointestinal tract of the human host. Eggs are excreted in the stools, and motile tapeworm segments can also be expelled from the bowels. In humans, the pork tapeworm causes invasive disease affecting soft tissues and the brain (cysticercosis). People of all ages and races and both genders are susceptible to infection, which is acquired by eating larvae-infected undercooked pork meat or by ingesting food contaminated with pork tapeworm eggs. |
| *Diphyllobothrium latum* | In North America, fish tapeworm infections have been previously reported in fish from the Great Lakes. There are six *Diphyllobothrium* species known to reside in Alaskan lakes and rivers, and some saltwater species may also be seen in North America. *Diphyllobothrium* infections are not species specific, and widespread reports describe infection in North American fish-eating birds and mammals. Humans are a definitive host, and crustaceans followed by fish are intermediate hosts. The incidence in the United States has been declining recently. Pike, perch, and salmon are among the fish most commonly infected. Reports are commonly made of *D. latum* infection in humans residing in Europe, Africa, and the Far East. *Prepatent period: 3–5 weeks* *Lifespan: up to 25 years* *Length of worms: 4–10 m* | Adult tapeworms live in the gastrointestinal tract of the human host. Eggs are excreted in the stools, and motile tapeworm segments can also be expelled from the bowels. The fish tapeworm does not cause invasive disease but, because of its length and potential to interfere with vitamin B12 absorption, can cause a number of nonspecific symptoms. People of all ages and races and both genders are susceptible to infection, which is acquired by eating undercooked, infected fish flesh. People preparing fresh fish, implements used to prepare fish (e.g., knives and cutting boards), and raw or undercooked fish meals (e.g., sushi, sashimi, ceviche) may be associated with a higher risk of infection. |
| **Tapeworms With Humans as Inadvertent or Unnatural Hosts (Sexual Reproduction of the Parasite Normally Occurs in Another Animal Species)** | | |
| *Hymenolepis nana* | *H. nana* (dwarf tapeworm) is a cosmopolitan intestinal tapeworm usually infecting rodents, mice, or rats. The intermediate host, a beetle, is not required to complete its life cycle in definitive hosts. Ingestion of tapeworm eggs by a definitive host, including humans, can reestablish an adult tapeworm infection. *Prepatent period: 2–3 weeks* *Lifespan of infection: many years because of autoinfection.* *Length of worms: 2.5–4 cm* | Often associated with environments with poor sanitation, the dwarf tapeworm causes few clinical problems; signs and symptoms include nonspecific abdominal complaints, loosening of the stools, perianal irritation, and the possible presence of small motile segments visible in the stool or on undergarments. |
| *Dipylidium caninum* | *D. caninum* is a cosmopolitan tapeworm infection of dogs, with inadvertent human infections occurring through ingestion of the intermediate host, a flea that has fed on the tapeworm eggs contaminating the animal's fur or dog feces. Human infections have been reported in Europe, the Philippines, China, Japan, Argentina, and North America. *Prepatent period: 3–4 weeks* *Lifespan: Less than 1 year.* *Length of worms: 10–70 cm* | Adult tapeworms live in the gastrointestinal tract of the inadvertent human host, usually a child. Perianal irritation may occur, with the passage of motile segments of the tapeworm or small "grain of rice"-like motile segments that may be seen in the stools. The proglottids are motile when passed and may be mistaken for maggots or fly larvae. |
| *Hymenolepis diminuta* | The rat tapeworm requires a grain beetle as an intermediate host, so it is most common in grain-producing areas of the world or where grain or other dry foods are stored. Human infections are uncommon. *Prepatent period: 3 weeks* *Lifespan: less than 1 year* *Length of worms: 20–60 cm* | Often associated with environments with poor sanitation, the rat tapeworm rarely infects humans and causes few clinical problems; these include nonspecific abdominal complaints, loosening of the stools, and perianal irritation and the possible presence of small motile segments visible in the stool or on undergarments. |

Adapted from MacPherson DW: Cestodes: intestinal and extraintestinal tapeworm infections, including echinococcosis and cysticercosis. In Sanford CA, Pottinger PS, Jong EC, eds: *The travel and tropical medicine manual*, ed 5, Philadelphia, 2017, Elsevier, pp 564-573.

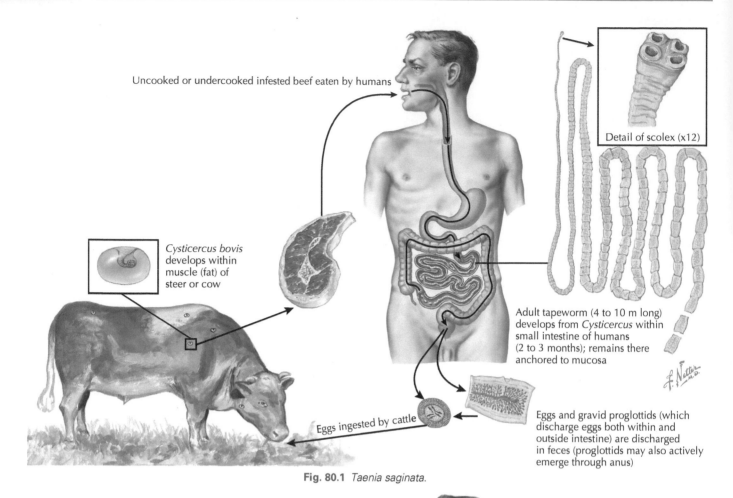

Uncooked or undercooked infested beef eaten by humans

Detail of scolex (x12)

*Cysticercus bovis* develops within muscle (fat) of steer or cow

Adult tapeworm (4 to 10 m long) develops from *Cysticercus* within small intestine of humans (2 to 3 months); remains there anchored to mucosa

Eggs ingested by cattle

Eggs and gravid proglottids (which discharge eggs both within and outside intestine) are discharged in feces (proglottids may also actively emerge through anus)

**Fig. 80.1** *Taenia saginata.*

Uncooked or undercooked infected fish containing plerocercoid is eaten by humans

Hyperchromic anemia indistinguishable from pernicious anemia develops in occasional cases

Egg expelled in feces; finds its way to water

Scolex

*Coracidium* emerges from egg via operculum

*Coracidium* eaten by cyclops (water flea); develops into procercoid (primary larva) within cyclops body

Cyclops eaten by fish; procercoid develops into plerocercoid (secondary larva) within fish

Laying orifice
Genital pore

Detail of proglottids

Adult tapeworm (2 to 10 m long) develops from plerocercoid in intestine of humans; lives there and lays eggs

**Fig. 80.2** *Diphyllobothrium latum.*

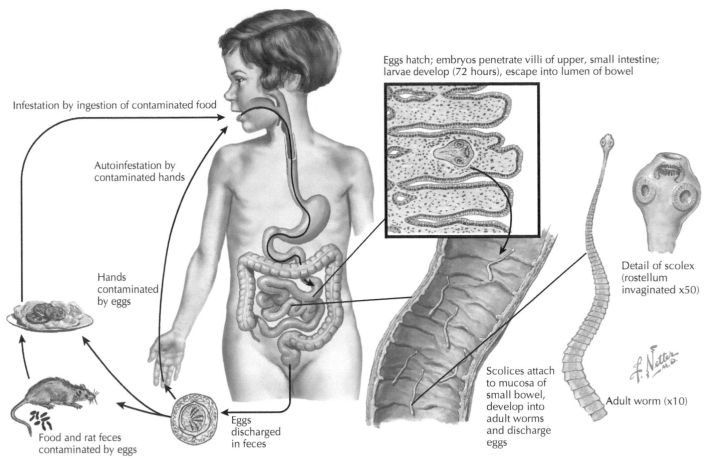

Eggs hatch; embryos penetrate villi of upper, small intestine; larvae develop (72 hours), escape into lumen of bowel

Infestation by ingestion of contaminated food

Autoinfestation by contaminated hands

Hands contaminated by eggs

Detail of scolex (rostellum invaginated x50)

Food and rat feces contaminated by eggs

Eggs discharged in feces

Scolices attach to mucosa of small bowel, develop into adult worms and discharge eggs

Adult worm (x10)

**Fig. 80.3** *Hymenolepis nana.*

rodent eats the infected beetle, the larvae mature to adults, and ultimately infective eggs are excreted into the environment, thus reestablishing an infection in another beetle. Neither the usual definitive hosts nor intermediate hosts are obligate requirements for the completion of this life cycle. Upon the ingestion of infected rodent feces, humans can serve as an inadvertent definitive host without the requirement of the intermediate beetle host, or humans can become reinfected by swallowing *H. nana* eggs passed in their own feces.

Intestinal infections with other animal tapeworms *D. caninum* or *H. diminuta* are sporadically detected in otherwise healthy persons. Infection occurs after the inadvertent ingestion of infected fleas *(D. caninum)* or grain beetles *(H. diminuta;* see Table 80.1).

Most human intestinal tapeworm infections are asymptomatic until a proglottid, a longer chain of proglottid segments (strobila), or the entire adult tapeworm is passed out of the bowel. Each of these living worm components can be motile; thus the presence of the tapeworm is often manifested by a patient's complaint of the sensation of wriggling in the perianal region or undergarments. Pernicious anemia caused by vitamin $B_{12}$ deficiency is a rare complication of fish tapeworm *(Diphyllobothrium)* infections. Complications of tissue invasion with larval forms of tapeworms are addressed elsewhere (see Chapter 81 and Chapter 85). Multiple adult tapeworm infections and infections with multiple species of intestinal tapeworms in a single individual at the same time are uncommon clinical phenomena.

## DIAGNOSIS AND MANAGEMENT

A tapeworm infection can often be diagnosed by the description given by the patient of passing a single segment or a chain of proglottids

per rectum and gross inspection of the specimen. A species-specific diagnosis requires examination of a specimen in a parasitology laboratory. Ova and parasite (O&P) stool examination can be used to detect both eggs and proglottids (segments) of tapeworms if present in the feces. *Taenia* species eggs cannot be distinguished as being pork or beef tapeworm eggs on routine laboratory examination. In these cases, the morphology of the uterine branches in a proglottid is diagnostic. The head (scolex) of each tapeworm species is also distinctive but rarely is an entire, intact worm available for laboratory examination. Serologic examination and other tests can confirm invasive pork tapeworm disease (see Chapter 81).

Except for *H. nana,* which can autoinfect its human host, all adult tapeworms have a limited lifespan in humans, although some can live for years. Specific management and treatment with antiparasitic drugs are indicated for a personal sense of well-being, as well as for public health considerations (Table 80.2).

## PREVENTION

In general, tapeworm infections can be prevented by food precautions, meticulous personal and environmental hygiene, and improved sanitary practices in animal husbandry and fish farming. At the individual level, persons should avoid eating raw, pickled, or undercooked meat and fish, especially in regions where food safety regulations are not likely to be strictly enforced. Food preparers should wash their hands before food preparation and again before eating, and avoid "snacking" on the unprocessed flesh during preparation. At the public health and farm industry level, human wastes should be collected, contained, and processed; untreated sewer sludge should not be spread on cattle

## TABLE 80.2   Treatment for Human Tapeworm Infections

| Parasite | Drug | Dosage |
|---|---|---|
| *Taenia saginata* | Praziquantel<br>Alternative: niclosamide | Single dose of 5–10 mg/kg<br>2.0 g once (50 mg/kg once) |
| *Taenia solium* (intestinal) | Praziquantel<br>Alternative: niclosamide | Single dose of 5–10 mg/kg<br>2.0 g once (50 mg/kg once) |
| *Diphyllobothrium latum* | Praziquantel<br>Alternative: niclosamide | Single dose of 5–10 mg/kg<br>2.0 g once (50 mg/kg once) |
| *Dipylidium caninum* | Praziquantel<br>Alternative: niclosamide | Single dose of 5–10 mg/kg<br>2.0 g once (50 mg/kg once) |
| *Hymenolepis nana* | Praziquantel<br>Alternative: nitazoxanide | *Adults and children:* Single dose of 25 mg/kg<br>500 mg × 3 days (1–3 years old: 100 mg bid × 3 days; 4–11 years old: 200 mg bid × 3 days) |
| *Hymenolepis diminuta* | Praziquantel or nitazoxanide | *Adults and children:* Single dose of 25 mg/kg<br>500 mg × 3 days (1–3 years old: 100 mg bid × 3 days; 4–11 years old: 200 mg bid × 3 days) |

*bid,* Twice per day.
Adapted from MacPherson DW: Cestodes: intestinal and extraintestinal tapeworm infections, including echinococcosis and cysticercosis. In Sanford CA, Pottinger PS, Jong EC, eds: *The travel and tropical medicine manual*, ed 5, Philadelphia, 2017, Elsevier, pp 564-573.

grazing lands, and piggeries should avoid feeding raw garbage to swine. Raw sewage should not be released into freshwater lakes or streams where copepods serving as intermediate hosts of the fish tapeworm can become infected and subsequently transmit the infection to fish, which in turn may be ingested by humans.

Cosmopolitan palates that cannot eschew sushi, sashimi, steak tartare, and other raw, smoked, pickled, or fermented meat and fish products, as well as enthusiastic home cooks who taste their culinary creations before adequate cooking, need to entertain the possibility of tapeworm disease. The use of gamma irradiation to process raw meat and seafood products may increase food safety in the future, especially in regions where the ingestion of raw, smoked, pickled, fermented, or undercooked meat and/or fish is culturally ingrained.

## EVIDENCE

Dixon MA, Braae UC, Winskill P, Walker M, Devleesschauwer B, Gabriël S, Basáñez MG. Strategies for tackling *Taenia solium* taeniosis/cysticercosis: A systematic review and comparison of transmission models, including an assessment of the wider Taeniidae family transmission models. *PLoS Negl Trop Dis* 2019;13(4):e0007301. doi: 10.1371/journal.pntd.0007301. eCollection 2019 Apr. PMID:30969966. *A review article on control means and methods for the prevention of pork tapeworm transmission.*

Rodriguez-Canul R, Argaez-Rodriguez F, Pacheco de la Gala D, et al. *Taenia solium* metacestode viability in infected pork after preparation with salt pickling or cooking methods common in Yucatan, Mexico. *J Food Prot* 2002;65:666-669. *A study showing that some traditional Mexican cooking methods may be sufficient to kill tapeworm larvae in infected pork.*

## ADDITIONAL RESOURCES

Cabello FC. Salmon aquaculture and transmission of the fish tapeworm. *Emerg Infect Dis* 2007;13:169-171. *Summary of epidemiologic studies associating human outbreaks of fish tapeworm in Brazil with salmon shipped from Chile, and implications for future outbreaks in other regions. Parasite destruction by various fish preparation methods is reviewed.*

Centers for Disease Control (Atlanta, USA). Alphabetical index of parasitic diseases. Available at: https://www.cdc.gov/parasites/az/index.html. *Authoritative resource for health care professionals.*

Desowitz RS. New Guinea tapeworms and Jewish grandmothers. In *New Guinea tapeworms and Jewish Grandmothers: tales of parasites and people.* New York, 1987, WW Norton and Company, pp 36-45. *Entertaining account of cultural practices and tapeworm transmission among two very different ethnic groups.*

International Consultative Group on Food Irradiation (ICGFI): Facts about food irradiation, Vienna, Austria, 1999, ICGFI, pp 1-45. Available at: www.iaea.org/icgfi. Accessed January 27, 2009. *Basic introduction to the process of food irradiation and its impact on food-borne infections, including parasites.*

MacPherson DW. Cestodes: intestinal and extra-intestinal tapeworm infections, including echinococcosis and cysticercosis. In Sanford CA, Pottinger PS, Jong EC, eds: *The travel and tropical medicine manual*, ed 5, Philadelphia, Elsevier, pp 564-573. *Aspects of clinical diagnosis and treatment of tapeworm infections presented in detail for the healthcare provider.*

World Health Organization. Food Safety (4 June 2019). Available at: https://www.who.int/news-room/fact-sheets/detail/food-safety. *Global health authority on food and safety, global burden of illness and death associated with food safety issues.*

# 81

# Cysticercosis

*Natasha S. Hochberg, Anna Cervantes-Arslanian, Hector H. Garcia*

## ABSTRACT

Cysticercosis is infection caused by the larval stage of the cestode *Taenia solium*, the pork tapeworm. The clinically significant helminthic invasion to central nervous system (CNS) structures and the resulting spectrum of neurologic illness define neurocysticercosis (NCC). With the more extensive use of computed tomography (CT) and magnetic resonance (MRI) neuroimaging, and the availability of accurate serologic testing, NCC has been increasingly diagnosed not only in low- and middle-income countries (LMICs) where pigs are raised, but also in high-income countries with large immigrant populations originating from endemic areas. NCC is presently recognized as the most common helminth infection of the CNS.

## CLINICAL VIGNETTE

A 40-year-old male farmer from Nepal was seen in the emergency department at a hospital in an urban center in the United States. The patient reported new-onset seizures as well as headache. An MRI was performed that demonstrated three colloidal parenchymal cystic lesions with some perilesional edema (Fig. 81.1); subsequent head CT showed no calcification and a fundoscopic examination was unremarkable. The patient was started on antiepileptic medication and treated with a combination of steroids, albendazole (ABZ) and praziquantel (PZQ). Serologic testing for neurocysticercosis was positive. Upon follow-up examination, one cyst had disappeared and the other two had calcified; the patient reported no further seizures.

COMMENT: NCC accounts for approximately one-third of seizures worldwide. Notably, the patient had both MRI and head CT to evaluate the lesions. Combination therapy was warranted due to the number of cysts.

## CLINICAL VIGNETTE

A 38-year-old Cape Verdean man presented with new-onset vertigo and 2 months of pulsating occipital headaches, associated with sensitivity to light and occasional vomiting. On examination, he had nystagmus with right gaze, left-sided ataxia, and gait instability with presence of Romberg sign. MRI of the brain demonstrated multi-loculated cysts suggestive of racemose NCC with basal arachnoiditis (Fig. 81.2). Mild hydrocephalus was seen with dilation of the temporal horns of the lateral ventricles. He was started on steroids and dual anti-helminthics. At 1 month, the lesions had improved, but several cystic components continued to enhance. Hydrocephalus was mild. Steroids were discontinued. The patient was started on methotrexate and continued on dual anti-helminthics until radiographic cyst resolution at 4 months.

COMMENT: Extra-parenchymal involvement is less common, but often more dangerous. Subarachnoid and intraventricular disease are potentially neurosurgical cases. Hydrocephalus may form due to cerebrospinal fluid (CSF) obstruction (in intraventricular disease) or from impaired CSF resorption of the arachnoid granules due to arachnoiditis in subarachnoid disease. Whereas intraventricular disease may often be amenable to endoscopic removal of the cysts en toto, subarachnoid disease is often managed medically with prolonged courses of antihelminthic treatment. Consideration may be given to steroid-sparing agents.

## ETIOLOGY/PATHOPHYSIOLOGY

*Taenia solium* is found worldwide; however, rates are highest where humans live in close contact with pigs, typically in rural villages. It is estimated that 50 million people are infected with cysticercosis and 50,000 die annually worldwide. Endemicity has been demonstrated in Latin America, Eastern Europe, Africa, India, China, and other Asian countries; 11 to 29 million people have cysticercosis in Latin America, alone. In highly endemic areas, more than one-third of individuals are seropositive and almost 20% have evidence of CNS calcifications.

NCC is the most important cause of acquired epilepsy in LMICs and is becoming more common in high-income countries owing to tapeworm carriers immigrating and traveling from endemic regions. In high-income countries, the incidence is highest in major urban centers with large immigrant populations. Between 1500 and 2000 individuals are hospitalized with cysticercosis in the United States every year; 2% of seizures in two urban US centers were due to NCC. In Western Europe, the estimated prevalence is 0.02% to 0.67%.

Pigs are the obligate intermediate host, and humans are the definitive host for *T. solium*. Humans become infected and develop the pork tapeworm infection, taeniasis, by ingesting cysticerci in undercooked contaminated pork. The adult tapeworm (2 to 7 m in length) lives in the human small intestine; stool shedding of eggs and proglottids is associated with few or no symptoms, and infected carriers are often undetected (Fig. 81.3). Humans (and pigs) develop cysticercosis after ingesting gravid proglottids or embryonated ova excreted from a human tapeworm carrier (i.e., fecal-oral transmission); autoinfection may occur when hand hygiene is suboptimal.

That pork ingestion or direct contact with pigs is not required for the transmission of human cysticercosis is exemplified by reported cases of the disease that occurred among members of an Orthodox Jewish community in the United States, as well as among vegetarians in India.

After ingestion, the oncospheres hatch, penetrate the wall of the intestine, and migrate to subcutaneous tissue, musculature, CNS, and other organs. After 60 to 70 days, they mature into cysticerci. When they invade the CNS, larvae encyst (vesicular or cystic stage) and evade or modulate the host immune response (often generating minimal inflammation). Inflammatory responses vary based upon the parasite stage and location; when present, the perilesional inflammation is transitory. Early on, this response is characterized by lymphocytes, plasma cells, and eosinophils. After months to years, the host mounts a more exuberant inflammatory response with perilesional macrophages, T cells, and natural killer (NK) cells. The parasite ultimately degenerates and dies (colloidal and granular stages)—there is astrocytic gliosis, and calcification eventually occurs (calcific stage). Perilesional gliosis may be epileptogenic in conjunction with organic damage and inflammatory response.

## CLINICAL PRESENTATION

NCC induces neurologic syndromes that vary from asymptomatic infection to sudden death, but seizures and hydrocephalus (manifested by headache) are the most important presentations. In endemic areas, approximately one-third of seizure cases are associated with NCC, and the incidence of epilepsy can be as high as 90 to 122 per 100,000 persons in LMICs. Seizures may be focal but more commonly are generalized, or focal with generalization. Cognitive impairment, psychiatric symptoms, and focal neurologic deficits are common. Individual clinical presentation and clinical course are contingent on the type, stage, location, and number of parasites in the CNS as well as the host's immune response.

### Parenchymal Disease

Active parenchymal cysticercosis is the most common disease presentation (Fig. 81.4). Individuals with viable intact (vesicular) cysts are frequently asymptomatic; most patients with symptoms have CT or MRI evidence of cyst degeneration (colloidal and granular) or

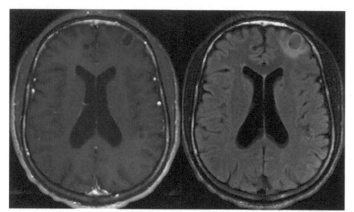

**Fig. 81.1** MRI demonstrates intraparenchymal cysts, one with pronounced perilesional edema.

inflammation (contrast enhancement or brain edema) surrounding the parasite.

Cysts can be single or multiple. There appears to be geographic variability in disease presentation. For example, in India and the United States, most patients have a single degenerating cysticercus, whereas in Latin America and China, individuals commonly have multiple cysts.

Cysticercotic encephalitis is the uncommon presence of large numbers of cysts and generalized cerebral edema associated with seizures and abnormal mental status.

Calcified lesions on CT are the hallmark of resolved parenchymal cysticercosis, but they may be clinically active. Patients usually experience seizures believed to be caused by disruption of calcified granulomas; perilesional inflammation may develop secondary to antigen exposure even from nonviable infection. Pericystic inflammation has been described as a predictor of seizure relapse.

### Extraparenchymal Disease

Extraparenchymal NCC can be found alone or in combination with parenchymal disease. Cysticerci lodged in the convexity of the cerebral hemispheres tend to be larger but produce similar clinical manifestations as intraparenchymal cysts. Cysticerci may also be inside the ventricles, the basal subarachnoid space, or the sylvian fissure. Intracranial hypertension results from mechanical factors or arachnoiditis.

Ventricular NCC is not uncommon, and cysts, which may be found in any ventricle and be difficult to detect by neuroimaging, may cause hydrocephalus by obstructing the normal flow of CSF. Clinical manifestations depend on the ventricle involved and degree of obstruction. It is critical to determine whether cysts are adherent to the lateral ventricle wall as this affects potential complications and management.

Active subarachnoid NCC occurs when cysts lodge outside of the brain. They tend to occupy the sylvian fissure or the basal cisterns, where more space is available. In those areas, cysts tend to form "racemose" clusters and grow larger with membrane expansion. They can induce a severe inflammatory response, which may be evident in CSF examination, and induce stroke secondary to vasculitis. Even inactive NCC of this area can cause CSF flow obstruction and hydrocephalus.

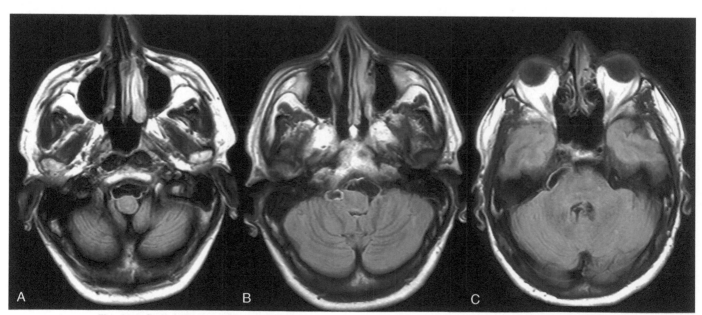

**Fig. 81.2** Brain MRI demonstrating multiloculated cysts within the pre-medullary (A), pre-pontine (B), right peri-mesencephalic (C), and cerebello-pontine angle (C) cisterns suggestive of racemose cysticercosis with basal arachnoiditis.

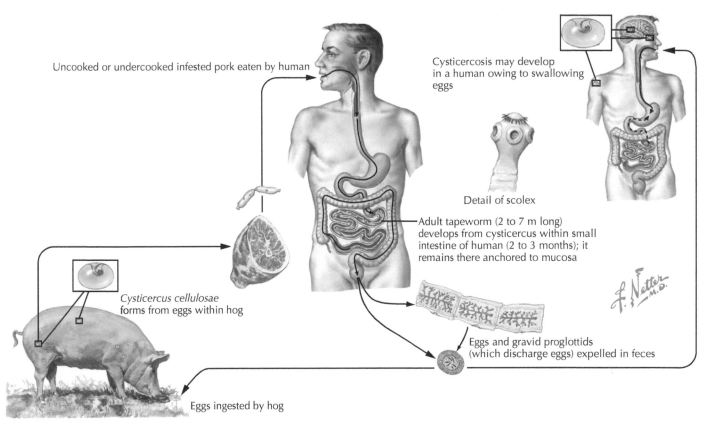

Uncooked or undercooked infested pork eaten by human

Cysticercosis may develop in a human owing to swallowing eggs

Detail of scolex

Adult tapeworm (2 to 7 m long) develops from cysticercus within small intestine of human (2 to 3 months); it remains there anchored to mucosa

Cysticercus cellulosae forms from eggs within hog

Eggs and gravid proglottids (which discharge eggs) expelled in feces

Eggs ingested by hog

**Fig. 81.3** The life cycle of *Taenia solium*.

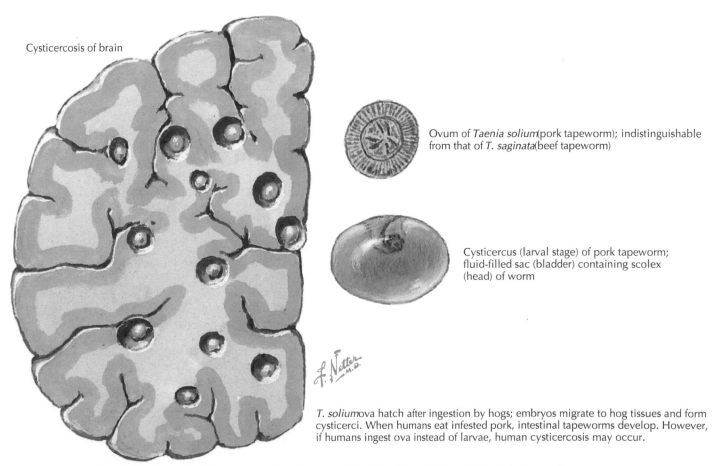

Cysticercosis of brain

Ovum of *Taenia solium* (pork tapeworm); indistinguishable from that of *T. saginata* (beef tapeworm)

Cysticercus (larval stage) of pork tapeworm; fluid-filled sac (bladder) containing scolex (head) of worm

*T. solium* ova hatch after ingestion by hogs; embryos migrate to hog tissues and form cysticerci. When humans eat infested pork, intestinal tapeworms develop. However, if humans ingest ova instead of larvae, human cysticercosis may occur.

**Fig. 81.4** Drawings of the ovum and cysticercus of *Taenia solium* and the pathologic features of neurocysticercosis.

## TABLE 81.1  Diagnostic Criteria

| | |
|---|---|
| Absolute criteria | • Histological demonstration of parasite from biopsy of brain or spinal cord<br>• Visualization of subretinal cysticercus<br>• Scolex within cystic lesion on neuroimaging |
| Neuroimaging criteria:<br>Major | • Cystic lesions without discernible scolex<br>• Enhancing lesions<br>• Multilobulated cystic lesions in subarachnoid space<br>• Typical parenchymal brain calcifications |
| Confirmative | • Resolution of lesions after cysticidal therapy<br>• Spontaneous resolution of single small enhancing lesions<br>• Migration of ventricular cysts on sequential neuroimaging |
| Minor | • Obstructive hydrocephalus or abnormal enhancement of basal leptomeninges |
| Clinical/exposure criteria:<br>Major | • Detection of anticysticercal antibodies or cysticercal antigens by immunodiagnostic tests<br>• Cysticercosis outside the CNS<br>• Household contact with *T. solium* |
| Minor | • Clinical manifestations suggestive of NCC<br>• Individuals coming from/living in endemic area |

Definitive diagnosis: One absolute criterion OR Two major neuroimaging criteria plus any clinical/exposure criteria OR One major and one confirmative neuroimaging criteria plus any clinical/exposure criteria OR One major neuroimaging criteria plus two clinical/exposure criteria (including at least one major clinical/exposure criterion) and exclusion of other pathologies producing similar neuroimaging.

Probable diagnosis: One major neuroimaging criteria plus any two clinical/exposure criteria OR One minor neuroimaging criteria plus at least one major clinical/exposure criteria.

*CNS*, Central nervous system; *NCC*, neurocysticercosis.
From Del Brutto OH et al, Revised diagnostic criteria for neurocysticercosis. *J Neurol Sci*. 2017;372:202-210.

## Other Manifestations

When located in the sella turcica, NCC mimics pituitary adenoma; spinal NCC causes level-related sensory and motor manifestations; and ophthalmic cysticercosis is usually retinal or vitreous. Cysts may develop undetected in other organs and in muscle.

## DIAGNOSTIC APPROACH

In endemic regions, seizures and intracranial hypertension are suggestive, and NCC should be high on the list of differential diagnoses. Providers should utilize the revised criteria to make either a definitive or probable NCC diagnosis; these include a set of neuroimaging, clinical/exposure, and absolute measures (Table 81.1). Brain imaging can be performed using CT or MRI, and each imaging technique has advantages and disadvantages; when possible, both modalities should be used. CT is more frequently available in LMICs and is less expensive (which is important, considering that NCC is a disease of poverty); however, poor-quality images from old CT machines are of limited utility. MRI, in turn, gives better images and images in different planes. Definition of small lesions, intraventricular lesions, and lesions close to the bone is much better with MRI than with CT. MRI's diverse imaging protocols also allow a better definition of the perilesional inflammation. Intraventricular and cisternal cysts are best imaged using volumetric-balanced, steady-state gradient echo sequences. As a drawback, MRI imaging of calcifications is suboptimal.

Serologic diagnosis of cysticercosis is most helpful when imaging is inconclusive. Antibody detection with enzyme-linked immunoelectro-transfer blot (EITB) assay, with purified glycoprotein antigens, is used most frequently due to higher sensitivity (98% sensitivity in patients with >1 viable cyst with specificity of nearly 100%); sensitivity drops to 70% in patients with a single degenerating parasite. The CDC lentil lectin-bound glycoproteins (LLGP)-EITB is available in the United States, Canada, and Peru. Other versions of the EITB assay are marketed in Europe, and used for diagnostic or research purposes in Mexico, Brazil, and India, among other developing countries. Enzyme-linked immunosorbent assays (ELISAs) have lower sensitivity and frequent cross-reactivity with related cestodes.

Antigen detection assays seem better at differentiating infections with viable parasites and are useful for monitoring disease evolution in cases of basal subarachnoid NCC.

Several molecular tests, including CSF testing via PCR or genomic sequencing, have been used for diagnosis of subarachnoid NCC, and cell-free *T. solium* DNA has been detected in urine. These have not been adopted for routine use as of yet.

## TREATMENT

There are two key points in NCC management: (1) symptomatic management is crucial, and (2) it should be approached according to the type of NCC.

The principles of NCC treatment involve appropriate use of symptomatic medication to control seizures, headache, or intracranial hypertension, followed by a decision on the specific measures targeted to destroy or inactivate the parasitic lesions. Fundoscopic examination is recommended prior to initiation of treatment.

Symptomatic treatment includes management of intracranial hypertension with steroids and hyperosmolar therapy (mannitol or hypertonic saline). In cases in which NCC obstructs CSF flow, CSF diversion via shunting or surgical removal of large cysts and cyst masses may be required. Other common therapies include analgesics and antiepileptic drug therapy to manage seizures. Seizures associated with NCC usually respond well to first-line antiepileptic drugs. Levetiracetam has the advantage of not interacting with antiparasitic drugs, although carbamazepine and phenytoin are sometimes used.

Treatment with cysticidal antiparasitic drugs is of benefit in most cases of NCC with viable (vesicular) or degenerating (colloidal) cysts. The regimen of choice for one to two parenchymal cysts is ABZ 15 mg/kg/day total (divided BID, total maximum dose 1200 mg) for 10 to 14 days. Multiple viable cysts are better treated with dual antiparasitics with ABZ (at the same dose) and PZQ 50 mg/kg/day for 10 to 14 days. Steroids are given concomitantly with antihelminthic therapy to decrease the inflammation resulting from exposure to antigens released from degenerating cysts. Of note, subarachnoid NCC may require antiparasitic drug treatment for long periods and repeated courses. Steroid-sparing agents such as methotrexate and etanercept may be considered. On the other hand, in patients with cysticercotic encephalitis, the use of antiparasitic medications is contraindicated because it could boost the inflammatory response and lead to fatal intracranial hypertension.

Surgery may be required in NCC, most frequently CSF shunting to control hydrocephalus. Neuroendoscopy is the approach of choice for management of intraventricular cysts that are nonadherent. Open surgery is rarely utilized to excise large cysts or cyst clumps causing mass effect, or for excision of fourth ventricular cysts.

## FUTURE DIRECTIONS

Gaps remain in our understanding of NCC and merit attention. These include a better understanding of the biology of the parasite and the mechanisms of parasite degeneration. Although the genome has been published, and data are emerging on the evolution of the parasite, genetic studies of strains may shed light on regional differences (e.g., why single lesions are more commonly found in India). Novel studies of human host-mediated immune responses such as RNA sequencing and omics studies have started to be done; future studies are needed to improve knowledge of antigenic response, disease pathogenesis, and mediators of inflammation. These studies will help explain the heterogeneity of the clinical response. Ultimately, these data will allow us to identify novel therapeutic targets and ideally prevent complications including seizures.

From a public health perspective, strategies to address the potential eradication of T. solium taeniasis/cysticercosis are paramount. Elimination in northern Peru through a combined intervention of screening, treatment, and porcine vaccination, and in Europe by sanitation efforts suggests that the disease can be eradicated globally. Improved animal husbandry and porcine meat inspection are effective but difficult to implement in LMICs. Better detection and treatment of tapeworm carriers could help interrupt human transmission, but ultimately, a combined platform of porcine chemotherapy and vaccination, and human diagnosis and treatment will be needed. Human vaccination may play a role.

## EVIDENCE

Abba K, Ramaratnam S, Ranganathan LN. Anthelmintics for people with neurocysticercosis. *Cochrane Database Syst Rev* 2010;2010(3):CD000215. *Randomized controlled trials comparing ABZ with placebo, no anthelmintic, or another anthelmintic regimen in subjects with NCC were reviewed to assess effectiveness and safety. In adults with viable lesions, ABZ may reduce the number of lesions, and in children with nonviable lesions, ABZ substantially lowered seizure recurrence (vs. no treatment).*

Bruno E, Bartoloni A, Zammarchi L, et al and the COHEMI Project Study Group. Epilepsy and neurocysticercosis in Latin America: A systematic review and meta-analysis. *PLoS Negl Trop Dis* 2013;7(10):e2480. *This study analyzed data on prevalence of lifetime epilepsy, active epilepsy, incidence, mortality, treatment gap and NCC proportion among people with epilepsy in Latin America and provides valuable data on prevalence to guide public health interventions.*

Coyle CM, Mahanty S, Zunt JR, et al. Neurocysticercosis: Neglected but not forgotten. *PLoS Negl Trop Dis* 2012;6(5):e1500. *This paper analyzed the prevalence of NCC and seizures due to NCC in different regions of the world.*

Del Brutto OH, Roos KL, Coffey CS, García HH. Meta-analysis: Cysticidal drugs for neurocysticercosis: albendazole and praziquantel. *Ann Intern Med* 2006;145:43-51. *This meta-analysis of randomized trials shows that ABZ or PZQ results in better resolution of colloidal and vesicular cysticerci, lower risk for recurrence of seizures in patients with colloidal cysticerci, and fewer generalized seizures in patients with vesicular cysticerci. The evidence favors ABZ over PZQ; larger courses of ABZ may be needed for treatment of patients with more than a few cystic lesions.*

Fleury A, Cardenas G, Adalid-Peralta L, et al. Immunopathology in Taenia solium neurocysticercosis. *Parasite Immunol* 2016;38:147-157. *This review paper describes the pathophysiology of NCC and the immune response variations in the different clinical stages of disease.*

Garcia HH, Gonzales I, Lescano AG, et al. Efficacy of combined antiparasitic therapy with praziquantel and albendazole for neurocysticercosis: a double-blind, randomised controlled trial. *Lancet Infect Dis* 2014;14(8):687-695. *This study demonstrated that the combination of ABZ plus PZQ increased parasiticidal effect in patients with multiple cysticercal cysts over single antihelminthic therapy.*

Garcia HH, Gonzalez AE, Tsang VC, et al. Cysticercosis Working Group in Peru. Elimination of Taenia solium transmission in northern Peru. *N Engl J Med* 2016;374:2335-2344. *This pivotal study demonstrated the effectiveness of interventions to eliminate T. solium in this region of Peru using mass screening and treatment (of humans and pigs) with and without vaccination followed by mass treatment of humans and mass treatment and vaccination of pigs.*

O'Keefe KA, Eberhard ML, Shafir SC, et al. Cysticercosis-related hospitalizations in the United States, 1998–2011. *Am J Trop Med Hyg* 2015;92:354-359. *This paper describes how many patients with cysticercosis are admitted to US hospitals and emphasizes the importance of considering this diagnosis in nonendemic countries.*

Rangel-Castilla L, Serpa JA, Gopinath SP, et al. Contemporary neurosurgical approaches to neurocysticercosis. *Am J Trop Med Hyg* 2009;280:373-378. *This article presents a retrospective analysis of the outcomes for NCC patients that had neurosurgical evaluations. Of 31 patients, 29 were treated with a variety of neurosurgical procedures (shunts, craniotomy, and endoscopy); neuroendoscopy seemed to be associated with a higher success rate.*

## ADDITIONAL RESOURCES

Centers for Disease Control and Prevention (CDC): Parasites - Cysticercosis. https://www.cdc.gov/parasites/cysticercosis/index.html. *Portal to multiple NCC resources on epidemiology, clinical signs, diagnosis, treatment, public health considerations, patient education materials, and more.*

Del Brutto OH, Nash TE, White AC Jr, et al. Revised diagnostic criteria for neurocysticercosis. *J Neurol Sci* 2017;372:202. *These revised NCC diagnostic criteria are applicable across a wide range of settings.*

Garcia HH, Gonzalez AE, Gilman RH for the Cysticercosis Working Group in Peru. Taenia solium cysticercosis and its impact in neurological disease. *Clinical Microbiology Reviews* 2020;33(3):e00085-19. *This review provides a comprehensive look at current literature on NCC.*

Schantz PM, Moore AC, Munoz JL. Neurocysticercosis in an Orthodox Jewish community in New York City. *N Engl J Med* 1992;327:692-695. *A fascinating account of NCC transmission in a US community that does not eat pork.*

White AC Jr, Coyle CM, Rajshekhar V, et al. Diagnosis and treatment of Neurocysticercosis: 2017 Clinical Practice Guidelines by the Infectious Diseases Society of America (IDSA) and the American Society of Tropical Medicine and Hygiene (ASTMH). *Clin Infect Dis* 2018;66(8):1159-1163. *These guidelines address diagnosis and management of patients with different forms of NCC. Subdivided into sections on parenchymal disease, single enhancing lesions, calcified lesions, ventricular and subarachnoid NCC, the guidelines present a comprehensive approach with graded evidence.*

# 82

# Food-Borne Trematodes: Liver, Lung, and Intestinal Flukes

*Johnnie A. Yates, Elaine C. Jong*

## ABSTRACT

Food-borne trematodes (FBTs) are a group of liver, lung, and intestinal flukes that typically infect humans through the ingestion of freshwater fish, crustaceans, and plants contaminated by encysted forms of the parasites. FBTs are prevalent throughout the world, with countries in Asia bearing the highest burden of infection. Transmission of FBTs is fostered by local culinary practices of consuming raw or undercooked fish and aquatic plants, the presence of freshwater snails that serve as intermediate hosts in the parasite life cycle, and contamination of surface water by untreated fecal waste from infected humans and mammalian reservoir hosts. In addition to occurring in residents of endemic regions, FBTs may also be found in those living in nonendemic areas who immigrated from countries where FBTs are prevalent, among international travelers who consume foods containing encysted larvae, or in those exposed to contaminated food imported from endemic regions. Acute fluke infections can manifest with gastrointestinal or pulmonary symptoms and eosinophilia, while chronic infections may be relatively asymptomatic. Two liver flukes are strongly associated with cholangiocarcinoma, a bile duct cancer. This chapter focuses on the FBTs of greatest medical importance: the liver flukes *Clonorchis sinensis*, *Opisthorchis viverrini*, *Fasciola hepatica* and *Fasciola gigantica*, the lung fluke *Paragonimus* spp., and the intestinal fluke *Fasciolopsis buski*.

## CLINICAL VIGNETTE

A 66-year-old woman with a history of chronic hepatitis B presented with intermittent epigastric and right upper quadrant pain that started several weeks after a trip to Vietnam to visit relatives. She denied fevers but reported anorexia and was noted to have lost 15 pounds. Physical examination revealed mild epigastric and right upper quadrant tenderness but no hepatomegaly. Laboratory evaluation was notable for leukocytosis (white blood cell count 20,200/μL) with 45% eosinophilia (absolute eosinophil count of 9090). Her aspartate aminotransferase and gamma glutamyl transferase levels were elevated, but the rest of her liver tests were normal. Stool examination was negative for ova and parasites. Right upper quadrant ultrasound revealed multiple ill-defined lesions in the liver. Magnetic resonance imaging of the abdomen found several areas of T2 hyperintensity with T1 hypointensity in the right lobe of the liver, diffuse gallbladder wall thickening, pericholecystic fluid, and no evidence of biliary ductal dilatation. Biopsy of a liver lesion demonstrated microabscesses, eosinophils, and Charcot-Leyden crystals. Serology for *Fasciola hepatica* performed at the Centers for Disease Control and Prevention was positive. She received triclabendazole 10 mg/kg as a single dose and her symptoms resolved within 2 weeks of treatment. Her eosinophilia resolved several weeks later, and a follow-up liver ultrasound 8 months after treatment was normal.

## GEOGRAPHIC DISTRIBUTION

According to the World Health Organization (WHO), more than 56 million people worldwide are infected with FBTs and approximately 750 million people are at risk of infection. The actual burden of disease is higher because of underdiagnosis and underreporting, and FBTs have been considered among the most neglected of the world's neglected tropical diseases (NTDs). FBTs have a global distribution but the greatest burden of disease occurs in Asia, with an estimated 12 million people infected with *Clonorchis sinensis* in China and 8 million with *Opisthorchis viverrini* in Thailand. *F. hepatica* is present in six continents while *Fasciola gigantica* is found predominantly in Asia and Africa. Multiple *Paragonimus* species are found throughout Asia, Africa, and the Americas (e.g., *Paragonimus kellicotti* is endemic in the Mississippi River basin of the United States). FBTs are transmitted mostly in rural agricultural communities, but their global dispersal is broadening due to the growth of aquaculture, increased globalization of the world's food supply, and increased human migration.

## LIFE CYCLE

Humans and animals become infected with FBTs when they ingest raw or undercooked freshwater fish, crustaceans, or plants that are contaminated with the encysted metacercariae of the flukes. *Fasciola* infections may also be transmitted through contaminated water. Humans serve as definitive hosts of FBTs, meaning that the parasites become sexually mature adults within the body of the human host. Adult liver and intestinal flukes produce ova (eggs) that exit the body in feces, while the ova of lung flukes may be expectorated in the sputum or swallowed and then excreted in the stools.

The life cycle of the intestinal fluke *Heterophyes*, a fish-borne trematode, is shown in Fig. 82.1 and serves as a model for understanding the life cycle of other fish-borne flukes such as *C. sinensis*, *O. viverrini*, and *Metagonimus yokogawai*. After contamination of freshwater environments by feces from infected mammalian hosts, *Heterophyes* eggs are ingested by suitable snail hosts (the first intermediate host) and hatch, with the emergence of parasite forms called *miracidia*. After going through further developmental stages in the snail gut, free-swimming parasite forms called *cercariae* are released by the snails into the water. The cercariae attach to and penetrate fish (the second intermediate host), transforming into encysted metacercariae in the flesh of the fish. The metacercariae are the infectious stage for mammalian hosts, and after infected fish are eaten by humans or other mammals, the *Heterophyes* metacercariae excyst and develop into adults in the small intestine, where they mature to become adult flukes.

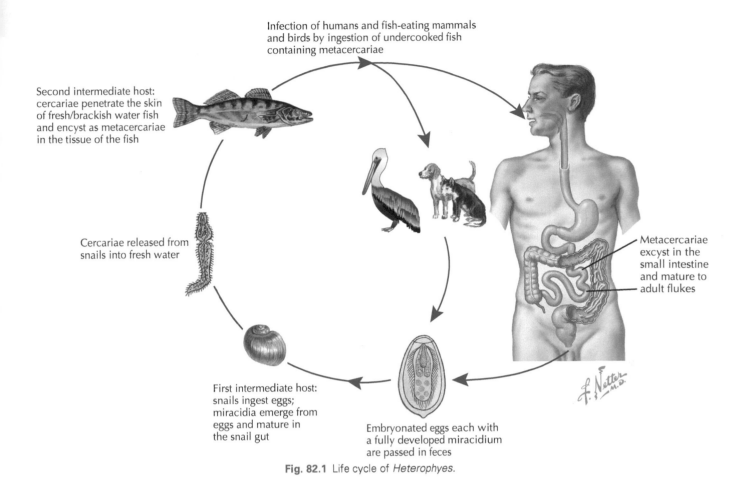

**Fig. 82.1** Life cycle of *Heterophyes*.

After ingestion of fish containing encysted metacercariae of the liver flukes *C. sinensis* and *O. viverrini*, the parasites excyst and the larvae migrate through the duodenum, the ampulla of Vater, and the extrahepatic bile ducts, eventually reaching the intrahepatic bile ducts, where they mature into adults. Adult *Clonorchis* and *Opisthorchis* flukes can survive in the human host for more than 20 years.

In the life cycle of the liver flukes *F. hepatica* and *F. gigantica*, cercariae shed by snails attach to freshwater plants (e.g., watercress) and become encysted metacercariae. After ingestion, *Fasciola* larvae excyst in the duodenum, enter the intestinal wall to reach the peritoneal cavity, penetrate Glisson's capsule, migrate through the liver parenchyma, and ultimately reach the bile ducts where the flukes mature and live for years.

In *F. buski* infections, the larvae excyst and then attach to the walls of the duodenum and jejunum, where they remain firmly attached as they develop into adults. Compared with other FBTs, *F. buski* has a relatively short lifespan (1 year) but adults can produce more than 20,000 eggs per day.

Freshwater crabs and crayfish serve as the second intermediate host of the lung fluke *Paragonimus*. After ingestion, metacercariae excyst in the small intestine and the larvae penetrate the intestinal wall, migrate through the peritoneal cavity, pass through the diaphragm into the pleural space, and then migrate into the lung parenchyma where the parasites mature into adults and produce eggs. The eggs are then expectorated in the sputum or swallowed and passed in the stool. Ectopic *Paragonimus* infections can result in the migration of the parasite into the brain or subcutaneous tissue.

## CLINICAL PRESENTATION

Among the numerous FBTs that cause animal infections, about 100 species are known to infect humans. The features of eight FBTs of medical and public health importance are summarized in Table 82.1. The acute phase of infection (when trematode larvae are migrating to their target organ) typically lasts for several weeks or months after exposure, and the clinical manifestations during this phase can range from asymptomatic illness to severe disease depending on the inoculum of metacercariae ingested. Symptoms of acute liver fluke infections include abdominal pain, fever, nausea, anorexia, weakness, and weight loss. Hepatomegaly and biliary obstruction may occur with heavy infections. Lung fluke infections can manifest with cough, hemoptysis (or rust-colored sputum), wheezing, fever, and chest pain. Although *Fasciolopsis buski* infections are often asymptomatic, heavy infections may cause severe diarrhea, abdominal pain, and malabsorption.

Peripheral eosinophilia is a common finding during the acute phase of FBT infections. Liver transaminases and alkaline phosphatase levels may also be elevated during this period. Radiographic findings of liver fluke infections are nonspecific but may include hepatic hypodensities, linear tracks, abscesses, and biliary abnormalities (e.g., gallbladder sludge, bile duct dilatation). The combination of marked eosinophilia and migratory tracks on liver imaging is highly suggestive of fascioliasis in the appropriate clinical/epidemiologic setting. Lung nodules or cysts, patchy consolidation, linear opacities, and pleural effusions are commonly seen on chest radiographs of patients with *Paragonimus* infections.

## TABLE 82.1   Selected Food-Borne Trematodes of Medical Importance

| Common Name | Genus and Species | Animal Reservoir Hosts | Source of Human Infection | Location in Human Body | Disease Manifestations |
|---|---|---|---|---|---|
| Liver flukes | *Fasciola gigantica* | Cattle, buffalo | Freshwater plants | Liver and biliary system | *Fascioliasis:* abdominal pain, anorexia and weight loss, mild intermittent fever, hepatospleno-megaly, jaundice, biliary abnormalities, necrotic lesions in hepatic tissue, fibrosis of biliary ducts. |
| | *Fasciola hepatica* | Sheep | Freshwater plants, espe-cially watercress | Liver and biliary system | |
| | *Clonorchis sinensis* | Dogs and cats | Freshwater fish | Liver, biliary system, and pancreatic duct | *Clonorchiasis:* anorexia, indigestion, abdominal pain, weakness and weight loss, gastrointes-tinal bleeding, formation of gallstones; invasion of pancreatic duct, chol-angitis, cholecystitis, and gallstones in severe infections. Increased risk of cholangiocarcinoma. |
| | *Opisthorchis viverrini* | Dogs, cats, and pigs | Freshwater fish, especially grass carp | Biliary system | *Opisthorchiasis:* similar to clonorchiasis; enlarged gallbladder, cholecys-titis, cholangitis, liver abscess, and gallstones in chronic heavy infections. Increased risk of cholan-giocarcinoma. |
| Lung flukes | *Paragonimus* species complex | Pigs, dogs, and a wide variety of felines | Freshwater crabs and crayfish | Pleural cavity and lungs; occasional brain invasion | *Paragonimiasis:* chest pain, cough with rust-colored sputum, fatigue, fever, focal hemorrhagic pneumonia, granuloma formation and fibrotic encapsulation in the lungs; possible larval migration to ectopic sites (e.g., brain, abdomen, groin, skin, heart). Mis-diagnosis of pulmonary tuberculosis or lung cancer. |
| Intestinal flukes | *Fasciolopsis buski* | Farm animals such as pigs, dogs and cats | Freshwater plants: water caltrops, water chestnuts, bamboo shoots | Small intestine | *Fasciolopsiasis:* epigastric pain, facial edema, urti-carial skin lesions, nausea and vomiting, diarrhea. |
| | *Heterophyes* | Dogs, cats, and fish-eating birds | Freshwater fish, especially mullet | Mucosa of small intestine | *Heterophyiasis:* abdominal pain, diarrhea, lethargy. |
| | *Metagonimus yokogawai* | Dogs and cats | Freshwater fish | Mucosa of small intestine | *Metagonimiasis:* similar to heterophyiasis. |

Compiled from "Foodborne Trematode Infections in Asia, Report of a Joint WHO/FAO Workshop," World Health Organization (WHO) Regional Office for the Western Pacific, 2004. Available at: https://apps.who.int/iris/bitstream/handle/10665/208026/RS_2002_GE_40_VTN_eng.pdf?sequence=1.

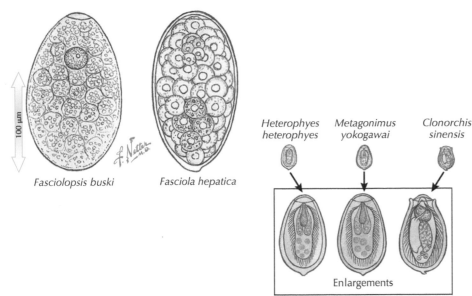

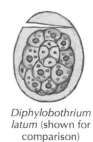

Fasciolopsis buski   Fasciola hepatica

Heterophyes heterophyes   Metagonimus yokogawai   Clonorchis sinensis

Enlargements

Paragonimus westermani   Diphylobothrium latum (shown for comparison)

**Fig. 82.2** Ova of selected food-borne trematodes.

Chronic liver fluke infections may be asymptomatic, or they may result in nonspecific abdominal symptoms. Cholelithiasis, recurrent cholecystitis, and cholangitis can also occur. The heavier parasite burdens seen among persons living in endemic regions are usually the result of repeated exposures, and such individuals may eventually manifest a variety of symptoms and end-organ disease due to the chronic inflammation and fibrosis induced by the flukes and their eggs. Some flukes may live in infected human hosts for more than a decade, and immigrants who were infected in their country of origin and subsequently moved to a nonendemic area may experience disease manifestations many years after resettlement.

Two liver flukes—*C. sinensis* and *O. viverrini*—have been categorized as class 1 carcinogens by the International Agency for Research on Cancer because of their strong association with cholangiocarcinoma, a bile duct malignancy with an extremely poor prognosis. Direct damage to the bile duct epithelium, chronic inflammation, parasite excretory-secretory products, and environmental and host factors are all thought to play a role in the pathogenesis of liver fluke-associated cholangiocarcinoma. *Fasciola* infections do not increase the risk of liver or bile duct cancer. *Paragonimus* species are not associated with malignancy, but symptoms such as hemoptysis and fatigue, along with radiographic findings such as lung nodules and pleural effusions, may result in lung fluke infections being mistaken for cancer or tuberculosis.

## DIAGNOSTIC APPROACH

The diagnosis of FBTs is based on finding and identifying trematode eggs in stool samples (and/or sputum samples in the case of *Paragonimus* infections). It is important to note that only adult flukes produce eggs. Therefore, stool specimens collected during the acute phase (sometimes referred to as the *prepatent period*, or the time from infection until the parasite develops into a sexually mature adult) will not harbor any eggs. The density of infection, as well as intermittent shedding of eggs, can affect the sensitivity of stool tests. In some cases of chronic infection, few (if any) eggs may be passed. Thus multiple specimens for ova and parasites may need to be examined if a trematode infection is suspected. Morphologic diagnosis can be challenging, as the eggs of one species may not be easily differentiated from the eggs

of another species based on morphology (e.g., *F. hepatica* vs. *F. buski*), and some of the eggs are relatively small (Fig. 82.2).

Serologic testing is more sensitive than stool microscopy for the diagnosis of FBTs and is the diagnostic method of choice during the acute phase of infection. Serology is also helpful in ectopic infections since the eggs cannot exit the gastrointestinal tract if the parasites reach their adult stage outside of their preferred target organ. However, serologic tests may not be readily available in many communities, and specimens may need to be sent to reference or research laboratories in order to obtain a definitive diagnosis.

Biopsies of liver or lung lesions are not usually necessary but may be indicated in cases of diagnostic uncertainty. Histopathologic findings may include microabscesses, granulomas, eosinophilic infiltrates, Charcot-Leyden crystals, or adult flukes or eggs. In some cases, adult liver flukes can be retrieved during procedures such as endoscopic retrograde pancreatography (ERCP).

## TREATMENT

Prompt treatment with anthelmintic medications may prevent some of the serious sequelae of FBT infections. Praziquantel (PZQ) is the drug of choice for the treatment of most flukes, with the exception of fascioliasis. Triclabendazole (TCZ) is used to treat *Fasciola* infections and was approved by the US Food and Drug Administration (FDA) in 2019 for this indication. In addition, TCZ may be used as an alternative for the treatment of paragonimiasis. Treatment regimens are listed in Table 82.2.

PZQ is also used in the treatment of schistosomiasis (a blood fluke infection) and has activity against cestodes (tapeworms). Therefore in areas that are co-endemic for these infections, central nervous system (CNS) and ocular schistosomiasis and/or cysticercosis should be ruled out prior to treatment with PZQ because of the risk of inducing inflammation and local hypersensitivity reactions from dying parasites in the CNS or eye.

Adverse effects of PZQ and TCZ include abdominal pain, nausea, headache, dizziness, and urticaria. Safety in pregnancy for either drug has not been established by the FDA, although the WHO does not restrict the use of PZQ in pregnant or breastfeeding women.

### TABLE 82.2    Drug Regimens for Treatment of Food-Borne Trematodes

| Trematode | Praziquantel, 600-mg Tablet: Adults and Children Over 4 Years of Age | Triclabendazole, 250-mg Tablet: Adults and Children Over 4 Years of Age |
|---|---|---|
| Liver flukes (excluding fascioliasis) | 25 mg/kg by mouth three times a day for 2–3 consecutive days, *or* 40 mg/kg by mouth as a single dose | Praziquantel is the drug of choice |
| Liver flukes (for fascioliasis) | Praziquantel efficacy is considered unsatisfactory for this infection | 10 mg/kg by mouth as a single dose |
| Lung flukes | 25 mg/kg by mouth three times a day for 2–3 consecutive days | 20 mg/kg by mouth given in two divided doses in 1 day |
| Intestinal flukes | 25 mg/kg by mouth as a single dose | Praziquantel is the drug of choice |

Data from WHO Foodborne Trematodiases Fact Sheet. Accessed September 29, 2019. Available at: https://www.who.int/news-room/fact-sheets/detail/foodborne-trematodiases.

## PREVENTION AND CONTROL

The risk of FBT infections can be reduced by thoroughly cooking freshwater fish, crustaceans, and plants before being eaten. Freezing at −20°C for 7 days kills encysted metacercariae. Populations at risk can be educated about the benefits of cooking aquatic products and should be informed that smoking and pickling raw fish or crustaceans do not destroy encysted parasites. However, changing long-standing cultural preferences and culinary practices is difficult.

Global control of FBTs requires a multipronged approach. Detection and treatment of newly acquired FBT infections in humans can prevent the morbidity and mortality associated with chronic infections and can help interrupt transmission through contaminated human feces. Similarly, chemotherapy of infected livestock can contribute to decreased environmental contamination with infected feces.

On a larger scale, agricultural communities in endemic areas would benefit from improved practices for managing human and animal waste and should avoid using wastewater to irrigate or fertilize crops and aquaculture farms. Inspection and regulation of exported foods may influence the techniques used in commercial aquaculture ventures but have little impact on rural farmers raising fish or vegetables for local consumption.

In regions that are highly endemic for *C. sinensis* or *O. viverrini*, the development of evidence-based guidelines for cholangiocarcinoma screening in patients infected with these parasites may help increase early detection of this lethal malignancy.

## EVIDENCE

Bui TD, Doanh PN, Saegerman C, et al. Current status of fasciolosis in Vietnam: An update and perspectives. *J Helminthol* 2016;90:511-522. *Review of the epidemiology of fascioliasis in Vietnam, noting that the central region of the country is highly endemic and F. gigantica is the main species.*

Furst T, Keiser J, Utzinger J. Global burden of human food-borne trematodiasis. A systematic review and meta-analysis. *Lancet Infect Dis* 2012;12:210-221. *WHO-commissioned assessment of the global burden of FBTs expressed in disability-adjusted life years (DALYs).*

Xia J, Jiang S, Peng H. Association between liver fluke infection and hepatobiliary pathological changes: A systematic review and meta-analysis *PLOS One* 2015;10(7):e0132673. Doi:10.1371/journal.pone.0132673. *Analysis of the evidence supporting the role of liver fluke infections in hepatobiliary disease, including cholangiocarcinoma.*

Yoshida A, Doanh PN, Maruyama H. *Paragonimus* and paragonimiasis in Asia: An update. *Acta Tropica* 2019;199:105074. *Review of the epidemiology of* Paragonimus *infections in Asia. Notes the re-emergence of paragonimiasis in Japan since the 1980s, a trend attributable to consumption of the raw meat of wild boars (a paratenic host).*

Yossepowitch O, Gotesman T, Assous M, et al. Opisthorchiasis from Imported raw fish. *Emerg Infect Dis* 2004;10:2122-2126. *Interesting account of a familial outbreak of liver fluke infection in Israel transmitted by illegally imported smoked carp from Siberia.*

## ADDITIONAL RESOURCES

Diaz JH. Paragonimiasis acquired in the United States: Native and nonnative species. *Clin Microbiol Rev* 2013;26(3):493-504. *Summary of imported and autochthonous* Paragonimus *infections in the United States.*

Keiser J, Utzinger J. Foodborne trematodiases. *Clin Microbiol Rev* 2009;22(3):466-483. *Broad review of food-borne trematode infections in humans.*

Mas-Coma S, Bargues MD, Valero MA. Human fascioliasis infection sources, their diversity, incidence factors, analytical methods and prevention measures. *Parasitol* 2018;145:1665-1699. *Compilation of the various sources of* Fasciola *infections, with color photographs of plants and traditional dishes.*

Qian M, Utzinger J, Keiser J, et al. Clonorchiasis. *Lancet* 2016;387:800-810. *Historical, epidemiologic and clinical review of Clonorchis sinensis, with recommendations for research priorities.*

Sripa B, Echaubard P. Prospects and challenges toward sustainable liver fluke control. *Trends Parasitol* 2017;3(10):799-812. *Discussion of the complexities of* Opisthorchis *control programs and the need to incorporate social-ecological approaches.*

# 83

# Schistosomiasis

Elaine C. Jong

## ABSTRACT

Schistosomiasis (bilharzia) is a disease caused by infection with parasitic flat-worms known as *blood flukes*, belonging to the genus *Schistosoma*. According to the World Health Organization (WHO), schistosomiasis transmission has been reported from 78 countries, and it is estimated that more than 200 million people living in tropical endemic areas are infected. Chronic schistosomiasis causes significant morbidity and premature mortality, and negatively affects both individual health and socioeconomic development in poor and rural communities that lack access to safe drinking water and adequate sanitation.

The infection is acquired through direct skin contact with water inhabited by freshwater snails (intermediate hosts), shedding infective larvae into the water. The infective larvae, called *cercariae*, penetrate the intact wet skin of human hosts, mature in the vasculature of the liver, and migrate to the mesenteric (*Schistosoma mansoni* and *Schistosoma japonicum*) or pelvic (*Schistosoma haematobium*) veins. The female worms lay eggs that are excreted in the stool or urine. Acute disease manifestations include rash, fever, abdominal pain, and bloody stools or urine; chronic infection is characterized by an inflammatory immune response to eggs retained in the tissues (granuloma formation and fibrosis), leading to intestinal, hepato-splenic, or urogenital disease.

The burden of infection is directly related to the period of time persons living in endemic regions are immersed in or wading in bodies of water containing infected snails. Established infections do not stimulate protective host immunity, so new schistosome infections may be acquired during every instance of contact with infected-snail-contaminated water. The snails tend to live among vegetation close to the shoreline and to shed infective cercariae during the middle of the day when the sun is shining brightly. This is the time of day when residents in endemic areas are most likely to be in the water—fishing, boating, farming, bathing, washing clothes, gathering water, swimming, and engaging in other water activities. Males in the population seem to acquire heavier parasite burdens from childhood onward and experience more severe disease manifestations than females, ascribed to longer periods of water exposure of males through occupational and recreational activities. Visitors to endemic areas have acquired schistosomiasis while living and working among resident populations or while participating in recreational activities such as swimming, snorkeling, and kayaking in contaminated bodies of freshwater. When travelers return home to temperate nonendemic regions, acute disease manifestations may not be recognized at first when they occur.

## CLINICAL VIGNETTE

"A 30-year-old male patient was evacuated from Namibia and evaluated at a US medical center because of a 2-week history of headaches, unilateral (left) vision loss, and one episode of loss of consciousness consistent with a seizure. The patient had been a Peace Corps Volunteer for 2 years in Tunisia and was serving an additional 2 years in Namibia. He had no history of recreational freshwater exposure in Tunisia or Namibia. However, he had snorkeled during a 2-day period at Cape Maclear in Lake Malawi [about 3 months before presentation]. A physical exam and white blood cell count (including eosinophil count) were normal. Computed tomography (CT) and magnetic resonance imaging (MRI) scans detected an enhancing left parietal lesion with extensive edema. Open brain biopsy was performed because the lesion was initially presumed to be a meningioma. The biopsy specimen demonstrated a granulomatous abscess containing parasite eggs that had peripheral spines consistent with schistosome eggs. *S. haematobium* eggs were identified in his urinary sediment but not in his stool. Schistosomiasis serology using the Falcon assay screening test-enzyme linked immunosorbent assay (FAST-ELISA) was positive at CDC. A confirmatory immunoblot was positive for *S. hematobium* antibody but not *S. mansoni* antibody. He was treated with praziquantel (PZQ; 60 mg/kg body weight per day, orally in three divided doses) for schistosomiasis, phenytoin for his seizure disorder, and dexamethasone for the cerebral edema." At a medical follow-up 3 months afterward, "his symptoms had resolved and a CT scan documented continuing improvement. He returned to Namibia to complete his Peace Corps service." *Reported by Wolfe M, Parenti D, Pollner J, Kobrine A, Schwartz, A. Schistosomiasis in U.S. Peace Corps Volunteers—Malawi, 1992. MMWR, July 30, 1993. 42(29):565-570. Available at: http://www.cdc.gov. Accessed November 13, 2019.

COMMENT: This historical case report of schistosomiasis in a Peace Corps volunteer illustrates several clinical points: presentation of schistosomiasis as a space-occupying lesion in the brain, infection in a person with unbeknownst exposure to infected freshwater snails in Lake Malawi in sub-Saharan Africa, onset of clinical signs and symptoms months after the presumed infectious episode, appropriate medical examination and parasite diagnostic tests, and management and treatment of schistosomiasis involving the central nervous system.

## GEOGRAPHIC DISTRIBUTION

Schistosomiasis is prevalent in tropical and subtropical areas of Africa, the Middle East, South America, Southeast Asia, and the southern provinces of the People's Republic of China. Surveillance data are incomplete but indicate that more than 90% of infected persons live in sub-Saharan Africa. In the past, the parasite was also transmitted in some islands in the Caribbean, the Philippines, and in Japan, but control programs appear to have been successful in eradicating the disease in these areas (Fig. 83.1). Three species of *Schistosoma* are associated with the majority of human disease: *Schistosoma mansoni*, *Schistosoma japonicum*, and *Schistosoma haematobium*, and humans serve as the definitive host where sexual reproduction of the parasite takes place. Other schistosome species have been recognized less commonly as agents of intestinal schistosomiasis in humans, reported only from specific areas and having limited transmission: *S. mekongi* in Southeast Asia, and *S. guineensis* and related *S. intercalatum* in Africa. Each *Schistosoma* species has a specific snail intermediate host necessary for the completion of asexual life-cycle stages (Table 83.1). The geographic range of disease transmission is actually determined by the geographic

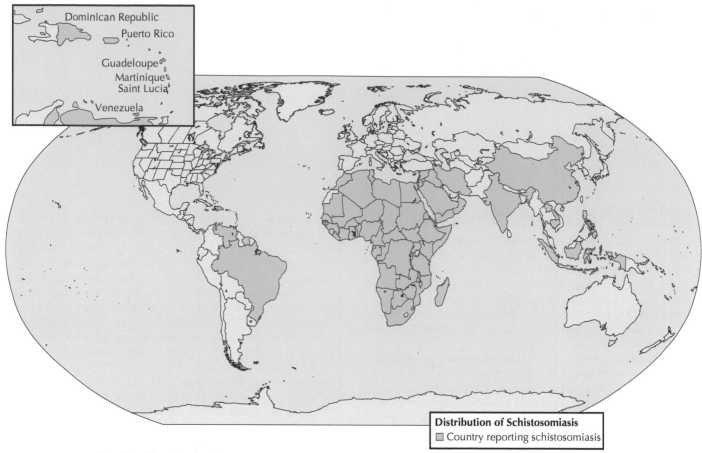

**Fig. 83.1** Map showing the global distribution of schistosomiasis. (Adapted from Centers for Disease Control and Prevention [CDC]: *Health information for international travel 2020*, Atlanta: U.S. Department of Health and Human Services, Public Health Service, 2019 Map 4-11. Available at: https://wwwnc.cdc.gov/travel/yellowbook/2020/travel-related-infectious diseases/schistosomiasis#5591.)

**TABLE 83.1  Parasite Species and Geographic Distribution of Schistosomiasis**

| Disease Description | Species | Snail Intermediate Host | Geographic Distribution |
|---|---|---|---|
| Intestinal schistosomiasis | *Schistosoma mansoni* | *Biomphalaria* species | Africa, the Middle East, the Caribbean, Brazil, Venezuela, and Suriname |
| | *Schistosoma japonicum* | *Oncomelania* species | China, Indonesia, the Philippines |
| | *Schistosoma mekongi* | *Neotricula* species | Several districts of Cambodia and the Lao People's Democratic Republic |
| | *Schistosoma guineensis* and related *S. intercalatum* | *Bulinus* species | Rain forest areas of central Africa |
| Urogenital schistosomiasis | *Schistosoma haematobium* | *Bulinus* species | Africa, the Middle East, Corsica (France) |

Adapted from World Health Organization (WHO): Schistosomiasis Fact Sheet of 2 March 2020. Available online: http://www.who.int/news-room/fact-sheets/detail/schistosomiasis. Accessed November 29, 2020.

distribution of the requisite snails, and expansion of areas where schistosomiasis is transmitted has been associated with new habitats for snails created by the development of water resources: dams, irrigation waterways, streams, and ponds, and the concurrent migration of infected human populations into the new habitats. Snails spread to new areas not only by migration along contiguous waterways but also by introduction into new bodies of freshwater on the feet of aquatic birds.

## LIFE CYCLE

Schistosomiasis is transmitted to humans by infective parasite larvae called *cercariae* released into freshwater by infected aquatic snails.

The cercariae are free-swimming and capable of penetrating intact wet human skin. Upon skin penetration, the cercaria sheds its motile forked tail section, and the tail-less body, called a *schistosomulum*, accesses the circulation and migrates to the hepatic vasculature, where maturation takes place. After the schistosomula mature into male and female adults, the adults pair up, mate, and migrate to the mesenteric veins of the intestine (intestinal and hepatic schistosomiasis) (Fig. 83.2), or to the venous plexus of the urinary bladder (urogenital schistosomiasis) (Fig. 83.3). Eggs are deposited on the peritoneal surface of the intestine or bladder, and the eggs work their way through the walls of the intestine or bladder into the viscus cavity, and exit the body in the fecal stream or in the urine, respectively.

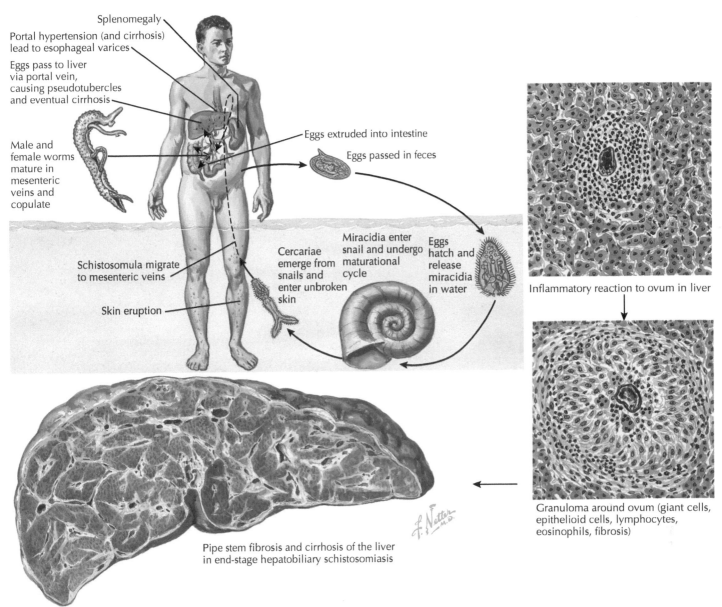

Splenomegaly

Portal hypertension (and cirrhosis) lead to esophageal varices

Eggs pass to liver via portal vein, causing pseudotubercles and eventual cirrhosis

Male and female worms mature in mesenteric veins and copulate

Eggs extruded into intestine

Eggs passed in feces

Schistosomula migrate to mesenteric veins

Skin eruption

Cercariae emerge from snails and enter unbroken skin

Miracidia enter snail and undergo maturational cycle

Eggs hatch and release miracidia in water

Inflammatory reaction to ovum in liver

Granuloma around ovum (giant cells, epithelioid cells, lymphocytes, eosinophils, fibrosis)

Pipe stem fibrosis and cirrhosis of the liver in end-stage hepatobiliary schistosomiasis

**Fig. 83.2** Life cycle of *Schistosoma mansoni* and pathogenesis of hepatic schistosomiasis.

The transmission cycle begins anew when untreated human feces and urine containing parasite ova (eggs) contaminate the freshwater aquatic environment. Under appropriate conditions of light and temperature, the schistosome eggs burst open in the water, releasing free-swimming parasite larvae called *miracidia*. The miracidia are infective only for the species-specific snail host. A week or so after the snails become infected, they will release cercariae, the parasite larval stages that are infective for humans, starting the cycle all over again.

## CLINICAL PRESENTATION

*Acute schistosomiasis:* Within hours after cercarial penetration of the skin, with transformation of the cercarial stage parasites into schistosomula, some individuals develop a pruritic skin rash called cercarial dermatitis (also known as "swimmer's itch"), representing each point of parasite entry. The rash may last for a few days, and itching can be relieved by antihistamines.

An asymptomatic incubation period then follows lasting approximately 4 to 8 weeks, during which time the schistosomula migrate to

the hepatic circulation, mature to adult male and female worms, mate, and migrate against the direction of portal blood flow to the mesenteric veins or the venous plexus surrounding the bladder, depending on species. After copulation, the female worm migrates to the terminal venules and deposits eggs onto the peritoneal surface of the intestines or the bladder. With the aid of intrinsic mechanical movements of the egg and egg secretions, the eggs transit the walls of the viscus and break through the mucosa on the luminal side. Local tissue injury resulting from the passage of parasite eggs through the walls of the intestines into the bowel lumen is associated with abdominal pain and bloody stools in intestinal schistosomiasis, and passage of parasite eggs through the urinary bladder wall into the bladder lumen may result in grossly bloody urine in urogenital schistosomiasis. Light infections may be asymptomatic and go unnoticed.

Dispersion of *S. haematobium* eggs to ectopic locations outside the bladder may occur because of the complex anatomy of the vesicle venous plexus, which on occasion allows migratory worm pairs and loose eggs to gain access to other parts of the circulation. Eggs shunted to the systemic circulation may end up in ectopic locations such as

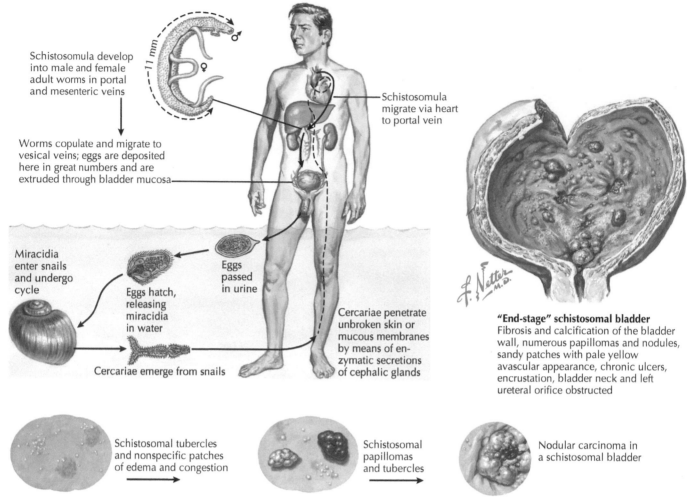

Schistosomula develop into male and female adult worms in portal and mesenteric veins

—11 mm—

Worms copulate and migrate to vesical veins; eggs are deposited here in great numbers and are extruded through bladder mucosa

Schistosomula migrate via heart to portal vein

Miracidia enter snails and undergo cycle

Eggs hatch, releasing miracidia in water

Eggs passed in urine

Cercariae emerge from snails

Cercariae penetrate unbroken skin or mucous membranes by means of enzymatic secretions of cephalic glands

f. Netter M.D.

**"End-stage" schistosomal bladder**
Fibrosis and calcification of the bladder wall, numerous papillomas and nodules, sandy patches with pale yellow avascular appearance, chronic ulcers, encrustation, bladder neck and left ureteral orifice obstructed

Schistosomal tubercles and nonspecific patches of edema and congestion

Schistosomal papillomas and tubercles

Nodular carcinoma in a schistosomal bladder

**Fig. 83.3** Life cycle of *Schistosoma haematobium* and schistosomiasis of the urinary bladder.

the brain (seizures), heart (acute myocarditis, heart block), spinal cord (transverse myelitis, pain, paralysis), and skin ("egg dermatitis" rash).

A severe illness may occur 4 to 8 weeks after initial infection, around the time of parasite oviposition, as described previously, which probably represents a pronounced hypersensitivity reaction to a variety of circulating parasite antigens. The illness is characterized in most cases by a high fever and may be accompanied by cough and urticaria. A peripheral blood eosinophilia is often present. This illness was first described in association with *S. japonicum* infections in the district of Katayama near Hiroshima, Japan, and was called *Katayama fever*; a similar illness may occur in acute schistosomiasis caused by *S. mansoni* and also has been reported recently in some cases of *S. haematobium* infection. Acute schistosomiasis may be misdiagnosed as malaria, as initial results of definitive diagnostic tests may be negative in both acute febrile illnesses.

***Chronic schistosomiasis:*** Adult schistosomes can live in humans for decades, persisting long after the initial infective episode(s) with prolonged production of eggs by the female worms. In intestinal schistosomiasis, the intestinal walls become scarred by egg granulomas over time, leading to further entrapment of parasite eggs in the tissues. Eggs that do not penetrate the walls are passively swept away in the direction of the portal blood flow towards the liver and become trapped in the portal triads, where granulomatous reactions around the eggs lead to fibrosis (see Fig. 83.2). Bridging "pipe-stem fibrosis" of the liver is the hallmark of chronic severe infection with

*S. mansoni*. Cirrhosis, portal hypertension, splenomegaly, and esophageal varices represent late stages of chronic hepatic schistosomiasis, and exsanguination from bleeding esophageal varices is a common cause of premature death in infected adults in their second or third decade of life.

In chronic urogenital schistosomiasis, egg granulomas in the bladder wall may eventually cause ureteral obstruction, leading to hydronephrosis and subsequent renal failure (see Fig. 83.3). Impaired bladder function leading to incomplete urinary voiding is accompanied by recurrent urinary tract infections. Chronic inflammation of the bladder wall caused by egg granulomas is linked epidemiologically to an increased risk of bladder cancer in *S. haematobium*–infected persons. The involvement of the female genital organs can result in dyspareunia and impaired fertility.

## DIAGNOSTIC APPROACH

The diagnosis of schistosomiasis is made by morphologic identification of the characteristic eggs in stool, urine, or tissue specimens (Fig. 83.4). The *S. mansoni* egg is relatively large and characterized by a prominent hook-like lateral spine. The *S. haematobium* egg is characterized by a prominent terminal spine. The *S. japonicum* egg is relatively small and has a small inconspicuous lateral knob; it may traverse smaller blood vessels than the eggs of other species before becoming lodged and trapped in end-organ tissues.

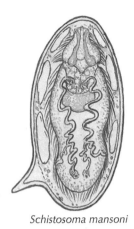
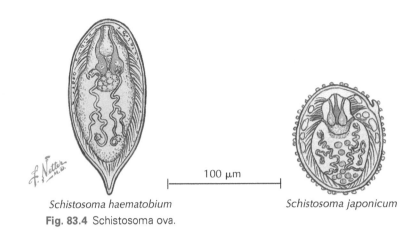

*Schistosoma mansoni*                *Schistosoma haematobium*                100 μm                *Schistosoma japonicum*

**Fig. 83.4** Schistosoma ova.

In light infections the ova and parasite examinations of stool and urine specimens may not yield the diagnostic forms, even if egg concentration methods are used: Kato-Katz for intestinal schistosomiasis and urine filtration for the urogenital schistosomiasis. Tests for schistosome circulating cathodic antigen (CCA) are more sensitive than stool examinations and can used for diagnosis in *S. mansoni* transmission areas. Molecular techniques such as loop-mediated isothermal amplification (LAMP) and the polymerase chain reaction (PCR) can provide high specificity and high sensitivity for the detection of schistosome components in infected patients, helping identify low-intensity and prepatent infections, but such diagnostic tests are not generally available at this time.

Serological and immunological tests may be useful in showing exposure to infection and the need for further medical evaluation for patients living in nonendemic or low transmission areas if at least 8 to 10 weeks have elapsed since presumed exposure to schistosomiasis: the FAST-ELISA based on *S. mansoni* adult microsomal antigen and Western blot for confirmation and speciation are available from the Centers for Disease Control and Prevention (CDC).

Another way to make the diagnosis when stool samples are negative in persons with a strong history of possible exposure to intestinal schistosomiasis is to identify the characteristic eggs of *S. mansoni* or *S. guineensis/S. intercalatum* in wet mounts of rectal mucosal snips. Schistosomiasis may also be diagnosed from tissue biopsy samples taken during colonoscopy, cystoscopy, or surgical procedures in more seriously ill patients. Typical findings seen during cystoscopy of patients with advanced genitourinary disease are shown in Fig. 83.3.

## TREATMENT

### Praziquantel

PZQ is the drug recommended by the WHO to treat all forms of schistosomiasis, both intestinal and urogenital schistosomiasis. However, PZQ is inactive against juvenile parasites that the host may be harboring at the time of treatment, and these may mature and start egg production following a round of chemotherapy—thus when PZQ is used for treatment of acute schistosomiasis, a second dose given approximately 3 months after the initial exposure is recommended to eradicate any remaining adult flukes.

Praziquantel is an oral drug that is manufactured as 600-mg tablets, scored into four segments of 150 mg each. An oral dose of 60 mg/kg/day given in three divided doses for 1 day, each dose separated by 4 to 6 hours, is indicated for the treatment of schistosomiasis in the official package insert. Adverse side effects are not serious but include nausea,

dizziness, and abdominal discomfort. PZQ may be used in patients 4 years of age and older, but safety in pregnant and lactating women has not been established, and caution is advised for use in persons with impaired renal function, as drug elimination from the body is mainly through renal excretion. PZQ is thought to interfere with calcium ion influx in the adult parasite, but the molecular basis of activity has not been completely defined.

Because of economic constraints in the countries most affected by endemic schistosomiasis, clinical studies have investigated alternate dosage regimens and duration of treatment for PZQ drug therapy. PZQ at a dose of 40 mg/kg given once for treatment of *S. haematobium* has been shown to be highly efficacious and is currently considered standard. Other treatment protocols employing the PZQ package insert dose of 60 mg/kg/day given in three divided doses for durations longer than 1 day have been investigated for the treatment of schistosome infections in focal areas where emerging parasite resistance to the drug is a concern.

If acute schistosomiasis involves the central nervous system, PZQ treatment may be deferred to avoid drug-induced disintegration of parasite eggs and worsening local inflammatory reactions. If respiratory manifestations of acute schistosomiasis are prominent, therapy with corticosteroids may be implemented prior to the initiation of antiparasitic treatment.

### Artemether

Artemether (ART) and artesunate are semisynthetic artemisinin (qinghaosu) derivatives that are powerful antimalarials used in artemisinin-based combination therapy (ACT) for the treatment of patients with severe malaria. However, in addition to its main use against malaria parasites, artemisinin is active against immature schistosome worms. This is in contrast to PZQ that kills mature schistosomes but not juveniles. Studies show that when an artemisinin derivative is used together with the highly effective PZQ as a combination therapy against all forms of schistosomiasis, individual cure rates are higher than with PZQ treatment alone. In a systemic review and meta-analysis of published reports on artemisinin-based therapies for the treatment of schistosomiasis, PZQ and an artemisinin derivative (ART or artesunate) were compared as agents for monotherapy and in combination therapy: patients treated with artesunate alone had significantly lower parasitological cure rates than those treated with PZQ alone. However, patients treated with PZQ combined with an artemisinin derivative showed a higher cure rate than with PZQ alone. The role of artemisinin derivatives in prophylaxis of schistosomiasis and important constraints are discussed below.

## PREVENTION AND CONTROL

Travelers can be instructed to take personal precautionary measures to prevent schistosomiasis infection, such as not swimming or wading in bodies of freshwater when visiting endemic areas, and toweling the skin off immediately if drops of water from contaminated sources are splashed onto bare skin. However, avoiding personal contact with contaminated water is impractical advice for residents who live in endemic areas under impoverished conditions: their livelihood and activities of daily living largely center around the local water supply.

A three-pronged approach is used to prevent and control schistosomiasis in endemic areas: (1) improved sanitation and hygiene to prevent raw sewage from contaminating bodies of freshwater; (2) environmental interventions to destroy snail habitats by use of herbicides and destruction of the snails themselves by the use of molluscicides; and (3) interval mass drug administration (MDA) programs targeting all local residents presumed to be infected in endemic areas.

Since 1984, the WHO has recommended PZQ as treatment against all forms of schistosomiasis and for morbidity control among infected populations through annual or biennial MDA programs. Although PZQ MDA is highly effective in curing acute infections and managing severe morbidity due to chronic disease, the PZQ MDA programs have not been effective in lowering schistosomiasis transmission rates in affected communities because transmission and reinfection may occur following treatment.

Artemisinin derivatives currently used against malaria in ACT are active against juvenile schistosome parasites, and artemisinin-enforced praziquantel treatment (APT) has transmission-blocking capability: artemisinin kills juvenile schistosomes in human infections that are not affected by PZQ, thus decreasing the potential for surviving juveniles to mature, mate, and lay eggs after a treatment is finished. Because of the risk of emerging artemisinin resistance, the implementation of APT combined regimens for schistosomiasis MDA will be limited to endemic countries where ACT is not needed for malaria care. APT MDA programs might be acceptable and beneficial in schistosomiasis endemic areas of North Africa (including Egypt), the Middle East, China, and Brazil; further studies are needed.

Research on human vaccines against schistosomiasis progresses slowly, and even if a vaccine candidate can be shown to elicit good protection, it will take many years for full validation, large-scale production, and release of a product. One candidate vaccine that has reached clinical-trials level is the SM14/GLA-SE schistosomiasis vaccine based on the fatty acid-binding protein (FABP) and formulated with the glucopyranosyl lipid A (GLA-SE) adjuvant. The vaccine was primarily developed against *S. mansoni* but cross-reacts with *S. haematobium* and *S. japonicum*.

## EVIDENCE

Bergquist R, Elmorshedy H. Artemether and praziquantel: origin, mode of action, impact, and suggested application for effective control of human schistosomiasis. *Trop Med Infect Dis* 2018;3(4):125. doi:10.3390/tropicalmed3040125. *A thorough consideration of the challenges encountered in the control of human schistosomiasis and the feasibility of employing artemether and praziquantel in mass drug administration programs in high-risk transmission areas of countries.*

Kabuyaya M, Chimbari MJ, Mukaratirwa S. Efficacy of praziquantel treatment regimens in pre-school and school aged children infected with schistosomiasis in sub-Saharan Africa: a systematic review. *Infect Dis Poverty* 2018;7:73. Doi: 10.1186/s4j0249-018-0448-x. *Efficacy of praziquantel treatment with repeated standard doses of 40 mg/kg was demonstrated against S. haematobium and S. mansoni infections in preschool and school-aged children. Co-infections of S. haematobium and S. mansoni may require a higher praziquantel dose in this population and further investigations are needed to formulate guidelines.*

Meltzer E, Artom G, Marva E, et al. Schistosomiasis among travelers: new aspects of an old disease. *Emerg Infect Dis* 2006;12:1696-1700. Available at: www.cdc.gov/eid. Accessed March 15, 2010. *Well-written analysis of the clinical presentation and disease parameters among 137 Israeli travelers, most returning from trips to sub-Saharan Africa, in whom the diagnosis of schistosomiasis was made; 42.5% of the patients were asymptomatic. Serologic tests were more helpful in making the diagnosis in 87.6% of patients with acute schistosomiasis than traditional ova and parasite examinations. Diagnosis was made in 26.9% of the patients by finding ova in submitted urine, semen, or stool specimens; these patients tended to have chronic schistosomiasis.*

Perez del Villar L, Burguillo FJ, Lopez-Aban J, Muro A. Systematic review and meta-analysis of artemisinin-based therapies for the treatment and prevention of schistosomiasis. *PLoS One* 2012;7(9):e45867. doi:10.1371/journal.pone.0045867. Epub 2012 Sep 21. *This meta-analysis confirms that artemisinin derivatives used in combination with praziquantel have the potential to increase the cure rates in schistosomiasis treatment but not artesunate alone.*

## ADDITIONAL RESOURCES

Bergquist R, Gray DJ. Schistosomiasis elimination: beginning of the end or a continued march on a trodden path. *(Editorial). Trop Med Infect Dis* 2019; 4:76. doi:10.3390/tropicalmed4020076. *A comprehensive and thoughtful review of efforts to eliminate human schistosomiasis over the past 40 years.*

World Health Organization (WHO): Schistosomiasis Fact Sheet of 20 February 2019. Available online: http://www.who.int/news-room/fact-sheets/detail/schistosomiasis (accessed 5 January 2020). *Useful portal for accessing current information on global epidemiology, treatment trials, and prevention and control projects.*

# 84

# Filarial Diseases

*Jan M. Agosti*

## ABSTRACT

The filarial parasites covered in this chapter constitute a group of tissue-dwelling filarial nematodes that persist in the human host for years, causing damage to the lymphatic system that leads to elephantiasis and genital hydroceles (in the case of lymphatic filariasis) or marked inflammatory reactions in the skin and eyes leading to blindness (in the case of onchocerciasis). Both diseases are designated neglected tropical diseases by the World Health Organization (WHO). Global elimination efforts are addressing this significant health burden, using annual or twice-annual mass drug administration of donated anthelmintic drugs in concert with vector-control measures. Significant progress has been made, particularly on onchocerciasis elimination from the Americas. A remaining challenge is to develop a strategy to use in areas of Africa co-endemic for *Loa loa* and lymphatic filariasis or onchocerciasis—where the existing drug regimens can precipitate significant adverse side effects due to rapid die-off of the high numbers of *L. loa* parasites. Another challenge is developing more effective macrofilaricidal treatments to kill adult worms in order to shorten the duration of mass drug administration elimination programs.

## CLINICAL VIGNETTE

A 23-year-old male from the Democratic Republic of Congo presented to the emergency room complaining of a worm in his eye. The patient had lived in rural Congo, bathed in lakes and rivers, and had friends with river blindness. For the past year, he had noted sporadic nontender swellings on his arms which resolved spontaneously. Past medical history was unremarkable, and he was taking no medications. The exam confirmed the presented of a 5-mm, moving worm in the scleral conjunctiva with mild conjunctival injection. There was mild photophobia without discharge. Ophthalmologic exam showed cornea was clear, no evidence of uveitis, and normal retina. There were no subcutaneous nodules or dermatitis. White blood cell count was normal, except for 43% eosinophils. Direct visualization of the ocular adult worm was considered diagnostic of *L. loa*; however, additional testing was required to evaluate for co-infection and microfilarial burden in order to guide treatment. Filarial IgG antibody level was positive; onchocercal IgG4, which was done to rule out co-infection with onchocerciasis, was negative. Thick smear of blood showed microfilaria with the characteristic *L. loa* sheath at the tail and a low level of microfilaremia. Since diethylcarbamazine (DEC), the drug of choice for treatment of *L. loa*, is contraindicated in patients with onchocerciasis due to the risk of severe ocular inflammation and resulting blindness, as a precaution, the Centers for Disease Control and Prevention (CDC) required pretreatment of the patient with ivermectin for potential onchocerciasis co-infection, before releasing the DEC for treatment of the *L. loa*.

## GEOGRAPHIC DISTRIBUTION AND MAGNITUDE OF DISEASE BURDEN

### Lymphatic Filariasis *(Wuchereria bancrofti, Brugia malayi, Brugia timori)*

Lymphatic filariasis is a leading cause of permanent and long-term disability worldwide. *W. bancrofti* is responsible for the vast majority of cases and is found throughout the tropics and in some subtropical areas worldwide (Fig. 84.1; Table 84.1). *B. malayi* is present in South, Southeast, and East Asia. *B. timori* is restricted to the island of Timor and its environs. The global burden of lymphatic filariasis is not well defined, but it is known to be endemic in 72 countries, placing 1 billion persons at risk, with the WHO most recently estimating 120 million people to be currently infected—2% of the world's population. Of these, 44 million have clinical manifestations such as lymphedema, elephantiasis, hydrocele, lymphangitis, chyluria, and renal disease. The remaining infected individuals often have subclinical lymphatic or renal injury. In addition to the disease burden, there is a significant social, psychological, and economic burden owing to social stigma, physical stigmata, and inability to work. The current estimates are that 4.6 million disability-adjusted life years (DALYs) are lost each year to this disease.

### Onchocerciasis *(Onchocerca volvulus)*

Onchocerciasis, commonly called *river blindness*, is caused by the filarial parasite *O. volvulus* and is one of the most common infectious causes of blindness worldwide, second only to trachoma. There are an estimated 25 million infected individuals in 30 countries, primarily in Africa but also Yemen and limited areas of the Americas, with 123 million living in endemic areas and at risk of infection (Fig. 84.2). Of those infected, an estimated 300,000 individuals are blind and another 800,000 are visually impaired. Each year, 1,990,000 DALYs are lost to onchocerciasis; of those, approximately 40% are a result of eye disease and 60% from severe skin manifestations. This disability affects not only the infected patients but also the children who escort blind individuals, preventing them from attending school. Fertile riverine areas needed for agricultural development are abandoned because of the risk of this disease, which has devastating socioeconomic consequences. Infections outside of endemic populations may occur in expatriates, usually only those living for a prolonged period in affected areas.

### Loiasis *(Loa loa)*

Loiasis occurs only in rainforest and swamp forest areas of Central and West Africa, as its distribution is primarily restricted by its transmission vector (see Fig. 84.2). Its primary significance is its co-endemicity in areas of Africa

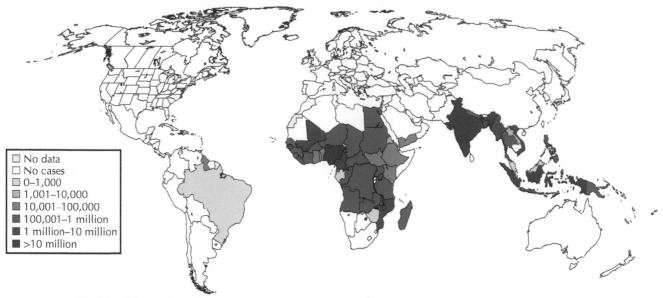

**Fig. 84.1** Epidemiology of lymphatic filariasis and presence of mass drug administration programs. (Reused from Our World In Data, a project of the Global Change Data Lab, a registered charity in England and Wales [Charity Number 1186433]. https://ourworldindata.org/grapher/prevalence-of-lymphatic-filariasis.)

**TABLE 84.1     Filarial Nematode Characteristics and Clinical Manifestations**

| Parasite Species (Disease Name) | Distribution | Vector | Adult Location | Microfilariae Location | Clinical Features |
|---|---|---|---|---|---|
| *Wuchereria bancrofti* (bancroftian lymphatic filariasis) | Tropical regions | *Anopheles, Aedes,* and *Culex* species | Lymphatics | Blood | Lymphangitis Elephantiasis Hydrocele Chyluria |
| *Brugia malayi* (brugian lymphatic filariasis) | South, East, and Southeast Asia | *Mansonia, Anopheles,* and *Aedes* species | Lymphatics | Blood | Lymphangitis Elephantiasis |
| *Brugia timori* (brugian lymphatic filariasis) | Indonesia | *Anopheles* species | Lymphatics | Blood | Lymphangitis Elephantiasis |
| *Onchocerca volvulus* (onchocerciasis, river blindness) | Africa, Central and South America | *Simulium* species | Skin | Skin | Dermatitis Subcutaneous nodules Eye lesions Visual impairment |
| *Loa loa* (loiasis) | Central and West Africa | *Chrysops* species | Connective tissue | Blood | Calabar swellings Subconjunctival eye worm |

with lymphatic filariasis and onchocerciasis, because patients with *L. loa* co-infection cannot be treated with the usual anti-microfilarial regimens used in mass drug administration programs owing to side effects from high numbers of dying *L. loa* microfilariae in the circulation and tissues.

## RISK FACTORS

### Lymphatic Filariasis *(Wuchereria bancrofti, Brugia malayi, Brugia timori)*

Risk factors for acquisition of the mosquito-borne filarial parasites *W. bancrofti, B. malayi,* and *B. timori* include living in endemic areas with exposure to a broad range of potential vectors that promote human-to-human spread (see Table 84.1). For lymphatic filariasis, the mosquito vectors are diverse and include *Anopheles* species, *Culex* species, *Aedes* species, and *Mansonia* species. Infants, children, and adults are exposed to mosquitoes carrying infective larvae (Fig. 84.3). There is

no natural animal reservoir for *W. bancrofti* or *B. timori*, but a very small percentage (under 5%) of domestic and wild animals such as monkeys and cats tested have been found to be infected with *B. malayi*. Urbanization in endemic areas may be contributing to an increased risk of lymphatic filariasis because infected and susceptible individuals may then live in close proximity to each other and to stagnant, contaminated water that provides a breeding ground for the vector. Irrigation projects may further increase risk.

### Onchocerciasis *(Onchocerca volvulus)*

Onchocerciasis is termed *river blindness* because the vector, the *Simulium* species black fly, breeds along rivers in fast-flowing water where its larvae can filter feed. Residents along rivers are at highest risk, but the fly has also been carried by monsoon winds for hundreds of kilometers, reinfecting areas that have previously eradicated the fly vector. There is no animal reservoir for onchocerciasis, and transmission

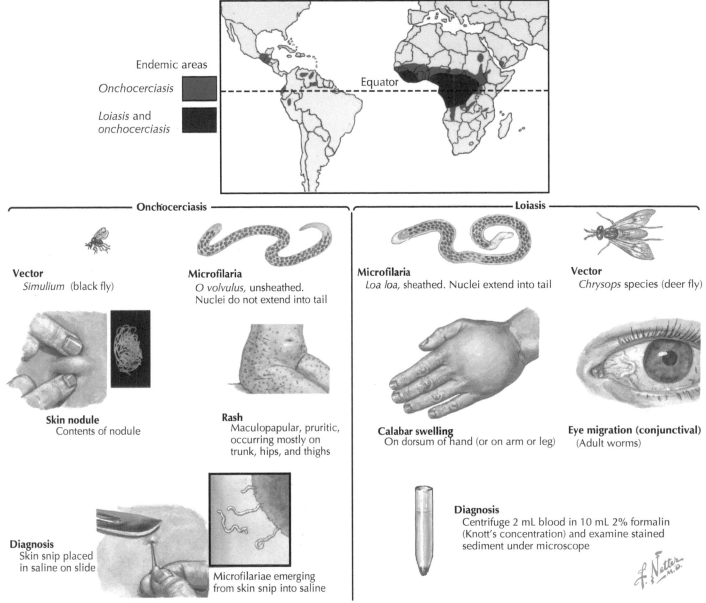

**Fig. 84.2** Filariasis: onchocerciasis and loiasis.

is human to human (see Fig. 84.4). The severity of ocular disease varies considerably among geographic zones, possibly based on the different vector-parasite complexes. The disease in the West African savanna is responsible for the most severe ocular disease; in the most affected villages, more than 10% of the population may be blind.

## Loiasis (Loa loa)

*L. loa* is transmitted by the day-biting tabanid deer flies of the genus *Chrysops*, which rest in the forest canopy and lay eggs in mud, so transmission is primarily in the rainy season. The flies become infected during a blood meal taken from an infected human and transmit the infective microfilariae (larvae) during subsequent human bites.

## CLINICAL PRESENTATION AND PATHOGENESIS

### Lymphatic Filariasis (Wuchereria bancrofti, Brugia malayi, Brugia timori)

Subclinical infection with the threadlike worms of *W. bancrofti*, *B. malayi*, and *B. timori* may persist for decades but is not benign because

there is significant subclinical damage to lymphatics, tissue, and kidneys (see Fig. 84.3). Ultrasound and lymphoscintigraphy have shown marked lymphatic dilatation with collateral channels even at this occult stage.

There are two distinct pathologic processes involved in progression: the first is acute dermatolymphangioadenitis (ADLA), in which there is cutaneous or subcutaneous inflammation with ascending lymphangitis and regional adenitis probably caused by secondary bacterial infection. There may be accompanying fever and chills. Recurrent episodes are an important cause of chronic lymphedema. The other is acute filarial lymphangitis (AFL), caused by an inflammatory response to the dying adult worms as a result of immunity or drug treatment, which manifests as a well-circumscribed nodule or lymphatic cord with or without lymphadenitis/lymphangitis and without fever. If lymphangitis is present, it may spread in a descending (centrifugal) manner. AFL rarely causes residual lymphedema, but these episodes damage the lymphatics and lead to recurrent ADLA.

The disease may progress to brawny lymphedema that can involve the limbs and breasts and may lead to hydroceles of the scrotum

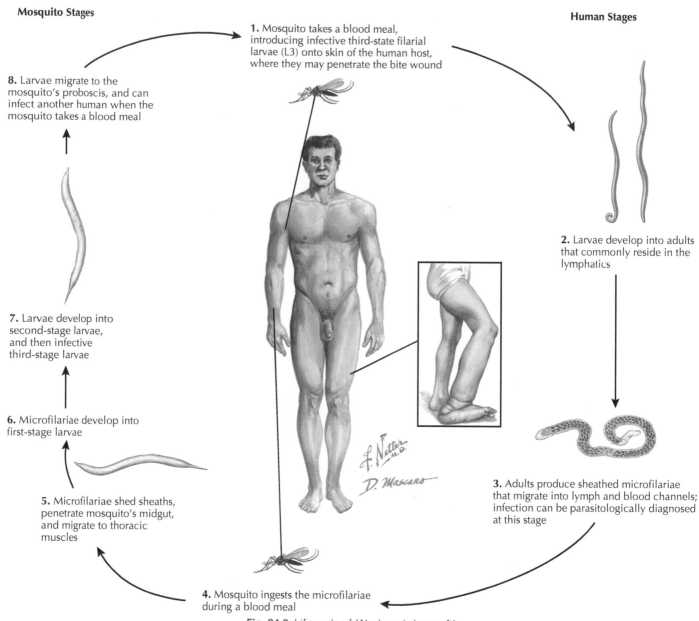

**Mosquito Stages**

**Human Stages**

**1.** Mosquito takes a blood meal, introducing infective third-state filarial larvae (L3) onto skin of the human host, where they may penetrate the bite wound

**8.** Larvae migrate to the mosquito's proboscis, and can infect another human when the mosquito takes a blood meal

**7.** Larvae develop into second-stage larvae, and then infective third-stage larvae

**6.** Microfilariae develop into first-stage larvae

**5.** Microfilariae shed sheaths, penetrate mosquito's midgut, and migrate to thoracic muscles

**2.** Larvae develop into adults that commonly reside in the lymphatics

**3.** Adults produce sheathed microfilariae that migrate into lymph and blood channels; infection can be parasitologically diagnosed at this stage

**4.** Mosquito ingests the microfilariae during a blood meal

**Fig. 84.3** Life cycle of *Wuchereria bancrofti.*

and vagina (Fig. 84.5). Further progression leads to the chronic state known as *elephantiasis*, in which the skin becomes thick, roughened, hyperkeratotic, and fissured (see Figs. 84.3 and 84.5). Radionuclide lymphoscintigraphic imaging shows lymphatic tortuosity, dermal backflow, obstruction, stasis, and poor regional lymph node visualization. Superinfection of these tissues becomes a problem at this stage. Hydroceles can become grossly enlarged to the degree that they become disabling. If there is obstruction of the retroperitoneal lymphatics, renal lymphatic pressure can increase to the point that the lymphatics rupture, causing chyluria. Hydroceles, genital lesions, chyluria, and elephantiasis of the upper leg are seen with bancroftian filariasis but not with *B. malayi* or *B. timori* infection, presumably because of localization of the given parasite within the body.

Renal manifestations such as hematuria, proteinuria, nephrotic syndrome, and glomerulonephritis have all been reported. Other reported manifestations include tropical pulmonary eosinophilia (paroxysmal

cough, fever, wheezing, and hypereosinophilia due to hyperresponsiveness to microfilariae), monoarthritis, polymyositis, urethral obstruction, fibrosing mediastinitis, and bladder pseudotumors.

## Onchocerciasis *(Onchocerca volvulus)*

Clinical manifestations of onchocerciasis include ocular lesions that can cause visual loss and skin lesions, including disfiguring skin changes and subcutaneous nodules (see Fig. 84.2). The adult worms typically entwine in subcutaneous nodules where they produce millions of microfilarial larvae that migrate to the skin, eyes, and other organs, where many die spontaneously, producing local tissue inflammation. Inflammation in the eyes leads to irreversible ocular lesions, resulting in impaired vision and eventual total blindness. Characteristic lesions include punctate keratitis (snowflake opacities), sclerosing keratitis, iridocyclitis, synechiae, optic nerve atrophy, and chorioretinitis.

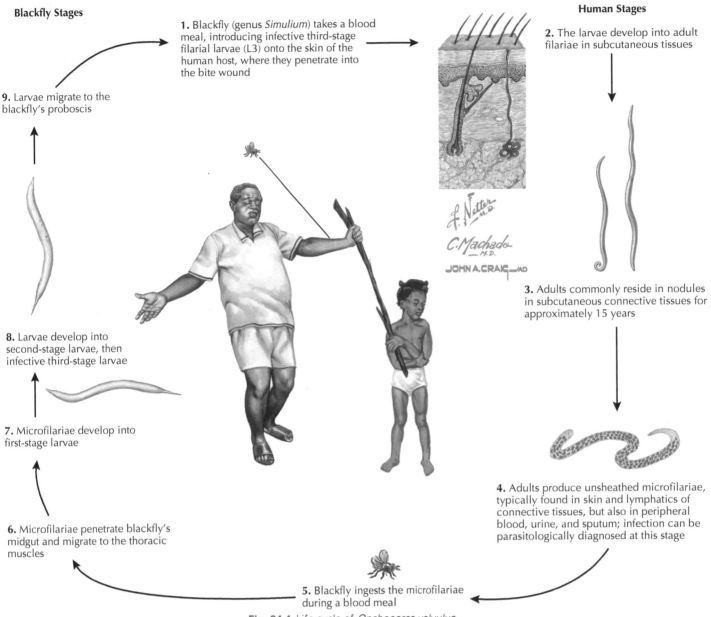

**Blackfly Stages**

**1.** Blackfly (genus *Simulium*) takes a blood meal, introducing infective third-stage filarial larvae (L3) onto the skin of the human host, where they penetrate into the bite wound

**Human Stages**

**2.** The larvae develop into adult filariae in subcutaneous tissues

**9.** Larvae migrate to the blackfly's proboscis

**3.** Adults commonly reside in nodules in subcutaneous connective tissues for approximately 15 years

**8.** Larvae develop into second-stage larvae, then infective third-stage larvae

**7.** Microfilariae develop into first-stage larvae

**6.** Microfilariae penetrate blackfly's midgut and migrate to the thoracic muscles

**4.** Adults produce unsheathed microfilariae, typically found in skin and lymphatics of connective tissues, but also in peripheral blood, urine, and sputum; infection can be parasitologically diagnosed at this stage

**5.** Blackfly ingests the microfilariae during a blood meal

Fig. 84.4 Life cycle of *Onchocerca volvulus*.

The death of microfilariae in the skin causes intense itching, papular dermatitis, atrophy, and patchy depigmentation of the skin—the last sometimes termed "leopard skin." Secondary infection from scratching is common; lymphadenitis is less common. Subcutaneous granulomas resulting from tissue reaction around adult worms are not symptomatic but are useful for rapid community surveys to identify disease burden. Nodding disease, epidemic epilepsy in children, is associated with onchocerciasis but causality is not definitive. Symptoms of onchocerciasis infection in travelers are generally limited to pruritus and rash that can be extensively delayed in presentation.

### Loiasis (Loa loa)

Loiasis is often present as an asymptomatic infection that is recognized only after the adult worm migrates across the subconjunctiva ("eye worm"). This is generally benign and should not be confused with the inflammatory response to intraocular microfilariae that leads to blindness in onchocerciasis. Another manifestation of loiasis is the Calabar

swelling, which is an evanescent localized skin lesion consisting of erythema and angioedema, particularly on the extremities (see Fig. 84.2). It may be accompanied by diffuse pruritus, even when the swelling is not present, and constitutional symptoms. Nephropathy, encephalopathy, and cardiomyopathy are rare manifestations. Nonresidents infected when visiting endemic areas may have more pronounced Calabar swellings and extreme eosinophilia.

## DIAGNOSTIC APPROACH

### Lymphatic Filariasis (Wuchereria bancrofti, Brugia malayi, Brugia timori)

The simplest method for detection is a thick blood film of capillary blood stained with Giemsa or hematoxylin and eosin, although it must be collected around midnight due to the periodicity of the microfilaremia. Concentration techniques improve the sensitivity of detection and include filtration of blood samples through a 3-micron

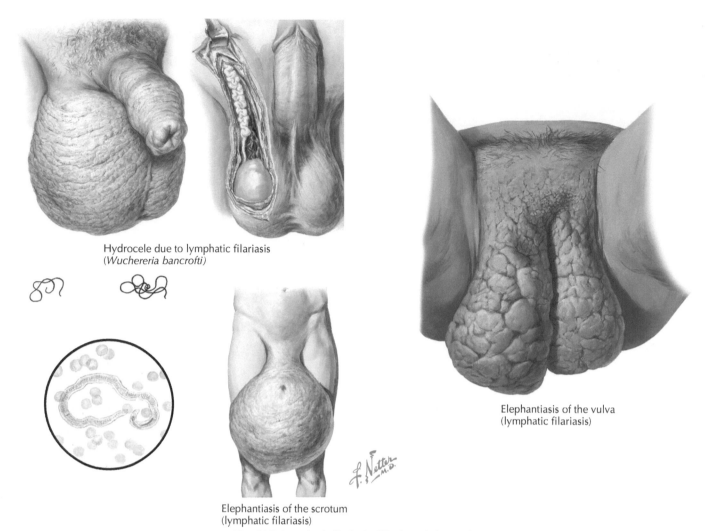

Hydrocele due to lymphatic filariasis
(*Wuchereria bancrofti*)

Elephantiasis of the vulva
(lymphatic filariasis)

Elephantiasis of the scrotum
(lymphatic filariasis)

**Fig. 84.5** Lymphatic filariasis *(Wuchereria bancrofti).*

pore polycarbonate filter or Knott's concentration technique with 2% formalin. Serum antifilarial IgG testing can be useful for diagnosis when microfilariae are not seen. This test is available through the US National Institutes of Health (NIH) or the CDC. Antifilarial immunoglobulin G4 (IgG4) antibody detection assays provide an index of exposure and can be used for early detection of resurgent transmission after mass drug administration programs have stopped.

For detecting *W. bancrofti* infections, lymphatic filarial antigen tests have replaced microfiliarial detection, because blood can be drawn at any time and because infection can be detected in patients without circulating microfilariae. The enzyme-linked immunosorbent assay (ELISA) based on the Og4C3 monoclonal antibody or the rapid immunochromatographic test (ICT) based on the monoclonal antibody AD12 is the preferred method for antigen detection. Polymerase chain reaction (PCR) methods for the detection of *W. bancrofti* or *Brugia* deoxyribonucleic acid (DNA) can also be used to detect microfilaremia effectively. Doppler ultrasonography can detect moving adult worms ("filarial dance sign") in the scrotum and breast, particularly. Since lymphedema may develop years after infection, laboratory tests may be negative in these patients. Because microfilariae are usually not detected in patients with tropical pulmonary eosinophilia, diagnosis requires epidemiologic risk and filarial antibody testing.

## Onchocerciasis *(Onchocerca volvulus)*

Skin snips are immersed in isotonic saline, and the microfilariae that have emerged after 0.5 to 24 hours are counted by microscopy. The sites for the skin snip are usually over the iliac crest, the scapula, and the lower extremities. Six snips provide the most diagnostic sensitivity. Microfilariae of *O. volvulus* are 270 to 320 microns long, unsheathed, and have a characteristic head and a pointed tail. They must be differentiated from the smaller skin-dwelling microfilariae of *Mansonella streptocerca* in Africa and *Mansonella ozzardi* in South America, as well as other rarer microfilariae. PCR of the skin can aid in diagnosis if the larvae are not visualized. Skin snips are limited in sensitivity in the early stages of infection or low-intensity infection. A patch test using topical DEC has also been developed but is not completely reliable.

Serologic detection of IgG4 antibody to the onchocerciasis recombinant antigen OV16 is sensitive and specific even in the rapid-format card test. Antibodies to OV16 often can be detected significantly before the appearance of microfilariae in skin snips. This is an exposure assay only and does not distinguish between prior and current infection. Determination of serum antifilarial antibody is available through the NIH or the CDC but can only determine prior exposure, not active disease, thereby limiting its utility in individuals from endemic areas. Microfilariae may be visible in the anterior chamber of the eye or the

| TABLE 84.2 | Drugs and Treatment Regimens for Patients With Filarial Nematodes | |
| --- | --- | --- |
| **Parasite** | **Treatment Regimens** | **Warnings** |
| *Wuchereria bancrofti, Brugia malayi, Brugia timori* | Diethylcarbamazine (DEC) 6 mg/kg orally (exclude co-infection with onchocerciasis or high-burden *Loa loa*). If onchocerciasis present, ivermectin 200–400 mcg/kg orally as a single dose. | DEC is unsafe in patients who have co-infection with onchocerciasis or loiasis. Mazzotti reactions (headache, fever, rash, pruritus, lymphadenopathy, lymphedema, occasionally postural hypotension) from dying microfilariae may be managed with antipyretics, analgesics, antihistamines, and, if necessary, systemic corticosteroids. Corticosteroid pretreatment and gradual dose introduction may be needed for heavy microfilaremia. |
| *Onchocerca volvulus* | Ivermectin 150–200 mcg/kg orally as a single dose (adults and children)[a]; or moxidectin 8 mg orally as a single dose (adults). Repeat every 3, 6, or 12 months as symptoms (pruritus, rash) recur. Consider following with doxycycline 200 mg orally daily for 6 weeks to attenuate adult worms. | If microfilariae are present in the eye, administer prednisone orally for 1 day before and 2 days after ivermectin. Mazzotti reactions (headache, fever, urticaria, pruritus, lymphadenopathy, arthralgias, edema, occasionally postural hypotension) from dying microfilariae may be managed with antipyretics, analgesics, antihistamines, and, if necessary, systemic corticosteroids. |
| *Loa loa* | DEC 8–10 mg/kg/day orally divided into three doses after food for 21 days; may require repeat treatment. Do not use in patients with microfilaremia greater than 8000/microliter (drawn at time of day for peak parasitemia) because of lysis of microfilariae leading to risk of encephalitis or death. | For patients with microfilaremia, strongly consider expert referral. Ivermectin, 200 mg orally twice daily for 21 days, results in slower lysis of microfilariae and so may be used to reduce parasite burden to less than 8000/microliter but may still result in encephalitis or death. |

[a]Except sick and infirm, children shorter than 90 cm height, pregnant and lactating women 1 week postpartum.

cornea on slit-lamp examination. Xenomonitoring for ongoing transmission is done by collecting pools of *Simulium* flies and screening with PCR for the larvae of *O. volvulus*. Humans are still used to attract flies to the monitoring traps, and because this potentially exposes them to disease, efforts are underway to develop an alternative attractant-baited flytrap to use for monitoring and vector control.

## Loiasis *(Loa loa)*

In order to diagnose loiasis, blood needs to be drawn usually between 10:00 a.m. and 2:00 p.m., owing to diurnal periodicity of the microfilariae in the blood. Giemsa-stained thick or thin blood smears for parasite visualization are specific, quantitative, and of low cost but are not feasible for use in community screening programs. Concentration techniques may improve the diagnostic yield. Quantification of parasite burden should be requested to guide treatment. PCR tests are specific and quantitative but costly. Serologic assays to detect antiparasitic antibodies are nonspecific, not quantitative, and only demonstrate prior exposure. Their utility may be limited to excluding infection. Identification of the adult worm upon physical examination or by a pathologist after excision is sufficient for diagnosis but must be followed by quantification of the microfilarial load to guide treatment.

## TREATMENT

### Lymphatic Filariasis *(Wuchereria bancrofti, Brugia malayi, Brugia timori)*

The drug of choice, DEC, can be obtained from the CDC under an investigational new drug protocol (see Table 84.2). Before treatment, coinfection with onchocerciasis must be excluded because of the risk of severe exacerbation of eye and skin disease (Mazzotti reaction). If onchocercal co-infection is present, ivermectin is the drug of choice. Coinfection with *L. loa* must be excluded because of the risk of serious

adverse reactions, including encephalopathy, renal failure, and death. The risk and severity of the adverse reactions are related to *L. loa* microfilarial density, leading some to avoid DEC if the microfilarial burden is greater than 8000/mL, and some sources even avoid it with a burden of greater than 2500/mL.

For ADLA, supportive treatment with rest, postural drainage, cold compresses, antipyretics, and analgesics is recommended. Treatment with antifilarial drugs is not recommended during the acute episode because it may provoke additional inflammatory response to the dying worms. After the acute episode has resolved, a single dose of two drugs—DEC (e.g., Hetrazan, Banocide, Notezine) and albendazole (e.g., Albenza, Eskazole, Zentel), or ivermectin (e.g., Mectizan, Stromectol) and albendazole—will kill the microfilariae and a proportion of the macrofilarial adults. Corticosteroid pretreatment may decrease the severity of adverse side effects associated with drug treatment of patients with heavy microfilaremia. Retreatment is usually required after 6 to 12 months, as indicated by blood smear or antigen testing. DEC will also treat intestinal *Ascaris* in the patient.

Once lymphatic filariasis is established, twice-daily washing of affected parts with soap and water, raising the affected limb at night, performing regular exercise of the limb to promote lymph flow, keeping the nails clean, wearing shoes, using antibiotic or antiseptic cream to treat small wounds or abrasions, and possibly wearing elastic bandages can slow the onset of elephantiasis and reverse some disease manifestations. Before surgical repair of hydroceles, treatment with one of the regimens in Table 84.2 is recommended. Further improvement in lymphatic pathology may be achieved with a 4- to 8-week course of doxycycline 200 mg daily, to decrease the survival and fertility of the adult worms.

Mass drug administration given to interrupt disease transmission in communities where there is a high prevalence of infection also helps individuals already infected by markedly reducing the incidence of filarial fevers, and one country reported a decrease in the incidence of hydrocele (Table 84.3).

## Onchocerciasis (Onchocerca volvulus)

A single dose of ivermectin 150 mcg/kg orally causes a rapid elimination of microfilaremia from the skin, which is sustained over several months until levels gradually rebound (see Table 84.2). Retreatment may be necessary at 3-, 6-, or 12-month intervals, since the drug kills the microfilariae but not the adult worms. Treatment should be continued for as long as there is evidence of continued infection, such as pruritus or evidence of microfilariae. Ivermectin treatment is contraindicated in pregnant women or children under the age of 5 years. Ivermectin is a potent microfilaricide but does not have a macrofilaricidal effect.

Some experts recommend following ivermectin with a 6-week course of doxycycline 200 mg orally daily since that kills Wolbachia, an endosymbiont rickettsial bacterium that is required for survival and fertility of adult female worms. There are limited data to suggest that the treatment of onchocerciasis with doxycycline in patients co-infected with L. loa is safe with Loa microfilarial loads greater than 8000/mL. DEC treatment is contraindicated in onchocerciasis due to the risk of Mazzotti reactions. Nodules on the cranium may be excised because they may increase the risk of blindness, but otherwise the excision of nodules does not provide any benefit.

Moxidectin has been approved by the US Food and Drug Administration (FDA) for treatment of onchocerciasis in individuals ages 12 and older. A single oral dose of 8 mg of moxidectin was superior to ivermectin 150 mcg/kg orally when skin microfilariae density was compared at 12 months post treatment, possibly due to moxidectin's longer half-life (20 to 43 days compared with ivermectin [<1 day]). Severe but reversible postural hypotension was observed. Moxidectin mass drug administration may offer a modality to shorten mass drug administration programs, particularly in areas with operational challenges such as conflict settings.

## Loiasis (Loa loa)

DEC rapidly eliminates microfilariae but must be used with caution when a high burden of microfilaremia is present (see Table 84.2). It has some inhibitory effect on adult worms but usually requires repeated therapy. Heavily microfilaremic individuals (>8000 microfilariae per milliliter of blood) should not be treated except with careful observation and possible management with corticosteroids. With burdens greater than 8000 microfilariae per milliliter of blood, symptomatic individuals should be treated with albendazole and doxycycline for 21 days to decrease the microfilarial burden to less than 8000 prior to DEC treatment. Albendazole can also be used for rescue treatment after two failed rounds of DEC.

## PROGNOSIS

### Lymphatic Filariasis (Wuchereria bancrofti, Brugia malayi, Brugia timori)

If disease is detected early, treatment can be effective at preventing disease progression. Once the infection is established, morbidity from chronic disease can result in lifelong disability.

### Onchocerciasis (Onchocerca volvulus)

Repeated ivermectin has been reported to cause regression of early lesions of the anterior segment of the eye, including iridocyclitis and sclerosing keratitis, and of optic nerve disease and visual field loss, but not of chorioretinitis. Visual loss from onchocerciasis is not reversible once established.

### Loiasis (Loa loa)

Calabar swellings are usually self-limited and last for a few hours to several days. They may recur at irregular intervals for years, even after the patient leaves the endemic area. Transocular migration of the worm does not result in impaired vision.

## PREVENTION AND CONTROL

### Lymphatic Filariasis (Wuchereria bancrofti, Brugia malayi, Brugia timori)

Travelers to endemic areas should use precautions against mosquito bites. The Global Alliance to Eliminate Lymphatic Filariasis and the World Health Assembly have established the goal of eliminating lymphatic filariasis disease. Lymphatic filariasis has been further designated as potentially eradicable by the International Task Force for Disease Eradication, but this will require sustained national and international commitment. Despite research, there is no filarial vaccine, so control efforts focus on interrupting transmission by mass drug administration programs and vector control through insecticides and bed nets. Long-lasting insecticide-treated bed nets have been shown to significantly decrease lymphatic filariasis microfilaremia in community-based studies of integration with malaria control.

The mainstay of control is annual community-wide mass drug administration programs, excluding children under age 2 years and pregnant and lactating women (see Table 84.3). In areas primarily outside of Africa where there is no co-endemicity with onchocerciasis, DEC 6 mg/kg and albendazole 400 mg are given together, each as a single dose annually. Where feasible, DEC-fortified salt as the only source of domestic salt for at least 6 months a year would be an alternative strategy to mass drug administration. In areas where there is co-endemicity with onchocerciasis, DEC is not used because of the possibility of severe ocular reactions. Instead, the regimen is albendazole 400 mg plus ivermectin 150 to 200 mcg/kg. Albendazole has the additional benefit of treating intestinal helminths Ascaris lumbricoides, Trichuris trichiura, and hookworm, with resultant benefits to growth in children. Ivermectin, in addition to treating Ascaris and Trichuris, is effective against scabies. These programs depend on donations of drugs from the pharmaceutical companies GlaxoSmithKline (albendazole) and Merck & Co. (ivermectin), as well as support from other donors.

Although the previously described regimens are effective at reducing microfilarial burden for a prolonged period, their effect on adult macrofilariae is less complete. Mass drug administration needs to be repeated annually for the lifespan of the adult worm, which can be 5 to 7 years for lymphatic filariasis. Higher doses and increased frequency of administration of ivermectin plus albendazole, or DEC plus albendazole, are being evaluated for their potential to decrease the survival of the adult macrofilariae and shorten the duration of mass drug administration programs. Ivermectin may cause severe reactions in areas co-endemic for L. loa and so should be used only with great caution in those areas. Defining a mass drug administration regimen for L. loa co-endemic areas is a current challenge.

Another potentially promising approach for a macrofilaricide is antibiotic therapy targeted against the Wolbachia rickettsial endosymbiont that the macrofilariae require for growth and development. Tetracycline, doxycycline, and rifamycin, which kill Wolbachia organisms, have been shown to cause the death of the Onchocerca adult worm and presumably kill adult lymphatic filariae as well. The challenge of these regimens is that they require weeks of treatment, which is difficult for mass drug administration programs, which usually rely on drugs that require only a single dose. L. loa do not contain Wolbachia organisms, so the anti-Wolbachia approach is of potential

**TABLE 84.3  Drug Recommendations and Approaches for Community Mass Drug Administration Control Programs**

| Treatment Regimens | Diethylcarbamazine Citrate (Dec) (50-mg or 100-mg Tablet) | Ivermectin[a] (3-mg Tablet) | Albendazole (400-mg Tablet; 100-mg/5 mL Oral Suspension) | Vector Control |
|---|---|---|---|---|
| Lymphatic filariasis, if not *Onchocerca* co-endemic | 6 mg/kg given as a single oral dose every 12 months. Used in combination with albendazole. | Effective in combination with albendazole, but drug donations not available unless onchocerciasis is coendemic. | 400 mg as a single oral dose for adults and children over 2 years of age; 200 mg as a single oral dose in children 1–2 years old. Repeat every 12 months. Used in combination with DEC. | Indoor residual spraying, insecticide-treated nets (for *Anopheles* species); breeding site reduction (for *Culex* species) |
| Lymphatic filariasis, if *Onchocerca* co-endemic | Not recommended because of adverse reactions from dying microfilariae, including potential blindness. | 150–200 mcg/kg as a single oral dose every 12 months. Used in combination with albendazole. | 400 mg as a single oral dose for adults and children over 2 years of age; 200 mg as a single oral dose in children 1–2 years old. Repeat every 12 months. Used in combination with ivermectin. | Indoor residual spraying, insecticide-treated nets (for *Anopheles* species) |
| Lymphatic filariasis, if *Onchocerca* and *Loa* co-endemic | Not recommended because of adverse reactions from dying microfilariae, including potential blindness and encephalitis. | Use only with extreme caution owing to the potential for adverse reactions from dying microfilariae, including potential microhemorrhagic encephalitis. | 400 mg as a single oral dose for adults and children over 2 years of age; 200 mg as a single oral dose in children 1–2 years old. Repeat every 12 months. | Indoor residual spraying, insecticide-treated nets (for *Anopheles* species) |
| Onchocerciasis | Not recommended because of adverse reactions from dying microfilariae, including potential blindness. | 150–200 mcg/kg orally given as a single dose. Retreat every 6–12 months for lifespan of adult worm. | Not applicable | Aerial spraying |

[a]Except sick and infirm, children shorter than 90 cm height, pregnant and lactating women 1 week postpartum.

use in co-endemic areas. Other potential macrofilaricides being evaluated include oxaboroles.

## Onchocerciasis *(Onchocerca volvulus)*

Travelers to endemic areas should avoid blackfly habitats, such as fast-flowing rivers and streams. The best prevention efforts include personal protection measures against biting insects. This includes wearing insect repellant such as *N,N*-diethyl-meta-toluamide (DEET) on exposed skin, wearing long sleeves and long pants during the day when blackflies bite, and wearing permethrin-treated clothing.

Mass drug administration of ivermectin prevents the eye and skin disease and may even interrupt transmission, depending on the frequency of treatment and the geographic extent of the distribution program. With ivermectin, the mass drug administration regimen kills primarily the microfilariae and not the adult worm; therefore the treatment must be continued until the adult worms die of natural causes. The estimated lifespan of the adult onchocercal worm is 14 to 17 years. DEC is considered contraindicated in onchocerciasis because of treatment-related ocular side effects.

The distinction between control and elimination programs of mass drug administration is important. In the former, ivermectin distribution will likely need to continue indefinitely because transmission persists. Sustainability under this strategic approach is critical, and therefore it is important to integrate with other control activities such as vitamin A distribution, treatment of intestinal helminths, and malaria bed net distribution. In the case of elimination, ivermectin is used more intensively (semiannually or quarterly) so that it can

eventually be halted when surveillance shows the parasite population has disappeared. Development of a potent macrofilaricide would greatly improve the feasibility of elimination because it would decrease the need to conduct mass drug administration annually for the entire lifespan of the adult worm, 13 to 17 years.

Significant progress has been made toward elimination in the Americas, where transmission has been interrupted or suppressed in most foci through semiannual ivermectin treatment of at least 85% of the eligible population and vector control. Colombia was the first of the affected countries to achieve countrywide interruption of transmission of the parasite. Active eye disease attributable to onchocerciasis (defined as a greater than 1% prevalence of microfilariae in the cornea or anterior chamber of the eye) was found only in Brazil and Venezuela; there has been no incident onchocercal blindness in the region since 1995. A Pan American Health Organization resolution in 2008 called for elimination of the ocular disease and interruption of transmission by 2012. With four of the six originally endemic American countries now WHO verified as having eliminated onchocerciasis transmission, and 95% of ivermectin treatments in the region halted, the regional focus is now on the remaining active transmission zone, called the Yanomami Area, on the border between Venezuela and Brazil. Both countries have difficult political climates that hinder the elimination task in this remote and relatively neglected region. As with other elimination efforts, "the final inch" is often the most difficult task.

In the Americas, onchocerciasis is found in limited foci; migration of infected human and fly populations is not a major problem, and most black fly species are inefficient vectors. In contrast, onchocerciasis in

Africa covers extensive areas and is associated with large human and fly population migrations, as well as very efficient black fly vectors. Black flies can travel hundreds of kilometers on monsoon and harmattan winds. Incompatibility in savannah and forest vector-parasite complexes seems to work to limit the infection. Vector control—aerial spraying of black fly breeding sites with larvicides—was effectively applied in the Onchocerciasis Control Program in West Africa from 1974 to 2002 with strong results, interrupting transmission in most of the program areas and maintaining this result for up to 12 years with no additional control effort in an area that had been sprayed for 14 years.

Ivermectin has been effective at controlling onchocerciasis as a public health problem in all areas; however, its potential for interruption of transmission is more promising in hypoendemic and meso-endemic areas. The African Programme for Onchocerciasis Control, which now supports control in endemic countries of Africa, uses donated ivermectin as its principal control tool in communities determined to be at high risk by rapid epidemiological mapping of onchocerciasis (REMO; identifies communities with an onchocercal nodule prevalence ≥20%) and Geographic Information Systems (GISs). Their strategy of community-directed treatment with ivermectin empowers communities to make their own decisions regarding the distribution process, such as selection and remuneration of distributors, time of distribution, method of distribution, and so on. This approach has now also recently been demonstrated to be effective at interrupting transmission in focal areas of Africa.

There is no documented resistance to ivermectin in onchocerciasis, although high-grade resistance emerged in intestinal trichostrongylid nematodes of sheep, cattle, and goats but not horses and dogs. An observation from Ghana of human poor responders (defined as microfilariae counts >10 microfilariae per snip after nine or more rounds of ivermectin) is being followed.

Indirect evidence for the needed duration of control programs comes from Guinea Bissau where civil unrest resulted in the suspension of a quarterly ivermectin treatment program in place for 5 years. Despite prevalence levels having dropped to near zero, recrudescence occurred, and it is likely that transmission will resume if interventions are halted before the entire duration of the lifespan of the adult *O. volvulus* (about 14 to 17 years).

## Loiasis *(Loa loa)*

The disease burden from loiasis is not significant enough to merit an elimination program, so its primary significance is that its coendemicity impedes the use of ivermectin needed for control programs directed toward lymphatic filariasis and onchocerciasis in Africa. Because of the profound levels of microfilaremia in loiasis and the side effects from rapid microfilarial death, the use of ivermectin should be considered contraindicated in areas of significant co-infection with *L. loa*. Residents of areas co-endemic for onchocerciasis or lymphatic filariasis who receive ivermectin as part of mass drug administration programs are at increased risk of adverse events, including encephalopathy, coma, and death, particularly if microfilarial burden is greater than 8000 microfilariae per microliter. The pathogenesis includes microfilarial emboli and microvascular hemorrhages. Subconjunctival hemorrhage is considered an early sign of treatment reaction. Corticosteroids do not prevent cerebral reactions and are even harmful once a reaction has occurred.

For individuals, DEC 300 mg orally once a week can be used to prevent infection. Using insect repellents (including permethrin-impregnated clothing) and covering the extremities may reduce the number of bites from infected flies. The flies are day biting, so insecticide-treated bed nets do not help, but other vector-control measures are being explored.

# EVIDENCE

Abegunde AT, Ahuja RM, Okafor NJ. Doxycycline plus ivermectin versus ivermectin alone for treatment of patients with onchocerciasis. *Cochrane Database Syst Rev* 2016(1):CD011146-CD011146. https://doi: 10.1002/14651858.CD011146.pub2. *Systematic review of literature on adding doxycycline to ivermectin treatment of onchocerciasis.*

Addiss D, Gamble CL, Garner P, Gelband H, Ejere HOD, Critchley JA. Albendazole for lymphatic filariasis. *Cochrane Database Syst Rev* 2005(4). https://doi: 10.1002/14651858.CD003753.pub3. *Systematic review of literature on treatment of lymphatic filariasis with albendazole.*

Barsoum RS. Tropical parasitic nephropathies. *Nephrology Dialysis Transplantation.* 1999;14(suppl_3):79-91. https://doi: 10.1093/ndt/14.suppl_3.79. *Review of the pathogenesis of renal disease from a variety of parasitic causes.*

Burri H, Loutan L, Kumaraswami V, Vijayasekaran V. Skin changes in chronic lymphatic filariasis. *Transactions of the Royal Society of Tropical Medicine and Hygiene.* 1996;90(6):671-674. https://doi.org/10.1016/S0035-9203(96)90428-9. *Case series with careful delineation and photos of the cutaneous changes of chronic lymphatic filariasis.*

Colebunders R, Nelson Siewe FJ, Hotterbeekx A. Onchocerciasis-associated epilepsy, an additional reason for strengthening onchocerciasis elimination programs. *Trends in Parasitology* 2018;34(3):208-216. https://doi.org/10.1016/j.pt.2017.11.009. *Review of Nodding disease and association with onchocerciasis.*

Cupp EW, Cupp MS. Impact of ivermectin community-level treatments on elimination of adult Onchocerca volvulus when individuals receive multiple treatments per year. *The American Journal of Tropical Medicine and Hygiene* 2005;73(6):1159-1161. https://doi.org/10.4269/ajtmh.2005.73.1159. *A reanalysis of several published reports to show that ivermectin monotherapy four times annually may have a macrofilaricidal effect.*

Hoerauf A, Pfarr K, Mand S, Debrah AY, Specht S. Filariasis in Africa—treatment challenges and prospects. *Clin Microbiol Infect* 2011;17(7):977-985. https://doi: 10.1111/j.1469-0691.2011.03586.x. *Review of filariasis elimination in Africa.*

Keating J, Yukich JO, Mollenkopf S, Tediosi F. Lymphatic filariasis and onchocerciasis prevention, treatment, and control costs across diverse settings: A systematic review. *Acta Tropica* 2014;135:86-95. https://doi.org/10.1016/j.actatropica.2014.03.017. *Systematic review of lymphatic filariasis and onchocerciasis eradication approaches.*

NTD. The World Health Organization 2030 goals for onchocerciasis: Insights and perspectives from mathematical modelling: NTD Modelling Consortium Onchocerciasis Group. *Gates Open Res* 2019;3:1545. https://doi: 10.12688/gatesopenres.13067.1. *Strategic review of elimination strategies with new treatment modalities.*

Opoku NO, Bakajika DK, Kanza EM, Howard H, Mambandu GL, Nyathirombo A, et al. Single dose moxidectin versus ivermectin for Onchocerca volvulus infection in Ghana, Liberia, and the Democratic Republic of the Congo: a randomised, controlled, double-blind phase 3 trial. *The Lancet* 2018;392(10154):1207-1216. https://doi: 10.1016/s0140-6736(17)32844-1. *Phase 3 clinical trial results comparing moxidectin to ivermectin for treatment of onchocerciasis.*

Pani SP, Yuvaraj J, Vanamail P, Dhanda V, Michael E, Grenfell BT, et al. Episodic adenolymphangitis and lymphoedema in patients with bancroftian filariasis. *Transactions of The Royal Society of Tropical Medicine and Hygiene* 1995;89(1):72-74. https://doi: 10.1016/0035-9203(95)90666-5. *Age-specific data on the frequency and duration of episodic adenolymphangitis (ADL) in patients with defined grades of lymphedema in bancroftian filariasis.*

Shenoy RK. Clinical and pathological aspects of filarial lymphedema and its management. *Korean J Parasitol* 2008;46(3):119-125. https://doi: 10.3347/kjp.2008.46.3.119. *Short description of clinical measures to ameliorate lymphedema from the treatment unit in India.*

Taylor MJ, Hoerauf A, Bockarie M. Lymphatic filariasis and onchocerciasis. *The Lancet* 2010;376(9747):1175-1185. https://doi.org/10.1016/S0140-6736(10)60586-7. *Review of filarial diseases.*

Taylor MJ, Makunde WH, McGarry HF, Turner JD, Mand S, Hoerauf A. Macrofilaricidal activity after doxycycline treatment of Wuchereria bancrofti: a double-blind, randomised placebo-controlled trial. *The Lancet* 2005;365(9477):2116-2121. https://doi.org/10.1016/S0140-6736(05)66591-9. *Controlled clinical trial results evaluating macrofilaricidal activity of doxycycline.*

Turner JD, Tendongfor N, Esum M, Johnston KL, Langley RS, Ford L, et al. Macrofilaricidal activity after doxycycline only treatment of Onchocerca volvulus in an area of loa loa co-endemicity: a randomized controlled trial. *PLOS Neglected Tropical Diseases* 2010;4(4):e660. https://doi:10.1371/journal.pntd.0000660. *Report of a randomized controlled clinical trial comparing the use of doxycycline, ivermectin, or the combination to kill Onchocerca macrofilaria by killing the Wolbachia endosymbiont necessary for Onchocerca growth and reproduction.*

## ADDITIONAL RESOURCES

Carter Center: Lymphatic Filariasis Elimination Program. Available at: www.cartercenter.org/health/lf/index.html. *Lymphatic filariasis fact sheet.*

Centers for Disease Control and Prevention (CDC): Lymphatic filariasis. Available at: www.cdc.gov/parasites/lymphaticfilariasis/index.html. *Lymphatic filariasis fact sheet and information for health professionals.*

Centers for Disease Control and Prevention (CDC): River blindness (onchocerciasis). Available at: www.cdc.gov/ncidod/dpd/parasites/onchocerciasis/factsht_onchocerciasis.htm. *Onchocerciasis fact sheet and information for health professionals.*

Centers for Disease Control and Prevention (CDC): Yellow Book; Health information for international travel. Available at: https://wwwnc.cdc.gov/travel/page/yellowbook-home. *Official publication of the CDC with advice on travel health issues, including vaccination.*

Fitzpatrick C, Nwankwo U, Lenk E, et al. An investment case for ending neglected tropical diseases. In K Holmes, S Bertozzi, B Bloom, P Jha, eds: *Disease Control Priorities* ed 3, Volume 6, Major Infectious Diseases. Washington, DC: World Bank, pp 411-432. Available at: www.dcp-3.org. *Summary review of Global Burden of Disease efforts with regard to neglected tropical disease elimination, including lymphatic filariasis and onchocerciasis, with emphasis on disease burden, intervention effectiveness, cost, and cost effectiveness.*

Ganesh B: Lymphatic Filariasis.www.pitt.edu/~super7/31011-32001/31201-31211.ppt. *PowerPoint presentation on lymphatic filariasis with clinical illustrations.*

Global Alliance to Eliminate Lymphatic Filariasis: Global Alliance website. Available at: www.filariasis.org. *Web portal for the global coordinating body for lymphatic filariasis elimination.*

Global Network: Onchocerciasis. Available at: http://globalnetwork.org/about-ntds/factsheets/onchocerciasis. *Onchocerciasis fact sheet.*

Global Programme to Eliminate Lymphatic Filariasis: progress report 2017. *Wkly Epidemiol Rec* 2018;93:589-604. Available at https://apps.who.int/iris/bitstream/handle/10665/275719/WER9344.pdf. *WHO update on global elimination efforts for lymphatic filariasis.*

Lymphatic Filariasis Support Centre (LFSC), Liverpool School of Tropical Medicine (LSTM): Filariasis.net website, 2006. Available at: https://www.lstmed.ac.uk/research/topics/lymphatic-filariasis. *Web portal on lymphatic filariasis.*

Melrose WD. Lymphatic filariasis: new insights into an old disease. *Int J Parasitol* 2002;32:947-960. *Recommended review of the pathogenesis and new aspects of the disease.*

Onchocerciasis Elimination Program for the Americas (OEPA). https://www.cartercenter.org/health/river_blindness/oepa.html. *Current status of onchocerciasis elimination in the Americas.*

Ottesen EA: Filariasis. In Cohen J, Powderly W, Opal S eds: *Infectious diseases*, ed 4, St Louis, 2016, Elsevier. *Useful and concise, clinically oriented summary of filarial diseases including treatment recommendations summarized and extensive references.*

World Health Organization (WHO): African Programme for Onchocerciasis Control. Available at: www.who.int/blindness/partnerships/APOC/en. *Web portal for WHO activities and supportive information on onchocerciasis.*

World Health Organization (WHO): Lymphatic filariasis. Available at: https://www.who.int/lymphatic_filariasis/en/. *Lymphatic filariasis fact sheet.*

World Health Organization (WHO): Onchocerciasis (river blindness). Available at: https://www.who.int/onchocerciasis/en/. *Onchocerciasis fact sheet.*

# 85

# Echinococcosis: Cystic and Alveolar Disease

*Christina M. Coyle*

## ABSTRACT

Within the genus *Echinococcus*, there are four species recognized: *Echinococcus granulosus*, *Echinococcus multilocularis*, *Echinococcus vogeli*, and *Echinococcus oligarthrus*. The larval cestodes of all four species can develop in the human host and can cause various forms of hydatid disease. A fifth species, *Echinococcus shiquicus*, a tapeworm of Tibetan foxes, has recently been described, but no infections have been reported in humans. Recent mitochondrial DNA studies have identified *Echinococcus felidis* as a distinct species. The adult worm resides in the African lion and the larval form is believed to occur in wild ungulates. To date, no cases have been reported in humans. This chapter focuses on disease caused by cystic echinococcosis (CE) caused by *E. granulosus* and alveolar echinococcosis (AE) caused by *E. multilocularis*.

*E. granulosus* is made up of a number of biologically and genetically distinct entities that have been referred to as *strains* or *subspecies*. Classic CE is caused by the adult worm, *E. granulosus*, which resides in the jejunum of canids (definitive hosts) and produces eggs that are passed in the stool. Eggs ingested by cows, sheep, moose, caribou, or humans (intermediate hosts) liberate an embryo in the duodenum, which passes through the intestinal mucosa to enter the portal circulation. Over 85% of the embryos are filtered by the liver and lungs, where they lodge and develop into hydatid cysts.

AE disease results from infection by *E. multilocularis*. Transmission to humans is usually through the accidental ingestion of parasite eggs shed by a variety of canids (fox, coyote, wolf, and dog) that had previously eaten an infected intermediate host, generally a rodent. In humans, the metacestode (larval) form develops in the liver, proliferating indefinitely by exogenous budding; it invades the surrounding tissue, mimicking a malignancy. The radiographic appearance, diagnosis, and management of AE are quite distinct from those of CE.

## INFECTION WITH *ECHINOCOCCUS GRANULOSUS* (CYSTIC ECHINOCOCCOSIS, OR HYDATID CYST)

### Life Cycle

*E. granulosus* is a small tapeworm of a wide variety of canids (e.g., domestic dogs, foxes, wolves, and dingoes), which serve as the definitive host. The adult tapeworm ranges in size (2–12 mm in length) and lives in the gut of the definitive host, attached by the scolex to the mucosa of the small bowel. The mature worm has an average of three proglottids: immature, mature, and gravid (Fig. 85.1). The terminal gravid proglottid contains eggs that are released into the feces. The eggs contain the true larval or hexacanth stage, which is accidently ingested by the intermediate host. Cows, sheep, moose, caribou, and humans can act as the intermediate host. Once an egg has been ingested by a suitable intermediate host, the oncosphere is released from its protective coat and penetrates the intestinal wall, using its larval hooks. The majority of oncospheres are deposited in the liver after migration through the portal circulation; those that escape the hepatic filter will enter the pulmonary circulation, where they may be trapped. A small number escape the pulmonary sieve and are distributed systemically. Once deposited in an organ, the oncosphere begins to develop into the metacestode form or an echinococcal cyst, which is a unilocular fluid-filled cyst containing the tapeworm larva (also known as a hydatid cyst). The cyst is lined by an endocyst, which is parasitic in origin and consists of an inner germinal layer surrounded by an acellular laminated membrane. Small vesicles, called brood capsules, bud internally from the germinal layer and develop asexually into protoscolices. The cycle is completed when the cyst is ingested by a canine carnivore; protoscolices are released into the gut of the canid, where they form new adult worms (see Fig. 85.1).

### Epidemiology

The distribution of *E. granulosus* is worldwide, with only a few areas such as Iceland, Ireland, and Greenland free of autochthonous human CE. The greatest prevalence of CE occurs in countries of the temperate zones, including South America, the entire Mediterranean littoral, the southern and central areas of the former Soviet Union, Central Asia, Australia, and areas of Africa. Most cases in the United States and central and western Europe are imported. In many parts of the world, CE is considered an emerging disease. For example, CE has been on a dramatic increase in the former Soviet Union and Eastern Europe in recent years (see Fig. 45.5).

Consistently, the highest prevalence is found among populations involved with sheep raising. People of both sexes and all ages appear susceptible. Socioeconomic and cultural characteristics are among the best-defined risk factors for human infection: dogs living closely with people, uncontrolled slaughter of livestock, and unsanitary living conditions. There may be as many as 40,000 tapeworms in a heavily infected dog, and each tapeworm sheds approximately 1000 eggs every 2 weeks. Dogs infected with *Echinococcus* pass eggs in their stools, which adhere to hairs around the anus and around the muzzle. Eggs are accidently ingested through intimate contact, usually with children. In addition, soil and vegetables can become contaminated, providing another route of infection. When dogs at risk for infection are maintained close to the family home, all members of the family may be exposed.

### Clinical Manifestations

CE may occur in persons younger than 1 year to over 75 years of age. In areas of endemic infection, most hospital cases are recorded among those 21 to 40 years of age, but severe morbidity may also occur in younger individuals.

The incubation of human hydatid disease is highly variable and often prolonged for several years. Clinically, CE often remains asymptomatic over long periods of time until symptoms appear, and at that point the disease may already be far advanced. On the other hand,

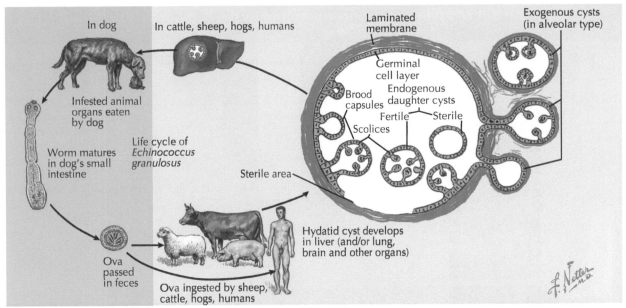

**Fig. 85.1** Life cycle of *Echinococcus granulosus*, with detail showing development of endogenous daughter cysts from a primary cyst in the liver. Although the life cycle of *Echinococcus multilocularis* is not shown, the formation of exogenous cysts by evagination of the primary cyst wall is illustrated to show the difference in cyst formation (the hydatid daughter cysts lack the pericyst layer, whereas the alveolar cysts retain this membrane layer).

many individuals will remain asymptomatic, and cysts are frequently observed as incidental findings when imaging is performed for another clinical reason. The liver is the most common site of cyst localization (65%), followed by the lungs (25%). Cysts are less frequently seen in the spleen, kidneys, heart, bone, and central nervous system. The clinical manifestations of CE are diverse and determined by the site of the cyst, its size, and its stage. Intracerebral, cardiac, bone, and ocular cysts may be very small (<2 cm) when they cause symptoms. In contrast, lung and liver cysts may grow asymptomatically to very large sizes (>40 cm) for many years before clinical signs develop. Most asymptomatic liver hydatid cases (75%) remain symptom free for more than 10 years, regardless of cyst size or type. Single cysts are most common; however, 20% to 40% of patients have multiple cysts or multiple organ involvement. Therefore all patients with hepatic cysts need a full radiographic evaluation of the pelvis and lung for extrahepatic cysts. In secondary echinococcosis, new cysts develop from released protoscolices after spontaneous or trauma-induced cyst rupture or during surgical treatment.

## Hepatic Disease

Owing to the distensible nature of the liver, hepatic cysts may grow for years before becoming symptomatic. Solitary cysts are most common, and the right lobe is affected more often than the left lobe. The size of the cyst is variable, ranging from 1 to 15 cm. Hepatic cysts can cause pain in the upper abdominal region secondary to pressure from an enlarging cyst. Within the liver, the maturing cyst form of *E. granulosus* consists of an internal germinal layer surrounded by an acellular, parasite-derived laminated layer; together they form the endocyst. The endocyst is surrounded by the pericyst, or adventitial layer, and is largely derived from host tissue. By 3 weeks of development, the cyst is approximately 250 μm in size, and by 20 weeks, it is about 1 cm in size. The central cavity of an *E. granulosus* cyst is filled with hydatid fluid, which is similar in composition to the interstitial fluid. Intracystic pressure increases as the cyst grows and reaches maturity.

As the cyst grows, cells bud from the germinal layer. They develop stalks, becoming "brood" capsules in the central cystic cavity. From the wall of the brood capsule, a protoscolex develops. The protoscolex is able to develop into a daughter or secondary cyst if accidentally spilled or into a mature worm in the definitive host if ingested. As the primary cyst matures the brood capsule detaches from the germinal layer and ultimately will degenerate, releasing protoscolices into the hydatid cyst fluid, forming the "hydatid sand" characteristically found in older cysts. Daughter cysts, a pathognomonic feature of *Echinococcus*, resemble the primary cyst with the exception that the pericyst is not present (see Figs. 85.1 and 45.2).

Complications of hepatic echinococcosis can occur with devastating consequences if not recognized and managed properly by the clinician. Communication with and rupture of hepatic cysts into the biliary tree are well described and can result in cholangitis and cholestasis. With frank intrabiliary rupture, the daughter vesicle and germinative membrane pass into the main biliary ducts, resulting in obstructive jaundice in some cases. In these instances, patients experience epigastric pain and right-upper-quadrant pain. A cystobiliary fistula should be considered when there is acute or intermittent pain mimicking biliary colic, and these patients should be managed surgically. If cysts are left untreated, cholangitis can occur, with resultant bacterial superinfection of the cyst cavity and abscess formation. It is important to establish the diagnosis of a cystobiliary communication preoperatively because, owing to the risk of sclerosing the biliary tree, sclerosing agents should be avoided. Less commonly, portal hypertension can complicate hepatic CE by extrinsic compression of the liver or by obstruction of the inferior vena cava and hepatic outflow tract. Rarely, parasitic emboli may result from rupture of a cyst into the hepatic vein or inferior cava. Rupture or leakage of the cyst into the peritoneal cavity can result in acute or intermittent allergic manifestations and secondary CE. Mild to severe anaphylactoid reactions (and occasionally death) may follow the sudden massive release of cyst fluid.

Two commonly used imaging-based classification systems correlate individual hepatic cyst stage with natural progression of hepatic

**TABLE 85.1     Treatment Modalities for Uncomplicated Cystic Echinococcosis Stratified by Cyst Stage**

| World Health Organization Classification, 2001 | Radiographic Image | CURRENT PRACTICE | | | |
|---|---|---|---|---|---|
| | | Surgery | Percutaneous Techniques | Medical Treatment | Suggested[a] |
| CE1 | Hepatic unilocular anechoic cystic lesion with a double sign (ultrasound) | Practiced | Practiced | Practiced | PAIR + ABZ if >5 cm ABZ alone if <5 cm |
| CE2 | Multiseptated, "rosette-like" or "honeycomb" cyst in the liver (ultrasound) | Practiced | Rarely practiced | Practiced | Non-PAIR PT + ABZ Surgery + ABZ |
| CE3a | Hepatic cyst with detached membranes ("water lily" sign) (ultrasound) | Practiced | Practiced | Practiced | PAIR + ABZ if >5 cm ABZ alone if <5 cm |
| CE3b | Hepatic cyst showing daughter cysts in solid matrix (CT scan) | Practiced | Rarely practiced | Practiced | Non-PAIR PT + ABZ Surgery + ABZ |
| CE4 | Hepatic cyst showing heterogeneous hypoechoic or hyperechoic contents and no daughter cysts (ultrasound) | Regard as inactive; unless complicated, they should not be treated Watch and wait | | | |
| CE5 | Calcified hepatic cyst (CT scan) | | | | |

*ABZ*, Albendazole; *CE*, cystic echinococcus; *CT*, computed tomography; *PAIR*, percutaneous aspiration-injection-reaspiration; *PT*, percutaneous treatment.

[a]See text for discussion.

Image for CE1 courtesy Prof. P. Kern, Division of Infection Diseases and Clinical Immunology, Comprehensive Infectious Diseases Center, University Hospitals, Ulm, Germany; images for CE2 and CE3b courtesy Prof. C. Coyle, Department of Medicine, Division of Infectious Diseases, Albert Einstein College of Medicine, New York; image for CE3a courtesy Prof. E. Brunetti, Division of Infectious and Tropical Diseases, University of Pavia, IRCCS S. Matteo Hospital Foundation, Pavia, Italy; images for CE4 and CE5 courtesy Prof. B. Gottstein, Institute of Parasitology, University of Bern, Bern, Switzerland.

cysts and have therapeutic implications. They are the World Health Organization (WHO) Informal Working Group on Echinococcosis (IWGE) and the Gharbi classification systems; both are very similar, and this chapter refers to the WHO-IWGE classification system. The WHO-IWGE standard classification allows for a natural grouping of cysts: active CE1 and 2, CE3, and inactive (CE4 and CE5). Table 85.1 summarizes the stratified approach to treatment based on cyst stages. CE1 is a unilocular anechoic cystic lesion with a double sign. CE2 is a multiseptated, "rosette-like" "honeycomb" cyst. Both CE1 and CE2 are considered active, usually fertile cysts containing viable protoscolices. CE3a is a cyst with detached membranes ("water lily" sign), whereas CE3b has daughter cysts in a solid matrix. A cyst with heterogeneous hypoechoic or hyperechoic contents and no daughter cysts is the CE4 stage, and calcified cysts are considered CE5. CE types 4 and 5, which are inactive, have normally lost their fertility and are degenerative. When the physician is considering what the best modality of treatment for a patient with hepatic CE would be, he or she must take into consideration the location, size, and staging of the cyst.

## Pulmonary Disease

Pulmonary hydatid disease has been reported in up to 30% of cases of hydatidosis in some series and can be either primary or secondary.

Uncomplicated lung cysts rarely produce symptoms and are usually found incidentally at imaging. In patients who are symptomatic, chest pain with cough, dyspnea, and hemoptysis are often the presenting symptoms. Fever is less common and usually seen in cases of bacterial superinfection. The hydatid cyst can occur anywhere in the lung, but it settles more often on the right side and has a predilection for the lower lobes. It generally is a single cyst, and in up to 25% of cases there is a coexisting hepatic cyst. Pulmonary cysts may be multiple and/or bilateral in approximately 30% of cases. Typically, cysts grow at an average rate of 1 cm in diameter per year. However, growth rates of up to 5 cm/year have been noted. Because the period of initial cyst growth is frequently asymptomatic, pulmonary hydatid disease most frequently manifests in the second to third decade of life. Cysts may be entirely asymptomatic in 75% of cases at initial detection, and asymptomatic pulmonary cysts are more likely to be found in areas of the world where chest radiographs are taken for mass tuberculosis screening programs.

Pulmonary hydatid cysts have a two-layer wall; the outer layer (exocyst), white and fragile, is made up of concentric sheets of hyaline, and the inner layer is the germinal layer, or the endocyst. Unlike hepatic cysts, daughter cysts are rarely seen. The content is liquid, with some solid elements that constitute the "hydatid sand," made up of hooklets and scolices. In "closed" cysts the fluid resembles water. It contains antigenic elements that are responsible for the anaphylactic phenomenon that may appear when the cyst ruptures.

As with hepatic lesions, the adventitia (pericyst) is the result of the inflammatory response of the organ in which the parasite settles. It consists of three layers: an inner layer, which is smooth and glossy; a middle layer, which is of a fibrous nature; and an outer layer, with active inflammation. Atelectasis is seen around the adventitia, but more extensive alterations such as bronchiectasis and interstitial sclerosis can be seen.

When pulmonary cysts rupture into neighboring bronchi with formation of cystobronchial fistulas, a typical "salty water and grape skin" expectoration may occur. Cysts may also become infected in this setting. A large pulmonary cyst can cause stenosis and occlusion of the bronchi. Pleural effusion may develop in 30% of the cases and may occur secondary to cyst rupture into the pleural space. Rupture can lead to severe complications, such as massive hemoptysis and tension pneumothorax.

In contrast to perforation into a bronchus, rupture of a hydatid cyst into the pleural cavity usually causes pneumothorax, pleural effusion, or empyema. Cyst rupture into the pleural cavity can also result in tension pneumothorax. Rupture of a cyst into the pleural cavity or into the bronchial tree may also lead to secondary larval spread or to allergic and anaphylactic reaction. Although extrapulmonary hydatid cyst rupture may cause fatal anaphylaxis, the incidence of this phenomenon is low in association with pulmonary cysts.

The typical radiographic appearance of pulmonary hydatid disease is that of one or more homogeneous round or oval masses with smooth borders surrounded by normal lung tissue (Fig. 85.2). Large cysts can shift the mediastinum, induce a pleural reaction, or cause atelectasis of the surrounding parenchyma. Pulmonary cysts rarely calcify, and the contents of the cyst are usually homogeneous, with a density close to that of water. The introduction of air between the pericyst and exocyst produces the appearance on imaging of a thin layer around the exocyst, which is referred to as the *crescent*, or *meniscus*, *sign*. If the ruptured cyst communicates with the tracheobronchial tree, evacuation of the contents of the cyst results in an air-fluid level. The endocyst may appear to float in the remaining fluid, producing a characteristic radiographic feature known as the *water-lily sign*.

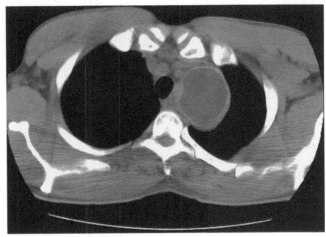

Fig. 85.2 Computed tomographic appearance of pulmonary cystic echinococcosis. (Courtesy Prof. C. Coyle, Department of Medicine, Division of Infectious Diseases, Albert Einstein College of Medicine, New York.)

## Bone Disease

Hydatid disease of the bone is rare and accounts for only 0.5% to 4% of echinococcal disease in humans. The spine is the most common site, accounting for 50% of cases, followed by the pelvis and hip. Other osseous sites for infection include the femur, tibia, fibula, ribs, scapula, clavicle, and tarsal bones. Patients experience pain, swelling, and occasionally pathologic fracture when the condition occurs in the long bones or neural compromise when it occurs in the spine.

In cases involving the spine, radiographs show pedicle erosion and loss of vertebral body height. Plain radiographic findings include multilocular osteolysis and reactive sclerosis of a honeycomb nature. Computed tomography (CT) demonstrates erosions of the body, pedicle, and lamina of the cancellous bone without a subperiosteal reaction. Magnetic resonance imaging (MRI) is superior to CT in evaluating the extent of disease preoperatively and can also be helpful for evaluating recurrence. Successful treatment of vertebral hydatidosis represents a challenge because of its invasive features. Surgery is the treatment of choice, but adjuvant anthelmintic chemotherapy is essential to control the disease locally, avoid systemic spread, and prevent recurrences. Wide surgical excision is particularly difficult to achieve in the spine and the pelvis, but medical therapy alone is not appropriate.

## Hydatid Disease of the Brain

Hydatid cysts of the brain are uncommon, accounting for 2% to 3.6% of all intracranial space-occupying lesions in endemic areas. Children are much more frequently affected than adults; 50% to 93% of intracranial cysts are found in children younger than 17 years of age. Associated extracranial cysts are common, making a thorough radiologic evaluation of the patient, with chest radiographs, and abdominal ultrasound or CT mandatory. Furthermore, cardiac ultrasound is indicated, especially in children, when the occurrence of primary cerebral hydatid cysts may imply a communication between the right and left sides of the heart.

Cerebral hydatidosis manifests differently in children and adults. Signs of increased intracranial pressure with papilledema dominate in the younger age group, whereas focal findings such as hemiparesis, speech disorders, and hemianopsia, sometimes associated with epileptic seizures, are more prevalent in the older age group.

Cerebral hydatid cysts are usually unilocular on neuroimaging. They are most often supratentorially localized in the distribution of the terminal branches of the middle cerebral artery, usually temporoparietooccipitally. The definitive treatment is surgical removal.

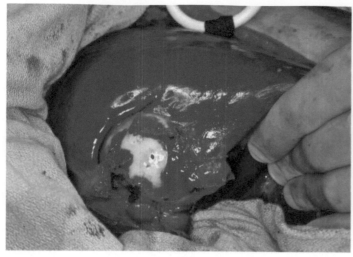

Exposure of large hepatic cyst wall

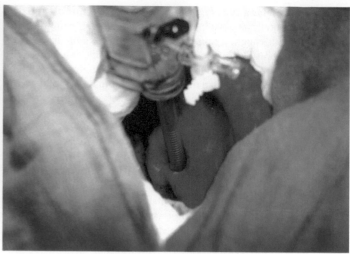

Aspiration of cyst contents

Residual hepatic cavity with deflated cyst

Excised cyst capsule

**Fig. 85.3** Surgical management of hepatic cystic echinococcosis. (Courtesy Prof. C. Coyle, Department of Medicine, Division of Infectious Diseases, Albert Einstein College of Medicine, New York.)

## Diagnosis

The diagnosis of CE in individual patients is based on identification of cyst structures by imaging techniques, predominantly ultrasound and CT, and confirmation by detection of specific serum antibodies on immunodiagnostic tests. For clinical practice it should be noted that confirmation via enzyme-linked immunosorbent assay (ELISA) using crude hydatid cyst fluid ranges from 85% to 98% for liver cysts, 50% to 60% for lung cysts, and 90% to 100% for multiple organ cysts. Specificity is limited by cross-reactions because of other cestode infections, some helminths, malignancy, cirrhosis, and presence of anti-P1 antibodies. Confirmatory tests must be used (arc-5 test; antigen B [AgB] 8 kDa/12 kDa subunits or EgAgB8/1 immunoblotting) in cases that are doubtful. Cysts in the brain, bone, or eye and calcified cysts often induce no or low antibody responses. In routine laboratory practice, at least two different tests are used to improve accuracy.

## Treatment

Four therapeutic modalities exist to treat hepatic CE: chemotherapy, surgery, percutaneous drainage, and "watch and wait." For many years, surgery was the only treatment available for CE. Percutaneous drainage (consisting of percutaneous aspiration-injection-reaspiration [PAIR]) of hepatic hydatid cysts is now accepted as an alternative to surgery in the appropriate group of patients. Medical therapy is usually used in conjunction with surgery and PAIR but can be used alone if the patient is not a candidate for either surgery or PAIR. The "watch and wait" method has been employed successfully for inactive cysts.

## Surgery

Surgery has been the mainstay of therapy. In uncomplicated CE1 and CE3a liver cysts, surgery is increasingly being replaced by alternative therapies. In patients with complicated cysts (rupture, cystobiliary or most cases of cystobronchial fistulas, compression of vital organs and vessels, hemorrhage, secondary bacterial infection), surgery remains the treatment of choice (see Chapter 45).

Surgical procedures can be divided into conservative and radical. Conservative procedures aim at sterilization and evacuation of cyst content, including the hydatid membrane (hydatidectomy), and partial removal of the cyst. The cyst is punctured and partially aspirated and then a scolicidal agent is injected, followed by total aspiration (Fig. 85.3). The risks are anaphylactic shock and chemical cholangitis or alveolar or bronchial damage if the cyst communicates with the biliary or bronchial tree, respectively. Other risks include spillage and secondary CE, with relapse rates of up to 20% being reported after surgery for liver cysts and up to 11.3% for lung cysts. After partial removal of the cyst, a residual cavity remains, which is at risk of secondary bacterial infection and abscess formation. The goal of a radical procedure is to remove the cyst completely. This can be achieved with or without hepatic or lung resection. There is greater intraoperative risk with a radical procedure, but fewer postoperative complications and relapses. Conservative procedures are preferred in pulmonary hydatid disease.

Efforts to avoid spillage are imperative, and during surgery the surgical fields should be protected with pads soaked with scolicidal agents. At present, 20% hypertonic saline is recommended, and when injected into the cyst it should be in contact with the germinal layer for at least 15 minutes for optimal efficacy. To avoid the risk of chemically induced sclerosing cholangitis, the possibility of a cystobiliary fistula should be ruled out before introduction of hypertonic saline into a hepatic cyst. This can be done before surgery with magnetic resonance cholangiopancreatography (MRCP) or intraoperatively, by checking for bile staining in the fluid aspirated from the cyst. Similarly, the surgeon should avoid injecting a scolicidal agent into the bronchial tree if there is a cystobronchial fistula. Albendazole (ABZ) should be started at least 1 day to a week before surgery and then continued up to at least 1 to 3 months after surgery to prevent secondary CE and relapse. The length of adjunctive ABZ therapy has never been formally evaluated.

Surgery should be carefully evaluated and is the first choice for complicated cysts. In the liver, it is indicated for removal of large cysts (>10 cm), CE2 to CE3b cysts with multiple daughter cysts, cysts communicating with the biliary tree, and cysts exerting pressure on vital organs.

## Percutaneous Aspiration of Hepatic Cysts Under Ultrasonographic Guidance

Percutaneous treatments of CE are image-guided techniques aimed at evacuating cyst contents and inactivating the germinal membrane and protoscolices using a scolicidal agent. The technique consists of percutaneous aspiration using ultrasound (or CT) guidance, aspiration of the cyst contents, and reinstillation of a protoscolicidal agent that must remain in the cyst cavity for 15 minutes, and reaspiration of the cyst contents (PAIR). Prior to instillation of a protoscolicidal agent, the cyst fluid must be tested for bilirubin; injection of contrast medium into the cyst followed by radiologic imaging can verify if there is a cystobiliary communication. The presence of such a communication is an absolute contraindication to the placement of a scolicidal agent into the cavity as it can lead to sclerosing cholangitis and liver failure. ABZ should be given at least 4 hours before the procedure and continued for at least a month afterward. A Cochrane review found only two randomized trials examining PAIR in hepatic hydatid disease. One compared PAIR versus surgery and found that it was as effective as surgery with less adverse effects; another small trial reported that PAIR with or without ABZ was more effective than ABZ alone. The indication for undertaking PAIR is having CE1 and CE3a cysts with a diameter of 5 to 6 cm to less than 10 cm in diameter in which cystobiliary communication has been ruled out. PAIR is contraindicated for inaccessible or superficially located liver cysts, CE2 cysts and CE3b cysts, inactive or calcified cystic lesions, complicated cysts, and cysts with biliary communication. In addition, PAIR is also contraindicated in lung cysts.

## Chemotherapy

Mebendazole (MBZ) and ABZ have shown efficacy against CE. MBZ was the first benzimidazole (BMZ) carbamate agent found to have in vivo activity in hydatid disease. Both ABZ, 10 to 15 mg/kg body

weight per day in two divided doses, and MBZ, 40 to 50 mg/kg body weight per day in three divided doses, for 3 to 6 months have demonstrated efficacy. ABZ is preferred over MBZ as it is more active in vitro and has improved gastrointestinal absorption and bioavailability as well as reportedly better clinical results. As a result, ABZ has replaced MBZ in the treatment of echinococcosis. Administration of ABZ with fat-rich meals facilitates absorption and bioavailability. Adverse reactions (abdominal discomfort, alopecia, liver toxicity, agranulocystosis) are reversible upon discontinuation of the drug. Liver enzymes and a complete blood count need to be followed while patient is taking ABZ. Bone marrow suppression is a rare but serious complication. Experimental studies on ABZ have indicated that is teratogenic in rats and rabbits, therefore its use is contraindicated during pregnancy.

BMZs are indicated for inoperable patients with liver or lung CE, patients with multiple cysts in two or more organs, and patients with peritoneal cysts. Smaller cysts (<5 cm) with little adventitial reaction seem to respond best to medical therapy. Cysts with multiple daughter cysts (CE2) are more likely to recur with medical treatment, and older cysts with a thick or calcified surrounding adventitial reaction are likely to be refractory to treatment. With treatments for less than 3 months (i.e., one or two cycles), fewer patients respond and a slightly smaller proportion are cured. A systemic review which collected data from 711 treated patients with 1308 cysts from six countries using the WHO cyst classification found highly active CE1 cysts and those less than 6 cm had a better response to BMZ treatment. Highly active cysts, such as small CE1 and CE3a cysts, respond best to medical therapy alone, whereas stage CE2 and CE3b cysts respond poorly. The length of treatment with ABZ has not been established, but some experts believe that a 3-month course is minimum. The cysts should be followed by ultrasound for signs of involution such as solidification or decrease in size. At least 12 months should lapse before treatment is considered as a success.

The role of praziquantel in hydatid disease has not been defined, although there is some evidence to support a role for the use of praziquantel (40 mg/kg) once a week in combination with ABZ as chemotherapy before and after surgery or percutaneous procedures. Combined therapy is more protoscolicidal in this setting and may reduce the risk of disease recurrence and intraperitoneal seeding of infection that could develop via cyst rupture and spillage, but studies are small and nonrandomized. Currently there are no formal recommendations to support the addition of praziquantel to ABZ. MBZ and ABZ have been used in pulmonary hydatid disease. Cure rates are only 25% to 30%. Most experience with medical treatment of pulmonary hydatidosis is in children. In a 16-year retrospective of 36 children with pulmonary disease treated medically, complications were more likely in cysts with a mean diameter of 6 cm at the beginning of medical therapy. There are some suggestions that anthelmintics weaken the cyst wall and increase the risk of cyst rupture in pulmonary disease. Wen and Yank found a 77.3% incidence of cyst rupture in 21 patients with hydatid disease who were treated with ABZ.

## Watch and Wait

Ultrasound surveys have revealed CE4 and CE5 cysts which are consolidating and calcifying or have become calcified (completely inactive). If these lesions do not compromise organ functions or cause symptoms, then a "watch and wait" approach can be safely undertaken. If one chooses to "watch and wait" and leave the cysts untreated, the patient should be monitored with long-term ultrasonographic follow-up care. A period of 5 years minimum seems to be an adequate time frame.

CE is difficult to treat and cure, but a full knowledge of staging and complications can help the clinician choose the best treatment options for his or her patient.

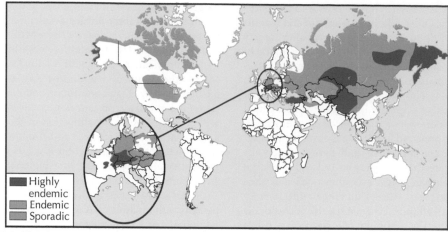

**Fig. 85.4** Global geographic distribution of alveolar echinococcosis. (Data from Torgerson PR, Keller K, Magnotta M, Ragland N: The global burden of alveolar echinococcosis, *PLoS Negl Trop Dis* 4:e722, 2010.)

## Prevention

The main risks for transmission of CE to humans are present in locales where sheep raising is predominant and *E. granulosus* is established in a dog-sheep-dog cycle, with dogs serving as the definitive host and the sheep serving as the intermediate host. The sheep are infected from eating grass and vegetation contaminated with feces (containing eggs) from infected wild canids or infected domestic dogs. During sheep slaughtering, dogs eat infected viscera of sheep and new infections may occur. Prevention of human disease is highly dependent on increased awareness among the shepherds and their family members, meat processers, and veterinarians and their willingness to change behaviors. Avoiding close personal contact with dogs and washing hands thoroughly after working with or handling animals is basic but difficult to do in the high-risk areas. Humans also should avoid ingestion of water and raw vegetables that may be fecally contaminated. Blocking access of dogs to slaughterhouses and periodically treating dogs with anthelmintic drugs will help to lower the prevalence of infection in the canine environment.

## INFECTION WITH *ECHINOCOCCUS MULTILOCULARIS* (ALVEOLAR ECHINOCOCCOSIS)

### Life Cycle

*E. multilocularis* is a microscopic tapeworm of canids (fox, coyote, wolf, or domestic dog and less frequently cat) characterized by its small size (length of up to 4.5 mm) that causes rare but serious infections of humans when the eggs are accidentally ingested. Transmission of *E. multilocularis* occurs in a sylvatic (natural) cycle when foxes or other wild canids—such as coyotes, raccoons, dogs, and wolves—or wild cats, which can serve as definitive hosts, eat infected rodents and other small mammals that serve as intermediate hosts, harboring the metacestode (larval) stage in the liver.

After an infected rodent is eaten, the larvae mature into adults in the gastrointestinal tract of the definitive host, and eggs are passed in the feces. Environmental contamination of soil, grass, berries, and other ground vegetation thus results. The ova are ingested then by foraging rodents, which serve as intermediate hosts, and the natural life cycle is completed when the rodents are eaten by their predators. Humans serve as accidental intermediate hosts and inadvertently become infected by hand-to-mouth transmission through close association with dogs or cats that have eaten infected rodents and are passing eggs. Some of the eggs passed in the fecal stream stick to the perianal fur and contaminate human hands. Accidental ingestion of eggs may also occur when humans drink water contaminated by fox feces or eat unwashed vegetables and berries from the ground.

Once the *E. multilocularis* eggs are ingested, embryos (oncospheres) are released, penetrate through the small intestinal mucosa, and are carried in the portal blood flow to the liver where the larvae (metacestodes) begin to develop. The larval mass develops into an alveolar lesion, infiltrating the liver parenchyma, destroying the surrounding tissue, and resembling a malignancy in both its appearance and behavior. There is contiguous extension of AE from the liver to other organs, and the infection may metastasize to the brain and lung by hematogenous routes.

## Epidemiology

AE has been reported in parts of central Europe, much of Russia, the Central Asian republics, western China, the northwestern portion of Canada, and western Alaska. The annual incidence in endemic areas of Europe has increased from a mean of 0.10 per 100,000 during 1993 to 2000 to a mean of 0.26 per 100,000 during 2001 to 2005. There is evidence of parasites spreading from endemic to previously nonendemic areas in North America and North Island, Hokkaido, Japan, principally because of the movement or relocation of the fox (Fig. 85.4). Hunters, trappers, and persons who work with fox fur are at risk of exposure. Hyperendemic foci have been described in some Eskimo villages of the North American tundra and in western China, where dogs feed on infected rodents.

## Clinical Manifestations

In humans, the initial phase of AE infection is usually asymptomatic. Estimates of the incubation period vary from less than 5 to up to 15 years. The reported ages of patients at diagnosis ranges from 5 to 89 years, with a mean of 45 plus or minus 15 years.

The metacestode stage of *E. multilocularis* is characterized by an alveolar structure composed of numerous small vesicles (<1 mm–3 cm in diameter). The lesions grow slowly and can grow to diameters of 15 to 20 cm. Most patients become symptomatic in the progressive phase, when the metacestode has infiltrated large parts of the liver. Symptoms include abdominal pain, jaundice, hepatomegaly, sometimes fever and anemia, weight loss, and pleural pain.

Mature lesions consist of a central necrotic cavity filled with white amorphous material that is covered with a thin peripheral layer of dense fibrous tissue. Focal areas of calcification occur, and extensive

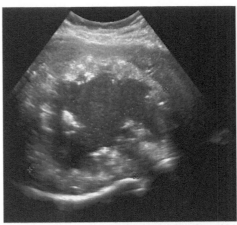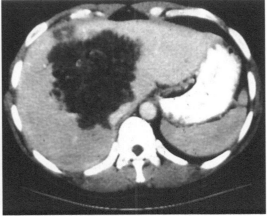

Conventional and active alveolar echinococcosis lesions in ultrasound (left) and CT images (right), showing typical findings in patients presenting with symptoms and requiring treatment

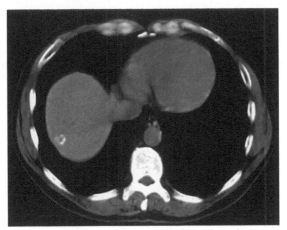

CT scan showing an inactive calcified alveolar echinococcosis lesion, typical of an "abortive" course of infection that is dying out spontaneously and will not require treatment

**Fig. 85.5** Alveolar echinococcosis lesions in the liver. (Ultrasound courtesy Prof. Peter Kern, Division of Infectious Diseases and Clinical Immunology, Comprehensive Infectious Diseases Center, University Hospitals, Ulm, Germany; CT scans courtesy Prof. Bruno Gottstein, Institute of Parasitology, University of Bern, Bern, Switzerland.)

infiltration by proliferating vesicles can be seen. Initially the metacestodes establish infections in the liver, but over time they can spread by extension to adjacent organs or hematogenously to distant sites (e.g., lungs, brain, bones).

The advanced stage is characterized by severe hepatic dysfunction, often associated with portal hypertension. The duration of the disease is variable from weeks to years. Mortality rates in untreated or inadequately treated AE patients can be high; in several published studies, the average survival rate 10 years after diagnosis was 29% (range 0% to 23%), and the survival rate after 15 years was 0%.

## Diagnosis

Ultrasonography is the method of choice for screening; it is usually complemented by CT, which detects the large number of lesions and characteristic calcifications (Fig. 85.5). Focal liver lesions are described as a mixed pattern of infiltration, necrosis ("pseudocysts") and calcification with multivesicular, honeycomb-like areas. MRI may facilitate the diagnosis in some cases.

The diagnosis can be confirmed by parasite identification in surgical or biopsy material. In most cases, histologic examination is sufficient, but recognition of characteristic structures in fine-needle biopsy specimens or calcified samples may be difficult or impossible. In these cases polymerase chain reaction (PCR) is used mainly for the direct detection of parasite nucleic acid in biologic specimens. The Em2plus-ELISA is a diagnostic test that employs a mixture of affinity-purified *E. multilocularis* metacestode antigens and has been used successfully for the serologic differentiation of AE and CE. The *E. multilocularis* protoscolex-derived Em18 antigen has also been used to differentiate AE and CE. ELISA and Western blot tests based on the Em18 show sensitivities and specificities in the ranges of 91% to 100% and 77% to 97%, respectively.

## Treatment

Surgery is the mainstay of treatment of AE, and radical resection of the lesion from the liver or affected organs is the goal of surgery. Radical surgery is possible in virtually all cases diagnosed at an early stage but in only 20% to 40% of the advanced cases. It may lead to complete cure, but resection is often incomplete because of diffuse and undetected parasite infiltration into host tissues. Therefore postsurgical chemotherapy should be carried out for at least 2 years with evaluation using positron emission tomography (PET) and MRI, with monitoring of the patients for a minimum of 10 years for possible recurrence.

For chemotherapy of AE in humans, MBZ is given at daily doses of 50 mg/kg in three divided doses or ABZ is given at daily doses of 15 mg/kg in two divided doses. Cyclic or continuous treatment is practiced. In a Swiss study, therapy for nonresectable AE with MBZ or ABZ resulted in a significantly increased 10-year survival rate.

Liver transplantation has been used successfully in otherwise terminal cases. For patients who have incomplete resection or undergo liver transplant chemotherapy a BMZ is recommended for many years to life.

## Prevention

The primary intervention for preventing AE is to interrupt the hand-to-mouth transmission of *E. multilocularis* eggs from animal to human. Hunters, trappers, and people who work with fox fur should wear protective gloves when handling animals and their byproducts. People should try to prevent their domestic dogs and cats from eating rodents that might be infected and should always wash their hands after petting, grooming, or handling animals with potential exposure to contaminated feces.

Environmental efforts to control the sylvatic cycle have included baiting wild foxes with bait containing scolicidal (anthelmintic) drugs. However, given the scale of the problem and the expense of such programs, this is likely to be practical only for targeted regions, such as urban areas where control is a high priority.

## EVIDENCE

Giorgio A, Tarantino L, de Stefano G, et al. Hydatid liver cyst: an 11-year experience of treatment with percutaneous aspiration and ethanol infection. *J Ultrasound Med* 2001;20:729-738. *Giorgio and others showed that PAIR of multivesiculated cysts does not allow complete healing (i.e., progression to stages CE4 or CE5).*

Hegglin D, Ward PI, Deplazes P. Anthelmintic baiting of foxes against urban contamination with *Echinococcus multilocularis*. *Emerg Infect Dis* 2003;9:1266-1272. *AE is not just a disease of hunters and trappers. This article presents the results of a study looking to reduce E. multilocularis egg contamination by foxes inhabiting urban areas intensively used by humans for recreational activities.*

Khuroo MS, Wani NA, Javid G, et al. Percutaneous drainage compared with surgery for hepatic hydatid cysts. *N Engl J Med* 1997;337:881-887. *This is one of two randomized clinical trials that are available. Khuroo and others found PAIR combined with periinterventional BMZ derivatives to be as effective as open surgical treatment, with fewer complications and lower cost.*

Velasco-Tirado V, Alonso-Sardon M, Lopez-Bernus A, et al. Medical treatment of cystic echinococcosis: systemic review and meta-analysis. *BMC Infectious Diseases.* 2018;18:306. Available at: https://doi.org/10.1186/s12879-018-3201-y. Accessed December 29, 2020. *A comprehensive review and meta-analysis of treatment options for cystic echinococcosis.*

## ADDITIONAL RESOURCES

Brunetti E, Kern P, Vuitton DA. Writing Panel for the WHO-IWGE: Expert consensus for the diagnosis and treatment of cystic and alveolar echinococcosis in humans. *Acta Tropica* 2010;114:1-16. *A consensus treatment article with recommendations for treatment of CE and AE using the Infectious Diseases Society of America grading system as a basis. An excellent resource for physicians managing echinococcus disease in their patients.*

Eckert J, Gemmell MA, Meslin F-X, Pawlowski ZS, eds. *WHO/OIE manual on echinococcosis in humans and animals: a public health problem of global concern.* Paris, 2001, World Organisation for Animal Health (Office International des Epizooties) and World Health Organization. Available at: www.oie.int and www.who.int. Accessed September 8, 2010. *A detailed and authoritative compendium on the global epidemiology, pathobiology, laboratory diagnosis, treatment, and control of Echinococcus infections.*

Junghanss T, Menezes da Silva A, Horton J, et al. Clinical management of cystic echinococcosis: state of the art, problems and perspectives. *Am J Trop Med Hyg* 2008;79:301-311. *A comprehensive approach to clinical management of CE, presenting a system for staging disease and correlating radiographic images—an essential guideline for healthcare providers seeing patients with CE.*

Nasseri-Moghaddam S, et al. Percutaneous needle aspiration, injection, and re-aspiration with or without benzimidazole coverage for uncomplicated hepatic hydatid cysts. *Cochrane Database Syst Rev* 2011(1): p CD003623 *A comprehensive review examining PAIR in hepatic hydatid disease.*

Santivanez S, Garcia H. Pulmonary cystic echinococcosis. *Curr Opin Pulm Med* 2010;16:257-261. *A review and update on pulmonary disease due to CE of the lung.*

Smego RA Jr, Sebanego P. Treatment options for hepatic cystic echinococcosis. *Int J Infect Dis* 2005;9:69-76. *A review of surgical and nonsurgical options to treat CE of the liver. PAIR appears to have greater clinical efficacy with lower rates of major and minor complications; surgery should be reserved for patients with difficult-to-manage cyst-biliary communication, obstruction, or hydatid cysts refractory to PAIR.*

Stojkovic M, et al. Treatment response of cystic echinococcosis to benzimidazoles: a systematic review. *PLos Negl Trop Dis* 2009;3(9):e524. *A systemic review examining the use of BMZs in echinococcus which collected data from six countries using the WHO classification found a strong between cyst activity and response to BMZ treatment.*

Stojkovic M, Mickan C, Weber TF, Junghanss T. Pitfalls in diagnosis and treatment of alveolar echinococcosis: a sentinel case series. *BMJ Open Gastro* 2015;2:e000036. doi:10.1136 *A retrospective case series which analyzes diagnostic and treatment data of patients with confirmed AE.*

Torgeson PR, Keller K, Magnotta M, Ragland N. The global burden of alveolar echinococcosis. *PLoS Negl Trop Dis* 2010;4:e722. *Estimation of the global incidence of AE by country based on a detailed review of published literature and data from other sources. The authors conclude that the global burden of AE is comparable to that of several other diseases in the neglected tropical disease category. AE is a particular problem in rural China on the Tibetan plateau.*

# 86

# Trichinellosis

*Zvi Shimoni, Paul Froom*

 **ABSTRACT**

Trichinellosis (trichinosis) is a parasitic infection caused by a roundworm of the *Trichinella* species. There are eight known species, and all are capable of causing human disease. The most common infections are caused by *Trichinella spiralis*, which can be found in pigs, rodents, horses, bears, and foxes. Another common species is *Trichinella nativa*, which infects humans after ingestion of infected bear or dog meat. Humans are incidental hosts. Reservoir hosts include carnivorous animals such as rats, mice, and foxes. Public heath efforts have decreased the risk of infection worldwide but there are still endemic rural areas. The diagnosis is by serology or muscle biopsy, but it can be made without diagnostic tests in the individual symptomatic patient who is part of a documented outbreak. Treatment is with antihelminthic drugs, and prednisone is used to decrease the risk of a systemic inflammatory reaction.

## CLINICAL VIGNETTE

A 34-year old male who had eaten partially cooked wild boar developed abdominal pain and diarrhea 2 days later. Nine days later he presented to the emergency room with severe muscle pain but was afebrile. He was one of 30 people at a party, all of whom became ill, and most were hospitalized with similar symptoms. On physical exam, his legs and arms were severely tender to palpation, but he did not have periorbital edema. The white blood cell (WBC) count was $9.8 \times 10^9$/L, and there were 25% eosinophils on the differential count. Serum creatine phosphokinase (CPK) levels were four times the upper reference range. He was treated with albendazole 400 mg twice a day and discharged with a recommendation for continued therapy for 14 days. Seven days after discharge he was readmitted with a temperature of 39°C and recurrent myalgia. The CPK value was normal but the WBC was now $14.0 \times 10^9$/L with a differential eosinophil count of 35%. The patient was treated with prednisone 40 mg/d for 5 days and, after 12 hours after the beginning of treatment with prednisone, the fever and symptoms resolved. The albendazole was continued for a total of 14 days. The patient was asymptomatic on follow-up 30 days later. A serologic test for *Trichinella* was negative, but it was positive in 21 of the other patients who were hospitalized.

COMMENT: The serologic test is very specific but has a relatively low sensitivity, so that a negative test does not rule out the infection. The diagnosis was made because he had compatible symptoms, shared an epidemiologically implicated meal, and the serologic test was positive in most of those who ate the undercooked pork. The worsening symptoms after an initial response to antihelminthic drugs was found in about 50% of patients in one study, occurring as early as 4 and as late as 17 days after the beginning of treatment. The mechanism for the systemic inflammatory reaction is probably largely mediated by eosinophils. It appears prudent to add 5 days of treatment with prednisone to lessen the patient's reaction to the effects of antihelminthic drug treatment, but the evidence for that recommendation is not based on a randomized controlled trial.

## GEOGRAPHIC DISTRIBUTION

The disease has a worldwide geographic distribution, causing major public health problems in many parts of the world. *T. spiralis* is found primarily in the United States, South America, Europe, and South Asia, whereas *T. nativa*, when it is associated with ingestion of bear meat, is found more often in Canada and Northern Asia.

Trichinellosis is common in Thailand and in low- and middle-income countries (LMICs), where many outbreaks are reported each year as a result of consumption of undercooked pork or wild animals. Although the disease is rare in the Western world, outbreaks are reported in Europe and the United States, especially among immigrants who continue to eat undercooked meat or when there are breakdowns of veterinary services. Most infections today are still caused by the ingestion of wild or domestic pork, but cases have been reported after ingestion of infected meat from other animals including horses, bears, walruses, rodents, dogs, lions, panthers, and crocodiles. Game animals as sources of infection have increased greatly in both LMICs and high-income countries; in the United States and other Western countries game animal–associated cases now exceed pork-associated cases, which are rare.

## LIFE CYCLE

The life cycle includes enteral and parenteral phases (Fig. 86.1).

### Enteral Phase

Humans are infected after eating encysted *Trichinella* larvae present in raw or inadequately cooked meat. The cyst walls are digested by acid-pepsin digestion in the stomach; larvae are then released and pass into the small intestine. The larvae invade the small intestine's epithelial wall, molt four times, and develop into adult worms. The males die after copulation, whereas the females produce 500 to 1500 newborn larvae, which are deposited into the mucosa of the duodenal wall over a 2- to 3-week period before being expelled in the fecal stream.

### Parenteral Phase

Newborn larvae enter the bloodstream and seed various organs including myocardium, lungs, brain, pancreas, and lymph nodes, but only the larvae that invade the skeletal muscle survive. The individual muscle fibers invaded by the *Trichinella* larvae show degeneration and necrosis and heavy infiltration with lymphocytes and eosinophils. The larvae are encysted within a few weeks, and the cyst walls may calcify over time.

## CLINICAL FEATURES

Many infections are thought to be asymptomatic, yet attack rates in outbreaks can approach 100%. Factors affecting the manifestation

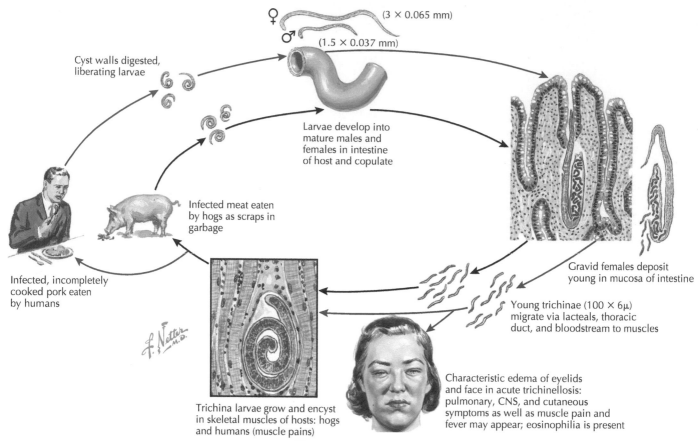

Cyst walls digested, liberating larvae

(3 × 0.065 mm)

(1.5 × 0.037 mm)

Larvae develop into mature males and females in intestine of host and copulate

Infected meat eaten by hogs as scraps in garbage

Infected, incompletely cooked pork eaten by humans

Gravid females deposit young in mucosa of intestine

Young trichinae (100 × 6μ) migrate via lacteals, thoracic duct, and bloodstream to muscles

Trichina larvae grow and encyst in skeletal muscles of hosts: hogs and humans (muscle pains)

Characteristic edema of eyelids and face in acute trichinellosis: pulmonary, CNS, and cutaneous symptoms as well as muscle pain and fever may appear; eosinophilia is present

**Fig. 86.1** Trichinellosis.

of clinical disease are the number of living larvae ingested, species involved, and host factors including age, sex, ethnic group, and immune status. There are two distinct phases. During the first week after ingestion, symptoms are consistent with those of gastroenteritis from other causes; they include diarrhea, vomiting, and abdominal discomfort. This acute enteric phase can be delayed if the infective dose is low. The second, parenteral phase corresponds to the invasion of skeletal muscles and occurs 8 to 42 days after infected meat is eaten. The most common symptom at this stage is muscular pain, and the most distinctive physical finding is periorbital edema. Systemic symptoms such as fever, headache, and rashes are common. Rarely, death may occur from myocarditis, encephalitis, or pneumonitis.

## DIAGNOSTIC APPROACH

In the enteric phase, a history of ingesting raw or undercooked meat is helpful in differentiating parasitic infection from other causes of gastroenteritis. Still, the diagnosis of an individual case in the enteric phase and early into the parenteral phase can be difficult because diagnostic antibody test results are commonly not positive until 3 to 5 weeks after infection. Also, the sensitivity of a muscle biopsy to identify diagnostic encysted *Trichinella* larvae in infected muscle will be dependent on the degree of muscle involvement and choice of the area for biopsy.

Serologic tests (enzyme immunoassays [EIAs]) for *Trichinella* using excretory-secretory (ES) antigens are usually available from commercial clinical laboratories as well as through the public health department. During the parenteral phase, even with negative *Trichinella* antibody test results, a febrile illness accompanied by headache, periorbital and facial edema, and myalgias is suggestive of trichinellosis in persons with a compatible food history. Elevated blood levels of

muscle enzymes such as CPK as well as elevated eosinophil counts in the peripheral blood increase the probability even further.

In an outbreak, however, the most important diagnostic information is a history of ingestion of undercooked or raw meat with multiple symptomatic patients. Because of the relative insensitivity of the serologic tests for diagnosis of acute trichinellosis, the following case definition is used by the US Centers for Disease Control and Prevention (CDC):

1. A positive muscle biopsy or positive serologic test result in a patient with symptoms compatible with trichinellosis.

or

2. At least one person with a positive serologic test result or muscle biopsy; associated cases are defined by compatible symptoms in persons who shared the epidemiologically implicated meal or implicated meat product. This is important because 50% or more of infected persons early in the course of the disease will not have positive serology.

## TREATMENT

Although treatment is still controversial, most experts recommend albendazole 400 mg orally twice a day for 8 to 14 days as standard therapy and concomitant prednisone 40 to 60 mg for an unspecified number of days. There are other, alternative antiparasitic drugs that can be substituted for albendazole. Prednisone was added to recommended therapeutic regimens after an observational study reported that prednisone is effective in preventing and treating systemic symptoms that occur after treatment with albendazole. The optimum dose of prednisone is uncertain, but 1 mg/kg for 5 days without tapering is probably sufficient. Aggressive supportive treatment and perhaps higher doses of steroids are warranted in those with life-threatening complications such as myocarditis, encephalitis, or pneumonitis.

## PREVENTION

The two ways to prevent the disease are first, to prevent infected meat from reaching the consumer, and second, to educate people on the importance of proper handling and cooking of potentially high-risk meats. Commercial producers of pork are required, in many countries, not to feed swine garbage (which could contain infected meat scraps) and to prevent swine ingestion of small mammals potentially infected with *Trichinella*, including rodents, raccoon, skunks, and opossums. Commercial producers of pork products such as hams, sausage, and jerky are required to follow brining, smoking, and cooking guidelines that will kill *Trichinella* larvae in pork. However, even with effective commercial standards, the risk of trichinellosis remains with meat from swine raised on small noncommercial farms, specialty meat products (sausage, jerky, and so on) from small-scale meat processing concerns, and wild game meat, often home butchered. Some experts advise that preventive efforts should now emphasize testing of free-ranging and backyard pigs.

Cooking is the most reliable method of destroying *Trichinella* in any type of meat, and a temperature of 170°F (77°C) is above the thermal death point. This temperature is usually achieved if the meat is no longer pink, but the most accurate way of measuring temperature during cooking is with a rapid-read cooking thermometer. *T. spiralis* in pork can be killed at a temperature of 160°F (71°C) for 2 minutes (Table 86.1). Microwaving meat is not recommended as a method of cooking to eradicate parasites because uniform temperatures are not obtained throughout the meat. *T. spiralis* in pork cuts that are less than 6 in thick are killed by freezing at −20°F (−29°C) for 6 days (see Table 86.1). However, freezing may not kill *Trichinella* larvae in game meats, which are relatively resistant to low temperatures.

## EVIDENCE

Marucci G, Tonanzi D, Cherchi S, et al. Proficiency testing to detect Trichinella larvae in meat in the European Union. *Vet Parasitol* 2016;231:145-149. *The authors demonstrated that testing for Trichinella larvae in meat is very reliable.*

Pozio E. Searching for Trichinella: not all pigs are created equal. *Trends Parasitol* 2014;30:4-11. *Excellent review on the distribution of trichinella in pigs worldwide, and the effectiveness of prevention in raising pigs in a controlled environment. Efforts now should emphasize testing of free-ranging and backyard pigs.*

Rostami A, Gamble HR, Dupouy-Camet J, Khazan H, Bruschi F. Meat sources of infection for outbreaks of human trichinellosis. *Food Microbiol* 2017;64:65-71. *Excellent summary of meat sources of infection for outbreaks of human trichinellosis other than pork including wild boar, horse meat, dog meat, bear meat, and other unusual outbreaks.*

Shimoni Z, Klein Z, Weiner P, et al. The use of prednisone in the treatment of trichinellosis. *Isr Med Assoc J* 2007;9:537-539. *Description of clinical and laboratory data in 30 cases associated with ingestion of pork, and comparison of clinical outcomes when the patients were treated with antiparasitic drugs plus prednisone or antiparasitic drugs alone.*

### TABLE 86.1 Consumer Processing Guidelines to Kill *Trichinella spiralis* Larvae in Pork

| Meat | Process | Temperature[a] Fahrenheit | Temperature[b] Centigrade | Duration |
|------|---------|-----------|-----------|----------|
| Fresh pork | Cooking | 160°F | 71°C | 2 minutes |
| Fresh pork | Cooking | 140°F | 60°C | 6 minutes |
| Fresh pork[c] | Freezing | −20°F | −29°C | 6 days |
| Fresh pork[c] | Freezing | −10°F | −23°C | 10 days |
| Fresh pork[c] | Freezing | −5°F | −17°C | 20 days |

[a]Refers to minimal internal temperature achieved.
[b]Guidelines for killing *Trichinella* species in pork may not apply to *Trichinella* larvae in other meats, which may be more resistant to both cooking temperatures and freezing temperatures effective for processing pork.
[c]Applies to pork cut less than 6 in thick.

Uspensky A, Bukina L, Odoevskaya I, Movsesyan S, Voronin M. The epidemiology of trichinellosis in the Arctic territories of a Far Eastern District of the Russian Federation. *J Helminthol* 2019;93:42-49. *There are still areas in the world at high risk for trichinellosis. A seroprevalence of 24.3% was detected in 259 people tested by an enzyme-linked immunosorbent assay (ELISA). The highest prevalence was detected among people who consumed traditional local foods made from the meat of marine mammals.*

Wilson NO, Hall RL, Montgomery SP, Jones JL. Trichinellosis surveillance—United States, 2008-2012. *MMWR Surveill Summ* 2015;64(1):1-8. *Comprehensive review of recent epidemiology, laboratory diagnosis, and control measures for trichinellosis in the United States. Reported human cases of trichinellosis associated with consumption of wild game meat, mostly from bear meat, now outnumber cases associated with pork in the United States.*

## ADDITIONAL RESOURCES

Centers for Disease Control and Prevention (CDC): Trichinellosis. www.cdc.gov/parasites/trichinellosis. August 8, 2012 (Accessed 09/12/19). *Web portal for information on trichinellosis, including epidemiology, disease, diagnosis and treatment, prevention and control; includes patient education materials and resources for health care professionals.*

Gilbert DN, Moellering RC, Eliopoulos GM, Sande MA, eds. *The Stanford guide to antimicrobial therapy*, 6th ed. Hyde Park, VT, 2019. *Antimicrobial Therapy, Clinical guidelines for treatment of trichinellosis.*

Gottstein B, Pozio E, Nöckler K. Epidemiology, diagnosis, treatment, and control of trichinellosis. *Clin Microbiol Rev* 2009;22(1):127-145. *Extensive comprehensive review of trichinellosis.*

Kazura JW. Tissue nematodes, including trichinellosis, dracunculiasis, filariasis, loiasis, and onchocerciasis. In Bennett JE, Dolin R, Blaser MJ (eds): *Mandell, Douglas, and Bennett's Principles and Practice of Infectious Diseases*, 8th ed. Philadelphia, Elsevier, 2015, pp. 3208-3215. *A clinical consideration of tissue nematodes.*

Shimoni Z, Froom P. Uncertainties in diagnosis, treatment and prevention of trichinellosis. *Expert Rev Anti Infect Ther* 2015;13:1279-1288. *An extensive general review covers the various aspects of diagnosis, treatment and prevention of trichinellosis.*

# 87

# Chagas Disease

*Louis M. Weiss, Fabiana Simão Machado*

## ABSTRACT

Chagas disease is caused by infection with the protozoan parasite *Trypanosoma cruzi*. It is endemic in many areas of Latin America, where it remains an important cause of morbidity and mortality. It is a significant cause of heart disease and gastrointestinal (GI) dysfunction, such as megacolon and megaesophagus. There has been increased immigration from endemic areas to North America and Europe, where Chagas disease is recognized with increased frequency as an imported infection. Chagas disease is also an opportunistic infection in immunocompromised individuals, including those with human immunodeficiency virus (HIV) infection and acquired immunodeficiency syndrome (AIDS). The diagnosis of acute infection is based on finding trypomastigotes in the blood, and chronic Chagas disease is suggested by serologic testing. The antiparasitic drugs benznidazole and nifurtimox are used to treat acute infection, but antiparasitic treatment in the chronic phase is not uniformly recommended. The management of chronic chagasic cardiomyopathy is similar to that of cardiomyopathies from other causes. The management of the GI complications generally involves both medical and surgical interventions.

## CLINICAL VIGNETTE #1

The patient, a 48-year-old woman, had immigrated to the United States from Brazil 10 years earlier. She grew up in a rural area of Brazil but had lived in a city prior to immigration. She presented to her primary care physician complaining of difficulty swallowing, saying that this was getting progressively worse. In addition, on review of symptoms, she was also noted to have a 10-year history of constipation and had been using laxatives regularly for the preceding 5 years. She reported no history of heart disease, and her electrocardiogram (ECG) was normal. Imaging (see Fig. 87.5) revealed the presence of gastrointestinal manifestations of Chagas disease; megaesophagus (shown by a barium swallow) and megacolon with retained barium due to a radiographic study that had been done 7 months previously (revealed by a flat plate of the abdomen).

## CLINICAL VIGNETTE #2

A 65-year-old male who had immigrated to the United States from Bolivia 25 years earlier presented to a cardiologist for increasing shortness of breath. Chest x-ray demonstrated an enlarged heart (see Fig. 87.3). On physical examination the patient was found to have congestive heart failure (CHF) with an ejection fraction of 35%. ECG demonstrated a right bundle branch block, and an echocardiogram demonstrated the presence of an apical aneurysm. A blood smear was done, on which no parasites were seen. Serology for *Trypanosoma cruzi* was performed and proved to be positive.

## GEOGRAPHIC DISTRIBUTION AND EPIDEMIOLOGY

Chagas disease is present in Latin America (Mexico, Central and South America) with the exception of the Caribbean countries. Vector-borne transmission of the *T. cruzi* parasite usually occurs in individuals living in primitive houses in areas where the sylvatic cycle is active. Living quarters are often invaded by infected vectors (triatome or reduviid bugs), which become domiciliary and feed at night on people, dogs, and other mammals that live in and around the household. The parasite has a complex life cycle (Fig. 87.1). During a blood meal from an infected mammalian host, the insect vector ingests blood-form trypomastigotes, which undergo transformation; after 3 to 4 weeks infective, nondividing metacyclic trypomastigotes are present in the hindgut of the vector. These forms are deposited with the feces of the vector during subsequent blood meals. Transmission to the new host occurs when the parasite-laden feces contaminate oral or nasal mucous membranes, the conjunctiva, or insect-bite wounds. When trypomastigotes (Fig. 87.2A) enter a host cell, they transform into intracellular amastigotes, which multiply by binary fission (see Fig. 87.2B) and then transform to trypomastigotes, which are released as the host cell ruptures. Trypomastigotes infect adjacent uninfected cells or disseminate through the lymphatics and the bloodstream (see Fig. 87.2A) to infect new cells in distant tissues. Although any nucleated mammalian cell can be parasitized, the cells of the cardiovascular, reticuloendothelial, nervous, and muscle systems as well as adipose tissue appear to be favored. Other modes of transmission include blood transfusion, organ donation, congenital transmission, breast milk, ingestion of contaminated food or drink, and laboratory accidents.

In recent years there has been increased immigration of infected, usually asymptomatic, individuals from endemic areas to nonendemic areas such as North America and Europe; thus Chagas disease is being recognized with increasing frequency worldwide. In the United States there have been a handful of autochthonous cases of Chagas disease. Immigration into the United States of potentially chronically infected individuals has led to the screening of blood donors to identify people who are asymptomatic but have the potential to transmit the infection via blood transfusion.

## CLINICAL AND LABORATORY MANIFESTATIONS

Individuals who are serologically positive for Chagas disease usually do not recall having an acute infection. In addition, they often do not know that they are chronically infected with *T. cruzi*. This is because vector-borne acute Chagas disease is usually mild. After an incubation period of 1 to 2 weeks, a newly infected individual may develop fever, chills, nausea, vomiting, diarrhea, rash, and meningeal irritation. A raised inflammatory lesion at the site of parasite entry (a chagoma), unilateral periorbital edema (Romaña sign), conjunctivitis,

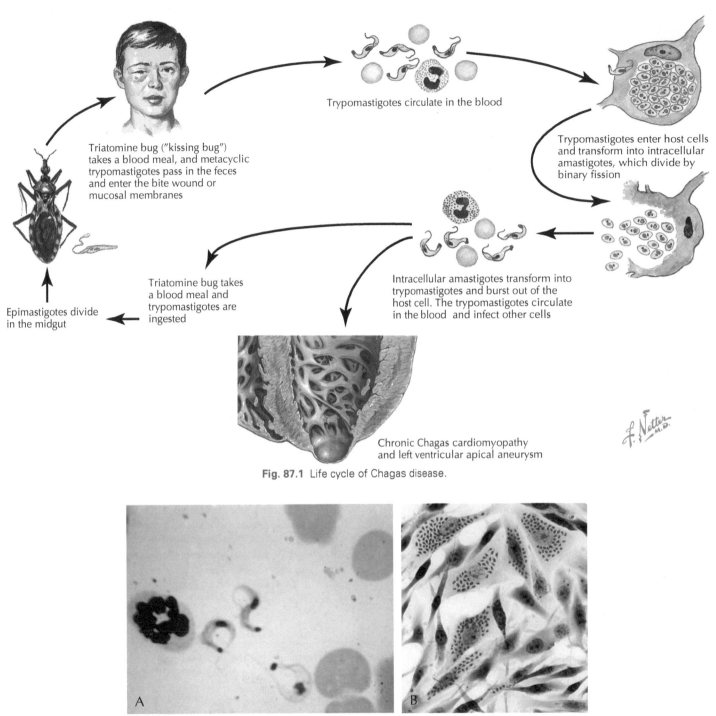

Fig. 87.1 Life cycle of Chagas disease.

**Triatomine bug ("kissing bug") takes a blood meal, and metacyclic trypomastigotes pass in the feces and enter the bite wound or mucosal membranes**

**Trypomastigotes circulate in the blood**

**Trypomastigotes enter host cells and transform into intracellular amastigotes, which divide by binary fission**

**Epimastigotes divide in the midgut**

**Triatomine bug takes a blood meal and trypomastigotes are ingested**

**Intracellular amastigotes transform into trypomastigotes and burst out of the host cell. The trypomastigotes circulate in the blood and infect other cells**

Chronic Chagas cardiomyopathy and left ventricular apical aneurysm

Fig. 87.2 (A) Trypomastigotes of *Trypanosoma cruzi* in a peripheral blood smear. Image of trypomastigotes. (B) Intracellular amastigotes of *Trypanosoma cruzi* in a culture of myoblasts. (A, From Zaiman H: *Pictorial presentation of parasites*, American Society of Tropical Medicine and Hygiene.)

lymphadenopathy, and hepatosplenomegaly have been observed in acute infection. During acute infection, anemia, thrombocytopenia, and elevated liver enzymes may be present. Blood-form trypomastigotes can be observed in wet preparations of blood and cerebrospinal fluid in many patients. Results of serologic tests for *T. cruzi*–specific immunoglobulin G (IgG) are often negative during acute infection.

Myocarditis, cardiomegaly, vasculitis, and CHF develop in a small percentage of acutely infected patients. The presence of arrhythmias is usually a poor prognostic finding. The mortality rate of acutely naturally infected patients, often children, is less than 2%, and the common

mode of death is acute myocarditis and/or meningoencephalitis. In most patients an immune response develops, the parasitemia wanes, and signs and symptoms resolve completely in 2 to 4 months. These individuals then enter the indeterminate phase of infection, which is characterized by a positive serology but the absence of clinical manifestations. This phase may last from months to an entire lifetime, and, as noted, most chronically infected persons never develop clinical manifestations attributable to the persistence of the parasites.

Most experts believe that 15% to 30% of infected individuals develop chronic Chagas disease. Chronic Chagas disease is usually

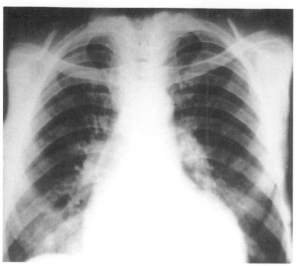

**Fig. 87.3** Pathologic findings of an enlarged heart on radiograph of patient with Chagas disease. (Reused with permission from Barrett MP et al: The trypanosomiases. *Lancet.* 362[9394]:1469-80, 2003, Figure 7.)

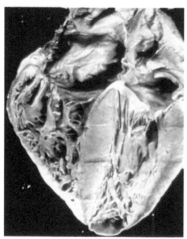

**Fig. 87.4** Autopsy specimen of a heart that is dilated and hypertrophied and has an apical aneurysm. (Courtesy Armed Forces Institute of Pathology.)

divided into cardiac and GI manifestations. The cardiac manifestations usually predominate, although in some geographic areas GI alterations are most important. The GI manifestations of chronic Chagas disease are part of the so-called *megasyndrome* complex.

Chronic chagasic heart disease may manifest insidiously as CHF or abruptly with arrhythmias and/or thromboembolic events such as stroke. Dilated congestive cardiomyopathy is an important manifestation and usually occurs years or even decades after a person first became infected (Fig. 87.3). Apical aneurysm of the left ventricle is one of the hallmarks of chronic chagasic cardiomyopathy and is observed by cardiac imaging and at autopsy (Fig. 87.4). Histology of cardiac tissue from these patients reveals chronic inflammation and fibrosis. The destruction of conduction tissue results in atrioventricular and intraventricular conduction abnormalities. The most common ECG abnormality is a right bundle branch block, which may also be associated with an anterior fascicular block. Conduction defects may necessitate the placement of a pacemaker. Increases in levels of brain natriuretic peptide (BNP) have been demonstrated to be important in the evaluation of patients with chronic chagasic heart disease. Echocardiography and cardiac magnetic resonance imaging are often useful in assessing the severity of disease.

The GI dysfunction that may accompany chronic Chagas disease is likely the result of injury to the enteric nervous system. Affected individuals may develop dilatation of portions of the GI tract. Although megacolon and megaesophagus are most common, megastomach, megaduodenum, megajejunum, megagallbladder, and megacholedochus have all been described (Fig. 87.5). Other associated dysfunctions include achalasia and aspiration pneumonia, disturbances of gastric emptying, alterations in intestinal transit, and motility disorders of the colon and gallbladder. Patients may develop GI and/or cardiac dysfunction or both. Imaging studies of the GI tract by methods such as computed tomography and ultrasound as well as pressure and motility studies are useful in assessing the extent of damage. These should be performed in conjunction with radiologists and GI specialists.

## LABORATORY DIAGNOSIS OF CHAGAS DISEASE

The diagnosis of acute *T. cruzi* infection is usually made by the detection of parasites in wet mounts of blood or cerebrospinal fluid and in Giemsa-stained slides (see Fig. 87.2A). Testing for anti–*T. cruzi* IgM antibodies is not useful. Inoculation of blood samples into a special medium or into mice may be required to demonstrate the parasite, but these culture methods lack sensitivity because parasites may not be seen for several weeks. Parasites may at times be observed in other sites, such as pericardial fluid, bone marrow, brain, skin, and lymph nodes. If acute Chagas disease is suspected in an immunocompromised patient, an examination of other samples may be useful. Polymerase chain reaction (PCR) technology is thought to be the most sensitive method for detecting acute *T. cruzi* infection but is not widely available. PCR-based tests have been used in the diagnosis of congenital Chagas disease immediately after birth.

The diagnosis of chronic Chagas disease is usually based on detecting specific antibodies. Several serologic assays are employed, including indirect immunofluorescence (IFA) and enzyme-linked immunosorbent assay (ELISA). Serologic assays are used widely for clinical diagnosis and for screening of donated blood as well as in epidemiologic studies. An immunoprecipitation assay based on iodinated *T. cruzi* proteins is specific and sensitive and is being used as a confirmatory assay to test all donor samples that are positive in the screening test.

## CLINICAL MANAGEMENT

The treatment of Chagas disease must be considered at two levels: parasite-specific therapy and adjunctive therapy for the management of the clinical manifestations. There is as yet no ideal treatment for *T. cruzi*. Two drugs are available: nifurtimox (Lampit, Bayer 2502) and benznidazole (Exeltis, USA7-1051). Benznidazole was produced until 2003 by Roche (Rochagan and Radanil), when its manufacture was transferred to a pharmaceutical company linked to the Brazilian government, the Pernambuco State Pharmaceutical Laboratory (Lafepe). The drugs are available from the Centers for Disease Control and Prevention (CDC) Drug Service (www.cdc.gov/ncidod/srp/drugs/drug-service.html [404-639-3670] and Exeltis [877-303-7181] respectively). The drugs must be taken for extended periods and have severe side effects. These drugs appear to reduce the severity of acute infection, and the general consensus is that parasitologic cure is achieved in approximately 70% of persons with acute infection with a full course of either drug, but there are no large studies that support these data. The cure rate is thought to decrease as a function of the length of time for which patients have been infected, and perhaps less than 10% of individuals with long-standing chronic infection can be cured.

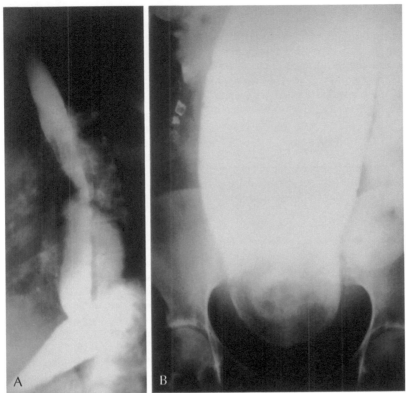

**Fig. 87.5** Gastrointestinal manifestations of Chagas disease. A 48-year-old woman came to the United States 10 years previously. She complained of difficulty swallowing and constipation. (A) Barium swallow revealed a megaesophagus. (B) A flat plate of the abdomen revealed megacolon and retained barium. The patient had undergone a barium enema 7 months previously. (From Tanowitz HB, Simon D, Gumprecht JP, et al: Gastrointestinal manifestations of Chagas' disease. In Rustgi VK, ed: *Gastrointestinal infections in the tropics*, Basel, 1990, Karger, p 64.)

There are no data at this time to state that treatment with either drug is beneficial in individuals with long-standing infections. Some experts recommend antiparasitic treatment only for patients with acute and congenital infections and for chronically infected children and young adults. Antiparasitic therapy for adults assumed to have long-standing infections is not generally recommended regardless of clinical status, although many do get treatment. For example, the question often arises as to whether to use antiparasitic therapy in individuals who are seropositive in a blood donation screen. Regardless of age, a seropositive person should be removed from the blood donation pool, and some have suggested that individuals 50 years of age or younger should be given the option of treatment. Allopurinol and several antifungal azoles have been shown to have anti–*T. cruzi* activity in vitro and in animal models. There are active research programs to develop new agents for therapy. In this context, several *T. cruzi* molecules are targets for drug design based on receptors and structures including cruzain, nitroreductase, and trans-sialidase. There are ongoing clinical trials of two drug therapies that combine benznidazole with a second drug (such as fosravuconazole) to look for more effective therapeutic regimens. An additional area of active research involves the identification of biomarkers for disease progression and cure.

The medical management of patients with chagasic heart disease is similar to the management of patients with CHF and cardiomyopathy from other causes and should be instituted in conjunction with cardiologists who specialize in CHF. Digoxin is useful in the treatment of CHF but should be used with caution in individuals with extensive fibrosis so as to avoid conduction disturbances. Angiotensin-converting enzyme (ACE) inhibitors and β-adrenergic blockers have also been successfully employed. Individuals with conduction disturbances may require the placement of a pacemaker. The use of anticoagulation therapy in patients with chagasic cardiomyopathy is controversial and should not be undertaken without consultation with a cardiologist.

Persons with severe chronic chagasic heart disease with dilatated cardiomyopathy and severe CHF may benefit from heart transplantation. A major concern in the heart transplant recipient would be the consequences of long-term immunosuppressive therapy after transplant—including the possible reactivation of *T. cruzi* infection from other tissue sites in the body. Stem cell transplantation is currently being evaluated in patients with severe CHF and chagasic heart disease.

Individuals with GI disturbances such as megaesophagus and megacolon are initially managed conservatively but may require surgery. These patients should be managed with GI specialists and surgeons familiar with these conditions.

## PREVENTION AND CONTROL

Improved housing, socioeconomic standards, and vector control have all been instrumental in reducing the spread of Chagas disease. The screening of blood and organ donors has also played an important role in reducing transmission.

## SUMMARY

Chagas disease should be considered in any individual who comes from an endemic area and manifests cardiac or GI dysfunction, and a serology test for Chagas disease should be obtained. If the serology is positive, the patient should be referred to an infectious disease or tropical disease specialist. Blood donors with a positive serology for Chagas disease should also be evaluated clinically for consideration of possible treatment.

## ACKNOWLEDGMENTS

The authors would like to acknowledge the work of our dear departed colleague Herbert B. Tanowitz on the previous edition chapter.

## ADDITIONAL RESOURCES

Bern C, Montgomery SP, Herwaldt BL, et al. Evaluation and treatment of Chagas disease in the United States: a systematic review. *JAMA* 2007; 298:2171-2181. *A detailed description of the epidemiology of Chagas disease in the United States, and the approach to diagnosis and treatment when Chagas disease is suspected in immigrants or returned travelers from endemic areas.*

Biolo A, Ribeiro AL, Clausell N. Chagas cardiomyopathy—where do we stand after a hundred years? *Prog Cardiovasc Dis* 2010;52:300-316. *Clinical perspective on Chagas cardiomyopathy and its management.*

Centers for Disease Control and Prevention (CDC): What happens to blood donors who test positive for Chagas disease? Available at: www.cdc.gov/chagas. Accessed May 18, 2010. *Fact sheet that provides information for healthcare providers and blood donors and CDC contacts for physician consultation, testing, and treatment. The CDC estimates that 300,000 or more T. cruzi–infected individuals of Hispanic origin currently live in the United States. Only approximately 11% of Chagas-seropositive blood donors or their physicians have contacted the CDC for consultation regarding treatment, according to the fact sheet.*

Echeverria LE, Morillo CA. American trypanosomiasis (Chagas disease). *Infect Dis Clin North Am* 2019;33(1):119-134. doi: 10.1016/j.idc.2018.10.015. *A useful review of Chagas disease epidemiology, diagnosis, and management.*

Gascon J, Bern C, Pinazo MJ. Chagas disease in Spain, the United States and other non-endemic countries. *Acta Trop* 2010;115:22-27. *Current epidemiology of Chagas disease in nonendemic countries.*

Nunes MCP, Beaton A, Acquatella H, et al; American Heart Association Rheumatic Fever, Endocarditis and Kawasaki Disease Committee of the Council on Cardiovascular Disease in the Young; Council on Cardiovascular and Stroke Nursing; and Stroke Council. Chagas cardiomyopathy: an update of current clinical knowledge and management: a scientific statement from the American Heart Association. *Circulation* 2018;138(12):e169-e209. doi: 10.1161/CIR.0000000000000599. *Consensus statement on management of Chagas Disease in the United States.*

Tanowitz HB, Machado FS, Jelicks LA, et al. Perspectives on *Trypanosoma cruzi*–induced heart disease (Chagas disease). *Prog Cardiovasc Dis* 2009; 51:524-539. *A useful review on the pathogenesis and pathology of* T. cruzi–induced heart disease.

# Emerging Infectious Diseases and Pandemics

*M. Patricia Joyce*

# 88

# Introduction to Emerging Infectious Diseases and Pandemics

*M. Patricia Joyce*

This section on emerging infectious diseases (EIDs) highlights new or reemerging infectious diseases that have been recognized over the past two decades and pose significant hazards to human health because of either rapidly increasing incidence in human populations or a growing geographic range. During the time period covered, a robust increase in international air travel, further development of the global food market, and continued growth in commercial shipping were all factors that contributed to the possibility that remote outbreaks in any part of the world can rapidly spread to become a global concern. Previously rare localized viral diseases have emerged to become international threats in a world where a traveler might cross the planet in less than one or two days, well before a short incubation period and the recognition of any illness.

Many EIDs are zoonoses or infections transmitted from animals to man: transmissions result from increased human encroachment on natural animal habitats, environmental changes that lead to increased insect vector populations, human consumption out of necessity or preference for wild or exotic animals, or even changes in animal husbandry practices. If a pathogenic agent is capable of cross-species infection from animal to human, further genetic mutations may facilitate person-to-person transmission and easy dissemination in large human populations. Other EIDs have existed in relatively isolated biological niches for some time; these emerge as significant threats when they are accidentally or purposely introduced into new environments, or engineered into new forms suitable for use as bioweapons.

Reemergent infectious diseases include familiar human pathogens, such as tuberculosis (TB) and viral influenza (flu). TB was controlled in the past by antimicrobial treatment and infection control measures, but development of strains resistant to multiple first-line drugs challenges present treatment and control measures. Flu vaccines and antiviral drugs have been used to prevent large-scale flu outbreaks with moderate success, but novel flu strains (e.g., avian influenza, swine influenza) could emerge and become capable of initiating global pandemics.

While recognized as a scourge of mankind since antiquity, TB remains a major cause of death and disability even to this day. As much as one-fourth of the world's population remains infected, although most TB disease is of the latent type. In a modern world of immunocompromised hosts living with HIV or receiving various chemotherapies for cancer or immunologic diseases, TB remains a threat to reactivate and be further transmitted. While excellent prophylaxis and treatment of sensitive TB is available in developed countries, access to affordable and effective treatment remains problematic for much of the world. Drug-resistant strains may be selected by insufficient or inappropriate treatment regimens, further causing difficulty in successful recovery. The emergence of multidrug-resistant and extensively drug-resistant strains is now well recognized in certain localities and presents major economic and public health challenges. Thus an ancient infection reemerges as a continuing threat.

Anthrax, caused by *Bacillus anthracis*, is a significant agricultural and veterinary disease worldwide with major economic implications for farmers and livestock producers, but fortunately human disease remains uncommon. Anthrax can present in humans as cutaneous disease, resulting from skin contact with spores from infected animals, animal products such as wool or hides, or soil. Gastrointestinal disease may result from the consumption of undercooked meat of infected animals, or ingestion of aerosolized spores. Inhalation anthrax occurs from breathing in spores aerosolized during processing of spore-contaminated animal products, but is unusual in modern textile (wool) and leather manufacturing. Unfortunately, anthrax spores have been used as a biologic weapon over the past century in several countries, and an anthrax mass incident remains a current risk.

Tularemia disease in humans is primarily associated with environmental exposure to animal reservoirs that include rabbits, muskrats, squirrels, and beavers, but may be transmitted by inhalation, ingestion, and insect bites by certain ticks and flies as well. Presenting as various ulceroglandular syndromes that progress to life-threatening septicemia and pneumonia, tularemia remains easily transmissible and also has potential as a biowarfare agent.

The flavivirus Zika emerged in the last two decades from being an infrequent mosquito-borne pathogen first isolated in 1947, in monkeys from the Zika Forest in Uganda, to become a cause of major outbreaks worldwide, even unto remote islands of the Pacific. Zika virus is transmitted via mosquitos, as well as by human sexual intercourse. While most adult human Zika disease has been mild, severe neurologic complications including Guillain-Barré syndrome have been reported. Tragically a profound syndrome of congenital microcephaly has been found to occur following Zika infection during pregnancy. Prevention of Zika transmission remains limited to avoidance of mosquito bites as well as avoidance of sexual intercourse by infected patients.

The rat pathogen *Angiostrongylus cantonensis* or Rat Lungworm spread beyond originally recognized foci in China to other countries in the Pacific, South and Southeast Asia, South America, the Caribbean, Africa, and the southeastern United States. Previously considered a veterinary disease, this zoonosis is now recognized to cause life-threatening disease in humans, which may include eosinophilic meningitis, encephalitis, cranial nerve palsies, radiculomyelitis, and ocular invasion. While treatment of the nematode infection is available, survivors may have profound and chronic neurologic disability. Human infection results from the accidental or deliberate ingestion of living infected snails (intermediate hosts) containing the larval nematode parasite, or fresh fruits and leafy green vegetables contaminated with infected snail mucus. Analysis of a recent outbreak of neuroangiostrongyliasis among humans in Hawaii has suggested that many cases worldwide are probably unrecognized or underreported.

Ebola virus disease outbreaks emerged in Africa (the first outbreak was identified in Congo in 1976), but subsequent outbreaks occurred in Central Africa, and most recently in West Africa in 2014 to 2016. The primary outbreaks occurred in endemic rural areas outside of major metropolitan areas, and index cases were thought to be associated with butchering bushmeat—but secondary cases occurred among people exposed to highly infectious secretions while providing care to infected or deceased patients. Initially presenting as a flu-like illness, the infection progresses to a febrile gastroenteritis with large volume losses, and the main transmission is probably fecal-oral. When EID occurs in resource-poor communities, health care personnel, drugs, supplies, and infrastructure to detect, treat, contain, and control community spread of the disease are often lacking. As happened in the recent West African Ebola virus outbreak, international medical relief workers responding to the health crisis sometimes become infected themselves: international spread of the infection became possible when infected relief workers were evacuated, and in turn required medical care and laboratory services in their home countries.

The emergence of novel pathogens such as SARS-CoV, MERS-CoV, and SARS-CoV-2 (COVID-19) coronaviruses may have been facilitated by mutations in these RNA viruses in animal reservoirs that facilitated cross-species infection. While transmission of these viruses originally may have occurred among humans working with, or exposed to, infected animals through aerosols or ingestion, these viruses have shown themselves to be highly infectious agents for human-to-human transmission, causing local and regional outbreaks and leading to international outbreaks though intercontinental travel of infected presymptomatic individuals.

Influenza viruses have infected man, birds, pigs, and other animals for millennia, with occurrence of severe pandemics in the historical past. Although medications and vaccines against influenza A and B are available, continuing viral mutations have made perennial transmission problematic. The concluding chapter in this section will review the mechanisms of genetic drift and genetic shift among influenza viruses and explore the potential for novel influenza viruses to emerge, along with the implications for global public health.

In an ever-changing landscape of threats from emerging and reemerging infectious diseases, knowledge of the current issues and a realistic expectation that future outbreaks, epidemics, and pandemics caused by EIDs will occur should motivate public health and infectious diseases experts—indeed, all health care providers and facilities—to participate in comprehensive strategic planning and preparation in communities, regions, nationwide, and globally.

## ADDITIONAL RESOURCES

Ellwanger JH, Kaminski VL, Chies JAB. Emerging infectious disease prevention: Where should we invest our resources and efforts? *J Infect Public Health* 2019;12(3):313-316. *A concise but accurate analysis of pathogen emergence from animal populations to human sentinels to the general human population, discussing targets for resource deployment to control emerging infectious diseases.*

Morens DM, Folkers GK, Fauci AS. The challenge of emerging and reemerging infectious diseases. *Nature* 2004;430(6996):242-249. *Outstanding review article on the burden of infectious disease, both traditional and emerging, with an in-depth discussion of the toll in human deaths and economic costs.*

Nii-Trebi NI. Emerging and neglected infectious diseases: Insights, advances, and challenges. *Biomed Res Int.* 2017;2017:5245021. *This is a thorough review article on global infectious diseases, both neglected and emerging, and summarizes current priorities and public health responses internationally. The author provides an in-depth description of past major infection control efforts and identifies areas for future advances in addressing emerging infectious disease threats.*

Paules CI, Eisinger RW, Marston HD, Fauci AS. What recent history has taught us about responding to emerging infectious disease threats. *Ann Intern Med* 2017;167(11):805-811. *The authors summarize the historical US governments efforts up to 2017 to provide pandemic response and planning. This is an important article to describe the need for preparation to meet pandemic needs before the threats emerge.*

# 89

# Multidrug-Resistant Tuberculosis

*Jonathan M. Wortham, Angela Starks, Sapna Bamrah Morris*

## ABSTRACT

Tuberculosis (TB) is an important public health problem. According to the World Health Organization (WHO), approximately 10 million persons developed TB during 2018, and approximately one-fourth of the world's population, more than 2 billion persons, are infected with *Mycobacterium tuberculosis*, the causative agent of TB. Worldwide, TB is the leading cause of death among persons living with human immunodeficiency virus (HIV) infection and the most common cause of death overall attributable to a single infectious agent. Nonetheless, TB incidence rates and prevalence rates of *M. tuberculosis* infection vary widely, with the majority of both occurring among persons living in a relatively small number of countries, often in resource limited settings. Drug-susceptible TB is curable through established ≤9 months treatment regimens and preventable by providing specific treatment to persons with *M. tuberculosis* infection. However, the emergence of multidrug-resistant (MDR) TB threatens progress toward curing cases, preventing morbidity for persons, and preserving the effectiveness of first-line treatment regimens for societies. Morbidity, both treatment- and disease-associated, is much more common among patients with drug-resistant TB. Although effective treatment for the majority of forms of drug-resistant TB exists, access to necessary medications is often limited or nonexistent in resource-limited settings where the incidence of drug-resistant TB is highest. To neutralize the dual threats of TB and drug resistance, new and stronger initiatives for improving diagnostic capacity, finding better therapeutics, and developing stronger medical and public health systems are needed.

## CLINICAL VIGNETTE

A 55-year-old woman with a history of well-controlled diabetes mellitus type 2 presented with unexplained weight loss, fatigue, and a productive cough. This patient had been born in a country with a high incidence of TB and had moved to the United States approximately 5 years before presenting to care with these symptoms. The patient reported that her cough had progressively worsened throughout the previous weeks. Clinical examination revealed that her temperature was 39.7°C, blood pressure 114/72 mm Hg, pulse 80 beats per minute, and respiratory rate 18 breaths per minute; she weighed 40.9 kg, approximately 9 kg less than at her previous health care visit 6 months prior. Her finger-stick glucose reading was 160 mg/dL.

This patient had received a diagnosis of tuberculosis (TB) 10 years earlier, which had been treated with a regimen of two medications for 6 months; thereafter her symptoms resolved. Although she reported no known recent contact with patients who had active TB disease, she did return to her country of birth for approximately 1 month each year. A chest radiograph revealed a right-upper-lobe infiltrate. Sputum was collected in the clinic and then again during the next 2 days while she was at home. The three samples were acid-fast bacillus (AFB) smear-negative, smear-positive, and smear-positive, respectively. The patient was started on the first-line TB treatment regimen, which comprised isoniazid, rifampin, pyrazinamide, and ethambutol.

During further interviews, the patient verified that she had been having symptoms consistent with TB for approximately 2 months before her most recent visit to her physician and reported that she had recently lived with extended family to care for an ailing relative. She continued to work throughout this time at her administrative job. She sat in a cubicle alone but had four coworkers in close proximity.

After approximately 6 weeks of therapy, her condition was very similar to that upon her initial presentation. She reported feeling slightly better after initially taking her medication, but her condition had worsened during the past week. She continued to have fevers and reported a decreased appetite. The patient's family and work contacts were all identified and tested for TB infection; several had positive test results, demonstrating evidence of infection from TB exposure. Some of the contacts were started on treatment for latent TB infection (LTBI) with rifampin.

The patient was initially referred to the public health department to initiate directly observed therapy and case management; however, she wanted to maintain continuity of care and have her private physician communicate with the TB clinic. Upon further review of the patient's medical chart, her sputum samples remained smear-positive for AFB; one of the sputum samples was then sent for nucleic acid amplification testing, which also included molecular evaluation for resistance. Results from this test demonstrated evidence of rifampin resistance. Culture isolates were sent for additional molecular detection of drug resistance and growth-based drug susceptibility testing by agar proportion. Both methods demonstrated resistance to both isoniazid and rifampin. After confirming that additional resistance was neither present nor acquired, the patient's treatment regimen was expanded to an all-oral regimen with five second-line anti-TB medications. The treatment course was continued for 12 to 18 months after culture conversion. At that point the patient's direct contacts had to be reevaluated and restarted on treatment for MDR LTBI with fluoroquinolone-based therapy.

COMMENT: It is important to have an index of suspicion for drug-resistant TB based on risk factors. In this case, a previous TB diagnosis, inadequate TB treatment (i.e., with fewer than three medications), and a lack of response to initial treatment for this episode increased clinical suspicion for MDR TB. For patients with such risk factors, treating clinicians should request molecular testing for drug resistance, culture, and phenotypic drug-susceptibility testing (DST) during the initial TB diagnostic evaluation. These results should guide the selection of an effective therapeutic regimen that minimizes medication-related toxicity. Consultation with an expert in the management of drug-resistant TB is always advisable when MDR TB is being managed. Prompt diagnosis not only benefits the patient with MDR TB but can also guide management of the patient's contacts. Because neither isoniazid nor rifampin will prevent progression to TB disease among persons with LTBI resulting from MDR organisms, the drug-susceptibility profile should also guide the choice of LTBI treatment regimen.

## EPIDEMIOLOGY

Globally, TB is a common infectious disease and the most common cause of death from a single infectious agent. TB caused an estimated 1.5 million deaths during 2018; while most TB deaths occurred among people without HIV, TB was the most common cause of death among persons living with HIV infection. Of the estimated 10 million TB cases diagnosed during 2018, approximately 500,000, or 5%, were MDR, meaning that they were resistant to isoniazid and rifampin, the two most important medications in first-line treatment regimens. The incidence of drug-resistant TB varies widely; the World Health Organization (WHO) estimates that 90% of MDR disease occurs among persons living in 30 countries with a high MDR TB burden (Fig. 89.1) Comparatively, drug resistance is relatively rare in the United States. In recent years, of the approximately 10,000 TB cases reported annually, 8% to 10% were isoniazid-resistant; 1% to 2% were MDR; and fewer than 10 cases (<1%) were extensively drug-resistant (XDR), meaning that they were resistant to isoniazid, rifampin, fluoroquinolone, and an injectable agent (i.e., capreomycin, amikacin, or kanamycin).

## BACKGROUND

TB is caused by one of a group of related acid-fast bacilli called the *M. tuberculosis* complex. Although *M. tuberculosis* and *M. bovis* are the most common causes of human disease, the bacille Calmette-Guérin strains of *M. bovis, M. africanum, M. caprae, M. microti, M. canetti,* and *M. pinnipedii* are other disease-causing organisms in the *M. tuberculosis* complex. These mycobacteria are typically transmitted from person to person through airborne droplet nuclei 1 to 5 μm in diameter. Contaminated particles are produced when a person with pulmonary or laryngeal TB coughs, speaks, or sings. Transmission occurs when another person inhales these contaminated droplet nuclei.

The majority of persons who inhale *M. tuberculosis*–containing droplet nuclei do not develop TB disease; macrophages either kill the mycobacteria and eliminate them from the body entirely or sequester them into granulomas in the lungs or elsewhere in the body; this condition of pathogen–immune system stalemate is often referred to as *latent TB infection* (LTBI). However, among the minority of persons who do not eliminate or sequester the bacilli after initial exposure, the mycobacteria cause *primary TB disease* in the lungs or travel through the pulmonary lymphatic system and disseminate through the bloodstream to infect almost any organ (e.g., the lymph nodes, larynx, brain [TB meningitis], spine [Pott disease], other bones and joints, the abdomen, or the genitourinary system). In less common circumstances, the site of infection can be outside of the lungs (e.g., gastrointestinal after the ingestion of contaminated food or cutaneous after direct exposure to infected tissues) and cause disease confined to the site of infection or similarly spread through lymphatic and hematogenous spread to other body sites, including the lungs themselves (when the lungs are seeded with bacilli through hematogenous spread, this sometimes results in the radiographic finding of a "miliary pattern").

In addition to developing disease after primary infection, disease can also occur months or years after initial infection and development of LTBI, an occurrence often referred to as *reactivation of* or *progression to TB disease*. The majority of TB disease cases diagnosed in the United States are attributed to reactivation disease. As in the case of disease following primary infection, the lungs are by far the most common site of reactivation disease. However, disease can occur at almost any organ structure.

The risk for developing TB disease, either primary or reactivation, is higher among young children (i.e., those <5 years of age), persons with immunocompromising conditions (e.g., HIV infection, certain hematologic conditions, or other specific malignancies), persons taking certain immunosuppressing medications (e.g., multiple cancer chemotherapeutic agents, antirejection medications for organ transplantation, tumor necrosis factor-α antagonists, and >15 mg of prednisone or equivalent daily for >4 weeks), persons with certain medical conditions (e.g., diabetes mellitus or silicosis), persons who have had certain surgical procedures (e.g., gastrectomy or jejunoileostomy), and persons who are malnourished (i.e., ≥10% below ideal body weight).

In the United States, TB disease is initially treated with a combination of four first-line medications (i.e., isoniazid, rifampin, pyrazinamide, and ethambutol) to prevent drug resistance. Drug resistance is thought to be a consequence of spontaneous mutations in genes associated with specific molecular targets that results in phenotypic resistance to these drugs either by modifying the drug target, preventing prodrug conversion, or altering efflux mechanisms. Random mutations that confer resistance to any single medication are thought to be rare, occurring with an approximate frequency of 1 in $10^6$ to 1 in $10^8$ cases (Table 89.1). Combination therapy prevents drug resistance, even for patients with high disease burdens (e.g., with cavitary disease), because random simultaneous mutations conferring resistance to three or four medications is extremely unlikely.

As in the case of drug-susceptible TB, drug-resistant TB can also be transmitted from person to person by contaminated respiratory droplets. Patients who were initially infected with already drug-resistant organisms are referred to as having *primary drug resistance*. Patients can also develop secondary or acquired drug resistance if the treatment regimen is inadequate because of too few effective medications or nonadherence with a correctly prescribed regimen. If a patient with initially drug-susceptible TB receives only isoniazid, the initially dominant population of isoniazid-susceptible bacteria will be killed, leaving only the isoniazid-resistant mycobacteria to proliferate and become the dominant population. If that patient, now with isoniazid-resistant TB, subsequently receives rifampin alone, the rifampin-susceptible organisms in the remaining population of mycobacteria will be killed, leaving the isoniazid-resistant, rifampin-resistant bacteria to proliferate. By using both drugs simultaneously, this resistance-inducing selection pressure is minimized (Fig. 89.2). Ethambutol is used as part of the standard initial regimen because in most cases the drug susceptibility profile of the patient's infection is unknown and ethambutol's bacteriostatic effect can slow the proliferation of unidentified resistant strains. Ethambutol can be discontinued if drug susceptibility studies have demonstrated susceptibility to the first-line drugs; for culture-negative TB that is presumably susceptible to all first-line medications, ethambutol is discontinued after 2 months of treatment. Pyrazinamide is included in the initial 2-month treatment phase because it has been shown to shorten the necessary overall length of treatment from 9 months to 6 months; this might be because pyrazinamide is active against mycobacteria in slightly acidic environments such as the intracellular space of macrophages, which is one of the places where mycobacteria can evade the normal host immune response (see Fig. 89.2). Patients can also acquire resistance to second-line TB medications, especially if therapy for drug-resistant TB is inadequate. Acquired drug resistance can also emerge, despite adherence to therapy, if drug levels are subtherapeutic, which can occur in instances of malabsorption or inadequate dosing.

## CLINICAL FEATURES AND DIAGNOSTIC APPROACH

*M. tuberculosis* infections, regardless of drug-resistance patterns, have similar clinical presentations. Patients with LTBI—those who are infected with *M. tuberculosis* but do not have clinically evident TB disease—have no symptoms or signs specific to this condition; nonetheless, such patients will often but not always have positive results for an interferon-γ release assay (IGRA) and/or a tuberculin skin test

**A. Percentage of previously treated TB cases with MDR/RR-TBª**

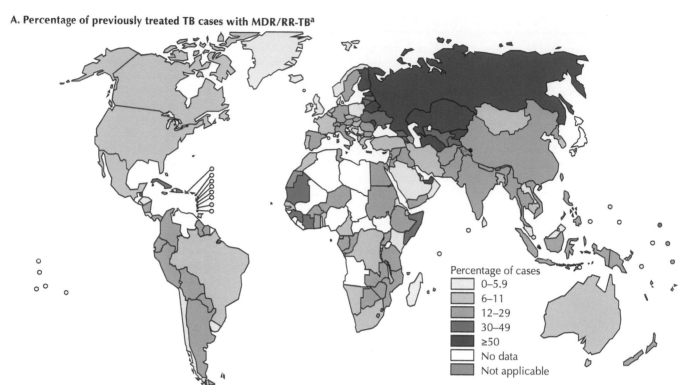

ª Percentages are based on the most recent data point for countries with representative data from 2004 to 2019. Model-based estimates for countries with data before 2004 are not shown. MDR-TB is a subset of RR-TB.

**B. Estimated incidence of MDR/RR-TBª in 2018, for countries with at least 1000 incident cases**

ª MDR-TB is a subset of RR-TB.

**Fig. 89.1** Percentage of previously treated TB cases with MDR/RR-Tbª and estimated incidence of MDR/RR-TBª in 2018 for countries with at least 1000 incident cases. Because isolated resistance to rifampin resistance is extremely rare, rifampin resistance often indicates MDR disease. *RR-TB*, Rifampin-resistant TB; *TB*, tuberculosis. (Reused with permission from Global tuberculosis report 2019, https://www.who.int/tb/publications/global_report/en/, Figures 3.31 and 3.32, page 60. https://apps.who.int/iris/bitstream/handle/10665/329368/9789241565714-eng.pdf?ua=1).

## TABLE 89.1 Resistance Mutations and Mechanisms of Action for Antituberculosis Drugs

| Agent | Mechanism | Genes Commonly Associated With Resistance |
|---|---|---|
| Isoniazid | Inhibits mycolic acid synthesis | katG<br>inhA<br>Others |
| Rifampin | Inhibits transcription (DNA-dependent RNA polymerase) | rpoB |
| Pyrazinamide | Unclear but could affect membrane energetics or fatty acid biosynthesis | pncA |
| Ethambutol | Inhibits mycolic acid synthesis | embB |
| Streptomycin | Inhibits protein synthesis | rpsL<br>rrs |
| Amikacin and kanamycin | Inhibits protein synthesis | rrs |
| Ethionamide | Inhibits mycolic acid synthesis | inhA |
| Fluoroquinolones | Inhibits DNA synthesis (DNA-gyrase) | gyrA |

(TST), both of which are tests for *M. tuberculosis* infection. Although LTBI treatment, also called *TB preventive therapy*, is highly effective for preventing drug-susceptible TB disease, no tests are available that differentiate persons with drug-susceptible LTBI from those with drug-resistant forms. In the United States, the LTBI treatment regimens preferred by the Centers for Disease Control and Prevention (CDC) include rifampin or another rifamycin (e.g., rifapentine) with or without isoniazid, although daily isoniazid monotherapy for 6 or 9 months is also commonly used and is still considered acceptable, but not preferred. Globally, the majority of regimens feature isoniazid. Because of resistance to isoniazid and rifamycins, these drugs are not acceptable for treating persons latently infected with MDR organisms. Current guidance recommends fluoroquinolone-based regimens for preventing TB disease in persons infected with MDR organisms that are presumed to be susceptible to fluoroquinolones.

Clinical presentations of drug-resistant TB disease are also largely indistinguishable from drug-susceptible TB, at least initially. Although TB disease can affect almost any organ structure, pulmonary disease is the most common, affecting more than 70% of patients. Presenting symptoms often include cough that is often prolonged for 2 or more weeks, fever, night sweats, fatigue, anorexia, and unexplained weight loss. Chest pain; dyspnea, especially on exertion; and coughing up blood (hemoptysis) are also common features of pulmonary disease. The presenting symptoms of extrapulmonary disease often include constitutional symptoms (e.g., fever, unexplained weight loss) along with symptoms specific to the affected organ structure (e.g., swollen lymph nodes, abdominal pain, mental status changes in patients with TB meningitis). As many as one-third of patients with extrapulmonary disease also have pulmonary disease.

Because symptomatic presentations of drug-resistant TB are similar to or even indistinguishable from drug-susceptible TB, eliciting specific epidemiologic and clinical risk factors for drug resistance during diagnostic evaluations can facilitate more prompt testing for and diagnoses of drug-resistant TB. These epidemiologic risk factors include:

- Exposure to a person with drug-resistant TB
- Exposure to or residence in an area or setting where drug resistance is highly prevalent

Clinical risk factors include:

- Previous receipt of an inadequate TB treatment regimen
- Poor or no improvement despite adherence to a first-line treatment regimen
- Diagnosis after treatment for latent infection when signs or symptoms of disease were present
- Prolonged use of fluoroquinolones or injectable agents (e.g., aminoglycosides) for prolonged periods before TB was considered, often for nonspecific pulmonary disease

Physical examination should be performed during all diagnostic evaluations for *M. tuberculosis* infection. Findings of TB disease include cachexia, cervical or regional lymphadenopathy, abnormal findings on auscultation of the lungs related to consolidation or pleural effusions, hepatosplenomegaly, abdominal tenderness related to peritonitis, spinal tenderness related to spinal TB (i.e., Pott disease), meningeal signs, and neurologic deficits.

In the United States, diagnostic evaluations for LTBI and TB disease, regardless of drug-susceptibility patterns, usually include a test for *M. tuberculosis* infection, either an IGRA or a TST. Because LTBI prevalence and TB disease incidence are relatively low in the United States, clinical guidelines for this setting recommend testing only persons who have specific epidemiologic risk factors for LTBI or TB disease, such as exposure to a person with infectious TB or previous residence in countries or settings where TB disease is common; also to be tested are those persons having signs or symptoms consistent with disease. These tests can be helpful for identifying asymptomatic persons with LTBI who can benefit from treatment to prevent progression to TB disease in this low-incidence setting. Although tests for *M. tuberculosis* infection can also assist with TB disease diagnoses, negative results do not exclude TB disease. Because these tests depend on the immune reaction to mycobacterial antigens, they have lower sensitivity for LTBI and TB disease among certain persons with strong risk factors for progression to disease (e.g., persons taking immunocompromising medications, those with other immunocompromising medical conditions, children <5 years of age, and those living with HIV infection). Tests can also be negative in the setting of overwhelming mycobacterial sepsis caused by immunosuppression. In resource-limited settings, tests for *M. tuberculosis* infection are often unavailable or only rarely used in settings where the prevalence of LTBI and incidence of TB disease are high.

When available, chest radiography should also be performed during all diagnostic evaluations for *M. tuberculosis* infection and TB disease. Although nonspecific, single or multiple areas of upper-lobe consolidation, upper-lobe fibronodular opacities, cavitation, mediastinal or hilar adenopathy, and pleural effusions are all radiographic signs consistent with TB disease. Computed tomography (CT) can be helpful in evaluating extrapulmonary TB and further characterizing pleural abnormalities, pulmonary infiltrates, cavitation, and lymphadenopathy among patients with pulmonary disease. Magnetic resonance imaging (MRI) with contrast can also be helpful in diagnostic evaluations for pulmonary and extrapulmonary disease; ring-enhancing lesions in the brain, lymph nodes, and other tissues are consistent with but not specific for TB disease affecting those organ structures. Although helpful for further characterizing disease, availability of CT, MRI, and even chest radiography can be unpredictable in resource-limited settings where the disease burden is often high.

Laboratory services are vital for diagnosing TB disease, including sputum smear microscopy as a key part of diagnostic evaluations for pulmonary TB. In pulmonary TB, acid-fast bacilli are often observed after smear and culture of sputa from affected patients. Although only approximately 50% of patients with diagnosed TB disease in the United States have positive sputum smear results, those results are usually available within 24 hours. Therefore, when positive, these results can assist

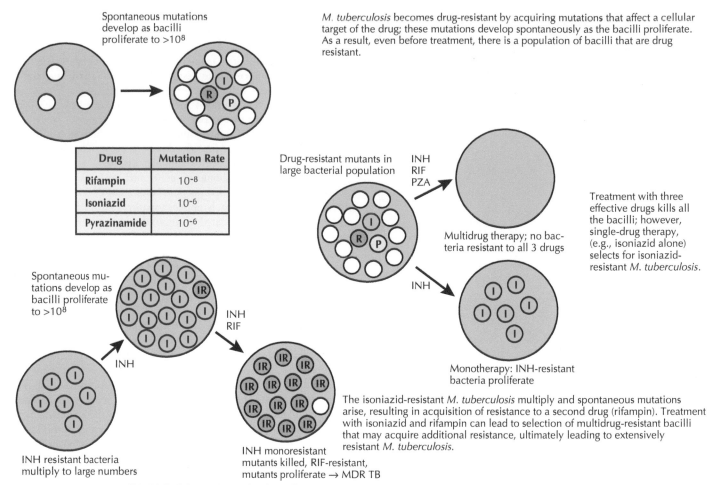

Spontaneous mutations develop as bacilli proliferate to >10⁸

*M. tuberculosis* becomes drug-resistant by acquiring mutations that affect a cellular target of the drug; these mutations develop spontaneously as the bacilli proliferate. As a result, even before treatment, there is a population of bacilli that are drug resistant.

| Drug | Mutation Rate |
|------|---------------|
| Rifampin | $10^{-8}$ |
| Isoniazid | $10^{-6}$ |
| Pyrazinamide | $10^{-6}$ |

Drug-resistant mutants in large bacterial population

INH
RIF
PZA

Multidrug therapy; no bacteria resistant to all 3 drugs

Treatment with three effective drugs kills all the bacilli; however, single-drug therapy, (e.g., isoniazid alone) selects for isoniazid-resistant *M. tuberculosis*.

Spontaneous mutations develop as bacilli proliferate to >10⁸

INH

INH
RIF

INH

Monotherapy: INH-resistant bacteria proliferate

INH resistant bacteria multiply to large numbers

INH monoresistant mutants killed, RIF-resistant, mutants proliferate → MDR TB

The isoniazid-resistant *M. tuberculosis* multiply and spontaneous mutations arise, resulting in acquisition of resistance to a second drug (rifampin). Treatment with isoniazid and rifampin can lead to selection of multidrug-resistant bacilli that may acquire additional resistance, ultimately leading to extensively resistant *M. tuberculosis*.

**Fig. 89.2** Schematic mechanism of emergence of drug resistance in *Mycobacterium tuberculosis*.

with rapid diagnoses and measuring response to therapy. Current guidance recommends performing nucleic acid amplification tests (NAATs) on at least one specimen from patients undergoing a diagnostic evaluation for tuberculosis; NAATs on sputum are 70% sensitive for diagnosing pulmonary TB. However, because initial sputum smear—even with the added sensitivity of NAATs—is relatively insensitive for diagnosing tuberculosis, empiric therapy for TB should be strongly considered when clinical suspicion for pulmonary disease is high, sputum smear results are negative, and alternative diagnoses have been excluded. NAATs are much more specific for TB disease compared with sputum smear, which may detect nontuberculous mycobacteria. Sputum culture, which can require as many as 6 to 8 weeks, should be performed when available. In the United States, more than 70% of patients with diagnosed TB disease will have positive cultures; identification of growth in culture should be performed to confirm infection with one of the organisms in the *M. tuberculosis* complex after positive culture results. Culture is particularly important for diagnosing drug-resistant pulmonary TB because growth-based DST can be performed on the organisms isolated from culture. Smear and culture of other fluids (e.g., cerebrospinal fluid and peritoneal fluid) often facilitates diagnosis of extrapulmonary TB disease. Unfortunately, as with radiography, laboratory capacity for performing culture is often limited or nonexistent in resource-limited settings.

After culture of *M. tuberculosis* isolates, drug resistance is determined through DST using antituberculosis drugs in liquid or solid media. DST is the standard of practice for cases of culture-confirmed TB disease in the United States. Therefore testing for resistance to first-line medications in the United States is often performed in commercially available automated liquid culture systems to facilitate more prompt diagnoses of drug-resistant disease. Solid media is more commonly used to test for resistance to second-line medications. Capacity for this kind of testing, often referred to as *phenotypic (or growth-based) DST*, is often limited or nonexistent in resource-limited settings. Molecular methods for determining drug-resistance patterns are becoming more prevalent; although specific procedures differ among laboratories, these methods typically rely on nucleic acid amplification to evaluate the mycobacterial DNA for mutations potentially associated with drug resistance. Some molecular methods combine NAATs for identifying *M. tuberculosis* complex in sputum specimens with tests to evaluate for mutations associated with resistance to specific drugs, especially rifampin. Because *M. tuberculosis* isolates that are solely resistant to rifampin are extremely rare, rifampin resistance often indicates MDR disease. These molecular methods can assist in promptly diagnosing TB and can offer results within 1 day of laboratory receipt; moreover, multiple platforms have been adapted for use in resource-limited settings. These tests also can enable more prompt development of treatment plans for drug-resistant TB. Although molecular methods for determining drug resistance are extremely valuable and some laboratories are reducing traditional phenotypic DST with primary use of advanced molecular methods such as whole genome sequencing, interpretation of results depends on identifying a mutation that has been characterized with respect to drug resistance. Because not all genetic mechanisms or mutations have been characterized, ongoing capacity for phenotypic DST is needed.

In the absence of phenotypic and molecular testing, drug resistance is usually diagnosed on the basis of the lack of response to first-line therapy (i.e., treatment for drug-susceptible TB). Typically, in drug-susceptible disease, symptoms related to TB disease improve; fewer acid-fast bacilli are visible by sputum smear microscopy; and eventually but not immediately, chest radiograph findings related to TB disease improve. Clinicians should always consider drug resistance when cultures remain positive after 2 months of treatment or patients do not improve after 2 months of first-line TB treatment, particularly in the context of epidemiologic or clinical risk factors for drug resistance.

## DIFFERENTIAL DIAGNOSIS

Although drug resistance, including multidrug resistance, should be considered in patients with specific risk factors and those who are not improving with first-line treatments, nonadherence to medications and subtherapeutic drug levels should also be considered in the differential diagnosis for drug-resistant TB. Because symptoms of TB disease can be nonspecific, differential diagnosis for pulmonary TB, regardless of drug-resistance profile, also includes community-acquired pneumonia caused by viral or bacterial pathogens. Furthermore, fungal pathogens and nontuberculous mycobacteria can have similar clinical presentations. Neoplastic processes (e.g., lung cancer) and autoimmune diseases should also be considered as possibilities. Sarcoidosis—another disease with similar clinical features—deserves special mention because, like pulmonary TB, it is a granulomatous process. Although sarcoidosis is classically associated with noncaseating granulomas and TB disease is associated with caseating granulomas, TB disease should be excluded by a thorough diagnostic evaluation before a sarcoidosis diagnosis is made. Extrapulmonary disease has a broad differential diagnosis that depends on the organ structure involved and the clinical presentation.

## MANAGEMENT AND THERAPY

In the United States and other areas where the rates of antituberculosis drug resistance are low to moderate, standard treatment of *fully drug-susceptible TB* consists of an initial daily regimen of isoniazid, rifampin, pyrazinamide, and ethambutol. Therefore patients with diagnosed TB are often started on this four-drug regimen while specimens are undergoing laboratory testing to confirm drug susceptibilities. After 8 weeks of therapy, pyrazinamide and ethambutol are discontinued and isoniazid and rifampin are continued, thus completing 6 to 9 months (26 to 38 weeks) of therapy.

Patients who have drug-resistant TB are often initiated on these inadequate treatment regimens while specimens are being examined in the laboratory. Because of the implications for both the patient and public health, providers should have an increased index of suspicion for drug resistance among patients with epidemiologic or clinical risk factors such as residence in areas with high rates of drug-resistant TB or previous treatment for TB (especially with less than three medications). Treatment for drug-resistant TB is more complex and requires a balance of choosing a regimen that maximizes efficacy while minimizing potential adverse events. Because resistance testing has not been as available in resource-limited countries, global trials of MDR-TB treatment have recently focused on evaluating entire regimens administered to patients suspected of having drug-resistant TB. In the United States, laboratory capacity for quickly identifying resistance and drug availability has allowed providers to design individualized regimens based on a patients' specific susceptibility pattern.

If drug resistance is suspected, diagnostic laboratory studies can be expedited, expanded regimens can be initiated, and patients can be promptly isolated to prevent further transmission.

Substantial changes have occurred in the approach to treatment of drug-resistant TB. Previously, providers had to create a five-drug regimen often including toxic medications that were difficult for patients to tolerate; these sometimes led to irreversible adverse effects (Table 89.2). New medications introduced during the last 5 years have substantially improved providers' ability to treat patients without the same risk. Both US-based and WHO guidelines now emphasize using all oral regimens for treating drug-resistant TB, and these have been associated with better treatment outcomes.

Overarching principles for treating drug-resistant TB include the following:

- Consultation should be requested with an expert in TB when suspicion for or confirmation of drug-resistant TB exists. In the United States, TB experts can be located through local health department TB control programs (https://www.cdc.gov/tb/links/tboffices.htm), through TB Centers of Excellence for Training, Education, and Medical Consultation supported by the Centers for Disease Control and Prevention (CDC) (http://www.cdc.gov/tb/education/rtmc/default.htm) and through international MDR-TB expert groups (e.g., the Global TB Network).
- Molecular testing should be obtained for the rapid detection of mutations associated with resistance.
- Regimens should include only drugs to which the patient's isolate has been documented or where a high likelihood of susceptibility exists. Patients should not be treated with medications to which their isolate is resistant.
- A regimen of five or more oral medications should be initiated (Table 89.3). After 5 to 7 months with documented clinical response (demonstrated by conversion of sputum cultures to negative and clinical improvement), a continuation phase of four or more oral medications should be continued for a total of 15 to 21 months after culture conversion.
- Treatment response should be monitored clinically, radiographically, and bacteriologically, with cultures obtained at least monthly for pulmonary TB. If cultures remain positive after 3 months of treatment, DST should be repeated.
- Evidence is mounting for the use of therapeutic drug monitoring where available, particularly in the case of delayed clinical response to treatment and to limit toxicity or adverse events.
- Patients should be asked and educated about possible adverse effects at each visit. The toxicities and poor tolerability of drugs used to treat drug-resistant TB are well established, and all adverse effects should be thoroughly investigated and ameliorated (see Table 89.2).
- Patient-centered case management should be used to help patients understand their diagnosis, understand and participate in their treatment selection, and discuss potential barriers to treatment completion and achieving a cure.
- In the event of progressive resistance or intolerance, providers should never add or change less than 2 drugs in a failing regimen.

The American Thoracic Society/CDC/Infectious Diseases Society of America drug-resistant treatment guidelines outline the best strategies for constructing regimens for treating patients with drug-resistant TB.

### Regimens for Single-Drug Resistance

For isolated isoniazid resistance, treatment options include the following:
- Rifampin, pyrazinamide, ethambutol, and fluoroquinolone for 6 months
  *or*
- Rifampin, pyrazinamide, and ethambutol for 9 months

A randomized controlled trial demonstrated equivalent outcomes with use of either regimen, but the fluoroquinolone arm had slightly

## TABLE 89.2    Adverse Effects of Antituberculosis Drugs

| System | Effect | TYPICAL CAUSATIVE AGENT | | Management Option |
| | | First Line | Second Line | |
| --- | --- | --- | --- | --- |
| Gastrointestinal | Gastritis, nausea, vomiting | Pyrazinamide Rifampin Isoniazid | Ethionamide Para-aminosalicylic Fluoroquinolone Linezolid | Consuming a small predose snack or meal<br>Proton pump inhibitor<br>If severe, discontinuation |
| | Liver injury | Pyrazinamide Isoniazid Rifampin | Ethionamide Bedaquiline | Interrupt therapy if transaminases >3 times upper limit of normal with symptoms or >5 times upper limit of normal without symptoms<br>Follow resolution with challenge[a] |
| Central nervous system | Paresthesia | Isoniazid | Ethionamide | Increase pyridoxine dose |
| | Asthenia | Isoniazid | Any drug | Reassurance<br>Time dose for later in the day |
| | Depression | — | Cycloserine | Antidepressant medication<br>Discontinuation if severe |
| | Sleep disturbance, agitation, tremulousness | — | Fluoroquinolone | Discontinuation if severe |
| | Development or worsening of movement disorder | Isoniazid | — | Discontinuation if severe |
| | Seizure | Isoniazid | Cycloserine Ethionamide | High-dose pyridoxine, antiseizure medication, discontinuation |
| | Vision change | Ethambutol | — | Interruption and ophthalmologic examination with observation for recovery |
| | Hearing or vestibular change | — | Streptomycin Amikacin Kanamycin Capreomycin | Discontinuation<br>Can progress despite interruption and can be irreversible<br>Consider a different injectable agent if treatment options are limited |
| | Psychosis | Isoniazid | Cycloserine | Discontinuation and consult a psychiatrist |
| | Optic neuropathy | — | Linezolid | Discontinuation |
| Renal | Uremia | — | Streptomycin | Discontinuation |
| | Interstitial nephritis | Rifampin Isoniazid | — | Discontinuation |
| Skin | Noninflammatory pruritus | Isoniazid | — | Attempt treatment with antihistamine |
| | Urticaria | Any drug | Any drug | Attempt treatment with antihistamine |
| | Acneiform rash | Isoniazid Rifampin | Cycloserine | Topical therapy<br>Discontinuation if severe |
| | Maculopapular rash | Any drug | Any drug | Discontinue, consider rechallenge after resolution |
| Endocrine | Hypothyroidism | — | Ethionamide | Attempt treatment with levothyroxine |
| Hematologic | Myelosuppression | — | Linezolid | Discontinue, consider rechallenge after resolution, if feasible |
| Cardiovascular | Heart rhythm changes (prolonged QT interval) | — | Bedaquiline | Discontinue |

[a]Patients should be counseled to stop taking medication immediately upon experiencing symptoms (vomiting or jaundice) and seek medical care.

better efficacy and similar safety. For isolated pyrazinamide resistance (or for any course of treatment using only isoniazid and rifampin), therapy should be extended to 9 months.

Resistance to rifampin alone is rare; the treatment approach for when it is present has been evolving over recent years. If susceptibility to isoniazid can be confirmed, treatment would include isoniazid, fluoroquinolone, pyrazinamide, and ethambutol for 6 to 9 months. If either extensive disease is present or concern exists regarding further resistance, a regimen similar to one designed for the treatment of MDR-TB might be used.

## Regimens for Multidrug-Resistant Tuberculosis

Treating MDR TB involves prolonged use of multiple medications that are associated with frequent adverse events (see Table 89.2). Clinical trials in this area are limited, and treatment recommendations are derived from observational studies and expert opinion. The American Thoracic Society/CDC/Infectious Diseases Society of America drug-resistant treatment guidelines were published in 2019; they were based on scientific evidence that included individual patient data meta-analyses (IPDMA). The IPDMA analyses reviewed data from more than 12,000 patients who had received MDR TB treatment and examined

## TABLE 89.3  Choosing Medications for Treating Multidrug-Resistant Tuberculosis

- Build a regimen **using five or more drugs to which the isolate is susceptible** (or has low likelihood of resistance), preferably with drugs that have not previously been used to treat the patient.
- Choice of drugs is contingent on capacity for monitoring for substantial adverse effects, patient comorbidities, and preferences or values (choices therefore subject to program and patient safety limitations).
- For children with TB disease who are contacts of source persons with infectious MDR-TB, if no isolate is obtained from the child, the source patient's isolate drug-susceptibility test result should be used.
- **TB expert medical consultation is recommended [best practice statement].**

| Step | Drug |
|---|---|
| 1. Choose one later-generation fluoroquinolone: | **Levofloxacin** <br> **Moxifloxacin** |
| 2. Choose both of these prioritized drugs: | **Bedaquiline** <br> **Linezolid** |
| 3. Choose both of these prioritized drugs: | **Clofazimine** <br> **Cycloserine** |
| 4. If a regimen cannot be assembled with at least five effective oral drugs **and the isolate is susceptible**, use one of these injectable agents[a]: | **Amikacin** <br> **Streptomycin** |
| 5. If needed, use these first- or second-line drugs only when **the isolate is susceptible**[b]: | **Pyrazinamide** <br> **Ethambutol** <br> **Delamanid**[c] |
| 6. If limited options and cannot assemble a regimen of at least five effective drugs cannot be assembled with those listed previously, consider use of the following drugs: | **Ethionamide or prothionamide**[d] <br> **Imipenem-cilastin/clavulanate or meropenam/clavulanate**[e] <br> ***para*-aminosalicylic acid**[f] <br> **High-dose isoniazid**[g] |
| **Drugs no longer recommended for inclusion in MDR-TB regimens:** | **Capreomycin and kanamycin** <br> **Amoxicillin or clavulanate when used without a carbapenem** <br> **Azithromycin and clarithromycin** |

[a]Amikacin and streptomycin should be used only when the patient's isolate is susceptible to these drugs. Because of their toxicity, these drugs should be reserved for when more effective or less toxic therapies cannot be assembled to achieve a total of at least five effective drugs.

[b]Patient preferences in terms of the harms and benefits associated with injectables (the use of which is no longer obligatory), the capacity for monitoring for substantial adverse effects, drug-drug interactions, and patient comorbidities should be considered in selecting step 5 agents over injectables. Ethambutol and pyrazinamide had mixed or marginal performance on outcomes assessed in the individual patient data meta-analyses (IPDMA); however, certain experts might prefer these drugs over injectable agents for building a regimen of at least five effective oral drugs.

[c]Data regarding dosing and safety of delamanid are available for children at least 3 years of age.

[d]Mutations in the *inhA* region of *M. tuberculosis* can confer resistance to ethionamide/prothionamide as well as to isoniazid at low concentrations. In this situation, ethionamide/prothionamide might not be the best choice of a second-line drug unless the isolate has been demonstrated to be susceptible with in vitro testing.

[e]Divided daily intravenous dosing limits feasibility. Optimal duration of use is not defined.

[f]Fair or poor tolerability and low performance. However, adverse effects are reported to be less common among children.

[g]Data not reviewed in the IPDMA. However, high-dose isoniazid can be considered in the presence of mutations with *inhA* promoter mutations but not with *katG* mutations, which result in inhibition of catalase activity and the development of high-level resistance (resistance at 1.0 mg/mL on solid media) to isoniazid that cannot be safely and effectively overcome by increasing the dose.

*MDR*, Multidrug-resistant. *TB*, tuberculosis.

Reused with permission from Nahid P, Mase SR, Migliori GB, et al: Treatment of drug-resistant tuberculosis: an official ATS/CDC/ERS/IDSA clinical practice guideline, *Am J Respir Crit Care Med*, 200:e93–e142, 2019.

the impact of each of the medications used. On the basis of those data, the following strategy has been recommended for building an effective regimen for treating patients with MDR TB (see Table 89.3):

- Bedaquiline, fluoroquinolones (levofloxacin or moxifloxacin), and linezolid should be included in the treatment regimen.
- Clofazimine should be included in the treatment regimen if possible, although it is often difficult to obtain.
- Cycloserine is recommended for inclusion in the treatment regimen.
- Pyrazinamide is recommended for treating patients with either MDR TB or single drug–resistant TB.
- Ethambutol should be included in a regimen for treating patients with MDR TB only when more effective drugs cannot

be assembled to achieve a total of at least five effective drugs in the regimen.

For selected oral drugs previously included in regimens for treating patients with MDR-TB, the following are recommended:

- Amoxicillin-clavulanate should *not* be included in treatment regimens except when the patient is receiving a carbapenem, in which case the inclusion of clavulanate is necessary.
- The macrolides azithromycin and clarithromycin should *not* be included in treatment regimens.
- Ethionamide or prothionamide should be included in the treatment regimen only if newer and more effective drugs are unavailable for constructing a regimen with at least five effective drugs.

- Para-aminosalicylic acid is no longer recommended to be included in the treatment regimen if newer and more effective drugs are available for constructing a regimen with at least five effective drugs.

## Emerging Treatments and Considerations for Their Use

New medications and regimens are being studied for use in treating drug-resistant TB, primarily in high-incidence countries outside the United States. In 2019, pretomanid, a new drug, was approved by the US Food and Drug Administration (FDA) for use in treating patients with XDR TB or nonresponsive or treatment-intolerant patients with MDR TB; pretomanid is used in combination with bedaquiline and linezolid. In one phase III clinical trial, the initial regimen of once-daily pretomanid 200 mg–bedaquiline 400 mg–linezolid 1200 mg for 2 weeks followed by once-daily pretomanid 200 mg–bedaquiline 200 mg–linezolid 1200 mg for a total of 26 weeks had 90% efficacy among 109 patients with XDR TB or nonresponsive or treatment-intolerant patients with MDR TB. The trial was stopped early because of the favorable outcomes; however, it documented high rates of adverse events, mostly attributed to linezolid. Thus a trial has begun comparing different initial doses of linezolid, for which clinical results are expected in 2021 or 2022.

Delamanid has been studied in multiple oral regimens for treating drug-resistant TB and has been met with success. Although delamanid does not have FDA approval for use in the United States (as of spring 2020), it is available on a case-by-case basis through compassionate use programs.

### Adjunctive Corticosteroids

For TB involving the pericardium, meninges, or central nervous system, systemic corticosteroids are routinely recommended during the initial 1 or 2 months of therapy; corticosteroids can also be considered if an inflammatory mass compromises vital organ structures. Although the use of corticosteroids should be limited to conditions for which objective evidence of benefit exists, systemic corticosteroids may be given at appropriate dosages in certain clinical circumstances when concurrent, effective TB treatment is being administered.

### Role of Surgery

Guidelines for the treatment of drug-resistant TB recommend elective partial lung resection (e.g., lobectomy or wedge resection) as an adjunct to effective chemotherapy for patients with MDR TB and localized disease plus adequate pulmonary capacity. These complicated clinical decisions should be made in conjunction with experts both in TB treatment and surgical intervention of this type and typically when clinical judgment is supported by bacteriologic and radiographic data that equate to increased risk for relapse or treatment failure with medical regimen alone.

### Case Management and Monitoring

Case management of TB typically includes assessing the safety of drug therapy through toxicity or adverse effect monitoring, monitoring response to treatment through clinical measures (e.g., weight, symptoms, bacteriology, and radiographic imaging), and adherence to treatment until completion through the use of patient-centered case management strategies, especially directly observed therapy (DOT). No published randomized or comparative studies are available that report drug-resistant TB patient outcomes with and without patient-centered case management. However, recent meta-analyses and other studies have reported benefits associated with patient-centered case management interventions relative to TB patient outcomes.

The Curry International TB Center publishes a drug-resistant TB survival guide with thorough descriptions of likely adverse events and monitoring schedules and tools. Examples of the tools include "drug-o-grams" that align clinical details with test results and treatment course as well as laboratory, bacteriology, and other toxicity-monitoring flow sheets and checklists for care plans. Such tools ensure timely drug toxicity monitoring and the provision of examinations required for assessing a patient's treatment response.

## SPECIAL POPULATIONS

### Persons Living With Human Immunodeficiency Virus Infection or Acquired Immunodeficiency Syndrome

An HIV test should be a standard component of every baseline evaluation for TB among all patients not previously known to be coinfected. A comprehensive review of the management of coinfection with MDR TB and HIV is beyond the scope of this chapter. However, certain key principles should be noted in evaluating or treating any HIV-infected person for TB, as follows:

- Presentation of pulmonary TB can be atypical, particularly with severe immunosuppression (e.g., CD4$^+$ cell count <200/mm$^3$).
- Time to treatment in antiretroviral-naïve patients has been studied extensively and the CD4$^+$ cell count should be taken into consideration. Current clinical practice guidelines recommend starting TB treatment first, followed by early initiation (at 2 to 8 weeks, depending on the patient's CD4$^+$ cell count at TB diagnosis) of antiretroviral therapy.
- On standard chest radiography, hilar and mediastinal adenopathy, lower zone opacities, and diffuse micronodular (miliary) opacities are more common among severely immunocompromised patients.
- Sputum smears for acid-fast bacilli are more likely to be negative.
- Extrapulmonary disease is more common.
- Rifampin resistance may be present among patients previously treated with rifabutin for prophylaxis of *Mycobacterium avium* complex and among those previously treated for TB with intermittent rifamycin administration.
- The toxicities and interactions associated with antiretroviral therapy and prophylaxis for opportunistic infections may overlap with those associated with antituberculosis drugs.
- Immune reconstitution inflammatory syndrome (IRIS), caused by an exuberant immune response to mycobacterial antigens, can occur during TB treatment, particularly after initiation of antiretroviral medications. Immune reconstitution is most often characterized by the return of constitutional symptoms and apparent worsening of TB disease. Although this syndrome can occur among persons not infected with HIV, it is observed most commonly weeks after the initiation of antiretroviral therapy among severely immunocompromised HIV-infected persons being treated for TB. Corticosteroids have an important role in treating IRIS.

### Pregnant Persons

Untreated MDR TB during pregnancy can be associated with adverse maternal and fetal outcomes; however, data are limited regarding preferred regimens. Evidence supports treatment of MDR TB during pregnancy, including prescriptions for second-line drugs. Per the FDA, the majority of second-line drugs are pregnancy category C (i.e., in the absence of well-controlled human studies, but adverse effects in animal reproduction studies, potential benefits may warrant use of drug in pregnant women despite risk) except bedaquiline and meropenem, which are classified as category B (i.e., in the absence of well-controlled human studies but no demonstrated adverse effects in animal reproduction studies), and aminoglycosides, which are category D (i.e., in the context of evidence for human fetal risk based on adverse reaction data from investigational or marketing experience, but potential benefits may warrant use of the drug in certain clinical situations). Despite low cure rates reported in the scientific literature, benefits of treatment to mother, child, and the community outweigh the harms. Because a preferred regimen has not been identified, treating these patients in consultation with a TB expert is of utmost importance. The majority of MDR-TB experts avoid aminoglycosides and ethionamide for pregnant women if alternative agents can be used in an effective treatment regimen.

## PUBLIC HEALTH CONSIDERATIONS

### Primary Prevention

Efforts to promptly identify and treat patients with contagious TB disease are foundational to public health prevention and control efforts. Diagnosis and treatment of drug-susceptible TB disease reduces morbidity and prevents drug resistance because treatment with too few medications can lead to drug resistance. In the United States, treatment costs for MDR TB disease can be more than 10 times the costs of treating drug-susceptible disease; treating XDR TB disease is even more expensive. Treatment also preserves the efficacy of first-line treatment regimens for public health TB control programs. Therefore, to facilitate treatment completion, DOT (i.e., treatment that includes a public health worker witnessing each medication dose, either in person or electronically by mobile device) is almost universal in the United States and increasingly common worldwide for patients with TB disease.

Additionally, in the United States, persons with infectious TB are commonly isolated to prevent M. tuberculosis transmission to others. Patients with drug-susceptible TB are often isolated until they have received at least 2 weeks of treatment, have improvement in symptoms, and have improvement in laboratory measures of disease (e.g., decrease in the degree of sputum-smear positivity). Experts are more stringent regarding patients with diagnosed, infectious MDR TB; such patients are often released from isolation only upon negative culture results for drug-resistant M. tuberculosis from the sputum. Patients with extrapulmonary disease do not require isolation as long as infectious forms of TB (e.g., pulmonary or laryngeal disease) have been excluded after a rigorous diagnostic evaluation.

### Contact Investigations

In the United States and other high-resource settings, contact investigations (i.e., efforts to systematically identify persons exposed to infectious disease) are also fundamental strategies for the prevention and control of TB. Contacts are subsequently evaluated for LTBI and TB disease. Household contacts are at relatively high risk for becoming infected with M. tuberculosis; approximately 20% to 30% are expected to have M. tuberculosis infection, and 1% are expected to have TB disease. These investigations are particularly important in the context of drug resistance. Because the most common treatment regimens will not effectively treat patients with presumed MDR-LTBI, management of contacts depends on knowing drug susceptibility results for the source case or cases. If the presumed source case is susceptible to fluoroquinolones, a 6- to 12-month fluoroquinolone-based regimen, possibly with the addition of a second drug, can be used for LTBI treatment for contacts presumed to be infected with MDR organisms. Experts also recommend clinical monitoring of contacts for MDR TB; to promptly diagnose MDR TB, this clinical monitoring should include regular queries about symptoms of TB disease and chest radiography at specified intervals. Clinical monitoring should always be performed in collaboration with the responsible local public health agency.

### Infection Control and Prevention

M. tuberculosis transmission can also occur in healthcare facilities and other congregate settings (e.g., jails, prisons, homeless overnight facilities, and long-term-care facilities). Such facilities should have specific infection control plans for preventing M. tuberculosis transmission; additionally, any facility that cares for persons with TB disease should have infection control plans. These plans should be informed by facility-specific assessments of risk for M. tuberculosis transmission; these risk assessments should consider factors such as TB incidence in the community and whether persons with infectious TB are expected to enter the facility. Administrative controls, which are designed to reduce the risk for exposures to infectious TB, are the foundation of any infection control plan. Administrative controls include designating a point person to coordinate infection control activities, facilitating prompt diagnosis of TB disease among workers and clients, and measures designed to facilitate prompt contact investigations when cases of TB disease are diagnosed. Engineering controls, also commonly referred to as environmental controls, form a secondary layer of protection; those efforts use ventilation to prevent transmission by diluting or eliminating contaminated droplet nuclei from facilities. Finally, personal protective equipment (e.g., fit-tested N-95 or better respirators) provides a tertiary layer of protection from infectious TB. However, overreliance on personal protective equipment at the expense of administrative and engineering controls can lead to a false sense of security.

## FUTURE DIRECTIONS

Substantial barriers still exist in controlling and preventing TB. However, regardless of drug susceptibility, MDR TB presents special challenges. Diagnostic tests have limited sensitivity for LTBI and TB disease overall, and certain tests—even chest radiography—are not widely available in resource-limited settings. Furthermore, these diagnostic tests do not discriminate between drug-susceptible and drug-resistant disease. Although phenotypic and molecular tests are widely available for diagnosing MDR TB in the United States, testing capacity is still limited or unavailable in certain settings where the incidence of MDR disease is high. Ensuring treatment completion is also a challenge; many TB programs globally do not have the medications or other resources necessary for ensuring adherence with curative treatment regimens for the 18 to 24 months often required to treat MDR TB properly. Finally, the HIV epidemic has facilitated transmission of MDR TB, particularly in sub-Saharan Africa.

Public health activities help to control and prevent drug-resistant TB. All require strong medical and public health infrastructures. First, more universal efforts in diagnosing and properly treating TB to cure would help prevent acquired drug resistance. Diagnosing TB disease is often difficult because symptoms are often nonspecific. Therefore it is important to ensure universal, competent TB care in medical systems and to introduce, when available, high-sensitivity diagnostics that can improve the promptness of TB diagnoses. Expansion of molecular tests for diagnosing drug-resistant TB would facilitate more prompt introduction of curative therapy for affected patients. After diagnosis, treating TB to cure requires consistent, adequate supplies of necessary medications. Therefore, to prevent transmission and further drug resistance, patients need uninterrupted supplies during treatment. Ensuring adherence, which is just as essential as medication procurement, often requires DOT. DOT expansion requires a trained public workforce. A safe, efficacious vaccine that protects against the acquisition of LTBI and TB disease would be a game changer for control and prevention. Although all of these interventions require sustained investments for research and programmatic activities, more widespread and well-designed TB prevention and control activities could save millions of lives.

## EVIDENCE

Ahuja SD, Ashkin D, Avendano M, et al. Multidrug resistant pulmonary tuberculosis treatment regimens and patient outcomes: an individual patient data meta-analysis of 9,153 patients. *PLoS Med* 2012;9:e1001300. *This is a meta-analysis that informed guidance for management of drug-resistant TB. Treatment success was better with a higher number of drugs, prothionamide or ethionamide, and fluoroquinolones.*

Conradie F, Diacon AH, Ngubane N, et al. Treatment of highly drug-resistant pulmonary tuberculosis. *N Engl J Med* 2020;382:893-902. *This article documents the success of a 3-drug, all-oral regimen of pretomanid, linezolid, and bedaquiline for treatment of XDR-TB or treatment intolerant/non-responsive MDR-TB.*

Food and Drug Administration (US) (FDA). Approval package; pretomanid 200 mg in the treatment of extensively drug resistant TB. Washington, DC, 2019, FDA. *This is the approval package for pretomanid, a drug approved in 2019 for treatment of treatment of adults with extensively drug resistant (XDR), treatment-intolerant or nonresponsive multidrug-resistant (MDR) TB.*

Fregonese F, Ahuja SD, Akkerman OW, et al. Comparison of different treatments for isoniazid-resistant tuberculosis: an individual patient data meta-analysis. *Lancet Respir Med* 2018;6:265-275. *A comparison of different regimens for isoniazid resistant tuberculosis. For isoniazid-resistant disease, addition of a fluoroquinolone was associated with better treatment success.*

Marks SM, Mase SR, Morris SB. Systematic review, meta-analysis, and cost-effectiveness of treatment of latent tuberculosis to reduce progression to multidrug-resistant tuberculosis. *Clin Infect Dis* 2017;64:1670-1677. *This is a systematic review that examined whether treating LTBI from MDR organisms is effective. While it found lower incidence among people treated for presumed MDR LTBI, patients with ethambutol-based regimens stopped treatment sooner.*

## ADDITIONAL RESOURCES

Centers for Disease Control and Prevention (CDC). *Reported tuberculosis in the United States, 2018*. Atlanta, 2019, US Department of Health and Human Services, CDC. *This report presents TB surveillance data and incidence in the United States.*

Centers for Disease Control and Prevention (CDC). Updated guidelines for the use of nucleic acid amplification tests in the diagnosis of tuberculosis. *MMWR Morb Mortal Wkly Rep* 2009;58(1):7-10. *These are guidelines for using nucleic acid amplifications tests for diagnosing tuberculosis.*

Curry International Tuberculosis Center and California Department of Public Health. *Drug-resistant tuberculosis: a survival guide for clinicians*, 3rd ed. San Francisco, 2016, University of California at San Francisco. *This is an excellent manual for clinical management of drug-resistant TB.*

Jensen PA, Lambert LA, Iademarco MF, Ridzon R. Guidelines for preventing the transmission of Mycobacterium tuberculosis in healthcare settings, 2005. *MMWR Recomm Rep* 2005;54(No. RR-17):1-141. *These are CDC guidelines for preventing TB transmission within healthcare facilities. In the United States, the principles articulated in these guidelines form the foundation for infection control plans in many healthcare facilities.*

Nahid P, Dorman SE, Alipanah N, Barry PM, Brozek JL, Cattamanchi A. Executive summary: official American Thoracic Society/Centers for Disease Control and Prevention/Infectious Diseases Society of America clinical practice guidelines: treatment of drug-susceptible tuberculosis. *Clin Infect Dis* 2016;63:853-867. *These are comprehensive guidelines focused on treatment of drug-susceptible TB in the United States and other high-resource, low-incidence settings.*

Nahid P, Mase SR, Migliori GB, et al. Treatment of drug-resistant tuberculosis: an official ATS/CDC/ERS/IDSA clinical practice guideline. *Am J Respir Crit Care Med* 2019;200:e93-e142. *These are comprehensive guidelines focused on treatment of drug-resistant TB in the United States and other high-resource, low-incidence settings.*

Panel on Opportunistic Infections in Adults and Adolescents with HIV. *Guidelines for the prevention and treatment of opportunistic infections in adults and adolescents with HIV: recommendations from the Centers for Disease Control and Prevention, the National Institutes of Health, and the HIV Medicine Association of the Infectious Diseases Society of America.* AIDSInfo, updated 2019. *These are guidelines for prevention of opportunistic infections in people living with HIV. Guidance for testing and treatment of TB disease and LTBI is included.*

Shah NS, Westenhouse J, Lowenthal P, et al. The California multidrug-resistant tuberculosis consult service: a partnership of state and local programs. *Public Health Action* 2018;8:7-13. *This is a case series detailing some of the challenging clinical issues that drug-resistant tuberculosis presents.*

World Health Organization (WHO). *Global tuberculosis report 2019*. Geneva, 2019, WHO. *This is a comprehensive report about tuberculosis incidence across the world.*

World Health Organization (WHO). *WHO consolidated guidelines on drug-resistant tuberculosis treatment*. Geneva, 2019, WHO. *These are comprehensive guidelines for the treatment of drug-resistant TB for a global context.*

# 90

# Anthrax

David L. Saunders, Maryam Keshtkar-Jahromi, Phillip R. Pittman

## ABSTRACT

Anthrax, caused by *Bacillus anthracis*, is a devastating disease for humans and animals, and one of the most important bioweapons. Though rare in the United States, natural anthrax exposure occurs sporadically worldwide. Anthrax most often manifests as cutaneous or gastrointestinal (GI) disease, with inhalation anthrax (IA) being associated with bioterrorism. Once diagnosed, anthrax must be aggressively treated with an antimicrobial regimen of fluoroquinolones, tetracycline, imipenem or some other penicillin, and supportive ICU care. Anthrax immune globulin is available from the US Centers for Disease Control (CDC) as an adjunct for IA treatment, as well as commercially available monoclonal antibodies. A licensed vaccine is available, and a new vaccine is approaching licensure. Prognosis for survival is relatively high for treated cutaneous (<1% mortality when treated) and GI anthrax (0% to 29% mortality). IA, even when treated early and aggressively, has a high mortality rate (45% in the 2001 "Amerithrax" attacks). Prevention and mitigation of this disease depend on the control of anthrax in animals, restriction or decontamination of animal products, and preparedness for rapid public health response to bioterrorism events.

## CLINICAL VIGNETTE

Three senior officials fall ill the night before a major diplomatic event at a foreign embassy, including the ambassador. The following morning, a major treaty governing a lucrative rare-metals extraction contract is scheduled to be signed. All three report fever, chills, headache, and myalgia, with one reporting GI distress. The ambassador's condition deteriorates overnight, and by morning he is barely responsive with profuse vomiting and diarrhea. The other two have increasing GI distress and at least one has mild ascites. Three other individuals have fallen ill as well. The event is cancelled, and an emergency response team brought in from the host country. Local law enforcement is notified that a member of the kitchen staff has not reported to work in 3 days. One of the other food handlers is noted to have a pruritic, painless papule on the left hand.

As the day progresses, the ambassador and one other staff member deteriorate and lapse into a coma. Evacuation to a local hospital reveals both to be in septic shock, with severe abdominal tenderness diffusely. Both are started on life support and empiric broad-spectrum antibiotics, with activity against potential enteric pathogens. The conditions of the others continue to deteriorate over the next 48 hours.

COMMENT: A high index of suspicion is necessary to diagnose acute anthrax, particularly from non-classical forms. In this case, GI anthrax might be confused with other enteric pathogens commonly associated with food poisoning. The severity of the presentation, timing around a sensitive diplomatic event, and the epidemiologic patterns observed should all raise suspicion for a deliberate bioterrorist attack. This is the most likely setting where inhalational and GI anthrax are likely to be observed. Diagnosis can be difficult and may be delayed. Rapidly assessing the situation and implementing appropriate empiric treatment are essential to saving lives.

## DISEASE BURDEN AND RISK FACTORS

Anthrax is a globally distributed zoonotic disease, primarily affecting livestock and wild herbivores, though more common today in Africa, Asia, and Central America. Sporadic animal outbreaks and rare human cases continue to occur elsewhere. Occasional *B. anthracis* outbreaks have been associated with war and drought conditions throughout history, starting with description of plagues appearing in the Bible's Old Testament. Most recently, an extended outbreak in the late 1970s killed more than 10,000 individuals in Zimbabwe following years of war. Modern-day anthrax occurs primarily in grazing animals including sheep, goats, and other domesticated species, as well as hippos, rhinos, and other wild herbivores. Sporadic human cases largely result from contact with infected animal carcasses causing the cutaneous form, or ingestion of contaminated meat leading to GI anthrax. Whether disease spread is primarily from anthrax spores persisting in the soil or from infected animal carcasses remains unsolved. Aerobic conditions in decaying tissues are thought to promote spore formation, and spores are known to persist in soil for decades, particularly in dry conditions when risk of transmission increases. More recently, outbreaks among intravenous drug users—with both cutaneous and septicemic anthrax from infected needles and heroin—have been reported in Scotland and Norway.

## ETIOLOGY/PATHOPHYSIOLOGY

*B. anthracis* is a rather large, aerobic, non-motile, gram-positive rod 1 to 2 microns in length and often appearing in chains of two to three organisms. On sheep's blood agar, thick, flat 2 to 5 mm gray-white colonies appear after incubation at 37 degrees for 15 to 24 hours. It grows best with carbon-dioxide enrichment, and is lysed by an anthrax-specific, gamma-delta bacteriophage (Fig. 90.1). It was the first causative organism identified for a bacterial disease by Robert Koch in 1877. Anthrax sporulates in settings of stress to include increasing oxygen tension, particularly in animal carcasses, which then contaminate the soil where they may survive for decades—particularly in dry conditions.

Anthrax spores enter the body primarily by three routes: skin abrasions, ingestion, and inhalation. A rare form of entry has been identified in Europe by drug injection, usually heroin. Once anthrax spores enter the body, they are phagocytosed by macrophages, wherein they germinate into vegetative bacteria or bacilli, released from macrophages into regional lymphatics where they multiply and enter into the bloodstream causing massive septicemia.

The vegetative form expresses potent virulence factors, including two binary exotoxins and a capsule coded in two plasmids, pXO1 and pXO2. The resulting toxemia can cause organ failures and host death. The plasmid pXO1 codes for three factors: protective antigen (PA),

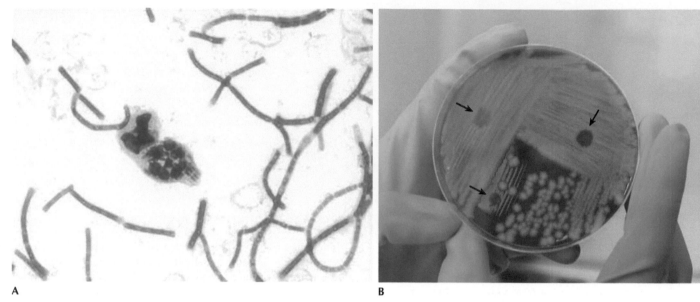

**Fig. 90.1** In vitro and clinical images of *Bacillus anthracis*. (A) Gram stain of infected guinea pig blood smear demonstrating intracellular bacilli chains next to a polymorphonuclear leukocyte. (Courtesy of Susan Welkos, PhD, Division of Bacteriology, US Army Medical Research Institute of Infectious Diseases, Fort Detrick, Maryland.) (B) Isolated colonies of *Bacillus anthracis* on sheep blood agar plate. The plate has been treated with an organism-specific gamma-phage causing cell lysis *(arrows)*. (Courtesy of Bret K Purcell, PhD, MD, Division of Bacteriology, US Army Medical Research Institute of Infectious Diseases and the Defense Threat Reduction Agency/Threat Agent Detection and Response Program, National Center for Disease Control, Tbilisi, Georgia, 2005.)

lethal factor, and edema factor, that produce the two binary exotoxins. The two binary exotoxins are *lethal toxin (LT)*, consisting of PA and lethal factor, and *edema toxin (ET)*, consisting of PA and edema factor. PA binds to a receptor on target cells where it is subsequently cleaved and forms a heptamer. PA heptamer binds three molecules of LT and ET and transports them into the interior of the cell where these toxins exert their biologic actions (Fig. 90.2).

LT is lethal to macrophages causing lysis and release of toxic cytokines including interleukin-1, tumor necrosis factor-α, and others. LT stimulates apoptosis in endothelial cells causing vascular leak syndromes, while inhibiting multiple normal immune cell line responses. In addition to causing local edema, ET impairs multiple normal immune cell responses including phagocytosis, cytokine production by macrophages, and dendritic cell function.

The plasmid, pXO2, the smaller of the two plasmids, carries the genes that produce the poly-glutamyl capsule. Historically, the capsule is noted to inhibit phagocytosis of the bacilli, whereas the exotoxins are involved in inhibiting the host's innate and acquired immune responses against infection. Both pXO1 and pXO2 are required for virulence; deleting one or both results in attenuation of *B. anthracis*.

The relative dose of anthrax spore exposure and local immunologic interactions appear to influence the course of disease and may explain, at least in part, the far higher lethality of inhalational disease. Cutaneous infection appears to be rapidly walled off, and largely limited to local infection, though systemic spread can sometimes occur if untreated. Spores are not thought to germinate in the alveolar spaces following initial exposure, but are rather translocated to regional hilar lymph nodes where they vegetate, encapsulate with accompanying virulence factor expression, and rapidly multiply. This leads to massive tissue damage and rapid systemic spread leading to septic shock (Fig. 90.3).

*Bacillus cereus*, normally associated with GI food poisoning, has also been found in rare instances to cause human disease similar to anthrax. This suggests possible recombination in the environment, and may further complicate diagnostic efforts.

## CLINICAL PRESENTATION

The major clinical presentations of *B. anthracis* infection are shown in Fig. 90.4. Though occurring in some agro-industrial settings in the past, such as wool sorters and tanners, with up to 100,000 cases estimated worldwide in 1958, the most serious inhalational form of anthrax is exceedingly rare today following improvements in industrial hygiene. There were only 400 anthrax cases, largely cutaneous, over the second half of the 20th century in the United States. Naturally occurring anthrax is now exceedingly rare in the United States, with only one case per year reported on average in the last 2 decades. These occur predominantly in those working with animal products including hides, drums, wool, hair, and the meat from infected animals. IA is at present most likely to occur from deliberate bioterrorist aerosol attack. The most striking example were the "Amerithrax" attacks in 2001 in Washington, D.C., which caused 11 cutaneous and 11 inhalational cases. Five of the 11 inhalational patients died, despite aggressive intensive care and appropriate antibiotic therapy. An accidental release of weapons-grade anthrax from a factory in Sverdlovsk, Russia in 1979, is believed to have caused more than 100 human deaths and killed livestock downwind of the factory. Though 96 cases with 64 deaths were eventually confirmed officially, the incident was covered up for years by Soviet authorities. Anthrax was used by Germany against enemy livestock in World War I. Japan used anthrax as an antipersonnel weapon in Manchuria as well as in World War II.

A small amount of *B. anthracis*, estimated at inhalation of 8 to 10,000 spores, may be fatal. Covert experiments with the spores of similar but nonpathogenic organisms demonstrated the ease of wide dissemination of potentially lethal anthrax spore concentrations using *Serratia marcescens* across the San Francisco Bay area in 1950, and *Bacillus globigii* in the New York City subway system in 1966. Because of its aerosol infectivity at very low doses, environmental persistence, lethality, and relative ease of manufacture and dissemination, *B. anthracis* is categorized as a Tier 1 bioterrorism agent by the US CDC,

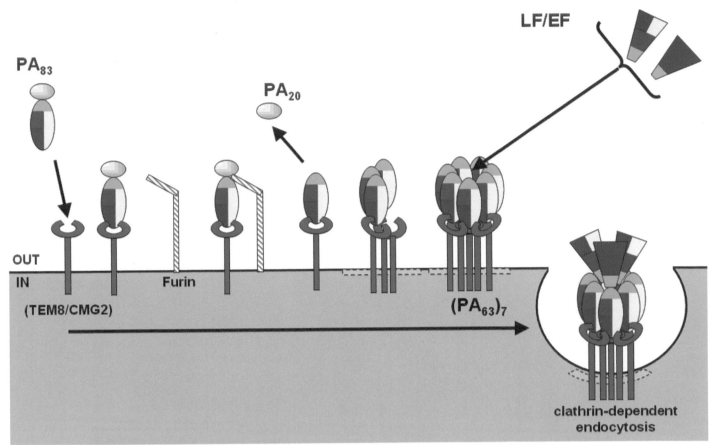

**Fig. 90.2** The different steps of anthrax toxins entry, and their inhibition by antibodies. Various steps of anthrax toxins entry. PA83 binds to its cell receptors and is processed by furin on the cell surface. PA20 is released and PA63 remains attached to the receptor. Heptamerization of PA63 induces the formation of LF/EF binding site. The toxin complex is then endocytosed. (Reused with permission from Froude JW, Thullier P, Pelat T. Antibodies against anthrax: mechanisms of action and clinical applications. *Toxins [Basel]*. 2011; 3[11]:1433-52, Figure 2A.)

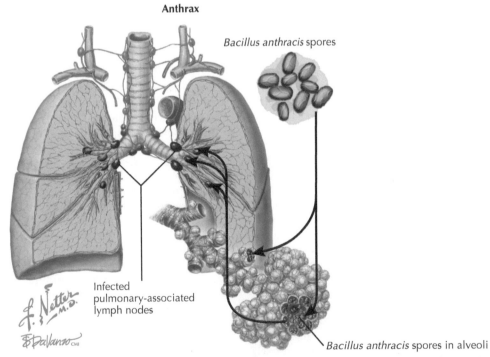

**Fig. 90.3** Pathophysiology of inhalation anthrax is thought to be caused by translocation of *B. anthracis* spores via draining lymphatics to mediastinal and peribronchial lymph nodes where they germinate into pathogenic bacilli. This leads to a rapid and severe hemorrhagic mediastinitis with development of substernal chest pain and mediastinal widening seen on chest x-ray.

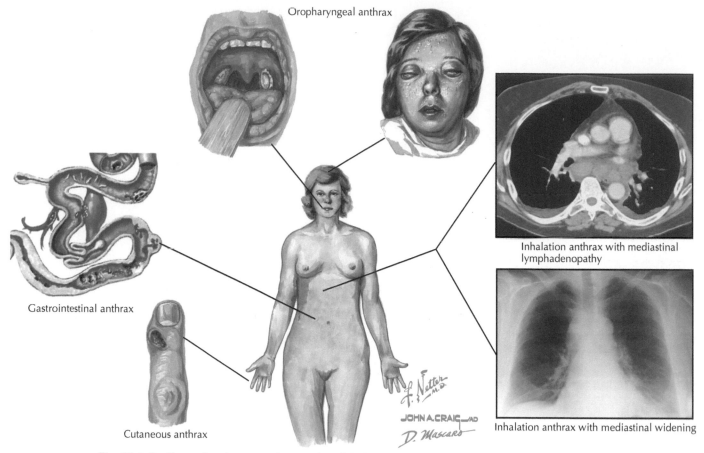

Oropharyngeal anthrax

Gastrointestinal anthrax

Cutaneous anthrax

Inhalation anthrax with mediastinal lymphadenopathy

Inhalation anthrax with mediastinal widening

**Fig. 90.4** *Bacillus anthracis* causes three major clinical syndromes to include cutaneous, oropharyngeal-gastrointestinal, and inhalational forms. Cutaneous anthrax is most common with a self-limited edematous ulcer forming on the skin of areas in contact with infectious spores. Oral and GI anthrax result from consuming contaminated animal products, most commonly meat, and feature ulcerative disease of the oropharynx and GI tract. Inhalation anthrax is the most severe form, and most commonly associated with deliberate biological attack through inhaling anthrax spores. The inhalational form begins with protean systemic symptoms, which rapidly progress to mediastinitis with widening seen on imaging studies, septic shock, meningitis, and death if untreated.

as well as a Category A select agent. Laboratory workers who handle anthrax must work in biosafety level 3 (BSL-3) containment conditions to limit the risk of exposure. Modern industrial hygiene practices have greatly lowered risks for workers handling animal products.

## Cutaneous Anthrax

Cutaneous anthrax is most commonly seen in nature and results from local infection with anthrax spores. The spores germinate in the skin, often with regional adenitis, as organisms are taken up by local lymphatics. Lesions typically begin as a painless and/or pruritic papule 1 to 12 days after infection, emerging most commonly within 1 week. Peripheral lesions may appear as well, along with regional lymphangitis, systemic symptoms to include headache, malaise, and fever. Vesicular lesions will typically develop a central vesicle, which ruptures and ulcerates. The lesion progresses with development of black eschar, which persists for 2 to 3 weeks, leaving a scar that may require cosmetic reconstruction. Fig. 90.5 shows characteristic lesion with black eschar and surrounding edema. In rare cases, cutaneous disease may progress to systemic involvement, which should be treated in the same manner as IA. It is important not to assume that cutaneous disease arose solely from skin exposure, and patients should be monitored for systemic disease, particularly if etiology of skin lesions is unclear.

## Inhalation Anthrax

While it can occur in industrial settings, and did so in the past, improvements in industrial hygiene have rendered this a rare event. As such, IA is by far most likely to result from deliberate exposure to anthrax spores through bioterrorism and/or military use. Anthrax spores are not thought to be pathogenic in and of themselves in lung tissue, but rather are taken up and migrate to regional lymph nodes where they germinate and cause massive and rapid tissue destruction.

Initial onset of disease may be mild and relatively nonspecific and there is sometimes a hiatus period before systemic symptoms become apparent. Acute disease progresses within 1 to 6 days from the time of initial exposure with nonspecific symptoms of fever, malaise, myalgia, and headache. Substernal chest pain, dyspnea, and cyanosis may develop rapidly followed by circulatory collapse and death as the organism rapidly invades hilar structures and disseminates. Clinical inhalational disease features massive hemorrhagic mediastinitis due to replication within mediastinal nodes (Fig. 90.6). This in turn leads to sanguineous pleural effusions and spread within the lung parenchyma via lymphatic channels. It is capable of causing a severe necrotizing pneumonia, as well as meningitis. Cerebrospinal fluid should be obtained as soon as possible to rule out meningitis, as it requires additional antibiotic therapy capable of crossing the blood-brain barrier.

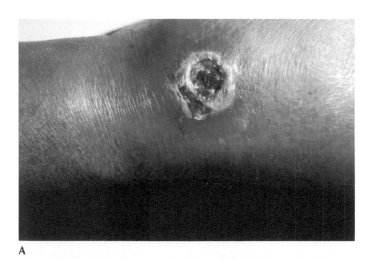

**Fig. 90.5** Characteristic lesions of (A) cutaneous anthrax with black eschar over an ulcerated base; and (B) significant surrounding edema (Photographs courtesy of the Centers for Disease Control and Prevention-Public Health Image Library, A, Image ID# 20334, CDC/F. Marc LaForce, MD, https://phil.cdc.gov/Details.aspx?pid=20334; B, Image ID# 19826, ArchilNavdarashvili, Georgia [Republic], https://phil.cdc.gov/Details.aspx?pid=19826.)

In the absence of immediate treatment with intravenous antibiotics and intensive supportive care, the inhaled organism is quickly disseminated throughout the body leading to overwhelming infection, sepsis, meningitis, and typically rapid death from septic shock within 24 to 36 hours. Even when treated in a timely fashion, mortality is thought to be quite high. Only 6 of 11 IA patients survived the Amerithrax attacks of 2001, perhaps the best-known example. It should be noted that delayed presentations have been documented, including the Sverdlovsk bioweapons leak where patients presented as late as 2 to 3 months after initial exposure. This is thought to be due to persistence of viable spores in the lung capable of germinating in delayed fashion.

### Gastrointestinal Anthrax

GI anthrax is less common than cutaneous anthrax, and can progress to very serious systemic disease requiring treatment similar to that for IA—and progressing just as rapidly within 1 to 6 days. The most common cause is consumption of infected animal products. Ulcerative lesions occur throughout the GI tract, with significant edema. In addition to intestinal infection, GI anthrax can involve the oropharynx with pseudomembranous lesions, progressing to ulceration and swelling from airway edema leading to respiratory compromise. GI anthrax is more likely than the cutaneous form to progress to systemic involvement, including development of meningitis.

### Anthrax Following Intravenous Drug Injection

Intravenous drug injection has been identified as a route of *B. anthracis* infection after an outbreak was described in intravenous drug users in Scotland in 2011. The injection site is usually the initial site of entry and can quickly spread systemically. Initial symptoms can include redness and swelling at the injection site, followed by shock and multiple organ failure due to rapid bacteremia. This can be rapidly fatal if not treated aggressively. Treatment usually requires wound debridement in addition to appropriate antibiotic therapy and sepsis management.

## DIFFERENTIAL DIAGNOSIS

Differential diagnosis is broad and based on clinical presentation of anthrax (Table 90.1). In an appropriate epidemiologic setting, the presence of an eschar with extensive edema disproportionate to the size of the lesion and the presence of gram-positive rods and few polymorphonuclear leukocytes on the Gram stain are strongly suggestive of cutaneous anthrax. Confirmatory testing should be performed on specimens from patients being evaluated for IA, including patients with a known exposure or clear epidemiologic link to exposure who are presenting with symptoms suggestive of IA. Because the window of opportunity for successful treatment is narrow, it is important to distinguish potential IA cases from more common disorders, such as community-acquired pneumonia (CAP), influenza, and influenza-like illnesses (ILIs). GI anthrax involvement is very hard to diagnose and may even manifest as sepsis syndrome. Oropharyngeal involvement can quickly lead to airway compromise and should be differentiated from other bacterial infections. Lower GI involvement can lead to perforation and bacteremia as the initial presentation.

## DIAGNOSTIC APPROACH

Clinical specimens must be taken before starting antibiotics. Tests should be collected based on the source of bacterial entry and clinical presentation in the setting of clinical suspicion (see Table 90.1). Diagnosis can be made by detecting antibodies or toxins in blood or testing directly to detect *B. anthracis* in clinical specimens (blood, urine, wound swabs, spinal fluid, or respiratory secretions) through culture or polymerase chain reaction (PCR). PCR can be used to confirm the presence of virulence factor genes (capsule and toxin), and whether an isolate represents virulent *B. anthracis*. Serologic tests have limited utility in severe anthrax cases as formation of antibodies lags behind clinical syndromes. Acutely ill patients require rapid diagnosis to start treatment and improve outcomes, and such treatment should be started empirically in suspected cases. Serology may be useful as a late confirmatory diagnostic tool in milder forms.

In a patient suspected of having cutaneous anthrax, clinical specimens should be taken from the edge of an eschar, ulcer base or fluid from unopened vesicles depending on the stage of disease. It is always recommended to collect two swabs, one for Gram stain and culture, the second for PCR assay. Full-thickness punch biopsy of a papule in 10% formalin is suggested to be submitted for histopathology and immunohistochemical staining. A punch biopsy specimen should also be evaluated for the presence of bacteria by Gram stain, culture, and PCR.

IA should be considered in any severe CAP presenting with significant shortness of breath, pallor, hypoxemia, and rapid progression of symptoms associated with mediastinal widening and/or pleural effusion, or consolidation in the setting of suspected exposure. Sputum, bronchial washings, and pleural fluids should be sent for Gram stain, bacterial culture, and PCR. Any pleural and/or bronchial biopsy specimen should be sent for Gram stain and immunohistochemistry. Some patients may have meningeal signs and should have cerebrospinal fluid evaluated for Gram stain, culture, and PCR. Blood cultures should be performed in all patients suspected of having IA. During the initial work-up, an extra tube of blood should be drawn and held as an acute serum sample and assayed together with a later convalescent serum sample: a ≥fourfold rise in anthrax antibody titer confirms the diagnosis.

Diagnostic capability is limited for GI anthrax. Bacteria are difficult to identify in stool, though oropharyngeal lesions swabs, ascites fluid, blood, and rectal swabs should be sent for Gram stain, culture, and PCR.

---

### ✴ CLINICAL VIGNETTE

The following day, one of the victims from the previous scenario expires from overwhelming septic shock. While two of the victims respond to empiric antibiotics, one complains of chest pain and is found to have mediastinal widening on plain chest x-ray. Autopsy reveals hemorrhagic mediastinitis in addition to bowel necrosis. Gram-positive rods are seen on histology of infected tissues. Empiric antibiotics are switched to dual therapy, with imipenem and ciprofloxacin given to the surviving victims. Airborne precautions are implemented.

COMMENT: Mediastinal widening in the current scenario is indicative of IA infection; it is likely that victims here were exposed via the GI tract. While the empiric antibiotics chosen are appropriate for likely GI anthrax, the possibility of IA argues for the addition of linezolid until meningeal involvement can be ruled out by cerebrospinal fluid analysis. Autopsy should generally be avoided in suspected anthrax cases, but in this instance, given an unclear diagnosis and the need to act rapidly, it was warranted. All personnel performing invasive procedures should wear personal protective equipment (PPE). Airborne precautions are not otherwise required for anthrax, which is not generally transmitted from person to person. Contact precautions are sufficient.

---

## TREATMENT

After initial diagnostic specimens are obtained, there should be no delays in initiating therapy for suspected anthrax infection, as these may prove fatal. Table 90.2 provides an overview of current recommendations for acute treatment and postexposure prophylaxis (PEP) for survivors and suspected exposures. With the exception of cutaneous anthrax without systemic involvement, antibiotic therapy should be instituted immediately, pending laboratory confirmation. All patients suspected of having systemic anthrax should be admitted to the hospital and closely monitored hemodynamically by telemetry, with continuous pulse oximetry, and monitoring kidney and liver function, electrolytes, and coagulation status as well. Patients who develop any headache or confusion should undergo brain imaging and lumbar puncture. Chest imaging and echocardiogram should be performed based on daily clinical evaluation. *B. anthracis* is susceptible to penicillin, chloramphenicol, tetracycline, erythromycin, streptomycin, and fluoroquinolones, but it is resistant to cephalosporin and trimethoprim-sulfamethoxazole. Considering *B. anthracis* may be carrying beta-lactamase genes, beta lactams are not recommended as first-line treatment due to developing resistance during treatment.

IA is a medical emergency and is the most severe anthrax syndrome. Treatment with at least three intravenous antibiotics (ciprofloxacin, meropenem, and linezolid) should be initiated for at least 14 days. These patients should also receive general sepsis supportive care including fluids, vasopressors, blood products in addition to consideration for steroids, anthrax monoclonal antibodies, and/or anthrax immune globulin. Drainage of pleural fluid, ascites, and/or pericardial fluid should be considered to reduce toxin levels and has been shown to improve survival. If anthrax meningitis has been ruled out definitively with lumbar puncture and CSF analysis, the antibiotic regimen can be reduced to ciprofloxacin and linezolid or clindamycin. Because viable anthrax spores have been found to persist in lung tissue for up to 60 days, both prolonged antibiotics and postexposure vaccination should be administered.

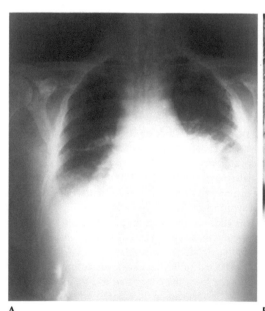 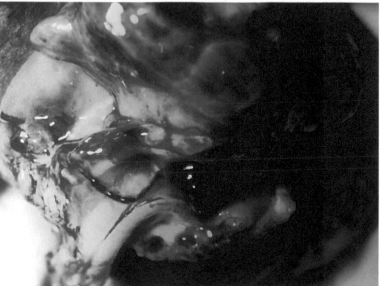

**A** **B**

**Fig. 90.6** (A) Chest radiograph showing a widened mediastinum due to inhalation anthrax, caused by the bacterium, *Bacillus anthracis*. It was taken 22 hours before this patient's death. (B) An oblique view of a gross pathologic lung sample from a chimpanzee that contracted fatal inhalation anthrax. This pathologic specimen demonstrates a markedly dilated, beaded lymphatic vessel connecting the hemorrhagic nodes on the right with smaller mediastinal nodes on the left. (Photographs courtesy of the Centers for Disease Control and Prevention-Public Health Image Library, A, Image ID# 1118, CDC/Dr. Philip S. Brachman, https://phil.cdc.gov/Details.aspx?pid=1118; B, Image ID# 5149, CDC/U.S. Army Photograph/Arthur E. Kaye, https://phil.cdc.gov/Details.aspx?pid=5149.))

Raxibacumab and obiltoxaximab are US Food and Drug Administration (FDA)-approved monoclonal antibodies, which inhibit binding of PA and translocation of anthrax toxins. They are considered adjunctive agents with antitoxin effects and should be considered in PEP or treatment of systemic anthrax—with the greatest benefit when used early during the course of disease. Raxibacumab is infused at 40 mg/kg IV over 2 hours, 15 minutes, while obiltoxaximab is infused at 16 mg/kg over 90 minutes. Both require premedication with diphenhydramine, up to 1 hour in advance. Anthrax immunoglobulin is derived from the plasma of vaccinated individuals, neutralizes anthrax toxins, and is recommended in the early stages of systemic infection. These products are held in the US Strategic National Stockpile and available through CDC in emergencies. Corticosteroids should be considered as an adjunctive therapy for treatment of mediastinitis, edema in the head and neck area, meningitis, or severe septic shock.

Localized uncomplicated **cutaneous anthrax** is treated with oral antibiotics for 7 to 10 days. Cutaneous anthrax with significant edema, especially on head and neck area, or any clinical evidence of systemic involvement should be treated as if IA. New guidelines recommend 60 days of antibiotic therapy based on assumption that cutaneous exposure may be linked to inhalational exposure. Cutaneous lesions without systemic infection should never be debrided surgically due to increased risk of systemic spread. Anthrax lesions following intravenous drug use with evidence of systemic involvement and bacteremia require surgical debridement.

**GI anthrax** can present as GI bleeding and this places the patient at a high risk for complications, including intestinal perforation, peritonitis, or mediastinitis. GI anthrax is generally treated similarly to IA due to the risk of bacteremia and systemic involvement.

## ✳ CLINICAL VIGNETTE

Given an apparent deliberate anthrax attack, the emergency response team moves swiftly to secure and clean the scene after taking forensic samples. The victims are isolated, and contact precautions are implemented for all patients and health care personnel. All patients, responders, embassy personnel, and health care staff are given an initial dose of the anthrax vaccine. Two more victims succumb. Two in critical condition have the monoclonal antibody raxibacumab added to triple antibiotic therapy.

COMMENT: Because of the persistence of spores in the environment, any potential areas affected should be thoroughly decontaminated by responders wearing appropriate PPE. Bleach solutions are sufficient for this purpose. The anthrax vaccine should be given in a compressed three-dose series at 0, 2, and 4 weeks for all exposed. Oral antibiotics should be administered to those potentially exposed in addition to the vaccine. Antibiotics should be continued for a total of 60 days for both those with suspected exposures and confirmed infections, as spores may continue to germinate in the lymphatics this late after exposure.

## PREVENTION

It should be noted that given the paucity of clinical data, important unknowns remain regarding the effectiveness of the various prophylactic measures, though their safety has been well described. Given the gravity of the disease, pre-exposure prevention should be an important consideration in anyone potentially at risk for anthrax. Antibiotics are currently not recommended for pre-exposure prophylaxis (PrEP), but are recommended for PEP in conjunction with the vaccine (see Table 90.2).

**TABLE 90.1    Anthrax Clinical Syndromes, Differential Diagnosis, Clinical Specimens, and Diagnostic Approach**

| Anthrax Clinical Syndrome | Differential Diagnosis | Clinical Specimens | Diagnostic Tests |
|---|---|---|---|
| Cutaneous anthrax | Bacterial skin infections | Skin lesion swab | Bacterial culture and Gram stain |
| | Cat scratch disease | Skin punch biopsy | Polymerase chain reaction (PCR) |
| | Spider bite | | Immunohistochemistry |
| | Rickettsial pox | | |
| | Syphilis | | |
| | Ecthyma gangrenosum | | |
| | Ulceroglandular tularemia | | |
| | Tularemia | | |
| | Tropical ulcer | | |
| | Plague | | |
| | Typhus | | |
| | Glanders | | |
| | Rat-bite fever | | |
| | Leprosy | | |
| | Cutaneous tuberculosis | | |
| | Atypical mycobacteria | | |
| | Varicella zoster | | |
| | Orf and milkers' nodules | | |
| | Herpes simplex | | |
| | Vaccinia | | |
| | Aspergillosis | | |
| | Mucormycosis | | |
| | Cutaneous leishmaniosis | | |
| | Drug induced necrosis (Warfarin) | | |
| | Antiphospholipid syndrome | | |
| | Skin malignancies | | |
| | Eczema | | |
| | Trauma | | |
| | Burns | | |

**TABLE 90.1    Anthrax Clinical Syndromes, Differential Diagnosis, Clinical Specimens, and Diagnostic Approach—cont'd**

| Anthrax Clinical Syndrome | Differential Diagnosis | Clinical Specimens | Diagnostic Tests |
|---|---|---|---|
| Inhalation anthrax | Community-acquired pneumonia<br>Influenza<br>Respiratory syncytial virus<br>Pneumonic plague<br>Pneumonic tularemia<br>Q fever<br>Psittacosis<br>Legionnaires' disease<br>Histoplasmosis<br>Coccidioidomycosis<br>Malignancy<br>Bacterial meningitis<br>Sepsis<br>Bacteremia<br>Encephalitis | Blood<br>Urine<br>Pleural fluid<br>Bronchial washing, respiratory secretions<br>Lung biopsy<br>Pleural biopsies<br>Pleural fluid<br>Cerebrospinal fluid<br>Skin biopsy<br>Ascitic fluid<br>Stool | Bacterial culture and Gram stain<br>PCR for detecting bacteria; PCR for detecting toxin or antibodies<br>Acute and convalescent serology against *B. anthracis* protective antigen (Enzyme-linked Immunosorbent Assay [ELISA], Western blot)<br>Immunohistochemistry |
| Gastrointestinal tract | Ulceroglandular tularemia<br>Bubonic plague<br>Foodborne illness<br>Sepsis and bacteremia<br>GI bleeding<br>Peritonitis<br>Acute appendicitis<br>Ruptured viscus<br>Diverticulitis<br>Dysentery<br>Necrotizing enteritis caused by Clostridium perfringens<br>Oropharyngeal (Vincent angina, Ludwig angina)<br>Streptococcal pharyngitis<br>Para pharyngeal abscess | Blood<br>Urine<br>Stool<br>Intraluminal lesions<br>Nasopharyngeal specimens<br>Ascites fluid | Bacterial culture and Gram stain<br>PCR for detecting bacteria<br>PCR for detecting toxin or antibodies<br>Acute and convalescent serology against *B. anthracis* protective antigen (ELISA, Western blot)<br>Immunohistochemistry |
| Intravenous drug users' anthrax | Sepsis<br>Cellulitis<br>Necrotizing fasciitis | Blood<br>Urine<br>Skin lesions<br>Cerebrospinal fluid | Bacterial culture and Gram stain<br>PCR for detecting bacteria<br>PCR for detecting toxin<br>Acute and convalescent serology against *B. anthracis* protective antigen (ELISA, Western blot) |

**TABLE 90.2    Treatment and Postexposure Prophylaxis Guidelines for Anthrax**

| Patient Category | Acute Infection Treatment | Postexposure Prophylaxis (PEP)[a] | Postexposure Vaccination |
|---|---|---|---|
| **Inhalation Anthrax (IA)** | | | |
| Immunocompetent, meningitis not ruled out | IV treatment with 3 antibiotics with good CNS penetration[b] for 2–3 weeks until stable<br>Then complete postexposure prophylaxis (PEP) with oral regimen | PEP regimen[a] for 42 to 60 days<br>Consider monoclonal antibodies[c] | 0, 2, and 4 weeks |
| Immunocompetent, meningitis ruled out | IV treatment with at least 2 antibiotics[d] 2–3 weeks until stable, then PEP[b] | Same as above | 0, 2, and 4 weeks |
| Immunocompromised (malignancy, HIV, immunosuppressive use such as high-dose corticosteroids >2 weeks, radiation therapy) | Same as above | PEP regimen[a] for at least 60 days<br>Consider monoclonal antibodies[c] | 0, 2, and 4 weeks |

## TABLE 90.2  Treatment and Postexposure Prophylaxis Guidelines for Anthrax—cont'd

| Patient Category | Acute Infection Treatment | Postexposure Prophylaxis (PEP)[a] | Postexposure Vaccination |
|---|---|---|---|
| Partially completed prior series of vaccinations | Same as above | Antimicrobial prophylaxis should be continued until 28 days after the first or 14 days after the last vaccine dose, whichever occurs later. | Complete scheduled vaccine series |
| Received complete PrEP vaccination series (at least 3 primary doses) | Same as above | If their most recent vaccine was <6 months ago, no antimicrobials is indicated. If their most recent vaccine was ≥6 months ago, antimicrobial prophylaxis until 14 days after the vaccine dose are indicated. | If their most recent vaccine was <6 months ago, no further action required. If their most recent vaccine was ≥6 months ago, an immediate vaccine dose is required. |
| **Gastrointestinal or oropharyngeal anthrax** | Same regimen for IA; corticosteroids for airway compromise | Follow inhalation PEP guidance if inhalation exposure not ruled out | Follow inhalation PEP guidance if inhalation exposure not ruled out |
| **Cutaneous anthrax with known cutaneous exposure without systemic involvement** | Oral ciprofloxacin, doxycycline, Levofloxacin, Moxifloxacin, clindamycin or penicillin (if susceptible) for 7–10 days | Follow inhalation PEP guidance if inhalation exposure not ruled out | Follow inhalation PEP guidance if inhalation exposure not ruled out |

[a]Recommended antibiotics include ciprofloxacin 500 mg po bid or doxycycline 100 mg po bid for adults. Amoxicillin 500 mg po tid can be substituted for those intolerant of quinolones, to be continued for 42 days after initiation of the vaccine series or for 14 days after the last dose, whichever is later (should not exceed 60 days).
[b]If meningitis is confirmed or not yet ruled out, therapy should be with at least three antibiotics including ciprofloxacin 400 mg IV q8h, meropenem 2 g IV q8h, and linezolid 600 mg IV q12h unless contraindications exist.
[c]Monoclonal antibodies against protective antigen (PA) should be used when these postexposure therapies are not available or are not appropriate. NOTE: The polyclonal Anthrax Immune Globulin Intravenous (AIGIV) is NOT recommended as part of PEP as it has been shown in an animal model to interfere with response to postexposure vaccination.
[d]If meningitis has been ruled out, antibiotics should include ciprofloxacin and linezolid or clindamycin 900 mg IV q8h.

Anthrax vaccine adsorbed (AVA) (BioThrax) is the only FDA-licensed vaccine for PrEP and PEP. The 2019 CDC ACIP (Advisory Committee on Immunization Practices) guideline recommends AVA to be administered to the following categories of high-risk people to prevent infection (PrEP): members of the US military deployed to areas designated by Department of Defense (DoD) as being at high risk for exposure, laboratory workers, and high-risk professionals (farmers, veterinarians, as well as livestock). The FDA has approved the vaccine for adults aged 18 to 65 years to be given intramuscularly (IM) (0.5 mL) at 0, 1, and 6 months for primary series. A single-dose booster is recommended at 6 and 12 months after the primary series, and then annually thereafter. For individuals who received the primary vaccine series plus two initial boosters, and not at high risk for exposure to B. anthracis, boosters can be given every 3 years instead of annually.

To prevent infection after suspected or known exposure to aerosolized B. anthracis spores (PEP), the ACIP recommends AVA (0.5 mL subcutaneously [SC] at 0, 2, and 4 weeks) in conjunction with a course of antimicrobial therapy (see Table 90.2). The SC route is recommended when the vaccine is given for PEP due to higher antibody concentrations shown by week 4. However, it can be given IM in a large-scale emergency response due to limitations in material supply or personnel that may delay vaccination of the exposed population. In case of vaccine supply shortages, two vaccine doses (0 and 2 to 4 weeks) or half-dose vaccinations in a three-dose primary series (0, 2, and 4 weeks) are recommended to be able to vaccinate a higher number of exposed individuals.

While appropriate for treatment of acute IA, note that anthrax immune globulin may blunt vaccine response and is no longer favored for PEP.

AV7909 (AVA plus CpG 7909 adjuvant) is another anthrax vaccine currently in Phase 3 trial and has demonstrated potential use for PEP, though AV7907 is not FDA-approved and can only be used under FDA Emergency Use Authorization (EUA) in case of large-scale need. This is intended to be added to the Strategic National Stockpile. When using AVA or AV7909 for PEP, antibacterial prophylaxis should be continued for at least 2 weeks after the last dose of vaccine is administered.

## EVIDENCE

Bartlett JG, Inglesby TV Jr, Borio L. Management of anthrax. *Clin Infect Dis* 2002;35(7):851-8. *Excellent description of clinical management of anthrax infection.*

Bower WA, Schiffer J, Atmar RL, Keitel WA, Friedlander AM, Liu L, Yu Y, Stephens DS, Quinn CP, Hendricks K; ACIP Anthrax Vaccine Work Group. Use of anthrax vaccine in the United States: Recommendations of the Advisory Committee on Immunization Practices, 2019. *MMWR Recomm Rep* 2019;68(4):1-14. https://dx.doi.org/10.15585/mmwr.rr6804a1. *Updated recommendations on anthrax vaccine administration.*

FDA: Supplemental New Drug Approvals NDA 20-634/S-047, 20-635/S-051, 21-721/S-015. 2008. [Accessed March 14, 2011]. Available from http://www.fda.gov/downloads/Drugs/EmergencyPreparedness/BioterrorismandDrugPreparedness/UCM133682.pdf. *Notification from the FDA regarding the approval of the use of levofloxacin for postexposure prophylaxis of inhalation anthrax.*

Grunow R, Verbeek L, Jacob D, Holzmann T, et. al. Injection anthrax—a new outbreak in heroin users. *Dtsch Arztebl Int* 2012;109(49):843-8. https://dx.doi.org/10.3238/arztebl.2012.0843. *Describes an uncommon but concerning route of infection.*

Guarner J, Jernigan JA, Shieh WJ, et al. Pathology and pathogenesis of bioterrorism-related inhalational anthrax. *Am J Pathol* 2003;163:701.

*Characterizes anthrax as a bioterrorism weapon and identifies the pathology and pathogenesis of the infection.*

Holty JE, Bravata DM, Liu H, et al. Systematic review: A century of inhalational anthrax cases from 1900 to 2005. *Ann Intern Med* 2006;144:270. *Anthrax review.*

Jernigan DB, Raghunathan PL, Bell BP, et al. Investigation of bioterrorism-related anthrax, United States, 2001: epidemiologic findings. *Emerg Infect Dis* 2002;8:1019-1028. *Important paper regarding the 2001 anthrax events in the United States.*

Meselson M, Guillemin J, Hugh-Jones M, et al. The Sverdlovsk anthrax outbreak of 1979. *Science* 1994;266:1202-1208. *Discussion of a classic outbreak of anthrax.*

Stern EJ, Uhde KB, Shadomy SV, Messonnier N. Conference report on public health and clinical guidelines for anthrax [conference summary]. *Emerg Infect Dis* 2008;14(4):e1. Available from http://www.cdc.gov/eid/content/14/4/e1.htm. *Summary of updated CDC recommendations for prophylaxis and treatment of anthrax.*

Walsh JJ, Pesik N, Quinn CP, et al. A case of naturally acquired inhalation anthrax: Clinical care and analyses of anti-protective antigen immunoglobulin G and lethal factor. *Clin Infect Dis* 2007;44:968. *Study involving a natural anthrax case with further analysis of the anthrax vaccine.*

Brachman PS. Inhalation anthrax. *Ann N Y Acad Sci* 1980;353:83-93. *A classic paper on inhalational anthrax.*

Brachman PS, Kaufmann AF: Anthrax. In Evans AS, Brachman PS, eds: *Bacterial infections of humans: epidemiology and control,* ed 3, New York, 1998, Plenum, pp 95-107. *An overview of the many aspects of anthrax disease, control, and epidemiology.*

*Guidance for Protecting Responders' Health During the First Week Following A Wide-Area Aerosol Anthrax Attack.* Department of Homeland Security; 2012. *Succinct recommendations for protecting first-responders following a deliberate anthrax attack.*

Quinn CP, Turnbull PCB: Anthrax. In Hausler WJ, Sussman M, eds: *Topley and Wilson's microbiology and microbial infection,* ed 9, London, 1998, Edward Arnold, p 799. *A very thorough anthrax overview.*

Sirisanthana T, Brown AE. Anthrax of the gastrointestinal tract. *Emerg Infect Dis* 2002;8:649-651. *An overview of GI anthrax.*

WHO guidelines approved by the Guidelines Review Committee. *Anthrax in Humans and Animals,* 4th ed. Geneva: World Health Organization; 2008. *International guidelines on anthrax clinical presentation, diagnosis, and management.*

## ADDITIONAL RESOURCES

Beatty ME, Ashford DA, Griffin PM, et al. Gastrointestinal anthrax: Review of the literature. *Arch Intern Med* 2003;163:2527-2531. *A review of articles on GI anthrax.*

Bozue J, Cote CK, Glass PJ, eds. *Medical aspects of biological warfare.* Office of the Surgeon General, Borden Institute, US Army Medical Department Center and School, Health Readiness Center of Excellence; 2018. *The definitive US military textbook on biological warfare agents, including basic science, epidemiology, and countermeasures.*

# 91

# Tularemia

Benjamin C. Pierson, Fernando B. Guerena, Phillip R. Pittman

 **ABSTRACT**

Tularemia is a zoonotic bacterial infection that occurs throughout the Northern hemisphere; it is caused by several subspecies of the gram-negative coccobacillus *Francisella tularensis*. A wide variety of animals can be infected with *F. tularensis*, and the bacteria can also be carried by insects. Human *F. tularensis* infection is acquired through contact with infected animals, by the bite of an infected insect, or through inhalation or ingestion of the bacteria, which are highly infectious. Thus tularemia is a human disease associated with outdoor activities, especially among hunters and trappers, and may be an occupational disease of landscape workers and gardeners. The severity of tularemia depends on the infecting subspecies of *F. tularensis*, the mode of disease transmission, and the size of the infecting inoculum. The spectrum of human *F. tularensis* ranges from mild, localized infection to life-threatening sepsis. Although *F. tularensis* is an uncommon human illness in the United States, because it is highly infectious, is easily transmissible, and causes severe or fatal illness, it is considered a potential agent of bioterrorism.

 **CLINICAL VIGNETTE**

A 45-year-old male presents to your clinic in rural Arkansas with the complaint of fevers, chills, myalgia, and malaise which came on abruptly 1 day ago. These systemic symptoms were preceded by the development of a "red bump" on the patient's right index finger that has now progressed to a painful blister. The patient's medical history is significant for type 2 diabetes mellitus and hypertension, well controlled with once daily doses of metformin and lisinopril. The patient works as a roofer, and is an avid outdoors enthusiast, reporting that he has been out hunting and trapping each weekend for the past month. He smokes half a pack a day of cigarettes, and partakes in limited social alcohol use, but denies illicit drug use and recent travel. He has no known drug allergies. The patient's vital signs include a temperature of 102.1°F, heart rate of 115, blood pressure of 110/80, and respiratory rate of 18, with room air pO2 96%. Physical examination reveals a generally unwell-looking middle-aged male, with a tender pustular lesion on the right index finger and right axillary lymphadenopathy to palpation, but no pulmonary or other systemic abnormalities are detected. Initial laboratory findings show leukocytosis, thrombocytopenia, as well as elevated AST, ALT, and creatinine. Gram stain of the patient's blood does not reveal any organisms; however, the *Francisella tularensis* IgG/IgM enzyme-linked immunosorbent assay (ELISA) test shows detectable IgM-class antibodies. Chest x-ray does not reveal any findings significant for a pneumonic syndrome. The patient was admitted to the hospital with a presumptive diagnosis of ulceroglandular tularemia and started on gentamicin at 5 mg/kg/day intravenously (IV). Due to the patient's elevated creatinine at admission, the drug was prescribed as three divided doses given every 8 h for a 10-day course. Further workup was sent to state laboratories for confirmatory testing including polymerase chain reaction (PCR), and the diagnosis of tularemia was confirmed. After 3 days of treatment, the patient had marked clinical improvement; however, the routine serum chemistry panel showed creatinine levels increased by 1.5-fold from time of admission. Given the concern for nephrotoxicity associated with gentamicin, the patient was transitioned to oral doxycycline 100 mg PO BID to complete 21 total days of treatment, and was discharged from the hospital on treatment day 7 after resolution of his acute kidney injury, with creatinine decreasing to his known pre-illness baseline, and continued clinical improvement, with strict return instructions for re-development of systemic infectious symptoms. The patient completed the course of antimicrobial therapy as an outpatient and made a full recovery.

COMMENT: Tularemia primarily occurs in rural settings and is a risk for hunters, trappers, and other outdoor enthusiasts who may come into contact with infected animals, their carcasses, or waste products; gardeners or landscapers may also be exposed to aerosols of contaminated soil. Most cases in the United States occur in the south central states including Arkansas, Missouri, and Oklahoma, but cases have been reported in every state except Hawaii. Very low inoculating doses of *F. tularensis* are capable of producing disease. The most feared complication of tularemia is pneumonic tularemia, occurring either due to inhalation of bacteria or by hematogenous spread from another site, which can rapidly progress to severe, multilobar pneumonia, acute respiratory distress syndrome, and death. If a diagnosis of tularemia is suspected, it is important to notify laboratory personnel so that they can use appropriate growth media for culture and ensure that correct biosafety measures are followed. While streptomycin has historically been the drug of· choice for tularemia, due to its limited availability it is rarely used, and gentamicin has been recommended for therapy in all adults, including pregnant women, and children.

## GEOGRAPHIC DISTRIBUTION AND MAGNITUDE OF DISEASE BURDEN

Tularemia, also known as *rabbit fever* or *deer fly fever*, is a zoonosis caused by a highly infectious, aerobic, gram-negative coccobacillus, *F. tularensis*. The natural reservoir for *F. tularensis* is small mammals such as rodents or rabbits. The bacterium is found throughout host animals in most of North America and Eurasia. In the United States, tularemia is most commonly caused by two subspecies of *F. tularensis*: *F. tularensis* subsp. *tularensis* (type A, which is subdivided into subtypes A1a, A1b, and A2), and *F. tularensis* subsp. *holarctica* (type B). In Europe and Eurasia, *F. tularensis* subsp. *holarctica* is the primary cause of tularemia. Human tularemia was first described in the United States in 1910 as "deer fly fever," and the causative agent (at that time known as *Bacterium tularense*) was identified after an outbreak of a plague-like illness of ground squirrels in Tulare County, California in 1911. In 1924, a United States Public Health Service physician, Edward Francis, identified *B. tularense* as the cause of human deer fly fever. To honor his contributions to the understanding of this organism, the bacterium was subsequently renamed *F. tularensis*.

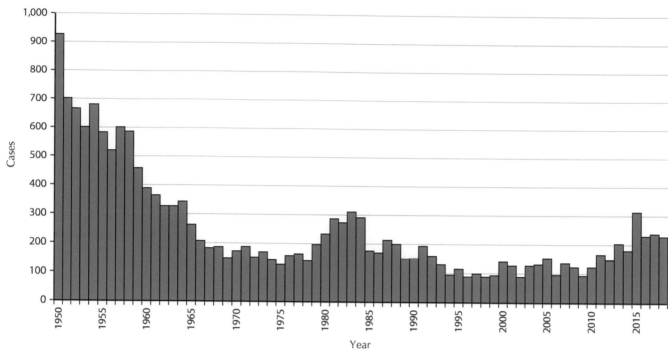

**Fig. 91.1** Reported tularemia, United States, 1950–2018. (From Centers for Disease Control and Prevention [CDC]: Tularemia statistics. Available at: https://www.cdc.gov/tularemia/statistics/index.html.)

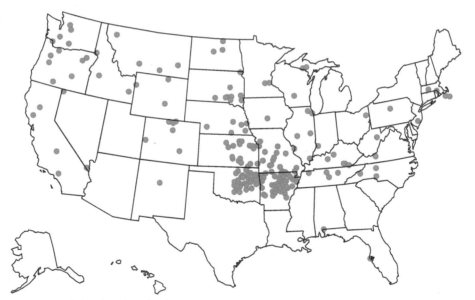

1 dot placed randomly within county of residence for each reported case

**Fig. 91.2** Reported cases of tularemia by county of residence, United States, 2018. (From Centers for Disease Control and Prevention [CDC]: Tularemia statistics. Available at: https://www.cdc.gov/tularemia/statistics/index.html.)

In the United States, human tularemia is rare but has been documented in every state except Hawaii. Tularemia was much more common in the early part of the 20th century than it is now (Fig. 91.1). Since 2000, most cases of tularemia have been reported from rural areas of the United States where infection of host animals is common (or enzootic), such as Arkansas, Kansas, Nebraska, and Missouri. In addition, many cases are reported from Martha's Vineyard, Massachusetts, where hunting clubs imported infected rabbits from enzootic areas in the 1920s and 1930s (Fig. 91.2). In 2018, the Centers for Disease Control and Prevention (CDC) received 229 reports of human tularemia, and from the years 2008 to 2018 there have been approximately 205 infections reported per year. Although the global incidence of tularemia has decreased markedly over the past 50 years, periodic outbreaks continue to occur, especially in Northern Europe and Eurasia. These outbreaks have been associated with drinking spring water, hunting, and other outdoor activities, and some have involved hundreds of cases of tularemia. Although rare in humans, tularemia occurs in a wide variety of animals and is maintained in an enzootic cycle with

*Francisella tularensis* infects many different animal species; transmission to humans in the United States occurs most commonly from contact with infected rabbits, rodents, or domestic cats.

Human infection can also occur from tick and deer fly bites or through ingestion of contaminated food or water.

**Fig. 91.3** Transmission of *Francisella tularensis*.

rodents and lagomorphs. Other animals, such as cats and nonhuman primates, may serve as incidental hosts. Outbreaks, or epizootics, of tularemia periodically occur in animal populations and may herald outbreaks of human disease.

## RISK FACTORS

The animals most commonly associated with transmission of tularemia to humans in the United States include lagomorphs (rabbits and hares) and rodents (voles, squirrels, muskrats, and beavers). Although animals are the primary reservoir of *F. tularensis*, the infection can also be transmitted by insect bites (especially those of ticks or deer flies) or by contact with bacteria in the environment. *Francisella tularemia tularensis* is a hardy organism that can survive for extended periods in water, mud, and frozen animal carcasses. Tularemia has been reported after skin or mucous membrane contact with contaminated animals or their environment, ingestion of contaminated food or water, and inhalation of aerosolized bacteria (Fig. 91.3). Activities associated with risk of tularemia include hunting, trapping, dressing, eating, or handling infected animals; activities that result in exposure to infected insects; farming or gardening with machinery that may aerosolize the carcasses of infected animals; and handling *F. tularensis* in a laboratory without appropriate personal protective equipment. Although inhalational tularemia does occur, the bacteria are not known to spread from person to person.

During and after World War II, several countries (including Japan, the former Soviet Union, and the United States) conducted research on the use of *F. tularensis* as a biologic weapon. More recently, most countries have suspended their biologic weapons research and destroyed their weapon stockpiles. However, as a highly infectious bacteria that can be easily mass-produced and aerosolized, with the potential to cause severe or fatal illness, *F. tularensis* continues to be designated a Category A Bioterrorism Agent by the CDC.

Recent CDC analyses of laboratory-documented human tularemia reported in the United States from 1964 through 2004 found that infections tend to be acquired during the warmer months (72% occurred from May through September) and are geographically diverse (type A1 infections cluster toward the eastern United States, type A2 infections occur toward the west, and type B is primarily clustered through the southern and central western areas of the country). Men constitute the majority of infections (74%), and younger age appears to be a risk for infections with type A, compared with type B (median age 38 vs. 50 years) (Fig. 91.4). Overall, 6% of infections occurred among persons with an immunocompromising condition (e.g., malignancy, organ transplant, or human immunodeficiency virus [HIV] infection). A source of infection was reported for 42% of the cases, and in this subgroup, direct animal contact accounted for about half of all infections (47% of type A and 53% of type B). Among persons with infection associated with animal contact, 53% of type A infections were attributed to lagomorphs and 30% to cats. No persons with type B infection reported lagomorph contact; those infections were linked to either rodents (33%) or cats (22%). Tularemia attributed to insects accounted for 44% of type A infections and 29% of type B infections; among infections resulting from insect bites, tick exposure (87%) was reported more commonly than contact with deer flies. A small proportion of types A (8%) and B (18%) infections were associated with multiple exposures or landscaping activities.

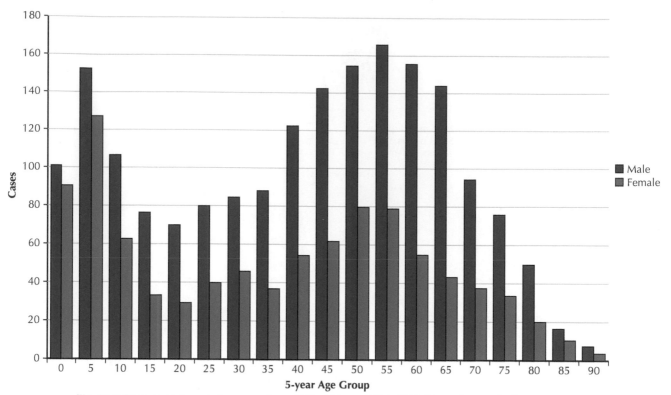

**Fig. 91.4** Reported tularemia by age and sex, United States, 2001–2018. (From Centers for Disease Control and Prevention [CDC]: Tularemia statistics. Available at: https://www.cdc.gov/tularemia/statistics/index.html.)

## CLINICAL FEATURES

The clinical presentation and severity of tularemia vary according to the anatomical portal of entry, the virulence of the *F. tularensis* strain involved, and the size of the bacterial inoculum; as few as 10 bacteria can cause human disease. The incubation period for tularemia is typically 3 to 5 days (range, 1 to 14 days). The clinical spectrum of tularemia is quite variable, ranging from localized infection to life-threatening systemic illness; forms include ulceroglandular, glandular, pneumonic, typhoidal, oculoglandular, pharyngeal, and septic syndromes (Fig. 91.5). After entry of *F. tularensis* via skin or mucous membranes, the bacteria disseminate to local lymph nodes and subsequently spread via the bloodstream to additional lymph nodes and organs, primarily the spleen, liver, kidney, lungs, and pleura. Regardless of the form of infection, untreated tularemia may become chronic, with fever, malaise, weight loss, and adenopathy lasting months.

Tularemia most commonly manifests as ulceroglandular or glandular disease (about 60% to 80% of infections) that follows entry of the bacteria into disrupted skin from contact with an infected animal or via the bite of an infected insect. A papule appears at the site of inoculation and, similar to other presentations of tularemia, the skin lesion is usually accompanied by the abrupt onset of symptoms of a systemic disease: chills, fever, headache, malaise, and myalgias. The skin lesion becomes pustular and tender with local lymphadenopathy, then ulcerates and may develop an eschar; the involved lymph nodes may become fluctuant and suppurate. In glandular tularemia, a less common form, a skin or mucosal lesion is not apparent.

About 15% of naturally infected persons have pneumonic tularemia, which may result from inhalation of bacteria or by hematogenous spread from another site. The occurrence of atypical infections, either in terms of clinical presentation or geographic location, and temporal clusters of pneumonic tularemia might indicate an act of deliberate infections. Typical signs and symptoms are similar to those of community-acquired pneumonia including fever, cough, and dyspnea. In early pneumonic tularemia, chest radiographic imaging may show nonspecific changes consistent with pneumonia such as diffuse peribronchial infiltrates that progress to patchy or lobar involvement with pleural effusions and perihilar lymphadenopathy. However, many patients demonstrate systemic findings alone and lack significant pulmonary involvement. In one study in which volunteers were intentionally exposed to aerosolized tularemia, up to 75% had no pulmonary signs or symptoms at the onset of illness. However, more serious complications can occur following pneumonic tularemia such as septic shock, acute respiratory distress syndrome, respiratory failure, and death. Less common presentations of tularemia include typhoidal (a systemic illness that does not involve specific organs or lymph nodes and lacks an obvious site of inoculation), oculoglandular (inoculation of the conjunctiva with mucosal ulceration and local lymphadenopathy), and oropharyngeal (ingestion of bacteria followed by oropharyngeal mucosal ulceration and local lymphadenopathy). Although rare, sepsis is the most serious form of tularemia, a systemic illness that may rapidly progress to shock, multisystem organ failure, and death.

Laboratory findings in tularemia are nonspecific and may include mild leukocytosis with normal differential hemoglobin and platelet counts are typically normal, the nonspecific markers of inflammation erythrocyte sedimentation rate and C-reactive protein are elevated, as well as abnormalities associated with multiorgan dysfunction, including elevated serum aminotransferases and creatine kinase.

## DIAGNOSTIC APPROACH

The relative low frequency of tularemia and the nonspecific findings associated with its many forms make its diagnosis challenging; a high index of suspicion is required, especially with typhoidal, pneumonic,

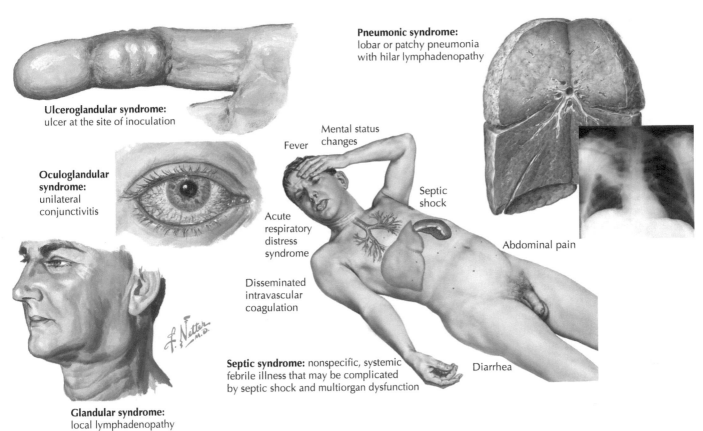

**Ulceroglandular syndrome:**
ulcer at the site of inoculation

**Oculoglandular syndrome:**
unilateral conjunctivitis

**Pneumonic syndrome:**
lobar or patchy pneumonia with hilar lymphadenopathy

Fever

Mental status changes

Septic shock

Acute respiratory distress syndrome

Abdominal pain

Disseminated intravascular coagulation

**Septic syndrome:** nonspecific, systemic febrile illness that may be complicated by septic shock and multiorgan dysfunction

Diarrhea

**Glandular syndrome:**
local lymphadenopathy

**Fig. 91.5** Syndromes associated with human tularemia.

or septic disease. Pneumonic tularemia may mimic other causes of atypical pneumonia (e.g., *Mycoplasma*, *Legionella*, or *Chlamydia* pneumonia, Q fever). Typhoidal tularemia can be indistinguishable from the innumerable causes of prolonged fever without an apparent source (e.g., invasive bacterial, mycobacterial, fungal, and parasitic infections; rheumatologic disorders; malignancies). An atypical presentation of ulceroglandular tularemia may suggest another localized bacterial or herpes virus infection, and glandular disease can be mistaken for a variety of bacterial, mycobacterial, fungal, parasitic, and viral infections that cause chronic fever and lymphadenopathy. Clinical suspicion may be heightened in enzootic areas when a patient has ulceroglandular tularemia, especially in the setting of an epizootic, but most clinicians in the United States lack familiarity with any presentation of tularemia. Factors that may signal a deliberate release of *F. tularensis* include clusters of severe, atypical pneumonia, or acute febrile illness that rapidly progress if untreated, or tularemia that occurs in urban areas, or among persons without obvious risk factors.

*F. tularensis* can be isolated from respiratory secretions, tissue biopsy specimens obtained from an ulcer, lymph node, or other affected tissue; fluid aspirates; tissue swabs or scrapings, or blood, although surprisingly, the organism is rarely isolated from blood. Isolation of *F. tularensis* from a clinical specimen is considered the gold standard for a diagnosis of tularemia. Clinical laboratories must be alerted that tularemia is suspected so that they can use appropriate growth media and proper biosafety measures. *F. tularensis* grows slowly in routine culture media and requires a cysteine-enriched growth medium to accelerate its growth. The bacteria are infrequently identified by Gram stain or isolated from routine cultures of skin or mucosal lesions, blood, sputum, respiratory secretions, or lymph node or pleural aspirates. Antimicrobial susceptibility testing should be performed only at an experienced microbiology laboratory.

Quantitative serum antibody testing for *F. tularensis* confirms most of the cases. A fourfold increase between acute and convalescent serum antibody titers taken 2 to 4 weeks apart is diagnostic. However, confirmation of tularemia by isolation of the organism or demonstration of a fourfold increase in serum antibody titers may not be timely enough for clinical or public health management of tularemia. Tularemia ELISA tests are commercially available and can detect IgM and IgG anti-*F. tularensis* antibodies with a short test turnaround time. IgM-class antibodies may become detectable as soon as 1 week after symptom onset, but due to possible persistence of IgM antibodies after acute disease resolution, lack of *F. tularensis* subspecies differentiation, and cross-reactivity with certain other infections, an IgM-positive result on an ELISA test is considered supportive but does not confirm the clinical diagnosis of tularemia while the confirmatory tests (culture and antibody titers) are pending. More specialized methods, available at local, state, and federal public health laboratories, include direct fluorescent antibody, PCR assays or immunohistochemical staining of clinical specimens; these methods can provide a rapid, presumptive diagnosis of tularemia, enabling appropriate patient management and public health interventions.

## CLINICAL MANAGEMENT AND DRUG TREATMENT

Rapid administration of appropriate antimicrobial therapy is the optimal treatment for tularemia. Although no controlled clinical trials have evaluated the duration of therapy required for cure or the efficacy of different antimicrobial regimens, the current standard of care is the use of aminoglycoside antibiotics streptomycin, gentamicin, and tobramycin as first-line agents, given their demonstrated highest rate of cure and lowest rate of relapse. Streptomycin is approved by the US Food and Drug Administration (FDA) for this indication, but has limited availability in the United States, limiting its use in the treatment of tularemia. Neither gentamicin nor tobramycin are FDA approved for the use of treating tularemia, but they have been used successfully for this purpose.

Recommended treatment regimens for adult patients with severe tularemia involve administering antimicrobials for at least 10 days: gentamicin or tobramycin is dosed at 5 mg/kg/day IV or intramuscularly (IM) once daily, with monitoring of serum drug levels and doses adjusted for renal insufficiency; if streptomycin is used, it is given at a dose of 1 g IM twice daily, not to exceed 2 g/day. For hematogenous meningitis, the use of streptomycin plus chloramphenicol is recommended. Aminoglycosides are ototoxic and nephrotoxic, and their dose must be adjusted according to the patient's body weight and creatinine clearance. The three aminoglycosides recommended for tularemia are FDA Pregnancy Risk and Lactation Category D. Treatment of severe tularemia in children also involves prompt administration of parenteral aminoglycosides with doses adjusted for age and weight, and a pediatric infectious disease specialist should be consulted if possible: streptomycin 15 mg/kg IM administered twice daily for a minimum of 10 days; or gentamicin 2.5 mg/kg IM or IV three times daily for a minimum of 10 days.

Tetracyclines and fluoroquinolones are alternate choices for treatment of tularemia. However, high relapse rates have been reported with both drug classes. As alternatives to an aminoglycoside, adults may be treated with doxycycline 100 mg IV or 500 mg by mouth (PO) twice daily for 14 to 21 days; or with ciprofloxacin 400 mg IV or 500 mg PO twice daily for 10 to 14 days. Children can be treated with ciprofloxacin 15 mg/kg IV or PO twice daily (maximum 800 mg/day) for a minimum duration of 10 days. More severe tularemia may require a longer course of treatment. Doxycycline is not recommended for use in children less than 8 years old. For postexposure prophylaxis (PEP) in events such as a deliberate release of tularemia or a laboratory incident, a 14-day course of doxycycline or ciprofloxacin is usually prescribed.

## PROGNOSIS

The prognosis of tularemia depends on several factors including the virulence of the *F. tularensis* strain, the mode of transmission, the clinical syndrome, the timeliness of diagnosis and antimicrobial treatment, and the overall health status of the affected patient. *F. tularensis* subtype A.I is the most virulent of the four subspecies with reported fatality rates as high as 35%. Inhalation of *F. tularensis* causes pneumonic tularemia, the most severe clinical disease. In ulceroglandular tularemia the infection appears to provoke a strong inflammatory response with localized clinical syndrome on the affected area. However, all types of localized tularemia can result in hematogenous dissemination to other organs such as lungs, heart, meninges, bones, liver, kidneys, and other organs with the development of sepsis, multiorgan failure, and death.

The timely diagnosis and the administration of antibiotics are the most important determinants of a good prognosis. Before the availability of antibiotics, fatality rates of 5% to 60% were reported for tularemia, depending on the *F. tularensis* strain and type of syndrome. With the availability of antibiotics, mortality is now reported as less than 2% overall but has been reported to be as high as 24% depending on the strain involved.

## PREVENTION AND CONTROL

Prevention of tularemia relies on minimizing the potential for contact with *F. tularensis*, especially in enzootic areas. Hunters, trappers, and others in contact with wild animals should wear gloves when handling these animals, especially rabbits, muskrats, prairie dogs, and other rodents. Game meat must be cooked thoroughly before consumption. Anyone engaged in outdoor activities should wear long sleeves, long pants, and long socks to prevent skin contact with ticks and deer flies. Ticks attached to skin need to be removed quickly with tweezers. Only drink treated water. Use Environmental Protection Agency (EPA)-registered insect repellents containing DEET (N,N-diethyl-*m*-toluamide), picaridin, IR3535, oil of lemon eucalyptus (OLE), para-menthane-diol (PMD), or 2-undecanone. When performing landscaping or gardening, do not mow over sick or dead animals, and consider using a mask when mowing to possibly reduce risk of inhalation.

Tularemia is recurrently reported as an occupational infection from research laboratories working with *F. tularensis*. However, it is also occasionally reported from clinical diagnostic laboratories. Biosafety level 2 (BSL-2) laboratory practices, equipment, and facilities are indicated for working with human or animal clinical specimens suspected of being infected with *F. tularensis*. All laboratory personnel must be informed in advance of tularemia being in the differential diagnosis. BSL-3 practices, equipment, and facilities are recommended for all manipulation of suspect cultures, animal necropsies, and experimental animal studies. *F. tularensis* is a select agent requiring registration with CDC and/or United States Department of Agriculture (USDA) for possession, use, storage and/or transfer.

### Immunization

Infection with tularemia is thought to confer lifelong immunity, although repeat infections have been reported. A variety of vaccines have been developed to prevent tularemia; however, currently there is no licensed human vaccine available for the civilian population in the United States. An investigational tularemia vaccine candidate, the attenuated live vaccine strain (LVS), has been administered in the United States since the 1950s under FDA Investigational New Drug (Vaccine) status and Institutional Review Board (IRB)-approved protocol and informed consent. The vaccine is offered to laboratory workers at risk for exposure to *F. tularensis*. A retrospective study showed the incidence of typhoidal tularemia fell from 5.70 to 0.27 cases/1000 employee-years; the incidence of ulceroglandular tularemia remained unchanged during the same 10-year time span, although the signs and symptoms of the latter were milder.

### Chemoprophylaxis

In the event of a deliberate release of *F. tularensis*, and laboratory or animal handling incidents resulting in a tularemia-related exposure, chemoprophylaxis may be effective in preventing the development of tularemia if given as early as possible within the incubation period. The Working Group on Civilian Biodefense in the United States recommends PEP treatment with oral doxycycline or ciprofloxacin for 14 days treatment for children and adults (including pregnant women). Chemoprophylaxis of persons in contact with someone who has tularemia is not indicated, as *F. tularensis* is not transmitted from person to person. There are no licensed antibiotics for pre-exposure chemoprophylaxis against *F. tularensis*.

### Infection Control

As noted earlier, tularemia is not transmitted from person to person; therefore in healthcare settings, standard precautions (frequent hand hygiene with soap and water or alcohol-based hand sanitizers; gown and eye protection [i.e., goggles or face shield] during activities that may generate splashes or aerosols of respiratory or other body fluids) alone are recommended by the CDC for infection control for patients infected with tularemia. *F. tularensis* may persist for extended periods of time in the environment; however, it is effectively inactivated by heat and standard disinfectants.

## Public Health Measures

Tularemia is listed as one of the 2020 national notifiable infectious diseases in the United States by the CDC. The current (2017) case definition of tularemia for public health reporting purposes consists of clinical and laboratory criteria. The clinical criteria include the previously described six types of clinical tularemia. The surveillance laboratory criteria are categorized as supportive (elevated but below fourfold increase in antibody titer), positive fluorescent assay or PCR, or confirmatory (isolation of *F. tularensis* from a clinical or autopsy specimen or fourfold or greater change in serum antibody titer). The prompt recognition of clusters of human or animal tularemia may prevent further cases of illness or identify an incident of deliberate release of *F. tularensis*. It is important that clinicians report any suspected cases of human tularemia to their local or state health department. Public health personnel can facilitate diagnostic and confirmatory laboratory testing and assist in the identification of potential exposures.

## ACKNOWLEDGMENT

The authors would like to acknowledge Dr. Jo Hoffman who contributed to the previous edition of this chapter.

## EVIDENCE

Burke DS. Immunization against tularemia: analysis of the effectiveness of live Francisella tularensis vaccine in prevention of laboratory-acquired tularemia. *J Infect Dis* 1977;135(1):55-60. *A retrospective analysis of the safety and effectiveness of a live, attenuated vaccine strain of tularemia to prevent or reduce the effect of tularemia disease.*

Dennis DT, Inglesby TV, Henderson DA, et al. Tularemia as a biological weapon: medical and public health management. *JAMA* 2001;285:2763-2773. *A comprehensive article on the history, clinical management, and public health response to tularemia when used as a biologic weapon.*

Feldman KA, Stiles-Enos D, Julian K, et al. Tularemia on Martha's Vineyard: seroprevalence and occupational risk. *Emerg Infect Dis* 2003;9:350-354. *The only reported outbreaks of pneumonic tularemia in the United States occurred on Martha's Vineyard, Massachusetts, in 1978 and 2000. This study evaluated the risk of exposure to* F. tularensis *among landscape employees.*

Kugeler KJ, Mead PS, Janusz AM, et al. Molecular epidemiology of *Francisella tularensis* in the United States. *Clin Infect Dis* 2009;48:863-870. *A CDC review of more than 500 cases of human and animal tularemia reported from 1964 to 2004, with an in-depth look at the epidemiology of subtypes of* F. tularensis *as determined by pulsed-field gel electrophoresis.*

Staples JE, Kubota KA, Chalcraft LG, et al. Epidemiologic and molecular analysis of human tularemia, United States, 1964-2004. *Emerg Infect Dis* 2006;12:1113-1118. *An earlier analysis of the same cases of tularemia as reported by Kugeler, with more focus on the epidemiology of human infections.*

Williams, MS, Baker MR, Guina T, Hewitt JA Lanning H, May JM, Fogtman B, Pittman PR. Retrospective analysis of pneumonic tularemia in Operation Whitecoat human subjects: disease progression and tetracycline efficacy. *Frontiers in Med* 2019;6:229. https://DOI: 10.3389/fmed.2019.00229. *A meta-analysis of a series of human studies including challenge studies with varying doses of live* F. tularensis Schu S4 *organisms.*

## ADDITIONAL RESOURCES

Adalja AA, Toner E, Inglesby TV. Clinical management of potential bioterrorism-related conditions. *NEJM* 2015;372:954-962. *Review of management of tularemia from a bioterrorism perspective.*

Centers for Disease Control and Prevention (CDC): Emergency preparedness and response: tularemia. Available at: www.bt.cdc.gov/agent/tularemia/index.asp. *CDC website for information on tularemia as a potential agent of bioterrorism. Accessed May 11, 2020.*

Centers for Disease Control and Prevention (CDC): Tularemia. Available at: www.cdc.gov/tularemia. *CDC website for tularemia and* F. tularensis. *Includes information for the public and public health and clinical professionals. Accessed May 11, 2020.*

*Francisella tularensis.* In U.S. Department of Health and Human Services, Public Health Service, Centers for Disease Control and Prevention, National Institutes of Health. Biosafety in microbiological and biomedical laboratories, ed 5, HHS Publication No. (CDC) 21-1112. *The 5th edition of the BMBL is an advisory document recommending best practices for the safe conduct of work in biomedical and clinical laboratories from a biosafety perspective.*

Hepburn MJ, Kijek TM, Sammons-Jackson W, et al. Tularemia. In Bozue J, Cote CK, Glass PJ, eds: *Medical aspects of biological warfare.* Office of The Surgeon General, Borden Institute, US Army Medical Department Center and School, Health Readiness Center of Excellence, Fort Sam Houston, Texas, 2018. *Infectious diseases textbook chapter with comprehensive information about all biodefense aspects of tularemia.*

Larson MA, Sayood K, Bartling AM, el al. Differentiation of Francisella tularensis subspecies and subtypes. *J Clin Microbiol* 2020;58:e01495-19. https://doi.org/10.1128/JCM.01495-19. Accessed May 11, 2020. *Review of subtypes of* F. tularensis *and their associated geographic distribution and relative virulence.*

Penn RL. *Francisella tularensis* (tularemia). In Bennett JE, Dolin R, Blaser MJ, eds: *Mandell, Douglas, and Bennett's principles and practice of infectious diseases,* ed 8, Philadelphia, 2014, Elsevier. *Infectious diseases textbook chapter with comprehensive information about all aspects of tularemia.*

Stevens DL, Bisno AL, Chambers HF, et. al. Practice guidelines for the diagnosis and management of skin and soft tissue infections: 2014 update by the Infectious Diseases Society of America. *Clin Infect Dis* 2014;59(2):e10-52. https://DOI: 10.1093/cid/ciu296. *These are updated recommendations on the management of skin and soft tissue infections from a panel of national experts convened by the Infectious Diseases Society of America.*

University of Minnesota Center for Infectious Disease Research and Policy (CIDRAP): Tularemia. Available at: www.cidrap.umn.edu/cidrap/content/bt/tularemia/index.html. *CIDRAP is a collaborative center of the University of Minnesota that focuses on emerging global challenges to public health, including preparedness and response to events such as bioterrorism and pandemic influenza. The site has extensive information about tularemia, including a comprehensive review of the subject that is updated frequently. Last accessed May 11, 2020.*

# 92

# Zika Virus Disease and Congenital Zika Virus Infection

*Susan L. Hills, Kate R. Woodworth*

 **ABSTRACT**

Zika virus (ZIKV) is a flavivirus primarily transmitted to humans by mosquitoes, mainly *Aedes aegypti*. Other commonly recognized transmission modes include sexual and intrauterine transmission. Following recognition of ZIKV transmission in Africa and Asia during the 1900s, outbreaks subsequently occurred in Pacific Island nations from 2007 and the Americas from 2015. Incidence subsequently declined, but available information suggests probable low-level ongoing transmission in parts of Africa, Asia, and the Americas with rare localized outbreaks. Most infections are asymptomatic or cause mild symptoms, with rash, fever, arthralgia, and/or conjunctival hyperemia commonly reported. Although severe disease is rare, Guillain-Barré syndrome (GBS) and other neurologic and ophthalmologic presentations have been reported. Congenital ZIKV infection can result in a variety of neonatal outcomes, ranging from asymptomatic infection to a severe and distinct pattern of anomalies. Among infants with severe outcomes, clinical findings most specific for congenital ZIKV infection include severe microcephaly with partially collapsed skull, cortical hypoplasia with abnormal gyral patterns, intracranial calcifications located between the cortex and subcortex, macular scarring with focal pigmentary retinal mottling, and arthrogryposis. Management of uncomplicated disease in adults and children is supportive. Pregnant women with laboratory evidence of possible infection should be monitored for adverse pregnancy outcomes. Infants with findings consistent with congenital ZIKV infection or born to mothers with laboratory evidence of possible ZIKV infection during pregnancy should have, in addition to standard evaluation, a head ultrasound and comprehensive ophthalmologic exam. No vaccine is available, and prevention relies on the avoidance of mosquito bites and sexual transmission of ZIKV.

## CLINICAL VIGNETTE

A 34-year-old woman presented to the emergency room in active labor at 38 weeks and 6 days of gestation. She had recently emigrated to the United States from Brazil, where she had received prior prenatal care. Early in her pregnancy she experienced a mild illness with rash and low-grade fever for which she did not seek medical care. Her last ultrasound, performed at 30 weeks' gestation, showed evidence of microcephaly and ventriculomegaly. She delivered her male infant by spontaneous vaginal delivery. He was born with Apgar scores of 7 and 9 at 1 and 5 minutes, respectively. Initial head circumference measurements were below the third percentile for sex and gestational age. Repeat head circumference measurements confirmed microcephaly. Weight and length were measured at the 25th and 10th percentiles, respectively. On examination, the infant was noted to have a small, irregularly shaped head and multiple contractures of both upper and lower extremities bilaterally. A head ultrasound was performed and revealed microcephaly, multiple subcortical calcifications, and ventriculomegaly. In addition, magnetic resonance imaging revealed polymicrogyria of the left frontal region and bilaterally in the temporal lobes. Given the maternal history and clinical findings in the infant consistent with congenital ZIKV infection, laboratory testing was performed. The infant's serum and urine were negative for ZIKV ribonucleic acid (RNA) by reverse transcription–polymerase chain reaction (RT-PCR) testing, but serum was positive for ZIKV IgM, with a ZIKV plaque reduction neutralization test (PRNT) titer of 1:1280 and dengue virus (DENV) PRNT titer of 1:20. Further evaluation included an ophthalmology exam that revealed left optic nerve atrophy. The infant passes an automated auditory brain-stem response hearing test. Based on clinical findings and laboratory test results, the infant was determined to have probable congenital ZIKV infection. The infant was referred to multiple specialists including neurology, ophthalmology, orthopedics, and early intervention services for further evaluation and management.

## ETIOLOGY/PATHOPHYSIOLOGY

ZIKV is a flavivirus in the family *Flaviviridae*. It is a single-stranded RNA virus related antigenically to dengue, yellow fever, Japanese encephalitis, West Nile, and St. Louis encephalitis viruses. There are two major ZIKV lineages, African and Asian.

Although infection with ZIKV most often results in asymptomatic infection or a mild systemic illness, in pregnant women ZIKV can cross the placenta, resulting in vertical transmission of the virus from the pregnant woman to her fetus. ZIKV is neurotropic in the fetus and causes direct damage to the developing brain. Severe structural brain abnormalities with intracranial volume loss can occur, resulting in collapse of the developing skull with overlapping cranial sutures, occipital bone prominence, and redundant scalp skin, a phenotype known as fetal brain disruption sequence.

## TRANSMISSION

ZIKV transmission among humans primarily involves a human-mosquito-human transmission cycle. The main vectors are *Aedes (Stegomyia)* spp. mosquitoes, primarily *Ae. aegypti*. These mosquitoes also are key vectors involved in the transmission of dengue, chikungunya, and yellow fever viruses. Another important mode of transmission is through the intrauterine route, which can result in congenital ZIKV infection. This can occur in any trimester of pregnancy, but the risk for severe outcomes is greatest when infection occurs during the first trimester. Sexual transmission is also well recognized, with male-to-female, male-to-male, and female-to-male sexual transmission having been documented. There are rare reports of laboratory-associated and intrapartum transmission, of probable transmission through platelet transfusion, and of possible transmission through breast feeding. One likely case of person-to-person transmission with an unknown transmission mode has been reported. The index patient had a uniquely high viral load, estimated to be about 100,000 times the normal level, and ultimately died. The case patient reported casual contact such as hugging and kissing the index patient and assisted hospital personnel in moving the patient after toileting but did not have direct contact with body

**Fig. 92.1** Countries and territories with history of Zika virus transmission. (Reused with permission from Hills SL, Fischer M, Petersen LR. Epidemiology of Zika Virus Infection. J Infect Dis. 2017 Dec 16;216(suppl_10):S868-S874, Fig. 1; doi: 10.1093/infdis/jix434. PMID: 29267914; PMCID: PMC5853392.)

fluids. Based on the transmission of other arboviruses through organ and tissue transplantation, transmission of ZIKV through these routes is theoretically possible but has not been reported.

## EPIDEMIOLOGY

ZIKV was first isolated in 1947 from a monkey in the Zika Forest in Uganda. For 60 years thereafter, only small numbers of human cases were identified in Africa and Asia. In 2007, a ZIKV disease outbreak was recognized in the Federated States of Micronesia, followed by an outbreak in 2013 in French Polynesia, with subsequent spread to other Pacific islands. Transmission was first recognized in the Americas in 2015 in Brazil, although retrospective analyses suggested earlier circulation of the virus in the region. Transmission was subsequently reported from much of South America, Central America, the Caribbean, and parts of North America (Fig. 92.1). In the continental United States, this included identified local transmission in Florida and southern Texas. The ZIKV outbreak in the Americas peaked during the first half of 2016, and incidence subsequently declined substantially. Although accurate and up-to-date epidemiologic data on ZIKV are limited, available information suggests probable ongoing low-level transmission in parts of Africa, Asia, and the Americas, with rare localized outbreaks. Transmission has likely been interrupted in some countries where ZIKV transmission had been recognized in the past, particularly geographically isolated places with small populations (e.g., small island nations). However, in areas with the appropriate vector, the potential for reemergence or reintroduction exists. As of 2020, transmission levels of DENV, which produces a clinically similar disease, are now far higher than levels of ZIKV in most areas.

## CLINICAL PRESENTATION

### Noncongenital Zika Virus Infection

Most ZIKV infections are asymptomatic or cause mild symptoms. For persons who develop symptomatic illness, the incubation period is 3 to 14 days. Common symptoms are rash (which may be pruritic), fever, arthralgia, conjunctival hyperemia, myalgia, and headache. Other reported symptoms include lethargy, vomiting, retro-orbital pain, lymphadenopathy, and edema of the extremities. Clinical illness typically resolves within a week. Severe disease requiring hospitalization and death following ZIKV infection are rare. The most commonly reported complication is GBS, with onset in most cases 5 to 10 days after the viral prodrome. Other neurologic or ophthalmologic presentations (e.g., meningoencephalitis, myelitis, acute maculopathy, uveitis) have occasionally been reported. Pregnant women have similar clinical presentations to nonpregnant persons, but when intrauterine transmission occurs, adverse pregnancy outcomes can include fetal loss or birth defects. Laboratory parameters in cases of ZIKV disease are most often normal, but reported abnormalities include leukopenia and elevated hepatic transaminases. In addition, ZIKV infection has been associated with severe thrombocytopenia, which can cause hemorrhage and death.

### Congenital Zika Virus Infection

Congenital ZIKV infection has been associated with a variety of neonatal outcomes, ranging from asymptomatic infection to a severe and distinct pattern of anomalies that has been referred to as congenital Zika syndrome. The clinical features most specific for congenital ZIKV infection include the following (Fig. 92.2):

1. Severe microcephaly with partially collapsed skull
2. Cortical hypoplasia with abnormal gyral patterns
3. Intracranial calcifications located between the cortex and subcortex
4. Macular scarring with focal pigmentary retinal mottling
5. Congenital contractures of major joints (arthrogryposis) associated with structural brain anomalies

However, the full spectrum of the associated phenotype, in particular the milder end of the spectrum, has not been well defined. Additional structural brain findings associated with congenital ZIKV infection include ventriculomegaly, corpus callosum agenesis, cerebellar

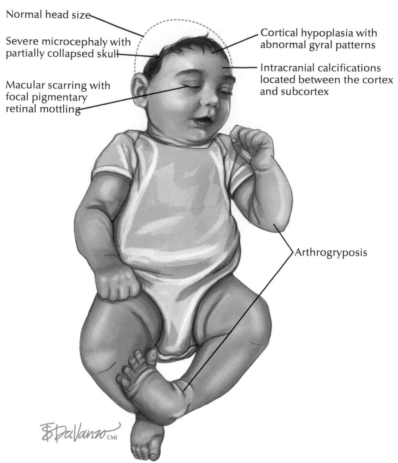

Normal head size

Severe microcephaly with partially collapsed skull

Macular scarring with focal pigmentary retinal mottling

Cortical hypoplasia with abnormal gyral patterns

Intracranial calcifications located between the cortex and subcortex

Arthrogryposis

**Fig. 92.2** Specific clinical findings in infants with congenital Zika virus infection.

---

### BOX 92.1    Clinical Findings in Congenital Zika Virus Infection

**Most Specific Clinical Findings**

- Severe microcephaly[a] with partially collapsed skull
- Cortical hypoplasia with abnormal gyral patterns
- Intracranial calcifications located between the cortex and subcortex
- Macular scarring with focal pigmentary retinal mottling
- Arthrogryposis associated with structural brain anomalies

**Other Clinical Findings**

- Other structural brain anomalies: Ventriculomegaly, corpus callosum agenesis, cerebellar hypoplasia, mild microcephaly, or postnatal-onset microcephaly
- Other structural eye anomalies: Microphthalmia, cataracts, chorioretinal atrophy, or optic nerve hypoplasia
- Neurologic sequelae: Hypertonia, dystonia, tremors, swallowing dysfunction, intellectual disability, hearing loss, or visual impairment

[a]Occipital frontal circumference greater than 3 standard deviations below the mean for age and sex.

---

hypoplasia, mild microcephaly, and postnatal-onset microcephaly (Box 92.1). Ophthalmologic findings include microphthalmia, cataracts, chorioretinal atrophy, and optic nerve hypoplasia. Neurologic sequelae likely associated with congenital ZIKV infection include hypertonia, dystonia, tremors, swallowing dysfunction, intellectual disability, sensorineural hearing loss, and visual impairment. Although surveillance and research studies have investigated the frequency of ZIKV-associated birth defects and neurodevelopmental abnormalities, differences in study approaches; laboratory definitions of ZIKV infection; the birth defects considered possibly, probably or definitely related; comprehensiveness and period of follow-up; and other factors make these rates hard to quantify. Efforts are ongoing to clarify the full spectrum of sequelae associated with ZIKV exposure in utero, including long-term neurodevelopmental outcomes.

## DIAGNOSTIC APPROACH

### Testing of Nonpregnant Persons

Based on a clinically compatible illness and travel to an area with risk of ZIKV transmission within the 2 weeks prior to illness onset (or sex with someone who has traveled to such an area), testing can be considered for persons with suspected ZIKV disease. Because ZIKV and DENV infections are often clinically indistinguishable and transmission of both viruses occurs in similar geographic areas, DENV testing should be performed concurrently. Nucleic acid amplification tests (NAATs) are the preferred method for laboratory testing to confirm infection. The US Centers for Disease Control and Prevention (CDC) recommends RT-PCR testing for ZIKV and DENV and states that this should be performed on serum collected within 7 days of symptom onset. Testing with ZIKV and DENV immunoglobulin M (IgM) antibody assays, with confirmatory testing using a PRNT to measure virus-specific neutralizing antibody titers, can be performed if serum samples are RT-PCR–negative or serum is collected more than 7 days after the onset of symptoms. However, antibody test results can be difficult to interpret because of cross-reactivity, which might preclude determination of which flavivirus is responsible for the person's recent infection. In addition, IgM antibodies can persist for months or even years after an infection, particularly for ZIKV, making determination of the timing of infection difficult, especially if a person frequently travels to or lives in an area with risk of transmission. To reduce the risk of false-positive test results, ZIKV testing should be requested only in the appropriate clinical and epidemiologic context; if the pretest probability of infection is low, there is a higher probability that an IgM-positive result will be a false positive. If indicated, ZIKV testing can be performed on other samples (e.g., RT-PCR or antibody testing of cerebrospinal fluid [CSF] or RT-PCR testing of urine). Depending on clinical symptoms and where the person is suspected of having been infected, consideration should be given to appropriate testing for other arboviral infections (e.g., chikungunya) and other possible etiologies of illness (e.g., rubella, measles, adenovirus, parvovirus, enterovirus, leptospirosis, malaria, rickettsiosis, and group A streptococcal infections). Persons with a laboratory-confirmed ZIKV infection are considered to have long-term, probably lifelong immunity.

## Testing of Pregnant Women

The CDC recommends that pregnant women with suspected ZIKV disease have diagnostic testing for ZIKV and DENV infection performed by RT-PCR and antibody testing on a serum specimen and ZIKV RT-PCR on a urine specimen. Similar considerations as for nonpregnant persons in interpretation of results and in consideration of alternate etiologies are relevant. In particular, if IgM is detected and a woman has had possible ZIKV exposure before the current pregnancy, it might not be possible to determine whether infection occurred during the current pregnancy. For asymptomatic pregnant women without ongoing risk for ZIKV infection, testing is not routinely recommended.

## Testing of Infants With Possible Congenital ZIKV Infection

Laboratory testing for ZIKV infection is recommended for all infants born to mothers with laboratory evidence of possible ZIKV infection during pregnancy. In this context, possible maternal ZIKV infection has a broad diagnostic definition and includes both molecular evidence of ZIKV infection by detection of ZIKV RNA in any maternal, placental or fetal sample, or possible serologic evidence of ZIKV infection (e.g., a positive or equivocal ZIKV IgM or DENV IgM result with evidence of ZIKV neutralizing antibodies by PRNT). Regardless of maternal testing. testing should also be performed for any infant with clinical findings consistent with congenital ZIKV infection if the mother had possible ZIKV exposure during pregnancy. Laboratory testing includes testing infant serum and urine by RT-PCR and serum by IgM antibody assay. RT-PCR and IgM antibody testing can also be performed on CSF if available. Infants with a positive RT-PCR test are considered to have confirmed congenital ZIKV infection. Those with nonnegative IgM testing but negative RT-PCR testing are considered to have probable congenital ZIKV infection. Testing should be performed as soon as possible after birth, as the period when ZIKV RNA and IgM antibodies can be detected in congenitally infected infants is likely limited. The differential diagnosis of congenital ZIKV infection includes both infectious and genetic etiologies, and all infants with suspected congenital ZIKV infection should have a comprehensive evaluation for alternate etiologies.

## CLINICAL MANAGEMENT

### Management of Zika Virus Disease and Its Complications

There is no specific antiviral treatment currently available for ZIKV disease. Management of uncomplicated disease in adults and children is supportive, including rest, fluids, antihistamines for pruritus, and acetaminophen for pain and fever. Because of the difficulty in clinically differentiating ZIKV and DENV infection, patients with suspected ZIKV infection also should be managed for possible DENV infection. Aspirin and other nonsteroidal antiinflammatory drugs (NSAIDs) should be avoided, to reduce the risk of hemorrhage, until dengue can be ruled out.

For patients who develop ZIKV-associated GBS, treatment options include intravenous immunoglobulin and plasmapheresis. For ZIKV-associated severe thrombocytopenia, the optimal management is not currently known; intravenous immunoglobulin has been associated with clinical improvement in some case reports, but it is unknown if this was due to the intervention or reflected the natural course of disease.

### Management of Pregnant Women With Laboratory Evidence of Possible Zika Virus Infection

Pregnant women with laboratory evidence of possible ZIKV infection should be evaluated and managed for possible adverse pregnancy outcomes, including fetal loss or serious birth defects. Prenatal care in the United States includes routine screening for fetal abnormalities at 18 to 22 weeks' gestation. Serial ultrasound monitoring can detect changes in fetal anatomy and growth patterns, but for the detection of congenital ZIKV-associated abnormalities, the sensitivity, specificity, and positive and negative predictive values of ultrasound are unknown. Limited data suggest that microcephaly can typically be detected at approximately 28 weeks' gestation, but questions remain about the optimal timing and frequency of ultrasounds. Data regarding the positive and negative predictive values and optimal timing for amniocentesis and its role in detection of congenital ZIKV infection are also unavailable. Nonetheless, if an indication exists for amniocentesis to evaluate abnormal prenatal findings, ZIKV NAAT testing should be considered. Given the limitations with these two screening modalities, the lack of effective interventions to prevent or treat congenital ZIKV infection, and the potential risks and costs, clinicians and patients should consider available guidance and then work together to decide on individualized evaluation plans. As more data become available, understanding of the role of prenatal ultrasound and amniocentesis should improve.

### Management of Infants With Possible Congenital Zika Virus Infection

In addition to standard evaluation of all newborns (i.e., comprehensive examination, anthropometrics, newborn hearing screening), infants with findings consistent with congenital ZIKV infection born to mothers with possible ZIKV exposure during pregnancy should have a head ultrasound and a comprehensive ophthalmologic exam before 1 month of age to evaluate for brain or eye abnormalities associated with congenital ZIKV infection. Referrals to appropriate specialists (e.g., infectious diseases, clinical genetics, neurology, ophthalmology) should be made depending on the clinical findings. Infants without clinical findings consistent with congenital ZIKV infection born to mothers with laboratory evidence of possible ZIKV infection during pregnancy should also undergo a head ultrasound and a comprehensive ophthalmologic exam, as some findings associated with congenital ZIKV infection might not be clinically apparent at birth. Long-term follow-up is important for any children born to mothers with possible ZIKV infection during pregnancy, regardless of findings at birth, to ensure early identification of any adverse neurodevelopmental outcomes and intervention for them. Infants without clinical findings consistent with congenital ZIKV infection whose mothers had possible ZIKV exposure in pregnancy but without laboratory evidence of infection do not require laboratory testing or additional evaluation beyond the standard of care for all newborns.

## PREVENTION AND CONTROL

Although the risk for travel-associated ZIKV disease has decreased compared with the risk when ZIKV was spreading and causing large outbreaks in the Americas, all persons should continue to take precautions to reduce risk for infection. Because of the greater potential implications of infection for pregnant women, they should not travel to an area with a ZIKV outbreak. If they are planning on traveling to an area with risk of transmission (based on past or current reports of ZIKV transmission), pregnant women should discuss their travel plans and the potential risks of ZIKV disease and other infectious diseases with a healthcare provider. Couples planning to conceive and to travel to an area with an outbreak or risk of transmission should also discuss plans with their healthcare providers and consider waiting to get pregnant for 2 months (if only the female partner travels) and 3 months (if the male partner travels).

Recommendations to prevent mosquito-borne transmission of ZIKV are the same as those for other *Aedes*-transmitted diseases such as dengue and chikungunya. As *A. aegypti* is an aggressive daytime biter, risk of infection is highest during the day, and the mosquitoes bite both indoors and outdoors. Measures to reduce risk of bites include wearing long sleeves and pants, using permethrin-treated clothing and gear, staying in accommodations with air conditioning or screens, and using insect repellants containing ingredients documented to be effective (e.g., N,N-diethyl-*m*-toluamide [DEET], picaridin, IR3535, oil of lemon eucalyptus, para-menthane-diol). Recommendations to prevent sexual transmission of ZIKV include using condoms or abstaining from sex during travel to an area with risk of transmission and for 3 months (if male traveler) or 2 months (if female traveler) after travel; if the partner of a pregnant woman travels, these measures should be for the duration of pregnancy. Travelers returning from an area with ZIKV transmission to an area without transmission but with an appropriate ZIKV vector should protect themselves from mosquito bites for 3 weeks to minimize the potential for spread of virus to the mosquito population and subsequent risk for local transmission of ZIKV.

In healthcare settings, workers should follow standard precautions when handling body fluids from patients with ZIKV infection. No specific additional precautions are needed. To reduce the risk for transfusion transmission of ZIKV, all blood donations in the United States are screened by ZIKV NAATs (or if platelets and plasma, can be subject to pathogen reduction technology). In addition, the US Food and Drug Administration (FDA) recommends blood donors with recent ZIKV infection be deferred for 120 days. Similarly, to reduce any risk of ZIKV transmission by human cells, tissues, or cellular and tissue-based products, the FDA has developed guidance for donations of these products. Living donors should be considered ineligible to donate if they have had within the past 6 months a diagnosis of ZIKV infection, travel to an area with risk for ZIKV transmission, or have had sex with anyone with either of these risk factors. Additionally, birth mothers are ineligible to donate umbilical cord blood, placenta, or other gestational tissues if they have had any of these risk factors during the pregnancy. Cadaveric donors are ineligible if there is history of ZIKV infection within the past 6 months.

## FUTURE DIRECTIONS

Many questions about noncongenital and congenital ZIKV infection and its treatment and prevention remain unanswered, and data from research studies and surveillance continue to be gathered. Work is also in progress to investigate improved diagnostic assays to strengthen the ability to accurately diagnose infections. Given the limitations of prevention and control measures, there has been substantial interest in

ZIKV vaccines, and at least nine candidate vaccines have been evaluated in phase I or II clinical trials. However, the currently low ZIKV transmission rates pose an obstacle for evaluating vaccine effectiveness.

## EVIDENCE

Counotte MJ, Meili KW, Taghavi K, et al. Zika virus infection as a cause of congenital brain abnormalities and Guillain-Barré syndrome: a living systematic review. *F1000Res* 2019;8: 1433. *Provides evidence for a causal relationship between ZIKV infection and adverse congenital outcomes and Guillain-Barré syndrome based on a review of up-to-date literature as of July 2019.*

de Araujo TVB, Ximenes RA, Miranda-Filho DB, et al. Association between microcephaly, Zika virus infection, and other risk factors in Brazil: final report of a case-control study. *Lancet Infect Dis* 2018;18:328-336. *Case control study investigating the association between microcephaly and congenital ZIKV infection.*

Duffy MR, Chen T, Hancock WT. Zika virus outbreak on Yap Island, Federated States of Micronesia. *N Engl J Med* 2009;360:2536-2543. *Documents the first large outbreak of ZIKV disease, describes common symptoms, and estimates frequency of asymptomatic infection.*

Krow-Lucal ER, Andrade MR, Cananea JNA, et al. Association and birth prevalence of microcephaly attributable to Zika virus infection among infants in Paraíba, Brazil, in 2015-16: a case-control study. *Lancet Child Adolesc Health* 2018;2:205-213. *Retrospective case-control investigation to assess the association of microcephaly and ZIKV.*

Melo A, Gama GL, Da Silva Junior RA, et al. Motor function in children with congenital Zika syndrome. *Dev Med Child Neurol* 2020;62(2):221-226, 2020. *Describes findings of evaluations of gross motor function among children with congenital Zika syndrome.*

Moore CA, Staples JE, Dobyns WB, et al. Characterizing the pattern of anomalies in congenital Zika syndrome for pediatric clinicians. *JAMA Pediatr* 2017;171:288-295. *Describes the distinct pattern of anomalies associated with congenital Zika syndrome.*

Rice ME, Galang RR, Roth NM, et al. Vital signs: Zika-associated birth defects and neurodevelopmental abnormalities possibly associated with congenital Zika virus infection—U.S. Territories and Freely Associated States, 2018. *MMWR Morb Mortal Wkly Rep* 2018;67:858-867. *Presents findings from children born in US territories and freely associated states to mothers with laboratory evidence of possible ZIKV infection during pregnancy who had follow-up care reported.*

Russell K, Hills SL, Oster A, et al. Male-to-female sexual transmission of Zika virus—United States, January–April 2016. *Clin Infect Dis* 2017;64(2):211-213. *Describes features of nine cases of male-to-female sexual transmission of ZIKV including information on timing of exposure and results of semen testing for two male travelers.*

## ADDITIONAL RESOURCES

Adebanjo T, Godfred-Cato S, Viens L, et al. Update: interim guidance for the diagnosis, evaluation, and management of infants with possible congenital Zika virus infection—United States, October 2017. *MMWR Morb Mortal Wkly Rep* 2017;66:1089-99. *US Centers for Disease Control and Prevention guidance for clinicians caring for infants with possible congenital ZIKV infection (up to date as of February 2020).*

Gregory CJ, Oduyebo T, Brault AC, et al. Modes of transmission of Zika virus. *J Infect Dis* 2017;216(Suppl 10):S875-S883. *A review of vector and non-vector-borne modes of transmission of ZIKV and interventions to reduce the risk of human infection through these routes.*

Hills SL, Marc F, Petersen LR. Epidemiology of Zika virus infection. *J Infect Dis* 2017;216(Suppl 10):S868-S874. *A review of the global epidemiology of ZIKV infection.*

Muñoz LS, Parra B, Pardo CA. Neurological implications of Zika virus infection in adults. *J Infect Dis* 2017;216(Suppl 10):S897-S905. *A review summarizing ZIKV-associated neurological complications in adults.*

Musso D, Ko AI, Baud D. Zika virus infection—after the pandemic. *N Engl J Med* 2019;381:15. *A review of ZIKV and ZIKV infection including information on epidemiology, transmission, disease presentations, diagnostic approaches, and clinical management.*

Sharp TM, Fischer M, Muñoz-Jordán JL, et al. Dengue and Zika virus diagnostic testing for patients with a clinically compatible illness and risk for infection with both viruses. *MMWR Recomm Rep* 2019;68:1-10. *US Centers for Disease Control guidance on Zika and dengue virus diagnostic testing for patients with a clinically compatible illness who recently traveled to or live in area with risk for transmission of these viruses, considering the current epidemiology of both viruses.*

## WEBSITES WITH USEFUL INFORMATION ON ZIKA VIRUS

General information: https://www.cdc.gov/zika/; https://www.who.int/emergencies/diseases/zika/en/; https://www.ecdc.europa.eu/en/zika-virus-disease.

Managing occupational exposures of healthcare personnel: https://www.cdc.gov/zika/hc-providers/infection-control/managing-occupational-exposures.html.

Pregnancy: https://www.cdc.gov/pregnancy/zika/index.html.

Prevention of exposure in healthcare settings: https://www.cdc.gov/zika/hc-providers/infection-control.html.

Prevention of sexual transmission: https://www.cdc.gov/zika/hc-providers/clinical-guidance/sexualtransmission.html; https://www.who.int/reproductivehealth/zika/prevention-guidelines-sexual-transmission-summary/en/.

Travel information and transmission map: https://wwwnc.cdc.gov/travel/page/zika-information.

U.S. Food and Drug Administration guidance on donor screening recommendations to reduce the risk of transmission of Zika virus by human cells, tissues, and cellular and tissue-based products: https://www.fda.gov/media/96528/download.

U.S. Food and Drug Administration guidance on measures to reduce the risk for Zika virus transmission by blood and blood components: https://www.fda.gov/media/99797/download.

# 93

# Neuroangiostrongyliasis (Rat Lungworm Disease)

*Vernon Ansdell*

## ABSTRACT

Neuroangiostrongyliasis (NAS), or rat lungworm disease (RLWD), is an important emerging infection spreading around the world. It is caused by *Angiostrongylus cantonensis,* a nematode parasite, and is the commonest parasitic cause of eosinophilic meningitis. Infection can also cause cranial nerve (CN) palsies, encephalitis, and radiculomyelitis. Another form of the disease, ocular angiostrongyliasis, is very rare but occurs when a larval worm invades the eye.

Initial symptoms are nonspecific, and an early diagnosis requires a very high index of suspicion. Somewhat later in the course of the illness there may be more specific symptoms, but even then they can be very unusual and confusing, and the diagnosis still requires a high index of suspicion. Many infections result in a benign illness with recovery within a few weeks following symptomatic and supportive care. It has become increasingly recognized, however, that some infections, particularly in certain parts of the world, may result in more severe illness with long-term neurologic deficits, significant disability, and death.

A presumptive diagnosis is made by finding evidence of an eosinophilic meningitis in a patient with a history of suggestive symptoms who is living in or has traveled to an endemic region. The diagnosis can be confirmed by finding larvae in the cerebrospinal fluid (CSF) or the eye, but this is rare. It can also be confirmed by real-time polymerase chain reaction (RT-PCR) testing for *A. cantonensis* DNA in the CSF. This important test became available recently but is currently available only in the United States and a few other countries. If PCR testing is available, antibody testing is not usually recommended for diagnosis in an acute illness. Except in very mild cases, treatment with high-dose corticosteroids is recommended for almost all cases. Treatment with anthelminthics has been controversial for decades because of theoretical concerns that it could provoke a severe inflammatory reaction in the brain or spinal cord. Increasing experience, however, suggests that early treatment (ideally within 2–3 weeks of infection) with anthelminthics such as albendazole combined with high-dose corticosteroids is safe and effective. It remains to be seen, however, if this approach will reduce the incidence of chronic neurologic sequelae, long-term disability, and death in more severe cases.

Prevention involves adequately cooking intermediate hosts such as slugs and snails or transport hosts such as freshwater shrimps, freshwater crabs, frogs, and monitor lizards. Green leafy vegetables and fruits should be thoroughly washed to remove any contaminating slugs, snails, or mucus.

## CLINICAL VIGNETTE

A 42-year-old organic farmer presented with a 2-day history of generalized headache. It was unusually severe, unlike any previous headaches he had had, and there was no associated fever. Past medical history was unremarkable. He lived on Hawaii Island in Hawaii and had no history of recent foreign travel. Physical examination by his primary care physician (PCP) was unremarkable, and he was diagnosed with tension-type headache. Ibuprofen and baclofen were prescribed.

Two days later he presented at a nearby urgent care clinic. His headache was worse, and he had developed shooting "electric shock" pains in his left arm and numbness in his right leg.

Physical examination was again unremarkable. His complete blood count (CBC) showed white blood cells (WBCs) 7800/mL, neutrophils 67%, lymphocytes 31%, and eosinophils 2%. A diagnosis of tension-type headache with anxiety was made and alprazolam was added to his drug regimen. Over the next 3 days his headache persisted, and the left arm pains became worse. In addition, he developed numbness in the left leg, and a sensation of "a thousand paper cuts" on the skin of his right leg.

Frustrated by worsening symptoms and lack of a specific diagnosis, he went to a nearby emergency room (ER). He was afebrile. Mild neck stiffness was noted but examination was otherwise unremarkable. His CBC showed WBCs 11,210/mL with 6% eosinophils. Computed tomography (CT) of the head was normal. He was told that he probably had stress headaches with anxiety, urged to follow up with his PCP, and referred to an outpatient neurologist.

By the following day the electric shock pains had become intolerable and were now described as like "ice picks." In addition, he had developed nausea, night sweats, and insomnia. He sought care in another ER. A friend had told him about RLWD, and he refused to leave the ER until he had had a lumbar puncture (LP). This was done and showed slightly turbid CSF. The opening pressure was 240 mm $H_2O$ (normal 60–200 mm $H_2O$). The CSF WBC count was 240/mL with 27% eosinophils. CSF protein was 49 mg/dL (normal 28–38 mg/dL).

Further history revealed that the patient often ate food directly from his farm and did not routinely examine or wash green leafy vegetables before consuming them. There was no history of intentionally eating raw slugs, snails, shrimp, or other unusual uncooked foods. He recollected that about 7 days prior to the onset of his headache he had experienced a few days of malaise with transient upper abdominal pain, nausea, and diarrhea associated with a cough and sore throat. Following his LP, a diagnosis of eosinophilic meningitis was made, with a presumptive diagnosis of NAS. CSF was sent for RT-PCR testing for *A. cantonensis* DNA. He was admitted to hospital and promptly started on treatment with high-dose corticosteroids and albendazole. The diagnosis of NAS was confirmed 3 days later with a positive PCR result.

COMMENT: There was no clear exposure history, but this patient may have been infected after unintentionally swallowing a slug hidden in unwashed salad vegetables from his farm. Symptoms prior to the onset of his headache may have been caused by migrating parasites on their way to the brain and spinal cord. He had a confusing array of very unusual migratory symptoms that led to a delay in diagnosis, fragmented care, and his resulting frustration. There were certain "red flags" that were not immediately recognized. These included an unprecedented severe headache and significant eosinophilia on his first ER visit; these could have prompted a more focused workup at an earlier stage of his illness. A detailed food history would have revealed that he had eaten unwashed salad vegetables directly from his farm, a significant risk factor for NAS. In addition, questioning regarding a prodromal illness might have prompted a more focused workup and resulted in an earlier diagnosis.

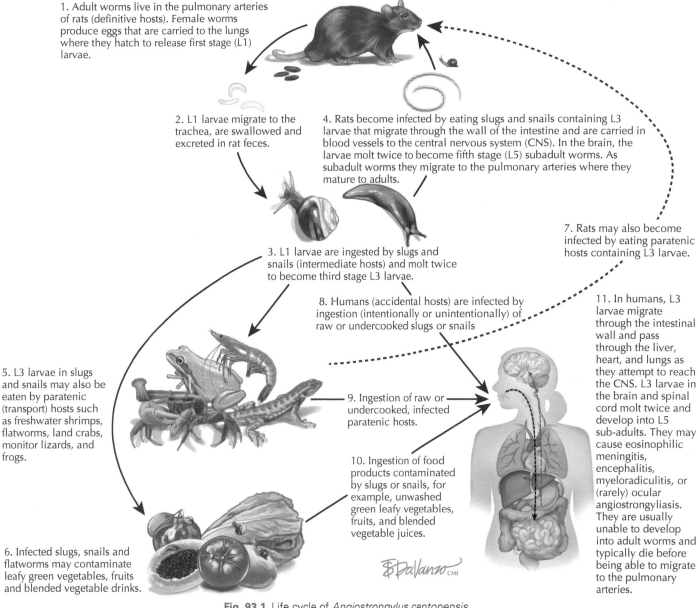

**1. Adult worms live in the pulmonary arteries of rats (definitive hosts). Female worms produce eggs that are carried to the lungs where they hatch to release first stage (L1) larvae.**

**2. L1 larvae migrate to the trachea, are swallowed and excreted in rat feces.**

**4. Rats become infected by eating slugs and snails containing L3 larvae that migrate through the wall of the intestine and are carried in blood vessels to the central nervous system (CNS). In the brain, the larvae molt twice to become fifth stage (L5) subadult worms. As subadult worms they migrate to the pulmonary arteries where they mature to adults.**

**7. Rats may also become infected by eating paratenic hosts containing L3 larvae.**

**3. L1 larvae are ingested by slugs and snails (intermediate hosts) and molt twice to become third stage L3 larvae.**

**8. Humans (accidental hosts) are infected by ingestion (intentionally or unintentionally) of raw or undercooked slugs or snails**

**11. In humans, L3 larvae migrate through the intestinal wall and pass through the liver, heart, and lungs as they attempt to reach the CNS. L3 larvae in the brain and spinal cord molt twice and develop into L5 sub-adults. They may cause eosinophilic meningitis, encephalitis, myeloradiculitis, or (rarely) ocular angiostrongyliasis. They are usually unable to develop into adult worms and typically die before being able to migrate to the pulmonary arteries.**

**5. L3 larvae in slugs and snails may also be eaten by paratenic (transport) hosts such as freshwater shrimps, flatworms, land crabs, monitor lizards, and frogs.**

**9. Ingestion of raw or undercooked, infected paratenic hosts.**

**10. Ingestion of food products contaminated by slugs or snails, for example, unwashed green leafy vegetables, fruits, and blended vegetable juices.**

**6. Infected slugs, snails and flatworms may contaminate leafy green vegetables, fruits and blended vegetable drinks.**

**Fig. 93.1** Life cycle of *Angiostrongylus cantonensis*.

## ETIOLOGY

NAS, or RLWD, is the result of infection by *A. cantonensis*, a nematode parasite. It is an important emerging disease that is spreading around the world. The natural life cycle of the parasite involves transmission between rats (definitive hosts), snails, and slugs (intermediate hosts). Humans are accidental hosts. Transport or paratenic hosts such as freshwater shrimps, flatworms, frogs, monitor lizards, land crabs, and centipedes may also be important in the life cycle (Fig. 93.1).

Humans may become infected in a variety of ways. For example: (1) Eating raw or undercooked intermediate or transport hosts in exotic foods such as "koi-hoi" (snails) in Thailand or Laos (Fig. 93.2) or eating raw shrimp dishes such as "tairo" in French Polynesia or "dancing" (live) shrimp in Thailand; (2) Swallowing raw snails or slugs as a wager or dare; (3) small children or the mentally challenged playing with and possibly swallowing slugs or snails; or (4) accidentally ingesting raw intermediate hosts such as small slugs or snails or paratenic hosts such as flatworms contaminating green leafy vegetables, fruits, or blended vegetable juices. The role of mucus containing infective larvae as a source of infection remains unclear.

## EPIDEMIOLOGY

*A. cantonensis* infection was first identified in rats in Canton (Guangzhou), China, in 1935 and the first case in humans was diagnosed in Taiwan in 1944 during World War II (WWII). In the chaotic conditions at the end of WWII and immediately afterwards there were large outbreaks of eosinophilic meningitis, subsequently recognized to have been caused by *A. cantonensis* infection in several Pacific islands.

Following WWII there was a significant increase in commercial shipping, with many of the ships carrying stowaway rats; this resulted in spread of the disease to other countries in the Pacific and South and Southeast Asia. Over the past 60 years the disease has spread beyond the traditional endemic areas to many of the Caribbean Islands, the southeastern United States, and countries such as Brazil and Ecuador in South America and African countries such as Egypt, Nigeria, Ivory Coast and South Africa. It has been identified in animals in the Canary Islands and Mallorca (Balearic Islands), suggesting that the disease may soon be seen in humans in areas such as southern Europe.

NAS is an important emerging disease. Over 2900 cases have been reported worldwide, but undoubtedly this is a significant underestimate,

**Fig. 93.2** Etiology includes eating raw or undercooked intermediate or transport hosts in exotic foods such as "koi-hoi" (snails).

because many cases are unrecognized or underreported. It is very likely that well over 10,000 cases have occurred worldwide since the disease was first identified in humans in 1944. Large outbreaks have been reported from China since 1997. There is a particularly high incidence of disease in northern and northeastern Thailand, where many cases are the result of eating raw or undercooked snails in dishes such as "koi-hoi" or raw or undercooked shrimp in dishes such as "dancing (live) shrimp." Typically both dishes are consumed with a variety of spices and local alcoholic drinks, none of which adequately prevent infection (see Fig. 93.2).

Increasing numbers of cases are being recognized in travelers from nonendemic areas to endemic areas. These cases are expected to increase as a result of factors such as worldwide spread of the disease, globalization of trade, climate change, increased travel to exotic areas, ignorance of the disease, and more adventurous eating habits in travelers (Box 93.1).

## CLINICAL PRESENTATION

Following infection, the earliest symptoms are very nonspecific. They result from migrating, neurotropic, third-stage (L3) larvae as they travel from the small intestine through the liver, heart, and lungs to the brain and spinal cord. Symptoms during this prodromal phase may include abdominal pain, nausea, vomiting, anorexia, diarrhea, malaise, myalgias, low-grade fever, headache, cough, sore throat, and dyspnea. Larvae in the skin may produce a rash and pruritus. Larvae

---

**BOX 93.1  Red Flags to Alert Clinicians to a Diagnosis of Neuroangiostrongyliasis**

- Unusual, confusing, occasionally migratory neurologic symptoms and signs.
- Allodynia with descriptions of pain such as "a thousand paper cuts," "severe sunburn" (without sun exposure), "lightning or electric shock sensations."
- New, unusually severe headache with or without peripheral blood eosinophilia.
- Meningitis and peripheral eosinophilia.
- History of travel to an endemic area.
- History of residence in an endemic area.
- History of ingestion of raw or undercooked slugs or snails.
- History of ingestion of raw or undercooked transport (paratenic) hosts (e.g., shrimps, land crabs, monitor lizards, flatworms, frogs, centipedes).
- History of ingestion of unwashed green leafy vegetables or fruits grown in an endemic area.
- History of an otherwise unexplained prodromal illness.

---

in the kidneys may cause microscopic hematuria. Unfortunately, as these are nonspecific signs and symptoms, it is very difficult to make a diagnosis of NAS during this phase of the illness.

Once the L3 larvae reach the central nervous system, more specific symptoms may appear. The most common manifestation of NAS is eosinophilic meningitis (EOM) or meningoencephalitis. In addition, there may be radiculomyelitis or CN involvement. Ocular angiostrongyliasis is very rare but occurs when a larva enters the eye.

The most common symptoms of EOM in adults are headache, neck stiffness, nausea, and vomiting. Headache is almost always present. It often comes on suddenly and can be severe. Significant fever is uncommon at this stage but may occur if there is severe encephalitis. Insomnia, night sweats, anorexia, and weight loss may also be present. If there is radiculomyelitis, there may be severe disabling nerve pain, patchy sensory loss, and dysesthesias sometimes described as "severe sunburn," "ice picks," "electric shocks," and a "thousand paper cuts." Crawling or burning sensations are also common. Photophobia, phonophobia, and heightened skin sensitivity (allodynia) may occur. Examples include the touch of clothing on the skin, which may be unbearable, or the breeze felt while driving with a car window rolled down. Muscle weakness and difficulty walking are common, and there may be bladder or bowel dysfunction. Focal neurologic findings that are migratory or often do not follow a dermatomal distribution are often present. Symptoms and signs may fluctuate and change over time, creating confusion and delaying the diagnosis.

Physical examination may reveal neck stiffness, papilledema, and impaired consciousness. CN involvement causes diplopia (CN VI), facial weakness (CN VII), tinnitus, and hearing loss (CN VIII). Deep tendon reflexes may be normal, increased, or decreased depending on the area involved in the brain or spinal cord.

Children tend to have less paresthesias but more fever, irritability, somnolence, lethargy, gastrointestinal symptoms (vomiting, anorexia, and abdominal pain), seizures, and limb weakness.

## DIAGNOSTIC APPROACH

NAS often presents with nonspecific, unusual, or bizarre symptoms. As a result, clinicians must have a high index of suspicion to make the diagnosis early in the course of the illness (see Box 93.1) Whenever possible, the goal is to make a *presumptive* diagnosis and start treatment with high-dose corticosteroids and albendazole within 2 weeks of infection.

### Lumbar Puncture

LP with measurement of the opening pressure and examination of CSF for eosinophils is an essential part of the evaluation of NAS. It is a low-risk procedure with additional therapeutic benefits including relief of headache (Fig. 93.3).

EOM is the hallmark of the disease. Traditionally it is defined as the presence of 10 or more eosinophils per milliliter of CSF or eosinophils accounting for 10% or greater of CSF WBCs or when there are at least 6 total WBCs/mL in the CSF. It is very important to make sure that the laboratory is staining the CSF for eosinophils. NAS is the commonest parasitic cause of EOM, but several other diseases may be responsible (Box 93.2). Eosinophil counts may be absent or low in the CSF in the early stages of disease. If the diagnosis of NAS remains a concern, then repeat examination within a few days may be indicated. If the diagnosis is strongly suspected, it may be reasonable to start treatment immediately with high-dose corticosteroids and albendazole.

A *presumptive* diagnosis of NAS usually requires all three of the following:
1. Characteristic symptoms and signs
2. CSF examination showing evidence of EOM
3. Exposure history, which may include residence in or travel to an endemic area

### Testing by Real-Time Polymerase Chain Reaction

The diagnosis is *confirmed* by finding *A. cantonensis* larvae in the CSF or in the eye, but this is rare. Larvae may be also be identified postmortem. The diagnosis may also be confirmed by detecting *A. cantonensis* DNA in the CSF via RT-PCR. This is a valuable, relatively new test, which may not be positive in the early stages of infection. CSF may therefore have to be reexamined in suspected cases. In that situation, treatment with high-dose corticosteroids and albendazole may be started without waiting for confirmation of the diagnosis. In countries where real-time PCR for *A. cantonensis* DNA in the CSF is available, it has become the test of choice for diagnosing NAS in the acute illness.

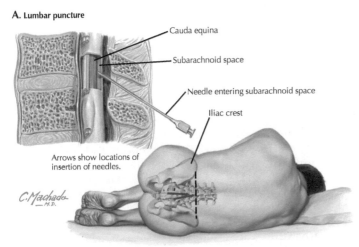
**A. Lumbar puncture**
Cauda equina
Subarachnoid space
Needle entering subarachnoid space
Iliac crest
Arrows show locations of insertion of needles.
C. Machado M.D.

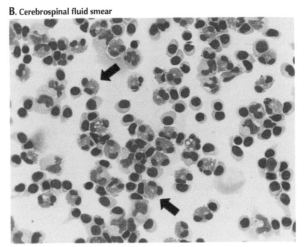
**B. Cerebrospinal fluid smear**

**Fig. 93.3** (A) Lumbar puncture with manometry to record the opening pressure. (B) Stained cerebrospinal fluid smear showing eosinophils in a patient with eosinophilic meningitis due to *Angiostrongylus cantonensis*. *Black arrows* indicate eosinophilic granulocytes with mostly bilobed, occasionally trilobed nuclei and normal eosinophilic granulation. (B, Reused with permission from Brummaier T, Bertschy S, Arn K, et al. A blind passenger: a rare case of documented seroconversion in an *Angiostrongylus cantonensis* induced eosinophilic meningitis in a traveler visiting friends and relatives. *Trop Dis Travel Med Vaccines* 2019;5:6.)

## BOX 93.2    Differential Diagnosis of Eosinophilic Meningitis

**Important Parasitic Infections**
- *Angiostrongylus cantonensis*
- *Gnathostoma spinigerum*
- *Baylisascaris procyonis*
- *Toxocara cani*
- Cysticercosis *(Taenia solium)*
- *Paragonimus* spp.
- Schistosomiasis *(Schistosoma japonicum)*
- Fascoliasis

**Important Nonparasitic Infections**
- Coccidiomycosis
- Cryptococcosis

**Important Noninfectious Causes**
- Hodgkin lymphoma
- Eosinophilic leukemia
- Intrathecal contrast material
- Ventriculoperitoneal shunts
- Idiopathic hypereosinophilic syndromes
- Drugs: nonsteroidal antiinflammatory drugs, ibuprofen, ciprofloxacin, trimethoprim-sulfamethoxazole, intraventricular gentamicin or vancomycin

## Eosinophils in the Blood

Elevated peripheral blood eosinophil counts greater than 500/mL are usually present at some point during the course of the illness.

## Imaging Studies

Magnetic resonance imaging (MRI) studies of the brain are valuable in the evaluation of patients with suspected NAS. They often help to exclude other causes for the patient's illness, and, unlike CT, there are several features that are suggestive albeit not diagnostic of NAS. These findings include leptomeningeal enhancement in contrast studies; increased signal intensity in the subcortical white matter on T2-weighted and FLAIR images; and nodular, linear, or hockey stick–like small lesions in white matter on gadolinium-enhanced T1 images. Characteristic worm tracks and microhemorrhagic lesions may be seen in countries where susceptibility weighted imaging (SWI) or gradient echo (GRE) is available. Brain MRI may be normal during the first 3 weeks of illness, but more sophisticated technology such as three-dimensional MRI enables an earlier diagnosis. In patients with myeloradicular signs and symptoms, focused MRI of the spine may also provide valuable information (Figs. 93.4 and 93.5).

## TREATMENT

### Role of Lumbar Puncture

LP is very important in the diagnosis of NAS, but it also has important therapeutic benefits. Removal of CSF reduces the intracranial pressure responsible for the severe headache and vomiting. In patients with persistent severe headache, repeat LP has been shown to provide significant short-term relief.

### High-Dose Corticosteroids

High-dose corticosteroids such as prednisolone or prednisone (e.g., 60 mg/day in divided doses in adults) or dexamethasone (e.g., 10 to 20 mg/day in divided doses in adults) have been shown to significantly improve specific clinical outcomes, as in the case of headache. It remains to be seen whether they can reduce other important outcomes such as long-term disability or death. Well-designed randomized double-blind placebo-controlled trials are urgently needed. High-dose corticosteroids are usually given for at least 2 weeks but may need to be tapered slowly over several weeks or even months to prevent recurrence of symptoms. In some situations, if there are recurrent symptoms, corticosteroids may need to be restarted.

One well-designed, double-blind, randomized, controlled trial from Thailand showed a statistically significant reduction in the duration of headache, analgesic use, and need for repeat LP in patients who were given high-dose corticosteroids.

Many experts give high-dose corticosteroids in almost all moderate and severe cases, recommending that they should be started as soon as a *presumptive* diagnosis has been made.

### Anthelminthics

The role of anthelminthics, such as albendazole, in the treatment of NAS has been controversial for decades because of theoretical concerns that the large-scale death of larvae could produce a severe inflammatory response in the brain and spinal cord. However, increasing clinical experience and studies in animals suggest that this concern may be unwarranted.

Key points on the treatment of NAS with albendazole are as follows:
- Albendazole is rapidly metabolized to the active metabolite albendazole sulfoxide. It crosses the blood-brain barrier and achieves high levels in the central nervous system.
- Albendazole is usually combined with high-dose corticosteroids to limit the inflammatory response to dead or dying parasites.
- Albendazole combined with high-dose corticosteroids has been used safely and apparently effectively for many years to treat humans living in endemic areas. Clinical experience from outbreaks and sporadic events in countries such as Thailand, Taiwan, mainland China, and Vietnam have been published.
- Animal studies in mice have shown albendazole to be effective and safe in the treatment of *A. cantonensis* infection. It appears that treatment must be given within the first 2 to 3 weeks after infection to be most effective.
- In the human brain and spinal cord, L3 larvae molt twice over approximately 2 to 3 weeks to become L5 immature adults. In the process they increase in volume 1000-fold. By treating patients as early as possible with an anthelminthic such as albendazole, L3 and L4 larvae are killed while they are relatively small. This could limit the damage caused during the period of larval migration through the CNS. In addition, a decreased larval worm mass could be assumed to produce less damage from an inflammatory response as they die in response to drug treatment.

The recommended dose of albendazole is 15 mg/kg/day in two divided doses taken by mouth for 2 weeks. Each dose should be taken with a fatty meal to improve absorption. Patients should be monitored carefully for clinical deterioration, including adverse drug reactions, while on treatment and after treatment is completed.

### Pain Management

Long term, a debilitating illness lasting for months or years is increasingly being recognized, particularly after severe disease. Pain management requires a multidisciplinary approach utilizing modalities such as primary care, physical therapy, occupational therapy, psychotherapy, and pain management specialists. In the acute phase of the illness, headache may be temporarily relieved by repeated LPs. Patients with persistent headache or severe disabling limb pain may not respond to simple analgesics such as acetaminophen or nonsteroidal

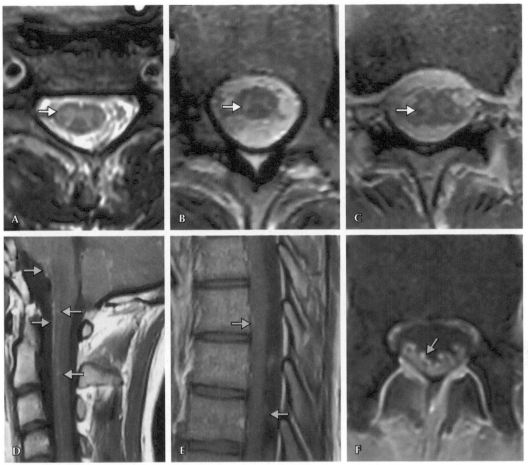

**Fig. 93.4** Magnetic resonance imaging of the spinal cord indicates meningomyelitis consistent with severe *Angiostrongylus* infection. Axial T2 sequences of spine at the cervical (A), midthoracic (B), and conus medullaris (C) levels show abnormally increased T2 signal of the central gray matter *(white arrows)*, indicating myelitis. Magnified sagittal T1 postcontrast sequences of the upper cervical and midthoracic spine highlight abnormal lepmeningeal enhancement of the cord with so-called sugar coating or zuckerguss appearance of leptomeninges *(blue arrows)* in conjunction with areas of abnormal central cord enhancement *(green arrows)* indicative of meningomyelitis (D and E). Axial T1 postcontrast of the cauda equina shows diffuse abnormal nerve root enhancement *(orange arrow;* F). (Reused with permission from McAuliffe L, Fortin Ensign S, Larson D, et al. Severe CNS angiostrongyliasis in a young marine: a case report and literature review. *Lancet Infect Dis.* 2019;19:e132-e142.)

antiinflammatory drugs; they may require short-term opioids during the acute phase of the illness. A variety of other options for symptom control (e.g., gabapentin, pregabalin, cannabinoids, ketamine, and acupuncture) appear to have limited value, although detailed studies of their use have not been done. Neuropathic pain can be very debilitating and difficult to treat.

## PREVENTION

Prevention involves adequately cooking intermediate hosts such as slugs and snails or transport hosts such as freshwater shrimps, frogs, and monitor lizards prior to ingestion or avoiding eating them altogether. Green leafy vegetables and fruits should also be thoroughly washed to remove any contaminating slugs or snails and residual mucus. Potentially contaminated beverages such as blended vegetable juices and kava should be carefully prepared to avoid any possibility of contamination.

## EVIDENCE

Chotmongkol V, Kittimongkolma S, Niwattayakul K, et al. Comparison of prednisolone plus albendazole with prednisolone alone for treatment of patients with eosinophilic meningitis. *Am J Trop Med Hyg* 2009;81:443-445. *An unblinded study that showed no significant benefit in the duration of headaches by adding albendazole to corticosteroids. Unfortunately, the albendazole group had a much longer duration of illness than the control group. which may have influenced the results.*

Chotmongkol V, Sawanyawisuth K, Thavornpitak Y, et al. Corticosteroid treatment of eosinophilic meningitis. *Clin Infect Dis* 2000;31:660-662. *An important, well-designed study that showed a 2-week course of high dose corticosteroids significantly reduced the duration of headache, analgesic use and need for repeat LPs in EOM.*

Hwang KP, Chen ER. Larvicidal effect of albendazole against Angiostrongylus cantonensis in mice. *Am J Trop Med Hyg* 1988;39:191-195. *A very important animal study that showed if albendazole is used for treatment of NAS it needs to be given in the first 2 to 3 weeks after infection to be most effective.*

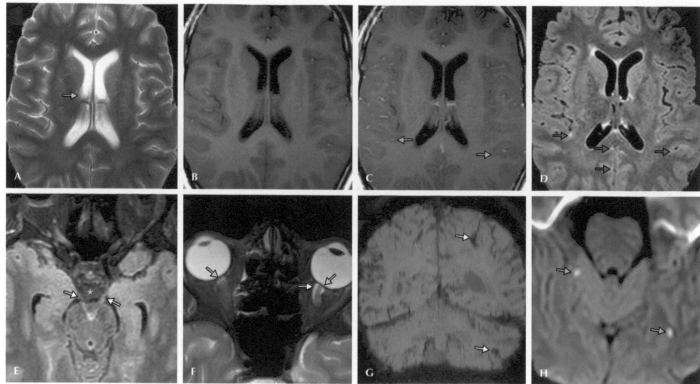

**Fig. 93.5** Magnetic resonance imaging of the brain reveals extensive disease involvement in severe angiostrongylus infection. Axial T2 image of the brain shows enlarged lateral ventricles *(green arrow)* out of proportion to the sulci (A) in conjunction with dilatation of the optic nerve sleeves *(white arrow)* and flattening of the lamina cribrosa *(green arrows;* F) implying raised intracranial pressure with resultant hydrocephalus and papilloedema. Axial T1 precontrast (B) and T1 postcontrast images (C) show areas of abnormal leptomeningeal enhancement, which are difficult to discern from normal background of pial vasculature. The highly sensitive T2 fluid-attenuated inversion recovery postcontrast sequence shows the overall extent of disease *(red arrows;* D). Magnified T2 fluid-attenuated inversion recovery postcontrast at the interpeduncular cistern shows abnormal enhancement of the oculomotor nerves *(white arrows)* as well as nonsuppression of cerebrospinal fluid signal in the suprasellar cistern *(asterisk;* E). Coronal susceptibility-weighted imaging highlights focal tract-like areas of abnormal blooming susceptibility, thought to represent hemorrhagic migratory tracks (G). Diffusion-weighted imaging shows punctate areas of restricted diffusion indicating ischemic changes, mediated by direct cytotoxic insult from the parasite or secondary to an inflammatory response to the primary infection (H). (Reused with permission from McAuliffe L, Fortin Ensign S, Larson D, et al. Severe CNS angiostrongyliasis in a young marine: a case report and literature review. *Lancet Infect Dis.* 2019;19:e132-e142.)

Jitpimolmard S, Sawanyawisuth K, Morakote N, et al. Albendazole therapy for eosinophilic meningitis caused by *Angiostrongylus cantonensis. Parasitol Res* 2007;100:1293-1296. *An important study that showed albendazole alone (without steroids) significantly reduced the duration of headache in the treatment of eosinophilic meningitis.*

Kliks MM, Kroenke K, Hardman JM. Eosinophilic radiculomyeloencephalitis: an angiostrongyliasis outbreak in American Samoa related to ingestion of Achatina fulica snails. *Am J Trop Med Hyg* 1982;31:1114-1122. *A fascinating, very detailed report of an outbreak of NAS in Korean fishermen in American Samoa.*

Lv S, Zhang Y, Chen SR, et al. Human angiostrongyliasis outbreak in Dali, China. *PLoS Negl Trop Dis* 2009;3(9):e520. http//doi.org.10.1371/journal.pntd.0000520. *A detailed report of an outbreak of NAS in 33 patients in Dali, China. The majority (26/33) of patients appeared to respond well and without significant side effects to treatment with albendazole and dexamethasone. The lead author is an expert on the global epidemiology of NAS.*

Sawanyawisuth K, Limpawattana P, Busaracome P, et al. A 1-week course of corticosteroids in the treatment of eosinophilic meningitis. *Am J Med* 2004;117:802-803. *This study assessed the effectiveness of a short, 1-week*

*course of high-dose corticosteroids. Fifteen percent of patients relapsed after steroids were discontinued and, in most cases, they needed to be restarted or further LPs were necessary to relieve headaches.*

Slom TJ, Cortese MM, Gerber SI, et al. An outbreak of eosinophilic meningitis caused by Angiostrongylus cantonensis in travelers returning from the Caribbean. *N Engl J Med* 2002;346:668-675. *One of the earliest reports of an outbreak of NAS in travelers. It highlighted the importance of NAS in travelers from nonendemic to endemic areas.*

## ADDITIONAL RESOURCES

Ansdell V, Kramer KJ, McMillan JK, et al. Guidelines for the diagnosis and treatment of neuroangiostrongyliasis: updated recommendations. *Parasitology* 2020;1–7. *Recently updated, evidence-based guidelines for the diagnosis and treatment of NAS.*

Ansdell VE, Wattanagoon Y. Angiostrongylus cantonensis in travelers: clinical manifestations, diagnosis, and treatment. *Curr Opin Infect Dis* 2018;31:399-408. *A detailed review of NAS in travelers. For many reasons, the number of cases in travelers is expected to increase in the future.*

Barratt J, Chan D, Sandaradura I, et al. Angiostrongylus cantonensis: a review of its distribution, molecular biology and clinical significance as a human pathogen. *Parasitology* 2016;143:1087-1118. *A comprehensive review of angiostrongyliasis. Particularly valuable sections on epidemiology, global distribution, pathophysiology, diagnosis, and treatment. Very well referenced.*

Eamsobhana P. Eosinophilic meningitis caused by Angiostrongylus cantonensis—a neglected disease with escalating importance. *Trop Biomed* 2014;31:569-578. *A comprehensive review of NAS from one of the most knowledgeable experts in the world on the disease.*

Johnston D, Marlena C, Dixon M, Elm J, et al. Review of Cases of Angiostrongyliasis in Hawaii, 2007–2017. *Am J Trop Med Hyg* 2019;101(3):608-616. *A detailed review of 82 cases of NAS diagnosed in Hawaii from 2007 to 2017.*

Martins YC, Tanowitz HB, Kazacos KR. Central nervous system manifestations of Angiostrongylus cantonensis infection. *Acta Tropica* 2015;141:46-53. *A detailed review of NAS with focus on the important neurologic features of the disease.*

Murphy GS, Johnson S. Clinical aspects of eosinophilic meningitis and meningoencephalitis caused by Angiostrongylus cantonensis, the rat lungworm. *Hawaii J Med Public Health* 2013;72(6 Suppl 2):35-40. *An important article that focuses on diagnosis and treatment and makes suggestions for further research.*

Prociv P, Turner M. Neuroangiostrongyliasis: the subarachnoid phase and its implications for anthelminthic therapy. *Am J Trop Med Hyg* 2018;98: 353-359. *An interesting article that argues early treatment with high-dose corticosteroids and anthelminthics may result in better outcomes. The authors emphasize the challenges involved in making the diagnosis early in the illness.*

Qvarnstrom Y, Xayavong M, da Silva ACA, et al. Real-time polymerase chain reaction detection of Angiostrongylus cantonensis DNA in cerebrospinal fluid from patients with eosinophilic meningitis. *Am J Trop Med Hyg* 2016; 94:176-181. *This review evaluated RT-PCR assay for the detection of A. cantonensis DNA in human CSF. The test may be particularly useful in the acute phase of the illness when antibody detection may give negative results.*

Sears WJ, Qvarnstrom Y, Dahlstrom E, et al. AcanR3990 qPCR: a novel, highly sensitive, bioinformatically-informed assay to detect Angiostrongylus cantonensis infections. *Clin Infect Dis* 2020;ciaa1791. *Important information on a new highly sensitive and specific assay for the diagnosis of NAS. It may enable diagnosis of NAS at an earlier stage in the illness and result in earlier treatment.*

Wang QP, Lai DH, Zhu XQ, et al. Human angiostrongyliasis. *Lancet Infect Dis* 2008;8:621-630. *A comprehensive review of NAS.*

Wang QP, Wu ZD, Wei J, et al. Human Angiostrongylus cantonensis: an update. *Eur J Clin Microbiol Infect Dis* 2012;31:389-395. *An updated review of NAS.*

# 94

# Ebola and Other Emerging Viral Hemorrhagic Fevers

*Elizabeth R. Schnaubelt, David M. Brett-Major*

## EBOLA VIRUS DISEASE

### Etiology and Transmission

The term *Ebola virus disease* most commonly refers to infection with Ebola Zaire virus (EBOV), the causative agent of the 2014–2016 West Africa Ebola virus disease epidemic, as well as several recent epidemics in the Democratic Republic of Congo. EBOV is one of several filoviruses, a diverse group of single-stranded negative-sense RNA viruses with a prominent linear component that, on electron microscopy, resembles a filament. Three other species of *Ebolavirus*—Sudan virus, Taï Forest virus, and Bundibugyo virus—have caused limited human outbreaks. Despite over 40 years of health emergencies resulting from the spread of Ebola viruses among humans, how the human chain of infections begins remains a matter of debate. In general, an initial zoonotic crossover event occurs; bats are thought to be involved either directly or indirectly via bats infecting livestock, or bushmeat (wild animals including primates) that are subsequently consumed by humans. Outbreaks increase in size by human-to-human transmission. This may occur explosively through contact with blood from a terminal patient who has bled, participation in burial rituals in which body washing is conducted, or exposure during healthcare delivery. In sustained outbreaks, the most common mode of human transmission is probably fecal-oral.

### Geographic Distribution and Burden

Since Belgian missionaries first identified an Ebola outbreak in Congo in 1976, the main burden of disease has been in Central Africa and, more recently, in West Africa. The 2014–2016 West Africa experience involved both the repatriation of international deployers to North America and Europe for care as well as some travel-associated outbreaks in those areas. There have been cases of laboratory-associated disease as well. As in the case of many emerging infectious diseases, communities with little infrastructure or healthcare services are the worst affected. Adding to that burden, stressed, resource-poor health systems suffer disproportionately from cases of nosocomial infection (Fig. 94.1). Sometimes a particular community practice is implicated in large transmission events. The most common association is with burial practices, when mourners from a broad geographic area may congregate and partake of ritual washings of a deceased.

### Public Health Importance

An outbreak of Ebola virus disease (EVD) may have striking impacts on health and wellness for communities that experience even a few cases. It is a disease with high mortality that engenders fear of both the consequences of the disease as well as the measures authorities may take to control it. Healthcare workers associated with early cases may be among the waves of secondary and tertiary infection from a sentinel case. This can have a compounding impact on both the availability of health services and a community's willingness to engage them. Downstream effects of this have included worse outcomes for the control of other diseases in the community where presentation for care or access to public health programming is important (e.g., HIV, tuberculosis, and malaria management; vaccine and antihelminth programming; and the management of noncommunicable disease).

### Risk Factors

Other than being in an area endemic for or experiencing an Ebola virus disease outbreak, risk factors for becoming infected are controversial. It seems prudent to avoid exploration of bat habitats as well as involvement in butchering bushmeat. However, once an outbreak is under way, activities such as caregiving for those who are ill, participating in traditional burial practices, having other close contact with ill or deceased persons, or even having repeated casual contact (as experienced among motorcycle taxi operators in Sierra Leone in 2014) all place persons at increased risk.

### Clinical Features and Manifestations

Occasionally a patient will present without fever and with nonspecific symptoms until later in the course of an illness. However,

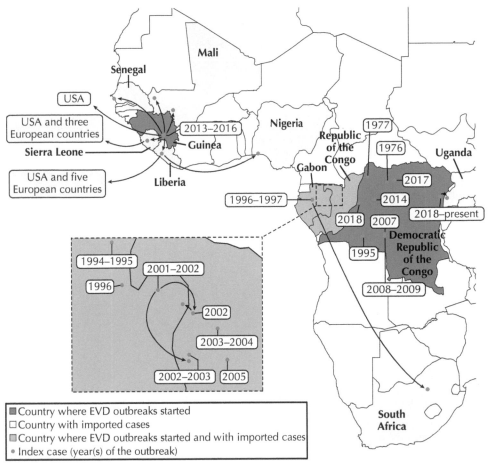

**Fig. 94.1** Ebola virus disease (EVD) outbreaks. (Reused with permission from Kuhn JH. Ebolavirus and Marburgvirus infections. In: Kasper DL, Fauci AS, Hauser SL, Longo DL, Jameson JL, Loscalzo J, eds. *Harrison's Principles of Internal Medicine.* Vol 2. 19th ed. Columbus, OH: McGraw-Hill Education; 2015:1323-1329.)

most patients experience an influenza-like illness before showing a febrile gastroenteritis that can be profound, sometimes with cholera-like volume loss of 4 to 10 L/day. Older patients with comorbidities may separately or concomitantly present with primary neurologic symptoms resembling an encephalitis. These patients have very high case fatality rates. Some patients present with a febrile arthralgia reminiscent of children recovering from meningococcal disease while having little mortality, as this syndrome may be associated with the immune response. Although it is called a viral hemorrhagic fever (VHF) and some bleeding may occur relatively early in the illness, most patients show no signs of hemorrhage or do so when they are experiencing fulminant disease just before death. Occasionally patients manifest a fine papular rash. Hiccups are common among those who are volume depleted. Secondary bacterial infection is a concern over the course of illness, as is concomitant or supervening malaria in endemic areas. Patients may manifest profound visceral pain, particularly in the mediastinal and epigastric areas (Fig. 94.2).

## Diagnostic Approach

The mainstay of diagnosis in Ebola virus disease is nucleic acid testing with real-time reverse transcription-polymerase chain reaction (RT-PCR). A variety of assays are available on a number of distributed molecular platforms, such as for tuberculosis. Patients may have false-negative results for 2 to 3 days after the onset of an influenza-like illness. Some serologic assays, including rapid tests, have been

developed, with utility for outbreak identification but to date less so for case management. Diagnostic laboratories in field settings are often not equipped for concomitant infection diagnosis or usual clinical laboratory studies for characterizing and managing a patient's sepsis. In such instances, other arrangements such as point-of-care clinical platforms may have to be considered.

## Clinical Management

As in most cases of sepsis, interventions focus both on the host and the pathogen. Host-based strategies include volume and electrolyte repletion, pain control, and consideration of concomitant medical challenges such as preexisting comorbidities and coinfection. Up to one-third of Ebola virus disease patients in malaria-endemic areas also have malaria on presentation. Bacterial superinfection also is a concern. Patients who develop pulmonary involvement signaled by hypoxia should be assessed for pneumonia and also considered for embolus and pregnancy-related complications.

A number of investigational new drugs have been assessed for therapy against EBOV, most recently in a study referred to as PALM, the Pamoja Tulinde Maisha trial, which means "Together Save Lives" in the Kiswahili language. In it, mixes of monoclonal antibody therapy performed better than earlier-generation formulations as well as remdesivir, a small-molecule candidate. However, the potential role of combination therapy and the importance of small-molecule therapy in the setting of CNS disease remains an open question. These medications are not readily available. When a potential Ebola virus disease

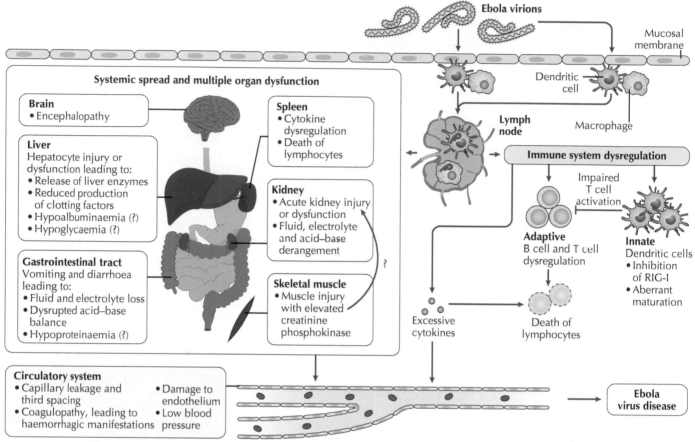

**Fig. 94.2** Conceptualized pathogenesis of Ebola virus disease. Ebola virus particles enter the body through dermal injuries (microscopic or macroscopic wounds) or via direct contact via mucosal membranes. Primary targets of infection are macrophages and dendritic cells. Infected macrophages and dendritic cells migrate to regional lymph nodes, producing progeny virions. Through suppression of intrinsic, innate, and adaptive immune responses, systemic distribution of progeny virions and infection of secondary target cells occur in almost all organs. Key organ-specific interactions occur in the gastrointestinal tract, liver, and spleen, with corresponding markers of organ injury or dysfunction that correlate with human disease outcome. The question marks indicate speculated manifestations. *RIG-I,* Antiviral innate immune response receptor RIG-I. (Reused with permission from Jacob ST, Crozier I, Fischer WA, et al. Ebola virus disease. *Nat Rev Dis Primers* 2020;6:13. https://doi.org/10.1038/s41572-020-0147-3, Figure 5.)

case is being managed, in addition to immediate and close coordination with public health authorities, national authorities for investigational drugs should also be contacted. In the United States, the dominant gatekeeper for such therapies is the National Institute for Allergy and Infectious Diseases at the National Institutes of Health.

## Prognosis

Overall case fatality rates vary depending upon the emergency and setting, from less than one in five in western critical care units among predominantly redeployed international medical personnel to anywhere from less than one-third to three-quarters or more in outbreaks among those residing in Africa. Pregnancy, older patients with cardiovascular risk factors, and very young children tend to experience very poor outcomes in outbreaks. Survivorship entails the risks of stigmatization as recovered patients seek to reenter life in home communities and be accepted. Furthermore, in some cases, survivors suffer from prolonged neurocognitive impacts, fatigue, muscle or joint pain, and a generally decreased functional status. Additionally, the virus may persist in immunoprivileged sites. Both persistent viral detection as well as reactivated systemic disease have been associated with the testes, eyes, and central nervous system.

## Prevention and Control

Vaccines now exist under both emergency use authorization and research trials for the prevention of Ebola virus disease, and there are investigational new drugs for postexposure prophylaxis as well. They must be sought through public health and regulatory authorities. In endemic areas, avoidance of bushmeat and good food handling practices as well as universal precautions in healthcare delivery and encouragement of safe burial practices all contribute to prevention of a sentinel case and amplification of disease transmission. The mainstay of control of an Ebola virus disease emergency is rapid, early, accessible case identification; isolation of the patient in conjunction with high-quality medical care; and aggressive yet respectful contact tracing and monitoring.

## OTHER EMERGING VIRAL HEMORRHAGIC FEVERS

EBOV is not the only VHF threat to patients and communities, even though it has been so common in the past few years that its data have dominated the conversations on risk management of the several other threats in this class. The viruses that cause VHFs are varied

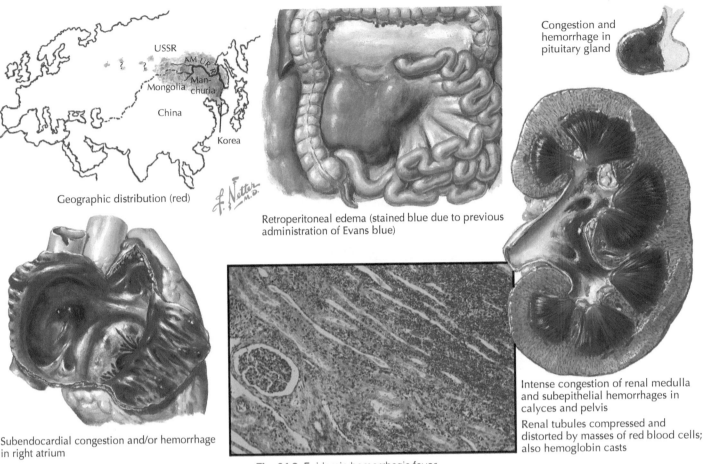

Geographic distribution (red)

Retroperitoneal edema (stained blue due to previous administration of Evans blue)

Congestion and hemorrhage in pituitary gland

Intense congestion of renal medulla and subepithelial hemorrhages in calyces and pelvis

Renal tubules compressed and distorted by masses of red blood cells; also hemoglobin casts

Subendocardial congestion and/or hemorrhage in right atrium

**Fig. 94.3** Epidemic hemorrhagic fever.

and have in common an impact on vascular integrity that may result in bleeding but most often leads to tissue injury, pathologic volume distribution, and sepsis frequently associated with high morbidity and mortality (Fig. 94.3). Diagnosis is most often made by RT-PCR or detection of virus-specific antibodies. We remain early in our understanding of how the viruses cause severe disease and the best way to prevent and mitigate their effects. Even their names—in clinical coding, viral taxonomy, and the literature—change frequently. Medical countermeasures approved by the US Food and Drug Administration (FDA), such as vaccines and therapeutics, are generally lacking. Early identification, isolation, and notification of infection control staff and health authorities of a sentinel case are critically important to limit potential outbreaks and negative public health consequences.

## Lassa

Lassa fever virus is an arenavirus endemic to West Africa, where case counts may rival the larger Ebola virus disease epidemics. The virus is transmitted by an agricultural, sometimes peridomestic pest, the multimammate rat. The range of the virus likely is increasing in Africa in concert with movement of the reservoir host (Fig. 94.4). Although aerosol exposure from rat urine used to be implicated as the primary mode of acquisition of the disease, more recent evidence suggests that food contamination from urine is a larger factor. The proximity to human habitation and prevalence of the virus among populations of this rodent results in increasing initiations of outbreaks. Lassa fever virus is thought to have a higher subclinical and mild case rate than EBOV, although the virus spreads and amplifies in people similarly by

■ Countries reporting endemic disease and substantial outbreaks of Lassa Fever
□ Countries reporting few cases, periodic isolation of virus, or serologic evidence of Lassa virus infection
□ Lassa Fever status unknown

**Fig. 94.4** Lassa fever outbreak distribution map. (Reused from Content source: Centers for Disease Control and Prevention, National Center for Emerging and Zoonotic Infectious Diseases [NCEZID], Division of High-Consequence Pathogens and Pathology [DHCPP], Viral Special Pathogens Branch [VSPB]. https://www.cdc.gov/vhf/lassa/outbreaks/index.html.)

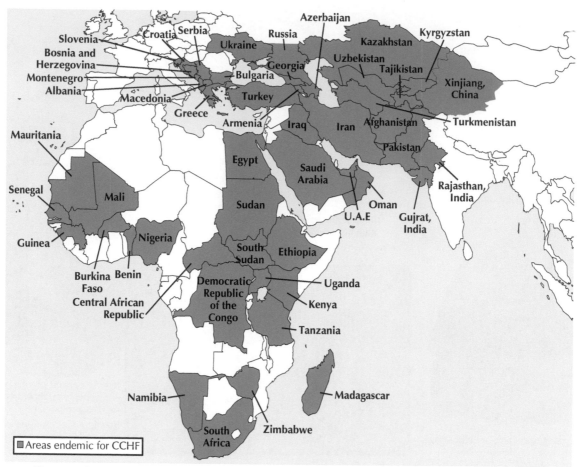

**Fig. 94.5** Crimean-Congo hemorrhagic fever distribution map. (Reused from Content source: Centers for Disease Control and Prevention, National Center for Emerging and Zoonotic Infectious Diseases [NCEZID], Division of High-Consequence Pathogens and Pathology [DHCPP], Viral Special Pathogens Branch [VSPB]. https://www.cdc.gov/vhf/crimean-congo/outbreaks/distribution-map.html.)

fecal-oral and burial practice routes. Pregnant patients are at particular risk for poor outcomes. Ribavirin is used in therapy. Recently there has been occasional concomitant use of favipiravir. At high doses, ribavirin has hematologic and renal implications, and current guidelines should be referenced when employing it. Sensorineural deafness is a common finding among survivors of Lassa fever.

## Marburg

Marburg virus also is a filovirus with a similar range to the Ebola viruses. Its routes of transmission are similar, although bat associations, particularly with outbreaks among miners, are stronger. Patient and community management approaches are similar, as well, though data on Marburg disease is less robust and medical countermeasures have not yet advanced for this disease.

## Crimean-Congo Hemorrhagic Fever

Crimean-Congo hemorrhagic fever (CCHF), a tick-borne disease, is endemic across a wide swath of Eurasia and more recently Africa (Fig. 94.5). Several serotypes are recognized implicating regions of spread. Cases often are associated with livestock which amplify the virus; the livestock are sometimes ill or die. Although cases that present to hospital may have a case fatality rate as high as two out of five, the mild and asymptomatic case rate may be high in some instances as well. Nosocomial infections have occurred from CCHF, and ribavirin has been used in both therapy and postexposure prophylaxis.

## Rift Valley Fever

Rift valley fever (RVF) is a culicine mosquito-borne threat associated with epizootic events as in the midst of sheep abortions, predominantly in East Africa (Fig. 94.6). In people, infection may occur due to exposure to body fluids of affected livestock during husbandry or bites from infected mosquitos. The attack rate among humans is not clear, although human-to-human transmission is not recognized. People do not seem to be an amplifying host for the mosquitos. Clinical features are often mild and nonspecific, with fever, myalgias, arthralgias, and headache. Up to 1 in 10 known cases may be severe, manifesting with retinitis, meningoencephalitis, or fulminant hepatitis with hemorrhage.

## New World Arenaviruses

Like Lassa fever viruses, New World arenaviruses may cause a VHF. They are associated with peridomestic or agricultural rodents. In some instances, such as for Argentine hemorrhagic fever (AHF)—for which the experience is most robust—convalescent plasma and ribavirin have been used in the management of acute cases. As in Lassa fever, there is increased interest in pursuing other drugs for management of the disease, such as favipiravir. This class of viruses appears to be on the move. Members of the group have been found in ticks in Florida, although natural transmission of disease outside of South America has not been observed. An imported case of AHF was recently reported from Brussels in a returned traveler from Argentina.

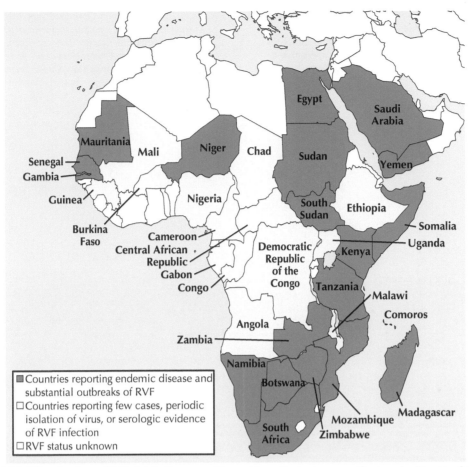

**Fig. 94.6** Rift valley fever *(RVF)* distribution map. (Reused from Content source: Centers for Disease Control and Prevention, National Center for Emerging and Zoonotic Infectious Diseases [NCEZID], Division of High-Consequence Pathogens and Pathology [DHCPP], Viral Special Pathogens Branch [VSPB]. https://www.cdc.gov/vhf/rvf/outbreaks/distribution-map.html.)

## Hantaviruses

Hantaan virus is the prototypical VHF among hantaviruses, transmitted by mice and associated with severe renal injury causing hemorrhagic fever with renal syndrome (HFRS). It is associated with the Korean peninsula, and shipping has brought it far and wide in isolated outbreaks. One such event occurred in a Maryland mall basement linked by utility conduit to Baltimore harbor. Ribavirin has been employed in its treatment. Sin Nombre virus is a hantavirus infection that has been reported in the southwestern United States; it causes a severe pulmonary syndrome known as hantavirus pulmonary syndrome (HPS), also rodent-linked but not a VHF.

## Other Viral Hemorrhagic Fevers

Elsewhere in this text, dengue (Chapter 70) and yellow fever (Chapter 64) are discussed. These too are VHFs, although with a broad impact meriting focused attention relevant to their management. The underlying principles are the same. All of the VHFs likely start as influenza-like illnesses. They progress to vascular injury with multiorgan involvement, possibly volume shifts, and in some cases more classic bleeding dysesthesias. Discerning VHFs from among the many causes of influenza-like illnesses is difficult early in the disease course. Consequently severe disease tends to be what first is recognized, hence the name it is given. Not all VHF are associated with outbreaks of human-to-human transmission. However, when the clinical syndrome is first recognized, all should trigger both public health reporting and safe and effective care, including an appropriately elevated infection prevention and control posture.

## ADDITIONAL RESOURCES

Brett-Major DM, Jacob ST, Jacquerioz FA, et al. Being ready to treat Ebola virus disease patients. *AJTMH* 2015;92(2):233-7. Available at https://www.ncbi.nlm.nih.gov/pmc/articles/PMC4347319/. *This paper addresses key operational considerations in clinical management that are implied but not always clear in guidelines for response to VHF emergencies.*

Centers for Disease Prevention and Control. Viral Hemorrhagic Fevers. https://www.cdc.gov/vhf/index.html. Accessed 10 March 2020. *The CDC website is well structured as a survey of VHF in this chapter and more.*

Escalera-Antezana JP, Rodriguez-Villena OJ, Arancibia-Alba AW, et al. Clinical features of fatal cases of Chapare virus hemorrhagic fever originating from rural La Paz, Bolivia, 2019: a cluster analysis [published online ahead of print, 2020 Feb 17]. *Travel Med Infect Dis* 2020;101589. doi:10.1016/j.tmaid.2020.101589. *Less is widely recognized about New World VHF in contrast to the broader acknowledgement of such threats in Africa. This focused paper on clinical aspects of a recent outbreak of an emerging VHF related to Lassa fever virus and Junin virus (the cause of Argentine hemorrhagic fever) provides a local context to these threats, and emphasizes the importance of universal precautions in usual IPC practice.*

National Emerging Special Pathogens Training and Education Center (NETEC). https://netec.org/. Accessed 09 July 2020. *A network of US referral isolation care facilities, NETEC provides a range of resources and consultative services in the case management and systems appropriate to management of VHF and other patients experiencing high consequence communicable disease events.*

World Health Organization. Optimized Supportive Care for Ebola Virus Disease. https://www.who.int/csr/resources/publications/optimized-supportive-care/en/. Accessed 10 March 2020. *A compilation of several lines of effort in consolidating best practice for the nontherapeutic medication management of Ebola virus disease, this reference has several tools applicable to many of the VHFs.*

# 95

# Severe Acute Respiratory Syndrome and Coronavirus Disease 2019

Benjamin C. Pierson, Anthony P. Cardile, Phillip R. Pittman

## ABSTRACT

Within the past 20 years, three novel coronaviruses have emerged from the animal kingdom to infect humans, with devastating disease and economic consequences. They were severe acute respiratory syndrome–coronavirus disease (SARS-CoV), Middle East respiratory syndrome (MERS), and SARS-CoV-2 (COVID-19). This chapter discusses SARS-CoV and introduces readers to the currently emerging SARS-CoV-2 (COVID-19). MERS is discussed in Chapter 96. Severe acute respiratory syndrome (SARS) first appeared in November 2002 and ultimately resulted in 8096 probable human infections and 774 deaths worldwide. By July 2003, the global outbreak had been declared over. A new coronavirus, SARS-associated coronavirus (SARS-CoV), was identified as the causative agent; this virus has been shown to have jumped from bats to humans, although initially palm civets and raccoon dogs were suspected to be the animal reservoirs. The global response to the outbreak was extensive. Within a short period, the pathogen had been identified, new diagnostic tests had been developed, surveillance systems were created, infection control and prevention measures were instituted, and transmission among humans was stopped. A few community transmission events occurred, including the Amoy Gardens housing complex incident, which is important because this outbreak provided evidence for airborne transmission. It was unclear if and when person-to-person SARS-CoV transmission would reappear. Procedures were established by public health organizations, including the World Health Organization (WHO) and the US Centers for Disease Control and Prevention (CDC), to help guide diagnosis, reporting, surveillance, and prevention of future pandemics.

## CLINICAL VIGNETTE

A 65-year-old man presented to the emergency department with the complaint of a 7-day history of fever, malaise, and myalgia that had considerably worsened within the last day to include dyspnea and a nonproductive cough. His medical history was significant for diabetes, for which he currently takes metformin 500 mg PO bid; his most recent hemoglobin A1c was 7.1. He reports having recently returned from a trip to China, during which he visited a relative in a hospital and went to several wild animal markets. His vital signs were significant for fever, with a temperature of 102.1°F, tachypnea, a respiratory rate of 22 breaths/min, and oxygen saturation on room air of 91%. Physical examination revealed a generally ill-appearing man with inspiratory crackles heard at the lung bases. Chest radiography revealed diffuse interstitial infiltrates. Laboratory evaluation was significant for an elevated lactate dehydrogenase (LDH) of 943 IU/L and leukopenia, with a white blood cell (WBC) count of $3.3 \times 10^9$/L. Initial evaluation of sputum showed mixed oral flora and scant WBCs on Gram stain, and no bacterial pathogens were isolated from culture; a respiratory panel polymerase chain reaction (PCR) assay did not reveal the etiology of the patient's symptoms. He was admitted to the hospital and treated as a case of community-acquired pneumonia; however, on hospital day 2 the patient decompensated further, requiring admission to the intensive care unit (ICU) and ventilatory support. Owing to the patient's travel history and lack of a definitive diagnosis, local health authorities considered SARS-CoV as a potential etiology. Reverse-transcriptase PCR (RT-PCR) analysis confirmed the diagnosis of SARS. Over a period of 2 weeks, the patient's clinical condition further deteriorated, ultimately resulting in the patient's death on hospital day 17. Robust contact tracing and quarantine practices were implemented by the local public health authorities and no further cases of SARS in the local community were identified.

## SEVERE ACUTE RESPIRATORY SYNDROME

### Geographic Distribution and Magnitude of Disease Burden

In 2002 and 2003, a previously unknown infectious agent caused a widespread, global outbreak of life-threatening respiratory infections. The disease emerged as an outbreak of atypical pneumonia of unknown etiology in November 2002 in Foshan City, Guangdong province, China. The illness was first officially reported to the WHO in February 2003. In mid-March 2003, the WHO issued an alert calling attention to several outbreaks of severe atypical pneumonia in Hong Kong, Hanoi, and Singapore. Many of the initial SARS infections were traced to a guest staying at a Hong Kong hotel, and global spread occurred quickly, with multiple outbreaks reported in China, Southeast Asia, Europe, and North America. The illness was labeled SARS by the WHO on March 15, 2003. By mid-April 2003, the causative agent had been identified as a novel coronavirus, SARS-CoV.

An unprecedented global outbreak response (including the implementation of surveillance systems, epidemiologic studies, appropriate infection control measures, and development of laboratory diagnostics) was swiftly initiated. In July 2003, the WHO announced that the SARS outbreak was over. Although a few laboratory-associated SARS infections were reported in Asia after that, no infections have been reported worldwide since early 2004. Almost all persons with SARS were reported from China, Hong Kong, Taiwan, Singapore, or Toronto; as of December 2003, the WHO had received reports of SARS from 29 countries and regions: this meant a total of 8096 persons with probable SARS, resulting in 774 deaths, and a case-fatality rate of 9.6%. In the United States, 8 infections were documented by laboratory testing, and an additional 19 probable infections were reported. The syndrome arose and vanished within several months, and it is unclear when or if SARS will reemerge.

### Coronaviruses

Coronaviruses are enveloped, single-stranded positive strand ribonucleic acid (RNA) viruses that infect a wide spectrum of mammals and birds. There are three coronavirus groups: groups I and II affect mammals and group III affects birds. In humans, coronaviruses are primarily associated with upper respiratory infections. Prior to 2003, only two coronaviruses were known to cause human disease, which typically presented as mild cold-like syndromes, with more severe disease in the elderly and immunosuppressed. During the 2002–2003 SARS outbreak, several strains of a coronavirus unrelated to previously described coronaviruses were identified and isolated from clinical samples, including respiratory secretions, urine, and autopsy tissues. This new virus was identified as a novel group II coronavirus and named *SARS-associated coronavirus.*

Studies of animal ecology and virus evolution have revealed that SARS-CoV–like viruses are present in a variety of animal species including bats and animals commonly traded at live animal markets in southern China (e.g., masked palm civet, raccoon dog, red fox). At this time, it is unclear if these animals are susceptible hosts or carriers of SARS-CoV, but recent research indicates that the masked palm civet probably served as intermediate host between bats and humans during the 2002–2003 outbreak. Sequence analyses of SARS-CoV have shown that SARS-CoV–like viruses that infect masked palm civets are very similar (genome identity >99.6%) to the SARS-CoV that infected humans in the 2002–2003 SARS outbreak, suggesting that the virus had only recently circulated in the masked palm civet. The increased prevalence of SARS-CoV immunoglobulin G (IgG) antibodies among animal traders compared with a control group (13% vs. 1% to 3%) further supports this theory. In addition, the absence of SARS-CoV antibodies in the general population without clinical evidence of SARS suggests that SARS-CoV did not circulate widely before the 2002–2003 SARS outbreak.

## Transmission

The estimated incubation period for SARS is 2 to 10 days (median 5 to 6 days). The virus is detected at low levels in respiratory secretions during the initial days after the onset of illness, and peak viral levels occur during the second week of illness (e.g., 10 days). The primary route of SARS-CoV transmission is via the respiratory tract; during close contact with an infected patient, respiratory droplets may come into contact with one's mucous membranes either directly or indirectly through contaminated fomites. Studies have determined that SARS-CoV can remain stable on environmental surfaces for several days, although the virus can easily be inactivated by disinfectants. The virus has been isolated from respiratory secretions, saliva, tears, urine, and stool. Viral shedding generally does not persist beyond 4 weeks except in stool, in which the virus can be detected by reverse transcription-polymerase chain reaction (RT-PCR) for longer than a month. Isolation of the virus more than a month after the onset of illness is rare. Virus detection in nasopharyngeal specimens using quantitative RT-PCR found that the level typically peaks during the second week of illness, often when severely ill patients are seeking medical care. As patients improve clinically and the viral load decreases, transmission of the virus also decreases. Unlike other respiratory viral infections, such as influenza, transmission before symptom onset has not been reported. During the SARS outbreak, transmission occurred primarily in hospitals, less so within households, and to an even lesser extent within communities.

## Risk Factors

The most significant risk factors for a diagnosis of SARS were related to increased exposure to SARS-CoV; one of the highest risk groups was healthcare workers; 21% of all reported SARS-CoV infections occurred in this population. Nosocomial transmission of SARS-CoV was common early in the outbreak but subsequently decreased significantly as a result of early diagnosis and the reinforcement of infection

control practices. Nosocomial spread is theorized to occur via aerosolization during patient procedures such as intubation and bronchoscopy. Transmission has also been documented on an airplane, in an apartment complex (probably secondary to faulty plumbing and aerosolization of fecal matter), and among laboratory workers handling SARS-CoV. Transmission of the virus has not been reported via food- or waterborne sources or from an infected patient whose fever had resolved more than 14 days previously.

As seen with other infectious diseases, environmental and host factors influence the risk of transmission. Although the virus was initially thought to be highly infectious, the rate of secondary transmission of SARS-CoV is estimated to be low to moderate. Transmission modeling studies have estimated that each patient will infect an average of three persons. However, some SARS-infected patients designated as "superspreaders" have been documented to have very high secondary transmission rates (infecting an average of 36 contacts [range 11 to 74]), a phenomenon not unique to SARS. Transmission of SARS-CoV by superspreaders primarily occurred in hospital settings and was associated with a greater number of close contacts, delayed diagnosis, older age, more severe illness, and poor infection control practices.

Risk factors for poor outcomes in those infected with SARS-CoV include advanced age, comorbid conditions (i.e., diabetes, chronic hepatitis B, and other immunosuppressive conditions), atypical symptoms, elevated serum LDH at the time of hospital admission, and the presence of high viral SARS-CoV viral loads in nasopharyngeal secretions.

## Clinical Features

Among symptomatic patients, all have one of the following three key symptoms: fever (80%), cough (84%), and shortness of breath (82%), with 45% of symptomatic patients having all three of these symptoms. However, severity of disease among SARS patients varies from asymptomatic infection to fatal acute respiratory distress syndrome (ARDS) (Fig. 95.1). Seroprevalence surveys have documented asymptomatic infection, especially among animal traders in Guangdong, China. Overall, however, asymptomatic or mild disease is relatively uncommon in SARS (<1%), which is different from COVID-19 infection.

SARS affects persons of all ages; however, most infections occur among adults (median age approximately 59 years). Infections among children, especially those younger than 12 years of age, are uncommon. The disease is considerably less severe among children than it is in adults, and the outcome is much more favorable. Infections during pregnancy have been documented, with an increased risk of spontaneous abortion, preterm labor, severe pulmonary disease, and death. No reports of perinatal transmission have been noted.

The initial symptoms of SARS are nonspecific and consistent with an influenza-like illness. A prodrome that includes fever, headache, chills, rigors, malaise, and myalgias occurs approximately 1 to 2 days after exposure (Fig. 95.2). Nearly all patients report fever (with temperatures frequently exceeding 101°F), which typically precedes other prodromal symptoms but can also occur after the prodrome. The elderly and those with a history of chronic comorbid conditions, such as diabetes mellitus or chronic renal failure, may have atypical presentations (e.g., lack of fever).

SARS primarily affects the pulmonary system, and respiratory symptoms (typically including nonproductive cough and shortness of breath) appear more often during the second week of illness. The SARS surface spike (S) glycoprotein binds to human angiotensin-converting enzyme 2, found in the lower respiratory tract rather than the upper respiratory tract; this explains the distribution of pulmonary symptoms. In one reported patient series, gastrointestinal symptoms, primarily diarrhea, were prominent (73%), with high-volume diarrhea occurring

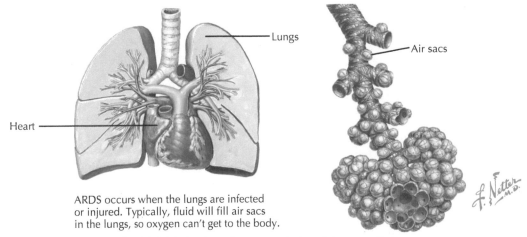

ARDS occurs when the lungs are infected or injured. Typically, fluid will fill air sacs in the lungs, so oxygen can't get to the body.

**Fig. 95.1** Mechanism of acute respiratory distress syndrome.

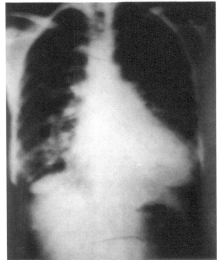

Early SARS symptoms are similar to other respiratory infections and include headache, chills, myalgia, and fever, typically seen in the first week following onset. This is followed by cough and shortness of breath, which typically appear in the second week

Chest radiographs can provide valuable information. Typical early findings include a ground glass appearance and focal opacities or consolidations in the peripheral lower lung fields, which often progress to bilateral patchy consolidations

**Fig. 95.2** Symptoms of severe acute respiratory syndrome.

in the second week of illness. In a separate report, diarrhea, nausea, and vomiting were less common (<25%). Mucus and blood in stool are uncommon, and the diarrhea is often self-limiting. Lymphadenopathy, rhinorrhea, sore throat, rash, and purpura are unusual.

In persons with SARS, inspiratory crackles at the lung bases and, less commonly, wheezing may be noted on auscultatory examination. Initially a consistent finding is the paucity of auscultatory findings relative to the degree of abnormality displayed on chest radiographs. By the second week of illness, clinical deterioration may occur, with pneumonia and hypoxemia necessitating hospitalization. During the 2002–2003 outbreak, respiratory failure and ARDS were the most common reasons for admission to an ICU. In several studies, approximately 20% to 30% of patients hospitalized with SARS were admitted to an ICU; about 75% of ICU patients required mechanical ventilation.

In up to 30% of patients, initial chest radiographic findings may be unremarkable or indistinguishable from those of other causes of infectious pneumonia. However, serial chest radiographs and high-resolution computed tomography (CT) may offer valuable information during the evaluation of a patient with suspected SARS, as

abnormalities appear in a large proportion of patients by days 7 to 10 of illness. Typical chest radiographs have a ground-glass appearance with focal opacities or consolidation in the peripheral lower lung fields; these focal findings often progress to bilateral patchy consolidation (see Fig. 95.2). Peripheral lung involvement was a very common finding in most case studies of the 2002–2003 outbreak; pulmonary cavitation, hilar lymphadenopathy, nodular infiltrates, and pleural effusion were unusual (Fig. 95.3).

Evidence of extrapulmonary dissemination of SARS-CoV can be found by laboratory and pathologic diagnostic methods. The virus has been detected in several extrapulmonary organs, including the gastrointestinal tract, kidneys, liver, and spleen. Studies to improve our understanding of SARS pathogenesis and the immune response are ongoing.

## Laboratory Findings

Common laboratory findings in patients with SARS include moderate lymphocytopenia with a low to normal WBC count (primarily caused by a decrease in T-cell lineages); mild thrombocytopenia; increased serum LDH (i.e., more than three to five times the upper

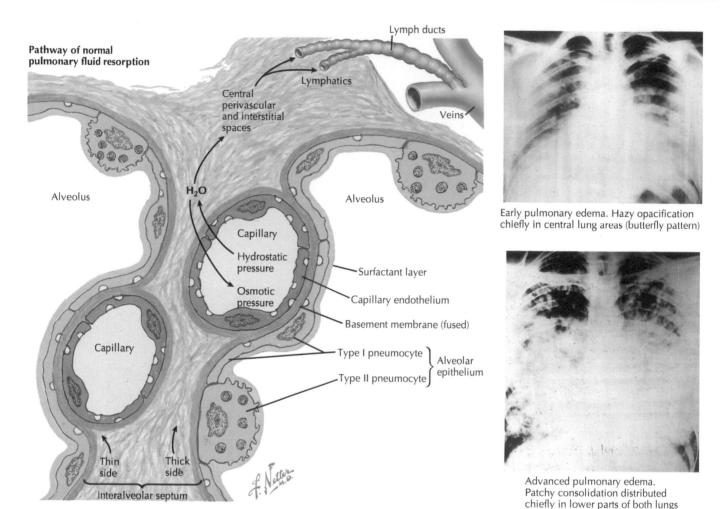

Pathway of normal pulmonary fluid resorption

**Fig. 95.3** Severe acute respiratory syndrome is a cause of noncardiac pulmonary edema. The infection causes the lung's capillaries to leak more fluid than normal into the air sacs (alveoli).

Early pulmonary edema. Hazy opacification chiefly in central lung areas (butterfly pattern)

Advanced pulmonary edema. Patchy consolidation distributed chiefly in lower parts of both lungs

limit of normal); and elevated serum hepatic transaminases. Decreased CD4 and CD8 T-cell counts can be significant: up to 30% of infected patients have a CD4 T-cell count of below 200/mm³. Elevated serum creatinine phosphokinase (noncardiac) and electrolyte abnormalities—including hypocalcemia, hypokalemia, hypomagnesemia, and hypophosphatemia—are also commonly seen in SARS.

## Diagnostic Approach

There has been no known transmission of SARS-CoV since 2004. Therefore new cases would have to emerge naturally from known reservoirs, a laboratory exposure incident, or via a deliberate release such as a bioterrorism event. According to CDC guidance, the diagnosis should be considered in those hospitalized with radiographically confirmed pneumonia or ARDS of unknown etiology and relevant epidemiologic risk factors. The risk factors identified in the 10 days prior to illness onset include at least one of the following:

- Travel to mainland China, Hong Kong, or Taiwan or close contact with an ill person with a history of recent travel to one of these areas
- Employment in an occupation associated with a risk for SARS-CoV exposure (e.g., healthcare worker with direct patient contact; worker in a laboratory that contains live SARS-CoV)
- Illness in association with a cluster of cases of atypical pneumonia without an alternative diagnosis

Laboratory testing should be coordinated with local health departments. Both SARS-CoV real-time RT-PCR and serum enzyme

immunoassays (EIAs) or immunofluorescence assays (IFAs) are available. Note that a biosafety level 3 (BSL-3) laboratory is required for other laboratory methods such as viral culture.

SARS-CoV RNA is most useful during the acute phase of disease and can be detected in multiple clinical specimens to include the upper/lower respiratory tract, blood, stool, and urine. Previous studies demonstrated that the virus can be most reliably detected during the second week of illness and that the use of samples from more than one body site was beneficial for confirmation. During the first week of clinical symptoms, the CDC recommends collecting a nasopharyngeal (NP) swab plus an oropharyngeal (OP) swab and a serum or plasma specimen. Previous studies demonstrated that the detection rates for viral RNA in plasma dropped to 25% at day 14 after fever onset. Thus, following the first week of symptoms, it is recommended to collect a NP swab plus OP swab and a stool specimen.

Serology may also be useful to confirm infection with SARS-CoV, and it is recommended to collect both acute and convalescent serum samples. Previous studies have demonstrated that serology is of low yield during the first week of illness, is positive in about half of patients during the second week of illness, and that seroconversion typically occurs after 4 weeks in most patients.

## Clinical Management and Drug Treatment

In the past, both ribavirin and systemic corticosteroids were widely utilized empirically for treatment. In addition, treatment with interferon, convalescent plasma, intravenous immunoglobulin (IVIG), and lopinavir/ritonavir was attempted. More recently, remdesivir has been demonstrated to have in vitro activity against SARS-CoV. In a systematic review of studies of SARS-CoV-1 therapeutics, most studies with ribavirin and corticosteroids were classified as inconclusive, with a number demonstrating possible harm. Study results for convalescent plasma, IVIG, interferons, and lopinavir/ritonavir were classified as inconclusive. Thus the mainstay of clinical management at this time is the provision of supportive care.

## Prognosis

Approximately 30% of patients with SARS improve clinically within 1 to 2 weeks after the onset of their illness, whereas 70% develop persistent fever and worsening respiratory symptoms often requiring hospitalization; some will be admitted to an ICU. The length of hospital stays has varied, but several studies have reported a median length of stay of approximately 2 weeks.

In hospitalized patients, common complications of SARS include cardiovascular abnormalities (e.g., hypotension, tachycardia) and hepatic dysfunction. Deep vein thrombosis (DVT) has been less common; however, in one Singapore case series, 30% of hospitalized patients had evidence of a DVT. Disseminated intravascular coagulation, acute renal failure, and neurologic and other complications are uncommon.

In the 2002–2003 SARS outbreak, poor outcome was associated with increasing age and the presence of comorbid conditions (e.g., diabetes mellitus, hypertension, cardiovascular disease, chronic renal disease). Overall crude mortality rates ranged from about 4% to 15%, with age-specific mortality rates highest among older adults. Mortality rates were higher in those aged 60 years or older (43%) as compared with those younger than 60 years of age (13%). Most SARS patients who survive have a complete, sometimes prolonged recovery, but some severely ill patients have reported decreased pulmonary function over the long term.

## Prevention and Control

In the absence of continued person-to-person SARS-CoV transmission worldwide, recommendations have been published by several public health organizations, including the WHO and CDC, to provide guidance on surveillance, clinical and laboratory evaluation, and reporting of suspected SARS infections. Being familiar with the clinical features of SARS-CoV disease, assessing travel history and exposure risk, and recognizing unusual clusters of unexplained pneumonia can help to maximize early detection. In the absence of person-to-person transmission, public health and healthcare personnel should be aware of specific settings that should raise suspicion for SARS-CoV infection. These situations include persons who are hospitalized for radiographically confirmed pneumonia or ARDS without an identifiable cause and who have at least one of the following three risk factors: travel to China, Hong Kong, or Taiwan or close contact with an ill person with a recent travel to one of these areas; employment in an occupation associated with a risk for SARS-CoV exposure (e.g., healthcare worker, worker in a laboratory that contains SARS-CoV); or illness in association with a cluster of cases of atypical pneumonia without an alternative diagnosis) in the 10 days before the onset of illness.

Clinicians evaluating patients who fit one of these three criteria should implement appropriate infection control measures, contact the local or state health department, and continue with a diagnostic evaluation. This evaluation should include testing for other respiratory pathogens (e.g., influenza, respiratory syncytial virus, *Streptococcus pneumoniae, Legionella* species). If no alternative diagnosis has been made after 72 hours or if a high index of suspicion for SARS exists, the clinician and health department should consider SARS-CoV

testing and contact the CDC for consultation. Currently laboratory-acquired SARS infection remains a possible scenario but a remote one, as adherence to strict biosafety and laboratory policies has significantly reduced this risk. The CDC is available for consultation and testing and has published guidelines for laboratory personnel working with SARS-CoV.

Globally, SARS has been designated as immediately reportable by the WHO International Health Regulations. The local or state health department should be promptly notified if a suspected SARS case is identified. Prompt case detection, implementation of infection control measures including patient isolation and standard and droplet precautions, and contact tracing have been shown to reduce transmission. Additional infection control measures should be instituted depending on the setting (e.g., healthcare, home, community).

Multiple vaccine candidates have been developed for the prevention of SARS; however, owing to the rapid resolution of the 2002–2003 outbreak, no studies to evaluate the efficacy of the vaccine candidates in human subjects could be completed. Several candidates showed promising results in animal studies, however, with most vaccine candidates targeting the surface S glycoprotein of SARS-CoV. The development of constructs for these vaccine candidates assisted in the expeditious development of vaccine candidates against other emerging coronavirus diseases, including the Middle East respiratory syndrome (MERS) and coronavirus disease 2019 (COVID-19).

One critical lesson learned from the 2002–2003 SARS outbreak is the apparent need for prompt collaboration and open communication among local, national, and international health agencies. Early diagnosis, timely reporting, the implementation of infection control measures, and continued research—including research regarding treatment and vaccine development—will help to identify and control possible future SARS outbreaks.

# CORONAVIRUS DISEASE 2019

## Geographic Distribution and Magnitude of Disease Burden

An outbreak of respiratory disease caused by a novel coronavirus was first detected in Wuhan City, Hubei Province, China, in December 2019. The outbreak was initially associated with an animal market in Wuhan. In January 2020, it was confirmed that the outbreak was caused by a novel beta coronavirus, initially termed "2019 novel coronavirus" (2019-nCoV). Subsequent genome sequencing would demonstrate that the novel coronavirus was in the same subgenus as SARS-CoV and a number of bat coronaviruses. Owing to rapid international spread of the virus, the WHO declared a public health emergency of international concern for the 2019-nCoV outbreak on January 30, 2020. On January 31, 2020, the US secretary of health and human services declared a public health emergency in the United States and announced new measures to prevent spread. The WHO subsequently named the disease coronavirus disease 2019 ("COVID-19") on February 11, 2020. The virus 2019-nCoV is now termed severe acute respiratory syndrome coronavirus 2 (SARS-CoV-2). It has been recommended that because of potential confusion with SARS-CoV, public communications should use the phrase "the virus that causes COVID-19."

SARS-CoV-2 has rapidly spread throughout the world on a scale not seen since the 1918 influenza pandemic. Since the beginning of January to the end of July 2020, cases of COVID-19 were reported from all WHO regions, namely the Americas, Europe, Southeast Asia, the eastern Mediterranean, Africa, and the western Pacific, with more than 72.5 million confirmed cases and over 1.6 million deaths due to the infection by late December 2020. The United States, India, Brazil, and Russia had the highest cumulative number of cases. SARS-CoV-2 has clearly demonstrated that emerging infectious diseases not only pose a serious challenge for the medical community but also have significant political and socioeconomic impacts.

## Approach to Diagnosis and Clinical Management

The medical and scientific communities are still learning about SARS-CoV-2, and information is changing rapidly. As a result, a very brief summary is provided here; it is limited to what was known at the time of writing. SARS-CoV-2 is globally distributed at this time and the local incidence varies owing to a multitude of factors. The primary risk of transmission stems from contact with an infected individual. Transmission most likely occurs person to person via respiratory droplets passing among close contacts (defined as being within 6 feet of an infected individual for 15 minutes or more). The virus may also be transmitted by direct contact with infectious secretions from an infected patient or during aerosol generating procedures in a medical setting. The contribution of airborne transmission in the community has yet to be determined. One concerning feature that has contributed to community spread of SARS-CoV-2 is that transmission can occur from both asymptomatic and presymptomatic individuals. Typically those exposed develop symptoms 4 to 5 days postexposure, but the incubation period can be as long as 14 days.

Symptoms of COVID-19 disease are nonspecific; they include fever or chills, cough, shortness of breath, fatigue, myalgias, headache, sore throat, and nasal congestion/rhinorrhea. Early reports suggest that loss of taste or smell is common; 50% of patients will report one or more gastrointestinal symptoms, with diarrhea (38%) and vomiting (13%) leading. Disease severity can range from asymptomatic to critical illness with multiorgan failure. Similar to SARS, risk factors for severe disease and increased mortality include older age, chronic medical conditions (e.g. cardiac disease, renal disease, diabetes), and immunosuppression. The disease course has generally been observed to be milder in pediatric than in adult patients. However, a minority of children who are actively/recently infected have been observed to develop a severe inflammatory syndrome with Kawasaki disease-like features (termed multisystem inflammatory syndrome in children [MIS-C] associated with COVID-19).

More data are needed to fully characterize laboratory findings in patients with COVID-19. Preliminary analysis shows that lymphopenia is found in 83% of hospitalized patients with COVID-19. Severe illness is associated with lymphopenia, neutrophilia, elevated serum alanine aminotransferase and aspartate aminotransferase, and elevated lactate dehydrogenase. Elevated D-dimer and lymphopenia have been associated with mortality.

After SARS-CoV-2 was initially recognized, RT-PCR assays for diagnosis were rapidly developed, with their production, distribution, and availability being variable throughout the world. SARS-CoV-2 RNA can be detected in the nasopharynx, oropharynx, lower respiratory secretions, blood, and stool. A variety of serological assays (antibody tests) are available, but immunity to the virus is poorly understood, thus limiting the clinical utility of such assays at this time. Antigen tests that detect fragments of proteins of the SARS-CoV-2 virus in patient samples have been developed, authorized, and are in clinical use for diagnosis. In December 2020, the first COVID-19 antigen test that can be used completely at home without a prescription received an emergency use authorization (EUA) from the United States (US) Food and Drug Administration (FDA). A mid-turbinate nasal swab is used by the Ellume COVID-19 Home Test, and a software application on a smartphone helps users perform the test and interpret results. The development of COVID-19 tests is rapidly advancing to expand access to COVID-19 testing; the FDA website should be consulted for details on molecular and antigen diagnostic tests that have FDA authorization and for information on new tests being developed.

The mainstay of management of COVID-19 patients is medical supportive care. A joint statement from the Infectious Diseases Society of America and the AABB (American Association of Blood Banks) clarifies the FDA EUA framework for use of high-titer units of COVID-19 convalescent plasma (CCP): CCP administered soon after hospital admission as part of a clinical trial may benefit patients by antibody-mediated pathogen neutralization; although antibody-dependent cellular toxicity and phagocytosis may also play a role. The use of CCP for nonhospitalized patients remains unclear.

Pharmacologic management of patients with COVID-19 is based on what is known about the pathogenesis of the infection and disease severity. As summarized in the US National Institutes of Health (NIH) COVID-19 Treatment Guidelines Panel, early in the course of infection, viral replication of the SARS-CoV-2 virus is occurring and the virus is disseminating to various organ sites. Later in the course of infection, disease is driven by an exaggerated immune or inflammatory response to the virus that leads to tissue damage. Using this framework to guide recommendations, the Panel recommends considering that outpatients who are at high risk of disease progression may benefit from receiving anti-SARS-CoV-2 monoclonal antibodies (bamlanivimab [EUA] or casirivimab plus imdevimab [EUA]) early in the course of infection.

The antiviral drug remdesivir is FDA-approved for use in hospitalized patients requiring supplemental oxygen but may be considered for all patients at high risk of disease progression. The combination of remdesivir plus dexamethasone is recommended for those requiring increasing amounts of supplemental oxygen. In the rare circumstances where corticosteroids (dexamethasone) cannot be used, the Panel recommends using baricitinib under EUA in combination with remdesivir for the treatment of COVID-19 in hospitalized, nonintubated patients who require oxygen supplementation. Baricitinib is an oral Janus kinase (JAK) inhibitor that may prevent cellular immune activation and inflammation and is FDA-approved to treat moderate to severe rheumatoid arthritis. Dexamethasone alone is recommended when combination therapy with remdesivir cannot be used or is not available, and dexamethasone alone is recommended for hospitalized patients requiring invasive mechanical ventilation or ECMO (extracorporeal membrane oxygenation). (See Fig. 95.4.) As all aspects of medical and pharmacologic management of COVID-19 are rapidly evolving, clinicians are strongly advised to access updated information and guidelines on a regular basis through accredited medical and public health websites. It is important to note that clinical treatment guidelines may greatly vary among regulatory agencies and professional societies of different countries.

## Disease Control and Prevention

Detailed discussion of infection control measures and public health management of COVID-19 cases is beyond the scope of this chapter and further information is available from the listings in Additional Resources. In December 2020, two vaccines received US FDA Emergency Use Authorization: the Pfizer-BioNTech COVID-19 vaccine (for use in individuals 16 years of age and older) and the Moderna COVID-19 vaccine (for use in individuals 18 years of age and older). Efficacy and safety studies of these vaccines in pediatric age groups and other special groups (e.g., pregnant women, immunosuppressed patients) are ongoing, and many other vaccine candidates are currently in clinical trials.

Doses and durations are listed in the footnote

| Disease Severity | Panel's Recommendations |
|---|---|
| **Not hospitalized, Mild to moderate COVID-19** | There are insufficient data to recommend either for or against any specific antiviral or antibod therapy. SARS-CoV-2 neutralizing antibodies (**bamlanivimab** or **casirivimab plus imdevimab**) are available through EUAs for outpatients who are at high risk of disease progression.[a] These EUAs do not authorize use in hospitalized patients.<br><br>**Dexamethasone** should not be used **(AIII)**. |
| **Hospitalized[a] but does not require supplemental oxygen** | **Dexamethasone** should not be used **(AIIa)**.<br><br>There are insufficient data to recommend either for or against the routine use of **remdesivir.** For patients at high risk of disease progression, the use of remdesivir may be appropriate. |
| **Hospitalized[a] and requires supplemental oxygen**<br><br>But does not require oxygen delivery through a high-flow device, noninvasive ventilation, invasive mechanical ventilation or ECMO | Use one of the following options:<br>• **Remdesivir**[b,c] (e.g., for patients who require minimal supplemental oxygen)**(BIIa)**<br>• **Dexamethasone**[d] **plus remdesivir**[b,c] (e.g., for patients who require increasing amounts of supplmetal oxygen)**(BIII)**[e,f]<br>• **Dexamethasone**[d] (e.g., when combination therapy with remdesivir cannot be used or is not available) **(BI)** |
| **Hospitalized[a] and requires oxygen delivery through a high-flow device or noninvasive ventilation** | Use one of the following options:<br>• **Dexamethasone**[d,f] **(AI)**<br>• **Dexamethasone**[d] **plus remdesivir**[b,c] **(BIII)**[e,f] |
| **Hospitalized[a] and requires invasive mechanical ventilation or ECMO** | **Dexamethasone**[d] **(AI)**[g] |

**Rating of recommendations:** A = Strong; B = Moderate; C = Optional
**Rating of evidence:** I = One or more randomized trials without major limitations; IIa = Other randomized trials or subgroup analyses of randomized trials; IIb = Nonrandomized trials or observational cohort studies; III = Expert opinion

[a] See the Panel's statements on the FDA EUAs for bamlanivimab and casirivimab plus imdevimab. These EUAs do not authorize use in hospitalized patients.
[b] The remdesivir dose is 200 mg IV for one dose, followed by 100 mg IV once daily for 4 days or until hospital discharge (unless the patient is in a health care setting that can provide acute care that is similar to inpatient hospital care). Treatment duration may be extended up to 10 days if there is no substantial clinical improvement by Day 5.
[c] For patients who are receiving remdesivir but progress to requiring oxygen through a high-flow device, noninvasive ventilation, invasive mechanical ventilation, or ECMO, remdesivir should be continued until the treatment course is completed.
[d] The dexamethasone dose is 6 mg IV or PO once daily for 10 days or until hospital discharge. If dexamethasone is not available, equivalent doses of other corticosteroids, such as prednisone, methylprednisone, or hydrocortisone, may be used. See the Corticosteroids section for more information.
[e] The combination of dexamethasone and remdesivir has not been studied in clinical trials.
[f] In the rare circumstances where corticosteroids cannot be used, baricitinib plus remdesivir can be used **(BIIa)**. The FDA has issued an EUA for baricitinib use in combination with remdesivir. The dose for baricitinib 4 mg PO once daily for 14 days or until hospital discharge.
[g] The combination of dexamethasone and remdesivir may be considered for patients who have recently been intubated **(CIII)**. Remdesivir alone **is not recommended**.

**Key:** ECMO = extracorporeal membrane oxygenation; EUA = Emergency Use Authorization; FDA = Food and Drug Administration; IV = intravenous; the Panel = the COVID-19 Treatment Guidelines Panel; PO = orally; SARS-CoV-2 = severe acute respiratory syndrome coronavirus 2

**Fig. 95.4** Pharmacologic management of patients with COVID-19 based on disease severity. (Reused from COVID-19 Treatment Guidelines Panel. Coronavirus Disease 2019 [COVID-19] Treatment Guidelines. National Institutes of Health. Available at https://www.covid19treatmentguidelines.nih.gov/. Accessed December 2020; Therapeutic Management of Patients with COVID-19, Figure 1, https://www.covid19treatmentguidelines.nih.gov/therapeutic-management/.)

## ACKNOWLEDGMENT

The previous edition of this chapter was written by Dr. Eileen Schneider.

## EVIDENCE

### SARS-CoV

Hui DSC, Zumla A. Severe acute respiratory syndrome: historical, epidemiologic, and clinical features. *Infect Dis Clin North Am* 2019;33(4):869-889. https://doi:10.1016/j.idc.2019.07.001. *This article provides a comprehensive review of SARS-CoV epidemiology and clinical features.*

Momattin H, Mohammed K, Zumla A, Memish ZA, Al-Tawfiq JA. Therapeutic options for Middle East respiratory syndrome coronavirus (MERS-CoV)—possible lessons from a systematic review of SARS-CoV therapy. *Int J Infect Dis* 2013;17(10):e792-e798. *This article provides a systematic review of SARS-CoV therapy.*

Ng EK, Hui DS, Chan KC, et al. Quantitative analysis and prognostic implication of SARS coronavirus RNA in the plasma and serum of patients with severe acute respiratory syndrome. *Clin Chem* 2003;49(12):1976-1980. *Describes kinetics of SARS-CoV RNA in serum/plasma samples.*

Peiris JS, Chu CM, Cheng VC, et al.; HKU/UCHSARS Study Group. Clinical progression and viral load in a community outbreak of coronavirus-associated SARS pneumonia: a prospective study. *Lancet* 2003;361(9371):1767-1772. *This study followed SARS patients managed with a standard treatment protocol of ribavirin and corticosteroids.*

Sheahan TP, Sims AC, Graham RL, et al. Broad-spectrum antiviral GS-5734 inhibits both epidemic and zoonotic coronaviruses. *Sci Transl Med* 2017;9(396):eaal3653. *This study demonstrates the in vitro activity of remdesivir against SARS.*

Shi Z, Hu Z. A review of studies on animal reservoirs of the SARS coronavirus. *Virus Res* 2008;133:74-87. *A comprehensive laboratory review of SARS-CoV–like viruses from several animal reservoirs (e.g., masked palm civet, bats) describing the possible evolutionary progression of SARS-CoV–like viruses in animals to SARS-CoV, which infected humans.*

Stockman LJ, Bellamy R, Garner P. SARS: systematic review of treatment effects. Version 2. *PLoS Med* 2006;3(9):e343. *Very useful systematic review of treatment of SARS.*

Tang P, Louie M, Richardson SE, et al.; Ontario Laboratory Working Group for the Rapid Diagnosis of Emerging Infections. Interpretation of diagnostic laboratory testsfor severe acute respiratory syndrome: the Toronto experience. *CMAJ* 2004;170(1):47-54. *Describes the diagnostic laboratory results of all SARS patients during an outbreak in Toronto.*

### SARS-CoV-2

Beigel JH, Tomashek KM, Dodd LE, et al.; ACTT-1Study Group Members. Remdesivir for the treatment of Covid-19—preliminary report. *N Engl J Med* 2020:NEJMoa2007764. *Study demonstrating potential benefit of remdesivir for COVID-19.*

Huang C, Wang Y, Li X, et al. Clinical features of patients infected with 2019 novel coronavirus in Wuhan, China. *Lancet* 2020;395(10223):497-506. *This is the first peer-reviewed publication describing in detail the clinical features of COVID-19.*

Rothe C, Schunk M, Sothmann P, et al. Transmission of 2019-nCoV infection from an asymptomatic contact in Germany. *N Engl J Med* 2020;382(10):970-971. *This is the first peer-reviewed publication demonstrating asymptomatic transmission of COVID-19.*

## ADDITIONAL RESOURCES

### SARS-CoV

Centers for Disease Control and Prevention (CDC), Centers for Disease Control and Prevention (CDC): Severe acute respiratory syndrome (SARS). Available at: https://www.cdc.gov/sars/index.html. Accessed July 17, 2020. *This site provides a wealth of information and guidance on SARS to include the document: "In the absence of SARS-CoV transmission worldwide: guidance for surveillance. Clinical and laboratory evaluation, and reporting version 2."*

Centers for Disease Control and Prevention (CDC), Centers for Disease Control and Prevention (CDC): Severe Acute Respiratory Syndrome (SARS) Infection Control. Available at: https://www.cdc.gov/sars/infection/index.html. Accessed July 23, 2020. This site provides infection control recommendations for different settings, including the hospital, home, and community. Implementation of infection control measures is critical for reducing transmission.

World Health Organization (WHO): SARS—how a global epidemic was stopped. Available at: https://apps.who.int/iris/bitstream/handle/10665/207501/9290612134_eng.pdf?sequence=1&isAllowed=y. Accessed July 27, 2020. *A detailed WHO summary document describing the historical events of the 2002–2003 SARS outbreak.*

World Health Organization (WHO): Severe acute respiratory syndrome. Available at: www.who.int/topics/sars/en. Accessed July 23, 2020. *WHO's primary website for SARS, with links to several helpful websites and documents.*

World Health Organization (WHO): WHO guidelines for the global surveillance of severe acute respiratory syndrome (SARS)—updated recommendations, October 2004. Available at: https://www.who.int/csr/resources/publications/WHO_CDS_CSR_ARO_2004_1/en/. Accessed July 27, 2020. *WHO guidelines for SARS, covering clinical laboratory criteria for global surveillance of SARS, guidance on how to approach suspected SARS patients during the interepidemic period, how to conduct global surveillance during an outbreak, and international reporting of a SARS case.*

### SARS-CoV-2

Centers for Disease Control and Prevention (CDC), Centers for Disease Control and Prevention (CDC): Coronavirus disease 2019 (COVID-19). Available at: https://www.cdc.gov/coronavirus/2019-nCoV/index.html. Accessed December 16, 2020. *This site provides a wealth of information and guidance on SARS-CoV-2 that is continuously updated.*

COVID-19 Treatment Guidelines Panel, National Institutes of Health: Therapeutic Management of Patients with COVID-19. Available at: https://www.covid19treatmentguidelines.nih.gov. Accessed December 15, 2020. *Comprehensive evidence-based guidelines along with the latest updates for therapeutic management of patients with COVID-19; detailed discussion of pharmacologic management based on different severities of disease.*

Infectious Diseases Society of America (IDSA): The COVID-19 Real-Time Learning Network. Available at: https://www.idsociety.org/covid-19-real-time-learning-network. Accessed December 17, 2020. *The Network "is supported by funds from the CDC, and brings together the latest clinical guidance, institutional protocols, clinical trial data, practice tools and resources from a variety of medical subspecialties around the world. Visit daily for must-have information that will help clinicians and decision-makers navigate the COVID-19 pandemic." This website is up-to-date and easy to navigate.*

U.S. Food and Drug Administration: Coronavirus Disease 2019 Testing Basics. Available at: https://www.fda.gov/consumers/consumer-updates/coronavirus-disease-2019-testing-basics. Accessed December 16, 2020. *Provides a useful overview of the multiple diagnostic and antibody tests for COVID-19 and summarized sample collection, result times, confirmation tests, and test interpretation.*

World Health Organization (WHO): Coronavirus Disease (COVID-19) Situation Dashboard/Map. Available at: https://www.who.int/emergencies/diseases/novel-coronavirus-2019/coronavirus-disease. Accessed December 17, 2020. *The WHO interactive dashboard/map provides the latest global numbers by country of COVID-19 cases on a daily basis. The main page of the WHO website provides access to global perspectives on COVID-19 vaccines, treatments, and tests, as well as public health guidance.*

# 96

# Middle East Respiratory Syndrome Coronavirus

*David L. Saunders, Anthony P. Cardile, Nathan K. Jansen*

## ABSTRACT

Unlike the coronaviruses causing severe acute respiratory syndromes, namely SARS-CoV and SARS-CoV-2 (the cause of Coronavirus Disease-2019 [COVID 19]), which developed rapid and aggressive person-to-person spread, the coronavirus that causes Middle Eastern respiratory syndrome (MERS-CoV) has yet to develop significant person-to-person spread. It remains a highly dangerous pathogen nonetheless. Roughly 2500 cases have been reported to date, with a 35% case fatality rate; however, actual rates of disease are uncertain, as up to 80% of cases may be asymptomatic. Bats are the major reservoir of MERS-CoV in nature, although camels appear to be the key reservoir host for human cases. This may have limited the spread of MERS-CoV, largely confining disease to the Arabian Peninsula. A high index of suspicion based on local epidemiology is essential to making the diagnosis. MERS-CoV has a predilection for DPP4 receptor–positive cells of the lower respiratory tract, in contrast to the ACE-2 receptor–positive cells in the upper respiratory tract used by SARS-CoV and SARS-CoV-2 for viral invasion. This difference in pathogenesis may also serve to limit the respiratory spread of MERS-CoV, although nosocomial transmission to health workers has been well documented. There are no confirmed or licensed effective treatments to date, although there are a few promising candidates. There are at least two experimental vaccines in clinical-stage development and multiple preclinical candidates. Supportive care, including intensive care unit (ICU) admission and mechanical ventilation, remains the mainstay of survival.

## CLINICAL VIGNETTE

A 56-year-old man fell ill in his home country a week after taking a pilgrimage to Mecca for the Hajj. His illness initially began with a mild cough but within a few days quickly progressed to fever and respiratory distress, prompting admission to a local hospital. He was initially diagnosed with acute pneumococcal pneumonia and treated empirically with antibiotics. The facility lacked an ICU or mechanical ventilation. The patient's condition worsened over 36 hours, at which point he deteriorated rapidly and died. No autopsy was performed. A week later, a healthcare worker who had helped to look after the patient fell ill with a mild respiratory illness, and missed a few days of work.

COMMENT: MERS-CoV can present as a range of illness severities from asymptomatic to severe with fulminant respiratory failure. Transmission has been focused in the Arabian Peninsula, likely limited by the geographic distribution of camels as the primary human reservoir host. As such, with a little over 2500 cases in 8 years, it has not emerged with nearly the severity of SARS-CoV-2 (COVID-19), which has become a global pandemic with more than half a million deaths over 6 months in 2020. However, visitors to the Arabian Peninsula have brought cases back to their home countries, even leading to extensive nosocomial and eventually person-to-person transmission, as in South Korea in 2015. Specific therapies remain in experimental clinical development. A high index of suspicion is needed to diagnose MERS-CoV.

## GEOGRAPHIC DISTRIBUTION AND MAGNITUDE OF DISEASE BURDEN

### Background

MERS is caused by a novel, highly pathogenic coronavirus strain termed MERS-CoV that first appeared in Jordan in 2012. Within 5 years more than 2000 cases had been reported, with approximately 600 deaths. Although the spread of MERS has not been nearly as dramatic or extensive as the current pandemic of COVID-19 (caused by SARS-CoV-2), there were documented cases in 21 countries within 3 years. To date, 2500 cases of MERS have been reported, with a 35% case fatality rate. Many of those who died had underlying medical conditions. The origins of nearly all human cases have been traced to the Arabian Peninsula. However, in 2015 there was a significant outbreak in South Korea, where more than 80 infections occurred within a few days and 186 over 3 months. This was thought to be due to a single "super spreader" event originating from an elderly man who had traveled to Saudi Arabia. The infection is believed to have spread beginning with the index case through healthcare workers at a network of hospitals.

### Coronaviruses

MERS-CoV is a *Betacoronavirus* in the *Coronaviridae* family. Coronaviruses are positive-sense single-stranded RNA viruses. MERS-CoV is thought to be more closely related to bat coronaviruses than to SARS-CoV. Although bats are believed to be the ultimate reservoir of MERS-CoV, the virus also infects camels, which appear to be the primary reservoir of human infections. The relatively limited geographic distribution of camels may be a factor that has limited the spread of MERS-CoV thus far compared with that of SARS-CoV and SARS-CoV-2. The virus appears to transmit asymptomatically among camels, and has been widespread in the herds studied. Sustained person-to-person transmission risk is thought to be low, although there has been documented nosocomial spread to healthcare workers from infected patients. The potential for mutation, allowing more efficient human-human transmission, remains a possibility. MERS-CoV has a relatively unique tropism for nonciliated bronchial epithelium, unlike most other respiratory viruses, which target ciliated epithelium. More detail on coronavirus infections can be found in Chapter 95 on severe acute respiratory syndrome coronaviruses (SARS and COVID-19).

### Transmission

Based on limited data from known transmissions, the estimated incubation period is 2 to 14 days (median 5 days). The time from clinically apparent symptoms to hospitalization is 4 to 5 days, with death occurring within 12 days for severe cases. It remains unclear as to whether MERS-CoV is transmitted primarily through contact or respiratory transmission, but sustained person-to-person transmission has not been documented. MERS-CoV was initially thought to use the angiotensin-converting enzyme 2 (ACE2) receptor to gain entry to bronchial epithelial cells, as in the case of SARS-CoV and SARS-CoV-2, but it is now thought to use the DPP4 receptor, which localizes to the lower respiratory tract. A relatively small population of lung cells are affected, supporting the hypothesis that a high infectious dose may be needed to cause infection, in contrast to SARS-CoV and SARS-CoV-2. The immune response to MERS-CoV infection appears to vary as well, with limited reports of poor outcomes associated with lack of a type-1 interferon response (interferon alpha), which in turn hindered a sufficient Th-1 immune response to clear the viral infection.

Details regarding viral transmissibility remain murky. As mentioned earlier, other viral respiratory pathogens may infect via receptor cells generally located in the upper respiratory tract, but MERS-CoV can infect via receptor cells located in the lower respiratory tract: this difference can have bearing on viral loads expelled by coughing and sneezing. It has yet to be established whether infections represent single zoonotic transmission with subsequent human transmission or multiple zoonotic

events in different locations. The virus was likely circulating undetected for 10 to 15 years prior to its discovery based on molecular clock analyses. At least seven separate zoonotic transmissions were reported prior to 2012. Although it is not thought to transmit effectively from person to person, the possibility remains that mutations could allow transformation into a more readily communicable respiratory illness.

The environmental stability of coronaviruses has been studied extensively. Although they are inactivated by heat and light, they can persist in the environment for significant periods of time, particularly in cold climates. MERS-CoV has been detected in and isolated from numerous sources including multiple bat species and dromedary camels but not sheep, goats, cattle, or other camels (Fig. 96.1). Undercooked camel meat or unpasteurized milk can be sources of spread. Camels can develop an upper respiratory infection similar to the common cold in humans, with viral shedding demonstrated in oral and nasal secretions and milk but not other fluids. This differs from viral shedding in humans, which occurs in both the upper and lower respiratory tracts and can persist for up to 1 month after infection.

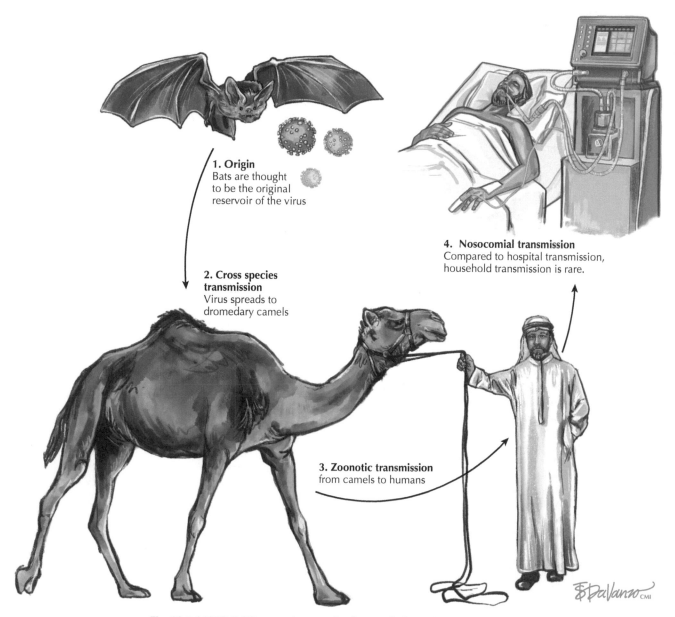

**1. Origin**
Bats are thought to be the original reservoir of the virus

**2. Cross species transmission**
Virus spreads to dromedary camels

**3. Zoonotic transmission**
from camels to humans

**4. Nosocomial transmission**
Compared to hospital transmission, household transmission is rare.

**Fig. 96.1** MERS-CoV has a unique mode of transmission to humans via camels.

## RISK FACTORS

Whereas the origins of outbreaks are zoonotic, transmission of the virus is usually reported in a healthcare or household setting. Sustained community transmission has not been described. Bats are thought to be the original reservoir of the virus, but animal-to-human transmissions have been shown only from dromedary camels. In a cohort of camel workers, activities including milking camels, cleaning farm equipment, and handling animal waste were associated with seropositivity for the virus. Nosocomial spread of the virus can also occur, leading to significant morbidity and mortality. Hospital outbreaks in Saudi Arabia and South Korea were highlighted by a primary initial case of a super spreader who infected multiple close contacts. Subsequent generations of cases had lower transmission rates than the index case, highlighting the importance of identifying the disease and the use of personal protective equipment in preventing further transmission. Compared with hospital transmission, household transmission is reported much less often. In a study from Saudi Arabia, only 12 out of 280 possible household contacts were identified to have secondary infection from 26 index patients, with mild or asymptomatic cases more likely in children. The risk factors for clinical progression include comorbid conditions such as heart disease and diabetes along with older age. Although comorbidities have been associated with infection, it is unclear whether these patients are predisposed to be infected or whether patients with comorbidities are more likely to present with severe illness.

## CLINICAL FEATURES

A diagnosis of MERS-CoV should be considered in anyone with lower respiratory tract symptoms such as cough and shortness of breath and especially with radiographic evidence of pneumonia who is living in the Arabian Peninsula or has recently travel from there. The median incubation period is around 5 to 7 days, but periods up to 13 days have been reported. Respiratory symptoms range from asymptomatic or mild to severe distress leading to death. Nonspecific symptoms such as malaise, sore throat, and fever can lead to lower respiratory symptoms within the first week of the illness's onset. The percentage of cases that are asymptomatic is unclear, and patients without comorbidities, especially children, are more likely to be reported as asymptomatic or mild cases. As in the case of other coronavirus infections, gastrointestinal complaints such as nausea and diarrhea are common. Acute renal failure is a frequent end-organ manifestation of the disease caused either by viral tropism for renal cells or sequelae of severe disease. Progression to acute respiratory distress syndrome (ARDS) is common, leading to high morbidity (see Figs. 95.1 and 95.2). Of the first 47 patients with MERS-CoV infection in Saudi Arabia, 42 required ICU-level care.

## LABORATORY/RADIOGRAPHIC FINDINGS

Laboratory findings for MERS-CoV infection are nonspecific and may include leukopenia, lymphopenia, thrombocytopenia and elevated lactate dehydrogenase levels. With severe disease, laboratory findings suggestive of end-organ dysfunction—such as elevated liver function tests, blood urea nitrogen, and creatinine—may be seen. Radiographic findings are also nonspecific and range from mild airspace disease to extensive bilateral abnormalities. Both interstitial and airspace involvement may be present, with subpleural and basilar involvement more generally reported. In a review of seven patients with confirmed disease who underwent computed tomography (CT) imaging, the majority showed extensive bilateral ground-glass opacities rather than consolidations.

## DIAGNOSTIC APPROACH

Individuals with compatible clinical features and an epidemiologic risk factor for MERS-CoV infection should be tested. The US Centers for Disease Control and Prevention (CDC) has established criteria for defining a person under investigation (PUI) for MERS-CoV:

- Severe illness (fever/pneumonia or ARDS) *plus* history of travel from countries in or near the Arabian Peninsula within 14 days before symptom onset *or* close contact with a symptomatic traveler who developed fever and acute respiratory illness within 14 days of traveling from countries in or near the Arabian Peninsula *or* a member of a cluster of patients with severe acute respiratory illness in which MERS-CoV is being evaluated
- Mild illness (fever and respiratory illness) *plus* a history of being in a healthcare facility within 14 days before symptom onset in the Arabian Peninsula in which recent healthcare-associated cases of MERS have been reported
- Fever or symptoms of respiratory illness *plus* close contact with a confirmed MERS case while the case was ill

The World Health Organization (WHO) criteria are similar to the CDC criteria and are referenced further on. Countries of concern in the Arabian Peninsula include Bahrain, Iraq, Iran, Israel/the West Bank/Gaza, Jordan, Kuwait, Lebanon, Oman, Qatar, Saudi Arabia, Syria, the United Arab Emirates, and Yemen. Close contact is defined as being within 6 ft of a confirmed case for a prolonged period of time while not wearing appropriate personal protective equipment (PPE) or having direct contact with infectious secretions of a confirmed cases while not wearing appropriate PPE. Other etiologies of community-acquired pneumonia should be ruled out, including *Streptococcus pneumoniae*, *Haemophilus influenzae* type b, *Legionella pneumophila*, and other recognized causes of bacterial pneumonia. Causes of viral pneumonia to be ruled out should include influenza, parainfluenza, adenovirus, rhinovirus, and respiratory syncytial virus. Potential patients who have traveled to endemic areas should also be queried about contact with camels (especially ill animals) and consumption of unpasteurized camel milk and undercooked camel meat.

Real-time reverse-transcriptase polymerase chain reaction (rRT-PCR) and serology assays are available for the diagnosis of MERS. It is recommended to collect multiple specimens from different body sites at different times following symptom onset. The available rRT-PCR assays target upstream of the E protein gene (upE), open reading frame 1a (ORF 1a), open reading frame 1b (ORF 1b), or the nucleocapsid (N) protein gene. In the United States the rRT-PCR assay is available at most state laboratories and is done via an emergency use authorization (EUA). There is no known cross reactivity between MERS-CoV and other human coronaviruses with the rRT-PCR assays per the WHO. MERS-CoV can be reliably detected in the respiratory tract, much less frequently in stool samples, and rarely in urine samples. It has been demonstrated that lower respiratory tract samples such as bronchoalveolar lavage (BAL) fluid and tracheal aspirates yield higher viral loads than upper respiratory samples, such as nasopharyngeal swabs.

Serology assays that have been developed for MERS include indirect fluorescent antibody (IFA) assay, enzyme-linked immunosorbent assay (ELISA), protein microarray, and serum neutralization. The CDC recommends that if a screening ELISA assays is positive for MERS, the result be confirmed via microneutralization assay. Most MERS patients seroconvert within the first week of diagnosis (the second week after the onset of symptoms), and IgM is not detected earlier than IgG. Interestingly, in one study all surviving patients but only approximately half of all fatal cases produced IgG and neutralizing antibodies.

When MERS-CoV is clinically suspected, it is recommended to collect three clinical specimens for RT-PCR testing: lower respiratory (sputum, BAL fluid, tracheal aspirate, pleural fluid), upper respiratory (oropharyngeal or a nasopharyngeal wash/aspirate), and serum. It is not recommended to test asymptomatic contacts routinely. The CDC recommends serology testing only for surveillance or investigational purposes and not diagnostics. The WHO recommends serology testing for defining a case for reporting under the International Health Regulations (IHR) or for serological surveys. Per WHO, if a patient has evidence of seroconversion in at least one screening assay and confirmation by a neutralization assay in samples taken at least 14 days apart, the individual is considered a confirmed case regardless of the RT-PCR assay result. Specific collection, storage, and shipping guidelines are noted in the references (further on) from the CDC and WHO. Most laboratory activities with MERS-CoV can be conducted in the BSL-2 in a class II BSC. However, a BSL-3 facility is required for animal studies, propagation in cell culture, and initial characterization of viral agents recovered in cultures of MERS-CoV specimens.

## CLINICAL MANAGEMENT AND DRUG TREATMENT

Given the initial protean manifestations of MERS-CoV, a low index of suspicion is needed to diagnose it. As in the case of other severe coronavirus infections, the spectrum of illness—ranging from asymptomatic infection to septic shock with multiorgan failure—remains incompletely defined. Contacts of PUI and confirmed cases should be thoroughly investigated and results reported to local health authorities. The mainstay of management is first and foremost to provide supportive care for acute respiratory illness and possible respiratory failure, generally in an ICU. Prevention and treatment of secondary bacterial infections are also essential.

Multiple therapeutic agents have been tried in limited studies, with unclear success rates. The most commonly used agent has been the antiviral ribavirin. Remdesivir, currently in development for Ebola and SARS-CoV-2 (COVID-19), was recently shown to reduce viral replication, lung damage, and disease severity in MERS-CoV in a nonhuman primate model. Lopinavir, ritonavir, and interferon beta (LPV/RTV-IFNb) in combination are currently being evaluated clinically. However, the combination of remdesivir and interferon-b has been shown to be more effective than LPV/RTV/IFNb in a mouse model.

Several monoclonal antibodies have been tested and found to have effective viral neutralization activity in vitro or in preclinical models against MERS-CoV infection. Many are aimed at the spike glycoprotein of MERS-CoV, which is similar to that of other coronaviruses. These have been isolated from cloned B cells recovered from survivors as well as camels, llamas, and rhesus macaques as well as created from artificially constructed libraries of antibody variants. Monoclonal antibody cocktails have been suggested to prevent the development of "escape mutants," which could theoretically develop resistance to antibody monotherapy. At least two candidates have entered early human clinical trials, although the small number of cases may make it difficult to test any developmental therapy.

Although monoclonal antibodies show promise therapeutically, vaccines will be required for large-scale prevention of disease. At least six vaccine approaches have been tried to date, including inactivated whole virus, live-attenuated virus, viral DNA, viral subunit, and viral vector vaccines. Nearly all effective vaccine candidates have targeted the spike protein, which is critical for viral cell invasion. Despite multiple vaccine candidates for MERS-CoV, given the relatively sporadic nature of the disease, clinical development has progressed slowly. A DNA vaccine (GLS-5300) has been tested in phase 1 clinical trials for MERS-CoV. DNA vaccines require electroporation of the surrounding skin following injection to allow immune cells to take up the DNA constructs. Two other vaccine candidates completing phase 1 clinical trials include MERS001, which uses the S protein in a chimpanzee adenovirus vector, and a modified vaccinia virus Ankara vaccine construct (MVA-MERS-S). The possibility of antibody-dependent enhancement (ADE) is a significant challenge for developing coronavirus vaccines. ADE results when nonneutralizing antibodies are produced following vaccination or natural infection and can cause subsequent infections to be worse, as ADE facilitates increased viral infectivity. Although it has not been demonstrated to date for MERS-CoV in the limited studies conducted, ADE has been shown for a number of other pathogenic viruses including SARS-CoV, dengue fever, and Ebola.

## PROGNOSIS

There is still a paucity of data on MERS-CoV and in most large available case series from the Middle East. Among confirmed cases, the mortality rate may be as high as 60%. In the MERS outbreak in South Korea, the morality rate was cited at approximately 20%. However, these estimates may be inaccurate, as there may have been mild or asymptomatic cases that went unrecognized. Individuals with chronic medical conditions and the elderly may face an increased risk of a poor outcome if they are infected with MERS-CoV. A systematic review of 637 MERS-CoV cases found that the following medical conditions were present (percent of patients): diabetes (50%), hypertension (50%), cardiovascular disease (30%), and obesity (16%). Some sources hypothesize that certain chronic diseases may increase DPP4 (the cellular receptor for MERS-CoV infection), which could play a role in disease susceptibility and severity. There are limited data in the literature regarding long-term outcomes following MERS infection. One study compared quality-of-life domains between patients with confirmed MERS-CoV infection and those who survived other severe respiratory diseases; there was no significant difference between the groups for the physical component or mental component summary scores. A more recent systematic review and meta-analysis found that lung function abnormalities, psychological impairment, and reduced exercise capacity were common among SARS and MERS survivors.

## PREVENTION AND CONTROL

The prevention and control of MERS-CoV outbreaks is best accomplished via the use of proper infection control and meticulous epidemiological methods. To prevent further community transmission, critical modalities include contact tracing, quarantine of close contacts, isolation/testing of symptomatic cases, and public education. Face masks or coverings may help to limit airborne disease transmission. The CDC recommends standard, contact, and airborne precautions in healthcare settings for the management of patients with known or suspected MERS-CoV infection. Proper infection control measures in hospital settings have been demonstrated to control MERS-CoV outbreaks.

## SUMMARY

Since 2012, MERS-CoV has continued to be a threat to persons in and traveling to the Arabian Peninsula, with high mortality (>30% with clinically apparent infection) but also high rates of asymptomatic disease (up to 80%). Because this disease tends to infect the lower respiratory tract in humans, large-scale person-person transmission has yet to occur, and it has not yet emerged in a way comparable to that of SARS-CoV-2, the virus causing COVID-19. There are no approved therapies, although multiple candidates remain in development.

# EVIDENCE

Alraddadi BM, Watson JT, Almarashi A, et al. Risk factors for primary Middle East respiratory syndrome coronavirus illness in humans, Saudi Arabia, 2014. *Emerg Infect Dis* 2016;22(1):49-55. *Describes risk factors for diagnosis and progression of disease.*

Assiri A, McGeer A, Perl TM, et al. Hospital outbreak of Middle East respiratory syndrome coronavirus. *N Engl J Med* 2013;369(5):407-416. *Provides one of the first description of nosocomial spread of MERS-CoV in Saudi Arabia.*

Badawi A, Ryoo SG. Prevalence of comorbidities in the Middle East respiratory syndrome coronavirus (MERS-CoV): a systematic review and meta-analysis. *Int J Infect Dis* 2016;49:129-133. *Studies comorbidities in over 600 cases of MERS-CoV.*

Drosten C, Meyer B, Muller MA, et al. Transmission of MERS-coronavirus in household contacts. *N Engl J Med* 2014;371(9):828-835. *One of the only reports on risk of transmission to household contact of confirmed cases.*

Faure E, Poissy J, Goffard A, Fournier C, Kipnis E, Titecat M, et al. Distinct immune response in two MERS-CoV-infected patients: can we go from bench to bedside? *PLoS ONE* 2014;9(2): e88716. https://doi.org/10.1371/journal.pone.0088716. *Summary of variances in immune response which may affect outcomes in MERS-CoV infection.*

Killerby ME, Biggs HM, Midgley CM, Gerber SI, Watson JT. (2020). Middle East respiratory syndrome coronavirus transmission. *Emerg Infect Dis* 2020;26(2):191-198. https://dx.doi.org/10.3201/eid2602.190697. *Recent summary of transmission dynamics for MERS-CoV.*

Kim KH, Tandi TE, Choi JW, Moon JM, Kim MS. Middle East respiratory syndrome coronavirus (MERS-CoV) outbreak in South Korea, 2015: epidemiology, characteristics and public health implications. *J Hosp Infect* 2017;95(2):207-213. *Excellent summary of the experience with MERS-CoV in South Korea.*

Memish ZA, Al-Tawfiq JA, Makhdoom HQ, Assiri A, Alhakeem RF, Albarrak A, Alsubaie S, Al-Rabeeah AA, Hajomar WH, Hussain R, Kheyami AM, Almutairi A, Azhar EI, Drosten C, Watson SJ, Kellam P, Cotten M, Zumla A. Respiratory tract samples, viral load, and genome fraction yield in patients with Middle East respiratory syndrome. *J Infect Dis* 2014;210(10):1590-4. *Demonstrates that lower respiratory samples are higher yield than upper respiratory specimens in MERS infected patients.*

Momattin H, Mohammed K, Zumla A, Memish ZA, Al-Tawfiq JA. Therapeutic options for Middle East respiratory syndrome coronavirus (MERS-CoV)– possible lessons from a systematic review of SARS-CoV therapy. *Int J Infect Dis* 2013;17(10):e792-8. *A recent review of limited therapeutic options that have been tested.*

Xu J, Jia W, Wang P, Zhang S, Shi X, Wang X, Zhang L. Antibodies and vaccines against Middle East respiratory syndrome coronavirus. *Emerg Microbes Infect* 2019;8(1):841-56. *Recent review of MERS-CoV vaccine and monoclonal antibody development.*

Yong CY, Ong HK, Yeap SK, Ho KL, Tan WS. Recent advances in the vaccine development against Middle East respiratory syndrome-coronavirus. *Front Microbiol* 2019;10:1781. *Excellent review of recent progress in vaccine development.*

# ADDITIONAL RESOURCES

Ahmed H, Patel K, Greenwood DC, Halpin S, Lewthwaite P, Salawu A, Eyre L, Breen A, O'Connor R, Jones A, Sivan M. Long-term clinical outcomes in survivors of severe acute respiratory syndrome and Middle East respiratory syndrome coronavirus outbreaks after hospitalisation or ICU admission: a systematic review and meta-analysis. *J Rehabil Med* 2020;52(5):jrm00063. *Analysis of long-term outcomes in SARS and MERS survivors.*

Ajlan AM, Ahyad RA, Jamjoom LG, Alharthy A, Madani TA. Middle East respiratory syndrome coronavirus (MERS-CoV) infection: chest CT findings. *AJR Am J Roentgenol* 2014;203(4):782-787. *Provides in-depth description of radiographic findings of confirmed MERS-CoV cases.*

Batawi S, Tarazan N, Al-Raddadi R, Al Qasim E, Sindi A, Al Johni S, Al-Hameed FM, Arabi YM, Uyeki TM, Alraddadi BM. Quality of life reported by survivors after hospitalization for Middle East respiratory syndrome (MERS). *Health Qual Life Outcomes* 2019;17(1):101. *A study in the literature outlining quality of life measures following MERS-CoV infection.*

Centers for Disease Control and Prevention (CDC): Middle East Respiratory Syndrome (MERS). Available at: https://www.cdc.gov/coronavirus/mers/index.html. Accessed July 20, 2020. *This site provides a wealth of information and guidelines on MERS-CoV.*

Drosten C, Meyer B, Muller MA, et al. Transmission of MERS-coronavirus in household contacts. *N Engl J Med* 2014;371(9):828-835. *One of the only reports on risk of transmission to household contact of confirmed cases.*

El Bushra HE, Al Arbash HA, Mohammed M, Abdalla O, Abdallah MN, Al-Mayahi ZK, Assiri AM, BinSaeed AA. Outcome of strict implementation of infection prevention control measures during an outbreak of Middle East respiratory syndrome. *Am J Infect Control* 2017;45(5):502-507. *Describes how MERS-CoV can be controlled in the hospital setting with proper infection control.*

Mackay IM, Arden KE. MERS coronavirus: diagnostics, epidemiology and transmission. *Virol J* 2015;12:222. *Provides overview of the epidemiologic features and transmission of the virus.*

Park JE, Jung S, Kim A, Park JE. MERS transmission and risk factors: a systematic review. *BMC Public Health* 2018;18(1):574. *Provides a large systematic review of risk factors associated with transmission.*

Sikkema RS, Farag E, Himatt S, et al. Risk factors for primary Middle East respiratory syndrome coronavirus infection in camel workers in Qatar during 2013-2014: a case-control study. *J Infect Dis* 2017;215(11):1702-1705. *One of the only descriptions of risk factors for animal-to-human transmission.*

World Health Organization (WHO): Middle East Respiratory Syndrome Coronavirus (MERS-CoV). Available at: https://www.who.int/emergencies/mers-cov/en/. Accessed July 20, 2020. *This site also provides a wealth of information and guidelines on MERS-CoV.*

# 97

# Influenza Virus: A Public Health Concern

*Maryam Keshtkar-Jahromi, Arthur C. Okwesili, Fernando B. Guerena, Phillip R. Pittman*

 **ABSTRACT**

Influenza is a disease that represents an important public health concern. The ability of influenza viruses to undergo genetic mutations and become the cause of global epidemics (pandemics) has been a constant worry to public health experts. Over the past century, four influenza pandemics (Spanish flu, 1918; Asian flu, 1957; Hong Kong flu, 1968; swine flu, 2009) have occurred. The 1918 pandemic was the most severe, when an estimated 50 million people died from the flu worldwide. The most recent pandemic occurred in 2009 to 2010, with the H1N1 influenza virus. The US Centers for Disease Control and Prevention (CDC) estimated that between 151,700 and 575,400 people died from this virus during the first year of the pandemic worldwide. Recently additional concerns have arisen to public health and bioterrorism officials regarding the use of influenza virus as a bioweapon. Advancements in biotechnology have allowed common microbes such as influenza virus to be genetically engineered to cause more severe symptoms, become resistant to current treatment modalities, and/or be crafted to mimic previous pandemics. Influenza virus has the ability to cause future devastating pandemics with significant public health impacts. We have learned from previous influenza pandemics and the recent COVID-19 pandemic that an efficient global preparedness plan to fight against large-scale outbreaks or pandemics is a necessity and requires national and international collaboration between public health agencies throughout the world.

## GEOGRAPHIC DISTRIBUTION AND SEASONALITY

Influenza viruses circulate among human populations worldwide, with yearly seasonal and geographic variations. The viruses that circulate among human populations are influenza A (H1N1 and H3N2) and B (Yamagata and Victoria) viruses. However, occasional infection and global spread in human populations occurs with avian and swine influenza viruses, including the highly pathogenic avian influenza A (H5N1 and H7N9) and the swine influenza A (H1N1, H1N2, and H3N2). Prior to the isolation and identification of the influenza virus, influenza pandemics were recognized by cases of the characteristic febrile respiratory syndrome comprising fever, chills, myalgia, malaise, headache, and upper respiratory symptoms such as rhinorrhea, cough, and pharyngitis that spread rapidly among human populations across geographic regions and countries. Global influenza surveillance reports indicate that human cases of influenza occur throughout the year. However, seasonal influenza epidemics peak mostly during the winter months in temperate areas of both hemispheres and during the rainy season in the tropics. Seasonal influenza occurs throughout the United States and its territories.

## EMERGENCE OF PANDEMIC INFLUENZA VIRUS STRAINS

It has been observed that pandemics occur when an animal influenza strain acquires the ability to cause human infection. An influenza virus strain can be circulating in animal species (birds or pigs) for years before it acquires the genetic capability to infect humans and subsequently cause sustained chains of human-to-human transmission. Transformation of an animal influenza strain into one capable of infecting humans may occur through either genetic reassortment or genetic mutations. Genetic reassortment is a process by which genes from animal and human influenza strains mix together to create a novel human-animal influenza reassortant (the result of the reassortment) virus (genetic shift) (Figs. 97.1 and 97.2). The novel reassortant virus may be capable of infecting humans, spreading rapidly, and causing severe infection; thus it would have the potential to cause a global epidemic (pandemic). Genetic mutations in animal or human influenza viruses are usually small cumulative changes in genes occurring continuously during viral replication that may result in significant alteration of viral surface antigens; occasionally mutations may occur that allow an animal virus to infect humans and transmit easily among them. The presence of a new (novel) influenza strain initially presents as small clusters of human infections that spread locally and then globally. Humans usually do not have any immunity against the new strain, so it can cause severe symptoms and spread worldwide, causing a pandemic with significant effects on human health. The primary risk factor for animal-to-human transmission is direct exposure to infected animals. The primary measure to prevent spread of these infections to humans is the control of influenza infections among animal and bird populations.

Many public health authorities question if there could be any predictive model for forecasting such outbreaks, but the answer is not yet fully known. The risk of such an outbreak with potential to cause a pandemic is multifactorial, depending on the extent of human-to-animal exposure in different populations, degree of influenza virus infection and gene reassortment in the animals, personal and social hygiene among individuals caring for or butchering animals, preexisting immunity in the human population exposed to the new virus, local and country-level long-term commitment to coordination between human and animal health authorities to control infections in animals, and successful implementation of an influenza pandemic preparedness plan. It is obvious that primary evidence of animal-to-human transmission is alarming and that sustained human-to-human transmission is the point at which regional, national, and international public health measures should be implemented to prevent the development of a pandemic.

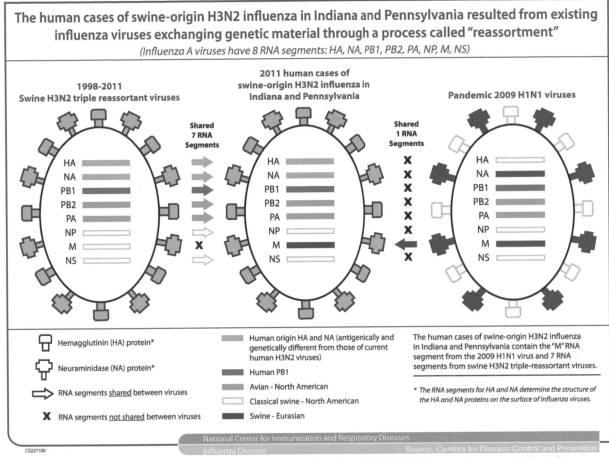

The human cases of swine-origin H3N2 influenza in Indiana and Pennsylvania resulted from existing influenza viruses exchanging genetic material through a process called "reassortment"
*(Influenza A viruses have 8 RNA segments: HA, NA, PB1, PB2, PA, NP, M, NS)*

**Fig. 97.1** This diagram depicts how the human cases of swine-origin H3N2 influenza virus, reported in Indiana and Pennsylvania in September 2011, resulted from the reassortment of two different influenza viruses. The diagram shows three influenza viruses placed side by side, with eight color-coded ribonucleic acid (RNA) segments inside each virus. Note: All influenza viruses contain eight RNA segments. These are labeled *HA* (hemagglutinin), *NA* (neuraminidase), *PB1, PB2, PA, NP, M,* and *NS*. (Reused courtesy of Douglas Jordan, MA, Centers for Disease Control and Prevention-Public Health Image Library, Atlanta, picture ID#1369 [https://phil.cdc.gov/Details.aspx?pid=13469].)

## PAST INFLUENZA PANDEMICS

The CDC defines an influenza pandemic as a global epidemic of a new influenza A virus. Although influenza pandemics have likely been occurring for centuries, only four have been fully recognized and confirmed by laboratory tests.

The 1918 influenza pandemic was caused by an A H1N1 influenza virus (likely avian) that spread worldwide at the end of World War I, from 1918 to 1919. The geographic origin of this virus is not known. It initially affected military populations in the United States and then spread to the rest of the population, causing about 675,000 deaths. The worldwide number of deaths was estimated to be approximately 50 million. Although the influenza A virus was not isolated until 1933, the entire genome of the 1918 pandemic influenza virus was sequenced from materials obtained from the exhumed remains of an American soldier, victim of the 1918 influenza, who had been buried in the Alaskan permafrost. The 1918 influenza was particularly virulent, with a high mortality among young adults that was not observed in subsequent influenza pandemics. The 1918 pandemic influenza in the United States occurred in two well described waves. The second wave was highly fatal and caused most of the deaths.

The 1957–1958 Asian influenza A H2N2 was the second pandemic recognized in history. It started in April 1957 from Guizhou in southern China, spread to Singapore and Hong Kong in May 1957, and then extended worldwide by November 1957. This H2N2 influenza virus had attack rates greater than 50% among children 5 to 19 years of age. It caused about 1.1 million deaths worldwide and about 116,000 deaths in the United States.

The 1968 Hong Kong influenza A H3N2 virus was first isolated in Hong Kong in July 1968. Increased excess mortality in the United States was attributed to influenza A H3N2 in the winter of 1969–1970. This influenza virus also disproportionately affected children, with attack rates greater than 40% among those 10 to 14 years of age. This influenza pandemic caused approximately 100,000 deaths in the United States and 1 million worldwide.

The fourth influenza pandemic started in the spring of 2009 and was first detected in the United States. It was characterized as a novel influenza virus A (H1N1) pdm09. Influenza A H1N1 viruses reemerged in 1977 in Tianjin, China, and have been circulating since then. However, the novel (H1N1) pdm09 was very different from the circulating H1N1 viruses. The (H1N1) pdm09 virus has circulated seasonally in the United States, causing significant morbidity and mortality. The CDC estimated that between 151,700 to 575,400 people died from this virus during the first year of the pandemic worldwide. The mortality caused by (H1N1) pdm09 was higher among people below 65 years of age as compared with the mortality due to seasonal influenza viruses,

## Genetic Evolution of H7N9 Virus in China, 2013

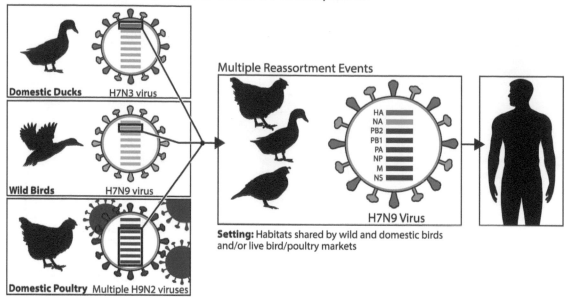

The eight genes of the H7N9 virus are closely related to avian influenza viruses found in domestic ducks, wild birds and domestic poultry in Asia. The virus likely emerged from "reassortment," a process in which two or more influenza viruses co-infect a single host and exchange genes. This can result in the creation of a new influenza virus. Experts think multiple reassortment events led to the creation of the H7N9 virus. These events may have occurred in habitats shared by wild and domestic birds and/or in live bird/poultry markets, where different species of birds are bought and sold for food. As the above diagram shows, the H7N9 virus likely obtained its HA (hemagglutinin) gene from domestic ducks, its NA (neuraminidase) gene from wild birds, and its six remaining genes from multiple related H9N2 influenza viruses in domestic poultry.

**Centers for Disease
Control and Prevention**
National Center for Immunization
and Respiratory Diseases

**Fig. 97.2** This diagram depicts the origins of the H7N9 virus from China and shows how the virus's genes came from other influenza viruses in birds. (Reused courtesy of Douglas Jordan, MA, Centers for Disease Control and Prevention-Public Health Image Library, Atlanta, picture ID# 15798 [https://phil.cdc.gov/Details.aspx?pid=15798].)

which is higher among people older than 65 years of age. However, the overall mortality estimates for (H1N1) pdm09 were lower than that due to previous pandemics. The United States mounted a complex, multifaceted, and long-term response to the pandemic and, on August 10, 2010, the WHO declared an end to the global 2009 H1N1 influenza pandemic. However, the (H1N1) pdm09 virus continues to circulate as a seasonal flu virus, causing illness, hospitalization, and deaths worldwide every year.

## INFLUENZA VIRUSES IN RECENT CIRCULATION

Influenza A viruses are the only influenza viruses historically known to cause pandemics. Novel influenza (influenza A viruses with subtypes that are different from those currently circulating) and variant influenza (influenza subtypes circulating in swine) pose public health concerns owing to their potential to cause pandemics. At the time of this writing, the circulating human influenza viruses include influenza A virus subtypes H3N2 and H1N1 as well as influenza B.

As discussed earlier, novel influenza A viruses of pandemic potential emerge through genetic reassortment, a process that can occur among avian, swine, and human influenza A viruses including the Eurasian swine influenza A virus. The last novel virus originating as a result of this process is the 2009 pandemic H1N1 virus, which is one of the currently circulating influenza A viruses.

Public health experts are routinely monitoring influenza activities, including systematic surveillance of avian and swine influenza

A virus for early warning of future influenza pandemics. In late June 2020, Chinese researchers reported the identification of a reassortant Eurasian avian-like (EA) H1N1 virus—with 2009 pandemic (pdm/09) and triple-reassortant-derived genes—termed the G4 genotype. The swine-based surveillance study conducted from 2011 to 2018 in China found that the genotype 4 (G4) EA H1N1 viruses have become the predominant genotype among swine influenza viruses since 2016. In addition, serological surveillance among swine farm workers in 10 Chinese provinces with high-density swine populations found that 10.4% (35 of 338) were positive for G4 EA H1N1, with a higher rate of 20.5% (9 of 44) among workers 18 to 35 years old; this suggested that the G4 EA H1N1 viruses had acquired increased human infectivity, thus raising public health concern for this virus's possible pandemic potential. However, there are no reports of human-to-human transmission of G4 EA H1N1 viruses to date. In the next several years, public health experts will continue to monitor G4 EA H1N1 and other strains of influenza A viruses, looking for further evidence of pandemic features such as human-to-human transmission or cases outside of endemic areas in China.

## INFLUENZA VIRUS AS A BIOWEAPON

Influenza is not considered a biological threat in the same way as traditional bioweapon agents, such as anthrax and smallpox, which have been weaponized in the past. However, advancements in biotechnology, especially in the area of genetic engineering of microbes, has likely made the alteration of common infections such influenza A more attractive to those who seek to produce bioweapons. A genetically manipulated influenza virus with increased virulence would possess characteristics similar to those of the viruses and hence diseases associated with the most concerning potential biological warfare or bioterrorism agents (CDC category A agents).

One may ask, therefore, what characteristics would make a genetically engineered influenza virus attractive as a bioweapon. First, influenza virus is easily accessible and available worldwide. It can be found in any of its natural reservoirs (human, avian, swine, etc.). In addition, influenza virus is not highly regulated by most health agencies worldwide. For instance, the US Federal Select Agents Program (US Department of Health and Human Services) lists only the reconstructed replication-competent forms of the 1918 pandemic influenza virus as a select agent. The US Department of Agriculture lists only the highly pathogenic strains of avian influenza virus as a select agent. All other strains of influenza are available for acquisition and can then be modified to become more virulent, resistant to current treatments modalities, and/or more easily transmitted from human to human. As such, when it comes to the accessibility of agents for the production of bioweapons, influenza is easy to obtain compared with typical select agent pathogens such as smallpox.

Second, influenza has the potential to be weaponized because of the relatively low cost and modest production requirements to create such a bioweapon. The methods to produce such a bioweapon are already available through open source literature as well as the dark web. In 2012, the National Science Advisory Board for Biosecurity tried to censor publication of two studies regarding the engineering of highly pathogenic avian influenza A (H5N1). It was recognized that the methodology presented in any of these studies could easily be recreated to construct bioweapon influenza viruses at very low cost.

Third, the delivery modes for a weaponized influenza strain are readily available. Biological agents can be deployed in a short period of time over large areas utilizing aerosol dispersion devices or other dissemination techniques. Terrorist organizations may also utilize members of their organizations as disease vectors by purposefully infecting them with communicable diseases and attempting to infect as many others as possible so as to attempt to initiate a pandemic. Although the incubation period for influenza is relatively short, it would still provide enough time for the perpetrators to escape notice before the effects of the weaponized influenza were seen.

Fourth, early detection of weaponized influenza is likely not possible. Weaponized influenza would probably present similarly to the early onset of most viral syndromes, with body aches, headaches, fever, weakness, nausea, cough, and so on, making it difficult to identify it quickly. Even with initial clusters of cases or outbreaks, test results would likely present as influenza A–positive. This would probably not trigger a robust response and investigation initially because the influenza virus occurs naturally and is always circulating within the human population.

Finally, a weaponized influenza is likely to be novel to humans. This means that the majority of the human population would lack specific immune protection to fight off the disease, potentially resulting in large numbers of casualties as well as creating fear and panic, which could disturb the norms of society. Judging from the impact of the COVID-19 pandemic on world economies, public health, medical infrastructures, and the world psyche, a weaponized influenza pandemic could be expected to cause similar devastation.

## PREVENTION AND CONTROL OF INFLUENZA

The influenza A viruses have been demonstrated to be persistent in human, animal, and avian reservoirs; genetically versatile; and capable of causing potentially severe disease that can spread rapidly in human populations. (See Chapter 8 for the clinical features, diagnostic approach, and management of influenza disease.) The development of effective influenza vaccines that elicit protective antibodies against the surface proteins of the virus are a primary measure for prevention of seasonal influenza. However, small mutations in the genes occur continually during viral replication and lead to changes in the surface proteins hemagglutinin (HA) and neuraminidase (NA). These changes accumulate over time and ultimately lead to viruses that are antigenically different from the viruses used to create prior influenza vaccines, a phenomenon called **antigenic drift**. Thus the strain composition of influenza vaccines is reviewed annually, months in advance of the upcoming influenza season, and the vaccines are reformulated as needed to cover the predominant influenza strains anticipated to be in circulation during the next influenza season based on data from the WHO influenza global surveillance network (see further on).

The pandemic potential of influenza A viruses has been demonstrated by the past influenza pandemics. During a pandemic caused by novel strain, the current seasonal vaccines will be insufficiently protective. New pandemic-strain vaccines usually cannot be produced until late in a pandemic, but they could still be of great benefit by reducing further transmission and also decreasing mortality in susceptible populations. In the interim—before new pandemic-strain vaccines become available and the optimal drug treatment and chemoprophylaxis can be determined—relatively simple measures to prevent transmission must be employed, such as quarantine of suspected or confirmed cases and individuals likely to have had significant exposure, use of facial masks, frequent hand washing, social distancing, avoidance of mass gatherings, and limitations on travel, among others. These measures are the same as those recommended for other respiratory pathogens transmitted among humans in a similar manner to influenza (e.g., SARS-CoV, SARS-CoV-2 [the agent causing COVID 19] and MERS-CoV).

## INFLUENZA PREPAREDNESS PLAN

National and international cooperation and collaboration are paramount for proper preparedness against pandemic influenza. The primary resource against influenza pandemics is a robust surveillance system followed by preparedness planning. Two pivotal events have advanced the modern global influenza surveillance systems. In 1952, the Global Influenza Surveillance and Response System was created by the WHO to monitor influenza viruses. In 1956, the CDC Influenza Branch was named a WHO collaborating center for the surveillance, epidemiology, and control of influenza. A committee review of the 2009 H1N1 influenza pandemic response resulted in 15 recommendations to the WHO and the member states. The recommendations included primarily the need to strengthen the WHO's capacity for sustained response, the revision of pandemic preparedness guidance, the establishment of a more extensive public health reserve workforce globally, and the pursuance of a comprehensive influenza research and evaluation program, among others.

Influenza surveillance programs established at local and country levels are the key tool that can help to monitor the number of cases and circulating strains in order to assess the risk of widespread infections compared with seasonal influenza. Country-level surveillance systems are designed to transfer information to collaborated international systems in order to implement timely preventive measures. Although global influenza surveillance and monitoring systems are much improved, it is always possible that the first outbreaks of a pandemic will not be detected or recognized. Public health authorities at local and country levels should have a plan to coordinate efforts and provide leadership across the country in case any alarming signal is detected by the surveillance systems. The goal is to collect, interpret, and disseminate information in order to monitor and assess situation before, during, and after pandemics. Unfortunately we have learned from the COVID-19 pandemic that many countries did not have a planned system or that their system was insufficiently prepared to stop a pandemic on a large scale. Moreover, to be able to implement intervention, national systems should continuously integrate with an international surveillance system and be monitored for sensitivity and accuracy in capturing information in a timely fashion.

In addition to establishing an efficient surveillance system, planning for surge capacity within local and regional health care systems (e.g., supplies of personal protection equipment [PPE] such as masks, gloves, gowns, antibiotics, antipyretics, hydration, oxygen, ventilation support, increased patient capacity in hospitals and intensive care units) is a key to providing medical care for additional patient loads. Lack of surge capacity was a huge problem during the COVID-19 pandemic in many countries, and a need for surge capacity should be anticipated in response to future global pandemics such as influenza.

Preparedness planning to prevent person-to-person transmission during an influenza pandemic would include a hierarchy of both individual and social measures that would be recommended or even mandated in a public health emergency. Individual measures might include self-quarantine, hand and respiratory hygiene, wearing a face mask outside the home, and social distancing. Social measures might include enforced quarantine; limitation of gatherings outside of living groups; reduced capacity/closure of commercial spaces (e.g., stores, restaurants, bars, gyms, and other businesses); limitation of the use of public spaces (e.g., parks, beaches, campgrounds); suspension of in-person classes, activities, and athletics at child care facilities, schools and colleges; limitation on travel, especially international travel; and cancelation of mass gatherings (e.g., religious services, theater, concerts, sports events, graduation ceremonies).

Planning for efficient modes of communication during pandemics is necessary to build public trust in the public health system, coordinate efficient use of limited resources, and provide relevant public health information to the public. Health education or health messaging for the lay public is a much larger challenge than most health crisis management teams realize, but effective communications are necessary to gain widespread acceptance of the individual and social measures meant to protect public health and safety and to minimize social and economic disruption.

Once animal sources or reservoirs have been identified, a plan to prevent animal-to-human transmission of pandemic influenza viruses must also be a part of preparedness planning, taking into account that the destruction of large numbers of birds, pigs, or other host animals may be necessary and that this will create severe hardship for the farmers and regions involved.

Countries should develop principles to guide national recommendations for the use of antivirals (for prophylaxis and treatment). National and international public health authorities should have a strategic global stockpile of antivirals and standard operating procedures for their rapid deployment. They should develop principles to guide national recommendations for the use of seasonal and pandemic vaccines. International organizations should support strain characterization as well as the development and distribution of vaccine prototype strains for possible vaccine production. They should also provide technical support, capacity building, and technology transfer for diagnostics, antiviral drugs, and influenza vaccines to resource-poor countries

and formulate mechanisms and guidelines to promote fair and equitable distribution of essential supplies. Key personnel should be regularly trained to be mobilized as part of a multisector expert response team for deployment to sites of animal or human influenza outbreaks of pandemic potential in case assistance is needed.

## EVIDENCE

Centers for Disease Control and Prevention, 2020 Centers for Disease Control and Prevention, National Center for Immunization and Respiratory Diseases (NCIRD). Influenza (Flu): Understanding Influenza Viruses. Available at: https://www.cdc.gov/flu/about/viruses/types.htm. Accessed on June, 30, 2020. *Discusses the types of influenza viruses and the currently circulating influenza virus in people.*

Christopher GW, Gerstein DM, Eitzen EM, Martin J. Chapter 1: Historical overview—from poisoned darts to pan-hazard preparedness. *Textbooks of Military Medicine: Medical Aspects of Biological Warfare*, 2nd ed. Office of the Surgeon General, Fort Sam Houston, TX, Borden Institute, 2018, pp. 10-26. *An overview of bioweapon, biological weapon convention, bio-crime and bioterrorism.*

Fineberg H. Pandemic preparedness and response-lessons from the H1N1 influenza of 2009. *N Engl J Med* 2014;370:1335-42. *The article discusses lessons learned from the most recent influenza pandemic in 2009.*

Madjid M, Lillibridge S, Mirhaji P, Casscells W. Influenza as bioweapon. *J R Soc Med* 2003;96:345-346. *Describes why an increased consideration of influenza as a bioweapon bioterror agent is warranted.*

Neumann G, Kawaoka Y. Predicting the next influenza pandemics. *J Infect Dis* 2019;219(Suppl 1):S14-S20. *Characterizes influenza as a potential public health concern for future pandemic.*

Qualls N, Levitt A, Kanade N, Wright-Jegede N, CDC Community Mitigation Guidelines Work Group, et al. Community mitigation guidelines to prevent pandemic influenza—United States, 2017. *MMWR Recomm Rep* 2017;66(1):1-32. *Updated guidelines from the Centers for Disease Control and Prevention for preventing influenza pandemics.*

Sun H, Xiao Y, Liu J, et al. Prevalent Eurasian avian-like H1N1 swine influenza virus with 2009 pandemic viral genes facilitating human infection. *Proc Natl Acad Sci U S A* 2020;117(29):17204-17210. *This article discusses the swine-based surveillance study that identified the new influenza A virus reassortant with potential to cause pandemic.*

Whitehouse CA, Beitzel B, Dembek ZF, Schmaljohn AL. Chapter 25: Emerging infectious disease and future threats. *Textbooks of Military Medicine: Medical Aspects of Biological Warfare*, 2nd ed. Office of the Surgeon General, Fort Sam Houston, TX, Borden Institute, 2018, pp. 657-666; 679-681. *An outstanding textbook with in-depth discussion of medical aspects of biological warfare, biological threat, and biodefense. It also discusses the emerging viral disease including different strains of influenza virus, and future bio-threats due to genetically engineered pathogens.*

## ADDITIONAL RESOURCES

Centers for Disease Control and Prevention. Influenza (Flu): Pandemic Influenza. Available at: https://www.cdc.gov/flu/pandemic-resources/index.htm. Accessed on July, 15, 2020. *Discusses the nature and history of past influenza pandemics and the lessons applicable to the prevention and mitigation of future pandemics.*

Coninx JK, Fukuda K, Harmanci H, Park K, Chamberland M, Niemi TC, Pluut E, Vivas C. Pandemic Influenza Preparedness and Response, A WHO Guidance Document, Global Influenza Programme. World Health Organization 2009. *Reprinted 2010. WHO guideline for influenza pandemic preparedness.*

U.S. Army Medical Research Institute of Infectious Disease (USAMRIID). *Medical Management of Biological Casualties Handbook*, 8th ed. Washington, DC, U.S. Government Publishing Office, 2014. *A handbook on the medical management of biological agents' casualties.*

Note: Page numbers followed by f, t, and b indicate figures, tables, and boxed material, respectively.